BERTRAND
W9-CDY-391

ELECTRONIC DEVICES AND CIRCUITS

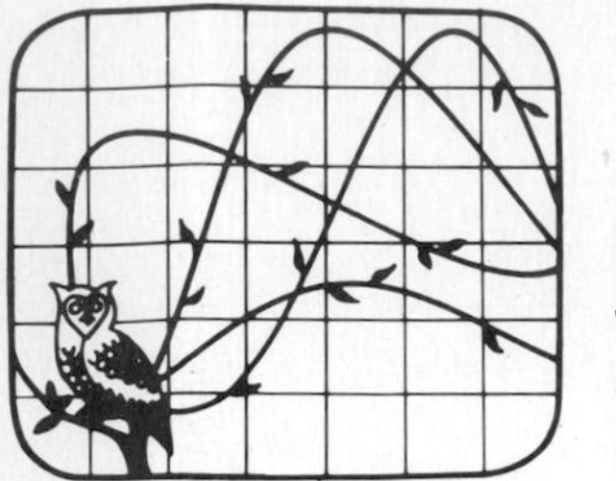

HRW
Series in
Electrical and
Computer Engineering

M. E. Van Valkenburg, Series Editor

P. R. Belanger, E. L. Adler, and N. C. Rumin INTRODUCTION TO CIRCUITS AND ELECTRONICS: AN INTEGRATED APPROACH
L. S. Bobrow FUNDAMENTALS OF ELECTRICAL ENGINEERING
L. S. Bobrow ELEMENTARY LINEAR CIRCUIT ANALYSIS
C. T. Chen LINEAR SYSTEM THEORY AND DESIGN
D. J. Comer DIGITAL LOGIC AND STATE MACHINE DESIGN
C. H. Durney, L. D. Harris, C. L. Alley ELECTRIC CIRCUITS: THEORY AND ENGINEERING APPLICATIONS
M. S. Ghausi ELECTRONIC DEVICES AND CIRCUITS: DISCRETE AND INTEGRATED
G. H. Hostetter, C. J. Savant, Jr., R. T. Stefani DESIGN OF FEEDBACK CONTROL SYSTEMS
S. Karni and W. J. Byatt MATHEMATICAL METHODS IN CONTINUOUS AND DISCRETE SYSTEMS
B. C. Kuo DIGITAL CONTROL SYSTEMS
B. P. Lathi MODERN DIGITAL AND ANALOG COMMUNICATION SYSTEMS
C. D. McGillem and G. R. Cooper CONTINUOUS AND DISCRETE SIGNAL AND SYSTEM ANALYSIS Second Edition
A. Papoulis CIRCUITS AND SYSTEMS: A MODERN APPROACH
S. E. Schwarz and W. G. Oldham ELECTRICAL ENGINEERING: AN INTRODUCTION
A. S. Sedra and K. C. Smith MICROELECTRONIC CIRCUITS
M. E. Van Valkenburg ANALOG FILTER DESIGN

ELECTRONIC DEVICES AND CIRCUITS

Discrete and Integrated

M. S. Ghausi
UNIVERSITY OF CALIFORNIA AT DAVIS

HOLT, RINEHART AND WINSTON
New York Chicago San Francisco
Philadelphia Montreal Toronto London
Sydney Tokyo Mexico City
Rio de Janeiro Madrid

Address correspondence to:
383 Madison Avenue, New York, NY 10017

Library of Congress Cataloging in Publication Data

Ghausi, Mohammed Shuaib.
Electronic devices and circuits.

Includes index.
1. Transistor circuits. 2. Transister amplifiers.
3. Digital electronics. I. Title.
TK7871.9.G42 1985 621.3815 30422 84-15783

ISBN 0-03-062481-9

Printed in the United States of America
Published simultaneously in Canada

5 6 7 8 016 9 8 7 6 5 4 3 2

CBS COLLEGE PUBLISHING
Holt, Rinehart and Winston
The Dryden Press
Saunders College Publishing

To my teachers and students

Contents

321

321

321

321

Chapter 7 FREQUENCY RESPONSE OF SINGLE-TRANSISTOR AMPLIFIERS 292

Chapter 8 FREQUENCY RESPONSE–MULTIPLE-TRANSISTOR AMPLIFIERS 351

Preface

Electronic Devices and Circuits emphasizes the understanding of basic design principles in electronic circuits. Currently, a large number of integrated-circuit (IC) packages are available both for analog and digital circuits. A circuit function can often be implemented, using the manufacturer's application guides alone. However, an understanding of the operation and properties of the devices and circuits within the package is very important for intelligent interfacing of these packages for the development of more efficient future generations of circuit design. This book therefore covers bipolar junction transistor (BJT), field-effect transistor (FET), and discrete and integrated circuits. A prerequisite for this book is a background in elementary circuit analysis, including a thorough understanding of loop and nodal analysis.

The book introduces the reader to circuit design using solid-state devices. Thus the terminal properties of the devices and their models are used, and the device physics is kept to a minimum. However, when appropriate, the physical relations governing the values of key parameters of the device are given without any derivation. Physical relations are given to show the basic limitations and also to motivate investigation of solid-state device theory later on. Both discrete- and integrated-circuit analysis and design methods are covered. Thus an understanding of the operation of diode, bipolar junction and field-effect transistors, and their applications are stressed. In order to [illegible] confusing the reader with regard to reference polarities, the commonly used *npn* BJT, and *n*-channel FET devices are covered with a passing mention of other types. The operation and application of devices for both small and large signals are stressed. For small-signal applications, the concept of equivalent circuit is expounded. Complicated models appropriate for computer-aided analysis and design are given; simple models, suitable for preliminary analysis and design, are utilized throughout the book. The latter is used extensively in order to provide insight into the analysis and design of electronic circuits. For large-signal applications, graphical analysis, using load line and the concept of the voltage transfer characteristic, is emphasized. The switching behavior of BJT and FET devices is covered at an early stage and integrated with the introductory analysis in order to provide full appreciation for the application of the devices in analog and digital circuits. An attempt has been made to balance theory and practice.

Chapters 1 through 6 cover transistor circuits at low frequencies. These chapters can be used in a first course, since they do not require a knowledge of complex frequency variable and frequency response. More material is included than can be covered in a first course, and thus some of the sections can be skipped. For Chapters 7 through 12, some knowledge of frequency response will be helpful. These chapters could be used in a second course. Again, all the material included in the book cannot be covered. Chapters 13 and 14 introduce the reader to the basics of digital circuits, namely, logic gates, flip-flops, clocks, and some of their applications in digital circuits and systems. The material in these chapters can be covered earlier or later depending on the reader's interest. If used in a first course, the detailed properties of the gates in Chapter 13 are to be omitted.

Chapter 1 covers diodes and their applications in electronic circuits, including large- and small-signal models, and zener diodes. Chapter 2 deals with bipolar junction transistors. The device *V-I* characteristics and transfer characteristic under large signals are discussed. Small-signal models for the BJT, including bias stabilization, are developed (these models are used in later chapters). Chapter 3 is concerned with field-effect transistors; these include junction field-effect transistors (JFET) and metal-oxide semiconductor (MOS) devices, both of the enhancement and depletion types. Their *V-I* characteristics, large-signal response, small-signal models, and biasing are covered. Chapters 4 and 5 cover low-frequency single- and multiple-transistor amplifiers including both BJT and FET devices. Differential and operational amplifiers, including Bi-FET and Bi-MOS op amps, are considered in some detail since op amps are used extensively in linear and nonlinear circuit applications. Chapter 6 examines power amplifiers, output stages, and thermal considerations.

Chapters 7 and 8 deal with the frequency response of single- and multiple-transistor amplifiers at low and high frequencies. The analysis and design of amplifiers utilize the small-signal circuits of BJT and FET devices. The concepts of dominant pole, bandwidth, and gain bandwidth are stressed in the design of such amplifiers. Chapter 9 presents a detailed coverage of operational amplifiers and their various applications in linear and nonlinear electronic cicuits. Ideal and practical op amps and their characteristics are discussed. Chapter 10 deals with the principles of feedback amplifiers. The advantages of negative feedback and the potential problem of instability are discussed in some detail. The chapter also includes op-amp frequency compensation and the analysis and design of feedback oscillators. Chapter 11 introduces tuned narrowband amplifiers with synchronously tuned and stagger-tuned design methods. Other types of frequency-selective circuits such as active RC and phase-locked loops are also covered. Chapter 12 introduces the concept of time-domain response in linear circuits. The interrelation between the frequency domain and time domain in amplifiers is given. The switching speeds of *pn* junction diodes and BJT and FET switches, which are important in digital circuits, are also included.

Chapter 13 serves as an introduction to digital circuits and integrated-circuit logic gates. Basic logic operations, logic gates, their properties, and interfacing are discussed. The logic families covered and compared are TTL, ECL, NMOS, CMOS, and I^2L. Chapter 14 covers latches, the different types of flip-flops, and their applications in registors and counters. The chapter concludes with a brief treatment of analog-to-digital (A/D) and digital-to-analog (D/A) conversion.

Appendixes A and D cover topics that are germaine to the book, and their coverage depends on the reader's choice. Appendix A deals with a quantitative treatment of semiconductor electronics and *pn* junction diodes. Some of this material may be covered with Chapter 1 if so desired; however, it is not necessary. Appendix B, which deals with the fabrication of integrated circuits, can be an optional topic. Appendix C is concerned with two-port network properties. Part of this appendix should be covered with Chapters 2 and 4. The rest of Appendix C should be covered before starting with Chapter 10. Appendix D deals with noise in bipolar transistors, field-effect transistors, and op amps. Appendix E includes a variety of typical manufacturer's data sheets from the discrete device to integrated-circuit analog and digital functional blocks, so that the reader can use actual data sheets and gain some knowledge of the practical range of parameter values of the various devices and circuits.

Each chapter is complemented with many illustrative examples to reinforce the material in the book and many exercises with answers to help the reader to master the material. Many problems are included at the end of each chapter for further practice assignments. Answers to selected problems are also given at the end of the book. A solutions manual for the instructor is available from the publisher.

This book was written at Oakland University in Michigan where the author was also the John F. Dodge professor of engineering at the time. The author wishes to acknowledge the many helpful suggestions offered by Professor Carroll Hill, who read the entire draft of the manuscript. I'm also grateful to Professor Hoda Zohdy who class-tested the first three chapters of the manuscript and provided some of the problems for these chapters. The author is also indebted to the following reviewers for their valuable comments at various stages of the manuscript:

Andrew Blanchard, University of Texas
Patricia Daniels, University of Washington
P. David Fischer, Michigan State University
Ronald Guentzler, Ohio Northern University
Stephen Haley, University of California at Davis
Michael Lightner, University of Colorado
Alan Marshak. Louisiana State University
Allen Nussbaum, University of Minnesota
Timothy Trick, University of Illinois
James Whelan, University of Southern California

The helpful comments and reviews provided by my editor at Holt, Rinehart and Winston, Deborah Moore, and series advisor Mac VanValkenburg are greatly appreciated. Paul Becker's enthusiasm and encouragement was a catalyst in undertaking this project.

Caroline Fauth and Sue Fightmaster did a fine job typing the manuscript and its several revisions. Special thanks are due to Mr. Zhong-xuan Zhou, who prepared the solutions manual.

My wife's understanding, patience, and moral support have been instrumental in the completion of this book.

M. S. Ghausi

ELECTRONIC DEVICES AND CIRCUITS

CHAPTER 1

Diodes and Circuit Applications

We begin our study of electronic devices and circuits with the diodes. The diode is the simplest nonlinear two terminal device, having a variety of applications in electronics. In this chapter we shall consider *pn*-junction diodes and some of their applications in signal processing including rectification, clipping, and other waveshaping functions. Graphical analysis and load lines are discussed in detail in this chapter. These techniques are important in the analysis of many nonlinear circuits and are applied to other electronic devices as well. The characteristics of an *ideal diode*, piecewise-linear approximation, transfer characteristic, and the concept of small-signal and large-signal models are also introduced.

This chapter utilizes the terminal properties of the diode. The physical operation of the *pn*-junction diode is discussed quantitatively in Appendix A in order to provide some understanding of the device physics. We shall use the terminal properties and *models* of the devices and their applications in electronic circuits in this text.

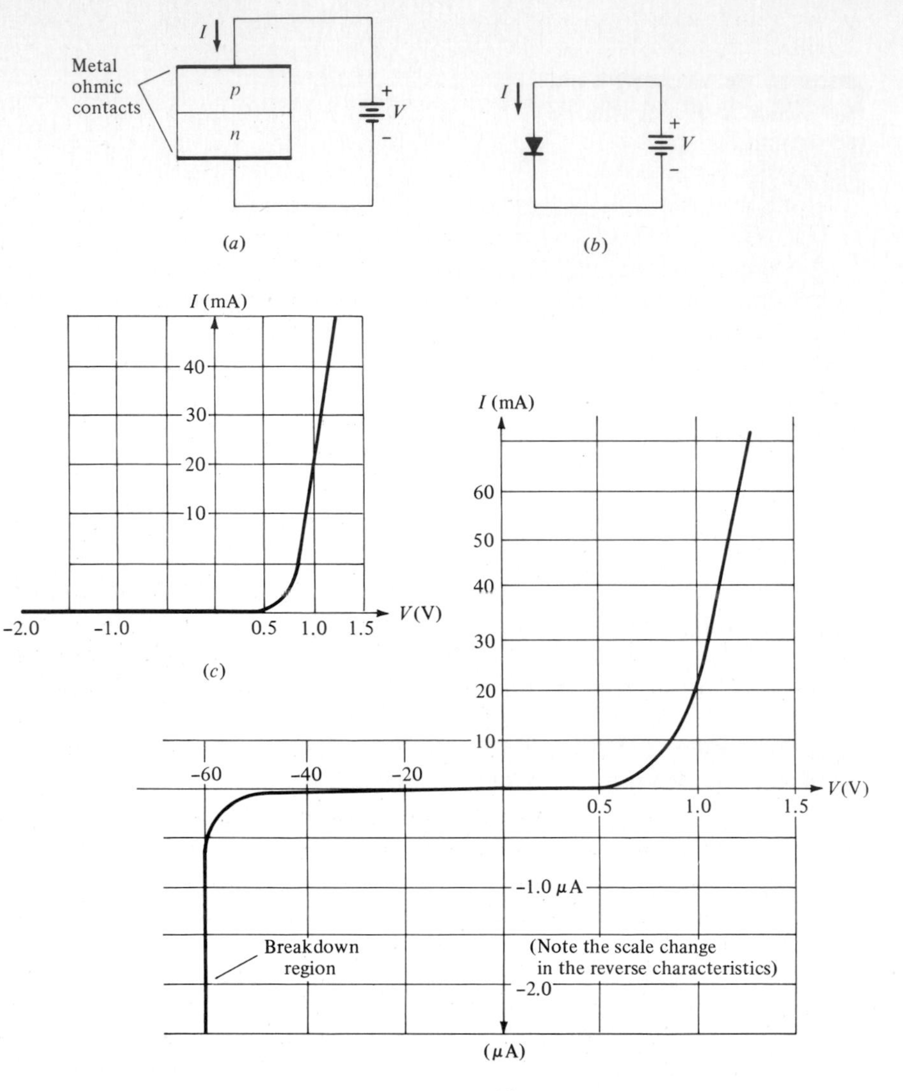

FIGURE 1.1 (*a*) A *pn* junction diode with applied voltage. (*b*) Diode symbol. (*c*) Typical diode characteristic in the forward region. (*d*) Typical diode characteristic in the forward and reverse regions (not to scale).

1.1 TERMINAL CHARACTERISTICS OF JUNCTION DIODES

Figure 1.1*a* is a schematic representation of a *pn*-junction. The *p* and the *n* terminals of the diode are sometimes called the *anode* and the *cathode* terminals, respectively (a carry-over from vacuum tubes). The letters *p* and *n* refer to the *p*-type and *n*-type semiconductors, respectively. Semiconductors are solid materials whose resistivities have values between those of conductors and insulators. They behave as conductors at very high temperatures and as insulators at very low temperatures. The most commonly used semiconductor in electronic devices is silicon (Si), which has a diamondlike structure with four valance electrons.

A pure semiconductor is referred to as an *intrinsic* semiconductor. In an intrinsic semiconductor the concentration of electrons, n, and holes, p, are equal. For Si, at room temperature $n = p = n_i = 1.5 \times 10^{10}$ holes or electrons per cm^3. The subscript i denotes intrinsic. A hole is the absence of an electron in a covalent bond. It behaves as if it were an independent mobile positive charge.

Semiconductors are often doped, i.e., carefully controlled impurities (e.g., 1 atom of impurity to 10^6 to 10^8 atoms of the host material) are added, so that either the holes or electrons dominate. Such semiconductors are referred to as *extrinsic* semiconductors. Usually electron donor atoms with five valence electrons, such as arsenic (As), or electron acceptor atoms with three valence electrons, such as gallium (Ga), are added to Si. In the former case electrons are the majority carriers and holes are the minority carriers, and the semiconductor is referred to as the n type. In the latter case holes are the majority carriers and electrons are the minority carriers, and the semiconductor is referred to as the p type. The concentration of impurity atoms is in the range 10^{14} to 10^{18} atoms per cm^3. The concentration of the holes and electrons can be found from $np = n_i^2$ and the charge neutrality condition. It should be noted that n_i is strongly temperature dependent.

A junction is formed by joining p-type and n-type semiconductors. Because of the nonuniform concentration of carriers, a gradient across the junction is created, which results in the diffusion of carriers; holes diffuse from the p to the n region, and electrons diffuse from the n to the p region. The result of the diffusion is to produce immobile ions of opposite charge on each side of the junction thus creating a *depletion region* or a *space-charge region*. As a consequence of this space charge, which is of opposite polarity on each side of the junction, an electric field is established across the junction. Since, at thermal equilibrium no net current can flow, the resultant electrostatic potential barrier counteracts the diffusion preventing a net flow of majority carriers to the other side.

When the junction is *forward-biased* (i.e., the positive terminal of the battery is connected to the p region), the potential barrier is reduced, and a large number of holes from the p region flow to the n region. Similarly, electrons from the n region flow to the p region, and a current flow results in *the direction of the arrow* shown on the symbol for a junction diode in Fig. 1.1*b*. When the diode is *reverse-biased* (i.e., the positive terminal of the battery is connected to the n region), the potential barrier is increased, and a very small number of holes in the n region flow into the p region, and vice versa. The current, therefore, is very small and in the direction opposite to the arrow. The basic physical mechanism of pn electronic conduction in junction diodes is quantitatively discussed in Appendix A.

The measured characteristic of a typical silicon (Si) junction diode at room temperature (20°C) is shown in Fig. 1.1*c*. Note the very small current (nearly zero) for negative voltage. Figure 1.1*d* shows an expanded scale for the reverse region showing the reverse voltage breakdown. The V-I relationship of a pn-junction diode, exclusive of the breakdown region, is given by

$$I = I_s(e^{(\lambda q/kT)V} - 1) = I_s(e^{\lambda V/V_t} - 1) \tag{1.1}$$

where V = applied bias with $V > 0$ being forward bias and $V < 0$ being reverse bias
q = electronic charge (1.60×10^{-19} C)
k = Boltzmann's constant (1.38×10^{-23} J/K)
T = absolute temperature K (K = 273 + °C)
λ = empirical scaling constant, which in a practical device lies between 0.5 and 1.0

$$V_t = \frac{kT}{q} \simeq 25 \text{ mV} \quad \text{at room temperature (290 K); } V_t \text{ is the thermal voltage} \tag{1.2}$$

For $V \gg V_t$: $$I \simeq I_s e^{\lambda V/V_t} \gg I_s \tag{1.3a}$$

For $V < -V_t$: $$I \simeq -I_s \tag{1.3b}$$

The simple theory, which describes the *V-I* characteristic of the diode in (1.1), ignores series resistance and thermal generation and recombination effects. From Eq. (1.3*a*) it is seen that for a forward-bias voltage larger than 25 mV, the current increases exponentially; for reverse-bias voltages, the current is constant and essentially equal to the saturation current I_s, provided the voltage is well below the breakdown voltage. In Appendix A it is shown in (A.39) that I_s is strongly temperature dependent. The reverse saturation current approximately doubles for every 6°C in silicon,* near room temperature. For the theoretical diode I_s is the order 10^{-15} A. However, due to generation, recombination, and other effects, the value of I_s for an actual Si diode at room temperature is typically 10^{-9} A.

Figure 1.2 shows the measured *V-I* characteristic of 1N4001 diode at room temperature. For this diode $I_s \simeq 10^{-9}$ A and $\lambda \simeq 0.58$. Note that the curve fits very well with the theoretical model if the empirical factor λ is used. For other values of λ the deviations can be significant since λ appears in the exponent. For diodes made from transistors such as in integrated circuits (Appendix B) the value of λ is almost unity. Note that the diode current is negligibly small for voltages smaller than 0.6 V. The diode is fully conducting at a voltage higher than this value. The voltage at which the diode is fully conducting is referred to as the *threshold voltage*. It is determined by the intersection of a linear line (which best fits the characteristic curve at high forward-bias values) with the voltage axis. The slope of this line is the inverse of the forward-bias resistance of the diode which is discussed later in Sec 1.9. In this particular diode the threshold voltage indicated by V_0 is equal to 0.74 V. The threshold voltage is also referred to as the *turn-on voltage* or *cut-in voltage*. The threshold voltage V_0 for Si diodes usually lies in the range from 0.5 to 0.9 V with a typical value of 0.7 V. The temperature variations of this voltage ranges from −1 to −3 mV/°C, with a typical value of −2 mV/°C near room temperature as shown in (A.45). In other

* We only consider Si since it is the predominant electronic semiconducting material used in integrated circuits. Other electronic semiconductors such as germanium (Ge) and gallium arsenide (GaAs) are also sometimes used for special applications. The latter is especially used for very-high-frequency requirements and light-emitting diodes.

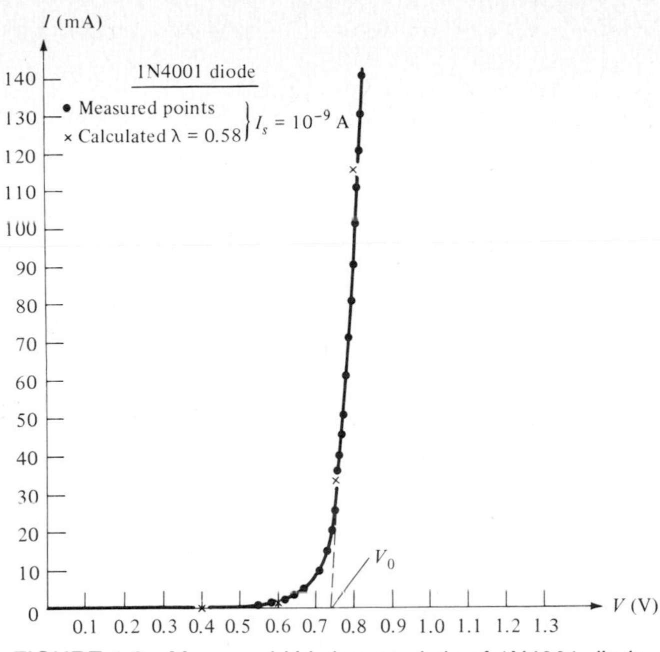

FIGURE 1.2 Measured *V-I* characteristic of 1N4001 diode.

words if $V_0 = 0.74$ V at room temperature, its value will decrease to 0.68 V for a 30°C rise in temperature.

From the exponential nature of the *V-I* characteristic for a forward-biased diode, it is seen that a relatively small voltage change beyond V_0 causes an appreciable change in the diode current.

EXAMPLE 1.1

The saturation current of a Si diode is $I_s = 10^{-9}$ A, and the empirical factor is $\lambda = 0.50$. Determine the following:

(a) The diode current for $V = 0.9$ V at room temperature (20°C)
(b) The threshold voltage at room temperature
(c) The above quantities if the temperature is raised to 62°C

From (1.3*a*), at room temperature, we have

$$I = 10^{-9}e^{0.50(0.9)/0.025} = 10^{-9}e^{18} = 65.6 \text{ mA}$$

From a plot of the *V-I* characteristic shown in Fig. 1.3 (at room temperature) we determine $V_0 = 0.86$ V.

For a 42°C rise in temperature I_s changes to

$$I_s(20° + \Delta T) \simeq I_s(20°)\ 2^{\Delta T/6°\text{C}} = 10^{-9}(2^7) = 1.28 \times 10^{-7} \text{ A}$$

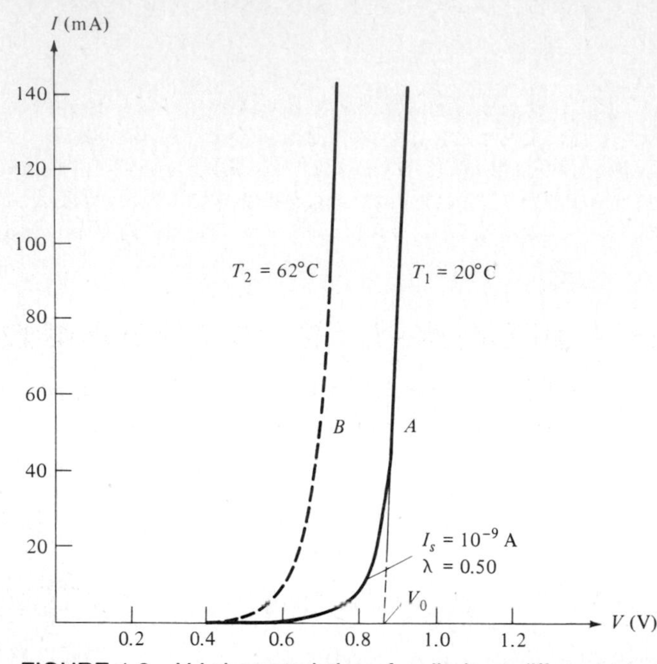

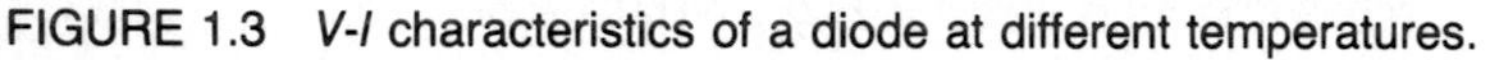
FIGURE 1.3 *V-I* characteristics of a diode at different temperatures.

and

$$I = 1.28 \times 10^{-7} e^{0.5(0.9)/0.027} = 1.28 \times 10^{-7} e^{16.7} = 2.29 \text{ A}$$

$$V_0(20° + \Delta T) \simeq V_0(20°) - 0.002\,\Delta T = 0.86 - 0.08 \simeq 0.78 \text{ V}$$

A plot of the *V-I* characteristics of this diode for the increased temperatures is shown in Fig. 1.3, curve *B*. Note the shift that is due to temperature increase.

Note that a small change in voltage can cause a large current change in the diodes. Specifically,

$$\frac{I_2}{I_1} = e^{\lambda(V_2 - V_1)/V_t} \tag{1.4a}$$

or

$$V_2 - V_1 = \frac{1}{\lambda} V_t \ln\frac{I_2}{I_1} = \frac{2.3V_t}{\lambda} \log\frac{I_2}{I_1} \tag{1.4b}$$

From Eq. (1.4*b*) a factor-of-10 change in current, i.e., $I_2 = 10I_1$, yields

$$V_2 - V_1 = \frac{2.3V_t}{\lambda} \simeq 0.115 \text{ V} \qquad (\text{for } \lambda = 0.5 \text{ and } V_t = 25 \text{ mV})$$

From (1.3*a*) we can write

$$V = \frac{V_t}{\lambda} \ln \frac{I}{I_s} = \frac{2.3V_t}{\lambda} \log \frac{I}{I_s} \tag{1.4c}$$

A plot of the curve described by (1.4*c*) is linear on semilog paper. This linear relationship usually holds for the range of the diode currents from 0.1 nA to about 1 mA. At high forward voltages ($V > 0.8$ V) other effects such as high-level injection effect and bulk ohmic effect take place, whereas at low currents ($I \leq 0.1$ nA) generation and recombination of electrons and holes take place (see Fig. A.4). From the linear plot (1.4*c*), on a logarithmic scale, one can determine the value of I_s.

EXERCISE 1.1

A Si diode has $I_s = 5$ nA and $V_0 = 0.75$ V at room temperature (300 K). Determine I_s and V_0 if the temperature is increased by (*a*) 20°C and (*b*) 50°C.

Ans (*a*) $I_s = 50.4$ nA, $V_0 = 0.71$ V; (*b*) $I_s = 1.61\ \mu$A, $V_0 = 0.65$ V

EXERCISE 1.2

For the 1N4001 diode, determine the change in voltage that will cause a 100-fold increase in the current.

Ans 0.2 V

EXERCISE 1.3

The measured *V-I* data of a certain diode are given as follows: $V = 0.50$ V, $I = 5\ \mu$A; $V = 0.55$ V, $I = 22\ \mu$A; $V = 0.60$ V, $I = 0.1$ mA; $V = 0.65$ V, $I = 0.3$ mA. Determine the value of I_s. (*Hint:* Plot the data on semilog paper.)

Ans $I_s = 5.4 \times 10^{-10}$ A

EXERCISE 1.4

Determine the *V-I* relation of the circuit in Fig. E1.4 if the diodes are identical. Assume $\lambda = 1$.

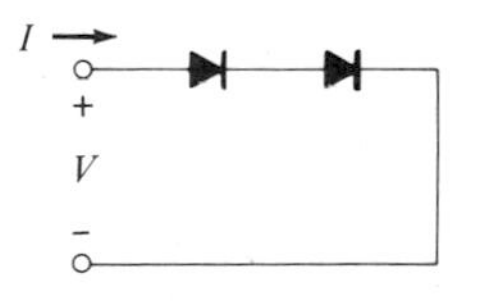

FIGURE E1.4

Ans $I = I_s(e^{V/2V_t} - 1)$

Because of the nonlinear nature of the *V-I* characteristics of diodes, circuit analysis for diode circuits with paper and pencil is not simple. In simple circuits, such as those using a single diode, the diode *V-I* characteristics can be used in a graphical analysis. If many diodes are present and the exponential nature of the *V-I* characteristics must be used, computer-aided analysis is mandated. However, if piecewise-linear approximation is used, the situation can be handled without too much difficulty. Before we delve into piecewise-linear analysis, we digress to review the Thevenin and Norton equivalent theorems in the next section.

1.2 THEVENIN EQUIVALENT NETWORK

The reader is assumed to have been exposed to this important theorem [1]. The Thevenin theorem is a powerful tool in simplifying analysis of complicated circuits. We shall briefly review it here since it is used extensively in this text.

Consider a general resistive circuit that includes a nonlinear two-terminal device. The network can be represented as shown in Fig. 1.4*a*, where *a* and *a'* denote the two terminals of the nonlinear device B, and A represents the rest of the network, which contains *only* linear circuit elements. The complicated *linear network* A can be replaced by its Thevenin equivalent circuit at terminals *a*, *a'* as shown in Fig. 1.4*b*. Note that the one-port network represented by B is *arbitrary* and can be linear or nonlinear. We emphasize that A *must* be linear and B is *arbitrary*. In Fig. 1.4*b*, $V_{eq} = V_{aa'}$ of the linear network A when the terminals *a* and *a'* are *open-circuited*. In Fig. 1.4*c*, $I_{eq} = I_{sc}$ is the current that flows from point *a* to *a'* if a wire is connected

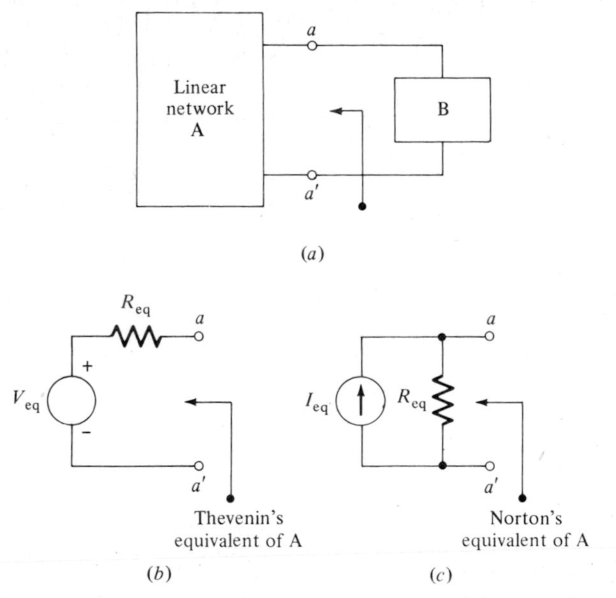

FIGURE 1.4 (*a*) Interconnection of an arbitrary nonlinear one-port B with a linear circuit A. (*b*) Thevenin's equivalent circuit. (*c*) Norton's equivalent circuit.

to *short-circuit* the terminals a and a' of network A. R_{eq} is simply determined from V_{eq} and I_{eq}; i.e., $R_{eq} = V_{eq}/I_{eq}$.

Figure 1.4*b* is called the *Thevenin equivalent representation* of the linear network A. Figure 1.4*c* is called the *Norton equivalent circuit* of network A. We shall illustrate the above through a few examples.

EXAMPLE 1.2

Consider the nonlinear circuit shown in Fig. 1.5*a*. The linear portion of the network as seen from terminals a and a' is to be replaced by the Thevenin equivalent circuit.

The open-circuit voltage across a, a' (i.e., the diode disconnected) is first found by simple analysis. Since the terminals a and a' are open-circuited, no current flows through the 25-Ω resistors, and $V_{ba'} = V_{aa'}$. Applying Kirchhoff's voltage law (KVL) around the loop yields

$$3 - 1 = I_1(50 + 50) \qquad \text{or} \qquad I_1 = 20 \text{ mA}$$

Hence $V_{ba'} = 1 + 0.02(50) = 2 \text{ V} = V_{aa'}$. Thus

$$V_{eq} = V_{aa'}\Big|_{\text{open circuit}} = 2 \text{ V}$$

Next, short-circuit the terminals a and a' by connecting a wire and determine the short-circuit current that flows from point a to a' through the wire. This

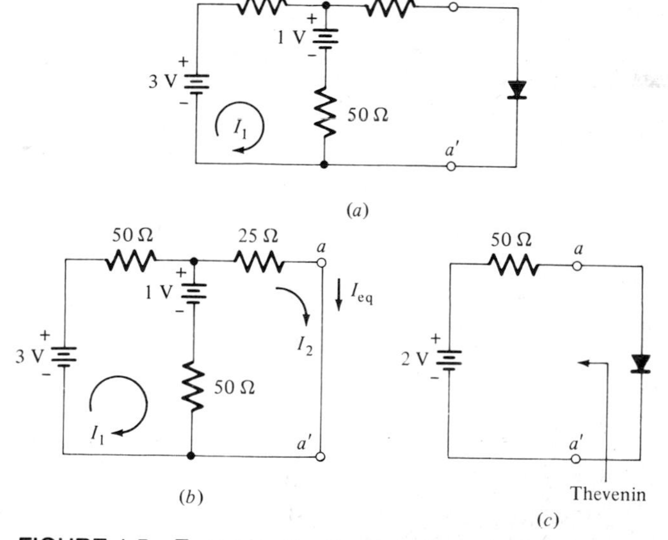

FIGURE 1.5 Example circuit with independent sources.

process yields a circuit shown in Fig. 1.5*b*. Analysis of this circuit is readily made by writing two loop equations, namely,

$$3 - 1 = I_1(50 + 50) - I_2(50)$$

$$1 + I_1(50) = I_2(50 + 25)$$

From the preceding equations we find $I_2 = 4$ mA or $I_{eq} = 4$ mA. Hence $R_{eq} = V_{eq}/I_{eq} = 2\text{ V}/4\text{ mA} = 50\ \Omega$. The Thevenin equivalent of Fig. 1.5*a* by looking to the left of terminals *a* and *a'* is shown in Fig. 1.5*c*.

Note that when there are *independent sources only* such as in Example 1.2 (two batteries present), one could set the voltage sources equal to zero (i.e., short-circuit each battery) and determine the impedance of the linear network at terminals *a*, *a'*, which is R_{eq}.

For this example, with *a* and *a'* open-circuited and the two batteries set to zero, the impedance looking into the terminals *a*, *a'* is

$$R_{eq} = 25 + 50 \| 50 = 50\ \Omega$$

(where $\|$ denotes in parallel with).

If independent current sources are present in the circuit, they are open-circuited for such calculations.

EXAMPLE 1.3

For the circuit shown in Fig. 1.6*a*, determine the Thevenin and the Norton equivalent circuits at the terminals *a*, *a'*.

Note that we have *only* independent sources in the circuit, and R_{eq} is determined by open-circuiting the current source and short-circuiting the voltage source. By inspection $R_{eq} = (2\text{ k}\Omega \| 2\text{ k}\Omega) + 1\text{ k}\Omega = 2\text{ k}\Omega$. To find V_{eq} we can use superposition since the network is linear. Recall that the superposition principle allows us to treat each source separately and determine the response due to each input, and the total response is the sum of the individual responses. Hence V_{eq} due to the current source is obtained by setting the voltage source equal to zero and determining the open-circuit voltage $V_{aa'}$. This process yields $V_{eq1} = 10^{-3}(10^3) = 1$ V. Similarly, V_{eq} due to the voltage source is obtained by open-circuiting the current source and determining the open-circuit voltage $V_{aa'}$. This process yields

$$v_{eq2} = -\left(\frac{2 \times 10^3}{2 \times 10^3 + 2 \times 10^3}\right) V_1 \sin \omega t = -\frac{V_1}{2} \sin \omega t$$

Note that a lowercase letter is used for an alternating-current (ac) signal. By superposition we have

$$V_{eq} = V_{eq1} + v_{eq2} = 1 - \frac{V_1}{2} \sin \omega t$$

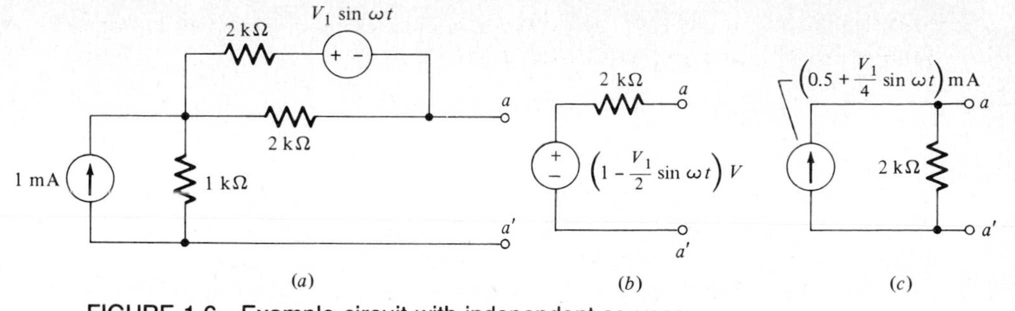

FIGURE 1.6 Example circuit with independent sources.

The resulting Thevenin and the Norton equivalent circuits are shown in Fig. 1.6*b* and *c*, respectively.

We shall next consider controlled sources and their representation and some examples, as they are often encountered in electronic circuits.

1.3 CONTROLLED SOURCES

There are two basic types of sources: *independent* sources and *dependent* or controlled sources. We have already considered independent sources. The graphic symbols and references for independent voltage and current sources are shown again in Fig. 1.7*a*. In this text, we shall use the polarity signs associated with the voltage sources. When only one polarity is indicated, it is with respect to ground (ground is *always* at zero voltage). We shall use arrows only to indicate the direction of current flow of the independent and dependent sources as in Fig. 1.7*a* and *b*.

The graphic symbols used in this book for dependent (controlled) voltage and current sources are shown in Fig. 1.7*b*. This convention of using different symbols is used to emphasize the difference between the dependent and independent sources.

The four *basic idealized* controlled-source representations are shown in Fig. 1.8. Note that these are basically three-terminal circuits. For the voltage-controlled voltage source (VCVS) in Fig. 1.8*a*, the input is a voltage, which may be dependent or

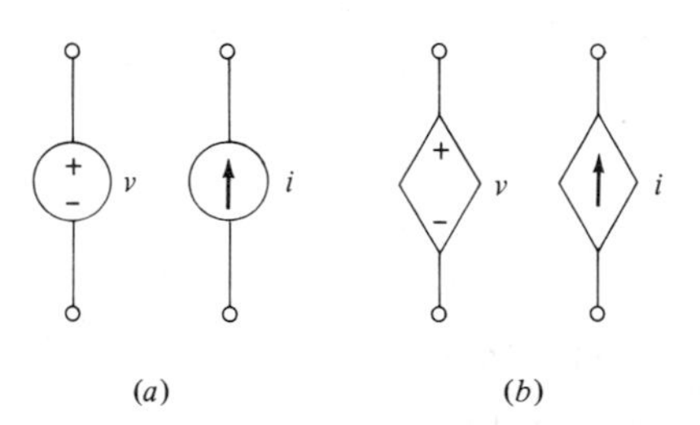

FIGURE 1.7 (a) Independent-voltage-source and independent-current-source representations. (b) Dependent, or controlled, voltage-source and current-source representations.

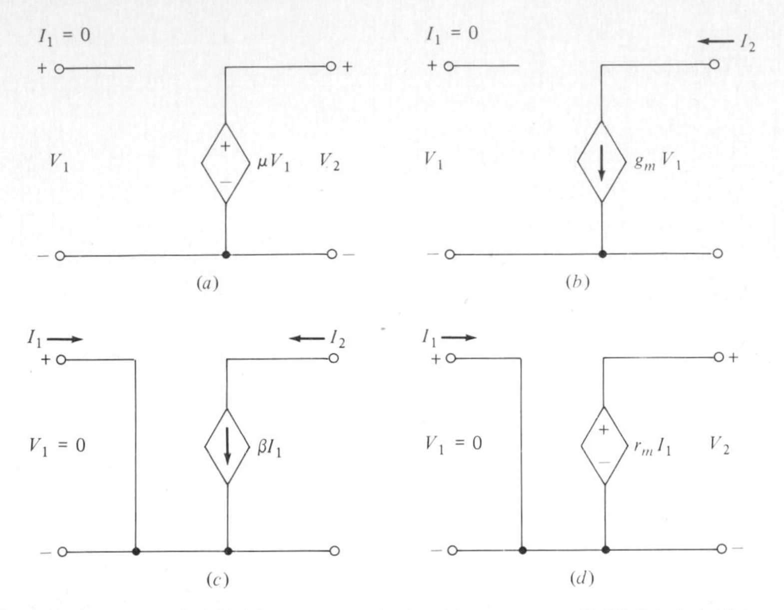

FIGURE 1.8 (*a*) Voltage-controlled voltage source (VCVS). (*b*) Voltage-controlled current source (VCCS). (*c*) Current-controlled current source (CCCS). (*d*) Current-controlled voltage source (CCVS).

independent, and the output is a dependent voltage source and $V_2 = \mu V_1$. For $\mu > 1$ we have an ideal voltage amplifier. The circuit in Fig. 1.8*b* is a voltage-controlled current source (VCCS). The circuit in Fig. 1.8*c* is a current-controlled current source (CCCS). Note that $I_2 = \beta I_1$, and for $\beta > 1$ we have an ideal current amplifier. Finally, in Fig. 1.8*d* we have a current-controlled voltage source (CCVS). The controlled-source elements are very important in electronic circuits and are basic constituents of electronic-device models. Although these models are idealizations, in conjunction with appropriate *RLC* elements they can adequately represent actual electronic devices in the linear operating mode. Controlled sources can be used to represent active as well as passive elements. For example, the *ideal transformer* in Fig. 1.9*a* may be represented by the controlled-source models shown in Fig. 1.9*b*. The *V-I* relations of an ideal transformer are given by

$$v_1 = \frac{n_1}{n_2} v_2 \tag{1.5a}$$

$$i_2 = \frac{n_1}{n_2} i_1 \tag{1.5b}$$

Equation (1.5) also describes the circuit in Fig. 1.9*b*. (superposition)

The short-circuiting and open-circuiting of sources *cannot be done* if there are controlled sources in the linear network. Controlled sources, as we shall see, occur

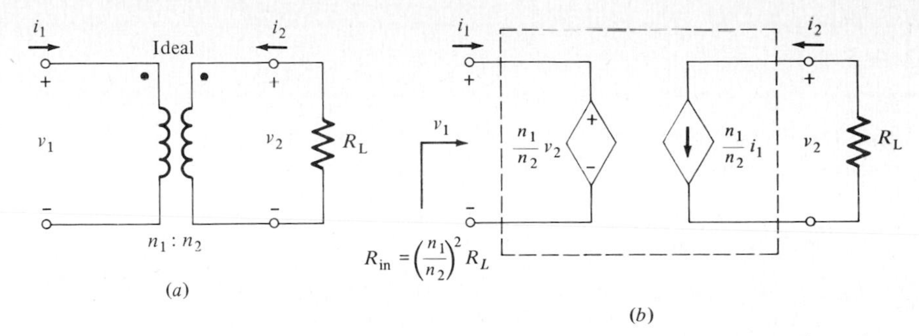

FIGURE 1.9 (*a*) An ideal transformer. (*b*) Controlled-source representation of an ideal transformer.

frequently in the analysis of electronic circuits. When controlled sources are present in the circuit, $R_{eq} = V_{eq}/I_{eq}$ is determined by the procedure given earlier: Determine $V_{eq} = V_{aa'}$ when the terminals a, a' are open-circuited; then determine I_{eq} the current from a to a' when the terminals a, a' are short-circuited.

EXAMPLE 1.3

Find the Thevenin equivalent for the circuit shown in Fig. 1.10*a*. Note that in this case we have a voltage-dependent current source.

When the terminals a, $a'$are open-circuited, we have

$$V_{eq} = V_{aa'}\Big|_{\text{open circuit } a,\, a'} = -g_m V_1 R_3 = -g_m R_3 \frac{R_2}{R_1 + R_2} V_s$$

$$I_{eq} = I_s \text{ (from } a \text{ to } a')\Big|_{\text{short circuit } a,\, a'} = -g_m V_1 = -g_m \frac{R_2}{R_1 + R_2} V_s$$

Hence

$$R_{eq} = \frac{V_{eq}}{I_{eq}} = R_3$$

as shown in Fig. 1.10*b*.

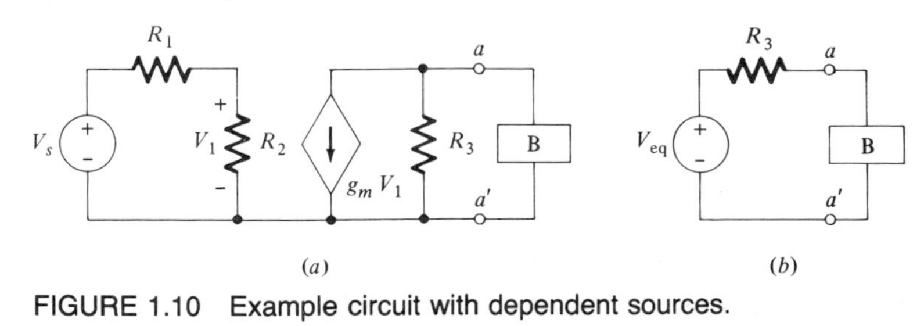

FIGURE 1.10 Example circuit with dependent sources.

EXAMPLE 1.4

Find the Thevenin equivalent for the circuit shown in Fig. 1.11*a*. From the circuit under open-circuit conditions, i.e., *a-a'* open, we have by inspection:

$$V_{aa'} = \frac{R_2}{R_1 + R_2} V_1 + (\alpha I_1)R_3$$

$$= \frac{R_2}{R_1 + R_2} V_1 + \frac{\alpha V_1}{R_1 + R_2} R_3 = V_1 \frac{R_2 + \alpha R_3}{R_1 + R_2} = V_{eq}$$

Under short-circuit conditions, i.e., *a-a'* short-circuited, we determine the current I_2 through *a-a'* which is I_{eq}. To find I_2, change the current source αI_1 and R_3 into a voltage source $\alpha I_1 R_3$ in series with R_3 as shown in Fig. 1.11*b* and write the two loop equations, namely:

$$V_1 = I_1 R_1 + (I_1 - I_2)R_2$$

$$\alpha I_1 R_3 + (I_1 - I_2)R_2 = R_3 I_2$$

From the above equations we solve for I_2, hence I_{eq}. The result is

$$I_2 = \frac{(\alpha R_3 + R_2)V_1}{R_1(R_2 + R_3) + R_2 R_3(1 - \alpha)} = I_{eq}$$

The Thevenin equivalent resistor R_{eq} is then determined from

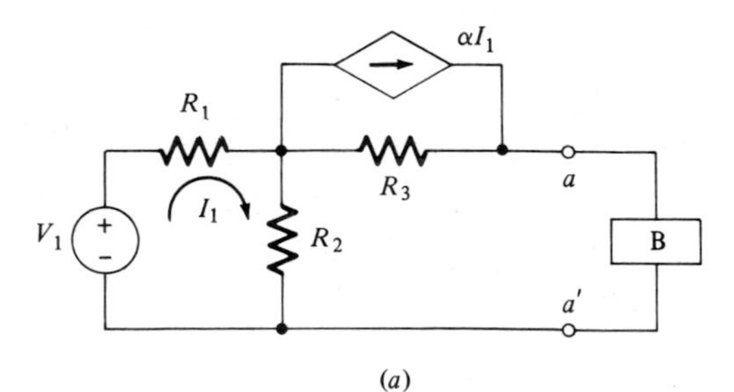

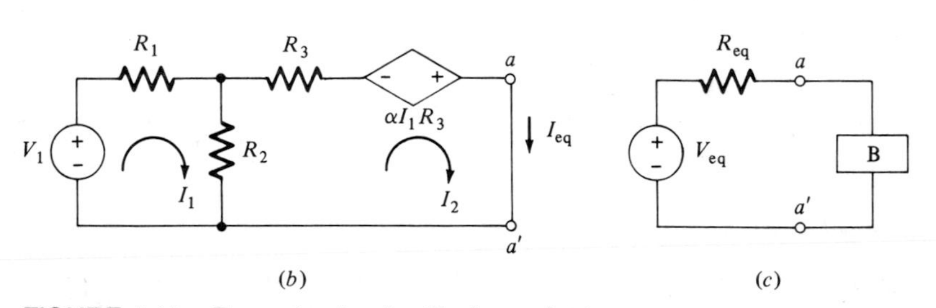

FIGURE 1.11 Example circuit with dependent sources.

$$R_{eq} = \frac{V_{eq}}{I_{eq}} = \frac{R_1(R_2 + R_3) + R_2R_3(1 - \alpha)}{R_1 + R_2}$$

Note that we cannot open-circuit the dependent current source, as it depends on the current I_1.

EXERCISE 1.5

Find the Thevenin equivalent circuit of Fig. E1.5. The voltage source is a sinusoidal signal $v_i = V_1 \sin \omega t$.

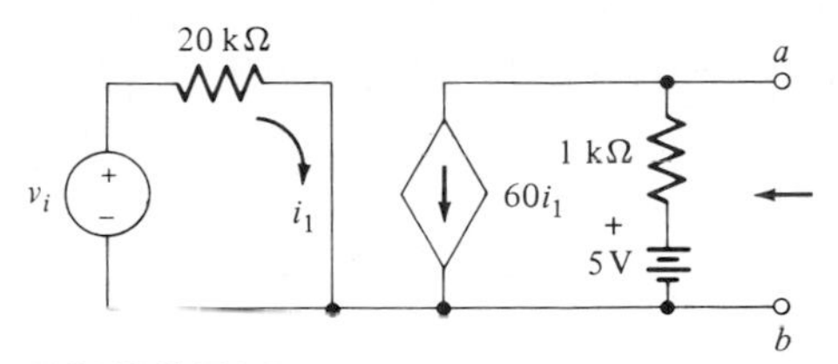

FIGURE E1.5

Ans $R_{eq} = 1\ \text{k}\Omega,\ v_{EQ} = 5 - 3V_1 \sin \omega t$

EXERCISE 1.6

For the circuit shown in Fig. E1.6, find the values of the Thevenin equivalent circuit as seen from the terminals *a*, *b*.

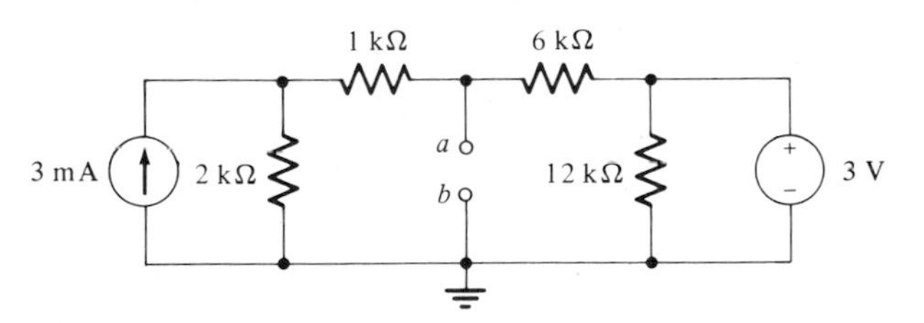

FIGURE E1.6

Ans $R_{eq} = 2\ \text{k}\Omega,\ V_{eq} = 5\ \text{V}$

1.4 GRAPHICAL ANALYSIS

Consider a circuit consisting of independent sources, resistors, and a one-port non-linear device such as a diode. This type of circuit can be readily analyzed graphically.

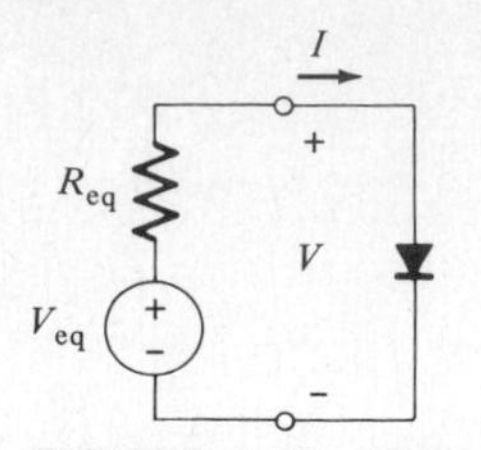

FIGURE 1.12 Thevenin equivalent of a resistive circuit with batteries and a single diode.

Pull out the diode terminals and represent the rest of the circuit across the diode by the Thevenin equivalent, as shown in Fig. 1.12. For the sake of clarity we shall consider separately the following three cases: a dc source, an ac source, and a combination of both. The equilibrium equations for the circuit are as follows: For the nonlinear diode,

$$I = I_s(e^{\lambda V/V_t} - 1) \tag{1.6}$$

for the Thevenin one-port,

$$V = V_{eq} - R_{eq}I \tag{1.7}$$

Note that (1.6) is simply the *V-I* characteristic of the diode and (1.7) is just the KVL equation. The solution of (1.6) and (1.7) is designated as I_q and V_q. In a simple case, such as the one shown in Fig. 1.12, the analysis is best performed graphically.

A. Constant (DC) Source

We shall consider now the analysis of the circuit in Fig. 1.5 utilizing the diode characteristic. Note that in this case we have already determined the Thevenin equivalent circuit with $V_{eq} = 2$ V and $R_{eq} = 50\ \Omega$. Let us assume that the diode with the *V-I* characteristic shown in Fig. 1.1*d* is used in the circuit.

The current through the diode and the voltage across the diode is determined as follows: The straight line corresponding to (1.7) is simply the equation of a straight line ($y = mx + b$) with $y = I$ and $x = V$ since $I = -V/R_{eq} + V_{eq}/R_{eq}$. In this case $m = -1/R_{eq}$ and $b = V_{eq}/R_{eq}$. The straight line is thus drawn on the same graph and scale as the device characteristic. The straight line is usually referred to as the *dc load line*. The intersection of the two lines yields the solution. The solution or the operating point is often called the *quiescent point Q*. In the above example, the quiescent point is $I_q \simeq 20$ mA, $V_q = 1.0$ V, as shown in Fig. 1.13.

B. Time-Varying (AC) Source

We shall next consider a time-varying source, as contrasted to the dc source of the previous case. The graphical technique can be applied in this case as well. When time-varying sources are present, the Thevenin equivalent voltage source will, of course, be time varying as well, as in Example 1.3. Consider the circuit in Fig. 1.12 where v_{eq} is time varying, such as a sinusoidal waveform. Let $v_{eq} = 1.75 \sin \omega t$ and $R_{eq} = 50\ \Omega$. Note that we use lowercase letters for ac signals.

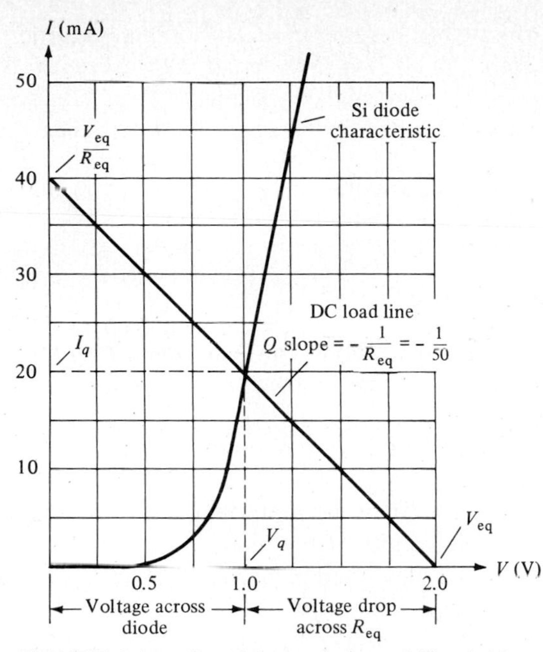

FIGURE 1.13 Graphical solution of Fig. 1.12.

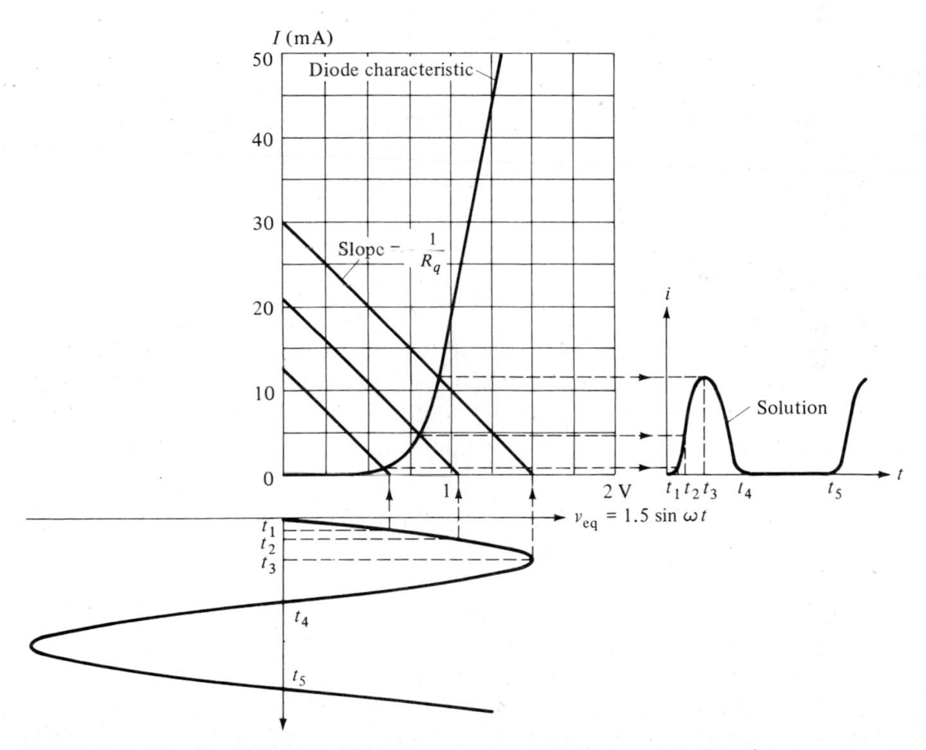

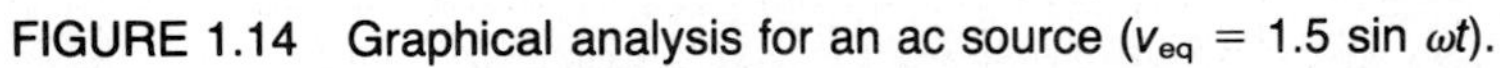
FIGURE 1.14 Graphical analysis for an ac source ($v_{eq} = 1.5 \sin \omega t$).

Graphical analysis is performed as before for several instants of time. At each instant of time the voltage value v_{eq} in (1.7) is known and the slope of the load line is fixed, namely, $-1/R_{eq}$. A point-by-point analysis at times $t_1, t_2, \ldots, t_n$ is shown in Fig. 1.14. Note that the solution $i(t)$ is heavily distorted in this nonlinear circuit for such a large signal.

The technique is, of course, applicable for any time-varying source not necessarily sinusoidal.

C. Simultaneous DC and AC Sources

Consider now the case where v_{EQ} is a superposition of an ac and a dc source; i.e., let $v_{EQ} = 1 + 0.75 \sin \omega t$. Note that a lowercase letter with capital letter subscript will be used for total ac and dc quantities.

The application of the technique for this case is similar and is illustrated in Fig. 1.15 without any further elaboration.

Note that in Figs. 1.14 and 1.15, $i(t)$ is not sinusoidal and is heavily distorted for such large signals due to the nonlinearity of the diode.

Graphical method of analysis is widely used in electronic circuits, if the analysis

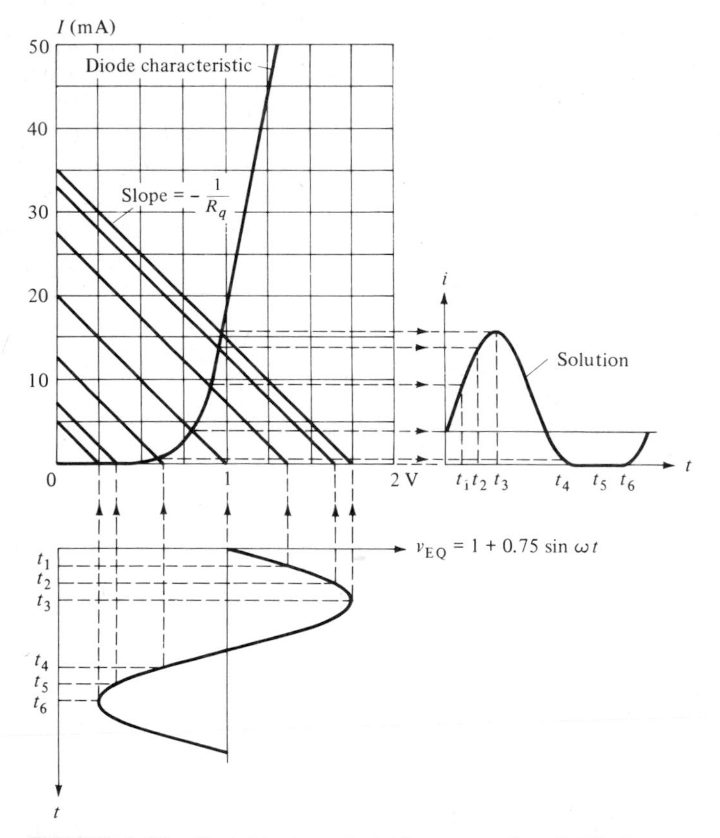

FIGURE 1.15 Graphical analysis for $v_{EQ} = 1 + 0.75 \sin \omega t$.

is done by hand, and we will have several occasions to encounter this method. The concepts of load line and Q point are thus fundamental and are applied for three-terminal devices in the next two chapters. It should be noted, however, that most analysis of large circuits is done using computer programs where load line analysis plays no point at all.

A general nonlinear circuit may alternatively be analyzed by piecewise-linear analysis. In piecewise-linear analysis, the nonlinear device is approximated by a piecewise-linear model using ideal diodes, and then linear analysis is used.

1.5 IDEAL DIODE

For piecewise-linear analysis, and in fact in many large-signal diode circuits applications it is convenient to introduce the *ideal diode,* as a first-order approximation.

The representation of the ideal diode and its characteristic is shown in Fig. 1.16*a*, *b*. (Note the distinction in the symbols, a *clear arrowhead* is used as a symbol for the *ideal diode*.)

The ideal diode is defined by the following equations:

$$I = 0 \qquad \text{for } V \leq 0 \tag{1.8a}$$

$$V = 0 \qquad \text{for } I \geq 0 \tag{1.8b}$$

From Eq. (1.8) it is seen that for a negative voltage, i.e., any reverse bias, the *ideal diode* reduces to an open circuit ($I = 0$) as shown in Fig. 1.16*c*. For a positive voltage, i.e., any forward bias, it is a closed circuit ($V = 0$) as shown in Fig. 1.16*d*. Thus once the *state* of the ideal diode is known, it can be replaced by either a *closed circuit* or an *open circuit*. Alternatively stated, the ideal diode can be represented by a switch which is OFF (open) for any reverse-biased voltage and ON (closed) for any forward-biased voltage. A combination of ideal diodes, resistors, and batteries provides us with

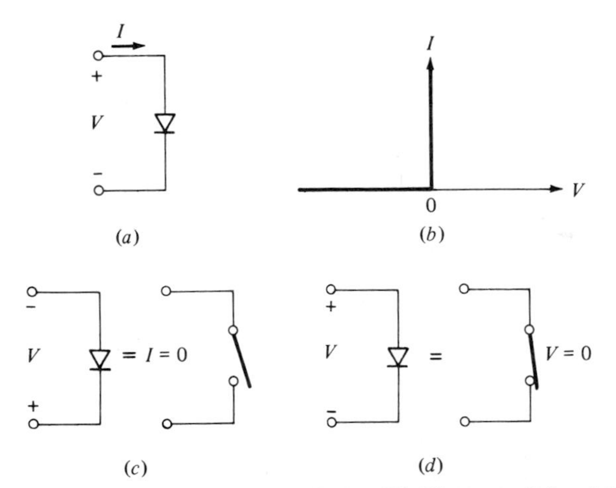

FIGURE 1.16 (*a*) Ideal diode. (*b*) Characteristic of the ideal diode. (*c*) Ideal diode with a reverse bias (an open switch). (*d*) Ideal diode with a forward bias (a closed switch).

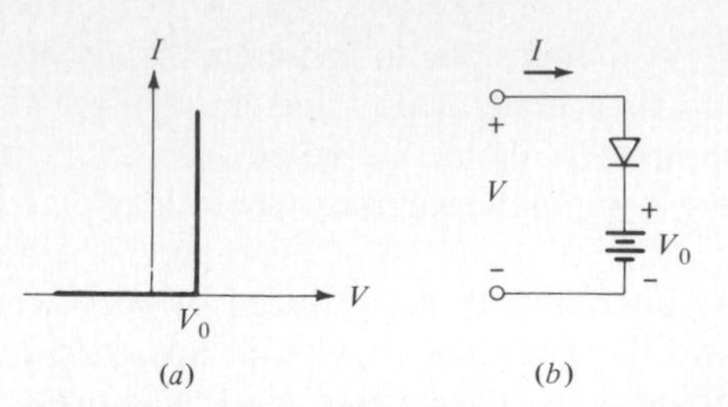

FIGURE 1.17 (*a*) An ideal diode with a voltage drop. (*b*) Circuit model.

the means of approximating nonlinear *V-I* characteristics with piecewise-linear *V-I* characteristics. Linear analysis can then be used for each section of the piecewise-linear approximation.

As an example consider the *V-I* characteristic shown in Fig. 1.17*a*. This *V-I* characteristic is modeled by an ideal diode and a battery as shown in Fig. 1.17*b*. This model may be viewed as a better approximation of an actual diode than that represented by the ideal diode. The model ignores the slope of the *V-I* characteristic and assumes that the voltage drop across the conducting diode is independent of the current flowing through thc diode.

EXERCISE 1.7

If a diode is connected between the terminals *a*, *b*, in Fig. E1.6, find the current through and the voltage across the diode if the diode is

(*a*) Assumed to be ideal.
(*b*) Represented by a constant-voltage-drop model.

Ans (*a*) $V_d = 0$, $I_d = 2.5$ mA; (*b*) $V_d = 0.7$ V, $I_d = 2.15$ mA

EXERCISE 1.8

The circuit shown in Fig. E1.6, with the diode connected at terminals *a*, *b*, can be equivalently drawn as shown in Fig. E1.8. Calculate the exact values of V_d and I_d if

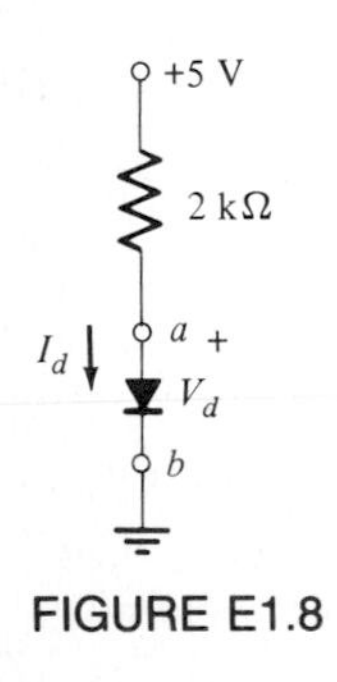

FIGURE E1.8

the diode has a voltage of 0.7 V at a current of 1 mA and $\lambda = 0.5$. [*Hint:* Use iteration in (1.4*b*) with $V_1 = 0.7$ V.]

Ans $V_d = 0.738$ V, $I_d = 2.13$ mA

EXERCISE 1.9

For the circuit shown in Fig. E1.9, assume identical diodes (of Exercise 1.8) and calculate the exact values of I_1 and V_{ab}.

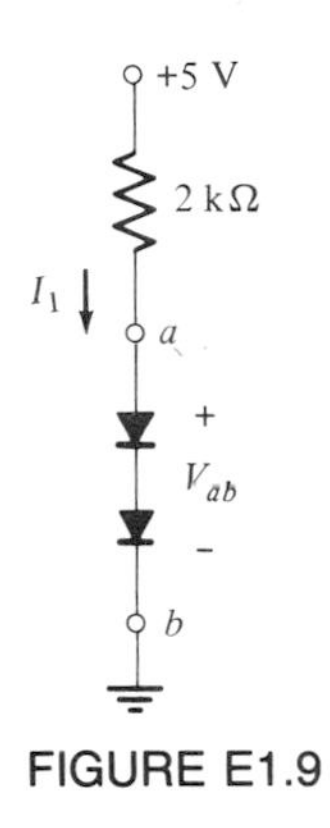

FIGURE E1.9

Ans $I_1 = 1.79$ mA, $V_{ab} = 1.425$ V

1.6 PIECEWISE-LINEAR APPROXIMATION

Piecewise-linear approximation is simply approximating a nonlinear *V-I* characteristic with several linear segments of *V-I* characteristics. The larger the number of segments, the more accurate is the piecewise-linear approximation.

A. Diode Large-Signal Model

The *V-I* characteristic of a junction diode may be approximated by a two-segment piecewise-linear characteristic, as shown in Fig. 1.18*a*. The circuit model is shown in Fig. 1.18*b*, where R_f represents the reciprocal of the slope of the *V-I* characteristic as shown. Notice that the addition of a resistor in the model (Fig. 1.18*b*) provides a better approximation to the *V-I* characteristic of an actual diode than the model in Fig. 1.17*b*. From the model in Fig. 1.18*b*, note that for $V \geq V_0$ the diode acts as a resistor R_f. For $V \leq V_0$ the diode is an open circuit. We shall refer to Fig. 1.18*b* as the *large-signal* piecewise-linear model of a *pn*-junction diode. For a typical diode $V_0 = 0.7$ V. If the *V-I* characteristic of the diode is available, V_0 and R_f of the diode can be readily determined.

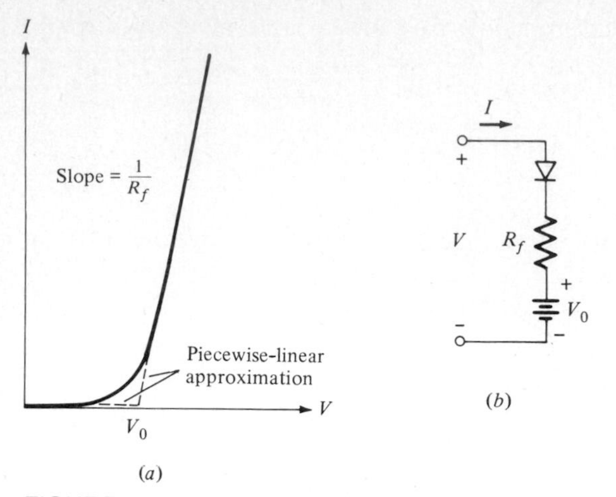

FIGURE 1.18 (*a*) An ideal diode with a voltage drop plus forward resistance. (*b*) Circuit model (large-signal model of a diode in the forward region).

EXAMPLE 1.5

For the circuit shown in Fig. 1.12, $V_{eq} = 2$ V and $R_{eq} = 50\ \Omega$. From the diode characteristic $V_0 \simeq 0.80$, $R_f \simeq 10\ \Omega$. Determine the current through and the voltage across the diode.

For $V \le V_0$, $I = 0$, and (1.7) is not satisfied. Hence, the solution must have $V > V_0$. For $V \ge V_0$ the diode is simply a resistor R_f, and we have to solve the two linear equations

$$I = \frac{1}{R_f}(V - V_0) = 0.10(V - 0.80) \tag{1.9a}$$

$$V = V_{eq} - R_{eq}I = 2 - 50I \tag{1.9b}$$

Solving for I and V gives the result $I = I_q = 20$ mA, $V = V_q = 1.0$ V.

In Sec. 1.4A we performed a graphical analysis to determine the Q point given the V-I characteristic and the Thevenin equivalent of the linear circuit. We shall determine the solution of the same problem using a piecewise-linear method of analysis.

B. Function Generation

For synthesis and generation of functions using resistors, ideal diodes, and batteries, simple piecewise-linear building blocks are used. A few of these are given in Fig. 1.19 for convenience. Using these blocks, a large class of driving-point functions can be generated. A *driving-point* function refers to the V-I relation of a one-port (two-terminal) circuit. We shall consider an example using these blocks.

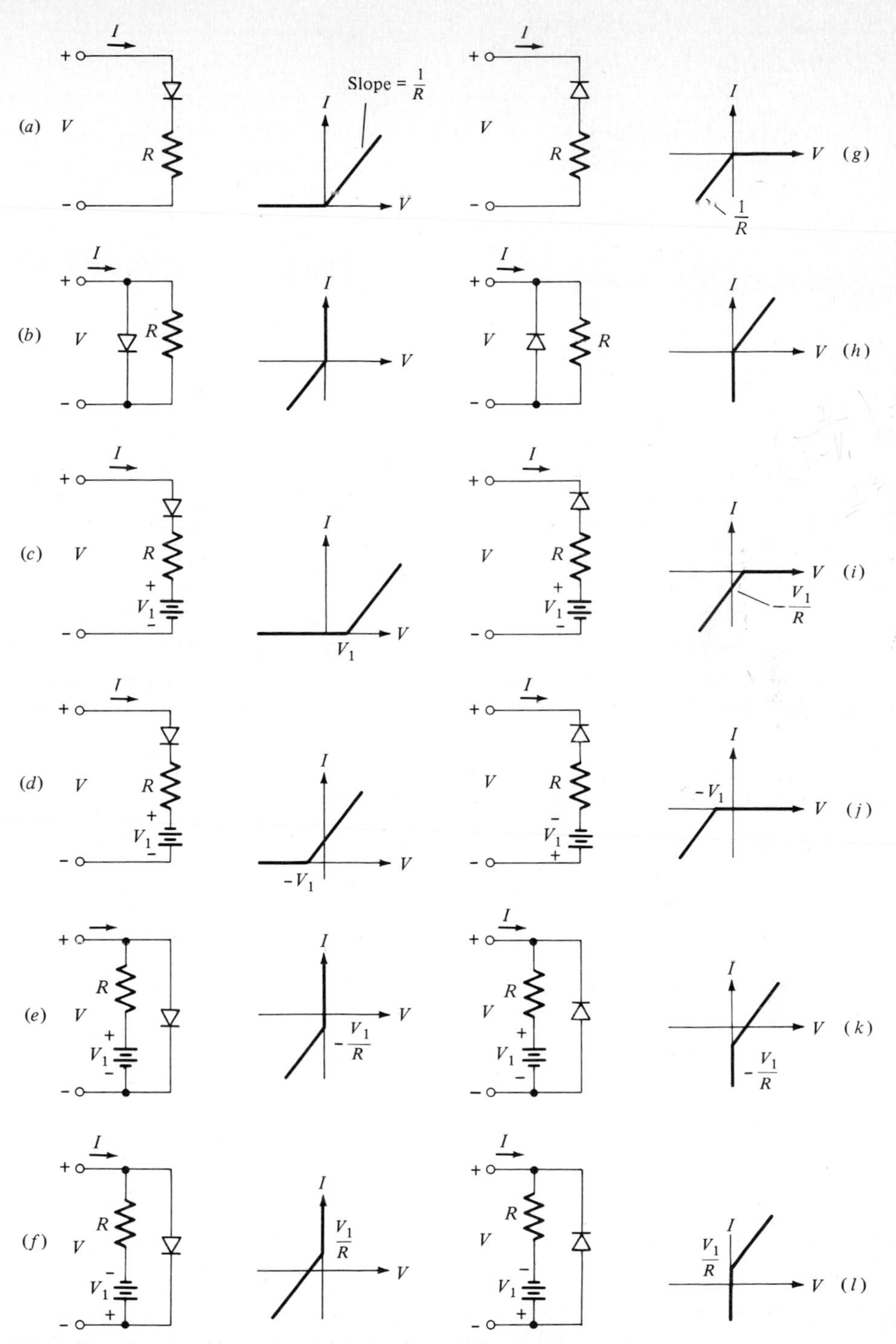

FIGURE 1.19 Various piecewise-linear building bocks.

EXAMPLE 1.6

Consider the nonlinear *V-I* characteristic shown in Fig. 1.20. This characteristic is to be approximated by three segments, as shown by the dashed lines on Fig. 1.20.

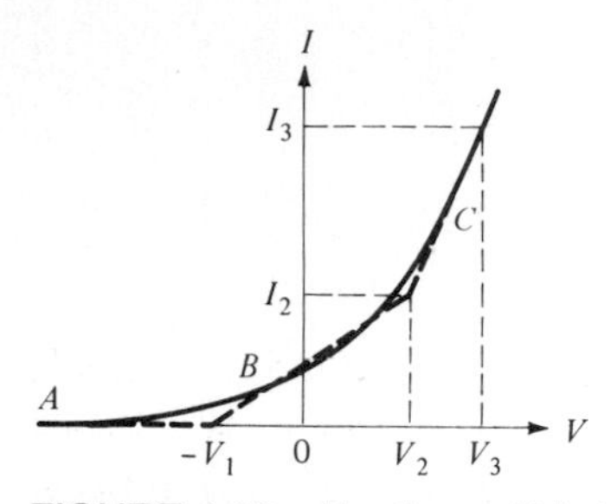

FIGURE 1.20 Nonlinear *V-I* characteristic with its piecewise-linear approximation for Example 1.6.

The model for Fig. 1.20 is obtained as follows: The approximation to segments *A* and *B* are obtained from Fig. 1.19*d* as shown in Fig. 1.21*a*. The *C* segment has a slope (conductance) larger than the *B* segment; hence the conductance must be increased for $V \geq V_2$, which can be done by adding a shunt conductance. The segment realization is readily accomplished by connecting the building block of Fig. 1.19*c* in shunt with the previous circuit. The resulting model is shown in Fig. 1.22*b*. The *V-I* characteristics of the circuit in Fig. 1.22*a* can be readily checked. For $V < -V_1$, D_1 and D_2 are open-circuited and $I = 0$. For $-V_1 < V < V_2$, D_1 is closed-circuited, and D_2 is open-circuited; hence $I = I_a$, and the *V-I* relation corresponds to that of segment *B*. For $V > V_2$, both diodes are short circuits; hence $I = I_a + I_b$, and the equivalent circuit corresponds to that of segment *C*.

Note that we have considered *monotonically increasing V-I* relationships, where the resulting resistors are all positive. It should further be noted that the

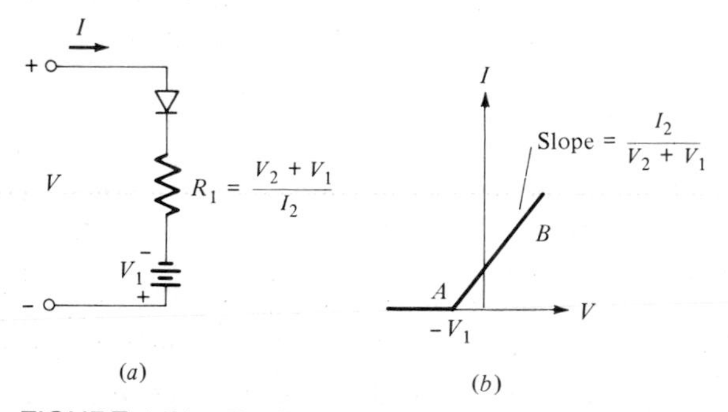

FIGURE 1.21 Realization of piecewise-linear segments *A* and *B*.

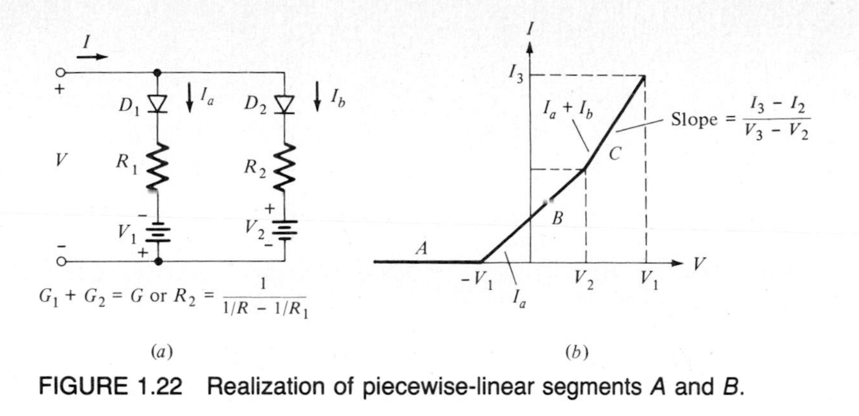

FIGURE 1.22 Realization of piecewise-linear segments *A* and *B*.

number of breakpoints corresponds to the number of diodes. In this example there are two breakpoints at V_1 and V_2, and hence two diodes are required.

EXERCISE 1.10

For the circuit shown in Fig. E1.10, determine the breakpoint (V_1, I_1) in the driving-point function.

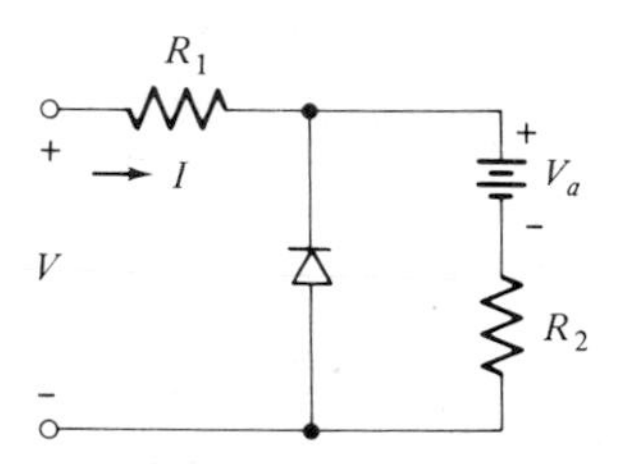

FIGURE E1.10

Ans $V_1 = -(R_1/R_2)V_a$, $I_1 = -V_a/R_2$

1.7 TRANSFER FUNCTION OF DIODE CIRCUITS

A voltage transfer function, which is a plot of the output voltage versus the input voltage, is very useful in determining the manner in which the input voltage controls the output voltage. For example, consider the circuit shown in Fig. 1.23*a*. If we assume an ideal diode, the transfer function is determined as follows. For $V_{in} \leq 0$, the diode is an open switch, and $V_{out} = 0$. For $V_{in} \geq 0$, the diode will conduct and is a closed switch, and $V_{out} = V_{in}$. The plot of V_{out} vs V_{in} for an ideal diode is then as shown in Fig. 1.23*b*. Similarly, if the diode is modeled as an ideal diode with a constant voltage drop V_0, the transfer function is shown in Fig. 1.23*c*. Finally, if the large-signal

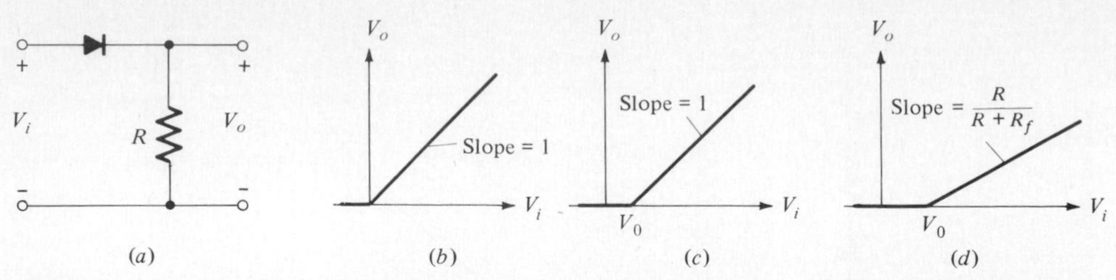

FIGURE 1.23 Transfer characteristic V_o vs V_i: (*a*) using an ideal diode model, (*b*) using an ideal diode with a battery model, (*c*) using an ideal diode plus a forward resistor model.

model of the diode (i.e., an ideal diode plus a battery V_0 and a resistor R_f) is used, the transfer function is shown in Fig. 1.23*d*. Once the transfer function is available, the output waveform can be determined for any given input waveform. Thus the transfer function is *waveform independent*. We shall illustrate the use of the voltage transfer function by the following example.

EXAMPLE 1.7

Determine the transfer function and the output waveform for the circuit shown in Fig. 1.24.

(a) Assume an ideal diode.
(b) Assume an ideal diode plus the cut-in voltage V_0.

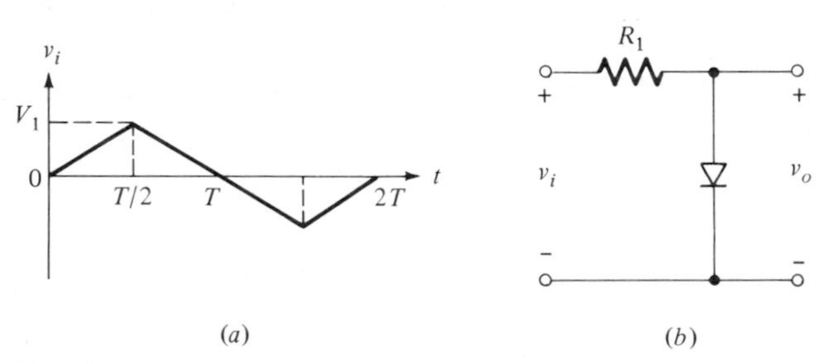

FIGURE 1.24 (*a*) Input waveform. (*b*) Example circuit.

For case (a), the transfer function is shown in Fig. 1.25*b*. The transfer function is obtained as follows: For $V_{in} \leq 0$, the diode is an open switch, and $V_{out} = V_{in}$. For $V_{in} \geq 0$, the diode conducts and is a closed switch; thus $V_{out} = 0$. The output waveform is then as shown in Fig. 1.25*c*.

For case (b) the transfer function is shown in Fig. 1.26*b*. It is obtained as follows: For $V_{in} = V_o$, the diode is an open switch and $V_{out} = V_{in}$. For $V_{in} \geq V_o$, the diode conducts and is a closed switch; thus $V_{out} = V_o$. The output waveform is then as shown in Fig. 1.26*c*.

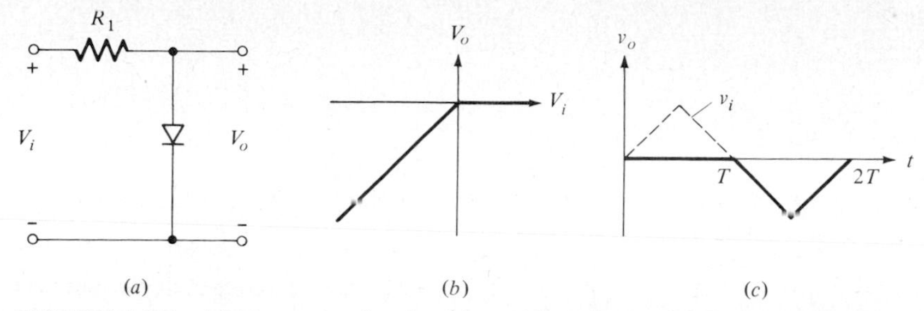

FIGURE 1.25 (*a*) Example circuit with an ideal diode. (*b*) Transfer function. (*c*) Output waveform.

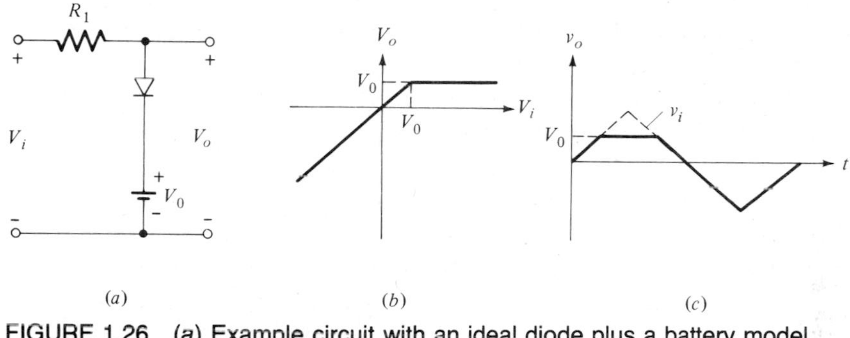

FIGURE 1.26 (*a*) Example circuit with an ideal diode plus a battery model. (*b*) Transfer function. (*c*) Output waveform.

We shall next consider some of the applications of Si diodes. Recall that V_o is typically 0.7 V, and we shall ignore V_o in the following section and thus assume ideal diodes. The inclusion of V_o and R_f of the model in Fig. 1.18 for a real diode should pose no problem. In certain cases where attention should be paid to these parameters we shall point them out.

EXERCISE 1.11

For the circuit shown in Fig. E1.11, the input voltage varies linearly from 0 to 100 V. Determine the output voltage.

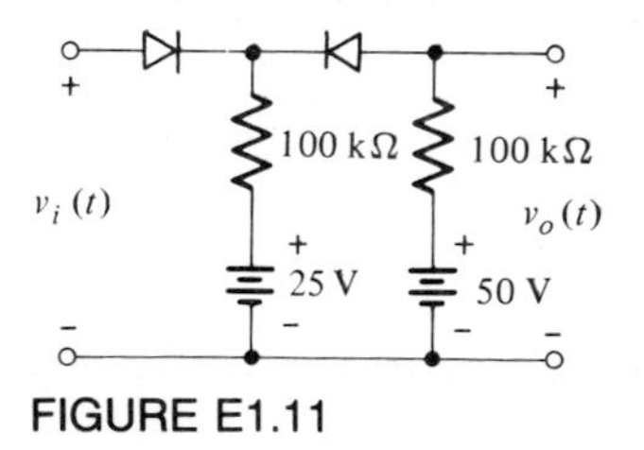

FIGURE E1.11

Ans $v_o(t) = 37.5$ V for $v_i(t) \leq 37.5$ V; $v_o(t) = v_i(t)$ for $37.5\text{ V} \leq v_i(t) \leq 50$ V; $v_o(t) = 50$ V for $v_i(t) \geq 50$ V

EXERCISE 1.12

For the circuit shown in Fig. E1.12, determine the voltage transfer characteristic.

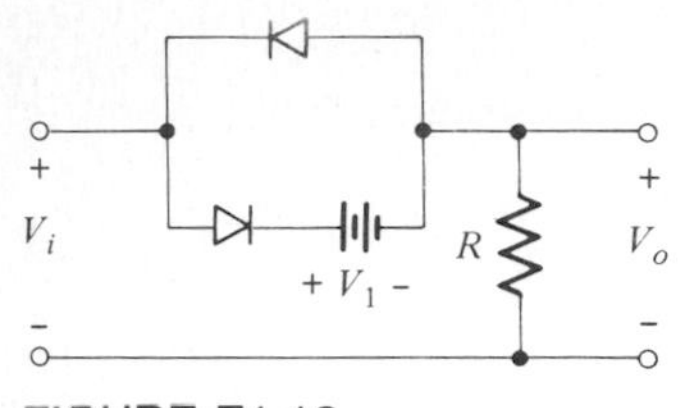

FIGURE E1.12

Ans $V_o = V_i$ for $V_i \leq 0$; $V_o = 0$ for $0 \leq V_i \leq V_1$; $V_o = V_i - V_1$ for $V_i \geq V_1$

EXERCISE 1.13

For the circuit shown in Fig. E1.13, determine the output voltage if $v_i(t) = 10 \sin \omega t$.

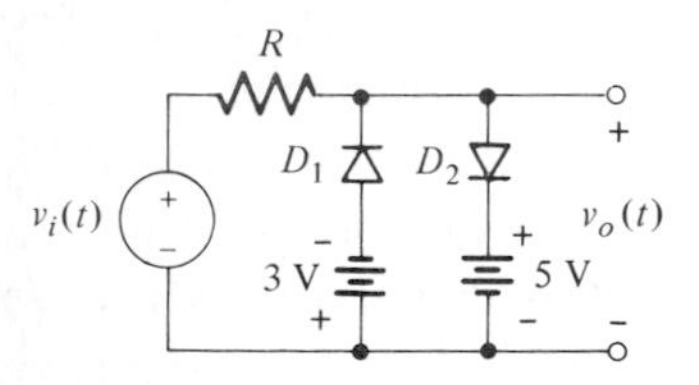

FIGURE E1.13

Ans $v_o(t) = 5$ V for $v_i \geq 5$ V; $v_o(t) = -3$ V for $v_i \leq -3$ V; $v_o(t) = v_i(t)$ all other times

1.8 DIODE APPLICATIONS

Diodes are used in a variety of applications, including rectifiers, detectors, function generators, clippers, waveform shapers, and logic gates. We shall discuss some of these applications in the folowing section.

A. Clipping or Limiting Circuit

The use of a diode in a waveshaping circuit is shown in Fig. 1.27. V_R is any arbitrary dc reference voltage. These circuits are also referred to as limiters since the amplitude of the output is limited to some particular reference-voltage level.

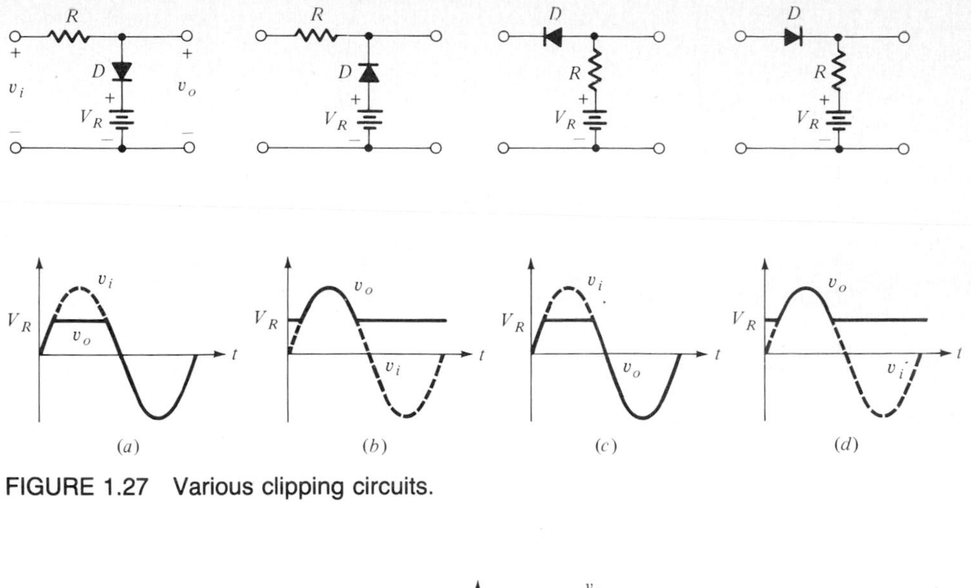

FIGURE 1.27 Various clipping circuits.

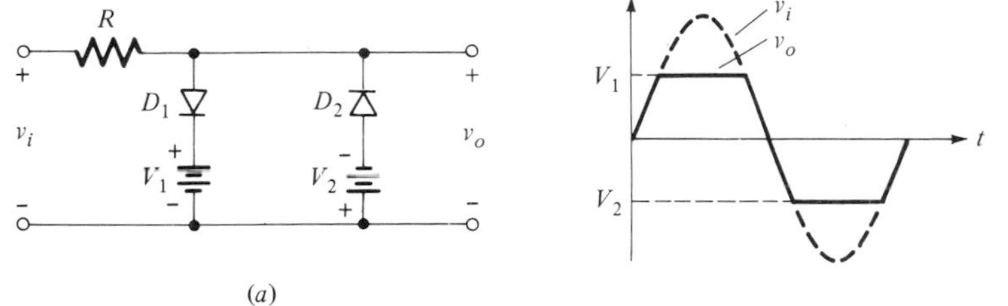

FIGURE 1.28 Double clipper: (*a*) schematic, (*b*) associated waveform.

Note that in Fig. 1.27*a*, for $v_i \geq V_R$ the diode is ON and hence close-circuited, and the output voltage is limited to V_R, i.e., $v_o = V_R$; for $v_i \leq V_R$ the diode is OFF and open-circuited, and hence $v_o = v_i$. For the circuit in Fig. 1.27*b*, for $v_i \geq V_R$ the diode if OFF and open-circuited, and hence $v_o = v_i$; for $v_i \leq V_R$ the diode is ON and short-circuited, and $v_o = V_R$. For $v_i \geq V_R$ the diode in Fig. 1.27*c* is OFF, while the one in Fig. 1.27*d* is ON, and the transfer functions are as shown in the figures.

Two diodes may be used to obtain double clipping as shown in Fig. 1.28. This technique provides an approximate square-wave generation from a sinusoidal waveform. (There are other techniques that result in better square-wave generators and will be considered in later chapters.)

B. Half-wave Rectifiers

A rectifier is a circuit that converts an ac signal into a unidirectional signal. Diodes are extensively used as rectifiers. A half-wave rectifier circuit and its associated input and output waveforms are shown in Fig. 1.29. Note that this circuit is basically a clipper circuit shown in Fig. 1.23 and discussed earlier.

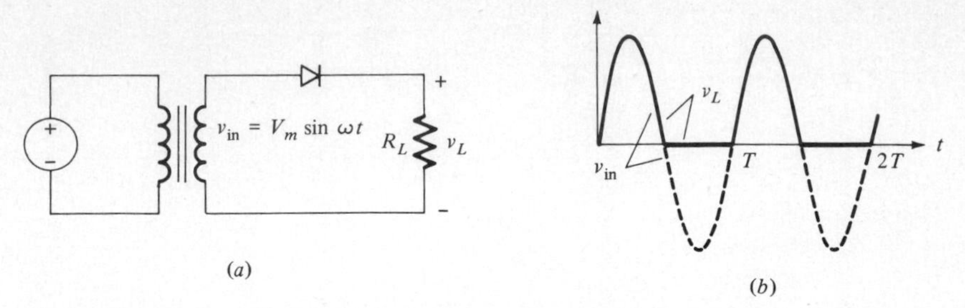

FIGURE 1.29 (*a*) Half-wave rectifier circuit. (*b*) Waveform of the half-wave rectifier.

We shall determine the average (dc) and the root mean square (rms) values of the output waveform of the half-wave rectifier.

The dc voltage V_{DC} is defined as

$$V_{DC} = \frac{1}{T}\int_0^T v(t)\,dt \tag{1.10}$$

We note from Fig. 1.29*b* that $v_0(t) = 0$ for $T/2 < t < T$; hence from (1.10) we have

$$V_{DC} = \frac{1}{T}\int_0^{T/2} V_m \sin \omega t = \frac{V_m}{\omega T}\left(-\cos \omega t \Big|_0^{T/2}\right)$$

$$= \frac{-V_m}{\omega T}\left(\cos\frac{\omega T}{2} - 1\right) \tag{1.11}$$

but

$$\omega = 2\pi f \qquad \text{and} \qquad f = \frac{1}{T} \tag{1.12}$$

Substitution of (1.12) into (1.11) yields

$$V_{DC} = \frac{V_m}{\pi} = 0.318 V_m \tag{1.13}$$

The root-mean-square (rms) value of a periodic waveform is defined by the following relation:

$$V_{rms} = \left[\frac{1}{T}\int_0^T v^2(t)\,dt\right]^{1/2} \tag{1.14}$$

For a sinusoidal voltage, $v = V_m \sin \omega t$, its value is

$$V_{rms} = \frac{V_m}{\sqrt{2}} \tag{1.15}$$

Thus the output V_{rms} for a half-wave rectifier with input $V_m \sin \omega t$ is

$$V_{rms} = \left[\frac{1}{T}\int_0^{T/2} (V_m \sin \omega t)^2\,dt\right]^{1/2} = \frac{1}{\sqrt{2}}\frac{V_m}{\sqrt{2}} = \frac{V_m}{2} \tag{1.16}$$

The rectification efficiency η is defined as

$$\eta = \frac{P_{dc}}{P_{ac}} \tag{1.17}$$

where P_{dc} is the dc power and P_{ac} is the ac power. For a half-wave rectifier in Fig. 1.29, from (1.13), (1.16), and (1.17) we find

$$\eta_{\text{half-wave}} = \frac{(V_m/\pi)^2/R_L}{(V_m^2)/4R_L} = \frac{4}{\pi^2} = 40.6 \text{ percent} \tag{1.18}$$

In an actual half-wave rectifier the efficiency is of course less than (1.18) because of the power loss in the resistance R_f of the actual diode.

C. Full-Wave Rectifiers

A full rectifier circuit using a center-tapped transformer is shown in Fig. 1.30. The center-tapped transformer is assumed to be ideal so that $v_1 = v_i = V_m \sin \omega t$. Note that each half of the transformer with its associated diode acts as a half-wave rectifier. The output voltage of the full-wave rectifier is shown in Fig. 1.30*b*. Instead of using a center-tapped transformer, we can use four diodes in order to obtain a full-wave rectifier. Such a circuit is shown in Fig. 1.31*a*. Note that the input signal v_i shown in Fig. 1.31*a* is the output of a supply transformer. A transformer is usually used in rectifiers for protection since it is useful for voltage changing.

Note that in Fig. 1.31*a* the current is steered by two diodes which conduct simultaneously, and the current can only flow from *a* to *b* through the load. In this case again both half-cycles of the sinusoidal input are rectified as shown in Fig. 1.31*b*, and the circuit is called a *bridge rectifier*.

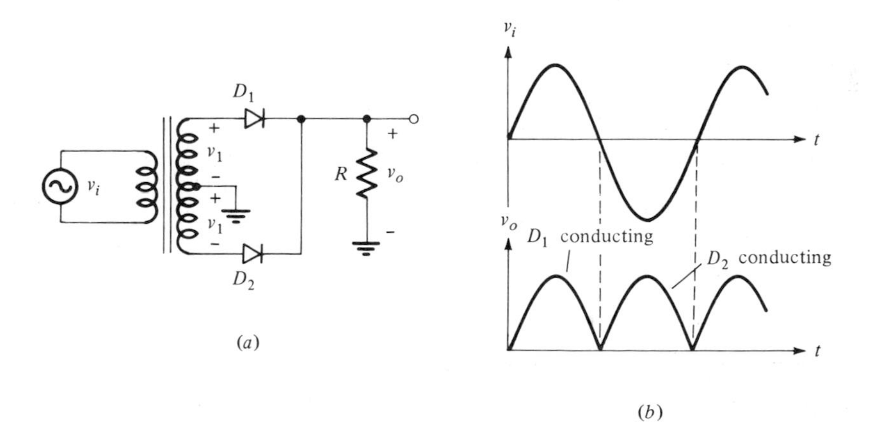

FIGURE 1.30 (*a*) Full-wave rectifier circuit. (*b*) Waveform of the full-wave rectifier.

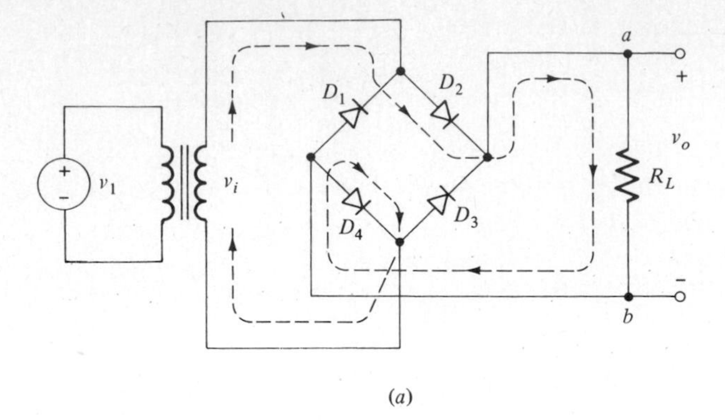

(a)

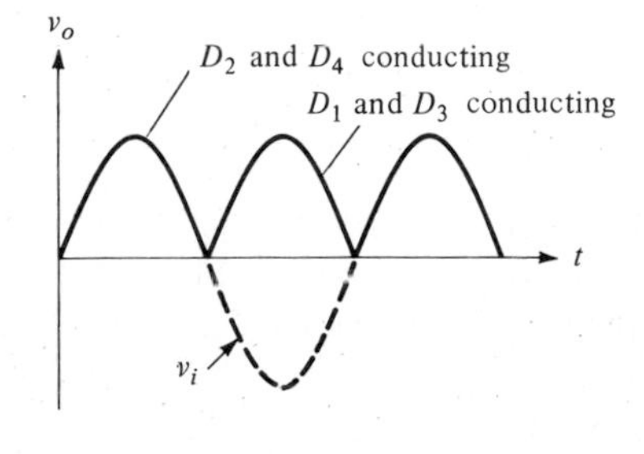

(b)

FIGURE 1.31 (a) Bridge rectifier circuit. (b) Waveform of the bridge rectifier circuit.

From the definitions of dc and rms voltages we see immediately that

$$(V_{dc})_{\text{full-wave}} = 2(V_{dc})_{\text{half-wave}} = \frac{2V_m}{\pi} \tag{1.19}$$

$$(V_{rms})_{\text{full-wave}} = \frac{V_m}{\sqrt{2}} \tag{1.20}$$

Thus the rectification efficiency of a full-wave rectifier is

$$\eta_{\text{full-wave}} = \frac{(2V_m/\pi)^2/R_L}{(V_m/\sqrt{2})^2/R_L} = \frac{8}{\pi^2} = 2\eta_{\text{half-wave}} \tag{1.21}$$

Again note that due to the reasons stated earlier, the actual rectification efficiency is smaller than this value.

D. Peak Detector

The peak detector operation is shown in Fig. 1.32. If $R = \infty$, i.e., a pure capacitive load, the voltage across C builds up and is equal to v_i until v_i reaches V_m. When v_i decreases, the capacitor voltage cannot decrease since the current must flow in the reverse direction through the diode. The diode current can only be in the forward

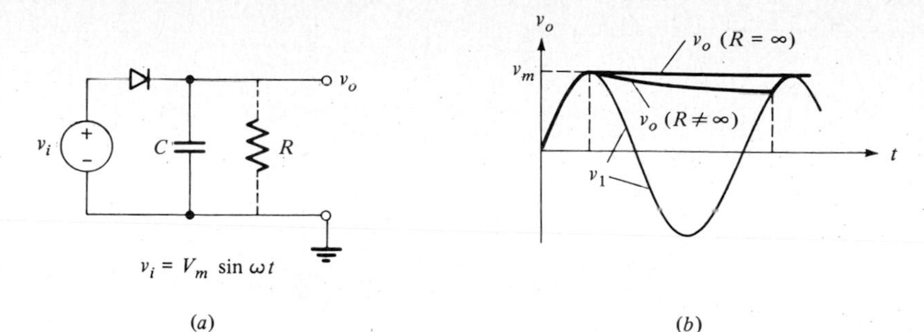

FIGURE 1.32 (*a*) Peak detector. (*b*) Associated waveform.

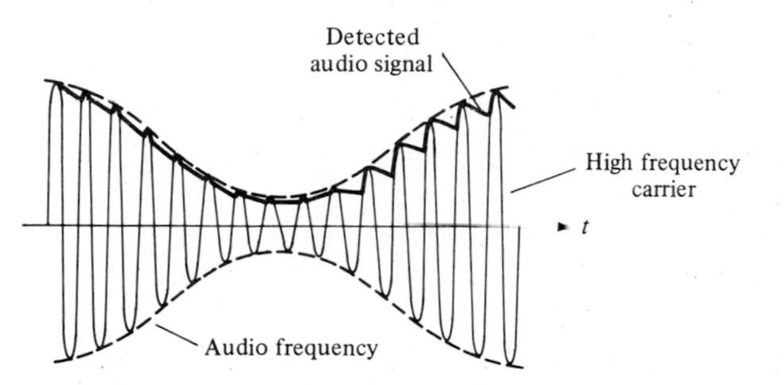

FIGURE 1.33 Detector (demodulator) waveform.

direction; hence v_i stays at the value of V_m as shown in Fig. 1.32*b*. Now consider a resistive load across the capacitor; i.e., $R \neq \infty$. In this case the capacitor can discharge since current can flow through the resistor. Recall that the capacitor voltage decays exponentially in a source-free *RC* circuit; i.e.,

$$v_0 = V_m e^{-t/RC} \tag{1.22}$$

The exponentially decaying voltage for $R \neq \infty$ is also shown in Fig. 1.32*b*.

The peak detector is used in amplitude-modulation (AM) radio to detect the audio signal as shown in Fig. 1.33. This process is usually known as *detection* or *demodulation*.

E. Power Supply Rectifiers and Filtering

Let us examine the full-wave rectifier circuit of Fig. 1.31, with the modification of placing a capacitor C across the load resistor R_L. The voltage across the *RC* load resistor is shown in Fig. 1.34. The output voltage between times t_1 and t_2 from (1.22) is given by

$$v_0 = V_m e^{-(t-t_1)/RC} \qquad t_1 \leq t \leq t_2 \tag{1.23}$$

The peak-to-peak ripple voltage is defined by

$$v_r = v_0(t_1) - v_0(t_2) = V_m(1 - e^{-(t_2-t_1)/RC}) \tag{1.24}$$

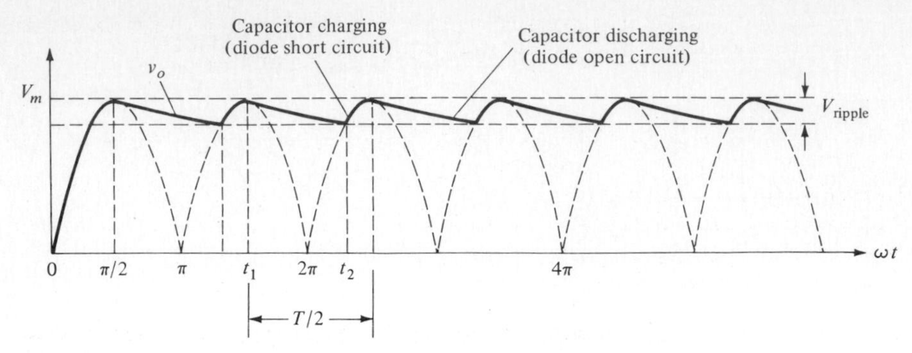

FIGURE 1.34 Full-wave rectified signal with filtering.

For large values of C such that $RC \gg (t_2 - t_1)$, we can use the well-known approximation

$$e^{-x} \simeq 1 - x \qquad \text{for } |x| \ll 1 \tag{1.25}$$

Hence from (1.24) and (1.25) we obtain the peak-to-peak ripple voltage v_r:

$$v_r(\text{peak to peak}) \simeq V_m \frac{t_2 - t_1}{RC} \tag{1.26}$$

The total nonconducting time, i.e., the time the capacitor is discharging, is $t_2 - t_1 \simeq T/2$. Since $T = 1/f_0$, where f_0 is the frequency of the input ac voltage, then

$$v_r\ (\text{peak to peak}) \simeq \frac{V_m}{2f_0RC} \qquad \text{for full-wave rectifier} \tag{1.27}$$

The dc component of the output voltage (see Fig. 1.34) is

$$V_{0,\text{dc}} = V_{\text{dc}} = V_m - \frac{1}{2}v_r = V_m\left(1 - \frac{1}{4f_0RC}\right) \tag{1.28}$$

For a half-wave rectifier, if a capacitor is placed across the rectifier circuit in Fig. 1.29*a*, the same procedure and equations apply except that $t_2 - t_1 \simeq T$; hence

$$v_r(\text{peak to peak}) \simeq \frac{V_m}{f_0RC} \qquad \text{for half-wave rectifier} \tag{1.29}$$

It should be noted that the diode currents are greatly in excess of average currents if the ripple voltage is a small fraction of V_m.

EXAMPLE 1.8

Consider a full-wave rectifier circuit of Fig. 1.31 with a capacitor C across R for filtering. Let the input signal be a 60-Hz sinusoid with a peak value

$V_m = 150$ V. If the load resistor $R = 10$ kΩ, find the value of the capacitance for a peak-to-peak ripple of 4 V. What is the dc voltage V_{dc}?

From (1.27): $$C = \frac{V_m}{2f_0Rv_r} = \frac{150}{120(10^4)(4)} = 31.2\ \mu\text{F}$$

From (1.28): $$V_{dc} = V_m - \tfrac{1}{2}v_r = 150 - 2 = 148\ \text{V}$$

F. Clamping or Level Restoring

A diode clamp circuit is shown in Fig. 1.35*a*. The input and output waveforms are shown in Fig. 1.35*b*. The capacitor is charged to the maximum negative value of the input. Since there is no resistance across C for it to discharge, the capacitor voltage remains charged to this value and remains constant. Thus the output voltage is given by

$$v_O = V_m + v_i = V_m(1 + \sin \omega t) \tag{1.30}$$

Note that output voltage never goes negative and is *clamped* between zero and $2V_m$.

G. Voltage Multiplier

The use of diodes as a voltage-doubling circuit is shown in Fig. 1.36*a* and *b*. The circuit in Fig. 1.36*a* is a half-wave voltage doubler, whereas the circuit in Fig. 1.36*b*

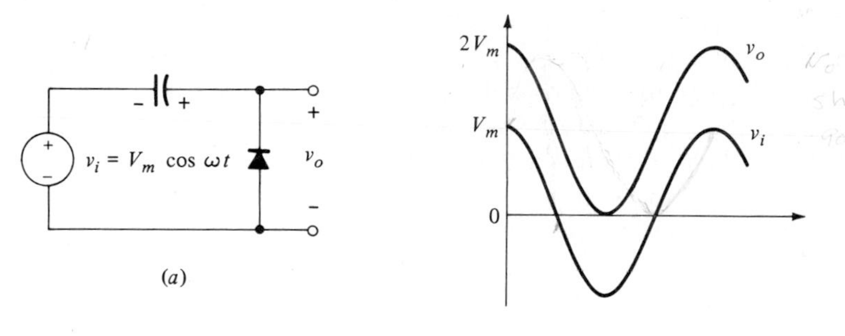

FIGURE 1.35 (*a*) Clamping circuit. (*b*) Associated waveform.

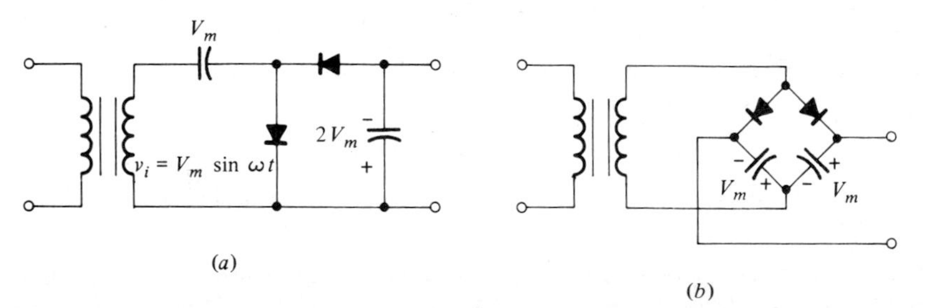

FIGURE 1.36 (*a*) Voltage doubler using half-wave rectifier circuit. (*b*) Voltage doubler using full-wave rectifier circuit.

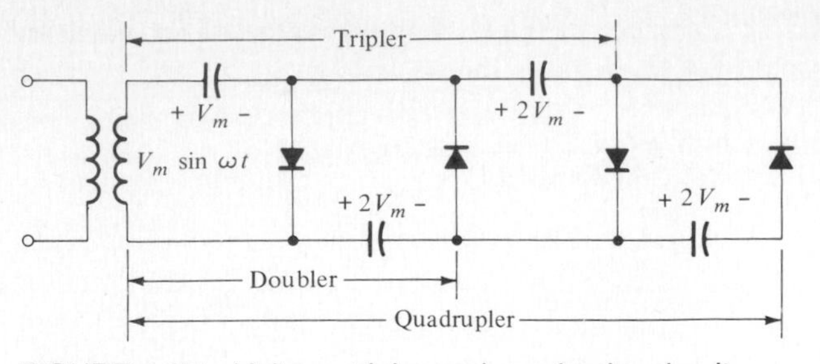

FIGURE 1.37 Voltage tripler and quadrupler circuit.

is a full-wave voltage doubler. The circuit delivers a dc voltage approximately equal to $2V_m$ at no load. Each of the two capacitors are alternately charged to the maximum value of the sinusoidal voltage. A voltage doubler can be viewed as a clamp followed by a peak detector. The extension of the voltage-doubler circuit to a voltage-tripler and -quadrupler circuit is shown in Fig. 1.37.

H. Diode Logic Gates

Diodes can also be used to function as logic gates, i.e., ON-OFF functions. Two such circuits are shown in Fig. 1.38*a* and *b*.

The circuits in Fig. 1.38*a* and *b* are referred to as AND and OR gates, respectively. The AND and the OR functions are discussed in detail in Chap. 13. Logic gates are the basic circuit components of a digital computer, and for reasons discussed in Chap. 13, more elaborate gates rather than diodes are always used.

In order to understand the operation of the circuits here, let us assume that all the signals are binary digital signals and that they assume only one of the two possible states (low and high), say, 0 and 5 V in this case. In Fig. 1.38*a*, the output voltage V_o becomes 5 V only if V_1 *and* V_2 are equal to 5 V. For example, if either of these voltages is zero, the corresponding diode is forward-biased, and hence V_o is short-circuited by the diode and becomes zero. In Fig. 1.38*b*, the output voltage V_o becomes 5 V if V_1 *or* V_2 is equal to 5 V. For example, if v_1 is 5 V, then D_1 is short-circuited, D_2 is reverse-biased and thus open-circuited, and V_o becomes 5 V. (In this circuit we also have the trivial case $V_1 = V_2 = V_o = 0$.)

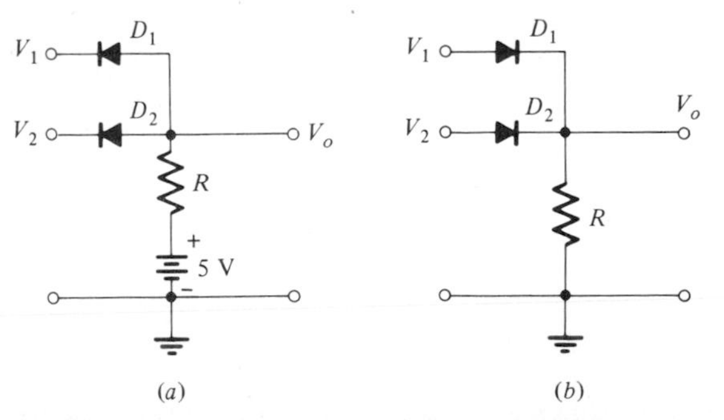

FIGURE 1.38 Diode logic gates: (*a*) AND gate and (*b*) OR gate.

EXERCISE 1.14

For the periodic waveform shown in Fig. E1.14, determine V_{DC} and V_{rms}.

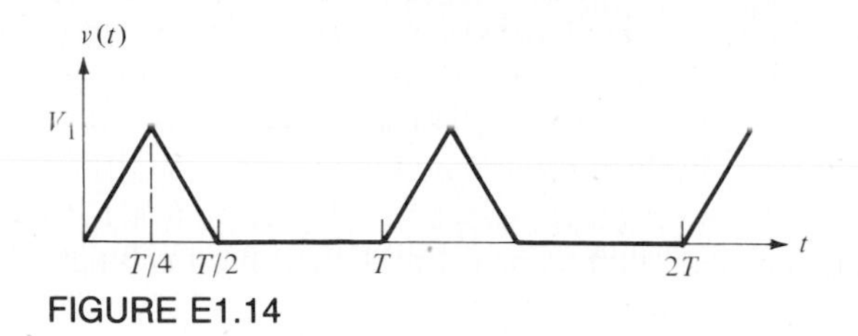

FIGURE E1.14

Ans $V_{DC} = V_1/4$, $V_{rms} = V_1/\sqrt{6}$

EXERCISE 1.15

For the half-wave rectifier circuit shown in Fig. E1.15, calculate C and V_m if the dc output voltage is 12 V with a peak-to-peak ripple of 0.1 V; use the constant-voltage-drop model ($V_o = 0.7$ V) for the diode.

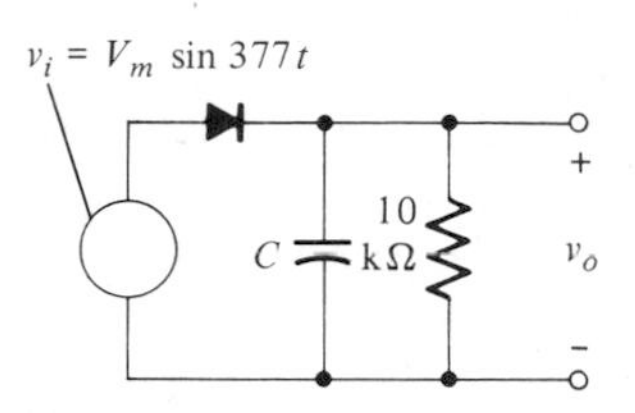

FIGURE E1.15

Ans $V_m = 12.75$ V, $C = 212\ \mu$F

EXERCISE 1.16

For the circuit shown in Fig. E1.16, determine V_o if (*a*) $V_1 = V_2 = 0$; (*b*) $V_1 = 0$, $V_2 = 2$ V; (*c*) $V_1 = 5$ V, $V_2 = 5$ V.

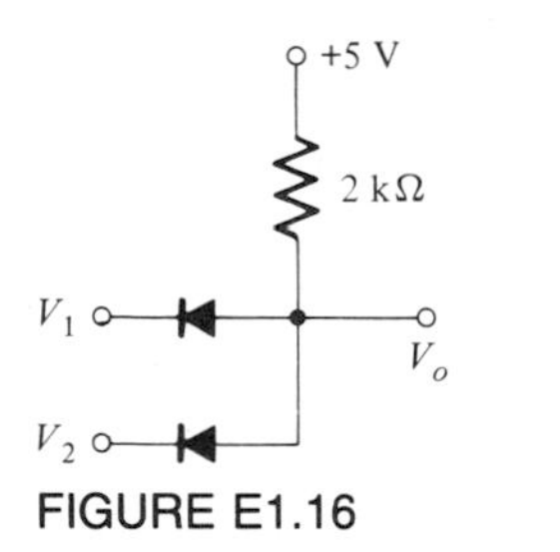

FIGURE E1.16

Ans (*a*) 0.7 V; (*b*) 0.7 V; (*c*) 5 V

1.9 SMALL-SIGNAL MODELS

Junction diodes are seldom operated solely in the linear region of the characteristic curve. We shall presently show that under small-signal operation at low frequencies, the diode can be modeled as a resistor. Small-signal models play a very important role in simplifying the analysis and design of electronic circuits. Small-signal models are widely used in the analysis of linear circuits employing bipolar transistors, field-effect transistors, etc. The general approach is to replace the nonlinear functions by their Taylor series expansions and retain only the linear parts under small-signal conditions. Thus the original nonlinear problem is replaced with a linear problem, which is much easier to solve. In other words we will have an approximate linear model (circuit) where all the powerful tools of linear systems such as superposition, Thevenin and Norton theorems, and transfer functions can be utilized.

For the simple circuit of a diode we have one nonlinear equation $i_D = f(v_D)$, where i_D and v_D are the total current through and the voltage across the diode. At low frequencies the diode has only one small-signal parameter, its dynamic resistance r_d. Next we show how to determine this resistance.

A. Low-Frequency Model

The circuit in Fig. 1.39*a* shows a diode biased by the dc voltage V_{DD} with the dc operating point at Q (i.e., I_q, V_q). A small-amplitude sinusoidal voltage is superimposed on V_{DD}, thereby varying V_q by an incremental amount. As the voltage across the diode changes to $V_q \pm V_m$ (with $V_m \ll V_q$), the operating point I_q and V_q changes to maximum limits of $I_q \pm \Delta I$ and $V_q \pm \Delta V$. These are illustrated in Fig. 1.39*b*, where the ac diode current is also shown. We are interested in ac signals only, i.e., sinusoidal

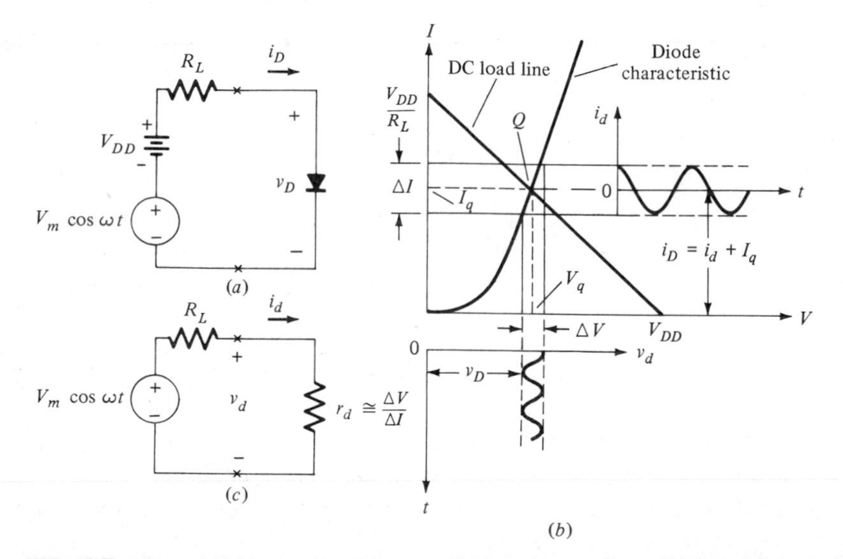

FIGURE 1.39 (*a*) Diode circuit for small-signal operation. (*b*) Load line and quiescent point. (*c*) Small-signal low-frequency model at the quiescent point.

signals of small amplitude such that the solution stays near the Q point. The diode characteristic near Q is approximated by a straight line, i.e., a linear approximation, and can be modeled by a dynamic resistance, r_d, namely,

$$g_d = \frac{\Delta I}{\Delta V} = \frac{1}{r_d} \tag{1.31}$$

Hence, the low-frequency model for the forward-biased junction diode is a resistance r_d, as shown in Fig. 1.39*c*. For small signals the value of the resistance depends on the *slope of the characteristic at the quiescent point*.

For the two-terminal device, namely, the diode, the *V-I* relationship is given by (1.1). The slope of the characteristic curve at point Q can be readily determined as follows:

$$I = I_s(e^{\lambda V/V_t} - 1) \simeq I_s e^{\lambda V/V_t} \tag{1.32}$$

$$\left.\frac{dI}{dV}\right|_Q = \left.\frac{\lambda}{V_t} I_s e^{\lambda V/V_t}\right|_{V_q} = \frac{\lambda}{V_t} I_q \tag{1.33}$$

From (1.33) we have

$$g_d = \frac{\lambda}{V_t} I_q = \frac{\lambda q}{kT} I_q = \frac{\lambda}{0.025} I_q \tag{1.34a}$$

$$r_d = g_d^{-1} = \frac{kT}{\lambda q I_q} = \frac{0.025}{\lambda I_q}\ \Omega \tag{1.34b}$$

EXAMPLE 1.9

Determine the small-signal dynamic resistance (r_d) for the 1N4001 diode of Example 1.1 at $I_q = 90$ mA and $I_q = 5$ mA. From (1.34*b*),

For $I_q = 90$ mA: $\quad r_d = \dfrac{0.025}{0.58(0.090)} = 0.48\ \Omega$

For $I_q = 5$ mA: $\quad r_d = \dfrac{0.025}{0.58(0.005)} = 8.6\ \Omega$

The value of r_d may also be derived analytically by making a Taylor series expansion of $i_D = f(v_D)$ near the operating point (I_q, V_q). We introduce the concept here and show the details because it is easier to understand it for a two-terminal device. The same concept is used in Chaps. 2 and 3 for different three-terminal devices such as bipolar transistors and field-effect transistors.

Consider the total diode voltage, which is the sum of the small-signal (ac) variation v_d and the dc quiescent point V_q. That is,

$$v_D = V_q + v_d \tag{1.35a}$$

In (1.35*a*) v_D varies as the sinusoidal voltage v_d varies (see Fig. 1.39*b*). Note that the Q point is where the circuit sits when the ac part is turned off, i.e., v_D at $Q = V_q$. Similarly,

$$i_D = I_q + i_d \tag{1.35b}$$

where I_q is the dc quiescent point and i_d is the small-signal ac current. The nonlinear *V-I* relationship of the diode in (1.1) can be expanded in a Taylor series near the point Q. Recall that the Taylor series expansion about a point $x = a$ is given by

$$f(x) = f(a) + \left.\frac{df}{dx}\right|_{x=a}(x - a) + \left.\frac{1}{2}\frac{d^2f}{dx^2}\right|_{x=a}(x - a)^2 + \cdots \tag{1.36}$$

Note that for x very near to a, i.e., $|x - a| = \epsilon \ll 1$, the nonlinear terms can be ignored and we have a linear approximation of the function near the point a. In our diode case we have $f(x) = f(v_D)$, and $x = a$ corresponds to the Q point, i.e., $v_D = V_q$. We can approximate (1.1) near the Q point as in (1.36), namely,

$$i_D = f(v_D)\Big|_Q + \left.\frac{\partial f}{\partial v_D}\right|_Q (v_D - V_q) + \text{nonlinear terms} \tag{1.37a}$$

where

$$i_D = I_s(e^{\lambda v_D/V_t} - 1) \simeq I_s e^{\lambda v_D/V_t} \tag{1.37b}$$

From (1.37)

$$i_D = i_D\Big|_Q + \frac{\lambda}{V_t} i_D\Big|_Q + \text{nonlinear terms} \tag{1.38a}$$

$$= I_s e^{\lambda V_q/V_t} + \left(\frac{\lambda}{V_t} I_s e^{\lambda V_q/V_t}\right) v_d + \text{nonlinear terms} \tag{1.38b}$$

From (1.35*b*) and (1.38*b*), for v_d very small,

$$I_q + i_d \simeq I_q + \left(\frac{\lambda}{V_t} I_q\right) v_d \tag{1.39}$$

or

$$i_d = \left(\frac{\lambda}{V_t} I_q\right) v_d \tag{1.40}$$

From (1.40) we obtain the small-signal diode model, which is a resistor, namely,

$$r_d = \frac{v_d}{i_d} = \frac{V_t}{\lambda I_q} = \frac{0.025}{\lambda I_q} \tag{1.41}$$

The circuit model is shown in Fig. 1.39*c*. Note that the value of r_d can also be obtained approximtely as $\Delta V/\Delta I$ *at* Q as shown in Fig. 1.39*b*.

Recall that

$$V_t = \frac{kT}{q} = 25 \text{ mV} \qquad \text{at room temperature}$$

The empirical scaling constant in (1.41) is λ, where $0.5 \leq \lambda \leq 1.0$. Note further that r_d can be varied by varying the dc current I_q. An example of varying r_d by varying the dc voltage (hence I_q), which is used in electronic circuits for volume control, is given in the following.

EXAMPLE 1.10

Consider the circuit shown in Fig. 1.40. The diode characteristic for the 1N4001 is shown in Fig. 1.2. For $V_m = 5$ mV, determine the ac voltage v_o at room temperature for $V_{DD} = 2$ V and 10 V.

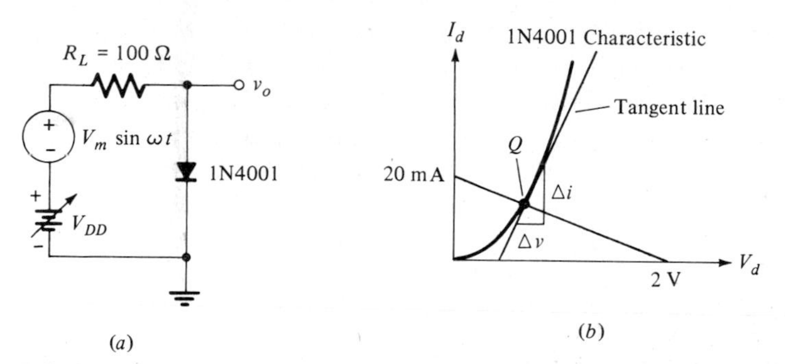

FIGURE 1.40 (a) Example circuit. (b) Load line for the circuit.

For $V_{DD} = 2$ V and $R_L = 100\ \Omega$ we draw the dc load line as shown in Fig. 1.40*b* and determine the Q point. In this case Q point is $I_q = 13$ mA, $V_q = 0.72$ V. The diode resistance may be found from (1.41), namely,

$$r_d = \frac{V_t}{\lambda I_q} = \frac{25 \text{ mV}}{0.58(13 \text{ mA})} = 3.31\ \Omega$$

Alternatively we may draw the tangent line to the characteristic curve at the point Q and determine the slope, namely,

$$r_d = \frac{\Delta V}{\Delta I} \simeq \frac{0.36}{0.1} = 3.6\ \Omega$$

In such calculations, of course, you can expect some deviations. For $r_d = 3.31\ \Omega$ we have

$$(v_o)_{ac} = \frac{3.31}{100 + 3.31} V_m \sin \omega t = 0.16 \times 10^{-3} \sin \omega t$$

The total voltage across the diode is then $0.72 + 0.16 \times 10^{-3} \sin \omega t$. Similarly, for $V_{DD} = 10$ V, $I_q \simeq 92$ mA and we have

$$r_d = \frac{25 \text{ mV}}{0.58(0.092)} = 0.48\ \Omega$$

and

$$(v_o)_{ac} = \frac{0.47}{100 + 0.47} 5 \times 10^{-3} \sin \omega t \simeq 0.23 \times 10^{-3} \sin \omega t$$

B. High-Frequency Model

We have already discussed the diffusion mechanism, which produces a net flow of carriers when the concentration is not uniform. At high frequencies, finite carrier transit time, and variable junction depletion region width must be taken into consideration. If this is done, the diode model becomes a shunt *RC* circuit as shown in Fig. 1.41*a*, where C_T is the total capacitance of a forward-biased *pn* junction. The capacitance C_T consists of the depletion-layer capacitance C_j and the carrier diffusion capacitance C_d. The depletion capacitance is nonlinear as given in (A.24), and the diffusion capacitance is proportional to the dc current. The high-frequency model for the forward region at the *Q* point (I_q, V_q) for a *pn*-junction diode is shown in Fig. 1.41*a*. The circuit elements are given by

$$r_d = \frac{V_t}{\lambda I_q} = \frac{0.025}{\lambda I_q} \tag{1.42}$$

$$C_j = K_1(V_0 - V_q)^{-m} \tag{1.43}$$

$$C_d = K_2 I_q = \frac{K_3}{r_d} \tag{1.44}$$

where K_1, K_2, K_3 are constants and $m = 0.33$ to 0.50. The diffusion capacitance C_d is usually larger than C_j. Note that V_0 is the potential barrier or the diffusion potential V_d determined by (A.18*c*). The range of values of C_d is typically from 10 to 100 pF, whereas for C_j it is in the range of 1 to 5 pF.

In the reverse region, the model for the *pn*-junction diode consists of a dynamic resistance r_c, which is very high ($10^6\ \Omega$ or larger), and the junction depletion layer capacitance C_j of the reverse-biased *pn* junction is denoted by C_c as shown in Fig. 1.41*b*. In this case we have

$$C_c = K_1(V_0 + V_q)^{-m} = C_j(0)\left(1 + \frac{V}{V_0}\right)^{-m} \tag{1.45}$$

Figure 1.41*c* shows a large-signal model, suitable for a computer-aided analysis and design (CAD). This model is particularly suitable for a commonly used CAD

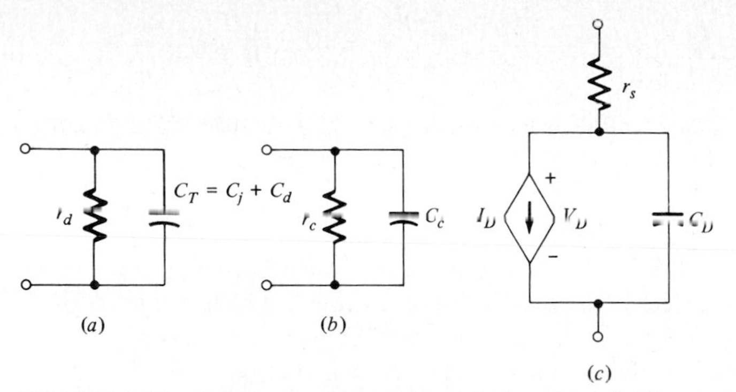

FIGURE 1.41 (*a*) Forward-biased small-signal model for a *pn*-junction diode. (*b*) Reverse-biased small-signal model for a *pn*-junction diode. (*c*) Large-signal model for *pn*-junction diode suitable for SPICE.

program called SPICE.* SPICE is an acronym for *s*imulation *p*rogram with *i*ntegrated *c*ircuits *e*mphasis. The program can be used for the diodes, bipolar transistors, and field-effect transistors, which are discussed in Chaps. 2 and 3. In Fig. 1.41*c*, the nonlinear *V-I* relationship of the diode is given by (1.1), and r_s is the series resistance. The capacitance ($C_D = dQ_D/dV_D$) is also a nonlinear function of the voltage.

EXERCISE 1.17

Determine the small-signal model (r_d) of the 1N4001 diode at room temperature for (*a*) $I_q = 0.1$ mA, (*b*) $I_q = 10$ mA, and (*c*) $I_q = 100$ mA.

Ans (*a*) 431 Ω; (*b*) 4.3 Ω; (*c*) 0.43 Ω

EXERCISE 1.18

For the circuit shown in Fig. E1.18, an 1N4001 diode is used. Determine $(V_0)_{dc}$, I_q, $(v_o)_{ac}$, and i_d.

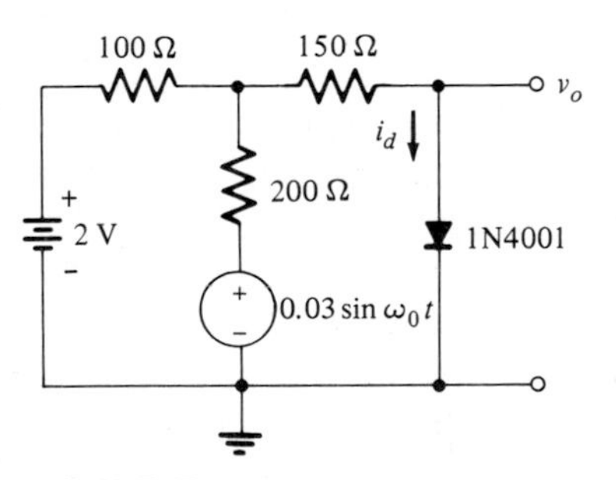

FIGURE E1.18

*SPICE was developed by Professor D. O. Pederson and his students at the University of California at Berkeley. For a discussion of SPICE see L. W. Nagel, *SPICE2, A Computer Program to Simulate Semiconductor Circuits,* Electronic Research Memorandum ERL-M520, University of California, Berkeley, May 1975.

Ans $(V_0)_{dc} = 0.65$ V, $I_q = 3.5$ mA, $(v_o)_{ac} = 5.5 \times 10^{-4} \sin \omega_0 t$, $i_d = 4.4 \times 10^{-5} \sin \omega_0 t$

1.10 BREAKDOWN (ZENER) DIODES

The breakdown voltages of a *pn* junction in the reverse-bias condition are mainly due to two mechanisms. One is the *avalanche breakdown,* which occurs at high voltages ($V \geq 10$ V). At high voltages, carriers acquire sufficient energy to create new electron-hole pairs and thus secondary carriers, which results in an avalanche breakdown. The other is the *zener breakdown* in highly doped diodes. This breakdown occurs at low voltages ($V < 10$ V) and can be controlled by controlling the concentration of impurities on each side of the junction. Regardless of the mechanism of breakdown, the breakdown diodes are usually referred to as zener diodes. Zener diodes find application in constant-voltage reference and regulator circuits.

The symbol and typical *V-I* characteristic of a low-voltage reference zener diode are shown in Fig. 1.42. The forward and the reverse characteristics are purposely shown to different scales. Note that the forward characteristic is similar to that of the conventional *pn*-junction diode. The reverse characteristic shows a breakdown voltage V_z, at which value it is almost independent of the diode current. A wide range of zener diodes are commercially available; values from 2 to 200 V are typical, with power

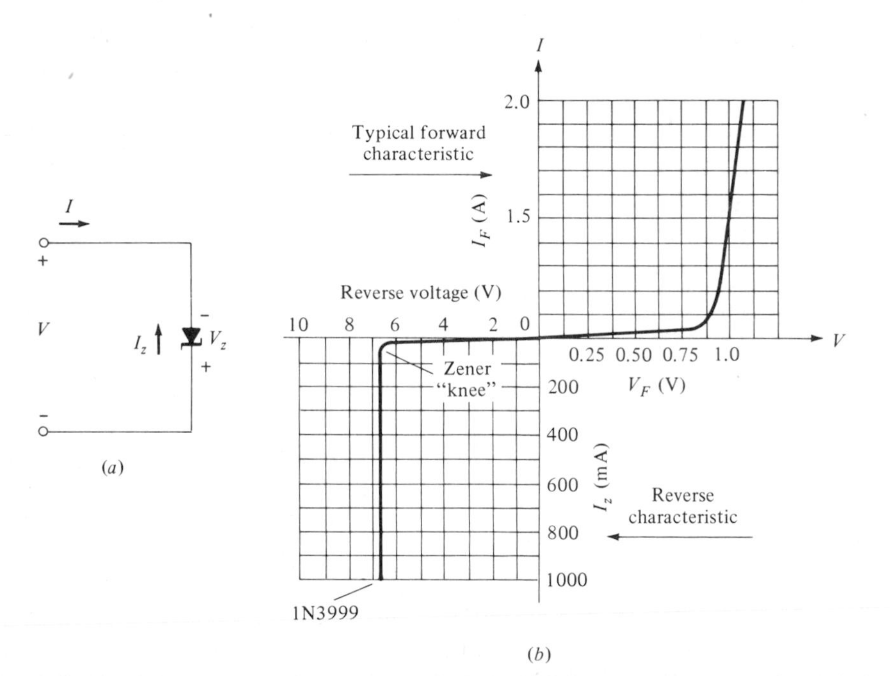

FIGURE 1.42 (*a*) Breakdown (zener) diode. (*b*) Typical breakdown (zener) diode *V-I* characteristic.

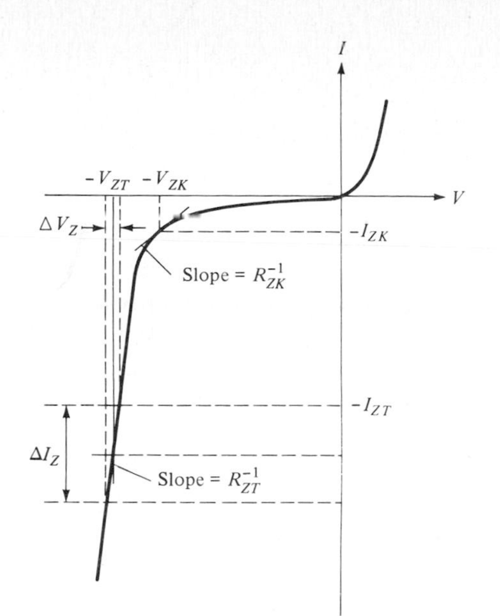

FIGURE 1.43 Zener-diode characteristic curve in the region of interest with key features.

ratings from a fraction of a watt to 100 W. In Appendix A the characteristics of 1N4728 to 1N4764 zener diodes are listed.

Since the reverse characteristic of a zener diode is the region of interest, let us examine this region more closely. A typical zener diode *V-I* characteristic, which exhibits the key parameters of the device, is shown in Fig. 1.43. I_{ZK} and V_{ZK} denote the zener "knee" current and voltage, I_{ZT} and V_{ZT} denote the test current and voltage. R_{ZK} and R_{ZT} are resistances defined by

$$R_{ZK} = \frac{V_{ZK}}{I_{ZK}} \tag{1.46a}$$

$$R_{ZT} = \frac{\Delta V_{ZT}}{\Delta I_{ZT}} \tag{1.46b}$$

R_{ZK} is much larger than R_{ZT}. Typical values of R_{ZK} for 1-W zener diodes are between 0.5 kΩ and 3 kΩ, whereas for R_{ZT} these are in the range of 5 Ω to 300 Ω.

A typical application of a zener diode for voltage regulation is shown in Fig. 1.44. The dc voltage V is unregulated and can vary, as can the load resistance. In spite of these variations, by proper design the voltage across the load can be held constant to the nominal zener value V_z. The operation of the circuit is as follows (assume the value of R is such that the operation is beyond the knee of the breakdown curve): If V changes, the voltage drop across R changes, since V_z is a constant. The change in the drop across R results in a change in current through R. This current flows through the zener diode, and the load current remains the same because the voltage across it has not changed. If the input voltage is constant and the load resistance is varied, the

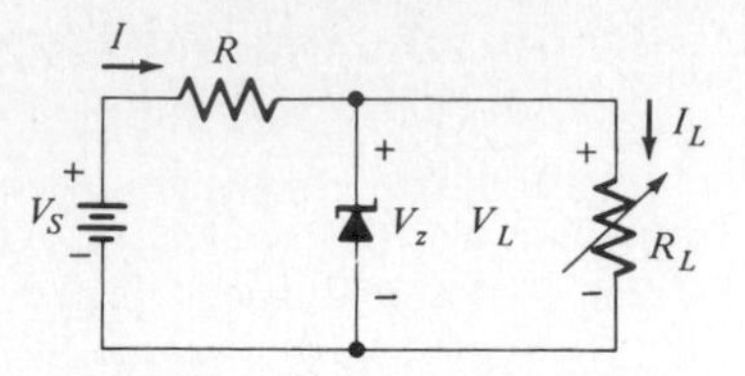

FIGURE 1.44 A simple voltage-regulator circuit.

variation in the load resistance causes a change in the load current. Since V and V_z are unchanged, the current variation of the load results only in changes in the zener diode current. Thus, the voltage across the load remains constant and equal to V_z in spite of these variations. It should be pointed out, however, that changes in temperature generally cause a change in the zener reference voltage. The temperature coefficient is about +2 mV/°C, which is similar but opposite in sign to that of a forward-conducting diode. For example, if the 1N3999-type zener diode with $V_Z \simeq 6.6$ V (shown in Fig. 1.42*b*) is connected in series with a 1N4001 diode with $V_0 \simeq 0.7$ V, the series combination with a voltage drop of about 7.3 V will exhibit a greatly reduced temperature coefficient.

The typical temperature coefficient of the zener diode is specified by the manufacturer. When the zener diode is operated at a much higher ambient temperature, the power dissipation of the device is reduced considerably. The power rating vs ambient temperature, known as a *derating curve,* is usually provided by the manufacturer (see Appendix E). The use of the derating curve for a diode is similar to that for a transistor and is discussed later in Sec. 6.7.

EXAMPLE 1.11

A 6.7-V zener diode is used in the circuit shown in Fig. 1.44. The zener is capable of dissipating 1.5 W at room temperature. The load current can vary from 30 to 200 mA. The minimum zener-diode current I_{zmin} to ensure the operation beyond the knee is 20 mA.

(a) Determine the value of R for a supply voltage of $V = 20$ V.
(b) For a fixed value of $R_L = 67\ \Omega$, what are the limits between which V can vary without loss of regulation?

(a) From the circuit we have the following relations:

$$I_z = I - I_L \qquad \text{and} \qquad R = \frac{V_s - V_z}{I}$$

From the above equations I_{zmin} occurs for I_{Lmax}; hence $I = (200 + 20) \times 10^{-3}$A. Thus

$$R = \frac{20 - 6.7}{0.22} = 60.45\ \Omega$$

For $I_L = 30$ mA, $I_Z = 220 - 30 = 190$ mA which is below the maximum I_Z. The range of the diode current is 20 and 224 mA (which is obtained from 1.5 W/6.7 V).

(b) At the minimum diode current $I_L = 6.7/67 = 100$ mA and $I = 100 + 20 = 120$ mA, and $V = 120 \times 10^{-3}(60.45) + 6.7 = 13.95$ V.

At the maximum diode current,

$$I = 100 + 224 = 324 \text{ mA}$$

$$V = 324 \times 10^{-3}(60.45) + 6.7 = 26.29 \text{ V}$$

Hence the output voltage will be constant at 6.7 V, and $I_L = 100$ mA in spite of the variations of the source between 13.95 and 26.29 V.

The circuit shown in Fig. 1.44 is called a *shunt zener regulator* since the zener diode is in shunt with the load. Voltage regulation is of importance in many applications. We want to maintain the voltage constant when the load varies from its minimum to maximum value. In other words, we want to ensure that the zener diode be in the ON state beyond the knee and under the maximum current value determined by the power dissipation rating and the ambient temperature.

EXERCISE 1.19

For the circuit shown in Fig. 1.44, a 1N4740 zener diode (25 mA $\leq I_z <$ 91 mA) is used. In this design R_L is fixed and R is to be chosen. Find R so that V_L remains at 10 V for 12 V $\leq V_s \leq$ 15 V, given that $R_L = 5$ kΩ. What is the maximum power dissipated by the diode?

Ans $R = 74.1$ Ω, 0.65 W

1.11 CONCLUDING REMARKS

This chapter dealt with the *V-I* characteristics of junction and zener diodes. The concepts of ideal diode, graphical analysis, load line, transfer characteristic, and circuit models were introduced. These concepts will also be used in Chaps. 2 and 3 when bipolar and field-effect transistors are discussed.

The circuit model shown in Fig. 1.17*b* will, in general, be used as a large-signal model of a Si diode (with $V_0 = 0.7$ V at room temperature) in subsequent chapters unless stated otherwise.

REFERENCES

1. L. S. Bobrow, *Elementary Linear Circuit Analysis,* Holt, New York, 1981 (for a review of basic circuit theorems).

2. J. Millman, *Microelectronics: Digital and Analog Circuits and Systems,* McGraw-Hill, New York, 1979.

3. D. Schilling and C. Belove, *Electronic Circuits, Discrete and Integrated,* McGraw-Hill, New York, 1979.

4. A. Sedra and K. Smith, *Microelectronic Circuits,* Holt, New York, 1982.

PROBLEMS

1.1 Find the resistance R_{ab} of the Si structure shown in Fig. P1.1, if the resistivity of the material is $\rho = 10^3\ \Omega \cdot \text{cm}$.

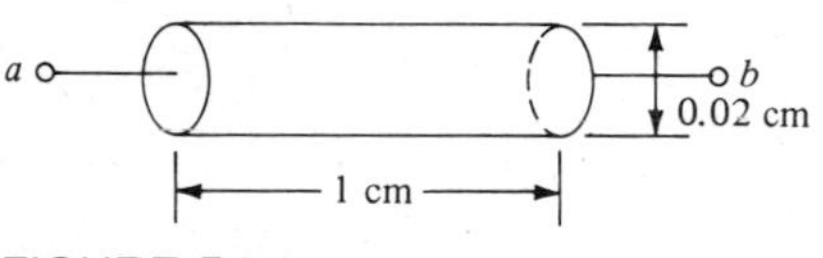

FIGURE P1.1

1.2 A copper wire has a diameter of 0.05 cm. What length of wire is needed to make a resistance equal to that in problem 1.1? Assume a conductivity of $5 \times 10^5\ (\Omega \cdot \text{cm})^{-1}$.

1.3 A silicon diode has $I_s = 5$ nA at room temperature (300 K). Find I_s at $T = 320$ K and $T = 350$ K.

1.4 For a Si diode $V_d = 0.8$ V and $I_d = 90$ mA at 20°C. If $V_0 = 0.7$ V at 20°C, determine V_0 at 100 and -50°C at the same value of I_d.

1.5 (*a*) Find the Thevenin equivalent circuit of Fig. P1.5.
(*b*) Find its Norton equivalent circuit.

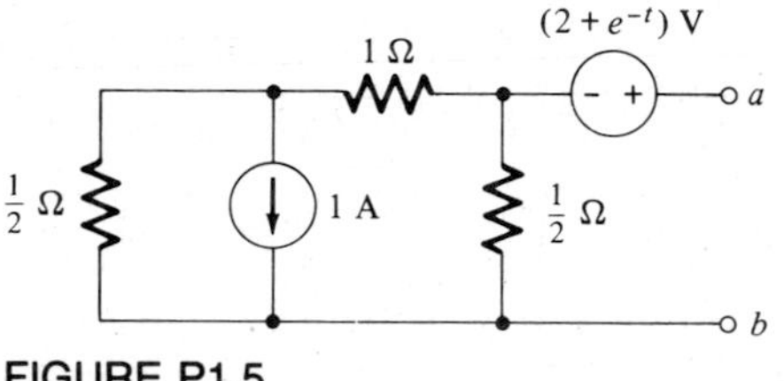

FIGURE P1.5

1.6 Find the Thevenin and the Norton equivalent circuits of Fig. P1.6.

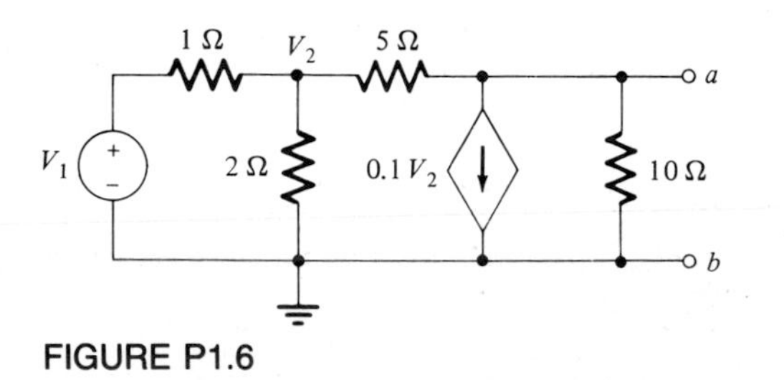

FIGURE P1.6

1.7 Find the Thevenin equivalent circuit of Fig. P1.7.

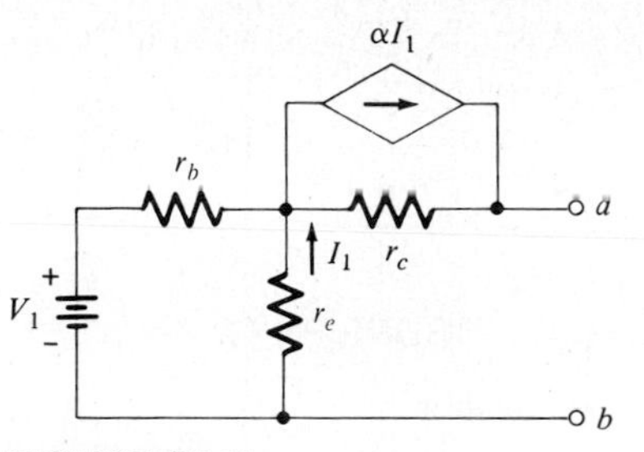

FIGURE P1.7

1.8 The V-I relationship of a Si diode is given in Fig. 1.2. If the diode is used as in Fig. P1.8, determine the voltage across and the current through the diode. (*Hint:* Use the Thevenin theorem and load line.)

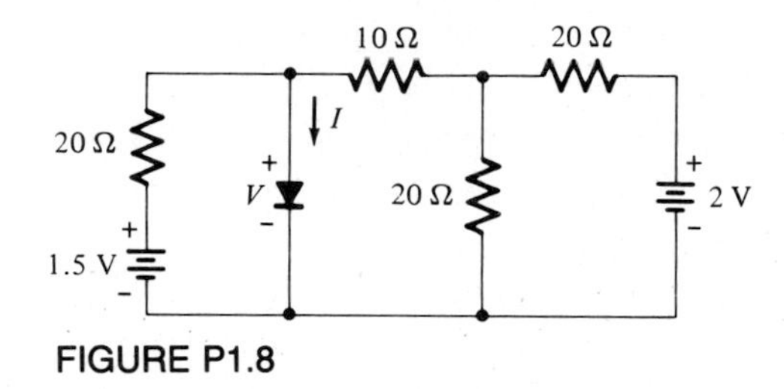

FIGURE P1.8

1.9 Repeat problem 1.8, but solve analytically, using a piecewise-linear approximation that includes V_0 and R_f.

1.10 For the circuit shown in Fig. P1.10*a*, draw the load line and comment on the operating point(s). The piecewise-linear V-I characteristic of the device $\mathcal{N}$ is as shown in Fig. P1.10*b*.

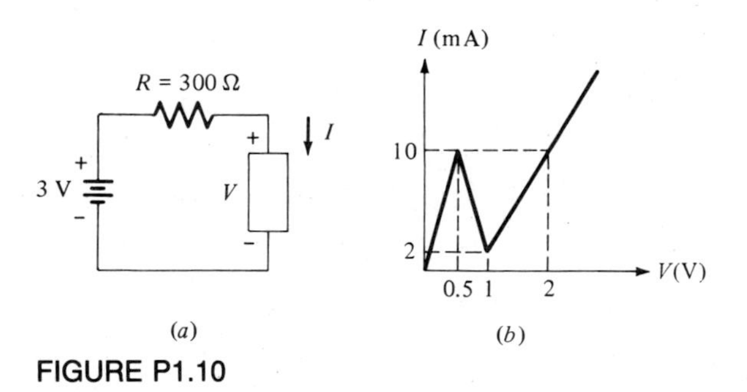

FIGURE P1.10

1.11 For the circuit shown in Fig. P1.11, determine the current I_x if the device V-I relation is given by

$$V_x = I_x^{1/3}$$

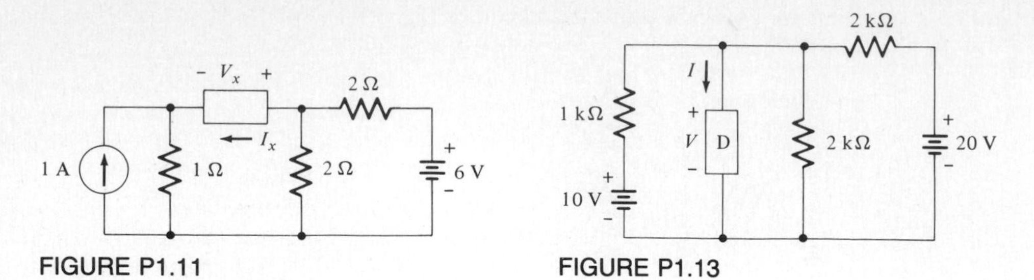

FIGURE P1.11

FIGURE P1.13

1.12 Repeat problem 1.11 for the device V-I relation

$$V_x = I_x^{1/2}$$

1.13 For the circuit shown in Fig. P1.13, determine the current through the nonlinear device D if the V-I relationship of the device is given by

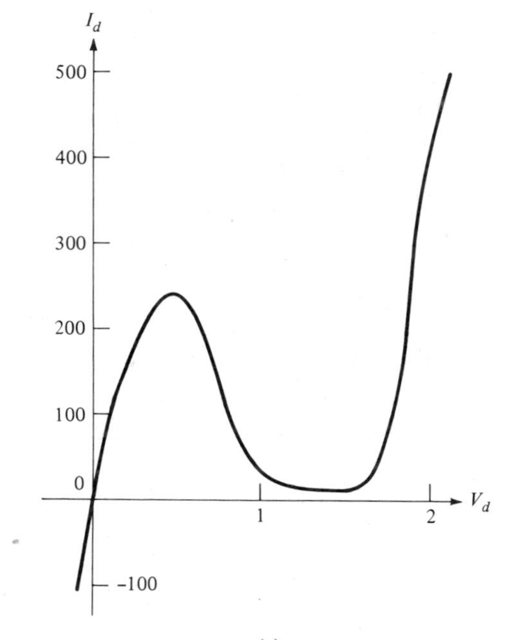

(a)

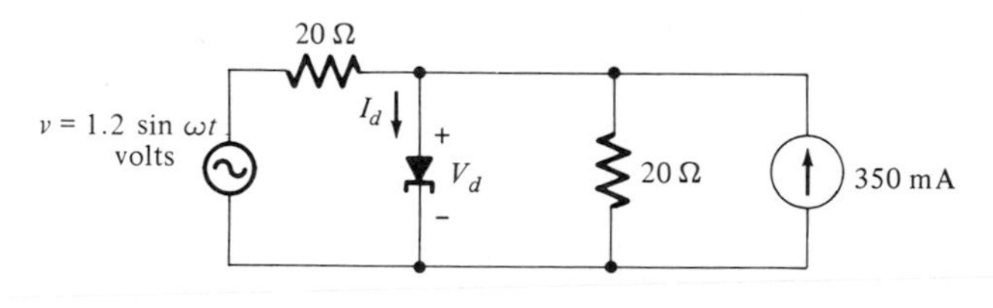

(b)

FIGURE P1.14

$$I = \begin{cases} 0.5V^{3/2} \text{ mA} & V > 0 \\ 0 & V \leq 0 \end{cases}$$

Graphical solution is acceptable.

1.14 A two-terminal device, known as *tunnel diode,* has the measured V-I characteristic shown in Fig. P1.14*a*. Calculate $v_d(t)$ for the circuit shown in Fig. P1.14*b*.

1.15 Show the piecewise-linear model for the ideal zener diode given in Fig. P1.15*a*. Repeat the above for an actual zener diode with the V-I characteristic *approximated* in Fig. P1.15*b*.

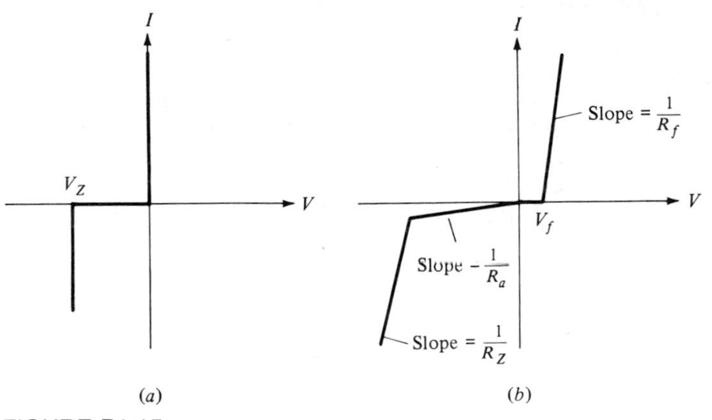

FIGURE P1.15

1.16 For the circuits shown in Fig. P1.16*a* and *b*, draw the transfer characteristic and sketch the output voltages in Fig. P1.16*a* if $v_i(t)$ is
(*a*) Sinusoidal 2 V peak to peak and $\omega = 10^4$ rad/s.
(*b*) Square wave 2 V peak to peak and a period $T = 10^{-2}$ s.

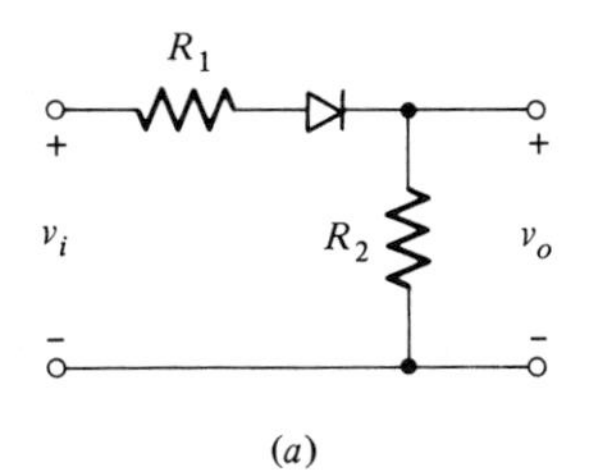

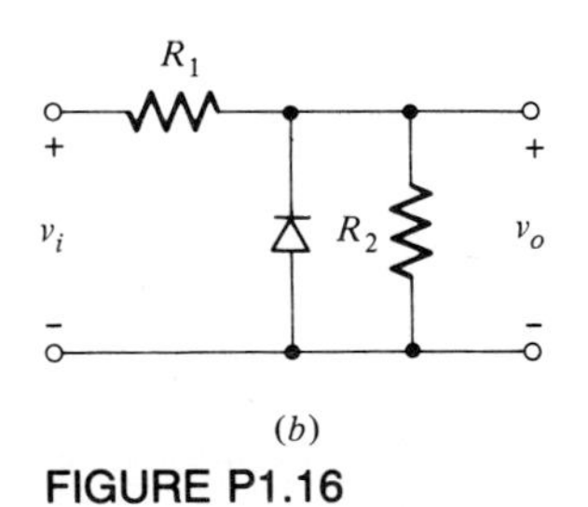

FIGURE P1.16

1.17 For the circuit shown in Fig. P1.17,
(*a*) Draw the transfer characteristic.
(*b*) Sketch $v_o(t)$ if $v_i(t)$ is as shown in Fig. P1.17*b*. Assume $V_Z = 5$ V.

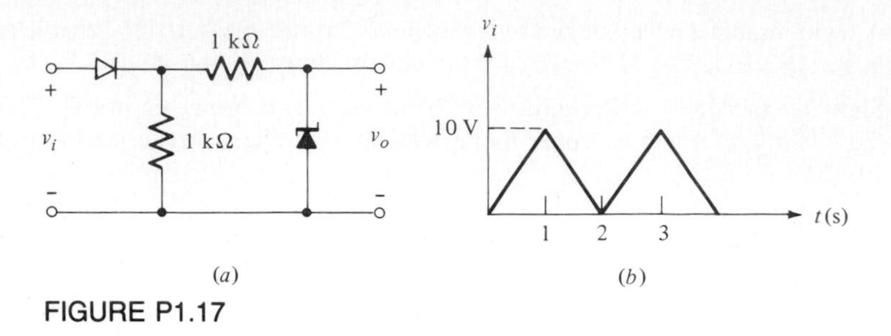

FIGURE P1.17

1.18 For the circuit shown in Fig. P1.18,
(*a*) Draw the transfer characteristic.
(*b*) Sketch the output waveform for the various input signals shown.

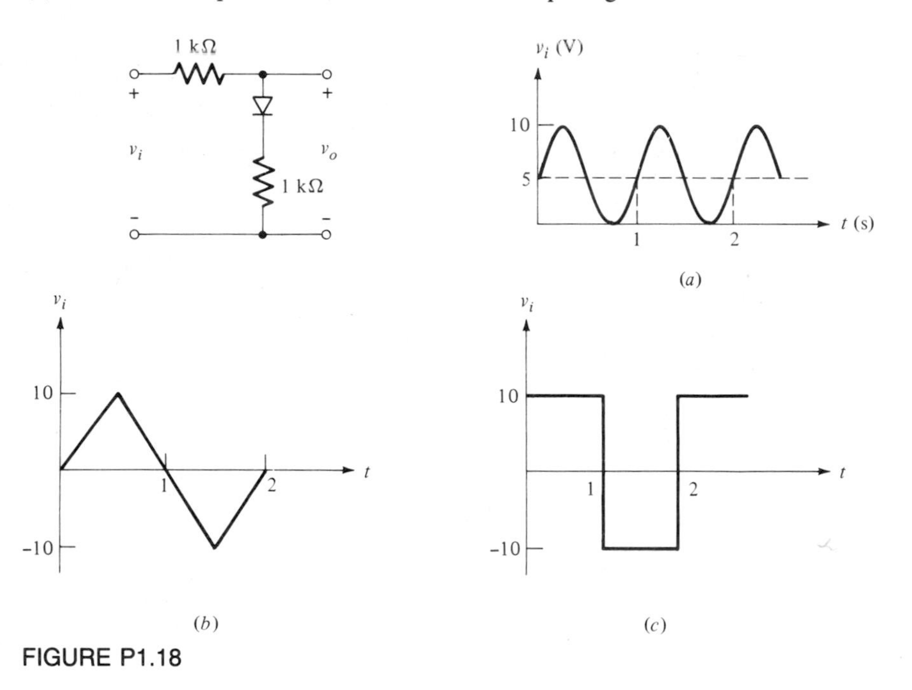

FIGURE P1.18

1.19 Repeat problem 1.18 with the diode reversed.

1.20 Repeat problem 1.18 for the input signal of Fig. 1.18*a* but consider an actual diode with characteristic as shown in Fig. 1.2. Use a piecewise-linear approximation with V_0 and R_f to approximate Fig. 1.2.

1.21 Graph the transfer characteristic of the circuits shown in Fig. P1.21*a* and *b*. The *V-I* characteristic of *each diode separately* is given in Fig. 1.2.

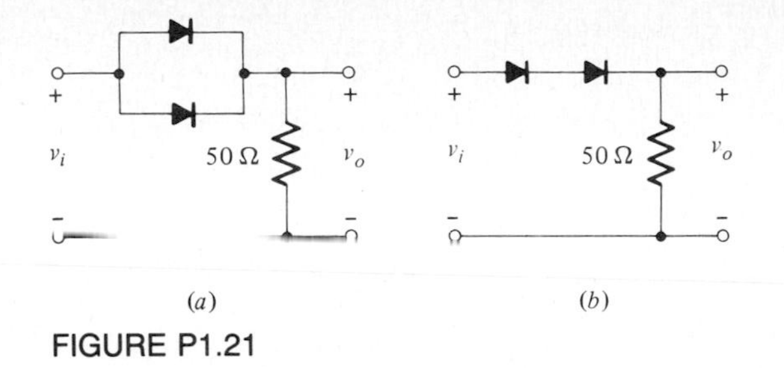

FIGURE P1.21

1.22 Repeat problem 1.21 for the circuits shown in Fig. P1.22*a* and *b*.

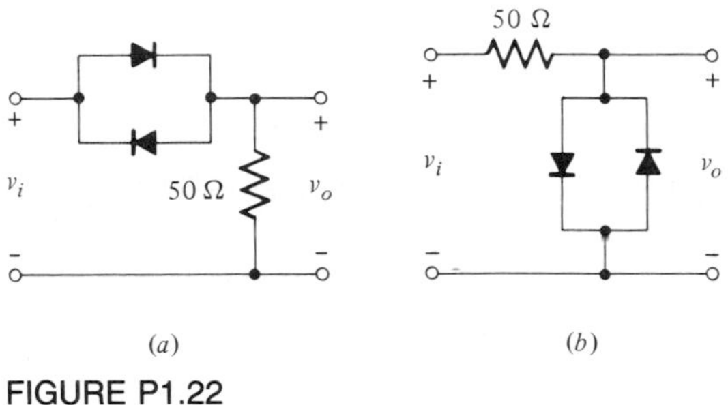

FIGURE P1.22

1.23 Sketch the transfer characteristic of the circuit shown in Fig. P1.23, given

$$V_Q = 0.7 \text{ V}, R_f = 50 \ \Omega, R_a = 10 \text{ k}\Omega, R_z = 5 \ \Omega, \text{ and } V_z = 5 \text{ V}.$$

(*Hint:* The one-port circuit at *a*-*b* is a piecewise-linear approximation of an actual zener diode.)

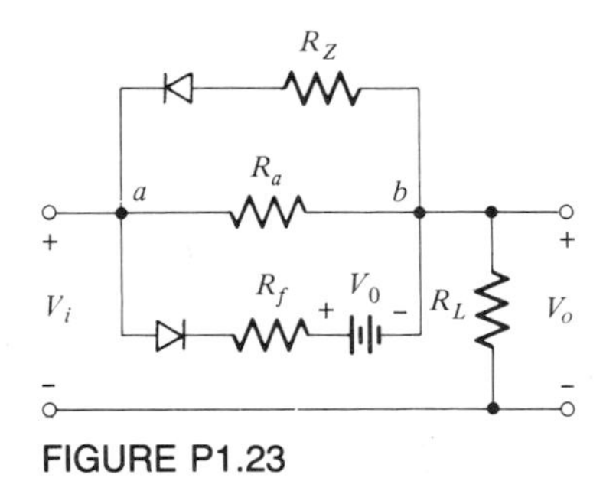

FIGURE P1.23

1.24 The input voltage to the circuit shown in Fig. P1.23 varies linearly with a slope of 1 V/s from 0 to 10 V. Sketch the output voltage to the same time scale as the input voltage. Note that ideal diodes are used.

1.25 (*a*) Show a circuit using ideal diodes, resistors, and batteries that will exhibit the transfer characteristic V_O vs V_i shown in Fig. P1.25*a*.
(*b*) Sketch the transfer characteristic of the circuit in Fig. P1.25*b*.

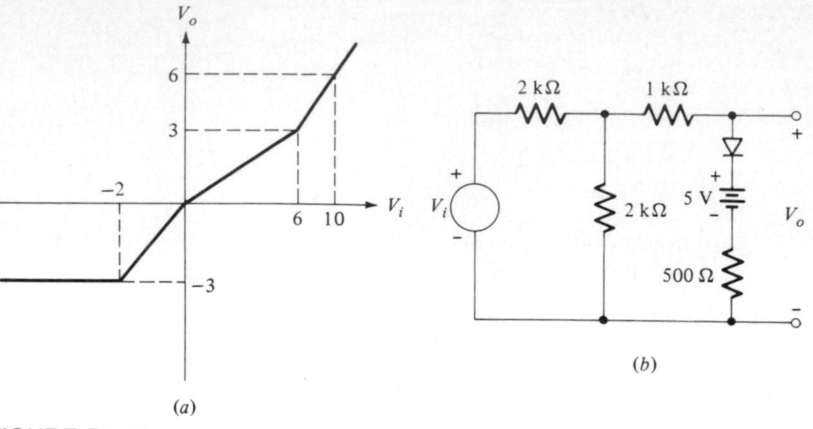

FIGURE P1.25

1.26 Consider a simple half-wave rectifier circuit shown in Fig. P1.26. Given $R = 100\ \Omega$ $C = 100\ \mu\text{F}$, $\omega = 2\pi(10^2)$ rad/s,
(a) Sketch the output voltage if D is an ideal diode.
(b) Sketch the output voltage if D is replaced by a zener diode with $V_z = -10$ V.

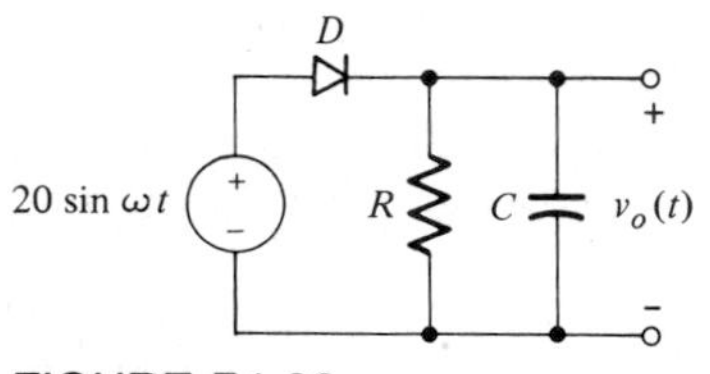

FIGURE P1.26

1.27 Assume that diodes D_1 and D_2 in Fig. P1.27 are defined by their piecewise-linear characteristics. D_1 and D_2 are Si diodes and their threshold voltage is $V_{01} = V_{02} = 0.7$ V. Their forward resistance is $R_{f1} = R_{f2} = 20\ \Omega$. Find the times at which D_1 and D_2 are triggered (i.e., start to conduct) as well as the currents i_1, i_2, and i_3 from $t = 0$ until the moment at which both diodes have been triggered.

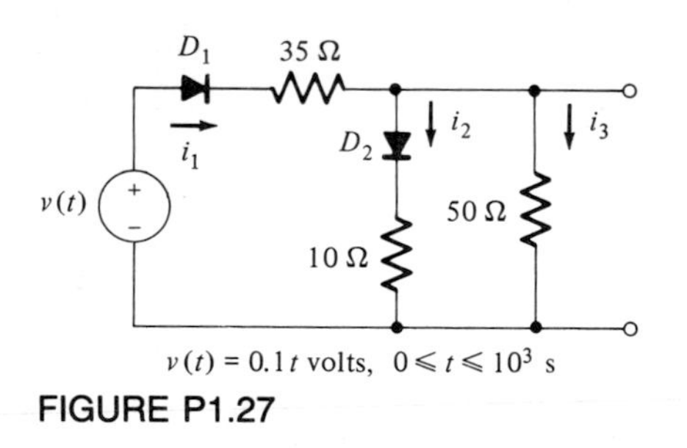

$v(t) = 0.1t$ volts, $0 \leqslant t \leqslant 10^3$ s

FIGURE P1.27

1.28 Sketch the transfer characteristic for the circuit shown in Fig. P1.28. If $v_i = 20 \sin 2\pi(10^3)t$, sketch $v_o(t)$ on the same scale and calculate V_{dc} and V_{rms} for the output voltage.

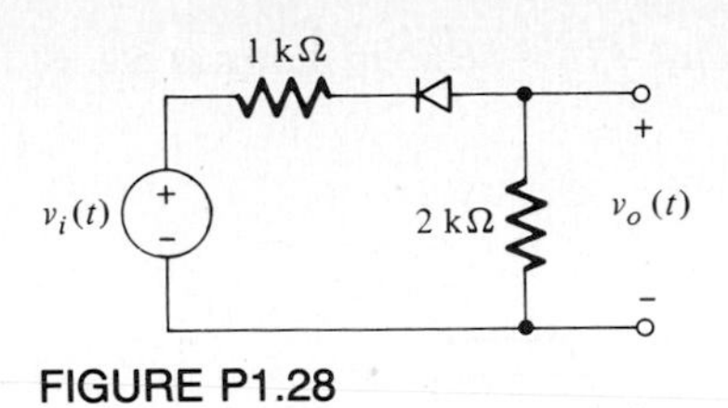

FIGURE P1.28

1.29 Find a realization, using ideal diodes, resistors, and batteries, for each transfer characteristic shown in Fig. P1.29.

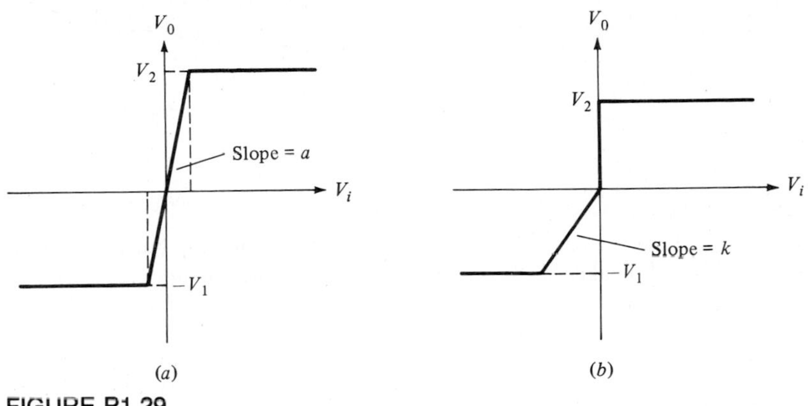

FIGURE P1.29

1.30 Sketch the driving-point characteristic I vs V for the circuit shown in Fig. P1.30.

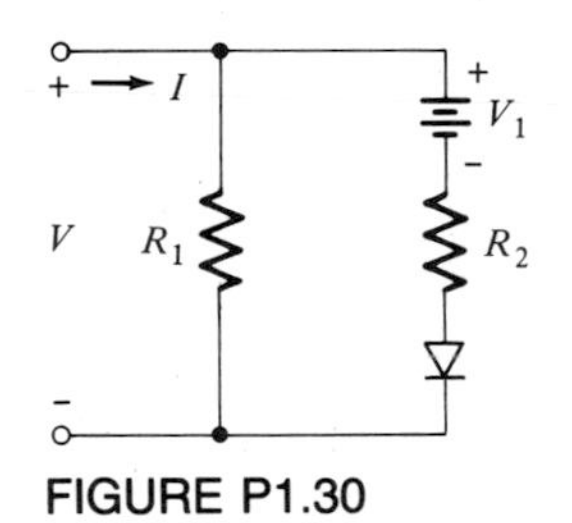

FIGURE P1.30

1.31 Sketch the driving-point I vs V for the circuits shown in Fig. P1.31a and b.

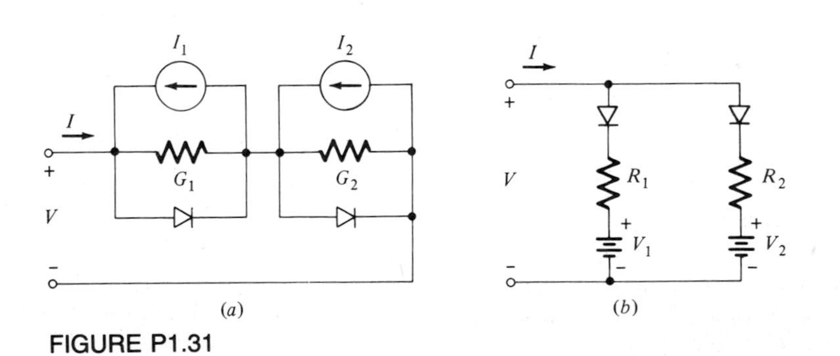

FIGURE P1.31

1.32 Find a realization for each driving-point V-I characteristic shown in Fig. P1.32. Use ideal diodes.

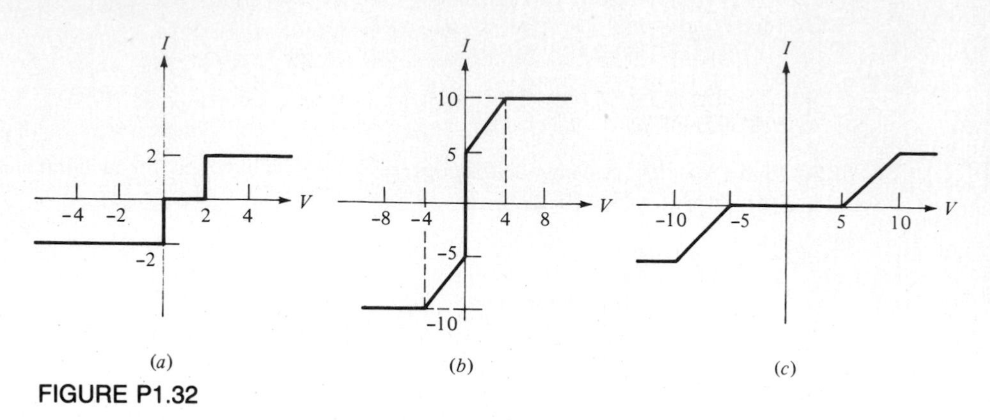

FIGURE P1.32

1.33 For the circuit shown in Fig. P1.33, determine the two breakpoints and sketch I vs V. (*Note:* At breakpoint the *current* through *and* the *voltage* across the diode are *simultaneously zero*.)

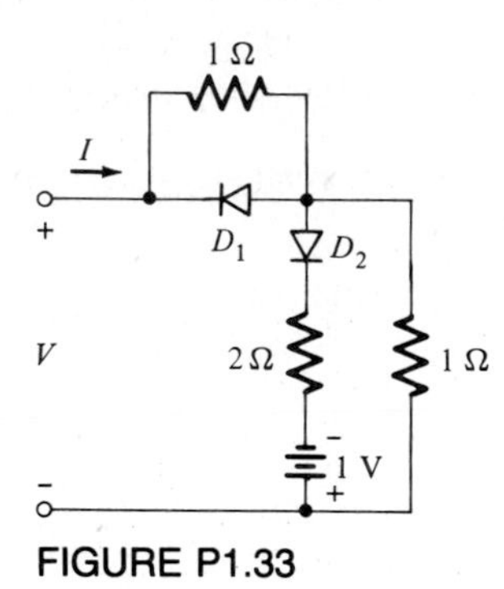

FIGURE P1.33

1.34 Determine V_{dc} and V_{rms} of the periodic waveforms shown in Fig. P1.34*a* and *b*.

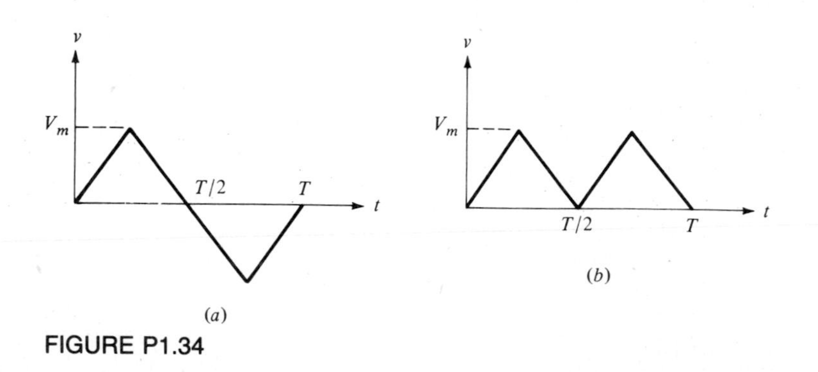

FIGURE P1.34

1.35 For the circuit shown in Fig. 1.29 use the large-signal model for the diode with $V_0 = 0.7$ V and $R_f = 50\ \Omega$.

(*a*) If $v_i(t) = 10 \sin \omega t$, sketch $v_o(t)$ on the same scale.

(*b*) If $v_i(t) = 10 + 5 \cos \omega t$, sketch $v_o(t)$ on the same scale.

(*c*) Determine V_{dc} and V_{rms} of the output voltage for each case.

1.36 The circuit shown in Fig. P1.36 is a full-wave rectifier.

(*a*) Assume ideal diodes with voltage drops $V_0 = 0.7$ V and sketch the output voltage $v_o(t)$.

(*b*) If a capacitor is connected in shunt across R, how is the voltage $v_o(t)$ affected? Sketch $v_o(t)$.

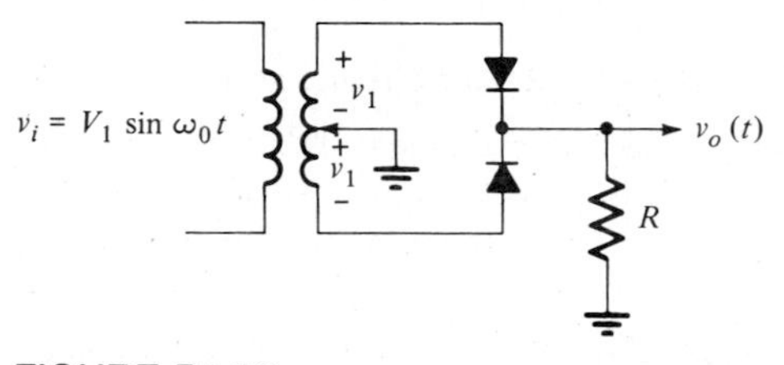

FIGURE P1.36

1.37 Repeat problem 1.36 for the full-wave rectifier circuit shown in Fig. 1.31*a* but assume actual diodes, i.e., $V_0 = 0.7$ V.

1.38 For the circuit in Fig. P1.38, sketch the response $v_o(t)$ for a pulse input $v_i(t)$ as shown.

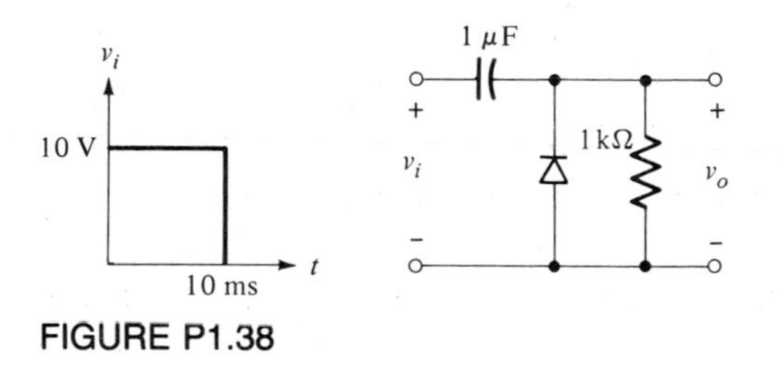

FIGURE P1.38

1.39 Sketch the driving-point characteristics, i.e., I vs V, for the circuit shown in Fig. P1.39. Assume $R_2 > R_1$.

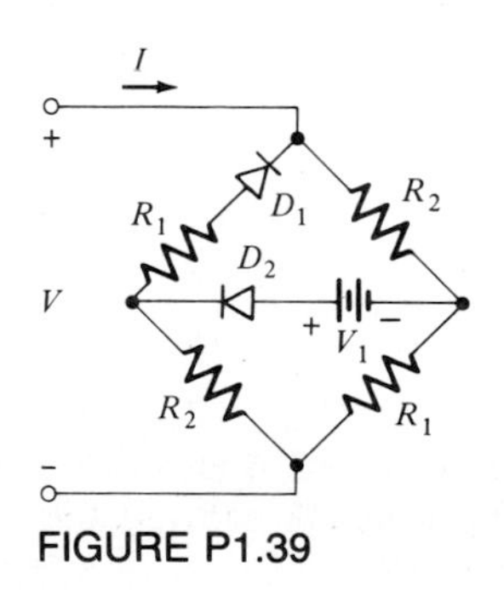

FIGURE P1.39

1.40 For the simple half-wave rectifier circuit shown in Fig. P1.40, the input is given by $v_i = V_m \sin \omega t$. Determine the dc output voltage and the rms value of the output voltage if the diode is modeled by a piecewise linear model with a voltage drop V_0 and a forward resistance R_f.

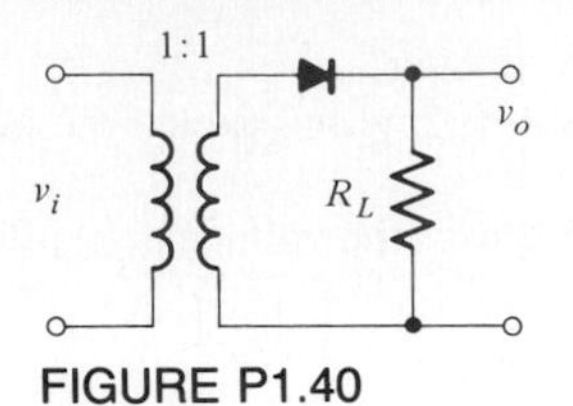

FIGURE P1.40

1.41 Repeat problem 1.40 for the full-wave rectifier circuit shown in Fig. P1.36.

1.42 The efficiency of rectification η_r is also sometimes defined as the ratio of dc *output power* to ac *input power*; determine η_r for the half-wave and full-wave rectifiers, assuming (*a*) ideal diodes and (*b*) piecewise-linear large-signal models for the diodes (i.e., include V_0 and R_f).

1.43 For the half-wave rectifier circuit with a smoothing capacitor shown in Fig. P1.43, the following are given:

$$v_i = 100 \sin 377t \qquad R = 100\ \text{k}\Omega \qquad C = 10\ \mu\text{F}$$

Find the dc output and the percent ripple.

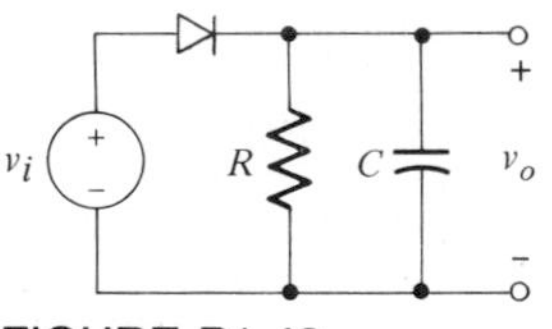

FIGURE P1.43

1.44 For the circuit shown in Fig. P1.44 the diode characteristic is given in Fig. 1.2. Find I_{Dq}, $(V_L)_{dc}$, r_d, i_d, and $(V_L)_{ac}$, analytically, using the diode large-signal and small-signal models. (*Hint:* $X_c = 1/\omega C \ll R_L$.)

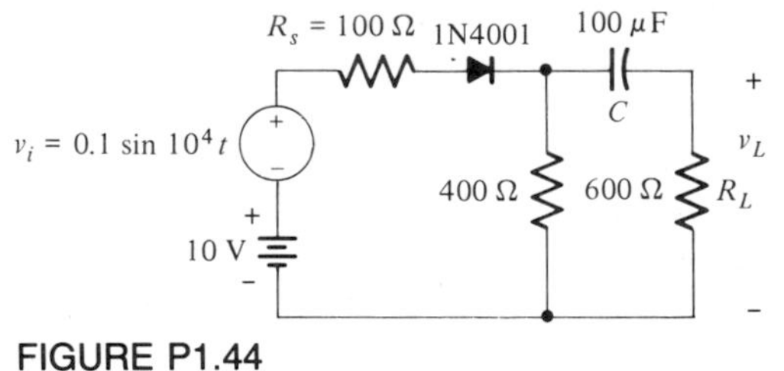

FIGURE P1.44

1.45 For the simple regulator circuit shown in Fig. P1.45, $V_z = 10$ V and 10 mA $\leq I_z \leq$ 200 mA. Determine the value of R_1 so that V_L is at 10 V with V_{dc} varying from 12 to 15 V and $R_L = 200\ \Omega$ with $\pm$ 10 percent allowed variations in the load.

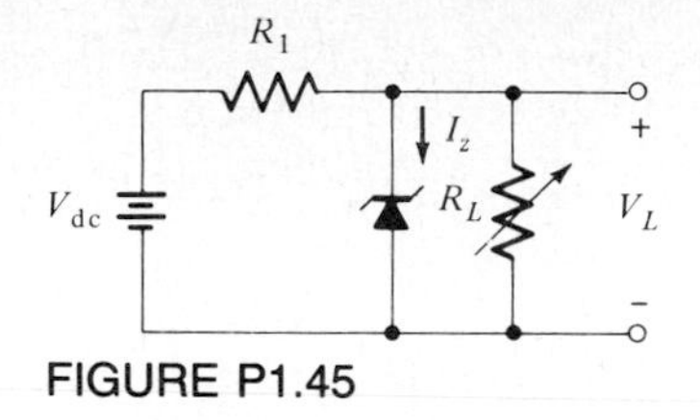

FIGURE P1.45

1.46 Determine the power dissipated in the zener diode in problem 1.45:
(*a*) When R_L is disconnected.
(*b*) When $R_L = R_{L\,min}$.

1.47 For the circuit in Fig. P1.45, if the load resistance varies from 25 to 75, determine the maximum value of R_1 to keep $V_L \simeq 10$ V. What is the maximum deviation in V_L if $R_z = 5\ \Omega$. Given that, $10\text{ mA} \leq I_z \leq 200\text{ mA}$ and $12\text{ V} \leq V_{in} \leq 15\text{ V}$.

1.48 For the zener regulator circuit shown in Fig. P1.45 the following are given: $V_L = V_z = 10$ V, $I_{z\,min} = 10$ mA, $V_i = 20$ V, and $R_S = 100\ \Omega$. The load resistor can vary from its minimum value to a maximum allowable value of 1 kΩ.
(*a*) Find $R_{L\,min}$.
(*b*) Find $I_{z\,max}$.

1.49 Consider the effect of R_z on the voltage regulator at the output voltage in problem 1.48 if $R_z = 5\ \Omega$. In other words, determine

$$\text{Voltage regulation} = \frac{V_{RL\,max} - V_{RL\,min}}{V_{RL\,min}} \times 100$$

1.50 Design the peak detector power supply circuit in Fig. P1.50; i.e., determine C and n such that the dc voltage across the load resistor is 50 V with less than 10 percent peak-to-peak ripple, given $v_i = 141 \sin 377t$.

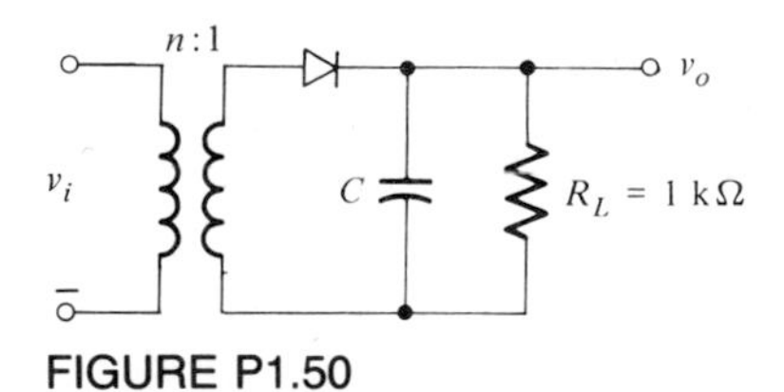

FIGURE P1.50

1.51 Design a full-wave power supply meeting the following specifications:
(1) No load output voltage, 24 V dc
(2) Full-load output voltage, ≥ 20 V dc at $I_L = 12$ A
(3) Full-load ripple, ≤ 1 V peak to peak at $I_L = 12$ A
The power source availability is 115 V rms, 60 Hz. Specify all the components needed and their values.

1.52 Sketch the transfer characteristic of the circuit shown in Fig. P1.52*a*. Use the idealized zener diode with a forward drop 0.7 V and breakdown voltage $-V_{z1}$ and $-V_{z2}$ (shown in Fig. P1.52*b*).

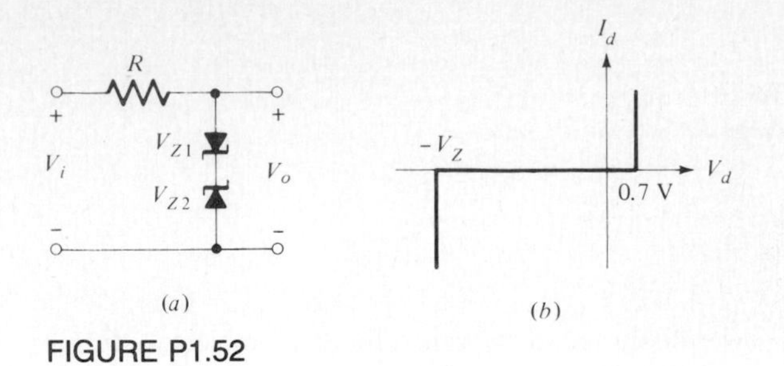

FIGURE P1.52

1.53 For the circuit shown in Fig. P1.53, assume identical diodes with $(V_0)_{on} = 0.7$ V at room temperature. Determine the voltage V_0 and the current in the diode D_2 for the following cases:
(a) $V_A = V_B = 0$ V
(b) $V_A = V_B = +5$ V
(c) $V_A = 0$, $V_B = +5$ V

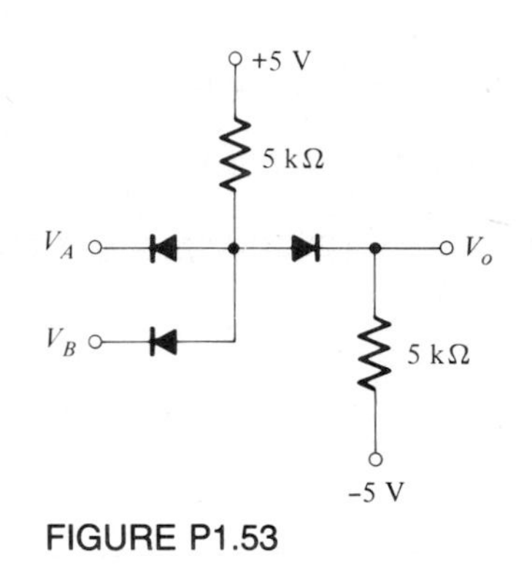

FIGURE P1.53

CHAPTER 2 Bipolar Junction Transistors

Bipolar junction transistors are very useful active devices that play an important role in today's electronic age, especially in wideband electronic amplifiers and fast digital circuitry, where they dominate the field.

A bipolar junction transistor (BJT) is comprised basically of two *pn* junctions connected back to back, and its operation depends on both electrons and holes; hence these devices are called *bipolar* junction transistors, or simply transistors. Because of yield and economy, usually one type of transistor, namely, *npn* is used in integrated circuits. The reason is that electrons have higher mobility than holes ($\mu_n/\mu_p = 2.6$ for Si at 300 K), which provides higher conductivity and better frequency performance. Complementary configurations, i.e., using both *pnp* and *npn* transistors, are also used for special circuit applications where they are needed.

In this chapter we discuss the *V-I* characteristics of BJTs in some detail. We also devote considerable effort toward developing circuit models for these devices, together with techniques for analyzing circuits containing these models. We will find that the type of models used (and the analysis techniques required) depend greatly on whether the signals that the transistor is handling are large or small. If the signals are very small, the operation of the transistor may be considered *linear,* and *small-signal analysis* may be used to predict many engineering quantities of interest. If the signals are large, the fact that the transistor is inherently a nonlinear device becomes important, and an entirely different analysis technique called *large-signal analysis* must be used. Both of these analysis techniques are widely used in areas of engineering other than electronics; in a sense, the transistor is simply an example nonlinear device to practice these methods on, although it is an extremely important example.

2.1 TRANSISTOR OPERATION

The essential parts of the physical structure of an integrated *npn* transistor are shown in Fig. 2.1, including the dominant region of transistor action. The BJT consists of a *p*-type semiconductor sandwiched between two *n*-type semiconductors (n^+ denotes heavily doped *n* type). The fabrication of integrated-circuit transistors is discussed in some detail in Appendix B, which the reader is encouraged to read in order to gain some appreciation for the material that follows.

The current-flow mechanism of the transistor can be described in terms of *pn*-junction theory, with some minor modification. Figure 2.2*a* shows a simplified schematic representation of the dominant region of transistor action of an *npn* transistor. I_E, I_B, and I_C are called the *emitter, base,* and *collector* currents, respectively, and are defined to be positive into the transistor in all cases. The actual direction of electron flow for the bias voltages applied is also shown. The circuit symbol and the actual direction of conventional current flow for the *npn* transistor is shown in Figure 2.2*b*. The three parts of the transistor are called the *emitter* (E), the *base* (B), and the *collector* (C), with the base region being very thin (of the order of 0.5 μm). The operation of a transistor can be explained best by considering the electron flow through the structure shown in Fig. 2.2*a*. Note that the *p*-type material (the base) is sandwiched between two *n*-type materials of different impurity concentrations (the emitter n^+ and the collector *n*). The opposite arrangement, i.e., *pnp* transistors such as lateral and

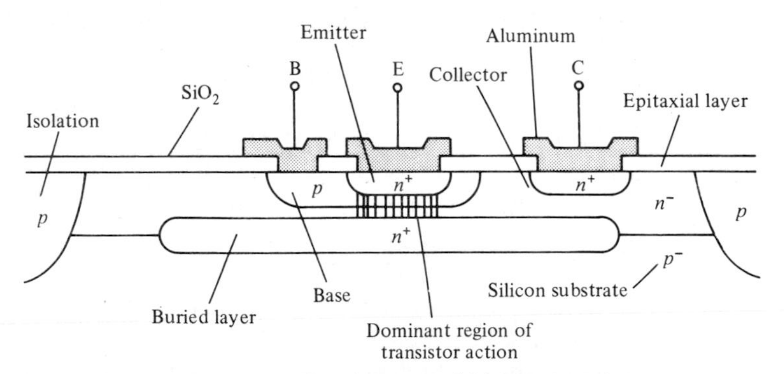

FIGURE 2.1 Structure of an integrated bipolar transistor.

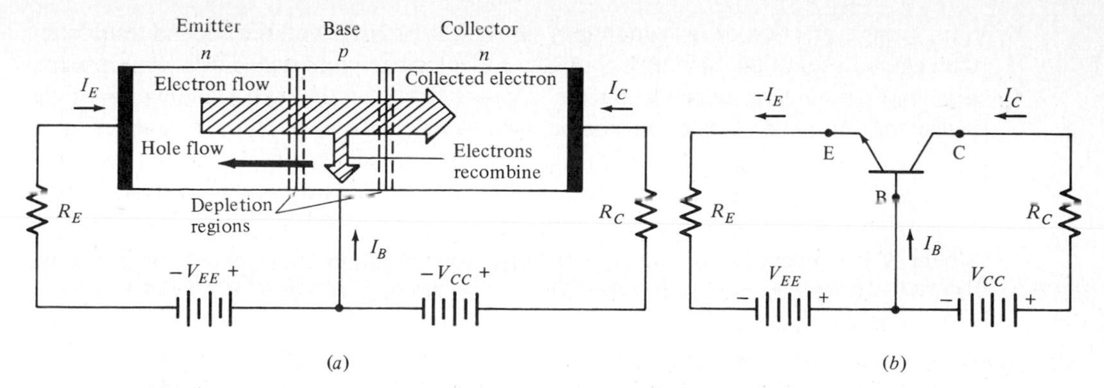

FIGURE 2.2 (*a*) Dominant region of transistor action enlarged with the associated junction voltages and terminal currents. (*b*) Circuit symbol with voltage and current directions.

substrate transistors, are also used in integrated circuits (see Appendix B). The following discussion applies equally to *pnp* transistors, but in order not to duplicate similar arguments, we shall consider only *npn* transistors. For *pnp* transistors, *everything* is reversed. *All* the voltages and currents have the opposite sign (see Fig. 2.3). Note that the actual direction of the emitter current is in the direction of the arrow.

Consider the biasing arrangement shown in Fig. 2.2, under normal mode of operation, where the emitter-base junction is forward-biased and the collector-base junction is reverse-biased. This is the usual biasing arrangement used in amplifiers. In this book the voltage V_{XY} is defined as positive when V_X is equal to or larger than V_Y. As a result of the forward bias, the emitter junction injects many majority carriers across the emitter junction (i.e., the base-to-emitter diode conducts), and electrons flow into the base region and holes into the emitter region, resulting in the emitter current. Most of the electrons injected into the base region are swept across the very thin base region by the rather large positive collector-base voltage and are "collected" by the collector. As a result of the finite transit time across the base, some recombination of electrons and holes takes place, and a small portion (1 percent or less) is lost in the base region. The collector current consists of two terms, the dominant

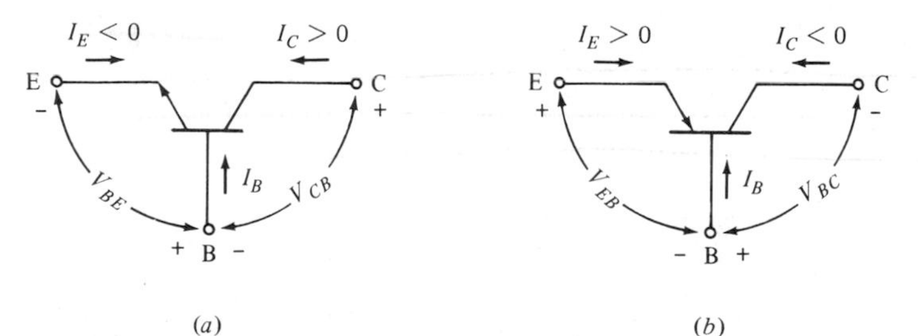

FIGURE 2.3 (*a*) *npn* transistor symbol. (*b*) *pnp* transistor symbol.

term being a fraction of I_E, which may be written as $\alpha_F I_E$, and the second term being the reverse saturation current I_{CO} of the collector-base junction diode. The proportionality constant α_F depends on the doping, construction, and dimensions of the transistor. An approximate expression used to calculate α_F at low frequencies is

$$\alpha_F \simeq 1 - \frac{1}{2}\left(\frac{W}{L_n}\right)^2 \tag{2.1}$$

where W is the base width and L_n is the diffusion length of the injected carriers in the base ($L_n = \sqrt{D_n \tau_n} \simeq 10$ μm for S_i) for *npn* transistors since $W/L_n \ll 1$, typical values of α_F range from 0.900 to 0.998. Because of the reverse bias across the collector junction, as shown in Fig. 2.2*a*, the reverse saturation current I_{CO} (which is very small) exists. It is of the order of nA (10^{-9}A) in silicon transistors at room temperature. Note that if the emitter junction is open-circuited, the collector current I_{CO} still flows. The collector current, according to the sign convention of Fig. 2.2, can then be written as

$$I_C = -\alpha_F I_E + I_{CO} \tag{2.2}$$

From Kirchhoff's current law (KCL), at the external node we have

$$I_E + I_B + I_C = 0 \tag{2.3}$$

The collector current, in terms of I_B and I_{CO}, is obtained by eliminating I_E in (2.3) and (2.2), to obtain

$$I_C = \frac{\alpha_F}{1 - \alpha_F} I_B + \frac{I_{CO}}{1 - \alpha_F} = \beta_F I_B + (\beta_F + 1) I_{CO} \tag{2.4}$$

where

$$\beta_F = \frac{\alpha_F}{1 - \alpha_F} \simeq \frac{1}{1 - \alpha_F} \quad \text{or} \quad \alpha_F = \frac{\beta_F}{1 + \beta_F} \simeq 1 - \frac{1}{\beta_F} \quad \text{for } \beta_F \gg 1$$

As we shall see the quantities α_F and $\alpha_F/(1 - \alpha_F)$ often appear in transistor circuit equations. The quantity $\alpha_F/(1 - \alpha_F)$ is commonly called β_F. A subscript 0, such as in α_0 and β_0, is usually used to denote dc quantities and F is implicitly understood. (Note that if $\alpha_0 = 0.98$, $\beta_0 \simeq 50$, and if $\alpha_0 = 0.995$, $\beta_0 \simeq 200$.) We note that α_F and β_F are sometimes defined as $\partial I_C/\partial I_E$ and $\partial I_C/\partial I_B$, respectively, and they are not exactly constants. From the above discussion, it is seen that current through the forward-biased emitter junction is transferred at essentially the same level through the reverse-biased collector junction—that is, I_C and $-I_E$ are very nearly equal, although I_C is always a little smaller in magnitude.

From (2.4), it should be clear that bipolar transistors are current-sensitive devices, i.e., basically these are current-controlled types in contrast to field-effect transistors (Chap. 3), which are voltage-controlled devices. The topic of current-sensitive and voltage-sensitive devices will be considered quantitatively and in more detail later on.

EXERCISE 2.1

For an *npn* transistor $W/L_n = 0.1$, determine α_F and β_F of the transistor.

Ans $\alpha_F = 0.995$, $\beta_F \simeq 200$

2.2 TRANSISTOR V-I RELATIONS

Mathematically, a transistor may be described in terms of two diodes *coupled* to each other. This is not unreasonable since a transistor consists of two *pn* junctions back to back, with the base region, which is common to both, *providing the coupling between input and output*.

The dc behavior of the *npn* transistor can be expressed mathematically in terms of coupled *pn*-junction diodes by the Ebers-Moll equations* [3, 4]

$$I_E = -I_{ES}\left(\exp\frac{V_{BE}}{V_t} - 1\right) + \alpha_R I_{CS}\left(\exp\frac{V_{BC}}{V_t} - 1\right) \tag{2.5a}$$

$$I_C = \alpha_F I_{ES}\left(\exp\frac{V_{BE}}{V_t} - 1\right) - I_{CS}\left(\exp\frac{V_{BC}}{V_t} - 1\right) \tag{2.5b}$$

where $V_t = kT/q$, and I_{CS} and I_{ES} are the collector and the emitter *saturation*-current values that are constant for a specific temperature. The polarities associated with (2.5) are shown in Fig. 2.9. Note that for the collector-base junction to be reverse-biased $V_{BC} < 0$, while for the emitter-base junction to be forward-biased $V_{BE} > 0$. These conditions correspond to the normal operation, which we have designated by α_F. For the reverse-active region, i.e., collector-base junction forward-biased and the emitter-base junction reverse-biased, we designate α_R, where $\alpha_R = \beta_R/(1 + \beta_R)$. Thus, the subscripts F and R designate forward and reverse conditions and are used for convenience. The Ebers-Moll equations have been shown to be true for any transistor regardless of geometry. The forward and the reverse α's are related by the following reciprocity relation:

$$\alpha_R I_{CS} = \alpha_F I_{ES} = I_S \tag{2.6}$$

where I_S is called the transistor *saturation* current. Of the four parameters α_R, α_F, I_{CS}, and I_{ES} (which, by definition, are all positive), only three are independent, and the fourth can be obtained from (2.6).

Two additional parameters, which are related to the above parameters, can also be defined. These two are I_{CO} and I_{EO}, the former having already been mentioned. The

---

*For *pnp* transistors all the polarities are reversed, i.e., $I_E = I_{ES}[\exp(V_{EB}/V_t) - 1] - \alpha_R I_{CS}[\exp(V_{CB}/V_t) - 1]$ and $I_C = -\alpha_F I_{ES}[\exp(V_{EB}/V_t) - 1] + I_{CS}[\exp(V_{CB}/V_t) - 1]$.

parameter I_{CO} is defined in terms of open-circuited conditions, i.e., the reverse collector current when $I_E = 0$ and V_{BC} is reverse-biased such that $\exp(V_{BC}/V_t) \ll 1$. Similarly, I_{EO} is the reverse emitter current that flows when $I_C = 0$ and $\exp(V_{BE}/V_t) \ll 1$. Note that both I_{CO} and I_{EO} are positive quantities. I_{CO} and I_{EO} may be obtained in terms of the other four parameters simply by applying the definitions. For example, to find I_{CO}, put $I_C = I_{CO}$ in (2.5*b*), set $I_E = 0$ in (2.5*a*), use the approximation $\exp(V_{BC}/V_t) \ll 1$, and eliminate $I_{ES}[\exp(V_{BE}/V_t) - 1]$ in (2.5*a*) and (2.5*b*) to get

$$I_{CO} = (1 - \alpha_F \alpha_R)I_{CS} \tag{2.7}$$

Similarly,

$$I_{EO} = (1 - \alpha_F \alpha_R)I_{ES} \tag{2.8}$$

The Ebers-Moll equations are important and useful in large-signal applications where the device nonlinearities must be considered. The equations apply for any bias or mode of operation. From (2.6), (2.7), and (2.8) we obtain

$$\alpha_F I_{EO} = \alpha_R I_{CO} \tag{2.9}$$

Since $\alpha_F > \alpha_R$, we have $I_{EO} < I_{CO}$.

EXERCISE 2.2

For the transistor circuit shown in Fig. E2.2, the following are given:

$$\alpha_F = 0.99 \qquad \alpha_R = 0.50 \qquad I_{ES} = 10^{-15}\text{ A} \qquad I_{CS} = 1.98 \times 10^{-15}\text{ A}$$

Determine the values I_C, I_B, and I_E at room temperature.

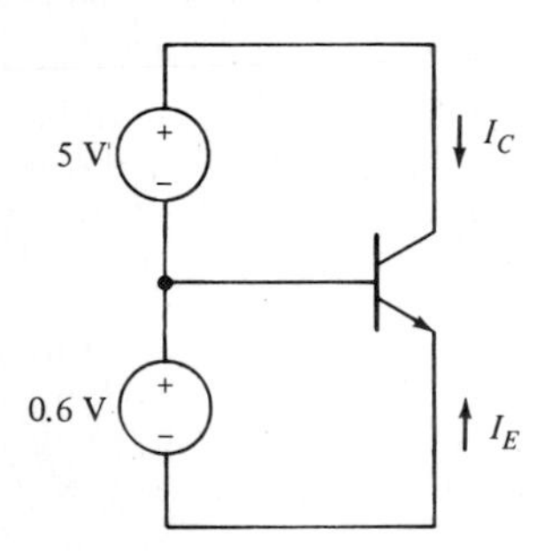

FIGURE E2.2

Ans $I_C = 2.62 \times 10^{-5}$ A, $I_E = -2.65 \times 10^{-5}$ A, $I_B = 0.03 \times 10^{-5}$ A

EXERCISE 2.3

An *npn* transistor has the following parameters: $\alpha_F = 0.99$, $I_{EO} = 2 \times 10^{-12}$ A,

$I_{CO} = 5 \times 10^{-12}$ A. Determine the collector current I_C at room temperature for $V_{BE} = 0.6$ V, $V_{BC} = -0.3$ V.

Ans 86.3 mA

2.3 CHARACTERISTICS OF BIPOLAR JUNCTION TRANSISTORS

The circuit *symbols* for *npn* and *pnp* transistors are shown in Fig. 2.3. Note that the emitter arrow in the symbol points in the actual direction of conventional (i.e., *not* electron flow) current flow from the emitter terminal in both types. In Fig. 2.4*a* (also Fig. 2.2*b*) the base is the common terminal between the input and the output, the circuit is described as being in the *common-base* (CB) configuration. Two other configurations, *common-emitter* (CE) and *common-collector* (CC), are also commonly used, and these are shown, without the biasing voltages, in Fig. 2.4*b* and *c*, respectively.

The *output* characteristics of a transistor in the CB connection is a plot of I_C vs V_{BC}, with I_E as a parameter. A typical plot of the output characteristic for an *npn* transistor is shown in Fig. 2.5. The mathematical relation is $I_C = f(V_{BC}, I_E)$. The relation is found by eliminating $\exp(V_{BE}/V_t) - 1$ in (2.5) and using (2.9) to obtain

$$I_C = -\alpha_F I_E - I_{CO}\left(\exp\frac{V_{BC}}{V_t} - 1\right) \tag{2.10}$$

The equation corresponding to (2.10) for the *pnp* transistor has all the polarities reversed.*

In the vast majority of applications, the CE configuration is used. (We shall show later that the CE configuration provides the largest power gain among the three

*For *pnp* transistors, $I_C = \alpha_F I_E + I_{CO}[\exp(V_{CB}/V_t) - 1]$.

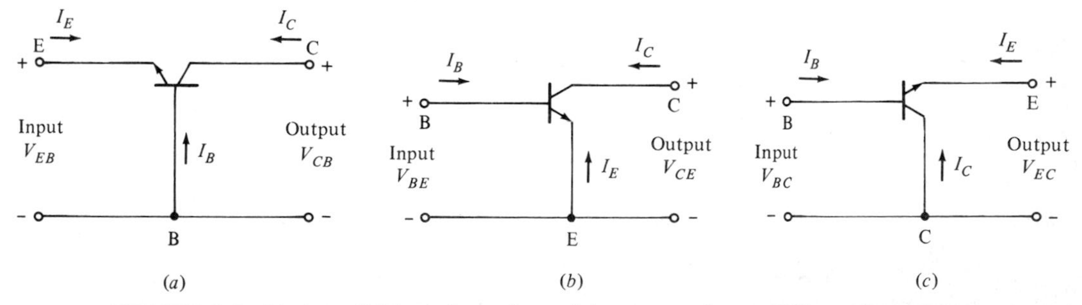

FIGURE 2.4 Various BJT configurations: (*a*) common-base (CB) configuration, (*b*) common-emitter (CE) configuration, and (*c*) common-collector (CC) configuration.

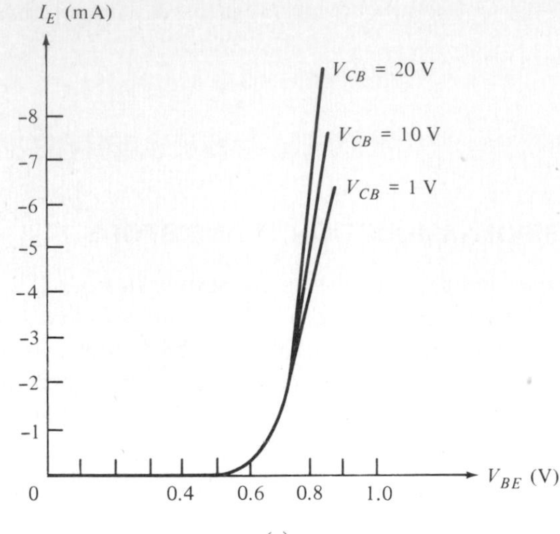

(*a*)

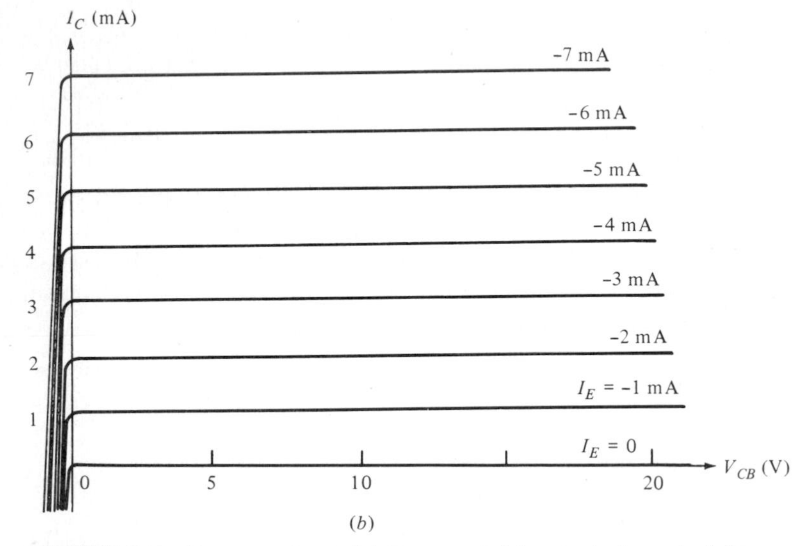

(*b*)

FIGURE 2.5 Common-base (*a*) input and (*b*) output characteristics.

configurations.) Consequently, manufacturers usually give the input and output *V-I* characteristics of the CE configuration. The measured input and output characteristics of a typical *npn* transistor in the CE configuration are shown in Fig. 2.6*a* and *b*, respectively.

The relation for the theoretical model in the CE configuration may be obtained from (2.5) and (2.3) by expressing $I_B = f(V_{BE}, V_{CE})$ and $I_C = f(I_B, V_{CE})$. The expressions become quite involved, difficult to interpret, and are less than useful. A useful form is obtained directly from (2.10) by expressing I_C in terms of I_B, V_{CE}, and V_{BE}.

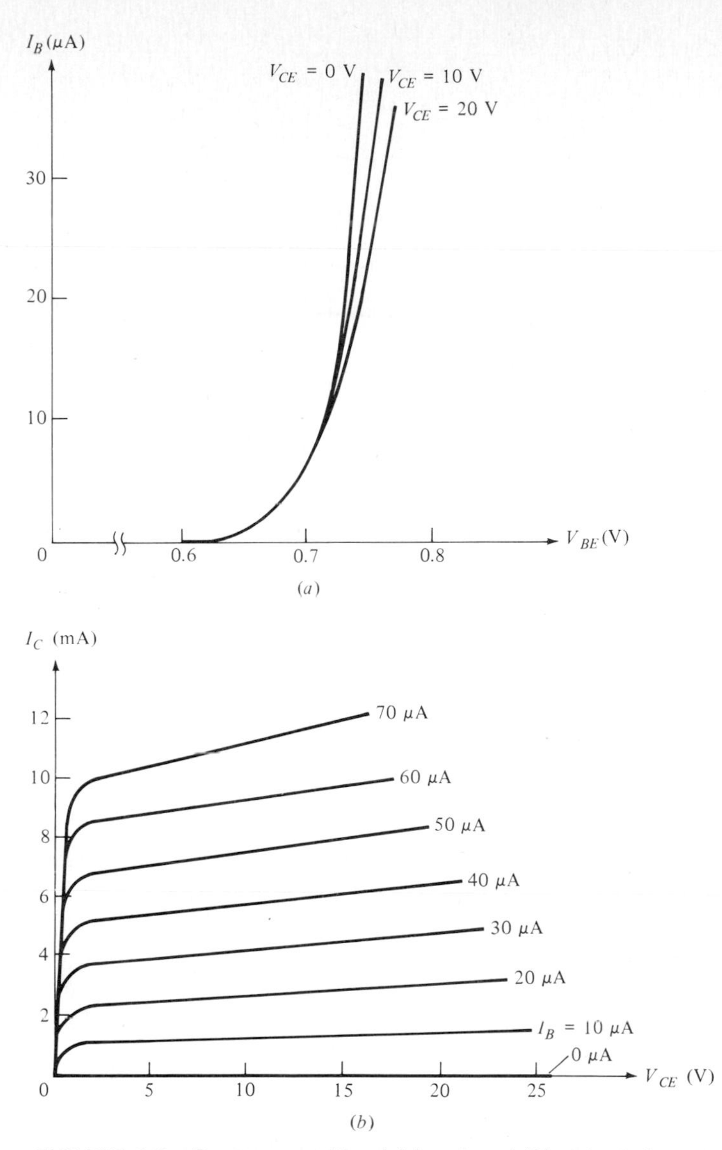

FIGURE 2.6 Common-emitter (*a*) input and (*b*) output characteristics.

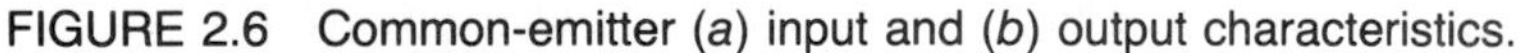

Since $I_E = -I_C - I_B$, we have

$$I_C = \frac{\alpha_F}{1 - \alpha_F} I_B - \frac{I_{CO}}{1 - \alpha_F}\left(\exp \frac{V_{BE} - V_{CE}}{V_t} - 1\right) \tag{2.11}$$

The output characteristic curve in Fig. 2.7 shows the entire region of operation including the *breakdown region,* where the collector currents get very large. The breakdown region results at high collector-base voltage and was not seen in Fig. 2.6*b* because the graph was not continued to high enough values of V_{CE}. As seen in Fig. 2.7,

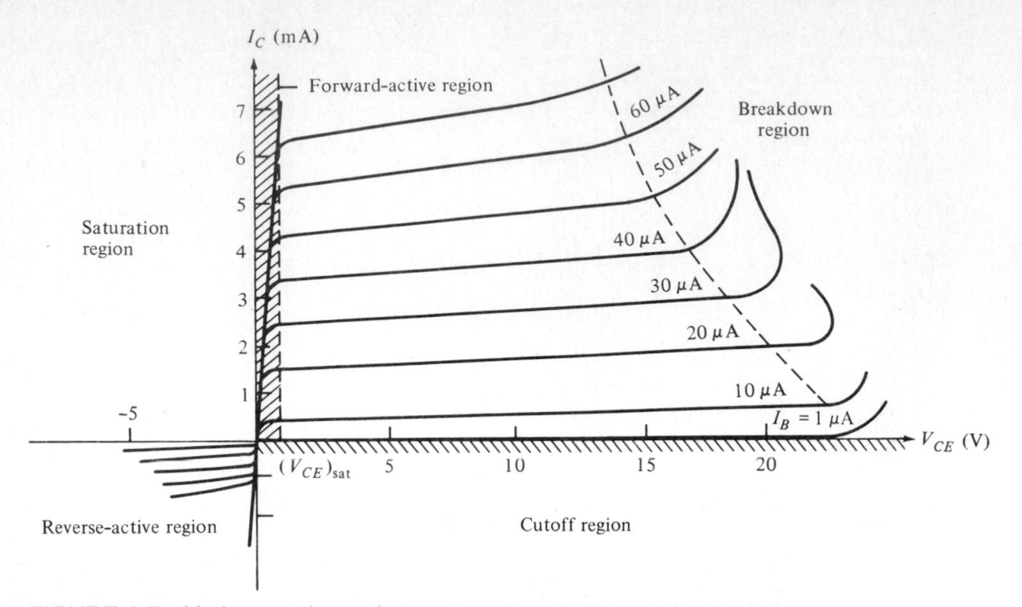

FIGURE 2.7 Various regions of operations in a BJT.

breakdown is caused by the onset avalanche multiplication of carriers in the collector junction. The primary carriers from the base region have sufficient energy to produce hole-electron pairs resulting in secondary carriers which contribute to the collector current. Hence, the maximum allowable voltage $V_{CE\max}$ must be kept well below this region.

For an *npn* transistor if $V_{CE} > (V_{BE} + 0.1 \text{ V})$, the exponential part of the second term in (2.11) is much less than 1, and (2.11) reduces to

$$I_C = \frac{\alpha_F}{1 - \alpha_F} I_B + \frac{I_{CO}}{1 - \alpha_F} = \beta_F I_B + (\beta_F + 1) I_{CO} \tag{2.12}$$

which is the same result as obtained qualitatively in (2.4).

2.4 REGIONS OF OPERATION IN BIPOLAR TRANSISTORS

The nonlinear *V-I* characteristics of the transistor describe the device under all bias voltages. In linear circuit applications, however, the transistor is not subjected to all the bias conditions. There are four possible bias conditions, depending on the state of the bias voltages of the two junctions. These four possible combinations of bias conditions define the region of operation of the transistor and are shown in Fig. 2.8. The four regions are:

1. The *forward-active region* corresponds to forward bias in the emitter-base junction and reverse bias in the collector-base junction. This is the normal transistor operation for amplification, and α_F and β_F characterize this region of operation.

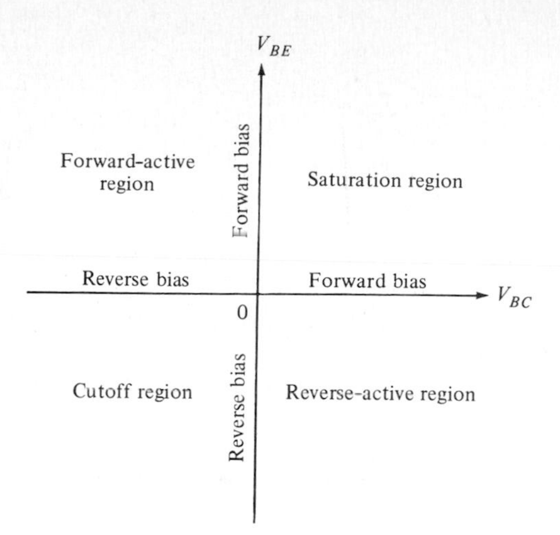

FIGURE 2.8 CE output characteristic for all regions of operation.

2. The *reverse-active region* corresponds to a reverse bias in the emitter-base junction and forward bias in the collector-base junction. The quantities α_R and β_R characterize this region. This region is rarely used.
3. The *cutoff region* corresponds to both junctions being reverse biased. Operation in this region is for switching application in the OFF mode, since the transistor acts like an open switch ($I_C \simeq 0$).
4. The *saturation region* corresponds to both junctions being forward-biased. Operation in this region is for switching application in the ON mode, since the transistor acts like a closed switch in this region ($V_{CE} \simeq 0$). The various regions of operation of an *npn* transistor output characteristic in the CE configuration is also shown in Fig. 2.7.

EXERCISE 2.4

Obtain a simple expression for the input and output characteristics of an *npn* transistor, in the CE configuration, operating in the normal-active mode. Plot the curves if the transistor has $\beta = 100$, $I_{ES} = 10^{-13}$ A.

Ans $I_B \simeq (I_{ES}/\beta)[\exp(V_{BE}/V_t)]$, $I_C \simeq \beta I_B$

2.5 LARGE-SIGNAL MODELS FOR BIPOLAR TRANSISTORS (EBERS-MOLL MODEL)

A transistor model for circuit applications called the *Ebers-Moll transistor model* is represented by the Ebers-Moll equations (2.5). The resulting model for a *npn* transistor is shown in Fig. 2.9. This model is also referred to as *the injection version* of the Ebers-Moll model. This model describes the dc behavior of the transistor.

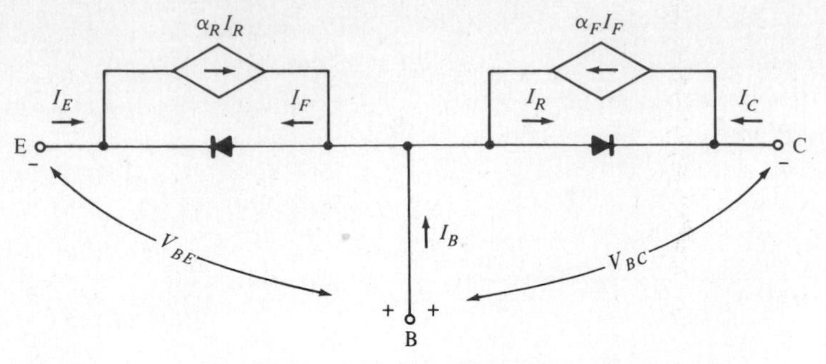

FIGURE 2.9 Ebers-Moll large-signal model for *npn* transistor in terms of the diode currents: $I_F = I_{ES}(e^{V_{BE}/V_t} - 1)$, $I_R = I_{CS}(e^{V_{BC}/V_t} - 1)$.

In most circuit analysis it is usually desirable to have the model in a form where the dependent sources are controlled by the *terminal* currents rather than by the *diode* currents. The alternative form is obtained as follows: One of the equations of the alternative form was obtained in (2.10). The other is obtained by eliminating $\exp(V_{BC}/V_t) - 1$ in (2.5) and then using (2.9), which is

$$I_E = \alpha_R I_C - I_{EO}\left(\exp\frac{V_{BE}}{V_t} - 1\right) \tag{2.13a}$$

$$I_C = \alpha_F I_E - I_{CO}\left(\exp\frac{V_{BC}}{V_t} - 1\right) \tag{2.13b}$$

The model corresponding to (2.13) is shown in Fig. 2.10. This model is referred to as the *transport version* of the Ebers-Moll model. This version is preferred in computer simulations [3]. Note that although the circuit *looks* the same, in Fig. (2.10) the dependent current sources are controlled by the *terminal* currents, not the diode currents. The model in Fig. 2.9 is in terms of short-circuited terminals, that is, when $V_{BC} = 0$, $I_E = -I_{ES}[\exp(V_{BE}/V_t) - 1]$, while the model in Fig. 2.10 is in terms of the open-circuited terminals, that is, when $I_C = 0$, $I_E = -I_{EO}[\exp(V_{BE}/V_t) - 1]$ and so on. The currents I_{ES} and I_{CS} are usually called *short-circuit saturation currents,* while I_{EO} and I_{CO} are called *open-circuit saturation currents.* These two models, which are

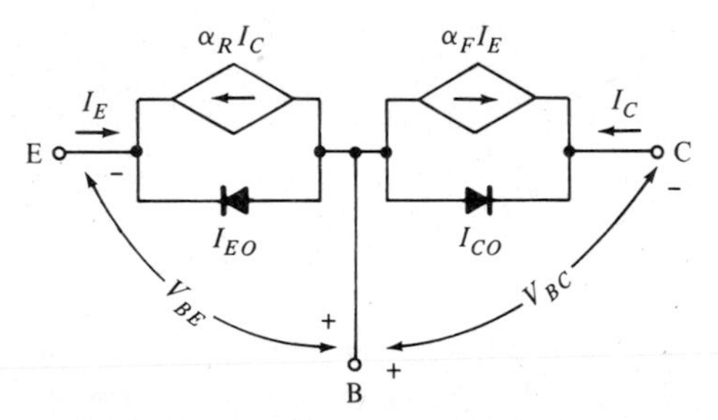

FIGURE 2.10 Ebers-Moll large-signal model for *npn* transistor in terms of terminal currents.

completely interchangeable, can be used for transistor operation in all four regions. As we might expect, if we restrict the operation of the transistor to one region only, then simplification in the model is possible. Let us examine such a simplified model for an *npn* transistor using the model in Fig. 2.10.

For the *forward-active* region, which is also referred to as the *normal* region, $V_{BC} < 0$ and $V_{BE} > 0$. For $|V_{BC}| \gg V_t (\simeq 25\text{ mV})$ (e.g., more negative than -0.1 V), the exponential part of the second term in (2.13*b*) is nearly zero, and it reduces to

$$I_C = -\alpha_F I_E + I_{CO} \simeq \beta_F I_B \qquad \text{for } I_{CO} \text{ negligible} \tag{2.14}$$

The model corresponding to the normal-active region is shown in Fig. 2.11*a*; note the disappearance of the collector diode caused by the fact that it has been assumed to be reverse-biased. I_{CO} is a small current of the order 10^{-9} A for Si at room temperature and increases for increasing temperature. The quantity α_F of a transistor in the active mode of operation is usually written as α_0, where the subscript F is omitted (it is understood since the reverse-active region is almost never used) and 0 is added to denote the dc value. The range of values of α_0 is from 0.900 to 0.998. The model of Fig. 2.11*a* can then be further approximated if we make the usual approximation of $I_{CO} \ll I_E$ and $\alpha_R \ll 1$; the result is shown in Fig. 2.11*b*.

The model for the *reverse-active* (also called *inverted-active*) mode is similar to the normal-active region except that the roles of the collector and emitter terminals are interchanged. In this mode, $V_{BE} < 0$ and $V_{BC} > 0$ for *npn* transistors. As mentioned earlier this region is almost never used.

In the *cutoff* region, *both* junctions are reverse biased. Thus, for an *npn* transistor, if $|V_{BE}|$ and $|V_{BC}|$ are both much greater than V_t, *all* the exponential terms in (2.13) can be ignored, and the equations reduce to

$$I_E = \left(\frac{1-\alpha_F}{1-\alpha_R\alpha_F}\right) I_{EO} = (1-\alpha_F) I_{ES} \tag{2.15a}$$

$$I_C = \left(\frac{1-\alpha_R}{1-\alpha_R\alpha_F}\right) I_{CO} = (1-\alpha_R) I_{CS} \tag{2.15b}$$

The model corresponding to (2.15) is shown in Fig. 2.12*a*. Since the values of I_{EO} and I_{CO} (I_{ES} and I_{CS}) are very small, the model can be further approximated as shown in Fig.

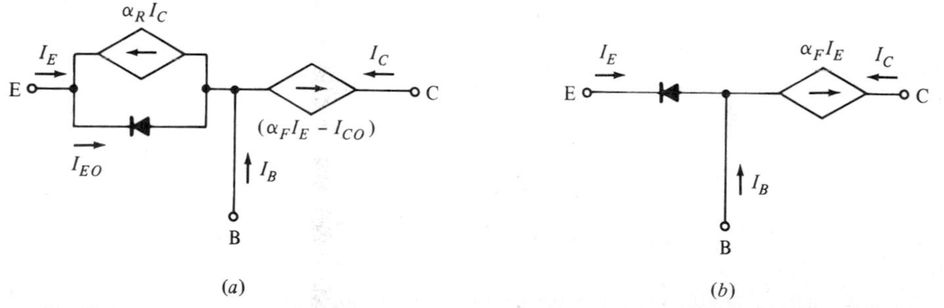

FIGURE 2.11 (*a*) Normal-active-region model. (*b*) Approximate model of (*a*).

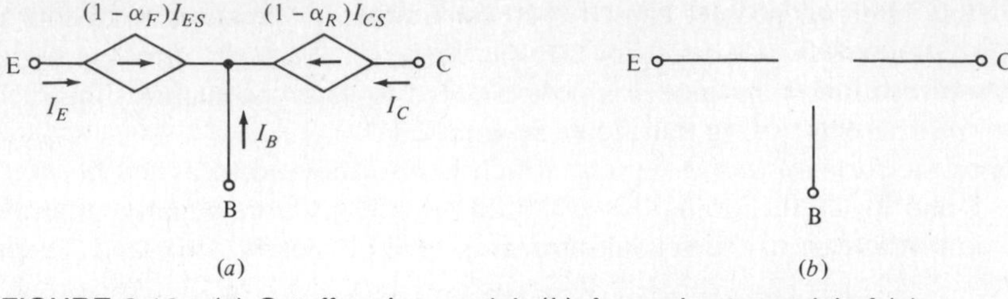

FIGURE 2.12 (a) Cutoff region model. (b) Approximate model of (a).

2.12*b*. A reverse voltage of the order of 0.1 V is usually sufficient to cut off the transistor. Note that in this mode, the transistor can be viewed as an OFF switch.

In the *saturation region* both junctions are forward-biased. In this case, the nonlinear model of Figs. 2.9 and 2.10 cannot be simplified if the exponential nature of the junction cannot be ignored. In many switching applications, however, the signals are large enough that the simplified model of Fig. 2.13*a* can be used for the saturation region for V_{BE}, $V_{CE} \gg V_t$. The saturation voltages, $(V_{BE})_{\text{sat}}$ and $(V_{CE})_{\text{sat}}$ are approximately 0.8 and 0.2 V, respectively. The emitter saturation voltage $(V_{CE})_{\text{sat}}$ can be obtained from the Ebers-Moll equations if we assume that V_{BE} and V_{BC} are both much larger than $V_t(\simeq 25\text{ mV})$ so that the -1 terms can be ignored as compared to the exponential terms in (2.5*a*) and (2.5*b*). Dividing I_C by I_B and noting that $V_{CE} = V_{BE} - V_{BC}$, with some manipulation (problem 2.12) we get

$$V_{CE} = V_t \ln \frac{1 + (I_C/I_B)(1 - \alpha_R)}{\alpha_R(1 - I_C/I_B\beta_F)} \tag{2.16a}$$

The edge of the saturation region is defined as the point where I_C is 90 percent of $\beta_F I_B$, namely, $I_C = 0.9\beta_F I_B$. Hence from (2.16*a*) we obtain

$$(V_{CE})_{\text{sat}} = V_t \ln \frac{1/\alpha_R + 0.9\beta_F/\beta_R}{1 - 0.90} \tag{2.16b}$$

As an example, for a silicon *npn* transistor if $\beta_F = 100$, $\beta_R = 1(\alpha_F = 0.99$, $\alpha_R = 0.5)$, we obtain

$$(V_{CE})_{\text{sat}} = 0.025 \ln \frac{2.0 + 0.9(100/1.0)}{1 - 0.90} = 0.025 \ln 920 \simeq 0.17\text{ V}$$

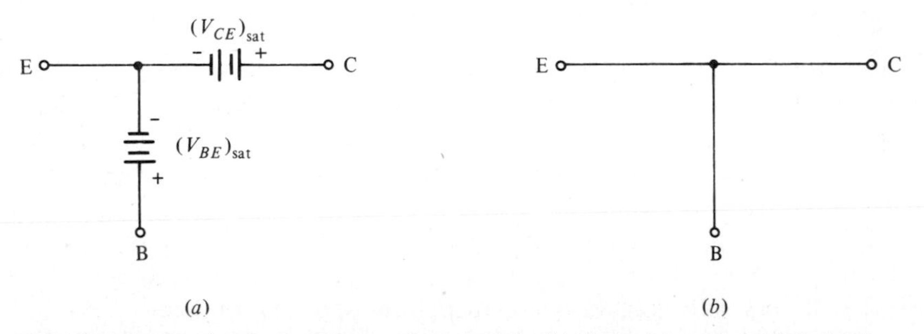

FIGURE 2.13 (a) Saturation region model. (b) Approximate model of (a).

The saturation voltages are temperature-dependent as discussed in the diode case. In fact, in some extreme cases the saturation voltages can also be ignored, in which case the model reduces to that of Fig. 2.13*b*. In this mode, the transistor can be considered as an ON switch. At saturation $I_B \geq I_C/\beta_0$, where $\beta_0 = \alpha_0/(1 - \alpha_0)$ as defined earlier.

In linear circuit applications, where the transistor is used as an amplifier, the operating conditions are confined to the forward-active region. In pulse and digital circuits, where the transistor is used as a switch, the transistor is ON when it is in the saturated region and OFF when it is in the cutoff region. During circuit operation, when the state of the transistor changes, the transistor will be driven between these two regions.

EXERCISE 2.5

For the circuit shown in Fig. E2.5 the transistor parameters are $\alpha_F = 0.99$ and $\alpha_R = 0.6$. Determine the voltage V_{CE}, and the currents I_B and I_C for (*a*) $V_{BB} = 4$ V, (*b*) $V_{BB} = 10$ V. Identify the mode of operation in each case.

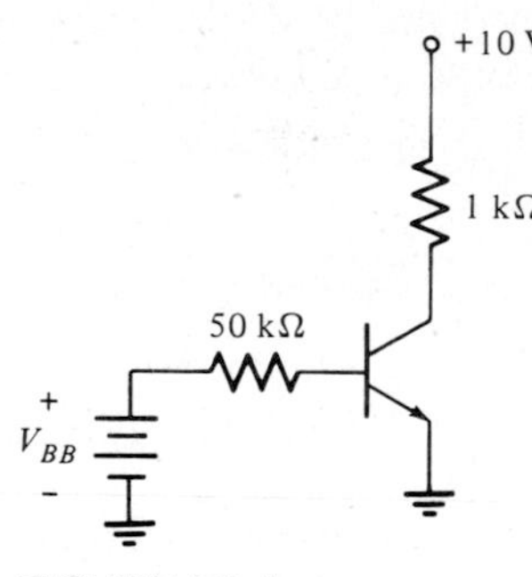

FIGURE E2.5

Ans (*a*) $V_{CE} = 3.4$ V, $I_B = 66$ μA, $I_C = 6.6$ mA, active mode; (*b*) $V_{CE} = (V_{CE})_{sat} = 0.16$ V, $I_B = 0.186$ mA, $I_C = 9.84$ mA, saturation mode

2.6 COMMON-EMITTER INVERTER

A simple *npn* common-emitter inverter circuit is shown in Fig. 2.14. As we shall see, this simple circuit exhibits the general properties of *amplifiers* as well. The input and output characteristics of the *npn* transistor is shown in Fig. 2.15*a* and *b*. The various regions of operation of the transistor are also shown on the characteristics for clarity of our discussion.

The dc transfer characteristic V_O vs $V_I (V_{CE}$ vs $V_{BE})$ of the circuit can be readily determined. For $V_I \leq 0$ the transistor is cut off so that $I_C = 0$ and $V_O = V_{CC}$ (point *A* in Fig. 2.15). Assume that V_I is increased from a value of zero up to ≤ 0.6 V. Since for these values of V_I, $V_{BE} \leq 0.6$ V, $I_B = 0$, and $I_C = 0$, we are still in the cutoff

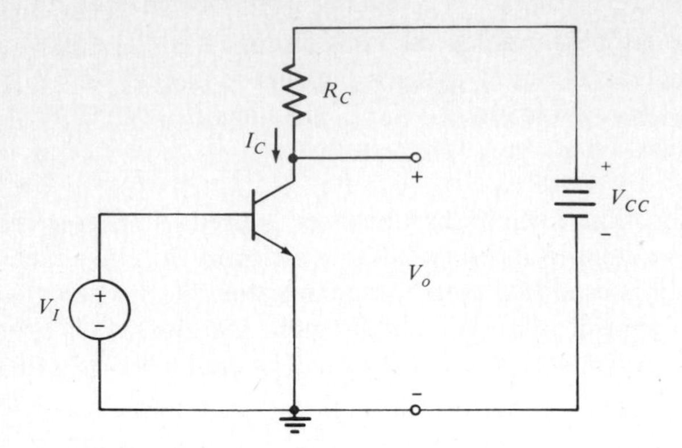

FIGURE 2.14 A simple inverter circuit.

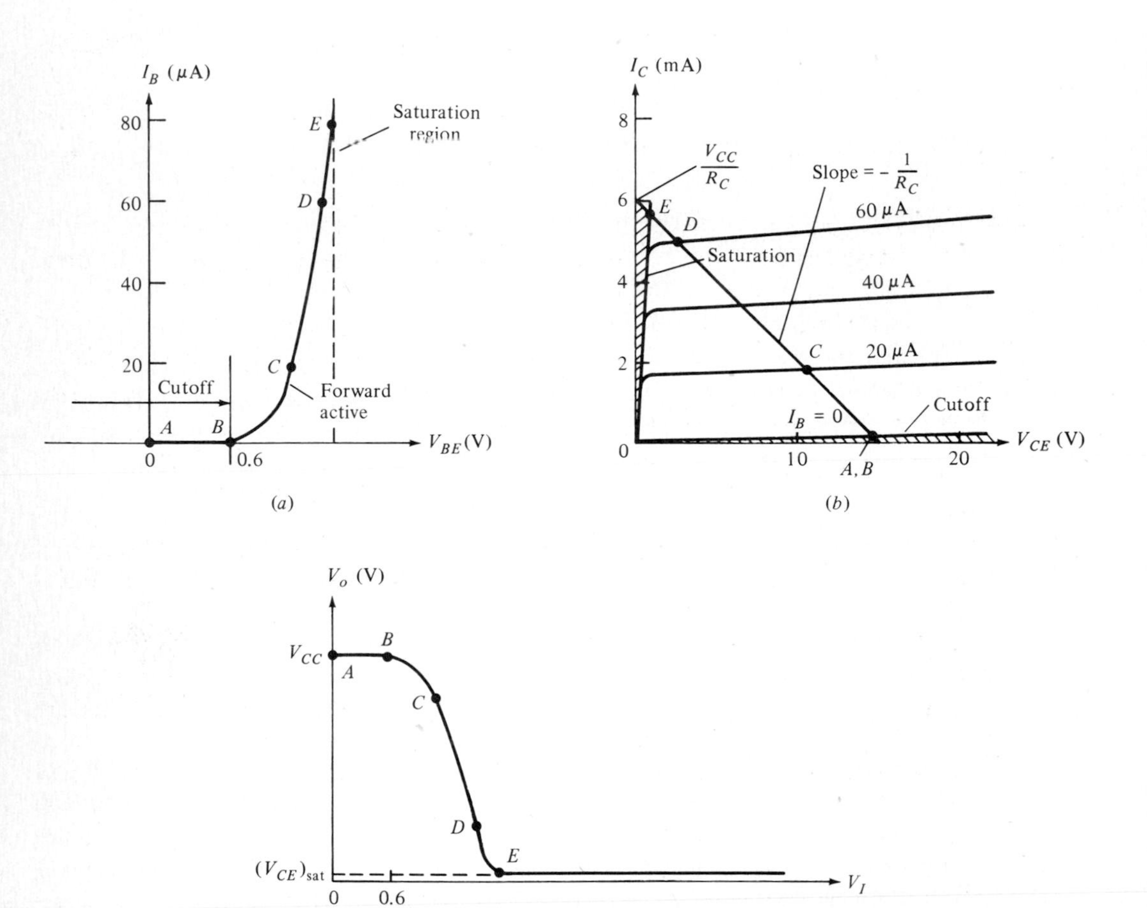

FIGURE 2.15 (*a*) Inverter input characteristic. (*b*) Inverter output characteristic with the load line. (*c*) Inverter transfer characteristic.

mode, and hence no current flows through R_C other than the leakage current I_{CO} (which is negligible), and therefore, $V_O = +V_{CC}$ (point B). For V_I exceeding the threshold voltage of 0.6 V, the transistor enters the forward-active region and from (2.14*b*)

$$I_C \simeq -\alpha_F I_E \simeq \beta_F I_B \tag{2.17a}$$

Since I_B increases exponentially with V_{BE} (Fig. 2.15*a*), so does I_C according to (2.16).

Writing a KVL around the output loop, we have

$$V_{CE} = V_{CC} - I_C R_C \tag{2.17b}$$

But $V_O = V_{CE}$, hence from (2.17) if the resistor R_C has a large value, the output voltage V_O drops very rapidly as I_C increases as shown in Fig. 2.15*c* (points C and D).

Precise prediction of the transfer characteristic may be done in the following way. In (2.17) I_C is a nonlinear function of V_{CE} and V_{BE} (hence I_B), so we rewrite (2.17) in functional form as

$$V_{CE} = V_{CC} - I_C(V_{CE}, I_B)R_C \tag{2.18a}$$

from which we get

$$I_C(V_{CE}, I_B) = -\left(\frac{1}{R_C}\right)V_{CE} + \frac{V_{CC}}{R_C} \tag{2.18b}$$

Note that (2.18) is just a reworked KVL which must be satisfied. However, it cannot be solved algebraically because we only know $I_C = f(V_{CE}, I_B)$ in graphical form. Hence we resort to a graphical method of solution.

The left-hand side of (2.18*b*) must equal the right-hand side. The right-hand side is the equation of a straight line in the I_C, V_{CE} plane and is called the *load line* since R_C is normally called *load resistor*. It has a slope of $-1/R_C$ and an intercept of V_{CC}/R_C as shown in Fig. 2.15*b*. The left-hand side is simply the output characteristic, i.e., the values of I_C for different values of V_{CE} and I_B. The point-by-point graphical solution is then indicated by points C and D. This graphical method of solution is quite generally useful in transistor circuit analysis and will often be used. In fact, it is the *only* general method that exists, and it is the *only* method that is capable of predicting a number of interesting amplifier phenomena—the most important of which is distortion.

The transfer characteristic is obtained by considering various values of the input voltage V_I. Note that $V_I = V_{BE}$, and for various values of V_{BE} we read the corresponding values of I_B. The intersection of the load line with these values of I_B determines the corresponding values V_{CE}, hence V_O.

Note that when the input voltage is near E, saturation occurs and $V_{CE} = (V_{CE})_{sat}$ ($\simeq 0.2$ V), and from this point on the output voltage, V_O remains constant as in Fig. 2.15*c*.

From Fig. 2.15*c* the switching behavior of the transistor under large-signal conditions is obvious: If $V_I \gg 0.6$, $V_O \simeq 0$; if $V_I \ll 0.6$ V, $V_O \simeq V_{CC}$. Note that V_O can be changed from V_{CC} to $(V_{CE})_{sat}$ by appropriate change in V_I. The reader should also notice that while the transistor is in the forward-active region, the graph of V_O vs V_I is *very steep,* so that a small change in input voltage can cause a very large change in the output voltage. This is called *voltage amplification*. Similarly, since in this region $I_C \simeq \beta_F I_B$, i.e., *current amplification* is also obtained: a small base current controls a much larger collector current. Thus under small-signal conditions in the forward-active region we get voltage amplification and current amplification and, hence, power amplification.

EXERCISE 2.6

For the circuit shown in Fig. 2.14, $V_{CC} = 20$ V, $R_C = 1$ kΩ. The transistors parameters are $\beta = 100$, $I_{ES} = 10^{-14}$ A.

(a) Determine the base current and the collector voltage if $V_I = 0.7$ V. (*Note:* $V_t = 25$ mV at room temperature.)
(b) If the transistor $(V_{CE})_{sat} = 0.3$ V, determine the minimum value of I_B that will saturate the transistor.

Ans (*a*) $I_B = 0.145$ mA, $V_{CE} = V_0 = 5.5$ V; (*b*) $I_B = 0.197$ mA

2.7 EMITTER-COUPLED INVERTER

In the simple inverter circuit of Fig. 2.14, the transistor switch does not turn off from saturation instantly. In Chap. 12 it is shown that the saturation process leads to storage time delay and thus lowers the speed of the transistor switch. One way to avoid this process is to connect a Schottky diode (see Appendix B) as a clamping circuit between the collector and the base terminals, thus preventing saturation and saturation delay time.

One circuit that is especially attractive for *high-speed switching* is shown in Fig. 2.16. This circuit, as described in Chaps. 12 and 13, avoids the saturation time of the transistor and has the fastest switching times. This inverter is usually called *emitter-coupled logic* (ECL) or sometimes *current-mode logic*. This circuit is capable of very fast switching speed less than 1 ns (10^{-9} s), and is used in very-high-speed integrated circuits (VHSIC). The circuit also finds wide application as an amplifier input stage in linear integrated circuits. Operational amplifiers utilize the emitter-coupled circuit as a differential amplifier. Because of the importance of the emitter-coupled circuit, its transfer characteristic is considered in the following.

We shall assume identical transistors with high current gain ($\beta_0 \geq 100$) so that $|I_C| \simeq |I_E|$. The resistor R_E is assumed to be very large so that R_E in conjunction with V_{EE} provides an equivalent current source.

For negative V_I the base-emitter junction of Q_2 is forward-biased and of Q_1 is reverse-biased; hence Q_1 is off and its collector voltage is at V_{CC}. The current I_E is solely from Q_2, and its collector voltage is at $V_{CC} - I_E R_C$. For V_I positive, the base-emitter junction of Q_1 is forward-biased and of Q_2 is reverse-biased, and thus the

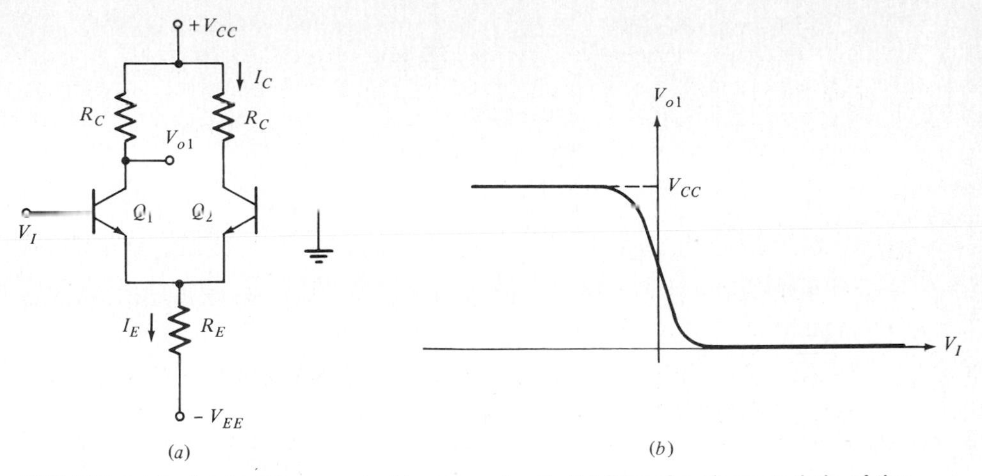

FIGURE 2.16 (a) Emitter-coupled inverter circuit. (b) Transfer characteristic of the circuit.

situation is reversed, and the collector voltage Q_1 is at

$$V_{01} = V_{CC} - I_C R_C \tag{2.19a}$$

But $I_C \simeq I_E \simeq V_{EE}/R_E$; hence

$$V_{01} = V_{CC} - \frac{V_{EE} R_C}{R_E} \tag{2.19b}$$

For $V_{CC} = V_{EE}$ and $R_C = R_E$ the transfer characteristic is shown in Fig. 2.16*b*. Note that the voltage levels are basically independent of transistor parameters and depend on the ratio of two resistors, which is desirable in integrated circuits.

More will be said about the circuit in Fig. 2.16 and its modifications and applications for linear and nonlinear circuits in later chapters.

EXAMPLE 2.1

For the circuit shown in Fig. 2.17*a*, the transistor input and output characteristics are given in Fig. 2.17*b* and *c*, respectively:

(a) Determine the Q points and β_0 of the transistor.
(b) Obtain the transfer characteristic V_{CE} vs V_{BE}.
(c) Determine graphically the ac voltage gain if an input signal voltage $v_{in} = 0.05 \sin \omega t$ is applied at the input as shown.
(d) Show the output signal $v_0(t)$ if $v_{in} = 0.5 \sin \omega t$.

From the input characteristic Fig. 2.17*b*, we obtain I_{Bq}, V_{BEq} (0.1 mA, 0.54 V), and from the output characteristic Fig. 2.17*c*, we obtain I_C, V_{CEq}

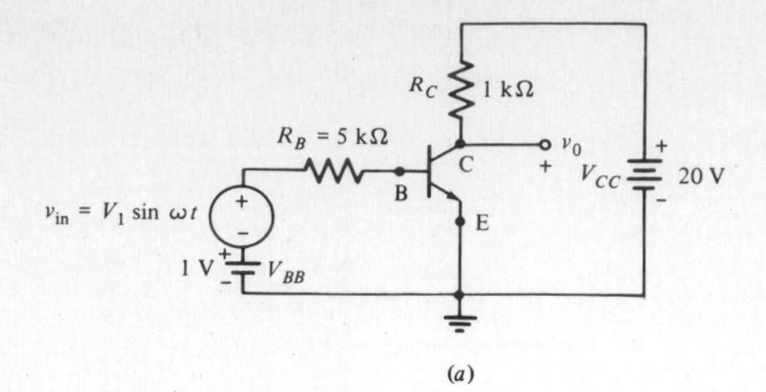

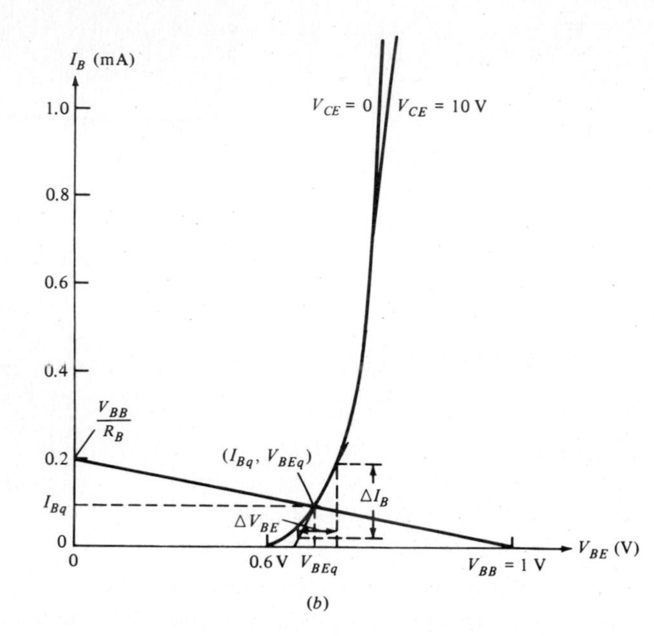

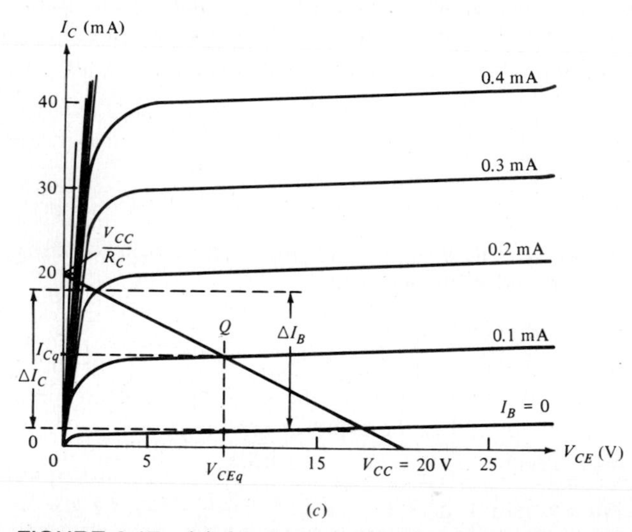

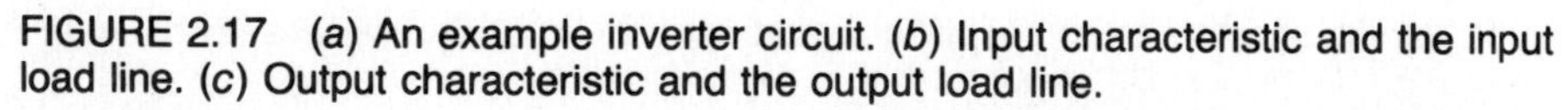

FIGURE 2.17 (*a*) An example inverter circuit. (*b*) Input characteristic and the input load line. (*c*) Output characteristic and the output load line.

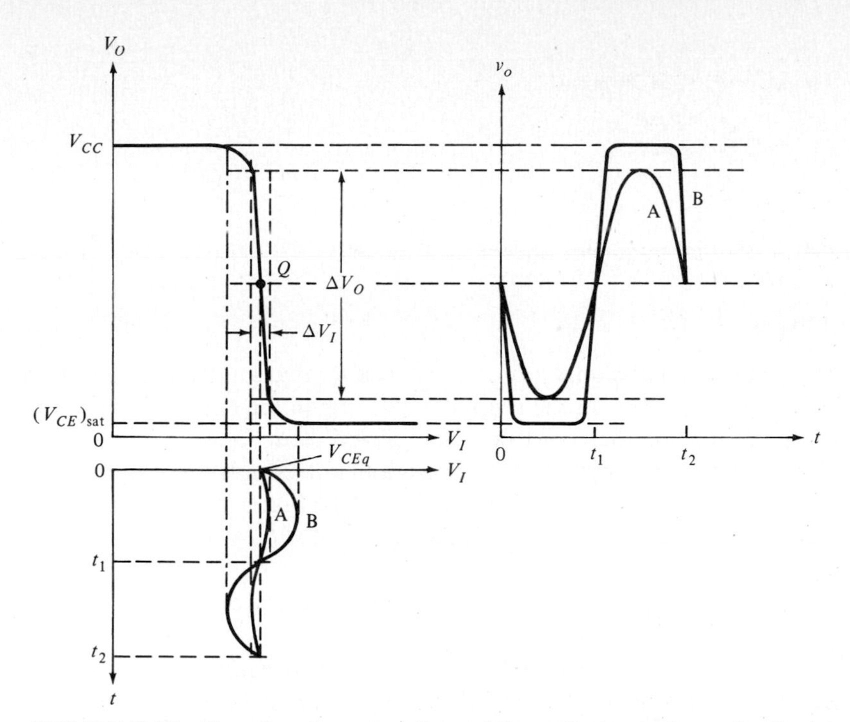

FIGURE 2.18 Transfer characteristic and the output waveforms for the input waveforms (one for small input signal and one for large input signal).

(10 mA, 9.8 V). From the output characteristic, near Q,

$$\beta_0 \simeq \frac{\Delta I_B}{\Delta I_C} = 100$$

Following the procedure discussed in Sec. 2.6, we obtain the transfer characteristic of the circuit as shown in Fig. 2.18. The ac voltage gain $v_0/v_{in} \simeq 18$. The output voltage for $v_{in} = 0.5 \sin \omega t$ is also shown in Fig. 2.18 (curve B). Note the distorted waveform of the output signal.

2.8 LOW-FREQUENCY SMALL-SIGNAL MODELS

Amplification of small signals is one of the most important functions in linear electronic circuits. When used as an amplifier, the transistor is usually biased (i.e., operated) in the linear portion of the normal-active region. The small-signal circuit is a model describing the operation of the device under small-signal conditions, i.e., the signal variation is near the quiescent Q point. We have already encountered the Q point in Chap. 1 and in Example 2.1. We assume in the following that the magnitude of the ac signal is small so that the operation of the circuit is near Q, and hence no distortion of the signal occurs.

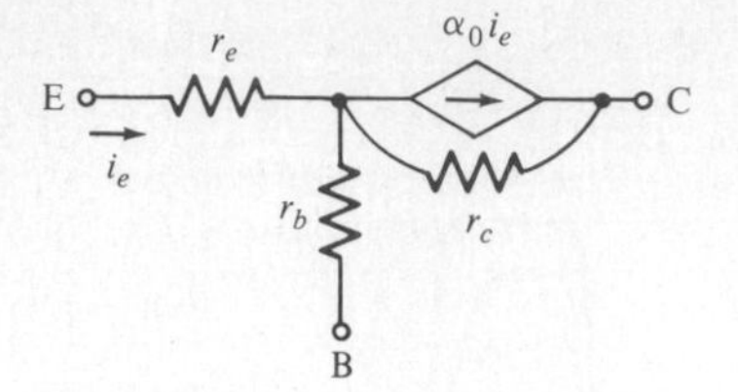

FIGURE 2.19 Low-frequency equivalent T model.

A. The Tee Model

The concept of a small-signal model developed for a *pn*-junction diode can be readily applied to the transistor. In the normal-active region the emitter-base junction diode is forward-biased and the collector-base junction diode is reverse-biased. For small-signal conditions, i.e., signals small enough that the operating region is near Q, the forward-biased diode is modeled simply by r_d (see Fig. 1.39), which we shall call r_e for a transistor. The value of the resistance r_e is given by (1.41)

$$r_e = \frac{kT}{\lambda q I_E} \simeq \frac{kT}{q I_E} = \frac{0.025}{I_E} \qquad \text{(at room temperature)} \tag{2.20}$$

The collector-base diode, which is reverse-biased, is equivalent (under small-signal conditions) to a leakage resistance, which we will call r_c. The value of r_c is very large (of the order of megaohms). Since current flows from the emitter to the collector through the base region, we must add the base-spreading resistance r_b in the base terminal as well as the controlled-current source to the model. The controlled-current source from (2.2) is simply $i_c = \alpha_0 i_e$. Note that lowercase i is used to denote ac signals, and α_0 denotes the value of α at low frequencies. The resulting model, known as the *equivalent Tee (T) model,* is shown in Fig. 2.19. Typical element values of the T model are $r_e = 10\ \Omega$ (for $I_C = 2.5$ mA), $r_b = 100\ \Omega$, $\alpha_0 = 0.99$, $r_c = 1\ \text{M}\Omega$.

These parameter values, of course, vary for different transistors and even from unit to unit. The model is, of course, valid for the transistor in any of the three configurations, namely, CB, CC, and CE configurations.

B. The Hybrid-π Model

A simple CE stage with its characteristics is shown in Fig. 2.17. The input and output characteristics of the CE transistor in the normal region are usually given by the manufacturer. It is assumed that by proper choice of the values of R_C and R_B the transistor is operating in the forward-active region at quiescent point Q. Biasing, i.e., the establishment of a proper operating point, is discussed in more detail in Sec. 2.10.

A bipolar transistor, as we have seen, is a nonlinear two-port (three-terminal) device. The nonlinear relations for the CE transistor characteristics may be written in the form:

$$v_{BE} = f(i_B, v_{CE}) \tag{2.21a}$$

$$i_C = g(i_B, v_{CE}) \tag{2.21b}$$

Taylor series for $f(x)$ @ c: $f(x) = f(c) + f'(c)(x-c) + \frac{f''(c)}{2!}(x-c)^2 + \cdots + \frac{f^{(n)}(c)}{n!}(x-c)^n \ldots$

where f and g are nonlinear functions and $i_C = i_c + I_{Cq}$, $v_{CE} = v_{ce} + V_{CEq}$, and so on; i_C is the total current ac plus dc, i_c is the ac current superimposed on the dc quiescent value of I_{Cq}, as shown in Fig. 2.20. Under small-signal conditions (i.e., $v_{ce} \ll V_{CEq}$, $i_c \ll I_{Cq}$), we want to obtain *analytically* a small-signal model for the transistor. We have already used this approach in obtaining a small-signal model for the diode in Chap. 1. In this case, however, we have *two* nonlinear functions of *two* variables to deal with. Expanding (2.21) in a two-dimensional Taylor series about the Q point, we get

c in above series = Q

$$v_{BE} = v_{BE}(I_{Bq}, V_{CEq}) + \left.\frac{\partial v_{BE}}{\partial i_B}\right|_Q (i_B - I_{Bq}) + \left.\frac{\partial v_{BE}}{\partial v_{CE}}\right|_Q (v_{CE} - V_{CEq}) + \cdots \tag{2.22a}$$

$$i_C = i_C(I_{Bq}, V_{CEq}) + \left.\frac{\partial i_C}{\partial i_B}\right|_Q (i_B - I_{Bq}) + \left.\frac{\partial i_C}{\partial v_{CE}}\right|_Q (v_{CE} - V_{CEq}) + \cdots \tag{2.22b}$$

We neglect the second- and higher-order terms (small-signal assumption, i.e., $i_B - i_{Bq}$ and $v_{CE} - V_{CEq}$ are small enough) and substitute the values of v_{BE} and i_C in (2.22). Note that

$$v_{BE} = v_{be} + V_{BEq} = v_{be} + v_{BE}(I_{Bq}, V_{CEq}) \tag{2.23a}$$

$$i_C = i_c + I_{cq} = i_c + i_C(I_{Bq}, V_{CEq}) \tag{2.23b}$$

Hence from (2.22) and (2.23) we have

$$v_{be} \simeq \left.\frac{\partial v_{BE}}{\partial i_B}\right|_Q (i_b) + \left.\frac{\partial v_{BE}}{\partial v_{CE}}\right|_Q (v_{ce}) \tag{2.24a}$$

$$i_c = \left.\frac{\partial i_C}{\partial i_B}\right|_Q (i_b) + \left.\frac{\partial i_C}{\partial v_{CE}}\right|_Q (v_{ce}) \tag{2.24b}$$

In (2.24*a*) the input voltage is a linear function of the input current and the output voltage, while in (2.24b) the output current is a linear function of the input current and

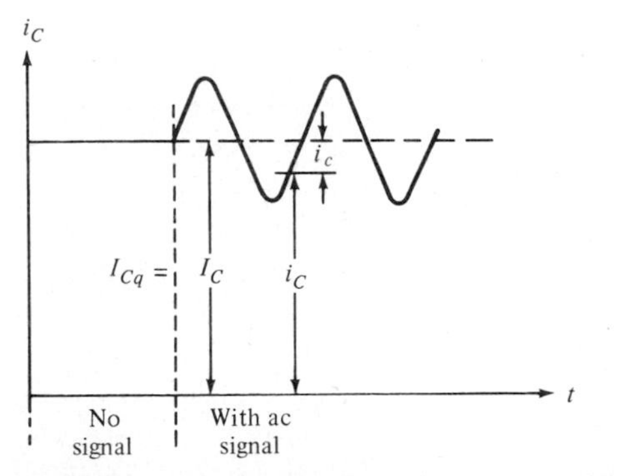

FIGURE 2.20 Illustration of ac, dc, and total currents for small-signal condition.

the output voltage. The partial derivatives in the linearized two-port description in (2.24) can be identified as the h parameters. (For a brief discussion of two-port theory the reader is referred to Appendix C.) Thus from (2.24) we can write*

$$v_{be} = h_{11} i_b + h_{12} v_{ce} \tag{2.25a}$$

$$i_c = h_{21} i_b + h_{22} v_{ce} \tag{2.25b}$$

where
$$h_{11} = h_{ie} = \left.\frac{v_{be}}{i_b}\right|_{v_{ce}=0} = \left.\frac{\partial v_{BE}}{\partial i_B}\right|_Q \simeq \left.\frac{\Delta v_{BE}}{\Delta i_B}\right|_{V_{CEq}} \tag{2.26a}$$

$$h_{12} = h_{re} = \left.\frac{v_{be}}{v_{ce}}\right|_{i_b=0} = \left.\frac{\partial v_{BE}}{\partial v_{CE}}\right|_Q \simeq \left.\frac{\Delta v_{BE}}{\Delta v_{CE}}\right|_{I_{Bq}} \tag{2.26b}$$

$$h_{21} = h_{fe} = \left.\frac{i_c}{i_b}\right|_{v_{ce}=0} = \left.\frac{\partial i_C}{\partial i_B}\right|_Q \simeq \left.\frac{\Delta i_C}{\Delta i_B}\right|_{V_{CEq}} \tag{2.26c}$$

$$h_{22} = h_{oe} = \left.\frac{i_c}{v_{ce}}\right|_{i_b=0} = \left.\frac{\partial i_c}{\partial v_{CE}}\right|_Q \simeq \left.\frac{\Delta i_C}{\Delta v_{CE}}\right|_{I_{Bq}} \tag{2.26d}$$

Note that I_{Bq}, V_{CEq} are the quiescent point values and Δ indicates incremental change around Q.

The h parameters are determined from the characteristic curves as follows: Consider the input and output characteristics shown in Fig. 2.17*b* and *c*, respectively. To determine h_{fe}, draw a vertical line through Q in the output characteristic curve. The ratio of the current increments in the neighborhood of Q from (2.26*c*) is then the required value of h_{fe}.

In this case

$$\left.\frac{\Delta I_C}{\Delta I_B}\right|_Q \simeq \frac{20 \times 10^{-3}}{0.2 \times 10^{-3}} = 100$$

This quantity is a very important small-signal parameter of the transistor, and as we have seen, it is usually denoted by β_0. It is to be noted that h_{fe} and β_0 are not exactly equal since, from (2.14) and neglecting I_{CO}, we have

$$I_C \simeq \frac{\alpha_0}{1 - \alpha_0} I_B = \beta_0 I_B \tag{2.27}$$

whereas

$$h_{fe} = \frac{\Delta i_C}{\Delta i_B} = \beta_0 + \frac{\Delta \beta_0}{\Delta i_B} I_B \tag{2.28}$$

---

*In the literature one often encounters the notation $v_{be} = h_{ie} i_b + h_{re} v_{ce}$ and $i_c = h_{fe} i_b + h_{oe} v_{ce}$, where the first subscripts, i, r, f, o denote the input, reverse, forward, and output, respectively, and the second subscript e denotes the common emitter configuration. We shall henceforth use this notation.

If the second term in (2.28) is much smaller relative to β_0, then they are equal, i.e., $h_{fe} \simeq \beta_0$. Usually, this is the case, and *we will assume it to hold throughout this text.*

The parameter h_{ie} is determined from (2.26*a*). Draw a tangent line at Q in the input characteristic, the inverse of the slope of the lines given the value of h_{ie}, namely,

$$h_{ie} = \left.\frac{\Delta V_{BE}}{\Delta I_B}\right|_Q \simeq \frac{0.11}{0.22} = 500\ \Omega$$

A more accurate method of determining h_{ie} is given in Sec. 2.9.

The parameter h_{re} is determined from the ratio of the voltage increments, from (2.26*b*), as read from a horizontal line drawn through Q at the input characteristics. However, as we see from the input characteristic v_{BE} is almost independent of v_{CE}, and hence $h_{re} \simeq 0$.

The parameter h_{oe} is determined from (2.26*d*) by drawing a tangent line through Q at the output characteristic and finding its slope. From the output characteristic we obtain

$$h_{oe} = \left.\frac{\Delta I_C}{\Delta V_{CE}}\right|_Q \simeq \frac{0.2 \times 10^3}{10} = 20\ \mu\text{mhos}$$

For the quiescent point $I_{Cq} = 10$ mA, $V_{CEq} = 9.8$ V, $I_{Bq} = 0.1$ mA, $V_{BEq} = 0.54$ V of the example, we have the following h parameters from the characteristic curves:

$$\begin{aligned} h_{ie} &\simeq 500\ \Omega \qquad & h_{re} &\simeq 0 \\ h_{fe} &\simeq 100 \qquad & h_{oe} &= 2.0 \times 10^{-6}\ \text{mho} \end{aligned} \tag{2.29}$$

Note that h_{ie} has the unit of *ohms* and h_{oe} is in *mhos*.* The other two parameters are h_{21}, a dimensionless *current ratio,* and h_{12}, a dimensionless *voltage ratio*.

---

*The symbol S which designates Siemens is also used as the dimension for reciprocal ohm, or mho, in some books.

EXAMPLE 2.2

For the amplifier circuit of Example 2.1, determine the output voltage using the h parameters if $v_{\text{in}} = 0.05 \sin \omega t$.

We have already determined the h parameters in (2.29). The two-port small-signal h parameters (see also Fig. C.8) and the associated circuitry from Fig. 2.17*a* is shown in Fig. 2.21. Note that in the small-signal ac model the batteries *do not* play any rôle. They serve only to establish the Q point and hence they appear only in large-signal model. Since $h_{re} \simeq 0$, the controlled-voltage source is zero. From the circuit we have

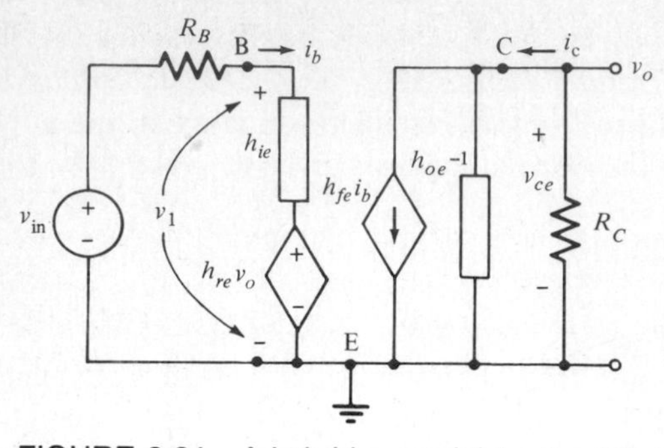

FIGURE 2.21 A hybrid-π model for the CE configuration.

$$v_O = -h_{fe}i_b(R_C \| h_{oe}^{-1}) = -100(10^3 \| 50 \times 10^3)i_b \simeq -10^5 i_b$$

But

$$i_b = \frac{v_{in}}{R_B + h_{ie}} = \frac{v_{in}}{5 \times 10^3 + 0.5 \times 10^3} = \frac{v_{in}}{5.5 \times 10^3}$$

Hence

$$v_O = -10^5\left(\frac{1}{5.5 \times 10^3}\right)v_{in} = -18.2(0.5 \sin \omega t) = 0.91 \sin \omega t$$

which is approximately the same result, namely, a voltage gain of $0.91/0.05 \simeq 18$. The parameters listed in (2.29) are typical for a variety of types of transistors. The characteristic curves vary appreciably with changes in the transistor's temperature environment; such data are often provided by the manufacturer. Consequently the transistor parameters vary with changes in the temperature as well as with Q-point variations. It is important, therefore, to stabilize the Q point in bipolar transistor circuits. This topic is discussed in the next section.

Typical variations of the h parameters for an *npn* silicon planar transistor as a function of the operating point and the temperature are shown in Fig. 2.22.

Equivalent Circuit

From (2.25), we may derive an *equivalent circuit* for the CE transistor. Viewing (2.25*a*) as a KVL equation (all terms are voltages) and (2.25*b*) as KCL (all terms are currents), we draw a circuit that has (2.25*a*) as its equation (see Fig. 2.23*a*). This method of analysis opens the door for us to use all the circuit analysis tools we have laboriously mastered in a previous circuits course.

The small-signal CE model for the linearized two-port in terms of the h parameters is shown in Fig. 2.23a. From the typical values given in (2.29), it is seen that h_{oe}^{-1} is a very large impedance, and if the load resistance R_C (in shunt with h_{oe}^{-1}) is small, that is, $R_C h_{oe} \leq 0.1$, then we may ignore h_{oe} in the model. Also, in (2.29) since $h_{re} \simeq 0$, we have $h_{re} v_{ce} \simeq 0$. In some cases we may leave h_{oe} in the model but ignore the controlled-voltage source at the input of the model. The approximation made when we ignore the voltage source provides considerable simplification in analysis and design. The fact that v_{ce} has a negligible effect on the input characteristic (Fig. 2.17b) also validates this approximation. In transistor literature, the input resistance h_{ie} is usually split into two series resistances, one indicating the base-spreading resistance r_b (the same r_b in the T model of Fig. 2.19), which is typically 100 Ω, and the other, denoted as r_{be}, as shown in Fig. 2.23b. Note that in the model $h_{re} v_{ce}$ is ignored, and $h_{fe} = \beta_0$ and $h_{oe}^{-1} = r_0$ is used. In the circuit, the symbol B′ is used to denote the internal base junction, which is not an accessible terminal. In a detailed analysis, or

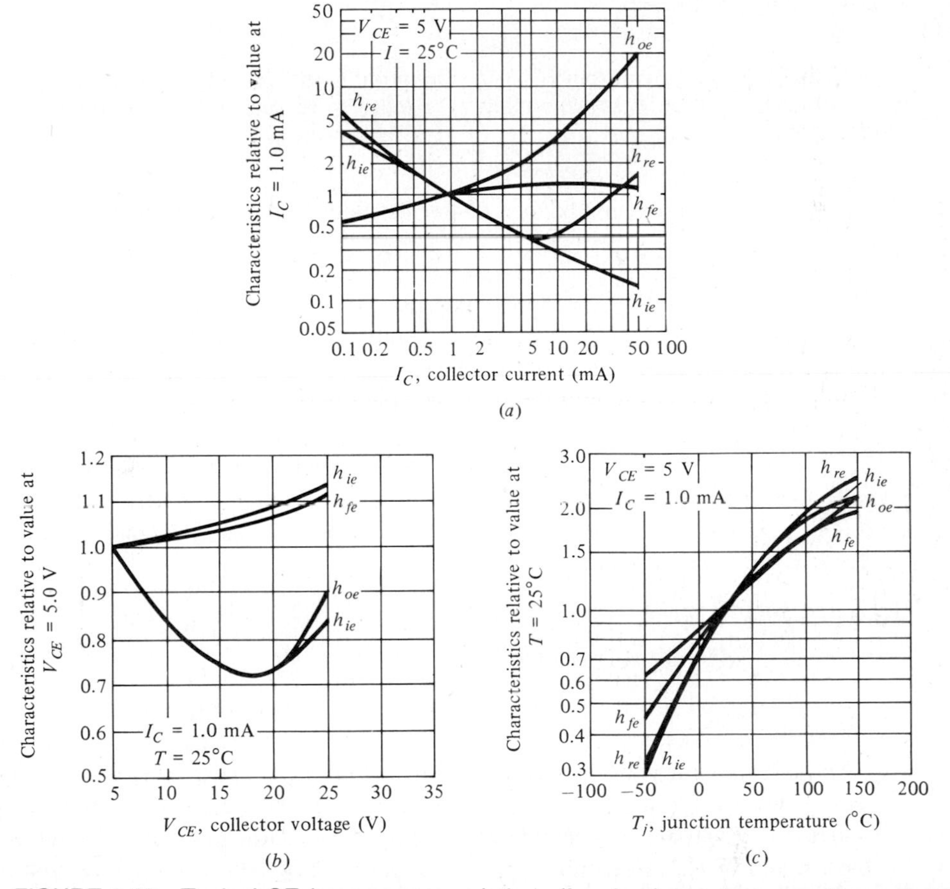

FIGURE 2.22 Typical CE h-parameter variations (for absolute values see Appendix E).

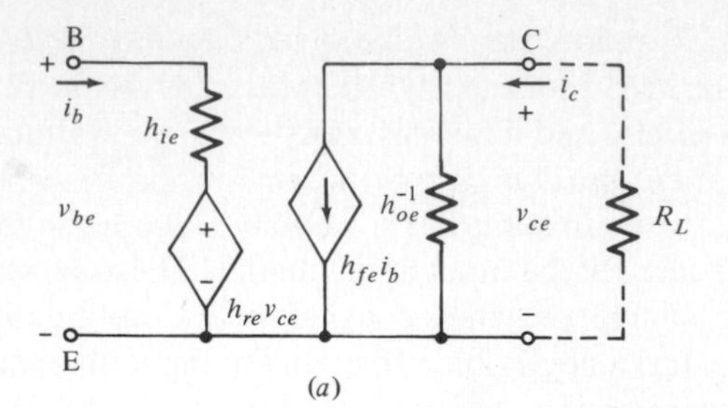

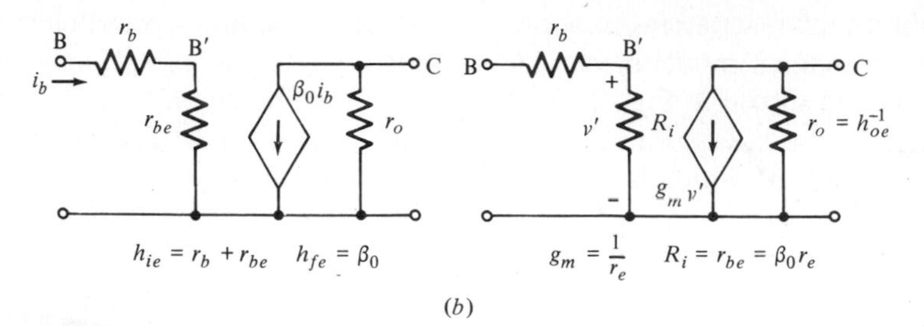

FIGURE 2.23 (*a*) Small-signal CE *h*-parameter model (exact). (*b*) Approximate model of (*a*) for $h_{re} \simeq 0$ or more accurately $h_{ie} \gg h_{fe}h_{re}R_L$. (*c*) Same circuit in terms of different variables.

by equating the input impedance of the T model to the model in Fig. 2.23*b*, we get

$$r_{be} = \beta_0 r_e = R_i \tag{2.30}$$

If we denote $i_b r_{be}$ by v', we have

$$v' = i_b R_i = i_b \beta_0 r_e \tag{2.31}$$

The current source $\beta_0 i_b$ at the output can then be written from (2.31) as v'/r_e or $g_m v'$, where $g_m = 1/r_e$.

The model of Fig. 2.23*b* can thus be represented as in Fig. 2.23*c*. The value of r_e depends on the operating point and is the same as in the T model (2.20). The model in Fig 2.23*c* is called the hybrid-π model and will be used extensively in this text. Typical parameters of the hybrid-π model are

$$\begin{aligned} &r_e = 10\ \Omega \qquad (\text{for } I_E = 2.5\text{ mA}) \qquad r_b = 100\ \Omega \\ &\beta_0 = 100 \qquad r_0 = 100\text{ k}\Omega \end{aligned} \tag{2.32}$$

The model shown in Fig. 2.23*c* is a reasonably accurate model of the transistor at low frequencies and may be used for small-signal transistor circuit analysis in any of the three configurations (CE, CB, CC) at low frequencies. The model in Fig. 2.23*c* can be further simplified, for the CB and CC configurations, under certain conditions, which are discussed in Chap. 4. If the simplified conditions cannot be met, then for the CB and CC configurations, the T model in Fig. 2.19 is more convenient than that of Fig. 2.23*c*. This is because when the circuit of Fig. 2.23*c* is rearranged to show the

CB and CC connections, the dependent sources will not be explicit functions of terminal voltages or current.

EXERCISE 2.7

The data sheet for the 2N2222A transistor is given in Appendix E. What are the h parameters of the transistor if $I_C = 10$ mA and $V_{CE} = 10$ V.

Ans $h_{ie} = 250\ \Omega$, $h_{fe} = 75$, $h_{oe} = 5 \times 10^{-6}$ mho, $h_{re} \simeq 0$

EXERCISE 2.8

For the circuit shown in Fig. E2.8, a 2N2222A transistor is used. If $v_i = 2 \times 10^{-3} \sin \omega t$, determine $v_O(t)$ and the corresponding ac voltage gain A_v.

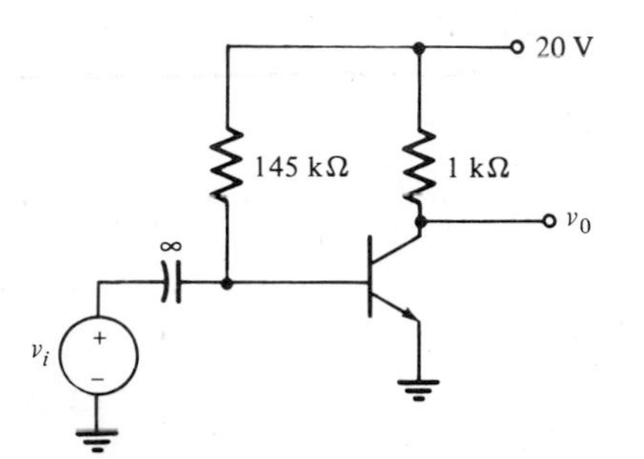

FIGURE E2.8

Ans $v_O(t) = 10 + 0.6 \sin \omega t$; $A_v = 300$

2.9 HIGH-FREQUENCY CIRCUIT MODELS FOR BIPOLAR TRANSISTORS*

When transistors are used to amplify high-frequency signals, the model for analysis and design must include the appropriate energy-storage elements (capacitances) and carrier transit time effects in the device in order to describe the frequency response characteristic of the transistor. In any model, one must make some compromise between accuracy and complexity. The simpler the model, the easier is the analysis and design, but it is less accurate. In order to obtain a realistic simple model, some *approximations must* be made that will provide a reasonable *compromise* between accuracy and complexity.

*The reader may wish to skip this section at this time and return to it when amplifier frequency response is considered.

A. The High-Frequency T Model

A high-frequency equivalent circuit which is applicable for transistors in all configurations and large- and small-signal conditions is shown in Fig. 2.24*a*. The model shown in Fig. 2.24*b* includes the series resistance. These models are *useful only* for a computer-aided analysis and design such as SPICE.* The Ebers-Moll model, as indicated earlier, includes all regions of operation of the transistor. The capacitances and the currents are nonlinear functions of the junction voltages.

For small-signal high-frequency applications we may use the T model of Fig. 2.19 with the associated capacitances of the *pn* junctions as shown in Fig. 2.25. In the model shown in Fig. 2.25, C_c is the junction capacitance of the reverse-biased collector-base junction. At high frequencies, the reactance of C_c is much smaller than r_c, so r_c is ignored in the high-frequency model. C_e consists of two parts: one is the junction capacitance of the forward-biased emitter-base junction, and the other is the diffusion capacitance (which is the dominant component).

The transit time effect is included in the so-called alpha cutoff frequency ω_α, namely,

$$\alpha(j\omega) \simeq \frac{\alpha_0}{1 + j\omega/\omega_\alpha} \tag{2.33}$$

where ω_α is the reciprocal of the transit time τ, namely,

$$\omega_\alpha = \frac{1}{\tau} = KV_t \frac{\mu_n}{W^2} \tag{2.34}$$

where K is a proportionality constant, usually in the range 3 to 15, μ_n is the mobility of electrons (1300 $\text{cm}^2/\text{V} \cdot \text{s}$ in Si), and W is the base width (typically 0.5 μm). For frequencies beyond ω_α the device is not useful, and hence (2.34) determines the *maximum frequency* of operation of the transistor.

---

*See the footnote on page 43.

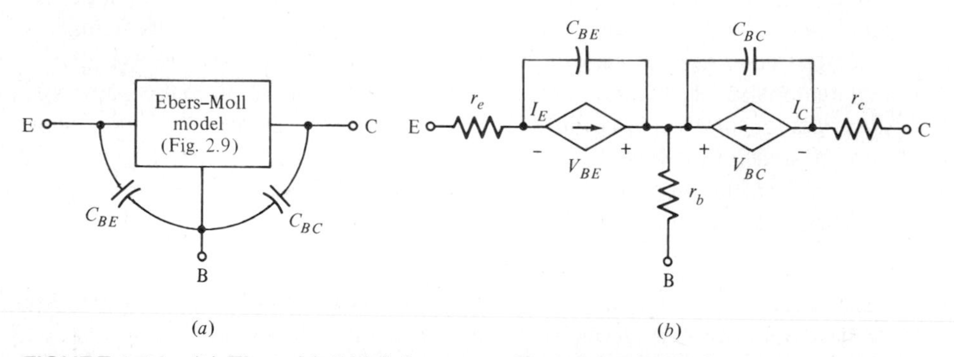

FIGURE 2.24 (*a*) Ebers-Moll high-frequency T model suitable for computer-aided analysis and design. (*b*) High-frequency model suitable for SPICE.

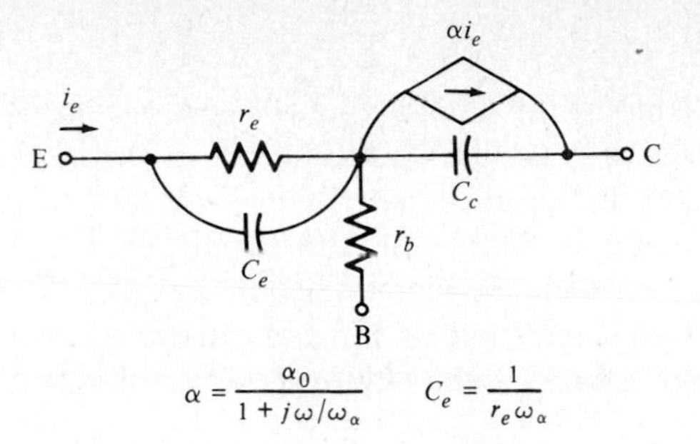

FIGURE 2.25 Simple high-frequency equivalent T model.

EXAMPLE 2.3

Determine the maximum frequency cutoff of an *npn* transistor, if $K = 5$ and $W = 0.5\ \mu$m.

From (2.34) we have

$$\omega_{\max} = \omega_\alpha = 5(25 \times 10^{-3})\left[\frac{1300}{(0.5 \times 10^{-4})^2}\right] = 6.5 \times 10^{10} \text{ rad/s}$$

$$f_{\max} = \frac{6.5}{2\pi} \times 10^{10} = 1.1 \times 10^{10} = 11 \text{ GHz}$$

The actual frequency cutoff of the transistor as an amplifier will naturally be smaller than ω_α, due to C_e, C_c, and parasitic capacitances. The value of C_e is given by [3]

$$C_e = \frac{1}{r_e \omega_\alpha} \tag{2.35}$$

A typical range of values for the elements of the T circuit in Fig. 2.25 is given below:

$$r_e = \frac{V_t}{I_E} \simeq 10\ \Omega \text{ (at } I_E = 25 \text{ mA)} \qquad \alpha_0 \simeq 0.95 - 0.99$$
$$r_b \simeq 20\text{–}100\ \Omega \qquad C_c = 1\text{–}20 \text{ pF} \qquad C_e \simeq 20\text{–}500 \text{ pF} \tag{2.36}$$

The lower values for capacitors listed in (2.36) correspond to high-frequency transistors.

Note that at low frequencies, i.e., when $\omega \ll \omega_\alpha$, the model of Fig. 2.25 reduces to that of Fig. 2.19.

B. The Hybrid-π High-Frequency Model

A complete hybrid-π high-frequency model for small-signal applications that is useful for computer-aided analysis and design is shown in Fig. 2.26. The model includes the series body resistance of the collector and emitter material as well as the substrate capacitance. For most applications, however, we may use the simplified model shown in Fig. 2.27. This model is very widely used in the literature and is called the *hybrid-π* equivalent circuit. Note that at low frequencies, where the capacitance reactances are so high they can be ignored, the model reduces to the low-frequency equivalent circuit shown in Fig. 2.23*c*.

A cutoff frequency f_β, usually associated with the CE configuration, is defined as:

$$h_{fe} = \frac{\beta_0}{1 + j(\omega/\omega_\beta)} = \beta(j\omega) \qquad \text{where } \omega_\beta = 2\pi f_\beta \tag{2.37}$$

The frequency response of $\alpha(j\omega)$ and $h_{fe}(j\omega) = \beta(j\omega)$ is shown in Fig. 2.28. Note that $\alpha(j\omega)$ is the frequency-dependent current gain of a CB transistor and $\beta(j\omega)$ is the short-circuited current gain of a CE transistor in Fig. 2.27. From the definition of ω_β in (2.37) and the equivalent circuit of Fig. 2.27 we obtain (problem 2.46)

$$C_\pi + C_c = \frac{1}{r_e \omega_T} \tag{2.38}$$

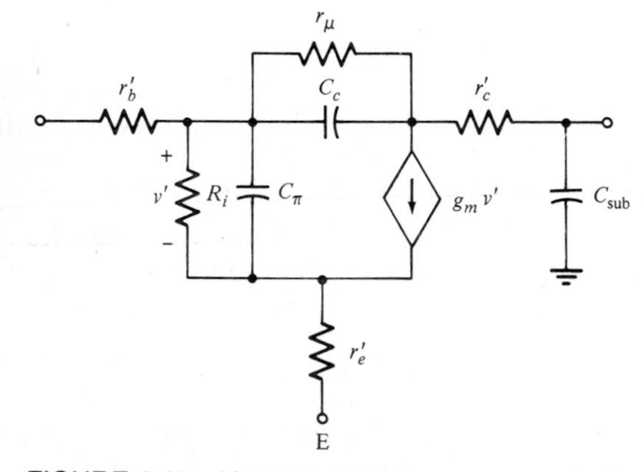

FIGURE 2.26 Hybrid-π high-frequency equivalent circuit suitable for computer-aided analysis and design.

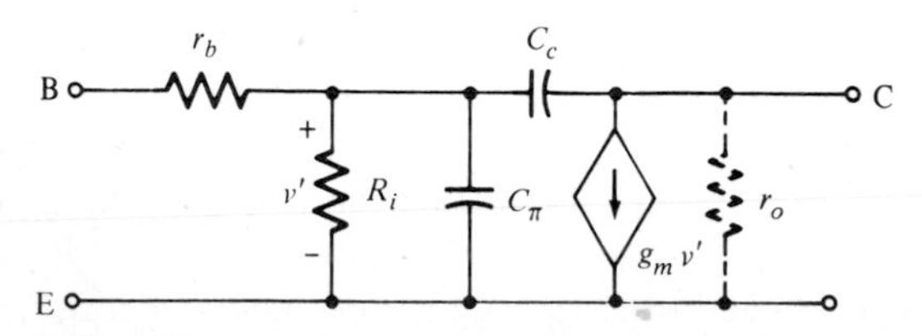

FIGURE 2.27 Simple high-frequency hybrid-π model: $C_c + C_\pi = 1/r_e \omega_T$, $R_i = \beta_0 r_e = \beta_0/g_m$.

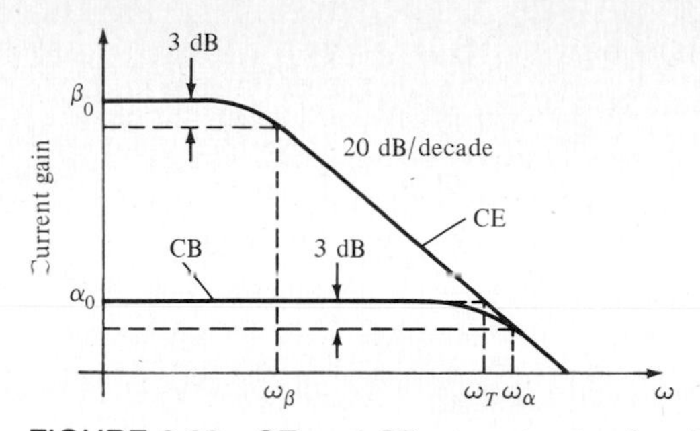

FIGURE 2.28 CE and CB current gains for short-circuited output.

where ω_T is defined as the radian frequency at which the short-circuit CE current gain is unity. From (2.37) and using the definition of ω_T, i.e., setting the magnitude of the gain equal to 1, we obtain

$$\omega_T \simeq \beta_0 \omega_\beta \qquad \text{or} \qquad f_T = \beta_0 f_\beta \tag{2.39}$$

The frequency f_T is a very important parameter of the transistor and is almost always given by the manufacturer. It is also sometimes called the *gain-bandwidth product* of the transistor. The frequencies f_T and f_α are related as one would expect. The relationship is approximately as follows:

$$f_T \simeq \frac{f_\alpha}{1 + m} \tag{2.40}$$

where m is a constant $0.2 \leq m \leq 1.2$. Typical variation of f_T vs I_e and V_{ce} is shown in Fig. 2.29. A typical range of values for the elements of the hybrid-π circuit is given in Table 2.1. The hybrid-π circuit parameters are readily determined from the manufacturer's data sheet. The value of f_T is either given directly, or it is found from the product of h_{fe} and f_1 at a high frequency ($f_1 \gg f_\beta$); this is a direct consequence of (2.37). The value of r_e is determined from 25 mV/I_e, and β_0 is determined from h_{fe} or

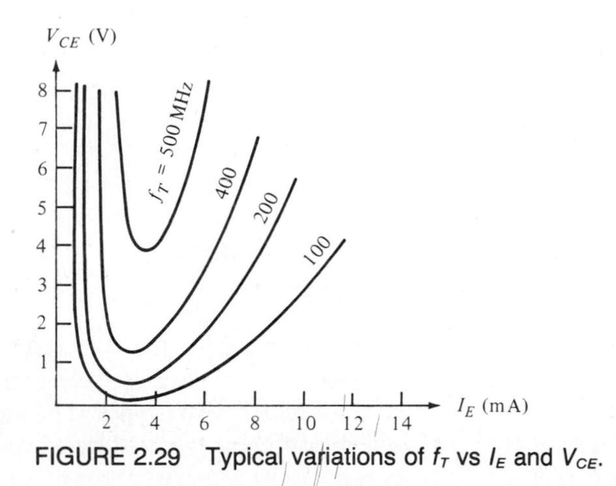

FIGURE 2.29 Typical variations of f_T vs I_E and V_{CE}.

TABLE 2.1 Element Values of the Hybrid-π Equivalent Circuit

$$g_m = \frac{1}{r_e} \simeq 40 I_E = 0.04 \text{ mho (for } I_E = 1 \text{ mA)}$$

$R_i = \beta_0 r_e$ $\qquad \beta_0 = 20 - 300$

$C_c = 1 - 20$ pF $\qquad r_b = 20 - 200\ \Omega$

$C_\pi \simeq 10 - 1000$ pF

$$r_0 = \frac{\kappa}{I_E} = 10 - 1000 \text{ k}\Omega,\ \kappa \text{ proportionality constant}$$

from h_{ie}/r_e (if $R_i \gg r_b$). The value of C_c is given by C_{obo} by the manufacturer. The value of r_b could be determined from the fact that $r_b \simeq h_{ie} - R_i$ at low frequencies. Since h_{ie} and R_i are usually both much larger than r_b, the difference between these two nearly equal quantities could lead to an erroneous or meaningless result. A better way is to find the real part of h_{ie} designated as $R_e(h_{ie})$ at $f \gg f_\beta$.

EXAMPLE 2.4

Determine the hybrid-π circuit parameters for the 2N2222 type BJT whose data sheet is included in Appendix E. The desired operating point is $I_{Cq} = 10$ mA, $V_{CE} = 10$ V.

From the data sheet using $I_{Cq} = 10$ mA, $V_{CEq} = 10$ V, we read the following:

$$f_T \geq 300 \text{ MHz} \qquad \beta_0 = h_{fe} \geq 75 \qquad r_e = 2.5\ \Omega$$

$$C_c \simeq C_{obo} \leq 8 \text{ pF} \qquad r_b = Re(h_{ie}) \leq 60\ \Omega \qquad r_0 = h_{oe}^{-1} \geq 40 \text{ k}\Omega$$

$$C_e \leq \frac{1}{r_e \omega_T} = 212 \text{ pF}$$

Note that we can alternatively determine $f_T = |h_{fe}|$ (100 MHz) $\geq 3(100) =$ 300 MHz, where the value of h_{fe} at high frequency is used.

It should be noted that transistor arrays are now available in a package. For example, the general-purpose *npn* transistor array data sheet for the 3086 type is shown in Appendix E. The array, supplied in a 14-lead package, consists of five general-purpose *npn* transistors on a common monolithic substrate. They may be used in discrete or integrated circuits.

The high-frequency hybrid-π model given in Fig. 2.27 for transistors is an approximate equivalent circuit which represents the best compromise between accuracy and circuit complexity. This model is valid for all transistor types. The element values of the model will of course be different for different types of transistors.

EXERCISE 2.9

Determine all the element values of the hybrid-π model in Fig. 2.27, for the 2N2222A transistor at $I_C = 10$ mA, $V_{CE} = 10$ V.

Ans $r_b = 60\ \Omega$, $R_i = 188\ \Omega$, $C_\pi = 204$ pF, $C_c = 8$ pF, $g_m = 0.4$ mho

2.10 BIPOLAR TRANSISTOR BIASING

For transistor amplifiers to function properly, we must ensure that the transistor Q point is in the normal-active region, i.e., the transistor is properly biased. The common-emitter output characteristic of a typical silicon transistor is shown in Fig. 2.30. The region beyond the area of safe operation is marked by shading. The limit of the safe operating region is the *maximum-dissipation hyperbola,* defined by

$$(P_D)_{\max} = V_{CE}I_C \tag{2.41}$$

where $V_{CE} \leq (V_C)_{\max}$ and $I_C \leq (I_C)_{\max}$. The limit is lowered in an increased-temperature environment. In low-power ($\leq$2 W) amplification we operate the transistor much below the dissipation hyperbola. Only in power amplifiers is the transistor operated close to the dissipation curve in order to deliver large power outputs. In such cases power transistors, which have power ratings from a few watts to over 100 W, are used, and large heat-sinks are needed. The design of power amplifiers are discussed in Chap. 6. Note that the power rating goes down for increasing temperatures (Sec. 6.7). In the following discussion, we consider low-power amplifiers ($P \leq 2$ W), i.e., operation is well below the power dissipation hyperbola.

A very simple biasing circuit is shown in Fig. 2.31 to illustrate some problems in biasing and the shift of operating point. The dc load line and the quiescent operating point are shown in Fig. 2.32. At room temperature, the quiescent point Q is at $I_{Cq} = 3.2$ mA, $V_{CEq} = 5.6$ V. If the temperature is increased to 75°C, however, the base current of the emitter junction diode will also increase, and the operating point

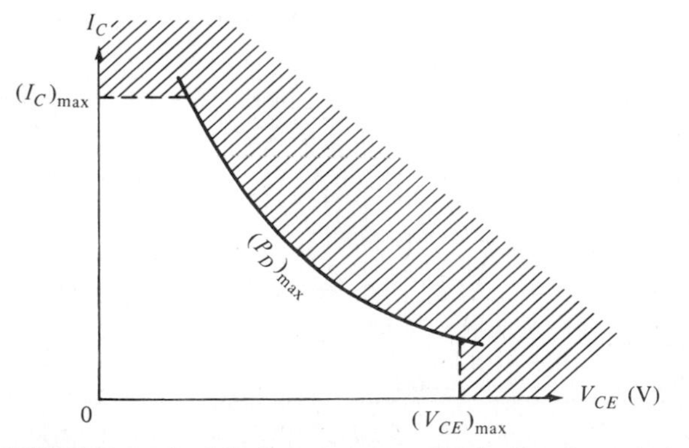

FIGURE 2.30 Maximum power dissipation hyperbola and the safe operating region.

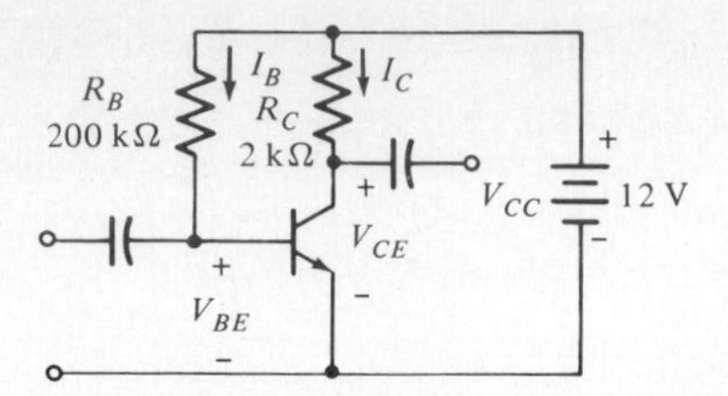

FIGURE 2.31 A fixed-bias circuit for *npn* transistors.

will shift to Q' as shown ($I'_{Cq} = 4.2$ mA, $V_{CEq'} = 3.6$ V); thus, the transistor parameters will be different. If the original operating point at 25°C had been at $I_B = 80\ \mu$A, the new operating point I_{cq} at 75°C would be in the saturation region—a very undesirable situation.

The aim of the biasing circuit is to establish the quiescent point and make it least sensitive to variations in transistor characteristics (which may result from temperature changes, transistor replacement, or other sources).

If the transistor is properly biased, the Q point will be fixed, and differences in transistor dc characteristics will not have a significant effect on circuit behavior.

We have seen in Fig. 2.22*c* that transistor characteristics are strongly temperature dependent. This can also be readily seen by examining the characteristic curves of transistors at different temperatures, as provided by the manufacturers. Further, the transistor parameters vary as the operating point changes. The question is how to choose and stabilize the operating point. The answer to this question depends on the circuit application. In the simple, low-power case considered here, the junction and ambient temperatures are essentially the same. This is the case when the supply voltage and the resistive load are such that the load line always falls well below the maximum power dissipation curve.

The transistor operating point is usually *selected* on the basis of optimal device performance in terms of gain (β_0), frequency response f_T, large dynamic swing of the

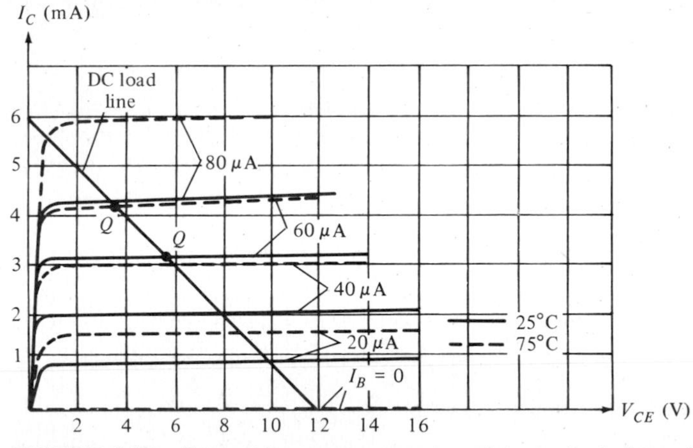

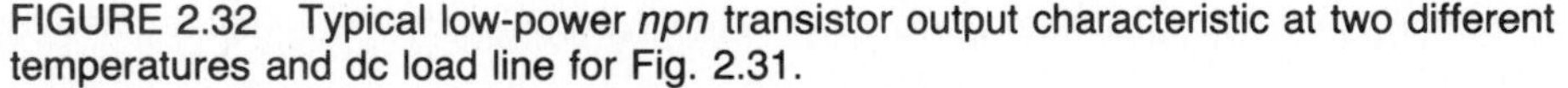
FIGURE 2.32 Typical low-power *npn* transistor output characteristic at two different temperatures and dc load line for Fig. 2.31.

output variable without distortion, and minimal noise considerations. Usually the manufacturer provides such operating-point data, and the *choice* of the Q point is not too difficult. In fact, it is often suggested by the manufacturer. Once the operating point is chosen, it must be *stabilized* by external circuits against temperature changes and transistor replacement. We shall next consider the shifts of the operating point and how to minimize this shift.

It should be emphasized at the outset that proper bias stabilization is that which maintains I_C and V_{CE} approximately constant against environmental changes and device replacement. A β_0 spread of $30 \leq \beta_0 \leq 300$ is not uncommon for the same type transistor. Note that by maintaining I_B constant we cannot achieve operating-point stability. For example, consider Fig. 2.32, which shows the common-emitter output characteristics of the same transistor for two different temperatures. (It could just as well have been the output characteristics of two different units of the same type transistors.) If I_B is maintained constant by external biasing, the operating point will move to a new point, as indicated in Fig. 2.32. In some cases, it is quite possible for such a shift to bring the transistor operating point near the nonlinear region or to bring it fully into the saturation region. We must design the circuit so that I_{Cq} and V_{CEq} are constant for various temperature conditions as well as to provide for interchangeability of transistors in assembly-line production.

The main sources of shift in the operating point of a transistor due to temperature changes are the variation of I_{CO}, V_{BE}, and β_0. Another important source is the variation of β_0 from unit to unit, which can vary by a factor of 10 among different specimens of the same type of transistor. I_{CO} approximately doubles for a 6°C increase in temperature above 25°C in silicon transistors. At room temperature, V_{BE} is approximately 0.7 V for Si and decreases approximately by 2 mV/°C for increasing temperature. The variation of β_0 is difficult to show analytically, but can be read directly from the CE output characteristics for various temperatures. β_0 roughly doubles for an increase of 80°C from its value at room temperature. Interchanging transistors, however, further complicates the problem, since variations can only be treated statistically, if such statistics are available (which they usually are not!).

In the active region, the relationship between I_C and I_B is given by (2.4):

$$I_C = (\beta_0 + 1)I_{CO} + \beta_0 I_B \tag{2.42}$$

From (2.42) a temperature increase changes I_{CO} and β_0 and hence the value of I_C for a fixed I_B. Typical values of β_0 and I_{CO} for Si transistor parameters at room temperature are given in Table 2.2. Also given are the *approximate* analytical expressions which

TABLE 2.2 Typical Values and Approximate Temperature Expressions for β_0, I_{CO}, and V_{BE} for Silicon Transistors

Parameters	Typical values at 25°C	Approximate values at different temperatures
$I_{CO}(25°C + \Delta T)$	1 nA	$[I_{CO}(25°C)](2)^{\Delta T/6°C}$
$\beta_0(25°C + \Delta T)$	50	$\beta_0(25°C)(1 + \Delta T/80°C)$
$V_{BE}(25°C + \Delta T)$	0.7 V	$(0.70 - 0.002\,\Delta T)$ V

may be used for a rough estimate of I_{CO}, β_0, and V_{BE} at different temperatures if more accurate data are not available.

A. Fixed Biasing Circuit

Let us now examine the simple biasing circuit of Fig. 2.31 known as *fixed biasing* circuit. As we shall show in this circuit I_B is fixed. The design is derived from the following equations. Note that for dc, the capacitors are open-circuited and no current flows through them. The collector-emitter and the base-emitter loop equations are

$$V_{CC} = V_{CE} + I_C R_C \tag{2.43}$$

$$V_{CC} = I_B R_B + V_{BE} \tag{2.44}$$

Normally, we design the biasing circuit so the operating point is nearly in the center of the active region, that is,

$$V_{CEq} \simeq \frac{V_{CC}}{2} \tag{2.45}$$

Hence, for a desired I_{Cq} and V_{CEq} we obtain the circuit parameters from (2.43), namely,

$$R_C = \frac{V_{CC} - V_{CEq}}{I_{Cq}} = \frac{V_{CC}}{2I_{Cq}} \tag{2.46}$$

from (2.42)

$$I_B = \frac{I_{Cq} - (\beta_0 + 1)I_{CO}}{\beta_0} \simeq \frac{I_{Cq}}{\beta_0} = \frac{V_{CC}}{2R_C\beta_0} \quad \text{for } I_{Cq} \gg \beta_0 I_{CO} \tag{2.47}$$

and, finally, from (2.44)

$$R_B = \frac{V_{CC} - V_{BE}}{I_B} = \frac{V_{CC} - 0.7}{I_B} \simeq \frac{V_{CC}}{I_B} \quad \text{for } V_{CC} \gg V_{BE} \tag{2.48}$$

We now show that the biasing circuit of Fig. 2.31 is not satisfactory for maintaining I_C and V_{CE} constant. This is due to (2.48), which maintains I_B = constant for a given V_{CC} and R_B. For example, if the transistor is replaced by another of the same type but with a β_0 twice as large as the first one (not an uncommon situation), the operating point will shift considerably, due to the change in β_0 as shown in the following example.

EXAMPLE 2.5

Consider the circuit of Fig. 2.31 for a Si transistor with the following numerical circuit parameters:

$$V_{CC} = 12 \text{ V} \qquad R_C = 1.2 \text{ k}\Omega \qquad R_B = 120 \text{ k}\Omega$$

The transistor parameters at room temperature (25°C) are $\beta_0 = 50$, $I_{CO} = 1$ nA, $V_{BE} = 0.7$ V.

(a) Determine the Q point at room temperature (25°C).
(b) Determine the Q point at room temperature but assume a replacement transistor with the same parameters except that $\beta_0 = 100$.
(c) Determine the Q point at 55°C with the original transistor.

The Q point as determined from (2.48), (2.47), and (2.43) is

$$I_B = \tfrac{12}{120} \times 10^{-3} = 0.1 \text{ mA} \qquad I_C \simeq \beta_0 I_B = 5 \text{ mA}$$

$$V_{CE} = 12 - 5(1.2) = 6 \text{ V}$$

For $\beta = 100$, we have

$$I_B = 0.1 \text{ mA} \qquad I_C \simeq 100(0.1) = 10 \text{ mA}$$

$$V_{CE} = 12 - 10(1.2) = 0$$

Obviously the new transistor will not function properly in the circuit; since $V_{CE} \simeq 0$, the transistor is in saturation.

Because of the temperature increase, the transistor parameters will change to $V_{BE} \simeq 0.64$ V, $\beta_0 \simeq 69$, $I_{CO} \simeq 32$ nA, as obtained from Table 2.2. The new operating point in this case is

$$I_B = 0.1 \text{ mA} \qquad I_C = 69(0.1) = 6.9 \text{ mA} \qquad V_{CE} = 12 - 6.9(1.2) \simeq 3.7 \text{ V}$$

Again notice a considerable shift in the operating point.

EXERCISE 2.10

For the circuit shown in Fig. E2.8, determine the dc currents I_B and I_C at room temperature if the transistor has (*a*) $\beta_0 = 75$, (*b*) $\beta_0 = 120$, (*c*) $\beta_0 = 200$.

Ans (*a*) $I_B = 0.133$ mA, $I_C = 9.98$ mA; (*b*) $I_B = 0.133$ mA, $I_C = 15.96$ mA; (*c*) $I_B = 0.133$ mA, $I_C \simeq 26.6$ mA

EXERCISE 2.11

For the circuit shown in Fig. E2.11, the following are given: $\beta_0 = 100, I_{CO} = 10^{-9}$ A, $(V_{BE})_{\text{on}} = 0.7$ V. Determine the Q point.

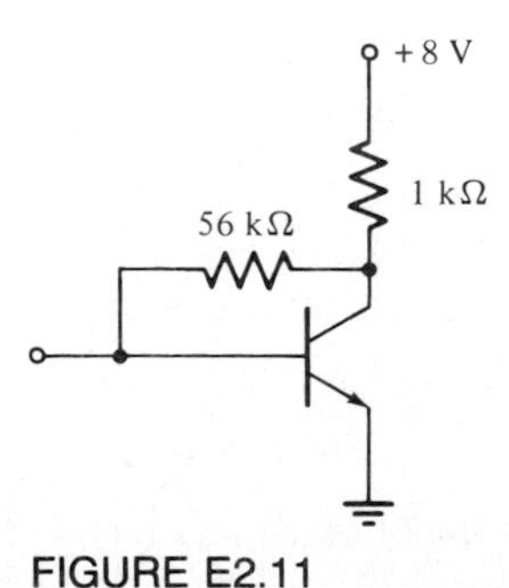

FIGURE E2.11

Ans $I_{Cq} = 4.65$ mA, $V_{CEq} = 3.3$ V

B. Self-biasing Circuit

The circuit shown in Fig. 2.33*a*, which is referred to as the *self-bias* or emitter-bias circuit, is one of the most frequently used biasing circuits for single-stage and *RC*-coupled stages in discrete transistor circuits. In this case, the addition of R_E in the emitter lead provides stabilization of the operating point. Physically, this improvement is provided as follows: If β_0 and/or I_{CO} is increased due to either interchangeability or temperature increase, then I_C increases. Since I_E is approximately the same as I_C, the increase in I_E causes an increase in the voltage drop across R_E, which decreases V_{BE}, hence a decrease in I_B. The decrease in I_B tends to decrease I_C, so a stabilization effect is achieved. This stabilization due to feedback is one of the important properties of negative feedback and is discussed further in Chap. 9.

If we replace the circuit to the left of the base-to-ground terminal by its Thevenin equivalent circuit, as shown in Fig. 2.33*b*, we obtain

$$V_{BB} = \frac{R_{b2}}{R_{b2} + R_{b1}} V_{cc} \tag{2.49a}$$

and

$$R_B = R_{b1} \| R_{b2} = \frac{R_{b1} R_{b2}}{R_{b1} + R_{b2}} \tag{2.49b}$$

Kirchhoff's voltage law applied for the base-emitter loop gives

$$V_{BE} = V_{BB} - I_E R_E - I_B R_B \tag{2.50}$$

For the approximate analysis, ignoring I_{CO} as compared to I_C, we have

$$I_C \simeq \beta_0 I_B \qquad \text{and} \qquad I_C \simeq \alpha_0 I_E \tag{2.51}$$

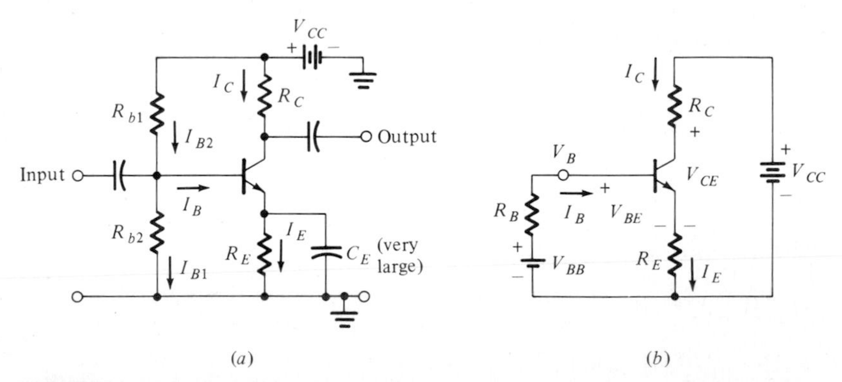

FIGURE 2.33 (*a*) Self-biasing circuit. (*b*) Equivalent representation of (*a*).

Hence from (2.50) and (2.51)

$$V_{BE} - V_{BB} = -(\beta_0 + 1)R_E I_B - I_B R_B \tag{2.52a}$$

or

$$I_B = \frac{V_{BB} - V_{BE}}{R_B + (\beta_0 + 1)R_E} \tag{2.52b}$$

KVL applied for the output loop gives

$$V_{CC} = V_{CE} + I_C R_C + I_E R_E$$

or

$$V_{CE} = V_{CC} - I_C\left(R_C + \frac{R_E}{\alpha_0}\right) \tag{2.53}$$

EXAMPLE 2.6

For the circuit shown in Fig. 2.33 let $V_{CC} = 20$ V, $R_C = 5$ kΩ, $R_E = 1$ kΩ, $R_{b1} = 20$ kΩ, $R_{b2} = 3$ kΩ. The transistor parameters are $\beta_0 = 100$, $V_{BE} = 0.7$ V, $I_{CO} = 1$ nA. Determine the Q point.
From (2.49*a*)

$$V_{BB} = \frac{3}{20 + 3}(20) \simeq 2.61 \text{ V}$$

From (2.49*b*)

$$R_B = (3 \| 20)\ k\Omega = 2.61 \text{ k}\Omega$$

From (2.52*b*)

$$I_B \simeq \frac{2.61 - 0.70}{2.61 \times 10^3 + 101 \times 10^3} = 18.4 \text{ mA}$$

Hence from (2.51) and (2.53),

$$I_{Cq} = 100(18.4 \times 10^{-6}) = 1.84 \text{ mA}$$

$$V_{CEq} = 20 - 1.84 \times 10^{-3}(6 \times 10^3) \simeq 11 \text{ V}$$

Bias Stabilization

As pointed out earlier, three parameters are the main cause of changes in the collector current; these are I_{CO}, V_{BE}, and β_0. The changes in these parameters for different temperatures may be either given by the manufacturer or can be estimated by Table 2.2.

A stability factor S can be used for each of these parameters as follows:

$$S_I = \frac{\partial I_C}{\partial I_{CO}} \simeq \frac{\Delta I_C}{\Delta I_{CO}} \tag{2.54a}$$

$$S_V = \frac{\partial I_C}{\partial V_{BE}} \simeq \frac{\Delta I_C}{\Delta V_{BE}} \tag{2.54b}$$

$$S_\beta = \frac{\partial I_C}{\partial \beta_0} \simeq \frac{\Delta I_C}{\Delta \beta_0} \tag{2.54c}$$

Thus the total change in I_C for small variations due to changes in I_{CO}, V_{BE}, and β_0 can be written to a first-order approximation as

$$I_C = S_I \, \Delta I_{CO} + S_V \, \Delta V_{BE} + S_\beta \, \Delta \beta_0 \tag{2.55}$$

Substituting the expression from (2.2) and (2.3), namely, $I_E = (1/\alpha_0)(I_C - I_{CO})$ and $I_B = I_E - I_C$, into (2.50) yields (note that the direction of I_E is changed in Fig. 2.33):

$$-V_{BB} + V_{BE} = -I_C\left[\frac{R_E}{\alpha_0} + \frac{R_B(1 - \alpha_0)}{\alpha_0}\right] + \frac{I_{CO}}{\alpha_0}(R_E + R_B) \tag{2.56}$$

From (2.56) and the various definitions of S in (2.54) we can determine the expression for each stability factor; namely, from (2.56) and (2.54a) we obtain

$$S_I = \frac{\partial I_C}{\partial I_{CO}} = \frac{1 + R_E/R_B}{(1 - \alpha_0) + R_E/R_B} \tag{2.57a}$$

or, in terms of β_0, (2.57a) can be written as

$$S_I = \frac{R_E + R_B}{R_E + R_B/\beta_0} \simeq 1 + \frac{R_B}{R_E} \qquad \text{for } R_E \gg \frac{R_B}{\beta_0} \tag{2.57b}$$

In (2.57b) if $R_B/R_E \rightarrow 0$, we have $S_I \rightarrow 1$, and it is essentially independent of β_0. If $R_B/R_E \gg \beta_0$ then $S_I \rightarrow \beta_0$. It is of course desirable to have S_I as small as possible, preferably 1. It should be noted, however, that choosing R_B as small as possible to make S_I small conflicts with the ac gain of the stage because the resistance R_B shunts the input of the amplifier. Usually R_B is such that it is much larger than the input impedance, that is, $R_B \gg r_b + \beta_0 r_e$. The requirement that R_E be as large as possible is limited by the available or specified voltage supply. Thus the design of a biasing circuit is a compromise between conflicting requirements. In certain cases, specified tight variations in I_{Cq} and V_{CEq} may have to be relaxed in order to achieve the design even with the biasing circuit of Fig. 2.33.

S_V is found from (2.54b) and (2.56), namely,

$$S_V = \frac{\partial I_C}{\partial V_{BE}} = \left[\frac{R_E}{\alpha_0} + \frac{R_B(1 - \alpha_0)}{\alpha_0}\right]^{-1} \tag{2.58a}$$

or in terms of β_0, (2.58a) can be written as

$$S_V \simeq \frac{-\beta_0}{R_B + \beta_0 R_E} \simeq \frac{-1}{R_E} \tag{2.58b}$$

A small value of S_V is desirable, and therefore from (2.58b) we want a large value of R_E—a requirement similar to that for small S_I.

S_β is obtained from (2.52) and (2.51) as follows:

$$I_{Cq} = \frac{\beta_0(V_{BB} - V_{BE})}{R_B + \beta_0 R_E} \quad \text{and} \quad \frac{I_{Cq2}}{I_{Cq1}} = \frac{\beta_{02}(R_B + \beta_{01}R_E)}{\beta_{01}(R_B + \beta_{02}R_E)} \tag{2.58c}$$

Subtracting unity from each side in (2.58*c*) and rearranging, we get

$$S_\beta = \frac{\Delta I_C}{\Delta \beta_0} = \frac{I_{Cq1}}{\beta_{01}} \frac{R_B}{R_B + \beta_{02}R_E} \tag{2.58d}$$

Note again that to make S_β small, we have to make R_E large—a similar requirement for making S_I and S_V small. Of these three parameter changes, the contributions due to the change in V_{BE} and β_0 are the largest as seen in Example 2.7. In other words, the effect of changes due to I_{CO} can be ignored in a Si transistor.

EXAMPLE 2.7

For Example 2.6 let the temperature change from 25 to 100°C, and determine the change in I_C due to the changes in I_{CO}, V_{BE}, and β_0 by using Table 2.2.

Consider the change in I_{CO}: $I_{CO}(25°\text{C}) = 1$ nA (given); $I_{CO}(100°\text{C}) \simeq 4\ \mu\text{A}$

$$\Delta I_C = S_I\,\Delta I_{CO} = \frac{100(1 + 2.61)}{100 + 2.61}(4 \times 10^{-6}) = 14.0\ \mu\text{A}$$

$$V_{BE}(25°\text{C}) = 0.7\ \text{V} \qquad V_{BE}(100°\text{C}) = 0.55$$

$$\Delta I_C = S_V\,\Delta V_{BE} = -\frac{100}{(20 + 100) \times 10^3}(-0.15) = 125\ \mu\text{A}$$

$$\beta_0(25°\text{C}) = 100\ \text{(given)} \qquad \beta_0(100°\text{C}) = 193$$

$$\Delta I_C = S_\beta\,\Delta\beta_0 = \frac{1.84 \times 10^{-3}}{100}\left(\frac{20}{20 + 193}\right)(193 - 100) = 169\ \mu\text{A}$$

Note that even though we have assumed a doubling of I_{CO} for every 6°C, the effect of I_{CO} is the smallest of the three.

It should be noted that if the changes are not small, then the change in the collector current can be obtained directly from

$$\Delta I_{Cq} = I_{Cq2}(I_{CO2}, V_{BE2}, \beta_{02}) - I_{Cq1}(I_{CO1}, V_{BE1}, \beta_{01}) \tag{2.58e}$$

From the expression of the stability factors it is clear that in order to minimize the changes in I_C the value of R_E should be as large as possible, consistent with other requirements. A large value of R_E will cause an ac signal loss unless it is bypassed by a large capacitance (>10 μF), which is used as shown in Fig. 2.33. The choice of the value of C_E depends on the desired low-frequency cutoff of the stage; this is discussed in Chap. 7.

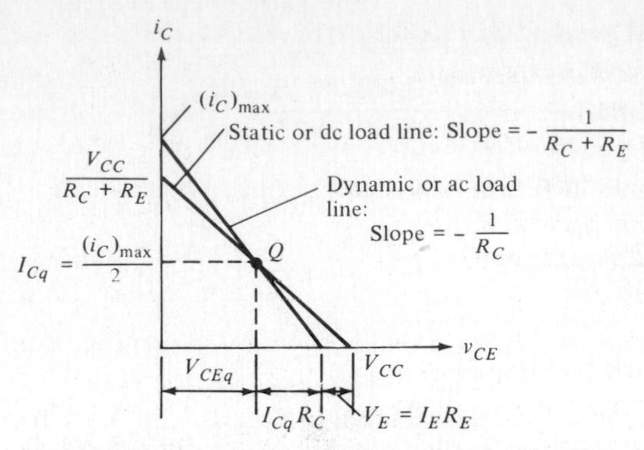

FIGURE 2.34 Static (dc) and dynamic (ac) load lines and the Q point.

The addition of a capacitor C_E across R_E introduces the concept of *ac load line*. We have already discussed the dc or static load line. Since at dc the capacitance C_E is open-circuited, the load line has a slope $-1/(R_C + R_E)$ as shown in Fig. 2.34. For ac signals, as we shall show in Chap. 7, the reactance of C_E is so small ($X_C = 1/\omega C_E < 2\ \Omega$ for $f = 10$ kHz and $C_E = 10\ \mu$F) that it effectively short-circuits R_E, and hence the ac load-line slope is given by $-1/R_C$, which is also shown in Fig. 2.34. We shall consider the ac load line again shortly.

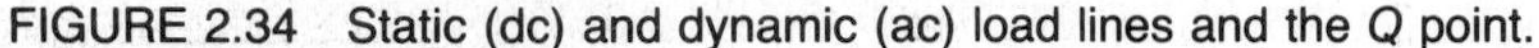

EXERCISE 2.12

For the circuit shown in Fig. E2.12, determine I_B, I_C, and V_{CE} if the transistor has $\beta_0 = 100$, $I_{CO} = 10$ nA, and $(V_{BE})_{on} = 0.7$ V.

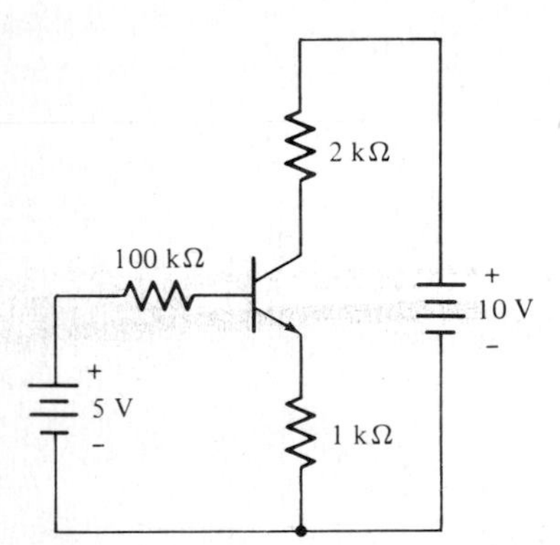

FIGURE E2.12

Ans $I_B = 21.4\ \mu$A, $I_C = 2.14$ mA, $V_{CE} = 3.56$ V

C. Design of the Biasing Circuit

If all the pertinent factors are included in the design, the design will not be simple. For a simple design, in order to obtain adequate stability, we usually choose

$R_B/R_E \leq (\beta_0)_{min}/5$. As a rule of thumb, the ratio R_B/R_E is generally chosen to be 5 or less; R_E is chosen according to $R_E \simeq V_{CC}/4I_C$.

EXAMPLE 2.8

Consider the design of the biasing circuit of Fig. 2.33 for a silicon transistor with the manufacturer's specifications at 25°C as follows: $(\beta_0)_{min} = 50$; $(\beta_0)_{max} = 150$; typical value for this transistor type is $\beta_0 = 100$. The supply voltage $V_{CC} = 20$ V. The desired Q point is $I_{Cq} = 5$ mA, $V_{CEq} = 5$ V.

For the collector-emitter loop, using Kirchhoff's voltage law, we have

$$V_{CC} = V_{CE} + I_C R_C + I_E R_E \simeq V_{CE} + I_C(R_C + R_E)$$

Substitution of numerical values yields

$$20 = 5 + 5 \times 10^{-3}(R_C + R_E) \qquad \text{or} \qquad R_C + R_E = 3\ \text{k}\Omega$$

We choose $R_E = V_{CC}/4I_C = 20/4(5 \times 10^3) = 1\ \text{k}\Omega$. For proper stability we choose $R_B/R_E = 5$; hence $R_B = 5\ \text{k}\Omega$. Now, neglecting I_{CO}, we have from (2.51)

$$I_C \simeq \beta_0 I_B \qquad \text{or} \qquad I_B = \frac{5 \times 10^{-3}}{100} = 50\ \mu\text{A}$$

V_{BB} is determined from (2.50), namely,

$$V_{BB} \simeq 0.7 + 5(1) + 0.05(5) = 5.95\ \text{V}$$

From (2.49) we can now solve for R_{b1} and R_{b2}, namely,

$$R_{b1} = \frac{R_B V_{CC}}{V_{BB}} \qquad \text{and} \qquad R_{b2} = \frac{R_{b1} V_{BB}}{V_{CC} - V_{BB}} = \frac{R_B V_{CC}}{V_{CC} - V_{BB}}$$

$$R_{b1} = (5 \times 10^3)\frac{20}{5.95} = 16.8\ \text{k}\Omega$$

$$R_{b2} = \frac{5(20)}{20 - 5.95} = 7.1\ \text{k}\Omega$$

We now determine the operating point for interchangeability, that is, for $(\beta_0)_{min}$ and $(\beta_0)_{max}$ rather than the typical value used in the design. For the $(\beta_0)_{min}$ case, the Q point from (2.52*b*), (2.51), and (2.53) is calculated to be

$$I_C = 4.7\ \text{mA} \qquad V_{CE} = 5.8\ \text{V}$$

For the $(\beta_0)_{max}$ case, the Q point will be at

$$I_C = 5.1\ \text{mA} \qquad V_{CE} = 4.7\ \text{V}$$

Hence we have adequate stability for interchangeability of transistors.

Next, let us determine the new Q point of the designed example for a temperature of 100°C, instead of 25°C. For the increased temperature the transistor parameters are approximately estimated by the rule in Table 2.2.

$$(\beta_0)_{\text{nominal}} \simeq 160 \qquad I_{CO} \simeq 0.2\ \mu\text{A} \qquad V_{BE} \simeq 0.55\ \text{V}$$

The analysis of the circuit with the new transistor parameters yields

$$I_C = 5.3\ \text{mA} \qquad V_{CE} = 4.1\ \text{V}$$

It is seen that β_0 variations, because of interchangeability, with one having a higher value of β_0 produce the same general effect as an increase in temperature. Thus, stabilizing for temperature variations also stabilizes unit-to-unit variations in β_0, but not necessarily to the same degree. The biasing circuit of Fig. 2.33 is, therefore, very useful, and the simple design method outlined above provides adequate stability for the operating point.

D. Maximum Symmetrical Swing

It should be noted that it is sometimes (but not always) desirable to locate the Q point so as to obtain a maximum symmetrical collector current swing. The maximum symmetrical swing criterion is normally appropriate for *output* stages of amplifiers, where large signals must be handled. To obtain the maximum symmetrical current or, alternatively stated, a large dynamic range, the Q point must *bisect the ac load line* as shown in Fig. 2.34.

The Q point for the maximum symmetrical swing [assuming that $(V_{CE})_{\text{sat}} \ll V_{CC}$] is determined as follows:

From (2.22), also Fig. 2.20, recall that

$$v_{CE} = v_{ce} + V_{CEq} \qquad i_C = i_c + I_{Cq}$$

The ac load line is given by

$$i_c = -\frac{1}{R_C} v_{ce} \tag{2.59a}$$

which can be written as

$$i_c = -\frac{1}{R_C}(v_{CE} - V_{CEq}) \tag{2.59b}$$

For $(i_C)_{\text{max}}$, $v_{CE} = 0$; hence

$$(i_C)_{\text{max}} = I_{Cq} + \frac{V_{CEq}}{R_C} \tag{2.60}$$

For *maximum symmetrical current,* we have

$$(i_C)_{\text{max}} = 2I_{Cq} \tag{2.61}$$

From (2.60) and (2.61) we obtain

$$I_{Cq} = \frac{V_{CEq}}{R_C} \tag{2.62}$$

Substituting (2.62) into the dc load-line equation, namely,

$$V_{CC} = V_{CEq} + I_{Cq}(R_C + R_E) \tag{2.63}$$

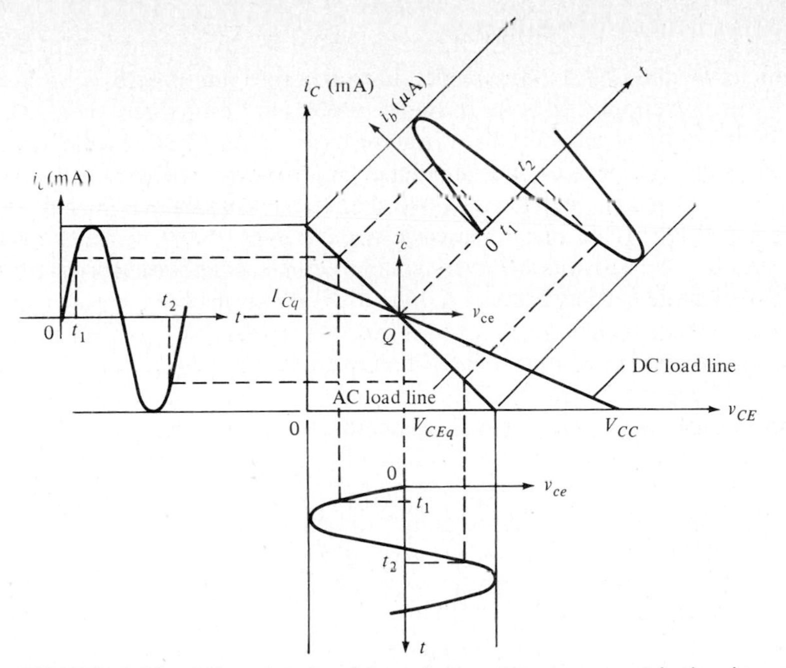

FIGURE 2.35 AC and dc load lines for maximum symmetrical swing.

yields

$$I_{Cq} = \frac{V_{CC}}{2R_C + R_E} \tag{2.64}$$

$$V_{CEq} = (I_{Cq})R_C = \frac{R_C}{2R_C + R_E}V_{CC} \tag{2.65}$$

Under the above condition we have linear operation over a maximum range of input signal. A graphical analysis and the various ac waveforms are shown in Fig. 2.35.

As a final note we mention that the maximum symmetrical swing is not always used for all applications. In class B power amplifiers discussed in Sec. 6.6, the quiescent point is intentionally designed near the cutoff so as to minimize power drain and increase the efficiency of the circuit. For the input stages, where the signals are usually very small, the Q point may be placed almost anywhere, *but you must then be very sure that it is stabilized.*

EXERCISE 2.13

For the circuit shown in Fig. 2.33a, the following are given: $V_{CC} = 20$ V, $R_C = 2$ kΩ, $R_E = 1$ kΩ. The transistor parameters are $\beta_0 = 100$, $I_{CO} \simeq 0$. Determine the values of R_{b1} and R_{b2} for a maximum symmetrical swing in I_C. What is the peak collector swing I_{cp}? Use $R_B = \beta_0 R_E/10$ for bias stability.

Ans $R_{b1} = 39.2$ kΩ, $R_{b2} = 13.5$ kΩ, $I_{cp} = 4$ mA

2.11 INTEGRATED-CIRCUIT (IC) BIASING

In addition to the circuit discussed above, there are other stabilized biasing circuits for discrete electronic circuits [see, for example, problem (2.32)] where feedback is used from the collector to the base terminal. In some cases, compensation techniques, using temperature-sensitive devices such as diodes, transistors, and thermistors (temperature-sensitive resistors), are also used. We shall not discuss these points in detail here except to point out that in integrated circuits the circuit of Fig. 2.33 is not suitable because large-valued resistors with very small tolerance cannot be fabricated; furthermore, if a very large capacitor across $R_E (C_E \geq 10\ \mu F)$ is needed, it must be provided externally. Further, in integrated circuits (see Appendix B) (1) close matching of active and passive devices over a large temperature range is achievable, (2) active elements are cheaper than passive elements, and (3) all the elements are fabricated on a very small chip, and hence the same temperature environment exists for the entire circuit. The design philosophy must take an altogether different viewpoint.

There are several useful biasing circuits given in the literature for the biasing of integrated circuits. We shall consider only two commonly used biasing arrangements in order to examine some of the basic concepts. Since the active and passive elements are fabricated in an identical manner in a monolithic structure, we can assume identical transistors and resistors over a wide temperature range. In the circuit of Fig. 2.36*a*, the two bases are driven from a common voltage node through equal resistances, so the base currents are the same:

$$I_{B1} = I_{B2} = I_B \tag{2.66}$$

Hence, the collector currents are matched for matched or identical transistors, and

$$I_{C1} = I_{C2} = I_C \tag{2.67}$$

The collector current is determined form Kirchhoff's voltage law,

$$V_{CC} = I_1 R_1 + (V_{CE})_1 \tag{2.68}$$

but

$$(V_{CE})_1 = V_{BE} \qquad \text{and} \qquad I_1 = I_{C1} + 2I_B \tag{2.69}$$

Solving (2.68) and (2.69) for I_C, we get

$$I_{C1} = I_{C2} = I_C = \frac{V_{CC} - V_{BE}}{R_1} - (2I_B) = \beta_0 I_B \tag{2.70}$$

If $V_{CC} \gg V_{BE}$ and $\beta_0 \gg 2$, then

$$I_C \simeq \frac{V_{CC}}{R_1} \tag{2.71}$$

Similarly,

$$(V_{CE})_2 = V_{CC} - I_C R_C \simeq V_{CC}\left(1 - \frac{R_C}{R_1}\right) \tag{2.72}$$

If R_C and R_1 are fabricated in the same diffusion step, the temperature sensitivity of

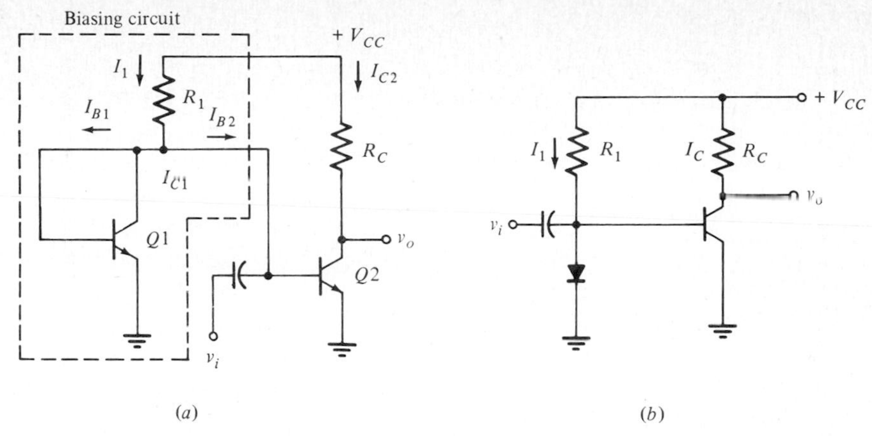

FIGURE 2.36 (*a*) IC biasing circuit. (*b*) Diode version of (*a*) (current mirror).

the ratio R_C/R_1 is very small. Usually $R_C = R_1/2$ so that

$$V_{CE} = \frac{V_{CC}}{2} \tag{2.73}$$

Note that the operating point is independent of temperature variation and depends only on the matching of the transistors. Since matching in integrated circuits can be achieved with very close tolerances, the circuit has a very well-stabilized quiescent point. For improved performance, small equal resistances (of the order of 100) are often inserted in series with the emitter leads of each transistor to obtain feedback. Note that no large bypass capacitors are required in this biasing.

Observe that Fig. 2.36*b* is an alternate representation of Fig. 2.36*a* in IC (i.e., transistor used as a diode). Since the transistor current I_C is approximately equal to I_1 (which is essentially the diode current as $I_{C1} \gg I_{B2}$), the circuit is also referred to as a *current repeater* or *current mirror* since from (2.69)

$$I_1 = I_{C1} + 2I_B = I_C + 2\frac{I_C}{\beta_0} \tag{2.74a}$$

$$I_C = \frac{\beta_0}{\beta_0 + 2}I_1 \simeq I_1\left(1 - \frac{2}{\beta_0}\right) \simeq I_1 \qquad \beta_0 \gg 2 \tag{2.74b}$$

It should be noted that an explicit assumption has been made that $\beta \gg 2$. If β is small, such as in lateral *pnp* transistors, then an additional transistor may be used to reduce the error. One such circuit is shown in Fig. 2.37. In this case it can be shown that (problem 2.45)

$$I_C = \frac{\beta_0^2 + 2\beta_0}{\beta_0^2 + 2\beta_0 + 2}I_1 \simeq I_1\left(1 - \frac{2}{\beta_0^2 + 2\beta_0}\right) \simeq I_1 \tag{2.75}$$

For $\beta_0 = 20$, in (2.74), I_C is within 9 percent of I_1; for (2.75) it is within 0.5 percent of I_1. For $\beta_0 = 100$, (2.74) is within 2 percent, and (2.75) is within 0.04 percent.

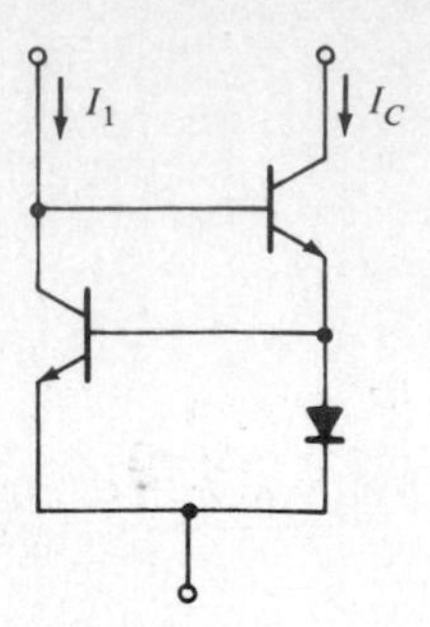

FIGURE 2.37 Improved current mirror circuit.

EXERCISE 2.14

For the circuit shown in Fig. E2.14, determine the currents I_1 and I_2 and the collector-to-ground voltage V_C. Assume $\beta_0 \gg 1$.

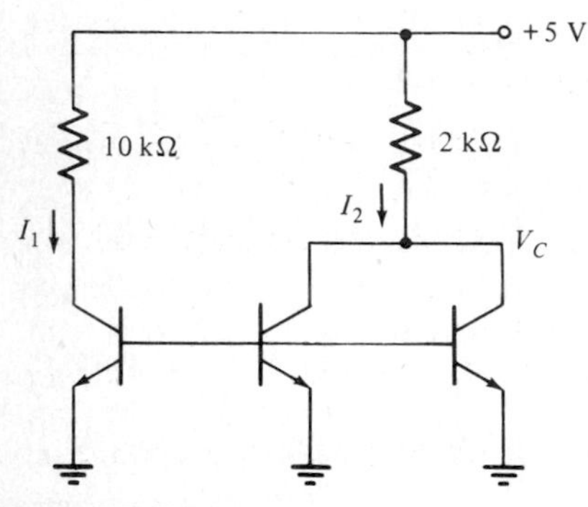

FIGURE E2.14

Ans $I_1 = 0.43$ mA, $I_2 = 0.86$ mA, $V_C = 3.28$ V

2.12 CONCLUDING REMARKS

The purpose of this chapter was to introduce the BJT, its *V-I* characteristics, and its behavior under large and small signals. Under large-signal applications the transistor is inherently a nonlinear device and graphical analysis or a computer-aided analysis, using the large-signal model, must be used. The common-emitter and emitter-coupled inverters are such examples and are used as switches in digital circuits. For small-signal applications, such as amplifiers, the transistor may be considered linear, and the small-signal models are used. Complicated models appropriate for computer-aided analysis and design are given; simple models, suitable for preliminary analysis and design, will be utilized throughout the text. The latter is used extensively in order to provide insight into the analysis and design of electronic circuits.

REFERENCES

1. C. A. Holt, *Electronic Circuits: Digital and Analog,* Wiley, New York, 1978.
2. D. Schilling and C. Belove, *Electronic Circuits—Discrete and Integrated,* McGraw-Hill, New York, 1979.
3. I. Getreu, *Modeling the Bipolar Transistor*, Tektronix, Portland, Oregon, 1976.
4. J. Millman, *Microelectronics: Digital and Analog Circuits and Systems,* McGraw-Hill, New York, 1979.

PROBLEMS

2.1 An *npn* transistor has dc voltage sources applied to it so as to forward bias the emitter-base junction with $V_{BE} = 0.6$ V and to reverse bias the collector-base junction with $V_{CB} = 3$ V. If the transistor parameters are $\alpha_F = 0.99$, $\alpha_R = 0.10$, and $I_{CS} = 10^{-13}$ A, determine the currents I_B, I_C, and I_E at room temperature, using the Ebers-Moll equations. What are the values of I_{ES}, I_{EO}, and I_{CO}?

2.2 Determine I_C, I_E, and I_B of a *pnp* transistor for $V_{EB} = 0.6$ V, $V_{BC} = 5$ V. The transistor parameters are $I_{CS} = 2 \times 10^{-14}$ A, $\alpha_F = 0.990$, $\alpha_R = 0.495$.

2.3 Plot the input characteristic I_B vs V_{BE} of an *npn* transistor operated in the active mode. Use the transistor parameters of problem 2.1 and let $V_{CB} \geq 3$ V.

2.4 The input and output characteristics of an *npn* transistor in the normal-active region are given approximately by

$$I_B = \frac{I_{ES}}{\beta_0}(e^{V_{BE}/V_t} - 1) \qquad I_C = \beta_0 I_B = -\alpha_0 I_E$$

(*a*) If the transistor parameters are $\beta_0 = 50$ and $I_{ES} = 10^{-14}$, plot the input characteristic.
(*b*) Plot the output characteristic for $I_B = 0.1$ μA, 1 μA, and 10 μA.

2.5 Derive the *V-I* relationships and plot the *V-I* characteristics for transistor-connected diodes if
(*a*) The collector-base terminals are short-circuited ($V_{BC} = 0$) as in Fig. P2.5*a*.
(*b*) The emitter terminal is open-circuited (i.e., $I_C = 0$) as in Fig. P2.5*b*.
Assume the transistor parameters as given in problem 2.1.

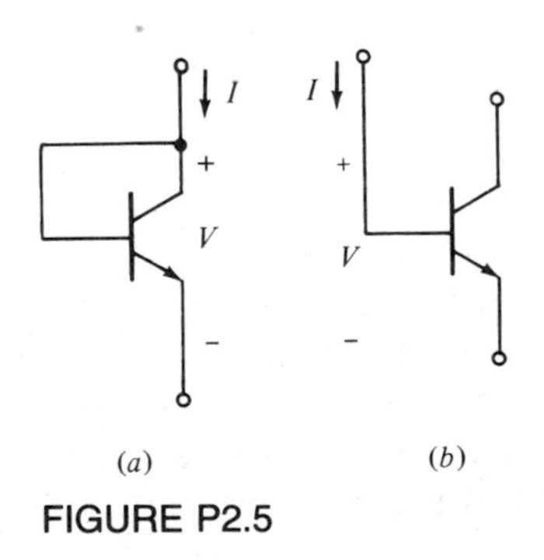

FIGURE P2.5

2.6 Plot I_C vs V_{CE} for $I_B = 10\ \mu$A and determine the saturation voltage $(V_{CE})_{sat}$ of the transistor in problem 2.1. (Plot only for $I_C > 0$, $V_{CE} > 0$.)

2.7 An *npn* transistor has the following parameters: $\alpha_F = 0.99$, $\alpha_R = 0.20$, $I_{ES} = 10^{-15}$ A.
(*a*) Plot the CE input characteristic for $V_{CE} = 5$ V.
(*b*) Plot the CE output characteristic for $I_B = 50\ \mu$A.

2.8 For the transistor in problem 2.7
(*a*) Plot the CB input characteristic for $V_{CB} = 10$ V.
(*b*) Plot the CB output characteristic for $I_E = 2$ mA.

2.9 A transistor may be falsely viewed as two diodes connected back to back as shown in Fig. P2.9. If the diodes obey the ideal-diode equation $I = I_s(e^{V/V_t} - 1)$ with $I_s = 10^{-12}$ A, plot:
(*a*) The input characteristic (I_1 vs V_{BE}) for $V_{CE} = 0$ and 0.2 V
(*b*) The output characteristic (I_2 vs V_{CE}) for $I_B = 10\ \mu$A

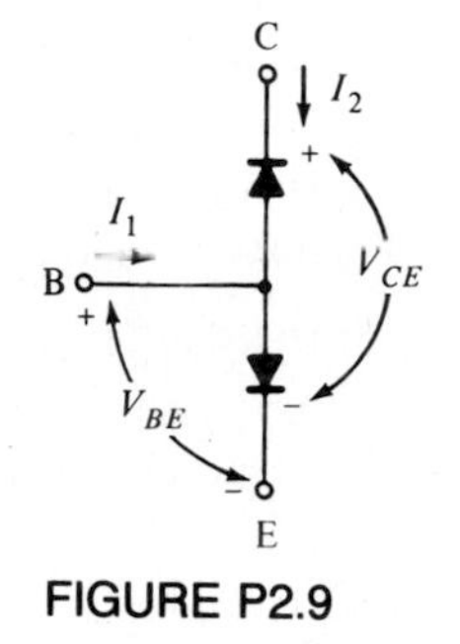

FIGURE P2.9

2.10 For the circuit shown in Fig. P2.10, determine the base voltage V_{BB} to saturate the transistor, given that $(V_{CE})_{sat} = 0.2$ V, $(V_{BE})_{sat} = 0.7$ V, and $\beta_0 = 50$.

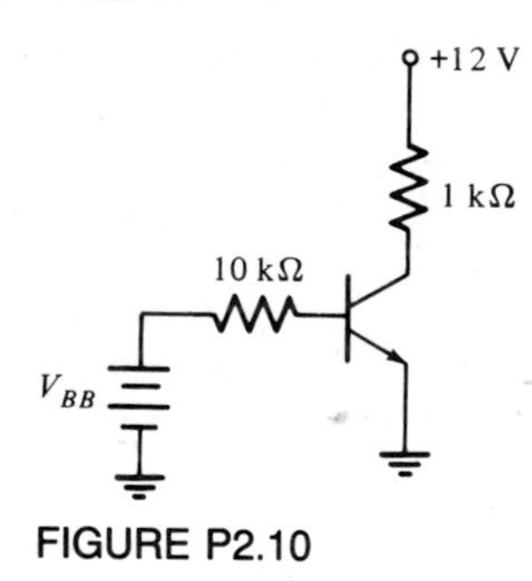

FIGURE P2.10

2.11 The input and output characteristics of a transistor are approximated crudely by the piecewise-linear method as shown in Fig. P2.11. Show a piecewise-linear model for the transistor using ideal diodes.

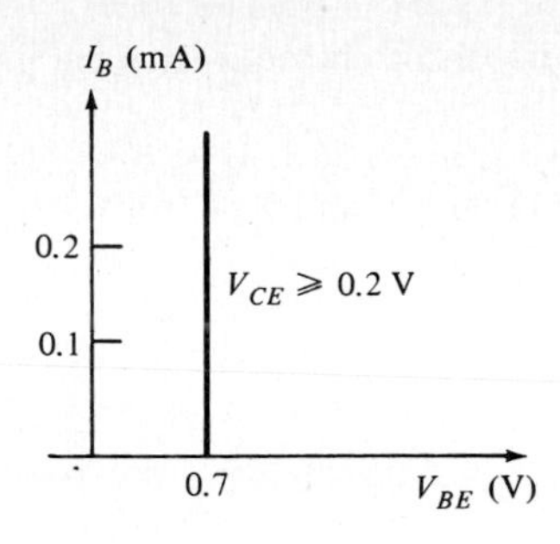

(a)

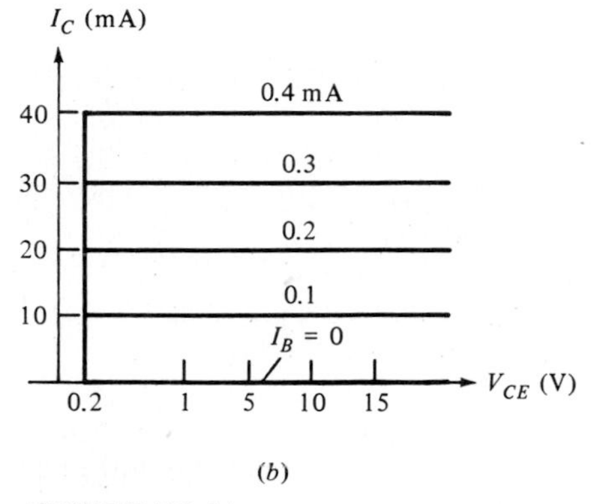

(b)

FIGURE P2.11

2.12 (a) For the transistor described in Fig. P2.11, give the values of β_0, α_0, $(V_{BE})_{on}$, $(V_{CE})_{sat}$, I_{CO}.

(b) Derive (2.16a).

2.13 Show a piecewise-linear model for the transistor with characteristics given in Fig. P2.13.

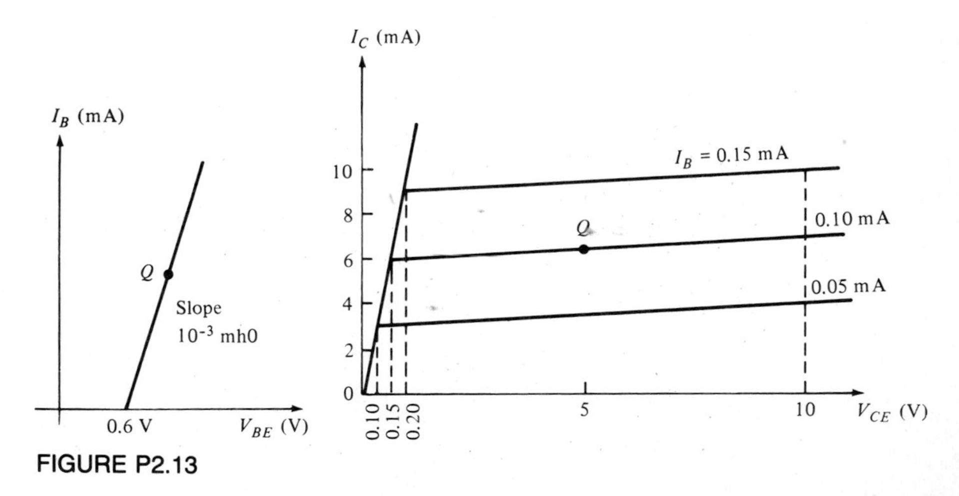

FIGURE P2.13

2.14 For the circuit shown in Fig. P2.14, the transistor characteristics are given as in Fig. 2.17*b* and *c*.
(*a*) Sketch the transfer characteristics V_O vs V_I if $-5\text{ V} < V_I < 5\text{ V}$.
(*b*) Sketch v_O if $v_I = V_B + V \sin \omega t$ with $V_B = 1$ V and $V = 0.1$ V.

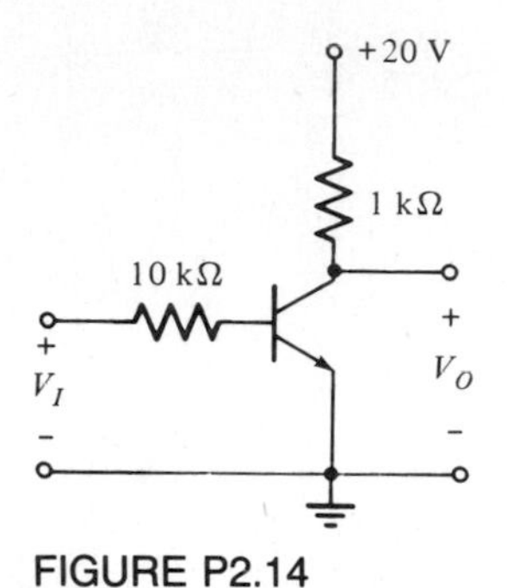

FIGURE P2.14

2.15 For problem 2.14*b*, determine the value of V_B for which the output voltage has the largest ac signal component. What are the corresponding amplitudes of the input and output ac signals? Show the load line.

2.16 For the circuit shown in Fig. P2.16, if $\beta_0 = 80$ and $V_{BE} \simeq 0.6$ V, $I_{CO} = 1$ nA.
(*a*) Sketch the transfer characteristic V_O vs V_I.
(*b*) Determine the collector and the emitter voltages for $V_I = -1, -5$ V.
(*c*) Repeat part (*b*) for $V_I = 1, 5$ V.

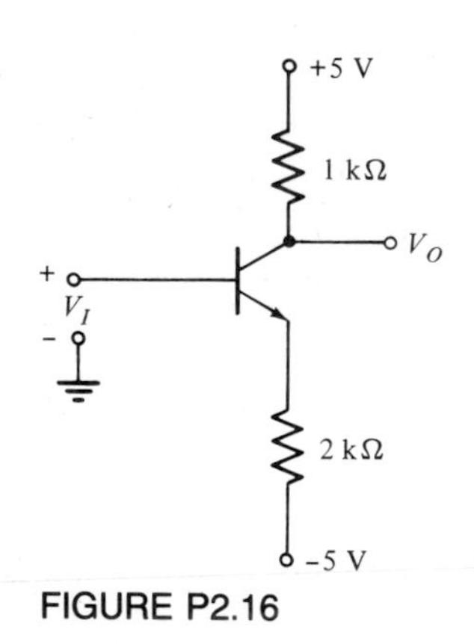

FIGURE P2.16

2.17 For the inverter circuit shown in Fig. P2.17, draw the transfer characteristics for $-5\text{ V} < V_I < 5\text{ V}$. The transistor characteristics are given in Fig. P2.13. Identify the value of V_I that brings the transistor at the edge of saturation.

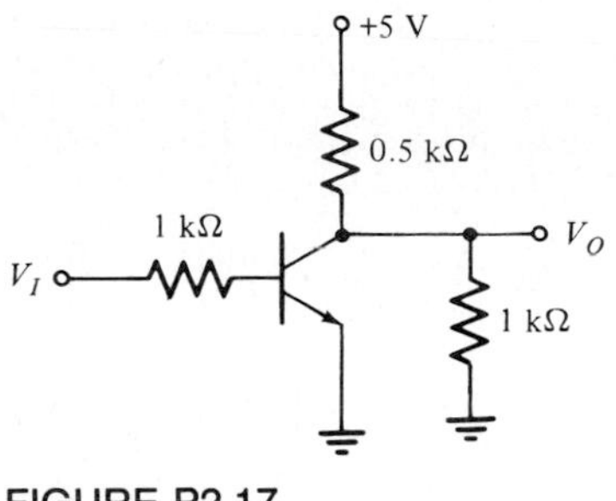

FIGURE P2.17

2.18 Repeat problem 2.17 if the transistor characteristics are given as in Fig. 2.17*b* and *c*.

2.19 For the circuit shown in Fig. P2.19, sketch the transfer characteristic, i.e., V_O vs V_I for all V_I. Assume the transistor to be open-circuited when in the cutoff mode and perfectly short-circuited when in the saturation mode. Derive the expression for the value of V_I that will saturate the transistor.

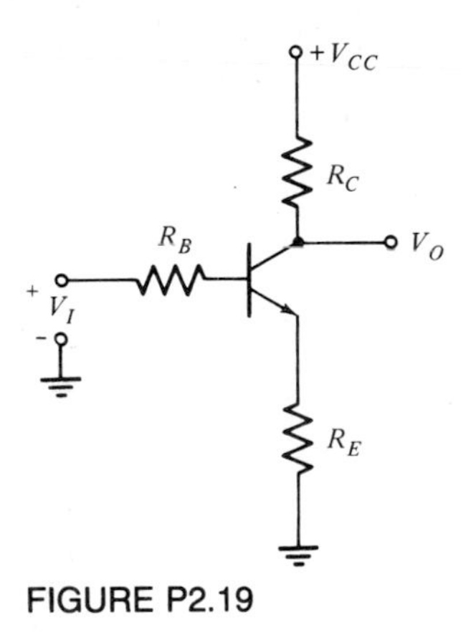

FIGURE P2.19

2.20 For the circuit in Fig. P2.20, the transistor parameters are $\beta_0 = 100$, $V_{BE} = 0.7$ V, and $(V_{CE})_{sat} = 0.2$ V.

(*a*) Sketch the transfer characteristics for $R_B = 10\ k\Omega$, $R_E = 1\ k\Omega$, $R_C = 2\ k\Omega$, $V_{CC} = 10$ V.
(*b*) What is the value of V_O for $V_I = 5$ V?

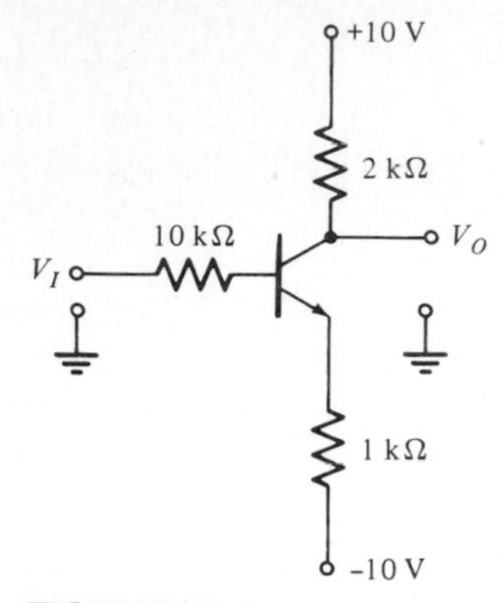

FIGURE P2.20

2.21 For the circuit of Fig. P2.21, determine the following:
(*a*) The Q point if $R_{b1} = 30\ k\Omega$ and $R_{b2} = 5\ k\Omega$, $\beta_0 = 100$, $I_{CO} \simeq 0$, and $(V_{BE})_{on} = 0.7$.
(*b*) The new Q point if the transistor is replaced with another unit with $\beta_0 = 50$ and all the other parameters remain unchanged.

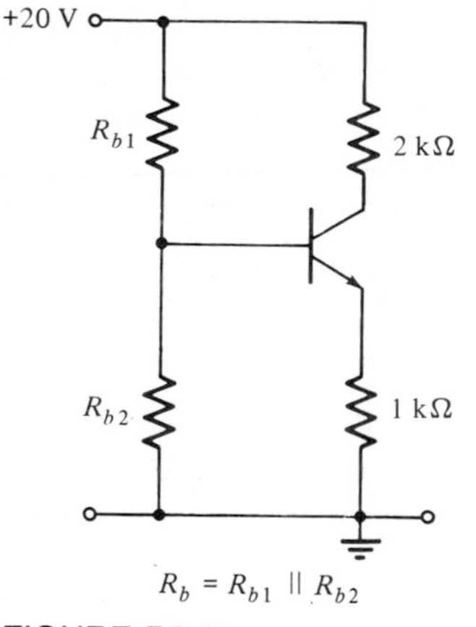

FIGURE P2.21

2.22 For the circuit in Fig. P2.21, determine the values of R_{b1} and R_{b2} so that the Q point is at $V_{CE} = 5$ V and $I_C = 1$ mA. The transistor parameters are $\beta_0 = 50$, $I_{CO} \simeq 1$ nA, $(V_{BE})_{on} = 0.7$ V. Determine the change in V_{CE} and I_C if the transistor β_0 varies from 20 to 100.

2.23 Determine the Q point of Fig P2.19 if $V_{CC} = 20$ V, $V_I = 1$ V, $R_C = R_E = 1$ kΩ, and $R_B = 5$ kΩ. Let $(V_{BE})_{\text{on}} = 0.7$ V, $\beta_0 = 100$, and $I_{CO} = 10$ nA.

2.24 Suppose that a nonlinear resistive two-port is characterized by

$$I_1 = 2V_1^2 + V_1V_2^2 \qquad I_2 = V_2(V_1 + V_1e^{-V_2})$$

Calculate the incremental admittance matrix about the operating point $V_1 = V_2 = 1$ V. Show the small-signal model.

2.25 The low-frequency V-I description of a certain hypothetical device is given by

$$V_I = 0.1e^{0.1I_1} \qquad V_2 = 0.1I_1^2(1 + e^{0.1I_2})$$

Calculate the incremental impedance matrix about the operating point $I_1 = 10$ A, $I_2 =$ 5 A. Show the small-signal model.

2.26 A bipolar transistor has a base width $W = 10^{-4}$ cm. Determine the maximum low-frequency common-base current gain α_0, the maximum common-emitter current gain β_0, and the alpha cutoff frequency ω_α, ω_β, and ω_T for the following:
(*a*) A *pnp* Si transistor with $L_p = 8 \times 10^{-4}$ cm
(*b*) A lateral *pnp* with $W \simeq 2 \times 10^{-4}$ cm, $L_p = 8 \times 10^{-4}$ cm
(*c*) An *npn* Si transistor with $L_n = 10^{-3}$ cm
Given that $\mu_n = 1300$ cm²/V · s, $\mu_p = 500$ cm²/V · s, $K = 10$, $m = 0.5$.

2.27 For the common-collector circuit shown in Fig. P2.27:
(*a*) If $R_E = 1$ kΩ determine the value of R_B so that $I_C = 5$ mA. What is the value of V_{CE} if the transistor $\beta = 20$?
(*b*) For the operating point determined in (*a*) what is the maximum undistorted ac signal for v_O? Show your sketch.

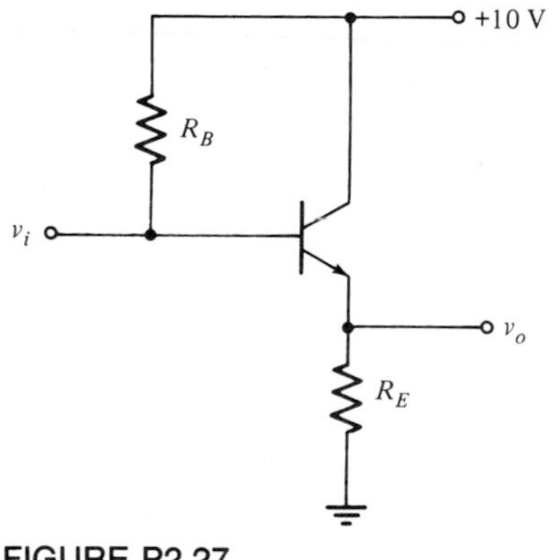

FIGURE P2.27

2.28 For the circuit shown in Fig. P2.28, sketch the transfer characteristics V_O vs V_I if $-5\text{ V} < V_I < 5$ V. The transistor parameters are given by $(V_{BE})_{\text{on}} = 0.7$ V, $(V_{CE})_{\text{sat}} =$ 0.2 V, $\beta_0 = 100$.

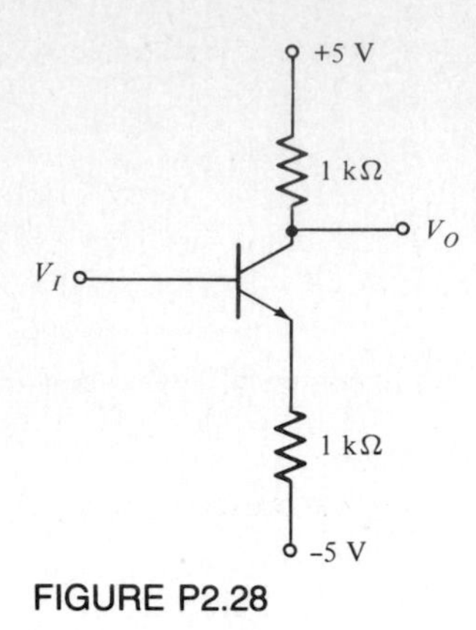

FIGURE P2.28

2.29 For the circuit shown in Fig. P2.29, determine the collector, emitter, and the base currents and V_{CE} for the following cases.
(*a*) The transistor $\beta_0 = 100$, $V_{BE} = 0.7$ V, $I_{CO} = 1$ nA
(*b*) The transistor $\beta_0 = 30$, $V_{BE} = 0.7$ V, $I_{CO} = 1$ nA
(*c*) The transistor $\beta_0 = 200$, $V_{BE} = 0.5$ V, $I_{CO} = 10$ nA

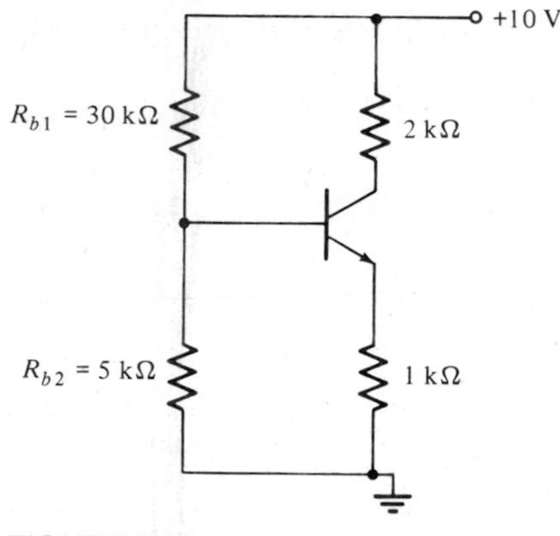

FIGURE P2.29

2.30 For the circuit shown in Fig. P2.29, if $R_{b2} = \infty$ and $\beta_0 = 10$, determine I_C and V_{CE}. For what value of β_0 will the transistor saturate?

2.31 For the circuit shown in Fig. P2.31, the desired Q point is at $I_C = 2$ mA, $V_{CE} = 5$ V. The transistor β variation from unit to unit ranges from 50 to 100, $V_{BE} \simeq 0.7$ V and is approximately the same for all units.
(*a*) Determine the values R_E, R_{b1}, and R_{b2} such that the quiescent current does not change by more than 20 percent.

(*b*) Determine the low-frequency ac current signal loss due to the biasing resistors R_{b1} and R_{b2}, for the designed circuit.

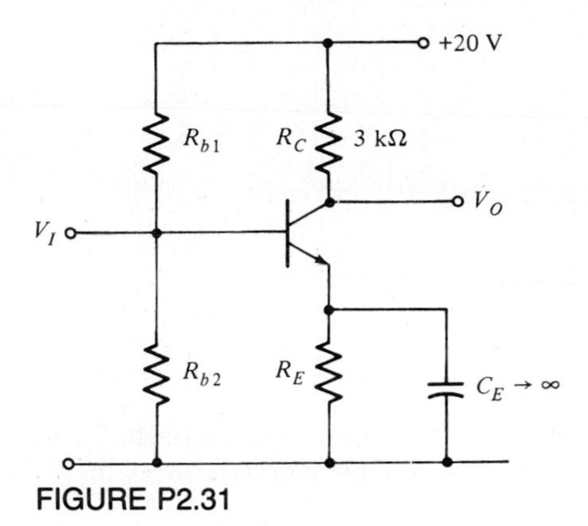

FIGURE P2.31

2.32 Consider the circuit shown in Fig. P2.32.
(*a*) If a Si transistor is used with $V_{BE} = 0.7$ V, $h_{FE} = 60$, and $I_{CO} \simeq 1$ nA, determine the quiescent point V_{CE}, I_C.
(*b*) If a Ge transistor is used with $V_{BE} = 0.2$ V, $h_{FE} = 60$, and $I_{CO} \simeq 5$ μA, what is V_{CE}, I_C?

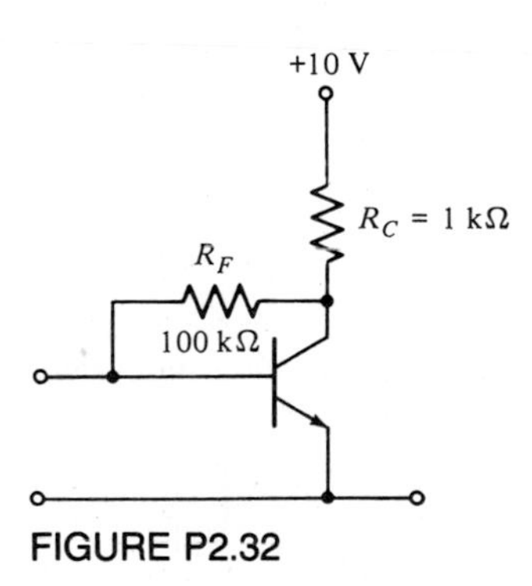

FIGURE P2.32

2.33 For the circuit shown in Fig. P2.33, determine the values of R_{b1} and R_{b2} such that a maximum symmetrical collector current swing is obtained. What is the quiescent point under this condition? The transistor has $\beta_0 = 50$, $I_{CO} = 1$ nA, and $V_{BE} \simeq 0.7$ V. Assume $R_B = 5R_E$.

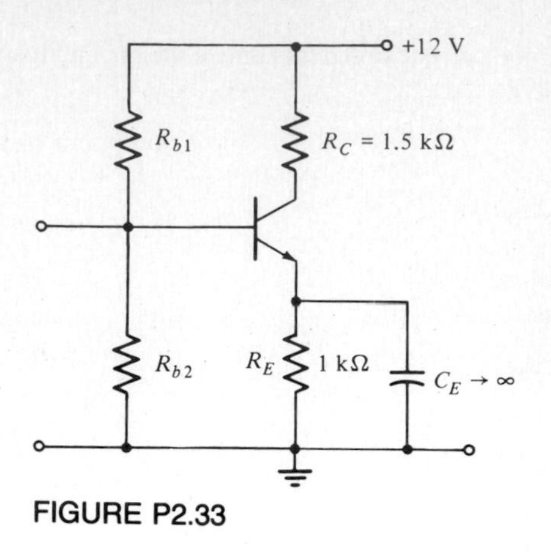

FIGURE P2.33

2.34 A Si transistor is used in the circuit shown in Fig. P2.34. Calculate I_C and V_{CE}. Determine the new operating point if another transistor of the same type, but with $\beta_0 = 100$, is used.

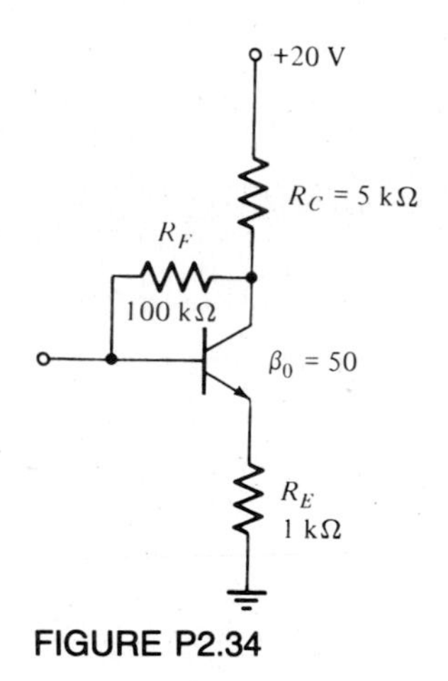

FIGURE P2.34

2.35 For the circuit shown in Fig. P2.35, calculate R_1 and R_2 for maximum possible undistorted output voltage swing. Plot $v_O(t)$ and $v_i(t)$ on the same scale. The transistor parameters are $\beta_0 = 100$, $V_{BE} = 0.7$ V, $I_{CO} = 1$ nA.

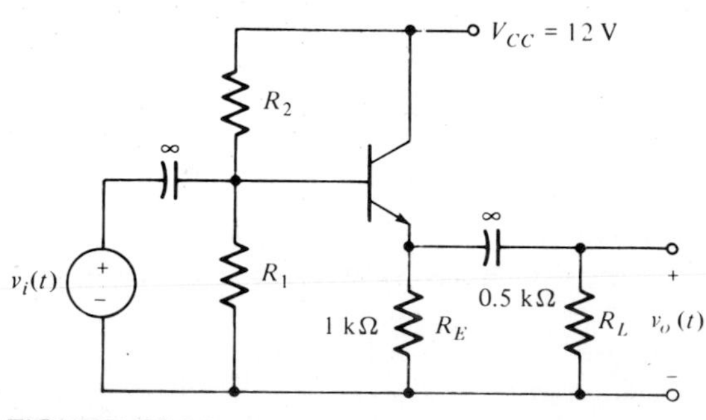

FIGURE P2.35

2.36 For the circuit shown in Fig. P2.36, the transistor characteristics are given in Fig. 2.18*a* and *b*. Sketch $v_i(t)$, $v_o(t)$ and determine the voltage gain of the amplifier. What is the Q point?

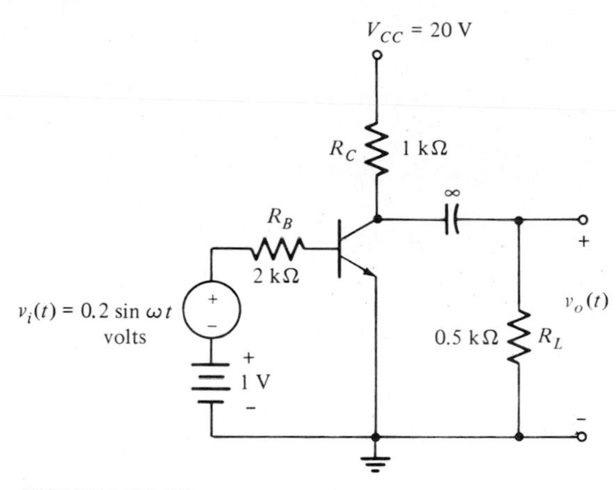

FIGURE P2.36

2.37 For the amplifier circuit shown in Fig. P2.37, the transistor input and output characteristics are given in Fig. 2.17*b* and *c*. Sketch $v_O(t)$ and determine the voltage gain of the amplifier.

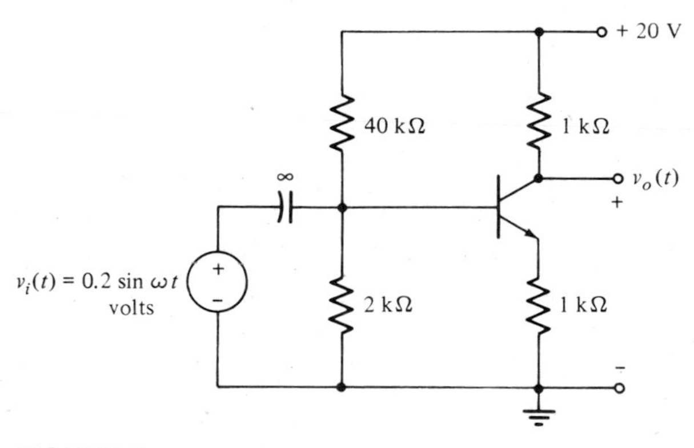

FIGURE P2.37

2.38 The piecewise-linear characteristics of a certain transistor in the CE configuration is given in Fig. P2.13. If the Q point is at I_{Cq} = 6.4 mA, V_{CEq} = 5 V, calculate the h parameters of the transistor.

2.39 Using the h parameters in problem 2.38, calculate the voltage gain and the current gain if the transistor is used as a two-port, with $R_L = R_S = 1$ kΩ.

2.40 Determine the high-frequency equivalent circuit parameters of the *npn* transistor type 2N2221a listed in Appendix E, at I_C = 10 mA, V_{CE} = 10 V.

2.41 Repeat problem 2.40 for the *pnp* transistor type 2N3251 given in Appendix E.

2.42 For the emitter-coupled circuit shown in Fig. P2.42 determine the collector current and voltage if the transistors are perfectly matched with $\beta_0 = 100$ and $V_{BE} = 0.7$ V.

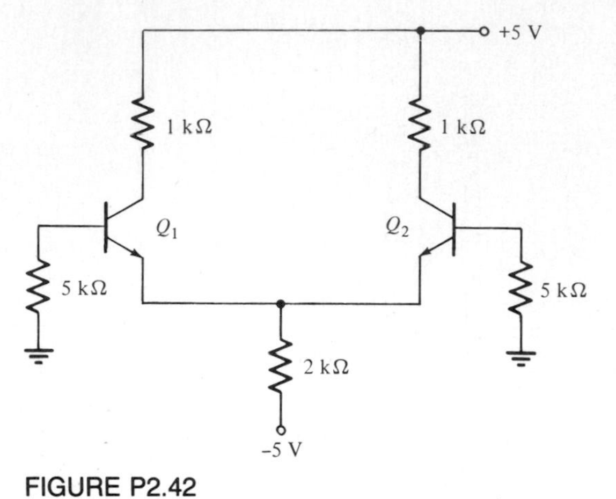

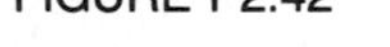
FIGURE P2.42

2.43 For the circuit shown in Fig. P2.43, assume identical transistors with $(V_{BE})_{\text{on}} = 0.7$ V and $(V_{CE})_{\text{sat}} = 0.2$ V and $\beta_0 = 50$. Determine the base and collector currents and voltages of each transistor. Note that the Q_2 transistor is normally ON.

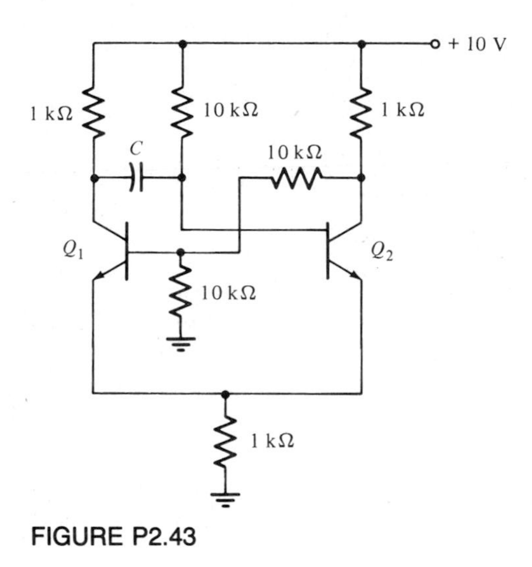

FIGURE P2.43

2.44 For the current mirror circuit of Fig. 2.37*b*, redrawn in Fig. P2.44, derive Eq. (2.74) by assuming $I_{b1} = I_{b2} = 1$ (normalized) and determine I_2/I_1.

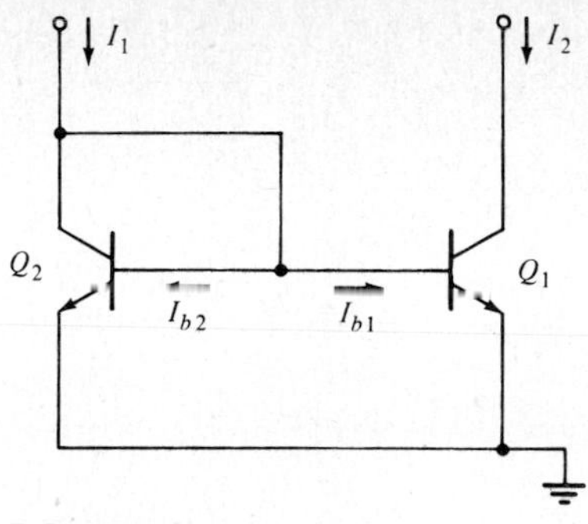

FIGURE P2.44

2.45 For the improved mirror circuit of Fig. 2.37, redrawn in Fig. P2.45, derive Eq. (2.75) by using the method of problem 2.44, namely, $I_{b2} = I_{b3} = 1$, and determine I_2/I_1. [Note that $I_b = I_c/\beta_0 = I_e/(\beta_0 + 1)$.]

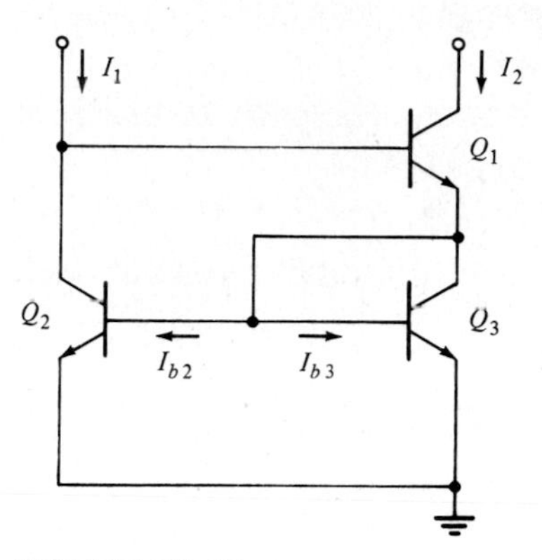

FIGURE P2.45

2.46 Derive (2.38).

2.47 Design the biasing circuit of Fig. 2.33 for maximum symmetrical swing, using an *npn* Si transistor. The supply voltage $V_{CC} = 20$ V, given that $40 \leq \beta_0 \leq 100$, $I_{CO} \simeq 1$ nA, and $V_{BE} \simeq 0.7$ V. Use $R_B \simeq 0.1(\beta_0)_{\min}R_E$, $R_C = 2R_E = 2$ kΩ.

CHAPTER 3 Field-Effect Transistors

The field-effect transistor (FET) is a semiconductor device whose operation depends on one type of carrier, the majority carrier, and is thus a unipolar device. The current in a FET is controlled by an electric field. FETs are three-terminal voltage-sensitive devices with extremely high input impedance. They are used as voltage-controlled sources in amplifier design as well as switches.

There are two basic types of FETs: one type is called the *junction field-effect transistor* (JFET, or simply FET), and the other type is called the *metal-oxide semiconductor field-effect transistor* (MOSFET, or simply MOS). There are two types of MOSFETs: one is called the *depletion type,* and the other is called the *enhancement type*. The MOSFET plays a dominant role in monolithic integrated circuits because it has a low fabrication cost, occupies less space, consumes less power, can be used as a resistor for very low voltages, and is less noisy than the BJT. These devices are used in medium-scale integration (MSI) and large-scale integration (LSI) and are instrumental in bringing the cost of microprocessors and computers to what it is today (see Appendix B for integrated-circuit fabrication and terminologies).

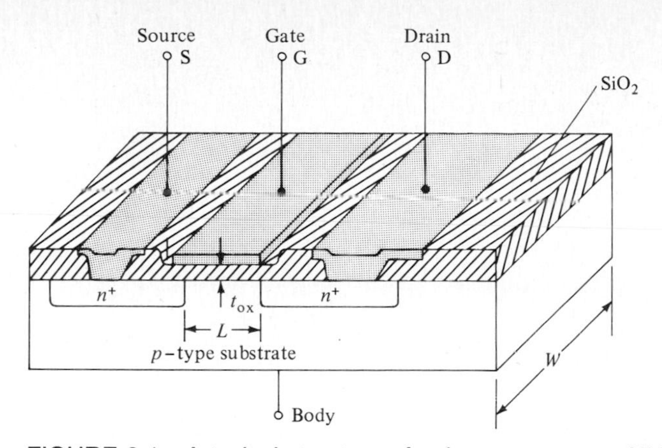

FIGURE 3.1 A typical structure of enhancement-type NMOS.

The physical structure of an enhancement-type *n*-channel MOSFET (NMOS) is shown in Fig. 3.1. *L* and *W* are the channel length and width, respectively. As we shall see, the *W*/*L* ratio is an important parameter. The dimensions of the devices are of the order of micrometers (1 μm = 10^{-6} m), and the total area occupied by an MOS device is of the order 50×10^{-8} cm^2, which is smaller than BJTs by about a factor of 5. The oxide layer is very thin, with thickness *t* of about 0.1 μm. The enhancement-type NMOS, shown in Fig. 3.1, and the complementary symmetry COS/MOS shown in Fig. 3.24 are among the dominant technologies in integrated circuits.

This chapter will consider the *V-I* characteristics of the JFET and MOSFET devices and their use as switches as well as small-signal amplifiers. Small-signal models and biasing are also covered in some detail.

3.1 JUNCTION FIELD-EFFECT TRANSISTORS (JFET)

The construction, simplified representation, and the graphic symbol for an *n*-channel JFET are shown in Fig. 3.2*a, b,* and *c,* respectively. The device consists of an *n*-type channel with two ohmic contacts, called the *source* and the *drain*, and two heavily doped *p*-type junctions with ohmic contacts, called the *gate*. The reason for the names source, drain, and gate will become apparent when we consider the operation of the device. FETs are also made with *p* channels and *n* gates. They operate in the same manner as *n* channels except that the polarities of all the voltages and currents are reversed for *p* channels. The symbol for a *p*-channel FET is the same as in Fig. 3.2*c* except that the direction of the arrow on the gate is reversed. To avoid unnecessary repetition, we will consider the *n* channel unless otherwise stated. We shall also show that the *n*-channel JFET has performance characteristics superior to the *p*-channel JFET, and hence it is a preferred choice.

The operation of the device is as follows. Let us assume that V_{GS} is zero so that the source and gate terminals are at ground potential, and V_{DS} is some dc voltage with the positive polarity on the drain, as indicated in Fig. 3.2*b*. Under these conditions, the *pn* junctions have zero potential difference, and essentially no current flows

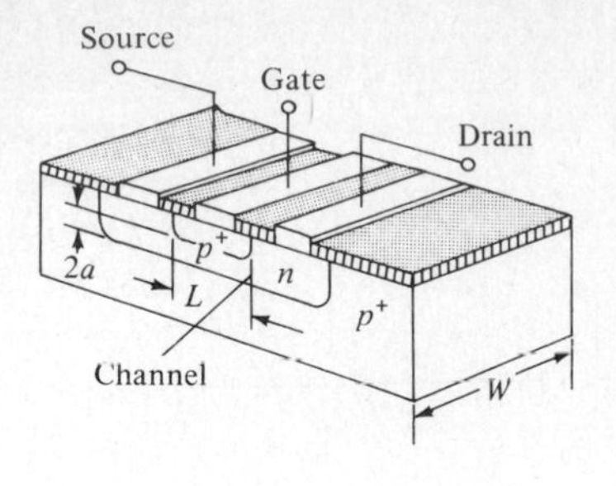

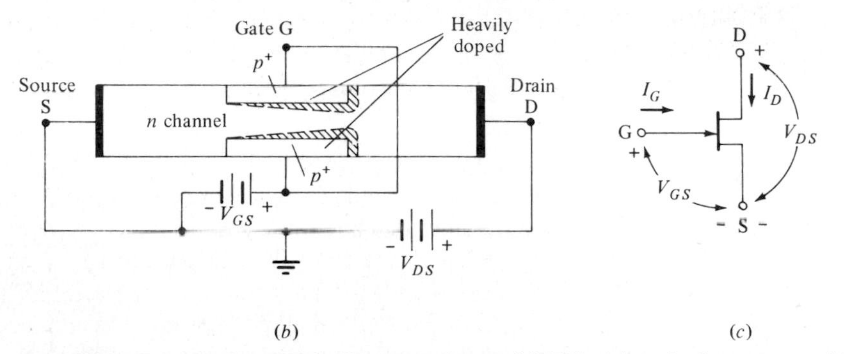

FIGURE 3.2 Junction field-effect transistors (JFET): (*a*) typical structure, (*b*) simplified representation, (*c*) symbol for *n*-channel JFET.

through the gate lead. The *n*-type material acts like a resistor for small values of V_{DS}, and electrons flow from the source to the drain. The initial amount of current is directly porportional to V_{DS} with the proportionality constant being the conductance of the *n*-type material. The *V-I* characteristics of JFET for small values of V_{DS} as shown in Fig. 3.3 are similar to those of a controlled resistor with the gate-to-source voltage playing the controlling effect. *The JFET is always operated with the gate junction reverse-biased, and thus the gate current is essentially zero*. This characteristic of the

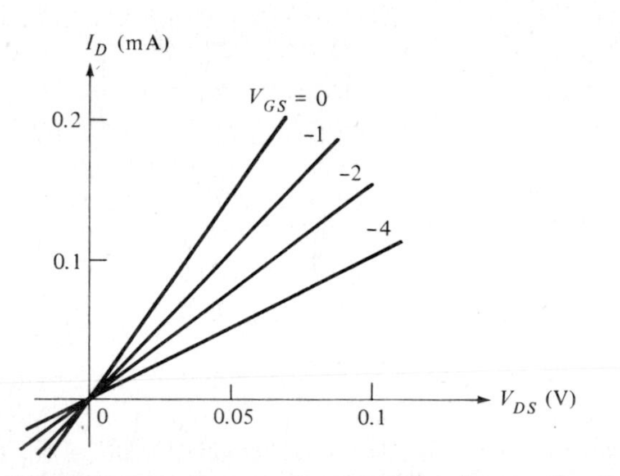

FIGURE 3.3 Output characteristics for small values of V_{DS} ($V_{DS} \ll V_{GS} \leq V_{PO}$).

device makes an FET usable as a resistor in integrated circuits at low voltage values of V_{DS}. At higher values of V_{DS} the V-I characteristic exhibits nonlinear behavior.

The operation of the device is qualitatively described as follows: as we increase the voltage V_{DS} from zero, the gate-drain junctions are increasingly reverse-biased, and as discussed in Appendix A, the depletion regions become wider. Note that the drain side of the junction has a wider depletion width because the reverse voltage is higher at the drain end than at the source end. The depletion region is shown by the dashed area in Fig. 3.4*a*. As the depletion region becomes wider, the channel becomes narrower, and thus the resistance increases, and the slope of the V-I characteristic becomes smaller. This is illustrated in Fig. 3.4*d*. The value of V_{DS} at which the channel pinches off is called V_{PO}; at this voltage the depletion regions on each side of the channel join together, and the channel connection between the source and the drain is pinched off, as shown in Fig. 3.4*b*. The pinch-off voltage is determined by the geometry and the physics of the device. For the n-channel JFET it is given by [1]

$$V_{PO} = \frac{qa^2N_d}{2K\epsilon_0} \tag{3.1}$$

where a = channel depth
q = electronic charge
K = relative dielectric constant (= 12 for Si)
$\epsilon_0 = (8.85 \times 10^{-14}$ F/Cm) = permittivity of free space
N_d = donor concentration of the n-type channel

In typical devices, V_{PO} ranges from 1 to 6 V.

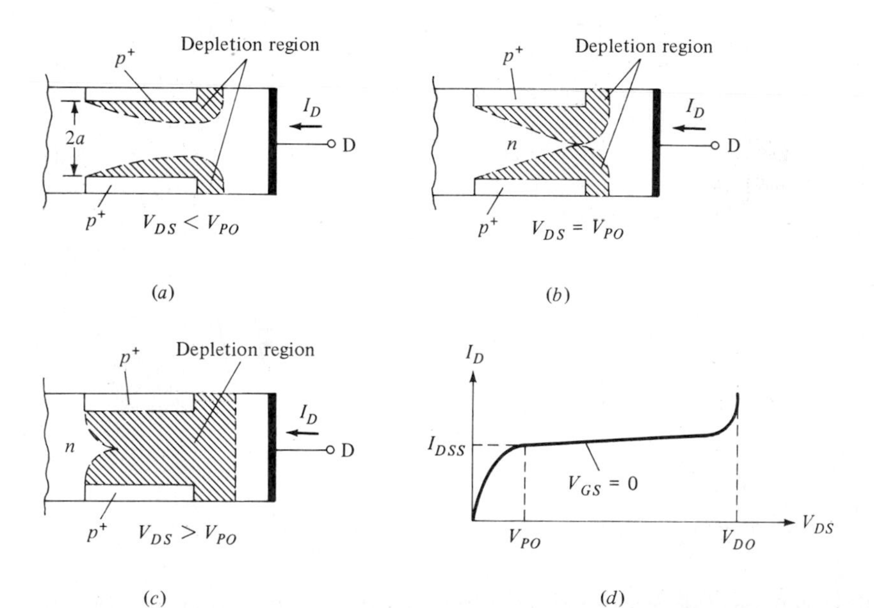

FIGURE 3.4 (*a*) Channel of JFET for $V_{DS} < V_{PO}$. (*b*) Channel of JFET for $V_{DS} = V_{PO}$. (*c*) Channel of JFET for $V_{DS} > V_{PO}$. (*d*) $I_D - V_{DS}$ curve for $V_{GS} = 0$.

EXAMPLE 3.1

Determine the pinch-off voltage for an n-channel JFET with $N_d = 10^{16}\text{cm}^{-3}$, $K = 12$, $a = 0.5\ \mu\text{m}$. From (3.1) V_{PO} is given by

$$V_{PO} = \frac{(1.6 \times 10^{-19})(5 \times 10^{-5})^2(10^{16})}{2(12)(8.85 \times 10^{14})} = 1.88\ \text{V}$$

The onset of pinch-off changes the physical process occurring in the device, and the current I_D remains almost constant and independent of V_{DS} for $V_{DS} > V_{PO}$. The value of I_D for $V_{GS} = 0$ and $V_{DS} = V_{PO}$ is denoted I_{DSS}, and this parameter is specified on the data sheets of JFETs. The current I_{DSS} is temperature dependent and decreases for increasing temperature. Increasing V_{DS} beyond the breakdown voltage causes avalanche breakdown. The breakdown voltage at zero gate voltage is denoted by V_{DO}. The breakdown voltage V_{DO} occurs at the drain end since the reverse voltage is highest at this end. The breakdown voltage is typically in the range of 20 to 50 V.

The measured output or drain characteristic and the transfer characteristic of a typical n-channel JFET are shown in Fig. 3.5. The reference directions are indicated in Fig. 3.2*c*. (For a p-channel JFET, the signs of all the voltages and currents are reversed.)

If V_{DS} is held constant and a gate voltage V_{GS} (which reverse biases the pn junctions in Fig. 3.2) is applied, the depletion region is increased, and thus the channel

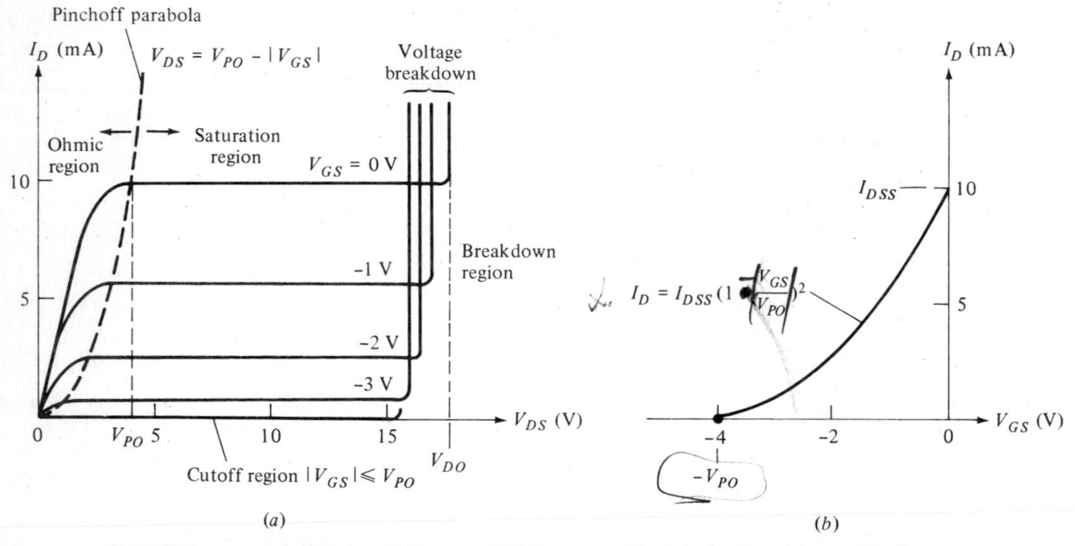

FIGURE 3.5 (*a*) Common-source JFET output (drain) characteristic. (*b*) Common-source JFET transfer characteristic.

resistance is increased, thereby decreasing I_D. The effect of increasing the magnitude of V_{GS} (i.e., making V_{GS} more negative) also reduces the value of V_{DS} at the threshold of pinch-off, as shown in Fig. 3.5*a*. The value of V_{DS} at the threshold of pinch-off is given approximately by

$$V_{DS}\,(\text{at pinch-off}) = V_{PO} + V_{GS} = V_P \tag{3.2}$$

Note that $V_{PO} > 0$ for *n*-channel JFET. In practice V_{GS} is never made more positive than 0.5 V in order to avoid gate-to-source current. The breakdown voltage V_B for an arbitrary V_{GS} is given by

$$V_B = V_{DO} + V_{GS} \tag{3.3}$$

Note that V_B is reduced as V_{GS} becomes more negative (Fig. 3.5*a*).

Note further that in Fig. 3.5*b* a small gate-voltage variation for V_{GS} near zero causes appreciable change in the drain current; thus, the FET is a voltage-sensitive device. As mentioned already, in integrated circuits FETs can be used as voltage-controlled resistors for $V_{DS} \ll V_{PO}$. When a FET is used as an amplifier, it is usually operated in the constant-current region (also called the *saturation** region). For large signals the current I_D can be changed from zero to a relatively large value; hence a FET can be operated as a switch under such conditions.

For analytic purposes, the *V-I* characteristics of FETs may be conveniently divided into three regions of operation not including the breakdown region. The three regions and their *V-I* relationships are:

1. *Region below pinch-off* (*ohmic region*): When the drain voltage is low and the channel is not pinched off, the *V-I* relationship is given by [1, 2]

$$I_D = I_{DSS}\left[2\left(1 + \frac{V_{GS}}{V_{PO}}\right)\frac{V_{DS}}{V_{PO}} - \left(\frac{V_{DS}}{V_{PO}}\right)^2\right] \qquad \text{for } 0 < V_{DS} \le V_{GS} + V_{PO} \tag{3.4}$$

For small values of V_{DS} we can approximate (3.4) as

$$I_D \simeq \frac{2I_{DSS}}{V_{PO}}\left(1 + \frac{V_{GS}}{V_{PO}}\right)V_{DS} \tag{3.5}$$

Note that (3.5) represents the linear *V-I* characteristic of JFET near the origin as shown in Fig. 3.3. The slopes of the lines are the linear conductances, namely,

$$r_{DS}^{-1} = \left.\frac{\partial I_D}{\partial V_{DS}}\right|_{V_{DS}} = \frac{2I_{DSS}}{V_{PO}}\left(1 + \frac{V_{GS}}{V_{PO}}\right) \tag{3.6}$$

Note that since V_{PO} is positive, V_{GS} is negative, and the resistance r_{DS} increases for increasing values of $|V_{GS}|$. For $V_{GS} = 0$, the ON resistance $r_{DS(\text{on})}$ in terms of the geometry and physics of the device is given by

---

*Note that the meaning of *saturation* in a FET is completely different from that of a BJT.

$$r_{DS(\text{on})} = \frac{L}{2qaW\mu_n N_D} \tag{3.7}$$

where the dimensions L, a, and W are shown in Fig. 3.2*a*. The values of $r_{DS(\text{on})}$ range from a few ohms to several hundred ohms for typical n-channel JFETs. Note that if the other parameters are kept constant, decreasing the W/L ratio increases the value of r_{DS}.

EXAMPLE 3.2

Calculate the ON resistance of a JFET with $N_D = 10^{16}\ \text{cm}^{-3}$, $a = 0.5\ \mu\text{m}$, $L = 20\ \mu\text{m}$, $W = 0.1$ cm, and $\mu_n = 1300\ \text{cm}^2/(\text{V} \cdot \text{s})$.

From (3.7) we have

$$r_{DS(\text{on})} = \frac{20 \times 10^{-4}}{2(1.6 \times 10^{-19})(0.5 \times 10^{-4})(0.1)(1.3 \times 10^{2})(10^{16})} = 961\ \Omega$$

2. *Region between pinch-off and breakdown (saturation region):* When the drain voltage is greater than the pinch-off value, the drain current is almost constant independent of V_{DS}. However, for $V_{DS} = V_{PO} + V_{GS}$, we have from (3.4)

$$I_D = I_{DSS}\left(1 + \frac{V_{GS}}{V_{PO}}\right)^2 \tag{3.8}$$

Equation (3.8) is called the *transfer characteristic* and is shown in Fig. 3.5*b*. Note that $V_{GS} < 0$ and V_{PO} is positive. Manufacturers specify the values of I_{DSS} and V_{PO} (see Appendix E). Thus the value of I_D for a given V_{GS} can be determined from (3.8). In Fig. 3.5*a* the pinch-off locus (parabola) separating the two boundaries is shown by the dashed line. This locus, i.e., the boundary between the linear and the saturation regions, is obtained from (3.8) with $V_{GS} = V_{DS} - V_{PO}$, namely,

$$I_D = I_{DSS}\left(\frac{V_{DS}}{V_{PO}}\right)^2 \tag{3.9}$$

3. *Cutoff region:* When the gate voltage is less than the pinch-off voltage, $|V_{GS}| < V_{PO}$, and thus $I_D = 0$.

EXAMPLE 3.3

Determine the values of I_D and V_{DS} for the JFET circuit shown in Fig. 3.6, given that $I_{DSS} = 5$ mA, $V_{PO} = 3$ V. Repeat for $R_D = 2\ \text{k}\Omega$.

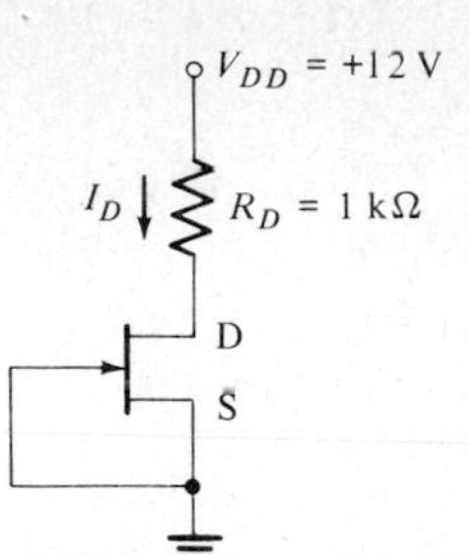

FIGURE 3.6 Example circuit.

We shall assume that the device is in the saturation region and then check the validity of our assumption. Note that $V_{GS} = 0$ in Fig. 3.6; hence from (3.8) we have

$$I_D = I_{DSS} = 5 \text{ mA}$$

The drain voltage is determined from

$$12 = I_D(10^3) + V_{DS} \qquad \text{or} \qquad V_{DS} = 7 \text{ V}$$

To check our assumption, we find $V_{PO} + V_{GS} = 3$ V, and since V_{DS} is larger than this value, our assumption is indeed correct; thus $I_D = 5$ mA, $V_{DS} = 7$ V.

If the value of R_D were 2 kΩ, then our assumption would be inaccurate since V_{DS} would be 2 V and less than V_{PO}. In this case we determine the values of I_D and V_{DS} from the ohmic region, namely, (3.4) and the circuit equation $V_{DD} = R_D I_D + V_{DS}$.

From the circuit we note that $V_{GS} = 0$, and from (3.4) with $V_{GS} = 0$ we have

$$I_D = 5\left[2\frac{V_{DS}}{3} - \left(\frac{V_{DS}}{3}\right)^2\right]$$

Also from the load-line equation:

$$12 = (2 \times 10^3)I_D + V_{DS}$$

Solving for V_{DS} from the above two equations, we get $V_{DS} = 2.4$ or 4.5. The second value is ruled out since it is larger than V_{PO}, and hence $V_{DS} = 2.4$ V and $I_D = 4.8$ mA.

The operation of a JFET as a switch is shown in Fig. 3.7*a*. When the input signal is at zero voltage level, the JFET is ON, as modeled in Fig. 3.7*b*, and v_o is at level V_1. When the input signal is at $-V_{PO}$, the JFET is OFF, as modeled in Fig. 3.7*c*, and v_o is at V_{DD}.

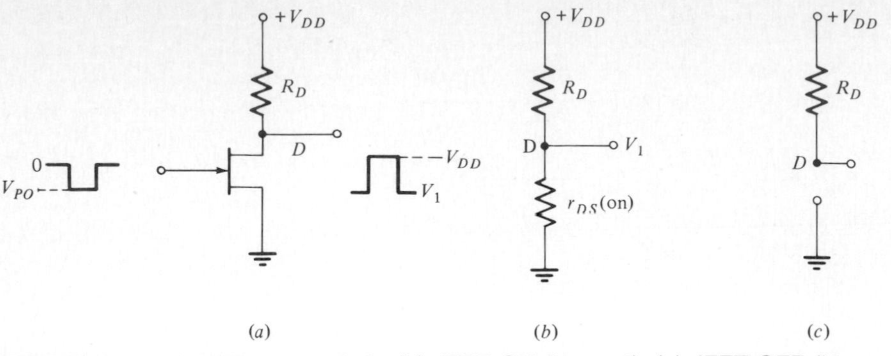

FIGURE 3.7 (*a*) JFET as a switch. (*b*) JFET ON ($V_{GS} = 0$). (*c*) JFET OFF ($V_{GS} = V_{PO}$).

EXERCISE 3.1

For an *n*-channel JFET the following are given: $V_{PO} = 4$ V and $I_{DSS} = 8$ mA. What is the minimum value of V_{DS} for the JFET to operate in pinch-off (*a*) for $V_{GS} = -3$ V and (*b*) for $V_{GS} = 0$.

Ans (*a*) 1 V, (*b*) 4 V

EXERCISE 3.2

For the JFET in Exercise 3.1 determine the resistance in the ohmic region for each case.

Ans (*a*) $r_{DS} = 1$ kΩ, (*b*) 250 Ω

The *p*-Channel JFET

The *p*-channel JFET circuit symbol is shown in Fig. 3.8*a*. Note that the direction of the arrow is reversed. The direction of the arrowhead therefore indicates whether the device is a *p* channel or an *n* channel. Furthermore observe the orientation of the S and the D terminals in Fig. 3.8*a* as compared with the *n*-channel JFET. With this orientation the *V-I* characteristics polarities are identical to those of the *n*-type *provided* that we replace V_{GS} with V_{SG} and V_{DS} with V_{SD}, as shown in Fig. 3.8*b* and *c*, respectively. In other words for the various regions we have the following conditions:

For the saturation region, $V_{SD} > V_{PO}$.
For the ohmic region, $V_{SD} < V_{PO}$.
For the boundary between the two regions, $V_{SD} = V_{PO} - |V_{SG}|$.

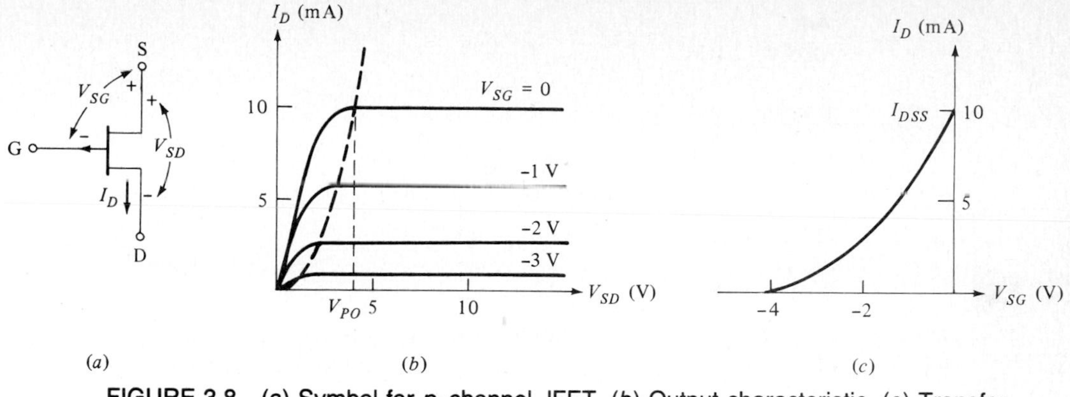

FIGURE 3.8 (*a*) Symbol for *p*-channel JFET. (*b*) Output characteristic. (*c*) Transfer characteristic.

EXERCISE 3.3

For a *p*-channel JFET the following are given: $V_{PO} = 4$ V and $I_{DSS} = 8$ mA. Find the value of I_D for $V_{SG} = -3$ V and $V_{SD} = 1$ V.

Ans 0.5 mA

3.2 METAL-OXIDE SEMICONDUCTOR FIELD-EFFECT TRANSISTORS (MOSFET)

There are two types of MOSFET: the depletion type and the enhancement type. The depletion type can be operated in the enhancement mode, but the reverse is not true. We shall consider each type separately in order to avoid confusion.

A. Depletion-Type MOSFET

The structure shown in Fig. 3.9*a* is an alternative way of realizing field-effect transistors. These structures consist of a lightly doped *p*-type substrate called the *body* into which two heavily doped *n*-type regions are diffused. In Fig. 3.9*a* a shallow channel of lightly doped *n*-type material is also formed between the source and the drain by diffusion, ion implantation, etc. The region between the source and the drain is covered by an oxide layer over which is deposited a metal plate that serves as a gate. In a MOSFET the gate is insulated from the semiconductor by the oxide layer; hence the name insulated-gate field-effect transistor (IGFET) is also sometimes used.

The circuit symbols for an *n*-channel depletion-type MOSFET, or simply NMOS, are shown in Fig. 3.9*b* and *c*. Sometimes, instead of the arrow, the letter *n* for *n* channel is used. When there is no arrow or any letter *n* or *p*, the *n* channel is

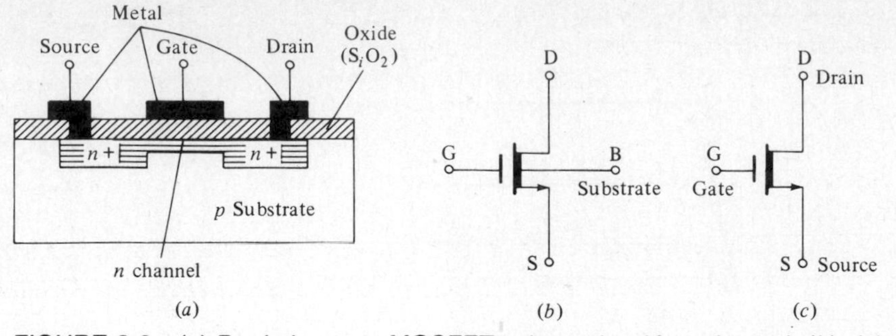

FIGURE 3.9 (*a*) Depletion-type MOSFET schematic with *n*-channel. (*b*), (*c*) NMOS-depletion-type circuit symbols (for PMOS reverse the direction of the arrow).

assumed, which is often the case in the literature. For a *p*-channel MOSFET or PMOS, the direction of the arrow is reversed. The operation of the MOSFET is basically similar to that of the JFET. In the structure of Fig. 3.9*a*, if a voltage is applied between the gate and the source, making the gate negative relative to the source, positive charges are induced in the channel through the SiO_2 of the gate capacitor. The induced positive charge causes a depletion of the majority carriers (electrons in the *n* channel), and thus the channel becomes less conductive.

If the gate is made sufficiently negative, the depletion region extends completely across the channel, and the channel cannot conduct current. This condition is the pinch-off and is generally within the same range of values as in the JFET. The *V-I* characteristics of the *depletion-mode* MOSFET are similar to those of JFET. Note that

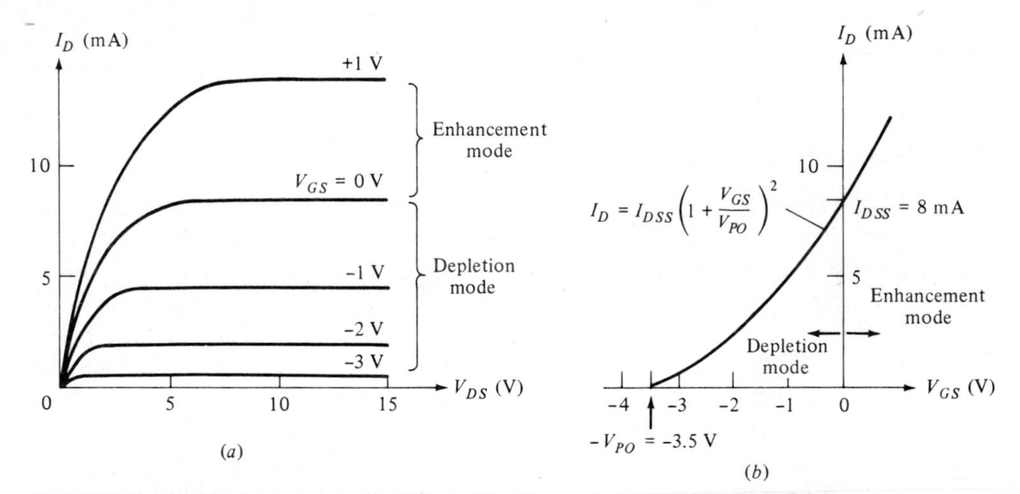

FIGURE 3.10 (*a*) Drain characteristics of *n*-channel depletion-type MOSFET. (*b*) Transfer characteristic.

the depletion-mode operation refers to negative values of V_{GS}; i.e., the gate is negative relative to the source.

If the gate voltage is made positive relative to the source in Fig. 3.9*a*, negative charges are induced in the channel through the gate capacitor, thus enhancing the concentration of the majority carriers in the channel. The conductivity of the channel is therefore increased, and the drain current can exceed I_{DSS}. The *enhancement-mode* operation refers to positive values of V_{GS} for an *n*-channel device.

Thus, a depletion-type MOSFET can be operated in the enhancement mode or the depletion mode. The measured *V-I* characteristics of a typical depletion-type MOSFET are shown in Fig. 3.10. Note that both enhancement- and depletion-mode characteristics are available in depletion-type MOSFETs as shown in Fig. 3.10. A comparison of the *V-I* characteristics between Fig. 3.10 and Fig. 3.5 indicates that the circuit operations of depletion type and JFET are similar and hence will not be elaborated any further.

EXERCISE 3.4

For a depeletion-type NMOS, $V_{PO} = 4$ V and $I_{DSS} = 8$ mA. Determine the minimum value of V_{DS} for the device to operate in the pinch-off if (*a*) $V_{GS} = 1$ V and (*b*) $V_{GS} = -1$ V.

Ans (*a*) 5 V, (*b*) 3 V

EXERCISE 3.5

For the circuit shown in Fig. E3.5, the depletion-type NMOS parameters are $V_{PO1} = 2V_{PO2} = 4$ V and $I_{DSS1} = I_{DSS2}$. Determine the value of V_{DS1}.

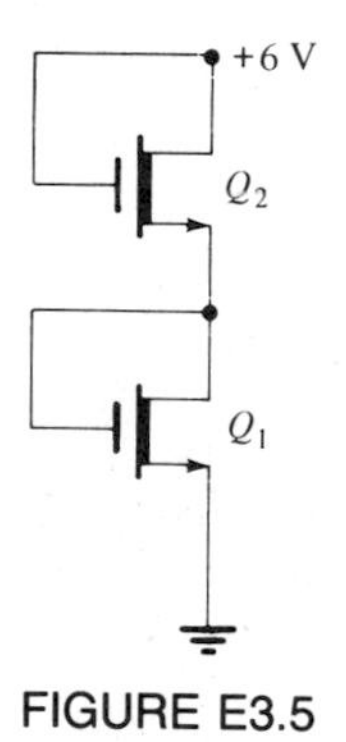

FIGURE E3.5

Ans 4 V

B. Enhancement-Type MOSFET

Another simple MOSFET structure shown *already* in Fig. 3.1 is redrawn in simple form in Fig. 3.11*a*. This structure is very common in integrated circuits because of its ease of fabrication, size, and very low power dissipation characteristics. In this case, no channel exists between the drain and the source if there is no applied voltage on the gate-to-source terminals as in Fig. 3.11*a*. This *n*-channel device is operated with a *positive gate-source voltage only*. Conduction occurs in a channel that is an *n*-type layer (formed by the accumulation of electrons on a *p*-type substrate) and is enhanced by the positive values of V_{DS}.

The circuit symbols for the enhancement NMOS device are shown in Fig. 3.12*a–c*. The symbol in Fig. 3.12*a* is for a nonstandard connection of the substrate. The substrate body B is usually connected to the source internally, and the terminal is not shown explicitly. We shall use the circuit symbol in Fig. 3.12*c* in this text. The normal current flow and voltage polarities of the NMOS and PMOS are shown in Fig. 3.12*c* and *d*, respectively. This type of MOSFET is useful *only* in the enhancement mode of operation, i.e., for the NMOS when the applied voltage makes the gate positive in relation to the source. When this situation exists, as in Fig. 3.11*b*, the

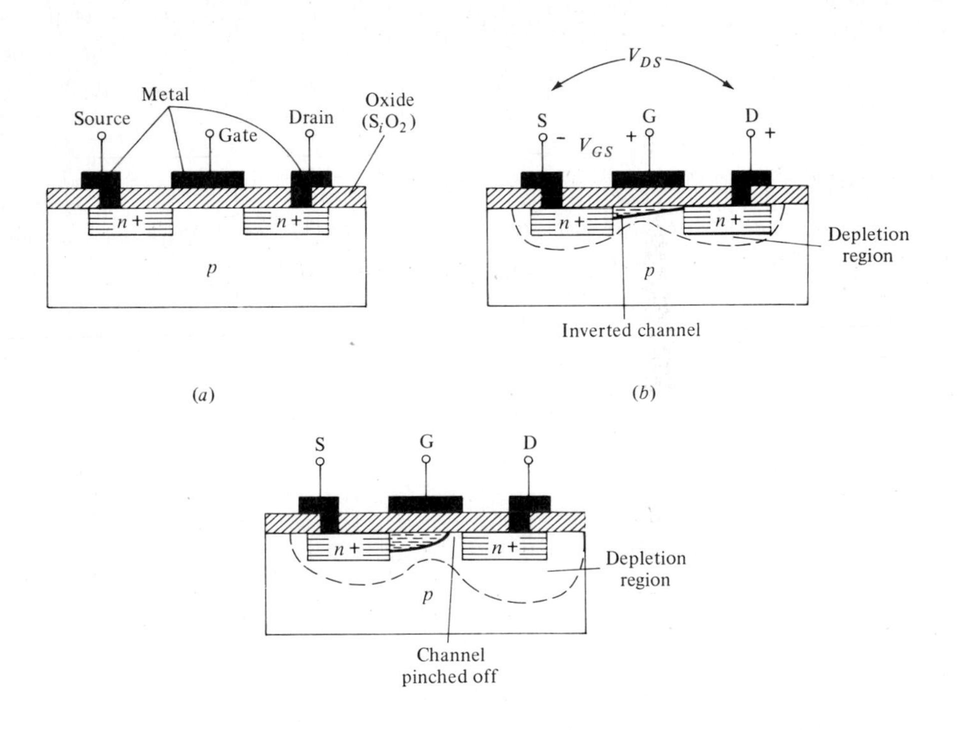

FIGURE 3.11 (*a*) Enhancement-type NMOS structure. (*b*) $V_{GS} > V_T$, $V_{DS} \leq V_{GS} - V_T$, and NMOS in Ohmic region. (*c*) $V_{GS} > V_T$, $V_{DS} \geq V_{GS} - V_T$, and NMOS saturated (beyond pinch-off).

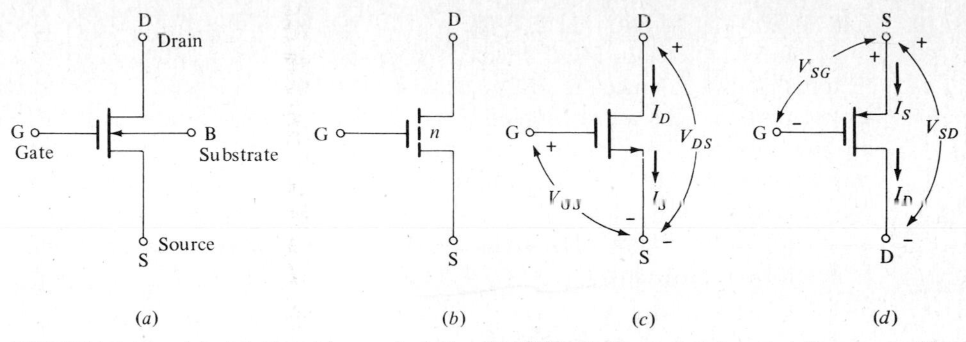

FIGURE 3.12 (a)–(c) Graphic symbols for NMOS with the current and voltage polarities in (c). (*d*) Graphic symbol and the current and voltage flow in a PMOS.

n-channel is induced, an inversion layer is formed, and the device can operate. The gate voltage at which the channel forms is called the *threshold voltage* V_T. The value of V_T depends upon the electrical properties of the substrate and oxide and the thickness of the oxide. The value of V_T for the device is specified by the manufacturer (see Appendix E). Typical values of V_T range from 1 to 5 V. Thus, when the gate-to-source potential V_{GS} is less than V_T, no channel exists, and the drain current is approximately zero. For higher values of V_{GS} (and $V_{GS} \geq V_T$), $V_{DS} \leq V_{GS} - V_T$, the NMOS is in the ohmic region as shown in Fig. 3.11*b*. In this case the density of electrons in the conduction channel will be fewer at the drain end and not constant throughout the channel. If we increase V_{DS} further such that $V_{DS} \geq V_{GS} - V_T$ and $V_{GS} \geq V_T$, the channel is pinched off as shown in Fig. 3.11*c*, and the device operates in the saturation region. Note that the quantity I_{DSS} has no meaning in the enhancement-type MOSFET because $I_D = 0$ when $V_{GS} = 0$ and is therefore not associated with this type of MOSFETs. The *V-I* characteristic of the enhancement-type MOSFET can also be divided into three regions.

The Ohmic Region

Here $V_{DS} \leq V_{GS} - V_T$ and the *V-I* characteristic is given by

$$I_D = K_n[2(V_{GS} - V_T)V_{DS} - V_{DS}^2] \tag{3.10a}$$

when

$$K_n = \frac{\mu_n \epsilon \epsilon_{\text{ox}}}{2t_{\text{ox}}} \frac{W}{L} = \frac{\mu_n C_0}{2}\left(\frac{W}{L}\right) \tag{3.10b}$$

where μ_n = surface mobility of electrons [$\mu_n = 800\ \text{cm}^2/(\text{V} \cdot \text{s})$]
ϵ = permittivity of free space ($= 8.85 \times 10^{-14}$ F/cm)
ϵ_{ox} = dielectric constant of SiO_2($\simeq$4)
t = thickness of the oxide

L is the channel length, and W is the width as shown in Fig. 3.1. The W/L ratio is an important parameter in MOS devices. In some cases this ratio (number) is indicated next to the device. Since the cost of an IC is related to the chip area, the designer would like to minimize the area occupied by the device on a chip. The size of MOS devices are minimized by making W/L close to unity. However, other requirements, such as power and speed, may dictate different values for W/L. Thus it is not uncommon to see W/L ratios as high as 20. For $t = 0.1$ μm the gate capacitance per unit area $C_0 = 3.5 \times 10^{-8}$ F/cm^2. The dividing locus between the saturation and ohmic regions is given by substituting $V_{DS} = V_{GS} - V_T$ in (3.10a), which is

$$I_D = K_n V_{DS}^2 = \frac{\mu_n C_0 W}{2L} V_{DS}^2 \qquad \textbf{(3.11)}$$

The locus described in (3.11) is shown by the dotted line in Fig. 3.13a.

The Saturation Region

Here $V_{DS} \geq V_{GS} - V_T$, and the current I_D is approximately constant as shown in Fig. 3.13a. The transfer characteristic is obtained by replacing V_{DS} by $V_{GS} - V_T$ in (3.11), which is

$$I_D = K_n(V_{GS} - V_T)^2 \qquad \textbf{(3.12)}$$

A plot of the transfer characteristic is shown in Fig. 3.13b.

The Cutoff Region

Here $V_{GS} < V_T$, and thus $I_D = 0$. The device is OFF in this region and is used in switching applications in this mode.

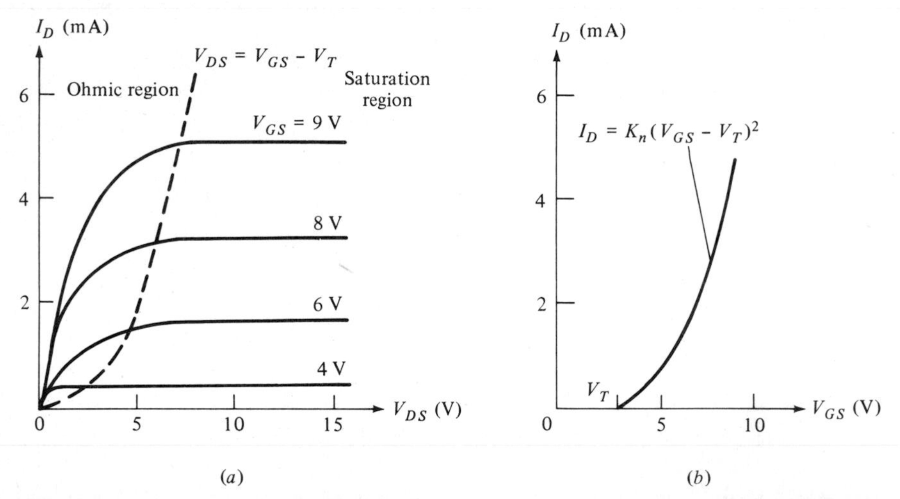

FIGURE 3.13 (a) Drain characteristics of n-channel enhancement-type MOSFET (NMOS). (b) Transfer characteristic.

EXAMPLE 3.4

For an NMOS structure shown in Fig. 3.1, the following are given: $W/L = 20$, $\mu_n = 800\ \text{cm}^2/(\text{V} \cdot \text{s})$, $V_T = 2$ V. Calculate I_D for $V_{GS} = 5$ V and the two values of $V_{DS} = 5$ V and 2 V. First, we determine

$$\frac{\mu_n C_O W}{2L} = \frac{800(3.5 \times 10^{-8})(20)}{2} = 2.80 \times 10^{-4}$$

For $V_T = 2$ V, $V_{GS} = 5$ V, and $V_{DS} = 5\ \text{V} > 5 - 2$ V, from the saturation region using (3.12), we obtain

$$I_D = 2.80 \times 10^{-4}(5 - 2)^2 = 2.52\ \text{mA}$$

For $V_{DS} = 2\ \text{V} < 5 - 2$ V, from the ohmic region (3.10*a*) we obtain

$$I_D = 2.80 \times 10^{-4}[2(3)(2) - (2)^2] = 2.24\ \text{mA}$$

It should be noted that the magnitude of the threshold voltage V_T decreases by about 2.5 mV/°C. On the other hand, the mobility of an electron in silicon decreases for increasing temperature, thus decreasing K_n and partially compensating for each other in the value of I_D in (3.12).

Finally note that we have assumed in the above that the source is connected to the substrate and both terminals are grounded. However, if the body (substrate) is at negative potential with respect to the source and the source at ground terminal as shown in Fig. 3.14*a*, the threshold voltage can be varied considerably as shown in Fig. 3.14*b*, where g_d is the output conductance. Thus by increasing V_{BS} from zero we can increase the value of V_T from its minimum value (which is at $V_{BS} = 0$) to more than double its minimum value.

It should be noted that for the PMOS to conduct current, $V_{SG} \geq |V_T|$. A commonly used configuration in integrated circuits utilizes both PMOS and NMOS devices and is called *complementary symmetry* MOS or CMOS, which is discussed in

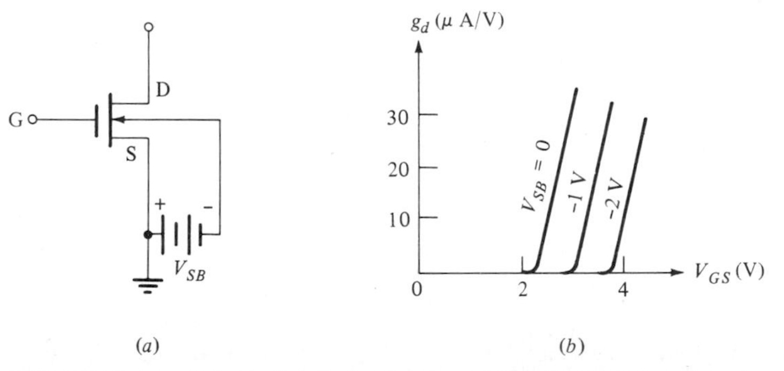

FIGURE 3.14 Effect of substrate bias on the threshold voltage.

Sec. 3.6. In order to distinguish between the NMOS and the PMOS, V_{Tn} and K_n are used for NMOS, and V_{Tp} and K_p are used for PMOS devices, i.e., the subscript n or p is added to K and V_T, whichever applies.

Finally, from the similarity of the V-I characteristics of JFET (Fig. 3.5), depletion-type MOSFET (Fig. 3.10), and the enhancement-type MOSFET (Fig. 3.13) we conclude that the same circuit model, as obtained in the next section, apply to all. The circuit parameters would naturally be different.

EXERCISE 3.6

For the circuit shown in Fig. E3.6 the enhancement-type NMOS parameters are $V_T = 2$ V and $K_n = 3 \times 10^{-4}$ A/V^2. Determine the values of I_D and V_{DS}.

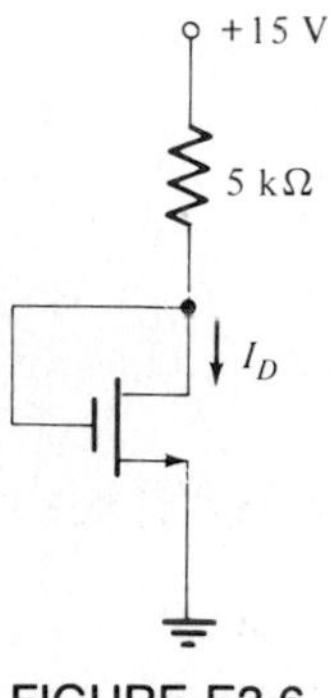

FIGURE E3.6

Ans $I_D = 2.07$ mA, $V_{DS} = V_{GS} = 4.63$ V

EXERCISE 3.7

Repeat Exercise 3.6 with the gate connected to the source as shown in Fig. E3.7.

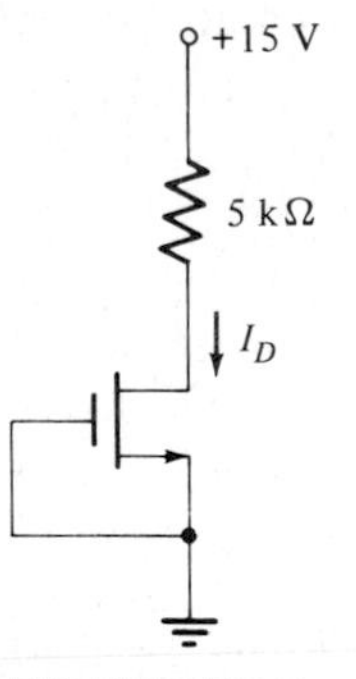

FIGURE E3.7

Ans $I_D = 0$, $V_{DS} = 15$ V

EXERCISE 3.8

For the circuit shown in Fig. E3.8 the parameters of the enhancement-type NMOS are $V_T = 3$ V, $K_n = 5 \times 10^{-3}$ A/V^2, and for the depletion type $I_{DSS} = 5$ mA. Find V_{DS1} and V_{DS2}.

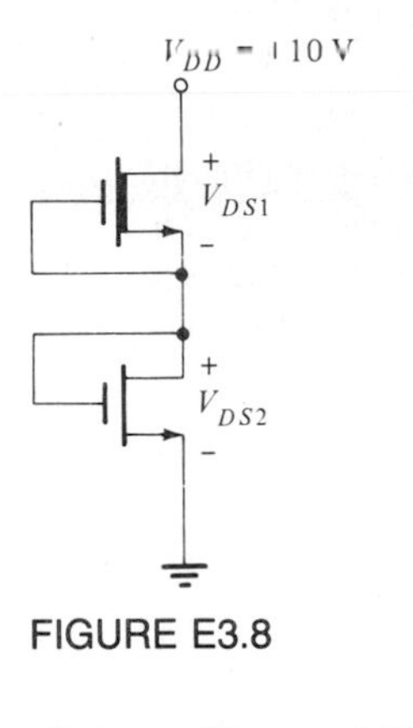

FIGURE E3.8

Ans $V_{DS1} = 6$ V, $V_{DS2} = 4$ V

3.3 COMPARISON OF JFET, MOS (NMOS AND PMOS), AND MESFET

As mentioned earlier the *V-I* characteristics of the three types of field-effect transistors are remarkably similar in spite of the differences in their constructions. From the circuit operation point of view (i.e., building circuit models) we do not have to worry about their differences except for differences in biasing for enhancement- or depletion-mode operation. For JFET and depletion *n*-type MOSFET the gate-to-source voltage is negative, whereas for the enhancement *n*-type MOSFET it is positive. Some of the differences among these are briefly considered below.

The JFET is only operated in the depletion mode, whereas the depletion-type MOS can be operated either in the depletion mode or enhancement mode. The enhancement-type MOS can be operated only in the enhancement mode. For enhancement-type MOS, increasing the gate voltage increases the conductivity, whereas in the depletion-type MOS and the JFET, increasing the gate voltage decreases the conductivity. In the MOS the leakage current is lower than the JFET because the gate is insulated. The enhancement-type MOS can also be viewed as a normally OFF device since for $V_{GS} = 0$, the drain current is zero. The depletion-type MOS and the JFET may be viewed as a normally ON device since for $V_{GS} = 0$, the drain current is nonzero and, in fact, substantial. The enhancement-type MOS therefore does not consume power under standby or static conditions. The low fabrication cost of the enhancement-type MOS and their other attractive properties have made MOS a more popular and commonly used technology choice as compared to JFET or the depletion-type MOS.

We have discussed the n-channel MOS field-effect transistor (NMOS) in the above. Of course, the p-channel MOS field-effect transistor (PMOS) is also possible. However, because of the following advantages, the NMOS is by far the dominant choice. First, recall from (Appendix A) that the hole and electron mobilities in silicon are $\mu_p = 500\ \text{cm}^2/(\text{V} \cdot \text{s})$ and $\mu_n = 1300\ \text{cm}^2/(\text{V} \cdot \text{s})$, respectively. In other words, the mobility of the n-type Si is more than twice that of the p type. Thus the ON resistance [see Eq. (3.7)] of an n-channel device of the same geometry is much smaller than that of the p channel under the same operating conditions. Furthermore K_n [see Eq. (3.10)] of an NMOS is higher than K_p of a PMOS, thus yielding higher gain. In order to achieve the same ON resistance, the p channel must therefore have more than twice the area of the n channel. The high mobility of electrons versus holes makes the frequency performance of the NMOS also superior to that of PMOS. In integrated circuits the reduction of device area on a chip is very important for large-scale integration. Also, higher packing density and smaller channel lengths of the devices make the n-channel MOS faster in switching applications than the p-channel MOS. The speed of the devices, which is governed by the transit time τ of the carriers, is determined approximately from the following equation [see also (3.44)]:

$$\text{Transit time } \tau = \frac{(\text{channel length})^2}{\text{mobility} \times \text{voltage}} \simeq \frac{L^2}{\mu_n (V_{DD}/2)} \tag{3.13}$$

Thus high mobility and small dimensions will improve the speed of the device. High voltage values will also improve the speed, but they will also increase power consumption and are therefore not desirable. It is for this reason that in switching applications, one speaks of a figure of merit for a device as the product of power and speed. This topic is discussed further in Sec. 12.12.

Another technology that is more suitable for high-frequency applications uses gallium arsenide (GaAs) in which the mobility of electrons is about 6 times that in silicon.

A shorter channel length can be realized by using a Schottky barrier instead of the pn gate junction (see Appendix B). The Schottky barrier is formed on the top surface, and the ohmic contacts form the drain and the source. This device is usually called the MESFET, which is derived from metal semiconductor field-effect transistors. These devices can be operated at very high frequencies, in the several gigahertz ($1\ \text{GHz} \simeq 10^9\ \text{Hz}$) range.

Currently NMOS is one of the dominant technologies in integrated circuits. PMOS is now used only in conjunction with NMOS to obtain the complementary symmetry COS/MOS or CMOS inverters. These inverters, as we shall consider in Sec. 3.5, have certain desirable properties (extremely low power dissipation and excellent temperature stability) that make them desirable in large-scale integrated circuits. In fact, CMOS technology is as dominant, if not more so, than NMOS in integrated circuits.

3.4 GRAPHICAL ANALYSIS

We have already used graphical analysis techniques in diodes and bipolar transistors. The basic technique is applied in the following for a FET circuit. Consider the simple

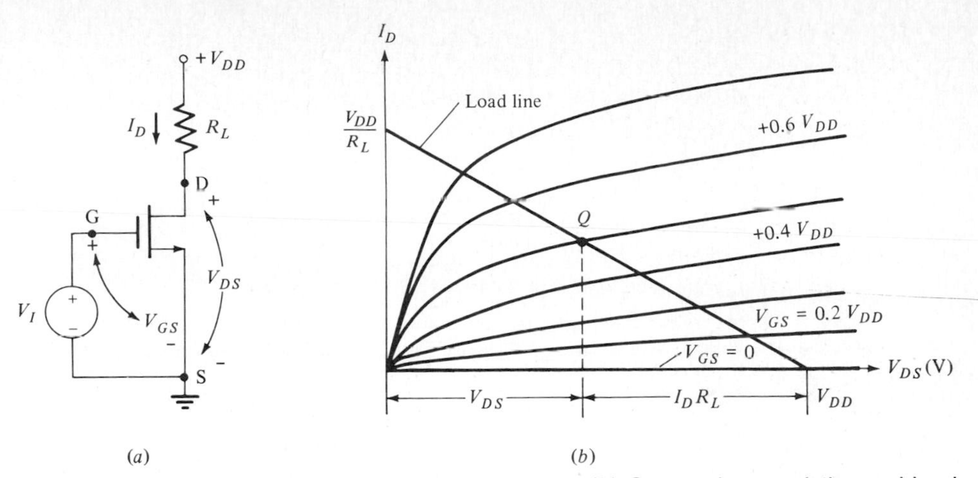

FIGURE 3.15 (*a*) NMOS enhancement-type circuit. (*b*) Output characteristics and load line.

NMOS enhancement-type FET circuit shown in Fig. 3.15. We shall obtain the transfer characteristics of the circuit, i.e., the output voltage V_O for applied input signal voltage V_I. At the input, noting that $I_G = 0$, we have

$$V_I = V_{GS} \tag{3.14}$$

At the output we have

$$V_{DD} = I_D R_L + V_{DS} \tag{3.15a}$$

$$V_O = V_{DS} \tag{3.15b}$$

The relationship between I_D and V_{DS} as well as the load line is shown in Fig. 3.15*b*. The transfer characteristic moving along the load line as $V_I(= V_{GS})$ is varied from zero to V_{DD} and is shown in Fig. 3.16. We note that the slope of the voltage transfer characteristic, which is the gain of the device in the linear region (near Q), is larger for increasing value of R_L. Note that the circuit can be operated under large signals as

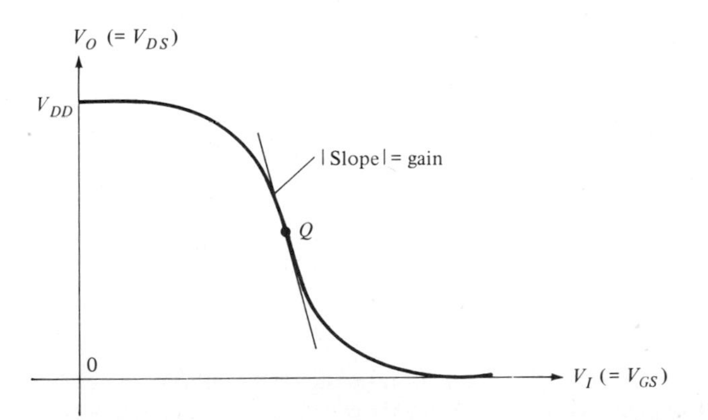

FIGURE 3.16 Transfer characteristics of the inverter circuit in Fig. 3.15.

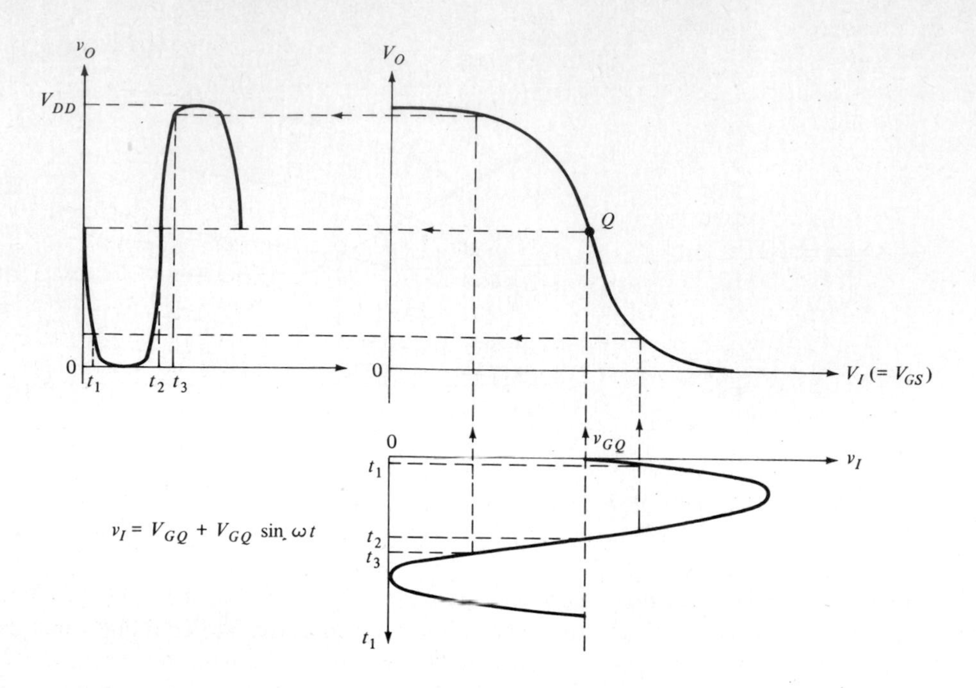

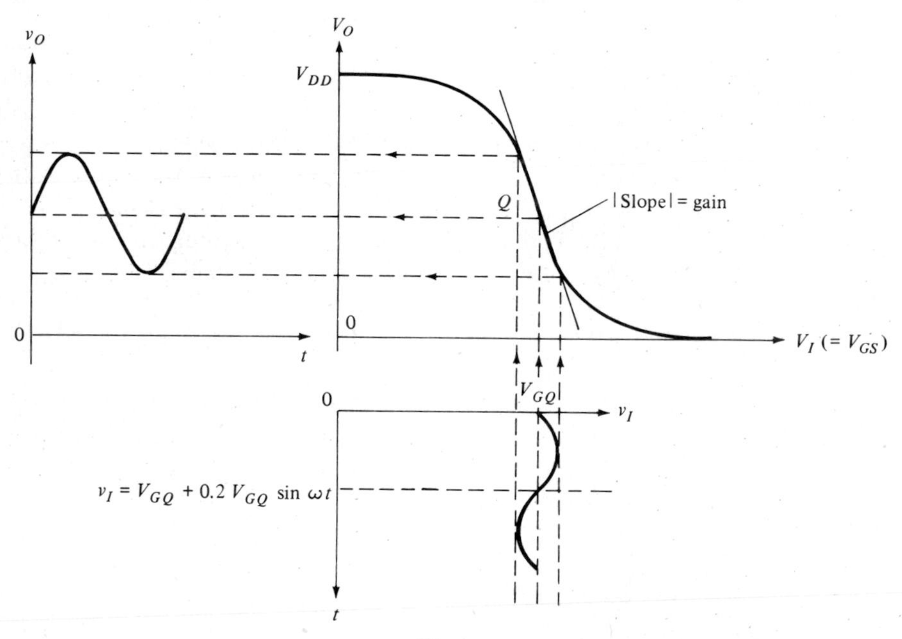

FIGURE 3.17 Output signal when the input is an ac signal superimposed on dc: (a) $v_I = V_{GQ} + V_{GQ} \sin \omega t$, (b) $v_I = V_{GQ} + 0.2V_{GQ} \sin \omega t$.

a switch between the levels of V_{DD} and 0. In other words, if $V_I = 0$, $V_O = V_{DD}$, and if $V_I = V_{DD}$, $V_O \simeq 0$. This complementing operation is used in digital and logic circuits.

A. DC and AC Signals Applied

Consider now the case where $v_I = V_{GQ} + V_1 \sin \omega t$, where V_{GQ} denotes the value of V_{GS} at the Q point. Let us first consider the case where $V_1 \simeq V_{GQ}$ as shown in Fig. 3.17*a*. In Fig. 3.17*a* notice the distortion in the output signal v_O. The output signal is clipped at the values 0 and V_{DD}. For small input signals $V_1 \ll V_{GQ}$ (say, $V_1 = 0.2V_{GQ}$), the operation is in the linear portion of the transfer characteristic, and thus an almost distortion-free amplification is achieved as shown in Fig. 3.17*b*.

B. Large-Signal Model

The large-signal model for the FET is clearly nonlinear. The transfer characteristic, i.e., I_D as a function of V_{DS} for different values of V_{GS}, is given separately for both JFET and MOSFET devices in Secs. 3.1 and 3.2. Graphical analysis may be used, for large-signal analysis, as done in the previous section. For analytical purposes the nonlinear characteristics of the devices mandate computer-aided analysis and design. For computer-aided analysis and design (CAD) we can then use elaborate accurate models, which include the nonlinearities and the body or substrate terminal junction diodes. One such model for MOS devices that is useful for a CAD program, such as SPICE,* is shown in Fig. 3.18. For low-frequency analysis, the capacitances are

*See footnote on page 43.

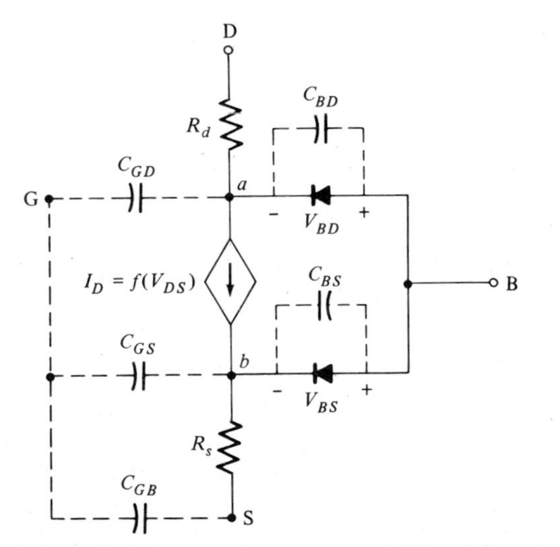

FIGURE 3.18 Large-signal model for enhancement-type MOSFET suitable for CAD (SPICE).

ignored. However, for high-frequency analysis, the capacitors (shown by dotted lines) must be included.

In the following sections, we shall use graphical analysis to determine the voltage transfer function of the NMOS and CMOS inverters. In other words we consider the switching behavior of these circuits under large-signal conditions.

3.5 THE NMOS INVERTER

Inverters find a variety of applications in switching and logic circuits. A basic inverter with a resistive load was shown in Fig. 3.15*a*. From the transfer characteristic of the circuit, shown in Fig. 3.16, it is readily seen that if the input V_I is a binary variable, i.e., it takes either of the two values 0 or V_{DD}, the output V_O will also be a binary with the value V_{DD} or 0, respectively. Thus, if we define V_{DD} as *logic 1* and 0 as *logic 0*, the output of the inverter produces an output that is the complement of its input. Inverters are also referred to as the NOT gates, which are discussed in detail in Chap. 13.

In integrated circuits there is no efficient way to implement resistors. In fact, in order to avoid using resistors, MOS devices are usually used as resistors. An inverter circuit that is commonly used in NMOS technology is shown in Fig. 3.19, where instead of a resistor, a depletion-type NMOS is utilized as the load. The *W/L* ratio of the load is usually made smaller than the *W/L* ratio of the driver in order to obtain a more desirable, steeper transfer characteristic. The voltage transfer characteristic V_O vs V_I of the inverter in Fig. 3.19 can be constructed graphically from the characteristics of the *driver* Q_1 (enhancement-type NMOS) and the *load* Q_2 (depletion-type NMOS). In Fig. 3.20*a* and *b*, we have the output *V-I* characteristics of the enhancement-type NMOS and the depletion-type NMOS, respectively. From the circuit in Fig. 3.19, we have the following relations:

$$V_I = V_{GS(\text{en})} \qquad \text{and} \qquad V_O = V_{DS(\text{en})} = V_{DD} - V_{DS(\text{dep})} \tag{3.16}$$

where (en) and (dep) denote the enhancement- and the depletion-type MOS transistors. Under steady-state conditions and with no current drawn at the output, the drain

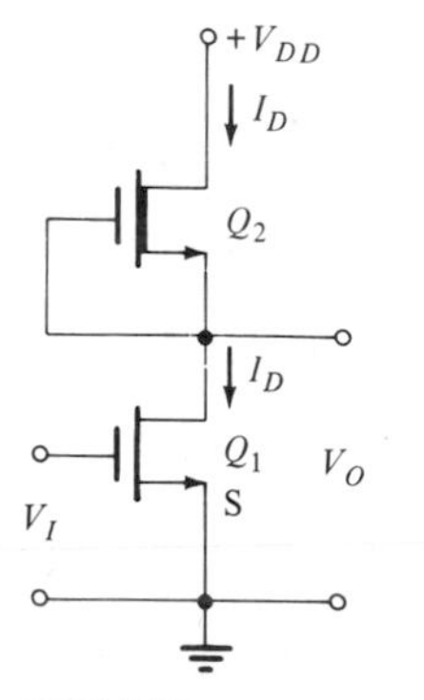

FIGURE 3.19 The NMOS inverter with depletion-type NMOS load.

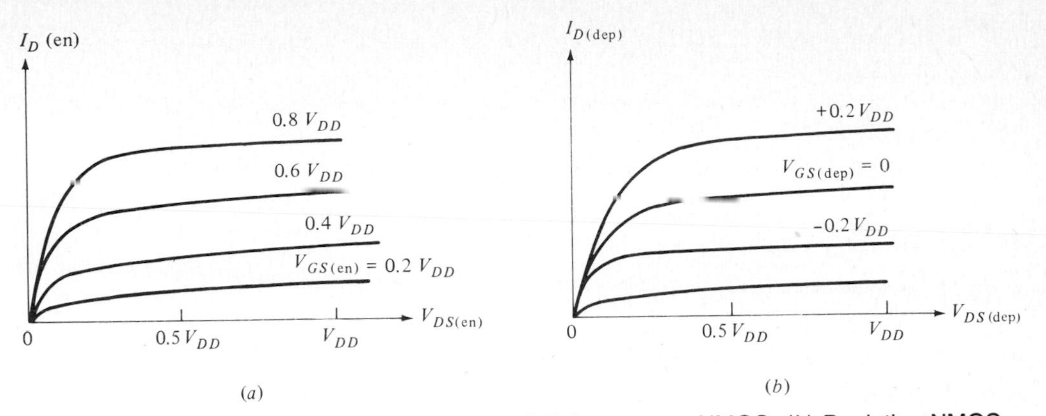

FIGURE 3.20 Output characteristics. (*a*) Enhancement NMOS. (*b*) Depletion NMOS.

currents are the same, namely,

$$I_{D(en)} = I_{D(dep)} = I_D \tag{3.17}$$

Note further from the circuit that $V_{GS(dep)} = 0$. From the *V-I* characteristics of the two types of NMOS devices shown in Fig. 3.20 and using the circuit constraints given in (3.16) and (3.17), we can now construct the transfer characteristic. Specifically, we superimpose plots of $I_{D(en)}$ vs $V_{DS(en)}$ and the plot of $I_{D(dep)}$ vs $V_{DS(dep)}$ (only for $V_{GS(dep)} = 0$ curve), as shown in Fig. 3.21*a*. Since the drain currents are the same, the intersection of the depletion MOS curve with those of the enhancement MOS curves yield V_O vs V_I as shown in Fig. 3.21*b*. Note that the transfer characteristic is similar to that of the basic inverter in Fig. 3.16.

For the NMOS inverters a useful quantity, which is usually used, is the ratio λ_r, defined as

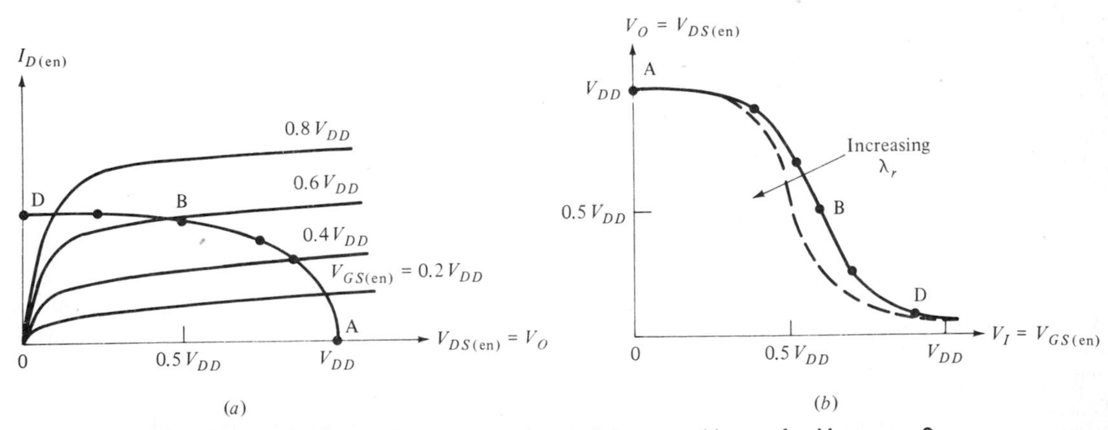

FIGURE 3.21 (*a*) Plots of $I_{D(en)}$ vs $V_{DS(en)}$ and $I_{D(dep)}$ vs $V_{DS(dep)}$ for $V_{GS(dep)} = 0$. (*b*) Transfer characteristic of the NMOS inverter.

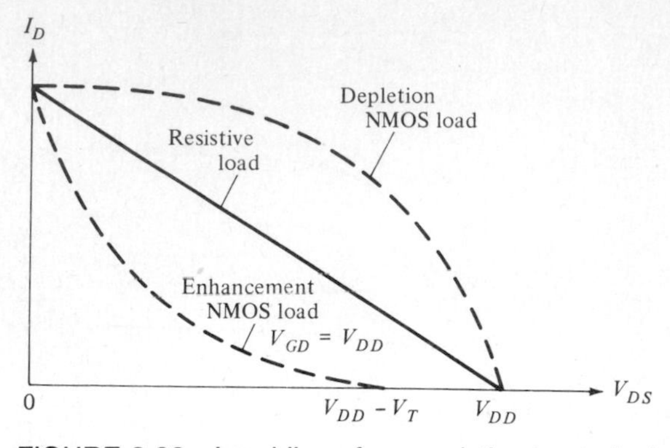

FIGURE 3.22 Load lines for a resistive load, depletion-type NMOS, and saturated enhancement NMOS.

$$\lambda_r = \frac{(W/L)_D}{(W/L)_L} \tag{3.18}$$

where the subscripts D and L denote the driver and the load, respectively. The larger the λ_r, the steeper is the slope of the transfer characteristic, as shown by the dotted line in Fig. 3.21*b*.

It should also be noted that an inverter using an enhancement NMOS load is also possible. The load line for the resistive load, depletion NMOS, and saturated ($V_{GD} = V_{DD}$) enhancement NMOS are compared in Fig. 3.22. The determination of the transfer characteristic for the inverter is left as an exercise in problem 3.13. Nonsaturated enhancement NMOS is also used as a load for an NMOS driver, in which case $V_{GG} > V_{DD} + V_T$ (see problem 3.14).

EXERCISE 3.9

For the circuit shown in Fig. 3.15*a*, with $V_{DD} = 20$ V and $R_L = 3$ kΩ, the NMOS characteristics are given in Fig. 3.13 (note that $V_T = 3$ V). Determine the maximum incremental voltage gain.

Ans -3

3.6 THE CMOS INVERTER

In the previous section we considered an inverter using both the enhancement- and depletion-type NMOS devices. With the advances of technology it is possible to fabricate NMOS and PMOS enhancement devices on the same chip. Such a circuit is called a *complementary symmetry MOS* (COS/MOS or CMOS for short). In order to fabricate such devices on the same chip some means must be provided for isolation of the devices as shown in Fig. 3.23. With the use of CMOS it is possible to design a circuit with essentially zero dc power dissipation. Power is dissipated only during the switching transients.

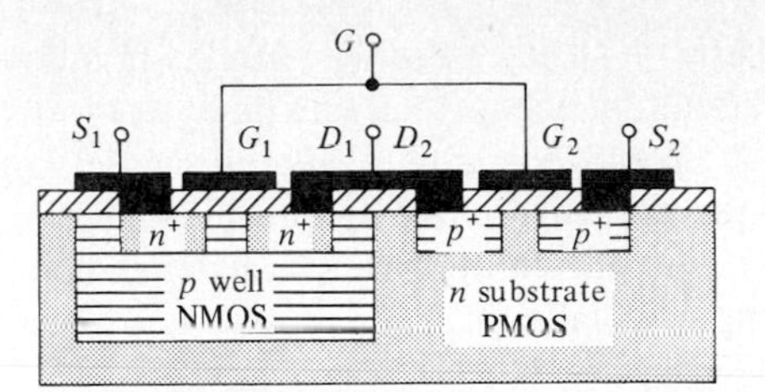

FIGURE 3.23 Construction of CMOS on a chip.

A CMOS inverter circuit is shown in Fig. 3.24*a*. In this circuit the PMOS, labeled Q_2, is used as a load to the NMOS driver, labeled Q_1. The transfer characteristic of each device is shown in Fig. 3.24*b*. The characteristic can be qualitatively verified as follows: From the circuit we note that when $V_I = 0$, the NMOS is cut off and $I_{D1} = 0$. For the PMOS, however, the gate-to-source voltage $V_{GS(p)} = -V_{DD}$. If V_{DD} is larger than V_{Tp}, where V_{Tp} is the threshold voltage of the PMOS, the PMOS will have an inversion channel with no current drawn at the output $I_{D1} = I_{D2} = I_D$. Since $I_D = 0$, we have $V_O = V_{DD}$. For $V_I = V_{DD}$, $V_{GS(p)} = 0$, and the PMOS is cut off and $I_{D2} = 0$. If V_{DD} is larger than V_{Tn}, where V_{Tn} is the threshold voltage of the NMOS, the

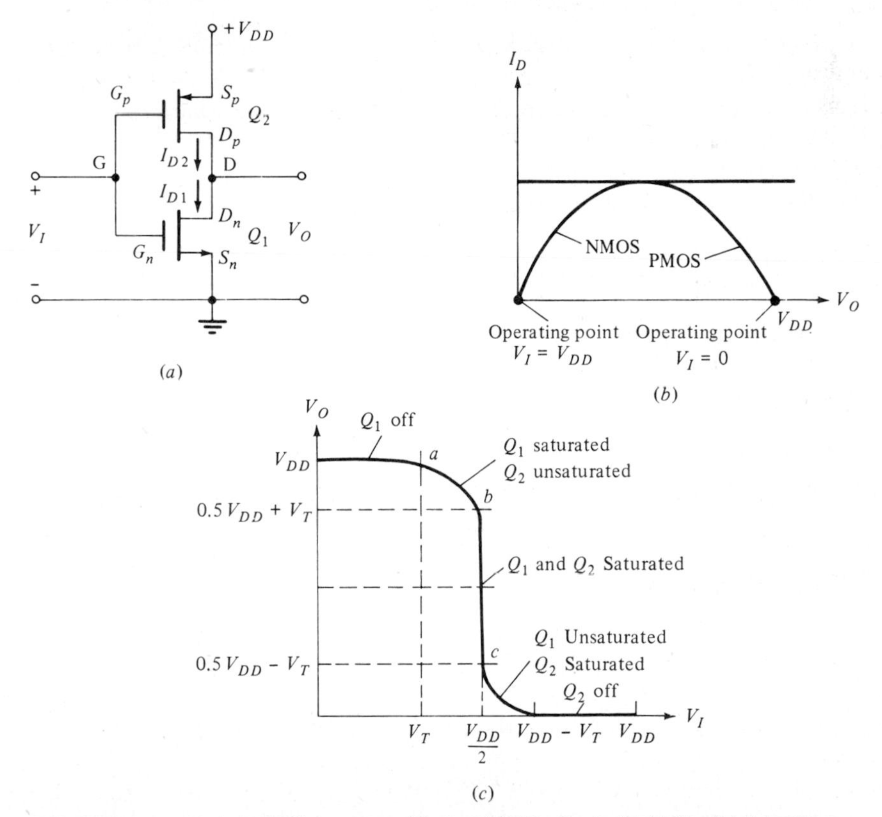

FIGURE 3.24 (*a*) CMOS inverter (Q_1 is NMOS; Q_2 is PMOS). (*b*) Load line. (*c*) Transfer characteristic of CMOS.

NMOS has an inversion channel. Since $I_D = 0$, we have $V_O = 0$. It should be noted that $I_D = 0$ for both levels of the logic circuit; hence virtually no dc power is dissipated in either state. It is for this reason that CMOS has an *extremely low* and negligible static power dissipation, which is the product of the leakage current and the supply voltage. Power dissipation occurs under dynamic conditions which depend on the load capacitance and the frequency of the input voltage ($P = C_L V_{DD}^2 f$), as discussed later in Sec. 13.9.

For the quantative derivation of the transfer characteristic of the CMOS inverter we shall make the following simplifying assumptions:

$$V_{Tn} = |V_{Tp}| = V_T \tag{3.19a}$$

$$K_n = K_p = K \tag{3.19b}$$

where K is the constant of the transfer function, e.g., K_n in (3.12) for the NMOS. Note that the NMOS and the PMOS need not be identical. The values of K_n and K_p can be made equal by making the W/L ratios inversely proportional to the mobility of holes and electrons, i.e.,

$$\left(\frac{W}{L}\right)_{\text{PMOS}} \simeq 2.5\left(\frac{W}{L}\right)_{\text{NMOS}} \tag{3.19c}$$

For small values of V_I, the NMOS Q_1 is cut off, and the PMOS Q_2 is unsaturated. No current will flow until $V_I = V_T$ and $V_O = V_{DD}$. For V_I slightly larger than V_T, Q_2 is still unsaturated, and Q_1 is saturated. The PMOS will become saturated when

$$|V_{DS(p)}| = V_{GS(p)} - V_T \tag{3.20}$$

From the inverter circuit we have

$$|V_{DS(p)}| = V_{DD} - V_O \tag{3.21a}$$

$$|V_{GS(p)}| = V_{DD} - V_I \tag{3.21b}$$

From (3.20) and (3.21) we obtain

$$V_O = V_I + V_T \tag{3.22}$$

Similarly, the NMOS will be saturated when

$$V_{DS(n)} = V_{GS(n)} - V_T \tag{3.23}$$

From the inverter circuit

$$V_{DS(n)} = V_O \qquad \text{and} \qquad V_{GS(n)} = V_I \tag{3.24}$$

From (3.23) and (3.24) we obtain

$$V_O = V_I - V_T \tag{3.25}$$

For both devices saturated, from (3.12), for the PMOS we have

$$I_{Dp} = K_p(V_{GS(p)} - V_T)^2 = K(V_{DD} - V_I - V_T)^2 \tag{3.26}$$

Similarly for the NMOS in saturation we have

$$I_{Dn} = K_n[V_{GS(n)} - V_T]^2 = K(V_I - V_T)^2 \tag{3.27}$$

With no current drawn at the output, the currents $I_{Dp} = I_{Dn}$. From equating (3.26) and (3.27) we obtain

$$V_I = \frac{V_{DD}}{2} \tag{3.28}$$

Substitution of (3.28) into (3.22) and (3.25) yields

$$V_O = 0.5V_{DD} + V_T \quad \text{when } Q_2 \text{ saturates} \tag{3.29a}$$

$$V_O = 0.5V_{DD} - V_T \quad \text{when } Q_1 \text{ saturates} \tag{3.29b}$$

In order to derive the complete expression for the transfer function, we now consider the case when the PMOS (Q_2) is nonsaturated and the NMOS (Q_1) is saturated. For the NMOS from (3.12) we have

$$I_D = K(V_{GS(n)} - V_T)^2 = K(V_I - V_T)^2 \tag{3.30}$$

For the PMOS from (3.10*a*)

$$I_D = K[2(V_{GS(p)} - V_T)V_{DS(p)} - V_{DS(p)}^2] \tag{3.31a}$$

$$= K[2(V_{DD} - V_I - V_T)(V_{DD} - V_O) - (V_{DD} - V_O)^2] \tag{3.31b}$$

We now equate (3.30) to (3.31*b*) and solve for V_O as a function of V_I, which yields

$$V_O = V_I + V_T + [(V_I + V_T)^2 + V_{DD}(V_{DD} - 2V_I - 2V_T) - (V_I - V_T)^2]^{1/2} \tag{3.32}$$

Equation (3.32) is the analytic description for the transfer characteristic. Note the symmetry of the transfer characteristic about $V_I = V_{DD}/2$. The reader can verify the result in (3.32) by considering Q_1 unsaturated and Q_2 saturated. The key features of the transfer characteristic, namely, the output voltages for various conditions of Q_1 and Q_2, are identified in the transfer characteristic in Fig. 3.24*b*. The experimental voltage transfer characteristics of a CMOS for three values of supply voltage and two extreme temperatures are shown in Fig. 3.25. Note the close agreement of Fig. 3.25 at low

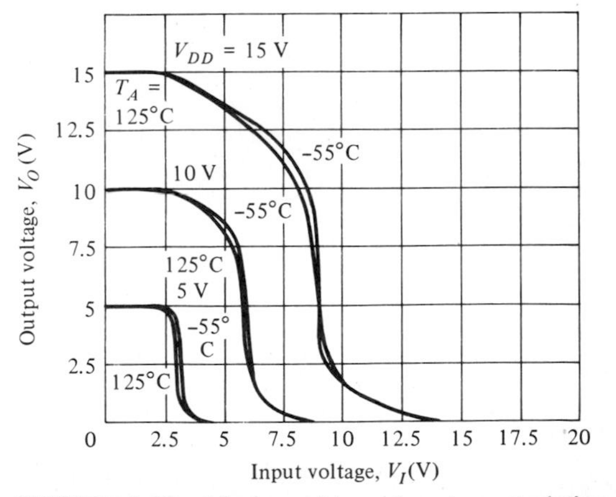

FIGURE 3.25 Measured transfer characteristics of CMOS for values of V_{DD} and two temperature extremes.

supply voltage with that of Fig. 3.24 and the excellent temperature stability at all values of V_{DD}.

EXERCISE 3.10

The NMOS and the PMOS currents vs the input voltage for a CMOS inverter is shown in Fig. E3.10. If $V_{DD} = 5$ V, $K_n = K_p = K = 0.1$ mA/V^2, and $V_{Tn} = -V_{TP} = |V_T| = 1$ V, determine the maximum current I_m.

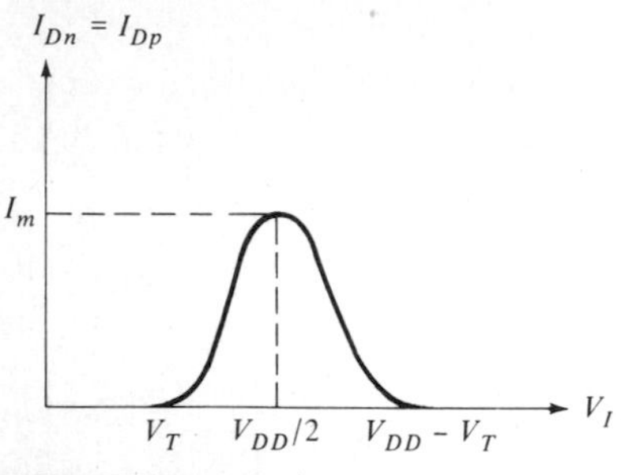

FIGURE E3.10

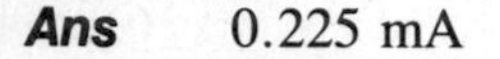
Ans 0.225 mA

The CMOS has a number of desirable properties that make it very attractive for digital circuitry and is thus widely used in integrated circuits. The major advantage is the extremely low power consumption of the gate under static conditions (of the order of 0.1 μW). The power dissipation occurs when the logic state is changed. The excellent temperature stability of the inverter as depicted in Fig. 3.25 is clearly evident. These inverters can thus be packed densely in an IC chip and used in MSI and LSI applications. (See Appendix B.)

3.7 FET SMALL-SIGNAL MODEL

When a FET is used as a small-signal amplifier, the device is operated in the saturation region, i.e., beyond pinch-off. A basic common-source JFET amplifier circuit is shown in Fig. 3.26. The quiescent point ($V_{DS} = 10$ V and $I_D = 5$ mA) and the load line for $V_{DD} = 20$ V and $R_D = 2$ kΩ are shown on the output characteristic curve. Recall that JFET and depletion-type MOSFET are biased so that V_{GS} is negative. For the enhancement-type MOSFET, V_{GS} must be positive and in fact $V_{GS} > V_T$.

We now wish to determine the low-frequency small-signal model around the operating point for the FET, as we did in Sec. 2.8 for bipolar transistors.

A. Low-Frequency FET Models

For the three-terminal FET devices we would like to obtain the two-port parameters. First we note that at the input port the gate current is zero; hence at the input port the

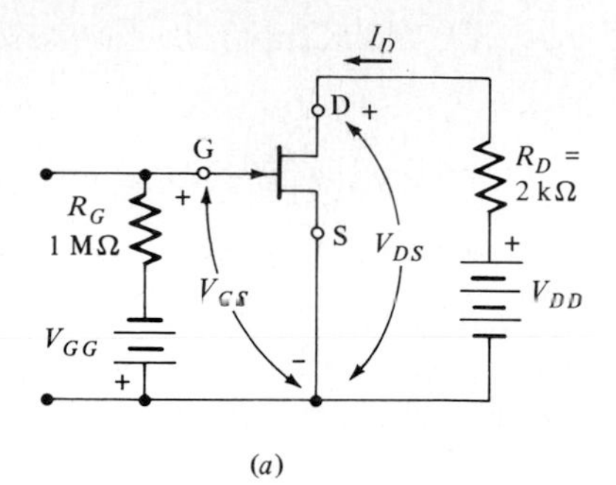

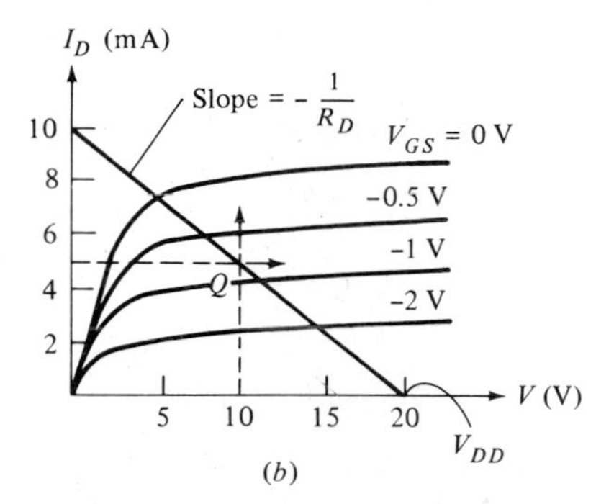

FIGURE 3.26 (*a*) Common-source JFET biasing circuit. (*b*) Output characteristics with load line and quiescent point.

gate to source terminals are open-circuited. At the output port we may write the dependence of I_D on V_{DS} and V_{GS} in the following functional form:

$$i_D = f(v_{GS}, v_{DS}) \tag{3.33}$$

where the symbols denote instantaneous quantities, e.g., $i_D = I_D + i_d$ and where I_D is the dc quiescent point, i_d is the ac signal, and i_D is the total current signal. From (3.33) it is seen that the output current is a function of the input and the output voltages. From Appendix C, we observe that a linearized relationship of i_2 $(= i_d)$ as a function of v_2 $(= v_{gs})$ leads directly to the two-port y parameters. We proceed as in (2.22), by obtaining a Taylor series expansion of (3.33) near the operating point Q. Thus we have

$$i_d \simeq \left.\frac{\partial i_D}{\partial v_{GS}}\right|_Q v_{gs} + \left.\frac{\partial i_D}{\partial v_{DS}}\right|_Q v_{ds} \tag{3.34}$$

In small-signal notation we write (3.34) as

$$i_d = y_{fs} v_{gs} + y_{os} v_{ds} \tag{3.35a}$$

Since we are considering low-frequency models, y_{fs} and y_{os} are simply conductances, and we shall rewrite (3.35*a*) as follows:

$$i_d = g_m v_{gs} + g_d v_{ds} \tag{3.35b}$$

where

$$g_m = y_{fs} = \left.\frac{i_d}{v_{gs}}\right|_{v_{ds}=0} = \left.\frac{\Delta i_D}{\Delta v_{GS}}\right|_{V_{DSq}} \quad (3.36a)$$

$$g_d = y_{os} = \left.\frac{i_d}{v_{ds}}\right|_{v_{gs}=0} = \left.\frac{\Delta i_D}{\Delta v_{DS}}\right|_{V_{GSq}} \quad (3.36b)$$

In (3.36) g_m is the transconductance, and g_d is the output conductance. The input current $i_g = 0$ for all values of v_{gs} and v_{ds} under small-signal conditions; hence y_{is} and y_{rs} are both identically zero. Using this fact and (3.35), we have a two-port model for an FET in the common-source (CS) configuration, as shown in Fig. 3.27*a*. *The model is, of course, valid for all types of FETs biased in the saturation region.* The element values, however, may be different in different types of FETs.

The current-controlled source may be changed to a voltage-controlled source by using the Thevenin equivalent form as shown in Fig. 3.27*b*. For such cases, it is convenient to define

$$\mu = \frac{g_m}{g_d} = g_m r_d \quad (3.37)$$

where μ is the amplification factor or gain of the device. It is the slope of the transfer function of the inverter in Fig. 3.16 for R_D very large. The value of $r_d = g_d^{-1}$ is determined from the manufacturer's data sheet from the value of y_{os} at low frequencies. It can also be obtained from the output characteristic at the Q point. From (3.36*b*) at the Q point we determine the slope of I_D vs V_{DS}, namely,

$$g_d = \left.\frac{\Delta i_D}{\Delta v_{DS}}\right|_{V_{GS}=\text{const}} = r_d^{-1} \quad (3.38)$$

The range of values of r_d are between 10 and 100 kΩ. The value of g_m is obtained from the manufacturer's specification of y_{fs} at low frequencies. The small-signal value of g_m for JFET and depletion-type MOSFET is determined from (3.8), namely,

7. * $$g_m = \frac{\partial I_D}{\partial V_{Gs}} = 2I_{DSS}\frac{1 + V_{GS}/V_{PO}}{V_{PO}} \quad (3.39)$$

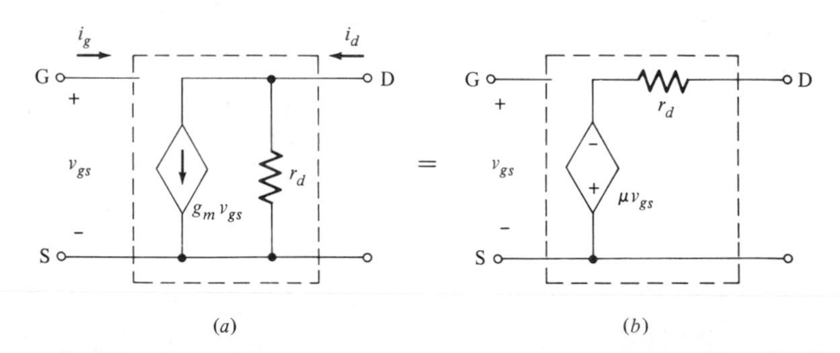

FIGURE 3.27 (*a*) Low-frequency small-signal model of an FET. (*b*) Equivalent representation.

For the enhancement-type MOSFET, it is determined from (3.12), namely,

$$g_m = \frac{\partial I_D}{\partial V_{GS}} = 2K_n(V_{GS} - V_T) = 2\sqrt{I_D K_n} \tag{3.40}$$

Recall that V_{GS} is negative in (3.39) and positive in (3.40). The reader should note the differences between the g_m for FET and BJT. In BJT $g_m = 1/r_e = 25 \text{ mV}/I_E$, which is dependent on the emitter current at Q point. In FETs g_m is dependent on the gate-to-source voltage V_{GS}. The range of values of g_m for FETs are usually between 10^{-4} to 10^{-2} mhos.

EXAMPLE 3.5

For the depletion-type NMOS-MN82 data sheet given in Appendix E, determine the element values of the small-signal model at the Q point: $V_{DS} = 15$ V and $I_D = 5$ mA

$$g_m = |y_{fs}| = 10 \times 10^{-3} \text{ mho} \qquad \text{at } f = 1 \text{ kHz}$$

$$g_d = |y_{os}| = 0.25 \times 10^{-3} \text{ mho} = r_d^{-1} \qquad \text{at } f = 1 \text{ kHz}$$

EXAMPLE 3.6

Determine the transconductance of an enhancement NMOS for $V_{GS} = 4$ V, $V_T = 2$ V, at the Q point $I_D = 5$ mA, $V_{DS} = 5$ V, given that $K_n = 5 \times 10^{-4}$ mA/V^2.

We note that the NMOS is in saturation since $V_{DS} > V_{GS} - V_T$. From (3.40) we obtain

$$g_m = 2(5 \times 10^{-4})(4 - 2) = 2 \times 10^{-3} \text{ mho}$$

EXERCISE 3.11

Determine the low-frequency small-signal parameters of a JFET with $V_{PO} = 5$ V, $I_{DSS} = 6$ mA for the following cases and comment on the mode of operation in each case. (*a*) $V_{GS} = -2$ V, $V_{DS} = 2$ V, (*b*) $V_{GS} = -2$ V, $V_{DS} = 4$ V.

Ans (*a*) $g_m = 9.6 \times 0^{-4}$ mho, $r_d = 2.08$ $k\Omega$, ohmic region; (*b*) $g_m = 1.44 \times 10^{-3}$ mho, $r_d = \infty$, saturation region

B. High-Frequency FET Models

At very high frequencies we cannot ignore the carrier transit time and the interterminal capacitances. For the JFETs a small-signal high-frequency equivalent circuit that may be used in a computer-aided analysis and design is shown in Fig. 3.28*a*. For the MOSFET the corresponding high-frequency model is shown in Fig. 3.28*b*. A simplified high-frequency model that is sufficiently accurate for most applications and is used for both JFET and MOSFET is shown in Fig. 3.29. We shall henceforth use the model of Fig. 3.29 known as the π model for high-frequency small-signal representation of FETs in this text. It should be noted, however, that the gate-to-source capacitance C_{gs} is usually much larger than the gate to drain capacitance C_{gd}. Typical values of C_{gs} vary from 10 pF for a low-frequency FET to less than 2 pF for a high-frequency FET. The gate-to-drain capacitance C_{gd} is typically in the range from 0.2 to 2 pF with the low value for a high-frequency MOSFET. Typical values of the parameters for JFETs and MOSFETs are shown in Table 3.1. Note that the values of C_{ds} and C_{gd} are comparable, and both are smaller than C_{gs}. We have included C_{gd} in the model since its value due to the Miller effect can be appreciable for a large gain (Miller effect is discussed in Sec. 7.7). Finally it should also be mentioned that for small MOS devices the values of the capacitances could be smaller by an order of magnitude than those listed in Table 3.1.

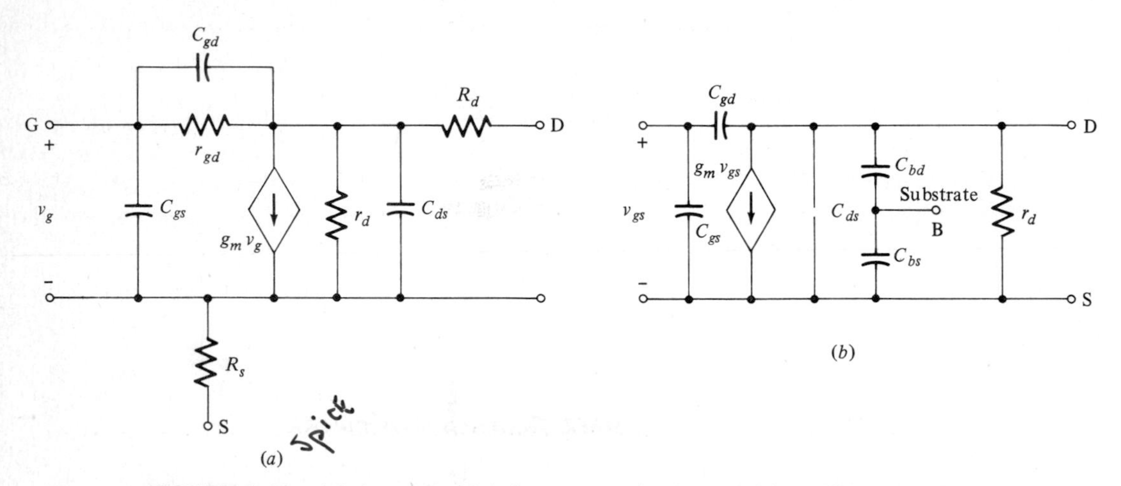

FIGURE 3.28 (*a*) Accurate high-frequency circuit model for JFET. (*b*) Accurate high-frequency circuit model for MOSFET.

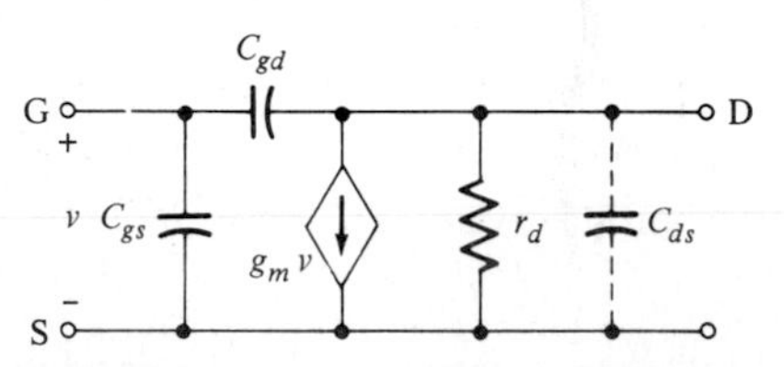

FIGURE 3.29 Simple high-frequency small-signal model of an FET ($C_{gs} \gg C_{gd}$; C_{ds} can usually be ignored).

TABLE 3.1 Typical Parameter Values for JFETs and MOSFETs

Parameter	JFET	MOSFET (both types)
$\|y_{fs}\| = g_m$ mhos	0.1×10^{-3}–10×10^{-3}	0.5×10^{-3}–10×10^{-3}
$\|y_{os}\| = g_d = r_d^{-1}$ mhos	10^{-6}–10^{-5}	10^{-5}–10^{-4}
$C_{iss} = C_{rss} + C_{gs}$ pF	2–10	2–10
$C_{rss} = C_{gd}$ pF	0.2–2	0.2–2
$C_{ds} = C_{oss} - C_{rss}$*		
$\mu = g_m r_d$		

*Is in the range of 0.2 to 2 pF and can usually be ignored.

The circuit element values can be readily obtained from manufacturers' data sheets. The data are usually given in terms of the y parameters. More specifically, the common-source short-circuit input capacitance C_{iss}, reverse transfer capacitance C_{rss}, and output capacitance C_{oss} are given (see Appendix E). From the high-frequency model of Fig. 3.28 and the manufacturer's y parameters we obtain

$$C_{gd} \simeq C_{rss} \qquad C_{gs} \simeq C_{iss} - C_{rss} \qquad C_{ds} \simeq C_{oss} - C_{rss} \tag{3.41}$$

EXAMPLE 3.7

Determine the high-frequency model circuit parameters of the depletion NMOS type MN82 given in Appendix E (for $V_{DS} = 15$ V and $I_D = 5$ mA).

From the data sheet at $f = 1$ MHz, we read

$$C_{rss} = 0.3 \text{ pF} \qquad \text{hence } C_{gd} = 0.3 \text{ pF}$$

$$C_{iss} - C_{rss} = 4 - 0.3 = 3.7 \text{ pF} \qquad \text{hence } C_{gs} = 3.7 \text{ pF}$$

$$C_{oss} - C_{rss} = 1.6 - 0.3 \qquad \text{hence } C_{ds} = 1.3 \text{ pF}$$

The values of g_m and r_d were already determined in Example 3.6.

In spite of the fact that the capacitances of FETs are small, as we shall see in Chaps. 7 and 8, the bipolar transistor has a higher gain-bandwidth product than the FET devices and is, therefore, used in high-speed (wideband) applications. This is mainly due to the fact that gain-bandwidth product is proportional to g_m/C, where g_m for BJTs are higher than that of FETs by as much as an order of magnitude or larger. For high-frequency applications, the load resistance is necessarily small, and capacitance C_{ds}, which is in shunt with R_L, can be ignored, just as in the bipolar transistor (BJT) model. Note that the FET model is similar to the BJT model *except* that r_b and R_i are not present in the FET model. *The input properties of the two devices are quite different;* in the FET the low-frequency input impedance is infinite, and it must, therefore, be driven by a voltage source. The input impedance of the BJT is moderate, and it can be driven by a voltage or a current source. The latter is most often used because the BJT is a current-controlled device.

Finally, it should be clear that the model we have shown in Fig. 3.28 can be used for FETs in any of the three configurations, common source (CS), common drain (CD), and common gate (CG), by merely reorienting the model. The properties of FETs in the various configurations are discussed in Chap. 4. Circuit applications of FETs are discussed in later chapters throughout the text.

C. Maximum Operating Frequency

The *maximum operating frequency*, or *cutoff frequency*, f_m is defined as the frequency at which the FET can no longer amplify the input signal. In other words the current through the input is equal to the dependent current source $g_m v_{gs}$. That is,

$$2\pi f_m (C_{gs} + C_{gd}) v_{gs} = g_m v_{gs} \tag{3.42a}$$

or

$$f_m = \frac{g_m}{2\pi(C_{gs} + C_{gd})} \tag{3.42b}$$

From (3.40) and (3.42) we have

$$f_m = \frac{2K_n(V_{GS} - V_T)}{2\pi(C_{gs} + C_{gd})} \tag{3.43}$$

If we assume $C_{gs} + C_{gd} = C_0 WL$, then from (3.10*b*) and (3.43) we obtain

$$f_m \simeq \frac{\mu_n(V_{GS} - V_T)}{2\pi L^2} \tag{3.44}$$

Notice that in order to maximize f_m, we must reduce the channel length and increase the mobility and the gate voltage. Compare (3.44) with (3.13). The limit in (3.44) applies to all FET devices. In practice the maximum frequency limit of FET is much lower than (3.44). The limit is determined by the speed with which the device can charge or discharge the load capacitance.

EXERCISE 3.12

(*a*) What is the maximum cutoff frequency of an NMOS if the channel length is $L = 10\ \mu\text{m}$, $V_T = 2$ V, and $V_{GS} = 3$ V. (*b*) Determine the same if $L = 2\ \mu\text{m}$.

Ans (*a*) $f_{max} \simeq 127$ MHz; (*b*) 3.17 GHz

3.8 DUAL-GATE FET

A dual-gate FET is an *n*-channel depletion-type MOSFET with two independent insulated gates as shown in Fig. 3.30*a*. These transistors have a series arrangement

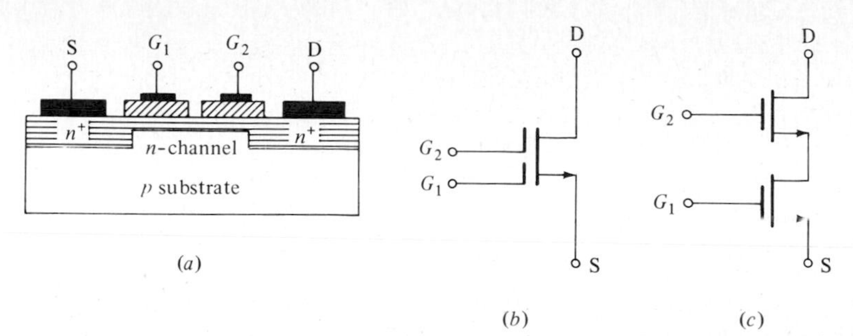

FIGURE 3.30 (*a*) Dual-gate MOS structure. (*b*) Circuit symbol. (*c*) Equivalent representation.

of two separate channels, each channel having an independent control gate. The circuit symbol for the dual-gate FET is shown in Fig. 3.30*b*. The dual-gate MOSFETs are used at very high frequencies (several hundred megahertz) for mixer, modulator, and multiplier functions. In normal operation the input signal is applied to gate 1, and the output is taken from the drain. The source terminal is grounded. The signal applied to gate 2 is used to modulate the input signal at gate 1. The *V-I* characteristics of a 3N205-type dual-gate FET is shown in Fig. 3.31*a* and *b*. In Fig. 3.31*a*, I_D vs V_{DS} for $V_{G2S} = 4$ V and V_{G1S} as a parameter is shown, while the characteristic in Fig. 3.31*b* shows I_D vs V_{G1S} for $V_{DS} = 15$ V and V_{G2S} as a parameter. Note that in this device $I_D = 0$ when $V_{G1S} = 2$ V or $V_{G2S} = -1$ V. The dual-gate FET exhibits a very small capacitance between the drain and gate 1 with typical values less than or equal to 0.01 pF. When gate 2 is at ac ground, the dual-gate FET can be represented as in Fig. 3.30*c*, which is a common-source, common-gate (CS-CG pair) arrangement. This arrangement is similar to CE-CB in BJT, which is referred to as the *cascode circuit*. (Analysis of the cascode and its effective low-capacitance property are discussed in Secs. 5.3 and 8.2, respectively.)

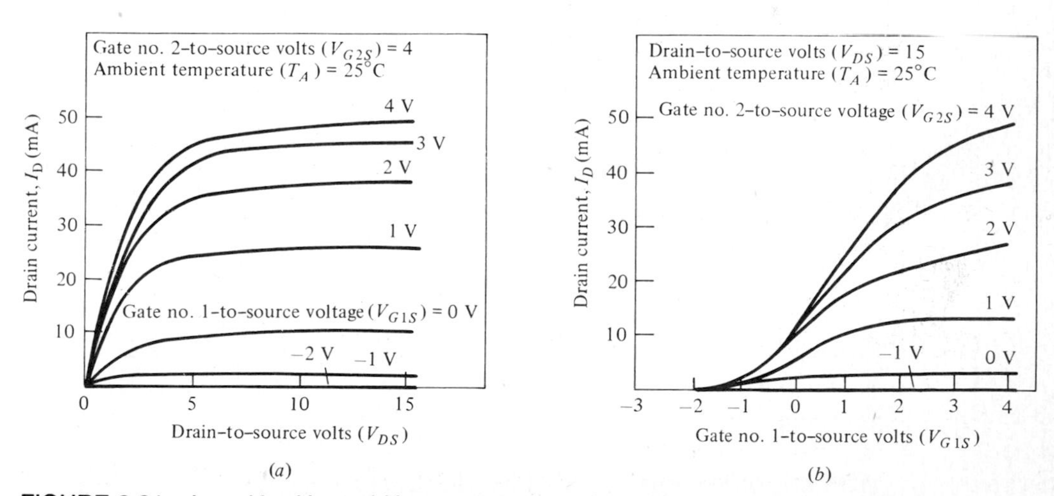

FIGURE 3.31 I_D vs V_{DS}, V_{G1} and V_{G2} as parameters. I_D vs V_{G1}, V_{G2} as a parameter.

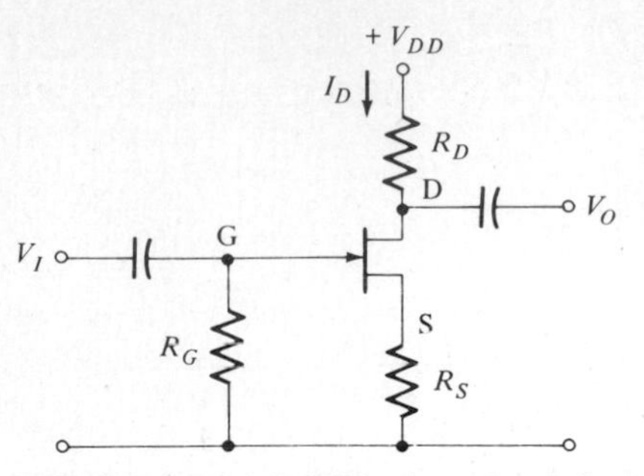

FIGURE 3.32 Self-biasing circuit for a (n-channel) JFET and depletion-type NMOS.

3.9 BIASING TECHNIQUES FOR FETS

The dc characteristics of FET devices are not as heavily temperature dependent as those of bipolar transistors. For example, in an NMOS device V_T decreases for an increase in temperature, while K_n in (3.12) decreases due to the decrease in mobility of the electron in silicon thus compensating for each other. An examination of the FET for various temperatures as given by the manufacturers will also readily show this fact. In fact, if the required output voltage swing can be obtained at the highest temperature, it can be virtually assured at low temperatures if the quiescent operating point is maintained fixed. The second simplifying factor in the design is that the input circuit draws no dc current. In FETs the gate current is usually very small (of the order of 0.1 μA in JFET and 1 nA in MOSFET) so that it can be entirely neglected in the design, which simplifies things enormously. Variability of the device from unit to unit, however, should be taken into consideration.

A. Biasing Circuit for JFET and Depletion-Type MOSFET

A biasing circuit for an n-channel JFET that is also applicable to depletion-type MOSFET devices is shown in Fig. 3.32. The selection of an operating point (I_D, V_{DS}) for an FET amplifier stage is usually suggested by the manufacturer. The design is best illustrated through a simple example.

EXAMPLE 3.8

Design the biasing of a JFET amplifier stage shown in Fig. 3.32 for the quiescent point $I_D = 2.0$ mA, $V_{DS} = 10$ V. The JFET parameters are $I_{DSS} = 8$ mA, $V_{PO} = 6$ V; the supply voltage $V_{DD} = 20$ V.

From (3.8) we have $2 = 8(1 + V_{GS}/V_{PO})^2$ which yields $V_{GS} = -3$ V. We choose an arbitrarily large value for R_G, say, 0.5 MΩ. This is done in order to provide a path for the leakage gate current. The voltage drop across R_G is $I_G R_G$, which is negligible as compared to V_{GS} since $I_G \leq 0.1$ μA (for depletion-type MOS it is of the order 1 nA). From KVL, at the input we have

$$V_{GS} = I_G R_G - I_D R_S \simeq -I_D R_S$$

$$R_S = \frac{-V_{GS}}{I_D} = \frac{3.0}{2.0 \text{ mA}} = 1.5 \text{ k}\Omega$$

The value of R_D is determined from KVL applied to the source-drain circuit,

$$V_{DD} = I_D R_D + V_{DS} + V_S$$

For a supply voltage $V_{DD} = 20$ V, the value of R_D is

$$R_D = \frac{V_{DD} - V_{DS} - V_S}{I_D} = \frac{20 - 10 - 1.5(2.0)}{2.0 \times 10^{-3}} = 3.5 \text{ k}\Omega$$

If the FET characteristic curve is available, such as in Fig. 3.5, we can determine $R_S + R_D$ by drawing the load line corresponding to the desired operating point Q and the supply voltage V_{DD}. The value of V_{GS} at the operating point determines R_S and hence R_D.

The V-I characteristics of a device can vary either by temperature changes or interchangeability from unit to unit. For example, consider the transfer characteristic at two different temperatures, two units of the same type, or the maximum and minimum values such as shown in Fig. 3.33. A common biasing circuit that is applicable to both JFET and depletion-type MOSFET is shown in Fig. 3.34a. The circuit is redrawn for simplicity in Fig. 3.34b. For the circuit in Fig. 3.34b we have

$$R_G = R_{g3} + R_{g2} \| R_{g1} \tag{3.45a}$$

$$V_G = \frac{R_{g2}}{R_{g1} + R_{g2}} V_{DD} \simeq \frac{R_{g2}}{R_{g1}} V_{DD} \qquad \text{for } R_{g1} \gg R_{g2} \tag{3.45b}$$

The input load line for this circuit is shown on the transfer characteristic curve in Fig. 3.33. Note that V_{GS} is negative. If the worst-case information is available such as $I_{DSS,\max}$, $V_{PO,\max}$, $I_{DSS,\min}$, $V_{PO,\min}$, the design of biasing circuit for a nominal Q point with allowable percent variation proceeds as in the following example.

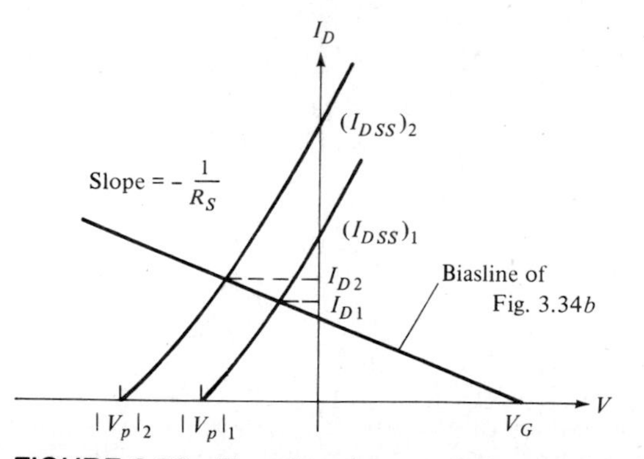

FIGURE 3.33 Transfer characteristic variations in JFETs.

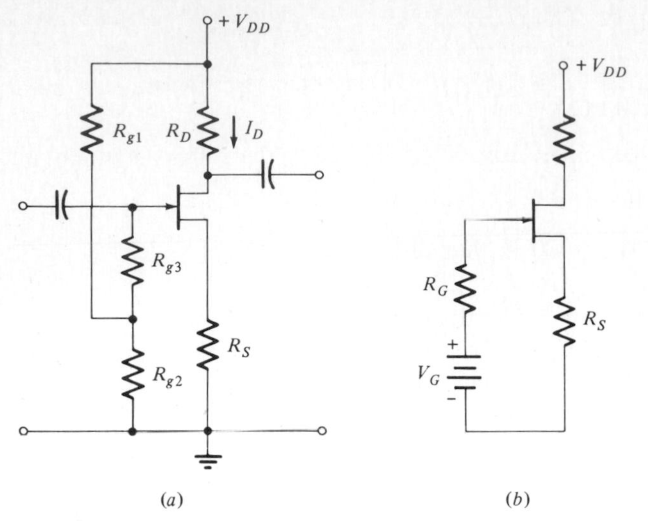

FIGURE 3.34 (*a*) Biasing circuit for JFET and depletion-type MOS. (*b*) Equivalent representation of the circuit.

EXAMPLE 3.9

Design the biasing circuit for a depletion-type MOS shown in Fig. 3.34. Given that $V_{PO,\max} = 6$ V, $V_{PO,\min} = 4$ V, $I_{DSS,\max} = 10$ mA, $I_{DSS,\min} = 8$ mA. The nominal operating point desired is $I_D = 5$ mA, $V_{DS} = 8$ V, with a 10 percent allowable variation in I_D. The supply voltage $V_{DD} = 20$ V. We shall determine the circuit element values corresponding to Fig. 3.34.

For $I_{D,\max} = 5.5$ mA and the maximum value of I_{DSS} from (3.8) we determine V_{GSM},

$$5.5 = 10\left(1 + \frac{V_{GSM}}{6}\right)^2 \qquad \text{or} \qquad V_{GSM} = -1.55 \text{ V}$$

For $I_{D,\min} = 4.5$ mA we obtain V_{GSm},

$$4.5 = 8\left(1 + \frac{V_{GS}}{4}\right)^2 \qquad \text{or} \qquad V_{GSm} = -1.0 \text{ V}$$

Hence $R_S = \Delta V_{GS}/\Delta I_D = 0.55/10^{-3} = 550\ \Omega$. From Fig. 3.34*b*, at the input we have the relation

$$V_G = V_{GS} + I_D R_S$$

But $V_{GS} = \frac{1}{2}(V_{GSM} + V_{GSm}) = -1.275$ V. Hence $V_G = -1.275 + 5 \times 10^{-3}(0.55 \times 10^3) = 1.475$ V.

We arbitrarily choose large values of R_{g3} and R_{g1} in order to avoid the loss of ac signal, say, $R_{g1} = R_{g3} = 0.5\ \text{M}\Omega$. From (3.45*b*) $R_{g2}/500K = 1.475/20$

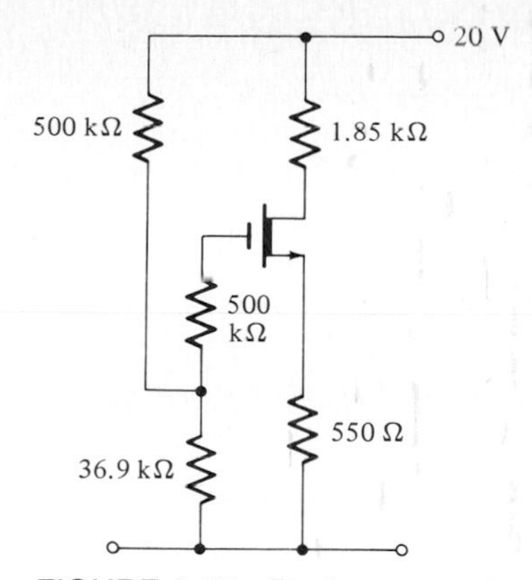

FIGURE 3.35 Design example circuit.

or $R_{g2} = 36.9$ kΩ. The value of R_D is determined from the output circuit in Fig. 3.34.

$$V_{DD} = I_D R_D + V_{DS} + I_D R_S = V_{DS} + I_D(R_D + R_S)$$

Hence $R_D + 550 = (20 - 8)/5 \times 10^{-3}$ or $R_D = 1.85$ kΩ. Note that the variation of V_{DS} for this design is given by $\Delta V_{DS} = (\Delta I_D)(R_S + R_D) = \pm 0.5 \times 10^{-3}(2.4 \times 10^3) = \pm 1.2$ V. In other words, the value of V_{DS} will be in the range of 6.8 to 9.2 V. Note that the minimum value V_{DS} is larger than $V_{PO,\max}$, and I_D is independent of V_{DS} in the saturation region. The designed circuit is shown in Fig. 3.35.

Note that the simple self-biasing circuit of Fig. 3.32 cannot be used for the enhancement-type MOSFET since this biasing circuit reverse-biases the gate.

B. Biasing Circuit for the Enhancement-Type MOSFET

The enhancement-type MOS biasing requirements are the following:

1. The gate voltage is of the same polarity as the drain voltage. For NMOS V_{GS} is positive and larger than V_T.
2. $V_{DS} \geq V_{GS} - V_T$ to ensure that the operating point is in the saturation region.

A simple circuit that achieves the above objectives is shown in Fig. 3.36*a*. As we shall show, this circuit is not desirable since the drain current variations can be unacceptably large.

In the manufacturing of FETs the devices obviously cannot be identical,and they vary from unit to unit. Even though the manufacturers do not always supply such information, variability from unit to unit exists, and the operating point shifts because the devices are not exactly interchangeable. For the same unit the temperature variations can also cause a shift in the device characteristic. A common biasing circuit that

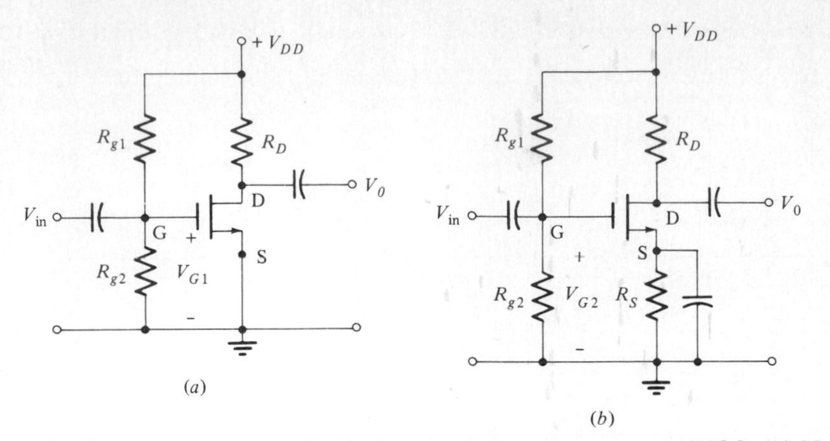

FIGURE 3.36 (*a*) A simple biasing circuit for enhancement NMOS. (*b*) Modified biasing circuit to improve stability of operating point.

circumvents the variations in the Q-point is shown in Fig. 3.36*b*. To see how the operating point can be stabilized despite interchangeability or variations due to temperature change, consider the modified FET biasing circuit shown in Fig. 3.36*b*, where R_S is added and plays the same role as the emitter resistor R_E in BJT biasing circuits, i.e., it provides feedback. Due to the nature of the manufacturing process, good control over V_T cannot be maintained. Within the same type of FET, these parameters may exhibit a spread of 3 to 1 in the above characteristics. For example, the transfer characteristics of two FETs of the same type might be as shown in Fig. 3.37. In order for I_D to be approximately the same for both FETs, the slope of the bias line must be considerably decreased (i.e., R_S is considerably increased). However, if R_S is made very high, the drain current will be very low for a given supply voltage, or the transistor might be biased in the ohmic region, which is undesirable.

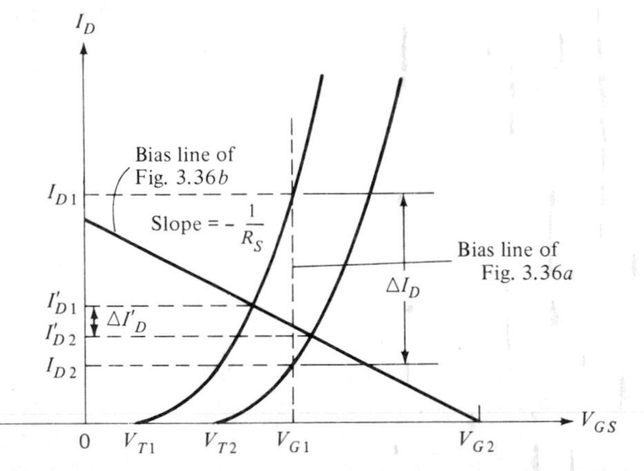

FIGURE 3.37 Bias line and quiescent drain current shift for Fig. 3.36*b* shown on the transfer characteristic.

The design equations for the modified biasing circuit of Fig. 3.36*b* are as follows. At the input circuit the equations are

$$V_{GS} = V_G - R_S I_D \qquad \text{or} \qquad I_D = \frac{V_G}{R_S} - \frac{1}{R_S} V_{GS} \tag{3.46a}$$

where

$$V_G = \frac{R_{g2}}{R_{g2} + R_{g1}} V_{DD} \tag{3.46b}$$

At the output circuit the equation is

$$V_{DD} = I_D R_D + V_{DS} + V_S \tag{3.47}$$

The bias line corresponding to (3.46*a*) is shown superimposed on Fig. 3.37. Since the slope of the bias line is small, the variation of the quiescent point is also relatively small from unit to unit. Note that in the absence of R_S as in Fig. 3.36*a*, a constant V_{GS} will produce substantially different drain currents for different FETs. The differences between the drain currents in the two cases are shown by ΔI_D and $\Delta I'_D$ for each circuit. Note the small difference $\Delta I'_D$ for the improved circuit of Fig. 3.36*b* vs ΔI_D for the circuit in Fig. 3.36*a*. The advantage of using the resistor R_S is obvious. We shall henceforth use the biasing scheme of Fig. 3.36*b* for the enhancement MOS transistors.

EXAMPLE 3.10

Design the biasing circuit of Fig. 3.36*b* for an enhancement MOS device which under interchangeability and/or temperature extremes has the transfer characteristic that falls between $K_{nA} = 0.4$ mA/V², $V_{TA} = 2$ V and $K_{nB} = 0.2$ mA/V², $V_{TB} = 3$ V. The extreme transfer curves for this device are shown in Fig. 3.38.

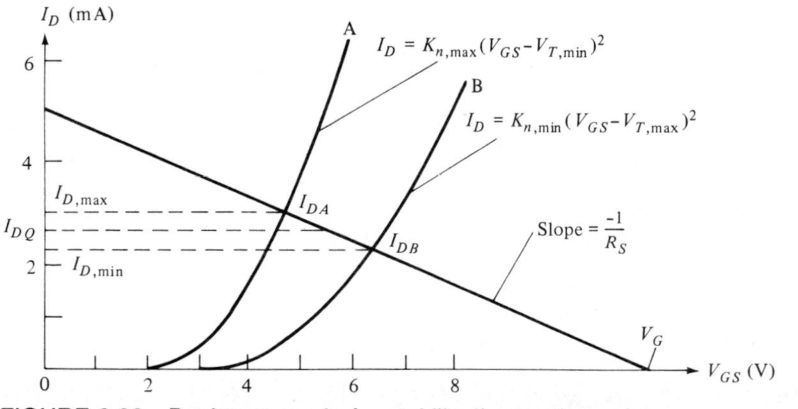

FIGURE 3.38 Design example for stabilized operating point.

The desired Q point is $I_D = 3$ mA, $V_{DS} = 6$ V, with an allowable 15 percent variation in I_D. The supply voltage $V_{DD} = 15$ V.

From (3.12) for case A, which yields $I_{D,\max}$, we have

$$3.45 = 0.4(V_{GSA} - 2)^2 \qquad \text{or} \qquad V_{GSA} = 4.94 \text{ V}$$

Similarly for case B, which yields $I_{D,\min}$, we have

$$2.55 = 0.2(V_{GSB} - 3)^2 \qquad \text{or} \qquad V_{GSB} = 6.57 \text{ V}$$

The slope and the intersection of the line connecting I_{DA} and I_{DB} determines the values of $-1/R_S$ and V_G, namely,

$$R_S = \frac{V_{GSB} - V_{GSA}}{I_{D,\max} - I_{D,\min}} = \frac{6.57 - 4.94}{0.9 \times 10^{-3}} = 1.8 \text{ k}\Omega$$

From (3.47) we determine the values of $R_S + R_D$, namely,

$$R_S + R_D = \frac{15 - 6}{3} \text{ k}\Omega = 3 \text{ k}\Omega$$

Hence, $R_D = 3 \text{ k}\Omega - 1.8 \text{ k}\Omega = 1.2 \text{ k}\Omega$. From (3.46$a$) using either the characteristic in A or B, we have

$$V_G = V_{GSA} + R_S I_{D,\max} = 4.94 + (1.8 \times 10^3)(3.45 \times 10^3) = 11.15 \text{ V}$$

If we choose $R_{g1} = 0.5$ MΩ, the value of R_{g2} is determined from

$$1 + \frac{R_{g1}}{R_{g2}} = \frac{V_{DD}}{V_G} \qquad \text{or} \qquad R_{g2} = 1.4 \text{ M}\Omega$$

The variation of V_{DS} for this design is given by

$$\Delta V_{DS} = \Delta I_D(R_S + R_D) = \pm 0.45 \times 10^{-3}(3 \times 10^3) = \pm 1.35 \text{ V}$$

In other words, V_{DS} of this design is in the range 4.65 to 7.35 V.

EXERCISE 3.13

For the biasing circuit of Fig. 3.36b, we wish to bias the device at $I_D = 1$ mA, $V_{DS} = 4$ V, with $V_{DD} = 10$ V. The NMOS parameters are given as $V_T = 1$ V, $K_n = 0.5$ mA/V^2. The gain requirements call for a value of $R_D = 5$ kΩ, and let $R_{g1} + R_{g2} = 1.5$ MΩ. Determine the circuit parameters.

Ans $R_S = 1$ kΩ, $R_{g2} = 0.51$ MΩ, $R_{g1} \simeq 0.99$ MΩ

EXERCISE 3.14

For the biasing circuit of Fig. 3.36b, the desired operating point is $I_D = 1$ mA, $V_{DS} = 6$ V. Given $V_{DD} = 12$ V, $R_D = 5$ kΩ and $R_{g1} + R_{g2} = 1.5$ mΩ. Determine R_S, R_{g1}, and R_{g2}. The V-I characteristic of the enhancement NMOS is given in Fig. E3.14.

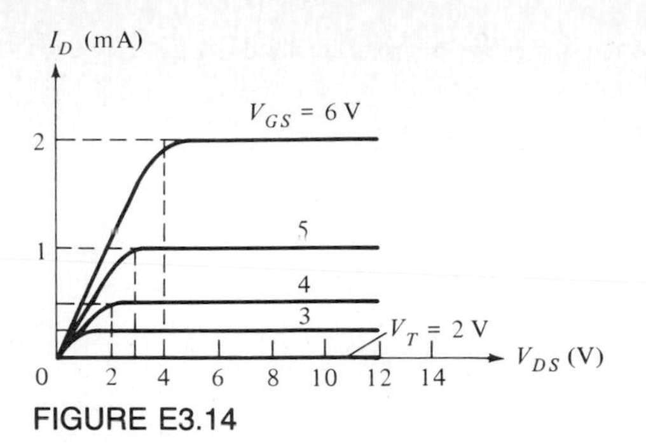

FIGURE E3.14

Ans $R_S = 1\ \text{k}\Omega,\ R_1 = R_2 = 750\ \text{k}\Omega$

3.10 CONCLUDING REMARKS

The objective of this chapter was to introduce the FET devices, their *V-I* characteristics and their applications as a linear amplifier and as a switch. There are basically two kinds of FET devices: the JFET and the MOSFET. MOS devices are of two types: the depletion type and the enhancement type. The depletion-type MOS devices are quite similar to the JFET and were therefore discussed briefly. The enhancement-type NMOS devices are commonly used as the driver of an inverter with a depletion-type NMOS load in integrated circuits. Among the inverters considered, the NMOS and the CMOS, which utilizes both NMOS and PMOS devices, inverters are the dominant choices for medium- and large-scale integrated circuits.

We also note that the MOS devices are used as voltage-controlled resistance elements over a portion of their *V-I* characteristics. The small-signal model of FET devices are simpler than those of BJTs, and their input impedances are extremely high and considered infinite for all practical purposes. This property of FET devices, for example, is utilized in op amps (short for operational amplifiers), discussed in Chap. 5.

REFERENCES

1. E. S. Yang, *Fundamentals of Semiconductor Devices,* McGraw-Hill, New York, 1978.
2. P. Richman, *MOS Field-Effect Transistors and Integrated Circuits*, Wiley, New York, 1973.
3. A. S. Sedra and K. C. Smith, *Microelectronic Circuits,* Holt, New York, 1982.
4. D. J. Hamilton and W. G. Howard, *Basic Integrated Circuit Engineering,* McGraw-Hill, New York, 1975.

PROBLEMS

3.1 Determine the ON resistance of an *n*-channel JFET if the dimensions of the JFET are given as follows: $W/L = 10$, $a = 1\ \mu\text{m}$, $N_d = 10^{16}\ \text{cm}^{-3}$.

3.2 The *V-I* characteristics of an enhancement-type NMOS are described by (3.10) and (3.12). Typical parameters of the NMOS are $\mu_n = 800\ C_m^2/V$, $\epsilon_{ox} = 4$, $\epsilon_0 = 8.85 \times 10^{-14}$ F/cm, $t_{ox} = 0.2\ \mu m$, $W/L = 10$, $V_T = 2$ V.
(*a*) Plot the output characteristic I_D vs V_{DS} for $V_{DS} > 0$ and $V_{GS} = 4$ V.
(*b*) Plot the transfer characteristic I_D vs V_{GS}.

3.3 Sketch the transfer characteristic V_O vs V_I of the basic inverter in Fig. 3.15 for $V_{DD} = 15$ V and $R_D = 3\ k\Omega$. The *V-I* characteristics of the NMOS transistor are those given in Fig. 3.13. (Note that $V_T = 3$ V.) Draw the load line and identify the values of V_O for $V_I = 1$ V and 3 V. Show the off and the saturation regions of the inverter.

3.4 Show that for small values of V_{DS} the resistance of an enhancement-type NMOS transistor is given by

$$r_D = \frac{L}{\mu_n W C_O}(V_{GS} - V_T)^{-1} \qquad \text{for } V_{DS} \ll V_{GS}$$

Determine the value of the resistance r_D for $V_{GS} = 3$ V if the NMOS parameters are given as in problem 3.2.

3.5 For the voltage-controlled linear attenuator circuit shown in Fig. P3.5, sketch V_O/V_I for different values of V_C if the NMOS is described by problem 3.2. Make reasonable approximations.

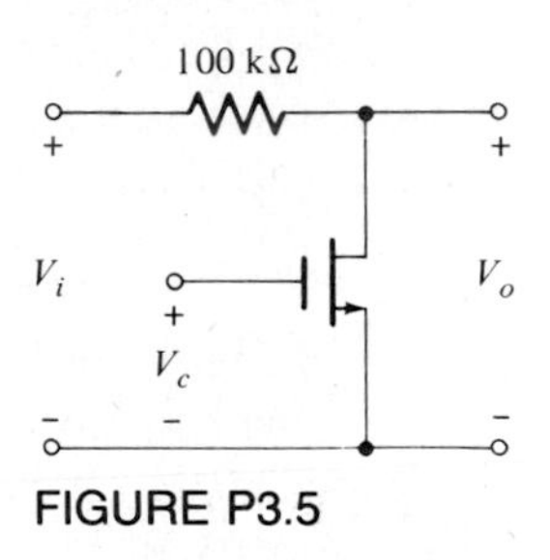

FIGURE P3.5

3.6 Plot the *V-I* relation of the NMOS shown in Fig. P3.6. The NMOS parameters are given in problem 3.2. Note that the transistor operates in the saturation region.

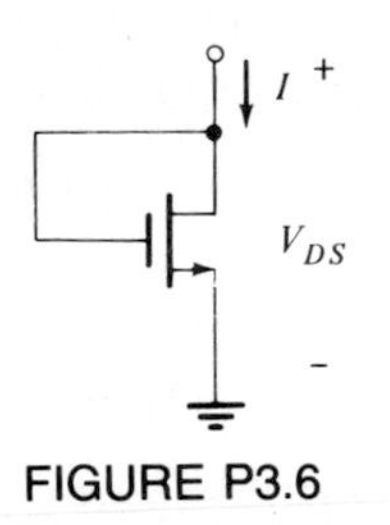

FIGURE P3.6

3.7 Determine I_D and V_{DS} of the JFET circuit shown in Fig. P3.7. The JFET parameters are $V_{PO} = 2$ V and $I_{DSS} = 5$ mA.

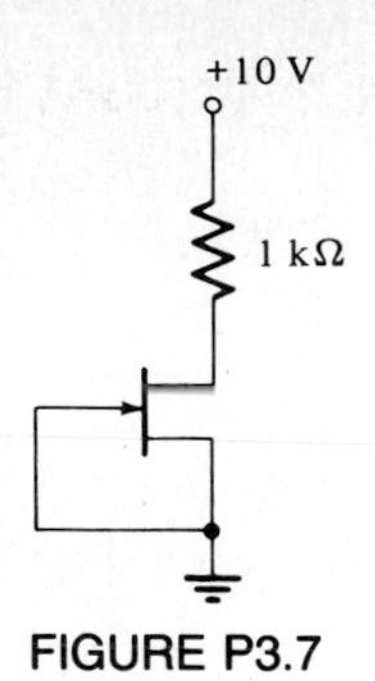

FIGURE P3.7

3.8 Repeat problem 3.7 for the circuit shown in Fig. P3.8. The JFET parameters are given in problem 3.7.

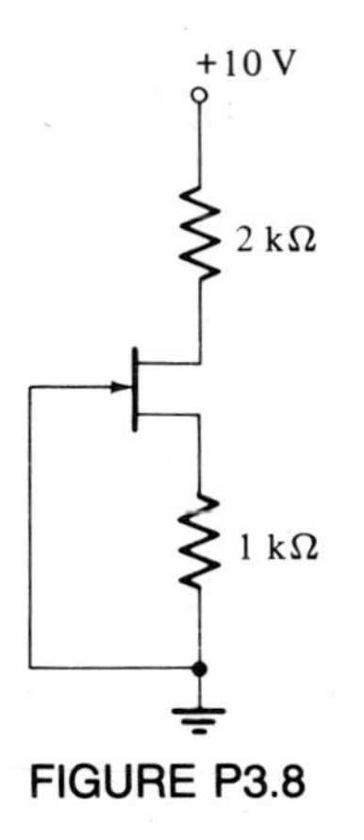

FIGURE P3.8

3.9 Show a piecewise-linear approximate model of the *V-I* characteristic in problem 3.6, using ideal diodes.

3.10 For the depletion-type MOSFET circuit shown in Fig. P3.10, assume that $I_{DSS} = 5$ mA and $V_{PO} = 2$ V. Find the values of the current I_D and the voltages V_{GS} and V_{DS}.

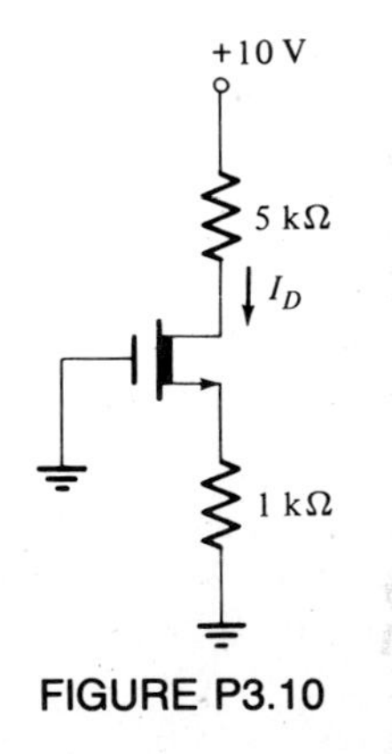

FIGURE P3.10

3.11 For the circuit shown in Fig. P3.11, determine the values of the drain voltage V_{DS2} and the drain current I_{D2}. The JFET parameters are $I_{DSS} = 5$ mA and $V_{PO} = 2$ V.

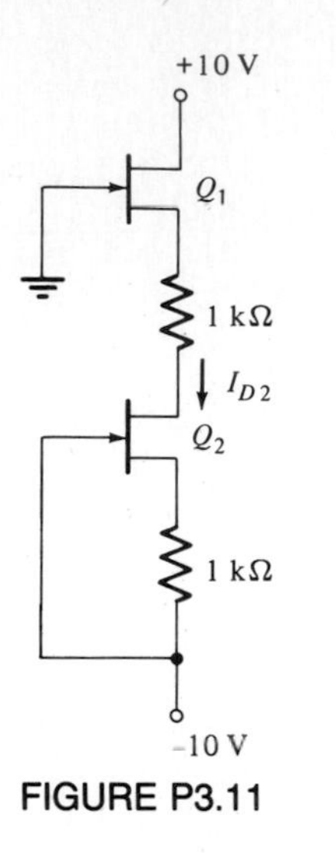

FIGURE P3.11

3.12 For the chopper circuit shown in Fig. P3.12, assume that $V_m \ll |V_P|$ and the control voltage is a square wave as shown. Sketch $v_o(t)$ if $r_{DS} \ll R_L$.

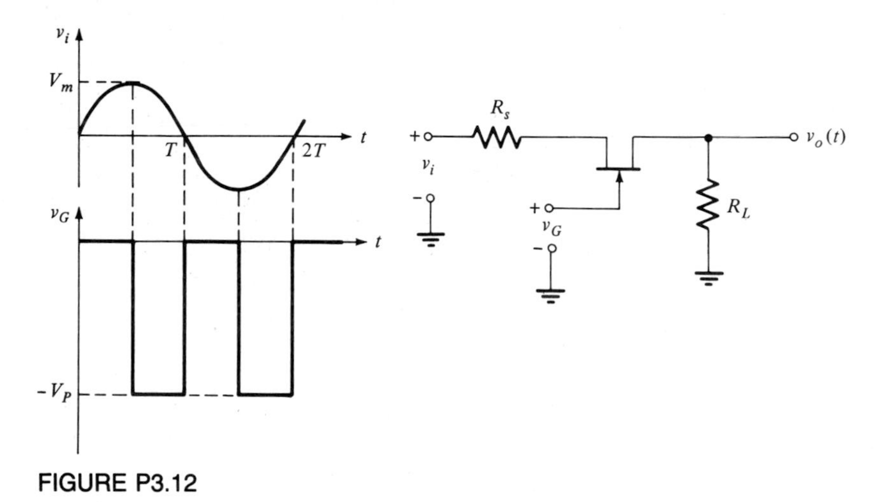

FIGURE P3.12

3.13 An enhancement NMOS driver with an enhancement NMOS saturated load is shown in Fig. P3.13. The *V-I* characteristics of an NMOS is shown in Fig. 3.13. Show the load line and sketch the voltage transfer characteristic V_O vs V_I for $V_{DD} = 10$ V, given $V_T = 3$ V and $K_n = 10^{-4}$ A/V^2.

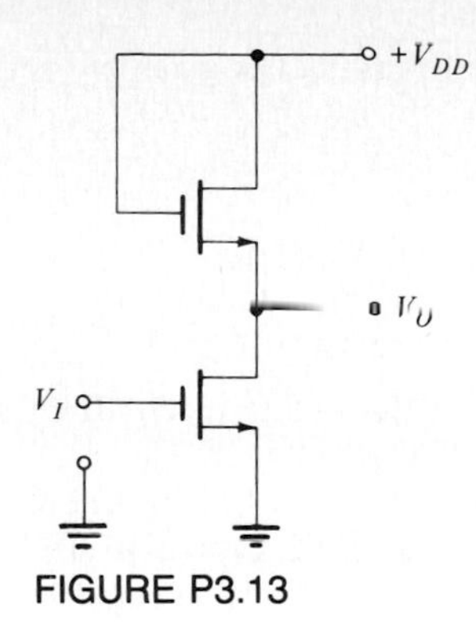

FIGURE P3.13

3.14 Repeat problem 3.13 for the circuit shown in Fig. P3.14 nonsaturated load if $V_{GG} = V_{DD} + V_T = 14$ V.

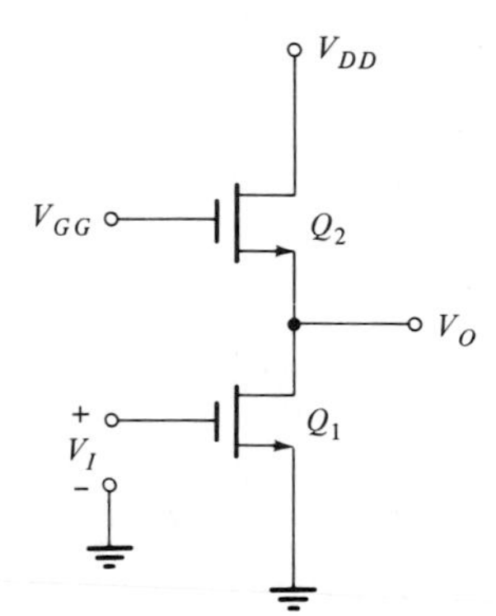

FIGURE P3.14

3.15 For the circuit in Fig. P3.15, calculate V_o:

(*a*) If the MOS parameters are identical, i.e., $K_n = 0.1$ mA/V^2 and $V_T = 2$ V

(*b*) If the threshold voltages are $V_{T1} = 2V_{T2} = 2$ V but $K_{n1} = K_{n2} = K_n$

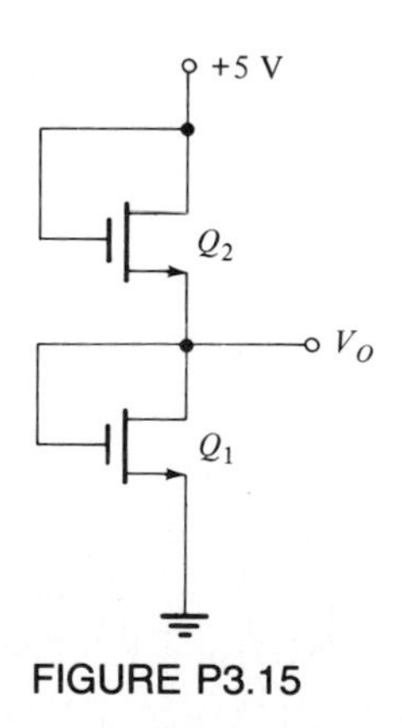

FIGURE P3.15

3.16 Show that the unity gain point of the CMOS in Fig. 3.23 is given by

$$V_I = \frac{V_I}{4} + \frac{3V_{DD}}{8}$$

3.17 For the CMOS inverter assume $V_{Tn} = V_{Tp} = V_T$ but $K_n \neq K_p$.
(*a*) Determine the value of V_I at which both Q_1 and Q_2 saturate.
(*b*) Determine the value of V_O at which each device saturates.

3.18 The parameters of a JFET are $I_{DSS} = 8$ mA, and $V_{PO} = 5.0$ V. The JFET stage is to be biased as in Fig. P3.18 at $I_D = 2$ mA, $V_{DS} = 8$ V.
(*a*) If the supply voltage V_{DD} is 20 V, determine the values of R_D and R_S.
(*b*) Find the voltage gain v_O/v_{in} at low frequencies if $g_d = 3 \times 10^{-4}$ mhos at the operating point.
(*c*) Determine the new operating point if the device parameters are changed to $I_{DSS} =$ 9 mA and $V_{PO} = 4.0$ V.

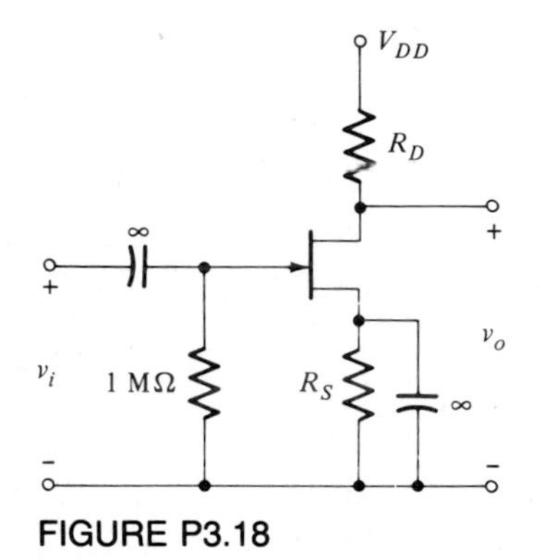

FIGURE P3.18

3.19 A feedback biasing scheme for MOSFETs is shown in Fig. P3.19 with $V_{DD} = 15$ V. If the device is an enhancement-type MOS with $K_n = 0.2 \times 10^{-3}$ A/V^2 and $V_T = 3$ V, design the circuit, i.e., determine R_D and R_G, so that the operating point is at $I_D =$ 5 mA. What is the new operating point if the device is replaced by another unit of the same type but with $K_n = 0.25 \times 10^{-3}$ A/V^2 and $V_T = 2.5$ V?

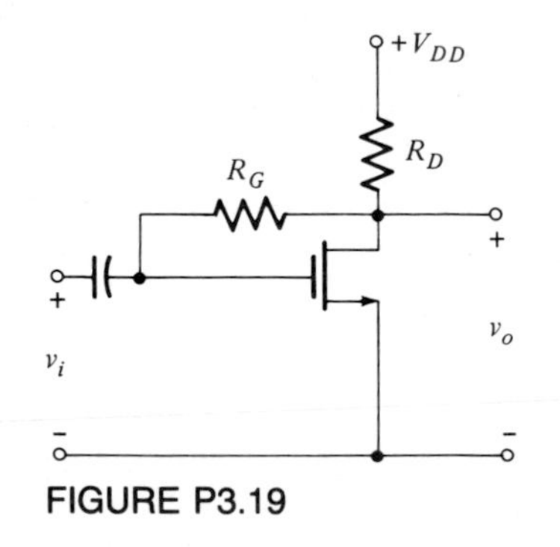

FIGURE P3.19

3.20 The *V-I* characteristic of a depletion-type MOSFET are shown in Fig. 3.10. Design the biasing circuit of Fig. 3.34 with $V_{DD} = 15$ V for an operating point $I_D = 4$ mA and $V_{DS} = 7$ V. Choose $R_{g1} = R_{g3} = 0.5$ MΩ, $R_S = 1$ kΩ.

3.21 A dual-gate FET can be used as a multiplier circuit as shown in Fig. P3.21. If a filter circuit is used at the output to filter out the dc, and the ω_1 and the ω_2 signals, show that $v_o = KV_1V_2 \sin \omega_1 t \sin \omega_2 t$. [*Hint:* Assume V_{DS} large enough so that the FET is in saturation and obtain a Taylor series expansion of $i_D = f(v_{G1S}, v_{G2S})$ about $V_{G1} = 0$, $V_{G2} = V_A$.]

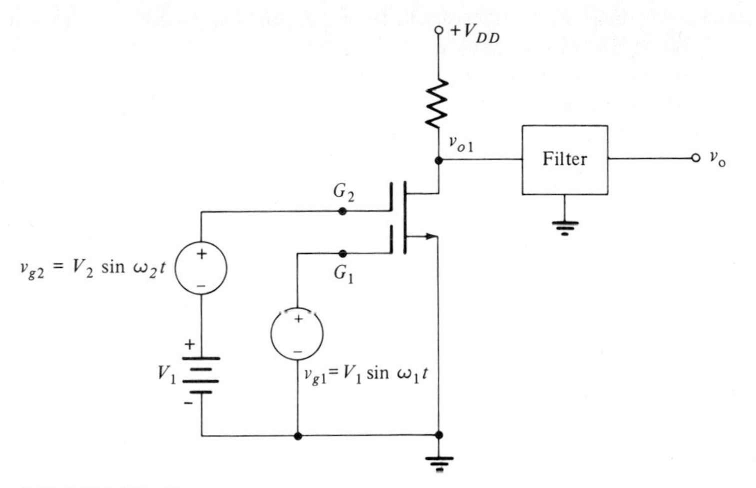

FIGURE P3.21

3.22 The biasing circuit shown in Fig. 3.36*b* is to be used for a depletion-type MOS amplifier. The desired operating point is $I_D = 4$ mA, $V_{DS} = 8$ V. The MOS transistor parameters are $I_{DSS} = 8$ mA, $V_{PO} = 5$ V. The supply voltage is $V_{DD} = 20$ V. For operating-point stability the value of R_S is chosen to be 2 kΩ.
(*a*) Determine the value of R_D.
(*b*) If $R_G = 1$ MΩ, determine the values of R_{g1} and R_{g2}.
(*c*) Find the new operating point if the MOS is replaced by another MOS of the same type but with $I_{DSS} = 10$ mA, $V_{PO} = 6$ V.

3.23 For the enhancement-type MOS amplifier stage shown in Fig. 3.36*b* the operating point is to be at $I_D = 4$ mA and $V_{DS} = 8$ V for $V_{DD} = 20$ V. The NMOS parameters are given as $K_n = 10^{-3}$ A/V^2, $V_T = 3$ V. At the operating point, the following data are given: at low frequencies $|y_{fs}| = 4000$ μmhos, $|y_{os}| = 0.1$ mmho, $C_{iss} = 12$ pF, $C_{rss} = 2$ pF, and $C_{oss} = 3$ pF.
(*a*) Determine the resistor values R_G and R_S, where $R_G = R_{g1} \parallel R_{g2}$.
(*b*) Show the incremental high-frequency model for the MOSFET and indicate all the parameters of the circuit model.

3.24 Determine the high-frequency circuit parameters of the π model for the JFET type JN51 given in Appendix E.

3.25 Determine the high-frequency circuit parameters for the enhancement MOS type MN82 given in Appendix E.

3.26 Estimate the maximum cutoff frequency of the following n-channel FET device if the channel length of the device is 1 μm and $V_{GS} - V_{GT}$ is $0.5V_{DD} = 5$ V. Bulk mobility of the electron $\mu_n = 1300\ \text{cm}^2/(\text{V} \cdot \text{s})$, and surface mobility of the electron $\mu_n = 800\ \text{cm}^2/(\text{V} \cdot \text{s})$.
(*a*) JFET
(*b*) MOS

3.27 Compare the maximum cutoff frequency of the n-channel JFET in problem 3.26 with the following:
(*a*) A p-channel JFET under the same conditions [$\mu_p = 500\ \text{cm}^2/(\text{V} \cdot \text{s})$].
(*b*) A MESFET with channel length 0.1 μm. The mobility of electron in GaAs is about 6 times that of Si.
(*c*) A BJT with base width 0.1 μm and $K = 5$.

CHAPTER

4

Single-Transistor (BJT and FET) Amplifiers

Amplification is one of the most important functions in electronic circuits. In this chapter we will discuss low-frequency small-signal amplification in a single-stage amplifier. Though amplifiers almost invariably comprise more than one stage, we will specifically consider a single-stage amplifier in this chapter to focus our attention on the salient characteristics of its performance; particularly the voltage gain, current gain, and input and output impedances, without concerning ourselves with interstage interactions and other problems that arise in a multistage amplifier. We shall include in our discussion both bipolar and field-effect transistors. The three possible configurations for each device are examined at low frequencies. The properties of each configuration are used in the analysis and design of multistage amplifiers in Chap. 5. Frequency response of the amplifiers is considered in Chaps. 7 and 8.

4.1 AMPLIFIER ANALYSIS IN GENERAL

Before we consider bipolar and field-effect transistors individually and in specific configurations, let us consider them in general as a two-port active network. Consider Fig. 4.1, where inside the box we may have a BJT or FET in any configuration. We consider low-frequency small-signal operation and models in this chapter. Hence all the elements are resistive (i.e., no reactive elements), and thus we use lowercase letters for the variables.

To begin the discussion in general terms, we consider the transistor as a two-port device, wherein a signal is fed in at an input port from a source having a source resistance R_S and is subsequently transmitted from the output port to a load resistance R_L. This arrangement, regardless of which terminal of the transistor is common to the input and output ports, is shown in Fig. 4.1. The two-port formalism discussed in Appendix C, which is applicable to any linear network, may now be used to obtain the various expressions for the network properties of the device. The input and output impedances and the current and voltage gain of the device as defined in Fig. 4.1 are

$$R_{\text{in}} = \frac{v_1}{i_1} \qquad R_o = \frac{v_2}{i_2} = \frac{v_2}{-i_o}$$
$$a_i = \frac{i_o}{i_1} \qquad a_v = \frac{v_2}{v_1} = \frac{v_o}{v_1} = \frac{i_o R_L}{i_1 R_{\text{in}}} \tag{4.1}$$

Note that in order to determine R_o, all independent sources are removed, the load is removed, the output port is excited with a voltage source v_2, and the value of the current i_2 is determined.

The power gain is given by

$$G_P = \frac{P_o}{P_i} = \frac{i_2^2 R_L}{i_1^2 R_{\text{in}}} = a_i^2 \frac{R_L}{R_{\text{in}}} = a_i a_v \tag{4.2}$$

Here a_i and a_v indicate the gains of the device, i.e., the current gain and voltage gains for *ideal current source* and *ideal voltage source,* respectively. Capital letters A_i and A_v are used for the actual gains such as v_o/v_{in} for finite values of R_S. In (4.1) and (4.2) any set of small-signal parameters may be used. A particular choice may be more convenient than the others because the parameters may be easily measured or may be obtained directly from the manufacturer's data sheets. For example, if the device

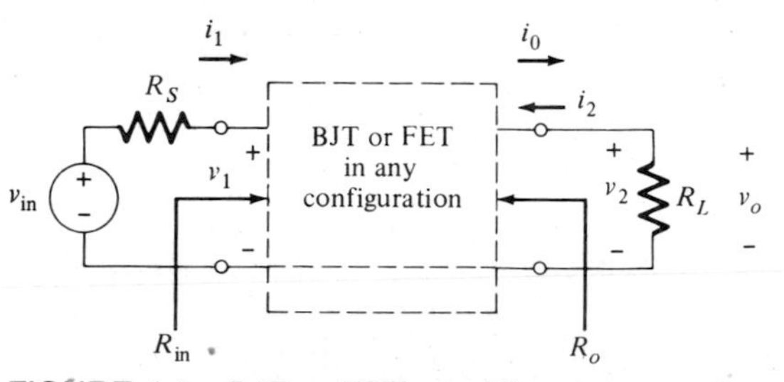

FIGURE 4.1 BJT or FET amplifier stage used as an active two-port circuit.

inside the box in Fig. 4.1 is a BJT, then we characterize it by the h parameters. If it is a FET, then the y parameters are most appropriate. Let us assume that we have a BJT; we then have the following matrix relation (note that we use lowercase variables since the circuits are purely resistive circuits under small-signal low-frequency conditions):

$$\begin{bmatrix} v_1 \\ i_2 \end{bmatrix} = \begin{bmatrix} h_i & h_r \\ h_f & h_o \end{bmatrix} \begin{bmatrix} i_1 \\ v_2 \end{bmatrix} \tag{4.3}$$

where an additional subscript (e, b, or c) may be used if the common terminal of the bipolar transistor is known (e.g., if the device is in the CE configuration, we would use h_{ie}, h_{re}, etc.). In FETs we may use the subscript s, g, or d, but for these devices, since they are *voltage-controlled devices,* either the y or g parameters are normally used. The y parameters are the preferred choice for FETs.

From (4.3) and the relations for the source and load circuits

$$-i_2 = G_L V_2 \qquad v_1 = v_{in} - i_1 R_S \tag{4.4}$$

we can obtain the network properties of the two-port device as discussed in Appendix C. Hence the signal transmission characteristics from a source to a load for the device in a given connection may be determined.

EXAMPLE 4.1

Let us assume that the transistor in Fig. 4.1 is in the CE connection. Suppose that at $I_C = 1$ mA, $V_{CE} = 5$ V, and the h parameters are given as follows:

$$h_{ie} = 1.4 \times 10^3\ \Omega \qquad h_{fe} = 50$$

$$h_{re} = 10^{-4} \qquad h_{oe} = 2 \times 10^{-5} \text{ mhos}$$

Let the source and load resistances be $R_L = R_S = 10^3\ \Omega$. We shall determine the input impedance R_{in}, output impedance R_o, current gain A_i, voltage gain A_v, and the power gain G_P of the circuit.

From the results derived in Appendix C and summarized in Table C.2, we have

$$R_{in} = h_{ie} - \frac{h_{re}h_{fe}}{h_{oe} + G_L} = 1.4 \times 10^3 - \frac{5 \times 10^{-3}}{2 \times 10^{-5} + 10^{-3}} \simeq 1.4 \text{ k}\Omega$$

$$R_o^{-1} = h_{oe} - \frac{h_{re}h_{fe}}{h_{ie} + R_S} = 2 \times 10^{-5} - \frac{5 \times 10^{-3}}{1.4 \times 10^3 + 10^3}$$

$$\simeq 1.8 \times 10^{-5} \text{ mho}$$

$$a_i = \frac{i_o}{i_1} = -\frac{i_2}{i_1} = -\frac{h_{fe}}{1 + h_{oe}R_L} = -\frac{50}{1 + 2 \times 10^{-2}} \simeq -49$$

Note that if we define $i_s = v_{in}/R_S$, then

$$A_i = \frac{i_o}{i_s} = -\frac{i_o}{i_1}\left(\frac{R_S}{R_S + R_{in}}\right) = -49\left(\frac{1}{1 + 1.4}\right) = -20.4$$

Similarly,

$$a_v = \frac{v_2}{v_1} = \frac{h_{fe}}{h_{re}h_{fe} - h_{ie}(G_L + h_{oe})}$$

$$= \frac{50}{10^{-4}(50) - 1.4 \times 10^3(10^{-3} + 2 \times 10^{-5})} \simeq -35.2$$

and the voltage gain v_o/v_{in} is given by

$$A_v = \frac{v_o}{v_{in}} = \frac{v_2}{v_1}\left(\frac{R_{in}}{R_{in} + R_S}\right) = -35.2\left(\frac{1.4}{1.4 + 1}\right) = -20.5$$

The minus sign signifies that the output signal is inverted with respect to the input. Finally, the power gain, since all parameters are real, is given by

$$G_P = \frac{P_o}{P_i} = -\frac{v_2 i_2}{v_1 i_1} = \left(\frac{v_2}{v_1}\right)^2 \frac{G_L}{G_{in}} = 35.2^2(10^{-3})(1.4) \times 10^3 = 1.73 \times 10^3$$

or

$$G_P = 10 \log \frac{P_o}{P_i} = 32.4 \text{ dB}$$

Note that in the above, once $R_{in}(= G_{in}^{-1})$ and i_o/i_1 are determined, we can also obtain from (4.2)

$$G_P = \frac{i_o}{i_1}\left(\frac{v_2}{v_1}\right) = \left(\frac{i_o}{i_1}\right)^2\left(\frac{R_L}{R_{in}}\right) = (-49)^2\frac{1}{1.4} = 1.72 \times 10^3$$

EXERCISE 4.1

The y parameters of a FET in the CS configuration are given as follows (all in mmho): $y_{is} = 0.01$, $y_{fs} = 10$, $y_{rs} = 10^{-3}$, $y_{os} = 0.1$. If $R_S = R_L = 1\ \text{k}\Omega$, determine the input impedance, the output impedance, and the voltage gain of the circuit.

Ans $R_{in} = 1\ \text{M}\Omega$, $R_o = 11.1\ \text{k}\Omega$, $A_v = -9.1$

We shall next consider BJT and FET in each of the three configurations individually in order to gain insight into their circuit properties.

4.2 THE COMMON-BASE (CB) CONFIGURATION

The transistor stage in the CB configuration is shown with the single-battery biasing circuit in Fig. 4.2. Capacitor C is a very large bypass capacitor, and its reactance at midband is essentially zero. Since the large value of the capacitor is of no concern in midband calculations, we label it $C \to \infty$, to indicate a short circuit at midband frequency.

To find the various properties of the transistor in the CB configuration, we shall use the small-signal equivalent for the transistor. We have already developed two models for bipolar transistor: the T model and the hybrid-π model. Let us consider both for the sake of completeness and comparison. We shall see that the simplified π model is most convenient for low-frequency gain and impedance calculations. The formula approach works better for computers.

A. Analysis via the T Model

The small-signal circuit of the CB stage using the T model corresponding to Fig. 4.2 is shown in Fig. 4.3. Recall that for small-signal ac calculations the supply voltage terminals are at ground potential; i.e., the batteries are *not* included as they only determine the Q point.

In order to find the network properties of the circuit, several approaches may be taken. Recall that bipolar transistors are current-controlled devices and, as discussed in Appendix C, the h or z parameters are most convenient. We may (1) find the h parameters of the CB stage directly or in terms of the CE parameters and then use the

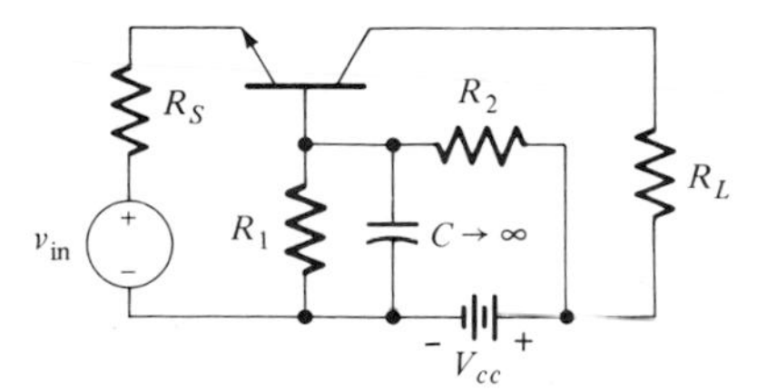

FIGURE 4.2 Single-stage CB transistor amplifier.

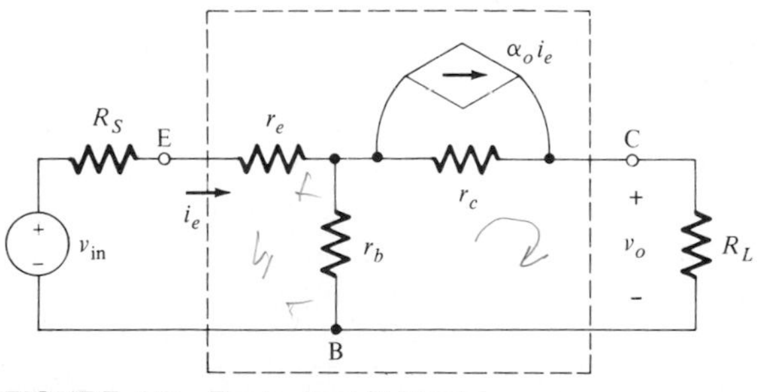

FIGURE 4.3 Equivalent T Model.

formulas in Appendix C to determine the midband properties or (2) use the z parameters directly and then use the formulas of Table C.2 in terms of the z parameters. By method (2), the z parameters of the CB configuration are

$$z_{12} = r_b \qquad z_{11} = r_e + r_b$$
$$z_{22} = r_c + r_b \qquad z_{21} = r_b + \alpha_0 r_c \tag{4.5}$$

$V_1 = Z_{11} I_1 + Z_{12} I_2$
$V_2 = Z_{21} I_1 + Z_{22} I_2$

The various network properties of the CB stage are then determined directly from (4.5) and the results of Table C.2:

$$Z_{\text{in}} = R_{\text{in}} = z_{11} - \frac{z_{12} z_{21}}{z_{22} + Z_L} = r_e + r_b\left(\frac{r_c(1 - \alpha_0) + R_L}{r_b + r_c + R_L}\right) \tag{4.6a}$$

$$\simeq r_e + r_b(1 - \alpha_0) \qquad (1 - \alpha_0)r_c \gg R_L \tag{4.6b}$$

$$Z_o = R_o = z_{11} - \left(\frac{z_{12} z_{21}}{z_{22} + Z_L}\right) = r_c - r_b \frac{\alpha_0 r_c - r_e - R_S}{r_e + r_b + R_S} \tag{4.7a}$$

$$\simeq r_c \qquad \text{for} \quad r_c \gg R_S \gg r_b \tag{4.7b}$$

$$\frac{i_o}{i_1} = \frac{z_{21}}{z_{22} + Z_L} = \frac{\alpha_0 r_c + r_b}{r_c + r_b + R_L} \simeq \alpha_0 \qquad R_L \ll r_c \tag{4.8}$$

$$\frac{v_o}{v_1} = \frac{z_{21} Z_L}{-z_{12} z_{21} + z_{11}(Z_L + z_{22})}$$

$$= \frac{R_L(r_b + \alpha_0 r_c)}{r_b[(1 - \alpha_0)r_c + r_e + R_L] + r_e(r_c + R_L)} \tag{4.9a}$$

$$\simeq \frac{\alpha_0 R_L}{r_e + r_b(1 - \alpha_0)} \qquad R_L \ll r_c \tag{4.9b}$$

Equation (4.9b) can be put in a more convenient form if we make use of

$$\beta_0 = \frac{\alpha_0}{1 - \alpha_0} \gg 1 \qquad \text{and} \qquad r_e = \frac{1}{g_m}$$

to get

$$\frac{v_o}{v_1} = \frac{\beta_0 R_L}{(\beta_0 + 1)r_e + r_b} \simeq \frac{R_L}{r_e} = g_m R_L \qquad \text{for } \beta_0 r_e \gg r_b \tag{4.9c}$$

It can be seen from (4.8) and (4.9) that the current gain of the device (i.e., the ratio of the output current to the input current for a current-source input) of a common-base stage is α_0, which is almost unity, and that the voltage gain of the device (i.e., the ratio of the output voltage to the input voltage for a voltage-source input) is approximately equal to $g_m R_L$. Note that for finite values of R, the voltage gain is

$$A_v = \frac{v_{\text{out}}}{v_{\text{in}}} = \left(\frac{v_o}{v_1}\right)\left(\frac{v_1}{v_{\text{in}}}\right) = \left(\frac{v_o}{v_1}\right)\left(\frac{R_{\text{in}}}{R_{\text{in}} + R_S}\right) \tag{4.10}$$

and the current gain is

$$A_i = \frac{i_o}{i_s} = \left(\frac{i_o}{i_1}\right)\left(\frac{i_1}{i_s}\right) = \left(\frac{i_o}{i_1}\right)\left(\frac{R_S}{R_S + R_{in}}\right) \tag{4.11}$$

where $i_s = v_{in}/R_S$.

The results in the form of (4.8) and (4.9) are quite useful for multistage cascaded amplifiers, as we shall demonstrate in Chap. 5.

B. Analysis via the Simplified π Model

The hybrid model of Fig. 2.23, which shall henceforth be referred to as the π model (see Fig. 2.27), can be used directly to obtain the circuit properties of the CB configuration. However it is more convenient to obtain the h parameters for the CB configuration from either the T model or the π model. From the T model, since the z parameters are known in (4.5), and the relationship between the h and z parameters (obtained in Appendix C) we have

$$h_{ib} = h_{11} = \frac{z_{11}z_{22} - z_{12}z_{21}}{z_{22}} = z_{11} - \frac{z_{12}z_{21}}{z_{22}}$$

$$= r_e + r_b - \frac{r_b(r_b + \alpha_0 r_c)}{r_c + r_b} \simeq r_e + r_b(1 - \alpha_0) \tag{4.12a}$$

$$h_{rb} = h_{12} = \frac{z_{12}}{z_{22}} = \frac{r_b}{r_b + r_c} \simeq \frac{r_b}{r_c} \tag{4.12b}$$

$$h_{fb} = h_{21} = -\frac{z_{21}}{z_{22}} = -\frac{r_b + \alpha_0 r_c}{r_c + r_b} \simeq -\alpha_0 \tag{4.12c}$$

$$h_{ob} = h_{22} = \frac{1}{z_{22}} = \frac{1}{r_c + r_b} \simeq \frac{1}{r_c} \tag{4.12d}$$

Typical T-parameter element values are listed below:

$$\alpha_0 = 0.98 \qquad r_e = \frac{25 \text{ mV}}{I_e \text{ mA}} = 25\ \Omega \qquad \text{at } I_e = 1 \text{ mA}$$

$$r_b = 200\ \Omega \qquad r_c = 2\ \text{M}\Omega \tag{4.13}$$

From the typical values we see that the π model of the CB configuration shown in Fig. 4.4*a* can be simplified to the one shown in Fig. 4.4*b*, where $h_{rb}(\simeq 10^{-4})$ and h_{ob}^{-1} ($\simeq$2 MΩ) are ignored. The model in Fig. 4.4*b* is referred to as the *simplified π model for the CB configuration.*

The simplified π model is very convenient to use in circuit analysis. Note that we can obtain all the above circuit properties directly without resorting to any formulas. This approach is recommended for most situations. From the circuit by inspection:

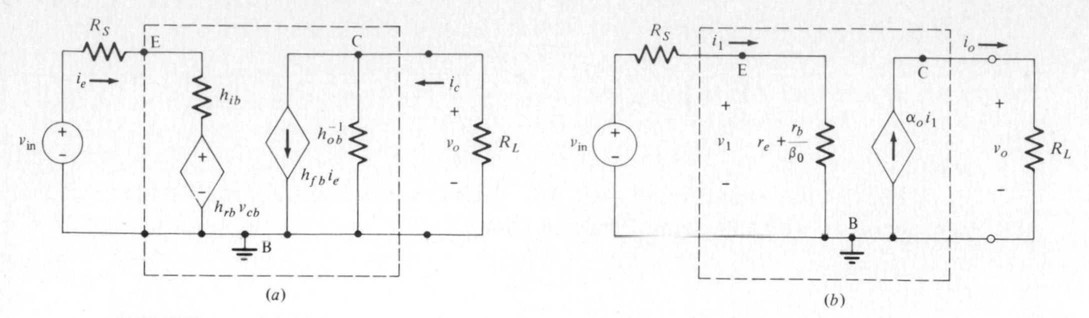

FIGURE 4.4 (*a*) Hybrid-π model. (*b*) Simplified model for the CB configuration.

$$R_{in} = r_e + \frac{r_b}{\beta_0} \qquad R_o \simeq \infty \tag{4.14}$$

$$\frac{i_o}{i_1} = \alpha_0 \tag{4.15}$$

Also

$$v_o = i_o R_L = (\alpha_0 i_1) R_L \qquad v_1 = i_1\left(r_e + \frac{r_b}{\beta_0}\right)$$

Hence

$$\frac{v_o}{v_1} = \frac{\alpha_0 R_L}{r_e + r_b/\beta_0} \simeq g_m R_L \tag{4.16}$$

EXAMPLE 4.2

For a BJT in the CB configuration determine R_i, R_o, A_i, and A_v. Assume the typical parameters in (4.13) and let $R_S = R_L = 1\ \text{k}\Omega$.

$$R_{in} = 25 + \frac{200}{50} = 29\ \Omega \qquad R_o = 2 \times 10^6 \to \infty$$

$$a_i \simeq \frac{i_o}{i_1} = 0.98 \qquad A_v = \frac{v_o}{v_{in}} \simeq \frac{i_o R_L}{i_o R_S} = 0.98$$

For the circuit $v_o/v_{in} \simeq 10^3/29 = 34$.

Note that the input impedance of a CB stage is *very low* and the output impedance is *very high*. As a result of this *tremendous mismatch* between the input and the output circuits, the voltage gain and the current gain of *cascaded CB* stages are less than unity, and no power gain can be obtained unless a transformer is used between

TABLE 4.1 Network Properties of Transistor Configurations for a Typical Transistor*

	CB	CE	CC (EF)
R_{in} (ohms)	29	1450	50k
R_o (ohms)	2M	10k	49
$a_i = \dfrac{i_o}{i_1}$	0.98	−50	50
$a_v = \dfrac{v_o}{v_1}$	34.0	−34	0.98
$G_P = \dfrac{v_o i_o}{v_1 i_1}$	33.0	1700	49
$A_v = \dfrac{v_o}{v_{in}} = A_i$ †	0.98	20	0.98

*For transistor parameters see (4.13).
†Since $R_S = R_L = 1\ \text{k}\Omega$.

the stages. This fact is readily seen in the above for $R_L = R_S$ and then finding both A_i and A_v are approximately equal to α_0, and hence $G_P \simeq \alpha_0^2 < 1$.

The CB stage is used to change the impedance levels in a circuit and to provide isolation for CE stages in tuned amplifiers. It is also used as a noninverting amplifier with a voltage gain. Sometimes it is used as a current-controlled current-source element. The comparison of the midband properties of a typical CB stage with those of the other configurations is shown in Table 4.1.

4.3 THE COMMON-EMITTER (CE) CONFIGURATION

A CE amplifier stage with the dc biasing circuitry is shown in Fig. 4.5. Biasing of the transistor has already been considered in Chap. 2. We shall assume then that the values of R_L, R_E, R_{b1}, and R_{b2} are already established. It is also assumed that the values of

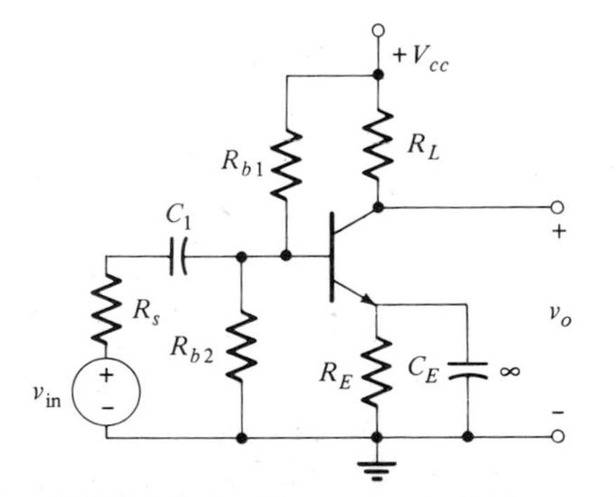

FIGURE 4.5 Single-stage CE transistor amplifier.

the capacitors are so large that they are essentially short circuits (i.e., the bypass capacitors are assumed to be infinite). The effect of these capacitors on the very low frequency response of the amplifier is considered in Chap. 7.

A. Analysis via the T Model

The equivalent circuit for Fig. 4.5 using a T model for the transistor is shown in Fig. 4.6. Note that $R_b = R_{b1} \parallel R_{b2}$ is connected from the base terminal to ground since the supply voltage terminals are at ground potential for ac calculations.

The z parameter of the CE configuration are

$$\begin{aligned} z_{12} &= r_e & z_{11} &= r_b + r_e \\ z_{22} &= r_c(1 - \alpha_0) + r_e & z_{21} &= -\alpha_0 r_c + r_e \end{aligned} \tag{4.17}$$

The circuit properties are then determined as in before, namely,

$$Z_{\text{in}} = R_{\text{in}} = r_b + r_c \frac{r_c + R_L}{r_e(1 - \alpha_0) + R_L + r_e} \tag{4.18a}$$

$$\simeq r_b + \frac{r_e}{1 - \alpha_0} \qquad R_L \ll r_c(1 - \alpha_0) \tag{4.18b}$$

$$Z_o = R_o = r_c(1 - \alpha_0) + r_e \frac{R_S + r_b + \alpha_0 r_c}{R_S + r_b + r_e} \tag{4.19a}$$

$$\simeq r_c\left[(1 - \alpha_0) + \frac{r_e}{R_S}\right] \qquad r_b \ll R_S \ll r_c \tag{4.19b}$$

$$\frac{i_o}{i_1} = -\frac{\alpha_0 r_c - r_e}{r_c(1 - \alpha_0) + r_e + R_L} \tag{4.20a}$$

$$\simeq -\frac{\alpha_0}{1 - \alpha_0} = -\beta_0 \tag{4.20b}$$

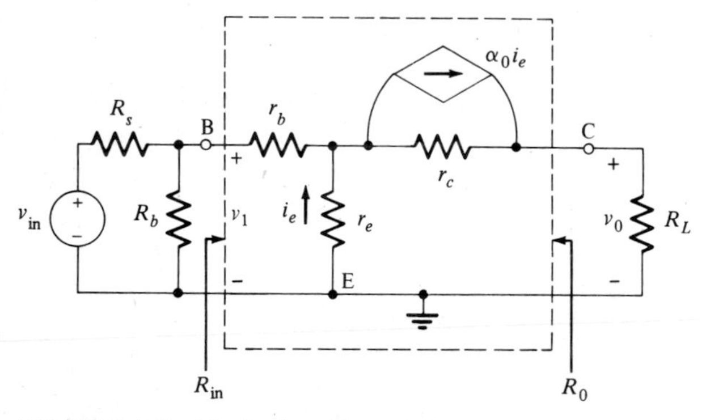

FIGURE 4.6 Equivalent T model.

$$\frac{v_o}{v_1} = -\frac{R_L(\alpha_0 r_c - r_e)}{r_b[r_c(1 - \alpha_0) + r_e + R_L] + r_e(r_c + R_L)} \tag{4.21a}$$

$$\simeq -\frac{\alpha_0 R_L}{r_e + r_b(1 - \alpha_0)} \qquad r_c(1 - \alpha_0) \gg R_L \tag{4.21b}$$

B. Analysis via the Simplified π Model

It is much easier and preferable to use the simplified π model of Fig. 4.7 for circuit analysis, especially in the CE configuration. The hybrid parameters are almost always given by the manufacturer (see Appendix E). One need not resort to formulas and can directly perform the analysis and obtain the desired information. We shall use and encourage this approach. From the circuit in Fig. 4.7, the input resistance is

$$R_{\text{in}} = r_b + R_i = r_b + \beta_0 r_e \tag{4.22}$$

The output resistance of the model is infinite ($R_o = \infty$) since the collector-current generator is ideal. The current gain a_i is readily determined as follows:

$$v' = R_i i_1 = \beta_0 r_e i_1$$

$$i_o = -g_m v' = -\frac{1}{r_e} v' = -\frac{1}{r_e} \beta_0 r_e i_1$$

hence

$$a_i = \frac{i_o}{i_1} = -\beta_0 \tag{4.23}$$

The voltage gain of the circuit is readily found from $a_i R_L / R_{\text{in}}$. Or directly from the circuit

$$v_o = i_o R_L = (-g_m v')R_L$$

$$v' = \frac{R_i}{R_i + r_b} v_1$$

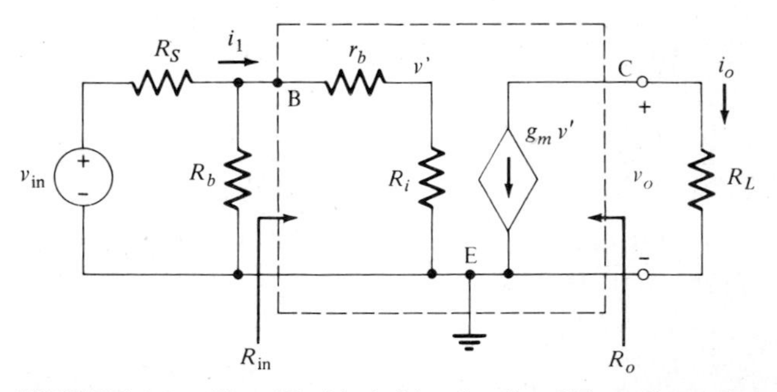

FIGURE 4.7 Simplified hybrid π for the CE configuration.

hence

$$a_v = \frac{v_o}{v_1} = -\frac{g_m R_i R_L}{R_i + r_b} = -\frac{\beta_0 R_L}{\beta_0 r_e + r_b} \tag{4.24}$$

Note that part of the current from the source ($i_s = v_{in}/R_S$) is lost in the biasing resistance R_b. Usually this resistance is much larger than the input resistance of the transistor, i.e., $R_b \gg R_{in}$. However, if this approximation does not hold, we have at the input the situation shown in Fig. 4.7. The current through the transistor then is given by the current divider rule:

$$i_1 = i_s\left(\frac{R_s'}{R_s' + R_{in}}\right) \tag{4.25}$$

where $R_s' = R_S \parallel R_b$ and $R_{in} = r_b + \beta_0 r_e$. The overall voltage gain v_o/v_{in} is found from the circuit as follows:

$$v_o = -g_m v' R_L$$

$$v' = R_i i_1 = \beta_0 r_e \left(\frac{V_{in}}{R_S}\right)\left(\frac{R_s'}{R_s' + R_{in}}\right)$$

Hence

$$A_v = \frac{v_o}{v_{in}} = -\frac{\beta_0 R_L R_s'}{R_S(R_s' + R_{in})} \tag{4.26}$$

EXAMPLE 4.3

For the CE amplifier stage shown in Fig. 4.5, determine the input impedance R_i, the output impedance R_o, the voltage gain v_o/v_{in}, and the current gain i_o/i_s, where $i_s = v_{in}/R_S$. The transistor parameters are $\beta_0 = 50$, $r_b = 200\ \Omega$, $r_e = 25\ \Omega$ (at $I_e = 1$ mA), $R_L = R_S = 1\ \text{k}\Omega$, $R_b = 15\ \text{k}\Omega$.

From the circuit of Fig. 4.7 we have by inspection $R_{in} = r_b + \beta_0 r_e = 200 + 50(25) = 1.45\ \text{k}\Omega$, $R_o = \infty$ [the actual value from (4.19*b*) is $R_o = 40\ \text{k}\Omega$ obtained from the T model]. Note that $R_b \gg R_{in}$ and its effect can be ignored in calculations of current and voltage gains.

$$a_i = \frac{i_o}{i_1} = -\beta_0 = -50$$

$$a_v = \frac{v_o}{v_1} = -\frac{50(10^3)}{50(25) + 200} = -34.5$$

$$A_v = \frac{v_o}{v_{in}} = -\frac{50(10^3)(10^3)}{10^3(10^3 + 1.45 \times 10^3)} \simeq -20.4$$

$$A_i = \frac{i_o}{i_s} = \frac{v_o/R_L}{v_{in}/R_S} = A_v = -20.4$$

From the results obtained in above it is seen that for CE configuration, the current gain is $-\beta_0$ for a current-source input, and the voltage gain is roughly $-g_m R_L$ for a voltage-source input. Note that in this configuration, both current and voltage gains of greater than unity can be achieved. The input impedance is medium and is essentially independent of the load impedance if R_L is not too high. The output impedance is *very* high and fairly independent of the source impedance. The mismatch of the input and output impedances of cascaded CE stages is the least of the three configurations (see Table 4.1), and power amplification can be obtained in this configuration without using transformers. This configuration, therefore, enjoys wide application and is the principal configuration in amplifier design. The midband properties of a CE stage as compared to the other configurations for a typical transistor are shown in Table 4.1.

EXERCISE 4.2

The T-parameter element values of a BJT at room temperature and at ($I_C = 10$ mA, $V_{CE} = 5$ V) are given as follows: $\alpha_0 = 0.99$, $r_b = 100\ \Omega$, $r_c = 3\ \text{M}\Omega$. Determine the hybrid-π model parameters of the transistor.

Ans $r_b = 100\ \Omega$, $\beta_0 = 100$, $r_e = g_m^{-1} = 2.5\ \Omega$, $R_i = 250\ \Omega$, $r_o = 30\ \text{k}\Omega$

EXERCISE 4.3

For the circuit shown in Fig. 4.5, $R_S = 10\ \text{k}\Omega$, $R_L = 1\ \text{k}\Omega$, and $R_{b1} \| R_{b2} = 20\ \text{k}\Omega$. Determine the voltage gain and the current gain of the circuit if the BJT parameters are given as in the above.

Ans $A_v = -9.5$, $A_i = -95$

4.4 THE COMMON-COLLECTOR (CC) CONFIGURATION

The transistor stage in the CC configuration with the biasing circuitry is shown in Fig. 4.8. As in the previous case we assume that C is large enough to be considered a short circuit and that R_b is also very large, i.e., $R_b \gg R_S$.

A. Analysis via the T Model

The equivalent circuit for Fig. 4.8, using the T model is shown in Fig. 4.9. The various network properties of this circuit are obtained in exactly the same manner as before; that is, we find the z or h parameters in terms of the T model (see problem 4.2) and then use Table C.2. The expressions are

$$R_{\text{in}} = r_b + r_c \frac{R_L + r_e}{(1 - \alpha_0)r_c + R_L + r_e} \tag{4.27a}$$

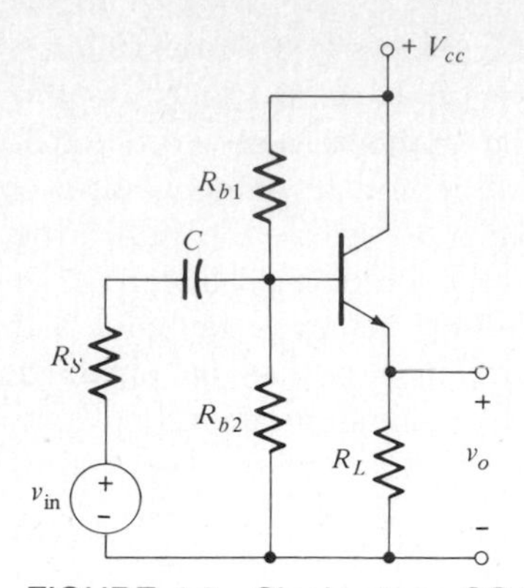

FIGURE 4.8 Single-stage CC transistor amplifier.

$$\simeq r_b + \frac{R_L + r_e}{1 - \alpha_0} \simeq \beta_0(R_L + r_e) \qquad (1 - \alpha_0)r_c \gg R_L \qquad \beta_0 \gg 1 \tag{4.27b}$$

$$R_o = r_e + \frac{(1 - \alpha_0)r_c(R_S + r_b)}{R_S + r_b + r_c} \tag{4.28a}$$

$$\simeq r_e + (R_S + r_b)(1 - \alpha_0) \qquad r_c \gg R_S + r_b \tag{4.28b}$$

$$a_i = \frac{i_o}{i_1} = \frac{r_c}{r_c(1 - \alpha_0) + R_L + r_e} \tag{4.29a}$$

$$\simeq \frac{1}{1 - \alpha_0} \simeq \beta_0 \qquad r_c(1 - \alpha_0) \gg (R_L + r_e) \tag{4.29b}$$

$$a_v = \frac{v_o}{v_1} = \frac{r_c R_L}{r_b[(1 - \alpha_0)r_c + R_L + r_e] + r_c(R_L + r_e)} \tag{4.30a}$$

$$\simeq \frac{R_L}{R_L + r_e} \simeq 1 \qquad R_L + r_e \gg r_b(1 - \alpha_0) \tag{4.30b}$$

Note that the input impedance is high and the output impedance *very* low. The current gain is about β_0, and the voltage gain is very close to unity. This connection

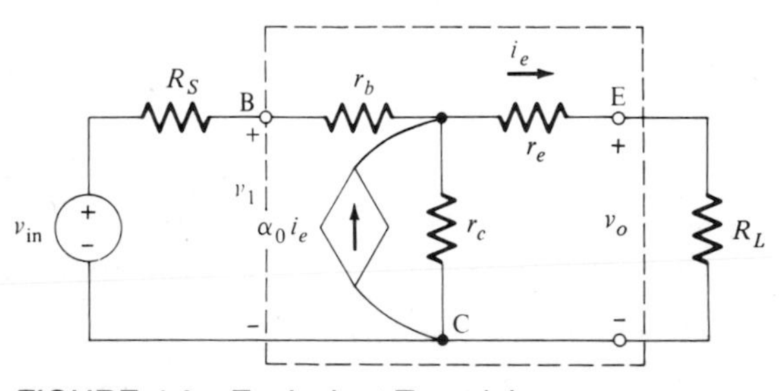

FIGURE 4.9 Equivalent T model.

is commonly called the *emitter follower*. The name emitter follower is used for the CC configuration because the output voltage is taken at the emitter, which follows the input voltage with almost no loss of signal, i.e., a voltage gain of unity.

B. Analysis via the Simplified π Model

The simplified CE model of Fig. 4.7 can be used for the transistor in the CC configuration by properly reorienting it as shown in Fig. 4.10. Note that $r_b + \beta_0 r_e = h_{ie}$ and $g_m v' = g_m i_b(\beta_0 r_c) = \beta_0 i_b$; hence the circuits in Fig. 4.10 and Fig. 4.6*b* are completely equivalent. We can now deal directly with the circuits without resorting to any formulas. From the circuit in Fig. 4.10 at the input of the transistor circuit we have

$$v_1 = i_b[h_{ie} + (\beta_0 + 1)R_L] = i_b[r_b + \beta_0 r_e + (\beta_0 + 1)R_L]$$

Hence

$$\begin{aligned} R_{\text{in}} &= \frac{v_1}{i_b} = r_b + \beta_0 r_e + (\beta_0 + 1)R_L \\ &\simeq \beta_0(r_e + R_L) \qquad \beta_0 \gg 1 \end{aligned} \tag{4.31}$$

At the output of the circuit we have

$$i_2 = -i_e = -(\beta_0 + 1)i_b$$

and

$$v_o = -i_b(R_S + h_{ie}) = -i_b(R_S + \beta_0 r_e + r_b)$$

Hence

$$R_o = \frac{v_o}{i_2} = \frac{R_S + h_{ie}}{\beta_0 + 1} \simeq r_e + \frac{R_S + r_b}{\beta_0} \qquad \text{for } \beta_0 \gg 1 \tag{4.32}$$

The current gain is simply i_e/i_b. From above we have

$$a_i = \frac{i_o}{i_b} = \beta_0 + 1 \simeq \beta_0 \qquad \text{for } \beta_0 \gg 1 \tag{4.33}$$

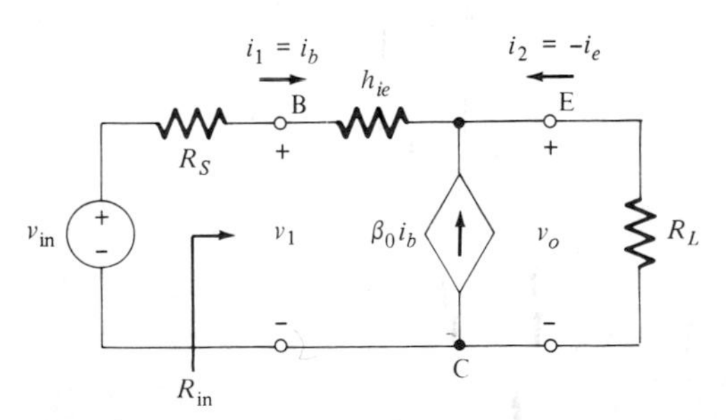

FIGURE 4.10 Simplified π model for the CC configuration.

The voltage gain is determined from

$$A_v = \frac{v_o}{v_{in}} = \frac{(\beta_0 + 1)R_L}{R_S + R_i + (\beta_0 + 1)R_L} \simeq \frac{R_L}{R_L + r_e} \simeq 1 \tag{4.34}$$

EXAMPLE 4.4

For the CC amplifier stage shown in Fig. 4.8, determine the input impedance, the output impedance, the voltage gain, and the current gain. The transistor parameters are as in Example 4.3.

$$R_{in} = \beta_0(r_e + R_L) = 50(25 + 10^3) = 50\ \text{k}\Omega$$

$$R_o = r_e + \frac{R_S + r_b}{\beta_0} = 25 + \frac{10^3 + 200}{50} = 49\ \Omega$$

$$a_i = \frac{i_e}{i_b} = \beta_0 = 50$$

Note that

$$\frac{i_o}{i_s} \simeq \frac{\beta_0 \times 10^3}{(1 + 50) \times 10^3} \simeq 0.98$$

$$\frac{v_o}{v_{in}} = \frac{10^3}{10^3 + 25} = 0.98$$

Note that i_s is defined as $i_s = v_{in}/R_S$.

From the above expressions and numerical values it is seen that for the emitter follower (CC configuration) the current gain, for a current-source input, is almost equal to β_0 and the voltage gain, for a voltage-source input, is almost unity. For the CC configuration, as we shall see, the input resistance is the highest ($R_{in} \simeq \beta_0 R_L$) and the output resistance the lowest ($R_o \simeq R_S/\beta_0$) among the three configurations. Because of these properties, the emitter follower is often used as a buffer stage. Since the output impedance of a CC stage is very low, it can also be used to drive a variable-load impedance. The midband properties of a CC configuration for a typical transistor as compared to the other configurations are summarized in Table 4.1.

EXERCISE 4.4

Determine the approximate input and output impedances of the BJT in Example 4.1 used in the CC configuration. Assume the same terminations.

Ans $R_{in} = 50\ \text{k}\Omega$, $R_o = 48\ \Omega$

4.5 COMPARISON OF BIPOLAR TRANSISTOR CONFIGURATIONS

We shall now compare the results of bipolar transistors in the three configurations. First the numerical results for a typical transistor will be compared and then the basic properties shall be summarized. Typical circuit element values for the equivalent-T model is given in (4.13). From the above, the derived quantities for the hybrid model are

$$h_{fe} = \beta_0 \simeq \frac{1}{1-\alpha_0} = \frac{1}{0.02} \simeq 50$$

$$h_{ie} = r_b + \beta_0 r_e = r_b + R_i = 1.45\ \text{k}\Omega \tag{4.35}$$

$$g_m = \frac{1}{r_e} = 0.04\ \text{mho} \qquad h_{oe}^{-1} = \frac{r_c}{\beta_0} = 40\ \text{k}\Omega \qquad h_{re} \simeq 0$$

The various network properties of the transistor described by (4.13) or (4.35) and embedded in a network as in Fig. 4.1 are shown in Table 4.1. The numerical values were obtained in the example for each configuration. For the sake of reference and convenience the simplified expressions for the three configurations are listed in Table 4.2. Since the manufacturers almost always provide typical values of h parameters in the CE configuration, we also listed the circuit properties of the various configurations in terms of these in Table 4.3. Let us examine the various quantities:

Input resistance: The input resistance of each configuration is different with CC having the highest and CB the lowest input resistance. The CE configuration has a medium input resistance.

Output resistance: The output resistance of CB and CE are very high and often can be assumed to be infinite. The CC output resistance is very low.

Current gain: The current gain of CE and CC are both high and equal to β_0, and for CB it is low and almost unity. Note that a phase reversal occurs in a CE stage.

Voltage gain: The voltage gains for CB and CE are high and of the same value,

TABLE 4.2 Small-Signal Network Properties of Bipolar Transistors*

	CB	CE	CC (EF)
R_{in}	$r_e + \dfrac{r_b}{\beta_0}$	$r_b + \beta_0 r_e$	$\beta_0(R_L + r_e)$
R_o	$r_c \to \infty$	$\dfrac{r_c}{\beta_0} \to \infty$	$r_e + \dfrac{R_S + r_b}{\beta_0}$
a_i	α_0	$-\beta_0$	β_0
a_v	$\dfrac{\alpha_0 R_L}{r_e + r_b/\beta_0}$	$-\dfrac{\alpha_0 R_L}{r_e + r_b/\beta_0}$	$\dfrac{R_L}{R_L + r_e}$

*$\beta_0 \gg 1$, $R_L \ll r_c$

TABLE 4.3 Approximate Relation* of Circuit Parameters for the Three Basic Configurations

	CB	CE	CC (EF)
R_{in}	$\frac{h_{ie}}{h_{fe}} = h_{ib}$	h_{ie}	$h_{ie} + h_{fe}R_L$
R_o	$h_{fe}h_{oe}^{-1} = h_{ob}^{-1}$	$\frac{1}{h_{oe}} > 10\text{ k}\Omega$	$\frac{h_{ie} + R_S}{h_{fe}}$
a_i	$\simeq 1 = -h_{fb}$	$-h_{fe}$	h_{fe}
a_v	$\frac{R_L}{h_{ib}}$	$\frac{h_{fe}R_L}{h_{ie}}$	$\simeq 1$

*Note that $h_{fe} = \beta_0 \gg 1$ is used in the relations.

and they both depend on the load resistance. The voltage gain of a CC stage is almost unity. Note the phase reversal of a CE stage.

Observe that the CE stage can provide both voltage and current gains and are widely used for amplification purposes. The other configurations are useful in different types of applications, e.g., as a buffer stage, the CC is used to prevent loading of the signal source, and the CB is used to match a very low impedance source. In other words, a CC is seen as a voltage source, and a CB is seen as a current source by the succeeding stage.

4.6 SMALL-SIGNAL PROPERTIES OF FET AMPLIFIERS

In Chap. 3 we obtained the small-signal model for FET amplifiers (Fig. 3.27). The small-signal model of FET is the same whether we have JFET or MOSFET and regardless of the MOS type, whether it is of the depletion type or enhancement type. In Fig. 4.11 we have shown the model again for convenience. Note that the model is remarkably simple and only two parameters are needed for low-frequency small-signal applications, namely, g_m (or y_{fs} as sometimes used) and r_d. The expressions for g_m are given in (3.39) and (3.40), and for $r_d = g_d^{-1}$ they are given in (3.38). Typical values of these parameters are

$$g_m = 2 \times 10^{-3}\text{ mho} \qquad r_d = g_d^{-1} = 20\text{ k}\Omega \tag{4.36}$$

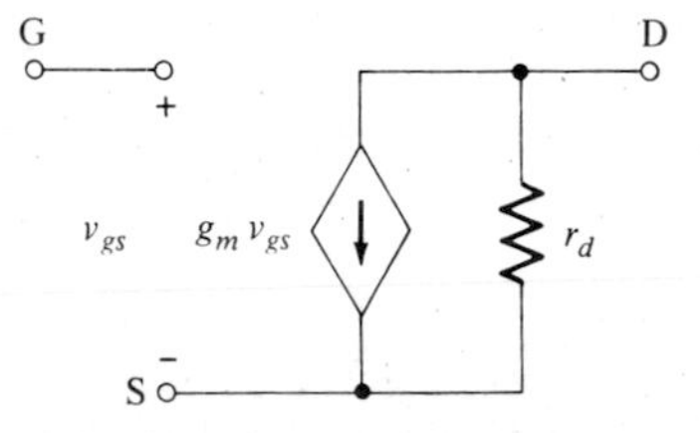

FIGURE 4.11 FET low-frequency model.

As pointed out in Chap. 3, the device parameters vary according to the type of the device and the operating point Q. We shall use the typical values only for the purposes of numerical calculations as we did for BJT devices. The range of values of r_d and g_m are listed in Table 3.1.

We shall next consider the properties of FET in the three configurations. Since the following discussion applies to both MOSFETs and JFETs, we shall only use FET henceforth. Recall that FETs are voltage-controlled devices, and hence the g or y parameters are suitable. The manufacturers almost always provide the values of the y parameters, from which g_m and g_d are found directly. Since the FET models are remarkably simple, we need not use the two-parameters description but instead use the circuits directly. That is precisely what we shall do in the following.

A. The Common-Gate Amplifier

The common-gate (CG) FET amplifier circuit and its equivalent circuit are shown in Fig. 4.12a and b, respectively. From the circuit,

$$i_2 = -\frac{v_o}{R_L} = (v_o - v_{\text{in}})\frac{1}{r_d} + g_m v_{gs} \tag{4.37}$$

Substitution of $v_{gs} = -v_{\text{in}}$ in (4.37) and solving for v_o/v_{in} yield

$$A_v = \frac{v_o}{v_{\text{in}}} = \frac{(\mu + 1)R_L}{r_d + R_L} \simeq g_m R_L \qquad \text{for } \mu \gg 1,\ r_d \gg R_L \tag{4.38}$$

where $\mu = g_m r_d \gg 1$. Similarly,

$$i_1 = v_{\text{in}}\frac{1}{R_G} + (v_{\text{in}} - v_o)\frac{1}{r_d} - g_m v_{gs} \tag{4.39}$$

From (4.37) and (4.39) and noting that $v_{gs} = -v_{\text{in}}$, we obtain the input and output resistance, namely,

$$R_i = \frac{v_i}{i_1} = \frac{R_L + r_d}{\mu + 1} \simeq \frac{1}{g_m} \qquad \text{for } r_d \gg R_L \tag{4.40}$$

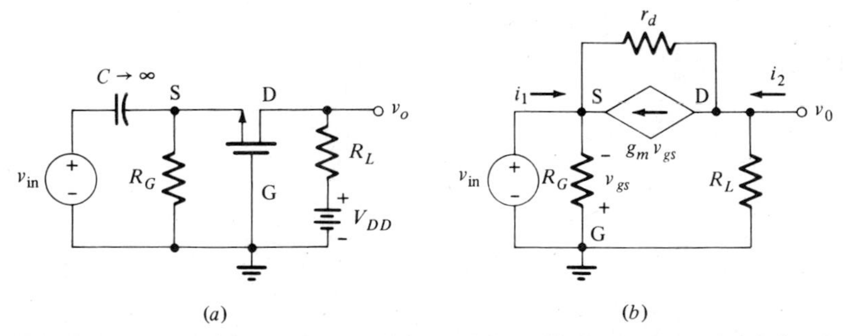

FIGURE 4.12 (a) Single-stage CG amplifier. (b) Equivalent circuit for the CG configuration.

and

$$R_o = \frac{v_o}{i_2} = r_d + (\mu + 1)R_G \simeq r_d + \mu R_G \tag{4.41}$$

Note that for the CG configuration the output resistance is very large and can be considered as infinite; the input resistance is relatively low. The voltage gain is dependent on R_L, and its maximum value is approximately μ. The CG configuration in FET is the counterpart of CB configuration in BJTs.

EXAMPLE 4.5

For the CG amplifier circuit shown in Fig. 4.12, the FET parameters are given in (4.36). Determine the input resistance R_i, the output resistance R_o, and the voltage gain v_o/v_{in} if $R_L = 2\ \text{k}\Omega$, $R_G = 1\ \text{M}\Omega$.

From (4.40) $R_i \simeq 1/g_m = 500\ \Omega$, $R_o \simeq 40\ \text{M}\Omega \simeq \infty$, $A_v = g_m R_L = (2 \times 10^{-3})(2 \times 10^3) = 4$.

B. The Common-Source Amplifier

A common-source (CS) FET amplifier with the dc biasing circuitry is shown in Fig. 4.13*a*. The small-signal equivalent is shown in Fig. 4.13*b*, with $R_G = R_{G1} \parallel R_{G2}$. From the circuit, by inspection we have

$$R_{\text{in}} = \infty \qquad R_o = r_d \tag{4.42}$$

The voltage gain is determined as

$$v_o = -g_m v (r_d \parallel R_L)$$

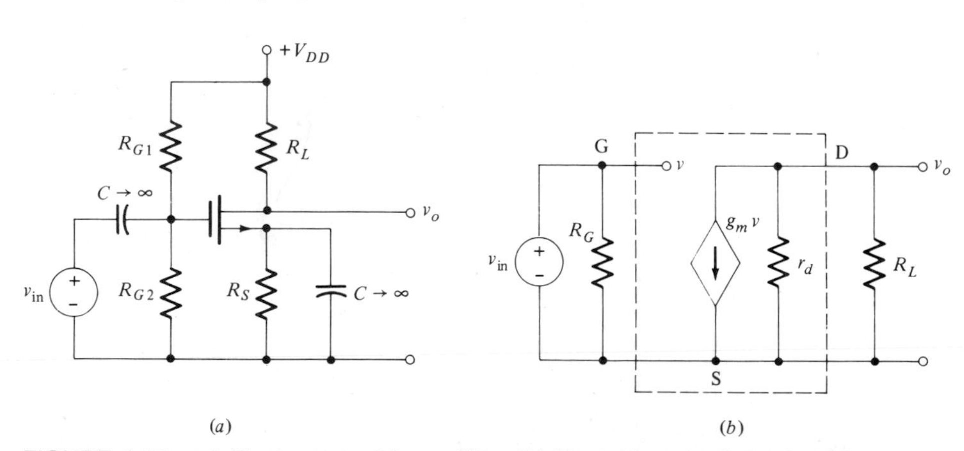

FIGURE 4.13 (*a*) Single-stage CS amplifier. (*b*) Equivalent circuit for the CS configuration.

Since $v = v_{\text{in}}$, we have

$$A_v = \frac{v_o}{v_{\text{in}}} = -g_m(r_d \parallel R_L) \simeq -g_m R_L \qquad \text{for } r_d \gg R_L \tag{4.43}$$

Note that the maximum voltage gain of a CS amplifier is given by

$$\left(\frac{v_o}{v_{\text{in}}}\right)_{\max} = -g_m r_d = \mu \gg 1$$

where μ is the amplification factor.

EXAMPLE 4.6

For the CS amplifier shown in Fig. 4.13, the FET parameters are given as in (4.36). Determine R_i, R_o, and A_v for $R_L = 2\ \text{k}\Omega$ and $R_G = 1\ \text{M}\Omega$. From (4.42), $R_i = \infty$, $R_o = r_d = 20\ \text{k}\Omega$. The voltage gain $A_v \simeq -g_m R_L = -2 \times 10^{-3}(2 \times 10^3) = -4$.

The CS configuration in FET is the counterpart of the CE configuration in BJT.

C. The Common-Drain Amplifier

A common-drain (CD) FET amplifier with the biasing circuitry is shown in Fig. 4.14*a*. The small-signal equivalent circuit is shown in Fig. 4.14*b*. Note that the controlled source and r_d are represented by its Thevenin equivalent circuit. From the circuit, by inspection $R_i = \infty$. The output resistance R_o is determined as follows:

$$v_o = -v_{gs} = \mu v_{gs} + r_d i_2 \tag{4.44}$$

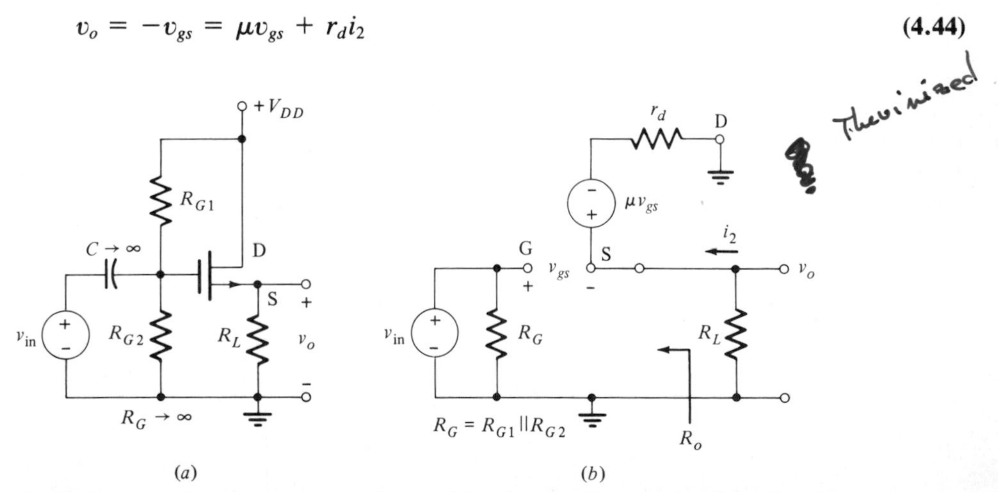

FIGURE 4.14 (*a*) Single-stage CD amplifier. (*b*) Equivalent circuit for the CD configuration.

Hence

$$R_o = \frac{v_o}{i_2} = \frac{r_d}{1 + \mu} \simeq \frac{r_d}{\mu} = \frac{1}{g_m} \qquad \text{for } \mu \gg 1 \tag{4.45}$$

The voltage gain of the circuit is found by solving for v_o/v_{in}. We have

$$v_{in} = v_o + v_{gs} \tag{4.46}$$

From (4.44) we have

$$v_o - \mu v_{gs} = i_2 r_d = -\left(\frac{v_o}{R_L}\right) r_d \tag{4.47}$$

From (4.46) and (4.47) we obtain

$$A_v = \frac{v_o}{v_{in}} = \frac{\mu R_L}{(\mu + 1)R_L + r_d} \simeq \frac{g_m R_L}{1 + g_m R_L} \tag{4.48}$$

For $g_m R_L \gg 1$ the voltage gain is close to unity. The CD configuration is therefore called the *source follower,* since the source voltage follows the input gate signal. The CS configuration in FET is the counterpart of the CC configuration in BJT.

EXAMPLE 4.7

For the CD amplifier shown in Fig. 4.14, FET parameters are given as in (4.36). Determine R_i, R_o, and the voltage gain for $R_L = 2\ \text{k}\Omega$, $R_G = 1\ \text{M}\Omega$.

By inspection $R_i = \infty$, $R_o = 1/g_m = 500\ \Omega$,

$$A_v = \frac{2 \times 10^{-3}(2 \times 10^3)}{1 + 2 \times 10^{-3}(2 \times 10^3)} = \frac{4}{5} = 0.80$$

EXERCISE 4.5

For the circuit shown in Fig. E4.5, determine the input impedance R_{in}. Assume $R_f \gg R_L$.

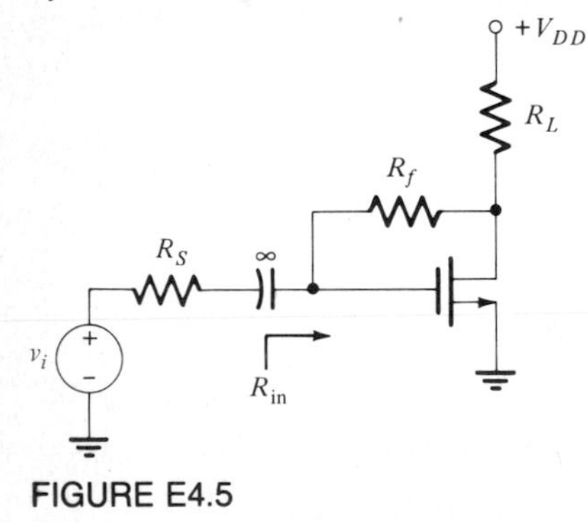

FIGURE E4.5

Ans $R_{in} = R_f/[1 + g_m(R_L \parallel r_d)]$

EXERCISE 4.6

For the CMOS circuit shown in Fig. E4.6, what is the output impedance and the maximum voltage gain?

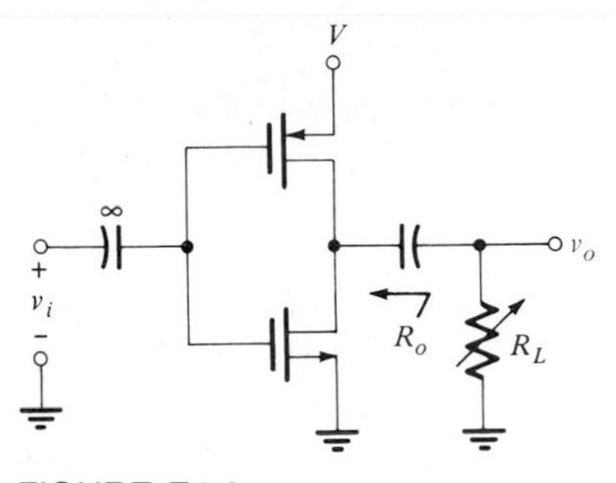

FIGURE E4.6

Ans $r_d/2$, $\mu = -g_m r_d$

4.7 COMPARISON OF FET CONFIGURATIONS

The various network properties of the FET, whose parameters are given in (4.36), are summarized in Table 4.4. For the sake of reference and convenience the simplified expressions are listed in Table 4.5.

An examination of the above quantities reveal the following:

Input resistance: Only the CG configuration has a noninfinite input resistance.

Output resistance: The output resistance of the CD configuration is the lowest, and that of the CG is the highest.

Voltage gain: The CD configuration provides a voltage gain of close to (but always less than) unity. The CS and CG configurations provide the same

TABLE 4.4 Network Properties of FET* for a Typical Unit

	CG	CS	CD (SF)
R_{in}	500 Ω	∞	∞
R_o	∞	20 kΩ	500 Ω
$A_v = \dfrac{v_o}{v_{in}}$	4	−4	0.80

*FET parameters are given in (4.36), namely, $g_m = 2 \times 10^{-3}$ mho, $r_d = 20$ kΩ, and $R_L = 2$ kΩ.

TABLE 4.5 Gain and Impedance Relations for FET Configurations

	CG	CS	CD (SF)
R_{in}	$\frac{1}{g_m}$	∞	∞
R_o	∞	r_d	$\frac{1}{g_m}$
$A_v = \frac{v_o}{v_{in}}$	$g_m R_L$	$-g_m R_L$	$\frac{g_m R_L}{1 + g_m R_L}$

voltage gain, while in the CS configuration polarity inversion is also obtained. Note that in FETs current gain is meaningless, especially in CS and CD configurations, since the input impedances of these circuits are infinite.

Finally note that the CS configuration is analogous to the CE amplifier, whereas the CD configuration is analogous to the CC configuration and the CG configuration is analogous to the CB configuration. The analogies are drawn from the voltage gain, polarity inversion property, and impedance characteristic of the devices. As in all analogies, however, we should be careful of their usage. For example, the input impedance of a CE stage is relatively medium, while that of a grounded source stage is infinite.

Before we conclude our discussion of single-stage FET amplifiers, we would like to note that in MOS technologies another MOS device, instead of a resistor, is used as the passive load element in order to reduce size and cost. Thus a common-source amplifier (enhancement-mode NMOS) may have either an enhancement-mode MOS load, a depletion-mode MOS load, or a PMOS load. The *V-I* characteristics of these interconnections and the voltage transfer characteristics were given in Fig. 3.22, Fig. 3.21*b*, and Fig. 3.24*c*, respectively. The voltage gain of these configurations are given by the ratio of the transductances $(-g_{mD}/g_{mL})$, where the subscripts D and L denote the driver (the common-source transistor) and the load, respectively. Since the transconductances are related to the geometry of the devices (W/L ratios), the maximum voltage gain of these amplifying stages are usually between 10 to 30 due to practical considerations. From (3.18) also recall that the larger the value of λ_r the larger is the gain of the amplifier.

4.8 PHASE-SPLITTER CIRCUITS

Two different circuits, one using a BJT and another using a FET that can be used as a phase-splitting circuit, are shown in Figs. 4.15 and 4.16, respectively. The above circuits provide the same gains for v_{o1} and v_{o2} but with different polarities. The dc voltages at the corresponding terminals are the same since essentially the same current flows through equal-valued resistors.

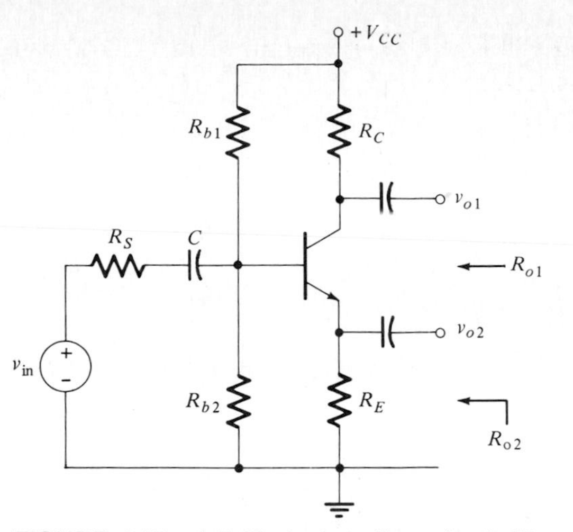

FIGURE 4.15 A BJT phase-splitter circuit ($R_E = \alpha_0 R_C$).

A. BJT Phase Splitter

Consider the BJT circuit shown in Fig. 4.15. The voltage v_{o2} is determined as if the circuit were an emitter follower. The input impedance of an emitter follower was determined in (4.31), namely,

$$R_{in} \simeq \beta_0(r_e + R_E) \simeq \beta_0 R_E \qquad \text{for } R_E \gg r_e \tag{4.49}$$

The base voltage of the circuit is then given by

$$v_b = \frac{(R_b \parallel R_{in})}{(R_b \parallel R_{in}) + R_S} v_{in} = \frac{R_b}{R_b + R_S} v_{in} \qquad \text{for } R_b \ll R_{in} \tag{4.50}$$

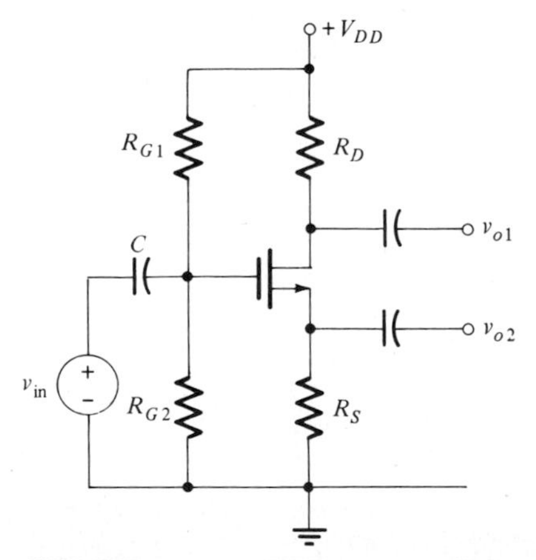

FIGURE 4.16 A FET phase-splitter circuit ($R_S = R_D$).

The voltage gain of an emitter follower was determined in (4.34), namely, for $R_C = R_E$,

$$\frac{v_{o2}}{v_b} = \frac{R_E}{R_E + r_e} \tag{4.51}$$

hence

$$\frac{v_{o2}}{v_{\text{in}}} = \frac{R_b}{R_b + R_S}\left(\frac{R_E}{R_E + r_e}\right) \tag{4.52}$$

Since $i_e = \alpha_0 i_e \cong i_c$, then the voltage at the collector, v_{o1}, is the same as the voltage at the emitter; hence

$$-v_{o1} \simeq v_{o2} = \frac{R_b}{R_b + R_S}\left(\frac{R_E}{R_E + r_e}\right) \tag{4.53}$$

As seen in (4.53) the two outputs are 180° out of phase with each other. The output impedances seen at the two terminals are vastly different and are determined in the following.

The output impedance, as seen from the emitter terminal, is that of an emitter-follower (CC), which is given by (4.32)

$$R_{o2} \simeq r_e + \frac{R_S + r_b}{\beta_0} \tag{4.54}$$

The output impedance as seen from the collector terminal is that of a CE configuration increased by the inclusion of R_E; hence it can be considered as infinite.

B. FET Phase Splitter

The analysis method for FET phase splitter is the same as in the BJT case. In this case if $R_D = R_S$, $v_{o1} = -v_{o2}$.

The Thevenin equivalent for the circuit for Fig. 4.16 as seen from terminals v_{o1} and v_{o2} to ground is shown in Fig. 4.17*a* and *b* (problem 4.26). The voltages for

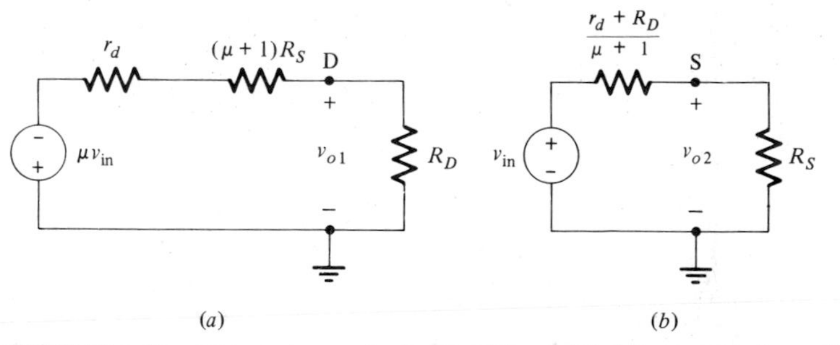

FIGURE 4.17 Thevenin equivalent of Fig. 4.16; (*a*) as seen from the D-to-ground terminals and (*b*) as seen from the S-to-ground terminals.

$R_D = R_S$ are given by

$$v_{o2} = -v_{o1} = \frac{\mu R_D v_{in}}{R_D + r_d + (\mu + 1)R_D} \tag{4.55}$$

Note that the equivalent resistance as seen from the drain (D) terminal is much higher than that seen from the source (S) terminal. The derivation of the Thevenin equivalent for the transistor in Fig. 4.15, using the simplified π model, is left as an exercise (problem 4.27).

EXAMPLE 4.8

For the FET phase-splitter circuit shown in Fig. 4.16, the FET parameters are $g_m = 2 \times 10^{-3}$ mho, $r_d = 20$ kΩ. Determine the voltages v_{o1} and v_{o2} if $v_{in} = V_m \sin \omega t$. What are the impedences as seen from the drain and source terminals? The circuit element values are $R_G = 1$ MΩ, $R_S = R_D = 5$ kΩ.

From (4.55) we obtain

$$v_{o2} = -v_{o1} = \frac{(40)(5 \times 10^3)v_{in}}{5 \times 10^3 + 20 \times 10^3 + 41(5 \times 10^3)} = 0.87V_m \sin \omega t$$

The impedance looking from the source terminal is

$$R_{o2} = \frac{r_d + R_D}{\mu + 1} = \frac{(20 + 5) \times 10^3}{41} = 610\ \Omega$$

The impedance looking from the drain terminal is

$$R_{o1} = r_d + (\mu + 1)R_S = 20 \times 10^3 + 41(5 \times 10^3) = 225\ \text{k}\Omega$$

EXERCISE 4.7

For the circuit shown in Fig. 4.15, $\beta_0 = 100$, what is the voltage gain v_{o1}/v_{in} if $R_E = 2R_C = 2$ kΩ, $R_{b1} \parallel R_{b2} = 20$ kΩ, and $R_S = 100\ \Omega$?

Ans -0.50

EXERCISE 4.8

For the circuit shown in Fig. 4.16, $\mu = 20$, $r_d = 20$ kΩ, what is the voltage gain v_{o1}/v_{in} if $R_S = 2R_D = 4$ kΩ and $R_{g1} \parallel R_{g2} = 1$ MΩ?

Ans -0.38

4.9 CONCLUDING REMARKS

This chapter considered the three configurations of BJT and FET amplifiers. Single-transistor amplifiers were considered in order to focus attention on the fundamental circuit properties of each configuration, namely, the input and output impedances and the current and voltage gains. It was also noted that in most MOS technologies a MOS is used as a passive load instead of a resistor. The small-signal circuit properties of BJT amplifiers summarized in Tables 4.2 and 4.3 and those of FET amplifiers listed in Table 4.5 will be utilized in the next chapter, where multiple transistors are considered.

REFERENCES

1. J. Millman, *Microelectronics: Digital and Analog Circuits and Systems,* McGraw-Hill, New York, 1979.
2. P. R. Gray and R. G. Meyer, *Analysis and Design of Analog Integrated Circuits,* Wiley, New York, 1977.
3. V. H. Grinich and H. G. Jackson, *Introduction to Integrated Circuits,* McGraw-Hill, New York, 1975.

PROBLEMS

4.1 Determine the common-emitter h parameters of the *npn* transistor type 2N2222 given in Appendix E, for $I_C = 1$ mA, $V_{CE} = 10$ V.

4.2 Determine the common-emitter h parameters of the *pnp* transistor type 2N3251 given in Appendix E, for $I_C = 1$ mA, $V_{CE} = 10$ V.

4.3 Show that the h parameters of the CB configuration in terms of the T-model parameters are given approximately by

$$h_{ib} \simeq r_e + \frac{r_b}{\beta_0} \qquad h_{ob} \simeq \frac{1}{r_c}$$

$$h_{fb} \simeq -\alpha_0 \qquad h_{rb} \simeq \frac{r_b}{r_c}$$

4.4 Find the h parameters of the *npn* type 2N2222 for the CB configuration.

4.5 Show that the h parameters of the CC configuration in terms of the T-model parameters are given approximately by

$$h_{ic} \simeq r_b + \beta_0 r_e \qquad h_{oc} \simeq \frac{\beta_0}{r_c}$$

$$h_{fc} \simeq -\beta_0 \qquad h_{rc} \simeq 1$$

4.6 Show that the T-model parameters in terms of the h parameters for the CE configuration are given by

$$\alpha_0 = \frac{h_{fe}}{1 + h_{fe}} \qquad r_c = \frac{1 + h_{fe}}{h_{oe}}$$

$$r_e = \frac{h_{re}}{h_{oe}} \qquad r_b = h_{ie} - \frac{h_{re}}{h_{oe}}(1 + h_{fe})$$

(*Hint:* The T-model parameters can be obtained in terms of the z parameters by inspection.) Determine the numerical values of the above for the transistor described in Example 4.1.

4.7 Determine the low-frequency y parameters of
(*a*) JFET type JN51 given in Appendix E.
(*b*) Depletion-type MOSFET MN82 given in Appendix E.

4.8 If the h parameters of the CB configuration are known, determine the h parameters of the CE configuration.

4.9 The transistor circuit shown in Fig. P4.9 is equivalent to a constant current source. Determine the Norton equivalent of the circuit looking into terminals a, b. What are the numerical values of the current and the parallel resistance if the transistor has the T-model parameters of (4.13), $R_1 = R_2 = 1$ kΩ and $R_3 = 2$ kΩ, $V = 5$ V?

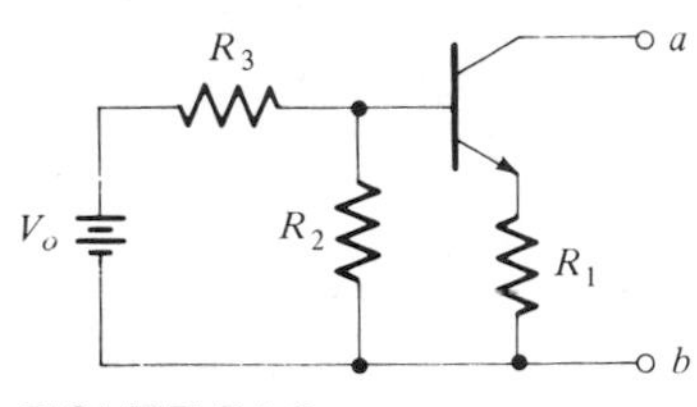

FIGURE P4.9

4.10 In an amplifier, it is sometimes useful to know a quantity referred to as the *voltage isolation* (VI), which is defined as follows:

$$\text{VI} = \frac{\text{voltage gain in forward direction}}{\text{voltage gain in reverse direction}}$$

For the BJT parameters given in (4.13), find the midband VI for the CE configurations for $R_L = R_S = 1$ kΩ.

4.11 Repeat problem 4.10 for the CB configuration.

4.12 Repeat problem 4.10 for the CC configuration.

4.13 For the BJT parameters given in (4.13), determine the maximum power gain G_P for the CE configurations. (*Hint:* See Appendix C.)

4.14 Repeat problem 4.13 for the CB configuration.

4.15 Repeat problem 4.13 for the CC configuration.

4.16 Determine the approximate y parameters of a CE stage from the low-frequency hybrid-π model in an effortless manner.

4.17 The transistor described in Example (4.1) has the following h parameters at $I_C = 5$ mA, $V_{CE} = 5$ V:

$$h_{ie} = 500\ \Omega \qquad h_{fe} = 60$$

$$h_{re} = 5 \times 10^{-5} \qquad h_{oe} = 5 \times 10^{-5}\ \text{mho}$$

For the same source and load terminations ($R_L = R_S = 1\ \text{k}\Omega$), determine the current, voltage, and power gain of the stage and compare them to those of Example 4.1.

4.18 If transformers are used at the input and output to obtain maximum power gain, determine the power gain and the transformer turns ratios if the transistor parameters are given in problem 4.17. Assume ideal transformers. The source and load terminations are $R_S = 50\ \Omega$ and $R_L = 1\ \text{k}\Omega$.

4.19 For the FET amplifier shown in Fig. P4.19, $g_m = 2 \times 10^{-3}$ mho, and $r_d = 20\ \text{k}\Omega$. Determine the voltage gain.

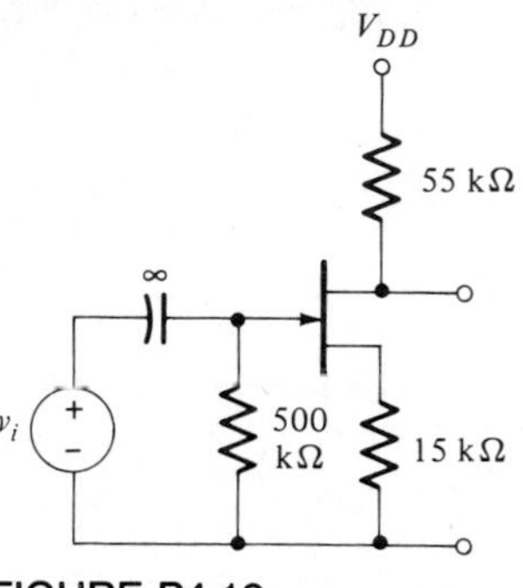

FIGURE P4.19

4.20 For the circuit shown in Fig. P4.20, determine

(*a*) $\dfrac{v_1}{v_s}$

(*b*) $\dfrac{v_2}{v_s}$

Use the same transistor parameters as in problem 4.17.

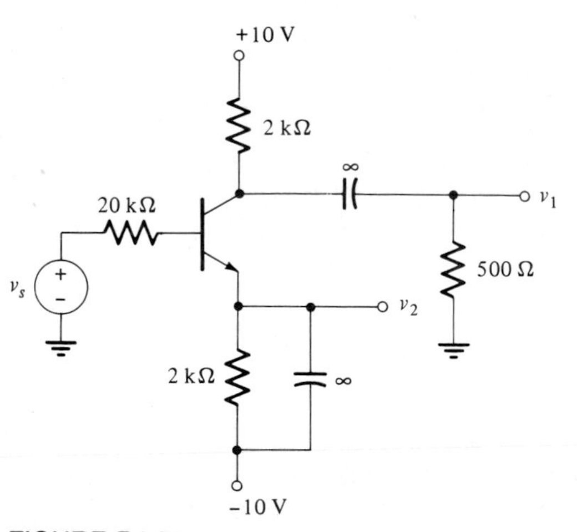

FIGURE P4.20

4.21 For the circuit shown in Fig. P4.21, determine
(*a*) The voltage gain v_o/v_i
(*b*) The input resistance R_i in an effortless manner
[*Hint:* r_e is changed to $r_e + 50\ \Omega$ in (4.18). Assume $\beta_0 = 100$.]

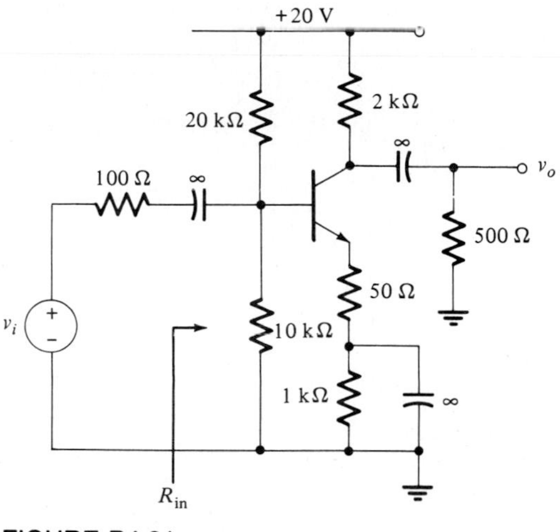

FIGURE P4.21

4.22 For the circuit shown in Fig. P4.22, $\beta_0 = 100$, $V_{BE} = 0.7$ V, $I_{CO} \cong 0$ (assume $r_b = 0$), determine
(*a*) The voltage gain v_o/v_i
(*b*) The input impedance R_{in}

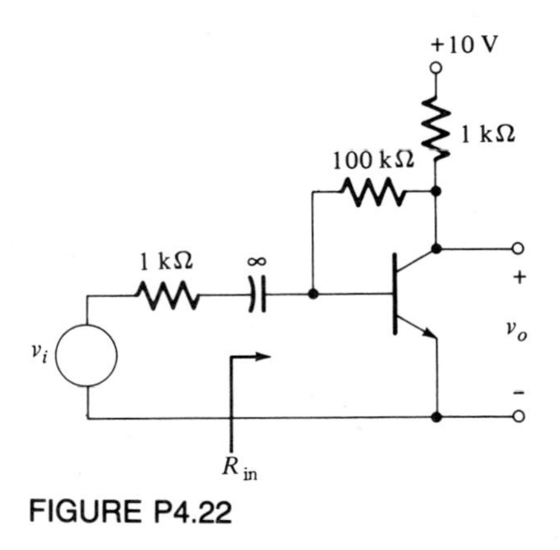

FIGURE P4.22

4.23 The MOSFET amplifier circuit shown in Fig. P4.23 is to be designed for a voltage gain of 10. The circuit is to be biased such that $I_D = 5$ mA, $V_{DS} = 5$ V, $V_{GS} = 3$ V. Determine the values of R_D and R_{G1}. Let $R_S = 0.5$ kΩ and $R_{G2} = 500$ kΩ.

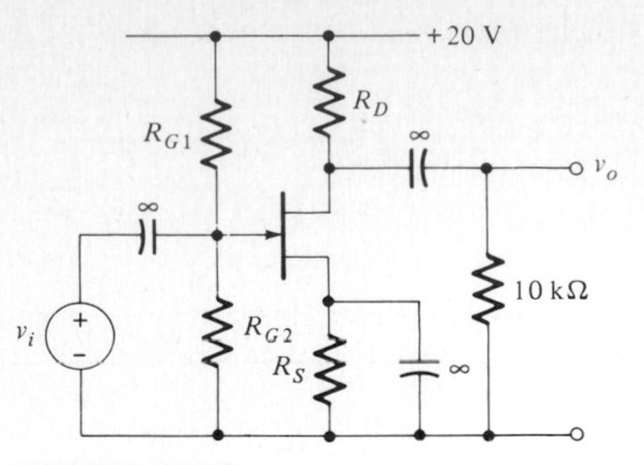

FIGURE P4.23

4.24 Determine the expression for the voltage gain and the input resistance of the FET circuit shown in Fig. P4.24. Assume $r_d \gg R_L$.

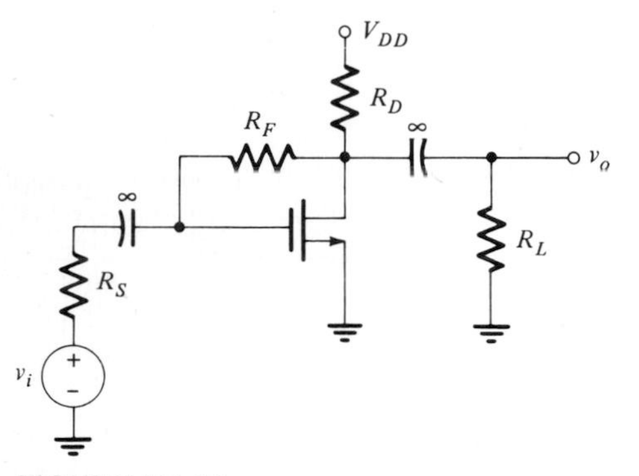

FIGURE P4.24

4.25 Determine the voltage gain of the circuit shown in Fig. P4.25.
(*a*) For $C_1 = \infty$
(*b*) For $C_1 = 0$

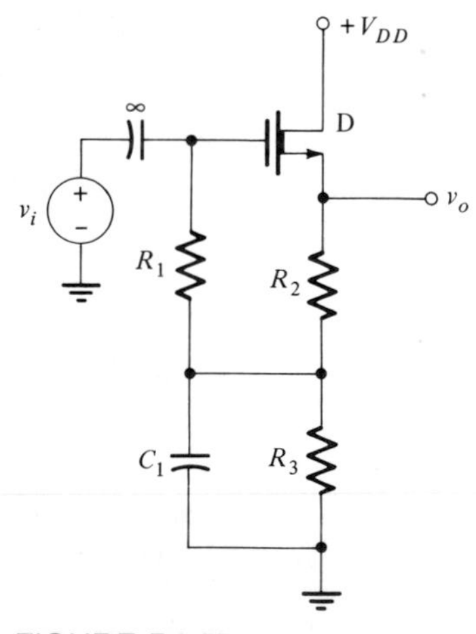

FIGURE P4.25

4.26 Show the details of arriving at the Thevenin equivalent circuits in Fig. 4.17.

4.27 Derive the Thevenin equivalent circuit of Fig. 4.15
(*a*) As seen from the collector-to-ground terminals v_{o1}
(*b*) As seen from the emitter-to-ground terminals v_{o2}

4.28 Derive the expression for the voltage gain v_{o1}/v_{in} for the circuit in Fig. 4.15 if $R_E \geq 10R_C$; assume $R_B \gg h_{ie}$.

4.29 Repeat problem 4.28 for Fig. 4.16 and assume $R_S \geq 10R_D$. (*Note:* The inequality conditions in problems 4.28 and 4.29 occur in differential amplifiers, in common-mode gain, as discussed in Chap. 5.)

4.30 Determine the input resistance of the BJT circuit in Fig. 4.15.

CHAPTER 5 Multiple-Transistor Amplifiers

In Chap. 4 we considered basic properties of single amplifier stages. Practical amplifiers seldom use single-stage amplifiers. In fact, in the construction of integrated circuits, a large number of transistors are used to achieve a number of objectives such as voltage gain, current gain, impedance level transformation, level shift, and current source. The basic properties of single-stage amplifiers obtained earlier can be used in the analysis and design of multiple-transistor circuits. However, in many applications combinations of two transistors are used as a compound circuit or subcircuit to take advantage of the sometimes quite extraordinary properties of some of these configurations. These subcircuits will be considered subsequently. They are the *Darlington configurations,* which consist of a cascade of CC-CC stages and a cascade of CC-CE stages; the CE-CB configuration, which is commonly called the *cascode configuration;* and the CC-CB pair, commonly called the *emitter-coupled pair*. The latter configuration, somewhat modified, is widely used in integrated circuits as a differential amplifier and will be considered in detail, including its characteristic for large-signal inputs, i.e., nonamplifier applications.

Combination of field-effect transistors and bipolar transistors (Bi-FETs), which takes advantage of the very large input properties of JFETs and the high gain capability of BJTs, is also considered. Cascaded amplifiers, using more than two transistors, are then considered. An important class of amplifiers is that which employs feedback. Feedback amplifiers are discussed in detail in Chap. 10.

In the subsequent sections the results obtained earlier in Tables 4.2 and 4.5 for BJTs and FETs will be used extensively.

5.1 THE DARLINGTON CONFIGURATIONS

The Darlington configuration consists of two transistors in a CC-CC or a CC-CE interconnection as shown in Fig. 5.1. In fact, some manufacturers package the Darlington pair as a composite device to obtain the so-called *super* β (i.e., β_0^2) composite transistors. The biasing current I_{dc} in Fig. 5.1 is used to accommodate a different dc emitter current of Q_1 and a dc base current of Q_2. In other words, with this flexibility we can have the same emitter currents for both transistors Q_1 and Q_2. If we do not provide for I_{dc}, the dc emitter current of Q_2 will be larger than the dc emitter current of Q_1 by a factor of β. Each of the configurations will be considered briefly:

A. The Common-Collector Cascade (CC-CC) Pair

The CC-CC pair of Fig. 5.1*a* is shown again in Fig. 5.2*a* with the pertinent quantities for analysis. The biasing circuitry is omitted for simplicity. The current gain of Fig. 5.2*a* is given by

$$\frac{i_o}{i_1} = \left(\frac{i_o}{i_{b2}}\right)\left(\frac{i_{b2}}{i_1}\right) \simeq \beta_{02}\beta_{01} \simeq \beta_0^2 \tag{5.1}$$

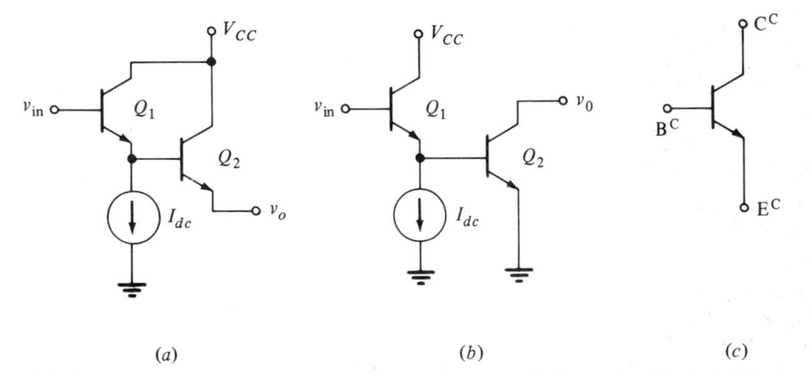

FIGURE 5.1 Darlington configurations: (*a*) CC-CC pair, (*b*) CC-CE pair, (*c*) Darlington composite transistor.

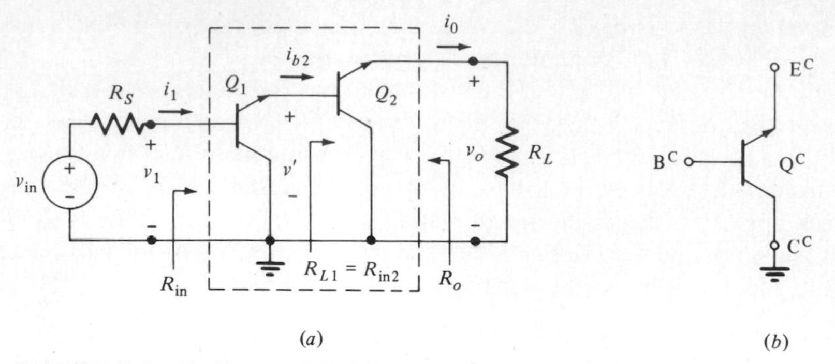

FIGURE 5.2 Darlington CC-CC configuration: (*a*) CC-CC cascade, (*b*) composite transistor.

Note that for simplicity we have assumed identical transistors $\beta_{01} = \beta_{02} = \beta_0$. This is a good assumption, since composite transistors are fabricated on the same chip and are nearly identical, much more so than can be obtained using discrete transistors.

The overall voltage gain is

$$\frac{v_o}{v_1} = \frac{v_o}{v'}\left(\frac{v'}{v_1}\right) \simeq \frac{R_L}{R_L + r_e}\left(\frac{R_{in2}}{R_{in2} + r_e}\right) \simeq 1 \tag{5.2}$$

The input impedance of a CC configuration is found by using the results in Chap. 4 (Table 4.2), namely,

$$R_{in} \simeq \beta_0(r_e + R_{L1}) \tag{5.3a}$$

where $R_{L1} \simeq R_{in2} \simeq \beta_0(r_e + R_L)$. Hence

$$R_{in} = \beta_0[r_e + \beta_0(r_e + R_L)] \simeq \beta_0^2(r_e + R_L) \simeq \infty \tag{5.3b}$$

Thus for small values of R_S, $v' = v_{in}$ and v_o/v_{in} is about unity. The output impedance is similarly found, namely,

$$R_o \simeq r_e + \frac{R_{o1} + r_b}{\beta_0} \tag{5.4a}$$

where $R_{o1} = r_e + (R_S + r_b)/\beta_0$. Hence

$$R_o = r_e + \frac{\beta_0(r_e + r_b) + R_S}{\beta_0^2} \simeq r_e \tag{5.4b}$$

Note that the biasing dc current source I_{dc} allows the bias currents (I_e) of Q_1 and (I_b) of Q_2 to be different. This biasing is accomplished by connecting a high resistance R_b from the emitter terminal of Q_1 to the supply voltage V_{CC} in Fig. 5.1*a*, thus providing a dc bias current, which can make the emitter quiescent current of Q_1 essentially independent of the base current of Q_2. Note that if R_b is connected from the emitter of Q_1 to ground it will rob some ac signal current and thus reduce the overall current gain to

$$\frac{i_o}{i_1} \simeq \beta_0^2\left(\frac{R_b}{R_b + R_{in2}}\right) \simeq \beta_0^2\left(\frac{R_b}{R_b + \beta_0(R_L + r_e)}\right) \tag{5.5}$$

The Darlington configuration can be represented as a single composite transistor as shown in Fig. 5.2*b*, where the superscript C denotes "composite." In order to illustrate the various properties, the same transistor parameters given in (4.13) will be used, namely,

$$\begin{aligned} \beta_0 &= 50 \qquad & r_e &= 25\ \Omega \qquad \text{at } I_e = 1\text{ mA} \\ r_b &= 200\ \Omega \qquad & r_c &= 2\ \text{M}\Omega \end{aligned} \tag{5.6}$$

EXAMPLE 5.1

Determine the input impedance R_i, the output impedance R_o, the current gain, and the voltage gain v_o/v_{in} of the CC-CC Darlington pair. The transistor parameters are given in (5.6), and $R_S = R_L = 1\ \text{k}\Omega$. From (5.3*b*)

$$R_{in} \simeq 50^2(10^3) = 2.5\ \text{M}\Omega$$

From (5.4*b*)

$$R_o = 25 + \frac{50(225) + 10^3}{50^2} = 30\ \Omega$$

From (5.1)

$$A_i = \frac{i_o}{i_1} = \beta_0^2 = 2500$$

From (5.2)

$$\frac{v_o}{v_i} = \frac{10^3}{10^3 + 25}\left(\frac{50 \times 10^3}{50 \times 10^3 + 25}\right) \simeq 1$$

Hence

$$A_v = \frac{v_o}{v_{in}} = \frac{v_o}{v_1}\left(\frac{R_{in}}{R_{in} + R_s}\right) = \frac{2.5 \times 10^6}{2.5 \times 10^6 + 10^3} \simeq 1$$

The circuit properties of the CC-CC configuration are listed in Table 5.1. Note that $R_{in} \simeq 2.5\ \text{M}\Omega$, $R_o \simeq 30\ \Omega$, $A_i \simeq \beta_0^2$, and $A_v \simeq 1$ and that these parameters describe the composite transistor.

TABLE 5.1 Network Properties of Some Two-Transistor Circuits

	CC-CC Darlington configurations	CC-CE Darlington configurations	CC-CB emitter-coupled	CE-CB cascode
R_{in}	$\beta_0^2(r_e + R_L) \simeq \infty$	$\beta_0^2 r_e \simeq \infty$	$2\beta_0 r_e$	$\beta_0 r_e$
R_{out}	r_e	∞	∞	∞
$a_i = \dfrac{i_o}{i_1}$	β_0^2	$-\beta_0^2$	$-\beta_0$	$-\beta_0$
$a_v = \dfrac{v_o}{v_i}$	1	$-\dfrac{R_L}{r_e} = -g_m R_L$	$-\dfrac{R_L}{2r_e} = \dfrac{1}{2} g_m R_L$	$-\dfrac{R_L}{r_e} = -g_m R_L$

Network properties of the CC-CC pair are listed in Table 5.1.

B. The Common-Collector, Common-Emitter (CC-CE) Pair

The CC-CE pair exclusive of biasing circuitry is shown in Fig. 5.3*a*. The composite transistor is shown in Fig. 5.3*b*. The current gain of the circuit in Fig. 5.3*a* is given by

$$\frac{i_o}{i_1} = \left(\frac{i_o}{i_{b2}}\right)\left(\frac{i_{b2}}{i_1}\right) = (-\beta_0)_2(\beta_0)_1 \simeq -\beta_0^2 \tag{5.7}$$

The voltage gain is

$$\frac{v_o}{v_1} = \frac{v_o}{v'}\left(\frac{v'}{v_1}\right) \simeq -\frac{\beta_0 R_L}{r_b + \beta_0 r_e}\left(\frac{R_{in2}}{R_{in2} + r_e}\right) \tag{5.8a}$$

where R_{in2} is the input resistance of a CE stage:

$$R_{in2} \simeq r_b + \beta_0 r_e = r_b + \frac{\beta_0}{g_m}$$

Note that $g_m = 1/r_e$, and since $\beta_0 r_e \gg r_b$, $R_{in2} \gg r_e$, and (5.8*a*) reduces to

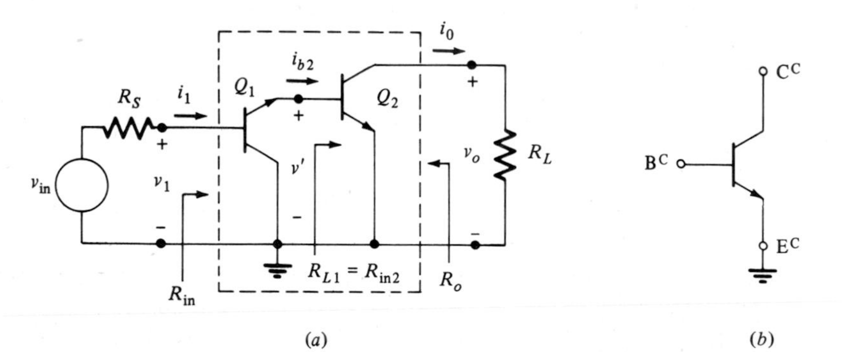

FIGURE 5.3 Darlington CC-CE configuration: (*a*) CC-CE cascade, (*b*) composite transistor.

$$\frac{v_o}{v_1} \simeq -\frac{R_L}{r_e} = -g_m R_L \tag{5.8b}$$

The input resistance of the circuit is that of a CC stage, namely,

$$R_{in} \simeq \beta_0(r_e + R_{L1}) = \beta_0[r_e + (r_b + \beta_0 r_e)] \simeq \beta_0^2 r_e \tag{5.9a}$$

The output resistance is simply that of a CE stage and thus is very large, i.e.,

$$R_o = \frac{r_c}{\beta_0} \tag{5.9b}$$

Note that the overall voltage gain is

$$\frac{v_o}{v_{in}} \simeq \frac{v_o}{v_1}\left(\frac{v_1}{v_{in}}\right) = -\frac{R_L}{r_e}\left(\frac{\beta_0^2 r_e}{\beta_0^2 r_e + R_S}\right) \simeq -\frac{R_L}{r_e} = -g_m R_L \tag{5.10}$$

The loss of signal due to R_b for the biasing current is negligible in this case since the input impedance of a CE stage is much smaller than the value of R_b. The circuit properties of the CC-CE pair are listed in Table 5.1.

EXAMPLE 5.2

Determine R_{in}, R_o, A_i, and A_v of the CC-CE pair if the transistor parameters are given in (5.6) and $R_S = R_L = 1\ \text{k}\Omega$. From (5.9*a*)

$$R_{in} \simeq 50^2(25) = 62.5\ \text{k}\Omega$$

From (5.9*b*)

$$R_o = \frac{r_c}{\beta_0} = \frac{2 \times 10^6}{50} = 40\ \text{k}\Omega$$

From (5.7)

$$A_i = \frac{i_o}{i_1} = -\beta_0^2 = -2500$$

From (5.10)

$$A_v = \frac{v_o}{v_{in}} = -\frac{10^3}{25} = -40$$

Note that in this case both the input and output impedances are very high; $R_{in} \simeq 62\ \text{k}\Omega$, $R_o \simeq 40\ \text{k}\Omega$, $A_i \simeq -\beta_0^2$, and $A_v = -40$.

Finally note that in the super β composite transistor of Fig. 5.1*c*, if C^C is grounded, we have the circuit in Fig. 5.2*b*, namely, the CC-CC combination. If E^C is grounded, we have the CC-CE combination shown in Fig. 5.3*b*. The circuit properties of the Darlington configuration are listed in Table 5.1. Note the super β property of these circuits. Of these two combinations the CC-CE combination is usually preferable because we obtain a voltage gain as well as current gain; hence we can achieve a higher power gain. We shall also show in Sec. 8.8 that the CC-CE combination has a better frequency response characteristic than the CC-CC combination.

EXERCISE 5.1

For the Darlington configuration (CC-CC), shown schematically in Fig. E5.1, determine the power gain if the transistor parameters are $\beta_0 = 100$, $r_e = 5\ \Omega$, $r_b = 100\ \Omega$.

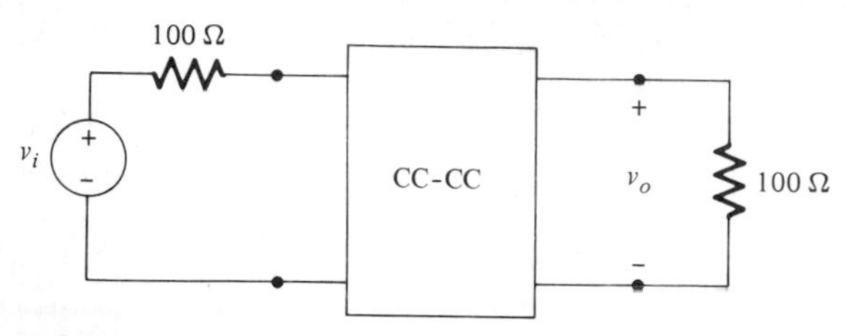

FIGURE E5.1

Ans 9.52×10^3 (39.8 dB)

EXERCISE 5.2

Repeat Exercise 5.1 for the CC-CE Darlington configuration.

Ans 1.6×10^5 (52 dB)

5.2 EMITTER-COUPLED AMPLIFIER (CC-CB)

An emitter-coupled amplifier circuit is shown in Fig. 5.4*a*. We will consider the circuit in a preliminary way to determine its basic properties. A more detailed analysis of the circuit, somewhat modified, is deferred to Sec. 5.6, where we consider differential amplifiers. The current gain of the circuit in Fig. 5.4*b*, for $R_E \gg R_{\text{in}2}$, is

$$\frac{i_o}{i_1} = \frac{i_o}{i_e}\left(\frac{i_e}{i_1}\right) = \alpha_0(-\beta_0) \simeq -\beta_0 \tag{5.11}$$

The voltage gain is given by

$$\frac{v_o}{v_{\text{in}}} = \frac{v_o}{v'}\left(\frac{v'}{v_1}\right)\left(\frac{v_1}{v_{\text{in}}}\right) \simeq \frac{R_L}{r_e}\left(\frac{R_{L1}}{R_{L1} + r_e}\right)\left(\frac{R_{\text{in}}}{R_{\text{in}} + R_S}\right) \tag{5.12}$$

Note that

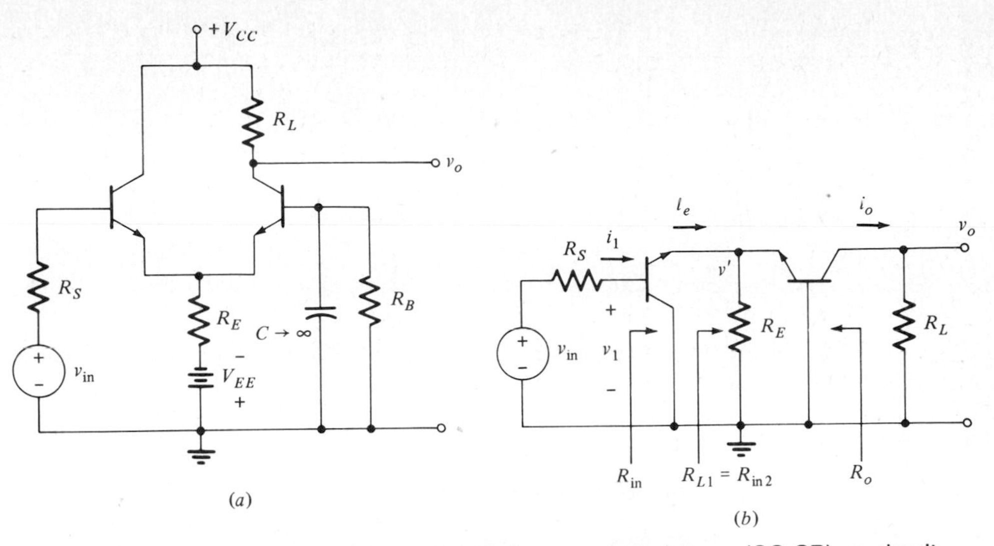

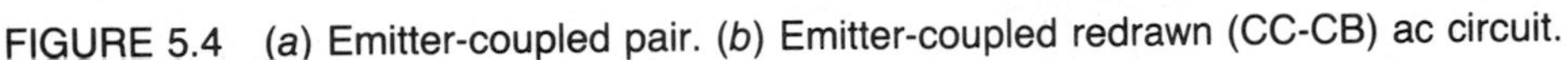
FIGURE 5.4 (*a*) Emitter-coupled pair. (*b*) Emitter-coupled redrawn (CC-CB) ac circuit.

$$R_{L1} = R_{\text{in}2} = r_e + \frac{r_b}{\beta_0} \simeq r_e$$

The input and output impedances of a CC-CB composite stage are given by

$$R_{\text{in}} \simeq \beta_0(R_{L1} + r_e) \simeq 2\beta_0 r_e \tag{5.13a}$$

$$R_o = r_c = h_{ob}^{-1} \tag{5.13b}$$

From (5.13*a*) and (5.12) we have

$$\frac{v_o}{v_{\text{in}}} = \frac{R_L}{r_e}\left(\frac{\beta_0 r_e}{2\beta_0 r_e + R_L}\right) \simeq \frac{\beta_0 R_L}{2\beta_0 r_e + R_S}$$

$$\simeq \frac{R_L}{2r_e} \simeq \frac{1}{2} g_m R_L \qquad \text{for } R_S \ll 2\beta_0 r_e$$

The input impedance is given by (5.13*a*) and the output impedance, R_o, is that of a CB stage (5.13*b*), hence R_o is almost infinite. The circuit properties of the CC-CB pair are listed in Table 5.1.

EXAMPLE 5.3

Determine R_{in}, R_o, A_i, and A_v of the CC-CB pair shown in Fig. 5.4*a*. The transistor parameters are given in (5.6), and let $R_S = R_L = 1\ \text{k}\Omega$, $R_E = 5\ \text{k}\Omega$. From (5.13*a*)

$$R_{\text{in}} = 2(50)(25) = 250\ \Omega$$

From (5.13*b*)

$$R_o = r_c = 2\ \text{M}\Omega$$

From (5.11)

$$A_i = -\beta_0 = -50$$

From (5.12*b*)

$$A_v = \frac{10^3}{2(25)} = 20$$

Note that the input impedance is small, $R_{\text{in}} = 250\ \Omega$, the output impedance is very large, $R_o = 2\ \text{m}\Omega$, and the current gain $A_i = -\beta_0$ and $A_v = 20$.

5.3 THE CASCODE CONFIGURATION (CE-CB)

The cascode configuration which is a combination of CE-CB in cascade, is shown in Fig. 5.5. The circuit can be similarly represented by a composite transistor. The composite equivalent transistor is shown in Fig. 5.5*b* and denoted by superscript "*e*" so as not to confuse it with the Darlington configuration. The gain and impedance calculations can be readily made as in the previous case with the aid of the results obtained in Table 4.2. However, in the cascode case we shall use the equivalent circuit approach, as it will demonstrate certain features of the circuit that will be helpful in later chapters.

The equivalent circuit of the cascode, using the simplified π models for the CE and CB transistors, is shown in Fig. 5.6. (Note that r_b is included with R_s', i.e., $R_s' = R_S + r_b$.)

From the circuit of Fig. 5.6 we have

$$\frac{i_o}{i_1} = \frac{i_o}{i_2}\left(\frac{i_2}{i_1}\right) = \frac{\alpha_0 i_2}{i_2}\left(\frac{-g_m v'}{i_1}\right) = \alpha_0\left(-\frac{g_m\beta_0}{g_m}\right) = -\alpha_0\beta_0 \tag{5.14}$$

To find the voltage gain, we have

$$v_o = (\alpha_0 i_2)R_L = -\alpha_0 g_m v' R_L$$

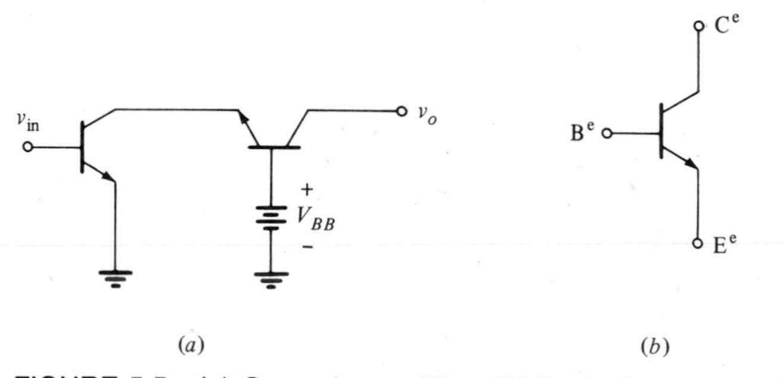

FIGURE 5.5 (*a*) Cascode amplifier. (*b*) Equivalent composite transistor.

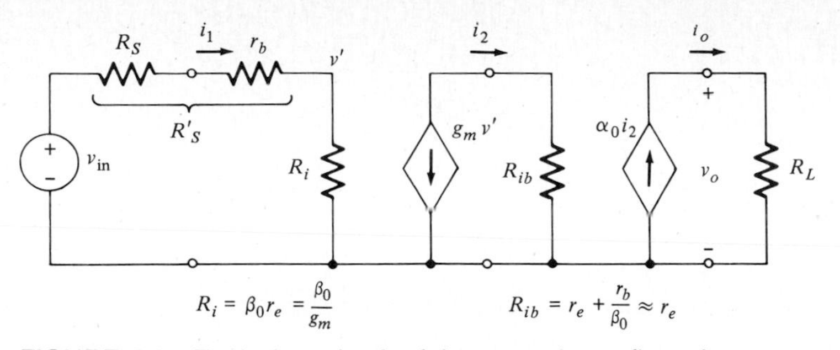

FIGURE 5.6 Equivalent circuit of the cascode configuration.

and

$$v' = \frac{R_i}{R_i + R_s'} v_{\text{in}} = \frac{\beta_0 r_e}{\beta_0 r_e + R_s'} v_{\text{in}}$$

Hence

$$\frac{v_o}{v_{\text{in}}} = -\alpha_0 g_m R_L \left(\frac{\beta_0 r_e}{\beta_0 r_e + R_s'} \right) \simeq -\frac{\beta_0 R_L}{\beta_0 r_e + R_s'} \tag{5.15}$$

The input and output resistances of a cascode are the input resistance of a CE and the output resistance of a CB stage, respectively, i.e.,

$$R_{\text{in}} \simeq \beta_0 r_e + r_b \qquad R_o = r_c \rightarrow \infty \tag{5.16}$$

It is interesting to note that the overall h parameters of the cascode (see Fig. 5.5b) are given by

$$\begin{aligned} h_{11}^e &\simeq h_{ie} & h_{12}^e &\simeq h_{re} h_{rb} \rightarrow 0 \\ h_{21}^e &\simeq h_{fe} & h_{22}^e &\simeq h_{ob} \rightarrow 0 \end{aligned} \tag{5.17}$$

The noninteractive (i.e., the reverse transmission $h_{12}^e \simeq 0$) nature of the cascode is especially useful at high frequencies (Sec. 8.3) and makes it especially attractive for isolating purposes such as in tuned amplifiers (Sec. 11.7). High-frequency behavior of this circuit is discussed in Sec. 8.3.

EXAMPLE 5.4

Determine R_{in}, R_o, A_i, A_v, and the overall h parameters of the equivalent cascode transistor shown in Fig. 5.5. The transistor parameters are given in (5.6) and $R_S = R_L = 1\ \text{k}\Omega$. From (5.16)

$$R_{\text{in}} = 50(25) + 25 = 1.45\ \text{k}\Omega \qquad R_o \simeq 2 \times 10^6\ \Omega$$

From (5.14)

$$A_i = -\beta_0 = -50$$

From (5.15)

$$A_v = -\frac{50(10^3)}{50(25) + 200 + 10^3} = -20.4$$

The overall h parameters of the cascode are determined from (5.17) and (4.12), namely,

$$h_{11}^e \simeq h_{ie} = 1.4 \times 10^3 \ \Omega$$

$$h_{12}^e \simeq h_{re}h_{rb} = 10^{-4}\left(\frac{200}{2 \times 10^6}\right) = 10^{-8} \simeq 0$$

$$h_{21}^e \simeq h_{fe} = 50$$

$$h_{22}^e \simeq h_{ob} = \frac{1}{2 \times 10^6} = 0.5 \times 10^{-6} \text{ mho}$$

Note that the reverse transmission (determined by h_{12}^e) is *truly negligible,* and the output impedance of the composite circuit is much larger than that of a CE stage.

The circuit properties of the cascode configuration are listed in Table 5.1.

EXERCISE 5.3

The cascode circuit, shown schematically in Fig. E5.3, is described by the following h parameters:

$$h_{11}^e = 600 \ \Omega \qquad h_{12}^e \simeq 0 \qquad h_{21}^e = 100 \qquad h_{22}^e \simeq 0$$

Determine R_i, R_o, A_i, and A_v for $R_S = R_L = 100 \ \Omega$.

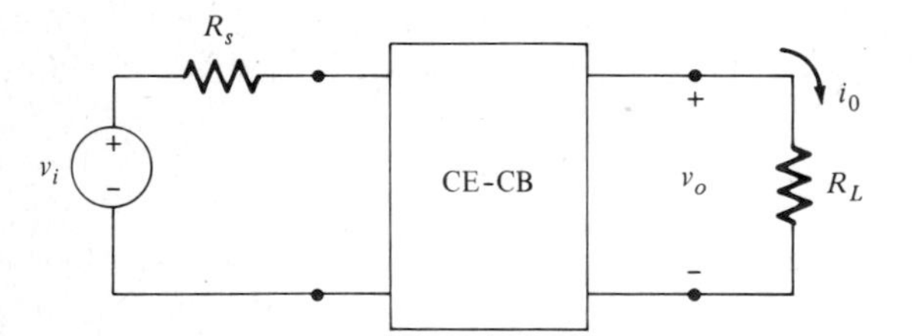

FIGURE E5.3

Ans $R_i = 600 \ \Omega$, $R_o \simeq \infty$, $A_i = -100$, $A_v = -14.3$

5.4 COMPARISON OF TWO-TRANSISTOR CIRCUITS

In Secs. 5.1 to 5.3 we considered some useful circuits comprised of two transistors. Each configuration has its own merit and application; however, we shall briefly compare these circuits in terms of their salient features. We shall consider low frequencies only, as their high-frequency performance characteristics are treated in Chap. 8.

Voltage gain, current gain, and input and output resistances of the various circuits are listed in Table 5.1. Note that v_1 is the input voltage source to the pair. If the voltage source v_{in} has an associated input source resistance, it should be incorporated in the analysis as discussed in the previous sections. Similarly the current gain is defined as i_o/i_1, where i_1 is the current input into the first transistor.

From Table 5.1, it is noted that the CC-CC pair has the highest input resistance and the lowest output resistance and is the compound equivalent of a CC stage. Note the super β performance of the Darlington configurations (β_0^2 in CC-CC and CC-CE configurations). The latter configuration also provides a phase reversal. The voltage gain is the highest for CC-CB and CE-CB configurations, and both configurations provide phase reversals. The use of an emitter-coupled pair in a differential amplifier configuration is extremely important in operational amplifiers and is discussed further in Sec. 5.7. The use of cascode amplifiers in tuned amplifier design is discussed in Chap. 11, and their low noise performance is discussed in Appendix D.

5.5 CMOS AMPLIFIER

A CMOS amplifier, exclusive of biasing circuitry, is shown in Fig. 5.7*a*. This circuit has already been discussed as an inverter in Sec. 3.6. Note that as an amplifier the Q point (assuming identical MOS) is at $V_{DD}/2$ (see Fig. 3.24). The equivalent circuit of the CMOS amplifier is shown in Fig. 5.7*b*. Note that the NMOS and PMOS devices are in parallel; hence the $2g_m$ and $r_d \parallel r_d = r_d/2$ in the equivalent circuit. The resistor R_G is shown dotted, as it is a high-valued resistor in the order of megaohms.

The voltage gain of the circuit is determined from

$$v_o = -2g_m v_{gs}(R_L \parallel 0.5r_d) \qquad v_{gs} \simeq v_i$$

Hence

$$A_v = \frac{v_o}{v_i} = -g_m\left(\frac{R_L r_d}{R_L + 0.5r_d}\right) \tag{5.18}$$

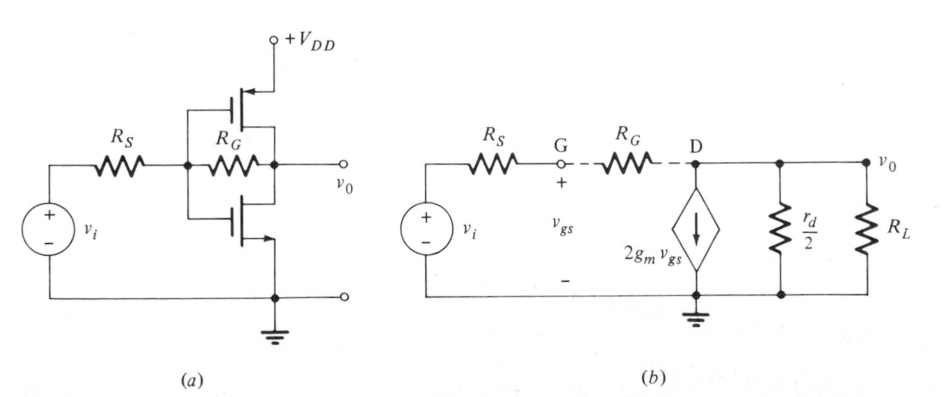

FIGURE 5.7 (*a*) CMOS amplifier. (*b*) Small-signal equivalent circuit of CMOS.

For example, if $g_m = 2 \times 10^{-3}$ mho, $r_d = 20$ kΩ, and $R_L = R_S = 1$ kΩ, we have

$$A_v = -2 \times 10^{-3}\left(\frac{1 \times 20}{1 + 10}\right) \times 10^3 = -36.4$$

One of the applications of CMOS in conjunction with crystals to obtain a crystal oscillator is given in Sec. 10.16.

5.6 MULTISTAGE CIRCUITS USING BOTH FET AND BIPOLAR TRANSISTORS

We have already seen that the input impedance of FET devices is extremely high. This property of FETs is often used to advantage as the *input circuit* for a voltage-source drive. The low-noise properties of FETs also make them attractive input circuits. The source-coupled JFETs are widely used in the differential input stage of an operational amplifier (op amp). Op amps are discussed in detail in Sec. 5.10. Many other combinations of FETs and BJTs are possible, and we shall consider three arrangements in the following sections.

B. Common-Source, Common-Base (CS-CB) Configuration

A CS-CB configuration, including the biasing circuitry is shown in Fig. 5.8*a*. The equivalent low-frequency small-signal model of the circuit is shown in Fig. 5.8*b*. The voltage gain is readily determined from the circuit. Note that the input resistance of a CB stage is $R_{ib} \ll r_d$, and for the FET stage $R_i \gg R_g$ (since the input resistance of a FET is larger than 10^{10} Ω); hence we have

$$v_o = i_o R_L = (\alpha_0 i_b) R_L \simeq \alpha_0(-g_{m1} v') R_L$$

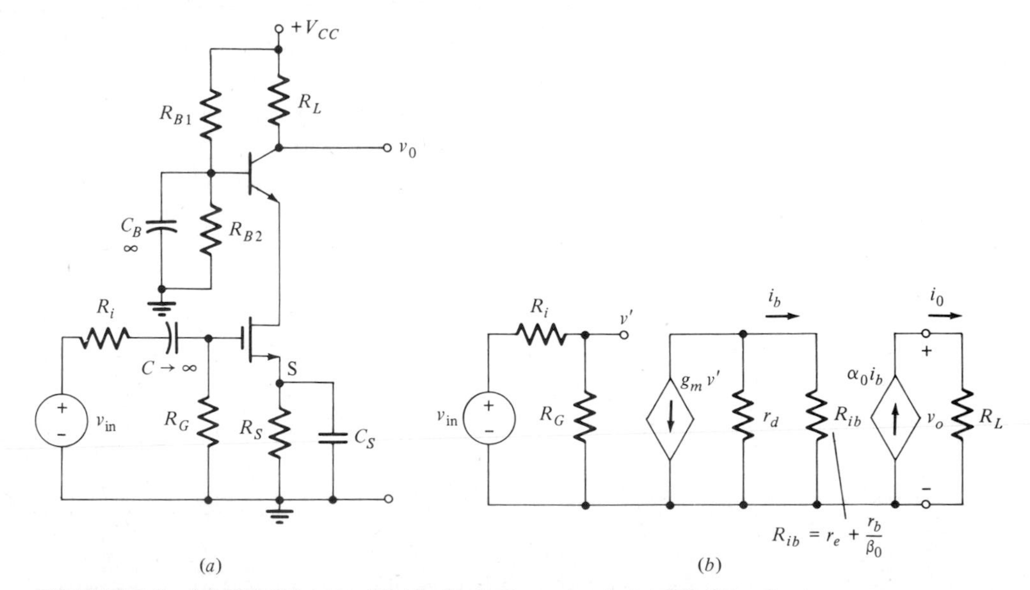

FIGURE 5.8 (*a*) CS-CB pair. (*b*) Equivalent ac circuit for CS-CB pair.

Thus

$$\frac{v_o}{v_{\text{in}}} = \frac{v_o}{v'}\left(\frac{v'}{v_{\text{in}}}\right) \simeq -\alpha_0 g_{m1} R_L \left(\frac{R_g}{R_g + R_i}\right) \simeq -g_{m1} R_L \tag{5.19}$$

Note that g_{m1} is the transconductance of the FET and is therefore designated with subscript 1. The reader should observe the similarities of the circuit with those of the bipolar cascode circuit. Note, however, that R_{in} of the FET is larger than that of BJT by several orders of magnitude and also observe that g_{m1} of the FET is usually much smaller than g_m of a BJT. The output impedance of the circuit is that of a CB stage.

B. Common-Drain, Common-Base (CD-CB) Configuration

The CD-CB configuration including the biasing circuitry is shown in Fig. 5.9*a*. This circuit is the hybrid equivalent of the emitter-coupled pair. The ac circuit is redrawn in Fig. 5.9*b* for the purpose of illustrating the following calculations. The voltage gain of the amplifier can be readily obtained by using the results of Chap. 4, as we did in the previous sections. First note that $R_G \gg R_S$ and $R_E \gg R_{ib} \simeq r_{e2}$, and thus $R_{L1} \simeq r_{e2}$. From Table 4.5 the voltage gain of a CD stage is

$$\frac{v'}{v_{\text{in}}} \simeq \frac{g_{m1} R_{L1}}{1 + g_{m1} R_{L1}} \simeq \frac{g_{m1} r_{e2}}{1 + g_{m1} r_{e2}}$$

and

$$\frac{v_o}{v'} \simeq g_{m2} R_L$$

Hence

$$\frac{v_o}{v_{\text{in}}} \simeq \frac{v_o}{v'}\left(\frac{v'}{v_{\text{in}}}\right) \simeq \frac{g_{m1} R_L}{1 + g_{m1} r_{e2}}$$

$$\simeq g_{m1} R_L \qquad \text{for } g_{m1} r_{e2} \ll 1 \tag{5.20}$$

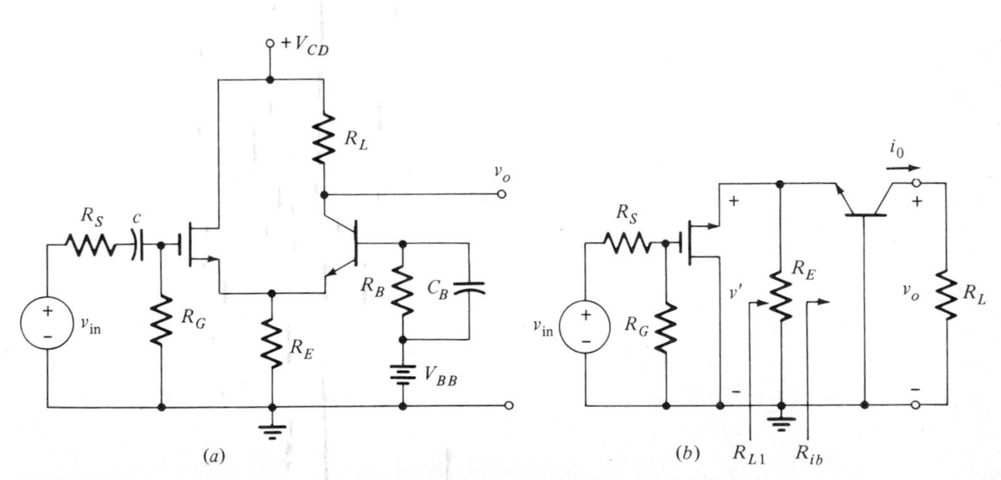

FIGURE 5.9 (*a*) CD-CB pair. (*b*) Equivalent ac circuit for CD-CB pair.

The output impedance of the circuit is that of a CB stage, namely, $R_o = r_c = h_{ob}^{-1}$. Note that the CD-CB cascade has the same voltage gain and impedance properties as the CS-CB cascade. In the CD-CB case we do not obtain polarity inversion, whereas in the CS-CB case we do.

C. Common-Source, Common-Emitter (CS-CE) Configuration

The complete CS-CE circuit and its equivalent are shown in Fig. 5.10*a* and *b*, respectively. Note that $R_G = R_{G1} \| R_{G2}$ and $R_B = R_{B1} \| R_{B2}$ are very large resistors. Although included in the circuit, they can be ignored. Also the resistor r_b of the CE stage is omitted for simplicity (its inclusion should pose no difficulty). The voltage gain (either using the results of Chap. 4 or using the equivalent circuit directly) can be readily determined. From the equivalent circuit in Fig. 5.10*b* we have

$$v_o = (-g_{m2}v')R_L$$

$$v' = -g_{m1}v_1R_{L1} \simeq -g_{m1}v_1(R_D \| R_i) \qquad \text{for } r_d \gg R_D,\ R_B \gg R_i$$

$$v_1 = \frac{R_G}{R_G + R_S}v_{in} \simeq v_{in}$$

since $R_G \gg R_S$. Hence

$$\frac{v_o}{v_{in}} = \frac{v_o}{v'}\left(\frac{v'}{v_1}\right)\left(\frac{v_1}{v_{in}}\right) \simeq -g_{m2}R_L(-g_{m1}R_{L1}) \simeq g_{m1}R_{L1}g_{m2}R_L \tag{5.21a}$$

Substitution for $R_{L1} = R_D \| R_i$, $R_i = \beta_0 r_{e2}$ and $g_{m2} = 1/r_{e2}$ in (5.21*a*) yields

$$\frac{v_o}{v_{in}} \simeq \beta_0 g_{m1}\left(\frac{R_LR_D}{R_D + \beta_0 r_{e2}}\right) \tag{5.21b}$$

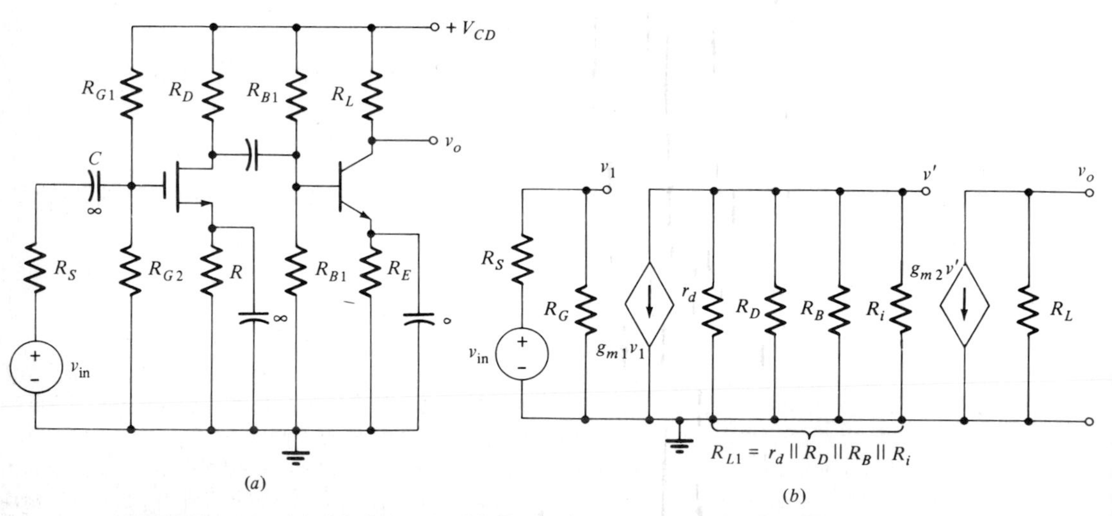

FIGURE 5.10 (*a*) CS-CE pair. (*b*) Equivalent ac circuit for CS-CE pair.

Note the distinction between g_{m1} and g_{m2} of the FET and the BJT, respectively. Observe that for this circuit the output impedance is that of a CE stage, which is very high.

EXAMPLE 5.5

For the circuit shown in Fig. 5.10*a*, the BJT parameters are as in (5.6), and the FET parameters are $g_m = 2 \times 10^{-3}$ mho and $r_d = 20$ kΩ. The other circuit parameters are $R_L = R_S = R_D = 1$ kΩ and $R_B = R_{B1} \| R_{B2} = 5$ kΩ. Determine the voltage gain of the circuit.

We calculate $R_{L1} = r_d \| R_D \| R_B \| R_i = (20 \| 1 \| 5 \| 1.25) \times 10^3 = 0.50$ kΩ. From (5.21*a*)

$$A_v = \frac{v_o}{v_{\text{in}}} = (2 \times 10^{-3})(0.5 \times 10^3)\left(\frac{1}{25}\right)(10^3) = 40$$

Note that the FET stage voltage gain is about unity, and the gain is provided by the BJT.

EXERCISE 5.4

For the circuit shown in Fig. 5.8*a*, the FET and BJT parameters are given as

FET:

$$g_m = 4 \times 10^{-3} \text{ mho} \qquad r_d = 10 \text{ k}\Omega$$

BJT:

$$g_m = 10^{-1} \text{ mho} \qquad \beta_0 = 100 \qquad r_b = 100\ \Omega$$

The source and the load resistances are 1 kΩ each. Determine the voltage gain, the input impedance, and the output impedance.

Ans $A_v = -4$, $R_{\text{in}} \simeq R_G$, $R_o \simeq \infty$

5.7 CASCADED AMPLIFIERS

We shall now consider cascaded amplifiers that use more than two transistors. A general cascaded amplifier is shown in Fig. 5.11. The individual two ports, indicated by boxes, may contain BJTs, FETs, or a combination of these. Note that the individual amplifying stages in Fig. 5.11 may be single stages or composite stages, which can be characterized as shown in Fig. 5.12. Note further that we are assuming that there is *negligible reverse transmission*, hence no interaction. The assumption of non-interaction is not always valid, especially at high frequencies. If the stages are

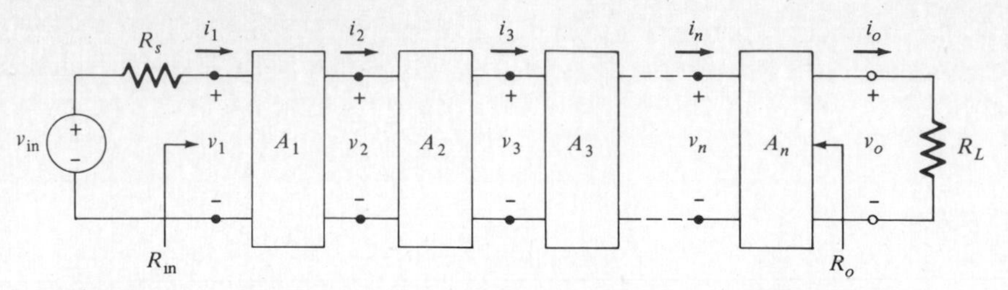

FIGURE 5.11 Schematic representation of cascaded amplifier stages.

interacting (i.e., coupled), we have to write the equilibrium equations for the entire system, which can be very complicated. Fortunately, however, most amplifying stages are only weakly interacting and can be represented approximately by the equivalent circuit of Fig. 5.12*b*.

The voltage and current gains, for noninteracting stages (Fig. 5.11), can be written as:

$$a_v = \frac{v_o}{v_1} = \frac{v_o}{v_n}\left(\frac{v_n}{v_{n-1}}\right)\cdots\left(\frac{v_3}{v_2}\right)\left(\frac{v_2}{v_1}\right) \tag{5.22a}$$

$$= \prod_{l=1}^{n} A_{vl} \tag{5.22b}$$

and

$$A_v = \frac{v_o}{v_{in}} = \left(\prod_{l=1}^{n} A_{vl}\right)\left(\frac{R_{in}}{R_{in} + R_S}\right) \tag{5.23}$$

where

$$A_{vl} = \frac{v_{l+1}}{v_l} = \frac{(R_i)_{l+1}}{(R_i)_{l+1} + (R_o)_l}\, a_l \simeq a_l \qquad \text{for } (R_i)_{l+1} \gg (R_o)_l$$

Similarly,

$$a_i = \frac{i_o}{i_1} = \frac{i_o}{i_n}\left(\frac{i_n}{i_{n-1}}\right)\cdots\left(\frac{i_3}{i_2}\right)\left(\frac{i_2}{i_1}\right) \tag{5.24a}$$

$$= \prod_{k=1}^{n} A_{ik} \tag{5.24b}$$

where

$$A_{ik} = \frac{i_{k+1}}{i_k}$$

$$i_1 = \frac{v_{in}}{R_S + (R_{in})}$$

Note that in cascaded amplifiers the *overall gain is the product of the individual stage gains*. If the circuit properties of the individual blocks are known, the voltage and current gains can be readily determined from (5.22) through (5.24).

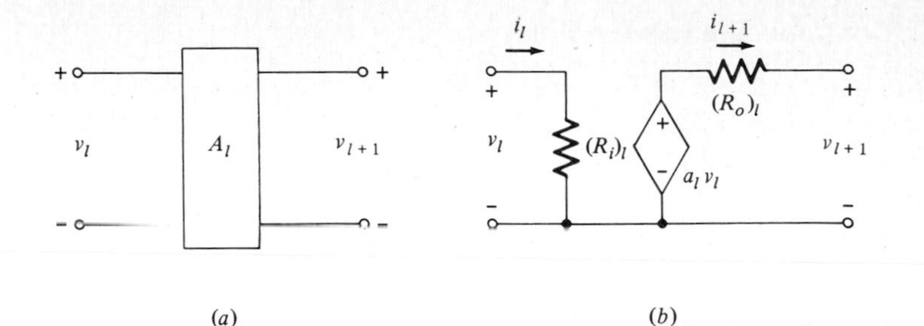

FIGURE 5.12 (a) Typical internal amplifier block. (b) Corresponding equivalent circuit (noninteracting $R_i \gg R_o$).

It should also be noticed that the input and the output stages are usually different, depending on the source and load requirements. For example, for a voltage-source drive we want the input stage to have a very large input impedance and for a current-source drive the input stage should have a very low input impedance. Noise considerations may dictate the use of an FET stage at the input. The input stage and the noise consideration are discussed separately in Appendix D. The output stage is usually determined by the load conditions and maximum signal swing without distortion. Output stages and power amplifiers are considered in Chap. 6.

EXERCISE 5.5

A multistage amplifier is shown schematically in Fig. E5.5. Determine the voltage gain, if the basic parameters of each amplifier, represented by Fig. 5.12, are given as follows:

$$A_1\text{: } a_1 = 20 \qquad R_i = 100\ \text{k}\Omega \qquad R_o = 500\ \Omega$$
$$A_2\text{: } a_2 = 25 \qquad R_i = 1\ \text{k}\Omega \qquad R_o = 200\ \Omega$$

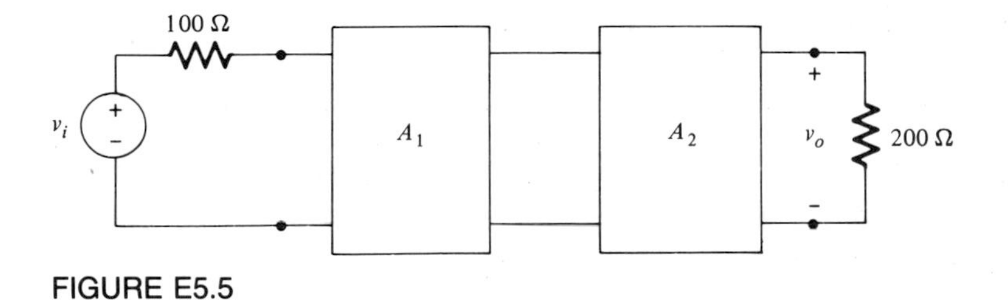

FIGURE E5.5

Ans 166.7

We shall next consider gain calculations for two three stage amplifiers: a cascade of three BJTs in the CE configuration and a cascade of 3 FETs in the CS configuration.

A. Cascaded BJT Common-Emitter Amplifiers

Consider the ac signal circuit of a three-stage BJT cascaded amplifier in the CE configuration as shown in Fig. 5.13*a*. (The complete circuit, which includes biasing, is shown in Fig. 8.14.) From the equivalent circuit in Fig. 5.13*b*, the voltage and current gains can be readily determined. Note that $R_i = \beta_0 r_e = \beta_0/g_m$, r_b is ignored for simplicity; the inclusion of r_b should pose no problem to the reader. The voltage gain is

$$A_v = \frac{v_o}{v_{\text{in}}} = \frac{v_o}{v_3}\left(\frac{v_3}{v_2}\right)\left(\frac{v_2}{v_1}\right)\left(\frac{v_1}{v_{\text{in}}}\right) \tag{5.25}$$

$$v_o = -g_{m3}\, v_3 R_{L3} = -g_{m3} R_L\, v_3$$

$$v_3 = -g_{m2}\, v_2 R_{L2} = -g_{m2}\,(R_3 \| R_{i3})\, v_2$$

$$v_2 = -g_{m1}\, v_1\,(R_{L1}) = -g_{m1}\,(R_2 \| R_{i2})\, v_1$$

$$v_1 = \frac{R_1 \| R_{i1}}{R_S + (R_1 \| R_{i1})}$$

From the above equations, assuming identical transistors, and

$$g_{m1} = g_{m2} = g_{m3} = g_m \qquad R_{i1} = R_{i2} = R_{i3} = R_i$$

we have

$$A_v = \frac{v_o}{v_{\text{in}}} = (-g_m)^3 R_L\,(R_3 \| R_i)(R_2 \| R_i)\left(\frac{R_1 R_i}{R_1 R_S + (R_1 + R_S)R_i}\right) \tag{5.26}$$

Similarly the current gain is found from

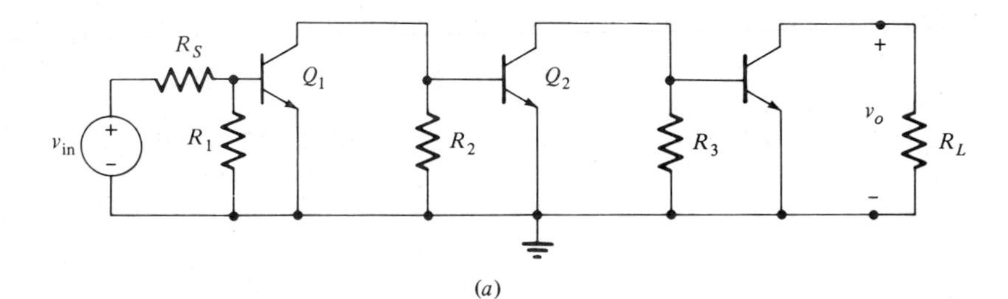

(*a*)

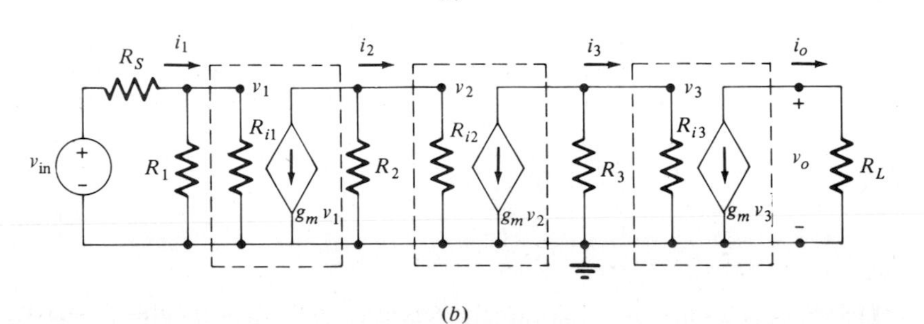

(*b*)

FIGURE 5.13 (*a*) CE cascaded stages (ac circuit). (*b*) Equivalent small-signal model of (*a*).

$$\frac{i_o}{i_1} = \frac{i_o}{i_3}\left(\frac{i_3}{i_2}\right)\left(\frac{i_2}{1}\right) \tag{5.27}$$

where $i_o = -g_m v_3 = -g_m(R_i i_3) = -g_m(\beta_0 r_e i_3) = -\beta_0 i_3$

$$i_3 = (-g_m v_2)\frac{R_3}{R_3 + R_i} = -g_m(i_2 R_i)\frac{R_3}{R_3 + R_i} = -\beta_0 \frac{R_3}{R_3 + R_i} i_2$$

$$i_2 = -(g_m v_1)\frac{R_2}{R_2 + R_i} = -\beta_0 \frac{R_2}{R_2 + R_i} i_1$$

Thus

$$\frac{i_o}{i_1} = (-\beta_0)\left(-\frac{\beta_0 R_3}{R_3 + R_i}\right)\left(-\beta_0 \frac{R_2}{R_2 + R_i}\right)$$

$$\frac{i_o}{i_1} = -\beta_0^3 \frac{R_2 R_3}{(R_2 + R_i)(R_3 + R_i)} \tag{5.28}$$

Note that

$$i_1 = \frac{v_{\text{in}}}{R_S} \frac{(R_1 \| R_S)}{(R_1 \| R_S) + R_i}$$

The signal loss due to the resistors R_1, R_2, and R_3 should be noted.

EXAMPLE 5.6

Determine the voltage and the current gains of the circuit in Fig. 5.13*a*. Assume the transistor parameters given in (5.6) and $R_S = R_1 = R_2 = R_3 = R_L = 1\ \text{k}\Omega$. We first calculate $g_m = 1/r_e = 0.04$ mho and $R_i = \beta_0 r_e = 1.25\ \text{k}\Omega$. Hence from (5.26)

$$A_v = \frac{v_o}{v_{\text{in}}} = -(0.04^3)(10^3)(1 \| 1.25) \times 10^3 (1 \| 1.25) \times 10^3 = 1.97 \times 10^4$$

Similarly,

$$A_i = \frac{i_o}{i_1} = -(50^3)\frac{1 + 1}{(1 + 1.25)(1 + 1.25)} = 4.94 \times 10^4$$

The resistors R_1, R_2, and R_3 are often designed to be smaller than in this example in order to get a wide bandwidth. Bandwidth determination and the gain-bandwidth trade-off are discussed in Chaps. 7 and 8.

B. Cascaded FET Common-Source Amplifiers

A three-stage cascaded CS amplifier exclusive of biasing circuitry is shown in Fig. 5.14*a*. (The complete circuit, which includes biasing, is shown in Fig. 8.10.) The equivalent circuit of Fig. 5.14*a* is shown in Fig. 5.14*b*. The voltage gain of the circuit

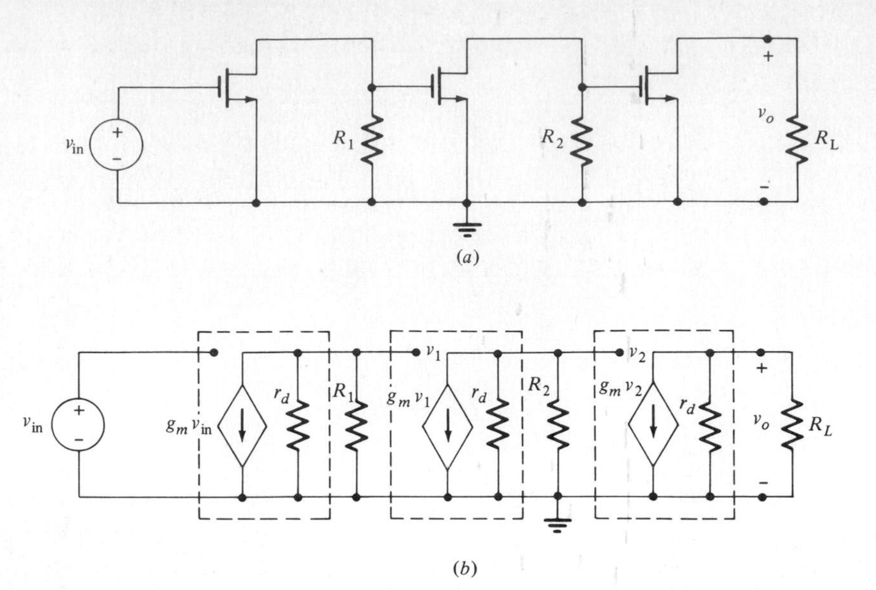

FIGURE 5.14 (*a*) CS cascaded stages (ac circuit). (*b*) Equivalent small-signal model of (*a*).

is given by

$$A_v = \frac{v_o}{v_{in}} = \frac{v_o}{v_2}\frac{v_2}{v_1}\frac{v_1}{v_{in}} \tag{5.29}$$

For identical FETs, i.e., $g_{mi} = g_m$ and $r_{di} = r_d$,

$$\begin{aligned} A_v &\simeq -g_m(R_L \parallel r_d)g_m(R_2 \parallel r_d)g_m(R_1 \parallel r_d) \\ &\simeq -g_m^3 R_L R_2 R_1 \qquad r_d \gg R_1,\ R_2,\ R_L \end{aligned} \tag{5.30}$$

EXAMPLE 5.7

For the cascaded FET amplifier shown in Fig. 5.14*a*, the FET parameters are $g_m = 2 \times 10^{-3}$ mho, $r_d = 20\ \text{k}\Omega$, and $R_1 = R_2 = R_L = 5\ \text{k}\Omega$. Determine the voltage gain of the amplifier.

From (5.30)

$$A_v = \frac{v_o}{v_{in}} = -[(2 \times 10^{-3})(5 \parallel 20) \times 10^3]^3 = 512$$

Note that the voltage gain of a FET is much lower than that of a BJT, in spite of the fact that higher-valued resistors are used. The smaller gain of FETs are due to the smaller value of g_m in FETs as compared to that of BJTs. A hybrid FET, BJT amplifier with the FET stage at the input is commonly used to take advantage of the desirable properties of each device.

Notice that in the above examples of cascaded amplifiers, calculations were made directly using the circuit models and the general formulas in (5.22) were not used. If the circuit properties are known, e.g., a_v, R_i, and R_O in Fig. 5.12*b*, we could use the results in (5.22). Such calculations are convenient for composite transistors.

It should be noted that there are a large class of multiple-transistor amplifiers that employ feedback. Feedback amplifiers are discussed in detail in Chap. 10.

5.8 DIFFERENTIAL AMPLIFIERS

A *differential amplifier* is also called a *difference amplifier,* and as the name implies, its function is to amplify the difference between two signals. The differential amplifier finds wide usage as a gain block in integrated circuits. Because of its balanced nature and symmetry, as we shall show, it can amplify very small signals. A differential amplifier usually requires a minimum number of external capacitors and can be used without bypass and coupling capacitors. Hence it is used for dc and ac amplification. The differential amplifier is the basic building block of operational amplifiers. Monolithic operational amplifiers are most widely used in analog integrated circuits and find a variety of applications. Operational amplifiers (op amps) are almost ideal gain blocks and are used in a very large number of applications. Because of the variety of applications of op amps in linear and nonlinear circuits, the op amp is discussed in detail in Chap. 9.

A. General Consideration

In order to illustrate the basic principles of operation of a differential amplifier, we will consider the schematic representation shown in Fig. 5.15. In an ideal differential amplifier the output voltage signal would be given by

$$v_o = A(v_1 - v_2) \tag{5.31}$$

where A is the voltage gain of the amplifier. In an actual differential amplifier the output signal is given by superposition as

$$v_o = A_1 v_1 + A_2 v_2 \tag{5.32}$$

where A_1 and A_2 depend on the difference and the sum of the signals. In *symmetrical* circuits it is convenient to talk about the in-phase signals, which are termed *common-mode* (CM) signals, and the difference or antiphase signals, which are called

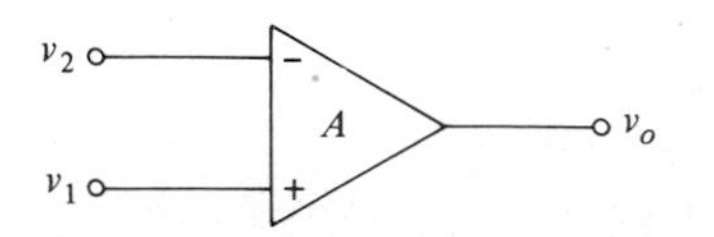

FIGURE 5.15 Schematic of a differential amplifier.

differential-mode (DM) signals. In accordance with this terminology, the CM signal v_c is defined by an average voltage, namely,

$$v_c = \tfrac{1}{2}(v_1 + v_2) \tag{5.33}$$

and the DM signal v_d is defined by *half* the difference, namely,

$$v_d = \tfrac{1}{2}(v_1 - v_2) \tag{5.34}$$

Then from (5.32), (5.33), and (5.34) we obtain

$$v_o = A_1(v_d + v_c) + A_2(v_c - v_d) \tag{5.35a}$$

$$v_o = A_d v_d\left(1 + \frac{v_c A_c}{v_d A_d}\right) \tag{5.35b}$$

where

$$\begin{aligned} A_d &= A_1 - A_2 \\ A_c &= A_1 + A_2 \end{aligned} \tag{5.36}$$

Note that A_d and A_c can be readily measured in a differential amplifier. For example, from (5.35*b*) we have

$$A_d = \left.\frac{v_o}{v_d}\right|_{v_c=0} \tag{5.37}$$

But $v_c = 0$ implies $v_1 = -v_2$. Hence we apply equal magnitude signals of opposite polarity and measure the output voltage. The resulting voltage gain for equal-magnitude antiphase input signals yields A_d. Similarly, from (5.35*b*) we have

$$v_o = A_c v_c\left(1 + \frac{A_d v_d}{A_c v_c}\right) \tag{5.38}$$

and

$$A_c = \left.\frac{v_0}{v_c}\right|_{v_d=0} \tag{5.39}$$

But $v_d = 0$ implies $v_1 = v_2$. Hence we apply equal-magnitude signals of the same polarity and measure the output voltage. The voltage gain of the equal-magnitude in-phase signal yields A_c. Generally, the desired signals in a differential amplifier are DM, and the undesired signals, such as those caused by temperature drift, are CM.

We now define a quantity given by most manufacturers called the *common-mode rejection ratio* (CMRR) as follows:

$$\text{CMRR} = \left|\frac{A_d}{A_c}\right| \tag{5.40}$$

The quantity CMRR is usually used as figure of merit for a differential amplifier. It is also sometimes referred to as the *discrimination factor* of a differential amplifier.

Ideally $A_c \simeq 0$. However, in most practical amplifiers A_c is nonzero but very small, whereas A_d is very large. Thus, CMRR is a very large number in a well-designed differential amplifier (of the order of $10^5 = 100$ dB, for a multistage differential amplifier). Substitution of (5.40) into (5.35*b*) yields

$$v_o = A_d v_d \left(1 + \frac{1}{\text{CMRR}} \frac{v_c}{v_d}\right) \simeq A_d v_d \tag{5.41}$$

Since CMRR is a very large number, and since for v_c and v_d of the same order of magnitude $(1/\text{CMRR})(v_c/v_d) \ll 1$, (5.41) is approximately equal to (5.31). From (5.31), (5.34), and (5.41) we have $A_d/2 = A$. Note that in a differential amplifier with $\text{CMRR} = 10^5$ a 1-μV differential input would give the same output as a 100-mV common-mode signal.

A basic differential amplifier circuit is shown in Fig. 5.16, which is an emitter-coupled amplifier. Due to the symmetry of the circuit, we can obtain considerable simplification if the transistors and resistors are assumed to be identical. We repeat that this is not a difficult condition to attain with integrated-circuit technology. Precise matching between the active and passive components is obtained because they are simultaneously fabricated in adjacent areas on a small chip. Because of this, the transistor pair also has a very good temperature-tracking characteristic. Thus the analysis of the circuit in Fig. 5.16 can be considerably simplified if we make use of its *symmetry*.

Let us consider a general symmetrical network N shown in Fig. 5.17*a*. If we apply equal signal voltages v_1 at both ports, the currents in the inside terminals i_a are equal due to symmetry; hence the total current (by superposition) is zero, and the terminals can be opened without affecting the circuit as shown in Fig. 5.17*b*. Now consider antiphase signals applied as shown in Fig. 5.17*c*. The voltage at wire *a* is, say, v_a due to $+v_1$ and $-v_a$ due to $-v_1$; hence by superposition the total voltage at *a* is zero, and we can ground these terminals without affecting the circuit.

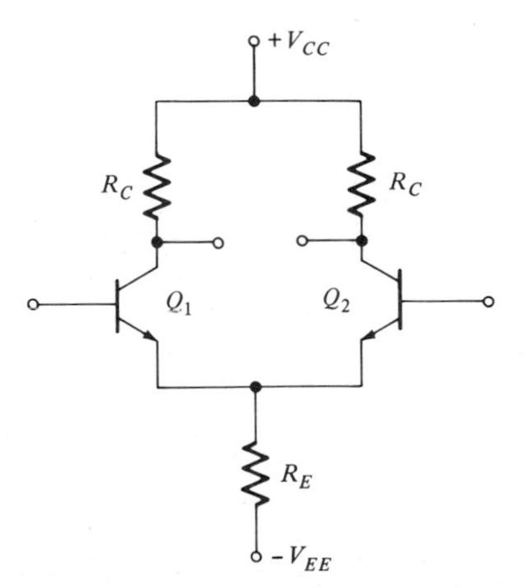

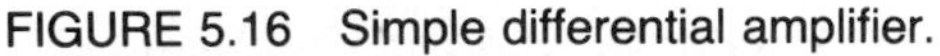
FIGURE 5.16 Simple differential amplifier.

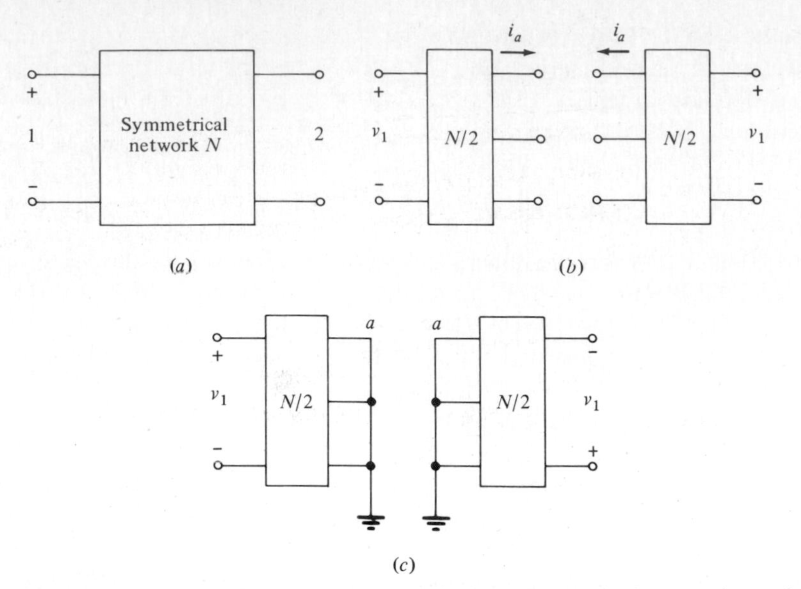

FIGURE 5.17 (a) General symmetrical linear network shown schematically. (b) Same circuit showing the half-network under in-phase equal excitations. (c) Same circuit under antiphase equal excitations.

EXERCISE 5.6

For the circuit shown in Fig. E5.6, determine the input impedance, R_i if (a) $v_2 = v_1$ or (b) $v_2 = -v_1$.

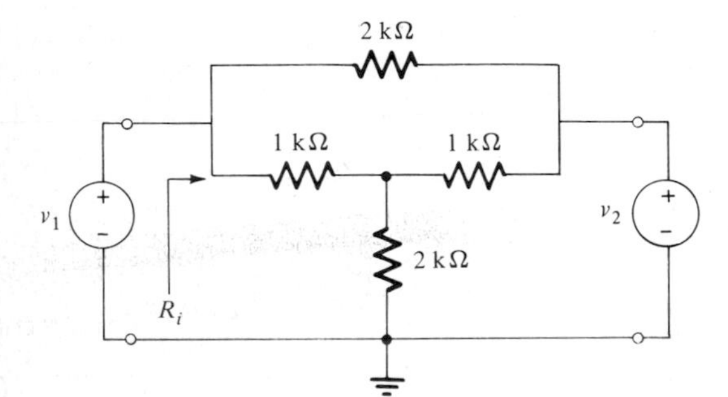

FIGURE E5.6

Ans (a) $R_i = 5\ \text{k}\Omega$, (b) $R_i = 0.5\ \text{k}\Omega$

B. Analysis of the Basic Differential Amplifier

Consider the basic differential amplifier redrawn in Fig. 5.18a, showing complete symmetry. We assume identical transistors and matched resistors. The ac half-circuit under common-mode and differential-mode signals are shown in Fig. 5.18*b* and *c*, respectively.

The differential-mode and common-mode voltage gains can be readily determined for Fig. 5.18*b* and *c*. Namely, for the differential-mode half-circuit we have the simple CE stage; hence its gain, as determined in Chap. 4, is given by

$$A_d = \frac{v_{o1}}{v_d/2} = \frac{-\beta_0 R_c}{\beta_0 r_e + r_b} \simeq \frac{-R_c}{r_e} \simeq -g_m R_c \qquad \text{for } \beta_0 r_e \gg r_b \tag{5.42}$$

The output voltage of a differential amplifier can be taken either differentially (i.e., $v_{o1} - v_{o2}$) or single ended (i.e., v_{o1}). For the single-ended case, from (5.42)

$$A_d = \frac{v_{o1}}{v_d} = -\frac{1}{2} g_m R_C \tag{5.43}$$

For the differential output,

$$A_d = \frac{v_{o1} - v_{o2}}{v_d} = -g_m R_C \tag{5.44}$$

Similarly for the common-mode case, the voltage gain is obtained from the half-circuit in Fig. 5.18*c*. We have already obtained the result for the phase splitter in (4.53); however, in this case $R_E > R_C$, and hence

$$A_c = \frac{v_{o1}}{v_c} \simeq -\frac{\alpha_0 R_C}{2R_E + r_e} \simeq -\frac{R_C}{2R_E} \tag{5.45}$$

If the output is taken differentially, $v_{o1} - v_{o2} = 0$; hence $A_c = 0$. From (5.43) and (5.45) the common-mode rejection ratio CMRR is

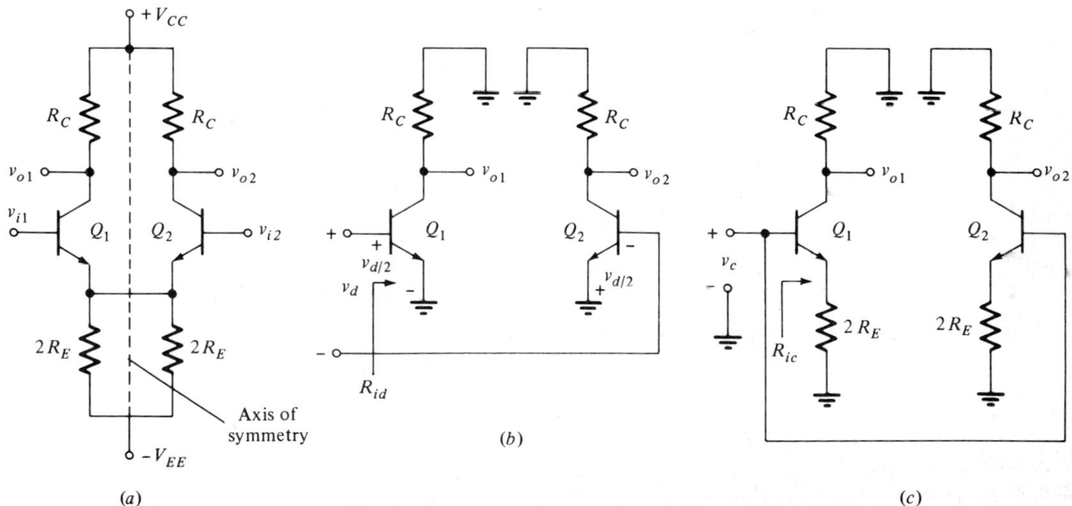

FIGURE 5.18 (*a*) Emitter-coupled differential pair. (*b*) Differential-mode circuit. (*c*) Common-mode circuit.

$$\text{CMRR} = \left|\frac{A_d}{A_c}\right| \simeq g_m R_E \tag{5.46a}$$

or expressed in decibels:

$$\text{CMRR} = 20 \log \left|\frac{A_d}{A_c}\right| \tag{5.46b}$$

As pointed out earlier, a large value of CMRR is desirable; hence the value of R_E must be made as large as possible.

The input differential resistance is the resistance seen by v_d in Fig. 5.18*b*. This is simply twice the input resistance of a CE stage, namely,

$$R_{id} = \frac{v_d}{i_b} = 2(r_b + \beta_0 r_e) \simeq 2\beta_0 r_e \tag{5.47}$$

The input common-mode resistance is the resistance seen by v_c in Fig. 5.18*b*. In this case it is simply

$$R_{ic} = \frac{v_c}{i_b} \simeq \frac{1}{2}\beta_0(2R_E) = \beta_0 R_E \tag{5.48}$$

EXAMPLE 5.8

Consider the basic differential circuit shown in Fig. 5.16. Let the transistor parameters be as in (5.6), and the circuit parameters as follows: $V_{CC} = -V_{EE} = 20$ V, $R_C = R_E = 10$ kΩ. Determine the collector currents, the single-ended, differential, and common-mode gains (A_d, A_c, and CMRR), and also R_{id} and R_{ic}. For simplicity and first-order approximation ignore r_b in the ac and (V_{BE}) on in the dc analysis.

From the circuit $V_{CE1} = V_{CE2} = V_{CC}/2$. The collector currents are determined from

$$I_c = \frac{V_{CC} - V_{CE}}{R_C} = \frac{20 - 10}{20 \times 10^3} = 0.5 \times 10^{-3} \text{ A}$$

since $I_e \simeq I_c$ and the total current through R_E is $2I_c$. For the ac circuit we use the results obtained in (5.43), (5.45), and (5.46), namely,

$$A_d = -\frac{1}{2}g_m R_C = -\frac{1}{2}\frac{I_e}{V_t}R_C = -\frac{1}{2}\left[\frac{0.5 \times 10^{-3}}{25 \times 10^{-3}}(20 \times 10^3)\right] = -200$$

$$A_c \simeq -\frac{R_C}{2R_E} = -\frac{1}{2}\frac{10^4}{10^4} \simeq -\frac{1}{2}$$

$$\text{CMRR} = \left|\frac{A_d}{A_c}\right| \simeq 400$$

$$\text{CMRR (dB)} = 20 \log 400 = 52 \text{ dB}$$

From (5.47)

$$R_{id} \simeq 2\beta_0 r_e = 2(50)\frac{25 \times 10^3}{0.5 \times 10^{-3}} = 5 \text{ k}\Omega$$

From (5.48)

$$R_{ic} \simeq \beta_0 R_E = 50(10 \times 10^3) = 500 \text{ k}\Omega$$

From the above expressions and analysis it is seen that in order to get a high CMRR, we must have high values for both R_E and g_m. The maximum value of R_E in Fig. 5.16 is limited by the supply voltage V_{EE} and the dc current, and the high value of g_m means high I_e, which is also limited by the supply voltage and the dc operating values. Note also from (5.47) and (5.42) that a high value of β_0 is clearly desirable to increase R_{id} and to a certain extent A_d. We shall next examine how to improve the circuit to get a better differential amplifier in terms of differential gain, CMRR, input and output properties.

C. Increase of CMRR

In order to make the CMRR as large as possible, from (5.46*a*) we must make the product $g_m R_E$ as large as possible. Since g_m is directly proportional to the dc current, the dc *current must also be stabilized.* As we saw earlier, the dc supply voltage V_{EE} and the dc operating point determine the value of R_E, and its upper limit is thus severely constrained. In practice a very high value of R_E is desirable. One circuit that achieves this is shown in Fig. 5.19.

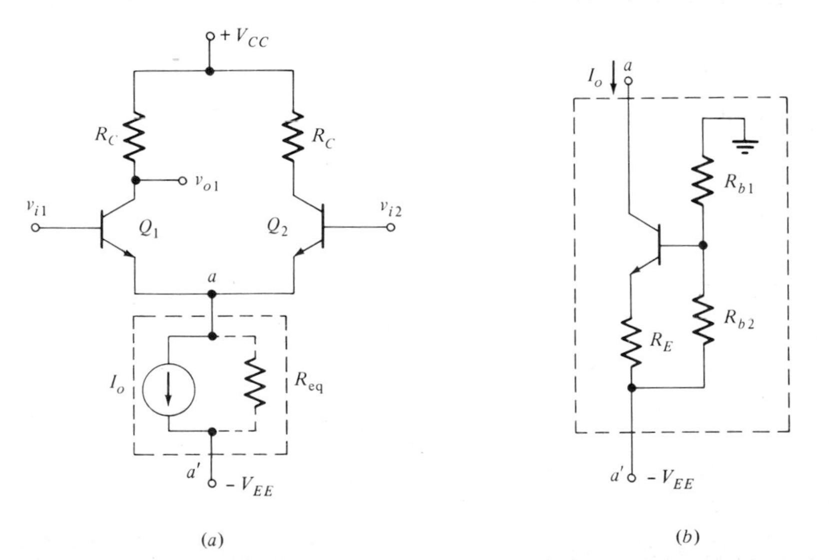

FIGURE 5.19 (*a*) Basic differential amplifier with large emitter resistor (R_{eq}). (*b*) Stabilized current source.

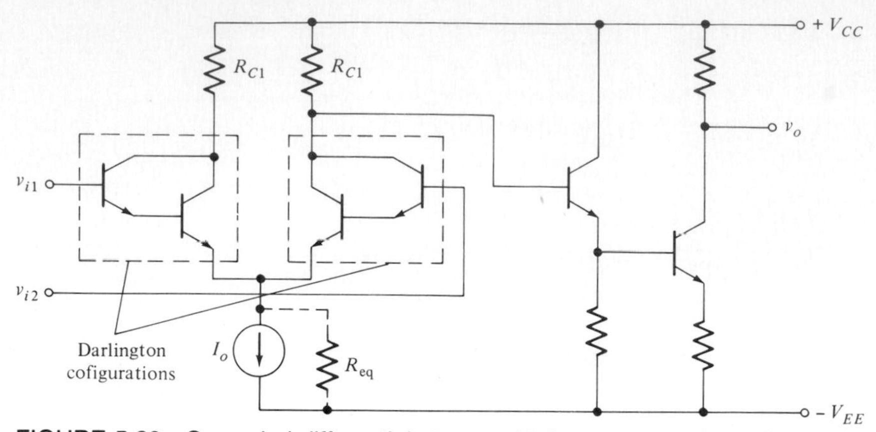

FIGURE 5.20 Cascaded differential stages with input stage using a Darlington configuration and the differential output converted to single-ended output.

For dc analysis the circuit is simply the stabilized discrete self-biasing circuit of Fig. 2.33, which keeps the variations of I_o small in spite of large variations in β_0 and temperature. Hence g_m is stabilized. From an ac point of view the Norton equivalent circuit is Fig. 5.19*b* as seen from terminals *a*, *a'* is simply a very large resistance (of the order of several megaohms, problem 4.9). Hence we can achieve a large value of R_E, which reduces A_C and increases CMRR.

As we have pointed out in Sec. 2.11, for IC circuits diode biasing schemes are used. Two such biasing circuits are the current mirror and the improved current mirror versions (Figs. 2.36 and 2.37). Since op amps are predominantly IC circuits, the current mirror versions are used.

D. Increase of R_{id} and A_d

The differential amplifier input resistance R_{id} and to a certain extent the gain A_d can be increased by increasing β_0 in (5.47) and (5.42). The increase in β_0 is achieved by using the super β Darlington composite configuration (see Fig. 5.1). From Table 5.1 it is seen that the Darlington configuration increases the input impedance significantly. The increase in A_d is achieved by cascading the basic emitter coupled pair with another stage of amplification as shown in Fig. 5.20.

5.9 SOURCE-COUPLED JFET PAIRS

Differential amplifier input stages must have an extremely high input impedance. Ideally it should be infinite. As we have already seen, FETs are ideal devices for voltage input signals as their input impedances are extremely high (larger than 10^{10} Ω). Thus source-coupled JFETs find application as input stages of an op amp. A simple *n*-channel JFET source-coupled pair is shown in Fig. 5.21*a*; of course, MOSFETs can also be used. When FETs are used in conjunction with BJTs to take advantage of the properties of each device, they are commonly referred to as Bi-FET and Bi-MOS.

The analysis of the circuit for the differential mode and common mode is similar to that of BJT and can readily be obtained from Fig. 5.21*b* and *c*, respectively.

For the differential-mode half-circuit we have a simple CS stage, which has already been analyzed in Sec. 4.6 The voltage gain of the differential-mode half-circuit is given by

$$\frac{v_{o1}}{v_d/2} = -g_m(R_D \parallel r_d) \simeq -g_m R_D \qquad \text{for } r_d \gg R_D$$

If the output is taken single ended,

$$A_d = \frac{v_{o1}}{v_d} = -\frac{1}{2} g_m(R_D \parallel r_d) \simeq -\frac{1}{2} g_m R_D \tag{5.49a}$$

If the output is taken differentially,

$$A_d = \frac{v_{o1} - v_{o2}}{v_d} = -g_m(R_D \parallel r_d) \simeq -g_m R_D \tag{5.49b}$$

Similarly, the voltage gain of the common-mode signal (Fig. 5.21*c*) was analyzed in Sec. 4.8 (see also Fig. 4.17*a*), which is

$$A_c = \frac{v_{o1}}{v_c} = -\frac{\mu R_D}{(\mu + 1)2R_S + r_d + R_D} \tag{5.50a}$$

$$\simeq \frac{g_m R_D}{1 + 2g_m R_S} \qquad \text{for } r_d \gg R_D,\ \mu = g_m r_d \gg 1 \tag{5.50b}$$

Of course, if the output is taken differentially, $v_{o1} - v_{o2} = 0$ (for identical JFETs),

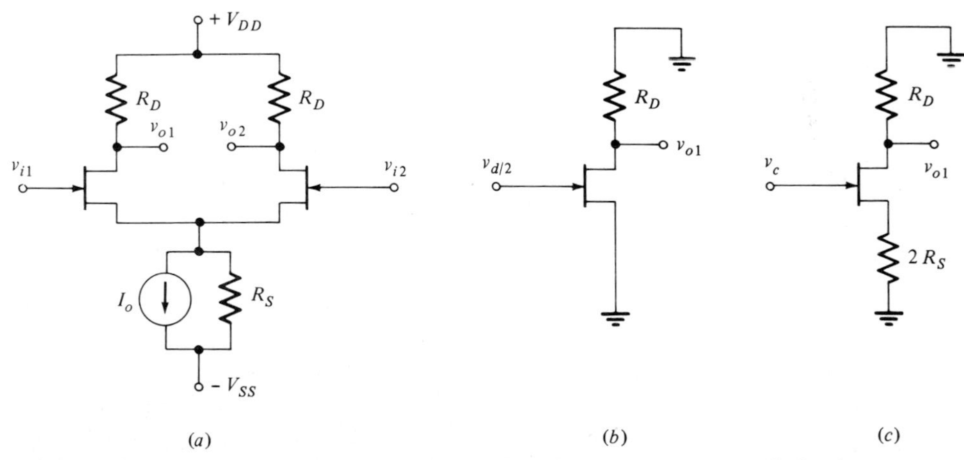

FIGURE 5.21 (*a*) Source-coupled JFET pair. (*b*) Differential-mode half-circuit. (*c*) Common-mode half-circuit.

and $A_c = 0$. The common-mode rejection ratio of the circuit, from (5.49*a*) and (5.50), is

$$\text{CMRR} = \left|\frac{A_d}{A_c}\right| = \frac{1}{2}(1 + 2g_m R_S) \simeq g_m R_S \qquad \text{for } 2g_m R_S \gg 1 \tag{5.51}$$

Note that the expression in (5.51) is similar to (5.46*a*). Note also that if the values of V_{PO}, I_{DSS}, and the Q point are given, then g_m is found from (3.39).

The input differential resistance R_{id} and common-mode resistance R_{ic} of a JFET is larger than 10^{10}, and hence R_{id} and $R_{ic} \simeq \infty$ for all practical purposes.

It is noted that for the same bias current the transconductance of FETs is much lower than that of BJTs. Hence, for equal resistance values, A_d and CMRR of a FET are much smaller than those of a BJT. The almost infinite input resistance of FETs makes them very attractive as the differential input circuits for op amps. These are then followed by BJTs to get high gain and CMRR. In fact, in Bi-FETs (e.g., LF351) this is precisely what is done.

EXAMPLE 5.9

Consider the JFET differential amplifier circuit shown in Fig. 5.21. The JFET small-signal parameters are $g_m = 2 \times 10^{-3}$ mho and $r_d = 20$ kΩ. The circuit parameters are $V_{DD} = -V_{SS} = 20$ V and $R_D = R_S = 10$ kΩ. Determine the single-ended differential gain A_d, the common-mode gain A_c, and CMRR. From (5.49*a*)

$$A_d = -\tfrac{1}{2}(2 \times 10^{-3})(10 \parallel 20) \times 10^3 = -6.67$$

From (5.50*a*)

$$A_c = -\frac{40(10^4)}{40(2 \times 10^4) + 2 \times 10^4 + 10^4} = -0.48$$

Hence

$$\text{CMRR} = \frac{6.67}{0.48} = 13.9 \qquad \text{or} \qquad 22.8 \text{ dB}$$

Notice that further stages of amplification are clearly needed.

5.10 THE OPERATIONAL AMPLIFIER

The operational amplifier (op amp) is the workhorse of linear integrated circuits. These amplifiers are nearly ideal high-gain blocks for many applications. The most widely accepted and available ones are monolithic op amps. These amplifiers are based on differential amplifiers. However, practical op amps have several

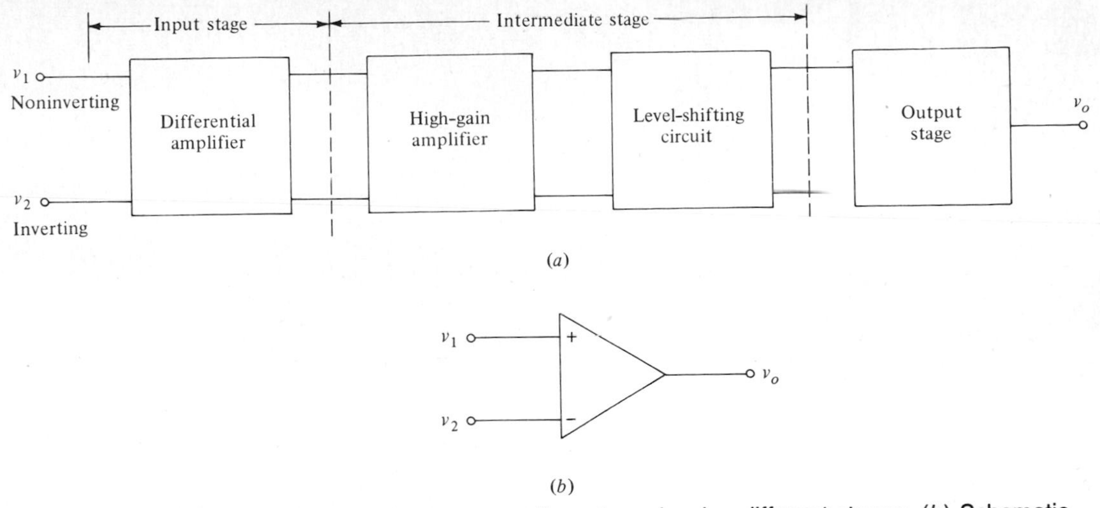

FIGURE 5.22 (*a*) Typical op-amp configuration, showing different stages. (*b*) Schematic representation of (*a*).

modifications in the circuit so as to make them useful as IC modules or the universal gain blocks. The basic constituents are shown in Fig. 5.22*a*. Figure 5.22*b* is commonly used as a simple representation of an op amp. The amplifier gain of Fig. 5.22*b* is henceforth expressed as

$$v_o = a_o(v_1 - v_2) \tag{5.52}$$

where a_o is the open loop gain of the op amp, which is the overall single-ended differential gain of Fig. 5.22*a*. The sign in (5.52) indicates the inverting and non-inverting terminals; e.g., if $v_1 = 0$, $v_o = -a_o v_2$ is a negative quantity.

In op amps since a_o is very large, 10^5 to 10^6 (100 to 120 dB), the differential input signals must be very small in order to prevent the amplifier from saturation. The large-signal characteristic of an op amp is discussed later in Sec. 5.11. Note also that input impedances of Bi-FET op amps are higher by several orders of magnitude than those of BJT op amps.

A. BJT Op Amps

We shall next examine the various blocks in Fig. 5.22*a*, namely, the level shifting and the output stage in BJT op amps.

Level Shifting

The reader will notice that the output of an op amp is single ended for a differential input. Since differential amplifiers are also used for dc amplifiers, the output level must be at zero dc voltage. The output level can be made to equal zero by using a

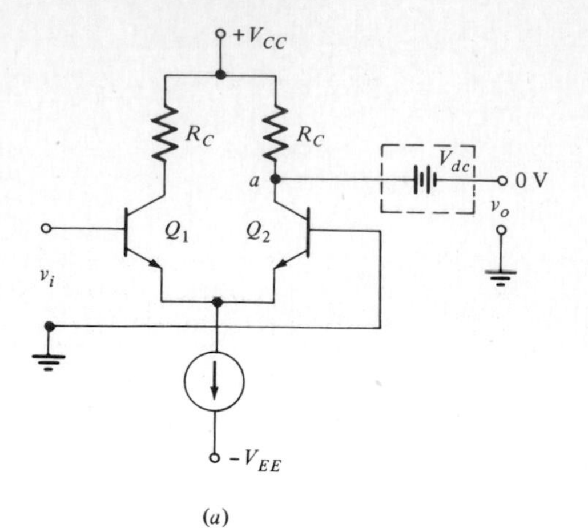

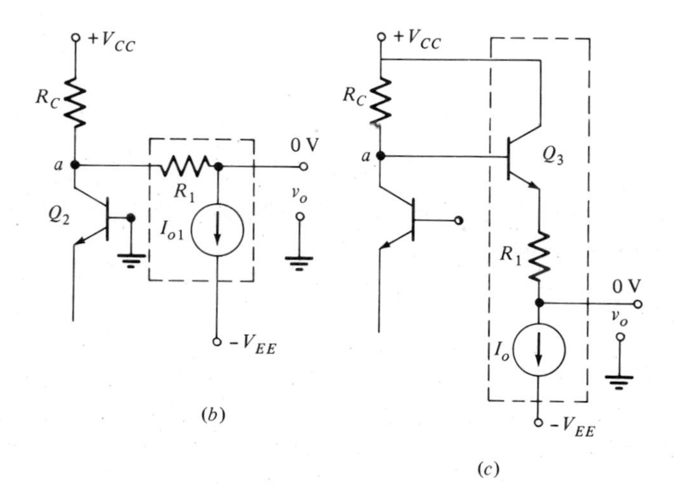

FIGURE 5.23 (*a*) Level-shifting circuits with a battery. (*b*) Level-shifting circuit with a current source. (*c*) Improved level-shifting network.

floating battery or, more practically, by using the offset circuit shown in Fig. 5.23*b* and *c*, respectively. Note that in Example 5.8, we found V_{CB} to be 10 V. Hence V_{dc} will be a 10-V battery as in Fig. 5.23*a*. Clearly this situation is not a desirable solution in IC. For the circuit in Fig. 5.23*b* and in Example 5.8, if we choose $I_{O1} = I_{C2} = 0.5$ mA, then

$$R_1 = \frac{V_{C2}}{I_{O1}} = \frac{10}{0.5 \times 10^{-3}} = 20 \text{ k}\Omega$$

It should also be noted, however, that the effective load resistance for Q_2 is now $R_{C2} \parallel R_1 = 10 \text{ k}\Omega$, which reduces the voltage gain by a factor of 2. The circuit in Fig.

5.23*c* achieves the level shifting via an emitter-follower circuit. Continuing Example 5.8, i.e., for $V_{C2} = 10$ V and $I_{O1} = 0.5$ mA, since the output terminal is to be zero voltage, we have

$$R_1 = \frac{V_{C2} - V_{BE}}{I_{O1}} = \frac{V_{C2} - 0.7}{I_{O1}} \simeq \frac{V_{C2}}{I_{O1}} = \frac{10}{0.5 \times 10^{-3}} = 20\ \text{k}\Omega$$

In this case negligible reduction in gain occurs since v_o/v_a is simply the voltage gain of an emitter-follower. The gain is unity, as the effective R_L of the emitter-follower is extremely high.

Output Stage

The main functions of an output stage are

1. To provide low output resistance
2. To provide large load current capability

An examination of the previous circuits thus far will reveal that the output resistance of the circuit is not low. A simple output stage for an op amp with complementary transistors is shown in Fig. 5.24. The output resistance of this stage, as we shall soon show, is very low. The large current capability of the circuit, large dynamic range, and other features and problems associated with the circuit are discussed in the next chapter.

We shall next consider the small-signal ac analysis of the 741 BJT op amp (see Appendix E) circuit shown in Fig. 5.25*a*. The circuit looks very complicated and at first glance appears to defy a paper-and-pencil analysis. We shall show that the circuit properties of the individual transistors (as obtained in Chap. 4 and this chapter) can be utilized to simplify the analysis and calculation of the basic properties of the op amp.

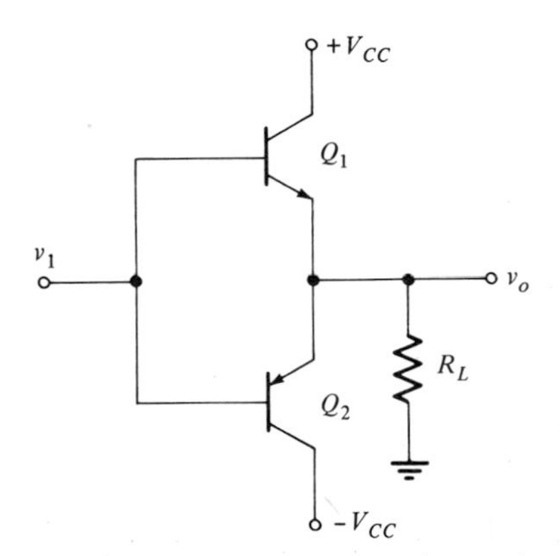

FIGURE 5.24 A commonly used output stage.

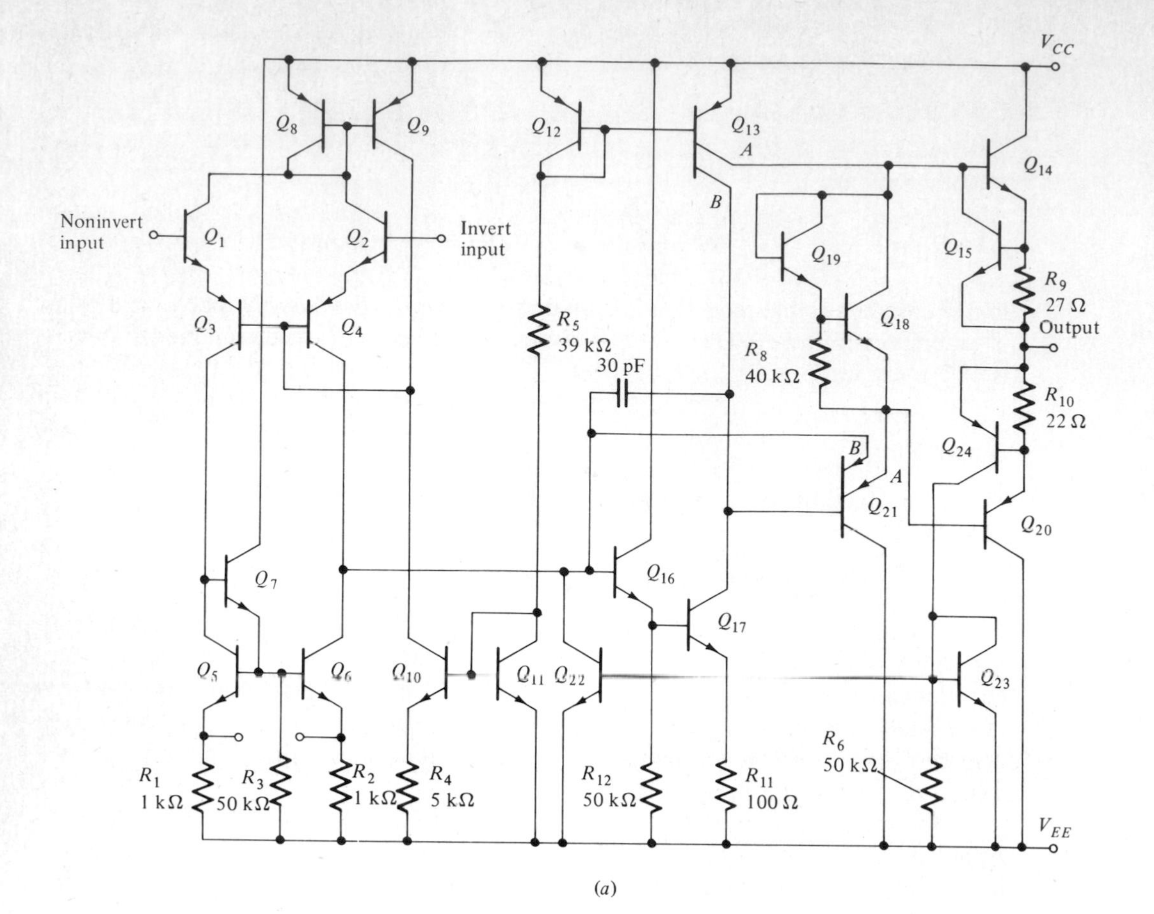

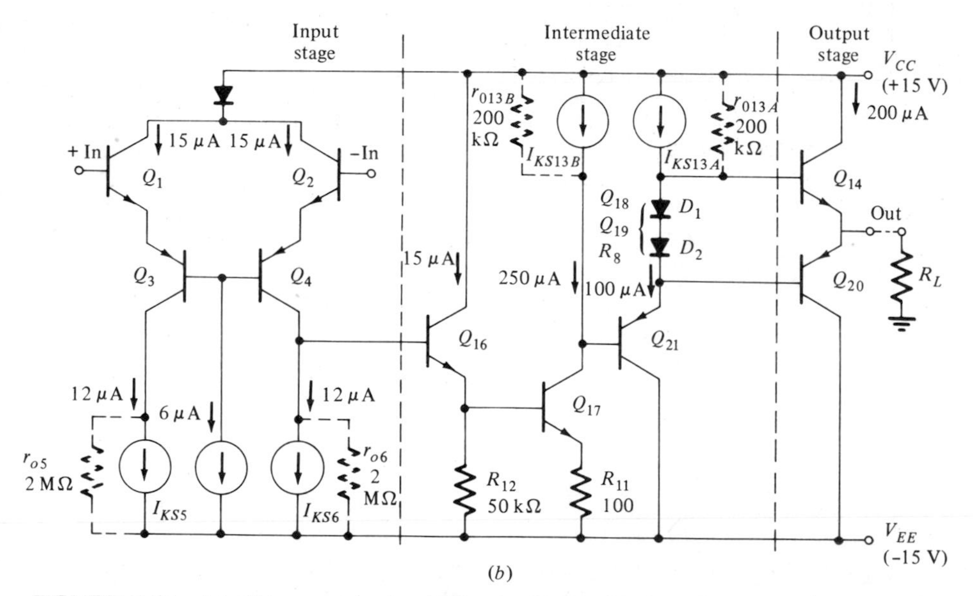

FIGURE 5.25 (*a*) 741 op-amp circuit (Appendix E). (*b*) Simplified version of (*a*) [2].

EXAMPLE 5.10

The following analysis utilizes the simplification and the numerical values of Ref. 3; namely, we shall assume $R_L = 2\ \text{k}\Omega$ and the following transistor parameters: all *npn* and *pnp* transistors have $\beta_0 = 200$ except for the *pnp* transistors at the input stage; i.e., for Q_3 and Q_4, $\beta_0 = 4$. To find the differential-mode (DM) gain A_d, the DM half-circuit is shown in Fig. 5.26. Notice that the input signal is $v_d/2$. The values of r_e for the indicated currents in Fig. 5.25*b* are given by (the subscript numbers correspond to the transistor Q's)

$$r_e = \frac{kT}{qI_e} = \frac{25\ \text{mV}}{I_e(\text{mA})}$$

Hence $r_{e17} = 100\ \Omega$, $r_{e4} = r_{e16} = r_{e2} = 1.67\ \text{k}\Omega$. (Note that $I_{c4} = 12\ \mu\text{A}$ and $\beta_{o4} = 4$; hence $I_{e4} = 15\ \mu\text{A}$.) The gain A_d of the ac circuit shown in Fig. 5.26 is determined from

$$\frac{v_{b20}}{v_{d/2}} = \left(\frac{v_{b20}}{v_{b17}}\frac{v_{b17}}{v_{b16}}\right)\left(\frac{v_{b16}}{v_{e4}}\frac{v_{e4}}{v_{d/2}}\right)$$

Using the properties of each stage, or Table 4.2, we can find the individual gains. Note that $r_{L4} \simeq (2 \parallel 5)\ \text{M}\Omega = 1.43\ \text{M}\Omega$. Hence the gain of a CC stage driving a CB stage is found from

$$\frac{v_{b16}}{v_{e4}} \simeq \frac{r_{L4}}{r_{e4}} = \frac{1.43 \times 10^6}{1 \times 67 \times 10^3} = 850$$

and

$$\frac{v_{e4}}{v_{d/2}} \simeq \frac{r_{L2}}{r_{L2} + r_{e2}} \simeq \frac{1.67}{1.67 + 1.67} = \frac{1}{2}$$

Hence for the input stage

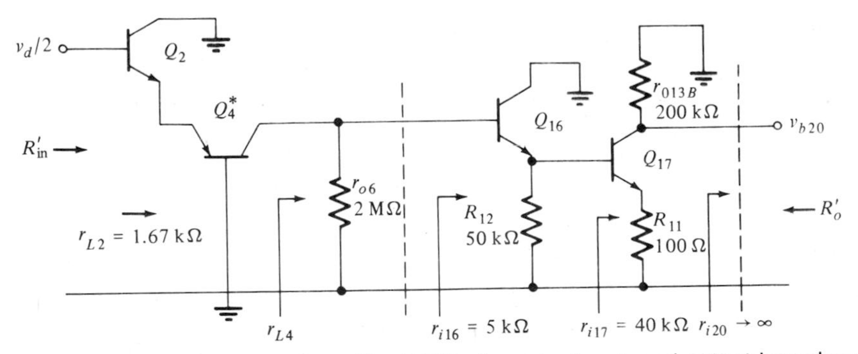

FIGURE 5.26 AC circuit from Fig. 5.25*b*, for gain, input, and output impedance calculations.

$$\frac{v_{b16}}{v_d} = \frac{1}{4}(850) \simeq 212$$

Similarly, for the second stage

$$\frac{v_{b17}}{v_{b16}} = \frac{r_{L16}}{r_{L16} + r_{e16}} = \frac{40 \parallel 50}{(40 \parallel 50) + 1.67} = 0.93$$

$$\frac{v_{b20}}{v_{b17}} \simeq -\frac{r_{o13B}}{r_{e17} + R_{11}} = -\frac{200 \times 10^3}{200} = -1000$$

Note that $v_{b21} = v_{b20}$ since Q_{21} is a CC stage with $R_{L20} \simeq \infty$. Hence the gain of the second stage is

$$\frac{v_{b20}}{v_{b16}} = -1000(0.93) = -930$$

Since there is a negligible voltage attenuation through the output complementary pair (see Chap. 6), the overall voltage gain A_d is

$$A_d = \frac{v_{b20}}{v_d} = -930(212) = -197{,}160 = a_o$$

The above gain is close to the typical value of $-200{,}000$ given in the data sheet. The input and output impedances of the circuit can be similarly found. The input impedance of a CC stage (Q_2) is

$$R'_{\text{in}} \simeq \beta_0(r_{L2} + r_{e2}) = 200(3.34 \times 10^3) = 0.668 \text{ M}\Omega$$

Hence the DM input resistance, which is twice the input resistance of the half-circuit, is given by

$$R_{\text{in}} = 2R'_{\text{in}} = 1.3 \text{ M}\Omega$$

The output resistance of the circuit is determined by the output stage. The following calculation will be easy to follow by referring to Fig. 5.25 and Table 4.2. For the emitter-follower (Q_{20}) we have

$$r_{O20} = \frac{r_{O22} \parallel r_{O13A} \parallel R_{i14}}{\beta_{20}} + r_{e20}$$

but

$$r_{O21} = \frac{r_{O17} + r_{O13B}}{\beta_{21}} + r_{e21} \simeq \frac{\beta_0^2(r_{e17} + 100) + 200 \times 10^3}{200} + 250$$

$$\simeq 1.25 \text{ k}\Omega$$

and

$$R_{i14} = \beta_{14} r_{e14} = 200\frac{25 \text{ mV}}{0.2 \text{ mA}} = 26 \text{ k}\Omega$$

hence

$$r_{o20} = \frac{1.25 \times 10^3 \parallel 200 \times 10^3 \parallel 26 \times 10^3}{200} + 125 \simeq 130\ \Omega$$

From the complete circuit Fig. 5.25*b*, we now add a 22-Ω resistor and note that the output resistances of Q_{14} and Q_{20} are in parallel, namely,

$$R_o = (130 + 22) \parallel (130 + 27) \simeq 76\ \Omega$$

which is very close to the typical value of 75 given by the manufacturer.

The manufacturer's specification of 741 is given in Appendix E. Typical values (for $R_L \geq 2$ kΩ) are given in Table 5.2. Notice that the results are in agreement with the manufacturer's specifications.

TABLE 5.2 Typical Parameters for BJT Op Amp (741) and for Bi-FET Op Amp (LF351)

	BJT (741)	Bi-FET (LF351)	MOSFETs* (NMOS and CMOS)
A_d	106 dB ($= 2 \times 10^5$)	100 dB ($= 10^5$)	60 dB ($= 2 \times 10^3$)
CMRR	95 dB	100 dB	70 dB
R_{in}	1 MΩ	10^{12} Ω	10^{12} Ω
R_o	75 Ω	75 Ω	1 kΩ

*Figures 5.28*a* and 5.29.

B. Bi-FET and Bi-MOS Op Amps

The complete data sheet of a Bi-FET (LF351) is given in Appendix E. The complete circuit and its simplified version are shown in Fig. 5.27*a* and *b*, respectively. From the circuit note that a Bi-FET op amp uses JFETs at the differential input stage followed by BJT stages, similar to the one in Fig. 5.25. The voltage gain at the input is smaller than in BJT, but the succeeding BJT stages provide the amplification. Notice that the output circuit of a Bi-FET (LF351) is almost identical to that of the BJT (741) op amp; hence the output properties are the same. The input impedance of a Bi-FET, however, is several orders of magnitude larger than that of a BJT op amp. Typical values of a Bi-FET (LF351) (for $R_L = 2$ kΩ) are given in Table 5.2.

It should also be noted that op amps with hybrid BJT-MOS are also available. Such op amps are called *Bi-MOS* and are commercially available, e.g., CA3130. The noise performance of Bi-FET op amps, however, are superior to that of Bi-MOS op amps.

C. All MOS Op Amps

In MOS technology both the NMOS and CMOS op amps are available. Figure 5.28*a* shows the circuit diagram of an all-MOS integrated NMOS op amp with internal

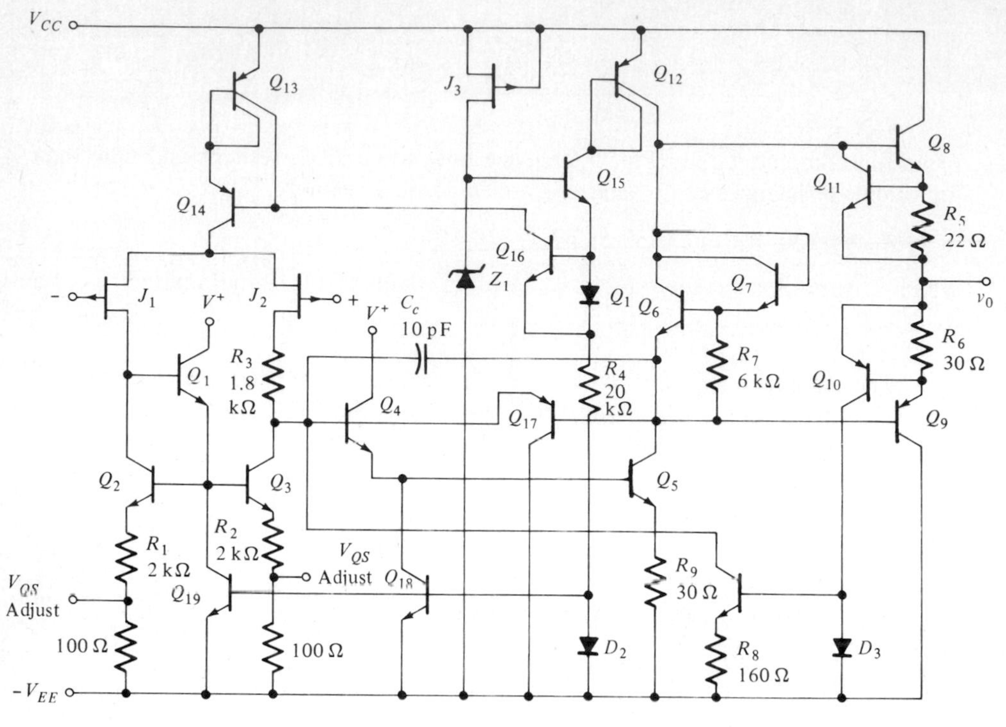

(a)

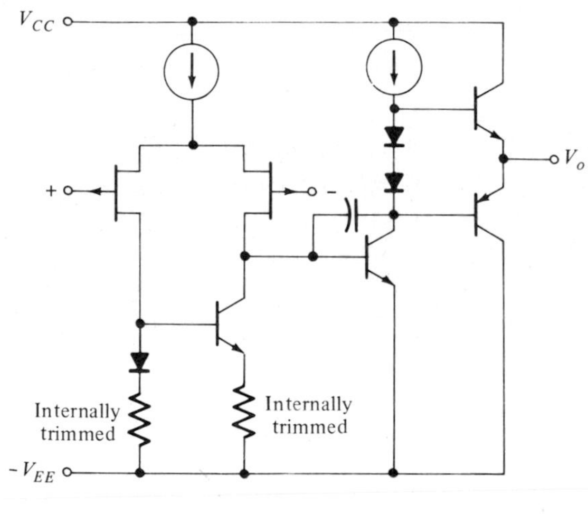

(b)

FIGURE 5.27 (*a*) Bi-FET (LF351) circuit (Appendix E). (*b*) Simplified version of (*a*).

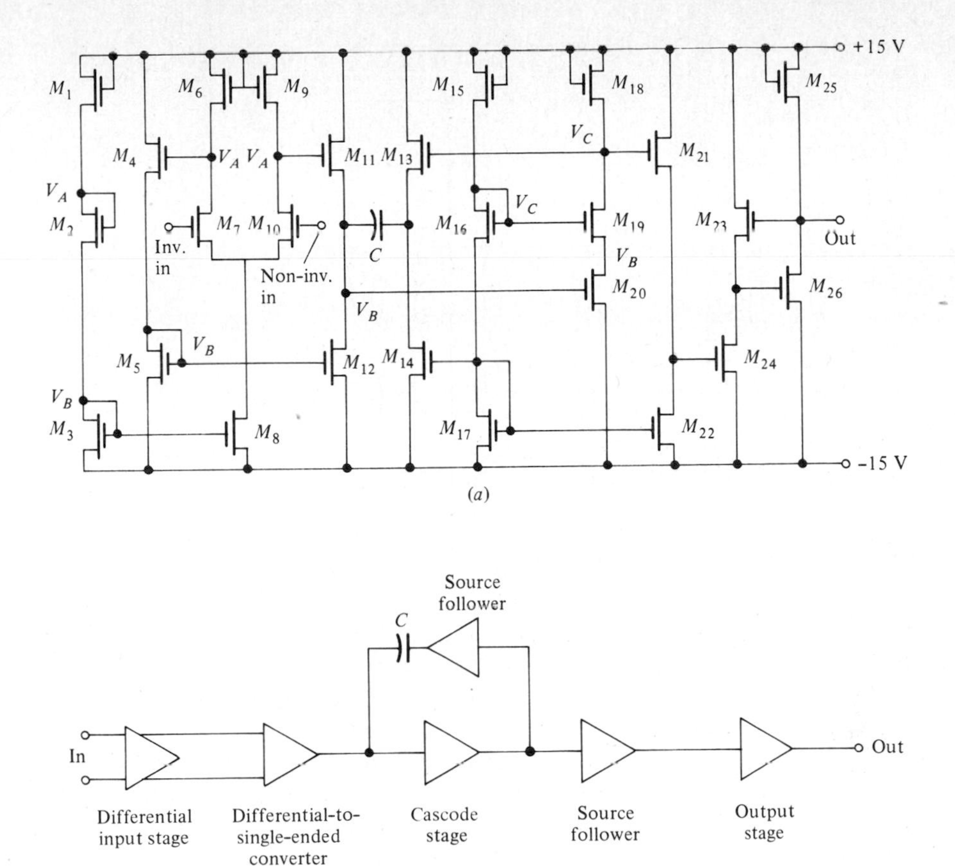

FIGURE 5.28 (*a*) Complete circuit diagram of a NMOS op amp. (*b*) Block diagram of the op amp.

compensation. The block diagram of the op amp is shown in Fig. 5.28*b*, where the individual functional blocks are identified.* The input stage is a source-coupled differential amplifier. The differential output signal is then applied to a differential single-ended converter. A cascode stage is then used for further amplification and isolation. The output of the driver is then fed to a source follower and the output stage. Typical calculations are carried out in the reference, and some of the results are listed in Table 5.2.

Figure 5.29 shows the schematic of a simple CMOS op amp[†] The M_5, M_6, M_9, M_{10}, and M_{11} FETs provide current-source biasing of the remaining circuitry. M_1

*See *Analog MOS Integrated Circuits*, edited by Paul R. Grey, D. A. Hodges, and R. W. Broderson, pp. 50–57, IEEE Press, New York, 1980.
[†]For CMOS op amp see ibid., 2–11.

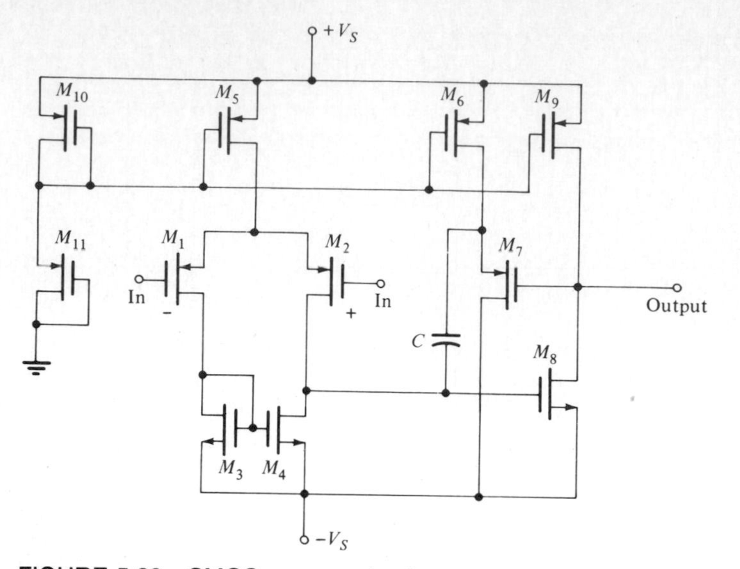

FIGURE 5.29 CMOS op-amp circuit.

through M_4 form an amplifier with differential input and single-ended output. M_8 provides the last amplifying stage. M_7 and C provide the compensation. Some of the performance parameters of the CMOS op amp are listed in Table 5.2. MOS op amps are used for special-purpose applications and in switched capacitor filters (see Sec. 9.7), where *they are compatible with LSI circuits*. They are about 5 times smaller in area than the BJT op amps and hence more economical in LSI circuitry. For general all-purpose applications, however, BJT op amps (e.g., 741) and the Bi-FET (e.g., LF351) are commonly used.

The reader can, by now appreciate the systematic approach in analyzing the low-frequency properties of a complicated circuit. Note that we have determined the basic properties of a complicated interconnected circuit such as the op amp by examining the basic properties of the individual or composite transistors such as the input and output resistances, the voltage, and/or the current gain as determined in Table 4.1. Other specifications, such as slew rate, offset voltage, and current, will be considered in Chap. 9, where we consider the op amp in detail. The gain-bandwidth product of this op amp is considered in Example 8.4.

5.11 THE IDEAL OP AMP

The circuit model for the op amp corresponding to Fig. 5.22 is shown in Fig. 5.30*a*, where a_o is the differential voltage gain, R_o is the output impedance, and R_i is the differential input impedance. We have already seen that for the 741 and the LF351 op amp, the voltage gain a_o is very high (10^5 to 10^6), R_i is very high ($R_i \simeq 1$ MΩ for 741, 10^{10} Ω for LF351), and R_o is very low (75 Ω).

We now introduce the concept of the *ideal op amp*. This concept simplifies the analysis and design of circuits utilizing op amps considerably, just as the concept of the ideal diode was very helpful in Chap. 1.

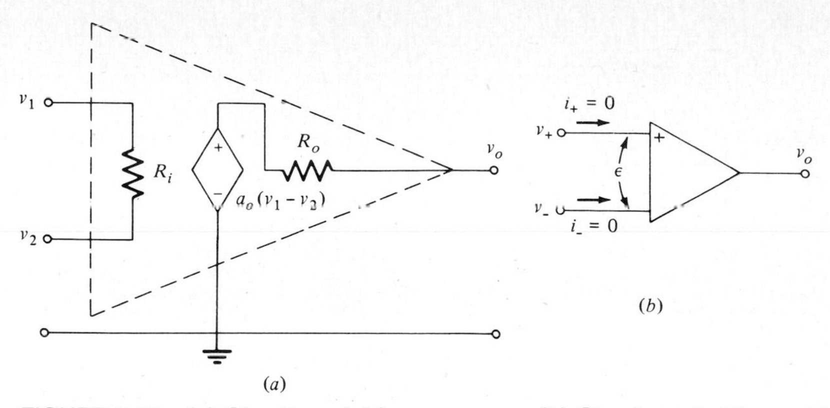

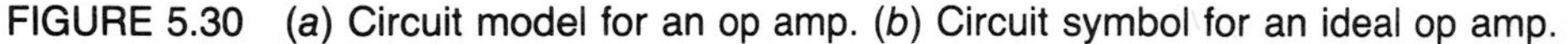

FIGURE 5.30 (*a*) Circuit model for an op amp. (*b*) Circuit symbol for an ideal op amp.

An ideal op amp is described by

$$v_o = a_o(v_+ - v_-) \tag{5.53}$$

where $a_o = \infty$, $R_o = 0$, and $R_i = \infty$, which implies $i_+ = i_- = 0$ and $\epsilon = v_+ - v_-$. The circuit symbol for an ideal op amp is shown in Fig. 5.30*b*. Note that for $\epsilon = 0$, this is referred to as a *virtual ground*. It should further be noted that $v_+ - v_-$ is at zero potential for the ideal op amp used in linear applications with negative feedback. However, v_+ or v_- need not be at zero potential. For nonlinear applications ϵ is not zero as discussed in Chap. 9.

Integrated-circuit op amps are nearly ideal and are widely used in electronic circuits. Inverting and noninverting voltage amplifiers and buffers are the simplest examples.

A. Inverting Amplifier

An inverting voltage amplifier is shown in Fig. 5.31*a*. Because of the virtual ground and the fact that $i = 0$, we have

$$v_i = R_s i_1 \qquad \text{and} \qquad v_o = -R_f i_1 \tag{5.54}$$

From (5.54) we obtain

$$\frac{v_o}{v_i} = -\frac{R_f}{R_s} \tag{5.55}$$

B. Noninverting Amplifier

A noninverting voltage amplifier is shown in Fig. 5.31*b*. Again, since $i = 0$ and $v_+ = v_-$, we have

$$v_- = \frac{R_2}{R_1 + R_2} v_o \qquad \text{and} \qquad v_+ = v_- = v_i \tag{5.56}$$

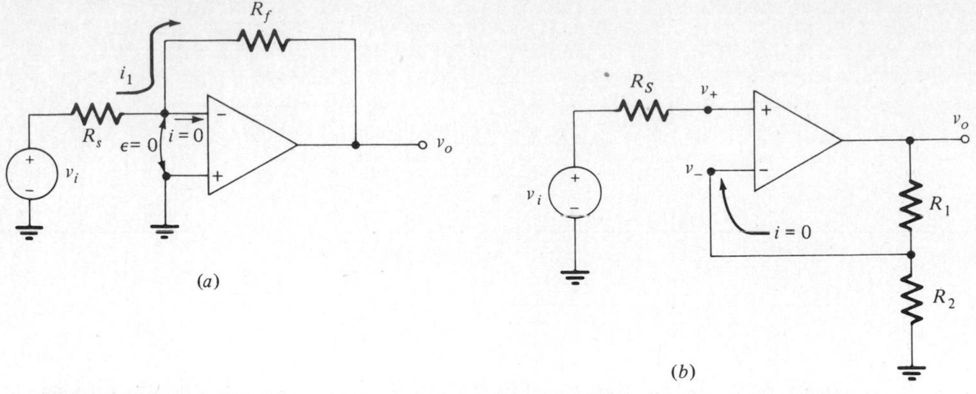

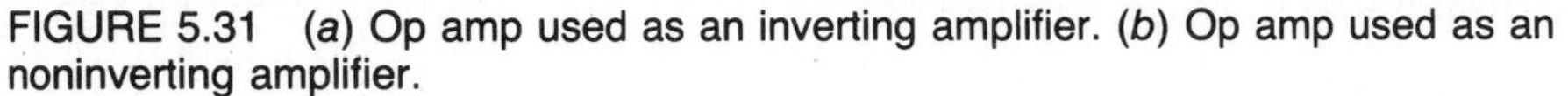

FIGURE 5.31 (*a*) Op amp used as an inverting amplifier. (*b*) Op amp used as an noninverting amplifier.

From (5.56) we obtain

$$\frac{v_o}{v_i} = \frac{R_1 + R_2}{R_2} = 1 + \frac{R_1}{R_2} \tag{5.57}$$

The results in (5.55) and (5.57) are very important and are frequently used in electronic circuit design.

EXERCISE 5.7

Design an amplifier using an op amp for a voltage of -50 if $R_s = 1\ \text{k}\Omega$.

Ans $R_f = 50\ \text{k}\Omega$

EXERCISE 5.8

Design an amplifier using an op amp for a voltage gain of $+10$. Use $R_2 = 1\ \text{k}\Omega$.

Ans $R_1 = 9\ \text{k}\Omega$

EXERCISE 5.9

The circuit shown in Fig. E5.9 is often used as a buffer. Determine v_o/v_i.

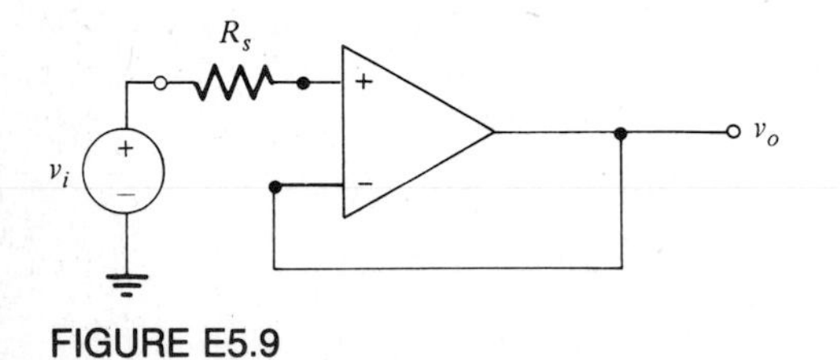

FIGURE E5.9

Ans Unity

EXERCISE 5.10

For the circuit shown in Fig. E5.10, determine the input resistance $R_{\text{in}} = V_1/I_1$.

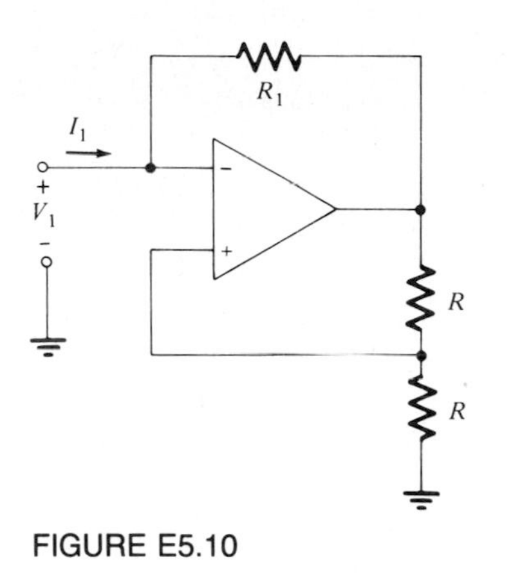

FIGURE E5.10

Ans $R_{\text{in}} = -R_1$

Further discussion of the op amps and their applications in linear and nonlinear circuits are given in Chap. 9.

5.12 LARGE-SIGNAL OPERATION OF THE DIFFERENTIAL PAIR

In the previous sections we considered small-signal properties of the differential pair and of the op amp. Since an op amp has a very large voltage gain ($>10^5$), it is evident that it does not take much of an input signal to saturate the transistors. We shall briefly consider then the so-called large-signal operation of the differential pair.

A. BJT Differential (Emitter-Coupled) Pair

The basic BJT differential pair is shown in Fig. 5.32. The transfer characteristic of the emitter-coupled pair was briefly discussed in Sec. 2.7 and is shown in Fig. 2.16. We shall consider it again in the following. For large-signal operation we must use the exponential *V-I* relationship of the transistor, namely,

$$I_{e1} = I_s e^{(V_{B1}-V_E)/V_t} \tag{5.58a}$$

$$I_{e2} = I_s e^{(V_{B2}-V_E)/V_t} \tag{5.58b}$$

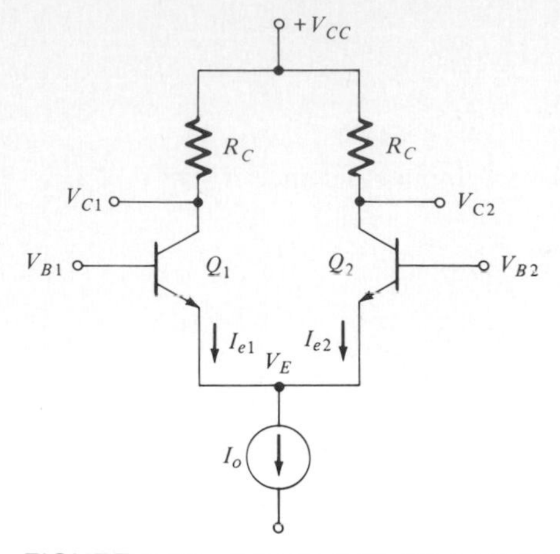

FIGURE 5.32 A basic BJT differential pair.

From the circuit we have

$$I_{e1} + I_{e2} = I_o \tag{5.59}$$

From (5.58) and (5.59) and noting that $I_{e1} = I_{e2}e^{(V_{B1}-V_{B2})/V_t}$, we have

$$I_{e1} = \frac{I_o}{1 + e^{(V_{B2}-V_{B1})/V_t}} \tag{5.60a}$$

$$I_{e2} = \frac{I_o}{1 + e^{(V_{B1}-V_{B2})/V_t}} \tag{5.60b}$$

Assuming identical transistors and $\alpha_0 \simeq 1$ (i.e., $I_c \simeq I_e$), the collector voltage is given by

$$V_{C1} = V_{CC} - I_{e1}R_C = V_{CC} - \frac{I_oR_C}{1 + e^{(V_{B2} - V_{B1})/V_t}} \tag{5.61}$$

Note that from (5.61) if $|V_{B2} - V_{B1}| \gg V_t$, I_{e1} is either zero or I_o depending on the polarity of $V_{B2} - V_{B1}$. For $I_oR_C = 2V_{CC}$, the transfer characteristic for both V_{C1} and V_{C2} vs $(V_{B2} - V_{B1})$ is shown in Fig. 5.33. Of course, if I_o is smaller than $2V_{CC}/R_C$ (say, $I_oR_C = V_{CC}$), then V_{C1} will swing from $+V_{CC}$ to zero for a very small change, i.e., the difference of the input signals. Note that the differential pair responds only to the *difference voltage* $V_{B1} - V_{B2}$. For the common mode $V_{B1} = V_{B2}$, and the current divides equally between the two transistors irrespective of their values.

Finally it should also be noted that the range of linear operation as shown in Fig. 5.33 is very small, namely, $|V_{B1} - V_{B2}| < 5V_t$ ($V_t = 25$ mV). Our discussions in the previous sections focused on this region. In fact, if $|V_{B1} - V_{B2}| \geq 5V_t$, the current I_o will flow entirely in Q_1 or Q_2, depending on the sign difference between $V_{B1} - V_{B2}$. Note that if the differential output of Fig. 5.32 is fed as input to another differential pair, the saturation effect will further be increased. Such circuits are used as comparators. Nonlinear operation of the op amp and its applications are discussed in Chaps. 9 and 13.

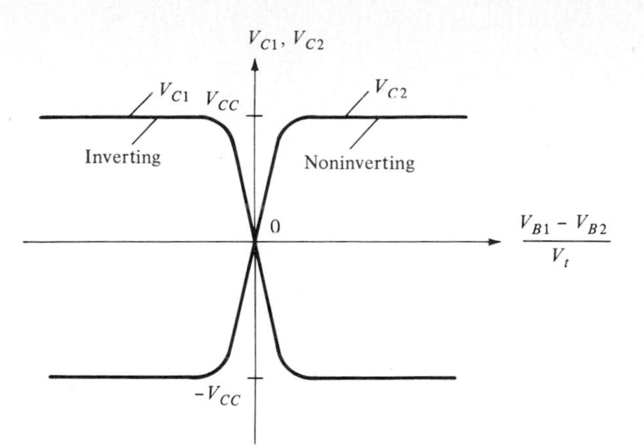

FIGURE 5.33 Transfer characteristic of Fig. 5.32.

EXERCISE 5.11

In Fig. 5.32 the circuit parameters are $V_{CC} = 6$ V, $R_C = 3$ kΩ, and $I_o = 5$ mA. The transistors are identical with $\beta_0 = 100$. Determine the values of V_{C1} for the following values of $V_{B1} - V_{B2}$: (*a*) −1 V, (*b*) − 10 mV, (*c*) 10 mV, (*d*) 1 V.

Ans (*a*) 6 V; (*b*) −0.02 V; (*c*) −2.98 V; (*d*) −9 V

B. FET Differential (Source-Coupled) Pair

The basic JFET differential pair is shown in Fig. 5.34. The nonlinear *V-I* characteristics of FETs were discussed in Chap. 3. For a JFET in the pinch-off region, from (3.8) we have

$$I_D = I_{DSS}\left(1 + \frac{V_{GS}}{V_{PO}}\right)^2 \tag{5.62}$$

Hence

$$I_{D1} = I_{DSS}\left(1 + \frac{V_{G1} - V_S}{V_{PO}}\right)^2 \tag{5.63a}$$

$$I_{D2} = I_{DSS}\left(1 + \frac{V_{G2} - V_S}{V_{PO}}\right)^2 \tag{5.63b}$$

From (5.63) we obtain

$$\sqrt{I_{D1}} - \sqrt{I_{D2}} = \sqrt{I_{DSS}}\,\frac{V_{G1} - V_{G2}}{V_{PO}} \tag{5.64}$$

From the circuit we have

$$I_{D1} + I_{D2} = I_o \tag{5.65}$$

From (5.64) and (5.65) we obtain

$$I_{D1} = \frac{I_o}{2} + \frac{I_o}{2}\frac{V_{G1} - V_{G2}}{V_{PO}}\sqrt{\frac{2I_{DSS}}{I_o} - \left(\frac{V_{G1} - V_{G2}}{V_{PO}}\right)^2\left(\frac{I_{DSS}}{I_o}\right)^2} \tag{5.66a}$$

$$I_{D2} = \frac{I_o}{2} - \frac{I_o}{2}\frac{V_{G1} - V_{G2}}{V_{PO}}\sqrt{\frac{2I_{DSS}}{I_o} - \left(\frac{V_{G1} - V_{G2}}{V_{PO}}\right)^2\left(\frac{I_{DSS}}{I_o}\right)^2} \tag{5.66b}$$

A plot of the transfer characteristic, i.e., V_{D1} and V_{D2} vs $(V_{G1} - V_{G2})/V_{PO}$ would be similar to the one shown in Fig. 5.33.

For small-signal conditions, i.e., linear operation, we assume in (5.66) that the nonlinear term is negligibly small, i.e.,

$$\left|\frac{V_{G1} - V_{G2}}{V_{PO}}\right| \ll \sqrt{\frac{2I_o}{I_{DSS}}} \tag{5.67}$$

Our discussion in the previous section was focused on this region. For linear and nonlinear operations we must limit I_o so that it is less or at most equal to I_{DSS}, in order not to forward-bias the gate-channel junction. The differential voltage that will drive the current I_o entirely through Q_1 or Q_2 is determined from (5.66), namely, I_{D1} or I_{D2} equals I_o for

$$\left|\frac{V_{G1} - V_{G2}}{V_{PO}}\right| = \sqrt{\frac{I_o}{I_{DSS}}} \tag{5.68}$$

In other words, if $(V_{G1} - V_{G2})/V_{PO}$ is positive and equal to the value given in (5.68), then Q_1 in ON and Q_2 is OFF.

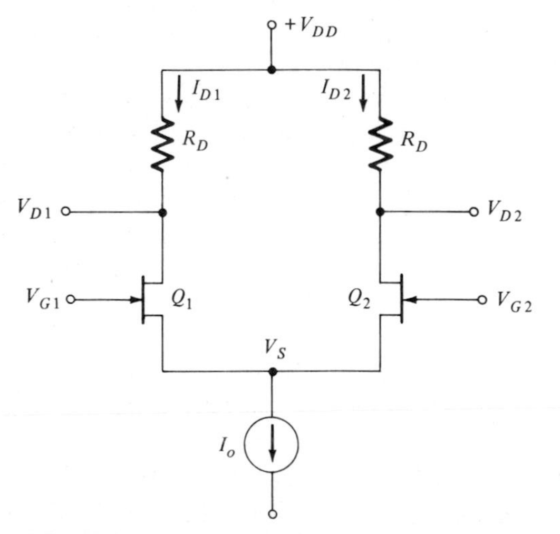

FIGURE 5.34 A basic JFET differential pair.

EXAMPLE 5.11

For the basic JFET differential pair shown in Fig. 5.34, the JFET parameters are $I_{DSS} = 5$ mA and $V_{PO} = 3$ V. The circuit parameters are $V_{DD} = 10$ V, $I_o = 2$ mA, and $R_D = 5$ kΩ.

(a) Determine the drain currents if $V_{G1} - V_{G2} = 0.3$ V.
(b) For what values of $V_{G1} - V_{G2}$ will the circuit be in the linear region.
(c) Determine the value of $V_{G1} - V_{G2}$ that will drive the current I_o entirely in Q_1.

The currents from (5.66) are

$$I_{D1}, I_{D2} = \frac{2}{2} \pm \frac{2}{2}\frac{0.3}{3}\sqrt{2\left(\frac{5}{2}\right) - \left(\frac{0.3}{3}\right)^2\left(\frac{5}{2}\right)^2} = 1.22, 0.78$$

For the linear region from (5.67),

$$V_{G1} - V_{G2} \ll 3 \times \sqrt{\tfrac{4}{5}} = 2.7 \qquad \text{or} \qquad |V_{G1} - V_{G2}| \leq 0.27$$

From (5.68) we have

$$V_{G1} - V_{G2} = 3\sqrt{\tfrac{2}{5}} = 1.90 \text{ V}$$

In other words if $V_{G1} = V_{G2} + 1.90$ V, Q_1 is ON and Q_2 is OFF.

5.13 CONCLUDING REMARKS

The purpose of this chapter was to consider multiple-transistor amplifiers. Some of the basic configurations discussed were the Darlington or super β, the cascode, and the differential pair circuits. The differential pair in BJT circuits is the emitter-coupled configuration, whereas in FET circuits it is the source-coupled configuration. These configurations are extremely important and are used in integrated op amp circuits. Op amps were discussed in some detail as they are important functional blocks in integrated circuits. A large variety of op amps are now available from the manufacturers. We have singled out for a detailed analysis the commonly used 741 BJT op amp and the LF351 Bi-FET op amp. Most of the op amps are configured similarly; namely, they consist of a differential input stage followed by a differential-to-single-ended converter, followed by an intermediate stage for further amplification and level shifting, which is then followed by an output stage.

Note that the op amps saturate if their linear region of operation, which is very small, is exceeded. This chapter also considered the nonlinear region of operation of the differential pair as it is widely used in nonlinear applications of the op amps. Because of the tremendous importance of op amp and their wide range of applications in electronic circuits, Chap. 9 revisits op amp and their applications in linear and nonlinear signal processing.

REFERENCES

1. P. R. Gray and R. G. Meyer, *Analysis and Design of Analog Integrated Circuits,* Wiley, New York, 1977.
2. A. Sedra and K. C. Smith, *Microelectronic Circuits,* Holt, New York, 1982.
3. V. H. Grinich and H. G. Jackson, *Introduction to Integrated Circuits,* McGraw-Hill, New York, 1975.

PROBLEMS

5.1 For the circuit shown in Fig. P5.1, the transistors are identical with $\beta_{01} = \beta_{02} = 100$. Determine the following at room temperature:
(*a*) Quiescent points for the transistors.
(*b*) The voltage gain of the amplifier. Assume that $r_b \ll R_i$.
(*Hint:* Assume $V_{BE} = 0.7$ V and $I_C \gg I_B$ for each transistor.)

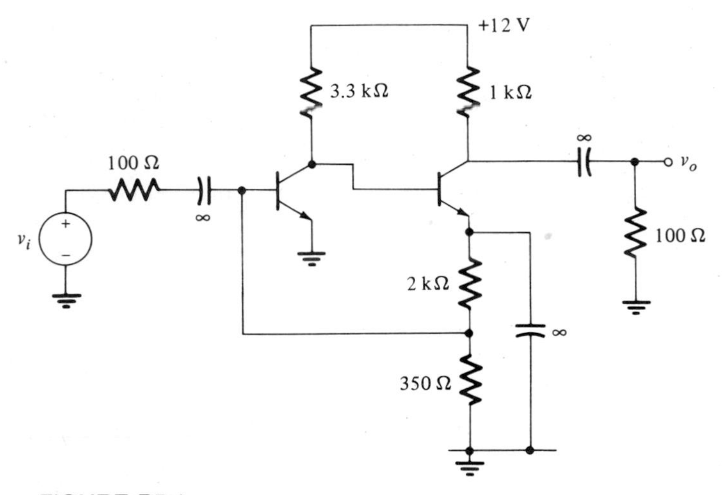

FIGURE P5.1

5.2 Determine the voltage gain of the two-stage FET amplifier shown in Fig. P5.2. The JFET parameters are given as follows: $r_d = 50$ kΩ and $g_m = 2 \times 10^{-3}$ mho.

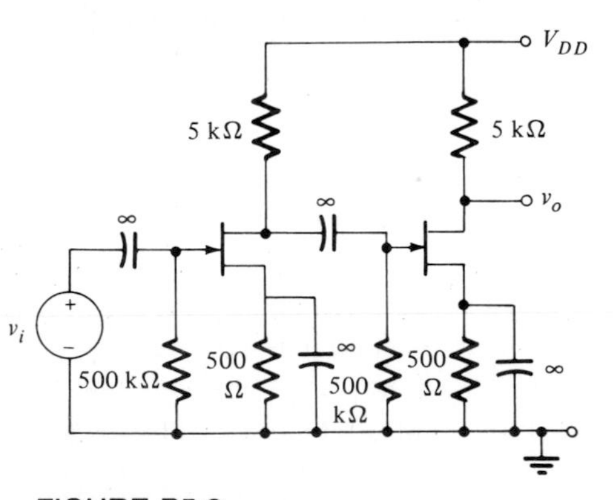

FIGURE P5.2

5.3 Repeat problem 5.2 assuming MOSFETs with the following parameters: $r_d = 20\ \text{k}\Omega$ and $g_m = 3 \times 10^{-3}$ mho.

5.4 An emitter-coupled pair (CC-CB pair) is shown in Fig. P5.4. Determine the approximate gain of the amplifier. For simplicity, assume that the transistor parameters are the same; i.e., $\beta_{01} = \beta_{02} = 100$, $r_{e1} = r_{e2} = 13\ \Omega$, $r_b = 0$. Let $R_L = R_E = 1\ \text{k}\Omega$ and $R_S = 50\ \Omega$.

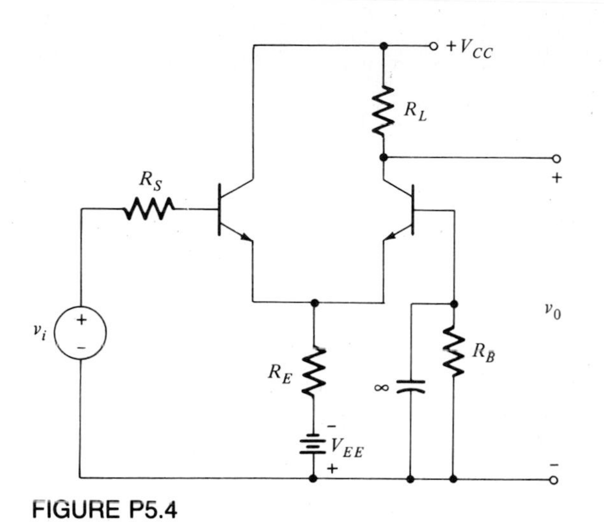

FIGURE P5.4

5.5 For the cascode amplifier shown in Fig. 5.6, $R_S = R_L = 10\ \text{k}\Omega$. The transistor parameters are the same as in problem 5.4. Determine the voltage gain of the amplifier.

5.6 Show that the circuit in Fig. P5.6 approximately realizes an ideal voltage-controlled current source, given that $r_e \ll R \ll r_c$. (*Hint:* Show that $y_{ij} \simeq 0$ except for y_{21}.)

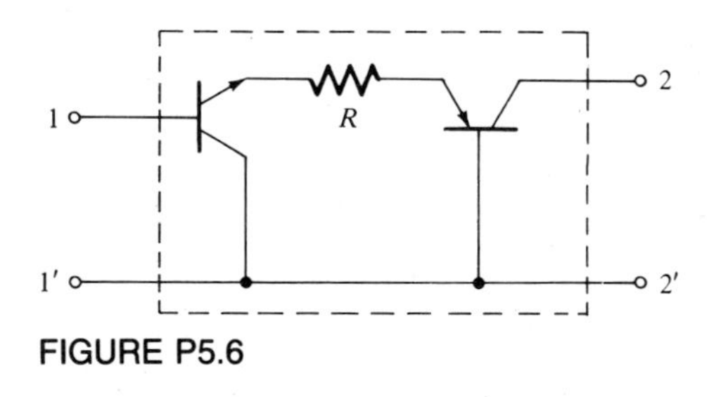

FIGURE P5.6

5.7 A hybrid cascode is shown in Fig. P5.7. Determine the voltage gain of the amplifier if the transistor parameters are

BJT: $\beta_0 = 50$ $\quad r_b = 100\ \Omega$

FET: $g_m = 2 \times 10^{-3}$ mho $\quad r_d = 50\ \text{k}\Omega$

$I_C = I_D = -1$ mA

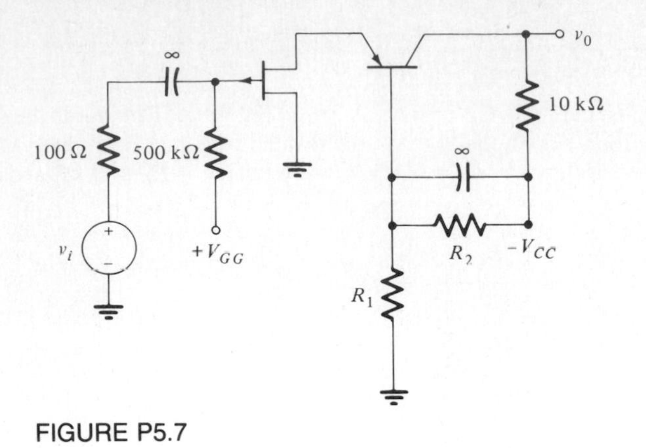

FIGURE P5.7

5.8 For a CC-CC pair the source and load terminators are $R_S = R_L = 1\ \text{k}\Omega$. Determine the voltage gain, the current gain, and the power gain of the circuit if the transistor parameters are $\beta_0 = 100$, $r_e = 25\ \Omega$, $r_b = 0$, and $r_c = 1\ \text{M}\Omega$.

5.9 For problem 5.8, determine the values of R_S and R_L that will yield maximum power gain. What is the value of the maximum power gain G_P?

5.10 For a CC-CE pair, the source and load terminations are $R_S = R_L = 1\ \text{k}\Omega$. Determine the voltage gain, the current gain, and the power gain of the circuit if the transistor parameters are $\beta_0 = 100$, $r_e = 25\ \Omega$, and $r_c = 1\ \text{M}\Omega$.

5.11 For problem 5.10, determine the values of R_S and R_L for maximum power gain. What is the value of the maximum power gain?

5.12 For a CE-CE pair, the source and the load terminations are $R_S = R_L = 1\ \text{k}\Omega$. Determine voltage gain, the current gain, and the power gain of the circuit if the transistor parameters are $\beta_0 = 100$, $r_e = 25\ \Omega$, $r_b = 0$, and $r_c = 1\ \text{M}\Omega$.

5.13 For problem 5.12, determine the values of R_S and R_L for maximum power gain. What is the value of the maximum power gain?

5.14 The h parameters of a certain transistor in the CE configuration is given by

$$h_{ie} = 1\ \text{k}\Omega \qquad h_{fe} = 100$$

$$h_{re} = 10^{-4} \qquad h_{oe} = 10^{-5}\ \text{mho}$$

Determine the voltage gain and the current gain of the circuit if the transistors are connected in cascode with $R_S = R_L = 1\ \text{k}\Omega$.

5.15 For the hybrid CD-CB pair shown in Fig. 5.9, determine the voltage gain if $R_S = 100\ \Omega$, $R_L = 10\ \text{k}\Omega$, and $R_E = 1\ \text{k}\Omega$. Assume the following transistor parameters:
BJT:

$$\beta_0 = 50 \qquad r_b = 100\ \Omega \qquad r_e = 10\ \Omega$$

MOSFET:

$$g_m = 4 \times 10^{-3}\ \text{mho} \qquad r_d = 20\ \text{k}\Omega$$

5.16 For the hybrid CS-CE pair shown in Fig. 5.10, determine the voltage gain if $R_S = 100\ \Omega$, $R_L = 10\ \text{k}\Omega$, and $R_D = R_B = 5\ \text{k}\Omega$. Assume the transistor parameters of problem 5.15.

5.17 Determine the voltage gain and the current gain of a three-stage cascaded CE amplifier in Fig. 5.13 if the transistor parameters are those given in problem 5.15. Let $R_S = R_L = 1\ \text{k}\Omega$ and $R_1 = R_2 = R_3 = 500\ \Omega$.

5.18 Repeat problem 5.17 if the transistor parameters are given by the h parameters as in problem 5.14. You may approximate $h_{re} \simeq 0$.

5.19 Determine the voltage gain of a three-stage cascaded CS FET amplifier in Fig. 5.14 if the FET parameters are given as in problem 5.15. Let $R_S = 100\ \Omega$, $R_L = 1\ \text{k}\Omega$, and $R_1 = R_2 = 10\ \text{k}\Omega$.

5.20 A two-port amplifier, modeled by its equivalent circuit in Fig. 5.12b, has the following circuit properties: $R_i = 0.1\ \text{M}\Omega$, $R_o = 100\ \Omega$, voltage gain $= 10^3$. For a source and load termination of $R_S = R_L = 50\ \Omega$, determine the overall voltage and current gains of the amplifier if two such amplifiers are connected in cascade as shown in Fig. 5.11.

5.21 A symmetrical two-port circuit with crosswires is shown in Fig. P5.21a.
(*a*) Show that for equal in-phase excitations (common mode) the circuit can be simplified as in Fig. P5.21b.
(*b*) Show that for equal antiphase excitations the circuit can be simplified as in Fig. P5.21c, *provided* the circuit is completely symmetrical, i.e., it has vertical and horizontal symmetry.

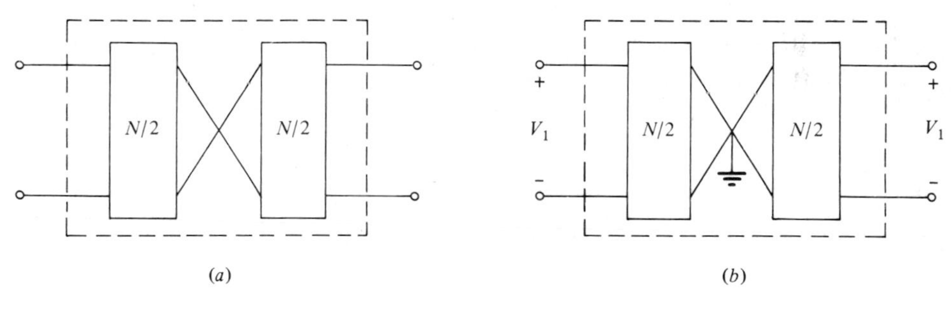

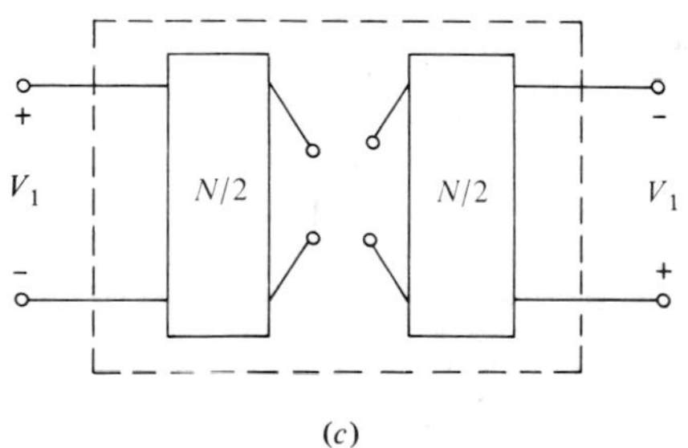

FIGURE P5.21

5.22 For a *completely symmetrical* two-port network, we define the open-circuit half-section driving point impedance as

$$Z_{och} = \left.\frac{V_1}{I_1}\right|_{V_1=V_2}$$

and the short-circuit half-section driving point impedance as

$$Z_{sch} = \left.\frac{V_1}{I_1}\right|_{V_1=-V_2}$$

where V_1, I_1, etc., are the conventional two-port voltage and current quantities.

(*a*) Show that $Z_{och} = z_{11} + z_{12}$ and $Z_{sch} = z_{11} - z_{12}$.

(*b*) Use the results in (*a*) to find the *z* parameters of the symmetrical two-port lattice network shown in Fig. P5.22.

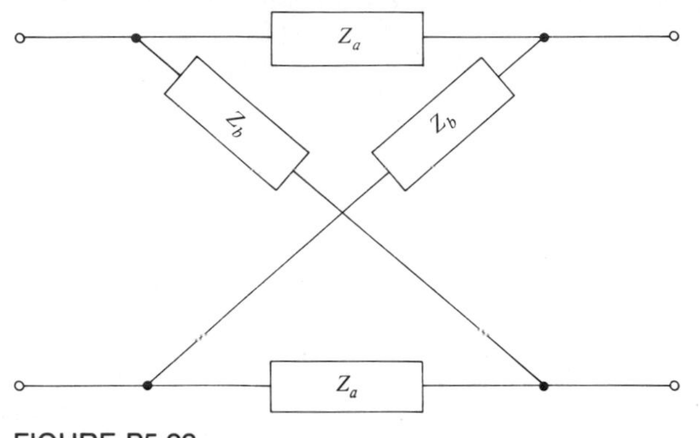

FIGURE P5.22

5.23 Find the two-port *y* parameters of the bridged-T networks shown in Fig. P5.23*a* and *b*. Use the concept of symmetry and half-symmetrical networks.

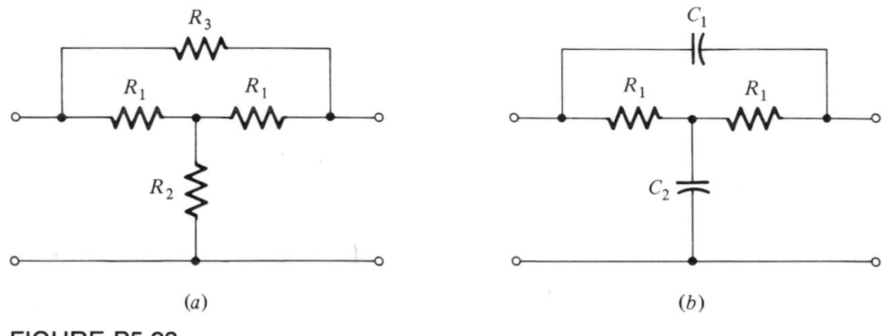

FIGURE P5.23

5.24 In dc amplifiers and integrated circuits, it is often of importance to provide a level shifting network without incurring a loss in the ac signal. Such a network is shown in Fig. P5.24, where the dc value of V_{C1} can be shifted to $V_{C3} = 0$ without a loss in ac gain.

(*a*) Determine the values of R_B such that V_{C3} is at zero potential with respect to ground. List the quiescent points of each transistor. Assume $\beta_0 = 100$ for each transistor.

(*b*) Show that the ac voltage gain from the collector of Q_1 to the collector of Q_3 is approximately unity and determine the voltage gain v_o/v_i. (*Hint:* Use the formulas in Table 4.2.)

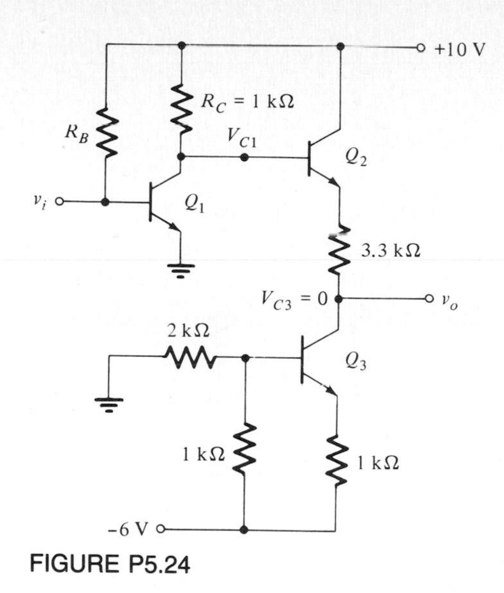

FIGURE P5.24

5.25 For the integrated-circuit difference amplifier shown in Fig. P5.25, find the quiescent points and the common-mode rejection ratio. Assume $\beta_{01} = \beta_{02} = 50$, $\beta_{03} = 100$, and $r_c \simeq 1\ \text{M}\Omega$. Note that $Z_{aa} \simeq r_c$.

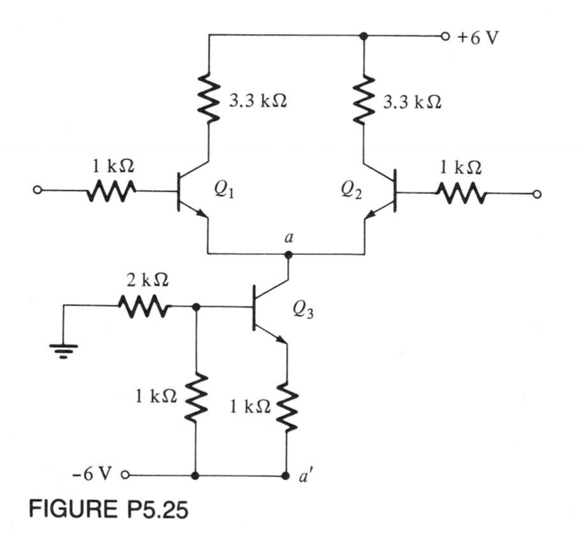

FIGURE P5.25

5.26 For the circuit shown in Fig. P5.26, assume matched transistors.
(*a*) Find the dc voltages at nodes 1 and 2, assume $V_{BE} = 0.7$ V and $\beta_0 = 100$.
(*b*) Determine the ac voltage gain v_o/v_i (assume $r_b = 0$).
(*c*) Find the expression for the voltage gain and the input impedance. Assume h_{oe}^{-1} and h_{re} to be zero.

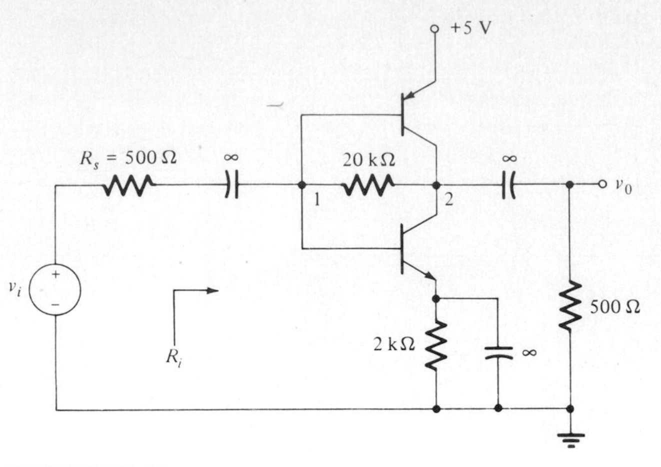

FIGURE P5.26

5.27 A simple JFET differential amplifier circuit is shown in Fig. P5.27. Assume identical JFETs with $g_m = 2 \times 10^{-3}$ mho and $r_d = 50$ kΩ. Find the CMRR.

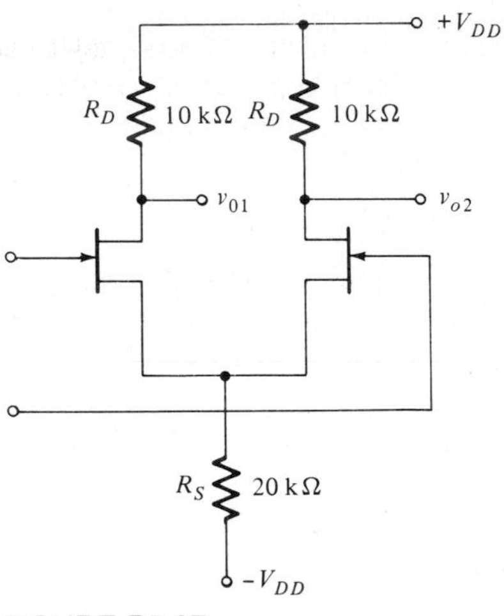

FIGURE P5.27

5.28 Determine CMRR and R_{id} of the circuit shown in Fig. P5.25 if the Q_1 and Q_2 are replaced by the Darlington configuration as shown in Fig. 5.1*b*. The circuit parameters are the same as in problem 5.25.

5.29 For the differential amplifier circuit shown in Fig. P5.29, the BJT parameters are $\beta_{01} = \beta_{02} = 100$. Determine the following for $I_o = 5$ mA.
(*a*) The single-ended differential-mode gain
(*b*) The single-ended common-mode gain
(*c*) The CMRR
(*d*) The differential- and common-mode input resistances R_{id} and R_{ic}.

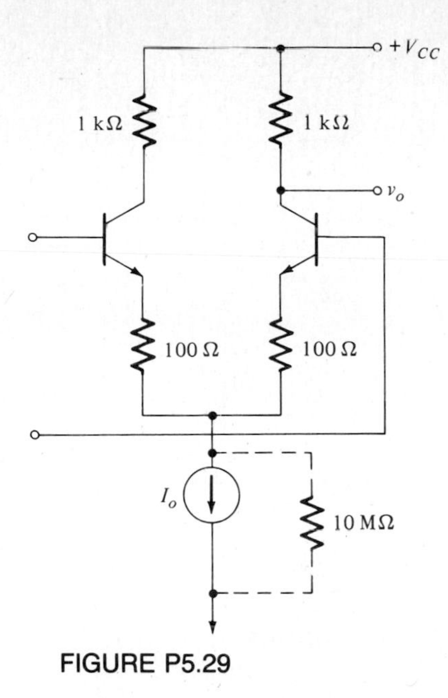

FIGURE P5.29

5.30 For a JFET differential pair shown in Fig. P5.30, determine the single-ended differential gain A_d, the common-mode gain A_c, and the CMRR. The JFET parameters are V_{PO} = 3 V, I_{DSS} = 4 mA, and V_{DD} = 20 V. (*Hint:* Determine I_D and g_m.)

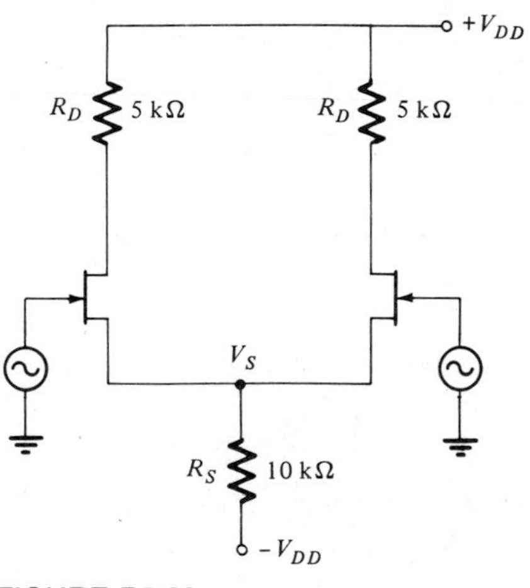

FIGURE P5.30

5.31 For problem 5.30, plot the drain currents vs the differential input voltage $V_d = |V_{G1} - V_{G2}| \leq V_{PO}$ for $I_D = 0.5 I_{DSS}$ and $I_D = I_{DSS}$. Determine A_d for each case. Is there a linear range of operation in the latter case?

5.32 For the circuit shown in Fig. P5.32, determine V_{C2}, V_{C3}, I_{C1}, I_{C2}, and I_{C3}.

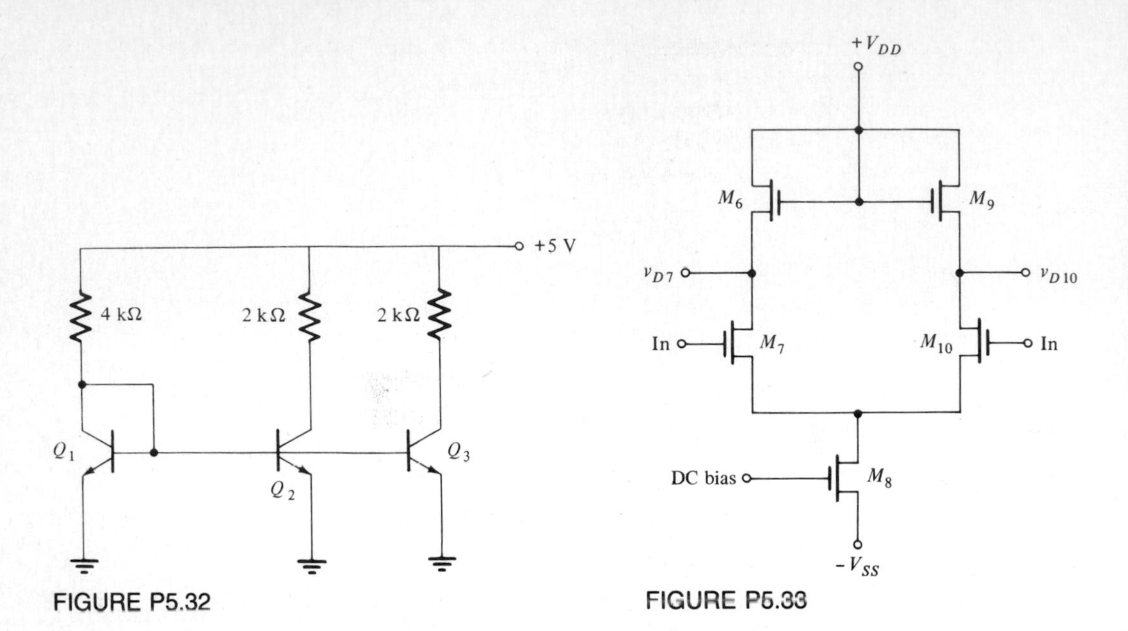

FIGURE P5.32

FIGURE P5.33

5.33 For the input stage shown in Fig. P5.33, M_7 and M_{10} MOS devices are in their saturation region. Show that the differential- and the common-mode gains are given by

$$A_d = \frac{g_{m7}}{g_{m6}} \qquad A_c = \frac{1}{2r_{o8}g_{m6}}$$

where r_{o8} is the output resistance of M_8.

5.34 For the circuit shown in Fig. P5.34, determine the input impedance R_{in}. Assume ideal op amps. (*Hint:* $V_1 = V_3 = V_4$, and start with $I_5 = 1$ A.)

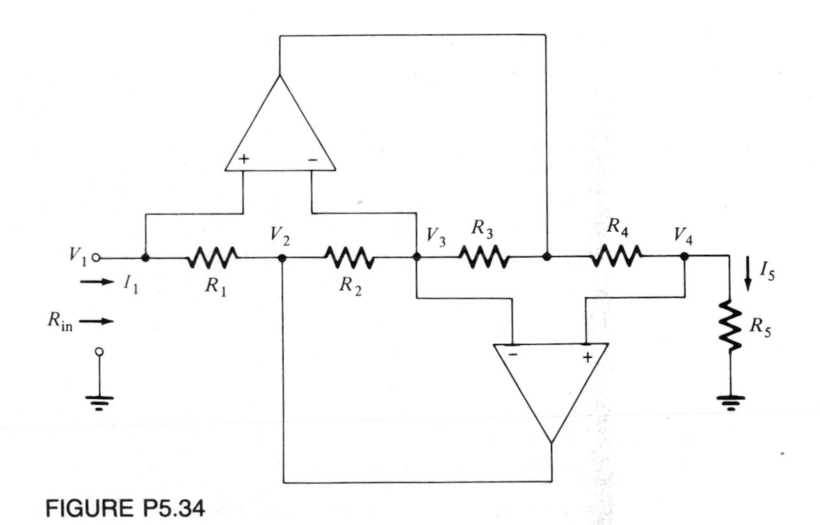

FIGURE P5.34

CHAPTER 6

Output Stages and Power Amplifiers

Thus far we have considered small-signal amplifiers, where the input and intermediate stages were used to obtain a large voltage gain or current gain. The operation of the devices were in the linear range so that the use of small-signal models were valid. However, the stages were not required to provide appreciable amounts of power. Since the operation of the amplifying stages was in the linear active region, distortion was negligible. The objective of the output stage is to deliver large-signal power to a load with minimum distortion. The output impedance of the output stage usually must be very low so that low values of load resistance can be driven without a loss in voltage gain.

In this chapter we will consider output stages for audio-frequency power amplifiers. The requirements for power amplifiers are quite different from those of small-signal low-power amplifiers. The output stage must drive a loudspeaker, a servomotor, or other low-impedance loads. In such cases appreciable amounts of power are required at the output stage. Other requirements are to keep the signal distortion low and the dc power requirements as small as possible. In other words, we wish to furnish the required power to the load as economically as possible while meeting the design specifications.

6.1 THE EMITTER FOLLOWER

The emitter-follower circuit has already been discussed in Sec. 4.4. It was noted that the output impedance of an emitter follower is very low, its input impedance is very high, and the voltage gain is almost unity for a large value of load resistance. (See Table 4.2.)

The basic emitter-follower circuit is shown in Fig. 6.1*a*. It is noted that different values of load resistance will have an effect on gain and the dc current of Q_1. Furthermore, the peak-to-peak voltage swing is less than V_{CC}. This circuit is used as an output stage for high-frequency ($\approx$MHz) low-power ($\leq$1 W) amplifiers.

A modified version of the circuit is shown in Fig. 6.1*b* where a current source is used to provide the high load impedance for Q_1 and a peak-to-peak swing voltage of larger than V_{CC}. The voltage gain in Fig. 6.1*b* is almost unity even though R_L can be a small load resistance of the order of 100 Ω. The transfer characteristics of the circuit in Fig. 6.1*b* are shown in Fig. 6.1*c*. In Fig. 6.1*b* note that

$$V_o = I_o R_L \tag{6.1a}$$

and for $I_o = I_{C2} = I_R = (V_{CC} - V_{BE})/R_1$, Q_1 is OFF **(6.1*b*)**

A distortion (clipping) will occur for small values of R_L, i.e., if $I_R R_L < V_{CC}$. Hence the value of R_L must be larger than a critical value determined by $-V_{CC}$ and I_R, namely,

$$R_L \geq R_1 \tag{6.2}$$

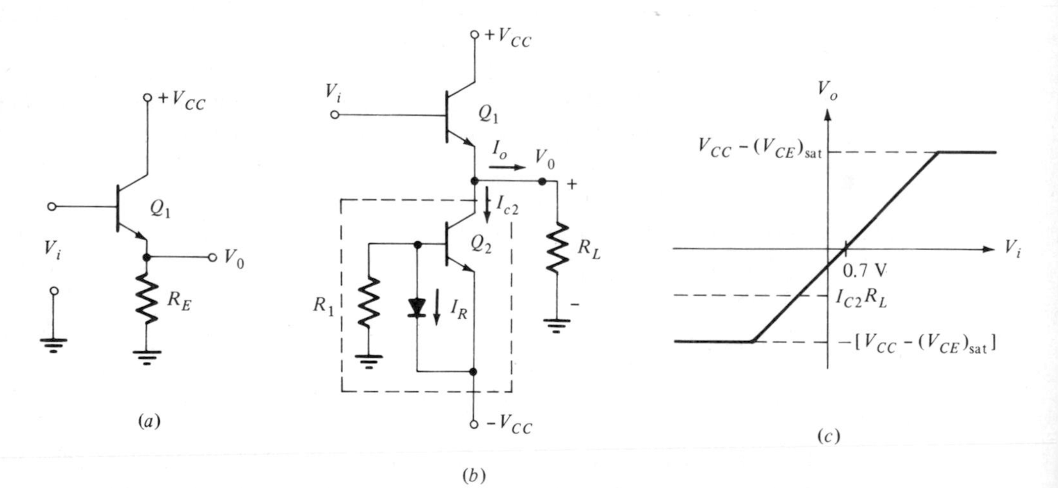

FIGURE 6.1 (*a*) Basic emitter follower. (*b*) Emitter-follower output stage with current source. (*c*) Transfer characteristic of (*b*).

If (6.2) holds the maximum voltage swing (peak to peak) is approximately $2[V_{CC} - (V_{CE})_{sat}]$. Thus to avoid clipping (see Fig. 6.1*c*), the condition in (6.2) must hold. For R_L smaller than that given by (6.2), the maximum peak voltage with no clipping is given by $R_L(V_{CC} - V_{BE})/R_1$. In many situations, such as power amplifiers, the value of the load resistor is very low (e.g., and 8-Ω loudspeaker); therefore, alternate circuitry must be explored. A very useful output stage with a peak-to-peak voltage of approximately $2V_{CC}$ even for small values of R_L is discussed in Sec. 6.6.

EXERCISE 6.1

For the circuit shown in Fig. 6.1*b*, the circuit parameters are $V_{CC} = 5$ V, $R_1 = 2$ kΩ, and $R_L = 1$ kΩ. The transistor parameters are $(V_{CE})_{sat} = 0.2$ V and $(V_{BE})_{on} = 0.7$ V. Determine the maximum unclipped output voltage and current swings. What is maximum average output power?

Ans $V_{om} = 2.15$ V, $I_{om} = 2.15$ mA, $P_{av} = 2.31$ mW

6.2 POWER AMPLIFIERS AND CLASSIFICATION

For power amplifiers the signal swing are often very large so as to be able to deliver maximum power to the load, and the load resistance is usually very low. Since the nonlinear performance of a power amplifier cannot be represented by a linear model, analysis must be performed graphically using the characteristics of the amplifying device. Simplifying assumptions are usually made to make the analysis and design tractable. We will consider such an analysis for audio-frequency power amplifiers, where the frequency range is below 20 kHz. In particular, we shall discuss distortion calculations and the efficiency of conversion of power amplifiers in the various modes of operations.

Power amplifiers are classified according to the load current flow during the sinusoidal input signal. The various classes are illustrated by the output circuit current in Fig. 6.2 (the collector current). It could just as well be the drain current of a FET.

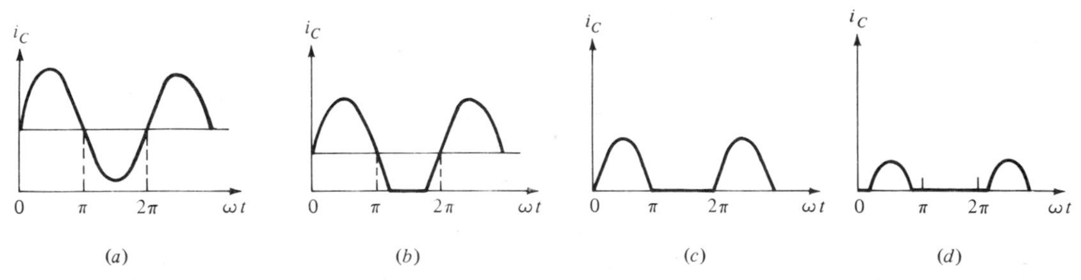

FIGURE 6.2 (*a*) Class A, current flows for the entire period. (*b*) Class AB, current flows for more than half the period. (*c*) Class B, current flows for half the period. (*d*) Class C, current flows for less than half the period.

In Fig. 6.2*a*, a current flows during the entire cycle of operation, and it is called *class A operation*. In class AB, Fig. 6.2*b*, current flows for more than a half-cycle but not for a full cycle. Class B corresponds to the case where current flows for one-half of full cycle as shown in Fig. 6.2*c*. In class C current flows for less than one-half cycle as shown in Fig. 6.2*d*. From Fig. 6.2 it appears that for low distortion only class A can be used. This is true for a *single* transistor, but as we shall show in the subsequent sections, class AB and class B stages can be made to operate essentially in a linear manner by using push-pull or a complementary symmetry arrangement. Class C amplifiers are normally used only at radio frequencies, where the nonlinear distortion can be filtered out by tuned resonant circuits. We shall consider class A and class B bipolar amplifiers in the following, as they are commonly used in audio-frequency power amplifiers.

6.3 CLASS A POWER AMPLIFIERS

When a single device is used in the output stage for linear amplification, the operation must be in class A mode. We shall consider two such amplifier configurations: the first case without transformers and the second case with transformers.

A. Class A Amplifiers without Transformers

A simple class A amplifier, using a transistor, is shown in Fig. 6.3. The operating-point bias conditions are such that the operation is not outside the hyperbola of maximum power dissipation. The limitations on the operating region of the transistor are shown in Fig. 6.4. Note that the maximum power dissipation of a transistor depends on the ambient temperature. As temperature increases, this value decreases (see Sec. 6.7). The maximum average power that a transistor can dissipate depends on the type and construction of the transistor. The maximum power is limited by the temperature of the collector-to-base junction. The junction temperature can rise because of environmental temperature change or self-heating due to collector junction power dissipation. Collector junction power dissipation raises the junction temperature, which in turn increases the collector current, leading to a further increase in power dissipation. This phenomenon (which is referred to as *thermal runaway*) must be avoided at all costs; otherwise, the transistor will be permanently damaged. In the design of power amplifiers, the transistor maximum collector dissipation is usually lowered for a case temperature greater than room temperature. This is called *derating* and is discussed later in Sec. 6.7.

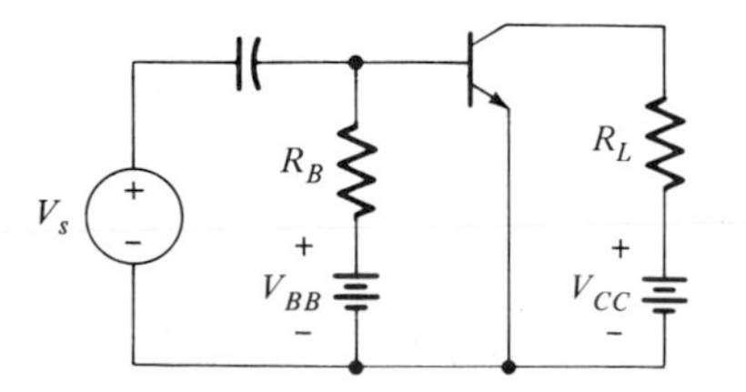

FIGURE 6.3 Simple class A power amplifier.

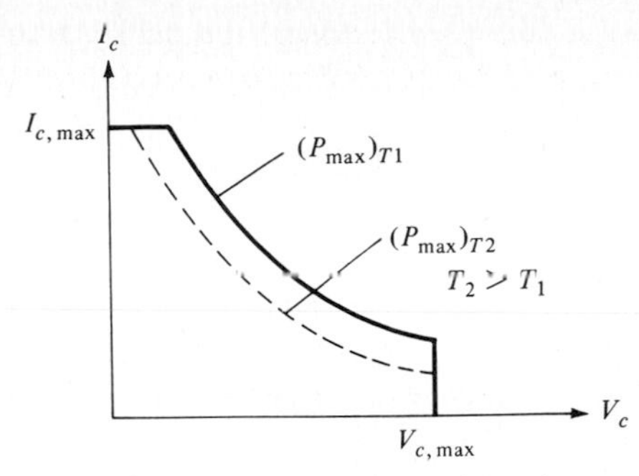

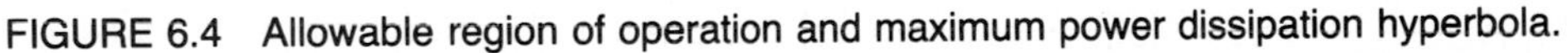

FIGURE 6.4 Allowable region of operation and maximum power dissipation hyperbola.

Assume the output characteristics and the load line for a transistor amplifier as shown in Fig. 6.5. Initially we assume that the output characteristics are equidistant for equal increments of input signal, so that the output signals will be sinusoidal for sinusoidal excitation. In other words, we assume a linear dynamic transfer characteristic, i.e., linear input-output relationship.

In this case nonlinear distortion is negligible, and the various power calculations are straightforward. The dc power required from the power supply is

$$P_{dc} = V_{CC} I_c \tag{6.3}$$

The ac output power is

$$P_{ac} = \frac{I_p}{\sqrt{2}} \frac{V_p}{\sqrt{2}} = \frac{I_p^2 R_L}{2} \tag{6.4}$$

where V_p and I_p are the peak values of the output circuit voltage and current, respectively. Under the linear assumption, the root-mean-square (rms) values of the load

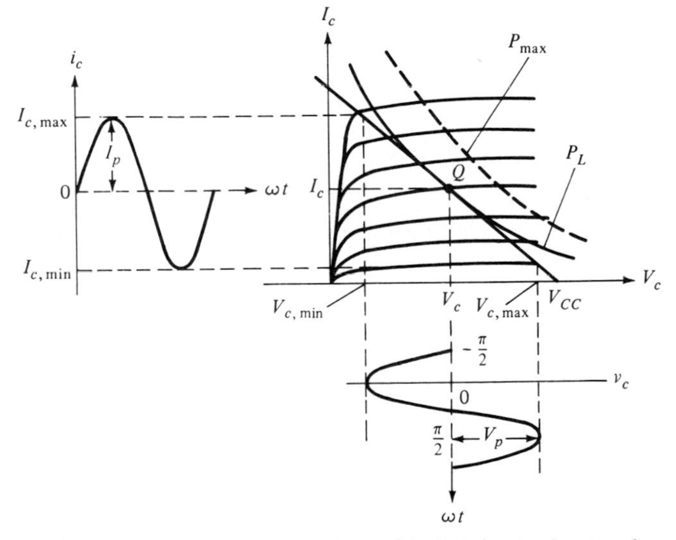

FIGURE 6.5 Collector current and voltage waveforms of a class A power amplifier.

current and voltage from the characteristic curve, in terms of the peak values, are

$$\frac{I_p}{\sqrt{2}} = \frac{1}{\sqrt{2}} \frac{I_{c,\max} - I_{c,\min}}{2} \tag{6.5}$$

$$\frac{V_p}{\sqrt{2}} = \frac{1}{\sqrt{2}} \frac{V_{c,\max} - V_{c,\min}}{2} \tag{6.6}$$

Hence

$$P_{ac} = \frac{I_p V_p}{2} = \frac{(V_{c,\max} - V_{c,\min})(I_{c,\max} - I_{c,\min})}{8} \tag{6.7}$$

We define the conversion efficiency of the dc power to the total ac power by

$$\eta = \frac{\text{ac power delivered to load}}{\text{dc power delivered from power supply}} \times 100\% = \frac{P_{ac}}{P_{dc}} \times 100\% \tag{6.8}$$

From (6.3), (6.7), and (6.8)

$$\eta = \frac{(V_{c,\max} - V_{c,\min})(I_{c,\max} - I_{c,\min})}{8(V_{CC} I_c)} \tag{6.9}$$

Under ideal conditions for maximum efficiency of class A power amplifiers (i.e., we drive to the extremes of the load line),

$$V_{c,\min} = 0 \qquad I_{c,\min} = 0 \qquad V_{c,\max} = V_{CC} \qquad I_{c,\max} = 2I_c$$

and (6.9) reduces to

$$\eta_m = \frac{P_{ac}}{P_{dc}} = \frac{I_p V_p}{2(V_{CC} I_c)} \simeq \frac{V_{c,\max} I_{c,\max}}{4V_{c,\max} I_{c,\max}} = 25\% \tag{6.10}$$

Hence the *maximum theoretical efficiency* for class A operation is 25 percent. The actual conversion efficiency is, of course, less than 25, since all the ac power is not delivered to the load if the device low-frequency parameters are included. Perhaps more important is that these limits cannot be met exactly.

The maximum collector power dissipation in a transistor, which sets the lowest bounds on the transistor power rating, is also a very important consideration in the design of power amplifiers.

The transistor collector dissipation P_C is given by

$$P_C = P_{dc} - P_L \tag{6.11}$$

where P_{dc} is the supplied power and is equal to $V_{CC} I_c$, and P_L is the output power to the load.

A figure of merit F_p for power amplifiers is defined as

$$F_p = \frac{\text{maximum collector dissipation}}{\text{maximum output power}} = \frac{P_{CM}}{P_{LM}} \tag{6.12}$$

For the resistive-coupled circuit of Fig. 6.3 we have

$$P_{LM} = \frac{I_p^2 R_L}{2} = \frac{V_{CC}^2}{8R_L} \tag{6.13}$$

The maximum collector dissipation occurs when P_L is minimum. Hence

$$P_{CM} = \frac{V_{CC} I_c}{2} - 0 = \frac{V_{CC}^2}{4R_L} \tag{6.14}$$

Thus from (6.11), (6.13), and (6.14) we obtain

$$F_p = \frac{V_{CC}^2/4R_L}{V_{CC}^2/8R_L} = 2 \tag{6.15}$$

Thus if the maximum output load power is specified as 10 W, the collector must be able to dissipate *at least* 20 W. This is the major undesirable feature of class A amplifiers, since it normally requires the use of an extremely large and massive heat sink for the transistor.

B. Transformer-Coupled Amplifiers

The efficiency of conversion can be improved considerably by preventing the quiescent current from flowing into the load. In the case of the series-fed amplifier shown in Fig. 6.3, the flow of the quiescent current is the cause of power loss and hence poor efficiency. Often it is also undesirable to pass the dc component of the current through the output device. (Consider the effect on your speakers of a steady-state 10-A current flowing!) Therefore, a more desirable circuit for the output stage is that shown in Fig. 6.6. The transformer at the output circuit also provides an impedance match in order to transfer maximum power to the load. The load, such as the impedance of the voice coil of a loudspeaker, is usually very low, e.g., 4 to 15 Ω. Sometimes the input circuit also utilizes a transformer for matching the output impedance of the driver to the input impedance of the output stage.

Recall that for an ideal transformer, the voltage-current relations are

$$V_1 = \frac{n_1}{n_2} V_2 \quad I_1 = \frac{n_2}{n_1} I_2 \tag{6.16}$$

where the subscript 1 indicates the primary winding and the subscript 2 indicates the secondary winding (current, voltage, and turns). From (6.16) and Fig. 6.6,

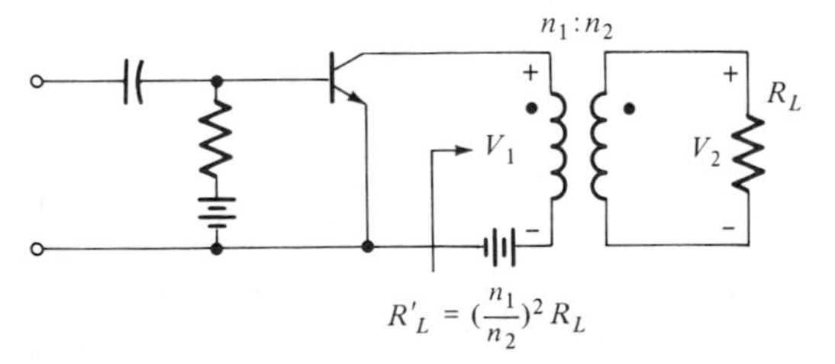

FIGURE 6.6 Transformer-coupled class A power amplifier.

$$R_L' = \frac{V_1}{I_1} = \left(\frac{n_1}{n_2}\right)^2 \frac{V_2}{I_2} = \left(\frac{n_1}{n_2}\right)^2 R_L \quad \textbf{(6.17)}$$

Of course, in a more complete analysis of efficiency and power loss, one must include the other parameters of the transformer, such as the core loss and the primary and secondary resistances.

For a transformer-coupled load, the static (dc) and dynamic (ac) load lines, discussed in Sec. 2.10, are shown in Fig. 6.7. The static load line is almost vertical, because of the very small primary resistance of the transformer. The maximum efficiency of a transformer-coupled stage is readily determined as follows. The maximum peak-to-peak voltage swing $2V_p = V_{\max} = 2V_{CC}$. Also, the peak-to-peak current swing $2I_p = I_{c,\max} = 2I_c$. The maximum ac power is then given by

$$(P_{ac})_{\max} = \frac{V_p I_p}{2} = \frac{V_{CC} I_c}{2} \quad \textbf{(6.18)}$$

The maximum efficiency is

$$\eta_m = \frac{(P_{ac})_{\max}}{P_{dc}} = \frac{V_{CC} I_c/2}{V_{CC} I_c} = 50\% \quad \textbf{(6.19)}$$

Hence the maximum efficiency of a class A stage is *doubled* by using transformer coupling to the load. Note also that for the same size swings the transformer-coupled case requires only $V_{CC}/2$ of the series-fed case.

The figure of merit of transformer-coupled class A amplifiers is the same as that of the series-fed case. This is readily seen to be the case since

$$P_{LM} = \frac{I_p^2 R_L'}{2} = \frac{V_{CC}^2}{2R_L'} \quad \textbf{(6.20)}$$

$$P_{CM} = V_{CC} I_c - 0 = \frac{V_{CC}^2}{R_L'} \quad \textbf{(6.21)}$$

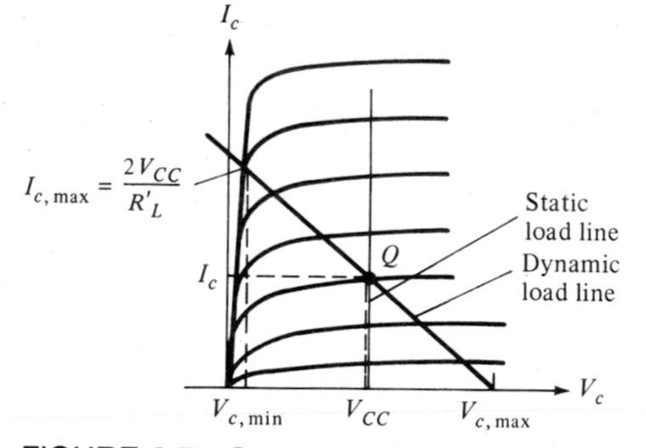

FIGURE 6.7 Static and dynamic load lines of a transformer-coupled class A power amplifier.

and

$$F_p = \frac{P_{CM}}{P_{LM}} = 2 \tag{6.22}$$

EXAMPLE 6.1

Consider the following specifications for the design of a class A power output stage. The audio output power to the load is to be 2 W. The output load resistance is 4 Ω, and the available supply voltage is 12 V. Design the class A amplifier for high efficiency.

Since the load resistance is very low, and the efficiency of a transformer-coupled stage is better than that of the series-fed class A amplifier, we shall use a transformer-coupled class A amplifier. The transistor to be chosen must have the following ratings:

$$(V_{CE})_{\max} > 2V_{CC} = 2(12) = 24 \text{ V}$$

From (6.22) the collector power dissipation must be at least twice the power output delivered to the transformer primary. Thus

$$P_C \geq 2(2) = 4 \text{ W}$$

There are many commercially available transistors that meet these specifications, and in addition have frequency capabilities suitable for audio frequencies. For the selected transistor we plot the 4-W dissipation hyperbola on the typical characteristic curve of the transistor. A load line is then drawn tangent to the hyperbola passing through the quiescent point $V_{CE} = 12$ V, as shown in Fig. 6.8. Hence the value of the quiescent current $I = \frac{4}{12} = 0.3$ A is known.

From the slope of the load line we determine the ac load resistance R_L for the transistor. For this example,

$$R_L' = \frac{12}{0.33} = 36 \ \Omega$$

The required turns ratio of the transformer is then found from

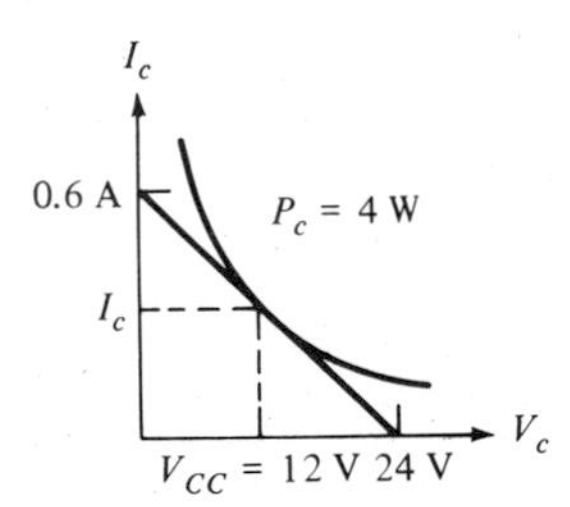

FIGURE 6.8 Load line of the design example.

$$R_L' = \left(\frac{n_1}{n_2}\right)^2 R_L \rightarrow \frac{n_1}{n_2} = \sqrt{\frac{R_L'}{R_L}} = \sqrt{\frac{36}{4}} = 3.0\ \Omega$$

This preliminary design is then experimentally tested and modified, as necessary, to achieve minimum distortion, maximum efficiency, etc.

EXERCISE 6.2

For Exercise 6.1

(*a*) Determine the efficiency of the circuit.
(*b*) If R_L is free to be chosen, what value of R_L will yield the maximum efficiency and what is the corresponding maximum efficiency?

Ans (*a*) $\eta = 10.74\%$; (*b*) $R_L = 2.23\ \text{k}\Omega$, $\eta_{\max} = 24\%$

6.4 DISTORTION CHARACTERIZATION AND CALCULATION

In the previous section we assumed a perfectly linear device, i.e., we assumed a linear dynamic transfer characteristic. However, we have seen in Chaps. 2 and 3 that the device transfer characteristics are nonlinear. For large-signal swings around Q, the nonlinear portion of the transfer characteristics is utilized, and a linear assumption can cause significant error in the analysis. If the device dynamic curve is nonlinear over the operating range, the waveform of the output differs from that of the input and is hence distorted. This type of distortion is known as *nonlinear* or *amplitude distortion*. A linear assumption of the dynamic curve $i = gv$, of course, cannot reveal this distortion. For example, the input characteristics of a bipolar transistor is as shown in Fig. 6.9. For field-effect transistors the transfer characteristics are described in (3.8)

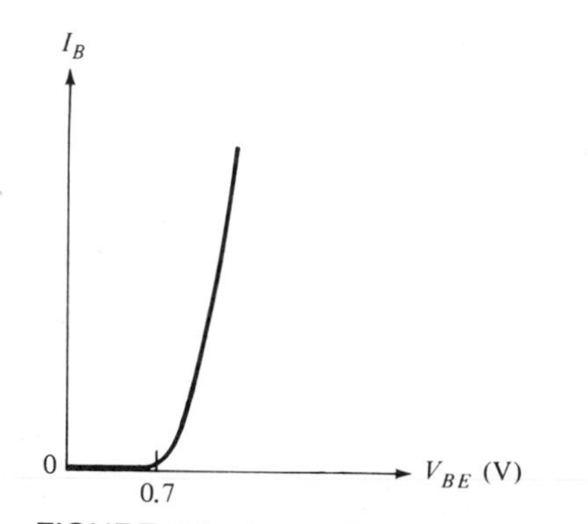

FIGURE 6.9 Input characteristic of a CE stage.

and (3.12) (which is a square law function). Thus, instead of assuming a linear dynamic characteristic, we express the characteristic curve with respect to the quiescent point Q by a power series of the form

$$y - y_0 = g_1 x + g_2 x^2 + g_3 x^3 + \cdots \tag{6.23}$$

where y is the output current and x is the input voltage. We use x and y so as to keep generality of the treatment and thus avoid considering a specific device BJT or FET. In some cases the simple expression $y - y_0 = g_1 x + g_2 x^2$ is acceptable; e.g., for FETs the transfer characteristic has a square law behavior, and hence $g_1 = 0$. This fact is borne out by (3.8) and (3.12); i.e., the drain current I_D is a function of the gate-to-source voltage (V_{GS}) square. It is for this reason that FET is desirable for low-distortion power amplifiers.

In order to characterize the distortion, we have the graphical waveforms, and thus graphical Fourier analysis may be performed. Since the output curve is an even function of time (see Fig. 6.10), the Fourier series expansion of (6.23) can be expressed as a series of cosine terms only:

$$y - y_0 = a_0 + a_1 \cos \omega t + a_2 \cos 2\omega t + a_3 \cos 3\omega t + \cdots \tag{6.24}$$

The Fourier components must now be determined graphically. The graphical Fourier analysis, in conjunction with Fig. 6.10, is illustrated in the following. A very accurate Fourier series expansion can readily be made by using a computer. However, to illustrate the method by a simple approximate technique, we assume a truncated Fourier series with five terms, up to $a_4 \cos 4\omega t$. The input is assumed to be a sinusoidal voltage signal:

$$x = x_m \cos \omega t \tag{6.25}$$

To evaluate the five components, the values of y at five different values of x are needed. We choose these to be as follows:

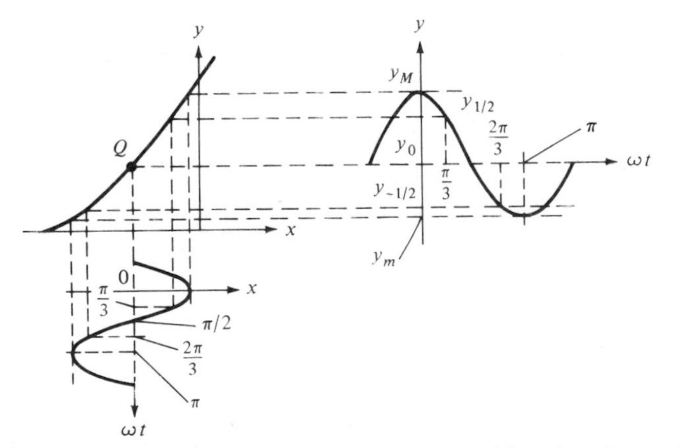

FIGURE 6.10 Transfer characteristics and input and output signals for distortion calculations (e.g., for JFET, $x = V_{GS}$ and $y = I_D$).

At $\omega t = 0$: $y = y_M$

At $\omega t = \frac{\pi}{3}$: $y = y_{1/2}$

At $\omega t = \frac{\pi}{2}$: $y = y_0$

At $\omega t = \frac{2\pi}{3}$: $y = y_{-1/2}$

At $\omega t = \pi$: $y = y_m$

From (6.25) and the following equation for the output current,

$$y = \sum_{n=0}^{4} a_n \cos n\omega t \tag{6.26}$$

We then solve for the values of a_n, which results in (see problem 6.6)

$$\begin{aligned}
a_0 &= \tfrac{1}{6}(y_M + 2y_{1/2} + 2y_{-1/2} + y_m) - y_0 \\
a_1 &= \tfrac{1}{3}(y_M + y_{1/2} - y_{-1/2} - y_m) \\
a_2 &= \tfrac{1}{4}(y_M - 2y_0 + y_m) \\
a_3 &= \tfrac{1}{6}(y_M - 2y_{1/2} + 2y_{-1/2} - y_m) \\
a_4 &= \tfrac{1}{12}(y_M - 4y_{1/2} + 6y_0 - 4y_{-1/2} + y_m)
\end{aligned} \tag{6.27}$$

If there is no distortion, $y_0 = (y_M + y_m)/2 = (y_{1/2} + y_{-1/2})/2$, and $y_M = y_m = 2(y_{1/2} - y_{-1/2})$. Thus $a_0 = a_2 = a_3 = a_4 = 0$, and we have a perfectly sinusoidal output signal. We define the *nth harmonic distortion* as the ratio of the nth harmonic amplitude to the fundamental component:

$$D_n = \left| \frac{a}{a_1} \right| \tag{6.28}$$

If the distortion is not negligible, the total output power can be expressed as

$$P_{ac} = (a_1^2 + a_2^2 + a_3^2 + \cdots)\frac{R_L}{2} = P_1(1 + D_2^2 + D_3^2 + \cdots) \tag{6.29}$$

where $P_1 = \frac{1}{2}a_1^2 R_L$ is the output power due to the fundamental component. We can rewrite (6.29) as

$$P_{ac} = (1 + D^2)P_1 \tag{6.30}$$

where D is called the *total distortion:*

$$D = \sqrt{D_2^2 + D_3^2 + D_4^2 + \cdots} \tag{6.31}$$

For a total distortion of 10 percent, which is usually more than can be tolerated, the total output power is only 1 percent higher than that contributed by the fundamental component. For a good high-fidelity system, the distortion is usually limited to less than 1 percent.

EXERCISE 6.3

The collector-to-emitter voltage of a distorted waveform is approximately described by

$$v_{CE} = V_{CEq} + V_o + V_1 \cos \omega t + V_2 \cos 2\omega t$$

If the measured values of the voltage waveform are $(V_{CE})_{\min} = 1$ V, $(V_{CE})_{\max} = 20$ V, and $V_{CEq} = 10$ V, determine the amount of second harmonic distortion D_2.

Ans $D_2 = 2.63\%$

6.5 CLASS B PUSH-PULL POWER AMPLIFIERS

When two devices are used in the class B push-pull configuration shown in Fig. 6.11, the collector dissipation can be reduced from that of a single-ended stage, and the efficiency can be considerably improved. The operation of the circuit in class B mode, assuming identical ideal transistors, is as follows. The input transformer supplies base currents virtually distortion free. In bipolar transistors because of the nonlinearity of the input *I-V* characteristic in Fig. 6.9, however, the load current will be distorted near the zero crossings, as shown in Fig. 6.12 by the dashed curve. This distortion is usually referred to as *crossover distortion*. It arises because $i_b = 0$ for $V_{BE} \leq 0.7$ V (hence $i_c = 0$). This operation in Fig. 6.12 is, therefore, strictly speaking the class AB mode. However, the break voltage of the $I_b - V_{BE}$ curve is very small (0.7 V) so that the operation is considered essentially in class B mode.

The load line for a single transistor of the class B push-pull circuit is shown in Fig. 6.13, where $R_L' = (n_1/n_2)^2 R_L$. The dc current for each transistor is found from the Fourier expansion. For example, from Fig. 6.12*b* the dc current for Q_1 is

$$(I_{dc})_1 = \frac{1}{T}\int_0^T i_{c1}\,dt = \frac{\omega}{2\pi}\int_0^{\pi} I_p \sin \omega t \; dt = \frac{I_p}{\pi} \tag{6.32}$$

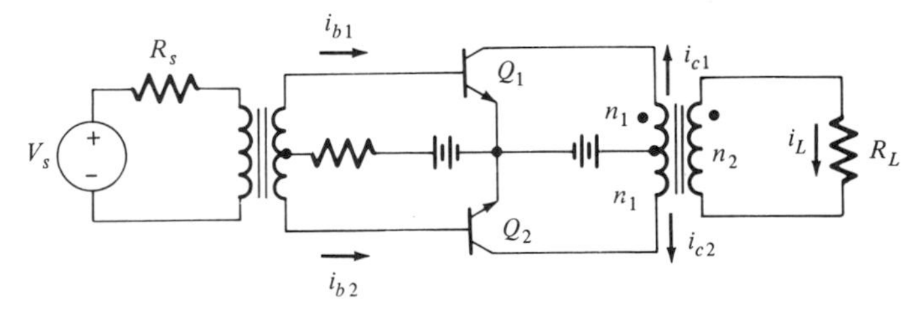

FIGURE 6.11 Class B push-pull amplifiers.

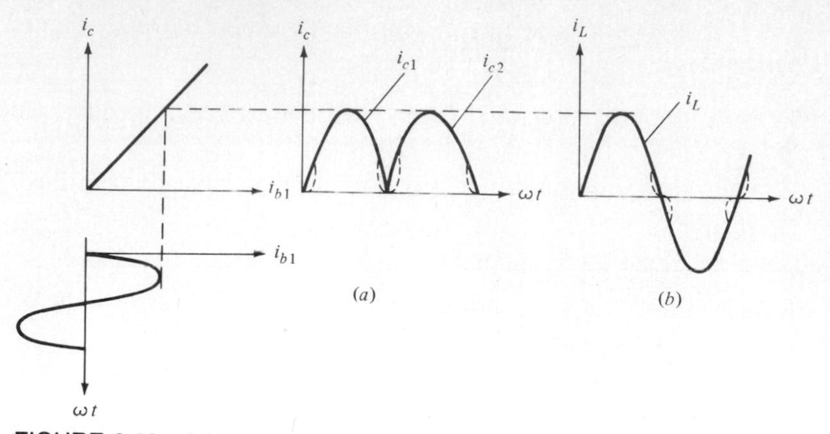

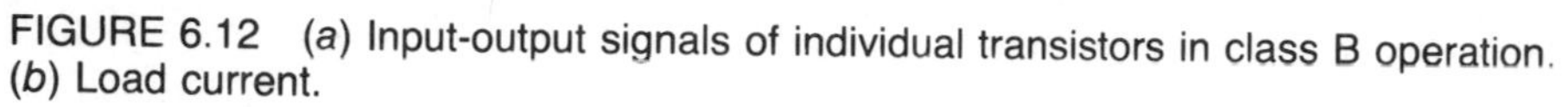

FIGURE 6.12 (*a*) Input-output signals of individual transistors in class B operation. (*b*) Load current.

Thus the dc input power is

$$P_{dc} = I_{dc}V_{CC} = \frac{2}{\pi}I_pV_{CC} \tag{6.33}$$

The factor of 2 is used to account for the dc current of both transistors. For maximum output power swing we have

$$(P_{ac})_{max} = \frac{I_p^2R_L'}{2} = \frac{I_pV_{CC}}{2} \tag{6.34}$$

Thus

$$\eta = \frac{V_{CC}I_p/2}{(2/\pi)I_pV_{CC}} = \frac{\pi}{4} = 78.5\% \tag{6.35}$$

The maximum theoretical efficiency of a transformer-coupled push-pull class in a linear amplifier is 78.5 percent. Note that the maximum efficiency is much higher than that of class A operation.

The average collector power dissipation for both transistors is given by

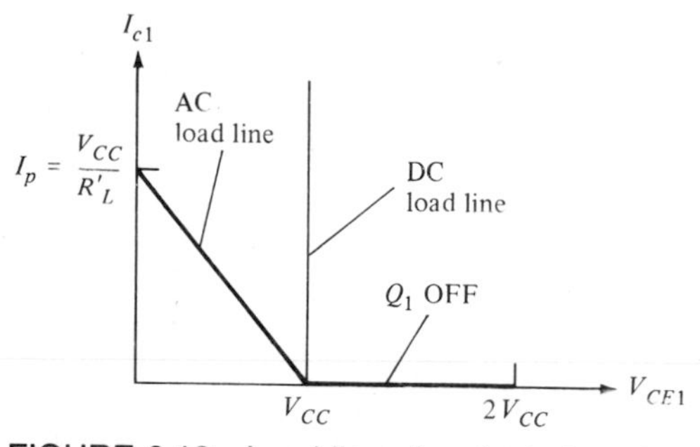

FIGURE 6.13 Load line of a single transistor in push-pull circuit.

$$2P_C = P_{dc} - P_L = \frac{2}{\pi} I_p V_{CC} - \frac{I_p^2 R_L'}{2} \tag{6.36}$$

To find the maximum power dissipation, P_{CM}, we differentiate P_C with respect to I_p and set the resulting expression equal to zero, which yields

$$I_p = \frac{2}{\pi} \frac{V_{CC}}{R_L'} \tag{6.37}$$

From (6.37) and (6.36) we obtain

$$2P_{CM} = \frac{2V_{CC}^2}{\pi^2 R_L'} \tag{6.38}$$

The maximum output power P_{LM} is

$$P_{LM} = (P_{ac})_{max} = \frac{I_p^2 R_L'}{2} = \frac{V_{CC}^2}{2R_L'} \tag{6.39}$$

The figure of merit for each transistor is

$$\frac{P_{CM}}{P_{LM}} = \frac{V_{CC}^2/\pi^2 R_L'}{V_{CC}^2/2R_L'} = \frac{2}{\pi^2} \simeq \frac{1}{5} \tag{6.40}$$

The improvement in comparison with class A amplifiers is a factor of 10. In other words, the class B push-pull amplifier operating at the ideal maximum efficiency may be designed for a power output approximately 5 times the power dissipation rating of the individual transistors. This is indeed quite an advantage. In addition, much smaller heat sinks are required.

EXAMPLE 6.2

Consider the same specifications as in Example (6.1) of the class A amplifier. We shall now consider the use of a class B push-pull circuit.

For the specified output load power of 2 W and the supply voltage of 12 V, from (6.39) we have

$$R_L' = \frac{V_{CC}^2}{2P_{LM}} = \frac{12^2}{2(2)} = 36$$

The peak current

$$I_p = \frac{V_{CC}}{R_L'} = \frac{12}{36} = 0.33 \text{ A}$$

The collector dissipation of each transistor in this case is

$$P_C \geq \tfrac{1}{5}(2) = 0.4 \text{ W}$$

Thus transistors with 2- or 1-W ratings, which could not be used in the previous example, can be used in this case and, in fact, have an extra margin of safety. The turns ratio is found from

$$\frac{n_1}{n_2} = \sqrt{\frac{R_L'}{R_L}} = \sqrt{\frac{36}{4}} = 3.0$$

Similarly, if the output impedance of the driver stage is known, the input coupling transformer turns ratio is selected such that thc input impedance of the power amplifier is matched to that of the driver.

6.6 COMPLEMENTARY (PUSH-PULL) OUTPUT STAGE

In class A operation power dissipation occurs in the absence of input signal. For class B operation, as we saw, the efficiency is much higher, and no power dissipation occurs with no input signal. By using complementary pairs (i.e., *npn* and *pnp* transistors), one can avoid altogether the use of transformers. A typical output stage is shown in Fig. 6.14*a*. The principle of operation of the circuit is described by the transfer characteristics in Fig. 6.14*b*. For $|V_i| \leq 0.7$ V both Q_1 and Q_2 are off. When V_i is larger than $(V_{BE})_{on}$ ($\simeq$0.7 V), Q_1 is forward-biased and acts as an emitter follower, with a gain of almost unity. For V_i larger than $V_{CC} - (V_{CE})_{sat}$, Q_1 saturates. For $V_i \geq 0$, Q_2 is OFF. The same situation holds true for negative V_i, in which case the roles of Q_1 and Q_2 are reversed. Note that the load resistor is connected to the emitter terminals of *both* transistors. Therefore, when either one is ON, it acts as an emitter follower. The dead zone, for $V_i \leq 0.7$ V, where *neither* transistor is ON and $V_0 = 0$, is the cause of the crossover distortion in push-pull output stages as demonstrated in Fig. 6.15. To avoid or ameliorate the crossover distortion, several solutions are commonly used. One is the use of an op amp and feedback as shown in Fig. 6.16. The op amp is

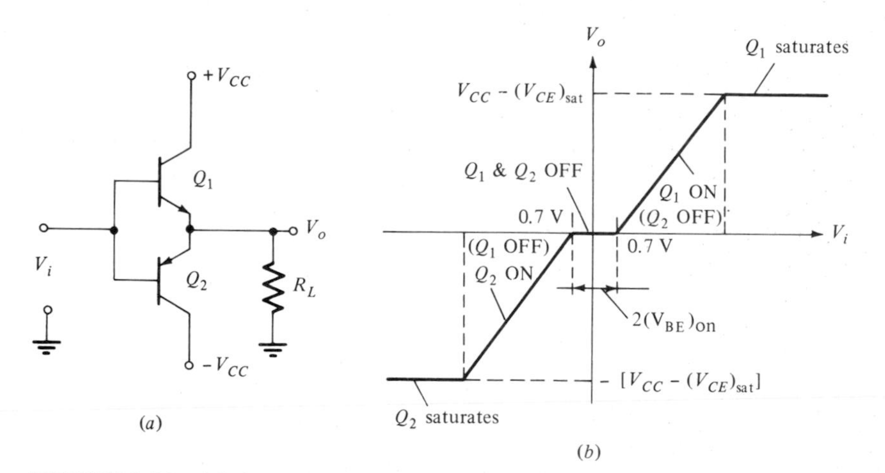

FIGURE 6.14 (*a*) A complementary output stage. (*b*) Its transfer characteristic.

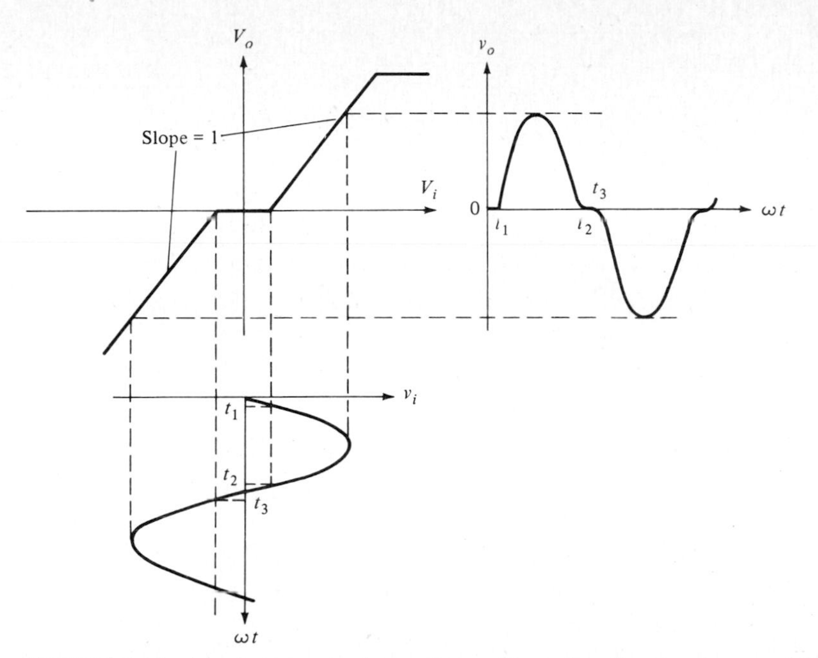

FIGURE 6.15 Input and output waveforms illustrating the dead-zone distortion.

connected in the unity-gain mode. Whenever V_0 and V_i differ by $0.7\ \mathrm{V}/a_o$, where $a_o \geq 10^5$, either of the transistors Q_1 or Q_2 will be ON depending on the sign of $V_i - V_o$. Note that in this case the dead zone is reduced to less than 7 μV on either side, which is entirely negligible, at least at low frequencies. This advantage of negative feedback to reduce distortion should be noted.

The alternate approach, which is used in most integrated-circuit design, is to bias Q_1 and Q_2 such that both are just barely in the active region when $V_i = 0$, as shown in Fig. 6.17*a*. The circuit is also sometimes called a *class AB output stage*, since as

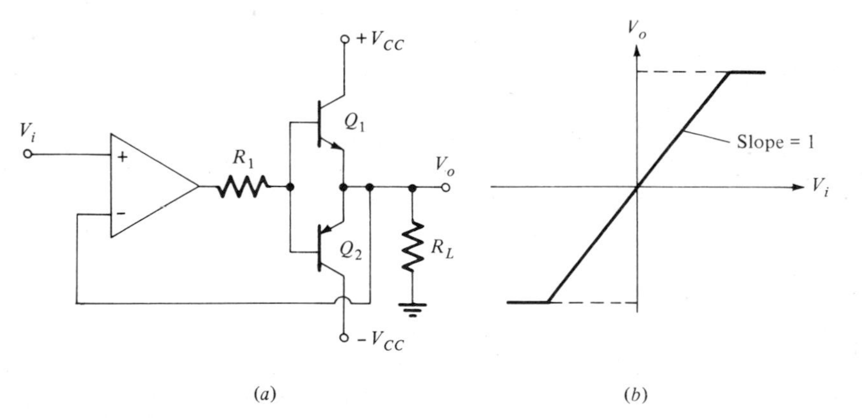

FIGURE 6.16 (*a*) Dead-zone distortion elimination through feedback. (*b*) Its transfer characteristic.

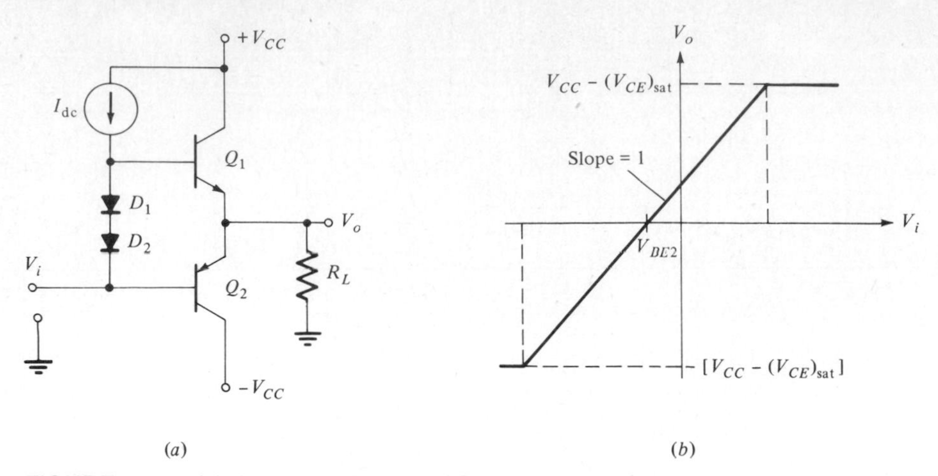

FIGURE 6.17 (*a*) A totem pole class AB output stage. (*b*) Its transfer characteristic.

in class A, the devices are always on and, as in class B, the devices are biased at *much smaller* current than the peak load current. The transfer characteristic of the circuit is shown in Fig. 6.17b. Note that this circuit is used for the op-amp output circuit in Fig. 5.25b. The load line of a single transistor is the same as shown in Fig. 6.13, where $(V_{CE})_{sat}$ is ignored in comparison with V_{CC}.

To find the efficiency of the complementary output stage, from (6.33)

$$P_{dc} = I_{dc}V_{CC} = \frac{2}{\pi} I_p V_{CC} \tag{6.41}$$

Similarly, as in (6.34), the maximum ac power is

$$(P_{ac})_{max} = \frac{1}{2}\frac{[V_{CC} - (V_{CE})_{sat}]^2}{R_L} \simeq \frac{1}{2}\frac{V_{CC}^2}{R_L} = \frac{1}{2} I_p V_{CC} \tag{6.42}$$

Hence the maximum efficiency is

$$\eta_{max} = \frac{(P_{ac})_{max}}{P_{dc}} \simeq \frac{\pi}{4} = 78.5\% \tag{6.43}$$

Again we note that for class B operation the standby power is zero, as the quiescent point Q is at $V_{CE} = V_{CC}$ and $I_c = 0$. From (6.40) it is seen that if we wish to deliver 20 W in a class B push-pull amplifier, then each transistor must be capable of dissipating only 4 W.

EXAMPLE 6.3

For a sinusoidal input, if the desired maximum peak-to-peak voltage at the load is 26 V and $R_L = 100\ \Omega$, find the efficiency and the average power dissipated by both transistors of the complementary class B output stage if $|V_{CC}| = 15$ V.

From (6.32) the average supply current $I_{dc} = V_p/\pi R_L$ and

$$P_{dc} = \frac{13}{100\pi}(2V_{CC}) = \frac{13(30)}{\pi(100)} = 1.242 \text{ W}$$

From (6.34)

$$P_{ac} = P_L = \frac{1}{2}\frac{13^2}{100} = 0.845 \text{ W}$$

Hence

$$\eta = \frac{P_{ac}}{P_{dc}} = 68.1\%$$

The average power dissipated by both transistors is

$$P_{av} = P_{dc} - P_{ac} = 397 \text{ mW}$$

Of course each transistor dissipates one-half of P_{av}, namely, 198 mW.

EXERCISE 6.4

For the circuit shown in Fig. 6.14*a* the circuit parameters are $V_{CC} = 10$ V and $R_L = 100\ \Omega$. If the transistor $(V_{CE})_{sat} = 0.2$ V, determine the average power delivered to the load for a maximum sinusoidal output voltage. What is the efficiency for this case?

Ans $P_{av} = 480$ mW, $\eta = 76.7\%$

6.7 ENVIRONMENTAL AND THERMAL CONSIDERATIONS

Power transistors dissipate a large amount of power, which generates heat and causes the collector junction temperature to rise. The temperature rise must be kept within a safe, acceptable limit in order not to damage the transistor. The maximum allowable collector junction temperature of Si power transistor is in the range of 150 to 200°C. Manufacturers' data sheets usually specify the maximum operating junction temperature range, the junction-to-case thermal resistance θ_{JC}, and the case-to-ambient thermal resistance θ_{CA}. The thermal resistance is related to the power dissipation P_D as

$$T_J - T_C = \theta_{JC} P_D \tag{6.44}$$

where θ_{JC} is in degrees Celsius per watt.

Similarly

$$T_C - T_A = \theta_{CA} P_D \tag{6.45}$$

The electrical analog of a thermal system expressed by (6.44) and (6.45) is shown in Fig. 6.18, where P_D, θ, and T are analogous to constant current, source resistance, and

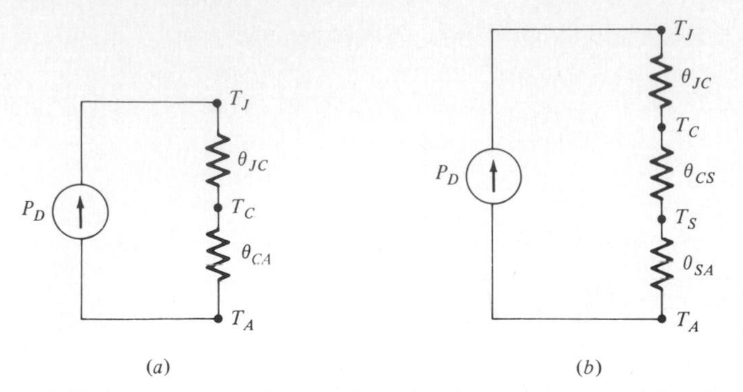

FIGURE 6.18 Electrical analog of a thermal system: (*a*) without heat sink and (*b*) with heat sink.

voltage, respectively. Note that θ denotes the thermal resistance between two points at which temperature measurements are taken. Usually the device is mounted on some form of a *heat sink,* such as mounting it on a chassis or some other metal surface. The metal heat conduction helps lower the junction temperature. The case-to-sink thermal resistance is denoted by θ_{CS}. Using the analogy in Fig. 6.18 (including the heat sink), we have

$$T_J - T_A = P_D(\theta_{JC} + \theta_{CS} + \theta_{SA}) \tag{6.46}$$

or

$$T_J - T_A = P_D(\theta_{JA}) \tag{6.47a}$$

where

$$\theta_{JA} = \theta_{JC} + \theta_{CS} + \theta_{SA} \tag{6.47b}$$

Note that if there is no heat sink, then

$$\theta_{JA} = \theta_{JC} + \theta_{CA} \tag{6.48}$$

Since the transistor case temperature cannot be held at ambient temperature, manufacturers provide a power-temperature derating curve as shown in Fig. 6.19,

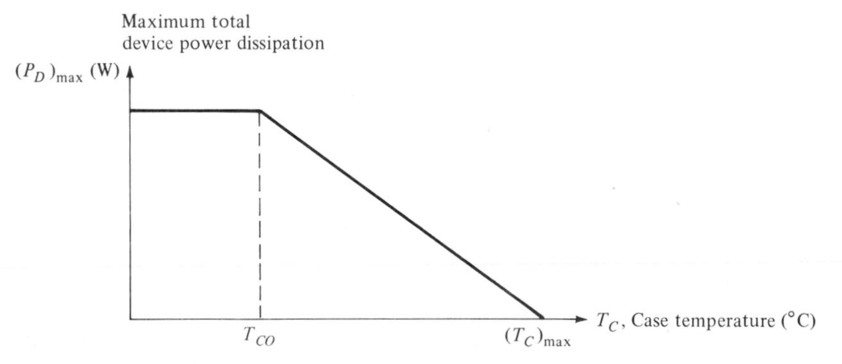

FIGURE 6.19 Power derating curve.

where T_{CO} is the case temperature at which derating begins. Note that T_{CO} is not necessarily 25°C and $(T_C)_{max}$ is in the range 150 to 200°C. Since derating is linear, the derating factor in watts per degree Celsius plus some other necessary parameters shown in Fig. 6.19 may alternatively be specified.

EXAMPLE 6.4

Determine the maximum dissipation at a case temperature of 100°C for a 70-W Si transistor (rated at 30°C). The derating factor that begins at 30°C is given as 0.6 W/°C.

$$(P_D)_{max} = 70 \text{ W} \qquad \text{at } 30°\text{C}$$

At 100°C,

$$(100 - 30)(0.6) = 42 \text{ W}$$

Hence $(P_D)_{max} = 18$ W at 100°C.

The thermal resistance from junction to free-air ambient, without heat sink, is typically

$$\theta_{JA} = 50°\text{C/W} \tag{6.49}$$

From (6.48) a silicon transistor operating in an ambient temperature of 25°C would exceed the maximum allowable junction temperature of 200°C for a power dissipation of 3 W. Hence, in order to be able to dissipate large amounts of power, we *must* use a heat sink, which will lower the thermal resistance considerably. The thermal resistance of heat sinks varies depending on the construction and mounting of the device, and θ_{SA} can be as low as 1°C/W. Typical values in (6.47b) are listed below:

$$\theta_{SA} \simeq 3°\text{C/W} \qquad \theta_{CS} \simeq 1°\text{C/W} \qquad \theta_{JC} \simeq 1°\text{C/W} \tag{6.50}$$

Hence from (6.47), with heat sink, we have

$$\theta_{JA} \simeq 3 + 1 + 1 = 5°\text{C/W}$$

Thus for every watt of power dissipation we have an increase of only 5°C instead of 50°C without a heat sink.

EXAMPLE 6.5

A power amplifier with 15-W output is operating at an efficiency of 68 percent. The maximum allowable junction temperature is 150°C. If the ambient temperature is 25°C, determine the maximum tolerable heat-sink thermal resistance. Use the typical values of $\theta_{CS} \simeq \theta_{JC} = 1°\text{C/W}$.

The power dissipated in the device is found from

$$P_D = \frac{15}{0.68} - 15 = 7.06 \text{ W}$$

From (6.46)

$$150 - 25 = 7.06(1 + 1 + \theta_{SA})$$

which gives $\theta_{SA} = 15.71°\text{C/W}$ maximum. Hence any heat sink with a thermal resistance below this value will be acceptable.

EXERCISE 6.5

A power transistor is rated 120 W at 30°C and has $\theta_{JC} = 0.7°\text{C/W}$. If the transistor is operated with a heat sink, $\theta_{SA} = 2°\text{C/W}$, determine the maximum power that can be dissipated at an ambient temperature of 50°C if $(T_J)_{\text{max}} = 200°\text{C}$ and $\theta_{CS} = 0.8°\text{C/W}$.

Ans 42.9 W

6.8 POWER FIELD-EFFECT TRANSISTORS (VMOS)

Thus far in this chapter we have concentrated on bipolar transistors. We have discussed FET devices in Chap. 3 and have also indicated that these devices fabricated in CMOS are ideal for low-power switching applications (see Sec. 3.5). New types of FET power transistors are available that find application in high-power circuits. These are the so called MOS POWER FETs or VMOS. These devices combine the advantages of the power bipolar transistors with those of the MOS. There are basically two types; one is a high-voltage double-diffused planar structure called DMOS, and the other is a low-voltage V-groove structure commonly called VMOS. A cross-sectional view of a VMOS is shown in Fig. 6.20*a*. This device is an *n*-channel enhancement MOSFET, but the fabrication is such that drain-to-source current flows vertically, hence the name VMOS. Recall that in the conventional FET, current flows horizontally from source to drain. For the *n*-channel enhancement device, the gate voltage must be positive with respect to the source. Under this condition the conducting channel is formed on both sides of the *p* body. Electrons flow from the source through the two channels to the n^- expitaxial and to the n^+ substrate and hence to the drain terminal. The channel length of a VMOS is shorter than that of the NMOS. The width-to-length ratio of VMOS is thus larger than the NMOS, hence I_D is increased, which in turn improves g_m. The DMOS, shown in Fig. 6.20*b*, resembles planar topology, but the drain current path penetrates the n^- epitaxy and the n^+ substrate to the backside contact. The current path is exactly the same as in the VMOS structure.

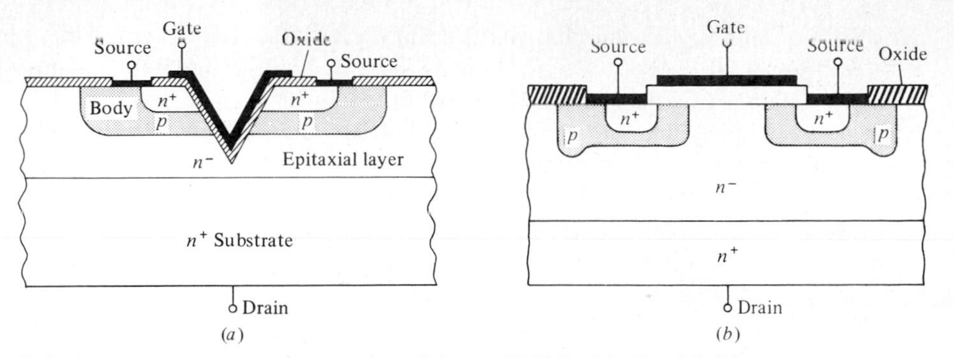

FIGURE 6.20 Cross-sectional view (*a*) of a VMOS; (*b*) of a DMOS.

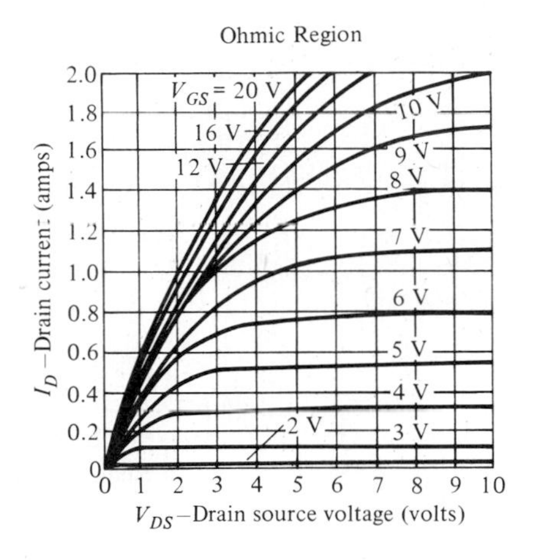

FIGURE 6.21 Output characteristics of a typical VMOS 2N6657 (Courtesy Siliconix Inc.).

Most of the commercially available power MOS devices fall into one of the two categories—VMOS or DMOS. The *V-I* characteristic of a typical VMOS is shown in Fig. 6.21. The VMOS is able to handle a larger current per unit area than the planar devices and it allows better packing density on the chip. The VMOS devices also have some other advantages such as low leakage current, high switching speed, and no thermal problems. The VMOS devices are unsymmetrical, so that the source and the drain terminals cannot be interchanged. Since complementary VMOS transistors are not available (to date), these devices cannot be used in class B push-pull operation. This is a drawback, since the efficiency of class A operation is much lower than that of class B.

6.9 CONCLUDING REMARKS

This chapter dealt with power amplifiers and output stages. The basic requirement for the output stage is to deliver the required signal power to the load with acceptable minimum signal distortion. Several output stages were considered in this chapter.

Class A amplifiers, using the emitter follower, achieve very low distortion but with poor power efficiency. Class AB and class B amplifiers, using the push-pull or complementary output stage, achieve essentially linear amplification with a reasonable power efficiency and are commonly used. Class C power amplifiers which are used at radio frequencies in communication circuits are not discussed here.

Power transistors dissipate a large amount of power, which generates heat. Hence thermal consideration must be also included in the design as discussed in this chapter.

REFERENCES

1. D. Schilling and C. Belove, *Electronic Circuits—Discrete and Integrated,* 2d ed., McGraw-Hill, New York, 1979.
2. P. R. Gray and R. G. Meyer, *Analysis and Design of Analog Integrated Circuits,* Wiley, New York, 1977.
3. J. Millman, *Microelectronics, Digital and Analog Circuits and Systems,* McGraw-Hill, New York, 1979.

PROBLEMS

6.1 Design a class A transformer-coupled power amplifier to supply a power of 10 W to an 8-Ω load resistance. The supply voltage is 12 V. Use a stabilized biased scheme with $R_B = 5R_E = 5$ kΩ and show all the circuit element values if the transistor has $\beta_0 = 50$. Also specify the transistor power dissipation rating.

6.2 A transistor is to be used in a class A power amplifier. The load resistance is 5 Ω. The power transistor ratings are $P_{C,\max} = 10$ W, $(V_{CE})_{\text{sat}} = 1$ V, $(V_{CE})_{\max} = 60$ V.
(*a*) If no transformer is used, determine the maximum attainable voltage swing, the maximum power dissipated by the load, and the efficiency of the amplifier.
(*b*) Repeat (*a*) if transformer coupling is used, with $n = 2$.

6.3 For the emitter-follower circuit shown in Fig. P6.3, determine the largest peak-to-peak value of the sinusoidal output voltage. Assume the following transistor parameters: $(V_{BE})_{\text{on}} = 0.7$ V, $\beta_0 = 200$, $(V_{CE})_{\text{sat}} = 0.2$ V.

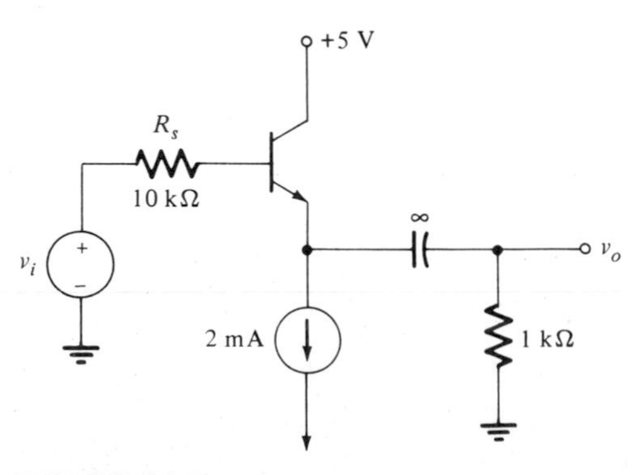

FIGURE P6.3

6.4 For a transformer-coupled class A power amplifier, the following specifications are given: the audio power to the load is to be 3 W, the load is a loudspeaker with an 8-Ω impedance, the available supply voltage is 12 V, the efficiency of the available transformers is 80 percent, and the driver stage has an output resistance of 1 kΩ. Specify the transistor ratings and the turns ratio of the input and output transformers if at the operating point $h_{ie} = 100\ \Omega$.

6.5 Show that for a class B push-pull amplifier, the power due to the fundamental component is given by

$$P_1 = \left[\frac{\sqrt{2}}{3}(I_{\max} + I_{1/2})\right]^2 \left(\frac{n_1}{n_2}\right)^2 R_L$$

where $I_{\max}$ and $I_{1/2}$ are the values of current at $\omega t = 0$ and $\omega t = \pi/3$.

6.6 (*a*) For FET devices show that in (6.24) $a_n = 0$ for $n \geq 3$.
(*b*) Derive (6.27).

6.7 A distorted output signal wave is approximated by

$$v_o = V_{CEq} + V_o + V_1 \cos \omega t + V_2 \cos 2\omega t$$

If the percent second harmonic distortion is defined as

$$D_2 = \frac{V_2}{V_1} \times 100\%$$

show that

$$D_2 = \left| \frac{\frac{1}{2}(V_{CEM} + V_{CEm}) - V_{CEq}}{V_{CEM} - V_{CEm}} \right| \times 100\%$$

where M and m denote maximum and minimum values, respectively. What is the percent distortion if

$$V_{CEM} = 18\text{ V},\ V_{CEm} = 1\text{ V, and } V_{CEq} = 10\text{ V?}$$

6.8 For distortion calculations, suppose we choose four points instead of five (as done in the text). If these four points are $\omega_0 t = 0,\ \pi/3,\ \pi/2$, and π and the corresponding values of y are $y_m,\ y_1,\ y_0,\ y_{-m}$, show that the coefficients of the Fourier series in this case are given by

$$a_0 = \frac{y_m + y_{-m}}{4} + \frac{y_0}{2}$$

$$a_1 = \frac{y_m}{4} - \frac{5}{12}y_{-m} + \frac{2}{3}y_1 - \frac{y_0}{2}$$

$$a_2 = \frac{y_m + y_{-m}}{4} - \frac{y_0}{2}$$

$$a_3 = \frac{y_m}{4} - \frac{y_m}{12} - \frac{2}{3}y_1 + \frac{y_0}{2}$$

6.9 For the push-pull circuit operating in class A, derive the expression for P_L, P_C, and η. Note that the effective load in this case is $(2n_1/n_2)^2 R_L$.

6.10 For the circuit shown in Fig. 6.1, let $I_{C2}R_L > V_{CC}$ and assume $(V_{CE})_{\text{sat}} \simeq 0$. Sketch the waveforms for voltage v_{ce1}, current i_{c1}, and power P_1 for the transistor Q_1 under full output conditions, i.e., $(v_{ce1})_{\text{peak}} = V_{CC}$. What is the average power dissipated in Q_1 under these conditions?

6.11 For the complementary output circuit shown in Fig. P6.11 the transistors are assumed identical with parameter values $(V_{BE})_{on} = 0.7$ V, $(V_{CE})_{sat} = 0.2$ V, and $\beta_0 = 100$. Determine the largest peak-to-peak value of the sinusoidal output before clipping occurs.

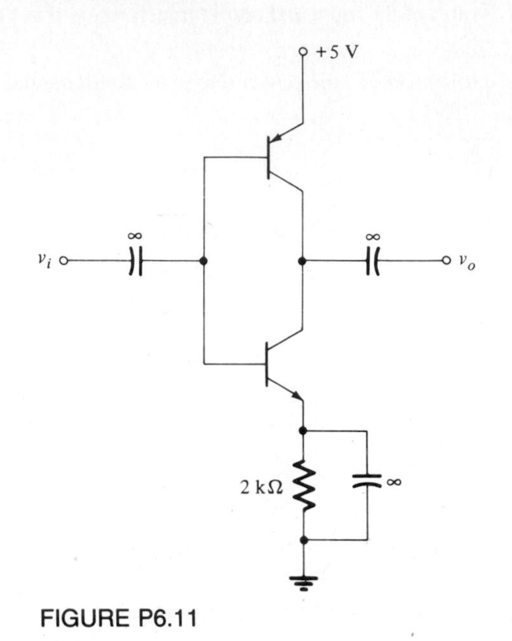

FIGURE P6.11

6.12 For the class B power amplifier circuit shown in Fig. P6.12, determine the maximum undistorted output peak voltage, the maximum input dc power, the output ac power, and the efficiency of the circuit.

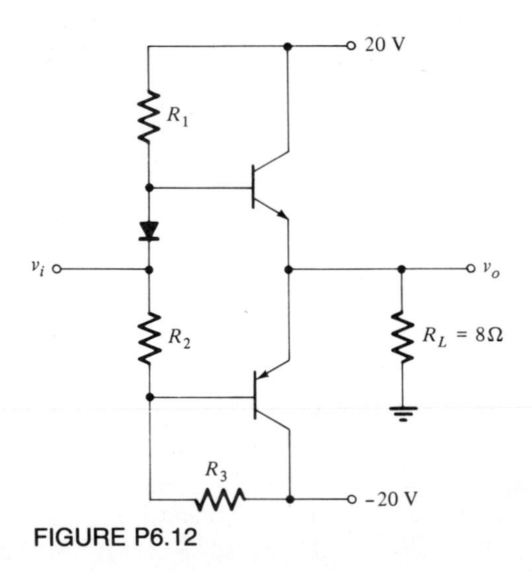

FIGURE P6.12

6.13 For the class B stage shown in Fig. 6.14, assume $(V_{CE})_{sat}$ and $(V_{BE})_{on}$ are both zero. Sketch the waveform for collector current, collector voltage, and the instantaneous power dissipation for one of the devices at maximum output.

6.14 For the circuit shown in Fig. P6.14, the transistors are assumed identical with the following parameters: $\beta_0 = 100$, $(V_{BE})_{on} = 0.7$ V, $(V_{CE})_{sat} \simeq 0$.
(*a*) Calculate the maximum positive and negative values of V_O if $R_L = 10$ kΩ.
(*b*) Repeat (*a*) for $R_L = 1$ kΩ.

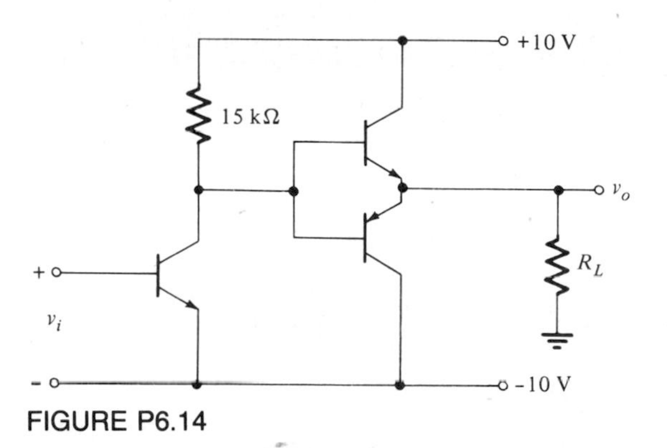

FIGURE P6.14

6.15 (*a*) For problem 6.14, what is the maximum value of v_O before clipping occurs in both cases?
(*b*) Determine the maximum power that can be delivered to the load R_L before clipping occurs in each case.

6.16 Determine the maximum power that can be handled by a 100-W power transistor operated with a heat sink $\theta_{SA} = 2$°C/W, $\theta_{JC} = \theta_{CS} = 1$°C/W at an ambient temperature of 40°C. The allowable junction temperature is 200°C.

6.17 Repeat problem 6.16 if no heat sink is used and the thermal resistance from junction to free-air ambient is $\theta_{JA} = 50$°C/W.

7 Frequency Response of Single-Transistor Amplifiers

Amplification is an important function in electronic circuits. We have already examined single-stage and multistage amplifiers and determined the voltage gain and current gain properties of BJTs and FETs. In the previous chapters we had ignored all the frequency-dependent circuit components. In other words, we used low-frequency models for the transistors and assumed very large capacitors for the bypass and coupling circuitry so that these can be treated as short-circuited at signal frequencies. This chapter will consider the frequency response of the amplifier circuit; i.e., the analysis and design will include the high-frequency equivalent circuit for the transistors and the effects of the bypass and coupling capacitors. Recall that when we considered "low-frequency" behavior of the devices in the previous chapters, it meant frequencies below which "high-frequency" effects take place, i.e., no shunt parasitic capacitances, no transit time effects, etc. In this chapter *low frequency* means frequencies at which the effects of the coupling and bypass capacitors are dominant and must be included.

This chapter will concentrate on the high frequency and very-low-frequency behavior of single-transistor amplifiers. At this stage, the reader is assumed to be familiar with Laplace transforms and the complex frequency s domain ($s = \sigma + j\omega$). We shall also make extensive use of poles and zeros and show how to simplify analysis and design using the concept of dominant poles and zeros. In a linear circuit frequency response and transient response are, of course, related. The latter is considered in detail in Chap. 12.

7.1 TYPICAL AMPLIFIER FREQUENCY CHARACTERISTICS

The normalized frequency response of a typical ac amplifier, such as that shown in Fig. 7.1*a*, is shown in Fig. 7.1*b*. This general type of response represents the frequency characteristics of FETs as well. In other words, although shown for a bipolar junction transistor, the frequency response could just as well represent the response of a field-effect transistor. The normalization scale, of course, depends on the device parameters and the external circuit elements.

The complete amplifier frequency response in Fig. 7.1*b* can be obtained from the circuit of Fig. 7.1*a* by replacing the device with its high-frequency small-signal equivalent circuit and then calculating the magnitude response of the desired gain function via computer aid. This calculation, if done by paper-and-pencil methods (i.e., noncomputer methods), is *very* involved and is not recommended. A simple approximate method for obtaining the frequency response of an amplifier is to consider the frequency response at various frequency ranges, namely, the low-frequency, midband, and high-frequency ranges. At low frequencies, say, $\omega \leq 10^4$ rad/s, the *shunt* capacitors of the high-frequency equivalent circuit of the device, which are of the order of 10 pF, have reactances $X_c = 1/\omega C > 10^7\ \Omega$. The shunt and parasitic capacitances are essentially open-circuited and can be ignored at low frequencies. Thus the circuit frequency behavior is determined by the bypass and coupling capacitors in this region. At high frequencies, say $\omega > 10^5$ rad/s, the bypass and coupling circuitry of the biasing circuit, which are of the order of 10 μF, have reactances $X_c = 1/\omega C < 1\ \Omega$. These capacitors are thus essentially short-circuited, and the

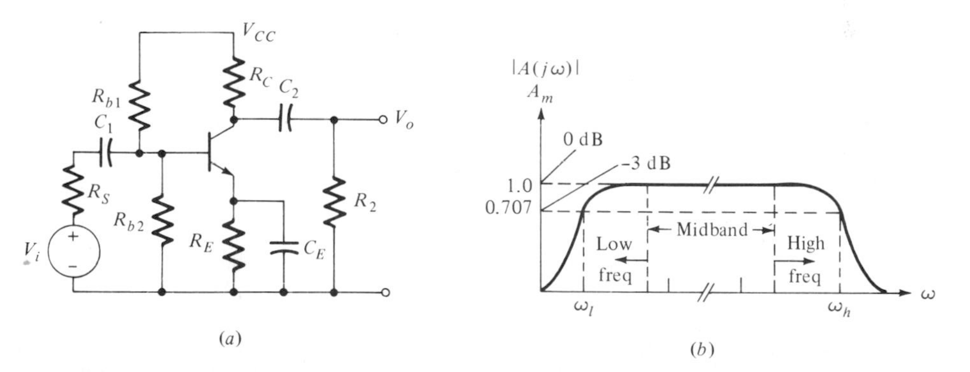

FIGURE 7.1 (*a*) Single-stage CE transistor amplifier. (*b*) Typical frequency response of the magnitude of the gain function.

circuit behavior is determined solely by the capacitors of the high-frequency transistor model and the parasitic capacitances.

At midband frequencies, say, $10^5 \geq \omega \geq 10^4$, the frequencies are low enough so that the shunt capacitors of the device and the parasitics can be approximated as an open circuit; at the same time, the frequencies are high enough so that the biasing and coupling capacitors can be approximated as a short circuit. We therefore have a flat gain in this region. The gain and impedance properties of FETs and BJTs discussed in Chaps. 4 and 5 are in this region. In the following we shall discuss the low- and the high-frequency response characteristics in detail.

In Fig. 7.1*b* the gain A corresponds to either voltage gain or current gain, ω_h is the high-frequency cutoff or upper 3-dB radian frequency, and ω_l is the low-frequency cutoff or lower 3-dB radian frequency. These frequencies are also called *half-power frequencies*. In other words, at $\omega = \omega_h$ and $\omega = \omega_l$, the power gain is lower than at its midband value by a factor of $\frac{1}{2}$, or alternatively stated, the voltage gain or current gain is lower than its midband value by a factor of $1/\sqrt{2}$. (Note that for a resistive load, power $P \alpha V^2$ or I^2 we have $10 \log \frac{1}{2} = 20 \log 1/\sqrt{2} = 3$ dB.) Usually ω_h is much larger than ω_l by several orders of magnitude. For example, in a video amplifier ω_h may be equal to $2\pi(10^7$ Hz), whereas ω_l may be $2\pi(10^2$ Hz). The amplifier bandwidth f_{3dB} or $\omega_{3dB} = 2\pi(f_{3dB})$ is defined as

$$f_{3dB} = f_h - f_l \simeq f_h \qquad \text{or} \qquad \omega_{3dB} = \omega_h - \omega_l \simeq \omega_h \tag{7.1}$$

The bandwidth of an amplifier is usually as large as the spectrum of the signals the amplifier has to amplify so that the frequency spectrum of the input signal is all amplified by the same gain; otherwise, distortion would result.

In view of the fact that the high-frequency (ω_h) and low-frequency (ω_l) cutoffs are separated by a wide range of frequencies (at which frequencies the gain is constant), considerable simplification results if they are treated separately. We can break up the frequency response into three regions: the low-frequency, the midband, and the high-frequency regions, as indicated in Fig. 7.1*b*. At midband the gain and impedances are constant, as in Chaps. 4 and 5.

The loss of gain in the low-frequency response region is due to the coupling and bypass capacitors of the biasing circuit and is unaffected by the high-frequency properties of the transistor. It has the high-pass (i.e., high frequencies are not attenuated) response characteristics shown in Fig. 7.2*a*. Note that because of capacitors C_1 and C_2, no signal is transmitted at dc, and at high frequencies the relatively large capacitances C_1, C_2, and C_E are essentially short-circuited. In a design the appropriate value of ω_l is obtained by the proper selection of values for these capacitors, as discussed in Sec. 7.4.

The high-frequency response and the attenuation are a result of the device capacitances as depicted in their high-frequency models and other parasitic and stray capacitances. The high-frequency response is unaffected by the coupling and bypass capacitors except that they do tend to introduce parasitic shunt capacitances into the circuit. In fact, at very high frequencies the lead inductances may also have to be included. The high-frequency response has a low-pass (i.e., low frequencies are not attenuated) characteristic, as shown in Fig. 7.2*b*.

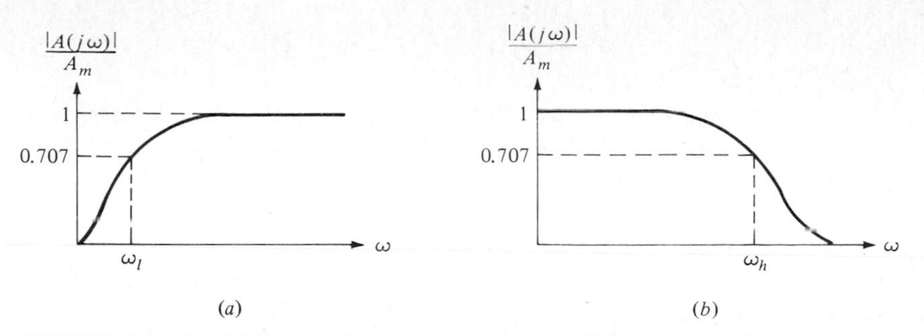

FIGURE 7.2 (*a*) High-pass frequency response. (*b*) Low-pass frequency response.

In the midband region, the frequencies are high enough so that the coupling and bypass capacitors are essentially short-circuited and low enough so that the device capacitances and parasitic capacitances can be neglected. The devices can thus be represented by their low-frequency, i.e., frequency-insensitive, models. In other words, at midband frequencies we have no reactive elements in the circuit, so that the gain is independent of the frequency. This was the situation in all cases considered in Chaps. 4 to 6. This chapter will examine, in detail, the gain functions and the frequency responses at both the low- and the high-frequency range.

7.2 FREQUENCY RESPONSE

The plot of magnitude and phase of a network function vs frequency is referred to as the *frequency response*. A network function may be a voltage gain function $A_v(s)$, a current gain function $A_i(s)$, a transfer impedance function $Z(s)$, or a transfer admittance function $Y(s)$. The frequency response information is very useful in the analysis and design of linear time-invariant networks. We shall briefly discuss the frequency response below. Consider a simple one-pole circuit shown in Fig. 7.3.

The voltage gain function of the circuit is given by

$$A_v(s) = \frac{V_o}{V_s} = -\frac{g_m R_L}{R_S(1/R_1 + sC_1)} = \frac{-g_m R_L R_1}{R_S(1 + sR_1C_1)} \tag{7.2}$$

where $R_1 = R_i \parallel R_S$. Equation (7.2) can be rewritten as

$$A_v(s) = \frac{K}{1 + s/p_1} \tag{7.3}$$

where K is the dc gain and p_1 is the pole of the network, namely,

$$K = -\frac{g_m R_L R_1}{R_S} \qquad \text{and} \qquad p_1 = \frac{1}{R_1C_1} \tag{7.4}$$

To find the magnitude response, we have

$$|A_v(j\omega)| = \left|\frac{K}{1 + j(\omega/p_1)}\right| = \frac{K}{\sqrt{1 + (\omega/p_1)^2}} \tag{7.5a}$$

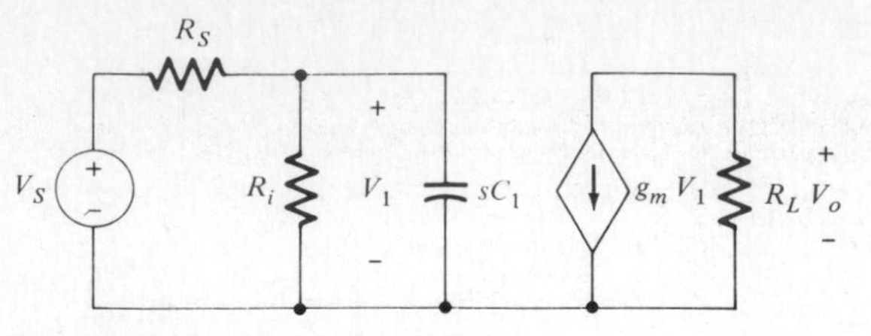

FIGURE 7.3 Example circuit.

A plot of log $|A_v(j\omega)|$ vs frequency, which is called the *magnitude response,* is shown in Fig. 7.4*a*. The phase function from (7.3) is

$$\arg A_v(j\omega) = \arg K - \tan^{-1}\frac{\omega}{p_1} = -\tan^{-1}\frac{\omega}{p_1} \tag{7.5b}$$

Note that at zero frequency $A_v(0) = K$ and at $\omega = p_1$ we have $|A_v(jp_1)| = K/\sqrt{2}$. In other words, $|A_v(jp_1)|$ is equal to $0.707A_v(0)$, or 3 dB lower than the dc value. *The 3-dB frequency of a one-pole circuit is thus* p_1. At the 3-dB frequency for this one-pole network function, the phase is $\arg A_v(jp_1) = -\tan^{-1}(p_1/p_1) = -45°$. The phase response of this circuit is shown in Fig. 7.4*b*.

Consider the simple *RC* circuits shown in Fig. 7.5*a* and *b*. The circuit in Fig. 7.5*a* has a shunt capacitance, and its transfer function is given by

$$\frac{V_2}{V_1} = \frac{1}{sR_1C_1 + 1} = \frac{1}{R_1C_1(s + 1/R_1C_1)} \tag{7.6a}$$

The frequency response of the circuit is similar to the one shown in Fig. 7.2*b*. Note that $\omega_{3\text{dB}} = 1/R_1C_1$, which corresponds to ω_h.

Similarly, for the circuit in Fig. 7.5*b* the transfer function is given by

$$\frac{V_2}{V_1} = \frac{R_2C_2s}{sR_2C_2 + 1} = \frac{s}{s + 1/R_2C_2} \tag{7.6b}$$

The frequency response of this circuit is similar to the one shown in Fig. 7.2*a*. The 3-dB frequency is $\omega_{3\text{dB}} = 1/R_2C_2$, which corresponds to ω_l.

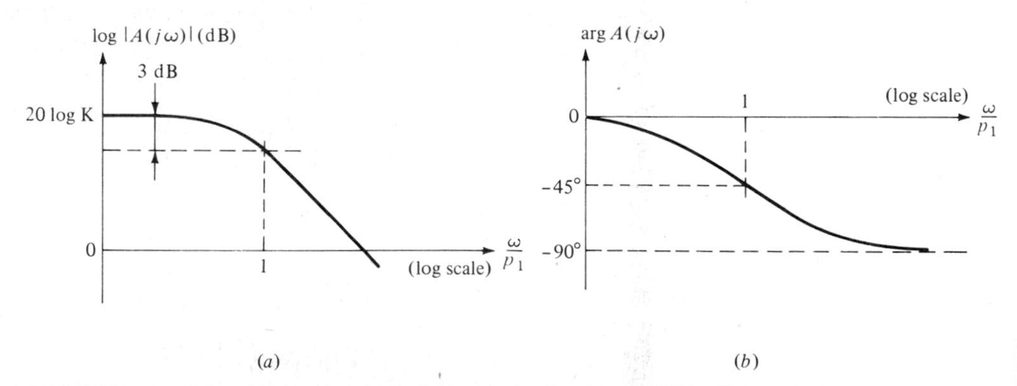

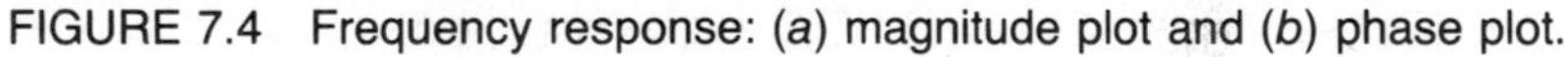
FIGURE 7.4 Frequency response: (*a*) magnitude plot and (*b*) phase plot.

EXAMPLE 7.1

Consider a combination of the circuits in Fig. 7.5*a* and *b* as shown in Fig. 7.5*c*. Let $R_1 = R_2 = 1\ \text{k}\Omega$, $C_1 = 100$ pF, $C_2 = 1\ \mu$F, and $K = 100$. Determine the midband gain, the low- and the high-frequency cutoffs.

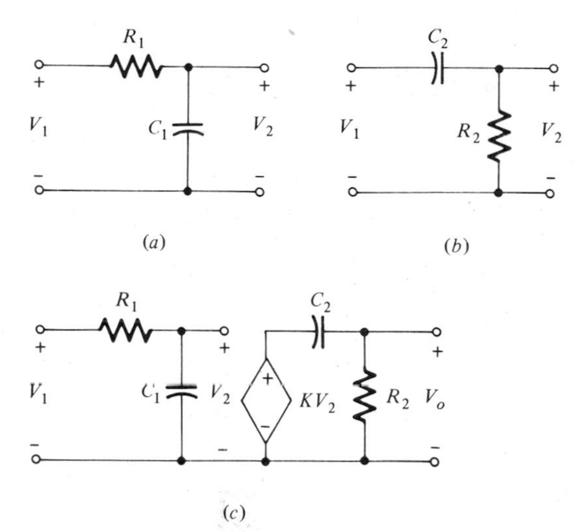

FIGURE 7.5 Example circuit: (*a*) low-pass circuit, (*b*) high-pass circuit, and (*c*) a combination of both.

The transfer function of the circuit in Fig. 7.5*c* is given by

$$A_v = \frac{V_o}{V_1} = \frac{KR_2C_2s}{(sR_2C_2 + 1)(sR_1C_1 + 1)} = \frac{Ks}{R_1C_1(s + 1/R_2C_2)(s + 1/R_1C_1)}$$

$$A_v(s) = \frac{10^9 s}{(s + 10^3)(s + 10^7)}$$

The above gain function has a frequency response similar to the one shown in Fig. 7.1*b*. Note that the midband gain $A_v = 100 = 40$ dB, $\omega_l = 10^3$ rad/s, and $\omega_h = 10^7$ rad/s.

We can determine the frequency response of the circuit in Fig. 7.5*c* in the low-frequency region by considering the shunt capacitor C_1 as an open circuit and calculating the frequency response due to the series capacitor C_2. Similarly, in the high-frequency region C_2 is short-circuited, and the frequency response is determined by C_1 alone.

EXERCISE 7.1

For the circuit shown in Fig. E7.1, $g_m = 5 \times 10^{-3}$ mho determine the midband gain, the low-frequency cutoff value, and the high-frequency cutoff value.

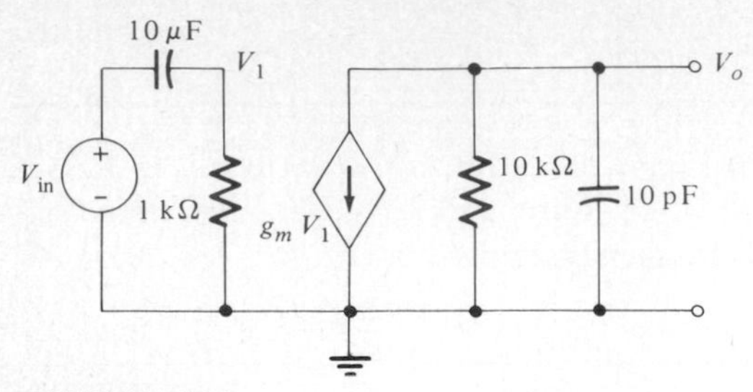

FIGURE E7.1

Ans $A_v = -50, f_l = 15.9$ Hz, $f_h = 1.6$ MHz

7.3 THE BODE PLOT

Knowledge of the magnitude and phase angle of a circuit at various frequencies as determined in the last section can be very useful in practice. In particular, you should recall that the response of a linear circuit to a sinusoidal excitation can be determined from the magnitude and phase angle of the circuit transfer function at the frequency of the sinusoid. Specifically, the output is a sinusoidal signal of the same frequency as the input with the amplitude of the output sinusoid equal to the product of the input amplitude and the magnitude of the transfer function at the frequency of the sinusoid. The phase of the output sinusoid is equal to the sum of the phase of the input and the phase of the transfer function at the input frequency.

Calculating the magnitude and the phase angle as was done in the last section, however, can be very tedious. It is desirable to have a simple technique for approximating the magnitude and phase angle of a given transfer function so that one can determine at least approximately how the circuit would respond to a particular sinusoidal excitation. The Bode plot provides such a simple approximation.

The Bode plot, or the asymptotic approximation, is essentially a break-point approximation. This method of approximation is best suited for real poles and zeros.

In general, the network function for a lumped linear time-invariant system can be written in factored form in terms of zeros and poles as

$$A(s) = K\frac{\Pi_i^l(1 + a_i s)\Pi_k^m(1 + b_k s + c_k s^2)}{s^p\Pi_r^q(1 + d_r s)\Pi_u^n(1 + e_u s + f_u s^2)} \tag{7.7}$$

where a, b, c, d, e, f, and K are real constants. The magnitude of $A(j\omega)$ in decibels from (7.7) is

$$\begin{aligned} 20 \log |A(j\omega)| = {} & 20 \log K + 20 \sum_i^l \log |1 + ja_i\omega| \\ & + 20 \sum_k^m \log |1 - c_k^2\omega + jb_k\omega| - 20p \log |\omega| \\ & - 20 \sum_r^q \log |1 + jd_r\omega| - 20 \sum_u^n \log |1 - f_0\omega^2 + je_u\omega| \end{aligned} \tag{7.8}$$

The phase of $A(j\omega)$ is

$$\arg A(j\omega) = \arg K + \sum_i^l \arg(1 + ja_i\omega) + \sum_k^m \arg(1 - c_k\omega^2 + jb_k\omega) - p\frac{\pi}{2} - \sum_r^q \arg(1 + jd_r\omega) - \sum_u^n \arg(1 - f_u\omega^2 + je_u\omega) \tag{7.9}$$

From (7.8) and (7.7) it is seen that we have the sum of four simple types of factors:

1. Constant term, K
2. Poles and zeros at the origin, $\pm j\omega$
3. Real poles and zeros, $\pm(1 + j\omega T)$
4. Complex poles and zeros, $\pm(1 + j\omega B - C\omega^2)$

Note that for the plus sign, $T = a_i$, $B = b_k$, and $C = c_k$ and for the minus sign, $T = d_r$, $B = e_u$, and $C = f_u$.

Each of the above factors is considered separately, and all are then added or subtracted, depending on whether the factors are due to the zeros or poles. The response is usually plotted on semilog paper with magnitude (in decibels) and phase on the linear scale and ω on the logarithmic scale. We shall now consider each factor separately.

1. The *constant term K* yields a constant magnitude (20 log K) and zero phase contribution.

2. The *pole and zeros at the origin,* for the magnitude response, represent terms that vary linearly with frequency when plotted on semilog paper, since

$$20 \log |(j\omega)^{\pm 1}| = \pm 20 \log \omega \text{ (in decibels)} \tag{7.10}$$

The slope of the straight line is obtained by taking the derivative of (7.10):

$$\frac{d}{d \log \omega}[\pm 20 \log \omega] = \pm 20 \text{ dB/decade} \tag{7.11}$$

Note that a unit change in log ω is equivalent to a decade change in frequency, i.e., 10 to 100, 1 to 10, etc., hence the term *per decade*. Sometimes we may wish to know this slope per octave, i.e., a change in frequency by a factor of 2, i.e., 1 to 2, or 4 to 8, etc. This may be readily determined, since for $\omega_2 = 2\omega_1$ we have

$$\log \frac{\omega_2}{\omega_1} = \log 2 = 0.301 \tag{7.12}$$

Hence, the slope per octave is given by

$$0.301(\pm 20 \text{ dB}) \simeq \pm 6 \text{ dB/octave} \tag{7.13}$$

It should be clear that if there are M more poles than zeros, then the asymptotic high-frequency slope is $-20M$ dB/decade, or $-6M$ dB/octave. In other words, beyond and above the *highest* frequency pole, the frequency rolloff will have the indicated slope. The phase shift for each extra pole or zero at the origin is minus or plus 90° since $\arg(\pm j\omega) = \pm\pi/2$.

3. The magnitude and phase terms due to the *real pole or zero* are

$$\pm 20 \log |1 + j\omega T| = \pm 20 \log \sqrt{1 + \omega^2 T^2} \tag{7.14}$$

$$\pm\arg(1 + j\omega T) = \pm\tan^{-1}\omega T \qquad (7.15)$$

To obtain the asymptotic behavior, we consider very low and very high frequencies; namely, for the magnitude

$$\omega \ll \frac{1}{T} \qquad 20 \log |1 + j\omega T| \simeq 20 \log 1 \simeq 0 \text{ dB} \qquad (7.16)$$

$$\omega \gg \frac{1}{T} \qquad 20 \log |1 + j\omega T| \simeq 20 \log \omega T \qquad (7.17)$$

Equations (7.16) and (7.17) represent two straight lines, one with a zero slope and the other with a slope of 20 dB/decade (or 6 dB/octave). The intersection of the low- and the high-frequency asymptotes yields a corner frequency, which is found by equating (7.16) and (7.17), yielding

$$20 \log \omega T = 20 \log 1 \qquad \text{or} \qquad \omega = \frac{1}{T} \qquad (7.18)$$

The frequency $\omega = 1/T$ is called the *break, or corner frequency*. The Bode plot for $\pm 20 \log |1 + j\omega T|$ and the actual magnitude response are shown in Fig. 7.6. The error between the asymptotic plot and the actual curve is maximum at the break frequency, with an error of ± 3 dB at $\omega = 1/T$.

The asymptotic phase behavior from $\pm\arg(1 + j\omega T)$ is readily seen to be as follows:

$$\omega T \gg 1 \qquad \pm\arg(1 + j\omega T) \simeq \pm\arg j\omega T = \pm 90^\circ \qquad (7.19)$$

$$\omega T \ll 1 \qquad \pm\arg(1 + j\omega T) \simeq \pm\arg 1 = 0^\circ \qquad (7.20)$$

At the break frequency the phase is $\pm\arg(1 + j1) = \pm 45^\circ$.

4. The magnitude response due to the *complex poles or zeros* is

$$\pm 20 \,|\log 1 + j\omega B - C\omega^2| \qquad (7.21)$$

The exact magnitude function is

$$\pm 20 \log \sqrt{(1 - C\omega^2)^2 + \omega^2 B^2} \qquad (7.22)$$

The low-frequency asymptote is again approximately $\pm 20 \log 1 = 0$ dB. The high-frequency asymptote is determined by the ω^2 term; namely, from (7.21)

$$\pm 20 \log C\omega^2 = \pm 40 \log \sqrt{C}\omega \qquad (7.23)$$

Equation (7.23) represents a straight line with a slope of ± 40 dB/decade (or ± 12 dB/octave). The intersection of the low- and the high-frequency asymptotes is given by

$$\pm 20 \log C\omega^2 = \pm 20 \log 1 \qquad \text{or} \qquad \omega = \frac{1}{\sqrt{C}} \qquad (7.24)$$

The error at the break frequency between the actual magnitude response and the asymptotic plot depends on the value of $B/2C$ (i.e., the real part of the poles or zeros).

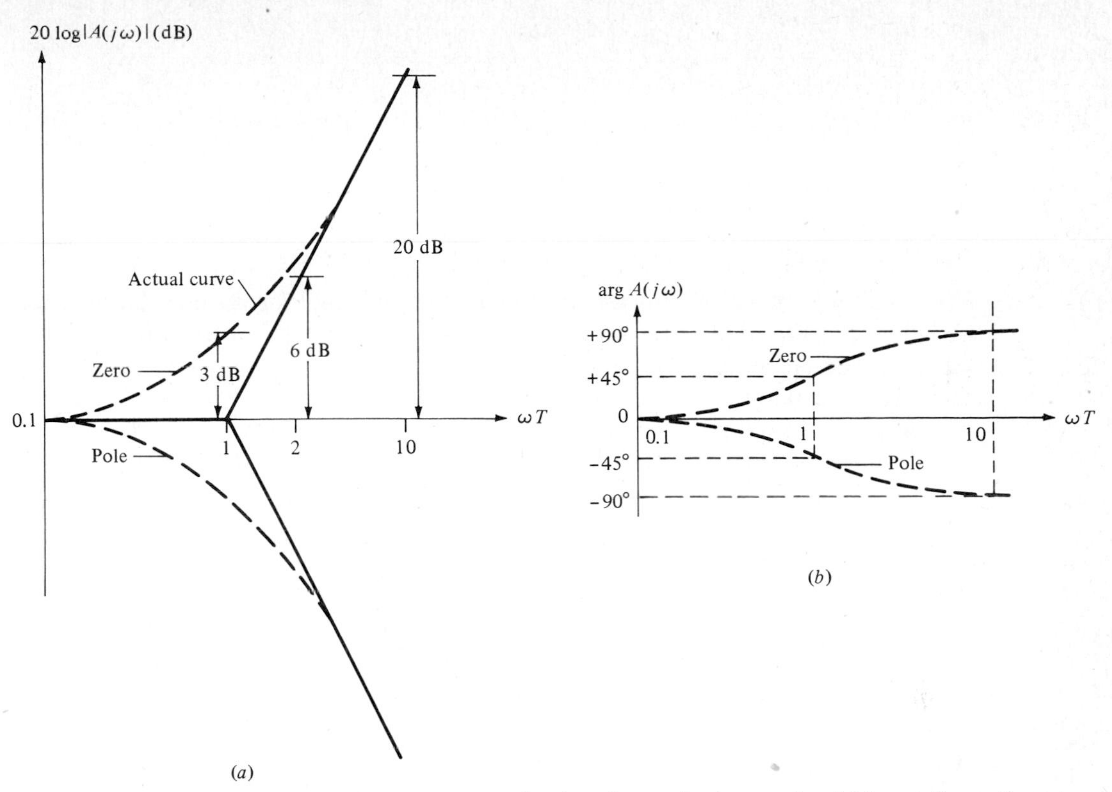

FIGURE 7.6 (*a*) Actual and asymptotic plot of magnitude vs ω for $A(s) = 1/(s + 1)$ and $(s + 1)$. (*b*) Phase response for $A(j\omega)$.

This is readily seen since the actual value of the magnitude from (7.22) at the break frequency is

$$\pm 20 \log \sqrt{(1 - C\omega^2)^2 + \omega^2 B^2}\,\Bigg|_{\omega=1/\sqrt{C}} = \pm 20 \log \frac{B}{\sqrt{C}} \tag{7.25}$$

Note that the Bode plot is *independent of the value of* B. Hence, from (7.25) it is seen that the closer the complex pole (zero) pairs are to the $j\omega$ axis, the larger the error. This is the reason why a Bode plot is not a good approximation for complex poles and zeros and is used mainly for real poles and zeros. For complex critical frequencies (poles and/or zeros), the error could be intolerable at the corner points.

The asymptotic phase behavior for the complex poles or zeros is readily determined as follows. The low-frequency asymptotic value of phase ($\omega \to 0$) is zero, while the high-frequency asymptotic value for the complex poles or zeros is

$$\pm \tan^{-1} \frac{B}{-C\omega}\Bigg|_{\omega \to \infty} = \pm 180° \tag{7.26}$$

For intermediate points there is no simplification except at the break frequencies. Consider a quadratic term as poles of a transfer function:

$$A(s) = \frac{1}{1 + sB + Cs^2} = \frac{1}{1 + (B/\sqrt{C})\sqrt{C}s + (\sqrt{C}s)^2} = \frac{1}{1 + 2\zeta s_n + s_n^2} \tag{7.27}$$

where $s_n = \sqrt{C}s$
$\zeta = B/2$
$C = \cos \phi$
$\phi =$ angle of pole from negative real axis

The actual magnitude and phase response and the Bode plot for various values of are shown in Fig. 7.7. Note that in Fig. 7.7*a* the Bode plot is independent of the value of ζ, and the break frequency is at $\omega_n = 1$.

To summarize, the Bode plot is logarithmic plot of $A(j\omega)$, and it has the advantage that product factors become additive. These plots are very useful for real poles and zeros, and they are also easy to sketch.

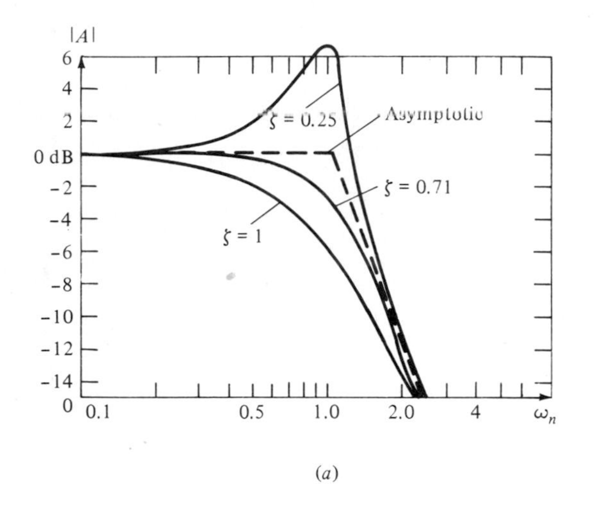

(*a*)

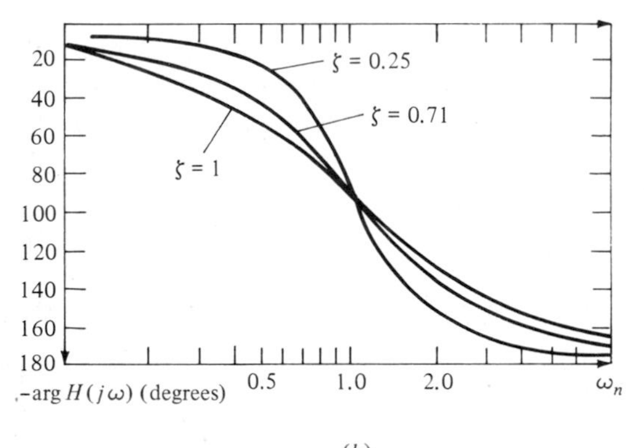

(*b*)

FIGURE 7.7 (*a*) Magnitude plot vs ω for $A(s) = 1/(1 + 2\zeta s_n + s_n^2)$. (*b*) Phase response for $A(j\omega)$.

EXAMPLE 7.2

Sketch the Bode plot for the following gain function:

$$A(s) = \frac{10^5(s + 5 \times 10^3)}{(s + 10^3)(s + 5 \times 10^4)}$$

Note that at $\omega = 0$, $A(0) = 10$ (or 20 log 10 = 20 dB). The break frequencies are at $\omega = 10^3$, 5×10^3, and 5×10^4. Thus the frequency rolls off at a slope of -20 dB/decade from the first break frequency (pole) until the second break frequency (zero). Because of the zero, we now have a $+20$ dB/decade contribution beyond the zero, causing an overall slope of zero, until the third break frequency (pole). Beyond this pole the frequency rolloff is again at a slope of -20 dB/decade (or equivalently, -6 dB/octave) as shown in Fig. 7.8*a*. The actual frequency response shown by the dotted line is also shown in Fig. 7.8*a*.

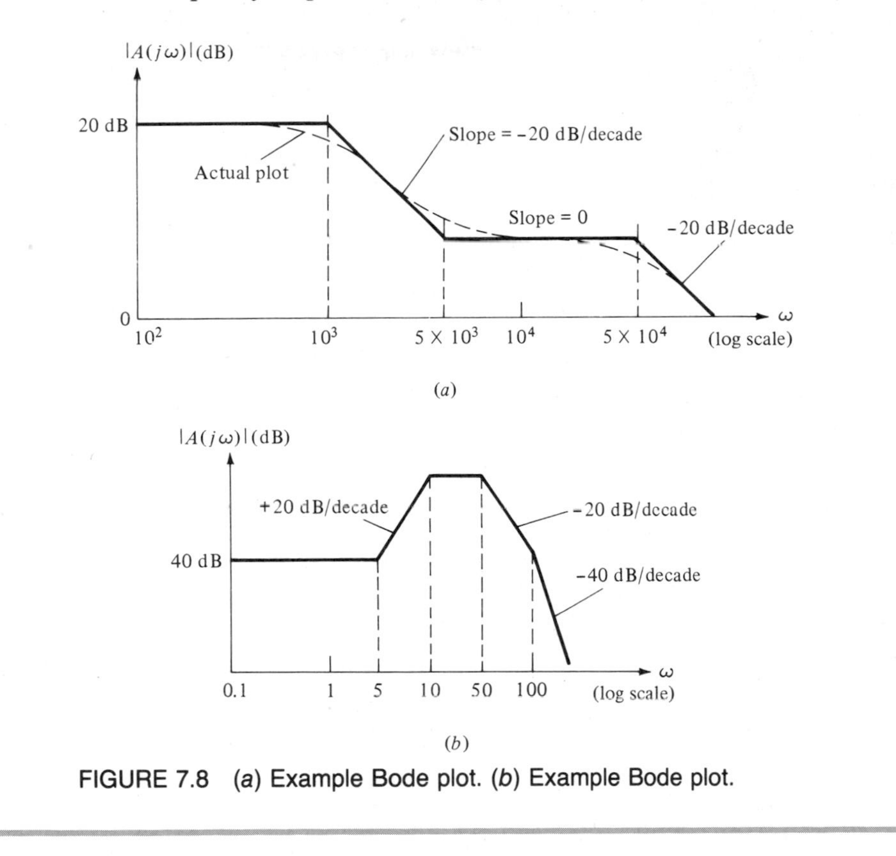

FIGURE 7.8 (*a*) Example Bode plot. (*b*) Example Bode plot.

EXAMPLE 7.3

We wish to sketch the Bode plot for the following gain function:

$$A(s) = -10^6 \frac{s+5}{(s+10)(s+50)(s+100)}$$

Note that at $\omega = 0$, the dc value is $A(0) = 10^2$. If we express the gain in decibels, we have $20 \log A(0) = 40$ dB. The Bode plot is shown in Fig. 7.8*b*.

EXERCISE 7.2

The following current gain function is given:

$$A_i(s) = 4 \times 10^{10} \frac{s+10}{(s+200)(s+2\times 10^4)^2}$$

Determine the midband gain and the low- and the high-frequency cutoffs and sketch the Bode plot.

Ans $A_i = 100$, $f_l = 31.8$ Hz, $f_h = 2.04$ kHz

these are correct

7.4 LOW-FREQUENCY RESPONSE OF BJT AMPLIFIERS DUE TO BIAS AND COUPLING CIRCUITRY

As discussed earlier, we now determine the effects of the bypass and coupling circuitry on the frequency response of the amplifier in Fig. 7.1*a* by considering the shunt capacitors of the high-frequency equivalent circuit (see Fig. 2.27) and the parasitic capacitances to be open-circuited at low frequencies.

The single-stage CE amplifier circuit at low frequencies is then as shown in Fig. 7.9. We are now interested in determining the low-frequency response of this circuit. Note that for low and midband frequencies we use the low-frequency model of the transistor, since at these frequencies the effects of the device capacitances are insignificant.

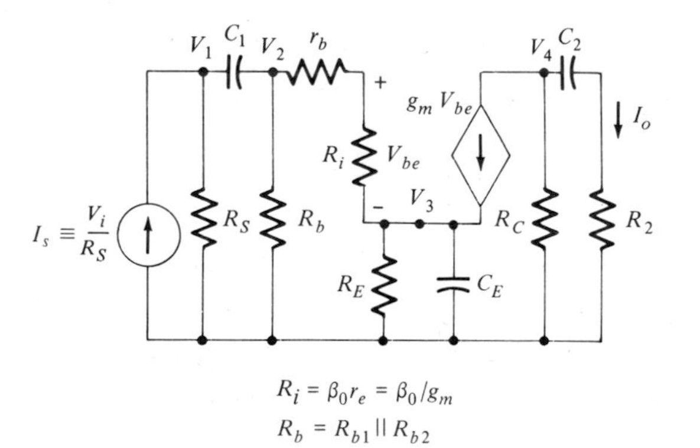

FIGURE 7.9 Low-frequency model of Fig. 7.1*a* ($R_b = R_{b1} \| R_{b2}$).

We can now determine the current gain function $A_i(s)$ for Fig. 7.9 by either nodal or loop analysis. The equilibrium equations of the circuit by nodal analysis in the transformed frequency domain are

$$I_s = V_1 G_S + (V_1 - V_2)sC_1 \tag{7.28a}$$

$$(V_1 - V_2)sC_1 = V_2 G_b + (V_2 - V_3)\frac{1}{r_b + R_i} \tag{7.28b}$$

$$(V_2 - V_3)\frac{1}{r_b + R_i} + g_m V_{be} = V_3(G_E + sC_E) \tag{7.28c}$$

$$g_m V_{be} + V_4 G_C + V_4 \frac{sC_2}{R_2 C_2 s + 1} = 0 \tag{7.28d}$$

Note that $G_1 = 1/R_S$, $G_E = 1/R_E$, $G_C = 1/R_C$, $g_m = 1/r_e$, and $R_i = \beta_0 r_e$ in the transistor circuit model, and

$$V_{be} = \frac{R_i}{r_b + R_i}(V_2 - V_3) \qquad \text{and} \qquad I_o = V_4 \frac{sC_2}{R_2 C_2 s + 1} \tag{7.29}$$

The solution of (7.28) after some manipulation is

$$A_i(s) = \frac{I_o}{I_s} = \frac{-a_m s^2(s + 1/R_E C_E)}{[s + 1/(R_2 + R_C)C_2](s^2 + a_2 s + a_1)} \tag{7.30}$$

where

$$a_2 = \frac{(1/C_1)(R_b + r_b + R_i) + [(R_b + R_S)/R_E C_E](R_i + \beta_0 R_E + r_b) + R_S R_b/R_E C_E}{R_b R_S + (R_b + R_S)(r_b + R_i)}$$

$$\simeq \frac{1}{C_1}\frac{1}{R_S + r_b + R_i} + \frac{1}{C_E}\frac{\beta_0}{R_S + r_b + R_i} \qquad R_b \gg R_i,\ R_S \tag{7.31a}$$

$$a_1 = \frac{R_b + R_i + \beta_0 R_E + r_b}{R_E C_E C_1[R_b R_1 + (R_b + R_S)(r_b + R_i)]}$$

$$\simeq \frac{R_b + \beta_0 R_E}{R_b R_E (R_S + r_b + R_i) C_E C_1} \qquad R_b \gg R_i,\ R_S \tag{7.31b}$$

$$a_m = \beta_0 \frac{R_S \parallel R_b}{(R_S \parallel R_b) + r_b + R_i}\frac{R_C}{R_C + R_2}$$

$$\simeq \beta_0 \frac{R_S}{R_S + R_i + r_b} \qquad R_b \gg R_S \qquad R_C \gg R_2 \tag{7.31c}$$

Note that in (7.31) we have used the approximation $\beta_0 \gg 1$, $\beta_0 R_E \gg r_b$.

Note that the gain function in (7.30) has a high-pass characteristic and that as $s \to \infty$, the magnitude of the current gain is equal to the midband value $-a_m$. In high-pass circuits the number of poles and zeros are the same. Stated differently, the order of the numerator and of the denominator is the same for the gain function of a high-pass circuit. From (7.30) it is also seen that the output coupling circuit (C_2) has no interaction with the emitter bypass capacitance (C_E) and the input coupling circuit (C_1), because R_2, C_2, and R_C do not appear anywhere in the other terms. The emitter bypass circuit and the input coupling circuit exhibit strong interactions as seen by their appearance in both terms a_1 and a_2. A numerical example will illustrate this point.

EXAMPLE 7.4

In the circuit of Fig. 7.1*a*, let the parameters be as follows: $R_C = 2\ \text{k}\Omega$, $R_b = R_{b1} \parallel R_{b2} = 6\ \text{k}\Omega$, $R_E = 2.7\ \text{k}\Omega$, $R_S = R_2 = 1\ \text{k}\Omega$, $C_1 = 2\ \mu\text{F}$, $C_2 = 10\ \mu\text{F}$, and $C_E = 50\ \mu\text{F}$. The transistor parameters are given as $\beta_0 = 50$, $g_m = 0.2$ mho, and $r_b = 100\ \Omega$. Determine the low-frequency gain function and the 3-dB cutoff frequency. Substitution of the numerical values in (7.30) yields

$$A_i(s) = -23.7 \frac{s^2(s + 7.4)}{(s + 33)(s + 1.14 \times 10^3)(s + 55)}$$

The pole-zero plot and the asymptotic plots are shown in Fig. 7.10*a*. The low-frequency cutoff point of this example is approximately equal to the farthest left pole, namely, $(1.14 \times 10^3)/2\pi$ Hz.

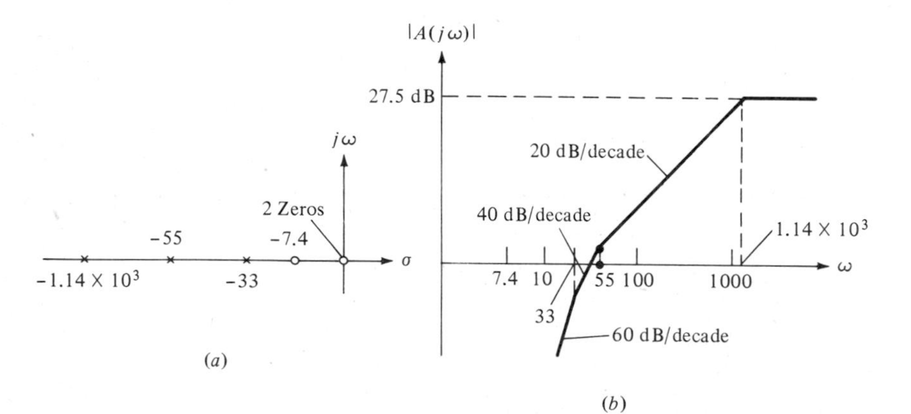

FIGURE 7.10 (*a*) Pole-zero plot of the gain function (not to scale). (*b*) Bode plot response of the gain function.

Note that in the above example, if we assume no interaction between the emitter bypass and input coupling circuits, the approximate location of the poles will be in considerable error. For example, if we assume C_E and C_2 to be perfect short circuits, the pole due to the input coupling circuit alone is determined from Fig. 7.11*a*. The pole of the circuit is given by

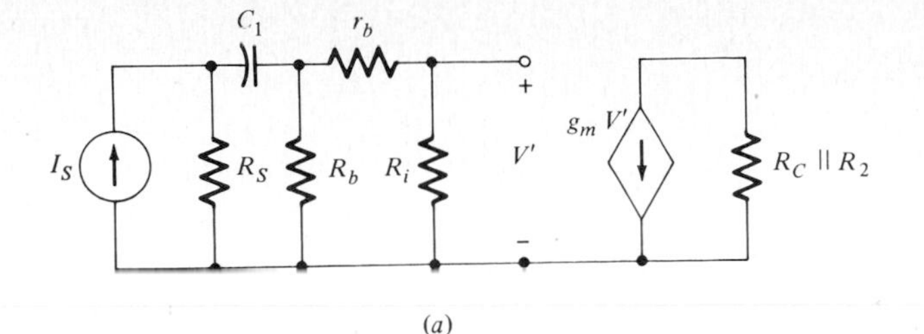

(a)

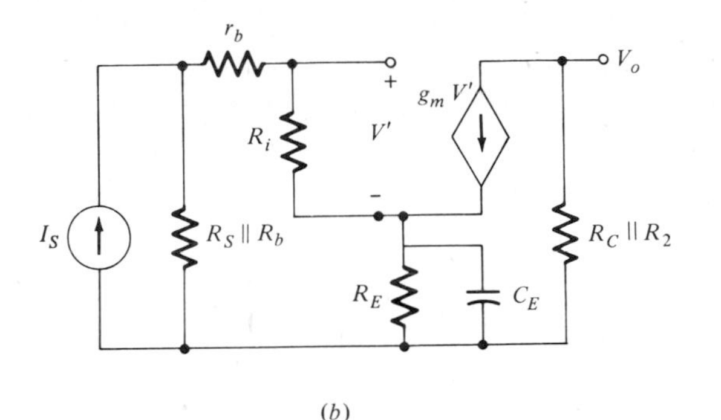

(b)

FIGURE 7.11 (a) Circuit to determine the pole due to C_1. (b) Circuit to determine the pole due to C_E.

$$P_a \text{ (due to } C_1 \text{ alone)} = \frac{1}{C_1[R_S + R_b \| (r_b + R_i)]} \tag{7.32a}$$

$$= -3.76 \times 10^2 \text{ rad/s (for this example)}$$

The pole due to C_E, assuming C_1 and C_2 are short circuits, is determined from Fig. 7.11*b* by short-circuiting C_1 and C_2. The pole of the circuit as determined by either nodal or mesh analysis is given by

$$P_b \text{ (due to } C_E \text{ alone)} = -\frac{1 + \beta_0 R_E/[(R_S \| R_b) + R_i + r_b]}{R_E C_E} \tag{7.32b}$$

$$= -8.31 \times 10^2 \text{ rad/s}$$

From (7.32*b*) it is seen that the resistance $R_S \| R_b$ has a strong influence on the poles due to C_E. Note further that the calculation of the pole locations by this method, that is, considering the pole due to each capacitance alone, is appreciably in error. The actual pole locations, as determined in Example 7.4, are at $p_1 = -55$ and $p_2 = -1.14 \times 10^3$. The effect of this interaction is to split the poles apart, as shown in Fig. 7.12. In a high-pass circuit, when the poles are on the negative real axis, the pole farthest to the left plays a dominant role in determining the low-frequency cutoff point of the amplifier. This is because the 3-dB cutoff frequency is the frequency at which the gain is 0.707 of its midband value. The midband value is reached asymptotically as the frequency increases. Thus, if $|p_2| \gg |p_1|, |p_c|$, where $|p_c|$ is the pole due to

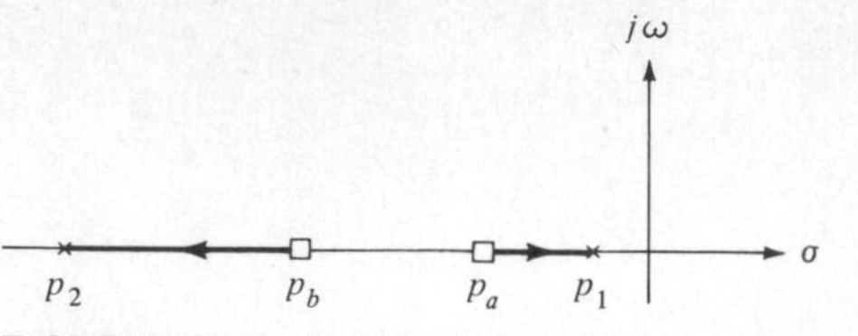

FIGURE 7.12 Pole splitting due to interaction.

C_c and $|p_2|$ and $|p_1|$ are the actual poles due to the combined effect of C_1 and C_E, then $\omega_1 \simeq |p_2|$.

In a design for a specified ω_l, we have three variables to choose, namely, C_1, C_2, and C_E. The external resistor values are already determined and known from the biasing design. In some cases C_2 may not be present, and even if present, it causes no problem since the output coupling circuit does not interact with the other circuits. Thus we consider the poles due to the combined effect of C_E and C_1. As we have pointed out (see Example 7.4), these poles are usually far enough apart so that one pole is much larger than the other. (In the following discussion, we use the negative of the values of the poles so that magnitude signs are not carried along.) In other words, we can approximate the actual roots by the following simple expressions:

$$
\begin{aligned}
s^2 + a_2 s + a_1 &= (s + p_2)(s + p_1) && \text{where } p_2 \gg p_1 \\
&\simeq (s + a_2)\left(s + \frac{a_1}{a_2}\right) && \text{where } a_2 \gg \frac{a_1}{a_2}
\end{aligned}
\tag{7.33}
$$

From (7.33) and (7.30) we thus have

$$
p_2 \simeq a_2 \simeq \frac{1}{R_S + r_b + R_i}\left(\frac{1}{C_1} + \frac{\beta_0}{C_E}\right) \tag{7.34a}
$$

$$
p_1 \simeq \frac{a_1}{a_2} \simeq \frac{R_b + \beta_0 R_E}{R_b R_E (C_E + \beta_0 C_1)} \tag{7.34b}
$$

The condition $p_2 \geq 10p_1$ (i.e., $p_2 \gg p_1$) enables us to use (7.33) and (7.34). In an analysis problem this condition can be readily checked.

In a design problem, however, the values of C_1 and C_E are unknown. If we wish to design on the basis of widely separated poles, we must meet the condition described above. Let us examine the condition $p_2/p_1 \geq 10$. From (7.34) we have

$$
\begin{aligned}
\frac{p_2}{p_1} &= \frac{(1/R_x)(C_E + \beta_0 C_1)/C_1 C_E}{(R_b + \beta_0 R_E)/R_b R_E (C_E + \beta_0 C_1)} \\
&= \frac{(1/R_x)(C_E + \beta_0 C_1)/C_1 C_E}{(1/R_p)\beta_0/(C_E + \beta_0 C_1)}
\end{aligned}
\tag{7.35}
$$

where $R_x = R_S + r_b + R_i$
$R_p = R_b \parallel \beta_0 R_E$

By setting $p_2/p_1 \geq 10$, we obtain from (7.35)

$$\frac{R_p(C_E + \beta_0 C_1)^2}{R_x \beta_0 C_1 C_E} \geq 10 \tag{7.36a}$$

or

$$\frac{R_p}{R_x}\left(\frac{C_E}{\beta_0 C_1} + 2 + \frac{\beta_0 C_1}{C_E}\right) \geq 10 \tag{7.36b}$$

The term in parentheses in (7.36*b*) is of the form $x + 2 + 1/x$, which has a minimum value of 4. Hence the most pessimistic condition is that $R_p/R_x \geq 2.5$. This guarantees the separation of poles by at least a factor of 10. The pole-separation condition, written explicitly in terms of the circuit parameters, is

$$R_b \| \beta_0 R_E \geq 2.5(R_S + r_b + R_i) \tag{7.37}$$

Hence, if (7.37) is met, no matter what the values of C_1 and C_E are, the poles are separated by at least a factor of 10. If R_S is not too large, this condition is often satisfied. In Example 7.4, where R_S is not small, we have $R_b \| \beta_0 R_E \simeq 6\ \text{k}\Omega$ and $R_S + r_b + R_i = 1.35\ \text{k}\Omega$, which clearly satisfies the condition in (7.37). For the numerical example it is seen that the poles are actually separated by a factor of 27.3. The poles as obtained by the approximate expressions (7.34*a*) and (7.34*b*) are

$$p_2 \simeq a_2 = 1.2 \times 10^3 \text{ rad/s} \qquad \text{and} \qquad p_1 \simeq \frac{a_1}{a_2} = 51 \text{ rad/s}$$

which clearly shows a very simple and fairly accurate method of predicting the pole locations due to the interaction of C_E and C_1. In fact, the error in calculation of this approximate method is always less than 10 percent.

In a design problem there is a great deal of freedom in the choice of C_1, C_2, and C_E for a given low-frequency cutoff ω_l. A simple design procedure is as follows.

First, ensure the satisfaction of the pole-separation condition (7.37). Under this condition, recall that no matter what the values of C_1 and C_E, the pole $p_2 \geq 10p_1$. Also if $p_c \leq p_2/10$, i.e., the pole due C_c is much smaller than the largest pole, which is identified as p_2, then $p_2 \simeq \omega_l$.

In an actual design, make $\omega_l \simeq p_2 \simeq a_2$ and arbitrarily select C_1 and C_E. The value of C_E is always larger than C_1 and C_2 since the effective reactance associated with it is small, as seen in (7.34*a*). Hence a good choice is

$$C_E = \beta_0 C_1 \tag{7.38a}$$

The value of C_1 is then determined from (7.34*a*) with $a_2 = \omega_l$:

$$C_1 = \frac{1}{\omega_l} \frac{2}{R_S + r_b + R_i} \tag{7.38b}$$

The pole p_1 is then obtained from (7.34*b*) and (7.38*a*), namely,

$$p_1 = \frac{R_b + \beta_0 R_E}{2\beta_0 R_b R_E C_1} \tag{7.39a}$$

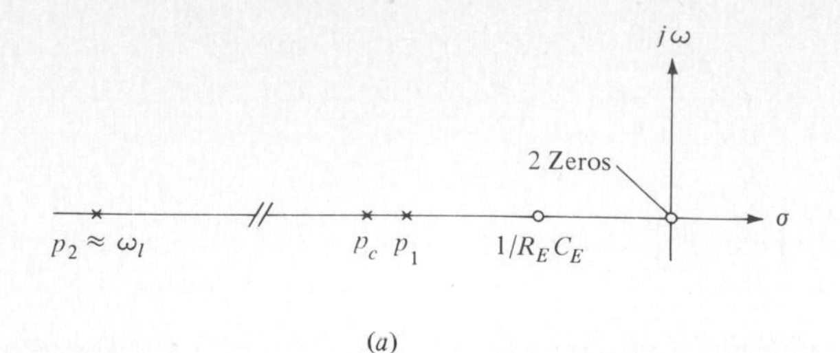

(a)

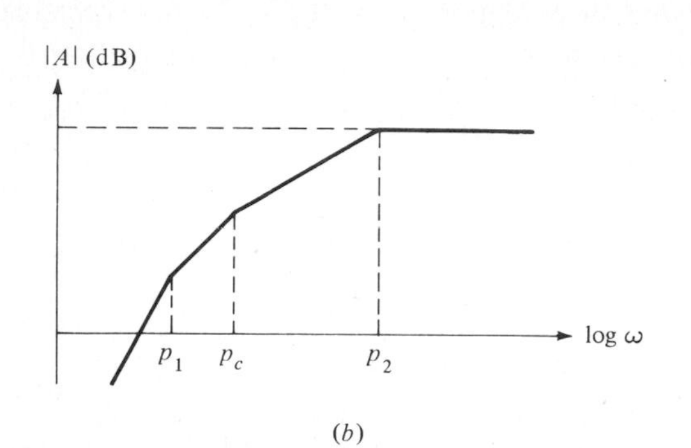

(b)

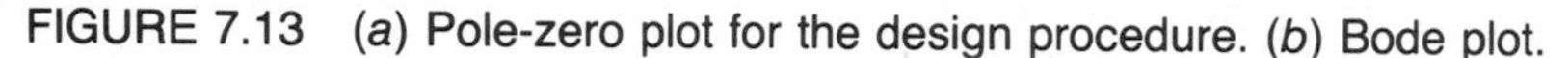
FIGURE 7.13 (a) Pole-zero plot for the design procedure. (b) Bode plot.

The output coupling circuit in this case must be chosen such that the output circuit pole $p_c \ll \omega_l$. Specifically, we may use the following relation:

$$p_c = \frac{1}{(R_2 + R_C)C_2} \leq \frac{\omega_l}{10}$$

or

$$C_2 \geq \frac{10}{\omega_l} \frac{1}{R_2 + R_c} \tag{7.39b}$$

The design equations are then given by (7.37), (7.38), and (7.39*b*). The pole-zero and Bode plots for this design method are shown in Fig. 7.13. The zero at $1/R_E C_E$ may or may not be larger than p_1, p_c depending on the element values. A design example using the above procedure is worked out in Sec. 7.14.

In integrated circuits the large values of the capacitors in the bypass and coupling circuitry are undesirable since they require a large area on the chip. Most integrated-circuit amplifiers employ different biasing schemes (see Fig. 2.38) and are often dc coupled in a differential configuration so that there is no low-frequency cutoff and dc signals are passed as well.

EXERCISE 7.3

For the BJT circuit shown in Fig. 7.1*a*, the circuit parameters are $R_S = 10\ \text{k}\Omega$, $R_2 = 1\ \text{k}\Omega$, $R_c = 2\ \text{k}\Omega$, $R_b = R_{b1} \parallel R_{b2} = 5\ \text{k}\Omega$, $R_E = 1\ \text{k}\Omega$, $V_{CC} = 20$ V, $C_1 =$

$C_2 = 10\ \mu F$, and $C_E = \infty$. The transistor $\beta_0 = 100$. Determine the lower 3-dB cutoff frequency of the current gain function and the midband value of the current gain.

Ans $f_1 = 5.5$ Hz, $A_i \simeq 67$

7.5 LOW-FREQUENCY RESPONSE OF FETS DUE TO BIAS AND COUPLING CIRCUITRY

Calculations of the low-frequency response due to biasing and coupling circuitry for FETs shown in Fig. 7.14*a* is similar to those for the bipolar transistor. For example, from the low-frequency equivalent circuit of the device, we have the circuit shown in Fig. 7.14*b*. The analysis and design considerations for low-frequency FETs are much simpler since there is no interaction between C_S and C_1.

Note that the input impedance of a FET is infinite, and we always apply a voltage-source input. Because of the open circuit due to the input impedance, the circuit divides into two noninteracting subcircuits. Thus, if we solve for the voltage gain function $V_o/V_i(s)$ in Fig. 7.14*b*, we will find that C_1 and C_S do not interact, and the pole due to each capacitor is determined separately and independent of the other. To provide adequate bypassing for the FET over the same frequency range as for the BJT, the value of the source bypass capacitor needed is much smaller than the emitter bypass capacitor, i.e., $C_E \gg C_S$, because of the drastically different values of g_m in BJTs and FETs. Usually the size of the capacitors is not so important in discrete circuits, in that we can use bigger capacitances at essentially no extra cost, but we can use a simple design because of the absence of interaction.

Thus the pole-zero locations due to the various circuits, considering them one at a time (i.e., considering one capacitance at a time and assuming perfect short circuits for the other capacitors), are determined by straightforward analysis as follows.

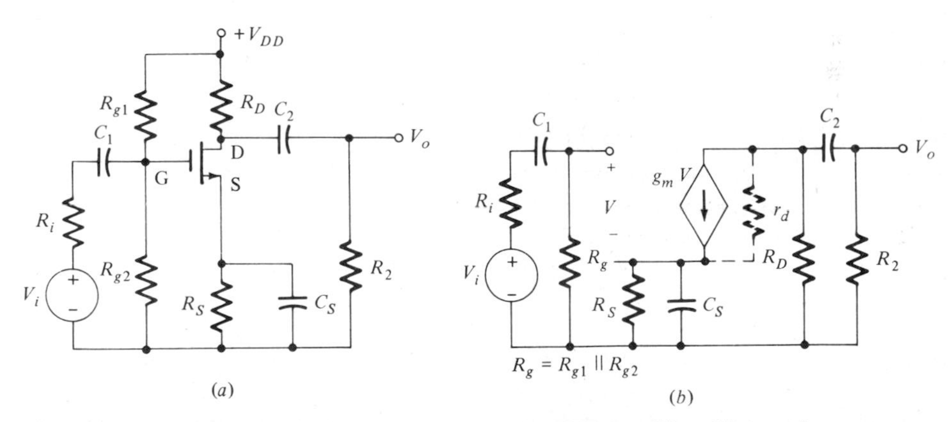

FIGURE 7.14 (*a*) Single-stage common-source FET amplifier. (*b*) Low-frequency small-signal equivalent circuit of (*a*).

We write the general gain expression due to each capacitor in the form

$$A_v(s) = A_m \frac{s + z_1}{s + p_1} \tag{7.40}$$

where $p_1 > z_1$, since the low-frequency response has a high-pass characteristic. Of course, z_1 can be equal to zero, as it is for the coupling capacitor circuits due to C_1 and C_2. The value of A_m is simply the midband gain and is readily determined by short-circuiting all the bypass and coupling capacitors. Thus, from Fig. 7.14*b*, the midband voltage gain A_m is

$$A_m = -g_m(r_d \parallel R_D \parallel R_2) \simeq -g_m(R_D \parallel R_2) \simeq -g_m R_L \tag{7.41}$$

where $R_L = R_D \parallel R_2$ and $r_d \gg R_L$. For the circuits the poles due to C_1 and C_2 individually can be written by inspection. The value of the zero is determined by making sure that (7.40) is correct at dc. For C_1 and C_2 it is obviously zero, since the signal path is open at dc and $A_v(0) = 0$.

Thus, for the drain-coupling circuit (C_2) the low-frequency voltage gain function is given by

$$\frac{V_o}{V_i} \simeq \frac{A_m s}{s + 1/(R_D + R_2)C_2} = \frac{A_m s}{s + p_d} \tag{7.42}$$

The voltage gain function of the source bypass circuit (assuming r_d very large) is given by

$$\frac{V_o}{V_i} \simeq \frac{A_m(s + 1/R_S C_S)}{s + (1 + g_m R_S)/R_S C_S} = \frac{A_m(s + z_s)}{s + p_s} \tag{7.43}$$

Finally, for the gate-coupling circuit we have

$$\frac{V_o}{V_i} = \frac{A_m s}{s + 1/(R_g + R_i)C_1} = \frac{A_m s}{s + p_g} \tag{7.44}$$

Note that in a single-stage BJT or FET amplifier we have three poles, two due to the coupling capacitors C_1 and C_2 and one due to the emitter (or source) bypass capacitance C_E (or C_S). In an analysis problem, if these poles are given or calculated from the circuit, the low-frequency cutoff can be determined as follows. We can write the low-frequency gain function of a FET or a BJT circuit as

$$A(s) = \frac{A_m s^2(s + z)}{(s + p_1)(s + p_2)(s + p_3)} \tag{7.45}$$

At the 3-dB cutoff frequency ω_l we have

$$|A(j\omega)| = \frac{A_m \omega_l^2(\omega_l^2 + z^2)^{1/2}}{[(\omega_l^2 + p_1^2)(\omega_l^2 + p_2^2)(\omega_l^2 + p_3^2)]^{1/2}} = \frac{A_m}{\sqrt{2}} \tag{7.46a}$$

From (7.46*a*), squaring both sides, we get

$$\begin{aligned} 2\omega_l^4(\omega_l^2 + z^2) &= (\omega_l^2 + p_1^2)(\omega_l^2 + p_2^2)(\omega_l^2 + p_3^2) \\ 2\omega_l^6 + 2\omega_l^4 z^2 &= \omega_l^6 + \omega_l^4(p_1^2 + p_2^2 + p_3^2) + \omega_l^2(\cdots) \end{aligned} \tag{7.46b}$$

Since the zero due to the bypass capacitor is always smaller than the pole associated with it and furthermore ω_l is larger than the largest pole, we can approximate (7.46*b*) as

$$\omega_l^6 \simeq \omega_l^4(p_1^2 + p_2^2 + p_3^2)$$

or

$$\omega_l \simeq \sqrt{p_1^2 + p_2^2 + p_3^2} \tag{7.46c}$$

where p_1, p_2, and p_3 are the three poles in (7.30) for a BJT or p_d, p_s, p_g in (7.42), (7.43), and (7.44) for a FET. Note that in the BJT design procedure we made $p_2 \geq 10p_1$ and $p_2 \geq 10p_c$, and hence from (7.46*c*) $\omega_l \simeq p_2$. For the FET design, since R_g is very large, $p_g \ll p_s, p_d$ for the same values of capacitors. Usually R_2 is also large such as in a cascaded amplifier stages, in which $R_2 = R_g$. For this case $p_d \ll p_s$ and $\omega_l \simeq p_s$. If the value of R_2 is small, then we can make use of (7.46*c*) in the design.

In an FET stage the source bypass capacitor C_S is usually the determining factor in establishing the low-frequency cutoff point. For the same low-frequency cutoff point, the value of C_S is much lower than that of the emitter bypass capacitor C_E of a bipolar transistor circuit because of the difference in the magnitudes of g_m in FETs and BJTs. The following example illustrates the calculation of ω_l for a single FET amplifier stage.

EXAMPLE 7.5

Consider an FET stage, such as shown in Fig. 7.14*a*, with the following circuit parameters: $R_D = 3.5$ kΩ, $R_2 = R_g = 1$ MΩ, $R_S = 1.5$ kΩ, $R_i = 50$ Ω, $C_1 = C_2 = C_S = 10$ μF. The FET parameters are $g_m = 2 \times 10^{-3}$ mhos and $r_d = 20$ kΩ. We wish to determine the approximate low-frequency cutoff point of the amplifier and the midband voltage gain.

The midband voltage gain for the circuit in Fig. 7.14*b* is obtained by short-circuiting the coupling and the bypass capacitors and then determining V_o/V_i. By doing this we obtain

$$\frac{V_o}{V_i} = \frac{R_g}{R_g + R_i}[-g_m(r_d \parallel R_D \parallel R_2)] \simeq -2 \times 10^{-3}(2.98) = -5.96$$

We determine the value of the poles due to each coupling and bypass circuit. From (7.42) the pole of the drain-coupling circuit alone, p_d, is given by

$$p_d = \frac{1}{(R_D + R_2)C_2} \simeq \frac{1}{(3.5 \times 10^3 + 10^6)(10 \times 10^{-6})} = 0.1 \text{ rad/s}$$

Since r_d is very large, i.e., $r_d \gg (R_D \parallel R_2)$, $1/g_m$, the pole due to the source bypass capacitance, p_s, from (7.43) is given by

$$p_s = \frac{1 + g_mR_S}{R_SC_S} = \frac{1 + 2 \times 10^{-3}1(1.5 \times 10^3)}{(1.5 \times 10^3)(10 \times 10^{-6})} = 2.66 \times 10^2 \text{ rad/s}$$

From (7.44) the pole due to the gate-coupling circuit p_g is given by

$$p_g \simeq \frac{1}{R_g C_1} = \frac{1}{10^6(10 \times 10^{-6})} = 0.1 \text{ rad/s}$$

Since $p_s \gg p_d$ and $p_s \gg p_g$, the low-frequency cutoff point is given by the pole of the source bypass circuit:

$$\omega_l = p_s = 2.66 \times 10^2 \text{ rad/s} \qquad \text{or} \qquad f_l = \frac{2.66 \times 10^2}{2\pi} = 42 \text{ Hz}$$

Note that for the same values of capacitors, the source bypass capacitance pole is larger than the other poles by three orders of magnitude. Note also that for the same cutoff frequency the value of C_E for a bipolar transistor is much larger than the value of C_S for an FET.

Note that if R_2 were specified as, say, 1.5 kΩ, then from (7.42) $p_d = 20$ rad/s, and ω_l would still be the same.

EXERCISE 7.4

For the FET circuit shown in Fig. 7.14*a*, the circuit parameters are $R_D = 2$ kΩ, $R_g = 1$ MΩ, $R_S = 2$ kΩ, $R_i = 100$ Ω, $R_2 = 2$ kΩ, $C_1 = C_2 = 10$ μF, and $C_S = 100$ μF. The FET parameters are $g_m = 4 \times 10^{-3}$ mho and $r_d = 20$ kΩ. Determine the midband voltage gain and the low-frequency cutoff frequency.

Ans $A_v = -4, f_l = 7.5$ Hz

7.6 HIGH-FREQUENCY RESPONSE CHARACTERISTICS OF BJTS

We shall now consider the high-frequency response of a single-stage CE amplifier shown in Fig. 7.1*a*. Recall that for high-frequency calculations the reactances due to the coupling and bypass capacitors are essentially short-circuited. Hence the only frequency-dependent parameters are the shunt capacitances of the device model (and the parasitic capacitances). The high-frequency circuit model, using the hybrid-π equivalent circuit for the transistor, is shown in Fig. 7.15*a*. Note that in this frequency range all the coupling and bypass capacitors are short-circuited. The frequency response of the circuit at high frequencies can now be determined by an analysis of the circuit in Fig. 7.15*a*. The algebra can be simplified if we replace the circuit to the left of terminals a, a' by its Norton equivalent circuit. This equivalence is shown in Fig. 7.15*b*. Nodal analysis of Fig. 7.15*b* yields the following equilibrium equations in the transformed domain:

$$\frac{R_s' V_i}{R_s(R_s' + r_b)} - V(G_t + sC_\pi) = (V - V_o)\,sC_c \qquad \textbf{(7.47a)}$$

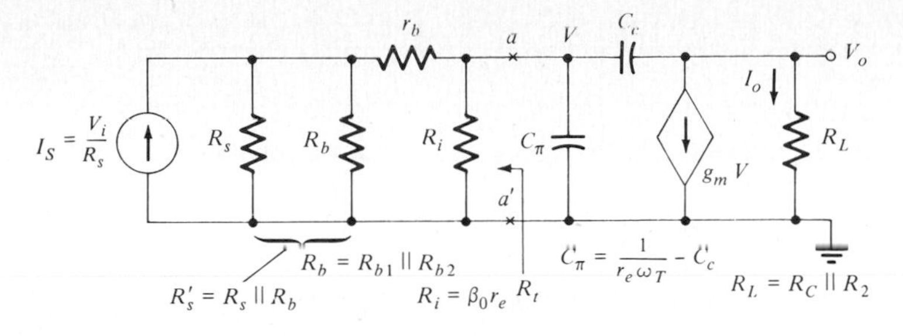

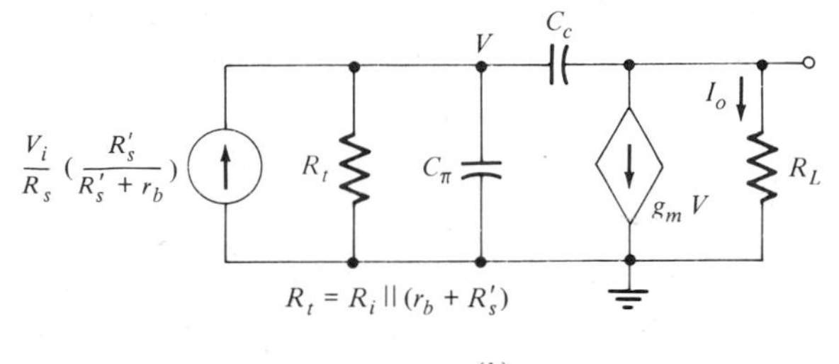

FIGURE 7.15 (*a*) High-frequency small-signal equivalent circuit of the CE transistor amplifier in Fig. 7.1*a*. (*b*) Simplified equivalent of Fig. 7.15*a*.

$$(V - V_o)\, sC_c = g_m V + V_o G_L \tag{7.47b}$$

If we eliminate V in (7.47) and solve for the voltage gain function, we obtain

$$\frac{V_o}{V_i} = \frac{R_s'(s - g_m/C_c)}{(R_s' + r_b)R_s C_\pi s^2 + s[(g_m + G_t)/C\pi + G_L(1/C_c + 1/C_\pi)] + G_t G_L/C_c C_\pi} \tag{7.48}$$

where $R_t = R_i \| (r_b + R_s')$ has been substituted in (7.48). Note also that $R_s' = R_s \| R_b$. The current gain function is obtained using the relations $I_S = V_i G_s = V_i/R_s$ and $I_o = V_o G_L$ in (7.48). Thus

$$\frac{I_o}{I_S} = \frac{R_s}{R_L}\frac{V_o}{V_{\text{in}}} = \frac{R_s'(s - g_m/C_c)/R_L(R_s' + r_b)C_\pi}{s^2 + s[(g_m + G_t)/C_\pi + G_L(1/C_c + 1/C_\pi)] + G_t G_L/C_c C_\pi} \tag{7.49}$$

In (7.49) we shall find again a wide separation between the two poles for typical transistor parameters. From Sec. 2.9 recall that since the model is not accurate near and beyond the frequency f_T, the expressions in (7.48) and (7.49) are valid for about $f \leq f_T/3$. We shall illustrate the analysis and determine the order of magnitudes of the poles and zeros for a typical transistor in Example 7.6 and then make use of the dominant-pole approximation concept in the preliminary design of transistor amplifiers.

EXAMPLE 7.6

Let the transistor parameters at the quiescent operating point ($I_C = 5$ mA, $V_{CE} = 5$ V) be as follows:

$$f_T = 400 \text{ MHz} \qquad r_b = 100\ \Omega \qquad C_c = 5 \text{ pF} \qquad \beta_0 = 50$$

The element values of the circuit in Fig. 7.1*a* are given as

$$R_b = R_{b1} \| R_{b2} = 6\text{ k}\Omega \qquad R_S = 1\text{ k}\Omega \qquad R_L = R_C \| R_2 = 500\ \Omega$$

Determine the pole-zero locations of the amplifier, plot the magnitude and phase response, and determine the current gain and bandwidth. From the above data we obtain

$$R_i = \beta_0 r_e = 250 \qquad C_\pi = \frac{1}{r_e \omega_T} - C_c = 74.6 \text{ pF}$$

Substitution of the numerical values in (7.49) yields

$$\frac{I_o}{I_S} = (2.4 \times 10^7)\frac{s - 4 \times 10^{10}}{s^2 + s(31.1 \times 10^8) + 2.67 \times 10^{16}}$$

For convenience we normalize* the frequency by a scale of 10^8 and rewrite the above expression as

$$\frac{I_o}{I_S} = 0.24\frac{s_n - 400}{s_n^2 + 31.1s_n + 2.67} = 0.24\frac{s_n - 400}{(s_n + 31.03)(s_n + 0.086)}$$

The pole-zero plot and the magnitude and phase responses of this example are shown in Fig. 7.16*a*, *b*, and *c*, respectively. The hybrid π in the figure indicates using the π model in this example. Remember that the above expression is valid for frequencies up to $f_T/3$, i.e., about 1.3×10^8 Hz because of the validity of the model. Note further that the magnitude of the zero is much

*Normalization is discussed later in Sec. 7.15.

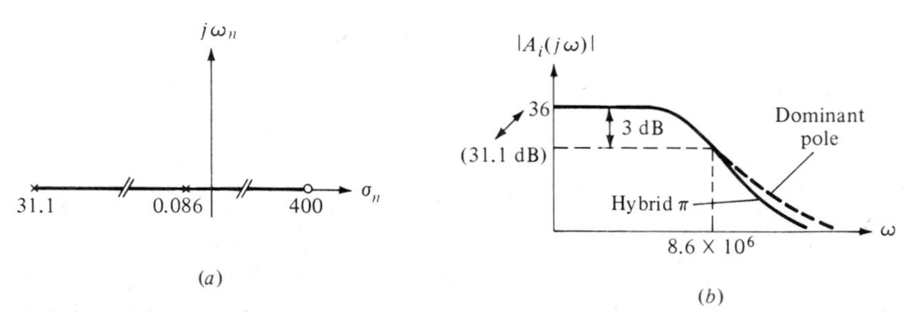

FIGURE 7.16 (*a*) Pole-zero plot of the example gain function. (*b*) Magnitude response of the example gain function (solid line). (*c*) Phase response of the example (solid line).

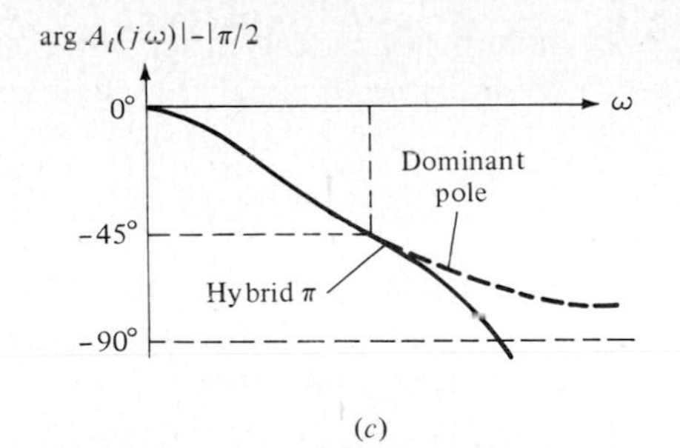

(c)

FIGURE 7.16 *(continued)*

larger than the poles and that the 3-dB bandwidth of this example from a Bode plot is $\omega_h = \omega_{3dB} \simeq 8.6 \times 10^6$ rad/s, which is the dominant pole of the circuit (i.e., in the low-pass frequency range of interest the smallest pole dominates the response as discussed in detail in Sec. 7.8.

The high-frequency 3-dB cutoff frequency is $f_{3dB} = \omega_{3dB}/2\pi = 1.37$ MHz. The low-frequency current gain is -36.

Recall that the low frequency here refers to the actual midband gain, since bypass and coupling circuitries are assumed to be short circuits and the frequency range starts from midband and extends to higher frequencies. This is understood throughout this chapter.

The current gain function in (7.49) can also be written as

$$A_i(s) = \frac{I_o}{I_S} = a_o \frac{b_1}{z_1} \frac{s - z_1}{s^2 + b_2 s + b_1} \tag{7.50}$$

where

$$a_o = \beta_0 \frac{R_S \| R_b}{R_i + r_b + R_S \| R_b} \tag{7.51a}$$

$$z_1 = \frac{g_m}{C_c} = \frac{1}{r_e C_c} \tag{7.51b}$$

$$b_2 = \frac{1}{C_\pi}(g_m + G_t) + \left(\frac{1}{C_c} + \frac{1}{C_\pi}\right) G_L \tag{7.51c}$$

$$b_1 = \frac{G_t G_L}{C_c C_\pi} \tag{7.51d}$$

Note that at $s = 0$, $A_i(0) = -a_o$, the midband current gain of a CE stage, and is the same as determined in Sec. 4.3. If a dominant-pole situation exists, then the two roots of (7.50) are far apart with the smaller pole the dominating factor in determining the high-frequency response of the amplifier. In this example, see also Fig. 7.16*b*, the pole at 8.6×10^6 rad/s is dominant. For two widely separated poles, as in (7.33), we have the condition $b_1/b_2 \ll b_2$ (note that we use b_i instead of a_i so as not to confuse the two situations), and (7.50) can be approximated as follows:

$$A_i(s) \simeq a_o \frac{b_1}{z_1} \frac{s - z_1}{(s + b_2)(s + b_1/b_2)} \tag{7.52}$$

Thus the dominant-pole expression (i.e., the smallest pole for high-frequency response) is b_1/b_2. It is further understood that $z_1 \gg b_1/b_2$, which is almost always the case since g_m/C_c has a very large value. Under these conditions (7.52) may be written in the simple approximate form

$$A_i(s) \simeq \frac{-a_o b_1}{b_2(s + b_1/b_2)} \tag{7.53}$$

The upper 3-dB radian frequency of the one-pole gain function is then given by the pole, namely,

$$\omega_h = \omega_{3\text{dB}} \simeq \frac{b_1}{b_2} = \frac{G_t}{C_\pi + C_c(1 + g_m R_L + R_L G_t)} \tag{7.54}$$

The reader is reminded that the expression in (7.54) is valid for frequencies up to $f_T/3$ due to the limitations imposed by the high-frequency hybrid-π model that has been utilized here. The magnitude and phase responses of (7.53), as compared to the more accurate expression in (7.50) for Example 7.6 are also shown by the dotted lines in Fig. 7.16*b* and *c*.

We may wonder at this time if one could make some simplifications in the circuit at the outset, such that the resultant gain function of the modified circuit is one pole, with the pole location given approximately by (7.54), and if this approximation can be achieved, what are the conditions and the limitations of the new model? As we would expect, this approximation can be achieved and is often applicable for gain and bandwidth calculations if we use the Miller theorem. We must be aware, of course, of the conditions under which it cannot be used. First, however, let us find out about the Miller theorem and its application in obtaining the unilateral approximation.

7.7 THE MILLER EFFECT AND THE UNILATERAL APPROXIMATION

We digress briefly to consider the Miller theorem, which is used in this chapter in order to obtain considerable simplification in circuit analysis and design.

Miller Theorem

Consider the circuit shown in Fig. 7.17*a* and let $V_2/V_1 = -A_o$. We shall show that Fig. 7.17*a* is completely equivalent to that of Fig. 7.17*b*. From Fig. 7.17*a* we have

$$I_1 = \frac{V_1 - V_2}{Z_1} = \frac{V_1(1 + A_o)}{Z_1} = \frac{V_1}{Z/(1 + A_o)} \tag{7.55a}$$

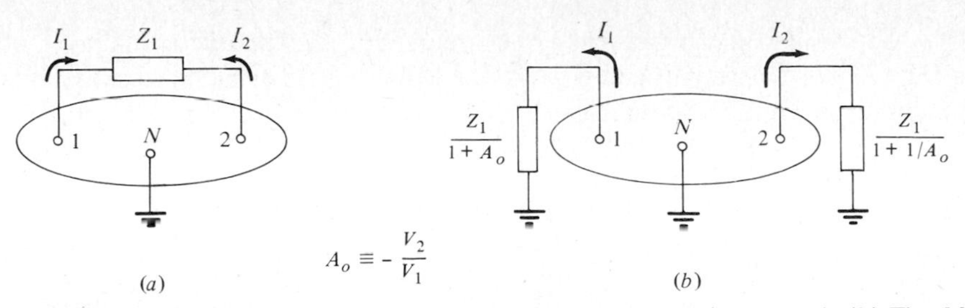

FIGURE 7.17 (*a*) A general linear circuit (with $a_o = -V_2/V_1$ known). (*b*) The Miller theorem applied to (*a*).

Similarly,

$$I_2 = \frac{V_2 - V_1}{Z_1} = \frac{V_2(1 + 1/A_o)}{Z_1} = \frac{V_2}{Z_1/(1 + 1/A_o)} \tag{7.55b}$$

For $A_o \gg 1$ note that the value of Z_1 is decreased considerably when reflected across the terminals 1 to N; thus if Z_1 is a resistor, the value of the resistor is *decreased* by A_o, and if it is a capacitor, its value is *increased* by a factor of A_o. It should be further noted that this theorem is useful only if A_o is known or can be found independently.

The apparent increase in the input capacitance of the circuit caused by the voltage gain of the amplifying device is known as the *Miller effect*. The effect of C_f at the output is negligible and henceforth ignored. The Miller effect may increase the effect of the feedback capacitor C_f by a *very* large factor, since A_o is approximately the midband voltage gain and can be very large. For example, in Fig. 7.18, if $C_1 = 90$ pF, $C_f = 5$ pF, and $A_o = 50$, the *total input capacitance* will be 345 pF. The application of the Miller theorem for BJTs and FETs in order to obtain considerable simplifications is discussed next. Needless to say, the dual of the Miller theorem can be similarly proved (problem 7.12) and applied to get certain other simplifications. Consider the circuit shown in Fig. 7.18*a*. We use general symbols so that we can apply the results obtained in this section to bipolar and field-effect transistors. Note that in the bipolar transistor case, R_i is included in Y_1; this admittance is not of direct concern here. Similarly, the output capacitance may be included in Y_L.

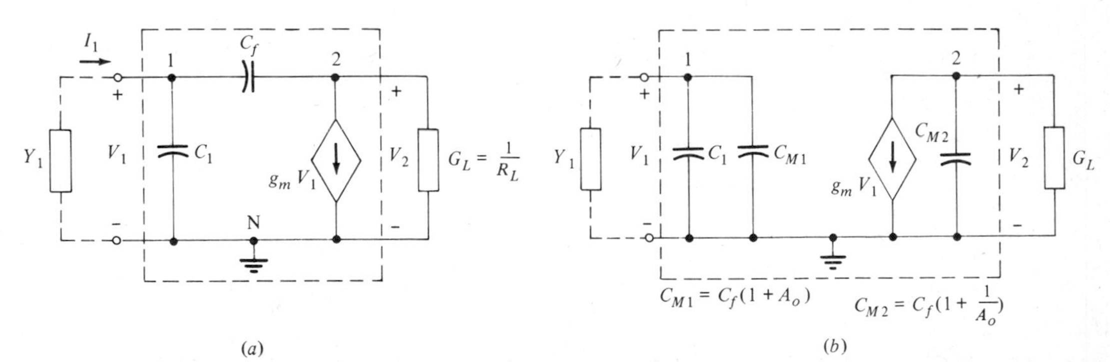

FIGURE 7.18 (*a*) A circuit that could represent BJT or FET. (*b*) Unilateral circuit after the application of the Miller theorem.

Applying Miller's theorem to the circuit in Fig. 7.18*a*, we have the circuit shown in Fig. 7.18*b*, where C_{M1} and C_{M2} are the Miller effect capacitances due to C_f. From the circuit in Fig. 7.18*b* we have

$$A_o = \frac{-V_2}{V_1} = g_m(sC_{M2} + G_L) \tag{7.56a}$$

If $G_L \gg |sC_{M2}|$, we can approximate (7.56*a*) as

$$A_o = \frac{g_m}{G_L} = g_m R_L \tag{7.56b}$$

If $A_o \gg 1$, we have

$$C_{M1} = C_f(1 + g_m R_L) \simeq g_m R_L C_f \tag{7.56c}$$

$$C_{M2} = C_f\left(1 + \frac{1}{g_m R_L}\right) \simeq C_f \tag{7.56d}$$

It should be noted from (7.56) that the above approximations imply

$$\omega C_f \ll g_m, G_L \tag{7.57}$$

which further constrains the range of frequencies for the applicability of the model.

Now applying Miller's theorem to the BJT and FET high-frequency equivalent circuits, we have the "unilateral" circuits shown in Figs. 7.19 and 7.20. Note that since the values of C_c and C_{gd} are usually small, our previous approximation (namely, $G_L \gg sC_{M2}$) implies that we can ignore their effects at the output. We shall henceforth use the circuit in Fig. 7.19*b* for the high-frequency BJT model and the circuit in Fig. 7.20*b* for the high-frequency FET model. The models will be referred to as the *unilateral models*, since signal transmission is in one direction, namely, from the source to the load. In some applications, such as in feedback amplifiers (Chap. 10) and tuned amplifiers with inductive load (Chap. 11), the role of the reverse transmission cannot altogether be ignored, and therefore these models should be used with care in those cases. Note that the inclusion of the output capacitances creates no difficulty and can be incorporated in the circuit if desired.

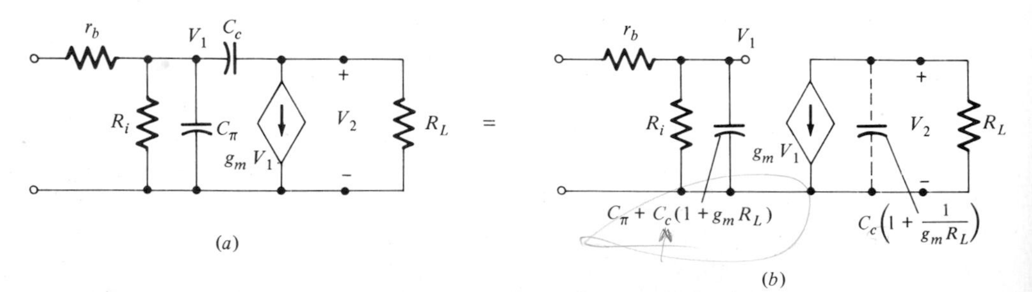

FIGURE 7.19 (*a*) Hybrid-π model of BJT. (*b*) Unilateral model of BJT.

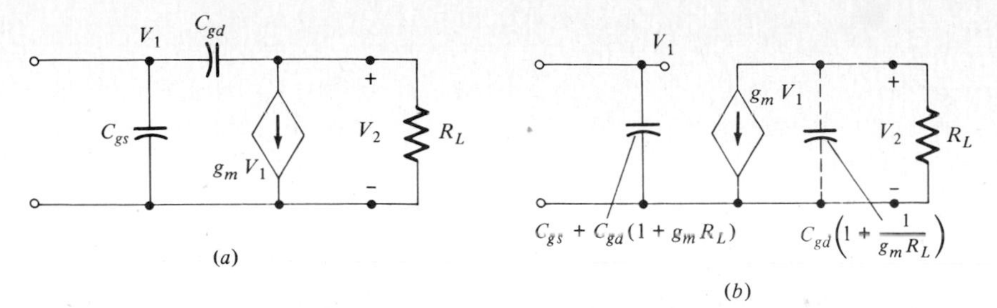

FIGURE 7.20 (*a*) Pi model of FET. (*b*) Unilateral model of FET.

An alternate approach, which of course yields the same results, is to obtain the input admittance of the circuit in Fig. 7.18*a*. This method yields

$$Y_{\text{in}} = \frac{I_1}{V_1} = s(C_1 + C_f) + \frac{g_m - sC_f}{G_L + sC_f} sC_f \tag{7.58}$$

If we assume that in the range of frequencies of interest

$$\omega C_f \ll g_m, G_L \tag{7.59}$$

then (7.59) can be approximated by

$$\frac{I_1}{V_1} \simeq s[C_1 + C_f(1 + g_m R_L)] = s[C_1 + C_f(1 + A_o)] \tag{7.60}$$

which is precisely what is shown in Fig. 7.18*b*.

We would like to emphasize that the decoupling of the input and output circuits is the most attractive simplifying feature of the equivalent circuits in Fig. 7.19*a* and Fig. 7.20*b* in analysis and design. With this model, which we shall henceforth refer to as the *unilateralized model*, the analysis and design of cascaded multistage low-pass amplifiers are simplified in that the overall circuit can be broken into single "noninteracting" stages, as will be seen later in the following sections.

The central theme in this section has been to make an approximation by removing C_f, i.e., decoupling the input and output circuits and yet including its effect at the input and on forward transmission. This approximation presents no serious problem in calculating the forward gain and the 3-dB bandwidth of low-pass amplifiers, provided the load resistance is not very large. In bandpass amplifiers, however, or when examining the stability of two ports or the stability of some feedback amplifiers, the effect of the output impedance and the reverse transmission cannot be ignored, and the unilateral model must not be used.

Finally, it should be repeated that in Fig. 7.18, if R_L is not a resistor but a general impedance $Z_L(s)$, the model can become complicated, as seen in (7.60) by replacing R_L with $Z_L(s)$. Hence, in order to preserve the simplicity of the analysis and design, we *shall henceforth consider Z_L as a pure resistance for the Miller effect* even though this may not be the case. Specifically, we *will use the models in Fig. 7.19b* for BJT, where $R_L = Z_L(0)$, i.e., the dc load for the stage and the model in Fig. 7.20*b* for FET in the same manner. *The Miller effect at the output shall be ignored.*

EXERCISE 7.5

For the circuit shown in Fig. E7.5, the transistor has $\beta_0 = 100$, $V_{BE} = 0.7$ V. Determine (*a*) the midband voltage gain, (*b*) the input resistance R_{in}, and (*c*) the 3-dB low-frequency cutoff f_1.

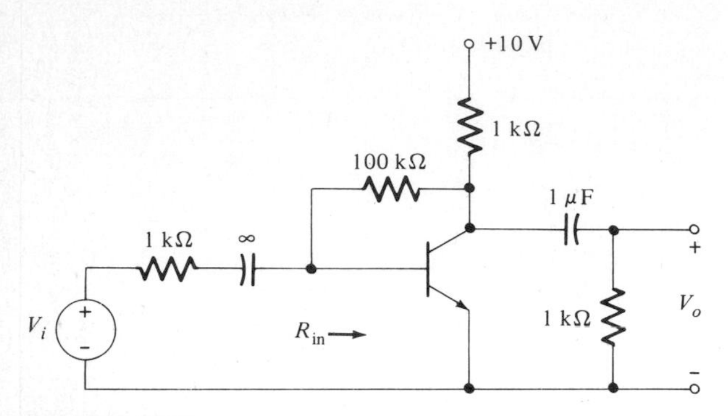

FIGURE E7.5

Ans $A_v = 24.8$, $R_{in} = 333\ \Omega$, $f_1 = 80$ Hz

EXERCISE 7.6

For the circuit shown in Fig. E7.6, determine the values of the input resistance and capacitance due to the Miller effect. What is the 3-dB cutoff frequency and the voltage gain V_o/V_i?

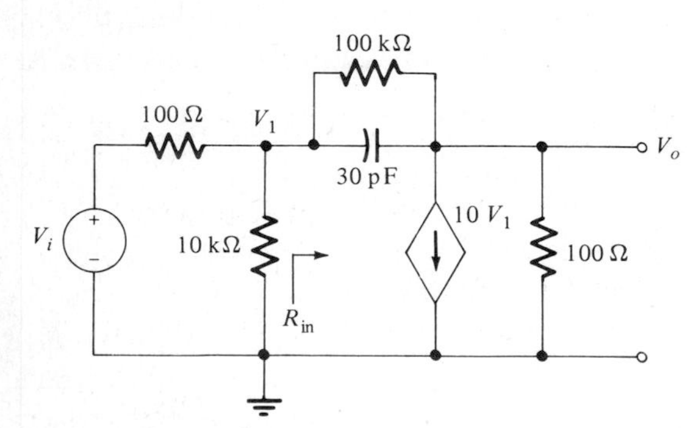

FIGURE E7.6

Ans $R_{in} = 100\ \Omega$, $C_{in} = 3 \times 10^4$ pF, $f_{3dB} = 104$ kHz, $A_v = 500$

7.8 GAIN AND BANDWIDTH CALCULATION FOR BJTS USING THE UNILATERAL MODEL

To bring clearly into focus the simplicity, limitations, and usefulness of the unilateral model, we consider the analysis of the single-stage resistively loaded amplifier of Sec.

7.5. For convenience we again show the unilateralized model of Fig. 7.15*a* in Fig. 7.21.

For the circuit of Fig. 7.21 and from (2.38) we have

$$C_\pi = \frac{1}{r_e \omega_T} - C_c \tag{7.61}$$

From Fig. 7.21*a* we define $C_i = C_\pi + C_c(1 + g_m R_L)$; hence

$$C_i = \frac{1}{r_e \omega_T} + g_m R_L C_c = \frac{D}{r_e \omega_T} \tag{7.62}$$

where

$$D = 1 + R_L C_c \omega_T \tag{7.63}$$

D is the degradation factor in the bandwidth due to the feedback effect of C_c. From the circuit of Fig. 7.21 it is a simple matter to write the gain function:

$$\frac{V_o}{V_{\text{in}}} = \frac{-g_m R_L R_s'}{R_s(R_s' + r_b)} \frac{1}{G_t + s[C_\pi + C_c(1 + g_m R_L)]} = \frac{-g_m R_L R_s'}{R_s(R_s' + r_b)} \frac{1}{G_t + sC_i} \tag{7.64}$$

The current gain function is

$$A_i = \frac{I_o}{I_{\text{in}}} = -\frac{g_m R_s'}{R_s' + r_b} \frac{1}{G_t + sC_i} = -\frac{g_m R_s'}{(R_s' + r_b)C_i} \frac{1}{s + (1/R_t C_i)} \tag{7.65}$$

Since the above gain function has only one pole, the 3-dB radian frequency is given by the pole, namely,

$$\omega_{3\text{dB}} = \frac{1}{R_t C_i} = \frac{G_t}{C_\pi + C_c(1 + g_m R_L)} \tag{7.66}$$

Comparison of (7.66) and (7.54) shows that the expressions are nearly the same if R_L is not too large; i.e., if $R_L G_t \ll (1 + g_m R_L)$, the expression in (7.54) reduces to the result of the unilateral model given in (7.66). *The unilateral model is thus very useful and sufficiently accurate for gain and bandwidth calculations in the analysis and design of low-pass amplifiers.*

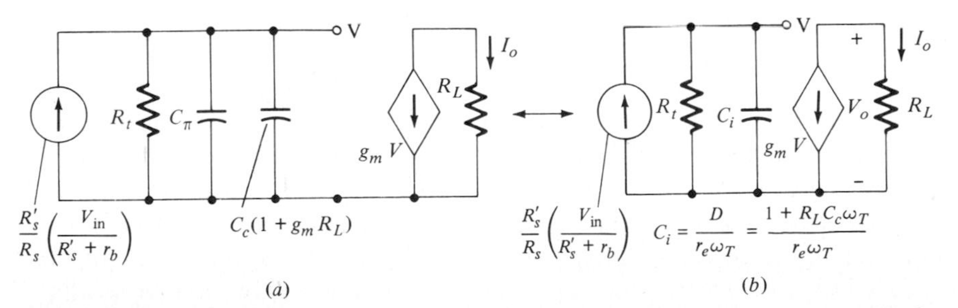

FIGURE 7.21 (*a*) Unilateral high-frequency equivalent circuit of Fig. 7.1*a*. (*b*) Same circuit with capacitors combined.

EXAMPLE 7.7

Let us now consider the previous example for the unilateral CE model of Fig. 7.21. Determine the current gain and the 3-dB bandwidth for the same circuit parameters as in Example 7.6.

From (7.65) the low-frequency current gain is given by

$$A_i(0) = \frac{I_o}{I_{\text{in}}} = \frac{-g_m R'_s}{G_t(R'_s + r_b)} = -36$$

The bandwidth is then determined from (7.66) to be

$$\omega_{3\text{dB}} = \frac{G_t}{C_\pi + C_c(1 + g_m R_L)} = 0.086 \times 10^8 \text{ rad/s}$$

Hence $f_{3\text{dB}} = (8.6 \times 10^6)/2\pi = 1.37$ MHz, which is the same result as was obtained using the hybrid-π model. However, notice the considerable simplification as a result of using the unilateral model.

For future reference, let us rewrite the expressions for gain and bandwidth in CE stages with a resistive load. Thus the gain and bandwidth are given by

$$A_v = \frac{V_o}{V_{\text{in}}} = -\beta_0 \frac{R_L}{R_1 + r_b + R_i} \tag{7.67}$$

$$A_i = \frac{I_o}{I_s} = -\beta_0 \frac{R_1}{R_1 + r_b + R_i} = -\beta_0 \frac{1}{1 + (r_b + R_i)/R_1} \tag{7.68}$$

$$\omega_{3\text{dB}} = \frac{\omega_\beta}{D} \frac{R_1 + r_b + R_i}{R_1 + r_b} = \frac{\omega_\beta}{1 + R_L C_c \omega_T} \frac{R_1 + r_b + R_i}{R_1 + r_b} \tag{7.69}$$

where $R_i = \beta_0 r_e$ and $\omega_\beta = \omega_T/\beta_0$, $R_1 = R_S \| R_b = R'_s$. The biasing resistance R_b is usually large enough to be ignored, i.e., $R_1 \simeq R_S$.

From (7.68) and (7.69), it is seen that when R_1 is very large, i.e., $R_1 \gg R_i + r_b$, the current gain approaches β_0, and the bandwidth approaches ω_β/D. If, in addition, $R_L \to 0$, $\omega_{3\text{dB}} \to \omega_\beta$, as we would expect.

The effect of R_1 and R_L on the gain and bandwidth of a CE stage can be seen in (7.68) and (7.69), as R_L increases, so does the degradation factor D, and thus the bandwidth decreases, whereas a decrease in R_L increases the bandwidth. Note also that for transistors with a high f_T and/or where the load resistance is not too small, the D factor is significant and cannot be ignored in bandwidth calculations. Lowering R_S increases the bandwidth and decreases the gain. For a very small R_S, the effect of r_b on the bandwidth will be pronounced, and due to inaccuracies in the modeling of r_b, bandwidth calculation under this condition may not give reliable results. Note the condition of validity of the unilateral model as given in (7.59), i.e., $R_L \omega_{3\text{dB}} C_c \ll 1$.

Thus the unilateral model should not be used in circuit applications where R_L is large, since the reverse transmission or the output impedance plays a dominant role. The model is inaccurate for these purposes, as was discussed in the previous section.

We emphasize the importance of the inclusion of r_b in bandwidth calculations for wideband amplifiers. It does not play an important role in midband gain calculations, and we often ignore it as was done in Chaps. 4 to 6, but we cannot ignore it in bandwidth calculations in (7.69), especially when R_S is small.

7.9 THE DOMINANT POLE-ZERO CONCEPT AND APPROXIMATION

We digress briefly to discuss the dominant pole-zero concept. This concept is very important when the network functions have widely separated pole-zero patterns and the analysis is to be performed by pencil and paper. In the analysis of electronic circuits, we may often not need to go into great detail, as approximate results may be satisfactory, at least in the initial design phase. If a detailed analysis is needed, computer methods of circuit analysis are used. A design based on the simple approximate methods can be very useful, however, as an initial iteration in a computer-aided design.

An approximation based on the dominant pole-zero concept simply ignores second-order effects due to far-removed poles and zeros of the low-pass network functions in high-frequency response calculations. In low-frequency response calculations, i.e., for the high-pass circuits, recall that the largest pole dominates the response as given by (7.46). The inclusion of only first-order effects yields considerable simplification in the analysis and design of electronic circuits. We have already used such an approximation in Secs. 7.4 and 7.5, where we ignored second-order effects due to the nondominant poles. In most cases, the inclusion of only dominant poles yields satisfactory results in determining the magnitude response, hence 3-dB calculations. When the role of the phase shift is important, however, such as in a *stability study of feedback amplifiers* (Chap. 10), one should be extremely careful. For example, the phase-shift contribution due to the nondominant poles and zeros may well cause instability, and ignoring them completely would give meaningless and disastrous results.

Let us first consider the magnitude approximation by ignoring the far removed, or the so-called nondominant, poles and zeros in high-frequency response calculations. For the sake of clarity, we consider the following simple case:

$$A(s) = \frac{1 + s/z_1}{(1 + s/p_1)(1 + s/p_1^*)(1 + s/p_2)} \tag{7.70}$$

Let p_1 and p_1^* be the dominant poles (* denotes the conjugate pole), as shown in Fig. 7.22, where $|p_2|, |z_1| \gg |p_1|$. The magnitude function from (7.70) is

$$A(j\omega) = \frac{\sqrt{1 + (\omega/z_1)^2}}{|(1 + j\omega/p_1)(1 + j\omega/p_1^*)|\sqrt{1 + (\omega/p_2)^2}} \tag{7.71}$$

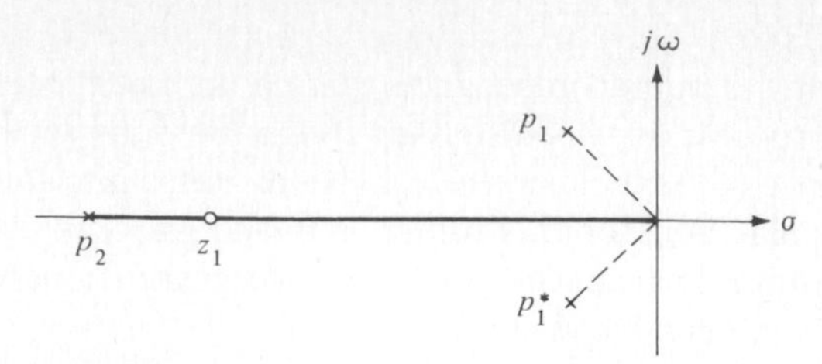

FIGURE 7.22 Pole-zero plot of (7.70).

For the range of frequencies where $|\omega/z_1| \leq \frac{1}{5}$ and $|\omega/p_2| < \frac{1}{5}$, (7.71) can be approximated as

$$|A(j\omega)| \simeq \frac{1}{|1 + j\omega/p_1|\,|1 + j\omega/p_1^*|} \tag{7.72}$$

Hence the magnitude function is essentially that of the dominant pair of poles in (7.72), neglecting the nondominant poles and zeros. In some cases, the nondominant poles may be complex. Note, however, that a pole (or a zero) as far apart as by a factor of 5 has a negligible effect on the magnitude response. For example, if

$$A(s) = \frac{A(0)p_1p_2}{(s + p_1)(s + p_2)} \qquad \text{with } p_2 = 5p_1 \tag{7.73}$$

then

$$|A(j\omega)| = \frac{A(0)}{|1 + j\omega/p_1|\,|1 + j\omega/5p_1|} \tag{7.74a}$$

$$= \frac{A(0)}{\{[1 + (\omega/p_1)^2][1 + \omega^2/25p_1^2]\}^{1/2}}$$

$$\simeq \frac{A(0)}{\sqrt{1 + (\omega/p_1)^2}} \qquad \text{for } \omega \leq p_1 \tag{7.74b}$$

The 3-dB bandwidth of (7.73) is obviously less than p_1; thus, at p_1 (i.e., a frequency slightly higher than the 3-dB frequency), the contribution of the farther pole p_2 to the magnitude response is less than 2 percent. This error decreases monotonically for decreasing frequencies. Note that the phase contribution due to the far pole p_2 is 25 percent, which is appreciable. Magnitude responses for various locations of nondominant poles and zeros are shown in Fig. 7.23. Also included is an approximate pole-zero cancellation case, which may be of interest. Note that the magnitude responses are essentially the same, while the phase responses are quite different.

The concept of dominant poles is of considerable help in obtaining simplified approximate expressions and can be used for both frequency and time-domain analysis (Chap. 12). For frequency-response calculations, focusing on dominant poles amounts to ignoring the far-removed poles in the gain function of high-frequency circuits and to ignoring the small-valued poles in the gain function of low-frequency circuits.

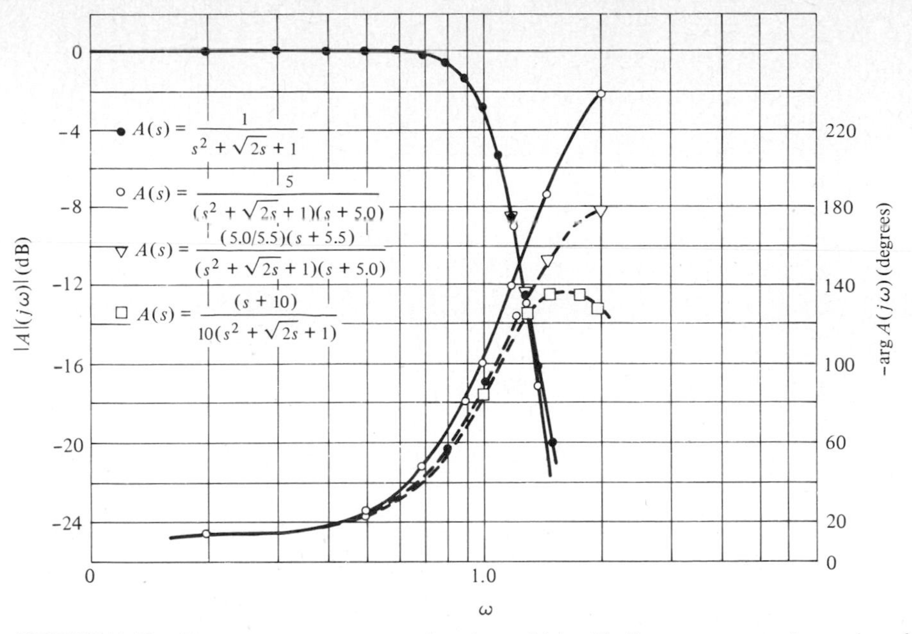

FIGURE 7.23 Frequency response of various $A(s)$ with the same complex pairs of dominant poles.

EXERCISE 7.7

The current gain function of a certain amplifier is given by

$$A_i(s) = 10^8 \frac{s + 10^7}{(s + 10^6)(s + 10^8)}$$

Determine the 3-dB cutoff frequency.

Ans $f_{3\text{dB}} = 0.16$ MHz

EXERCISE 7.8

Repeat Exercise 7.7 for the following:

$$a_v(s) = \frac{10^{13}(s + 3 \times 10^7)}{[s + (1 + j) \times 10^6][s + (1 - j) \times 10^6](s + 9 \times 10^6)}$$

Ans $f_{3\text{dB}} = 2.26 \times 10^5$ Hz

7.10 HIGH-FREQUENCY RESPONSE OF FETS

For the single-stage common-source FET amplifier, shown in Fig. 7.14a, the midband and high-frequency circuit model is shown in Fig. 7.24a. Using the Miller effect

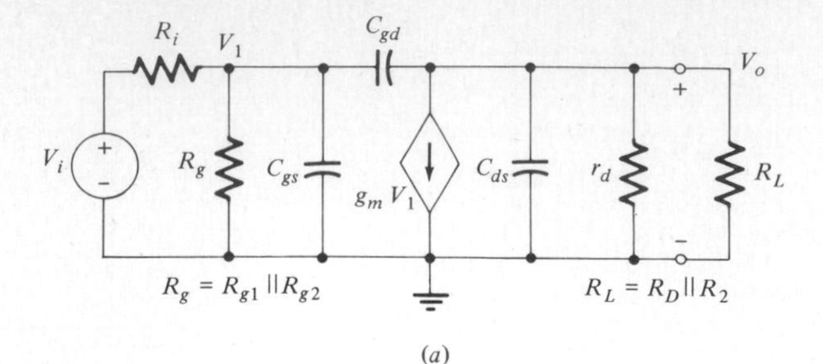

(a)

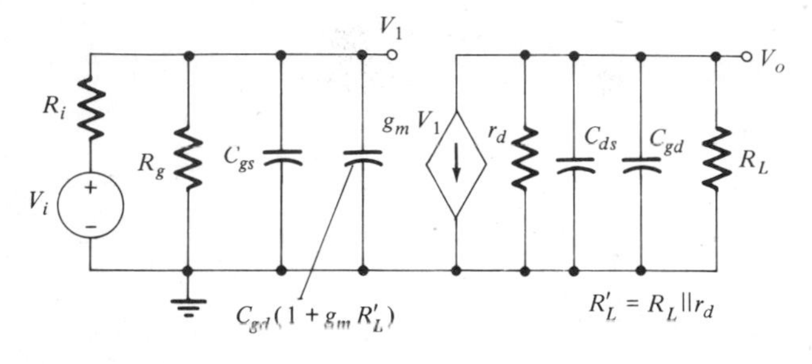

(b)

FIGURE 7.24 (*a*) High-frequency equivalent circuit of a single-stage common-source FET amplifier shown in Fig. 7.14*a*. (*b*) Simplified unilateral equivalent circuit of (*a*).

approximation, we have already obtained the unilateral model for FET in Fig. 7.20. Thus the unilateral model of FET including all the circuit parameters of Fig. 7.24*a* is shown in Fig. 7.24*b*. In FETs the input signal is always a voltage source since the FET low-frequency input impedance is extremely high ($>10^{10}\ \Omega$). Thus, for a voltage-source input (R_S very small) the input capacitance is in shunt with a very small resistance yielding a very large pole, which will be nondominant. The bandwidth in such cases is determined by the output circuit capacitances in shunt with R_L and r_d in Fig. 7.24*b*. In multistage amplifiers, however, the input capacitance of every interior and the final stage play a significant part in bandwidth calculations. This is due to the Miller effect since the effective input capacitance is increased by the voltage gain of the stage and this capacitance is in shunt with the load of the preceding stage, which may also be a high. This is discussed in detail in Chap. 8, where we consider multistage amplifiers.

The analysis of FETs is much simpler than those of BJTs because of the absence of r_b and R_i in the FET model. Let us briefly consider the single-stage FET amplifier in the CS configuration in Fig. 7.14*a* and its high-frequency model in Fig. 7.24*b*. For the circuit in Fig. 7.24*b* the voltage gain function is given by

$$A_v = \frac{V_o}{V_i} = -\frac{R_g}{R_g + R_i} \frac{g_m R'_L}{[1 + s(R_g \parallel R_i)C_1](1 + sR'_L C_2)}$$

$$\simeq -\frac{g_m R'_L}{(1 + sC_1 R_i)(1 + sC_2 R'_L)} \qquad \text{for } R_i \ll R_g \tag{7.75}$$

where $R_L' = R_L \parallel r_d$

$$C_1 = C_{gs} + C_{gd}(1 + g_m R_L')$$

$$C_2 = C_{ds} + C_{gd}$$

Note that for a voltage-source input, i.e., $R_i = 50\ \Omega$, the gain function is a one-pole circuit and the 3-dB bandwidth of the circuit is determined by the pole due to the output circuit.

EXAMPLE 7.8

For a single-stage FET amplifier circuit, shown in Fig. 7.14*a*, the circuit parameters are as given in Example 7.5, namely, $R_L = R_D \parallel R_2 = 3.5\ \text{k}\Omega$, $R_g = 1\ \text{M}\Omega$, $R_i = 50\ \Omega$. The FET parameters are given as $g_m = 2 \times 10^{-3}$ mho, $r_d = 20\ \text{k}\Omega$, $C_{gs} = 20$ pF, and $C_{gd} = C_{ds} = 2$ pF. Determine the voltage gain function, the midband voltage gain, and the high-frequency 3dB bandwidth.

From (7.75) we obtain

$$A_v = -\frac{2 \times 10^{-3}(3.5 \times 10^3)(20 \times 10^3)}{[1 + s(26.7 \times 10^{-12})(50)][1 + s(4 \times 10^{-12})(1.98 \times 10^3)]}$$

$$= -\frac{5.96}{[1 + s(1.33 \times 10^{-9})][1 + s(1.19 \times 10^{-8})]}$$

Thus $A_v(0)$, which is the midband voltage gain, is equal to -5.96, which is the same as in Example 7.5. The approximate 3dB bandwidth is determined by the dominant pole, namely,

$$\omega_{3\text{dB}} \simeq \frac{10^8}{1.19} = 0.84 \times 10^8 \text{ rad/s} \qquad \text{or} \qquad f_{3\text{dB}} = 13.38 \text{ MHz}$$

For high-frequency FET amplifiers the gain-bandwidth product may be considered as a figure of merit of the device. However, in cascaded CS stages the capacitance of an *interior stage* should be considered in such calculations, since the effective value of C_{gd} due to the Miller effect is added to C_{gs}. For R_L very large the maximum voltage gain of a FET is μ, and thus for an interior CS stage the total capacitance is $C_{gs} + \mu C_{gd}$, which is in shunt with r_d, yielding a gain-bandwidth product for an *interior stage* as

$$|A_v \omega_{3\text{dB}}| = \mu \frac{1}{r_d(C_{gs} + \mu C_{gd})} = \frac{g_m}{C_{gs} + \mu C_{gd}} \tag{7.76}$$

Note that gain and bandwidth can be traded, leaving their product unchanged. The maximum voltage gain that can be achieved in a single-stage amplifier occurs when $R_L = \infty$, and since R_L is in parallel with r_d, the maximum gain is $g_m r_d = \mu$. In the design of a single stage either gain or bandwidth can be specified; the other is

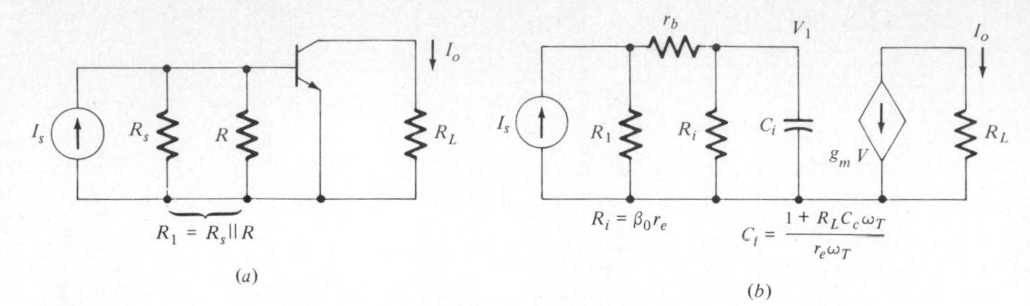

FIGURE 7.25 (a) Single-stage CE amplifier exclusive of dc biasing circuitry. (b) High-frequency small-signal equivalent circuit of (a).

determined by (7.75). Usually the gain, the bandwidth, the source, and the load impedance are specified. In such cases, multistage amplifiers are used. To improve the gain-bandwidth product capability of a stage, a compensating network can be used. Multistage FET amplifiers are discussed in Chap. 8. In multistage FET amplifiers the effect of C_{ds} is negligible, since it is in shunt with C_{gs} and $C_{gd}(1 + A)$ of the succeeding stage, which are both much larger than C_{ds}. *Henceforth C_{ds} will be ignored in the model.*

7.11 GAIN-BANDWIDTH RELATIONS IN BJTS

A single-stage CE transistor amplifier, exclusive of biasing circuitry, is shown in Fig. 7.25. We shall now show that the *gain-bandwidth product of a BJT stage is not constant*. This can readily be seen from the gain and bandwidth expressions in (7.68) and (7.69). For $R_1 \to 0$ the current gain approaches zero, while the bandwidth is finite. Thus the gain-bandwidth product (GBP) approaches zero at a frequency ω_1, namely, $\omega_1 = [(R_i + r_b)/r_b]\omega_\beta/D$. If the value of r_b could be made zero, the GBP would remain constant and equal to ω_T/D. Since r_b is never zero, it has a serious degradation effect on the GBP at high frequencies.

The current gain-bandwidth product, in general, is given by

$$\text{GBP} \simeq |A_i \omega_{3\text{dB}}| = \frac{\beta_0 \omega_\beta}{1 + R_L C_c \omega_T} \frac{R_1}{R_1 + r_b} = \frac{\omega_T}{D} \frac{R_1}{R_1 + r_b} \tag{7.77a}$$

Note that the voltage gain-bandwidth product is equal to

$$|A_v \omega_{3\text{dB}}| = \frac{\omega_T}{D} \frac{R_L}{R_1 + r_b} \tag{7.77b}$$

From (7.77*a*) it is seen that the GBP decreases with increasing R_L and decreasing R_1.

EXAMPLE 7.9

Consider the transistor parameters of Example 7.6, rewritten in the following for convenience:

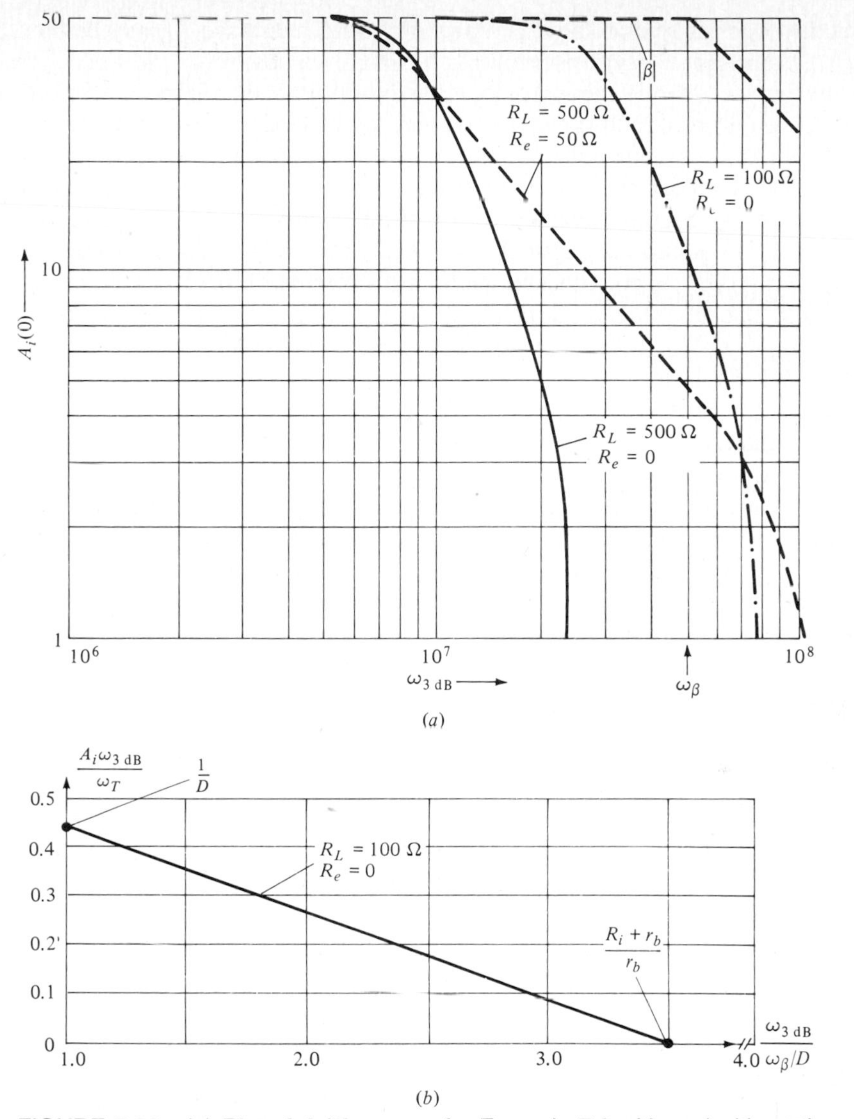

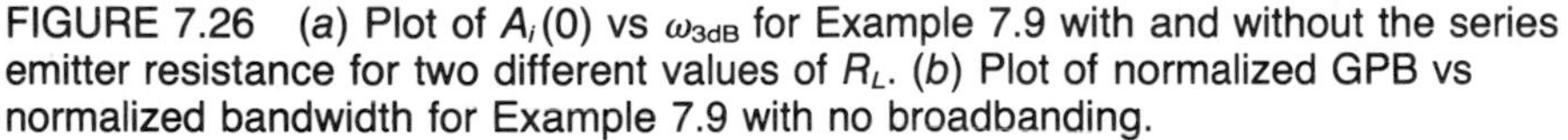

FIGURE 7.26 (a) Plot of $A_i(0)$ vs ω_{3dB} for Example 7.9 with and without the series emitter resistance for two different values of R_L. (b) Plot of normalized GPB vs normalized bandwidth for Example 7.9 with no broadbanding.

$$\omega_T = 2\pi(4 \times 10^8) \text{ rad/s} \qquad r_b = 100\ \Omega$$

$$C_c = 5 \text{ pF} \qquad \beta_0 = 50 \qquad g_m = 0.2 \text{ mho}$$

Let the input signal be from a current source. Plot the curernt gain vs bandwidth for various values of R_1 and two different values of R_L, namely, $R_L = 500\ \Omega$ and $R_L = 100\ \Omega$.

From Fig. 7.25 for each value of R_1 the bandwidth and the gain are determined from (7.66) and (7.65), and the corresponding plots for the two different load values are shown in Fig. 7.26*a*. Note that the actual source resistance of a practical current source and the effect of biasing resistance R_b may be included in R_1. From Fig. 7.26*a* it is seen that for small values of R_1, the gain is drastically reduced, while the increase in bandwidth is insignificant. Therefore, this type of amplifier circuit is very inefficient for wideband amplifiers, and thus broadbanding methods must be used in order to utilize the circuit efficiently. By "efficient" we mean that the gain-bandwidth product of the broadbanded stage in close to ω_T. Figure 7.26*b* shows the normalized GBP vs normalized ω_{3dB} for resistively broadbanded stages ($R_L = 100\ \Omega$, i.e., $D = 2.25$ in Example 7.9). It is seen that GBP is not constant and that it decreases for increasing bandwidth. It should also be pointed out that both β_0 and ω_T are functions of the operating point I_C. Usually the operating point is chosen such that these parameters are optimized. Even so, however, for a given transistor, the gain and bandwidth requirements of the amplifier often necessitate broadband multistage amplification.

From the previous section, (7.68) and (7.69), it is quite clear that if the bandwidth requirements of an amplifier stage are such that the current gain is less or equal to unity or $\omega_{3dB} \geq (\omega_\beta/D)[(r_b + R_i)/(r_b + r_e)]$, we cannot use the simple resistive circuit (Fig. 7.25) since at the equality sign, the gain of the stage will be approximately unity. For example, for the transistor parameters listed above and $R_L = 500\ \Omega$, suppose we wish to design the stage such that its bandwidth is $\omega_{3dB} = 2.5 \times 10^7$ rad/s. The quantity $(\omega_\beta/D)[(R_i + r_b)/(r_b + r_e)] = 2.31 \times 10^7$ rad/s, which is less than the required bandwidth. The circuit of Fig. 7.25 cannot meet this specification. Hence we must use some other sort of *broadbanding* to be able to meet this specification with the given transistor. The improvement in gain-bandwidth product at high frequencies $\omega \gg \omega_\beta/D$ is referred to as *broadbanding*. Let us now consider the use of an additional resistor in a feedback arrangement in order to improve the gain-bandwidth product at high frequencies. In addition to increasing the bandwidth, feedback has other desirable features, which are discussed in detail in Chap. 10.

7.12 BROADBANDING WITH A SERIES RESISTANCE IN THE EMITTER LEAD

For large bandwidth requirements, i.e., $\omega_{3dB} \gg \omega_\beta/D$, the addition of a small series feedback resistance (of the order of 50 Ω) at the emitter terminal, as shown in Fig.

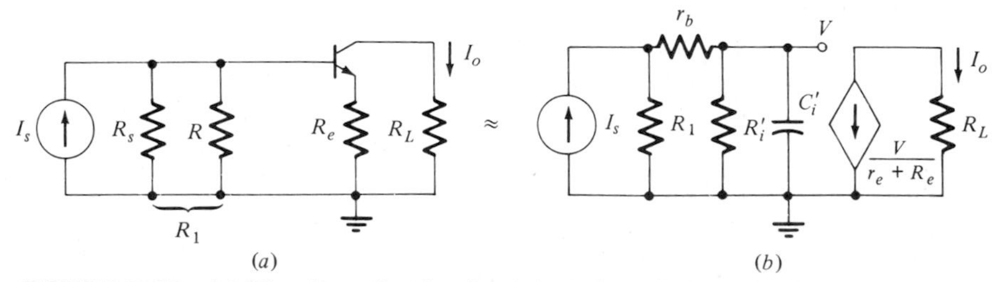

FIGURE 7.27 (*a*) Simple series-feedback broadbanded amplifier stage. (*b*) Small-signal high-frequency equivalent circuit of (*a*).

7.27a, can improve the gain-bandwidth product. Note that the effect of R_e is negligible at low frequencies since $R_e \ll R_E$. We shall show that in this arrangement the degrading effect of r_b on the bandwidth is small and that this type of amplifier has a much better GBP capability than the simple circuit of Fig. 7.25. Actually, the same effect (i.e., increasing r_e) can be achieved by decreasing the emitter current (since $r_e = V_t/I_e$). This approach leads to some complications, however, since β_0 and f_T are functions of the current and will not remain constant, as was seen in Fig. 2.22.

The circuit model of Fig. 7.27b is an approximation that includes only the dominant natural frequency of the circuit. This circuit is readily obtained by using the dual of the Miller theorem (problem 7.12).*

The circuit in Fig. 7.27b is the same as that in Fig. 7.25b, with r_e replaced by $r_e + R_e$ so that the expressions for the gain and bandwidth are also similar to (7.68) and (7.69). These are given by

$$A_i = \frac{I_o}{I_s} = -\beta_0 \frac{R_1}{R_1 + (r_b + R_i')} \simeq -\beta_0 \frac{1}{1 + \beta_0(R_e + r_e)/R_1} \tag{7.78a}$$

$$\omega_{3\text{dB}} = \frac{\omega_\beta(R_1 + r_b + R_i')}{D'(R_1 + r_b)} \simeq \frac{\omega_\beta}{1 + (R_L + R_e)C_c\omega_T} \frac{R_1 + R_i'}{R_1 + r_b} \tag{7.78b}$$

where $R_i' = \beta_0(r_e + R_e)$ and $D' = 1 + (R_L + R_e)C_c\omega_T$. The effectiveness of R_e in this series feedback arrangement in improving the GBP can best be illustrated by a numerical example.

---

*Another way of obtaining this circuit model is to consider the BJT equivalent-T model, where r_e is changed to $r_e + R_e$ in Fig. 2.25. Since the hybrid-π and the T models are equivalent, the new hybrid-π model will have r_e replaced by $r_e + R_e$. Note that for a better approximation of equivalency a small capacitor of value $C_e = 1/R_e\omega_\alpha$ is to be added across R_e. However, since this capacitance is very small, it is usually ignored.

EXAMPLE 7.10

Consider Example 7.9 where $R_L = 500\ \Omega$, $R_e = 50\ \Omega$, and a current source excitation is at the input. Plot the gains vs bandwidth for various values of R_1.

From the circuit shown in Fig. 7.27b and the gain and bandwidth expressions in (7.78a) and (7.78b) we have a plot A_i vs $\omega_{3\text{dB}}$ as shown in Fig. 7.26a. The figures are drawn on the same scale for comparison purposes. Note that beyond a certain bandwidth ($\omega_{3\text{dB}} \geq 9 \times 10^6$ rad/s in this example), the gain is higher for the same bandwidth using the additional series resistance R_e. For a bandwidth $\omega_{3\text{dB}} = 2.5 \times 10^7$ rad/s, which could not be realized by the simple circuit of Fig. 7.25, the gain using this broadbanding scheme is a little over 10.

The addition of R_e also provides a degree of freedom in the design of an amplifier stage for a specified bandwidth. In fact, there exists an optimum value for R_e when the load resistance and the 3-dB bandwidth are specified. This optimum value for a specified ω_{3dB} and R_L is given in problem 7.29. It has been found, in general, that $(R_e)_{opt}$ has a rather broad range and that its value is usually not critical; i.e., the loss of gain near the optimum value is usually not significant. In practice, a somewhat lower value of R_e than that obtained from (7.82) is desirable in cascaded stages, because a lower value of R_e lowers the D factor of the preceding stage, thus improving the overall gain-bandwidth product. Thus, for wideband stages, we shall use a convenient value of R_e, usually in the range $25 < R_e < 75\ \Omega$.

7.13 BROADBANDING WITH A RESISTANCE IN SHUNT FROM COLLECTOR TO BASE

Broadbanding is also achieved by placing a large resistor in shunt from collector to base, as shown in Fig. 7.28*a*. Biasing circuitry has been omitted for simplicity. This configuration is referred to as a *shunt-feedback stage,* as contrasted to the previous case, which is known as a *series-feedback* configuration. The effect of R_f in stabilizing the operating point has already been noted in problem 2.32. Here we examine the gain-bandwidth capability of the stage.

The familiar method of analysis can be used by replacing the device and using the hybrid model for the BJT as shown in Fig. 7.28. However, this approach is quite cumbersome and will *not* be used. An alternate approach is to use the Miller theorem.

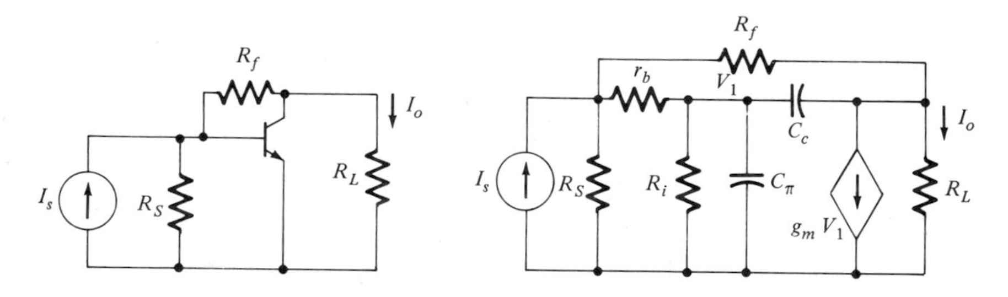

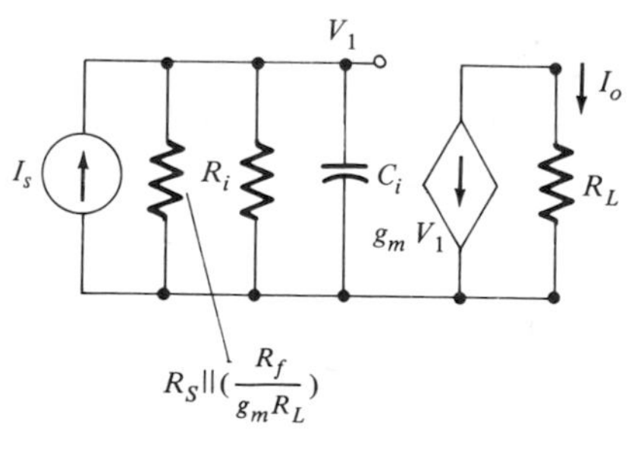

FIGURE 7.28 (*a*) Simple shunt-feedback broadbanded amplifier stage. (*b*) Small-signal high-frequency equivalent circuit of (*a*). (*c*) Simplified circuit with the Miller theorem applied.

If we make the practical assumption of $R_i \gg r_b$ and $R_f \gg R_L$, we have the simplified circuit shown in Fig. 7.28*c*. The approximate midband gain and bandwidth of the circuit is then given by

$$A_i(0) \simeq \frac{\beta_0}{1 + \beta_0 R_L/R_f} \tag{7.79a}$$

$$\text{for } R_f \gg R_i,\ R_L$$

$$f_{3\text{dB}} \simeq \frac{f_\beta}{D}(1 + \beta_0 R_L/R_f) \tag{7.79b}$$

EXAMPLE 7.11

For the shunt-feedback circuit of Fig. 7.28*a*, assume the same transistor parameters as Example 7.9. Consider a current-source input ($R_S \to \infty$) with R_L = 100 Ω and R_f = 1 kΩ. Determine the current gain and the 3-dB bandwidth of the amplifier.

From (7.79*a*) and (7.79*b*) and the transistor parameters we obtain

$$A_i(0) \simeq \frac{50}{1 + 50(0.1)} = 8.3 \qquad f_{3\text{dB}} \simeq \frac{8 \times 10^6}{2.25}(1 + 5) = 21 \text{ MHz}$$

Note that for the circuit in Fig. 7.25 for R_L = 100 Ω and a current gain of 8, the corresponding $f_{3\text{dB}}$ = 8.6 MHz. (This value could also be read from Fig. 7.26*a*.) The increase in bandwidth by a factor of almost 2.5 in the shunt-feedback circuit is to be noted. A more accurate computer-aided analysis of Fig. 7.28*b*, using the circuit parameters of this example, yields

$$A_i(0) = 7.7 \qquad f_{3\text{dB}} = 20 \text{ MHz}$$

Note that the simplified expressions of gain and bandwidth are reasonably accurate for a preliminary analysis and design purposes.

7.14 SIMPLE SINGLE-TRANSISTOR DESIGN EXAMPLE

Let us consider the complete design of a single-stage CE amplifier for a specified gain and frequency response. As mentioned earlier, amplifiers almost invariably use more than one stage. We could also use high-frequency op amps. However, we shall consider a design example illustrating the materials developed in the preceding sections. We assume that we have transistors available in stock, which we must use. The quiescent operating point suggested by the manufacturer for an optimum gain-frequency response is approximately I_C = 5 mA, V_{CE} = 5 V. At this operating point, the typical transistor parameters at room temperature are as follows:

$$\beta_0 = 100 \qquad f_T = 500 \text{ MHz} \qquad r_b = 100\ \Omega \qquad C_c = 4 \text{ pF}$$

For this particular design we also have the following specifications: the supply voltage is 20 V; the source and load terminations are R_S = 10 kΩ and R_L = 500 Ω, respectively.

We wish to achieve the following approximate performance parameters in the single-stage amplifier. The low- and high-frequency cutoff points, respectively, are to be $f_l \leq 1$ kHz and $f_h \simeq 5$ MHz. The stage must have a current gain of at least 20 dB (i.e., $A_i \geq 10$) with a one-pole rolloff near the upper band-edge frequency. The operating point is to be reasonably stable for normal environmental temperature changes and interchangeability of transistors.

We make a quick, rough check to see if the specifications can be met in a single stage. The maximum current gain for any single stage is β_0, which is 100 in this case. The GBP from the specification is $A_i f_{3dB} \geq 50$ MHz, which is less than f_T. At this point there is no obvious indication of not being able to meet the specifications. We now proceed with the design, starting with dc biasing, high-frequency design, and low-frequency design, respectively.

A. DC Biasing

From the biasing design procedure of Example 2.2 we have already determined the following circuit resistor values for reasonably stabilized biasing (note that we would usually select standard-size resistors closest to the calculated values):

$$R_C = 2\text{ k}\Omega \qquad R_E = 1\text{ k}\Omega$$

$$R_{b1} = 17\text{ k}\Omega \qquad R_{b2} = 7\text{ k}\Omega \qquad R_b = R_{b1} \| R_{b2} = 5\text{ k}\Omega$$

B. High-Frequency Design

First, we design the high-frequency circuit because the broadbanding resistor R_1 needed at the input affects the low-frequency design. Hence, the value of R_1 must be determined. This will become clear when we come to the low-frequency design. We calculate $(f_\beta/D)[(R_i + r_b)(R_b + r_e)] = 3.9 \times 10^6$ Hz a quantity that is less than the specified high-frequency cutoff. Hence, the simple circuit of Fig. 7.25 cannot do the job. We must, therefore, choose a series- or a shunt-feedback broadbanding scheme. Let us try the series-feedback configuration of Fig. 7.27*a* and arbitrarily choose $R_e = 35\ \Omega$, as recommended earlier.

From (7.78*b*) we determine the value of R_1 from

$$2\pi(5 \times 10^6) = \frac{2\pi(5 \times 10^6)}{7.28} \frac{R_1 + 4000}{R_1 + 100} \qquad \text{or} \qquad R_1 = 520\ \Omega$$

The current gain is determined from (7.78*a*):

$$A_i \simeq -100 \frac{1}{1 + 4000/500} = -11.5$$

which meets the desired specification. Hence the preliminary high-frequency design is complete.

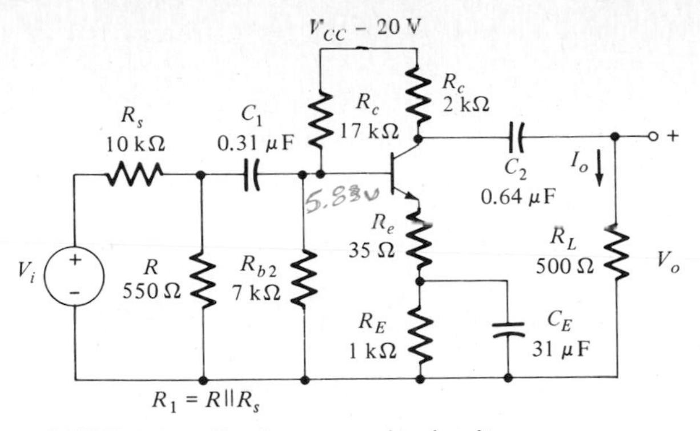

FIGURE 7.29 Design example circuit.

C. Low-Frequency Design

We proceed according to the design method given in Sec. 7.4. First we check the pole-separation condition as in (7.37). We find that $R_b \parallel \beta_0 R_E = 5$ kΩ is larger than $2.5(R_i + r_b + R_1) = 3.1$ kΩ, and thus the design method will be valid. Note that we have used $\beta_0 r_e$ and not $\beta_0(r_e + R_e)$ for R_i since for frequencies near or below the low-frequency cutoff point, the emitter Z_E due to R_E(and C_E) is much larger than the value of R_e. Since these impedances are in series, the value of R_e can be ignored in comparison with Z_E.

Thus from (7.38) and (7.39) we obtain

$$C_1 = \frac{10}{2\pi(10^3)(520 + 100 + 500)} = 0.31\ \mu\text{F}$$

$$C_E = 100C_1 = 31\ \mu\text{F}$$

$$C_2 = \frac{10}{2\pi(10^3)(2.5 \times 10^3)} = 0.64\ \mu\text{F}$$

In practice, standard-size capacitors are selected. Note the particularly large value of the emitter bypass capacitor C_E. Because of their relatively large values (>25 μF) vs physical size, electrolytic capacitors are normally used. Electrolytic capacitors have an associated series resistance in the range of 1 to 5. This small resistance can affect the gain and bandwidth performance of the stage, especially if R_e is zero. The circuit for this design is shown in Fig. 7.29.

EXERCISE 7.9

For the circuit shown in Fig. E7.9, determine the voltage gain and the upper and lower 3-dB cutoff frequencies given $g_m = 5 \times 10^{-3}$ mho, $r_d = 20$ kΩ, $C_{gs} = 20$ pF, and $C_{gd} = C_{ds} = 3$ pF.

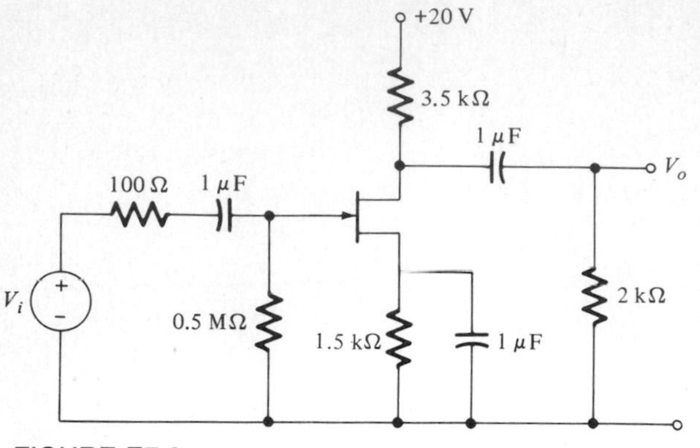

FIGURE E7.9

Ans $A_v = -6.0, f_l = 902$ Hz, $f_h = 19.1$ MHz

EXERCISE 7.10

For the circuit shown in Fig. E7.10, determine the upper and the lower 3-dB frequencies and the voltage gain given $\beta_0 = 100$, $r_b = 100\ \Omega$, $C_c = 5$ pF, and $\omega_T = 10^9$ rad/s.

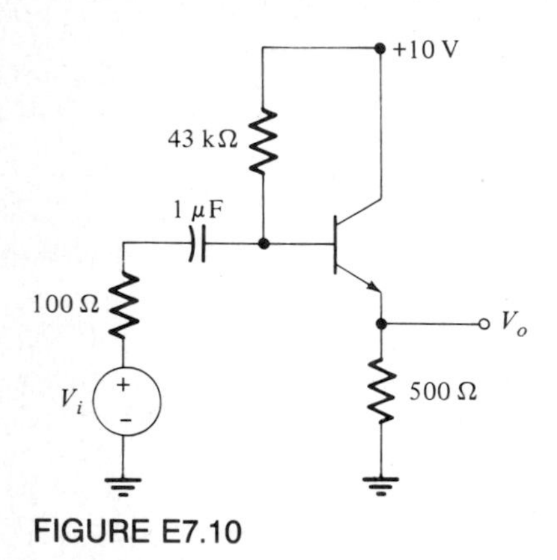

FIGURE E7.10

Ans $f_l = 6.8$ Hz, $f_h = 113.7$ MHz, $A_v \simeq 1$

Before we conclude this chapter, it is advisable to digress a little and consider circuit normalization. In the analysis and design of circuits using paper and pencil, it is often desirable to use a suitable change of scale in amplitude and frequency. Under an appropriate scale change, the tedium of computation with large numbers and powers of 10 is reduced to simple numerical operations. If a digital computer is used in the analysis, no advantage, in general, is accrued by normalization.

7.15 CIRCUIT NORMALIZATION (FREQUENCY AND MAGNITUDE SCALING)

A change in the frequency scale is referred to as *frequency normalization,* and a change in the scale of amplitude is called *resistance normalization*. If the radian frequency is normalized with respect to Ω_0 and the resistance is normalized with respect to R_0, the normalized element values, designated by the subscript n, are given by

$$s_n = \frac{s}{\Omega_0} \qquad R_n = \frac{R}{R_0}$$
$$L_n = \frac{L\Omega_0}{R_0} \qquad C_n = R_0\Omega_0 C \tag{7.80}$$

From (7.85) it is apparent that the normalized element values are dimensionless; hence dimensional analysis cannot be used as a check after normalization. Normalization should always be done with respect to known values of Ω_0 and R_0. The choice of R_0 and Ω_0 is usually obvious in a given problem.

EXAMPLE 7.12

Consider the simple circuit shown in Fig. 7.30*a*. The values of L and R are to be designed for a voltage gain with a maximally flat magnitude response; i.e.,* the poles are to be located as in Fig. 7.30*b*. Plot the frequency response for $\phi = 50°$ and $\phi = 30°$.

The voltage gain function of the circuit V_o/V_1 is given by

$$\frac{V_o}{V_1} = \frac{-g_m R}{s^2LC + sRC + 1} = \frac{-g_m R}{LC} \frac{1}{s^2 + s(R/L) + 1/LC}$$

The normalized transfer function using (7.80) is

$$\frac{V_o}{V_1} = \frac{-g_m R}{L_nC_n(s/\Omega_0)^2 + (s/\Omega_0)(R_n/L_n) + 1/L_nC_n}$$

$$A_V(s_n) = \frac{A_V(0)/L_nC_n}{s_n^2 + s_n(R_n/L_n) + 1/L_nC_n}$$

For two-pole locations as in Fig. 7.30*b* we have

$$A_V = \frac{A(0)}{s_n^2 + \sqrt{2}s_n + 1}$$

Equating the denominators from the above expressions, we have

$$\frac{R_n}{L_n} = \sqrt{2} \qquad \frac{1}{L_nC_n} = 1$$

*Maximally flat magnitude response is discussed in Sec. 8.10.

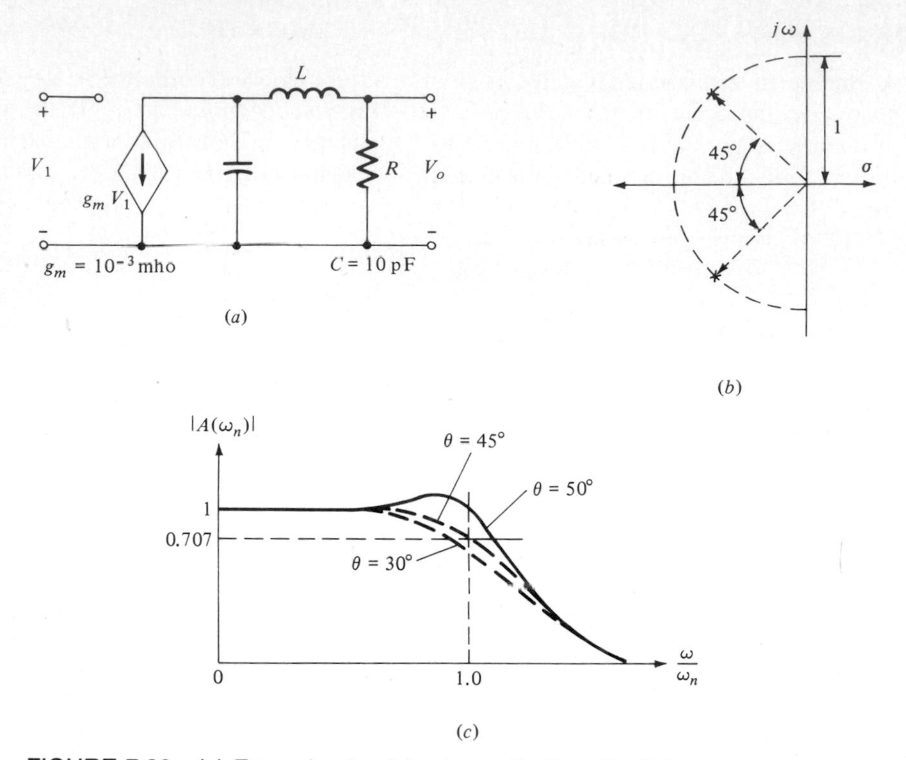

FIGURE 7.30 (*a*) Example circuit for normalization. (*b*) Pole-zero plot for maximally flat magnitude response. (*c*) Magnitude response for various pole angles θ.

We set $C_n = 1$; hence $L_n = 1$, and $R_n = \sqrt{2}$. One appropriate set of values of R_0 and Ω_0 is $R_0 = 1\text{ k}\Omega$, since g_m is known. From (7.80) $\Omega_0 = 10^8$, since C is known. Hence $L = 10\ \mu\text{H}$. The 3-dB bandwidth is $\omega_{3\text{dB}_n} = 1$. The actual 3-dB bandwidth is $\omega_{3\text{dB}} = 1/LC = 10^8$ rad/s. The gain of the circuit is $-g_m R = -\sqrt{2}$. The responses for $\phi = 50°$ and $\phi = 30°$ are shown in Fig. 7.30*c*. Note that for $\phi > 45°$ the frequency response has peaking.

EXERCISE 7.11

Normalize the frequency to $\Omega_0 = 10^6$ rad/s in Exercise 7.8, and determine the normalized 3-dB frequency and the actual 3-dB frequency.

Ans $(f_{3\text{dB}})_n = 0.226$, $f_{3\text{dB}} = 2.26 \times 10^5$ Hz.

7.16 CONCLUDING REMARKS

In this chapter the frequency response of single BJT and FET amplifiers was considered. Single-transistor amplifiers were considered in order to focus on the salient features of low- and high-frequency performance of the amplifiers. The use of the Miller theorem in obtaining a simple unilateral model for BJT and FET circuits was seen to be very useful. Simple models and approximate techniques were used in order to gain insight into the performance of the circuit.

The concept of dominant poles and zeros in the calculation of low and high cutoff frequencies were emphasized since these concepts will be used in Chap. 8 and the following chapters.

REFERENCES

1. D. Comer, *Modern Electronic Circuit Design,* Addison-Wesley, Reading, Mass., 1976.
2. D. Schilling and C. Belove, *Electronic Circuits: Discrete and Integrated,* McGraw-Hill, New York, 1979.
3. J. Millman, *Microelectronics: Digital and Analog Circuits and Systems,* McGraw-Hill, New York, 1979.

PROBLEMS

7.1 For the circuit shown in Fig. P7.1, determine C_1 so that the lower 3-dB frequency of the current-gain function is $f_l = 50$ Hz. Consider the following two cases:
(*a*) $R_e = 0$
(*b*) $R_e = 50\ \Omega$
(Assume $\beta_0 = 100$, $r_b = 100\ \Omega$, and $h_{ie} = 1\ \text{k}\Omega$.)

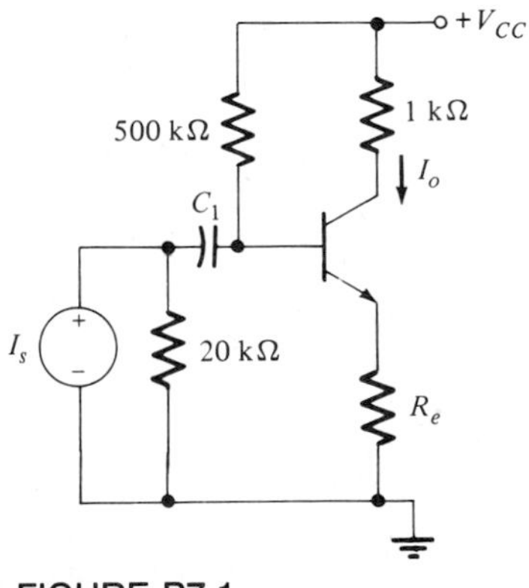

FIGURE P7.1

7.2 For the circuit shown in Fig. P7.2, determine the values of C_1 and C_E such that the low-frequency cutoff is at f_l = 100 Hz. Determine the midband voltage gain and the upper cutoff frequency f_h. Sketch V_o/V_{in} (asymptotic plot). Assume the following parameters:

$$\beta_0 = 50 \qquad r_b \simeq 0 \qquad f_T = 200 \text{ MHz} \qquad C_c = 6 \text{ pF}$$

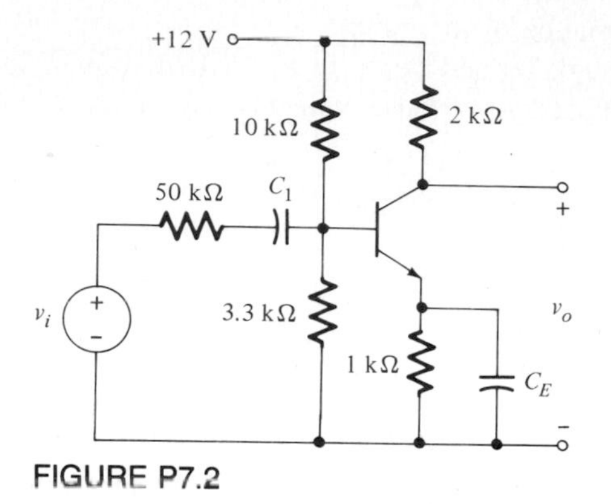

FIGURE P7.2

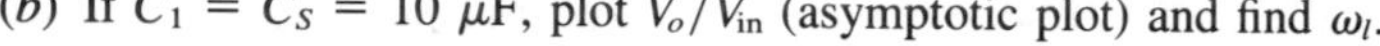

7.3 For the JFET amplifier shown in Fig. P7.3, $g_m = 2 \times 10^{-3}$ mho, and r_d = 20 kΩ. (*a*) Determine the complete low- and mid-frequency voltage gain function. (*b*) If $C_1 = C_S$ = 10 μF, plot V_o/V_{in} (asymptotic plot) and find ω_l.

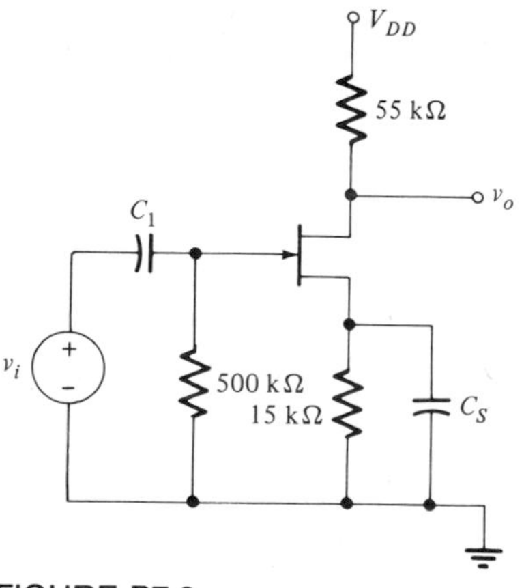

FIGURE P7.3

7.4 For the circuit shown in Fig. P7.3, if C_1 = 2 μF and $C_S = \infty$, determine the upper and lower 3-dB frequency of the voltage gain. Assume C_{gs} = 20 pF and $C_{ds} = C_{gd}$ = 2 pF.

7.5 For the circuit shown in Fig. P7.5, determine the lower 3-dB frequency of the voltage gain. Assume r_b = 0 and β_0 = 100.

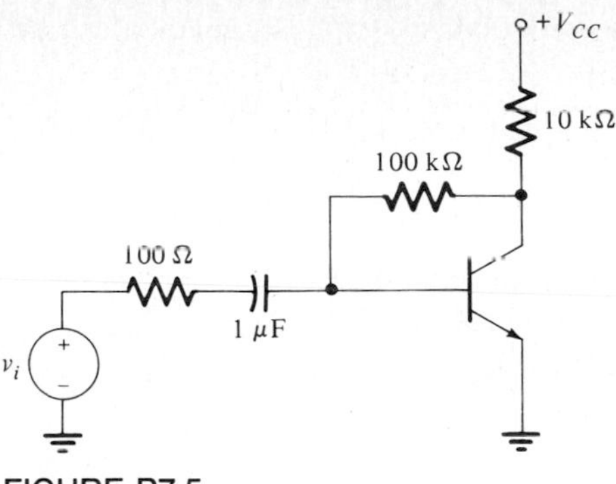

FIGURE P7.5

7.6 (*a*) Determine the lower 3-dB frequency and the voltage gain for the MOSFET circuit in Fig. P7.6. The FET parameters are given as $g_m = 4 \times 10^{-3}$ mho and $r_d = 20$ kΩ.
(*b*) Find R_{in} of the circuit using the Miller theorem.

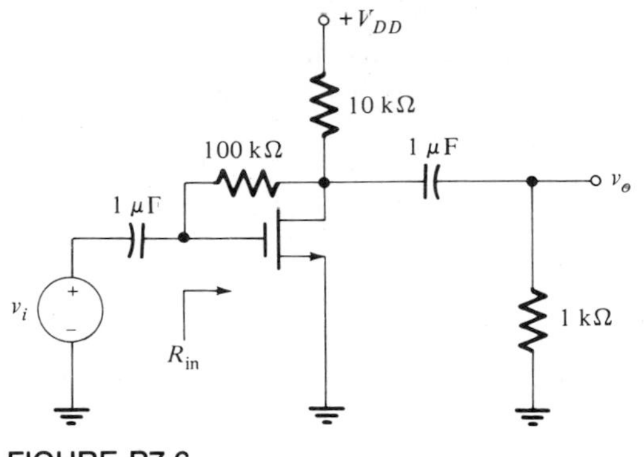

FIGURE P7.6

7.7 The low-frequency gain function of a transistor amplifier is given by

$$A_i = \frac{10^3 s_n^2 (s_n + 1)}{(s_n + 30)(s_n + 30)(s_n + 4)} \qquad s_n = \frac{s}{10^3}$$

Determine the lower 3-dB frequency f_l and sketch the asymptotic plot.

7.8 The high-frequency gain function of a certain amplifier is given by

$$A_i = \frac{10^4 (s_n + 50)}{(s_n + 25)(s_n + 55)(s_n + 5)}$$

where the frequency is normalized by a factor of 10^6, i.e., $s_n = s/10^6$.
(*a*) Determine the midband gain and the approximate 3-dB bandwidth of the amplifier.
(*b*) What is the approximate phase angle of A_i at $\omega_n = 5$?
(*c*) Sketch the asymptotic plot.

7.9 The high-frequency gain function of a cascaded FET amplifier is given by

$$A_v = \frac{10^3(s_n + 12)(s_n + 15)}{(s_n + 1)(s_n + 13)(s_n + 1.5)(s_n + 16)}$$

where $s_n = 10^6$. Determine the midband gain and the approximate 3-dB bandwidth of the amplifier.

7.10 Determine the exact and the approximate 3-dB bandwidths of an amplifier that has the following high- and low-frequency gain function:

(*a*) $A_i(s) = \dfrac{A_o}{(1 + s/p_o)^3}$

(*b*) $A_i(s) = \dfrac{A_o s^3}{(s + p_o)^3}$

7.11 For the circuit shown in Fig. 7.18*a*,
(*a*) Derive (7.58), but use Y_L instead of G_L, if the load impedance is a shunt LC circuit, i.e., $Y_L = sC + 1/sL$. Under what condition is the real part of Y_{in} negative?
(*b*) Repeat (*a*) for a series LC circuit, i.e., $Z_L = sL + 1/sC$. Use the Miller approximation in both cases and comment on your results.

7.12 (*a*) Verify the dual of Miller's theorem shown in Fig. P7.12*a*.
(*b*) Apply Miller's theorem to determine the low-frequency input and output impedances R_i and R_o in Fig. P7.12*b*.

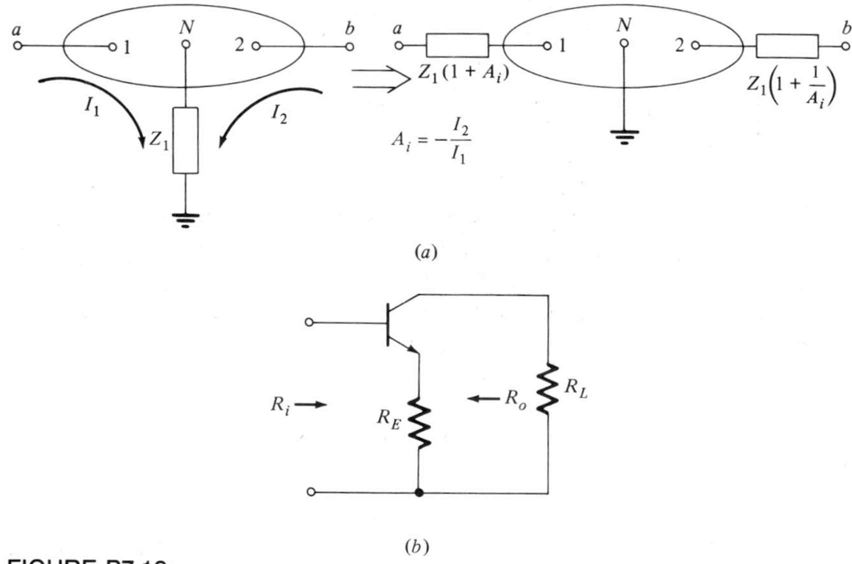

FIGURE P7.12

7.13 The complete gain function of an amplifier is given by

$$A_n(s) = K\frac{s(s + 100)}{(s + 10^3)(s + 3 \times 10^3)(s + 10^6)(s + 10^7)}$$

(*a*) Determine the low-frequency cutoff ω_l.
(*b*) Determine the high-frequency cutoff ω_h.
(*c*) Determine the value of K if $A_{\text{mid}} = 100$.

7.14 For the circuit in Fig. 7.29, let $R_e = 0$ and the derive expression for the current gain function. You may split it into a high-frequency and a low-frequency equivalent circuit. Determine ω_h, ω_l, and the current gain of the circuit. Assume $f_T = 500$ MHz, $C_c = 2$ pF, $\beta_0 = 100$, and $r_b = 0$.

7.15 Repeat problem 7.14 with $R_e = 50\ \Omega$.

7.16 For the circuit shown in Fig. P7.16, the FET parameters are $C_{gs} = 10$ pF, $C_{gd} = C_{ds} = 2$ pF, $g_m = 4 \times 10^{-3}$ mho, and $r_d = 20$ kΩ. Use the Miller approximation and find the voltage gain function V_o/V_{in} and $\omega_{3\text{dB}}$.

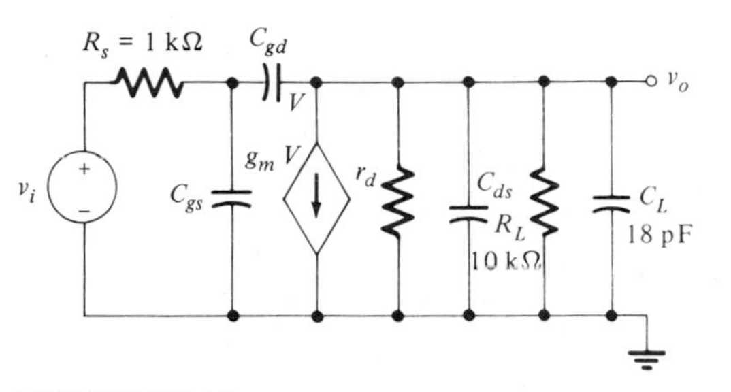

FIGURE P7.16

7.17 Repeat problem 7.16 and use 7.58 with $Y_L = G_L + sC_L$ instead of G_L. Note that C_L *is to be included,* since additional capacitance C_L is added at the output. Observe the pole-zero cancellation in your results.

7.18 The manufacturer of a certain type of transistor gives the following information at room temperature: at the operating point (which is at $I_E = 2.5$ mA, $V_{CE} = 5$ V), h_{fe} measured at 10 kHz is 100, h_{fe} measured at 50 MHz is 4, h_{ie} measured at 10 kHz is approximately 1 kΩ, and the real part of h_{ie} measured at 50 MHz is 50 Ω. For a current-source input and a load resistance of 1 kΩ, $f_{3\text{dB}} = 200$ kHz. Determine the values of g_m, r_e, r_b, β_0, C_π, C_c, R_i, f_T, and f_β for the transistor.

7.19 For the simple resistive stage shown in Fig. P7.19, the parameters are as given in problem 7.18.
(*a*) Find R_I for a current gain of 10 and determine the upper 3-dB frequency.
(*b*) Find R_I for a 3-dB bandwidth of 2 MHz and determine the current gain.

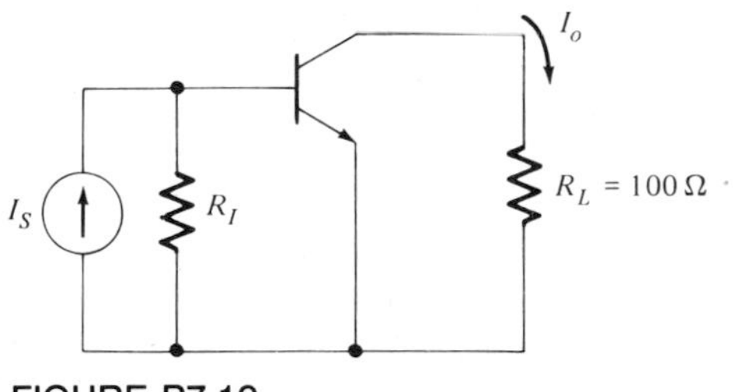

FIGURE P7.19

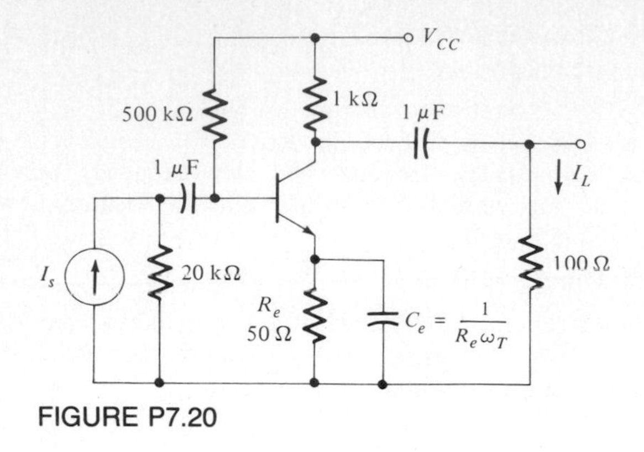

FIGURE P7.20

7.20 For the transistor circuit shown in Fig. P7.20, determine the midband current gain and the upper and lower 3-dB frequencies given $f_T = 300$ MHz, $C_c = 5$ pF, $\beta_0 = 100$, $r_b = 100\ \Omega$, and $r_e = 10\ \Omega$. (*Hint:* Figure 7.27*b* is applicable.)

7.21 For the base-compensated broadbanded stage shown in Fig. P7.21, the transistor parameters are as given in Example 7.6. For proper compensation, C_1 is chosen such that $R_1C_1 = R_iC_i$. Given that $R_S = R_L = 100\ \Omega$, perform the following:
(*a*) Find the expressions for $A_i(s)$ and $\omega_{3\text{dB}}$, and show that

$$\text{GBP} = A_i\omega_{3\text{dB}} = \frac{\omega_T}{D}\frac{R_S}{R_S + r_b}$$

(Note that $A_i = I_o/I_s$, where $I_s = v_{\text{in}}/R_S$.)
(*b*) Design the circuit for a bandwidth of $\omega_{3\text{dB}} = 8 \times 10^7$ rad/s. What are the gains $A_i(0)$ and $A_v(0)$?
(*c*) If $R_S = 0 = r_b$, can the unilateral model be used for the bandwidth calculation? Why?

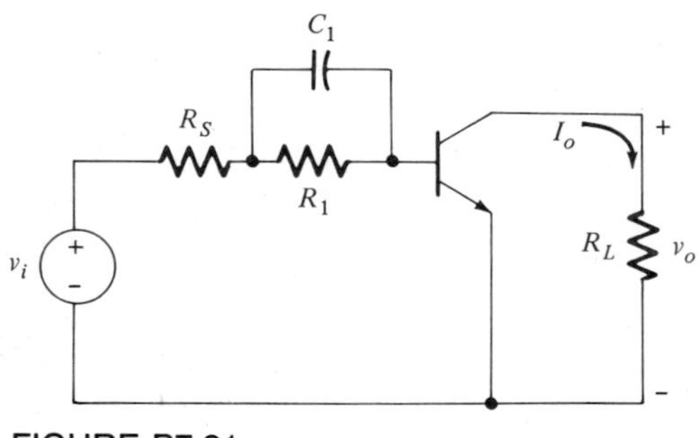

FIGURE P7.21

7.22 For the MOSFET source-follower circuit shown in Fig. P7.22, determine the following:
(*a*) the expression for the lower cutoff frequency f_l
(*b*) The expression for the midband voltage gain
(*c*) The expression for the higher 3-dB frequency f_h

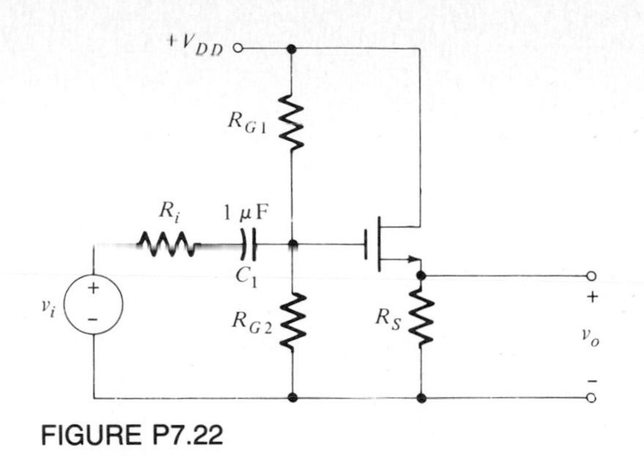

FIGURE P7.22

7.23 A single-stage MOSFET amplifier is shown in Fig. P7.23. The PMOS parameters and the circuit element values are $|y_{fs}| = 5$ mmho, $C_{iss} = 10$ pF, $C_{rss} = 2$ pF, $C_{oss} = 4$ pF, $r_d = 20$ kΩ, $R_i = 50$ Ω, $R_D = R_S = 2$ kΩ, $R_G = R_{G1} \| R_{G2} = 50$ kΩ, and $C_1 = C_2 = 1$ μF. Find

(*a*) The midband gain V_o/V_i and the lower 3-dB frequency ω_l of the amplifier.

(*b*) The higher 3-dB frequency ω_h of the amplifier.

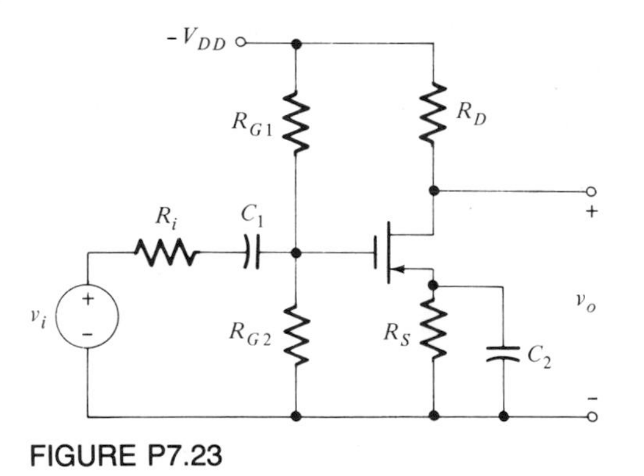

FIGURE P7.23

7.24 For the resistive shunt-feedback circuit of Fig. 7.28, $R_L = 500$ Ω and $R_f = 5$ kΩ. The transistor parameters are as given in Example 7.9 except that we assume $r_b = 0$. Determine the gain and the bandwidth of the circuit using the Miller approximation for R_f and C_c.

7.25 A CC stage with $R_L = 500$ Ω and $R_S = 100$ Ω is shown in Fig. P7.25.

(*a*) Determine the upper 3-dB bandwidth of the amplifier if the transistor parameters are as in Example 7.9. What is the midband gain of the circuit? (Use the hybrid-π model and suitable approximations, i.e., $C_c = 0$.)

(*b*) Determine the output impedance $Z_o(s)$ and show that it is inductive.

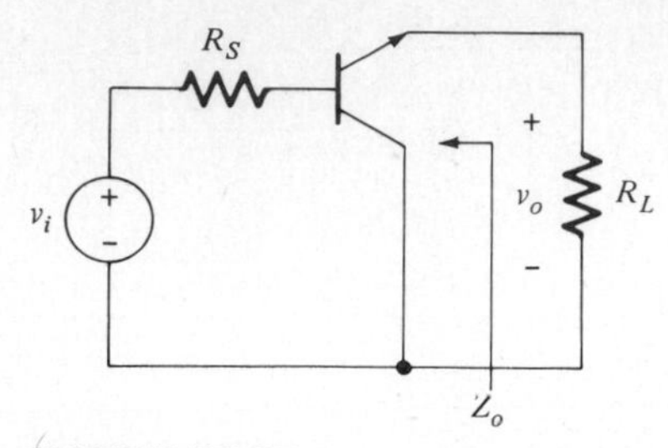

FIGURE P7.25

7.26 An emitter-follower circuit is shown in Fig. P7.26. The circuit is biased such that $I_e = 1$ mA, $V_{CE} = 5$ V. Determine the midband voltage gain and the upper and lower 3-dB frequencies of the amplifier if the transistor parameters are $f_T = 300$ MHz, $\beta_0 = 100$, $C_c = 2$ pF, $r_b = 100\ \Omega$, and $R_b = R_{b1} \parallel R_{b2} = 10$ kΩ.

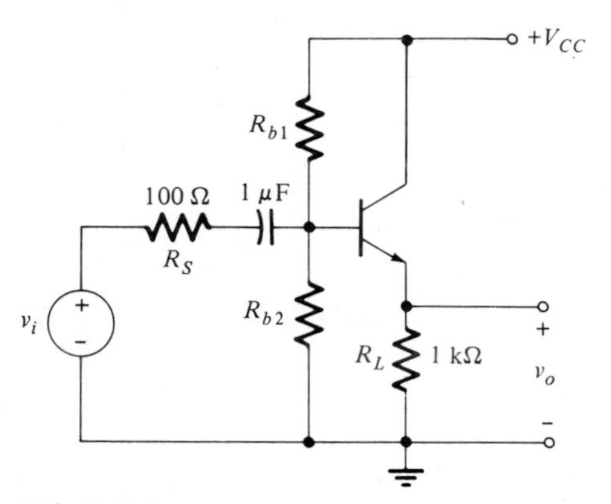

FIGURE P7.26

7.27 For the circuit shown in Fig. P7.27, determine the midband voltage gain and the upper 3-dB frequency. The transistor parameters are given as $\beta_0 = 100$, $C_c = 2$ pF, $r_b = 100\ \Omega$, and $f_T = 300$ MHz.

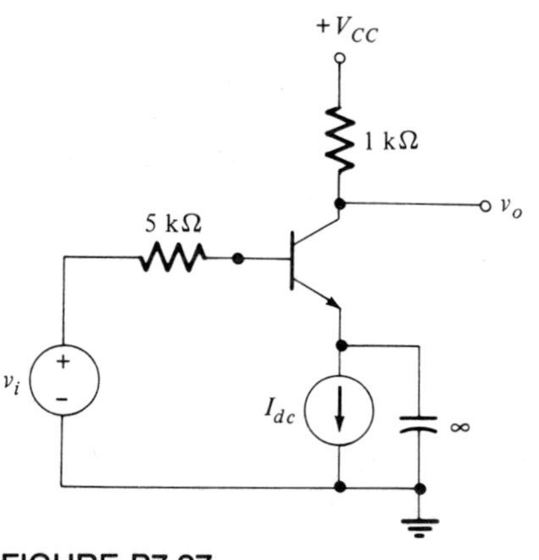

FIGURE P7.27

7.28 For the source-follower circuit shown in Fig. P7.28,

(*a*) Determine the midband voltage gain and the output impedance.

(*b*) Determine the upper and lower 3-dB frequencies of the circuit if the MOSFET parameters are $g_m = 3 \times 10^{-3}$ mhos, $r_d = 50\ \text{k}\Omega$, $C_{gs} = 10$ pF, and $C_{gd} = C_{ds} = 2$ pF.

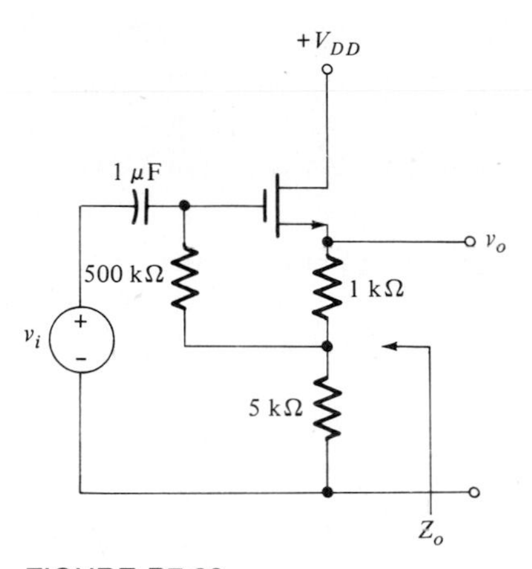

FIGURE P7.28

7.29 For the series-feedback circuit shown in Fig. 7.27*a*,

(*a*) Show that the optimum value of R_e for a specified $\omega_{3\text{dB}}$ and R_L is given by the following equation:

$$(R_e)^2_{\text{opt}} + K_1(R_e)_{\text{opt}} + K_2 = 0$$

where

$$K_1 = -\frac{2[r_b\omega_{3\text{dB}}(1 + R_LC_c\omega_T) - (\beta_0 r_e + r_b)\omega_\beta]}{\omega_T(\omega_{3\text{dB}}r_bC_c - 1)}$$

$$K_2 = \frac{-r_b\beta_0 r_e C_c\omega_T - r_b(1 + R_LC_c\omega_T) + r_b(\omega_{3\text{dB}}/\omega_\beta)(1 + R_LC_c^2\omega_T)}{C_c\omega_T(\omega_{3\text{dB}}r_bC_c - \beta_0)}$$

(*b*) For $R_L = 500\ \Omega$, $\omega_{3\text{dB}} = 4 \times 10^7$ rad/s and the same transistor parameters as in Example 7.9, determine the value of $(R_e)_{\text{opt}}$ and $A_i(0)$.

7.30 For the circuit shown in Fig. P7.30, the current gain function is given by

$$A_i(s) = \frac{-g_mR_1(1/LC_i)}{s^2 + s[(R_1 + r_b)/L + 1/R_iC_i] + (R_1 + r_b + R_i)(1/LC_iR_i)}$$

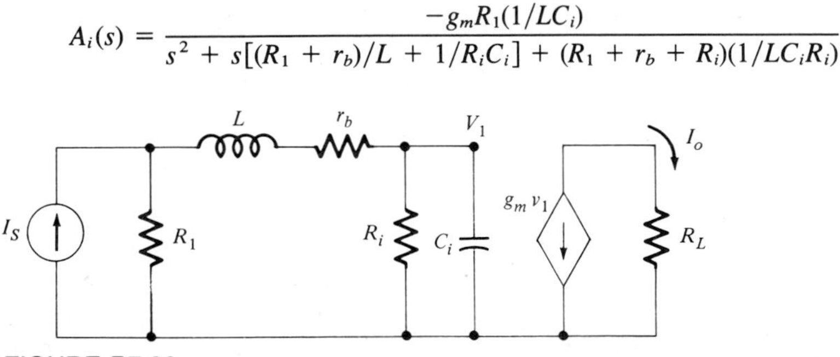

FIGURE P7.30

Given $R_L = 100\ \Omega$, $R_1 = 90\ \Omega$, $L = 2.13\ \mu H$, and the transistor parameters $f_T = 300$ MHz, $C_c = 4$ pF, $\beta_0 = 50$, $r_e = 5\ \Omega$, and $r_b = 50\ \Omega$,

(*a*) Determine $A_i(0)$ and ω_{3dB}. *Use normalization.*

(*b*) For $L = 0$ determine $A_i(0)$ and ω_{3dB}. Note that this is another example of broad-banding.

7.31 For the circuit shown in Fig. P7.31, assume an *ideal op amp*. Design the circuit for a low-frequency voltage gain of $A_V = 50$ and $\omega_{3dB} = 10^4$ rad/s.

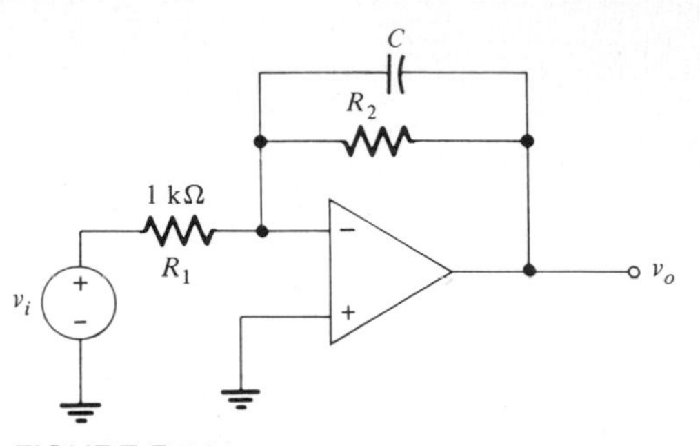

FIGURE P7.31

7.32 Design a single-stage CE transistor amplifier using a 2N2222 transistor (see Appendix E) to meet the following specifications: $A_i = 20, f_l \le 500$ Hz, $f_h \ge 2$ MHz, $R_L = 100\ \Omega$, $R_S = 1\ k\Omega$, $V_{CC} = 5$ V. Show your circuit. Use the data sheets given.

7.33 Design a single-stage CS amplifier using an enhancement-type NMOS (parameters given in Appendix E) for the following specifications: $A_V = 10, f_l \le 500$ Hz, $f_h \ge 5$ MHz, $R_L = 2\ k\Omega$, $R_S = 50\ \Omega$, $V_{DD} = 10$ V. Show your circuit. Use the data sheets given.

7.34 For the emitter-follower circuit shown in Fig. P7.26, show that the upper 3-dB frequency is given by

$$\omega_{3dB} = \frac{\omega_\beta}{D}\left[1 + \frac{\beta_0(R_L + r_e)}{R_S + r_b}\right]$$

where $D = 1 + R_L C_c \omega_T$. Determine the numerical value of ω_{3dB} if the BJT parameters are given as in problem 7.30; assume $R_B (= R_{b1} \parallel R_{b2})$ to be very large.

CHAPTER

8

Frequency Response Multiple-Transistor Amplifiers

In Chap. 7, we considered the frequency response of single-transistor amplifiers. As mentioned in Chap. 5, amplifiers seldom use a single transistor, and hence we must consider analysis and design of amplifiers utilizing more than one transistor. The concepts developed in the previous chapter, such as the unilateral (Miller) approximation, dominant poles and zeros, etc., will be extensively used in the following sections.

We start with the CE configuration for BJTs and the CS configuration for FETs, as these configurations are usually used as the main amplifying blocks. The cascode and the differential pair are considered next. The frequency performance of the op amp is determined from the differential mode as discussed in Sec. 8.4. The chapter concludes with multistage-transistor-amplifier design using BJT and FET devices and for specified magnitude response characteristics.

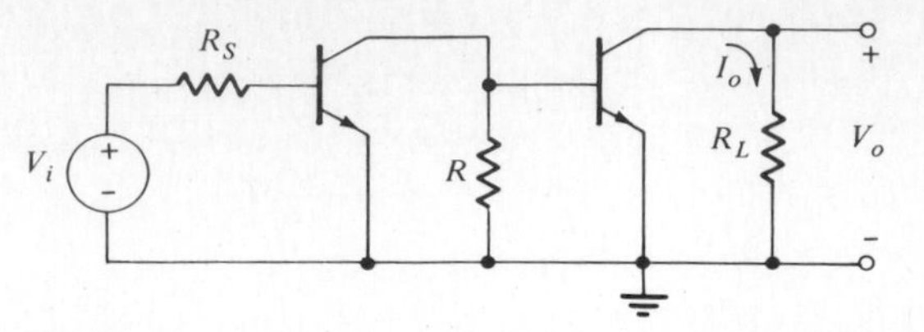

FIGURE 8.1 Signal circuit for a two-stage CE transistor amplifier.

8.1 *RC*-COUPLED CE-CE CASCADE

Consider a two-transistor CE-CE cascaded amplifier shown, exclusive of biasing circuitry, in Fig. 8.1. The hybrid-π model and the unilateral approximation for the hybrid-π model are shown in Fig. 8.2*a* and *b*, respectively. Note that the transistors are assumed to be identical. The unilateral approximate model for the second stage is simple, since the load for the second stage is purely resistive. As discussed in Chap. 7, from (7.61) we have

$$
\begin{aligned}
C_{i2} &= C_\pi + C_{M2} = \left(\frac{1}{r_e\omega_T} - C_c\right) + C_c(1 + g_m R_{L2}) \\
&= \frac{1 + R_{L2}C_c\omega_T}{r_e\omega_T} = \frac{D_2}{r_e\omega_T}
\end{aligned}
\tag{8.1}
$$

where $D_2 = 1 + R_{L2}C_c\omega_T$.

For the first stage, however, the load is not a pure resistance but an *RC* impedance as seen in Fig. 8.2. Hence, in order to *simplify analysis and design,* we consider

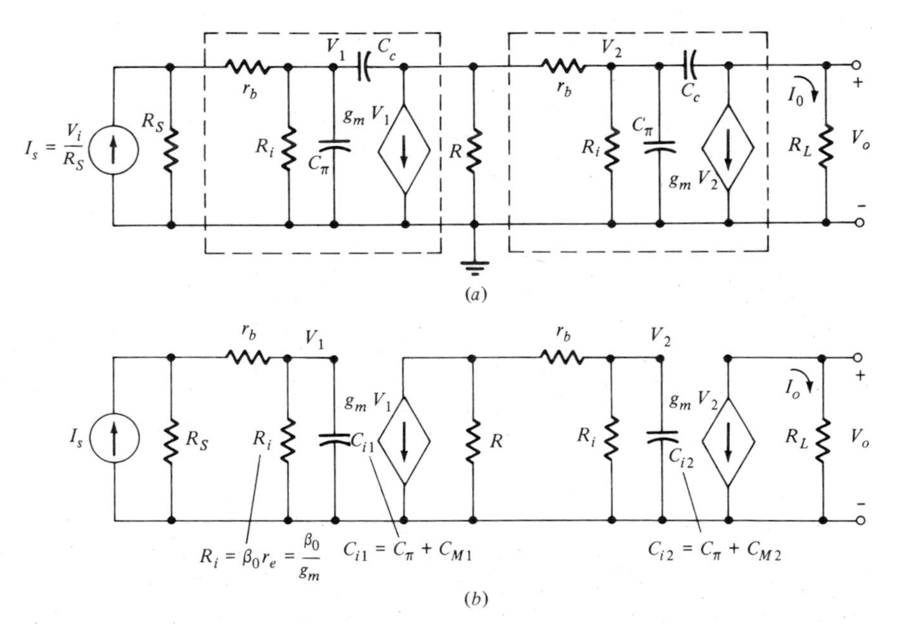

FIGURE 8.2 (*a*) Hybrid-π model. (*b*) Unilateral model, incorporating Miller effect.

only $Z_L(0)$, i.e., the dc load (resistance) seen by the first stage as pointed out and emphasized in the previous chapter. Bandwidth calculations based on $Z_L(0)$ will be slightly pessimistic since at the 3-dB frequency Z_L will be smaller than $Z_L(0)$. Thus, $R_{L1} = R \parallel (r_b + R_i)$ for the first stage. This approximation leads to the value of C_{i1} given by $C_{i1} = C_\pi + C_{M1}$, and as in (8.1) we have

$$C_{i1} = \frac{1 + R_{L1} C_c \omega_T}{r_e \omega_T} = \frac{D_1}{r_e \omega_T} \tag{8.2}$$

where $D_1 = 1 + R_{L1} C_c \omega_T$. Note that even for identical transistors the values of C_{i1} and C_{i2} are different, as they depend on the value of load resistance for each stage, which can be different, thus contributing different Miller-effect capacitances. From the circuit in Fig. 8.2*b*, we readily obtained the following relations:

$$I_o = -g_m V_2 \tag{8.3a}$$

$$V_2 = -\frac{\beta_0 g_m R}{sR_i C_{i2}(r_b + R) + R + R_i + r_b} V_1 \tag{8.3b}$$

$$V_1 = \frac{R_S R_i}{sR_i C_{i1}(r_b + R_S) + R_S + R_i + r_b} I_s \tag{8.3c}$$

and

$$A_I = \frac{I_o}{I_s} = \frac{I_o}{V_2}\frac{V_2}{V_1}\frac{V_1}{I_s} \tag{8.4a}$$

From (8.3) and some simplification we obtain

$$A_I = \frac{A_I(0) p_1 p_2}{(s + p_1)(s + p_2)} \tag{8.4b}$$

where

$$A_I(0) = \beta_0^2 \frac{R_S}{R_S + r_b + R_i} \frac{R}{R + r_b + R_i} \tag{8.5a}$$

$$p_1 = \frac{1}{R_i C_{i1}} \frac{R_S + r_b + R_i}{R_S + r_b} = \frac{\omega_T}{\beta_0 D_1} \frac{R_S + r_b + R_i}{R_S + r_b} \tag{8.5b}$$

$$p_2 = \frac{1}{R_i C_{i2}} \frac{R + r_b + R_i}{R + r_b} = \frac{\omega_T}{\beta_0 D_2} \frac{R + r_b + R_i}{R + r_b} \tag{8.5c}$$

An example will illustrate the gain and bandwidth calculations from (8.5).

EXAMPLE 8.1

Calculate the gain and the bandwidth of a two-stage cascaded CE-CE configuration shown in Fig. 8.1*a*. The BJT parameters are given as

$$f_T = 300 \text{ MHz} \qquad \beta_0 = 50 \qquad r_b = 50\ \Omega$$

$$C_c = 4 \text{ pF} \qquad r_e = \frac{1}{g_m} = 5\ \Omega$$

Let $R_S = R_L = R = 100\ \Omega$. From (8.5*a*) the current gain is given by

$$A_I(0) = 50^2\left(\frac{100}{100 + 50 + 250}\right)\left(\frac{100}{100 + 50 + 250}\right) = 156.25$$

The voltage gain is the same as the current gain, since $R_L = R_S$ in this example:

To find p_2 and p_1, we have to first determine D_2 and D_1. Note that we start from the last stage since its load resistance is known.

$$D_2 = 1 + R_{L2}C_c\omega_T = 1 + 100(4 \times 10^{-12})(2\pi \times 3 \times 10^8) = 1.75$$

$$D_1 = 1 + R_{L1}C_c\omega_T = 1 + [R \parallel (r_b + R_i)]C_c\omega_T = 1.57$$

Having determined the values of D_2 and D_1, we determine p_2 and p_1 from (8.5*c*) and (8.5*b*), namely,

$$p_2 = 2\pi\frac{3 \times 10^8}{50(1.75)}\left(\frac{400}{100}\right) = 2\pi(1.37 \times 10^7) \text{ rad/s}$$

$$p_1 = 2\pi\frac{3 \times 10^8}{50(1.57)}\left(\frac{400}{100}\right) = 2\pi(1.52 \times 10^7) \text{ rad/s}$$

The 3-dB bandwidth as calculated directly from (8.4*b*) by setting $|A_I(j\omega)| = A_I(0)/\sqrt{2}$, with the values of p_1 and p_2 given in above (*or* from Fig. 8.9, discussed later in Sec. 8.4), is $\omega_{3dB} = 2\pi(9.3 \times 10^6)$ rad/s or $f_{3dB} \simeq 9.3$ MHz.

8.2 *RC*-COUPLED CS-CS CASCADE

Analysis of a two-stage cascaded FET amplifier is similar. In fact, it is *much simpler* than for BJT because of the simpler circuit model for FET as compared to that of BJT. Consider a two-stage cascaded FET amplifier shown in Fig. 8.3*a* with its ac signal model in Fig. 8.3*b*. The π model and the unilateral approximation incorporating the Miller effect is shown in Fig. 8.3*c* and *d*, respectively. Note that the small value of C_{ds} is ignored and not shown in the model, since $C_{ds} \ll C_{gs}$. The inclusion of C_{ds}, if desired, is straightforward. Note further that for an FET amplifier the input is a voltage source and the input capacitance of the first stage yields a nondominant pole even though it includes the Miller-effect capacitance C_{M1}. Thus, basically we have a one-dominant-pole voltage gain function for a two-stage amplifier as the following example will illustrate.

The voltage gain function of the circuit in Fig. 8.3*c* is determined as follows:

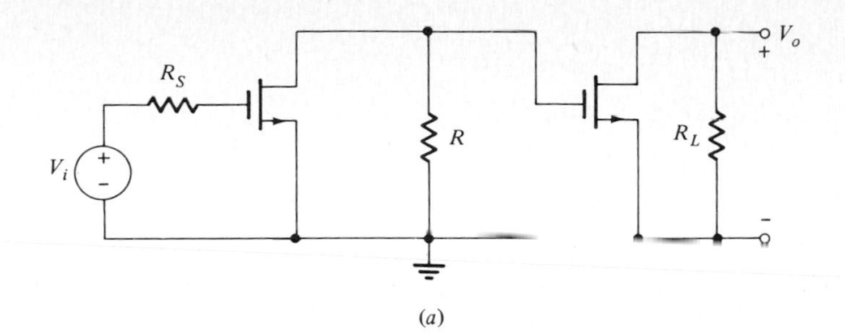

(a)

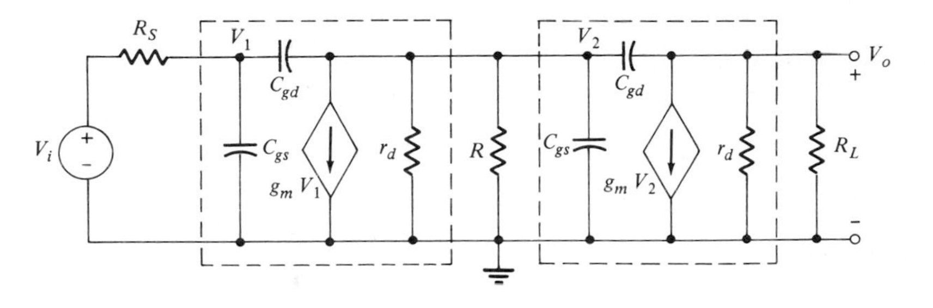

(b)

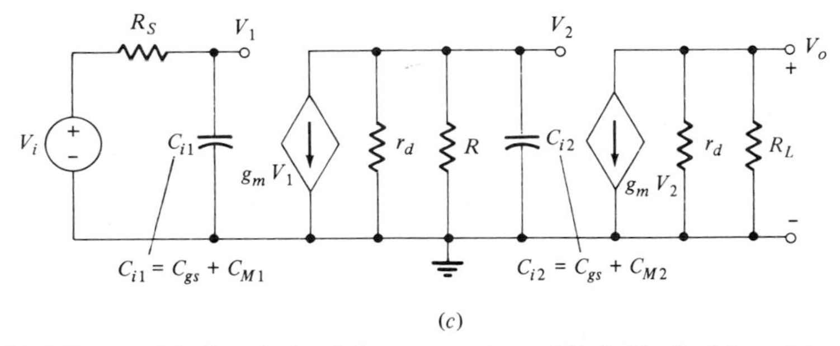

(c)

FIGURE 8.3 (a) Signal circuit for a two-stage CS, field-effect transistor amplifier. (b) Hybrid-π model. (c) Unilateral model.

$$V_o = -(g_m R_{L2})V_2 \qquad \text{where } R_{L2} = r_d \parallel R_L \tag{8.6a}$$

$$V_2 = \frac{-(g_m R_{L1})V_1}{1 + sR_{L1}C_{i2}} \qquad \text{where } R_{L1} = r_d \parallel R \tag{8.6b}$$

$$V_1 = \frac{V_i}{1 + sR_S C_{i1}} \tag{8.6c}$$

From (8.6) we can write the overall voltage gain function.

$$A_v = \frac{V_o}{V_i} = \frac{V_o}{V_2}\frac{V_2}{V_1}\frac{V_1}{V_i} = \frac{A_v(0)p_1p_2}{(s + p_1)(s + p_2)} \tag{8.7}$$

where

$$A_v(0) = (g_m R_{L1})(g_m R_{L2}) = g_m^2 R_{L1} R_{L2} \tag{8.8a}$$

$$p_1 = \frac{1}{R_S C_{i1}} \quad \text{and} \quad p_2 = \frac{1}{R_{L1} C_{i2}} \tag{8.8b}$$

EXAMPLE 8.2

Calculate the gain and the bandwidth of the two-stage FET amplifiers shown in Fig. 8.3*a*. The FET parameters are given as

$$g_m = 2 \times 10^{-3} \text{ mho} \qquad r_d = 20 \text{ k}\Omega \qquad C_{gs} = 10 \text{ pF} \qquad C_{gd} = 2 \text{ pF}$$

$$C_{ds} = 1 \text{ pF}$$

Let $R_S = 100\ \Omega$ and $R = R_L = 2\ k\Omega$. From (8.6) we have $R_{L2} = 20 \times 10^3 \parallel 2 \times 10^3 = R_{L1} = 1.82\ \text{k}\Omega$. Hence from (8.8*a*) we have

$$A_v(0) = g_m^2 R_{L1} R_{L2} = 4 \times 10^{-6}(1.82 \times 10^3)(1.82 \times 10^3) = 13.2$$

$$C_{i2} = C_{gs} + C_{M2} = C_{gs} + C_{gd}(1 + g_m R_{L1}) = 10 + 2(1 + 3.6) = 19.2 \text{ pF}$$

$$p_2 = \frac{1}{R_{L1} C_{i2}} = \frac{1}{(1.82 \times 10^3)(19.2 \times 10^{-12})} = 2.87 \times 10^7 \text{ rad/s}$$

$$p_1 = \frac{1}{R_S C_{i1}} = \frac{1}{100(19.2 \times 10^{-12})} = 5.2 \times 10^8 \text{ rad/s}$$

Note that $p_1 \gg p_2$ and the first pole is nondominant due to the small value of the source impedance. The 3-dB bandwidth of the amplifier is determined by the dominant pole, namely,

$$f_{3\text{dB}} = \frac{2.87 \times 10^7}{2\pi} = 4.57 \text{ MHz}$$

EXERCISE 8.1

For the circuit shown in Fig. E8.1, the device parameters are

JFET: $g_m = 2 \times 10^{-3}$ mho $\quad r_d = 10\ \text{k}\Omega \quad C_{gs} = 20$ pF $\quad C_{gd} = 2$ pF

BJT: $\beta_0 = 100 \quad r_e = 10\ \Omega \quad C_c = 5$ pF $\quad \omega_T = 10^9$ rad/s $\quad r_b \simeq 0$

Determine (*a*) the midband voltage gain, (*b*) the lower 3-dB frequency, and (*c*) the upper 3-dB frequency.

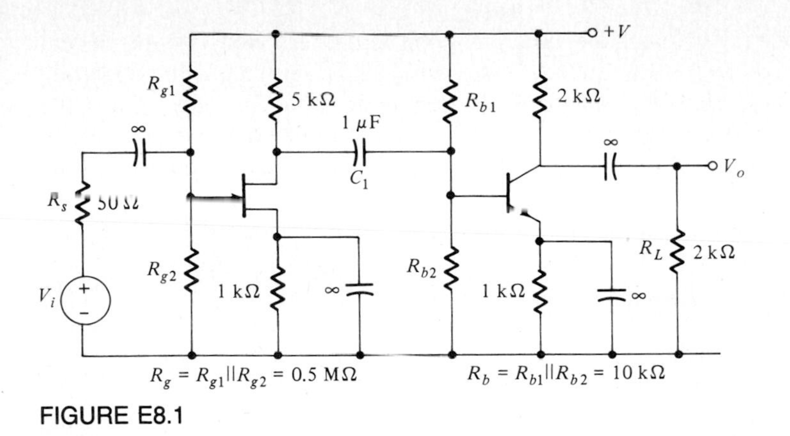

FIGURE E8.1

Ans $A_v = 142$, $f_l = 36.8$ Hz, $f_h = 0.37$ MHz

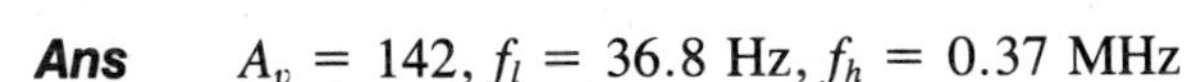

8.3 THE CASCODE (CE-CB) CIRCUIT

The midband properties of the cascode circuit were discussed in Sec. 5.3. We shall now consider its frequency response. A cascode circuit, exclusive of the dc biasing circuitry, is shown in Fig. 8.4*a*. Figure 8.4*b* shows the high-frequency equivalent

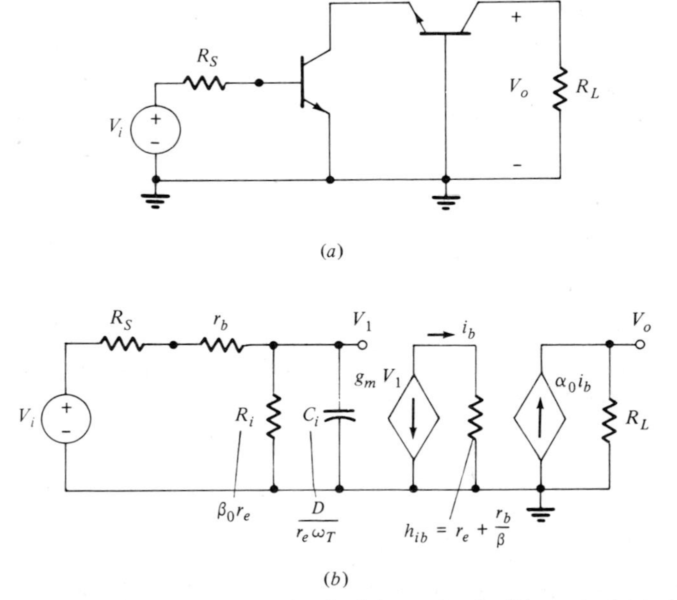

FIGURE 8.4 Cascode circuit: (*a*) ac circuit, (*b*) equivalent circuit.

circuit of the cascode (again using the unilateral approximation). Note that the current gain of a CB stage has its 3-dB frequency at ω_α, which is larger than ω_T and certainly much larger than ω_β, the approximate 3-dB frequency of a CE stage. Thus the resulting pole of the CB stage is larger by an order of magnitude than the pole due to the CE stage. Hence the nondominant pole due to CB is ignored at the outset without any significant error.

The midband voltage gain of the circuit can be readily obtained (5.15) as

$$A_v(0) = \frac{V_o}{V_i} \simeq -\beta_0 \frac{R_L}{R_S + B_0 r_e + r_b} \qquad \textbf{(8.9a)}$$

The 3-dB bandwidth of the circuit is given by the single pole of the circuit, namely

$$\omega_{3\text{dB}} = p_o = \frac{1}{C_i R_i \parallel (R_S + r_b)} = \frac{\omega_\beta}{D} \frac{R_i + R_S + r_b}{R_S + r_b} \qquad \textbf{(8.9b)}$$

The complete voltage gain function is given by

$$A_v(s) = \frac{A_v(0) p_o}{s + p_o} \qquad \textbf{(8.9c)}$$

To illustrate the use of this simplified approach consider the following example.

EXAMPLE 8.3

Determine the gain and the bandwidth of the cascode circuit shown in Fig. 8.4. The transistor parameters are given as

$$g_m = 0.4 \text{ mho} \qquad \beta_0 = 100$$

$$\omega_T = 4 \times 10^9 \text{ rad/s} \qquad C_c = 5 \text{ pF}$$

$$R_S = R_L = 200\ \Omega \qquad r_b = 20\ \Omega$$

From the above data we have $r_e = 2.5\ \Omega$ and $R_i = \beta_0 r_e = 250\ \Omega$. Also

$$D = 1 + \left(r_e + \frac{r_b}{\beta_0}\right) C_c \omega_T = 1 + 2.7(5 \times 10^{-12})(4 \times 10^9) \simeq 1.05$$

Note the low value of the D factor, which is due to the small value of the input resistance of the CB stage (which is the load resistance for the CE stage). As a result, the bandwidth of the CE stage is quite broad. From (8.9*a*) and (8.9*b*) we have

$$A_v = -100\left(\frac{200}{200 + 250 + 20}\right) = -43$$

$$\omega_{3\text{dB}} = \frac{4 \times 10^7}{1.05}\left(\frac{250 + 200 + 20}{220}\right) \simeq 8.14 \times 10^7 \text{ rad/s}$$

or

$$f_{3\text{dB}} \simeq 12.9 \text{ MHz}$$

A complete analysis of this circuit using the hybrid-π models for the transistors will yield three zeros and four poles. There is only one dominant pole and the computer aided analysis result* is $f_{3\text{dB}} = 12.9$ MHz. The results in this case are remarkably identical.

*See P. Gray and C. Searle, *Electronic Principles,* Wiley, New York, 1969, p. 528.

The reader should also notice that if a single-stage CE transistor is used with $R_L = 200\,\Omega$, the D-factor of the CE stage will be 5, and hence the bandwidth will be smaller by a factor of 5 than the cascode in Example 8.3. Note that in the cascode, the CE stage provides the current gain, and the CB stage provides the voltage gain, each to quite high frequencies for the configuration.

8.4 THE EMITTER-COUPLED (CC-CB) CIRCUIT

The emitter-coupled circuit was discussed in Secs. 5.2 and 5.7 in conjunction with op amp. Since this circuit is the basic element of an op amp, we shall analyze the frequency response of the circuit with respect to its differential-mode and common-mode half-circuits.

The emitter-coupled circuit is shown in Fig. 8.5. The differential-mode and the common-mode half-circuits with their equivalent circuits are shown in Fig. 8.6*a* and *b*, respectively. Notice that since R_E is very large in order to simulate a current source, we have to include the parasitic capacitance C_e across R_E. C_e is not a deliberately added emitter bypass capacitor, but it is nonetheless there and cannot be ignored since it is across a very large resistance.

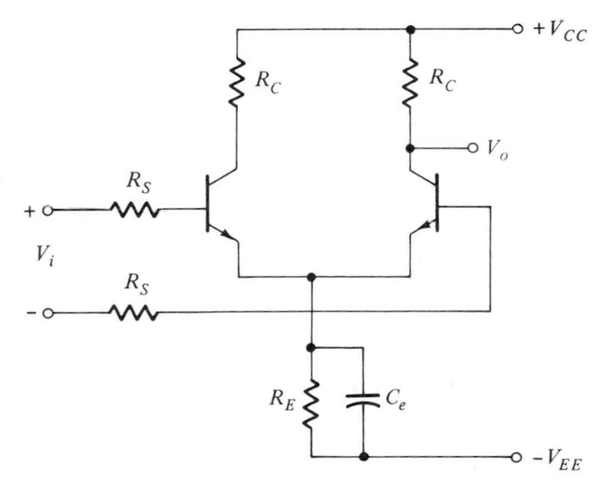

FIGURE 8.5 Differential amplifier circuit.

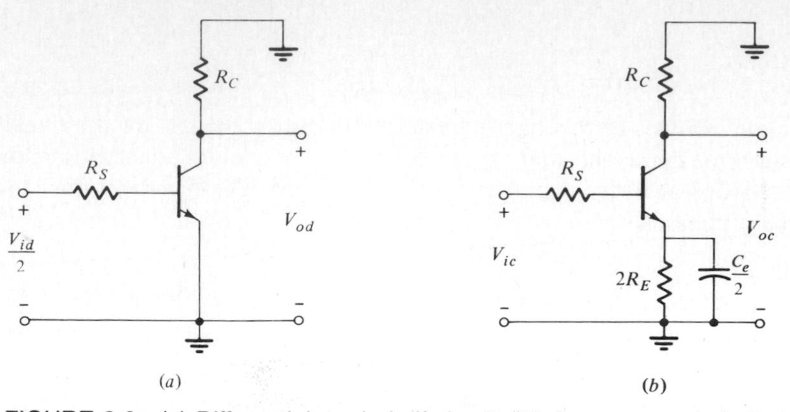

FIGURE 8.6 (*a*) Differential-mode half-circuit. (*b*) Common-mode half-circuit.

A. Differential Mode

The differential-mode half-circuit shown in Fig. 8.6*a* is simply a CE stage. Since the gain and bandwidth expressions of a CE stage are already determined in (7.68) and (7.69), we shall rewrite the results for convenience:

$$A_d = \frac{V_{od}}{V_{id}} = \frac{A_d(0)p_d}{s + p_d} \tag{8.10a}$$

where

$$A_d(0) = \frac{-\beta_o R_C/2}{R_S + r_b + \beta_0 r_e} \tag{8.10b}$$

$$p_d = \frac{\omega_\beta}{1 + R_C C_c \omega_T}\left(1 + \frac{\beta_0 r_e}{r_b + R_S}\right) \tag{8.10c}$$

Note that if R_C is very large, the hybrid-π model should be used for better accuracy. The Bode plot of (8.10*a*) is shown in Fig. 8.7*a*.

B. Common Mode

Since R_E is very large, the effect of the stray capacitance C_e cannot be ignored as the impedance due to C_e at the signal frequencies will be smaller than the parallel resistance R_E. The effect of this capacitance is to produce a zero at $1/R_E C_e$. Thus for a very high value of R_E we have

$$A_c = \frac{V_{oc}}{V_{ic}} \simeq -\frac{R_C}{Z_E} = -\frac{R_C}{2R_E}(1 + sR_E C_e) \tag{8.11}$$

Note that $A_c(0)$ is the same as in (5.45). An analysis of the circuit in Fig. 8.6*b* will show that there is a pole p_c in the transfer function. The value of p_c is unimportant, but it is much larger than p_d and the zero where $z_c = 1/R_E C_e$. [The expression for p_c is given in (7.32*b*), and note that in this case C_e is on the order of picofarads.] The Bode plot of A_c is shown in Fig. 8.7*b*.

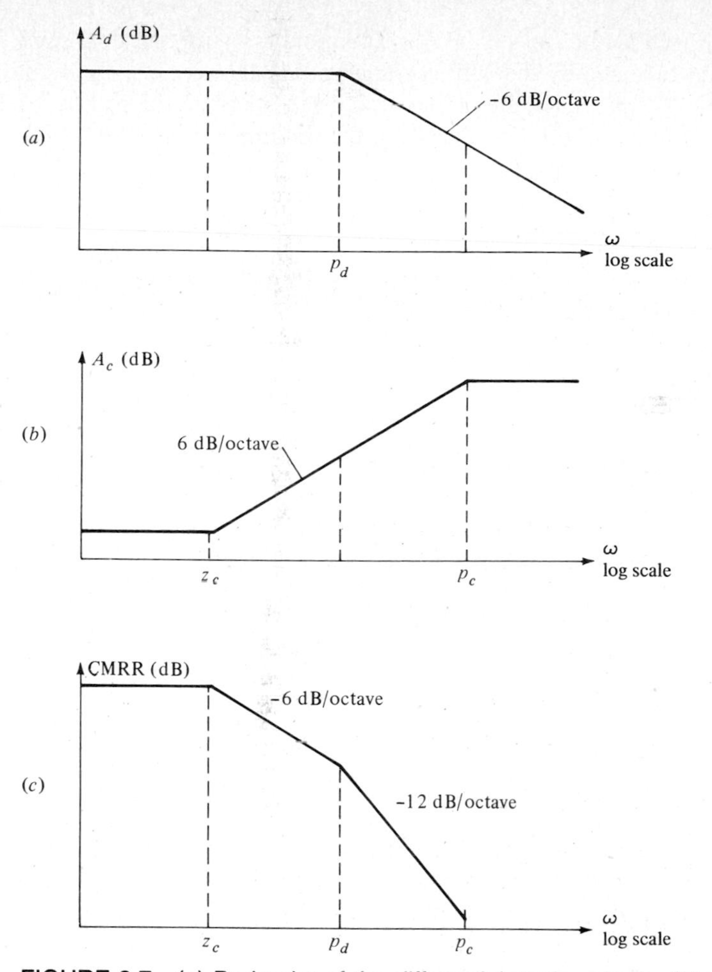

FIGURE 8.7 (*a*) Bode plot of the differential-mode gain A_d. (*b*) Bode plot of the common-mode gain A_c. (*c*) Bode plot of the common-mode rejection ratio CMRR.

The common-mode rejection ratio $|A_d/A_c|$ as a function of frequency is shown in the asymptotic plot in Fig. 8.7*c*. *Note that CMRR falls off as a function of frequency as shown in Fig. 8.7c.*

The 3-dB frequency of an op amp, such as the 741 in Fig. 5.25, is determined by a deliberately used internal compensating capacitance. This is because of stability reasons, as an op amp is seldom operated in an open-loop configuration. Chapter 10 will consider these matters.

EXAMPLE 8.4

Determine the 3-dB bandwidth of the 741 op amp shown in Fig. 5.25*a*. In Fig. 5.25*a*, the 30 pF from the base of Q_{16} to the collector of Q_{17} determines the

3-dB bandwidth (see also C_1 in Fig. B.1). This pole is *by far* the dominant pole, since when reflected by the Miller effect as in Example 5.2, $C_i \simeq 900$ (30 pF) $\simeq 27 \times 10^{-9}$ F. Since the effective resistance in parallel to this Miller capacitance was found to be $r_{L4} \simeq 1.4$ MΩ, the dominant pole, due to the compensation capacitor, p_D is at

$$p_D = \frac{1}{(1.4 \times 10^6)(27 \times 10^{-9})} \simeq 26.5 \text{ rad/s}$$

Hence

$$f_{3\text{dB}} \simeq \frac{26.5}{6.28} \simeq 4.2 \text{ Hz} \tag{8.12}$$

which is astoundingly low, yet we shall see in Chap. 10 that these op amps are routinely used in closed-loop feedback arrangement to amplify signals at frequencies up to several hundred kilohertz.

Since the gain of the amplifier is around 200,000, as determined in Example 5.10, the gain-bandwidth product of a 741 is thus approximately 1 MHz.

Notice that the compensating capacitor for a Bi-FET (LF351) in Fig. 5.27*a* is 10 pF, hence for the same resistance value of 1 MΩ and the same gain and thus the same Miller effect, the 3-dB frequency is about 3 to 4 times that of a 741 op amp.

EXERCISE 8.2

For the JFET differential pair shown in Fig. E8.2, determine (*a*) the differential-mode gain and its 3-dB frequency and (*b*) the common-mode rejection ratio. The device parameters are $g_m = 4 \times 10^{-3}$mho, $r_d = 50$ kΩ, $C_{gs} = 20$ pF, and $C_{ds} = C_{gd} = 3$ pF.

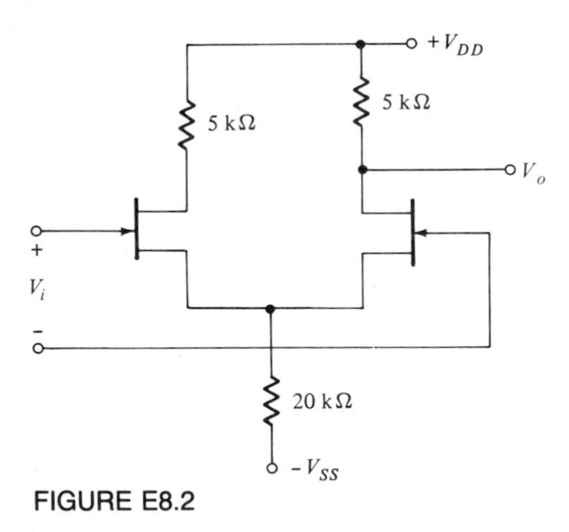

FIGURE E8.2

Ans (*a*) $A_d = 10, f_{3\text{dB}} = 5.3$ MHz; (*b*) CMRR $= 80(38$ dB)

8.5 DARLINGTON (CC-CE) CONFIGURATION

A configuration that is useful for wideband amplification is a cascade of CC-CE stages shown in Fig. 8.8. Biasing and coupling circuitry in this figure have been omitted. Since the output impedance of a CC stage is inductive in nature at high frequencies (problem 7.25), it interacts with the capacitive input impedance of the CE stage to provide complex poles and thus can lead to a broadbanding effect if the poles are properly placed.

In order to simplify the analysis and design, we neglect C_c of the CC stage. This omission is often permissible for wideband applications, since C_c is in shunt with $R_I + r_b$, which is a low resistance, and the resulting error is thus negligible in most cases. The high-frequency equivalent circuit for Fig. 8.8, with the above approximation, is shown in Fig. 8.9. The resistors R_{i1} and R_{i2} need not have the same value, thus providing some flexibility in the design. Even if the operating points are identical, the addition of a small external emitter resistance provides a degree of freedom, as was seen in the series-feedback stage. We therefore distinguish the parameters of each stage by appropriate subscripts.

We write the nodal equation for the circuit in the s domain and then determine the current gain function. After some manipulation and using the approximations $\beta_0 \gg 1$ and $\beta_0 r_e \gg r_b$, we obtain (note that we have used $r_{b1} = r_{b2} = r_b$ and $\omega_{T1} = \omega_{T2} = \omega_T$)

$$A_I = \frac{I_o}{I_s} = \frac{R_L}{(R_I + 2r_b)D}$$

$$\frac{s/\omega_T + 1}{\left(\dfrac{s^2}{\omega_T}\right) + \left(\dfrac{s}{\omega_T}\right)\left[\dfrac{R_I + \beta_0(r_b + r_{e1})}{\beta_0(R_I + 2r_b)} + \dfrac{R_I + \beta_0 r_{e2}}{\beta_0(R_I + 2r_b)D}\right] + \dfrac{r_{e2}}{D(R_I + 2r_b)}} \tag{8.13}$$

where $D = 1 + R_L C_c \omega_T$. Equation (8.13) can be rewritten for convenience in the normalized form.

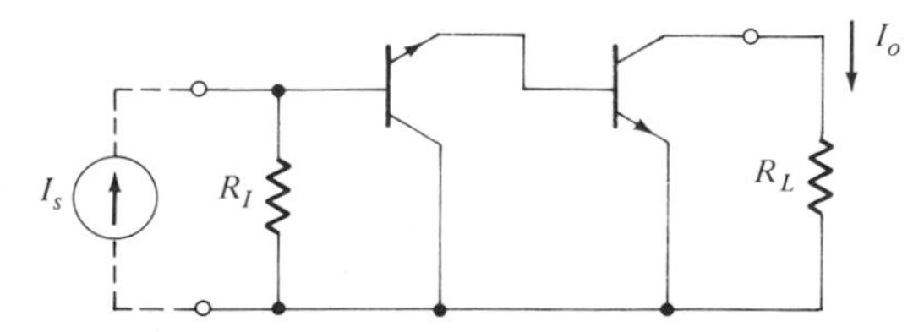

FIGURE 8.8 Amplifier in the CC-CE pair configuration.

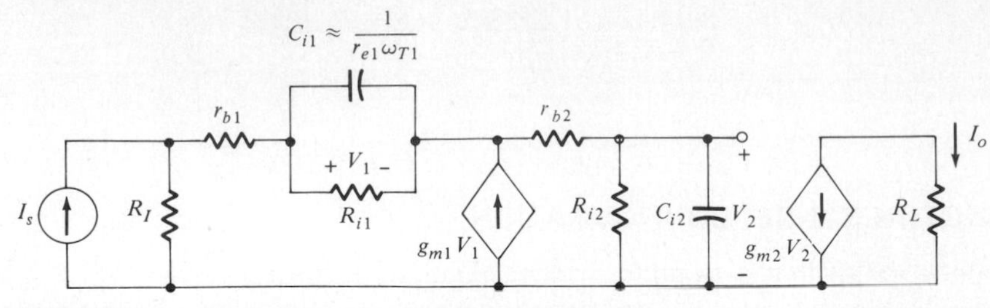

FIGURE 8.9 Equivalent circuit of the CC-CE pair.

$$A_I(s_n) = \frac{bA_I(0)(s_n + 1)}{s_n^2 + as_n + b} \simeq \frac{A_I(0)b}{s_n^2 + as_n + b} \tag{8.14a}$$

where

$$A_I(0) = \frac{R_I}{r_{e2}} \qquad s_n = \frac{s}{\omega_T} \tag{8.14b}$$

$$a = \frac{R_I + \beta_0(r_b + r_{e1})}{\beta_0(R_I + 2r_b)} + \frac{R_I + \beta_0 r_{e2}}{\beta_0(R_I + 2r_b)D} \tag{8.14c}$$

$$b = \frac{r_{e2}}{(R_I + 2r_b)D} \tag{8.14d}$$

Note that the zero is located at ω_T and is nondominant in most cases since $a, b, \ll 1$, as will be seen in the numerical example. The denominator poles will be complex if $a^2 < 4b$. An example will illustrate the gain-bandwidth capability of the CC-CE pair.

EXAMPLE 8.5

For the operating point and transistor parameters given in Example 8.1, we calculate the gain and the 3-dB bandwidth. Let $R_I = 500\ \Omega$ and $R_L = 100\ \Omega$. The current gain is given by (8.14*b*), that is,

$$A_I(0) = \frac{R_I}{r_{e2}} = \frac{500}{5} = 100$$

From (8.14*c*) and (8.14*d*) we determine the values of a and b

$$a = \frac{500 + 50(55)}{50(500 + 100)} + \frac{500 + 50(5)}{50(500 + 100)(1.7)} = 0.122$$

$$b = \frac{5}{(500 + 100)(1.7)} = 0.0049$$

Hence (8.14*a*) can be written as

$$A(s_n) \simeq \frac{0.49}{s_n^2 + 0.122s + 0.0049}$$

$$\simeq \frac{0.49}{(s_n + 0.061 + j0.034)(s_n + 0.061 - j0.034)}$$

Note that the zero is nondominant, since its value is much larger than the magnitude of the complex pole pair, i.e., $|p_1| = 0.07 \ll 1$, and can therefore be ignored for purposes of bandwidth calculation. Thus, the 3-dB bandwidth calculation from the above yields $(\omega_{3\text{dB}})_n = 0.055$; hence $\omega_{3\text{dB}} = 0.055\omega_T$ or $f_{3\text{dB}} =$ 16.5 MHz.

Note that *this bandwidth could not be realized by a cascade of resistive broadbanded stages*. A smaller value of R_l can achieve a two-pole, maximally flat magnitude response characteristic (see problem 8.10) with a wider bandwidth.

As a final comment on the circuit of Fig. 8.8, note that the current gain (8.14*b*) is dependent only on R_l and r_{e2}. If an external resistance R_{e2} is added such that $R_{e2} > r_{e2}$, then the current gain of the amplifier will be almost essentially the ratio of the two resistors and will be almost independent of the transistor parameters. This is a desirable feature, and the ratio feature is particularly attractive in the design of integrated circuits.

For convenience, and later reference, we have plotted in Fig. 8.10 the 3-dB bandwidth for a gain function with two poles, real or complex. For two real poles, if the smaller of the two poles is labeled p_1, the bandwidth is determined in terms of p_1 for any ratio $p_2/p_1 \leq 5$. For ratios larger than 5 the second pole is nondominant and $\omega_{3\text{dB}} = p_1$.

For a complex pole pair, if the magnitude is p_1 and the angle with the negative real axis is ψ, the 3-dB bandwidth can be determined for any $\psi \leq 45°$. Note that $\psi = 45°$ corresponds to a maximally flat magnitude function (discussed in Sec. 8.11), and for $\psi > 45°$ the magnitude response will exhibit peaking and is of no interest. Also $\psi = 30°$ corresponds to the linear-phase case (discussed in Sec. 12.3).

In Example 8.4 the normalized 3-dB frequency for the complex pole-pair corresponding to $\psi = 31°$ from Fig. 8.10 is $0.79p_1$; hence $\omega_{3\text{dB}} = 0.79(0.07)\omega_T = 2\pi(16.5$ MHz).

8.6 GAIN AND BANDWIDTH CALCULATIONS FOR AN *RC*-COUPLED FET MULTISTAGE AMPLIFIER

A three-stage, *RC*-coupled, FET amplifier is shown in Fig. 8.11*a*. Its signal circuit, which assumes a short-circuit for the bypass and coupling capacitors, is shown in Fig. 8.11*b*. The source and load terminations are also given as shown. Note that the *n*-channel FET devices are biased as described in Sec. 3.9, namely, the biasing point is $I_D = 2.0$ mA and $V_{DS} = 10$ V. We wish to calculate the midband voltage gain, and

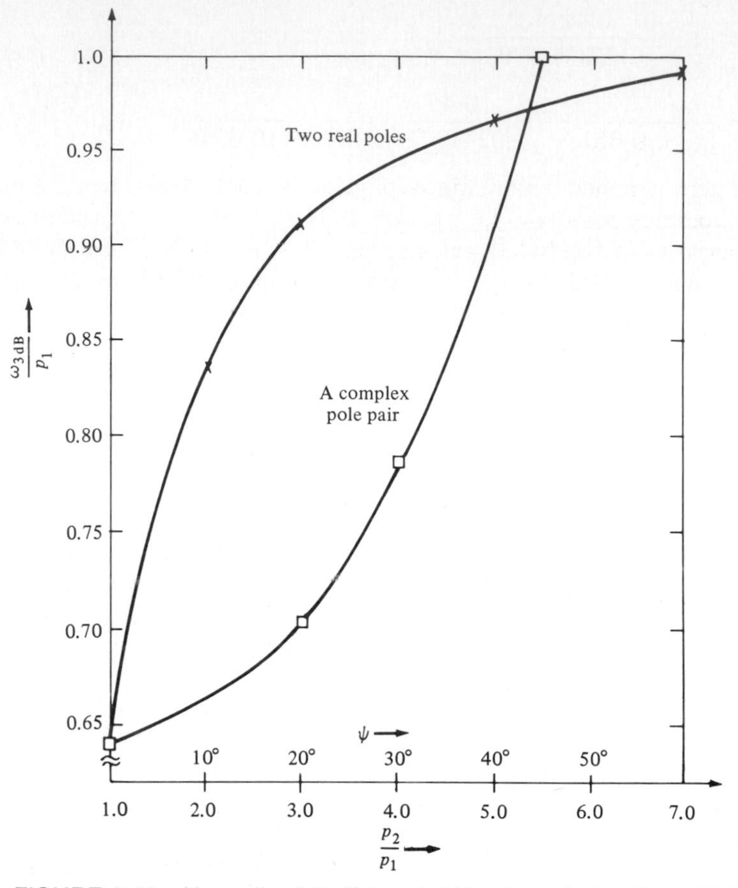

FIGURE 8.10 Normalized 3-dB bandwidth of a gain function with two real poles $|p_2| > |p_1|$ and with a complex pole pair having an angle ψ with the negative σ axis.

the upper and lower 3-dB cutoff frequencies of the amplifier. The FET parameters at the operating point are

$$g_m = 2 \times 10^{-3} \text{ mho} \qquad C_{gd} = 2 \text{ pF} \qquad C_{gs} = 10 \text{ pF}$$
$$r_d = 20 \text{ k}\Omega \qquad C_{ds} = 1 \text{ pF}$$

A. High-Frequency Response

For midband and high-frequency calculation the effect of bypass and coupling capacitors can be ignored since the capacitors are essentially short-circuited at mid-high frequencies (e.g., for this circuit at $f \simeq 10$ kHz, $1/\omega C \leq 2\ \Omega$). The midband and high-frequency equivalent circuit of the amplifier is shown in Fig. 8.11*c*. We can simplify the circuit further by using Miller's theorem as discussed previously. The simplified unilateral circuit is shown in Fig. 8.11*d*.

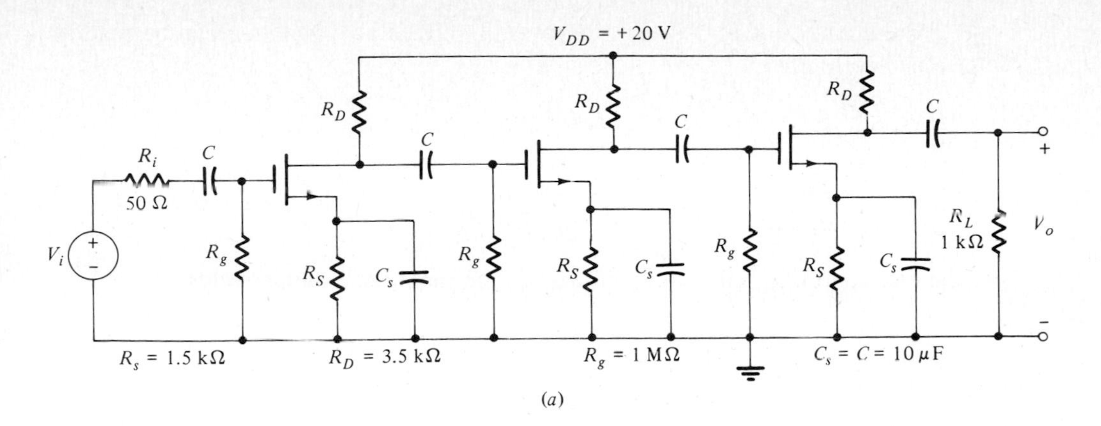

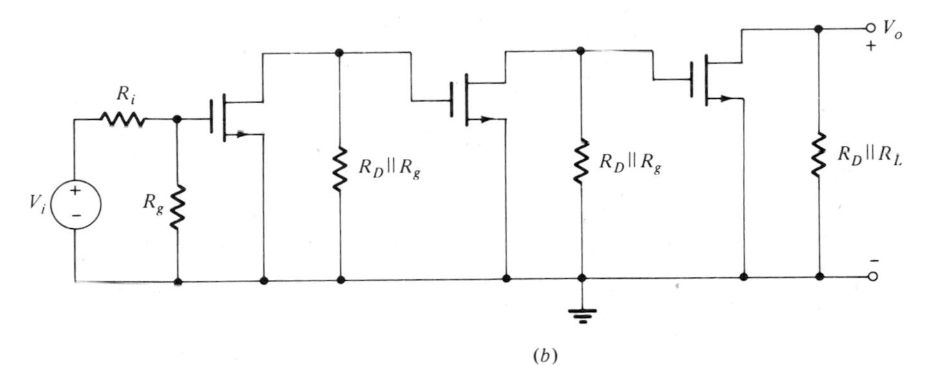

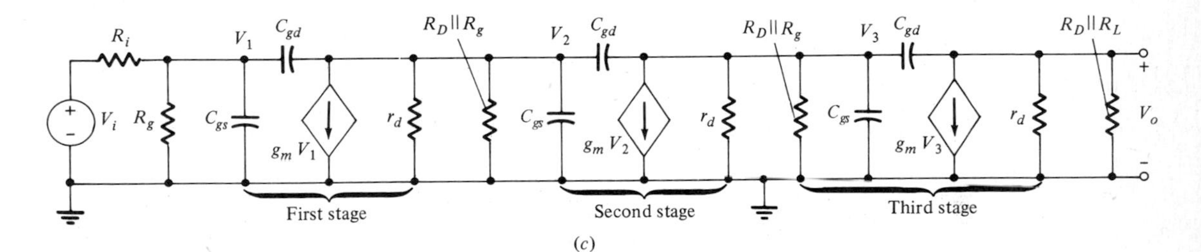

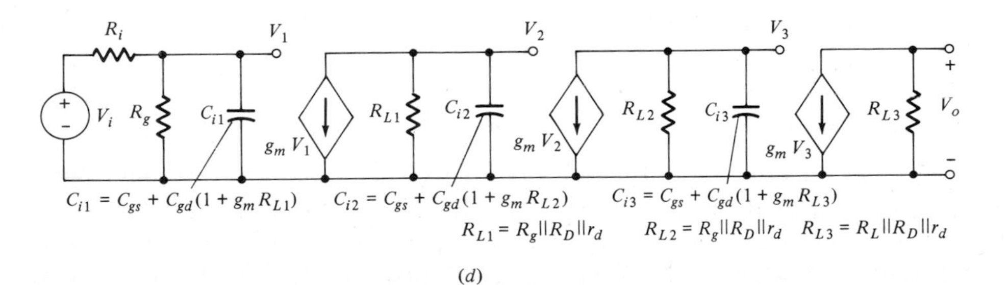

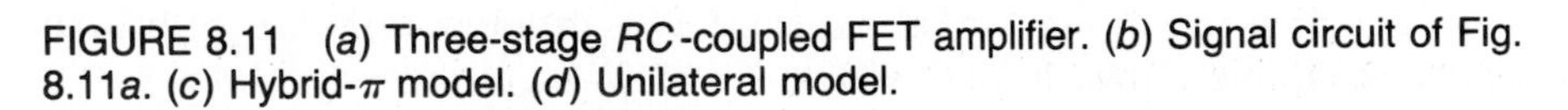

FIGURE 8.11 (*a*) Three-stage *RC*-coupled FET amplifier. (*b*) Signal circuit of Fig. 8.11*a*. (*c*) Hybrid-π model. (*d*) Unilateral model.

For the circuit of Fig. 8.11*d*, the overall voltage gain function can be readily determined as follows:

$$\frac{V_o}{V_3} = -g_m R_3 \qquad \frac{V_2}{V_1} = \frac{-g_m R_{L1}}{1 + sC_{i2}R_{L1}}$$
$$\frac{V_3}{V_2} = \frac{-g_m R_{L2}}{1 + sC_{i3}R_{L2}} \qquad \frac{V_1}{V_i} = \frac{R_g/(R_g \parallel R_i)}{1 + sC_{i1}(R_g \parallel R_i)} \tag{8.15}$$

For the circuit in Fig. 8.11*a*, substitution of the numerical values yields

$$\frac{V_o}{V_3} = -2 \times 10^{-3}(10^3 \parallel 3.5 \times 10^3) = -1.50$$

$$\frac{V_3}{V_2} = \frac{-5.97}{1 + s[15 \times 10^{-12}(2.98 \times 10^3)]} = \frac{-5.97}{1 + s/(2.27 \times 10^7)}$$

$$\frac{V_2}{V_1} = \frac{1}{1 + s[23.9 \times 10^{-12}(2.98 \times 10^3)]} = \frac{-5.97}{1 + s/(1.40 \times 10^7)}$$

$$\frac{V_1}{V_i} = \frac{1}{1 + s(23.9 \times 10^{-12})(50)} \frac{1}{1 + s/(8.37 \times 10^8)}$$

The overall midband voltage gain $A_v(0) = -(1.50)(5.97)^2 = -53$. Hence, the approximate overall gain function can be written as

$$\frac{V_o}{V_i} \simeq \frac{-53}{[1 + s/(2.27 \times 10^7)][1 + s/(1.40 \times 10^7)]}$$

Note that for a three-stage FET amplifier we have two dominant poles. The third pole due to the input circuit of the first stage can only become significant if R_i has a value similar to that of R_D. From the above, the 3-dB bandwidth can be readily calculated directly or using Fig. 8.10, with $p_1 = 1.40 \times 10^7$ and $p_2/p_1 \simeq 1.62$; we read $\omega_{3dB} = 0.79p_1 = 1.10 \times 10^7$ rad/s. Hence

$$f_{3dB} = \frac{\omega_{3dB}}{2\pi} = \frac{1.10 \times 10^7}{2\pi} \simeq 1.75 \text{ MHz}$$

For more than two dominant poles in the gain function, an approximate method of calculating ω_{3dB} is given in Sec. 8.7B.

B. Low-Frequency Cutoff

The approximate low-frequency cutoff point of the FET amplifier is readily determined using the method of Sec. 7.5. Since the parameters of the first and second stages are the same as those in Example 7.5, the low-frequency cutoff points of the first and second stages are $f_{l1} = f_{l2} = 42$ Hz [which is determined mainly by (7.43), the source bypass circuit]. For the third stage, since R_L is different from R_g, the pole due to the drain coupling circuit from (7.42) is

$$p\Big|_{\substack{\text{drain}\\\text{circuit}}} = \frac{1}{(3.5 \times 10^3 + 10^3)(10 \times 10^{-6})} \simeq 22 \text{ rad/s}$$

This is still much smaller than the pole due to the source bypass circuit, namely, $p_s = 2\pi(42) \simeq 264$ rad/s. Hence, the third stage will also have the same dominant pole as the other stages and the low-frequency cutoff $f_{l3} = 42$ Hz. For three stages, each having identical one-dominant-pole low-frequency response (remember that for a high-pass circuit the largest poles are dominant), it will be shown in Sec. 8.6 that the overall low-frequency cutoff of the amplifier for three identical dominant poles is about twice the value of the pole, namely,

$$f_l = \frac{42}{0.51} = 82 \text{ Hz}$$

To summarize, for the circuit of Fig. 8.11, the results are

$$A_v = -53 \qquad f_l = 82 \text{ Hz} \qquad f_h = 1.75 \text{ MHz}$$

Before we consider the analysis and design of multistage bipolar transistor amplifiers, it is expedient at this point to develop an approximate method of calculating the 3-dB cutoff frequency of an amplifier where there is a cluster of dominant poles in the network function. The cluster of dominant poles in the high-frequency gain function leads to a bandwidth shrinkage in the overall frequency response, i.e., a decrease in f_h. This factor must be taken into consideration in the design of an amplifier. For the low-frequency gain function the cluster leads to expansion, i.e., an increase in f_l as in the above.

8.7 THE EFFECT OF A NUMBER OF DOMINANT POLES ON THE BANDWIDTH AND FREQUENCY RESPONSE

In order to appreciate the effect of dominant pole-zero locations on the frequency response of an amplifier, we consider in this section simple cases where we have only a set of dominant real poles. The effect of complex poles and/or zeros is discussed later in Sec. 8.10.

A. Identical Real Poles: Low-Pass Case

Let the amplifier gain function be given by

$$A(s) = \frac{A_0 p_o^n}{(s + p_o)^n} \tag{8.16}$$

where A_0 is the overall midband gain. $A(s)$ is either a current gain or a voltage gain function, p_o is the dominant pole location of each stage, and n is the number of stages having the dominant pole at p_o. Note that (8.16) assumes identical poles, but it does not imply identical stages. The stages may have different load resistors, hence different Miller capacitances.

The overall upper 3-dB cutoff frequency or the 3-dB bandwidth of the amplifier can now be determined as follows. From the definition of the 3-dB radian frequency we have

$$A(\omega_{3dB}) = \frac{A_0}{|1 + j\omega_{3dB}/p_o|^n} = \frac{A_0}{\sqrt{2}} \tag{8.17a}$$

or

$$\left[1 + \left(\frac{\omega_{3dB}}{p_o}\right)^2\right]^{n/2} = \sqrt{2} \tag{8.17b}$$

Solving (8.17*b*), we get

$$\omega_{3dB} = p_o\sqrt{2^{1/n} - 1} = p_o S_n \tag{8.18}$$

where

$$S_n = \sqrt{2^{1/n} - 1} \tag{8.19}$$

The factor S_n, which is always less than unity, is called the *identical pole shrinkage factor*. This factor tells us by how much the overall bandwidth will be reduced (for a given number of stages) from that of the individual stage bandwidth, which is equal to p_o.

An approximate, but more convenient expression for S_n for n large, is obtained as follows. Since

$$2^{1/n} = \exp\frac{\ln 2}{n} = \exp\frac{0.693}{n} = 1 + \frac{0.693}{n} + \left(\frac{0.693}{n}\right)^2 + \cdots$$

$$\simeq 1 + \frac{0.693}{n}$$

$$S_n = \sqrt{2^{1/n} - 1} \simeq \sqrt{\frac{0.693}{n}} = \frac{0.83}{\sqrt{n}} \qquad n \geq 4 \tag{8.20}$$

Table 8.1 gives the value of S_n for n up to 5. Note that for $n \geq 4$ the approximate expression is accurate to within 5 percent and is much simpler to use. For $n < 4$ we must use the exact values given in Table 8.1.

TABLE 8.1 Real-Pole Shrinkage Factors

n	1	2	3	4	5
$S_n = \sqrt{2^{1/n} - 1}$	1	0.64	0.51	0.44	0.39
$S_n \simeq 0.83/\sqrt{n}$	—	—	—	0.42	0.37

EXAMPLE 8.6

The shrinkage factor is used in a design as follows: Suppose that the gain and bandwidth requirements of an amplifier are such that a three-stage *RC*-coupled amplifier must be used for the main amplifier. If the desired overall bandwidth is designated by $(\omega_{3dB})_o$, then the individual stages (assuming identical and one dominant pole per stage) must have a bandwidth $(\omega_{3dB})_i$

$$(\omega_{3dB})_i = \frac{(\omega_{3dB})_o}{0.51} = 1.96(\omega_{3dB})_o$$

where 0.51 is the shrinkage factor due to three identical stages. Note that if a cascade of stages is to be down 3 dB, then the individual stages must be down less than 3 dB. Therefore, they must have a greater 3-dB bandwidth.

B. Nonidentical Real Poles, Low-Pass Case

The poles of a cascaded amplifier, in general, may not be identical. Let the gain function of the amplifier be given by

$$A(s) = \frac{A_0}{(1 + s/p_1)(1 + s/p_2)(1 + s/p_3)\cdots} \tag{8.21}$$

This situation arises often in BJTs and FETs where due to the interaction of stages, the dominant poles are not identical even though identical stages are used. The approximate gain function of an *RC*-coupled transistor amplifier, will be of the form of (8.21).

The magnitude function for (8.21) is

$$|A(j\omega)| = \frac{A_0}{\{[1 + (\omega/p_1)^2][1 + (\omega/p_2)^2][1 + (\omega/p_3)^2]\cdots\}^{1/2}} \tag{8.22a}$$

where A_0 is the magnitude of the gain function. We equate (8.22) to $A_0/\sqrt{2}$ in order to obtain the 3-dB radian frequency, namely,

$$\frac{A_0}{2} = \left.\frac{A_0}{\{[1 + (\omega/p_1)^2][1 + (\omega/p_2)^2][1 + (\omega/p_3)^2]\cdots\}^{1/2}}\right|_{\omega=\omega_{3dB}} \tag{8.22b}$$

$$2 = \left[1 + \left(\frac{\omega_{3dB}}{p_1}\right)^2\right]\left[1 + \left(\frac{\omega_{3dB}}{p_2}\right)^2\right]\left[1 + \left(\frac{\omega_{3dB}}{p_3}\right)^2\right]\cdots \tag{8.23}$$

Since ω_{3dB} is less than any one of the poles, we can expand (8.23) and retain only the terms of the order ω_{3dB}^2, i.e.,

$$2 \simeq 1 + \omega_{3dB}^2\left(\frac{1}{p_1^2} + \frac{1}{p_2^2} + \frac{1}{p_3^2} + \cdots\right) \tag{8.24}$$

Hence

$$\omega_{3\text{dB}} \simeq \frac{1}{[(1/p_1)^2 + (1/p_2)^2 + (1/p_3)^2 + \cdots]^{1/2}} \tag{8.25}$$

From (8.25) it is noted that if any of the poles $p_i \geq 4p_1$, where the poles are numbered such that $p_1 < p_2 < \cdots < p_n$, its contribution in calculating the bandwidth can be neglected with a 6 percent error since $(1/p_1)^2(1 + 1/16) \simeq (1/p_1)^2$. For the worst case, where the poles are all identical, the error between (8.25) and (8.18) is about 20 percent. Hence the error will always be less than 20 percent by using (8.25).

C. High-Pass Case

In multistage amplifiers, the gain function also has a set of dominant poles due to the coupling and bypass capacitors of the various stages. In this case, if we assume identical dominant poles, the gain function can be written as

$$A(s) = \frac{A_0 s^n}{(s + p_1)^n} \tag{8.26}$$

where p_1 is the dominant pole (s) of the low-frequency circuits and A_o is the overall midband gain. In this case the low-frequency 3-dB cutoff radian frequency ω_l is determined from

$$|A(j\omega_l)| = \frac{A_0 \omega_l^n}{(p_1^2 + \omega_l^2)^{n/2}} = \frac{A_0}{\sqrt{2}} \tag{8.27}$$

Solution of (8.27) yields

$$\omega_l = \frac{p_1}{\sqrt{2^{1/n} - 1}} = \frac{p_1}{S_n} \tag{8.28a}$$

or using the more convenient expression for S_n from (8.20), we have

$$\omega_l(s) \simeq \frac{p_1\sqrt{n}}{0.83} = 1.2p_1\sqrt{n} \qquad n \geq 4 \tag{8.28b}$$

Note that since $S_n \leq 1$, $\omega_l \geq p_1$, and the equality sign obviously holds for $n = 1$ in (8.28*a*). For example, if in a three-stage amplifier each stage contributes one dominant pole (remember that for low-frequency response functions, the farthest pole from the origin is dominant), we then have

$$A_l(s) = \frac{A_0 s^3}{(s + p_1)^3} \tag{8.29}$$

Note that (8.29) could just as well describe the low-frequency gain function of a single stage where the three poles of the various coupling and bypass circuits are equal, and the zero is smaller than p_1 so that it is ignored. From (8.29) and (8.28), for $n = 3$, we obtain

$$\omega_l = \frac{p_1}{\sqrt{2^{1/3} - 1}} = \frac{p_1}{0.51} = 1.96p_1 \tag{8.30}$$

If the low-frequency circuit dominant poles are not identical, we can use the expression derived in (7.46c), namely,

$$\omega_l \sim (p_1^2 + p_2^2 + p_3^2 + \cdots)^{1/2} \tag{8.31}$$

where p_1, p_2, p_3, etc., are the pole locations of the low-frequency equivalent circuit of the amplifier. Again note that the approximation in (8.31) for the worst case, where $p_1 = p_2 = p_3 = p_i$, has an error of 20 percent. Hence the error by using (8.31) is always less than 20 percent. When the poles are identical we, of course, use (8.28) instead of (8.31).

EXERCISE 8.3

The gain-function of a certain amplifier is given by

$$A_v(s) = \frac{10^{21}s^3}{(s + 40)^3(s + 10^6)(s + 2 \times 10^6)(s + 10^7)}$$

Determine the midband gain and the low-frequency and the high-frequency cutoff values.

Ans 50, $f_l = 12.5$ Hz, $f_h = 132$ kHz

EXERCISE 8.4

Repeat Exercise 8.3 for the following gain function:

$$A_i(s) = \frac{10^{16}s^2}{(s + 5)(s + 40)(s + 10^7)^2}$$

Ans 100, $f_l = 6.37$ Hz, $f_h = 1.02$ MHz

EXERCISE 8.5

Determine the 3-dB bandwidth of the following normalized gain function:

$$A_v(s_n) = \frac{A_0}{s_n^2 + 3s_n + 3} \qquad s_n = \frac{s}{10^7}$$

(*Hint:* Use Fig. 8.10.)

Ans $f_{3dB} = 2.15$ MHz

8.8 GAIN AND FREQUENCY RESPONSE CALCULATIONS FOR *RC*-COUPLED CASCADED CE TRANSISTOR STAGES

We turn our attention now to the analysis of a cascaded, *RC*-coupled, CE multistage amplifier. In particular, we shall obtain expressions for the midband gain and the upper 3-dB cutoff frequency of such an amplifier. Familiarity with these expressions will be of aid when we consider design procedures for such amplifiers.

A simple three-stage cascaded *RC*-coupled CE transistor amplifier is shown in Fig. 8.12*a*. The circuit is redrawn in Fig. 8.12*b*, referred to as the signal circuit where all the low-frequency coupling and bypass capacitors are short-circuited, since they do not affect the high-frequency response of the amplifier. The collector resistances and the biasing resistances are absorbed in the interstage resistances R_1, R_2, R_3, i.e., $R_2 = R_b \parallel R_{C1}$. The equivalent circuit of the amplifier using the hybrid-π model is shown in Fig. 8.13*a*, which can be used in an accurate computer-aided analysis. However, for preliminary design purposes or quick paper-and-pencil calculation, the approximation based on the Miller theorem can be used to provide a reasonably accurate means of calculating the 3-dB bandwidth of an amplifier. We make the, by now familiar, approximation of using $Z_L(0)$ for the Miller effect as was done previously.

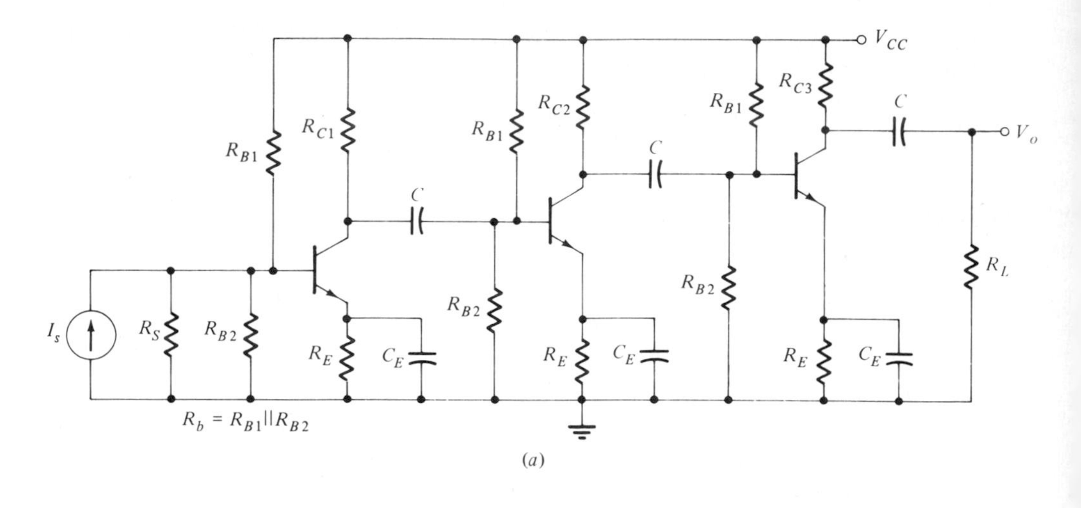

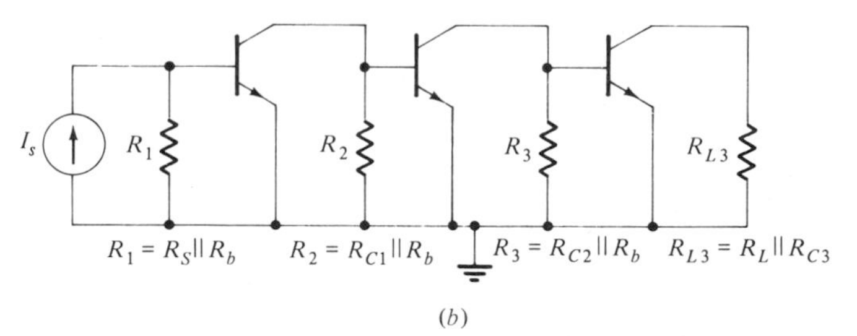

FIGURE 8.12 (*a*) Three-stage *RC*-coupled BJT amplifier. (*b*) Signal circuit of Fig. 8.12*a*.

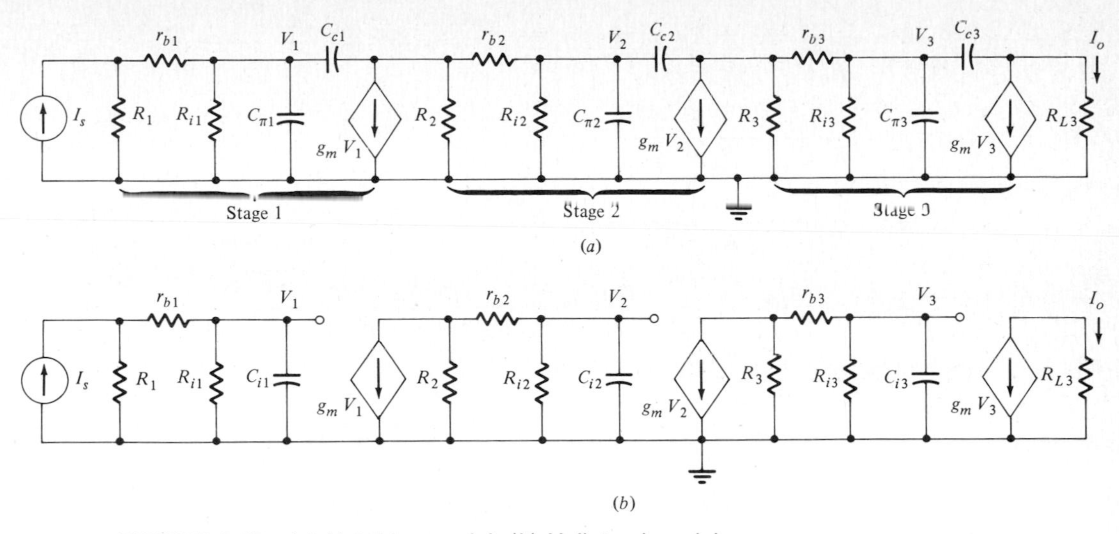

FIGURE 8.13 (*a*) Hybrid-π model. (*b*) Unilateral model.

Under this approximation the model of Fig. 8.13*a* reduces to that of Fig. 8.13*b*, where C_i are given by (8.1), namely,

$$C_{i3} = \frac{1 + R_{L3}C_{c3}\omega_{T3}}{r_{e3}\omega_{T3}} = \frac{D_3}{r_{e3}\omega_T} \tag{8.32a}$$

$$C_{i2} = \frac{1 + R_{L2}C_{c2}\omega_{T2}}{r_{e2}\omega_{T2}} = \frac{D_2}{r_{e2}\omega_{T2}} \tag{8.32b}$$

$$C_{i1} = \frac{1 + R_{L1}C_{c1}\omega_{T1}}{r_{e1}\omega_{T1}} = \frac{D_1}{r_{e1}\omega_T} \tag{8.32c}$$

$$\text{where } \begin{aligned} R_{L3} &= R_L \parallel R_C \\ R_{L2} &= R_C \parallel (r_{b3} + R_{i3}) \\ R_{L1} &= R_2 \parallel (r_{b2} + R_{i2}) \end{aligned} \tag{8.33}$$

and we have assumed that $R_{C3} = R_{C2} = R_{C1} = R_C$.

The current gain function of the circuit in Fig. 8.13*b* can now be easily determined to be

$$A_I = \frac{I_o}{I_s}(s) = \frac{I_o}{V_3}\frac{V_3}{V_2}\frac{V_2}{V_1}\frac{V_1}{I_s} \tag{8.34}$$

In (8.34) each ratio contributes one pole except I_o/V_3, which is equal to $-g_m$. Hence, from (8.34) and Fig. 8.13*b* we can write the gain function in the following convenient form:

$$A_I = \frac{I_o}{I_s} = \frac{A_I(0)p_1p_2p_3}{(s + p_1)(s + p_2)(s + p_3)} \tag{8.35}$$

where the poles p_i are determined as in (8.5), namely,

$$p_1 = \frac{1}{R_{i1}C_{i1}} \frac{R_1 + R_{i1} + r_{b1}}{R_1 + r_{b1}} \tag{8.36a}$$

$$p_2 = \frac{1}{R_{i2}C_{i2}} \frac{R_2 + R_{i2} + r_{b2}}{R_2 + r_{b2}} \tag{8.36b}$$

$$p_3 = \frac{1}{R_{i3}C_{i3}} \frac{R_3 + R_{i3} + r_{b3}}{R_3 + r_{b3}} \tag{8.36c}$$

$$A_I(0) = \frac{I_o}{I_s}(0) \simeq -\frac{\beta_{01}R_1}{R_1 + r_{b1} + R_{i1}} \frac{\beta_{02}R_2}{R_2 + r_{b2} + R_{i2}} \frac{\beta_{03}R_3}{R_3 + r_{b3} + R_{i3}} \tag{8.36d}$$

The midband voltage gain is given by

$$A_V(0) = \frac{V_o}{V_s} = \frac{R_L I_o}{R_s I_s} = \frac{R_L}{R_s} A_I(0)$$

The approximate 3-dB bandwidth is determined from (8.25).

EXAMPLE 8.7

Determine the midband gain and the upper 3-dB frequency of the *RC*-coupled, cascaded CE stages of the type shown in Fig. 8.12. Consider the following numerical example.

For simplicity, assume identical transistors with the identical operating points: $V_{CE} = 5$ V and $I_C = 5$ mA. The biasing resistor $R_{Ci} = 1$ kΩ, $R_b = 3.5$ kΩ. The transistor parameters at the operating point are as follows:

$$\omega_T = 300 \text{ MHz} \qquad r_b = 50\ \Omega$$

$$C_c = 4 \text{ pF} \qquad \beta_0 = 50$$

Let $R_S = R_L = 50$ Ω. From the above information, namely, $I_c = 5$ mA, we find

$$r_e = \frac{25 \text{ mV}}{5 \text{ mA}} = 5 = \frac{1}{g_m} \qquad R_i = \beta_0 r_e = 250\ \Omega \qquad \frac{1}{r_e \omega_T} = 1.06 \times 10^{-10} \text{ F}$$

Note that the interstage resistance values are

$$R_1 \simeq 50\ \Omega \qquad R_2 = R_3 = (1 \times 10^3) \| (3.5 \times 10^3) = 780\ \Omega$$

From (8.36*d*) the magnitude of the midband current and voltage gains are

$$A_I(0) = \frac{-(50)50}{50 + 50 + 250}\left(\frac{(50)780}{780 + 50 + 250}\right)\left(\frac{(50)780}{780 + 50 + 250}\right)$$

$$= 9320 = 79.4 \text{ dB}$$

$$A_V(0) = \frac{R_L}{R_S} A_I(0) = 9.32 \times 10^3 = 79.4 \text{ dB}$$

since $R_L = R_S$ in the example. The input capacitances of the three stages from (8.32) are

$$C_{i3} = (1.06 \times 10^{-10})D_3 = (1.06 \times 10^{-10})(1 + R_{L3}C_c\omega_T) = 1.44 \times 10^{-10}\ \text{F}$$

$$C_{i2} = (1.06 \times 10^{-10})D_2 = (1.06 \times 10^{-10})(1 + R_{L2}C_c\omega_T) = 2.76 \times 10^{-10}\ \text{F}$$

$$C_{i1} = (1.06 \times 10^{-10})D_1 = (1.06 \times 10^{-10})(1 + R_{L1}C_c\omega_T) = 2.76 \times 10^{-10}\ \text{F}$$

From (8.36) we obtain

$$p_3 = 3.61 \times 10^7\ \text{rad/s} \qquad p_2 = 1.88 \times 10^7\ \text{rad/s} \qquad p_1 = 5.08 \times 10^7\ \text{rad/s}$$

We normalize the frequency by a factor of 10^7 (i.e., $\Omega_0 = 10^7$ for convenience. The 3-dB bandwidth is then determined from (8.35) to be

$$(\omega_{3\text{dB}})_n \simeq \frac{1}{[(1/p_{1n})^2 + (1/p_{2n})^2 + (1/p_{3n})^2]^{1/2}}$$

$$\simeq \frac{1}{[(1/5.08)^2 + (1/1.88)^2 + (1/3.61)^2]^{1/2}} \simeq 1.59$$

Hence, the denormalized 3-dB radian frequency is $\omega_{3\text{dB}} = 1.59 \times 10^7$ rad/s or

$$f_{3\text{dB}} = \frac{1.59 \times 10^7}{2\pi} = 2.54\ \text{MHz}$$

8.9 DESIGN OF *RC*-COUPLED CE TRANSISTOR AMPLIFIERS FOR SPECIFIED GAIN AND BANDWIDTH

In a wideband multistage amplifier design, the circuit shown in Fig. 8.11*a* is not often used because of the inefficient gain-bandwidth product of the individual stages, as was demonstrated in Sec. 7.11. Usually a broadbanding scheme is utilized in the design, as discussed in the next section. Nonetheless, we will provide a design example for the circuit of Fig. 8.11 in order to illustrate the use of shrinkage factor and other ideas important in a design and furthermore to bring out the need for broadbanding techniques.

Let us assume that the proper operating point for obtaining high f_T, β_0, etc., for the available transistors is $V_{CE} = 5$ V and $I_C = 5$ mA. The parameters of the available transistors are assumed to be the same as those given in Example 8.1. The design requires an approximate bandwidth $f_{3\text{dB}} = 5$ MHz, and the overall voltage gain is to be larger than or approximately equal to 1000 (60 dB). The source and load resistances are $R_S = R_L = 100\ \Omega$.

First, we design the biasing circuit in accordance with the design procedure given in Sec. 2.10. For the operating point $V_{CE} = 5$ V, $I_C = 5$ mA and a supply voltage of 12 V, the following design values were obtained:

$$R_C = 1\ \text{k}\Omega \qquad R_E = 400\ \Omega, \qquad R_{b1} = 15\ \text{k}\Omega \qquad R_{b2} = 4.7\ \text{k}\Omega$$

The low-frequency cutoff is not specified; hence we may use any reasonably large values of C. If it were specified, we would use the design method in Sec. 7.3 and (8.31).

We then proceed with the design of the high-frequency circuitry. For the given transistor parameters, gain, and bandwidth specifications, if we use two stages, the individual stage gain must be greater than 30, and the individual stage bandwidth is about 7.7 MHz ($= f_{3dB}/0.64$), which is twice as large as $f_\beta/D = 3.5$ MHz. Since the GBP for simple resistive broadbanding decreases drastically for frequencies larger than f_β/D, we cannot realize the required specification with two stages when using the given transistors.* We can thus estimate that three stages will probably be sufficient.

From the specified bandwidth we now determine the gain and bandwidth of each of the individual stages when using three stages of amplification. For an amplifier with three identical-pole stages, the pole for each stage is given by (8.18); that is,

$$p_o = \frac{\omega_{3dB}}{S_n} = \frac{\omega_{3dB}}{\sqrt{2^{1/3} - 1}} = \frac{\omega_{3dB}}{0.51}$$

Since $\omega_{3dB} = 2\pi(5 \text{ MHz}) = 3.14 \times 10^7$ rad/s, the individual poles have the value of $p_o = 6.15 \times 10^7$ rad/s, as obtained from the above. Now from (8.36), starting with last stage since the D factor is known, we obtain

$$D_3 = 1 + R_L C_c \omega_T = 1.7 \qquad C_{i3} = \frac{D_3}{r_e \omega_T} = 1.82 \times 10^{-10} \text{ F}$$

$$R_3 = \frac{R_{i3}}{p_3 R_{i3} C_{i3} - 1} - r_b = \frac{250}{1.8} - 50 = 90\ \Omega$$

Also

$$R_{L2} = R_3 \parallel (r_b + R_i) = 90 \parallel 300 = 70\ \Omega \qquad D_2 = 1 + R_{L2} C_c \omega_T = 1.49$$

$$R_2 = \frac{R_{i2}}{p_2 R_{i2} C_{i2} - 1} - r_b = 123\ \Omega$$

and

$$R_{L1} = R_2 \parallel (r_b + R_i) = 123 \parallel 300 = 88\ \Omega \qquad D_1 = 1.62$$

$$R_1 = \frac{R_{i1}}{p_1 r_{i1} c_{i1} - 1} - r_b = 100\ \Omega$$

In this case, coincidentally, $R_1 = R_S$. If the specified value of R_S is larger than that of R_1, then a resistor R_a of value $R_b \parallel R_a \parallel R_S = R_1$ may be added in shunt with R_b (see

---

*The proof of this lies in the fact that if $p_1 = p_2 = \ = 2\pi(7.7 \times 10^6)$ rad/s for two identical amplifier stages, then from the individual stage bandwidth expression, $p_2 = (\omega_\beta/D) \times (R_l + R_i + r_b)/(R_l + r_b)$, which yields $R_l = 158\ \Omega$. The current gain of each of the individual stages, a_i is found from $a_i = \beta_0 R_l/(R_l + R_i + r_b)$, which is $a_i \simeq 17$. Therefore, since the product of the individual stage gains $17^2 < 1000$, two stages of amplification are insufficient to meet the design specifications.

Fig. 8.12*a*). Generally, however, the input stage may be altogether different, depending on what we wish to achieve at the input, in addition to realizing the gain and bandwidth requirements of the design specification. For example, noise considerations at the input are very important and may dictate a choice (see Appendix D). For the resistor values of this design and $R_L = R_S = 100\ \Omega$, the magnitude of the gain is

$$A_V = A_I = \beta_0^3 \frac{R_S}{R_S + r_b + R_i} \frac{R_2}{R_2 + r_b + R_i} \frac{R_3}{R_3 + r_b + R_i} = 2100 > 1000$$

Hence the design requirements are met. The overall voltage gain of this design example is given by

$$A_V = \frac{V_o}{V_s} = \frac{I_o R_L}{R_S I_s} = A_I \frac{R_L}{R_S} = 2100$$

since $R_L = R_S$. Now that we have met the gain specifications, the preliminary design is complete. The designed circuit is shown in Fig. 8.14, where R_C is split into low- and high-frequency collector load resistors by the use of a shunt capacitor. When discrete circuits are used, the preliminary design is experimentally tested and, if necessary, adjusted to exactly meet the design specifications. Note that since the low-frequency cutoff has not been specified, the values of the capacitors are arbitrary, say, all C's are 10 μF, and these values are not shown in Fig. 8.14.

Note that if the design specifications had called for the same gain, the same load termination, and $f_{3\text{dB}} = 10$ MHz, similar calculation yields the value of $R_3 = 4.5\ \Omega$. The current gain of the third stage would be 0.7, and the circuit would thus be useless for this and greater bandwidths. In fact, in simple stages of the type considered here, we would not be able to meet the specifications with the given transistors, unless another type of broadbanding were used. It is quite clear from this example that the transitor cascade amplifier circuit shown in Fig. 8.14 does not make efficient use of the gain-bandwidth product. We thus need broadbanding techniques, some of which were described in Secs. 7.12 and 7.13. We further explore, in the next section, some

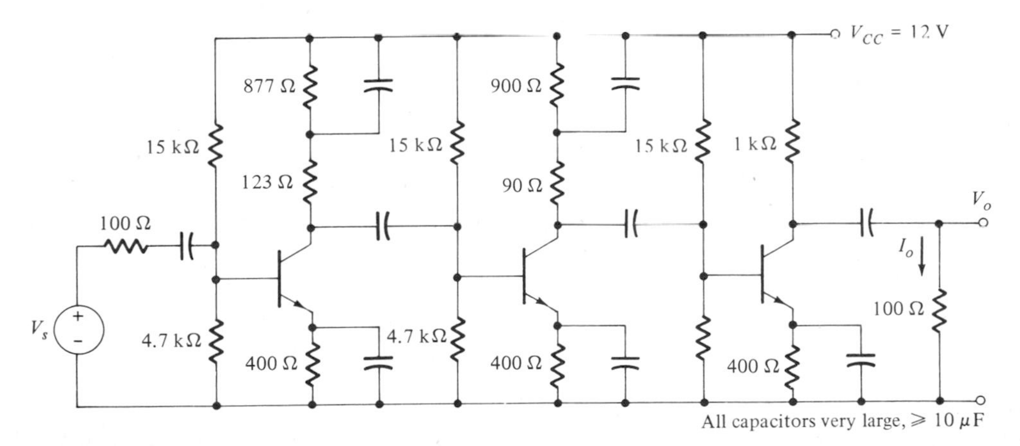

FIGURE 8.14 Amplifier of the design example.

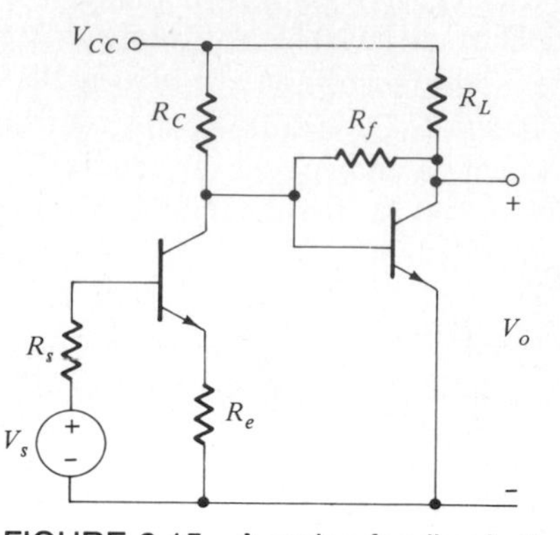

FIGURE 8.15 A series-feedback stage cascaded with a shunt-feedback stage.

of these broadbanded stages in conjunction with the multistage amplifier configuration.

8.10 MULTISTAGE BROADBAND TRANSISTOR AMPLIFIERS

In multistage amplifier design there are many factors to be taken into consideration. These factors vary, depending on the particular application of the amplifier. In this section we will briefly consider a few typical amplifier circuits and examine their gain-bandwidth capabilities.

Transistor amplifiers must often use feedback in order to achieve desensitivity, as well as broadbanded frequency response. Since feedback amplifiers are extremely useful, the topic is discussed in Chap. 10. A few such circuit arrangements are the so-called series-shunt pair (Fig. 10.16), shunt-series pair (Fig. 10.18), series-series triple (Fig. 10.19), and shunt-shunt triple (Fig. 10.20), which are commonly used in amplifier design. These are named in accordance with their feedback connections. The analysis and design of these circuits, using feedback techniques, are discussed in detail later in Chap. 10.

Figure 8.15 shows a series-feedback amplifier with a voltage-source input in *cascade* with a shunt-feedback amplifier. For a current-source input, the shunt-feedback stage would precede the series-feedback stage so as not to lose signal current. More stages may be cascaded, of course, as alternate shunt-series-shunt, etc., if the gain specifications cannot be met with two stages.

EXAMPLE 8.8

The circuit (Fig. 8.15) to be analyzed is redrawn for convenience in Fig. 8.16, wherein the circuit element values are indicated. The transistor parameters are the same as those in Example 8.1. Calculate the approximate gain and bandwidth of the circuit.

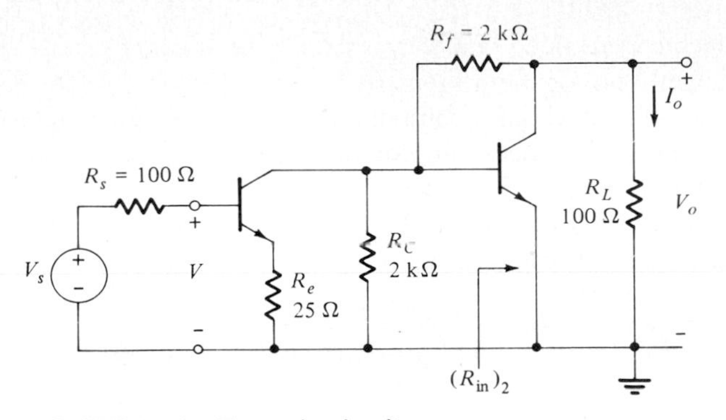

FIGURE 8.16 Example circuit.

The midband input impedance of a shunt-feedback stage, using the Miller approximation, is given by

$$(R_{in})_2 \simeq [G_f(1 + g_m R_L)]^{-1} \parallel (r_b + R_{i2})$$

$$= \frac{R_f}{1 + g_m R_L} \parallel (r_b + \beta_0 r_{e2}) \simeq 75\ \Omega$$

Since $(R_{in})_2$ is much smaller than $R_C \parallel R_0$ (where R_0 is the output impedance of stage 1, which is very high), we can consider the second stage to be driven by a current source and analyze the circuit for its gain and bandwidth separate of the first stage. The gain and bandwidth of the second stage may now be determined as in Sec. 7.13. Thus, without going into the details of the calculation, we find that the current gain and the bandwidth of the second stage are

$$A_{i2}(0) = 14 \qquad p_2 \simeq 2\pi(12\ \text{MHz})$$

The first stage may now be separately analyzed as a series-feedback configuration having a load resistance $(R_L)_1 = 75\ \Omega$. For this load value, the given transistor parameters, and $R_S = 100\ \Omega$, we can use the expressions in (7.78*a*) and (7.78*b*) to find the current gain and the bandwidth. Substitution of the numerical values into these expressions yields

$$A_{i1}(0) = \frac{I_1}{I_s} \simeq 3 \qquad p_1 = 2\pi(36\ \text{MHz})$$

Thus the overall current and voltage gains are given by the product, since the stages are approximately noninteracting.

$$A_I = A_{i1}A_{i2} = 42 = A_V$$

since $R_L = R_S$. The overall 3-dB bandwidth is determined from Fig. 8.10 or (8.25). For a p_2/p_1 ratio of 3 we read

$$\omega_{3\text{dB}} = 0.91(2\pi \times 12\ \text{MHz}) \qquad \text{or} \qquad f_{3\text{dB}} \simeq 11\ \text{MHz}$$

The above results were obtained in a very approximate manner just to illustrate the gain-bandwidth capability of configurations such as shown in Fig. 8.16. Note that a bandwidth of 10 MHz, with some gain, could not be realized using simple cascaded resistive broadbanded stages. Feedback amplifiers also can produce complex pole pairs, as contrasted with the cascaded CE stages, which yield real poles only. A mixed pair such as the CC-CE pair of Sec. 8.4 also yielded complex poles. Complex poles can often be designed to get flat response with large bandwidth.

8.11 DESIGN OF CASCADED FETS FOR SPECIFIED GAIN AND BANDWIDTH

The design of cascaded CS field-effect transistors, shown in Fig. 8.17*a*, for a specified overall voltage gain and 3-dB bandwidth, given the FET parameters and the source and load terminations, proceeds as follows: We first examine an interior stage for which the gain-bandwidth product for the maximum gain μ is given by (7.76), namely,

$$A_{vi}\omega_{3\text{dB}} = \frac{g_m}{C_{gs} + C_i} = \frac{g_m}{C_{gs} + C_{gd}(1 + \mu)} \tag{8.37}$$

For a reasonable gain per stage, say, $A_{vi} \geq 4$, the bandwidth is then limited by (8.37), namely,

$$\omega_{3\text{dB}} \leq \frac{g_m/4}{C_{gs} + C_{gd}(1 + 4)} \tag{8.38}$$

The design equations for the overall bandwidth and gain specifications are given by

$$\omega_{3\text{dB}} = \frac{p_o}{S_{n-1}} \tag{8.39}$$

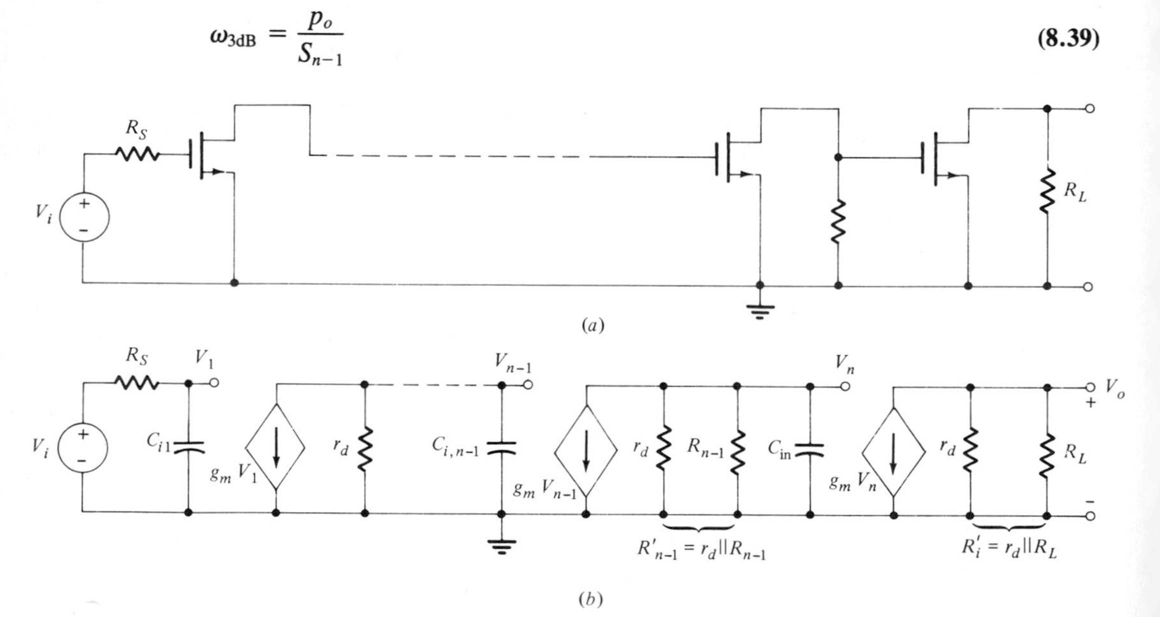

FIGURE 8.17 (*a*) A cascade of *n* stages for FET design. (*b*) Signal equivalent circuit for the design.

where S is the shrinkage factor (e.g., for $n = 4$, $S_3 = 0.51$), and

$$A_V = (g_m R_L')A_{vi}^{n-1} \tag{8.40}$$

where $R_L' = R_L \| r_d$ and n is the number of stages as shown in Fig. 8.17*b*. Note that $n - 1$ is used since the pole of the first stage (due to the voltage source input, i.e., R_S, is very small) is much larger than the interior stage poles. If the value of R_S is large, say, of the order of kiloOhms, then the pole due to the first stage cannot be ignored, and we use n in (8.39). The gain of the last stage is determined by the specified value of R_L. In the design, since R_L is specified, for identical poles the gains of each stage will not be identical.

EXAMPLE 8.9

Design a cascaded CS transistor for an overall voltage gain $A_v \geq 100$, $\omega_{3\text{dB}} = 2\pi(5 \text{ MHz})$ using FETs with the following parameters:

$$g_m = 2.5 \times 10^{-3} \text{ mho} \qquad r_d = 20 \text{ k}\Omega \qquad C_{gs} = 5 \text{ pF} \qquad C_{gd} = C_{ds} = 1 \text{ pF}$$

The terminations specified are $R_S = 100\ \Omega$, $R_L = 1\text{ k}\Omega$. We first check from (8.38) to see if the bandwidth specification is not too severe.

$$\frac{g_m/4}{C_{gs} + C_{gd}(1 + 4)} = \frac{2.5 \times 10^{-3}}{4(5 + 5) \times 10^{-12}} = 6.2 \times 10^7 \geq 2\pi(5 \times 10^6)$$

Since the condition in (8.38) is met, we go ahead and determine the number of stages from (8.40)

$$100 = (2.5 \times 10^{-3})(10^3)A_{vi}^{n-1} \qquad \text{or} \qquad A_{vi}^{n-1} = 40$$

For $n = 2$, it cannot be designed since $A_{vi} = 40 \simeq \mu$. For $n = 3$, $A_{vi} = \sqrt{40}$ or $A_{vi} = 6.3 < \mu$, so we can try it. The simplified equivalent circuit for Fig. 8.17*a* is shown in Fig. 8.17*b*. For the circuit $C_{i3} = C_{gs} + C_{gd}(1 + g_m R_L') = 5 + (1 + 2) = 8$ pF. From the pole

$$p_2 = \frac{\omega_{3\text{dB}}}{S_2} = \frac{2\pi(5\text{ MHz})}{0.64} = \frac{1}{R_2' C_{i2}}$$

we determine $R_2' = 2.55\text{ k}\Omega$ or $R_2 = 3\text{ k}\Omega$. At this point we notice that the gain of the second stage is $g_m R_2' = 6.4$, which is about the same as A_{vi}. Since the load resistance for the second stage is larger than R_L, the capacitance C_{i2} will be larger than C_{i3}, and R_1 will be smaller than R_{L2}; hence A_{v1} will be less than 6.3, and three stages would not do. We next try $n = 4$. For $n = 4$, from (8.40) $A_{vi}^3 = 40$ or $A_{vi} \simeq 3.4$. We now proceed with the design for four stages with $p_3 = p_2 = p_1 = p_o$. For $n = 4$, $S_3 = 0.51$ and $C_{i4} = 8$ pF. We have

$$p_o = \frac{\omega_{3\text{dB}}}{0.51} = \frac{1}{R_3 C_{i4}}$$

$$p_o = \frac{\omega_{3\text{dB}}}{0.51} = \frac{1}{R_3 C_{i4}}$$

which yields

$$R_3' = 2.03\ \text{k}\Omega \qquad \text{or} \qquad R_3 \simeq 2.3\ \text{k}\Omega$$

The gain of the third stage is now $g_m R_3' = 5.1 > 3.4$, and the design can be met. We next find $C_{i3} = 5 + (1 + 5.1) = 11.1$ pF, and

$$p_o = \frac{\omega_{3\text{dB}}}{0.51} = \frac{1}{R_2' C_{i3}}$$

which implies

$$R_2' = 1.46\ \text{k}\Omega \qquad \text{or} \qquad R_2 \simeq 1.47\ \text{k}\Omega$$

The gain of the second stage is $g_m R_2' = 3.6$.

For the second stage $C_{i2} = 5 + (1 + 3.6) = 9.6$ pF, and from

$$p_o = \frac{\omega_{3\text{dB}}}{0.51} = \frac{1}{R_1' C_{i2}}$$

we have

$$R_1' = 1.7\ \text{k}\Omega \qquad \text{or} \qquad R_1 \simeq 1.8\ \text{k}\Omega$$

The gain of the first stage is 4.25. The overall gain of the designed circuit is

$$A_v = 4.25(3.6)(5.1)(2.5) = 233 > 100$$

and hence the design is complete.

EXERCISE 8.6

The signal circuit, exclusive of biasing circuitry, of a three-stage amplifier is shown in Fig. E8.6: Design the circuit for an overall bandwidth of $\omega_{3\text{dB}} = 10^7$ rad/s. Determine the values of R_1, R_2, and the overall voltage gain. The FET parameters are $g_m = 3 \times 10^{-3}$ mho, $r_d = 10\ \text{k}\Omega$, $C_{gs} = 5$ pF, $C_{gd} = 2$ pF, and $C_{ds} \simeq 0$.

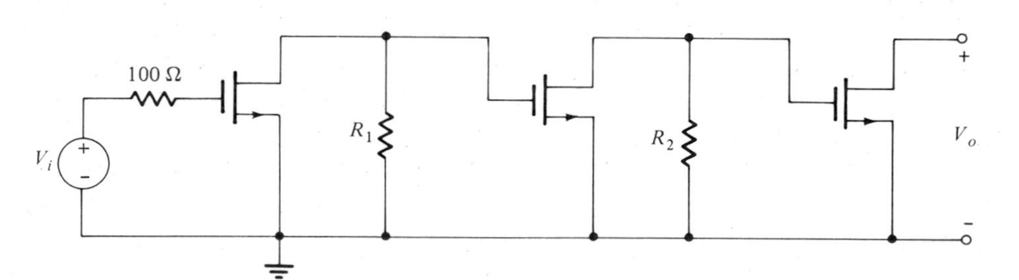

FIGURE E8.6

Ans $R_2 = 9.4\ \text{k}\Omega,\ R_1 = 1.5\ \text{k}\Omega,\ A_v = -120$

Before concluding this section, we would like to remind the readers of the availability of amplifier chips in integrated circuits. We have discussed general design principles and techniques that are utilized in the design of some of these amplifiers. It should be noted that special differential input and output amplifiers are available for high-frequency amplification in IC packages. These are the automatic-gain-control (AGC) types and non-AGC types. For the non-AGC types, typical examples are MC1553 and MC1733 types. For AGC amplifiers, typical examples are MC1550 and MC1590 types. These amplifiers can be used for frequencies up to 100 MHz. The characteristics of these as well as other types are given in the manufacturers' data books listed in Appendix E.

8.12 MAXIMALLY FLAT MAGNITUDE (BUTTERWORTH) FUNCTIONS

A class of all pole functions that exhibit maximally flat magnitude characteristic is known as *Butterworth functions*. The magnitude of the nth-order Butterworth functions is given by

$$|A(j\omega)|^2 = \frac{A_o^2}{1 + (\omega/\rho)^{2n}} \tag{8.41}$$

Note that the 3-dB bandwidth of (8.41) is given by ρ, since at $\omega = \rho$, we have

$$|A(j\omega)|^2 = \frac{A_0^2}{2} \tag{8.42}$$

Butterworth functions are all pole functions and can be written in the normalized form ($s_n = s/\rho$) as

$$A(s) = \frac{A_0}{s_n^n + a_{n-1}s_n^{n-1} + \cdots + a_1 s_n + 1} \tag{8.43}$$

where the coefficients are constrained such that (8.43) reduces to (8.41) for $s = j\omega$. The constraints can be readily determined by examining the magnitude function from (8.43), namely,

$$|A(j\omega)|^2 = \frac{A_0^2}{1 + b_1\omega_n^2 + b_2\omega_n^4 + \cdots + b_n\omega_n^{2n}} \tag{8.44}$$

where b_i in (8.44) are functions of a_i in (8.43). By long division near $\omega \rightarrow 0$ [and noting that $1/(1 + \epsilon) \simeq 1 - \epsilon$ for $\epsilon \ll 1$], (8.44) can be written as

$$|A(j\omega)|^2 = A_0^2\,(1 - b_1\omega_n^2 - b_2\omega_n^4 - \cdots - b_n\omega_n^{2n}) \tag{8.45}$$

TABLE 8.2 Butterworth Polynomials ($s_n = s/\rho$)

$s_n + 1$
$s_n^2 + \sqrt{2}s_n + 1$
$s_n^3 + 2s_n^2 + 2s_n + 1 = (s_n + 1)(s_n^2 + s_n + 1)$
$s_n^4 + 2.613s_n^3 + 3.414s_n^2 + 2.613s_n + 1 = (s_n^2 + 0.765s_n + 1)(s_n^2 + 1.848s_n + 1)$

Equation (8.45) is simply the Maclaurin series expansion of $|A(j\omega)|^2$ *near* ω equal to zero. In other words,

$$f(\omega^2) = |A(j\omega)|^2 = A_0^2 + f'(0)\omega + \frac{f''(0)}{2!}\omega^2 \times \frac{f'''(0)}{3!}\omega^3 + \cdots \tag{8.46}$$

By setting as many derivatives of $|A(j\omega)|$ equal to zero as possible, we obtain the maximal flatness, hence the name maximally flat magnitude function. Since the magnitude is an even function, the odd terms do not appear in (8.45), and from (8.46) and (8.45) we have

$$b_i = 0 \qquad i = 1, 2, \ldots, n - 1 \tag{8.47}$$

From (8.45) and (8.44) we have the expression given in (8.41), where $b_n = 1/\rho^{2n}$. The poles of (8.41) are located on a semicircle of radius ρ, which can be set equal to 1, in the normalized frequency plane, resulting in the following Butterworth polynomials:

$$n \text{ even:} \quad \prod_{k=1}^{n/2} [s_n^2 + (2\cos\theta_k)s_n + 1] \qquad \theta_k = \frac{(2k-1)\pi}{2n} \tag{8.48a}$$

$$n \text{ odd:} \quad (s_n + 1)\prod_{l=1}^{(n-1)/2} [s_n^2 + (2\cos\theta_l)s_n + 1] \qquad \theta_l = \frac{l\pi}{n} \tag{8.48b}$$

Table 8.2 lists the first four Butterworth polynomials. Pole locations of the low-order polynomials are shown in Fig. 8.18. The corresponding magnitude responses are shown in Fig. 8.19.

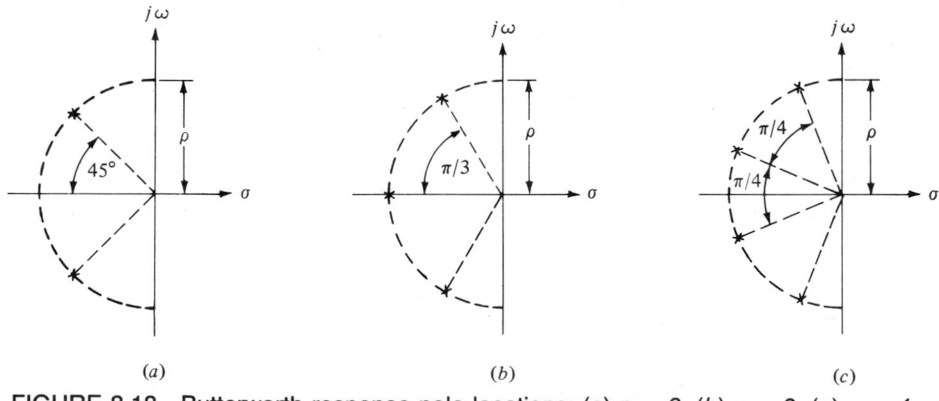

FIGURE 8.18 Butterworth response pole locations: (a) $n = 2$, (b) $n = 3$, (c) $n = 4$.

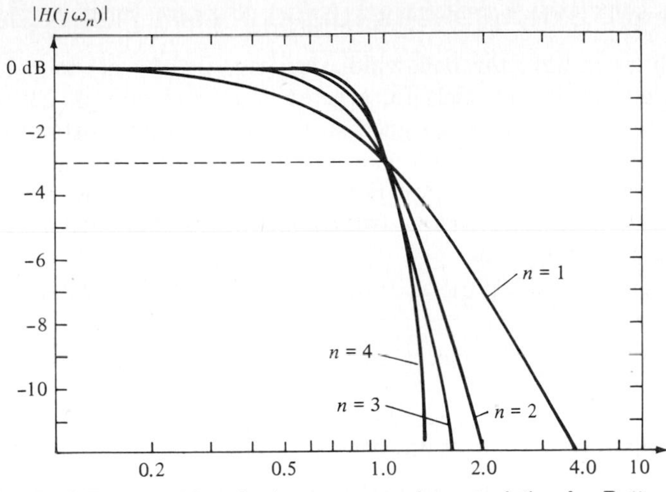

FIGURE 8.19 Magnitude response characteristics for Butterworth functions for $n = 1, 2, 3, 4$.

EXERCISE 8.7

For the circuit shown in Fig. E8.7, the amplifier is an ideal VCVS with a gain of K.

(a) Derive the expression for the voltage gain function.
(b) Design the circuit for a maximally flat magnitude response with a gain of 10 and $\omega_{3\text{dB}} = 2\pi(10^4)$ rad/s, given $R_1 = 1\ \text{k}\Omega$ and $C_1 = C_2$.

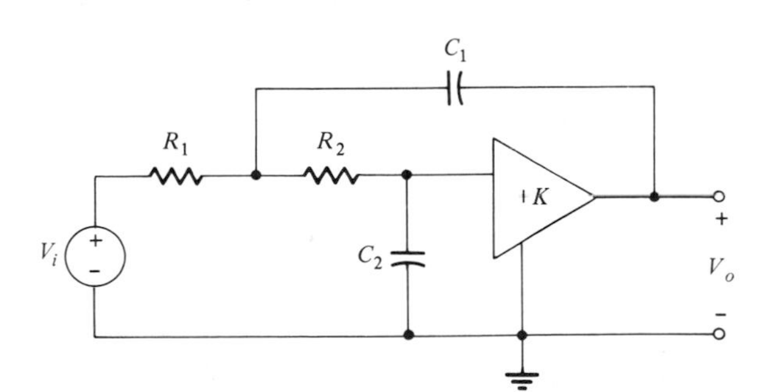

FIGURE E8.7

Ans (a) $A_v = \dfrac{V_o}{V_i} = \dfrac{K/R_1R_2C_1C_2}{s^2 + s\left[\dfrac{1}{R_1C_1} + \dfrac{1}{R_2C_1} + \dfrac{1}{R_2C_2}(1 - K)\right] + \dfrac{1}{R_1R_2C_1C_2}}$

(a) $K = 10$, $R_2 = 13.2\ \text{k}\Omega$, $C_1 = C_2 = 0.0044\ \mu\text{F}$.

8.13 STAGGER TUNING AND SYNCHRONOUS TUNING IN AMPLIFIER DESIGN

In wideband amplifier design, the bandwidth shrinkage effect can be reduced or eliminated by stagger tuning, and thus for a given gain level, a high GBP can be realized. A *stagger-tuned* circuit is designed such that the desired overall network function pole-zero pattern is broken into simple pole-zero locations, which are assigned to individual noninteracting stages. For example, an overall four-pole Butterworth function is achieved by a cascade of two noninteracting stages having complex pole pairs as indicated in Fig. 8.20.

$$A(s) = \frac{A_0\rho^4}{(s^2 + 0.765\rho s + \rho^2)(s^2 + 1.848\rho s + \rho^2)} \tag{8.49a}$$

$$= \frac{A_{01}\rho^2}{s^2 + 1.848\rho s + \rho^2}\,\frac{A_{02}\rho^2}{s^2 + 0.765\rho s + \rho^2} = A_1(s)A_2(s) \tag{8.49b}$$

This simple procedure cannot be used in transistor stages *where interaction occurs*. For op-amp stages this can readily be done as discussed in Chap. 9, and the order of the blocks are arbitrary. Analytically, for such cases stagger-tuned design is achieved by obtaining the overall gain function of the amplifier and then equating the coefficients of the overall gain-function with those of the polynomial resulting from the desired overall pole-zero pattern. Because approximate models are used for the transistors, however, calculated values frequently fail to achieve the design specifications, necessitating the varying of some of the values of the circuit elements on an experimental basis until the desired results are achieved. Such an alignment for interacting stages can be difficult, because varying one component may change many coefficients of the network function, and thus many circuit elements may have to be varied by trial and error to achieve the overall desired response. Nevertheless, it is commonly done as in TV alignment. In some applications, such as tuned amplifiers, the interaction is intentionally reduced by a deliberately introduced mismatch between the input and the output of the individual stages. The cascading of shunt-feedback

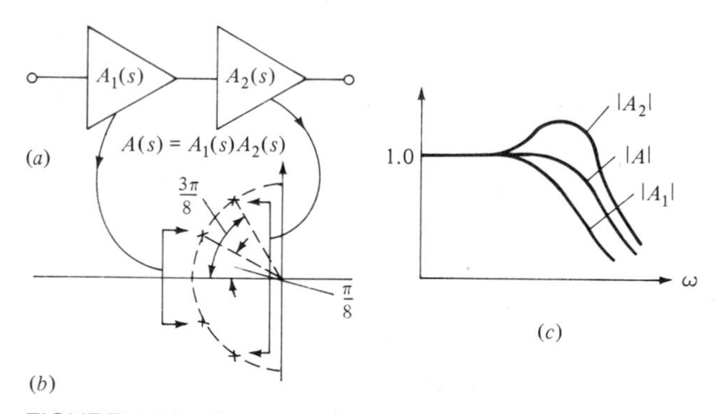

FIGURE 8.20 Stagger-tuned system with two noninteracting blocks. (*a*) Individual blocks in cascade. (*b*) Pole locations of the individual gain functions. (*c*) Magnitude response of the individual gain functions and the overall gain function.

amplifiers and series-feedback amplifiers, or vice versa, is another example of mismatch that reduces interaction between stages. The cascode circuit, as a unit, can be considered a noninteracting block, and thus stagger-tuned with another stage at its output.

Experimentally, the alignment of transistor amplifiers in a stage-by-stage manner to achieve the widest overall bandwidth with no frequency peaking in the overall gain function is a form of stagger tuning to obtain a maximally flat magnitude response. In such cases, the alignment is best performed by starting with the last stage and progressing to the preceding stage. The gain function of each unit (which may be single stage or multistage) is adjusted by varying around the preliminary design values so as to get the desired response from the unit. Usually the individual block or unit is arranged such that the interaction is small. As an example, consider the shunt-series feedback pair of Fig. 10.18 as a unit. It is shown in Chap. 10 that this circuit can be designed to obtain a complex pair of dominant poles. Its input impedance is shown to be low, whereas its output impedance is very high. If we cascade two such units, the interaction can be very small.

Suppose we wish to design the circuits to realize an approximate four-pole Butterworth response. We may initially design the circuits by ignoring the interaction and realize one complex pole pair with $\psi = \pi/8$, say, $A_1(s)$ in Fig. 8.20*a*, using one shunt-series pair and $A_2(s)$ to have $\psi = 3\pi/8$ by another shunt-series pair. The initial design may then be adjusted, starting with A_2 such that $A_2(s)$ and $A_1(s)$ in cascade yield the widest overall bandwidth with no frequency peaking in the overall response. For interacting stages, the sequence of realization can make some difference; i.e., in the above example, if the interaction is not very small, the results of assigning the complex pole pair with $\psi = \pi/8$ to A_1 or A_2 can give different results, because of the effect of the termination impedance in each stage. Only in the case of noninteracting stages is the assignment arbitrary. Op-amp-based gain functions are therefore ideal as noninteracting blocks due to their extremely high input and very low output impedances. Interaction between blocks using op amps are therefore negligible and thus facilitate design. This topic is discussed in detail in Sec. 9.5.

A *synchronously tuned* design is one in which the individual noninteracting stages have identical pole-zero patterns. In noninteracting stages the overall pole-zero pattern has a certain multiplicity, depending on the form of the gain function of the individual stage. If the interaction is very small, an identical pole-zero pattern may be assigned to the individual stage or unit. In all synchronously tuned circuits the effect of the shrinkage factor must be included in the design. The CE, cascaded, identical one-pole-per-stage design described in Sec. 8.1 is one example of synchronous tuning using single poles. In the case of shunt-series pairs, which were considered for illustrative purposes earlier in this section, if the individual units are assigned such that each unit has a two-pole maximally flat magnitude response, the resulting design is a synchronously tuned design. In such a case, both units will have complex poles with $\psi = \pi/4$ as shown in Fig. 8.21*a*.

In other words, the overall gain function is given by

$$A(s) = \frac{A_0 \rho^4}{(s^2 + \sqrt{2}\rho s + \rho^2)^2} \tag{8.50a}$$

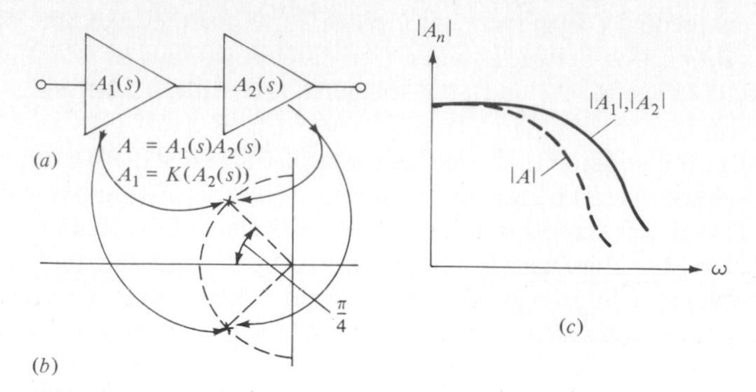

FIGURE 8.21 Synchronously tuned system with two noninteracting blocks. (*a*) Individual blocks in cascade. (*b*) Pole locations of the individual gain functions. (*c*) Magnitude response of the individual gain functions and the overall gain.

$$= \frac{A_{01}\rho^2}{s^2 + \sqrt{2}\rho s + \rho^2}\frac{A_{02}\rho^2}{s^2 + \sqrt{2}\rho s + \rho^2} \tag{8.50b}$$

Note that in this case the overall gain function does *not* have a maximally flat magnitude response characteristic in spite of the fact that the individual block response is a two-pole maximally flat magnitude function, as shown in Fig. 8.21*b*.

The shrinkage factor for this case is (see problem 8.25).

$$S = (\sqrt{2} - 1)^{1/4} = 0.804 \tag{8.51}$$

Hence the overall bandwidth will be smaller than the individual stage bandwidth by the above shrinkage factor.

In a synchronously tuned design the overall GBP is generally smaller than that of the staggered-tuned design. However, the advantage gained through using synchronous tuning is ease of alignment, i.e., each noninteractive stage or unit is aligned so as to have the same response characteristics as the other stages or units, and the alignment procedure is thus simpler than in the stagger-tuned case. In the above example, if the interaction is negligible, each unit is aligned to have the widest bandwidth with no frequency peaking (beginning with the last stage) in order to achieve a synchronously tuned alignment.

We mention again that op-amp-derived blocks, such as those discussed in Sec. 9.5 for noninteracting circuits to design high-order filters or amplifiers (provided that the 3-dB frequency of the system is much smaller than the gain-bandwidth product of the op amp), can be used.

8.14 CONCLUDING REMARKS

The purpose of this chapter was to determine, in a simple manner, the frequency response of multiple-transistor amplifiers. A number of two-stage amplifier configurations were considered for their frequency performance and gain capabilities. Some of these circuits are utilized as a functional block on their own, such as the cascode configuration, while others are used as a constituent of a more complex amplifier. In the latter category is the emitter-coupled (or the source-coupled) pair, which are used as the differential input stage of an op amp. It was also noted that the 3-dB bandwidth of the op amp without feedback is very low and is determined by the compensating capacitor. The input-output interaction of some combination can yield complex poles such as in the Darlington CC-CE configuration. An important class of multistage amplifiers is the feedback amplifier, which was mentioned in this chapter. However, a detailed treatment of feedback amplifiers is reserved for Chap. 10.

REFERENCES

1. P. R. Gray and R. G. Meyer, *Analysis and Design of Analog Integrated Circuits,* Wiley, New York, 1977.
2. D. J. Comer, *Modern Electronic Circuit Design,* Addison-Wesley, Reading, Mass., 1976.

PROBLEMS

8.1 For the circuit shown in Fig. P8.1, the transistors have the following parameters:

$$\beta_0 = 100 \qquad f_T = 400 \text{ MHz} \qquad C_c = 2 \text{ pF} \qquad r_b = 50\ \Omega \qquad r_e = 13\ \Omega$$

Determine the midband voltage gain and approximate 3-dB bandwidth of the amplifier. (Note that $I_{E1} \simeq I_{E2} = 2$ mA.)

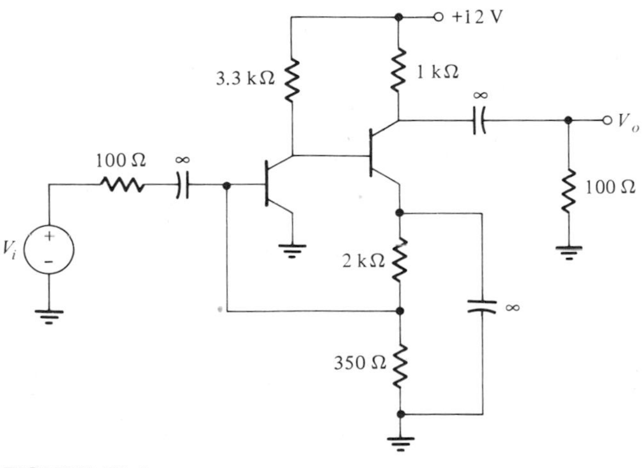

FIGURE P8.1

8.2 For the two-stage JFET amplifier shown in Fig. P8.2, determine the following:
(*a*) Midband voltage gain
(*b*) The upper and lower 3-dB frequencies. The FET parameters are

$$r_d = 10 \text{ k}\Omega \qquad C_{gs} = 6 \text{ pF} \qquad C_{ds} = 1 \text{ pF}$$

$$g_m = 4 \times 10^{-3} \text{ mho} \qquad C_{gd} = 2 \text{ pF}$$

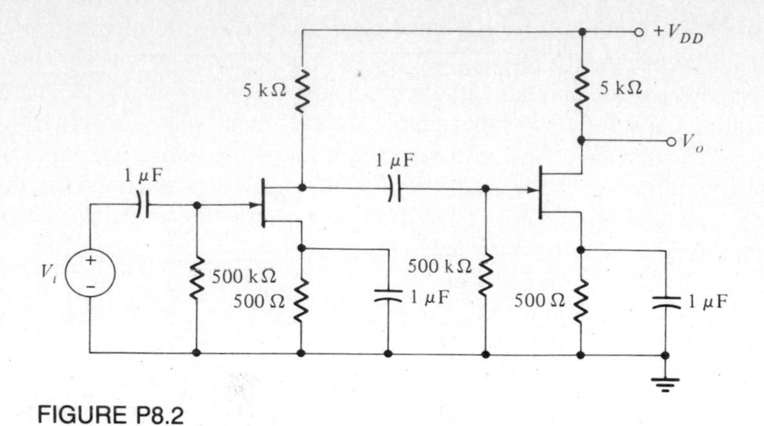

FIGURE P8.2

8.3 Repeat problem 8.2 assuming a MOSFET with the following parameters:

$$r_d = 20\ \text{k}\Omega \qquad C_{gs} = 6\ \text{pF} \qquad C_{ds} = 1\ \text{pF}$$

$$g_m = 4 \times 10^{-3}\ \text{mho} \qquad C_{gd} = 0.5\ \text{pF}$$

Why is the effect of C_{gd} dominant even though it has a smaller value than C_{ds}?

8.4 For the amplifier circuit shown in Fig. P8.4, determine the voltage gain and the input and output impedances at midband frequencies. What is the upper 3-dB frequency of this amplifier if the transistor parameters are given as in problem 8.1.

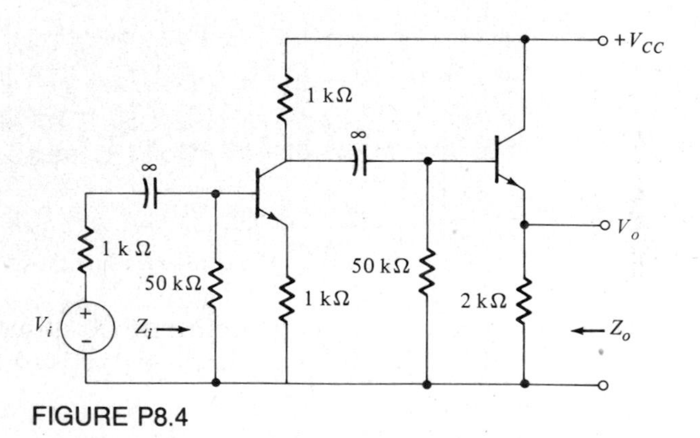

FIGURE P8.4

8.5 An amplifier has a voltage gain function given by

$$A_v(s) = \frac{8 \times 10^{12} s^2 (s + 3)}{(s + 5)(s + 10)(s + 200)(s + 10^5)(s + 2 \times 10^5)}$$

Determine the approximate upper and lower 3-dB frequencies and the midband gain of the amplifier. Sketch the asymptotic plot.

8.6 Three amplifiers, each with high input and low output impedances, are cascaded. The frequency response and voltage gain of each amplifier is given by

Amplifier no.	f_l, Hz	f_h, kHz	a_v
1	100	100	10
2	200	150	5
3	50	200	10

It is also known that near f_l and f_h the amplifiers have a ±20 dB/decade slope. Find the approximate *overall* gain, f_l, f_h, and the 3-dB bandwidth of the amplifier.

8.7 An amplifier has a gain function given by

$$A_i(s) = \frac{5 \times 10^{14} s^2 (s + 5)}{(s + 8)(s + 100)^2 (s^2 + \sqrt{3} \times 10^6 s + 10^{12})}$$

Determine the midband gain and the upper and lower 3-dB cutoff frequencies. (You may use Fig. 8.10.)

8.8 Estimate the f_T required to design an amplifier (with a minimum number of transistors) with a midband overall current gain of 5000 and a bandwidth $\omega_{3\text{dB}} = 2\pi \times 10^6$ rad/s. The available transistors have $\beta_0 \simeq 60$ and $C_c \simeq 5$ pF. For simplicity, assume the CE stages to be identical interior stages of the type in Fig 8.12*b*. Let $r_b = 0$ and $r_e = 5\ \Omega$. Show the circuit. What are the gain and the bandwidth of the circuit if $r_b = 100\ \Omega$.

8.9 An amplifier with an overall voltage gain of 500 uses three identical *RC*-coupled FET stages, i.e., $R_S = R_{L1} = R_{L2} = R_L$. Estimate the increase in bandwidth if a MOSFET instead of JFET is used, i.e., assume identical parameters except (C_{gd}) MOSFET $\simeq \frac{1}{2}(C_{gd})$ JFET. *For simplicity* let $C_{gs} = 5C_{gd}$ (ignore C_{ds}) and $r_d \gg R_L$.

8.10 Determine approximately the midband voltage gain and the upper 3-dB frequency of the amplifier shown in Fig. P8.10. Assume that the transistor parameters are the same as in problem 8.1, Let $R_L = R_E = 1\ \text{k}\Omega$ and $R_S = 50\ \Omega$.
(*Hint:* The CB stage bandwidth is very much larger than that of the CC stage. Thus, use the low-frequency *h* model for the CB stage. Note also that for the CC stage the capacitor C_c of the hybrid-π model can be ignored, since the load resistance for the stage is small and the voltage gain of a CC is about 1.)

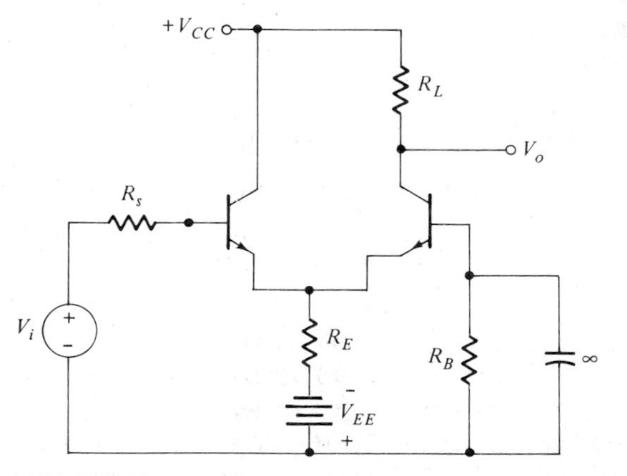

FIGURE P8.10

8.11 For the cascode amplifier shown in Fig. 8.4, $R_S = R_L = 10\ \text{k}\Omega$. The transistor parameters are as given in problem 8.1. Determine the voltage gain and the 3-dB bandwidth of the amplifier.

8.12 A cascode circuit using FETs are shown in the signal circuit in Fig. P8.12. Derive the expression for the voltage gain V_o/V_i and the 3-dB frequency. If the FET parameters are the same as in problem 8.3 and $R_S = 100\ \Omega$, $R_L = R = 1\ \text{k}\Omega$. What are the values of A_v and $f_{3\text{dB}}$?

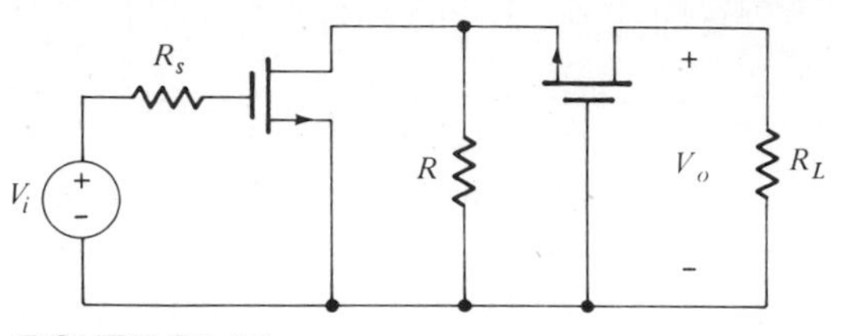

FIGURE P8.12

8.13 Design a transistor RC-coupled amplifier using the minimum number of stages to meet the following requirements:

$$A_V(0) = 2500 \qquad \omega_{3\text{dB}} = 2\pi(1\ \text{MHz}) \qquad R_L = 1\ \text{k}\Omega \qquad R_S = 100\ \Omega$$

The operating point is $V_{CE} = 5$ V and $I_C = 5$ mA. The transistor parameters at this quiescent point are as given in Example 8.7. For simplicity assume that $r_b = 0$ and ignore the biasing resistors. Show the designed signal circuit exclusive of biasing circuitry.

8.14 A hybrid cascode is shown in Fig. P8.14. Determine the midband voltage gain and the approximate 3-dB bandwidth of the amplifier if the transistor parameters are

BJT: $f_T = 100$ MHz $\quad \beta_0 = 50 \quad r_b = 100\ \Omega \quad r_e = 25\ \Omega$
JFET: $g_m = 2 \times 10^{-3}$ mho $\quad C_{gs} = 5$ pF $\quad C_{ds} = C_{gd} = 2$ pF

(*Hint:* The 3-dB cutoff frequency is much lower than f_T, so that the CB stage internal capacitance can be ignored.)

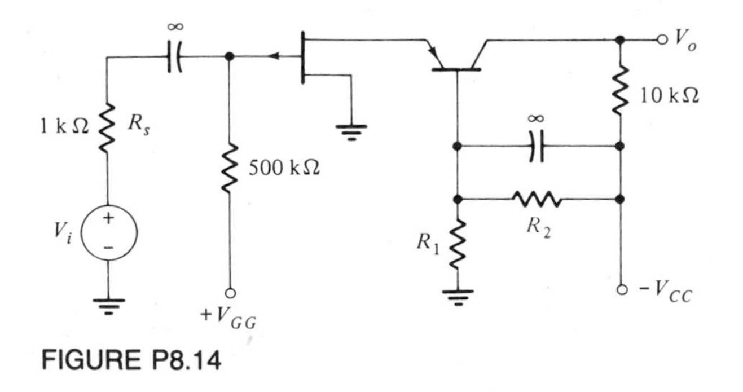

FIGURE P8.14

8.15 A transistor series-peaked equivalent circuit is shown in Fig. P8.15.
(*a*) Determine the current-gain function $A_i(s)$.
(*b*) If the transistor has the parameters given in Example 8.7 and $R_L = 100\ \Omega$, design the circuit (i.e., find R_1 and L) to obtain a maximally flat magnitude function with $\omega_{3dB} = 2\pi \times 10^7$ rad/s. What is the midband current gain of the circuit?

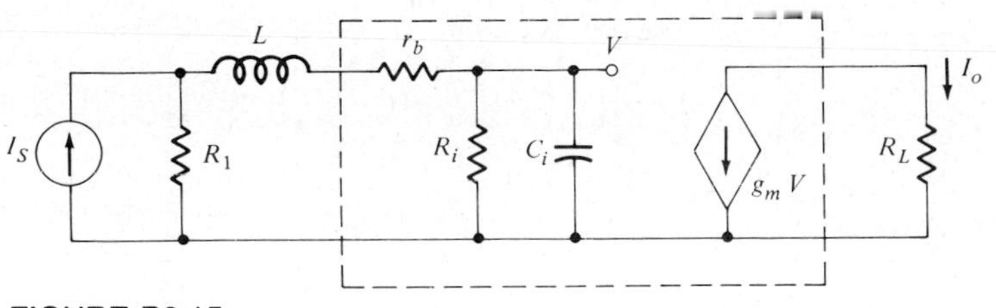

FIGURE P8.15

8.16 (*a*) Derive (8.12).
(*b*) Determining R_I for the CC-CE cascade of the text for a maximally flat magnitude response characteristic. Determine the midband gain and the 3-dB bandwidth for this case.

8.17 (*a*) Determine the shrinkage factor for a cascade of two stages, each stage having a two-pole maximally flat magnitude response.
(*b*) If the overall bandwidth is $\omega_{3dB} = 2\pi(1\text{ MHz})$, what is the individual stage bandwidth? Is the overall frequency response a maximally flat magnitude response?

8.18 For the circuit shown in Fig. P8.18, determine the midband gain and the upper 3-dB frequency if the MOSFET parameters are given as

$$g_m = 2 \times 10^{-3}\text{ mho} \qquad r_d = 20\text{ k}\Omega \qquad C_{gs} = 10\text{ pF} \qquad C_{dg} = C_{ds} = 2\text{ pF}$$

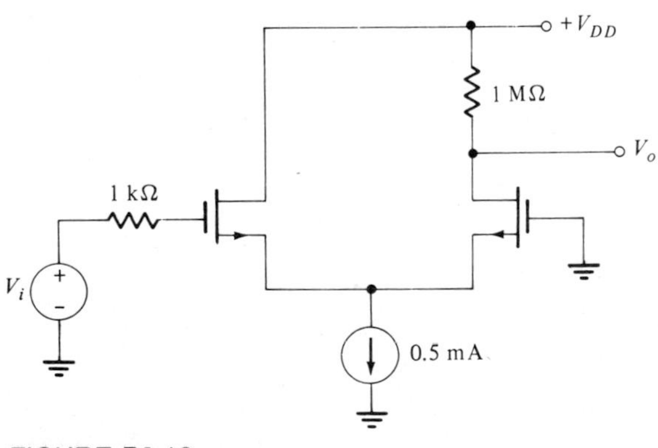

FIGURE P8.18

8.19 For the CMOS amplifier circuit shown in Fig. P8.19, determine the midband gain and the upper and lower 3-dB frequencies. For simplicity, assume the same transistor parameters as in problem 8.18.

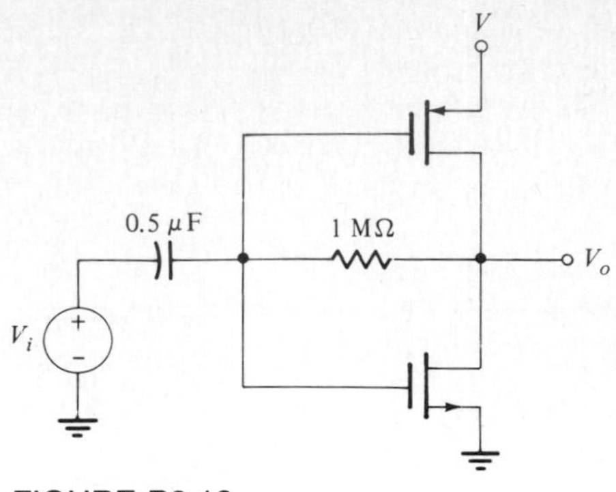

FIGURE P8.19

8.20 Design a cascaded CS FET amplifier using the minimum number of FETs to meet the following specifications:

$$A_v \geq 200 \qquad \omega_{3\text{dB}} \geq 2\pi(2\text{ MHz})$$

The following FET parameters are given: $g_m = 2 \times 10^{-3}$ mho, $r_d = 50$ kΩ, $C_{gs} = 10$ pF, $C_{gd} = 1$ pF, $R_S = 50$ Ω and $R_L = 1$ kΩ. Show the signal circuit of your design.

8.21 Determine the expressions for the common mode, differential mode, and CMRR of the basic FET op amp shown in Fig. P8.21. Determine the numerical values if the FET parameters are as given in problem 8.18.

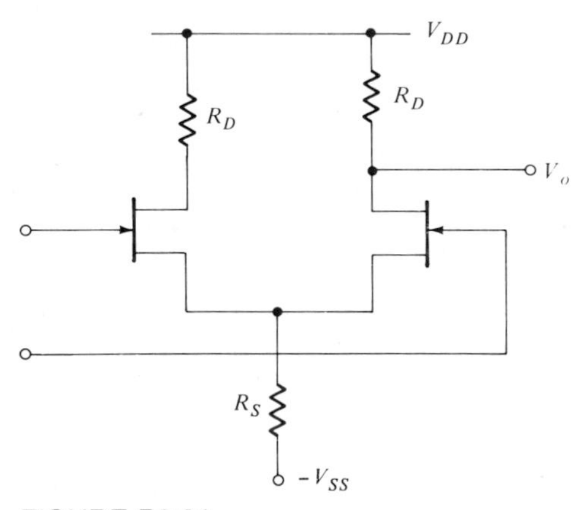

FIGURE P8.21

8.22 Design a cascode circuit using BJTs for a voltage gain of $A_v = 100$, $\omega_{3\text{dB}} = 2\pi(5\text{ MHz})$. Assume the transistor parameters of problem 8.1 and $R_L = R_S = 1$ kΩ.

8.23 Design a two-stage CE-CE cascade shown in Fig. P8.23 for the following specifications: $A_i \geq 100$, $f_{3\text{dB}} = 5$ MHz, $R_S = 10$ kΩ, $R_L = 100$ Ω. Show the ac high-frequency signal circuit only. Use the following BJT parameters: $\beta_0 = 100$, $f_T = 400$ MHz, $C_c = 2$ pF, $r_b = 50$ Ω, $r_e = 13$ Ω.

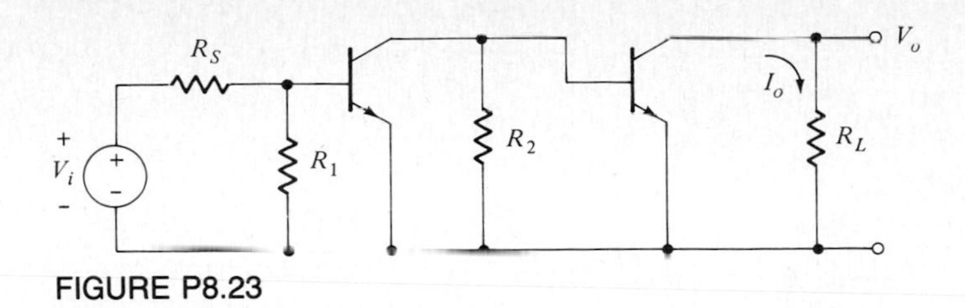

FIGURE P8.23

8.24 Repeat the design in problem 8.23 for the following specifications: $A_i = 100$ and $f_{3dB} \geq 5$ MHz. All other parameters remain the same.

8.25 Derive (8.51).

8.26 Design an amplifier, using 2N2222 BJTs (see Appendix E), for the following specifications: $A_v \simeq 50, f_l \simeq 100$ Hz, $f_h \simeq 1$ MHz, $R_{in} \simeq 50$ kΩ, $R_o \simeq 100$ Ω, and the output peak-to-peak signal voltage swing is to be 4 V. Assume a 20-V supply voltage, $R_S = 50$ Ω and $R_L = 10$ kΩ.
(*Hint:* Use a CC configuration at the input and the output.)

8.27 Design an amplifier, using an NMOS type MN82 (see Appendix E), for the following specifications: $A_V \simeq 50, f_l \simeq 1$ kHz, $f_h \simeq 1$ MHz, $R_{in} \simeq 1$ MΩ, $R_o \simeq 100$ Ω, and the output peak-to-peak signal swing is to be 4 V. Assume $R_S = 50$ Ω, $R_L = 10$ kΩ, and a supply voltage of 20 V. Note that the output stage can be different from a CS configuration.

CHAPTER 9

Operational Amplifiers and Applications

The monolithic operational amplifier (op amp) is one of the most important circuit blocks or modules in analog integrated circuits. Monolithic op amps are *small-size, low-cost, and highly reliable circuit modules*. Various types are available "off the shelf" at very low prices. The op amp has a tremendous variety of applications in linear and nonlinear electronics. We will consider in detail some of the op-amp applications in the processing of linear analog signals in this chapter. These include functional blocks, analog computation, and filters. Nonlinear applications will also be considered briefly so that the reader may appreciate the versatility of this important building block in the design of a large number and type of systems. Nonlinear applications include a variety of waveshaping circuits, comparator functions, and precision rectification. The chapter concludes with other practical considerations of op amps.

9.1 IDEAL OP AMP REVISITED

In Chap. 5 we discussed the internal configuration and properties of op amps. These are high-gain direct-coupled differential amplifiers utilizing emitter-coupled transistors at the input stage, shown in Fig. 5.25, or the Bi-FETs utilizing JFET at the input stage and BJT for the other stages, shown in Fig. 5.27. Two of the commonly used popular op amps are the 741 and the LF351 types.

The analysis of differential circuits as performed in Sec. 5.7 in terms of the differential-mode gain and common-mode gain, and the common-mode rejection ratio, and the input and the output impedances. We also determined the open-loop bandwidth in Sec 8.4.

The op amp has basically three terminals as shown symbolically in Fig. 9.1*a*. The subscripts p and n in Fig. 9.1*b*, correspond to positive and negative, i.e., noninverting and inverting terminals, respectively. The relation between the output and the input voltages is given by

$$V_o = a(V_2 - V_1) \tag{9.1a}$$

and

$$a(s) = \frac{a_0}{1 + s/p_o} \simeq \frac{a_0 p_o}{s} \qquad \text{for } \omega \gg p_o \tag{9.1b}$$

where $a_0 p_o$ is the gain-bandwidth product [typically 2π(1 MHz)] for the 741 and 351 op amps, a_0 is a very high voltage gain (typically 10^5 to 10^6), and p_o is the open-loop bandwidth [2π(5 Hz) to 2π(20 Hz)]. R_i is the input impedance, which is very high (typically $10^6\ \Omega$ for BJT and $10^{12}\ \Omega$ for Bi-FET), and R_o is the output impedance, which is low (50 to 100 Ω). An actual integrated-circuit op amp has many terminals

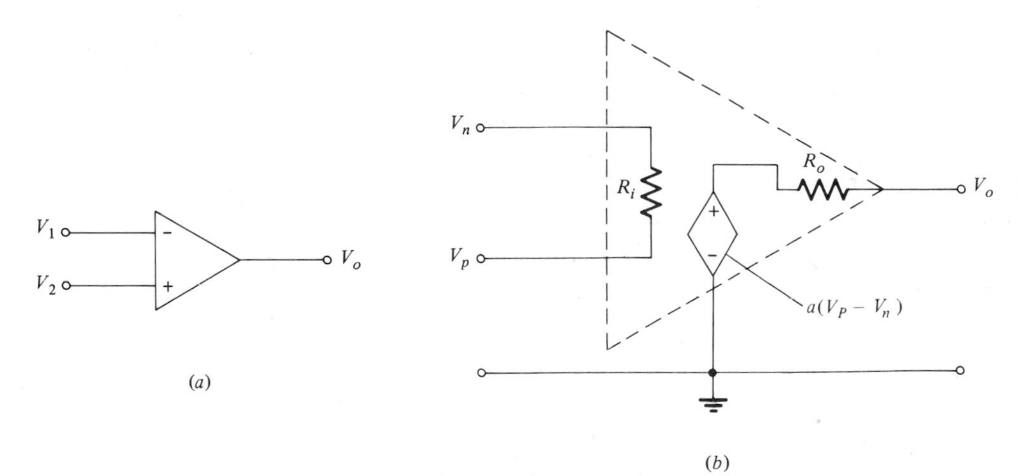

FIGURE 9.1 (*a*) Op-amp symbol showing the input and output terminals. (*b*) Op-amp circuit model.

that are used for the plus and minus dc power supplies, frequency compensation, null offset, etc., which will be discussed later in this chapter.

In order to be specific in the following discussion, we show the voltage reference points of Fig. 9.1*a* in Fig. 9.2*a* and consider time-domain analysis. Note that the op amp amplifies the *difference* between the voltage signals applied at its two input terminals. The circuit in Fig. 9.2*a* is shown in simplified form in Fig. 9.2*b*, which shall be henceforth called *ideal op amp*.

An ideal op amp, shown symbolically in Fig. 9.2*b*, was defined in (5.53) and is repeated in the following for convenience.

$$v_o = a_0(v_+ - v_-) \qquad \textbf{(9.2a)}$$

It has the following properties:

$$\begin{aligned} &\text{Voltage gain } a_0 = \infty \\ &\text{Bandwidth } p_o = \infty \\ &\text{Input resistance } R_i = \infty \\ &\text{Output resistance } R_o = 0 \\ &\text{Currents into the terminals } i_+ = 0 \qquad i_- = 0 \\ &\text{Potential difference } \epsilon = v_+ - v_- \end{aligned} \qquad \textbf{(9.2b)}$$

(Note that v_o is not zero and $\epsilon = 0$ only when negative feedback is used.) The subscript − and + signs indicate the inverting and noninverting terminals of the op amp, respectively. This means that a voltage signal applied at the minus (−) terminal will yield a negative voltage at the output, whereas at the plus (+) terminal it will yield a positive voltage at the output. The concept of an ideal op amp makes the analysis and design of circuits using op amps very simple, similar to the concept of an ideal

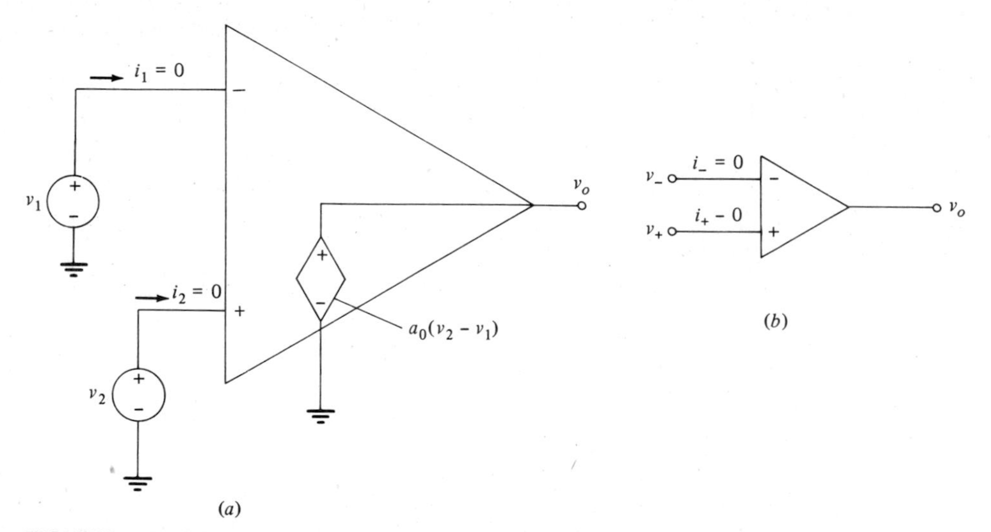

FIGURE 9.2 (*a*) Ideal op-amp circuit model showing reference voltages. (*b*) Ideal op-amp simplified model.

diode discussed in Chap. 1. It is interesting to note that the IC op amp has characteristics that are very close to the assumed ideal, and thus preliminary analysis assuming ideal op amps are satisfactory in most cases. The final analysis and design may include the actual op-amp circuit parameters if such an accuracy is warranted. In a number of applications, where the frequency characteristic of the op amp plays an important role, such as in feedback systems discussed in Chap. 10, the finite pole of an op amp must be considered as in Sec. 10.10.

9.2 BASIC CIRCUIT BLOCKS

In the following sections we will assume the op amp is ideal and obtain a number of useful circuit blocks. Op amps are invariably used in closed-loop configuration, i.e., feedback is applied around the op amp to reduce the gain and obtain other desired features. Negative feedback is used in the following circuits. (Feedback in discussed in Chap. 10.)

A. Inverting Amplifier

An inverting amplifier is shown in Fig. 9.3.* To determine the voltage gain of the circuit, we use Kirchhoff's voltage law. First recall from (9.2) that for an ideal op amp $i = 0$ and $v_- = 0$ since $v_+ = 0$ (this was referred to as a virtual ground in Chap. 5).

$$v_i = R_S i_s \tag{9.3a}$$

$$v_o = -R_f i_s \tag{9.3b}$$

---

*In a practical circuit there must be a dc resistance of value $R_S \parallel R_f$ in the plus lead for a BJT op amp (see Fig. 9.59) to compensate for the bias current. The bias current will cause a dc drop across R_S that will be amplified and give a very large offset voltage that may saturate the op amp. We ignore this consideration for the time being throughout this section. We shall, however, consider these practical considerations in Sec. 9.9. For a Bi-FET the input impedance is extremely high, and the parasitic shunt capacitance at the plus terminal to ground must be compensated for as shown in Fig. 9.60.

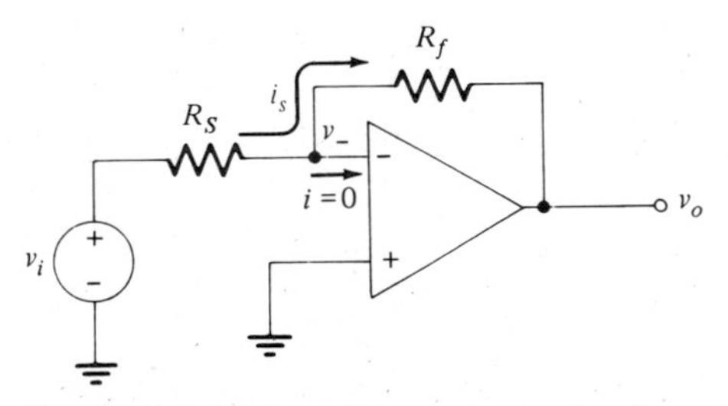

FIGURE 9.3 Inverting op-amp circuit.

From (9.3) we obtain

$$\frac{v_o}{v_i} = -\frac{R_f}{R_S} \tag{9.4}$$

The result in (9.4) is very important and will be frequently used in this chapter.

B. Noninverting Amplifier

A noninverting amplifier is shown in Fig. 9.4*a* and redrawn for convenience in Fig. 9.4*b*. Again, since $i = 0$, we have

$$v_- = \frac{R_2}{R_1 + R_2} v_o \tag{9.5}$$

but

$$v_- = v_+ = v_i$$

Hence

$$\frac{v_o}{v_i} = \frac{R_1 + R_2}{R_2} = 1 + \frac{R_1}{R_2} \tag{9.6}$$

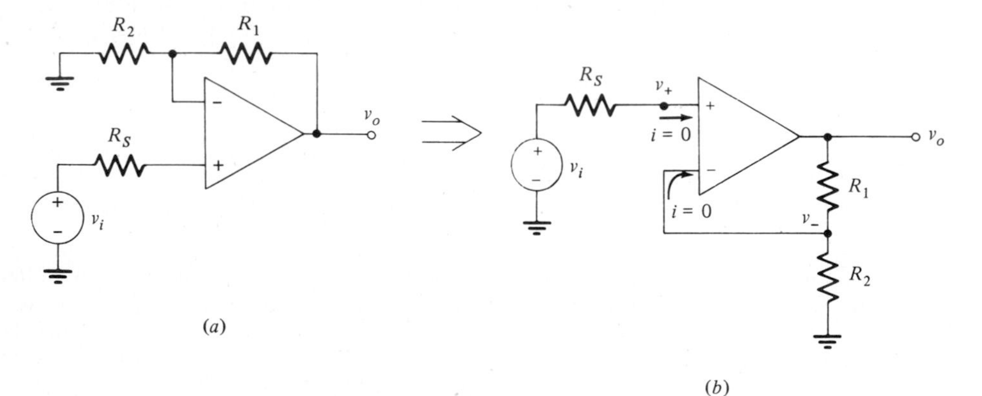

FIGURE 9.4 (*a*) Noninverting op-amp circuit. (*b*) Same circuit shown differently.

C. Voltage Follower or Unity Gain

The circuit for a voltage follower or a unity gain block is shown in Fig. 9.5*a*, redrawn in Fig. 9.5*b*. From Fig. 9.5*b*, we have

$$v_o = v_- \qquad \text{and} \qquad v_i = v_+ = v_-$$

Hence

$$\frac{v_o}{v_i} = 1 \tag{9.7}$$

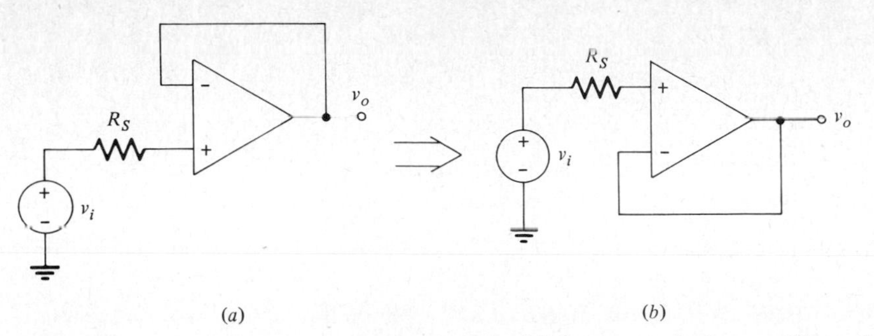

FIGURE 9.5 (*a*) Unity gain amplifier buffer. (*b*) Same circuit shown differently.

The voltage follower circuit is an *ideal buffer* (or an impedance transformer) since the input impedance in an actual unity gain op amp is almost infinite and its output impedance is almost zero for all practical purposes.

D. Current Sources

Op-amp current sources for a floating-load resistance R_L are shown in Fig. 9.6*a* and

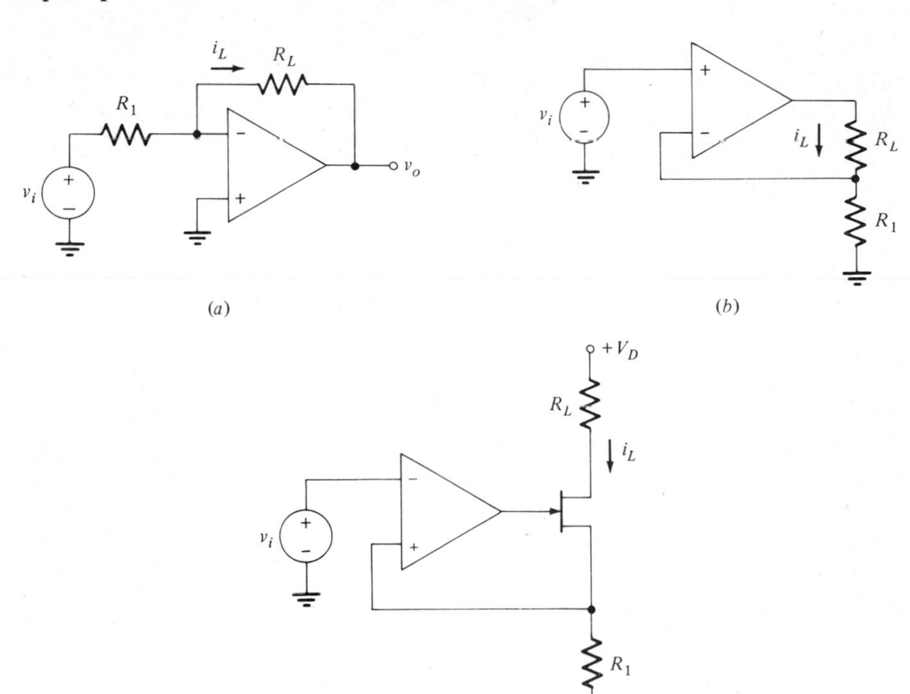

FIGURE 9.6 Op amp used as a (*a*) current source, (*b*) floating load, (*c*) nonfloating load.

b. In each case the load current is given by

$$i_L = \frac{v_i}{R_1} \tag{9.8}$$

The circuit shown in Fig. 9.6*c* avoids the floating load by incorporating a JFET. For this circuit the operating conditions require that the JFET gate is reverse-biased, and hence the relative operating levels of the circuit must ensure this condition.

E. Voltage-Controlled Voltage Source

The circuits in Fig. 9.7*a* is a voltage-controlled source (VCVS). This circuit has a very high input resistance and a very low output resistance and approximates the VCVS of Fig. 1.5*a*. Note that it is just a noninverting op amp of Fig. 9.4 with a fancy name. The simplified circuit representation of Fig. 9.7*a* is shown in Fig. 9.7*b*.

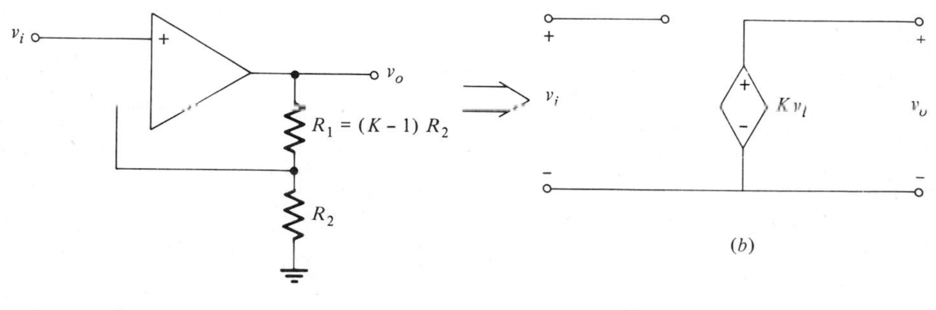

FIGURE 9.7 (*a*) Noninverting amplifier with gain *K*. (*b*) Voltage-controlled voltage source representation of (*a*).

EXERCISE 9.1

For the circuit shown in Fig. 9.3, a capacitor C is connected across R_f. Design the circuit for a voltage gain of -50 and a 3-dB bandwidth of $\omega_{3dB} = 10^4$ rad/s. Let $R_S = 1\ \text{k}\Omega$.

Ans $R_2 = 50\ \text{k}\Omega$, $C = 2 \times 10^{-9}$ F

EXERCISE 9.2

The circuit shown in Fig. 9.7 is to be used as a VCVS with a gain of 10. Determine the value of R_1 if $R_2 = 1\ \text{k}\Omega$.

Ans $R_1 = 9\ \text{k}\Omega$

EXERCISE 9.3

The input and output impedances of an actual op amp, used in a noninverting mode

TABLE 9.1 TYPICAL OP-AMP PARAMETERS

Circuit property	Approximate typical values: BJT (e.g., 741)	Bi-FET (e.g., LF351)
Input impedance (R_i)	1 MΩ	10^{12} Ω
Output impedance (R_o)	75 Ω	75 Ω
Open-loop gain (a_o)	2×10^5	10^5
Open-loop bandwidth ($f_{3dB} = p_o/2\pi$)	5 Hz	20 Hz
Unity gain bandwidth ($a_o f_{3dB}$)	1 MHz	2 MHz
Common-mode rejection ratio (CMRR)	95 dB	100 dB
Slew rate (SR)	0.7 V/μs	13 V/μs

as in Fig. 9.7, are given by

$$R_{in} = \frac{R_i a_0}{1 + R_1/R_2} \qquad R_{out} = \frac{R_o}{a_o}\left(1 + \frac{R_1}{R_2}\right)$$

(The derivation of these results are given in Chap. 10.) Determine the values of the input and output impedances of 741 if the op amp is designed for a gain of +10 (see Table 9.1).

Ans $R_{in} = 2 \times 10^{10}\ \Omega$, $R_o = 3.7 \times 10^{-3}\ \Omega$

9.3 ANALOG COMPUTATION

Op amps are used in analog computation, and the various mathematical operations are achieved as discussed in the following sections.

A. Inverter and Scale Changer

From (9.4) we see that the desired *scale change and the sign inversion* can be achieved by proper choice of R_f/R_S.

B. Adder (Summing Amplifier)

The circuit in Fig. 9.8, may be used to achieve summation. From the circuit, since $v_+ = 0$, we have $v_- = 0$ and

$$i_1 = \frac{v_1}{R_1} \qquad i_2 = \frac{v_2}{R_2} \qquad \cdots \qquad i_n = \frac{v_n}{R_n} \tag{9.9}$$

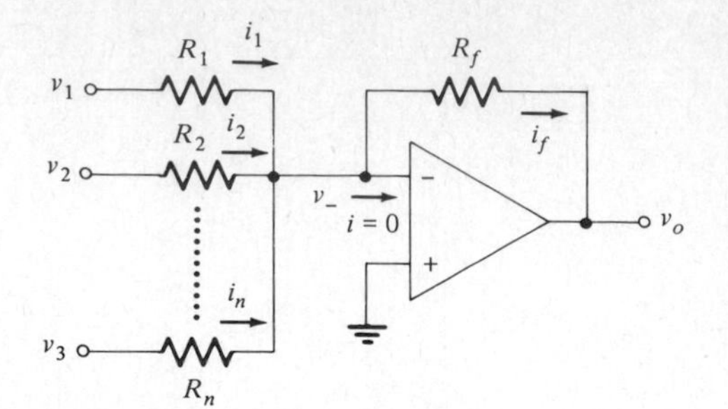

FIGURE 9.8 Op-amp adder.

also

$$v_o = -R_f i_f \qquad i_f = i_1 + i_2 + \cdots + i_n \tag{9.10}$$

From (9.9) and (9.10) we obtain

$$v_o = -\left(\frac{R_f}{R_1}v_1 + \frac{R_f}{R_2}v_2 + \cdots + \frac{R_J}{R_n}v_n\right) = -R_f \sum_{i=1}^{n} \frac{v_i}{R_i} \tag{9.11}$$

C. Integrator

The integrator circuit is shown in Fig. 9.9. This circuit yields an output voltage proportional to the integral of the input as follows:

$$\frac{v_i}{R} = i_R \qquad \text{and} \qquad i_C = -C\frac{dv_o}{dt} \tag{9.12}$$

But $i_R = i_C$; hence

$$v_o(t) = -\frac{1}{RC}\int_0^t v_i(\tau)\,d\tau + v_o(0) \tag{9.13}$$

Note that the specified initial conditions of a differential equation can be readily handled by placing a battery of appropriate value across C. An example of the analog computer simulation is given in the following.

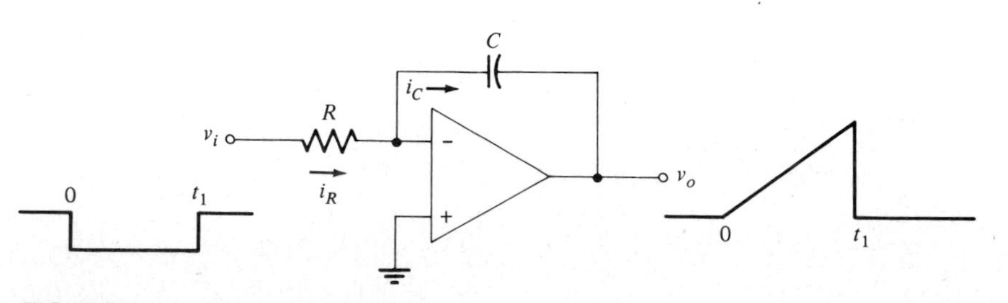

FIGURE 9.9 Op-amp integrator.

EXAMPLE 9.1

Consider a solution of the following second-order differential equation:

$$\frac{d^2x}{dt^2} + K_1\frac{dx}{dt} + K_0x + f(t) = 0$$

Assume zero initial conditions; i.e., $x(0) = 0$, $dx/dt(0) = 0$. We rewrite the above as follows:

$$\frac{d^2x}{dt^2} = -K_1\frac{dx}{dt} - K_0x - f(t)$$

We consider the highest-order term, d^2x/dt^2, and assume it is available. We next start drawing the circuit and use various outputs to generate d^2x/dt^2 as the output, which is the sum of three components in this case. This simulation is shown in Fig. 9.10. Now one integration by an operational amplifier of d^2x/dt^2 yields $-dx/dt$; another integration yields x. We provide one inverter for the sign change of dx/dt. The overall loop is then as shown in Fig. 9.11. Nonzero initial conditions can be handled readily as mentioned earlier.

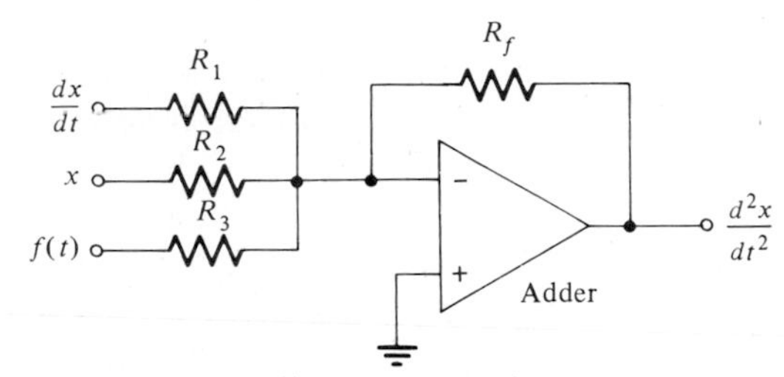

FIGURE 9.10 Example circuit, adder: $R_f/R_1 = K_1$; $R_f/R_2 = K_0$; $R_f/R_3 = 1$.

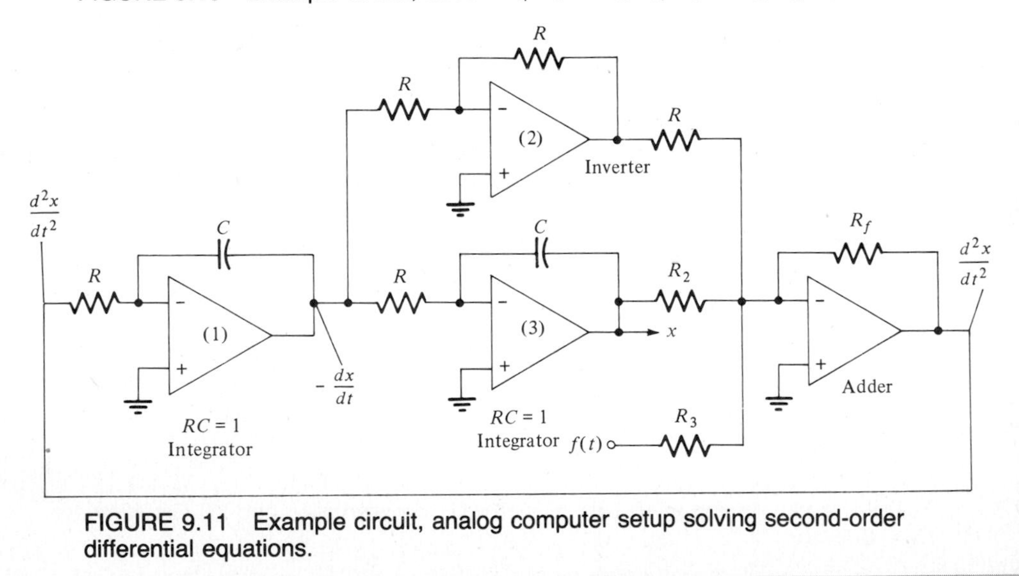

FIGURE 9.11 Example circuit, analog computer setup solving second-order differential equations.

An op amp differentiator is obtained by interchanging R and C in Fig. 9.9. However, differentiators are seldom used because of the following reasons:

1. In a differentiator a sudden change of input waveform may overload and saturate the amplifier.
2. The gain of a differentiator increases with frequency and thus is prone to instability or saturation.
3. It is more convenient to introduce initial conditions in an integrator.

The simulation of nonlinear differential equations requires the use of analog multipliers. In fact, analog multipliers have a wide variety of applications in communication circuits, and therefore, we shall discuss them in detail in the next section.

9.4 ANALOG MULTIPLIERS

Analog multipliers are important circuit modules that are used primarily for performing nonlinear operations such as multiplication, division, and raising variables to powers. The symbolic representation of a multiplier with its controlled-source model are shown in Fig. 9.12*a* and *b*, respectively.

An ideal single-ended analog multiplier is defined by

$$v_o = Kv_xv_y = 0.1v_xv_y \tag{9.14}$$

where v_o is the output voltage and v_x and v_y are input voltages. K is a scale factor, which is usually equal to 0.1 in most multiplier chips. The reason for this value is that for a 10-V full scale in v_o, v_x and v_y can each be 10-V full scale. The *ideal multiplier* is characterized by $v_o = 0$ when either $v_x = 0$ or $v_y = 0$ or both are zero.

Multipliers are classified according to the allowable polarities of the two inputs. If both the input polarities are allowed to be of one sign only, it is called a *one-quadrant multiplier*. If one of the inputs is positive or negative only with no restriction on the sign of the other input, it is called a *two-quadrant multiplier*. If both the inputs can assume positive and/or negative values, i.e., no restrictions on the sign of either inputs, it is called a *four-quadrant multiplier*. The latter is commonly used in most applications.

Multipliers, in addition to performing multiplication, can be used in other applications such as dividers and frequency multipliers. Integer power generation $y = \alpha x^2$ and $y = \beta x^3$ is obtained as shown in Fig. 9.13*a* and *b*.

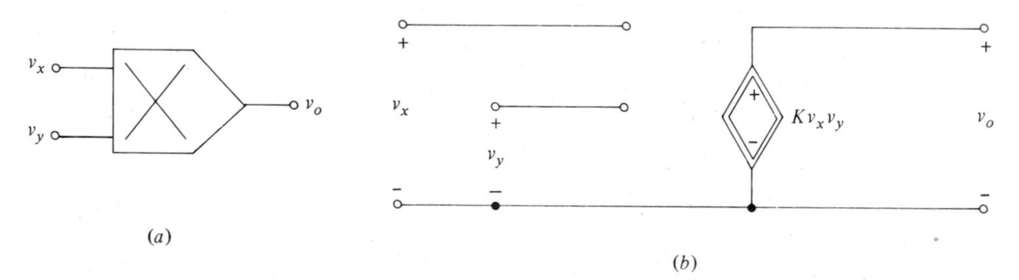

FIGURE 9.12 Analog multiplier: (*a*) symbolic representation and (*b*) circuit model ($K = 0.1$).

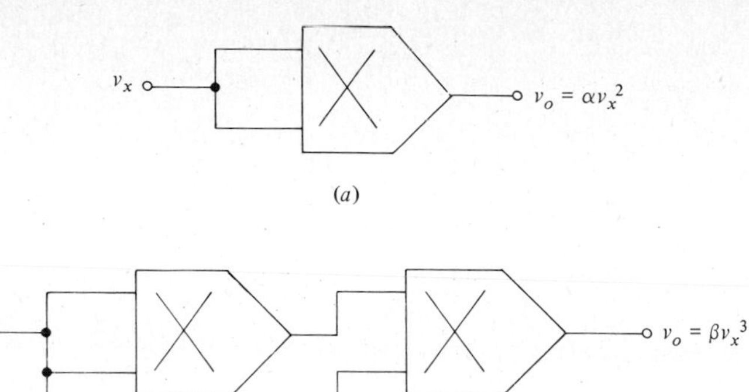

FIGURE 9.13 Variables raised to powers (a) $y = \alpha x^2$ and (b) $y = \beta x^3$.

EXAMPLE 9.2

A frequency doubler using a multiplier is obtained as follows: Consider the trigonometric identity

$$\sin^2 \omega t = \tfrac{1}{2}(1 - \cos 2\omega t)$$

Now if $v_x = v_y = V_m \sin \omega t$, the output is given by

$$v_o = KV_m^2 \sin^2 \omega t = \frac{KV_m^2}{2}(1 - \cos 2\omega t)$$

If the output is fed into a bandpass circuit of resonant frequency $\omega_0 = 2\omega$, the dc component can be filtered out, and the resulting circuit is a frequency doubler.

The use of multipliers to obtain a frequency tripler is left as an exercise in problem 9.7.

Divider circuits using multipliers and inverting and noninverting op amps are shown in Fig. 9.14*a* and *b*. (Note that the output of the multiplier must be of the same polarity as v_o to ensure negative feedback; hence $v_z > 0$.)

For the inverting op-amp circuit we have

$$K\frac{v_o v_z}{R} + \frac{v_x}{R} = 0 \tag{9.15a}$$

or

$$v_o = \frac{-1}{K}\frac{v_x}{v_z} \tag{9.15b}$$

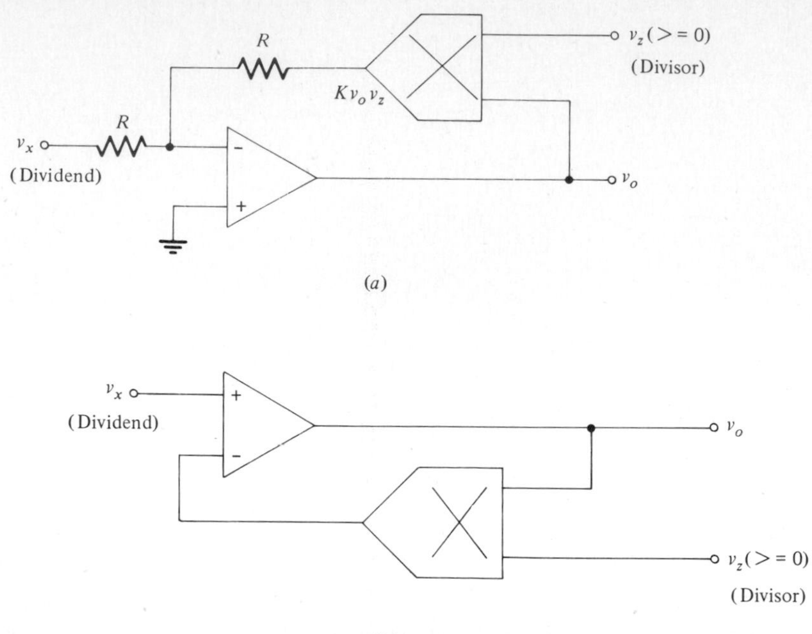

FIGURE 9.14 Division circuits (*a*) using inverting op amp and (*b*) using noninverting op amp.

Similarly, for the noninverting op-amp circuit we have

$$v_x = Kv_o v_z \tag{9.16a}$$

$$v_o = \frac{1}{K}\frac{v_x}{v_z} \tag{9.16b}$$

Note that the square root is readily obtained from the circuits in Fig. 9.14 by making $v_o = v_z$, i.e., connecting the v_z terminal to v_o. Under this condition, Eqs. (9.15*b*) and (9.16*b*) reduce to

$$v_o = \sqrt{\frac{-v_x}{K}} \tag{9.17a}$$

$$v_o = \sqrt{\frac{v_x}{K}} \tag{9.17b}$$

Note that $v_x < 0$ for (9.17*a*) and $v_x > 0$ for (9.17*b*), which implies the use of a one-quadrant multiplier.

The use of op amps in simulating a linear constant-coefficient differential equation was given in Sec. 9.3. With the availability of the multipliers that provide multiplication, division, and raising variables to powers we can now simulate nonlinear differential equations.

EXAMPLE 9.3

Show an analog simulation of the nonlinear differential equation (Van der Pol's)

$$\frac{d^2x}{dt^2} - \epsilon(1 - x^2)\frac{dx}{dt} + x = 0$$

where ϵ is a constant. We can rewrite the above equations as

$$\frac{d^2x}{dt^2} = \epsilon\frac{dx}{dt} - \epsilon x^2\frac{dx}{dt} - x$$

The solution requires two integrators and two multipliers as shown in Fig. 9.15.

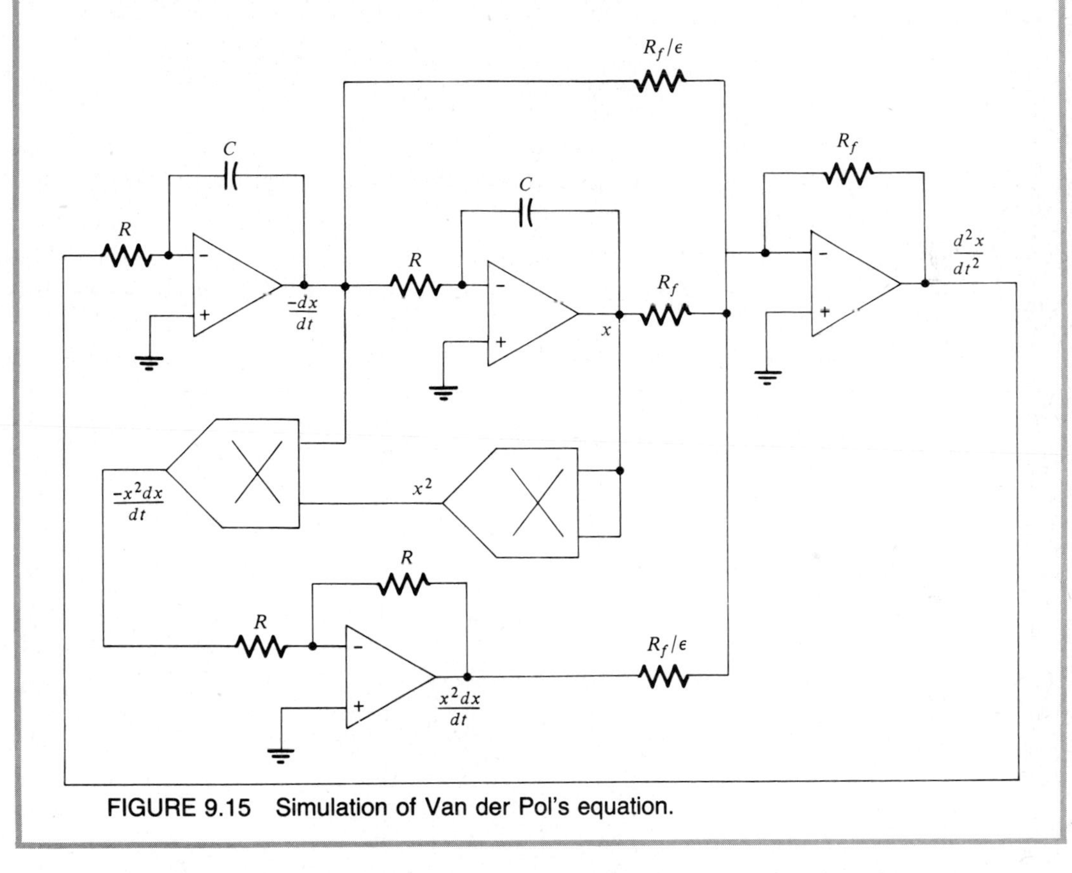

FIGURE 9.15 Simulation of Van der Pol's equation.

A. Analog Multiplier Circuit

A two-quadrant multiplier circuit is shown in Fig. 9.16*a*, which is essentially a gain-controlled amplifier that multiplies the input signal v_x with the external signal v_y

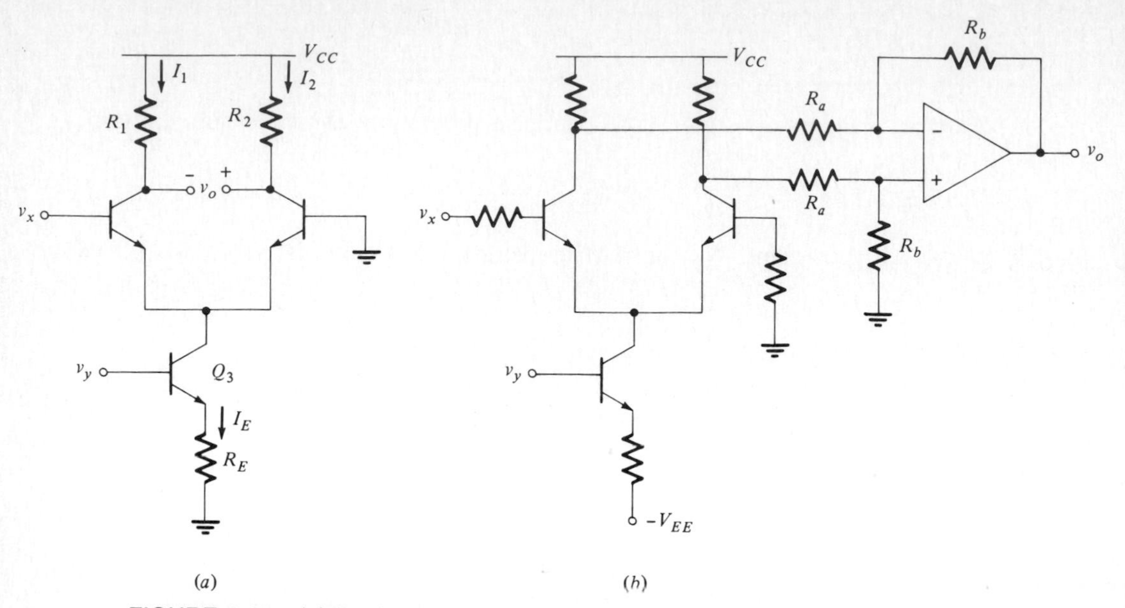

FIGURE 9.16 (*a*) Basic two quadrant multiplier circuit. (*b*) Practical version of (*a*).

to produce the output voltage v_o. The gain constant K is externally adjustable. The circuit makes use of the dependence of the transconductance of BJT on the emitter bias (the so-called variable transconductance method).

The principle of operation of the circuit is as follows: For small signals, such that $v_x \ll kT/q = V_t$, the differential output voltage v_o of the circuit is given by

$$v_o = g_m R_2 v_x \qquad \text{for } R_1 = R_2 \tag{9.18a}$$

where

$$g_m = \frac{qI_E}{kT} = \frac{I_E}{V_t} \tag{9.18b}$$

The transconductance g_m can be varied by varying v_y, because for $I_E R_E \gg V_{BE}$, we have

$$v_y \simeq I_E R_E \tag{9.19}$$

From (9.18) and (9.19) we obtain

$$v_o = \frac{R_2}{V_t R_E} v_x v_y \tag{9.20}$$

which is of the form given in (9.14). Note that the external gain controlling signal must be positive and larger than $V_{BE} \simeq 0.7$ V. The input signal v_x may be positive or negative and is attenuated to meet the small-signal condition as in Fig. 9.16*b*. The two-quadrant multiplier functional space is shown in Fig. 9.17.

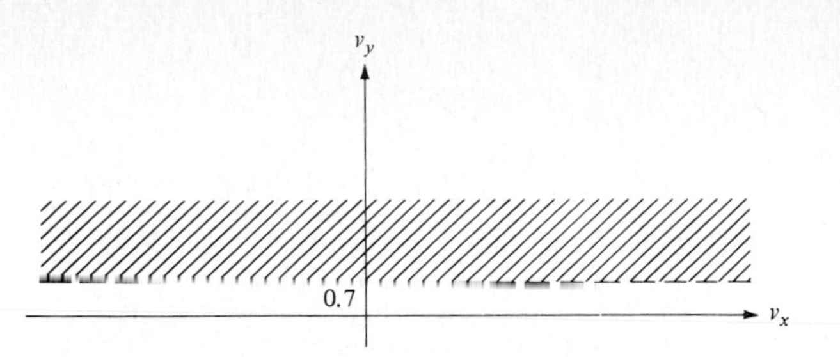

FIGURE 9.17 Region of operation of the circuit in Fig. 9.16.

B. Practical Considerations in Multipliers

In a practical multiplier the output is not given exactly by (9.14). In general, the output of a multiplier can be written as

$$v_o = Kv_xv_y \pm \epsilon_v \tag{9.21}$$

where ϵ_v is the error voltage. Actually, ϵ_v is of the form

$$\epsilon_v = K_xv_x + K_yv_y + K_o \tag{9.22}$$

and these constants are referred to as the x, y, and null-offset constants. Note that in a high-accuracy multiplier four separate adjustments (K, K_o, K_x, and K_y) are needed to set the multiplier gain constant K equal to 0.1/V. Manufacturers usually specify the multiplier accuracy by the worst-case error as a percentage of full-scale output voltage; i.e.,

$$\text{Accuracy} = \frac{|v_o - Kv_xv_y|}{\text{full-scale voltage}} \tag{9.23}$$

In practice, accuracy is measured with dc voltages applied at both the inputs at manufacturer's rated supply voltage.

For example, 1 percent full-scale accuracy with a ± 10 V output means that the actual output would be within ± 0.1 V of the desired level. In most applications using multipliers, temperature stability is also important. Temperature stability is usually given in terms of the temperature drift of the null-offset term K_o (mV/°C) and the gain constant K (ppm/°C).

A four-quadrant multiplier can be obtained from 2 two-quadrant multipliers and an op amp (problem 9.11). More practical four-quadrant multipliers are available by several IC manufacturers, e.g., the RCA CA3091D or the Motorola MC1595.

C. Amplitude Modulation

One way to transmit an audio signal in communication systems is via amplitude modulation (AM) using a high-frequency carrier wave. Analog multipliers can be used for such purposes as shown in Fig. 9.18, where one input is the modulating wave, the other input is the carrier wave, and the output is the modulated wave. Specifically, if

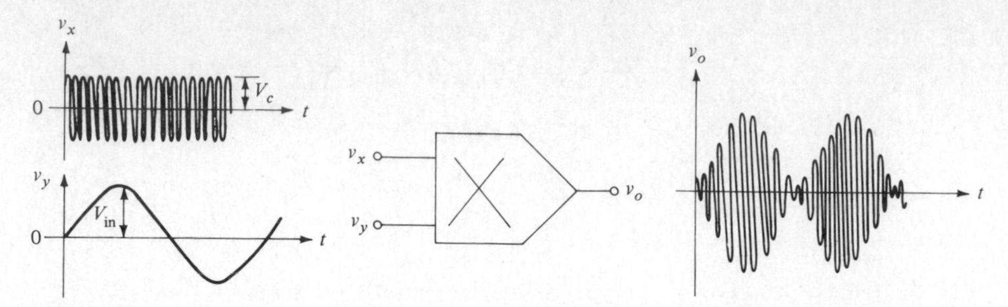

FIGURE 9.18 Analog multiplier used in modulation with input and output waveforms.

$v_x = V_c \sin \omega_c t$ and $v_y = V_m \sin \omega_m t$, then

$$v_o = \frac{v_x v_y}{10} = \frac{V_c V_m}{10}(\sin \omega_c t)(\sin \omega_m t) \tag{9.24}$$

The modulated output signal is shown in Fig. 9.18. The transmitted signal is recovered at the receiving end by a process called *demodulation* or *detection* (see Fig. 1.33). Using the trigonometric identity $\sin \alpha \sin \beta = \frac{1}{2}[\cos (\alpha - \beta) - \cos (\alpha + \beta)]$, we can write (9.24) as

$$v_o = \frac{V_c V_m}{20}[\cos (\omega_c - \omega_m)t - \cos (\omega_c + \omega_m)t] \tag{9.25}$$

Note that the output signal consists of the sum and difference of the modulating and carrier frequencies. Signals of the form of (9.25) are amplified via tuned amplifiers with center frequency at ω_c, and a bandwidth wide enough to pass the modulating audio frequencies. More is said about this in Chap. 11.

EXERCISE 9.4

Two sinusoidal voltages $v_x = V_0 \cos 10^5 t$ and $v_y = V_0 \cos (10^5 t + \phi)$ are applied to the analog multiplier, with $K = 0.1/\text{V}$, and the output of the multiplier is fed through a filter that eliminates the sinusoidal signal. Determine the value of ϕ if the output of the filter has (*a*) its maximum value of 1 V and (*b*) a value of 0.5 V.

Ans (*a*) $\phi = 0°$, (*b*) $\phi = 60°$

9.5 ACTIVE *RC* FILTERS

Filters are widely used for signal processing in electronic circuits. Active *RC* filters are the class of filters that utilize op amps as active elements in conjunction with passive element resistors and capacitors (*RC*), hence the name active *RC*. Since filters are used to affect the frequency behavior of the signal, we naturally deal with the frequency domain and hence poles and zeros of the transfer function. They may take any one of the following forms: low-pass, bandpass, high-pass, and band-elimination (bandstop) filters. The *idealized magnitude responses* of such filters are shown in Fig. 9.19. Of course, such idealized responses cannot be realized exactly by any physical system. However, they can be approximated to a satisfactory degree in most practical applications. We shall consider in this chapter, however, only the simple techniques in order to familiarize the reader with the design aspects of such filters. For example, the Butterworth transfer function to approximate the low-pass response of Fig. 9.19*a* was already discussed briefly in Chap. 8. We shall therefore consider the realization of simple functions using op amps in order to approximate the frequency selective behavior of Fig. 9.19. The op amps are assumed ideal.

Practical *RC* filters can be designed as a cascade of simple second-order networks. Thus, if we have a realization of a general biquadratic function (or biquad), namely, the ratio of two second-order polynomials

$$A(s) = \frac{V_o}{V_i} = \pm\frac{b_2s^2 + b_1s + b_0}{s^2 + a_1s + a_0} \tag{9.26}$$

then any general transfer function of higher order can be realized by cascaded noninteracting blocks with transfer functions given by (9.26). In (9.26) the denominator coefficients are all nonnegative constants for stability reasons (see Chap. 10). Since op amps are ideal noninteracting blocks, we can assume no interaction in cascading biquads. If the input-output mismatch between the blocks is not high enough, unity

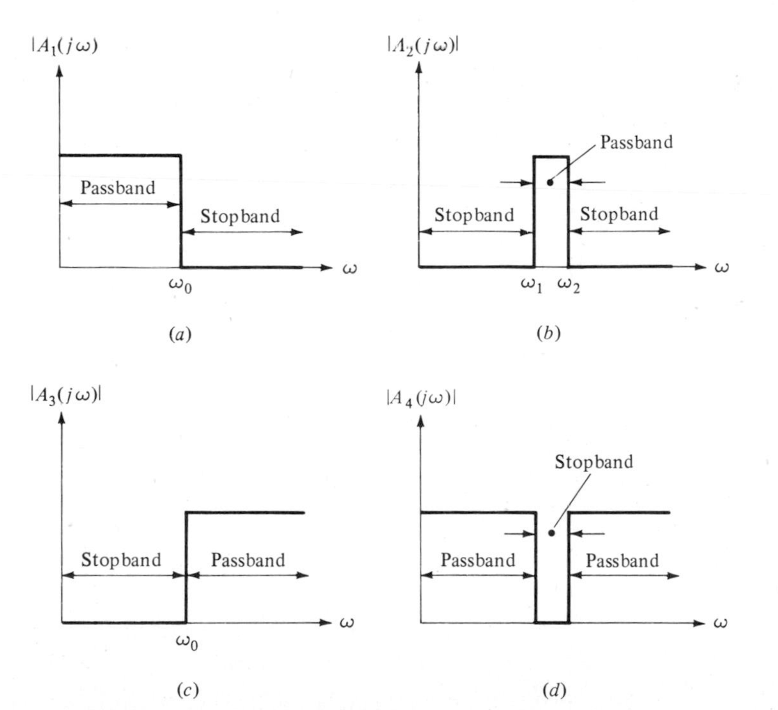

FIGURE 9.19 The various forms of ideal frequency-selective characteristics: (*a*) Low-pass filter, (*b*) bandpass filter, (*c*) high-pass filter, (*d*) band-elimination filter.

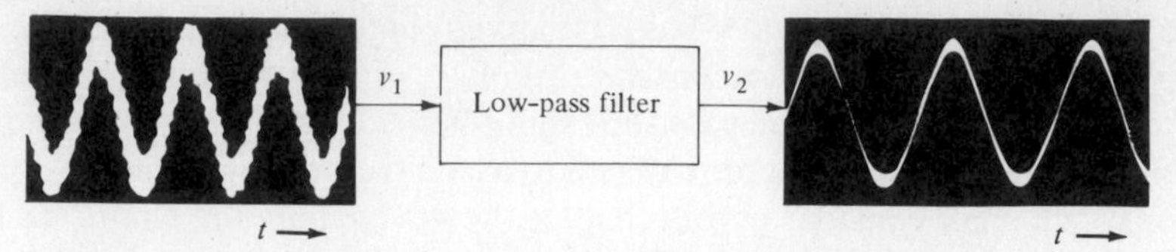

FIGURE 9.20 A low-pass filter application [5].

gain buffers can always be used. We shall next consider special cases of (9.26). The implementation of the general case in (9.26) will be discussed later in this section.

A. Low-Pass Sections

Low-pass circuits can be used to filter out high-frequency spurious signals as shown in Fig. 9.20. A high-order low-pass circuit can be designed by cascading several second-order low-pass sections.

Single-Pole Sections

For a high-order transfer function of odd order, the circuits shown in Fig. 9.21 can be used for the one-pole section. For the passive low-pass circuit in Fig. 9.21a, which

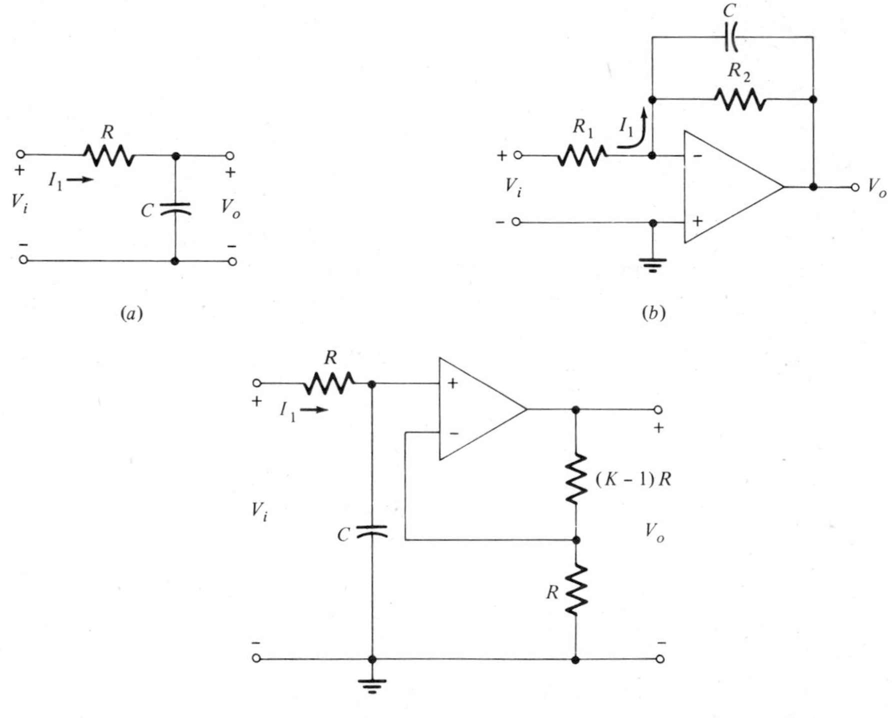

FIGURE 9.21 Single-pole low-pass section (a) using a passive network, (b) using an inverting op amp, and (c) using a noninverting op amp.

we have already discussed in Sec. 7.1, the transfer function is

$$\frac{V_o}{V_i} = \frac{1/RC}{s + 1/RC} \tag{9.27}$$

For a noninteracting one-pole section the active filter in Fig. 9.21*b* or *c* may be used. For the circuit of Fig. 9.21*b*, since $V_- = 0$ and $I = 0$, the gain function is obtained simply from

$$I_1 = \frac{V_i}{R_1} = -\frac{V_o}{R_2 \parallel 1/sC}$$

or

$$\frac{V_o}{V_i} = \frac{-1}{R_1C(s + 1/R_2C)} \tag{9.28}$$

For the circuit in Fig. 9.21*c* the transfer function is the same as in Fig. 9.20*a* but with a gain K since the op amp is used as a VCVS, as in Fig. 9.7*b*.

$$\frac{V_o}{V_i} = \frac{K/RC}{s + 1/RC} \tag{9.29}$$

Second-Order Low-Pass Sections

There are several circuits that provide a low-pass, two-pole transfer function of the form

$$A_v(s) = \frac{H_1}{s^2 + \alpha s + \beta} \tag{9.30}$$

The circuit shown in Fig. 9.22 is a simple circuit that can implement the gain function in (9.30). The voltage transfer function of the network in Fig. 9.22 is obtained by analysis, simply by replacing the noninverting op amp by a VCVS. The voltage across

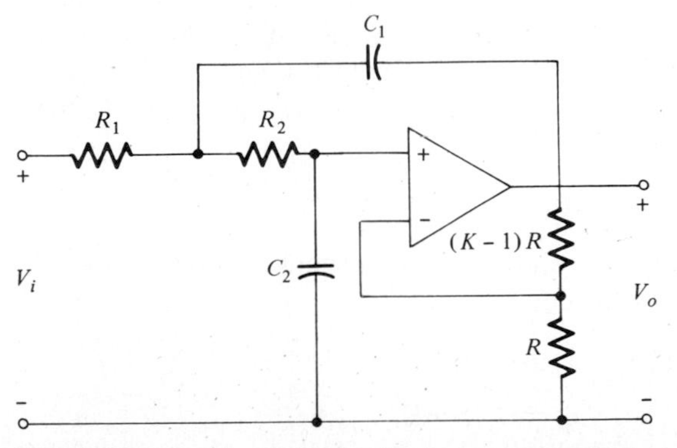

FIGURE 9.22 A second-order low-pass section.

C_2 is V_o/K, and no current is flowing into the op amp. By writing two loop equations and solving for V_o/V_i, we get

$$A_v = \frac{V_o}{V_i} = \frac{KG_1G_2S_1S_2}{s^2 + s(1/R_1C_1 + 1/R_2C_1 + 1/R_2C_2 - K/R_2C_2) + 1/R_1C_1R_2C_2} \quad \textbf{(9.31a)}$$

where $G_i = 1/R_i$ and $S_i = 1/C_i$, $i = 1, 2$.

Equation (9.31*a*) may be rewritten as

$$A_v = \frac{V_o}{V_i} = \frac{K/\tau_1\tau_2}{s^2 + s(1/\tau_1 + 1/\tau_2 - K/\tau_2 + 1/R_2C_1) + 1/\tau_1\tau_2} \quad \textbf{(9.31b)}$$

where $\tau_1 = R_1C_1$ and $\tau_2 = R_2C_2$. From (9.31*b*) and (9.30), by matching coefficients of (9.13*b*) with those of (9.30), we obtain

$$H_1 = \frac{K}{\tau_1\tau_2} \qquad \beta = \frac{1}{\tau_1\tau_2} \qquad \alpha = \frac{1}{\tau_1} + \frac{1}{\tau_2}(1 - K) + \frac{1}{R_2C_1} \quad \textbf{(9.32)}$$

There are five design variables: two R's, two C's, and the gain K. Since the number of unknowns is larger than the number of equations, two variables may be preselected to have convenient values, and the others are then determined from (9.32). Note that the source resistance specification can be incorporated in R_1.

EXAMPLE 9.4

Design a Butterworth two-pole filter using the circuit in Fig. 9.22 for a 3-dB bandwidth of $\omega = 2\omega(10^4)$rad/s. The source resistance is given as $R_S = 1\ \text{k}\Omega$.

For a two-pole Butterworth filter we have

$$A_v = \frac{V_o}{V_i} = \frac{K}{s^2 + \sqrt{2}(6.28 \times 10^4)s + (6.28 \times 10^4)^2}$$

We can match the coefficients of the above with those of (9.31) to solve for the element values. We shall use normalization, discussed in Sec. 7.15, in this example in order to familiarize the reader with it. Let us choose $R_{1n} = 1$ (normalized), which may be assigned to the source resistance, i.e., $R_S = R_{1n}$, and let $C_{1n} = 1$. Then from the specification and (9.31) by matching the coefficients, we obtain

$$K = 10 \qquad R_{2n} = 0.106 \qquad C_{2n} = 9.414$$

The actual element values corresponding to the normalization factors $R_n = 1\ \text{k}\Omega$ and $\Omega_n = 2\pi \times 10^4$ are

$$R_S = R_1 = 1\ \text{k}\Omega \qquad R_2 = 106\ \Omega \qquad K = 10$$

$$C_1 = 0.016\ \mu\text{F} \qquad C_2 = 0.15\ \mu\text{F}$$

B. Bandpass Sections

The application of bandpass filters in communication circuits is quite extensive, and one such example is shown in Fig. 11.1. In the ideal bandpass response, shown in Fig. 9.19*b*, the frequency response is attenuated beyond the bandpass at low and high frequencies. The bandpass gain function is given by

$$A_v(s) = \frac{H_2 s}{s^2 + a_1 s + a_o} \tag{9.33a}$$

Equation (9.33a) is usually expressed as

$$A_v(s) = \frac{H_2 s}{s^2 + (\omega_0/Q)s + \omega_0^2} = \frac{H_2 s}{s^2 + \omega_{3dB} s + \omega_0^2} \tag{9.33b}$$

In (9.33*b*) the center frequency ω_0 and the quality factor Q are put in prominence, where $Q = \omega_0/\omega_{3dB}$. These two parameters are very important in bandpass filters. In many applications a narrow bandwidth is desirable, i.e., $Q \geq 10$. Figure 9.23 shows a normalized bandpass response defining ω_0 and ω_{3dB}. Bandpass amplifiers are also called *tuned amplifiers*—this topic is treated in detail in Chap. 11.

The simplest and one of the most useful circuits for realizing a second-order bandpass response is shown in Fig. 9.24*a*. The circuit is also canonic in that it uses the minimum number of circuit elements. The voltage gain function of the circuit is readily obtained by noting that $V_- = 0$ since $V_+ = 0$. Thus from two loop equations we can solve for V_o/V_i, which yields

$$A(s) = \frac{V_o}{V_i} = \frac{-G_1 S_1 s}{s^2 + sG_2(S_1 + S_2) + G_1 S_1 G_2 S_2} \tag{9.34}$$

where $G_i = 1/R_i$ and $S_i = 1/C_i$, $i = 1, 2$. Again we have four degrees of freedom, i.e., four element values to choose in order to specify the pole position. We may narrow the choice by selecting the normalized values

$$S_{1n} = S_{2n} = 1 \qquad G_{2n} = 1 \qquad G_{1n} = \frac{1}{\gamma} \tag{9.35}$$

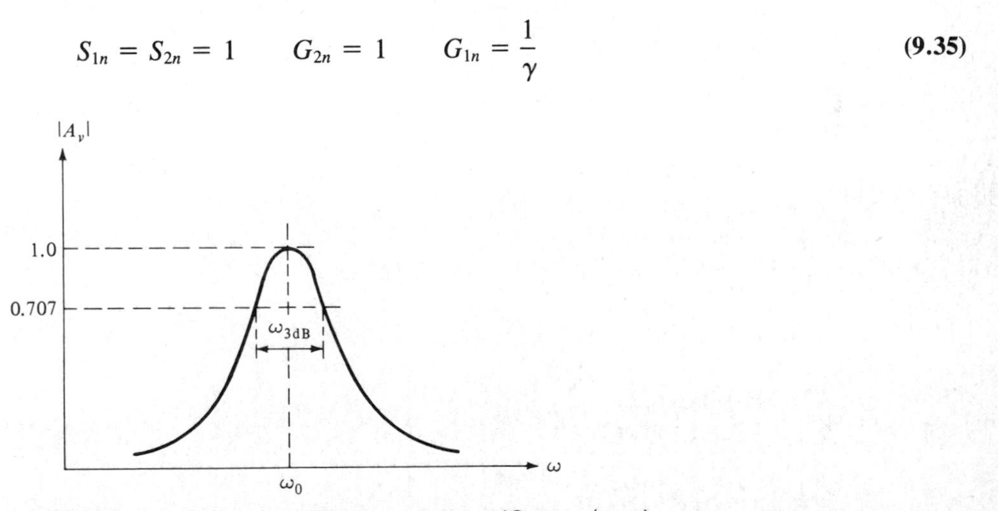

FIGURE 9.23 A typical bandpass response ($Q = \omega_o/\omega_{3dB}$).

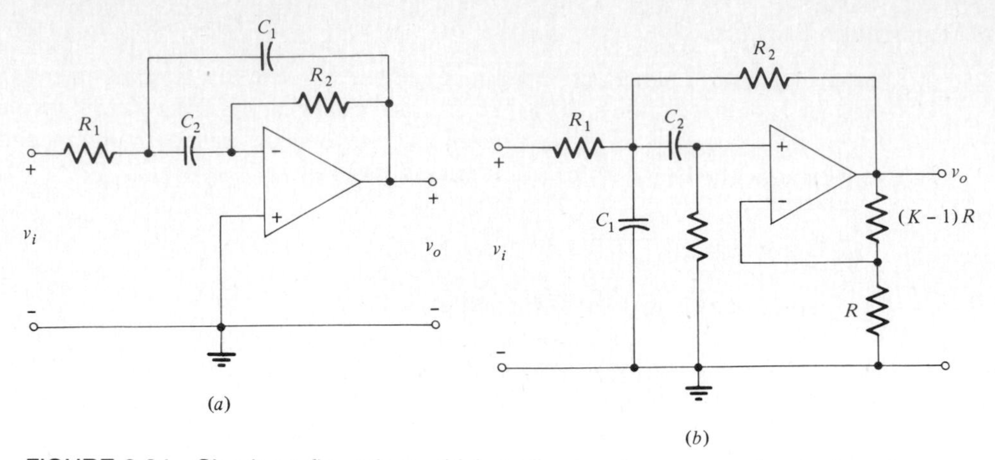

FIGURE 9.24 Circuit configurations which realize bandpass second-order transfer functions. (*a*) Canonical configuration using an inverting gain block. (*b*) Configuration using a noninverting gain block.

Substitution of the normalized values of (9.35) into (9.34) yields

$$A_v(s_n) = \frac{-s_n(1/\gamma)}{s_n^2 + 2s_n + 1/\gamma} \tag{9.36}$$

A comparison of (9.36) and (9.33*b*) yields

$$\omega_{\text{on}} = \frac{1}{\gamma} \quad \text{and} \quad Q = \frac{1}{2\sqrt{\gamma}} \tag{9.37}$$

Note that Q is unaffected by normalization, as it is the ratio of two frequencies. From (9.37) it is seen that a specified value of Q or ω_0 determines the other. These can be made independent of each other (i.e., specified independently) if we do not constrain G_{2n} or S_{2n} to be unity.

EXAMPLE 9.5

Design a bandpass section with $Q = 10$ and $\omega_0 = 2\pi(10^4)$ rad/s, using the circuit in Fig. 9.24*a*.

Let $C_1 = C_2 = C$. From a comparison of (9.33*b*) with (9.34) we have

$$\omega_0^2 = G_1S_1G_2S_2 = \frac{1}{R_1R_2C^2} = [2\pi(10^4)]^2$$

and

$$\frac{\omega_0}{Q} = G_2(2S_2) = \frac{2}{R_2C} = 2\pi(10^3)$$

Let $R_2 = 100\ \text{k}\Omega$ and from above we determine $R_1 = 250\ \Omega$ and $C_1 = C_2 = 0.0032\ \mu\text{F}$.

A circuit that utilizes a noninverting op amp is shown in Fig. 9.24*b*. The analysis and design of the circuit is similar and left as an exercise in problem 9.16.

C. High-Pass Section

High-pass filters are useful to filter out the undesired low-frequency components of the signal from the main high-frequency signals as shown in Fig. 9.25. The high-pass second-order gain function is given by

$$A_v(s) = \frac{H_3 s^2}{s^2 + b_1 s + b_0} \tag{9.38}$$

In (9.38) note that low frequencies are attenuated. A high-pass circuit using a non-inverting op amp is shown in Fig. 9.26. The voltage transfer function of Fig. 9.26 is found similar to the low-pass case and is given by

$$A_v = \frac{V_o}{V_i} = \frac{Ks^2}{s^2 + s(1/R_1C_1 - K/R_1C_1 + 1/R_2C_2 + 1/R_2C_1) + 1/R_1R_2C_1C_2} \tag{9.39a}$$

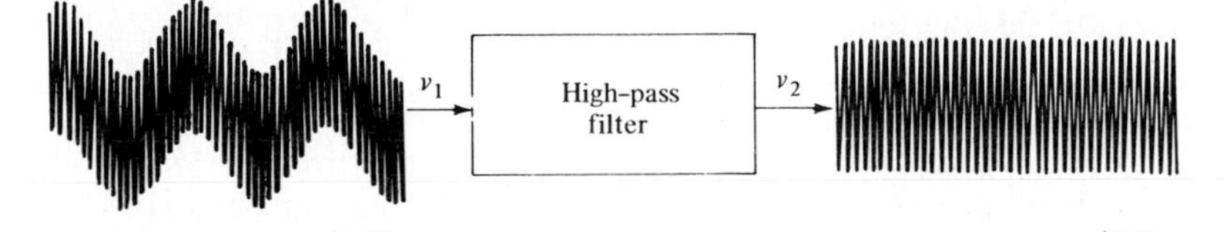

FIGURE 9.25 A high-pass filter application [5].

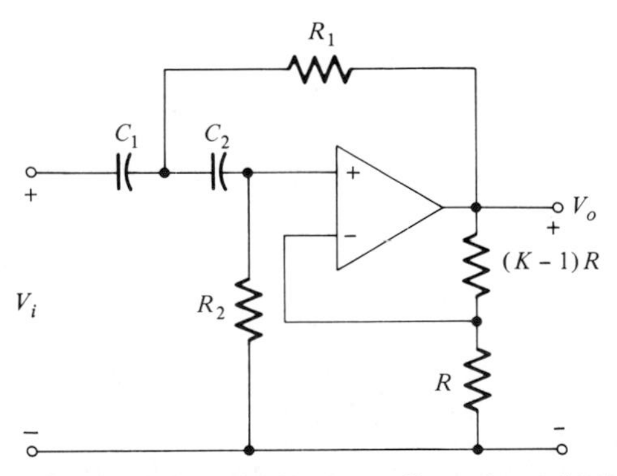

FIGURE 9.26 A circuit configuration which realizes a second-order high-pass transfer function.

Equation (9.39*a*) can be written as

$$A_v = \frac{Ks^2}{s^2 + s[(1-K)/\tau_1 + 1/\tau_2 + 1/R_2C_1] + 1/\tau_1\tau_2} \tag{9.39b}$$

where $\tau_1 = R_1C_1$ and $\tau_2 = R_2C_2$. The design equations as obtained by matching the coefficients are

$$H_3 = K \qquad b_0 = \frac{1}{\tau_1\tau_2} \qquad b_1 = \frac{1-K}{\tau_1} + \frac{1}{\tau_2} + \frac{1}{R_2C_1} \tag{9.40}$$

The design procedure is quite similar to the previous cases. Note that if the order of the high-pass filter is odd, a simple passive *RC* section shown in Fig. 9.27*a* or an active one-pole section of Fig. 9.27*b* may be used. It is to be noted that a high-pass circuit is obtained from a low-pass circuit by simply interchanging all *R*'s with *C*'s and vice versa.

D. Bandstop or Notch Section

Bandstop or notch filters are useful for rejecting undesired frequencies. One such application is shown in Fig. 9.28*a*. The ideal bandstop response is shown in Fig. 9.19*d*. The notch gain function is given by

$$A_v(s) = \frac{H_4(s^2 + \omega_z^2)}{s^2 + (\omega_p/Q)s + \omega_p^2} \tag{9.41}$$

A circuit that yields the function in (9.41) is shown in Fig. 9.29. The circuit will filter out completely the signal frequency ω_z from the input. The gain function of the circuit

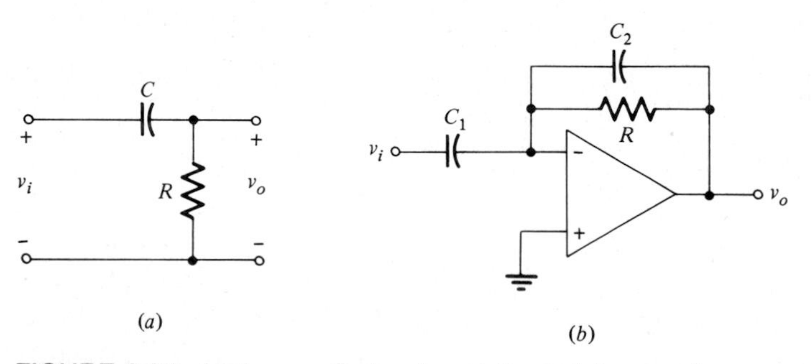

FIGURE 9.27 High-pass first-order sections: (*a*) a passive one-pole section, (*b*) an active one-pole section.

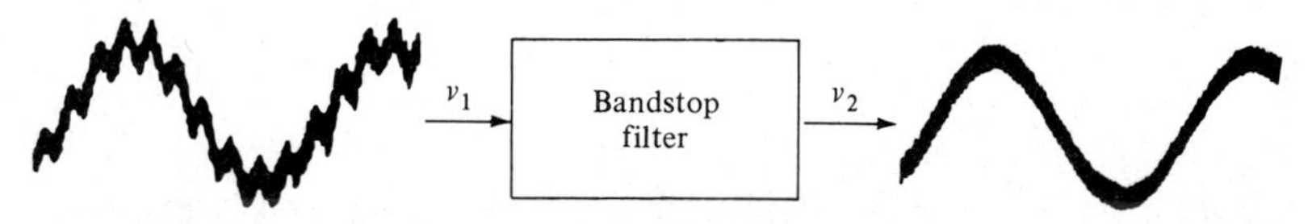

FIGURE 9.28 A bandstop filter application [5].

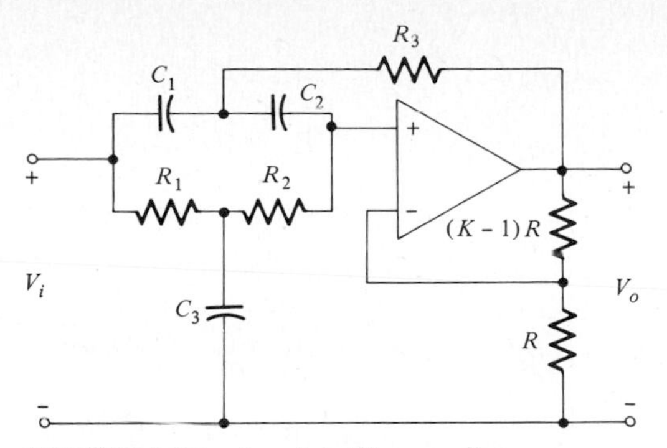

FIGURE 9.29 A notch filter section.

in Fig. 9.29 is obtained by replacing the op amp with a VCVS gain of K. Since no current can flow into the op amp, we can write loop equations and solve for V_o/V_i. The result is

$$A_v(s) = \frac{K(C_1C_2s^2 + G_1G_2)}{C_1C_2s^2 + [(C_1 + C_2)G_2 + C_2(G_1 + G_2)(1 - K)]s + G_1G_2} \tag{9.42}$$

where $C_3 = C_1 + C_2$

$$G_3 = G_1 + G_2$$

$$G_i = \frac{1}{R_i} \qquad i = 1, 2, 3$$

Design of the circuit for specific numerical values of R and C is straightforward and is left as an exercise in problem 9.22.

EXERCISE 9.5

The circuit in Fig. 9.29 is used to filter out an undesired signal frequency at 100 kHz. Determine the values of the capacitors if $C_1 = C_2$ and $R_1 = R_2 = 1\ \text{k}\Omega$.

Ans $C_1 = 1.6 \times 10^{-9}$ F

E. All-Pass Section

An all-pass section has a constant-magnitude gain function within the frequency range of interest, but the phase-shift varies. All-pass sections are useful in phase-shaping networks such as in delay equalization. A first-order all-pass section is shown in Fig. 9.30. From an analysis of the circuit we obtain

$$\frac{V_o}{V_i} = \frac{s - 1/RC}{s + 1/RC} \tag{9.43}$$

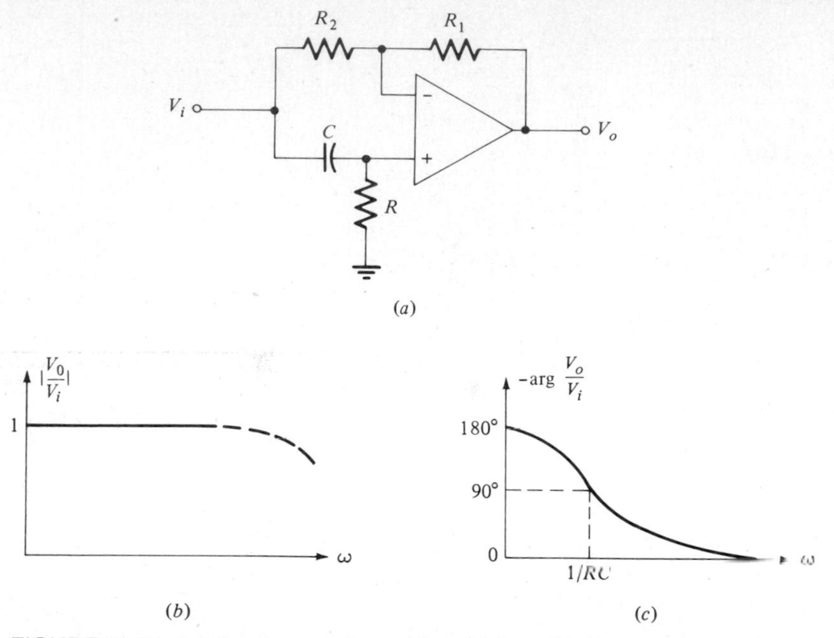

FIGURE 9.30 (*a*) An all-pass first-order section. (*b*) Magnitude response of (*a*). (*c*) Phase response of (*a*).

The magnitude of (9.43) is constant, but the phase function varies, namely,

$$\left|\frac{V_o}{V_i}\right| = 1 \qquad \arg\frac{V_o}{V_i} = \tan^{-1}(-\omega RC) - \tan^{-1}\omega RC = -2\tan^{-1}\omega RC \qquad \textbf{(9.44)}$$

A plot of the magnitude and phase in (9.44) is shown in Fig. 9.30*b* and *c*. Note that the magnitude response is shown dotted at higher frequencies since the amplifier frequency response and the circuit parasitics will cause the magnitude to decrease at very high frequencies.

Analysis and design of a second-order all-pass section is left as an exercise in problem 9.24.

F. General Single-Op-Amp Biquad

Circuit configurations are available that utilize both positive and negative feedback and that can realize a general biquad of the form (9.26), so that it can be used as a general biquad block. In fact, all four cases considered above are special cases of (9.26).

One circuit that may be used to realize a general biquadratic transfer function and, of course, can be used to realize special-case low-pass, high-pass, bandpass, all-pass, and band-reject functions is shown in Fig. 9.31. The general single-amplifier biquad (SAB) is useful as a block in modular design; i.e., it can be used as an all-purpose block. In fact, this block has been used in industry for voice-frequency filters. The hybrid IC fabrication of the circuit is shown in Fig. B.2. The transfer

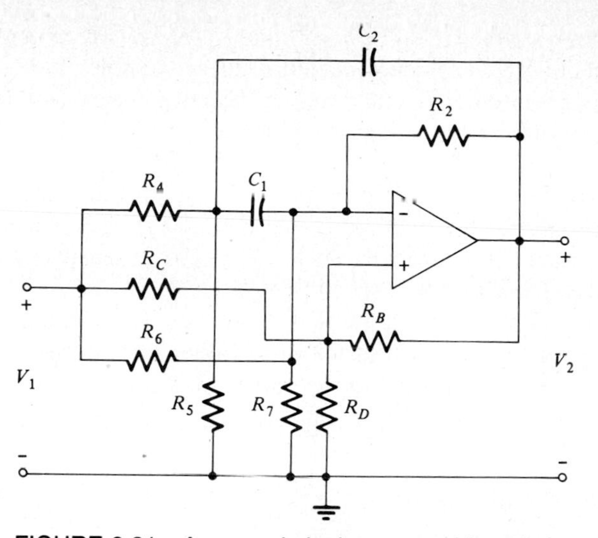

FIGURE 9.31 A general single op-amp biquad.

function of this circuit can be obtained and cast in the form of (9.26). The coefficients of (9.26) are related to the circuit-element values according to the following relations (problem 9.25):

$$b_2 = K_2 \tag{9.45a}$$

$$b_1 = \frac{K_2}{C_2}(G_1 + G_2 + G_3) + \frac{K_2}{C_1}(G_2 + G_3) - \left(1 + \frac{G_B}{G_A}\right)\left[\frac{K_1}{C_2}G_1 + K_3G_3\left(\frac{1}{C_1} + \frac{1}{C_2}\right)\right] \tag{9.45b}$$

$$b_0 = \frac{1}{C_1C_2}\left[K_2G_1(G_2 + G_3) - K_3G_1G_3\left(1 + \frac{G_B}{G_A}\right)\right] \tag{9.45c}$$

$$a_1 = \frac{C_1 + C_2}{C_1C_2}\left(G_2 - \frac{G_3G_B}{G_A}\right) - \frac{G_3G_B}{C_1G_A} \tag{9.45d}$$

$$a_o = \frac{G_1}{C_1C_2}\left(G_2 - G_3\frac{G_B}{G_A}\right) \tag{9.45e}$$

where

$$R_1 = R_4 \parallel R_5 \qquad K_1 = \frac{R_5}{R_4 + R_5}$$
$$R_A = R_C \parallel R_D \qquad K_2 = \frac{R_D}{R_C + R_D} \tag{9.46}$$
$$R_3 = R_6 \parallel R_7 \qquad K_3 = \frac{R_7}{R_6 + R_7}$$
$$G_i = \frac{1}{R_i}$$

It is noted that there are nine variables ($C_1, C_2, R_1, R_2, R_3, K_1, K_2, K_3, R_A/R_B$) and only five specifiable coefficients in (9.26). Therefore, four of these variables and either R_A or R_B can be independently prespecified by the designer. Usually, the capacitances are set to equal values, such as the normalized values $C_{1n} = C_{2n} = 1$; note that $R_B = (K - 1)R_A$.

Finally, for sensitivity reasons sometimes two or three op amps are used to realize a high-Q bandpass section. In fact, in some critical applications feedback configurations are used instead of cascaded blocks to improve sensitivity. These matters are treated in detail elsewhere.*

Active RC filters are suitable for hybrid IC fabrication. These filters are not suitable for monolithic technology because of the high values of C needed at voice frequencies and the tolerance requirements on the values of R and C. Active filters suitable for MOS technology are discussed in Sec. 9.6.

9.6 INDUCTANCE SIMULATION CIRCUITS

Since reasonable values of inductors cannot be fabricated in integrated circuits and inductors play an important role in LC filters, it is desirable to simulate inductors using BJTs or MOSFETs and RC elements. Op amps can be used to simulate an inductor. One such circuit, referred to as the *generalized impedance converter* (GIC), is shown in Fig. 9.32. The analysis of the circuit in Fig. 9.32 is straightforward if we use the properties of the ideal op amp with negative feedback. First note that $V_+ = V_- = V_1$; hence the three nodes have the same potential V_1. Next we find V_2 and V_3.

From the circuit $I_4 = I_5$; hence

$$(V_3 - V_1)\frac{1}{Z_4} = \frac{V_1}{Z_5} \quad \text{or} \quad V_3 = \left(\frac{Z_4}{Z_5} + 1\right)V_1 \tag{9.47a}$$

*M. S. Ghausi and K. R. Laker, *Modern Filter Design: Active RC and Switched Capacitor*, Prentice-Hall, Englewood Cliffs, N.J., 1981.

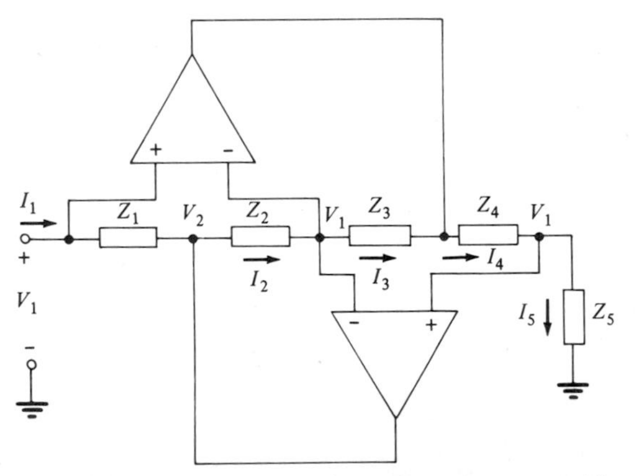

FIGURE 9.32 Basic configuration of a general impedance converter (GIC).

Similarly, $I_2 = I_3$, and

$$(V_2 - V_1)\frac{1}{Z_2} = (V_1 - V_3)\frac{1}{Z_3} \tag{9.47b}$$

From (9.47*a*) and (9.47*b*) we obtain

$$I_1 = \frac{V_1}{I_1} = \frac{Z_1 Z_3 Z_5}{Z_2 Z_4} \tag{9.48a}$$

Hence

$$Z_{\text{in}} = \frac{Z_1 Z_3}{Z_2 Z_4} Z_5 \tag{9.48b}$$

The result in (9.48) describes the generalized impedance converter property. Note that if we have

1. All $Z_i = R_i$ except $Z_4 = 1/sC$, then

$$Z_{\text{in}} = \left(\frac{R_1 R_3 R_5}{R_2} C\right) s \tag{9.49a}$$

2. All $Z_i = R_i$ except $Z_2 = 1/sC$, then

$$Z_{\text{in}} = \left(\frac{R_1 R_3 R_5}{R_4} C\right) s \tag{9.49b}$$

The choice in 1 with $R_3 = R_2$ and $R_5 = R_L$ yields

$$Z_{\text{in}} = (R_1 R_L C)s = sL \tag{9.50a}$$

where

$$L = R_1 R_L C \tag{9.50b}$$

The inductor simulation in 1, which is a special case of GIC, is shown in Fig. 9.33.

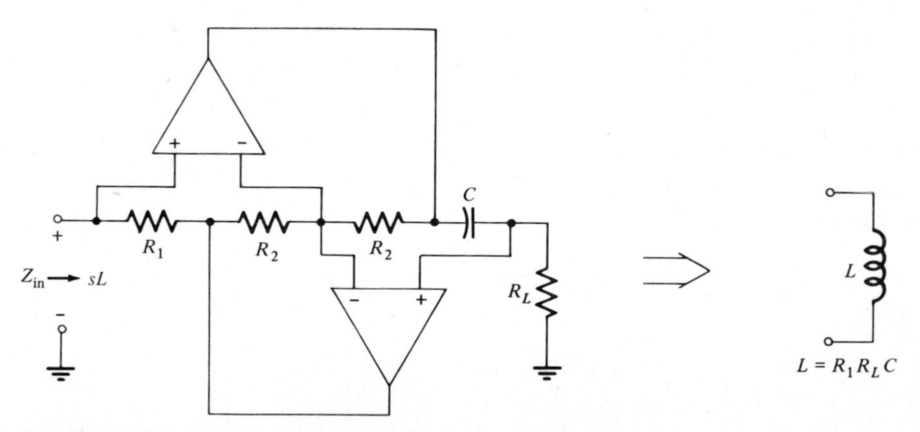

FIGURE 9.33 An inductor simulation circuit using GIC.

The use of inductors in LC filter design is very common. Thus LC filter design can be implemented using inductor simulation but without using inductors [5].

EXAMPLE 9.6

A second-order passive filter section is shown in Fig. 9.34a. Show an inductor simulation of this filter section with element values to give the same Q and ω_0.

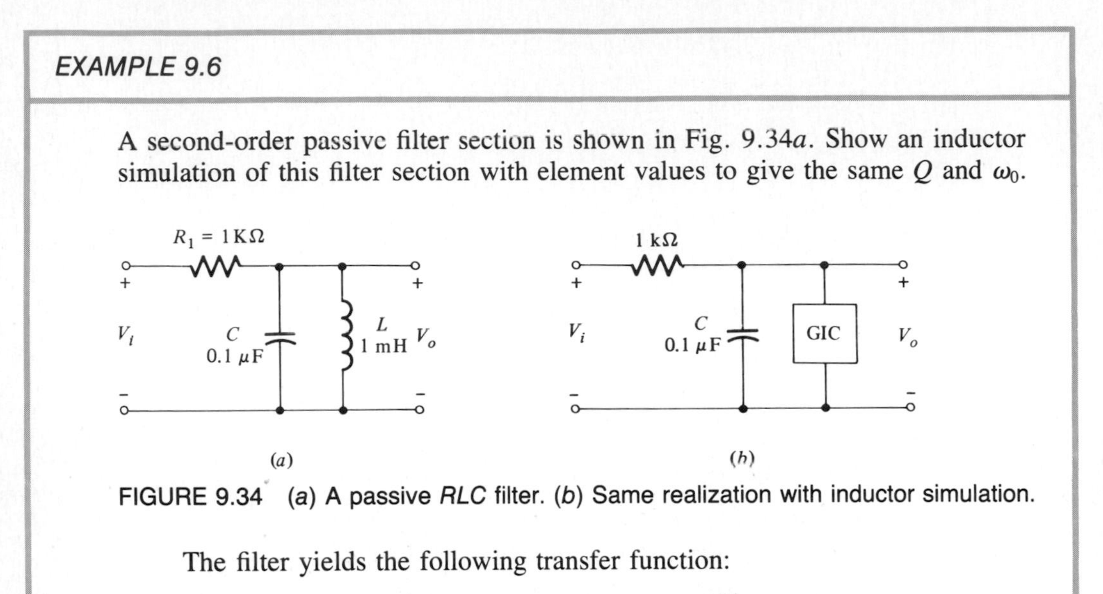

FIGURE 9.34 (a) A passive RLC filter. (b) Same realization with inductor simulation.

The filter yields the following transfer function:

$$\frac{V_o}{V_i}(s) = \frac{(1/R_1C_1)s}{s^2 + s(1/R_1C_1) + 1/LC_1} = \frac{10^4 s}{s^2 + s(10^4) + 10^{10}}$$

For the values indicated in the circuit $Q = 10$, $\omega_0 = 10^5$ rad/s. From (9.50) $L = R_1R_2C$, and a choice of $R_1 = R_2 = 1$ kΩ and $C = 10^{-9}$ F will yield $L = 1$ mH. The inductor simulated circuit is shown in Fig. 9.34b.

EXERCISE 9.6

For the circuit shown in Fig. 9.33, determine the value of the inductor if $R_1 = R_L = 1$ kΩ, $C = 100$ pF.

Ans $L = 100\ \mu$H

EXERCISE 9.7

For the GIC-derived bandpass section shown in Fig. E9.7,

(a) show that

$$\frac{V_o}{V_i} = \frac{2s(\omega_0/Q)}{s^2 + s(\omega_0/Q) + \omega_0^2}$$

where $\omega_0 = 1/RC$ and $R_2 = QR$.

(b) Design the circuit for $Q = 50$ and $\omega_0 = 10^5$ rad/s, given $R = 1\ \text{k}\Omega$. (*Hint:* Across the capacitor there is a shunt equivalent inductor, and note that $V_o/V_1 = 2$.)

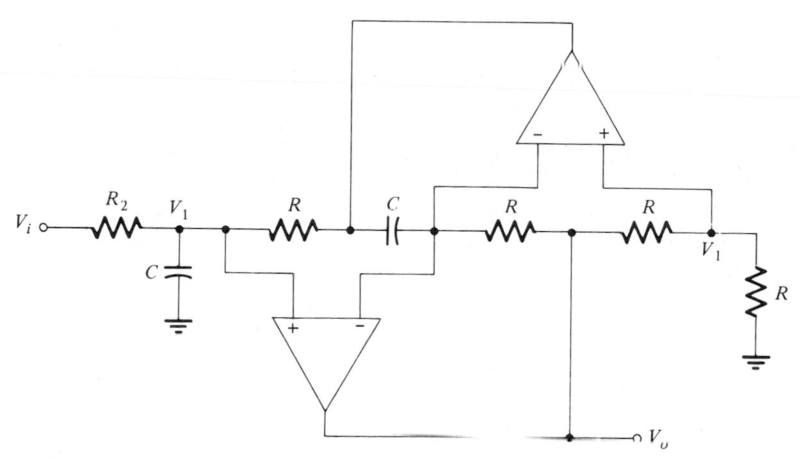

FIGURE E9.7

Ans $C = 10$ nF, $R_2 = 50\ \text{k}\Omega$

9.7 SWITCHED-CAPACITOR FILTERS

A switched capacitor (SC) in conjunction with MOS op amps can be designed to realize precision monolithic circuits to achieve frequency-selective filtering. These filters are becoming very popular since analog processing can be implemented with the digital processing on the same chip. Key features that make SC filters attractive are

1. Their suitability for monolithic filters.
2. Their suitability for MOS, LSI, VLSI technology.
3. MOS capacitors are nearly ideal and have very low dissipation factors and temperature coefficients ($\simeq 10$ ppm/°C), and capacitor ratios can be held to very tight tolerance (less than 0.1 percent).
4. Very low power consumption.
5. Very high Q filters can be designed readily.

We have already discussed MOSFET devices in Chap. 3. A MOSFET is shown in Fig. 9.35*a*. Recall that MOSFETs can be used as switches. Specifically, for $V_{GS} > V_T$ the switch is in the ON state, and when $V_{GS} < V_T$, the switch is in the OFF state. Enhancement-type MOS switches are normally OFF switches. We shall use the following idealized MOS switch for simplicity:

For $V_{GS} > V_T$, MOSFET is ON, and it is short-circuited.
For $V_{GS} < V_T$, MOSFET is OFF, and it is open-circuited.

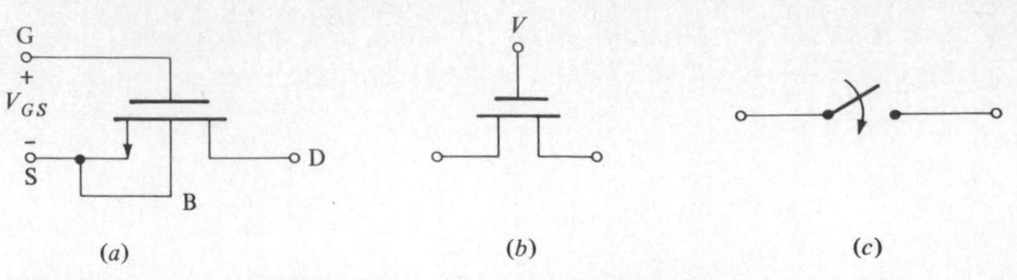

FIGURE 9.35 (*a*) MOSFET symbol. (*b*) Simplified model. (*c*) Switch representation.

The reader will recall that the MOS switch in the ON state has a resistance of the order of kiloohms, while in the OFF state the resistance is over several hundred megaohms, and the OFF/ON ratio of resistors is in the order of 10^5. Thus the idealization poses no problems. The effect of this ON resistance of the MOS switch gives rise to finite charging and discharging time for capacitors, but it is still negligible in the frequency range of operation of SC filters; for example, a 1-kΩ resistor charging a 10-pF capacitance has a time constant of 10 ns, which is much faster than the switching times of the SC circuits in these applications.

The MOS switch shown in Fig. 9.35*a* is then represented simply by Fig. 9.35*b* and further simplified as in Fig. 9.35*c*.

The voltage V_{GS} is generated by an external clock in a digital system. The *clock* is a periodic pulse train of period τ. The clock frequency $f_c = 1/\tau$ is usually less than 250 kHz. Typically the MOS switches are controlled by *two-phase, nonoverlapping clocks* as shown in Fig. 9.36. Note that when V_a is ON, V_b is OFF, and vice versa. The clocks have the same period and the upper and lower limits on the clock frequency are determined by other requirements. They are usually in the range 8 kHz $\leq f_c \leq$ 256 kHz. Note that 50 percent duty cycle shown in Fig. 9.36 is not necessary; it is for simplification purposes only. In practice it is somewhat lower in order to avoid overlapping. Figure 9.37*b* shows the equivalent circuit in terms of two single-pole,

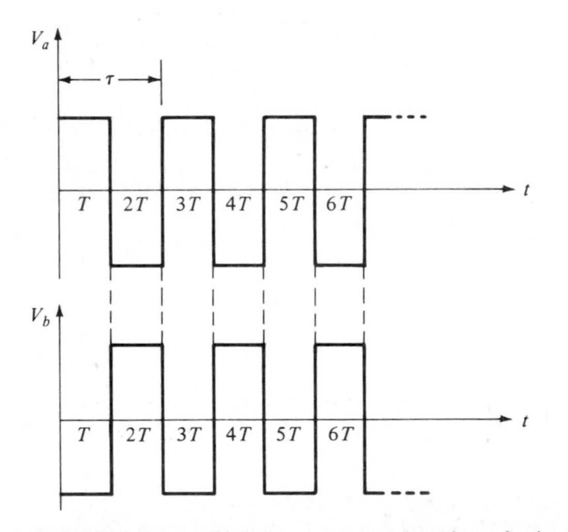

FIGURE 9.36 Biphase non-overlapping clocks V_a and V_b.

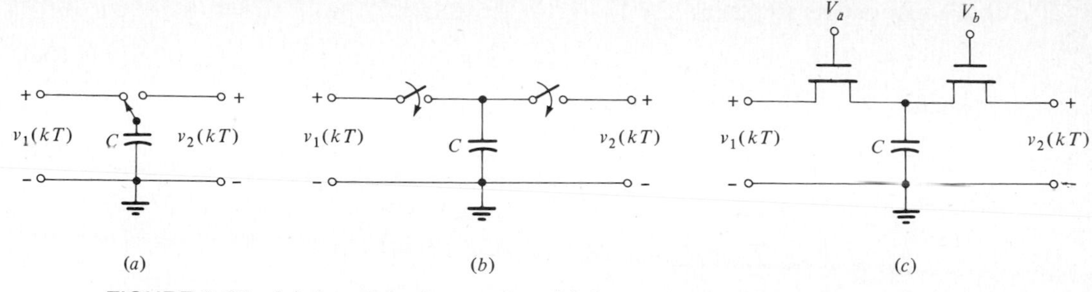

FIGURE 9.37 (*a*) A switched capacitor. (*b*) Same circuit with two switches. (*c*) MOS implementation of the switches.

single-throw (SPST) switches, and Fig. 9.37*c* shows a MOS implementation of the switches. The term *switched capacitor* is used because of the switching operation on a capacitor. Note that turning both switches OFF simultaneously does not affect the circuit, but turning both of them ON will cause improper circuit function. Note also that the input and output signals of the SC network change at discrete times *kT*.

A. The Switched Capacitor as a Resistor

Consider a capacitor switched ON and OFF by a clock as shown in Fig. 9.38b. We assume that the clock period is small enough so that voltage signals do not change appreciably during one period of the clock. That is, we assume instant charge and discharge for simplicity. The switching operation is then as follows: C is charged to v_1 by the input signal and discharged to v_2 at the output port in one period of a two-phase clock. There is a net transfer of charge from input to the output, and this change transfer in one period is given by

$$q = C(v_1 - v_2) \tag{9.51}$$

v_1 and v_2 remain essentially the same during this one period of the clock, enabling us to write from (9.51)

$$i = \frac{dq}{dt} \simeq \frac{\Delta q}{\Delta t} \simeq \frac{C(v_1 - v_2)}{\tau} \tag{9.52}$$

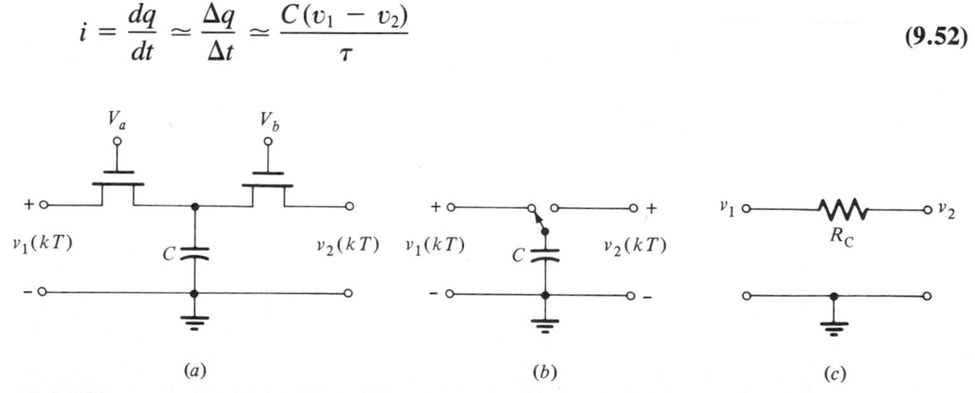

FIGURE 9.38 (*a*) Switched capacitor using MOS. (*b*) Basic switched capacitor. (*c*) Equivalent resistor of a switched capacitor.

We can rewrite (9.52) as

$$i = G_C(v_1 - v_2) \tag{9.53a}$$

where

$$G_C = C/\tau \tag{9.53b}$$

or

$$R_C = \frac{1}{G_C} = \frac{\tau}{C} = \frac{1}{f_c C} \tag{9.53c}$$

From (9.53) it is seen that the switched capacitor is equivalent to a resistor of value $1/Cf_c$ as shown in Fig. 9.38*c*. With this equivalence we can now replace the resistor of an active *RC* filter with a switched capacitor. Note that for a clock frequency in the range of 10 to 500 kHz, the values of the equivalent resistor for a capacitor of 1 pF lies in the range of 100 MΩ $\geq R \geq$ 2 MΩ. Remember that f_c must be much larger than the highest frequency of the signal to be filtered. For voice-frequency filters $f \leq 4$ kHz, and thus $f_c \geq 40$ kHz is high enough to guarantee the validity of our assumption of slowly varying signals.

B. Switched-Capacitor Integrator

In Fig. 9.39 we have an *RC* integrator with the transfer function given by

$$\frac{V_o}{V_i} = -\frac{Z_f}{R_1} = -\left(\frac{1}{R_1 C_f}\right)\frac{1}{s} \tag{9.54}$$

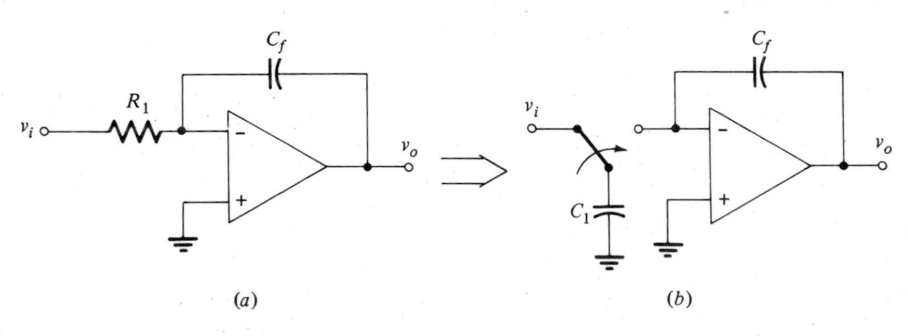

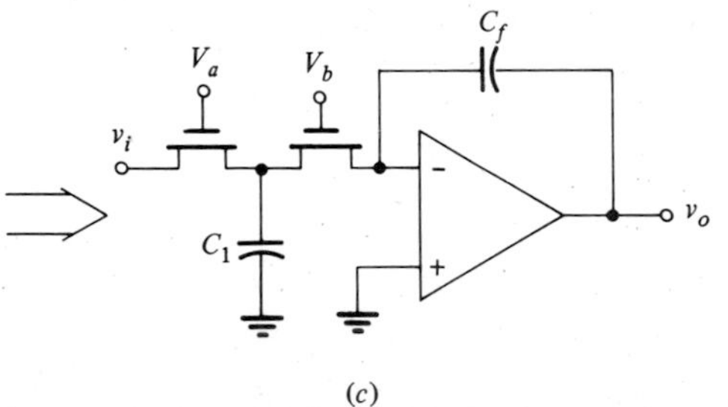

FIGURE 9.39 (*a*) An *RC* integrator. (*b*) A switched capacitor integrator. (*c*) MOS implementation of *b*.

By replacing the resistor with its equivalent SC value, we obtain

$$\frac{V_o}{V_i} = -\frac{f_c C_1}{C_f s} = -f_c \frac{C_1}{C_f} \frac{1}{s} \tag{9.55}$$

From (9.55) note that the transfer function depends on the clock frequency and *ratio of two capacitors*. We have already noted that this *ratio* can be held with tight tolerance in MOS technology.

C. Simple First- and Second-Order Circuits

A first-order low-pass circuit can be realized with a simple passive RC network as shown in Fig. 9.40*a*. Its equivalent SC network is shown in Fig. 9.40*b*. The transfer function is given by

$$\frac{V_o}{V_i} = \frac{1}{R_1 C_2 s + 1/R_1 C_2} \frac{1}{} = \frac{f_c C_1}{C_2} \frac{1}{s + f_c C_1/C_2} \tag{9.56}$$

Note that the pole at $f_c C_1/C_2$ depends on the clock frequency and ratio of two capacitors. A second-order bandpass section corresponding to Fig. 9.24*a* is shown in Fig. 9.41. The transfer function is given by

$$\frac{V_o}{V_i} = \frac{-f_c(C_a/C_1)s}{s^2 + sf_c(C_b/C_1 + C_b/C_2) + f_c^2(C_a/C_1)(C_b/C_2)} \tag{9.57}$$

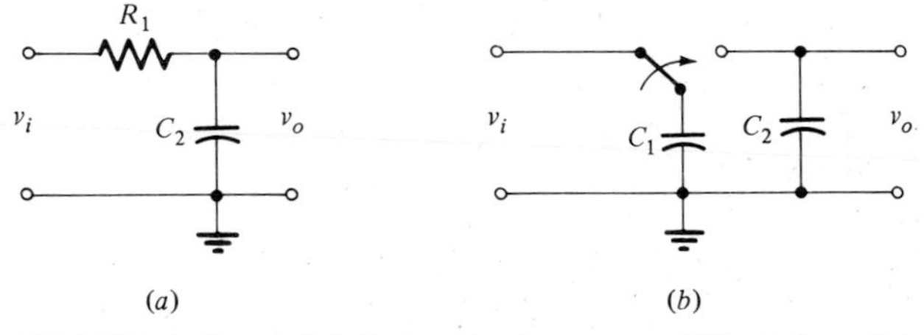

FIGURE 9.40 (*a*) A first-order low-pass RC section. (*b*) Equivalent of (*a*) with R_1 replaced by a switched capacitor C_1.

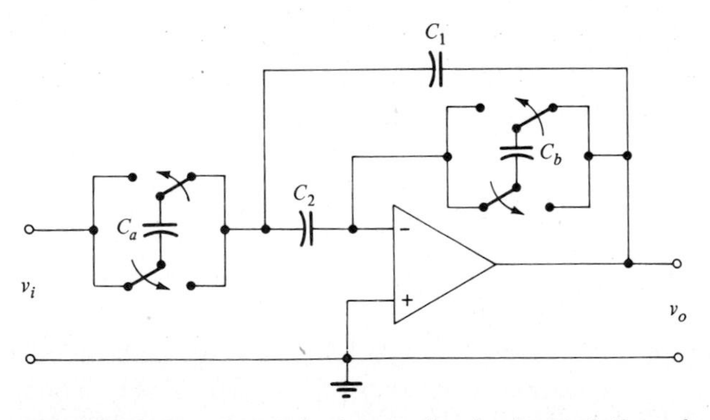

FIGURE 9.41 A switched capacitor implementation of a second-order bandpass section of Fig. 9.24.

Note again that the center frequency and the bandwidth depend on the clock frequency and the ratio of capacitors, both of which can be kept very accurate.

EXAMPLE 9.7

Design a bandpass, second-order function for a center frequency of $\omega_0 =$ 1 krad/s and $\omega_{3dB} = 100$ rad/s. Let the clock frequency be $f_c = 10$ kHz. We let $C_1 = C_2 = 10$ pF.

From (9.57) we have

$$\omega_{3dB} = f_c\left(\frac{C_b}{C_1} + \frac{C_b}{C_2}\right) \longrightarrow 10^2 = 10^4(2C_b)$$

$$\omega_0^2 = f_c^2\frac{C_a C_b}{C_1 C_2} \longrightarrow 10^6 = (10^4)^2 C_a C_b$$

From the above we get $C_b \simeq 0.05$ pF and $C_a = 20$ pF. In active SC filter design, since the values of the capacitors in the design are very small, circuit parasitic capacitances are very important and cannot be ignored. Hence parasitic-free biquads are used in practice.*

*See footnote on page 426.

Commercial switched capacitor filters using CMOS technology, such as MF10 universal monolothic dual-switched-capacitor filter, is available. The circuit can be configured to perform low-pass, bandpass, all-pass, high-pass, or notch functions. The circuit consists of two independent second-order sections in one package; thus a fourth-order filter design can be accomplished. An external clock and three to four resistors are needed for the various second-order filter sections. A sixth-order Butterworth filter is readily available in the MF6 filters.

EXERCISE 9.8

If the range of capacitor values in MOS switched-capacitor filters is $0.1 \text{ pF} < C < 10 \text{ pF}$, determine the range values of the equivalent resistors for (*a*) $f_c = 10$ kHz and (*b*) $f_c = 1$ MHz.

Ans (*a*) $10 \text{ M}\Omega < R < 1000 \text{ M}\Omega$, (*b*) $0.1 \text{ M}\Omega < R < 10 \text{ M}\Omega$

EXERCISE 9.9

Design a one-pole SC filter using the circuit in Fig. 9.21*b* for a voltage gain of -50 and a 3-dB bandwidth of 10^4 rad/s. Use a clock frequency of 100 kHz. The capacitor values must be less than or equal to 50 pF.

Ans $C = 10$ pF, $C_2 = 1$ pF, $C_1 = 50$ pF

D. Sample-and-Hold Circuit

Before we leave the SC topic, it is important to consider briefly the sample-and-hold (S/H) circuit that is useful in SC filters as well as in a number of applications, such as in analog-to-digital (A/D) signal processing. The sample-and-hold circuit is shown in Fig. 9.42. The function of the circuit, as the name implies, is to sample the input signal and hold it until the next sample. The unity gain block provides a buffer between C and a low-valued resistor (large-valued SC) to avoid charge leakage from C when the switch is open. The use and application of an S/H circuit are further discussed in Chap. 14 in conjunction with A/D converters.

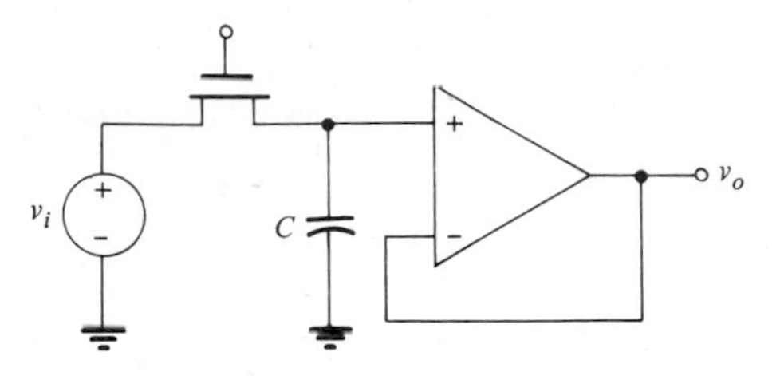

FIGURE 9.42 A sample-and-hold (S/H) circuit.

9.8 NONLINEAR OP-AMP APPLICATIONS

In this section we describe several applications of ideal op amps under large-signal conditions similar to the ideal-diode applications discussed in Chap. 1. To properly understand what follows, the reader should refer to the transfer characteristic of an ideal op amp shown in Fig. 9.43. This has been discussed in Chap. 5 (see Fig. 5.32). Note that since the voltage gain of an ideal op amp is infinite any potential difference between the input terminals will saturate the op amp and the output voltage will be a either $+V_{sat}$ or $-V_{sat}$ as indicated.

Op amps can be used with diodes to obtain superior rectification properties as follows: From Chap. 1, recall that the transfer characteristic of a diode has a cut-in voltage 0.7 V as shown in Fig. 9.44*b*. Now, if an op amp is used in the configuration shown in Fig. 9.45*a*, the transfer characteristic will be as shown in Fig. 9.45*b*. Thus *extremely* small signals can be rectified by this arrangement. In practice, since the gain of an op amp is about 10^6, an input voltage $v_i > 0.7/10^6 = 0.7\mu\text{V}$ can be rectified. Thus the op amp can be used to overcome the cut-in voltage of a diode, and when used in this manner, it is referred to as a *superdiode*. Note also that the slope of the transfer characteristic of the op-amp configuration is much closer to unity than that of a diode, which in Fig. 9.44*a* is $R/(R + R_f)$. We shall next consider some of the nonlinear applications of op amps.

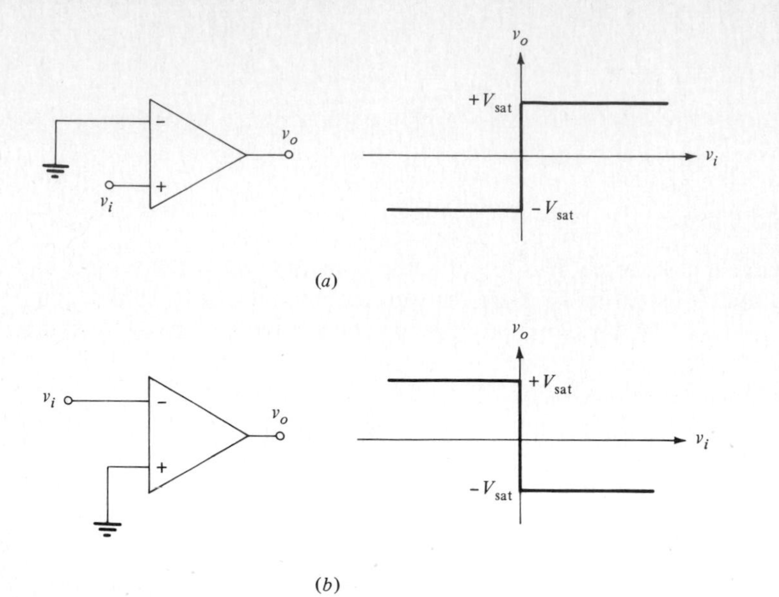

FIGURE 9.43 (a) Ideal op-amp noninverting characteristic driven into saturation. (b) Ideal op-amp inverting characteristic driven into saturation.

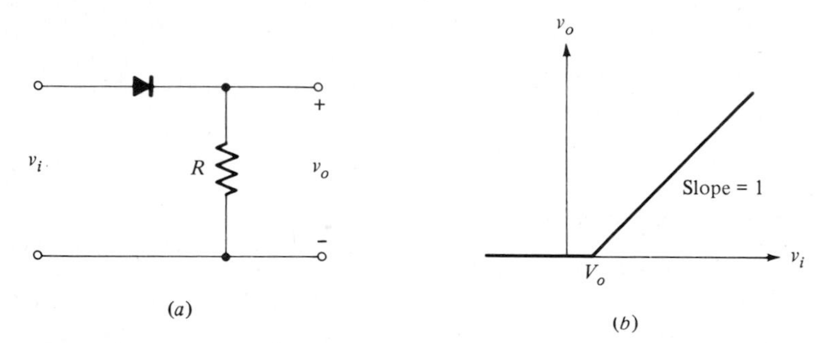

FIGURE 9.44 (a) A diode rectifier circuit. (b) Transfer characteristic of (a).

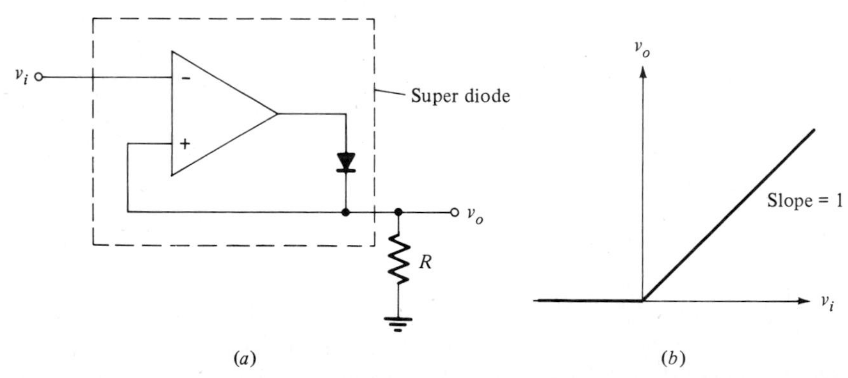

FIGURE 9.45 (a) An op amp used as a super diode. (b) Transfer characteristic of (a).

A. Precision Half-Wave Rectifier

The use of an op amp in the superdiode configuration to obtain precision half-wave rectifier is now obvious. The circuit and the waveforms are shown in Fig. 9.46. The operation is as follows: For $v_i > 0$, v_o is positive and the diode is ON (hence short-circuited), and the op amp is a voltage follower and $v_L = v_i$. For $v_i < 0$, $v_L = 0$ as shown in Fig. 9.46*b*. An improved version of the precision half-wave rectifier is shown in Fig. 9.46*c*. In this circuit diode D_2 keeps the feedback loop around the op amp closed when D_1 is OFF, and thus the op amp is prevented from saturation.

B. Precision Full-Wave Rectifier

The circuit of a precision full-wave rectifier is shown in Fig. 9.47*a*. The operation of the circuit is similar to the half-wave rectifier and the diode-bridge rectifier discussed earlier. The output voltage v_L, for $R_L = R_1$, is shown in Fig. 9.47*b*.

C. Clippers

If we connect a reference voltage V_o to R_L in the half-wave rectifier, we obtain the clipping circuit shown in Fig. 9.48*a*. The operation is identical except that for $v_i \geq V_o$, clipping occurs as shown in Fig. 9.48*b*.

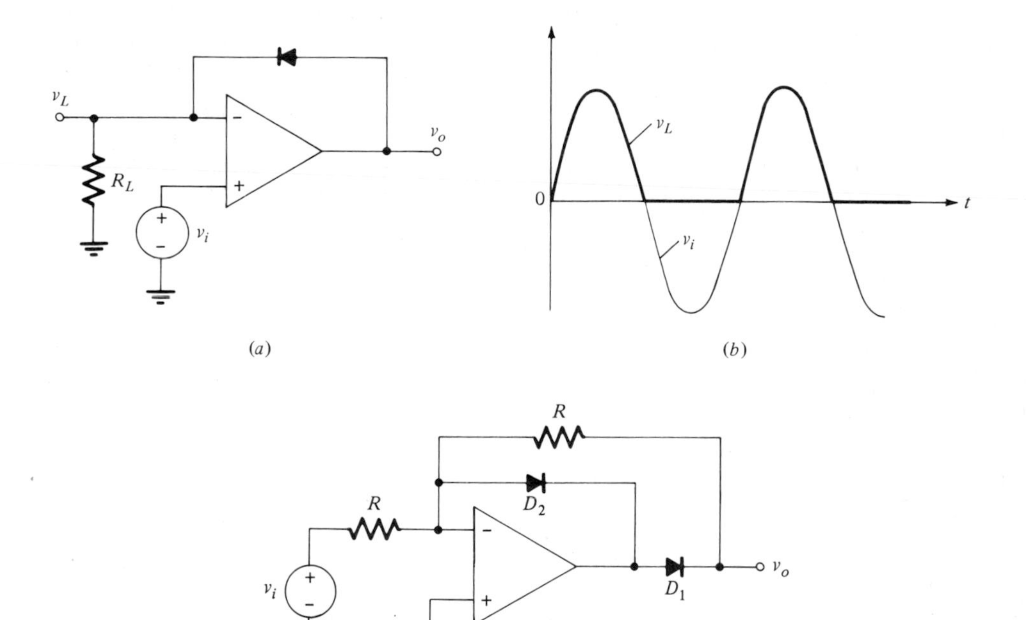

FIGURE 9.46 (*a*) An op-amp rectifier circuit. (*b*) Transfer characteristic of (*a*). (*c*) An improved precision half-wave rectifier circuit.

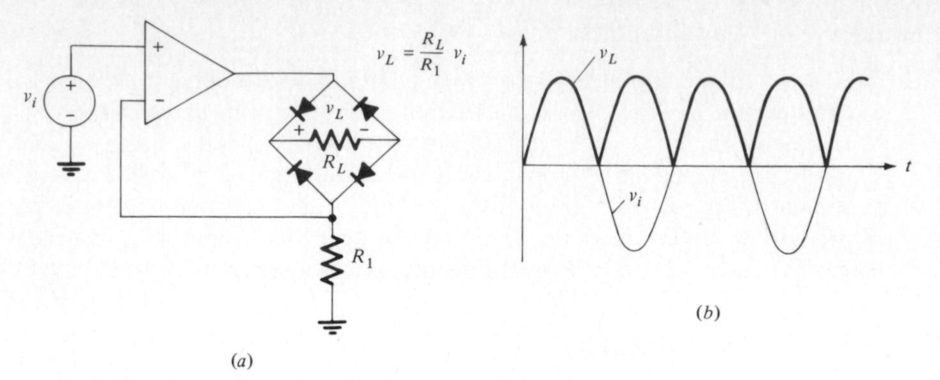

FIGURE 9.47 An op-amp full-wave rectifier circuit.

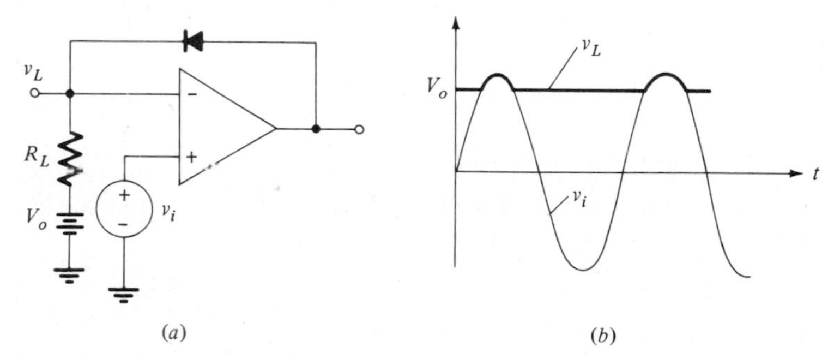

FIGURE 9.48 (*a*) A clipper circuit. (*b*) Associated waveforms.

Zener diodes may also be used with an op amp to obtain a voltage clipper or a limiter circuit as shown in Figs. 9.49 and 9.50.

EXERCISE 9.10

For the circuit shown in Fig. E9.10, determine and sketch the transfer characteristic. Assume an ideal zener diode with forward and reverse models of a battery 0.7 V and −5 V, respectively.

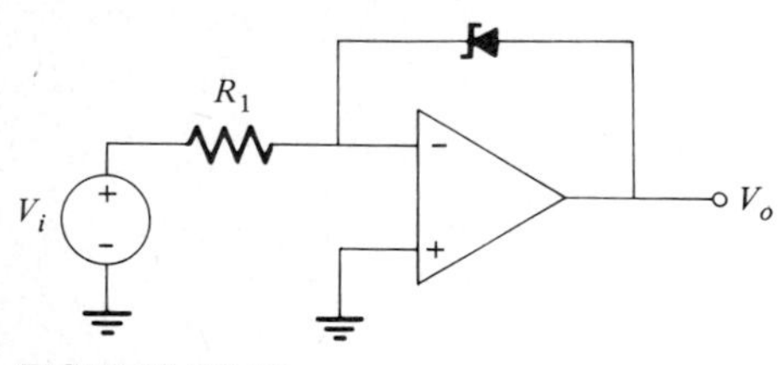

FIGURE E9.10

Ans $V_o = 0.7$ for $V_i < 0$, $V_o = -5$ V for $V_i > 0$

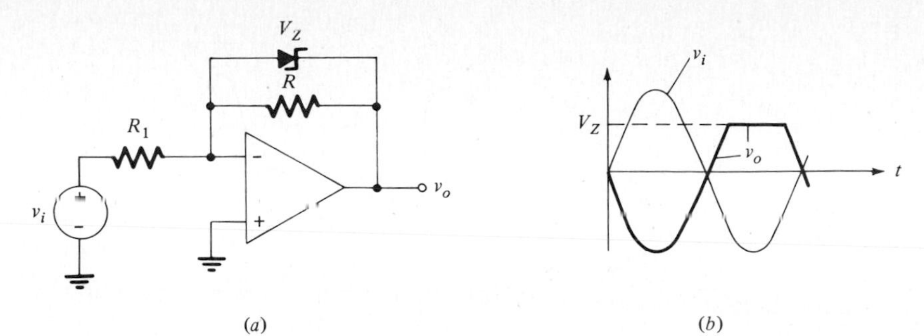

FIGURE 9.49 (*a*) A clipper circuit using a zener diode and an op amp. (*b*) Associated waveforms.

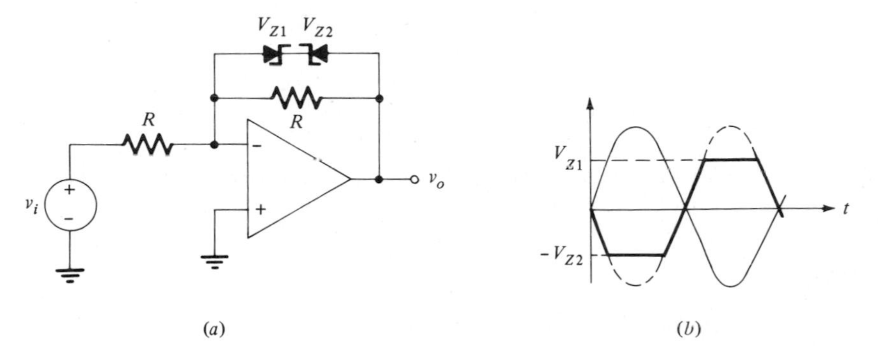

FIGURE 9.50 (*a*) A double-clipper (limiter) circuit. (*b*) Associated waveforms.

D. Active Peak Detector

The circuit shown in Fig. 9.51*a* may be used as a basic peak detector. The input and output waveforms are shown in Fig. 9.51*b*. The operation is simply described as follows: For $v_i > v_o$, v_a is positive, and the diode conducts. The capacitor C is charged through the diode to the value of the input voltage since the circuit is a voltage follower (unity gain). When $v_i < v_o$, the capacitor cannot discharge, and v_o is held at its maximum value.

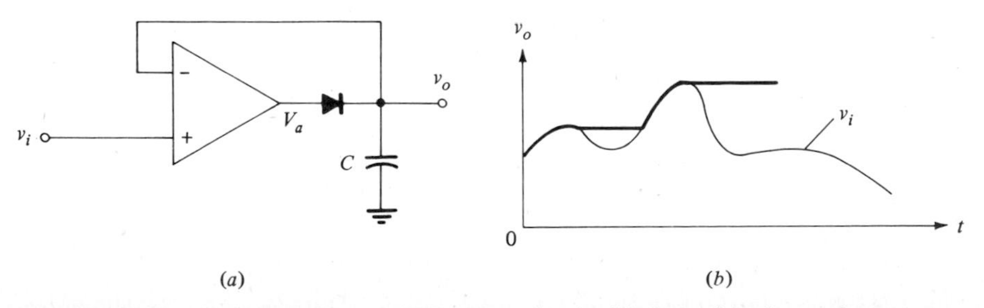

FIGURE 9.51 (*a*) An active peak-detector circuit. (*b*) Associated waveforms.

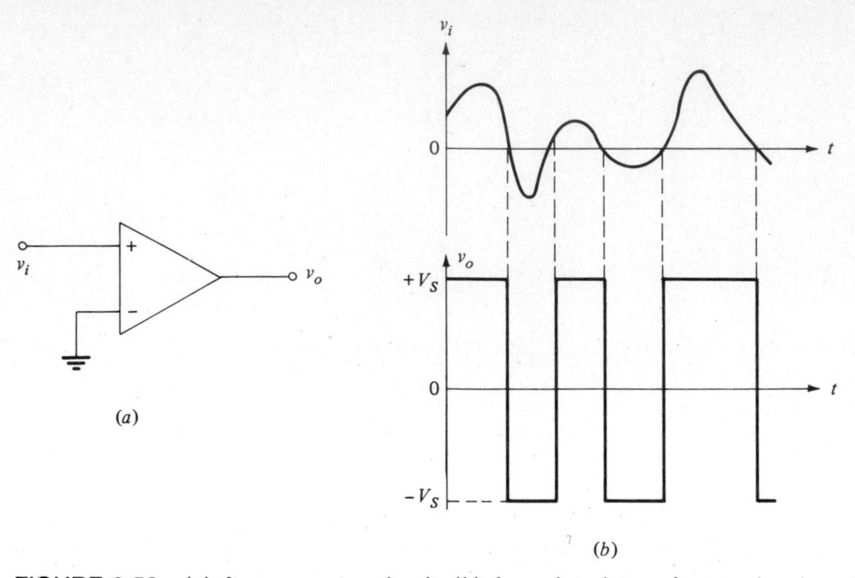

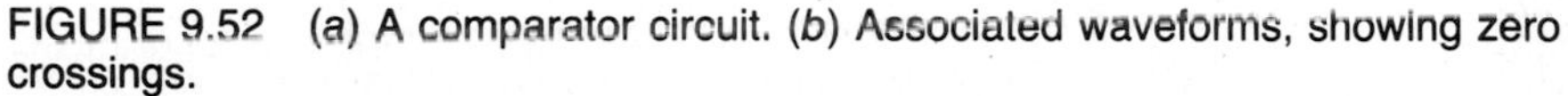

FIGURE 9.52 (*a*) A comparator circuit. (*b*) Associated waveforms, showing zero crossings.

E. Comparator

A comparator circuit with the input and output waveforms are shown in Fig. 9.52. The output voltage swings to plus or minus V_S depending on the input signal sign. This is expected as seen in Fig. 9.43.

A comparator is basically an op amp without any *negative feedback*. From Fig. 9.52 it is seen that a comparator circuit can be used to detect zero crossings. Note also that in Fig. 9.52*a*, if the negative terminal is connected to a reference voltage V_R, then v_i can be compared to that of V_R. A comparator is generally shown symbolically as in Fig. 9.53*a* with its transfer characteristic given in Fig. 9.53*b*.

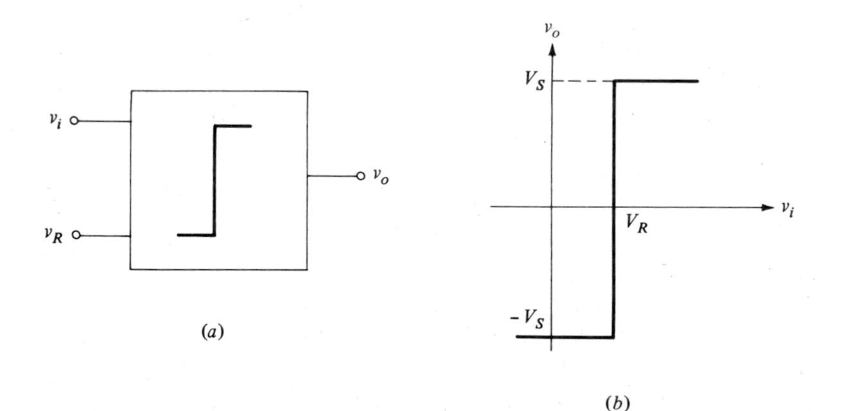

FIGURE 9.53 (*a*) A comparator circuit symbol. (*b*) Transfer characteristic of (*a*).

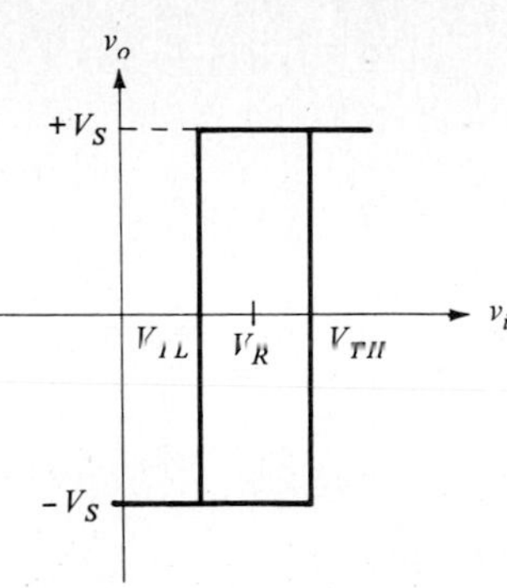

FIGURE 9.54 Transfer characteristic of comparator with positive feedback, showing hysteresis.

A comparator can also be designed to have two thresholds rather than one, which has just been described. The two threshold levels of a comparator are denoted by a high-threshold V_{TH} and a low-threshold V_{TL} around the reference voltage as shown in Fig. 9.54. Two such circuits that exhibit the hysteresis in Fig. 9.54 are shown in Fig. 9.55. These circuits are referred to as the *Schmitt trigger*, and due to the hysteresis in their characteristics, they exhibit memory. The circuits in Fig. 9.55*a* and *b* are a noninverting comparator with hysteresis and an inverting comparator with hysteresis, respectively.

The high and the low threshold levels can be determined as follows: Let us consider the noninverting Schmitt trigger and for simplicity the case of $V_R = 0$; that is, the negative terminal of the op amp in Fig. 9.55*a* is at ground potential. From the circuit the voltage v_+ at the positive input terminal of the op amp is given by

$$v_+ = v_o \frac{R_1}{R_1 + R_f} + v_i \frac{R_f}{R_1 + R_f} \tag{9.58}$$

For v_i positive and large enough to saturate, the op amp we have, $v_o = +V_S$, when v_+ is positive. The switching from $+V_S$ to $-V_S$ occurs when v_+ just passes zero in the

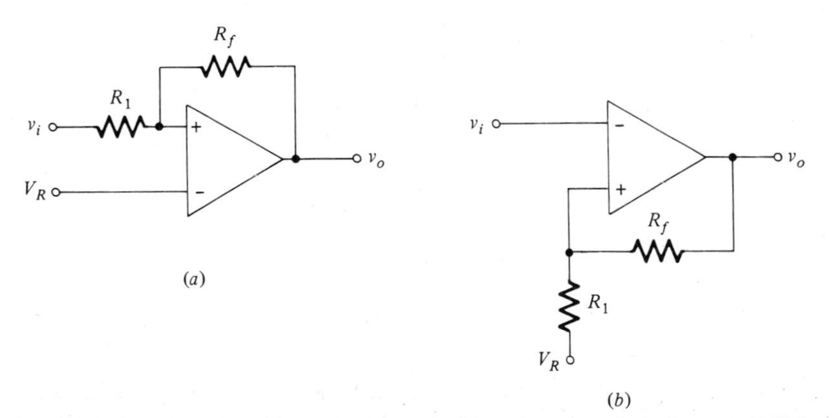

FIGURE 9.55 (*a*) An inverting Schmitt trigger. (*b*) A noninverting Schmitt trigger.

negative direction, and thus we can determine V_{TL} from (9.58) with $v_+ = 0$, $v_o = +V_S$, namely,

$$v_{i1} = V_{TL} = -V_S\frac{R_1}{R_f} \tag{9.59a}$$

Similarly, by increasing v_i for v_o to change from $-V_S$ to $+V_S$, we have

$$v_{i2} = V_{TH} = V_S\frac{R_1}{R_f} \tag{9.59b}$$

Of course, V_R does not have to be zero, and in general,

$$V_{TH} = V_R\left(1 + \frac{R_1}{R_f}\right) + V_S\frac{R_1}{R_f} \qquad V_{TL} = V_R\left(1 + \frac{R_1}{R_f}\right) - V_S\frac{R_1}{R_f} \tag{9.60}$$

For the inverting comparator the analysis is similar and is left as an exercise in problem 9.32. Comparators are widely used in digital circuits. We shall reserve further discussion of comparators for Chap. 13.

EXERCISE 9.11

For the inverting comparator circuit in Fig. 9.55*b*, the saturation voltage $V_S = \pm 10$ V, $R_1 = 2$ kΩ, and $R_f = 200$ kΩ. If $V_R = -2$ V, determine the two voltages at which the comparator will switch.

Ans $V_{TH} = -1.92$ V, $V_{TL} = -2.12$ V

F. Function Generator

The concave (or convex) *V-I* piecewise-linear functions generated as driving-point functions in Chap. 1 can be converted to transfer synthesis by connecting the circuit as in Fig. 9.56. From the circuit the *V-I* characteristics of the nonlinear resistor *N* is

$$V = f(I) \tag{9.61}$$

but

$$I = -I_1 = -\frac{V_i}{R_S}$$

hence

$$V = f\left(-\frac{V_i}{R_S}\right) = -Kf(V_i) \tag{9.62}$$

where *K* is a proportionality constant. This concept can be used to realize a piecewise-linear approximation to arbitrary concave or convex nonlinear curves.

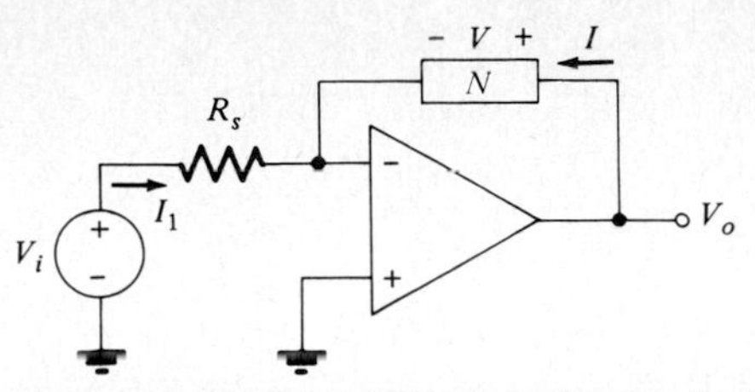

FIGURE 9.56 Transfer function realization circuit (a nonlinear curve approximator) with piecewise-linear models using diodes, resistors, and batteries in *N*.

EXAMPLE 9.8

Generate the function $y = e^x$ using a three-segment approximation. The driving-point realization for the three-segment piecewise-linear circuit using ideal diodes was done in Example 1.6 and is shown in Fig. 9.57*a*. The desired transfer function realization is then shown in Fig. 9.57*c*. Of course, if a higher accuracy is required, we may have to use a four-segment (or more) approximation.

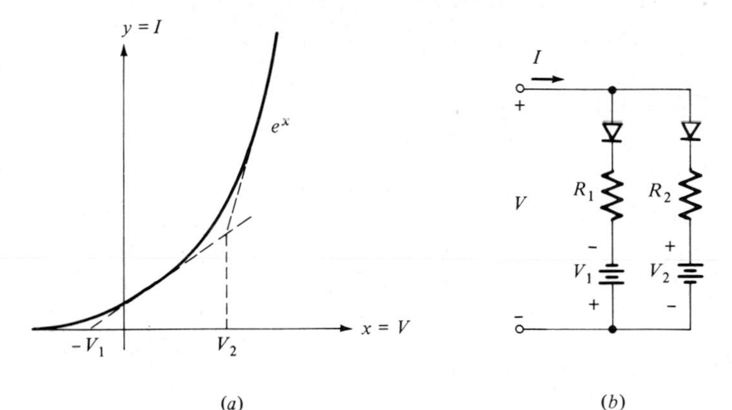

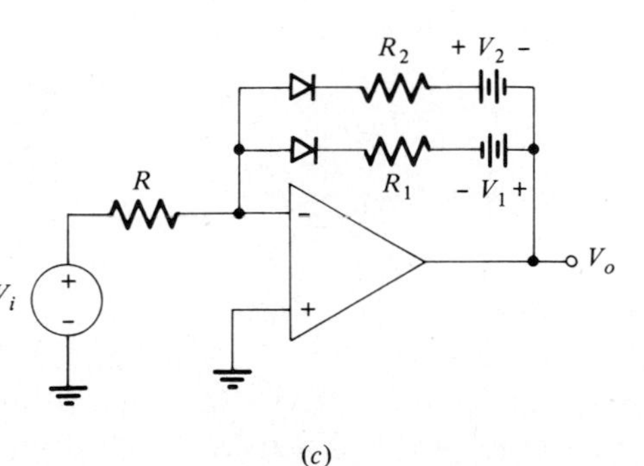

(c)

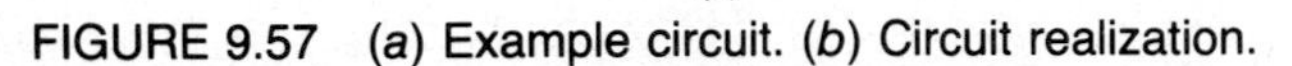

FIGURE 9.57 (*a*) Example circuit. (*b*) Circuit realization.

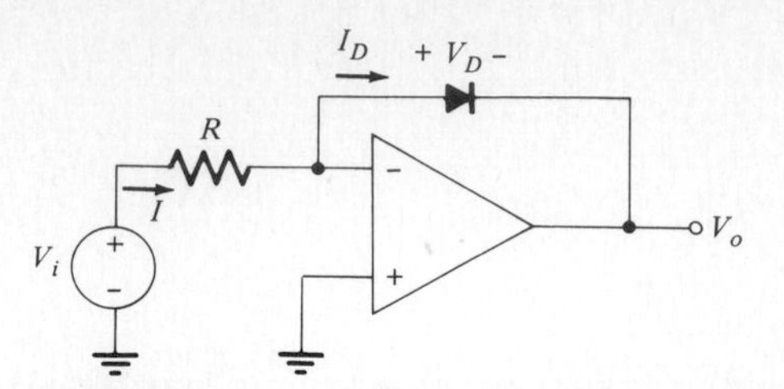

FIGURE 9.58 A logarithmic amplifier circuit.

G. Logarithmic Amplifier

Figure 9.58 shows a basic logarithmic amplifier circuit. The diode across the op amp is an exponential diode; i.e.,

$$I_D \simeq I_s e^{qV_D/kT} = I_s e^{V_D/V_T} \tag{9.63}$$

From this circuit we have

$$I = I_D = \frac{V_i}{R} \quad \text{and} \quad V_0 = -V_D \tag{9.64}$$

but

$$V_D = \frac{kT}{q} \ln \frac{I_D}{I_s} = V_t \ln \frac{I_D}{I_s} \tag{9.65}$$

From (9.64) and (9.65) we obtain

$$V_0 = -V_t \ln \frac{V_i}{IR} \tag{9.66}$$

The expression in (9.66) is logarithmic; note that it is valid only for V_i positive.

It should be mentioned that an *npn* CB transistor with the emitter terminal connected to the output of the op amp and the collector terminal connected to the inverting input terminal of the op amp can also be used as a logarithmic amplifier (problem 9.39). The expression for V_o is identical to (9.66). Again V_i must be positive for it to function properly.

9.9 NONIDEAL CHARACTERISTICS OF OP AMPS

In previous sections we have assumed an ideal op amp with properties given in (9.2*b*). Although IC op amps closely approximate the ideal characteristics, they nevertheless deviate from the idealizations depicted in (9.2*b*). For example, a BJT op amp such as 741, which we analyzed in Sec. 5.10, has the following properties: $R_i = 1\ \text{M}\Omega$, $R_o = 75\ \Omega$, $a_o \simeq 2 \times 10^5$. The frequency response of the op amp was discussed in Sec. 8.3. We have already observed that the 741 op amp is an internally compensated amplifier with the open-loop gain function, 3-dB frequency determined by the compensating capacitor (e.g., 30 pF in Fig. 5.25*a*). The 3-dB frequency was found to be

$f_{3dB} \simeq 5$ Hz. Hence the open-loop gain function of 741 can be written as given in (9.1*b*), namely,

$$a(s) = \frac{a_o}{1 + s/p_o} \tag{9.67}$$

with $a_o p_o = 2 \times 10^5(10\pi) = 2\pi(1 \text{ MHz})$.

Typical BJT (741), Bi-FET (LF351) op-amp parameters are listed in Table 9.1. The all-MOS (CMOS and NMOS) performances as op amps are similar to Bi-FET except that the open-loop gain is much lower and the output impedance is much higher (see Table 5.2). All MOS op amps, however, are smaller in size than the BJT and cost effective in LSI circuitry. For general-purpose applications BJT op amps or Bi-FETs are used. However, for some applications such as switched-capacitor filters or LSI circuitry, MOS op amps are used.

Note that the linear range of operation of an op amp is very small, i.e., $\epsilon = (V_+ - V_-) = V_o/a_o \rightarrow V_s/a_o \rightarrow 0$ and saturation can occur readily. In linear applications, saturation is to be avoided; consequently, op amps are seldom used in open-loop configuration. Invariably a resistor or *RC* network is added to avoid (or at least *control*) saturation. Compensation is needed in op amps with feedback because of stability reasons, as discussed in Chap. 10. Thus 741 and LF351 op amps are compensated internally to get a one-pole rolloff, i.e., 20 dB/decade rolloff.

In some op amps, such as 702, 709, and 1530 types, compensation is done externally (see Sec. 10.12). For these uncompensated amplifiers the dominant poles are at $f_1 \simeq 1$ MHz, $f_2 \simeq 5$ MHz. External compensation is used for these amplifiers to get the one-pole rolloff characteristics given by (9.67) in order to avoid instability, as discussed in Chap. 10. Thus *when the op-amp frequency characteristics are considered, we shall use (9.67),* as compensation will be provided externally if it is not done already internally.

When feedback is used around the op amp, the so-called closed-loop configuration such as the inverting and noninverting circuits in Figs. 9.3 and 9.4, the frequency response, and the input and output impedances are affected. These are considered in detail in Chap. 10.

EXERCISE 9.12

For the 741 op amp considered in Exercise 9.4, the 3-dB bandwidth is given by

$$\omega_{3dB} = p_o\left(1 + \frac{a_o}{1 + R_1/R_2}\right)$$

(The derivation is given in Chap. 10.)
Determine the bandwidth of the op amp for a voltage gain of 10.

Ans $f_{3dB} = 100$ kHz

EXERCISE 9.13

Repeat Exercise 9.12 for an LF351 op amp.

Ans $f_{3dB} = 200$ kHz

9.10 OTHER CONSIDERATIONS IN OP AMPS

Recall that an op amp is a differential circuit that uses emitter-coupled BJT or source-coupled FET devices at the input. In the analysis of op amps we have assumed perfect symmetry. In a practical situation perfect symmetry may not exist, and thus there will be some variations that can be adjusted. Some of the important ones are listed in the following sections.

A. Input Offset Voltage

The voltage input V_d that must be applied to make the output voltage V_o equal to zero is called the *input offset voltage*. Ideally when $V_d = 0$, $V_o = 0$, but in actual op amps some voltage input is required to make $V_o = 0$. This is done by a nulling adjustment (see offset null terminals in Appendix E). A typical value of the input offset voltage is 1 mV.

B. Input Offset and Input Bias Currents

Input offset current is the difference in the currents entering the two input terminals of an op amp when we make $V_o = 0$. Typically, in a BJT it is 50 nA; in a BiFET it is on the order of 50 pA.

Input bias current is the average of the two input currents at $V_o = 0$; typically, it is 100 nA in BJT and 1 nA in Bi-FET op amps.

These currents play a role that must be accounted for in a practical op-amp circuit, for example, in the inverting op-amp configuration shown in Fig. 9.3. We must provide a compensating resistor for the offset current in a BJT op amp as shown in Fig. 9.59. The same is true for other BJT op-amp circuits. See, for example, typical applications in Appendix E for 741; the compensating resistor is added in each case for integrator, clipper, differentiator, and filter. Thus, in practice, we *must* include the addition of this resistor.

In a BiFET the parasitic input capacitance C_1 (3 pF for LF351) interacts with feedback elements and creates a high-frequency pole, which is undesirable. To compensate, add C_2 such that $R_1C_1 = R_2C_2$ as shown in Fig. 9.60.

C. Unity Gain Bandwidth

This is the frequency at which the gain of the op amp is unity. For a one-pole op amp it is the value of $a_o f_{3dB}$; a typical value of 741 is approximately 1 MHz, which was calculated in Sec. 8.3. Thus, for an inverting amplifier in Fig. 9.3, for $R_f = R_S = 10$ kΩ the gain is unity, and the bandwidth is $f_{3dB} = 1$ MHz. For the one-pole roll off frequency response, gain can be traded with bandwidth. For example,

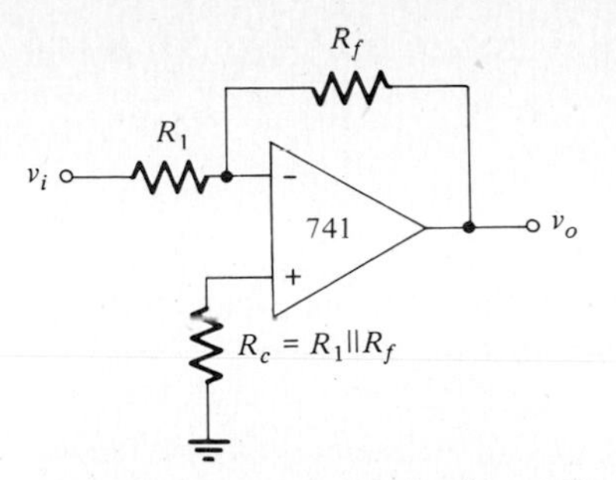

FIGURE 9.59 A practical inverting op-amp circuit using 741 op amp.

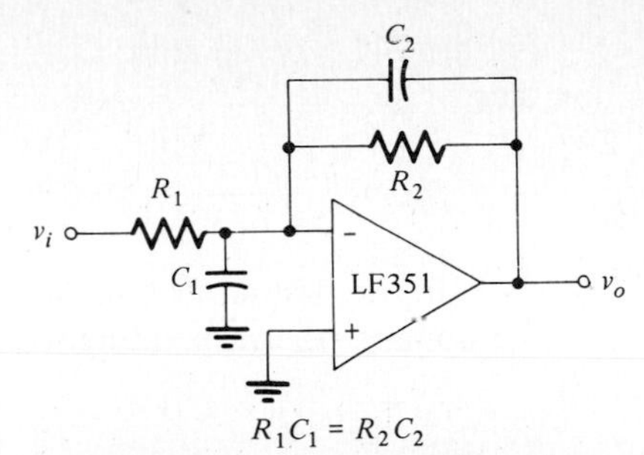

FIGURE 9.60 A practical inverting op-amp circuit using LF351 op amp.

for a voltage gain of 10 the 3-dB frequency is 100 kHz. If the frequency response characteristic is not compensated and it does not roll off at 20 dB/decade, more gain can be obtained at a certain frequency f_1 than the unity-gain-bandwidth product divided by f_1. The op amp, however, can be prone to instability as discussed in Chap. 10.

D. Slew Rate

Slew rate is basically a time-domain quantity and is discussed in Chap. 12. We shall discuss it here because it is an important consideration in op amps. Figure 9.61 shows a simplified model of the op amp, giving only the important quantities that relate to the slew rate. The two important quantities are the capacitor C_c and the current i_1, which is equal to the controlled source $G_m v_1$. From the circuit

$$i_1 = C_c \frac{dv_o}{dt} \tag{9.68}$$

slew rate (SR) is defined as

$$\text{SR} = \frac{dv_o}{dt} \tag{9.69}$$

From (9.68) and (9.69) we have

$$\text{SR} = \frac{i_1}{C_c} \tag{9.70}$$

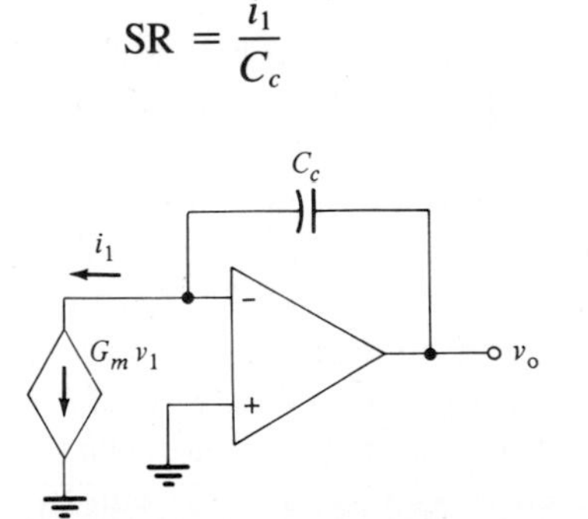

FIGURE 9.61 A simplified model of the op amp for the purpose of slew rate calculation.

For the 741 type op amp, $C_c = 30$ pF and $i_1 \simeq 20\ \mu$A (I_C through Q_6 in Fig. 5.25*a*); thus

$$SR = \frac{2 \times 10^{-5}}{30 \times 10^{-12}} \simeq 0.67\ \text{V}/\mu\text{s} \tag{9.71}$$

In the frequency domain, SR is related to the unity-gain-bandwidth product of the op amp. Corresponding to (9.68), in Fig. (9.61), we have

$$I_1 = G_m V_1 = C_c s V_o \tag{9.72}$$

The frequency at which the gain is unity, from (9.72) is given by

$$\frac{V_o}{V_1} = 1 = \frac{G_m}{\omega_\mu C_c} \tag{9.73}$$

Recall that the unity gain bandwidth ω_μ for the op amp is given by $a_o p_o$. In Ref. 2 it is shown that $G_m = 4V_t/I_1$ (where $V_t = 0.025$ V at room temperature). Hence from (9.70) and (9.73) with the appropriate substitutions for ω_μ and G_m, we have

$$SR = 4a_o p_o V_t \tag{9.74}$$

EXERCISE 9.14

Determine the SR of the 741-type op amp at room temperature if

$$a_o p_o = 2\pi(10^6)\ \text{rad/s}$$

Ans 0.63 V/μs

For the MOS op amp the slew rate is given by

$$SR = (V_{GS} - V_T)\,\omega_\mu \tag{9.75}$$

where ω_μ is the unity gain frequency of the MOS amplifier. The unity gain frequency $\omega_\mu = g_m/C_c$, where g_m is the conductance of the input NMOS and C_c is the compensation capacitor.

There are also other parameters of the op amp provided by the manufacturers, which are listed in Appendix E. The reader is referred to the references for further study.

There are many types of op amps available to the designer. We have considered the general-purpose 741 type only to be specific and also because they are popular. Other similar BJT op amps are 702, 709, and 1530 types. In the Bi-FET family, which uses FETs in the differential input stage and BJTs for the amplifying stages, we have used the popular LF300 family. For example, LF351 is the single Bi-FET, and LF353 is the dual (two op amps). Both 741 and LF351 are internally compensated; i.e., their frequency response has a one-pole rolloff given by (9.67). Bi-FETs and Bi-MOS op amps are gaining ground rapidly. Some manufacturers provide several op amps within one package, e.g., LM 348, which is a quad bipolar op amp with a cost of a nickle per op amp (at the time of writing this text), and the LF347, which is a quad version of LF351. Some are special for very high frequencies up to 300 MHz, e.g., AD380 type, and a high slew rate of up to 500 V/μs, and some are temperature compensated, either for bias current or for gain-bandwidth. Manufacturer's data books provide the characteristics and the details of their devices. Characteristics of the 741 and LF351 are given in Appendix E. For additional information other compensated op amps with their important parameters are also listed in Appendix E.

9.11 CONCLUDING REMARKS

The purpose of this chapter was to expose the reader to the variety of applications of the op amp. The concept of the ideal op amp is very important in the analysis and design of op-amp circuits. Analysis and design based on the ideal op amp is quite simple and since the IC op-amp characteristics are very close to the assumed ideal, the results of analysis will be in close agreement with the actual circuit performance.

Although we have examined a variety of linear and nonlinear applications of the op amp, the list of applications is by no means exhausted. Op amps are the workhorse of analog-system design. There are a large variety of op amps available, and a number of these such as the Bi-FET and Bi-MOS op amps take advantage of the almost infinite input impedance properties of the FET devices. Because of the volume production, manufacturers are constantly improving the performance and reducing the cost of their op amps. The cost and performance of op amps, therefore, dictate their choice in circuit applications wherever possible.

REFERENCES

1. R. G. Irvine, *Operational Amplifier—Characteristics and Applications,* Prentice-Hall, Englewood Cliffs, N.J., 1981.
2. A. Sedra and K. C. Smith, *Microelectronic Circuits,* Holt, New York, 1982.
3. J. K. Roberge, *Operational Amplifiers—Theory and Practice,* Wiley, New York, 1975.
4. J. T. Wait, L. P. Huelsman, and G. H. Korn, *Introduction to Operational Amplifier Theory and Applications,* McGraw-Hill, New York, 1975.
5. M. VanValkenburg, *Analog Active Filter Design,* Holt, New York, 1982.

PROBLEMS

9.1 For the inverting op-amp circuit shown in Fig. 9.3, if a_o is finite, show that

$$\frac{v_o}{v_i} = -\frac{R_f}{R_S[1 + (1/a_o)(1 + R_f/R_S)]}$$

9.2 For the simple integrator circuit shown in Fig. P9.2, sketch the output signal for the square-wave input signal. Show the amplitude and the time scale.

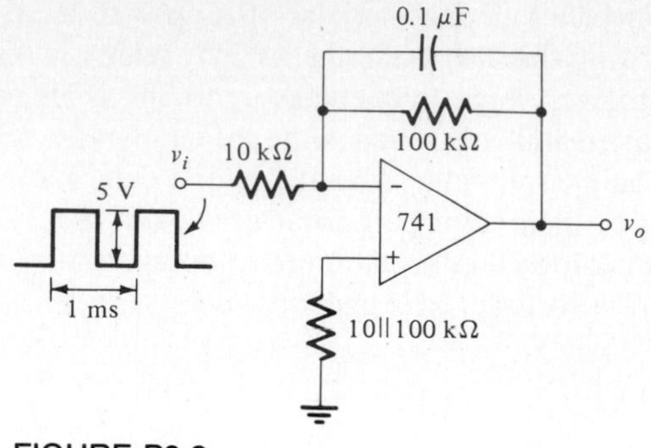

FIGURE P9.2

9.3 For the differentiator circuit shown in Fig. P9.3, derive v_o/v_i and sketch the output waveform for the given input waveform.

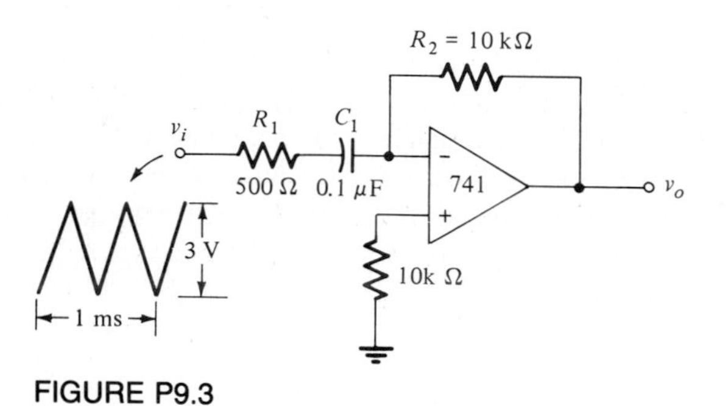

FIGURE P9.3

9.4 Show an analog computer simulation of the following ordinary linear differential equation

$$\frac{d^2x}{dt^2} + 2\frac{dx}{dt} + x = t$$

with the initial condition $x(0) = 1$ and $dx/dt\ (0) = 1$. Show the values of the circuit elements. Let $RC = 1$ ms.

9.5 Show the analog setup for the solution of the following system of differential equations and show the value of the circuit elements.

$$\frac{dx_1}{dt} = -5x_1 + x_2 \qquad \frac{dx_2}{dt} = 3x_1 - 3x_2$$

with

$$x_1(0) = 4 \text{ and } x_2(0) = 8.$$

9.6 Show a circuit implementation using an analog multiplier to detect the phase difference between two sinusoidal signals of the same frequency.

9.7 Show a circuit implementation of a frequency tripler using two multipliers with $K = 0.1/\text{V}$ and an op amp. (*Hint:* $\sin 3\theta = 3 \sin \theta - 4 \sin^3 \theta$.)

9.8 Determine the expression for the accuracy of a divider, using (9.15*a*) and (9.21). Assume that the resistors are trimmed so as to obtain precision accuracy, i.e., its ratio accuracy is much better than that of the multiplier.

9.9 Show a circuit implementation of the following:
(*a*) $y(t) = -2 \int x^2(t)\, dt$
(*b*) $y(t) = 2 \dfrac{d^2x}{dt^2} + 5x^2$

9.10 Show an analog simulation of the solution of the following nonlinear differential equation, which is known as the *Duffing equation:*

$$\frac{d^2x}{dt^2} + \omega^2 x + \beta x^3 = 0$$

Consider the special case of $\beta = -\omega^2/6$.

9.11 A four-quadrant transconductance multiplier circuit, shown in Fig. P9.11, provides $v_o = K v_x v_y$.
(*a*) Show that the circuit does perform the four-quadrant multiplication.
(*b*) Determine the expression for K.

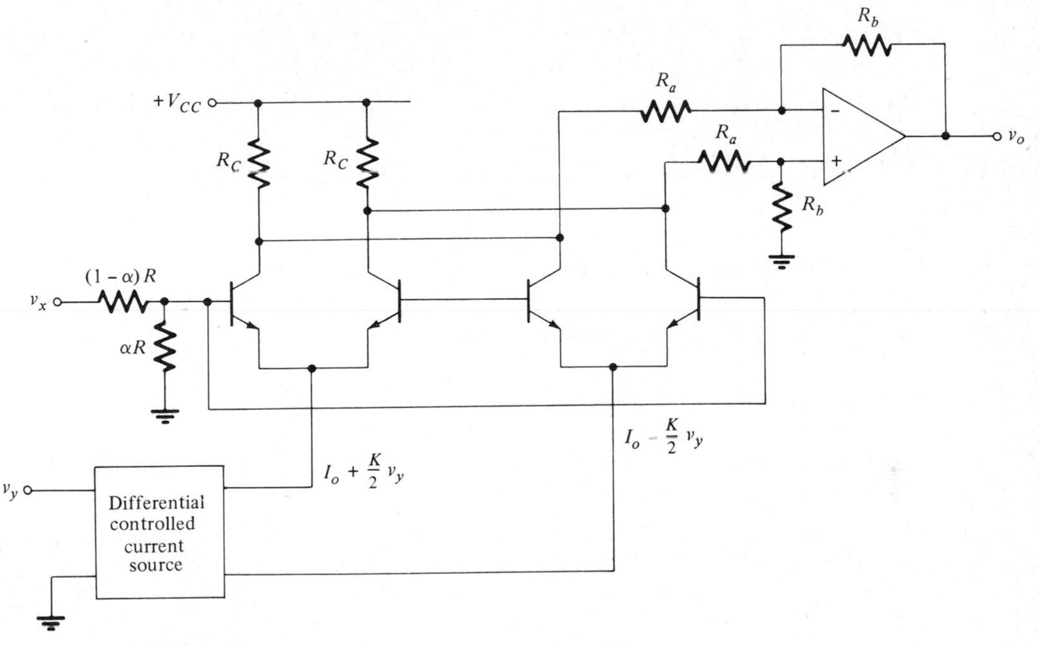

FIGURE P9.11

9.12 Design a circuit that has the following transfer function:

$$\frac{V_o}{V_i}(s) = \frac{3.95 \times 10^{10}}{[s + 2\pi(10^4)]^2}$$

Use only one-pole sections and equal-valued capacitors ($C_1 = C_2 = 0.01\ \mu\text{F}$). What is the overall bandwidth?

9.13 Derive (9.31*a*) and design the circuit for an identical two-pole response with 3-dB bandwidth $\omega_{3dB} = 2\pi(10^4)$ rad/s. Let $R_1C_1 = 10\ \mu$s.

9.14 Design the circuit of Fig. 9.22 for a two-pole maximally flat magnitude response with $\omega_{3dB} = 2\pi(10^4)$ rad/s. Assume $R_1C_1 = R_2C_2$, $K = 1$, and $C = 0.01\ \mu$F.

9.15 If two identical sections are cascaded and each section is designed as in problem 9.14, what is the overall bandwidth?

9.16 Derive the expression for the voltage gain function of the circuit in Fig. 9.24*b*. For $C_1 = C_2 = C = 0.01\ \mu$F and $R_1 = R_2 = R_3 = R$ design the circuit for $\omega_0 = 2\pi(10^3)$ rad/s and $Q = 10$.

9.17 Derive (9.34) and design the circuit for $Q = 10$ and $\omega_0 = 2\pi(10^3)$ rad/s. Let $C_1 = C_2 = 0.005\ \mu$F.

9.18 Design a low-pass three-pole Butterworth filter for $\omega_{3dB} = 2\pi(10^4)$ rad/s using a *minimum* number of op amps. The source impedance is 1 kΩ. Use equal values of capacitors $C = 0.01\ \mu$F whenever possible.

9.19 Repeat problem 9.18 using two op amps and with $R_S = 50\ \Omega$ and $C = 0.01\ \mu$F.

9.20 Repeat problem (9.18) for a four-pole Butterworth filter. Use Fig. 9.22 with $C_1 = 0.01\ \mu$F and $K = 2$ for the first section and $K = 1$ for the second section.

9.21 Design the high-pass section of Fig. 9.26 for a low-frequency cutoff of $2\pi(10^3)$ Hz with a maximally flat magnitude response. Use $R_1C_1 = R_2C_2$. Let $C_1 = 0.01\ \mu$F.

9.22 Derive the expression in (9.41) and design the circuit such that a frequency $f_0 = 10^3$ Hz is eliminated. Let $C_1 = C_2\ 0.01\ \mu$F and $R_1 = R_2$.

9.23 For the notch circuit shown in Fig. P9.23, determine the notch frequency if $C_1 = 0.01\ \mu$F. (*Hint:* Determine Z_{ab} first.)

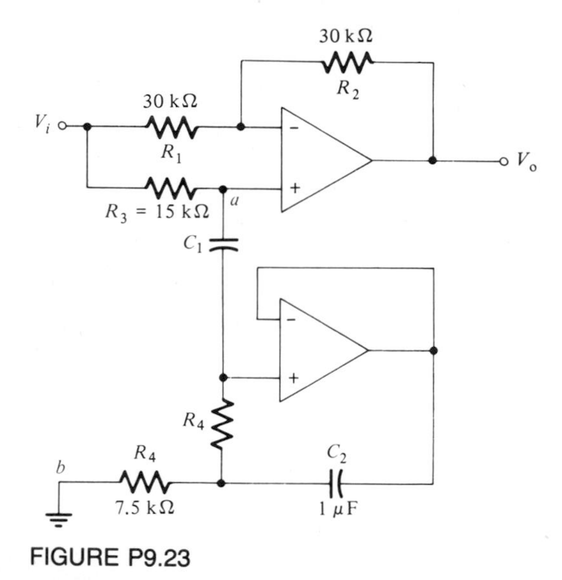

FIGURE P9.23

9.24 A second-order all-pass section is shown in Fig. P9.24. Derive the expression for the gain function if $R_4/R_3 = \frac{1}{4} R_2/R_1$, show that

$$\frac{V_o}{V_i}(s) = K \frac{s^2 - (1/Q)s + 1}{s^2 + (1/Q)s + 1}$$

where $K = -R_4/(R_3 + R_4)$.

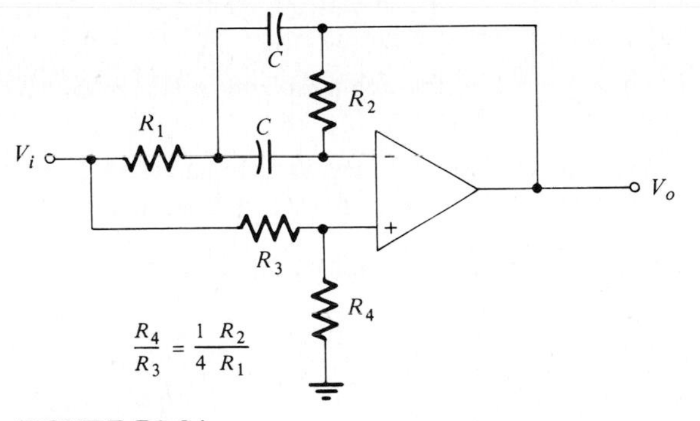

FIGURE P9.24

9.25 Derive (9.45).

9.26 Based on the basic input impedance relation given in (9.48) for a GIC,

(a) Show that the so-called frequency-dependent negative resistor (FDNR) can be obtained, namely,

$$Z_{\text{in}} = \frac{1}{s^2 D}\bigg|_{s=j\omega} = -\frac{1}{D\omega^2}$$

(b) Show a circuit realization using FDNR and RC that realizes the same voltage gain function as in Fig. 9.34a.

9.27 For the op-amp circuit shown in Fig. P9.27, the op amps are assumed to be identical with $a(s) = GB/s$. Determine the expression for the transfer function V_o/V_i.

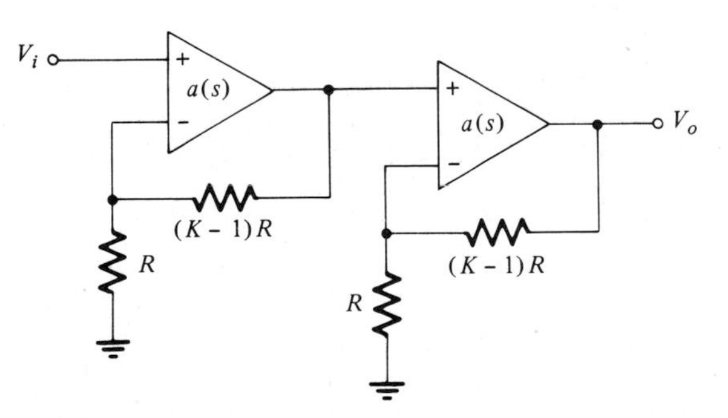

FIGURE P9.27

9.28 Design a switched-capacitor network to realize the following transfer function:

$$\frac{V_o}{V_i}(s) = \frac{s + 10^3}{s + 10^4}$$

Use a clock frequency of $f_c = 10^5$ Hz and equal-valued capacitors in the analog realization with $C = 1000$ pF.

9.29 Design a low-pass second-order SC filter for a bandwidth of $f_{3dB} = 10^3$ Hz, using a clock frequency of $f_c = 2 \times 10^4$ Hz. Use the analog realization in problem 9.13 with $C_1 = C_2 = 1000$ pF and $K = 1$.

9.30 Sketch the transfer characteristic of the circuit shown in Fig. 9.46. Sketch the output waveform for a sinusoidal input signal.

9.31 (*a*) Sketch the transfer characteristic for the circuit shown in Fig. P9.31. Assume $V_T = 0.7$ V for the forward characteristic of the zener diode. Show v_o for $v_i = V_m \sin \omega t$ with $V_m > |V_Z|$.
(*b*) Sketch the transfer characteristic v_o vs v_i for the precision rectifier shown in Fig. 9.47.

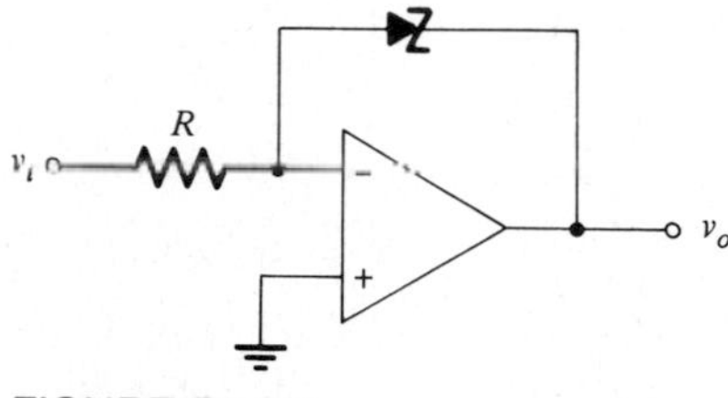

FIGURE P9.31

9.32 For the Schmitt trigger circuit shown in Fig. 9.55*b*,
(*a*) Show that the threshold voltages for $V_R = 0$ are given by

$$V_{TH} = \frac{R_1}{R_1 + R_f} V_s \qquad V_{TL} = \frac{-R_1}{R_1 + R_f} V_s$$

If $V_s = \pm 10$ V, $R_1 = 10$ kΩ, determine R_f for the Schmitt trigger to switch at +3 V.
(*b*) What is the switching voltage (center point between the two hysteresis voltages) if V_R is not zero?

9.33 Sketch the transfer characteristic of the circuits shown in Fig. P9.33. Assume $V_o = 0.7$ V and $V_z = -5$ V in each case, $V_1 = +3$ V.

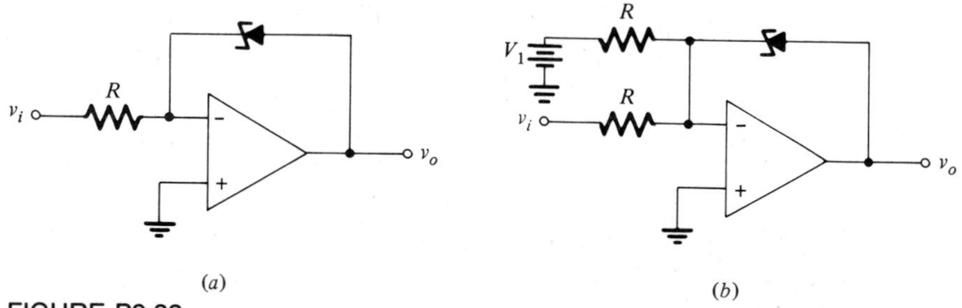

FIGURE P9.33

9.34 For the circuit shown in Fig. P9.34, determine v_o for $v_i = +5$ and -5 V.

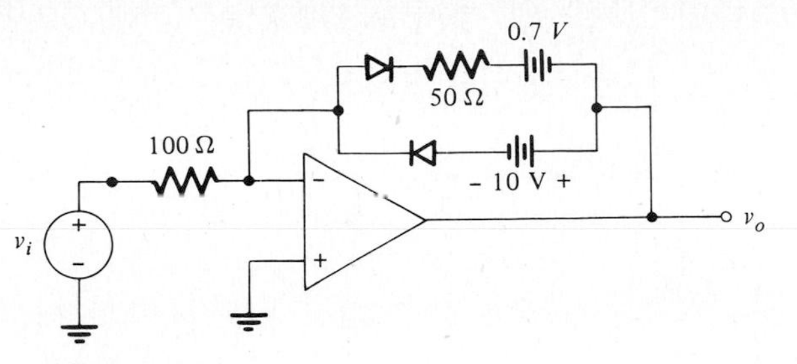

FIGURE P9.34

9.35 Sketch the transfer characteristic of Fig. 9.35 if $V_{z1} = V_{z2} = 3.3$ V and $V_o = 0.7$ V (note that the forward and the reverse characteristics of the zener diode are to be included).

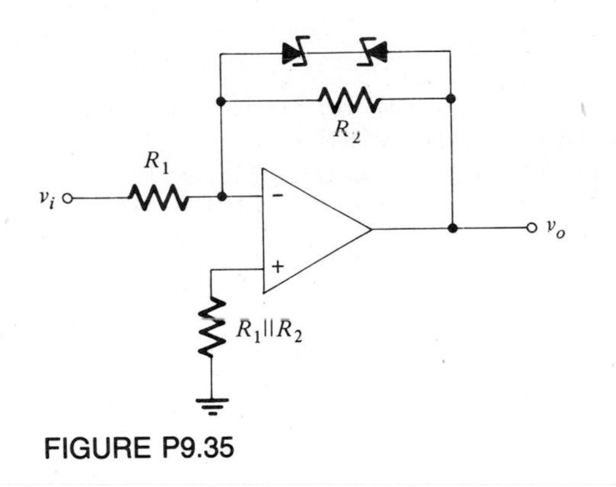

FIGURE P9.35

9.36 Using the circuit in Fig. 9.56, show that the network N achieves the transfer characteristics shown in Fig. P9.36.

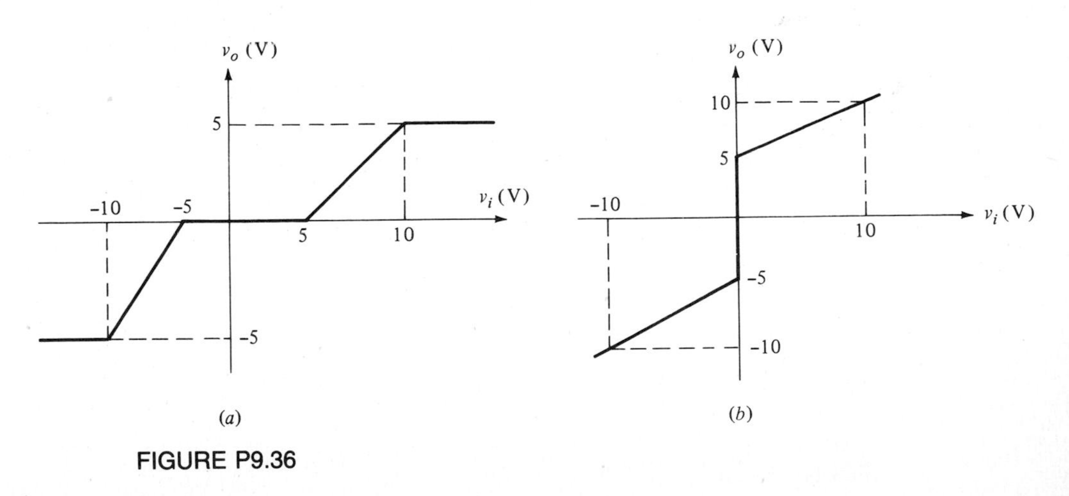

FIGURE P9.36

9.37 Sketch the transfer characteristic of the circuit shown in Fig. P9.37. Show how full-wave rectification is achieved. Assume ideal diodes.

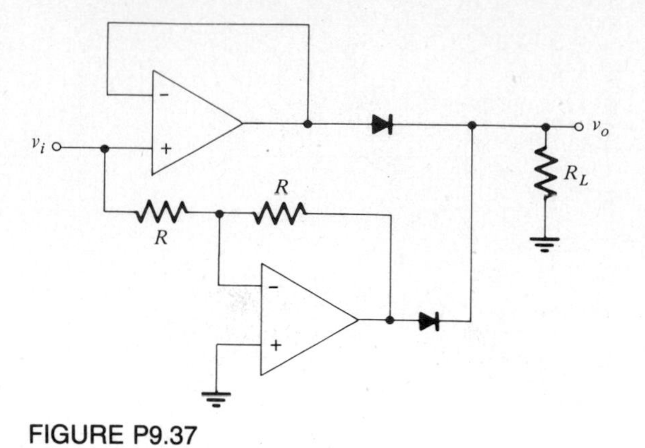

FIGURE P9.37

9.38 The circuit shown in Fig. P9.38 is a precision peak-detector circuit.
(*a*) Explain how the circuit performs peak detection.
(*b*) Why is this circuit preferable to the one shown in Fig. 9.51?

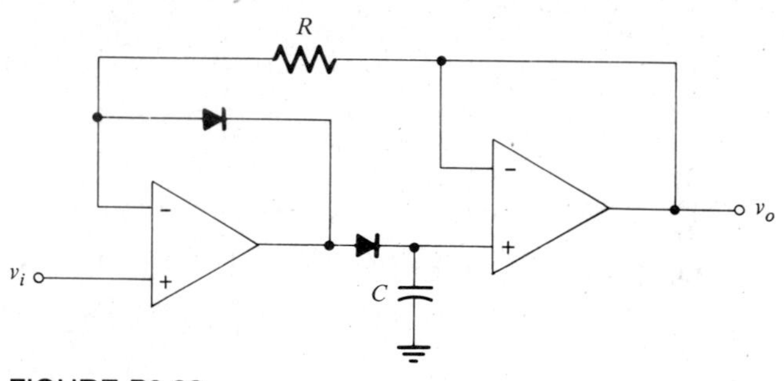

FIGURE P9.38

9.39 For the logarithmic amplifier circuit shown in Fig. P9.39, $I_c = I_s e^{V_{BE}/V_t}$. Show that $V_o = -V_t \ln(V_i/I_S R)$.

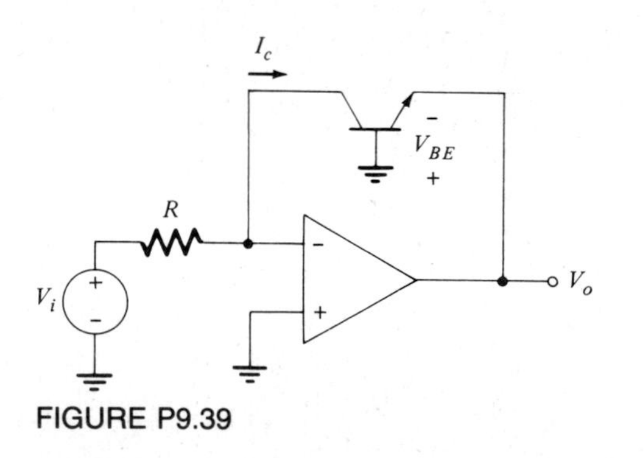

FIGURE P9.39

CHAPTER 10 Feedback Amplifiers and Oscillators

Feedback is present in all electronic circuits, intentionally or otherwise. In general, a system with feedback is one in which signal transmission exists in some manner from the output to the input. The nonunilateral nature of transistors is an example of unintentional internal feedback. External feedback is used deliberately in many systems to accrue some benefits. One of the many such examples in electronic circuits is the stabilization of the operating point in transistor amplifiers, as discussed in Chaps. 2 and 3. In fact, sensitivity requirements often necessitate the use of feedback, and most amplifiers are thus of the feedback type.

It should be mentioned at the outset that the conventional methods of analysis are applicable to all circuits, regardless of the presence and nature of feedback. An approximate method of analysis that focuses attention on feedback is very desirable, however, from the design viewpoint and considerable simplicity is thereby often achieved. The design can then be further refined and improved, if necessary, by more detailed analysis, preferably computer aided. It should be mentioned, however, that no one analysis method is the best for all situations.

In this chapter we consider the use of feedback in electronic circuits. Two-port concepts are used to present feedback, and it is therefore very important that the reader be familiar with the material in Appendix C. Op amps in feedback configurations are discussed in detail, in terms of interconnected two ports. We discuss some of the advantages resulting from the use of negative feedback and the potential disadvantage of feedback, which is the introduction of the potential problem of instability. For amplifiers, instability is, of course, to be avoided at all costs. Stability criteria are also briefly discussed including compensation. The discussion of feedback amplifiers will, of necessity, be introductory, since feedback control theory is a vast topic and cannot be adequately covered in one or two chapters.

Instability is *deliberately* used in some electronic circuits to perform certain circuit functions. Instability in such cases forms the basis of their operation, as in oscillators. In this chapter we also discuss, briefly, the linear analysis of sinusoidal feedback oscillators, including crystal oscillators.

10.1 BASIC FEEDBACK CONCEPTS AND DEFINITIONS

An *idealized* model of a single-loop feedback amplifier by means of a signal flow block diagram is shown in Fig. 10.1. The input and the output signals may be either current or voltage quantities. The idealizations are that the signals are transmitted only in the direction of the arrows. The electrical block diagram is shown in Fig. 10.1*b* and *c*. Note that in Fig. 10.1*b* the input quantities Φ are voltages, i.e., $\Phi_f = V_f$; hence Φ_a *must* be V_a. In Fig. 10.1*c*, $\Phi_f = I_f$; hence Φ_a *must be* I_a. The output quantities are $\Phi_2 = V_2$ in Fig. 10.1*b* and $\Phi_2 = I_o$ in Fig. 10.1*c*. We now define some terminologies that are used in the rest of this chapter.

The *feedback loop* is defined as the closed path consisting of the forward path, namely, the amplifier $\mathbf{a}(s)$, and the feedback path $\mathbf{f}(s)$.

The amplifier gain function $\mathbf{a}(s)$ is usually referred to as the *forward*, or sometimes *open-loop, transfer function*, i.e.,

$$\mathbf{a}(s) = \frac{\Phi_2}{\Phi_a} \tag{10.1}$$

Note that we use the symbol Φ instead of a V or I so as to have generality in our subsequent discussions; i.e., Φ_2 and/or Φ_1 may be current or voltage (in the transformed domain) depending on the circuit configurations. Also note that from Fig. 10.1, Φ_a is the sum of the input signal and the negative feedback signal. If Φ_a is current, then we use subscript i (i.e., $\mathbf{a}_i$), and if it is a voltage, we use $\mathbf{a}_v$ in (10.1) regardless of whether Φ_2 is current or voltage. Thus $\mathbf{a}$ can be a transconductance, transresistance, current gain, or voltage gain. From Fig. 10.1, we have

$$\Phi_a = \Phi_1 - \Phi_f \tag{10.2}$$

The *feedback function* $\mathbf{f}(s)$ is defined as

$$\mathbf{f}(s) = \frac{\Phi_f(s)}{\Phi_2(s)} \tag{10.3}$$

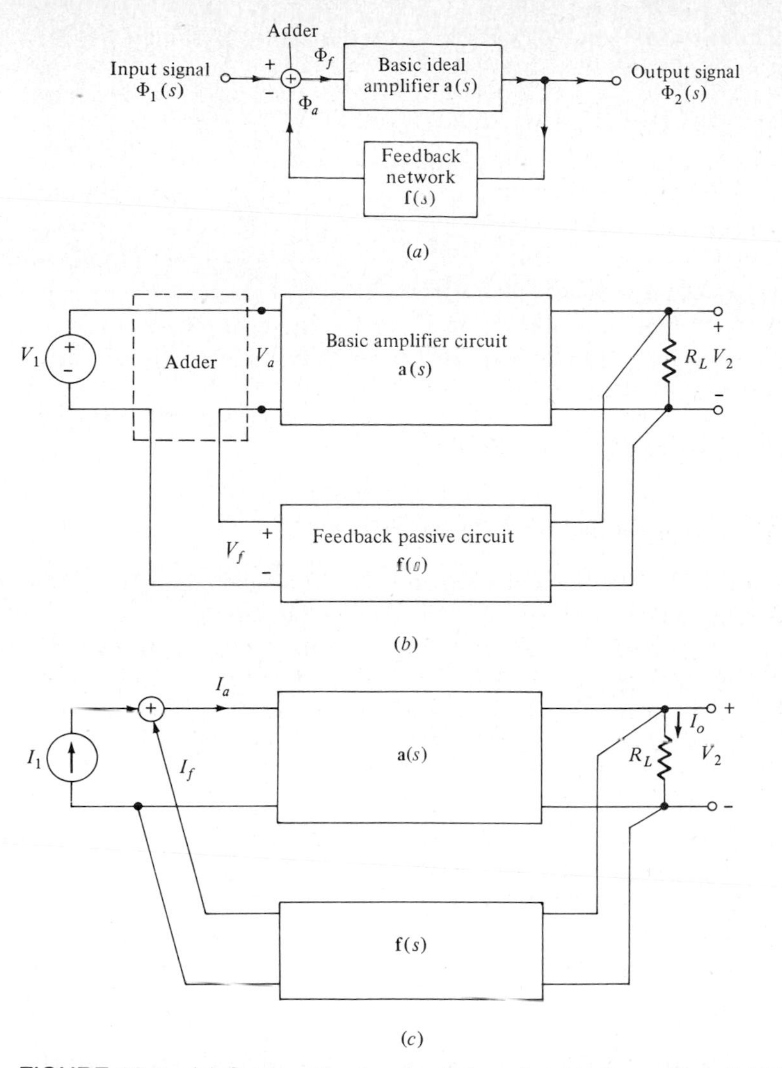

FIGURE 10.1 (*a*) Schematic of a single-loop feedback amplifier. (*b*) Electrical block diagram for a voltage feedback. (*c*) Electrical block diagram for a current feedback.

From (10.1) to (10.3) we solve for the gain function of the overall system, namely, for Φ_2/Φ_1, which we shall call $A(s)$.

$$A(s) = \frac{\Phi_2(s)}{\Phi_1(s)} = \frac{\mathbf{a}(s)}{1 + \mathbf{a}(s)\mathbf{f}(s)} \tag{10.4}$$

Note again that $A(s)$ need not be a voltage or a current gain; it will have the same dimension as of $\mathbf{a}(s)$. Equation (10.4) is very important and is usually referred to as the *basic feedback* equation. Many of the properties of feedback can be deduced from (10.4). In (10.4) A is referred to as the *closed-loop gain function*.

The *return ratio* (the *loop-transmission* function) or simply the *loop gain* is defined as

$$T(s) = \mathbf{a}(s)\mathbf{f}(s) \tag{10.5}$$

Note that if $|T(j\omega)| \gg 1$, then from (10.4) we have

$$A(s) \simeq \frac{1}{\mathbf{f}} \tag{10.6}$$

that is, for large values of loop gain the overall amplifier gain is given simply by the reciprocal of the feedback transfer function **f**. The feedback network is usually a passive network. The *return difference* $F(s)$, also sometimes referred to as the amount of feedback, is defined by

$$F(s) = 1 + T(s) = 1 + \mathbf{a}(s)\mathbf{f}(s) \tag{10.7}$$

Note that $F(s)$ is the denominator of $A(s)$. The equation

$$F(s) = 1 + \mathbf{a}(s)\mathbf{f}(s) = 0 \tag{10.8}$$

is usually called the *characteristic equation* of the system. Equation (10.8) is very important, since the roots of $F(s) = 0$ determine the poles of the closed loop or the natural frequencies of the system. Equation (10.8) is used for stability analysis of the system. For a stable system, the roots of $F(s)$ must be in the left half-plane, as discussed in Sec. 10.11.

The magnitude of $F(j\omega)$, which is usually expressed in decibels, namely,

$$20 \log |F(j\omega)| = 20 \log |1 + \mathbf{a}(j\omega)\mathbf{f}(j\omega)| \tag{10.9}$$

determines the amount of feedback. Note that $\mathbf{a}(s)\mathbf{f}(s)$, in general, is a complex quantity. *At low frequencies*, if $|1 + \mathbf{a}(j\omega)\mathbf{f}(j\omega)|$ is greater than unity, it is called *negative* (or degenerative) feedback, and if $|1 + \mathbf{a}(j\omega)\mathbf{f}(j\omega)|$ is smaller than unity, it is called *positive* (or regenerative) feedback. An unstable single-loop feedback amplifier may exhibit both negative and positive feedback, where the latter is the cause of instability, as discussed later in this chapter.

EXERCISE 10.1

For the simple circuits shown in Figs. 7.27 and 7.28, identify Φ_f as to whether it is a voltage or a current quantity.

Ans Φ_f = voltage across R_e; Φ_f = current through R_f

10.2 ADVANTAGES OF NEGATIVE FEEDBACK

Before we go into the analysis and design of feedback amplifiers, let us consider briefly the merits of negative feedback and the price paid for it. The main advantages are given in the following sections.

A. Improvement in Gain Sensitivity

The variation in gain a as a result of the active-device parameter changes, temperature changes, and *changes* in transistor current gain β is called *gain sensitivity*, and it can be reduced by feedback. The improvement in sensitivity is one of the most attractive features of negative feedback. To show this, consider the relation between the fractional (actually differential) change of $A(s)$ with feedback and the fractional change without feedback, as obtained by differentiating (10.5). We consider the low-frequency values in the following for simplicity, so that A, $\mathbf{f}$, and $\mathbf{a}$ are just real numbers and not functions of s. Differentiation of (10.4) yields

$$dA = \frac{1}{(1 + \mathbf{af})^2}\, d\mathbf{a} \tag{10.10a}$$

or

$$\left|\frac{dA}{A}\right| = \left|\frac{1}{(1 + \mathbf{af})}\right| \left|\frac{d\mathbf{a}}{\mathbf{a}}\right| \tag{10.10b}$$

In negative feedback $(1 + \mathbf{af}) > 1$; thus the closed-loop variation $|dA/A|$ is smaller than that of the open-loop variation $|d\mathbf{a}/\mathbf{a}|$. For example, in an amplifier with 26 dB of negative feedback, $\mathrm{F} = (1 + \mathbf{af}) = 20$, a 1 percent change in gain without feedback is reduced to a 0.05 percent change with feedback. [Note that for $|\mathbf{af}| \gg 1$, in (10.6) the closed-loop gain is essentially independent of $\mathbf{a}$.]

The sensitivity of A with respect to $\mathbf{a}$ is defined as

$$S_{\mathbf{a}}^{A} = \frac{dA/A}{d\mathbf{a}/\mathbf{a}} \tag{10.11}$$

Note that, in general, sensitivity is a function of s. The topic of sensitivity is discussed in Sec. 11.10 and will not be pursued here. From (10.10) and (10.11) at midband we have $S_{\mathbf{a}}^{A} = 1/(1 + \mathbf{af})$. The smaller the value of $S_{\mathbf{a}}^{A}$, the better the closed-loop gain sensitivity.

For large changes, which are often the case, one must use differences rather than the differentials, in which case, from (10.4).

$$\Delta A = A_2 - A_1 = \frac{a_2}{1 + a_2 f_2} - \frac{a_1}{1 + a_1 f_1} \tag{10.12a}$$

Multiply both sides by $1/A_1$ to get

$$\frac{\Delta A}{A_1} = \frac{a_2}{1 + a_2 f}\,\frac{1 + a_1 f}{a_1} - 1 = \frac{(a_2 - a_1)/a_1}{1 + a_2 f} = \frac{\Delta a/a_1}{1 + a_2 f} \tag{10.12b}$$

Thus, the fractional change from the original gain level is

$$\frac{\Delta A}{A_1} = \frac{\Delta a/a_1}{1 + a_2 f} = \frac{\Delta a/a_1}{1 + (a_1 + \Delta a)f} \tag{10.13}$$

For example, if $a_1 = 10^4$ and $f = 10^{-2}$, then $A_1 = 10^2$. If a changes by 100 percent, the change in A_1 will be less than 1 percent.

B. Reduction of Distortion

Distortion due to nonlinearities of the transfer characteristics of the output stages and of power amplifiers can also be reduced by feedback; this was shown in Sec. 6.6. However, noise signal distortion that appear at the input, i.e., signal-to-noise ratio, is *not* affected by feedback. Consider the feedback system shown in Fig. 10.2. We assume small signals and approximate linear operation so that superposition can be applied. Note that if the distortion appears at the output of the amplifier, the effect of the disturbing signal can be considerably reduced. For example, from Fig. 10.2, where v_d is the distortion signal and a_1 provides the main amplifications, we have

$$v_o = v_d \frac{a_2}{1 + a_1 a_2 f} + v_s \frac{a_1 a_2}{1 + a_1 a_2 f} \tag{10.14}$$

For $a_1 a_2 f \gg 1$, (10.14) reduces to

$$v_o \simeq v_s \frac{1}{f} + \frac{v_d}{a_1 f} \tag{10.15}$$

The disturbance is reduced by the gain between the input and the point of introduction of the disturbance. If $a_1 f \gg 1$, the effect of the disturbing signal is reduced considerably. Thus nonlinear distortion, since it occurs mainly in the output stage, can be considerably reduced by feedback. The noise signals introduced at the input stage, however, are most bothersome, and feedback cannot help in improving the signal-to-noise ratio.

C. Improvement in Bandwidth

Amplifier bandwidth can be increased by using feedback. We have already seen this property in Chap. 7 in conjunction with the shunt-feedback (Fig. 7.8) and series-feedback (Fig. 7.27) broadbanding. In fact, desirable transmission characteristics such as maximally flat magnitude response can be achieved in the design by employing the right amount of feedback in a multipole circuit. We shall consider these again in detail in this chapter.

D. Control over Impedance

The circuit input and output resistances can be vastly increased or decreased depending on whether current or voltage feedback is employed at the port. This property

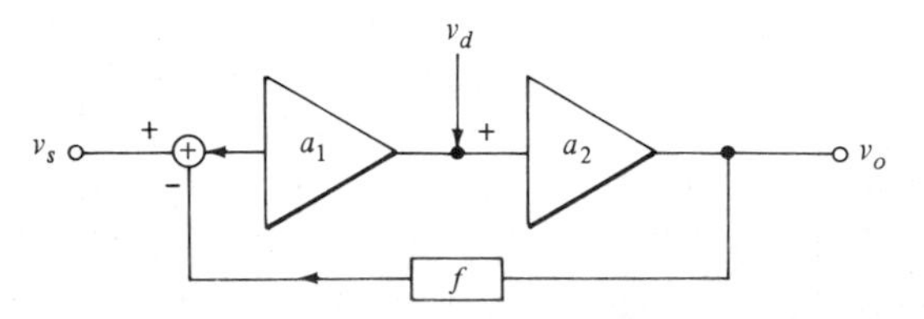

FIGURE 10.2 Feedback amplifier with the disturbing signal not appearng at the input.

has already been noted in conjunction with the Miller theorem and its dual. We shall examine further the effect of feedback on the input and output impedance of the circuit.

10.3 DISADVANTAGES OF NEGATIVE FEEDBACK

From the previous two sections it is quite clear that one of the disadvantages of negative feedback is the reduction of the overall gain of the amplifier. This is not a costly price to pay for the benefits derived. Further, since active devices are often inexpensive, the use of additional stages with feedback can readily take care of the specified closed-loop gain. In most cases, such as for op amps, the gain is much more than needed anyway.

Another troublesome and serious disadvantage is the potential introduction of the problem of oscillation into the system. This is possible since, in general, $\mathbf{a}(s)$ and possibly $\mathbf{f}(s)$ are frequency dependent, and if the loop transmission phase shift and attenuation characteristics are not properly controlled, the system may become unstable. Most often in the analysis and design we ignore the nondominant poles of the active device. One must be particularly careful if the design is based on the unilateral model for the transistors. For stability analysis, the unilateral model would not normally be used, and thus one must use a more complete model such as the hybrid-π circuit.

10.4 FEEDBACK CONFIGURATIONS AND CLASSIFICATIONS

In the single-loop feedback configuration shown in Fig. 10.1, the amplifier circuit $\mathbf{a}(s)$ is an active two-port, while the feedback network $\mathbf{f}(s)$ is usually a passive two-port network. The *idealization* of the assumed unilateral nature of $\mathbf{a}(s)$ and $\mathbf{f}(s)$, as depicted by the signal found in Fig. 10.1*a*, are never exactly met in practice, since amplifiers are never exactly unilateral and passive circuits are always bilateral. We must, therefore, make some reasonable assumptions in order to approximately achieve the situation in Fig. 10.1. Fortunately, the assumptions in Fig. 10.1 can be relaxed and satisfactorily met in many practical problems.

The two basic assumptions we make are the following:

1. The forward transmission through the feedback network is negligible in comparison with the forward transmission through the amplifier.
2. The reverse transmission through the amplifier is negligible in comparison with the reverse transmission through the feedback circuit.

The above approximations are very reasonable and often valid when the feedback path is through a passive circuit, which is almost always the case because of sensitivity reasons. Since the signal is amplified via the forward amplifier circuit and attenuated via the passive feedback circuit, the first assumption is thus quite valid. The second assumption is usually valid, since the amplifiers used in $\mathbf{a}(s)$ are approximately unilateral, while $\mathbf{f}(s)$ is bilateral; hence reverse transmission through the $\mathbf{a}$ circuit is negligible in comparison with that through the $\mathbf{f}$ circuit. If we further assume that the

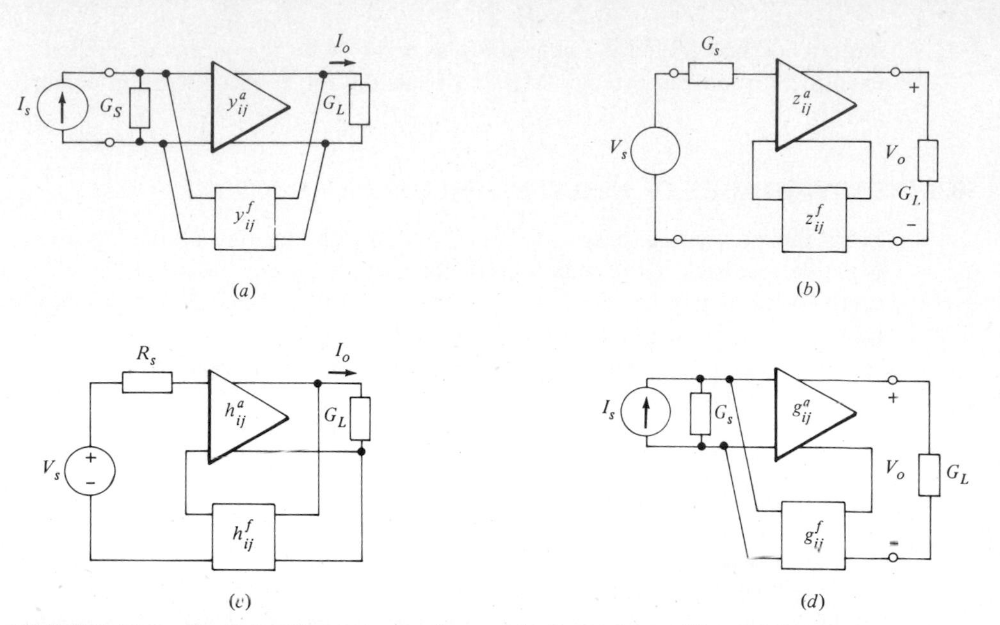

FIGURE 10.3 Basic feedback connections: (*a*) shunt-shunt, (*b*) series-series, (*c*) series-shunt, (*d*) shunt-series.

loading effects of the feedback network on the amplifier circuit are negligible, considerable simplification results in the initial design. The latter assumption is not necessary, but it is often made when the feedback parameter values are unknown.

Since we are usually dealing with two-port networks, let us consider the basic feedback connections. The four basic interconnections are shown in Fig. 10.3. These are the shunt-shunt, series-series, series-shunt, and shunt-series feedback configurations. Since the analysis of each of these is most conveniently carried out using a different set of small-signal parameters, a thorough review of Appendix C is appropriate at this point. For clarity,the $\mathbf{a}(s)$ blocks are shown as amplifier blocks with the appropriate two-port parameter descriptions. We use the two-port parameters in order to *derive* some basic relations; once the relations are obtained it is not necessary to use the two-port parameters in each case. Transistor feedback amplifier examples corresponding to these two-port diagrams were already shown in Fig. 8.15.

10.5 THE SHUNT-SHUNT CONFIGURATION

A composite two-port containing two parallel connected two-ports is shown in Fig. 10.4*a*. Since in the parallel connection, the y parameters are the most convenient to use, we shall make this choice in our analysis. The short-circuit admittance parameters of the amplifier and feedback circuits are designated by y_{ij}^a and y_{ij}^f, respectively.

From the two-port results (C.36) in Appendix C, the current gain function is given by

$$\frac{I_2}{I_1} = \frac{y_{21}G_L}{(y_{11} + G_S)(y_{22} + G_L) - y_{12}y_{21}} \tag{10.16}$$

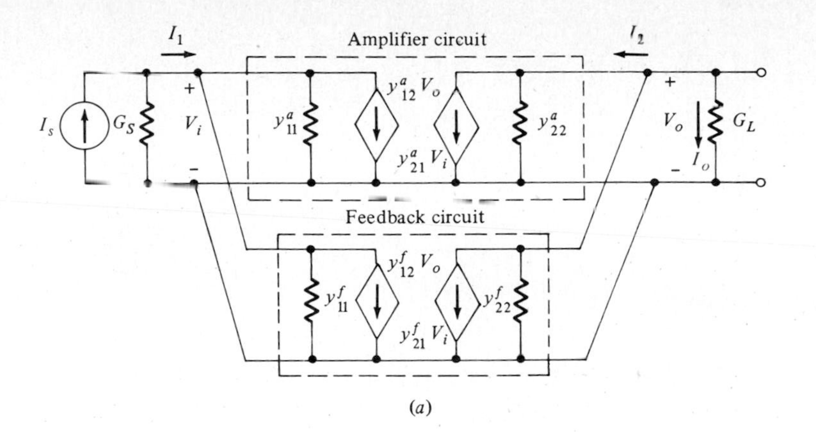

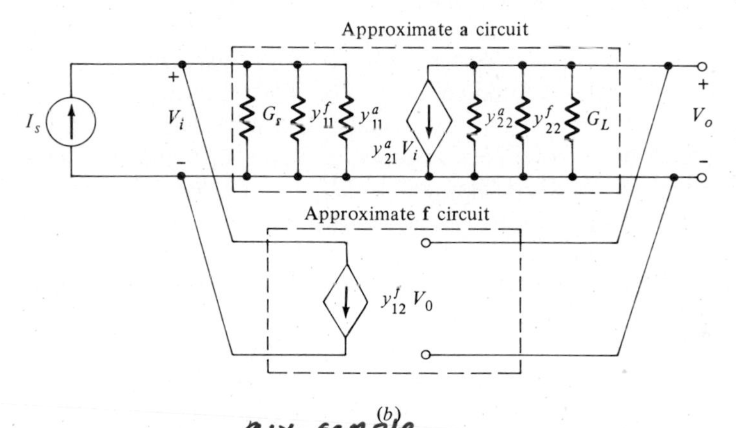

FIGURE 10.4 Shunt-shunt feedback configuration. (*a*) Complete two-port representation of the circuit. (*b*) Redrawn and simplified using the basic approximations.

In this case we have the y parameters of an interconnected two-port. Thus

$$\frac{I_o}{I_s} = \frac{-I_2}{I_1} = \frac{-y_{21}^T G_L}{(y_{11}^T + G_S)(y_{22}^T + G_L) - y_{12}^T y_{21}^T} \tag{10.17}$$

where $y_{ij}^T = y_{ij}^a + y_{ij}^f$. In order to express (10.17) in the form of the basic feedback equation (10.4), we rewrite (10.17) as

$$A_i = \frac{I_o}{I_s} = \frac{-y_{21}^T G_L/(y_{11}^T + G_S)(y_{22}^T + G_L)}{1 - [y_{12}^T y_{21}^T/(y_{11}^T + G_S)(y_{22}^T + G_L)]} \tag{10.18a}$$

We can rewrite (10.18*a*) as

$$\frac{I_o}{I_s} = \frac{-y_{21}^T G_L/(y_{11}^T + G_S)(y_{11}^T + G_L)}{1 - [(y_{12}^T/G_L)(y_{21}^T G_L)/(y_{11}^T + G_S)(y_{22}^T + G_L)]} \tag{10.18b}$$

From a comparison of (10.18*b*) and (10.4), we can at once identify the forward and feedback transfer functions as

$$\mathbf{a}_i(s) = \frac{-y_{21}^T G_L}{(y_{11}^T + G_S)(y_{22}^T + G_L)} \tag{10.19}$$

and

$$\mathbf{f}(s) = \frac{y_{12}^T}{G_L} \tag{10.20}$$

Therefore

$$T(s) = \mathbf{a}_i(s)\mathbf{f}(s) = -\frac{y_{12}^T y_{21}^T}{(y_{11}^T + G_S)(y_{22}^T + G_L)} \tag{10.21}$$

The subscript i is added to denote current addition at the input of Fig. 10.4. In the above equations *no* approximations are made. From (10.19) and (10.20), it is seen that both $\mathbf{a}(s)$ and $\mathbf{f}(s)$ depend on the y parameters of both two-port networks; i.e., the networks interact with each other. This situation does not simplify matters a whole lot. In order to *approximate* the idealized situation with as few assumptions as possible, we reexamine (10.18*b*) and incorporate the two basic assumptions stated in Sec. 10.4, which imply that $y_{21}^a \gg y_{21}^f$ and $y_{12}^a \ll y_{12}^f$. These approximations are quite reasonable as explained earlier and can provide considerable simplicity in the analysis and design as we shall soon see. It is also convenient to have the *feedback* **f** *dependent only on the* **f** *circuit*. The gain need not be a voltage ratio or a current ratio. It can be a gain or a transfer impedance or a transfer admittance in our formulation. However, once **f** is identified the dimension of **a** *must be* reciprocal of **f** since **af** is dimensionless. In this case the feedback signal is current, and because $\mathbf{f} = y_{12}^f$, we have a transresistance for **a**. Hence the approximate expression corresponding to (10.18*a*) is

$$A_z = \frac{V_o}{I_s} \simeq \frac{-y_{21}^a/(y_{11}^T + G_S)(y_{22}^T + G_L)}{1 - y_{12}^f y_{21}^a/(y_{11}^T + G_S)(y_{22}^T + G_L)} \tag{10.22}$$

Our identification of approximate $\mathbf{a}(s)$ and $\mathbf{f}(s)$ corresponding to (10.22) and (10.4) now becomes

$$\mathbf{a}_z(s) = \frac{-y_{21}^a}{(y_{11}^T + G_S)(y_{22}^T + G_L)} = \left.\frac{V_o}{I_s}\right|_{\mathbf{f}=0} \tag{10.23}$$

$$\mathbf{f}(s) = y_{12}^f \tag{10.24}$$

$$T(s) = \mathbf{a}(s)\mathbf{f}(s) = \frac{-y_{21}^a y_{12}^f}{(y_{11}^T + G_S)(y_{22}^T + G_L)} \tag{10.25}$$

Observe that the *input variable is current* I_s and the output variable for $\mathbf{a}_z$ is voltage V_o; hence this configuration achieves a transresistance, V_o/I_s, stabilization. We repeat that $\mathbf{f}(s)$ is an admittance function, y_{12}^f, in a shunt-shunt configuration and $\mathbf{a}(s)$ *must* be an impedance function since **af** is *dimensionless*. This is the reason for the subscript z in $\mathbf{a}_z$. Note that in (10.23) **a** depends on the feedback circuit, since $y_{11}^T = y_{11}^a + y_{11}^f$ and similarly $y_{22}^T = y_{22}^a + y_{22}^f$. The forward and feedback circuits corresponding to (10.23) and (10.24) are shown in Fig. 10.4*b*. Note that y_{11}^f and y_{22}^f are included in the **a** circuit. The circuit is redrawn in Fig. 10.4*b* to illustrate two points. First, G_S and

G_L are incorporated with the **a** circuit as they should since these must be there whether feedback is applied or not. Second, the two assumptions stated earlier, i.e., $y_{21}^a \gg y_{21}^f$ and $y_{12}^f \gg y_{12}^a$, are included in the circuit to simplify analysis.

The feedback parameter **f** in (10.24) depends *only* on the feedback circuit, which is a very desirable feature of this formulation. This approach, as we shall soon see, provides *considerable simplification*. If the values of the elements in the feedback network are known, such as in the analysis problem, one simply adds y_{11}^f and y_{22}^f with those of y_{11}^a and y_{22}^a as shown in Fig. 10.4*a*. This must be done, since (10.22) involves y_{11}^T and y_{22}^T. In a design problem, however, the feedback network values must usually be determined. In such cases, we make the *initial assumption of* $y_{11}^f \ll y_{11}^a$ and $y_{22}^f \ll y_{22}^a$ in the *preliminary design*. Then we reexamine our approximations and iterate, using the determined values for the feedback circuit, if necessary.

The input admittance of Fig. 10.4*a*, from (C.31) in Appendix C for the overall two-port, is

$$Y_{\text{in}} = (y_{11}^T + G_S) - \frac{y_{12}^T y_{21}^T}{y_{22}^T + G_L} = (y_{11}^T + G_S)\left(1 - \frac{y_{12}^T y_{21}^T}{(y_{22}^T + G_L)(y_{11}^T + G_S)}\right) \tag{10.26}$$

Incorporating the approximations such as in (10.22), we can rewrite (10.26) as

$$Y_{\text{in}} = (y_{11}^T + G_S)\left[1 - \frac{y_{12}^f y_{21}^a}{(y_{22}^T + G_L)(y_{11}^T + G_S)}\right] \tag{10.27}$$

From (10.27) and (10.25) or directly from (10.21) and (10.26), we obtain

$$Y_{\text{in}} = (y_{11}^T + G_S)[1 + T(s)] \tag{10.28}$$

The input impedance of the feedback amplifier would be what the source "sees." Since G_S is part of the source, so we have to subtract G_S from the result obtained in (10.28) to obtain the input admittance of the feedback amplifier. Similary, the output admittance is

$$Y_{\text{out}} = (y_{22}^T + G_L)[1 + T(s)] \tag{10.29}$$

The same statement concerning the source admittance is true for the output admittance. From (10.28) and (10.29) it is seen that shunt-shunt feedback increases the low-frequency input and output admittances [since $T(0) > 0$]; i.e., *the input and output impedances are decreased*. Note that in a *negative feedback amplifier we always have* $T(0) > 0$.

Consider as an example the inverting op-amp feedback amplifier circuit shown in Fig. 10.5*a*. We shall determine the circuit properties by the feedback method discussed in this section. From the **a** circuit (i.e., open loop in Fig. 10.5*c*) we find

$$V_o = -aV_i = -a(G_S + G_f)^{-1}I_s$$

Hence

$$\mathbf{a}_z = \frac{V_o}{I_s} = -\frac{a}{G_S + G_f} \tag{10.30}$$

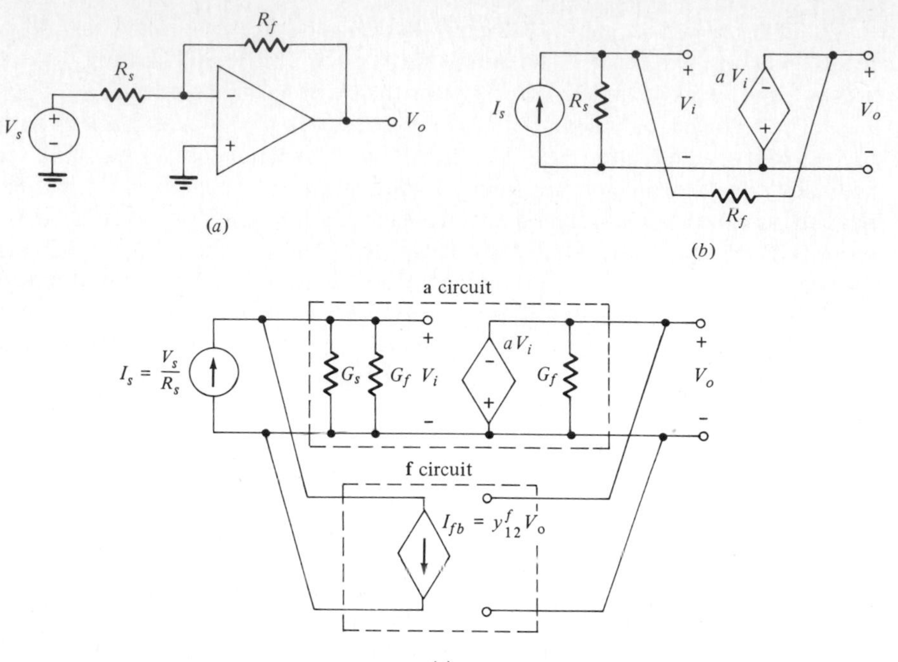

FIGURE 10.5 (*a*) An inverting op amp. (*b*) Equivalent circuit of (*a*). (*c*) The **a** circuit and the **f** circuit identified.

From the **f** circuit we have

$$\mathbf{f} = y_{12}^f = \frac{I_1}{V_2} = \frac{I_{fb}}{V_o} = -\frac{1}{R_f} = -G_f \tag{10.31}$$

thus

$$T = \mathbf{a}_z\mathbf{f} = \frac{aG_f}{G_S + G_f} \tag{10.32}$$

The overall gain (closed loop) is given by

$$A = \frac{\mathbf{a}}{1 + \mathbf{af}} \simeq \frac{1}{\mathbf{f}} \qquad \text{for } \mathbf{af} \gg 1 \tag{10.33}$$

Hence

$$\frac{V_o}{I_s} = -R_f$$

However, $I_s = V_s/R_S$; hence

$$\frac{V_o}{V_s} = -\frac{R_f}{R_S} \tag{10.34}$$

which is the same result as in (10.4). If we include the input and output resistances R_i and R_o of the op amp in the model in Fig. 10.5*b*, the values of these quantities as a result of the feedback can be readily determined from (10.28), (10.29), and (10.21).

From (10.28)

$$Y_{in} = (y_{11}^T + G_S)(1 + T) \tag{10.35}$$

where y_{11}^T is now given by $y_{11}^T = y_{11}^a + G_f = G_i + G_f$. Hence

$$G_{in} = (G_i + G_f + G_S)(1 + T) \tag{10.36}$$

Similarly

$$G_{out} = (G_o + G_f)(1 + T) \tag{10.37}$$

Note that G_S is usually not a part of G_{in} and hence must be separated.

EXAMPLE 10.1

As a numerical example, consider the 741 op amp, where $R_i = 1\ \text{M}\Omega$, $R_o = 75$, $a_o = 2 \times 10^5$. Let $R_S = 1\ \text{k}\Omega$ and $R_f = 100\ \text{k}\Omega$ in Fig. 10.5*a*. We shall find the gain, the input and the output resistances. From (10.34)

$$\frac{V_o}{V_s} = -\frac{100 \times 10^3}{10^3} = -100$$

From (10.32) the loop gain

$$T = \frac{(2 \times 10^5)(10^{-5})}{10^{-5} + 10^{-3}} \simeq 2 \times 10^3$$

From (10.36)

$$G_{in} = (10^{-6} + 10^{-5} + 10^{-3})(2 \times 10^3) \simeq 2 \text{ mhos} \qquad \text{or} \qquad R_{in} = 0.5\ \Omega$$

Note however that the input impedance seen from the voltage source is approximately equal to R_S. In other words, the source sees a resistance of about $0.5\ \Omega$ at *the input of the op amp.*

From (10.37) $G_{out} = (1.3 \times 10^{-2} + 10^{-5})(2 \times 10^3) = 26$ mhos, or $R_{out} \simeq 0.04\ \Omega$.

Note the *considerable reduction* in the values of the *input and output impedances* by the shunt-shunt feedback configuration.

If the amplifier $\mathbf{a}(s)$ is characterized by a simple pole, such as in 741 or LF351 or any other compensated op amp,

$$\mathbf{a}(s) = \frac{a_o}{1 + s/p_o} \tag{10.38}$$

Then

$$A = \frac{\mathbf{a}}{1 + \mathbf{af}} = \frac{a_o/(1 + s/p_o)}{1 + a_o \mathrm{f}_o/(1 + s/p_o)} \tag{10.39}$$

where the subscript o is added to **f** to indicate that it is constant with no poles and zeros. From (10.39) after simplification we obtain

$$A(s) = \frac{a_o/(1 + a_o \mathrm{f}_o)}{1 + s/p_o(1 + a_o \mathrm{f}_o)} \tag{10.40}$$

Note that the bandwidth of the amplifier is increased by the loop gain. For the 741 of Example 10.1, $p_o = 2\pi(5 \text{ Hz})$, the loop gain $T_o = 2 \times 10^3$, and the closed-loop bandwidth is

$$\begin{aligned}\omega_{3\text{dB}} &= p_o(1 + a_o \mathrm{f}_o) \simeq p_o T_o \\ &= 2\pi(5)(2 \times 10^3) = 2\pi(10^4) \text{ rad/s}\end{aligned}$$

Hence the 3-dB frequency is increased to 10 kHz. The Bode plot of the gain response with and without feedback is shown in Fig. 10.6.

Consider the simple single-stage shunt-feedback amplifier, employing negative feedback as shown in Fig. 10.7 (exclusive of biasing circuitry). We shall determine the gain and bandwidth of the circuit from the feedback viewpoint. We use the unilateral approximate model for the transistor as shown in Fig. 10.7*b* (see also Fig. 7.19*b*). Recall the limitation and caution required in the use of this model for high loop gain and stability analysis. In this case as we shall see, the loop gain is low, and the model can be used.

The **a** and the **f** circuits are shown in Fig. 10.8. Usually it is simple to find the expression of **a**(s) directly from the circuit. From Fig. 10.8*a*, letting $R_S' = R_S \parallel R_f$ and $R_f \gg R_L$, we have

$$\mathbf{a}_z(s) = \frac{V_o}{I_s} = -\frac{1}{r_e C_i} \frac{R_S'/(R_S' + r_b)R_L}{s + (1/R_i C_i)(R_S' + r_b + R_i)/(R_S' + r_b)} \tag{10.41}$$

We rewrite (10.41) for convenience as

$$\mathbf{a}_z(s) = -\frac{a_o p_o}{s + p_o} \tag{10.42}$$

where $a_o = \beta_0 \dfrac{R_S' R_L}{R_S' + r_b + R_i} \simeq -\beta_0 \dfrac{R_f R_L}{R_f + r_b + R_i}$ if $R_S \gg R_f \gg R_L$ **(10.43*a*)**

$$p_o = \frac{1}{R_i C_i} \frac{R_S' + r_b + R_i}{R_S' + r_b} \simeq \frac{1}{R_i C_i} \frac{R_f + r_b + R_i}{R_f + r_b} \tag{10.43b}$$

Also **f**(s), from (10.24) and Fig. 10.8*b*, is

$$\mathrm{f}_o = \frac{I_f}{V_o} = y_{12}^f = -G_f \tag{10.43c}$$

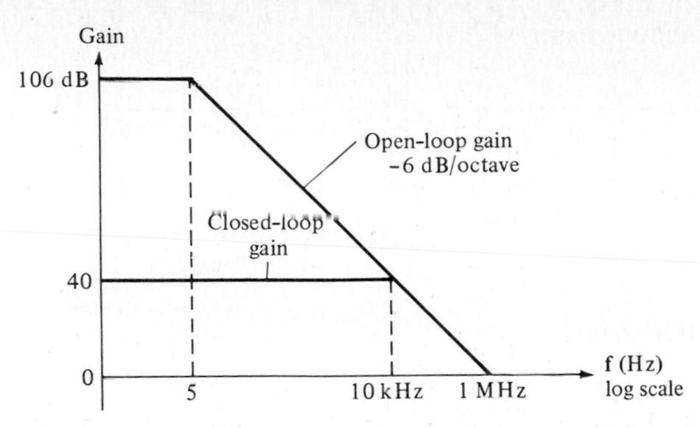

FIGURE 10.6 Bode plot of the open-loop and closed-loop one-pole gain functions.

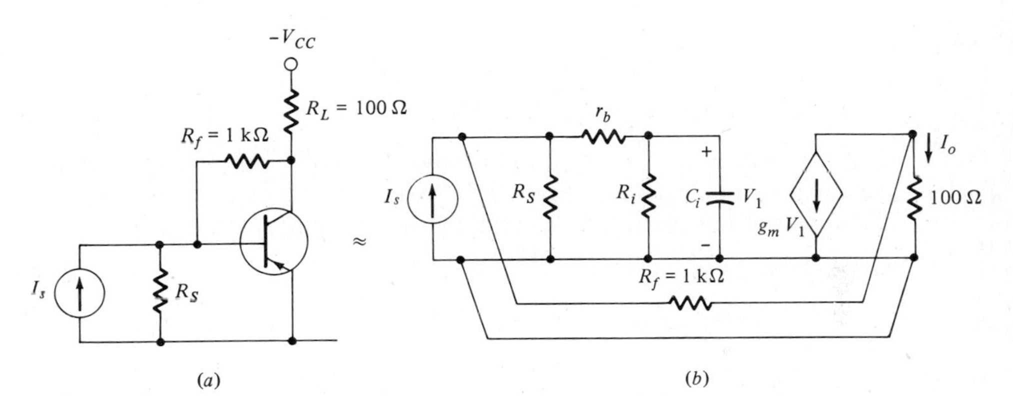

FIGURE 10.7 (*a*) Shunt-feedback amplifier for the example. (*b*) Equivalent circuit of (*a*).

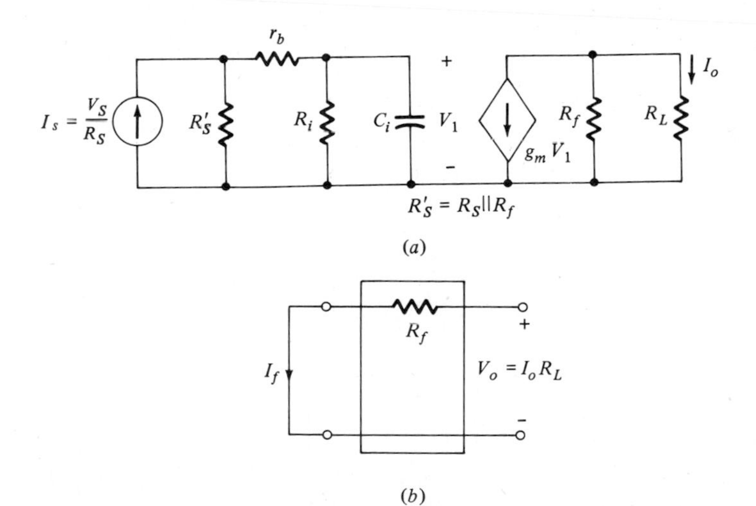

FIGURE 10.8 (*a*) The **a**(*s*) circuit for Fig. 10.7. (*b*) The **f**(*s*) circuit for Fig. 10.7.

The closed-loop gain function, using (10.4), is

$$A = \frac{\mathbf{a}}{1 + \mathbf{af}} = -\frac{a_o p_o/(s + p_o)}{1 + a_o p_o/(s + p_o)\mathrm{f}_o} = -\frac{a_o p_o}{s + p_o(1 + a_o \mathrm{f}_o)} \tag{10.43d}$$

Note that A has the same dimension as **a,** which in this case is $\mathbf{a}_z$, thus

$$A_z = \frac{V_o}{I_s}$$

hence

$$A_I = \frac{V_o/R_L}{I_s} = \frac{I_o}{I_s} \tag{10.44}$$

EXAMPLE 10.2

Determine the open-loop gain and bandwidth and the closed-loop gain and bandwidth of the circuit in Fig. 10.7. Assume a current source input. The transistor parameters are repeated in the following for convenience:

$$\omega_T = 2\pi(4 \times 10^8)\ \text{rad/s} \qquad \beta_0 = 50$$

$$C_c = 5\ \text{pF} \qquad r_b = 100\ \Omega \qquad r_e = 5\ \Omega$$

From the above, and noting that $R_L = 100\ \Omega$, we determine the following:

$$R_i = \beta_0 r_e = 50(5) = 250\ \Omega$$

$$C_i = \frac{D}{r_e\omega_T} = \frac{1 + R_L C_C \omega_T}{r_e \omega_T} = \frac{2.25}{5(2 \times 51 \times 10^9)} \simeq 180\ \text{pF}$$

From (10.43*a*)

$$a_o = -50\,\frac{(1 \times 10^3)(100)}{1 \times 10^3 + 0.35 \times 10^3} = -3700$$

From (10.43*b*)

$$p_o = \frac{1}{250(1.82 \times 10^{-10})}\,\frac{1.35 \times 10^3}{(1.1 \times 10^3)} = 2.70 \times 10^7\ \text{rad/s}$$

From (10.43*c*)

$$\mathrm{f}_o = -G_f = -10^{-3}\ \text{mho}$$

The loop gain $T(0) = a_o f_o = 3.7$. Note that a single-stage transistor feedback amplifier does not provide enough loop gain, and hence the amount of feedback is too small to be very effective. For this reason two or a three-stage amplifier or an op amp is usually used in feedback systems.

The closed-loop current gain and bandwidth are determined from (10.43d) and (10.44), namely,

$$A_z(0) = \frac{3.7 \times 10^3}{4.7} \qquad \text{and} \qquad A_I(0) = \frac{I_o}{I_s} \simeq -7.9$$

$$f_{3\text{dB}} = \frac{p_o}{2\pi}(1 + a_o f_o) = \frac{(2.7 \times 10^7)(4.7)}{2\pi} \simeq 21 \text{ MHz}$$

This is the example that was considered in Sec. 7.13.

EXERCISE 10.2

For the inverting op amp shown in Fig. 10.5a, $R_S = 5$ kΩ and $R_f = 50$ kΩ. Determine the approximate input impedance, the output impedance, the gain, and the bandwidth if a 741-type op amp is used. The open-loop bandwidth of 741 is 5 Hz.

Ans $R_{\text{in}} \simeq 0.25\ \Omega$, $R_o \simeq 0$, $A_V = -10$, $f_{3\text{dB}} = 100$ kHz

10.6 THE SERIES-SERIES CONFIGURATION

The complete series-series configuration is shown in Fig. 10.9a. For this configuration the z parameters are most appropriate, and based on two-port theory in Appendix C, we have

$$\frac{V_o}{V_s} = \frac{z_{21}^T R_L}{(z_{11}^T + R_S)(z_{22}^T + R_L) - z_{12}^T z_{21}^T} \tag{10.45}$$

Dividing both the numerator and the denominator by $(z_{11}^T + R_S)(z_{22}^T + R_L)$ and using the approximation $z_{21}^a \gg z_{21}^f$ and $z_{12}^f \gg z_{12}^a$, we have

$$A_y = \frac{I_o}{V_s} = \frac{z_{21}^a/(z_{11}^T + R_S)(z_{22}^T + R_L)}{1 - z_{12}^f z_{21}^a/(z_{11}^T + R_S)(z_{22}^T + R_L)} \tag{10.46}$$

The **a** and the **f** circuits are shown in Fig. 10.9b, where

$$\mathbf{a}_y(s) = \frac{z_{21}^a}{(z_{11}^T + R_S)(z_{22}^T + R_L)} = \left.\frac{I_o}{V_s}\right|_{\text{f}=0} \tag{10.47}$$

$$\mathbf{f}(s) = -z_{12}^f \tag{10.48}$$

$$T(s) = \mathbf{a}(s)\mathbf{f}(s) = \frac{-z_{21}^a z_{12}^f}{(z_{11}^T + R_S)(z_{22}^T + R_L)} \tag{10.49}$$

Note that the **a** circuit, unlike the idealized situation, includes z_{11}^f and z_{22}^f. As in the previous case, this addition does not pose a problem in the analysis. In the design, this effect may be ignored initially and then included as an iteration step if necessary. The

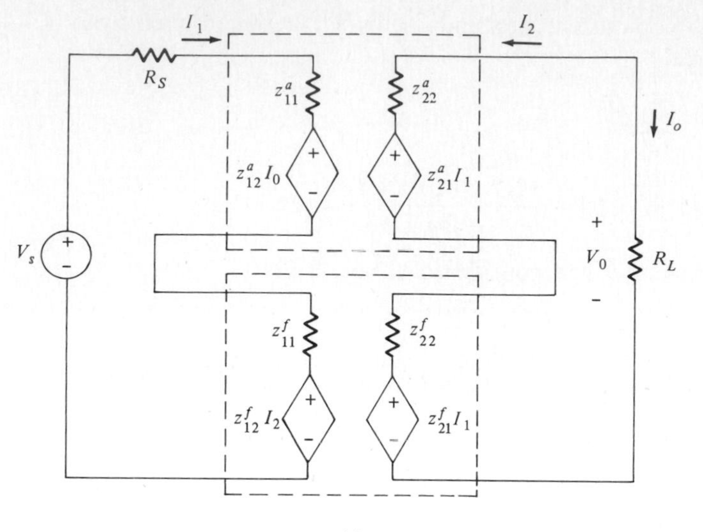

(*a*)

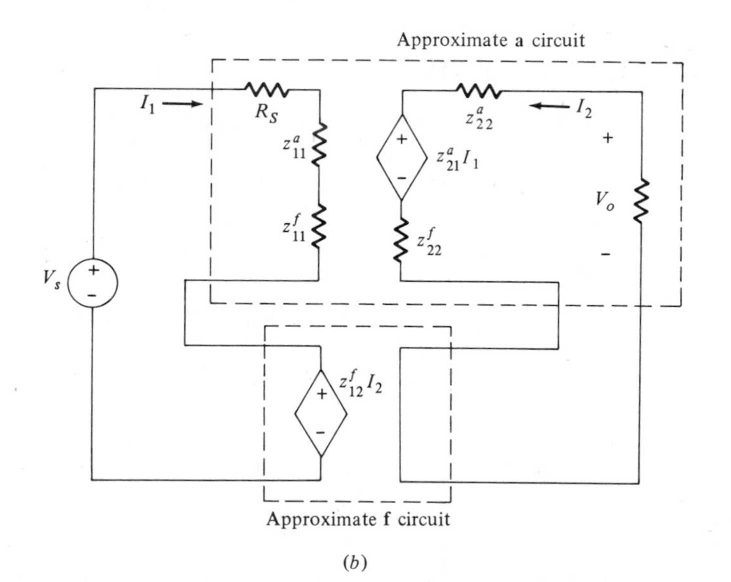

(*b*)

FIGURE 10.9 Series-series feedback configuration. (*a*) Series-series connected two ports. (*b*) Redrawn and simplified using the basic approximations.

f parameter depends only on the feedback circuit. Note that the *input variable is voltage* V_s and the output variable for $\mathbf{a}_y$ is current I_o; hence this configuration achieves stabilization in transconductance I_o/V_s. Note that since $\mathbf{f}(s)$ is an impedance function, $\mathbf{a}(s)$ *must* be an admittance function since **af** is dimensionless. Furthermore, note that because of negative feedback, $T(0) > 0$.

The input impedance of series-series feedback, using the result of Appendix C, for the overall two-port is

$$
\begin{aligned}
Z_{\text{in}} &= (z_{11}^T + R_S) - \frac{z_{12}^f z_{21}^a}{z_{22}^T + R_L} \\
&\simeq (z_{11}^T + R_S)\left[1 - \frac{z_{12}^f z_{21}^a}{(z_{22}^T + R_L)(z_{11}^T + R_S)}\right]
\end{aligned} \tag{10.50}
$$

From (10.49) and (10.50) we have

$$
Z_{\text{in}} = (z_{11}^T + R_S)[1 + T(s)] \tag{10.51}
$$

The input impedance of the feedback amplifier would be what the source sees. Since R_S is part of the source, we have to subtract R_S from the result obtained in (10.51) to obtain the input impedance of the feedback amplifier. Similarly,

$$
Z_o = (z_{22}^T + R_L)[1 + T(s)] \tag{10.52}
$$

and the output impedance is what R_L sees across its terminals. From (10.51) and (10.52), it is seen that the *series-series feedback connection increases the low-frequency input and output impedances* since $T(0) > 0$.

Consider the single-stage transistor feedback amplifier employing series negative feedback shown in Fig. 10.10*a*, exclusive of biasing circuitry. (Note that here we are simply examining the addition of an unbypassed emitter resistor discussed in Sec. 7.12.) We shall determine the overall gain and the bandwidth of the circuit from the feedback viewpoint. We use the unilateral approximate circuit model for the transistor, as shown in Fig. 10.10*b*. Note that the effective load resistor of the stage is not small; hence some error is introduced in the bandwidth calculation by the unilateral method. We use this simple model, however, in order to illustrate and emphasize the feedback technique. The **a** and the **f** circuits for this case are shown in Fig. 10.11. From Fig. 10.11, we have

$$
\mathbf{a}_y(s) = \frac{I_o}{V_s} = -\frac{g_m/C_i(R_f + R_S + r_b)}{s + (R_i + R_S + R_f + r_b)/R_i(R_f + R_S + r_b)C_i} \tag{10.53}
$$

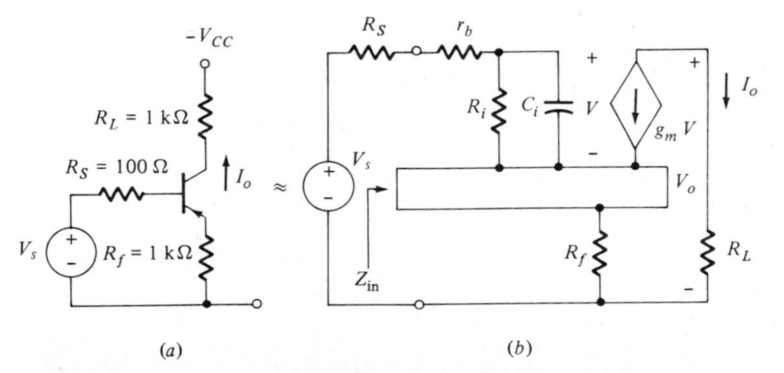

FIGURE 10.10 (*a*) Series-feedback amplifier for the example. (*b*) Equivalent circuit of (*a*).

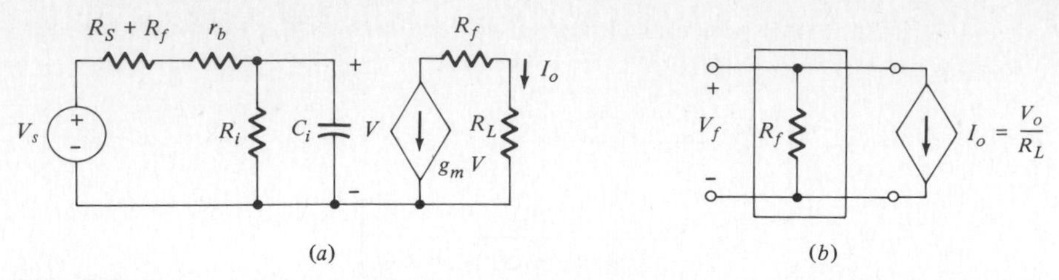

FIGURE 10.11 (*a*) The **a**(*s*) circuit for Fig. 10.10. (*b*) The **f**(*s*) circuit for Fig. 10.10.

For convenience, we rewrite (10.53) as

$$\mathbf{a}_y(s) = -a_o \frac{p_o}{s + p_o} \tag{10.54}$$

$$\text{where } a_o = \frac{\beta_0}{R_i + R_S + R_f + r_b} \tag{10.55a}$$

$$p_o = \frac{R_i + R_S + R_f + r_b}{R_i(R_f + R_S + r_b)C_i} \tag{10.55b}$$

Also, $\mathbf{f}(s)$ from (10.48) and Fig. 10.11*b* is

$$\mathbf{f}(s) = -z_{12}^f = -\frac{V_f}{I_o} = -R_f \tag{10.55c}$$

The closed-loop gain function using (10.55) is

$$A = \frac{\mathbf{a}}{1 + \mathbf{af}} = -\frac{a_o p_o/(s + p_o)}{1 + a_o p_o/(s + p_o)\mathrm{f}_o} = -\frac{a_o p_o}{s + p_o(1 + a_o \mathrm{f}_o)} \tag{10.56}$$

Note that if the closed-loop voltage gain is desired, we have to use $A_v = A_y R_L$ in (10.54).

The midband input impedance with feedback, in series with the voltage source, is

$$\begin{aligned}(R_{\text{in}})_f &= (r_b + R_i + R_f + R_S)(1 + a_o \mathrm{f}_o) \\ &= (r_b + R_i + R_f + R_S)\left(1 + \frac{g_m R_i R_f}{r_b + R_i + R_f + R_S}\right) \\ (R_{\text{in}})_f &= R_S + [r_b + R_i + (\beta_0 + 1)R_f]\end{aligned} \tag{10.57a}$$

Note that one usually considers R_S and R_{in} independent of each other; thus the input impedance in series with R_S is the one in the bracket. The output impedance is also increased by the amount of feedback and hence is very large, i.e.,

$$(R_{\text{out}})_f \simeq (1 + a_o \mathrm{f}_o)(R_L + r_o + R_f) \tag{10.57b}$$

where r_o is the output impedance of the CE transistor. Note that Fig. 10.10*b* is not valid for output impedance calculations at high frequencies and that the hybrid-π

model must be used for such calculations since it includes the feedback due to C_c and the value of r_o.

EXAMPLE 10.3

For the circuit shown in Fig. 10.10*a* we shall determine the gain, the bandwidth, and the input impedance. Assume the following transistor parameters:

$$\omega_T = 10^8 \text{ rad/s} \qquad \beta_0 = 200$$

$$C_c = 5 \text{ pF} \qquad r_b = 100\ \Omega \qquad r_e = 5\ \Omega$$

For the circuit element values of this example

$$R_i = \beta_0 r_e = 1\ \text{k}\Omega \qquad C_i = \frac{1 + (R_L + R_f)C_c\omega_T}{r_e\omega_T} = 4 \times 10^{-9}\ \text{F}$$

From (10.55*a*) and (10.55*b*) we obtain $-a_o = 91 \times 10^{-3}$ mho, $p_o = 7.9 \times 10^5$ rad/s, and $\mathrm{f}_o = -R_f = -1$ kΩ. Hence the midband loop transmission gain is

$$T(0) = \mathbf{a}(0)\mathbf{f}(0) = a_o \mathrm{f}_o = 91$$

Thus the closed-loop voltage gain and bandwidth are

$$A_v(0) = -\frac{a_o R_L}{1 + a_o \mathrm{f}_o} = -\frac{91}{92} = -0.99$$

$$\omega_{3\text{dB}} = p_o(1 + a_o\mathrm{f}_o) = (7.9 \times 10^5)(92) \simeq 7.3 \times 10^6 \text{ rad/s}$$

Note that the open-loop gain includes the value of R_f as in Fig. 10.9*b*. Notice also the increase in bandwidth and the decrease in gain as a result of feedback. From (10.57*a*)

$$(R_{\text{in}})_f = 100 + 10^3 + 200(10^3) \simeq 2 \times 10^5\ \Omega$$

Note that the source resistance sees a very large input resistance.

Finally, observe that we get similar results if we use the results in (7.78). In fact, for the series-shunt and shunt-series configuration where the emitter resistance also provides local feedback, it is much simpler to use the circuit model of Fig. 7.27*b*, when analyzing the **a** circuit, as will be done in the next two sections.

EXERCISE 10.3

For the circuit shown in Fig. 10.10, $R_f = 100\ \Omega$ and the other circuit parameters remain the same. Determine the voltage gain by using (*a*) the feedback method and (*b*) the method in Sec. 7.12. Use the transistor parameters of Example 10.3.

Ans (*a*) $A_V = -9.4$, (*b*) -9.5

EXERCISE 10.4

For the circuit shown in Fig. E10.4, the JFET parameters are given as $g_m = 5 \times 10^{-3}$ mho and $r_d = 20$ kΩ. Determine the feedback factor, the open-loop gain, and the closed-loop gain.

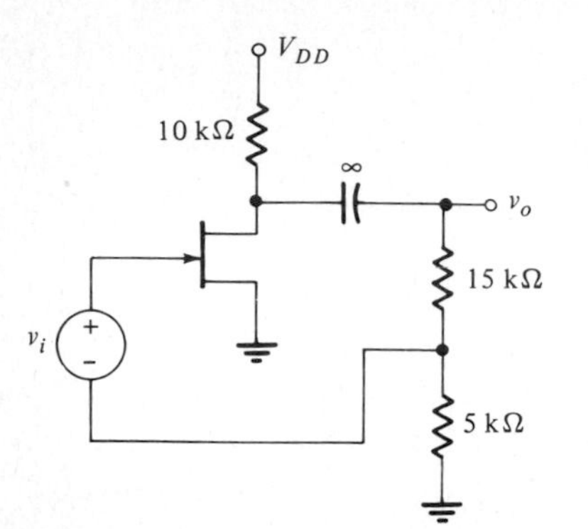

FIGURE E10.4

Ans $f_o = \frac{1}{4}$, $a_o = 25$, $A_v = -3.45$

10.7 THE SERIES-SHUNT CONFIGURATION

The series-shunt connection is shown in Fig. 10.12. For this configuration, since the input is in series and the output in shunt, the h parameters are the most convenient to use for analysis in terms of the two-port parameters.

The overall gain function in this case from Table C.2, from V_2/V_1 (incorporating R_S with h_{11} and G_L with h_{22}) is given by

$$A_v = \frac{V_o}{V_s} = \frac{-h_{21}^T}{(h_{11}^T + R_S)(h_{22}^T + G_L) - h_{12}^T h_{21}^T} \tag{10.58}$$

In order to express (10.58) in the form of (10.4) for identification purposes, the two basic assumptions to be used are $h_{21}^a \gg h_{21}^f$ and $h_{12}^a \ll h_{12}^f$. These approximations are depicted in Fig. 10.12*b*. We can rewrite (10.58) using the above approximations:

$$A_v \simeq \frac{-h_{21}^a/(h_{11}^T + R_S)(h_{22}^T + G_L)}{1 - (h_{21}^a h_{12}^f)/(h_{11}^T + R_S)(h_{22}^T + G_L)} \tag{10.59}$$

Note that in this case the feedback signal is a voltage, and we can use subscript v for the **a** circuit. From (10.59) and (10.4) we have

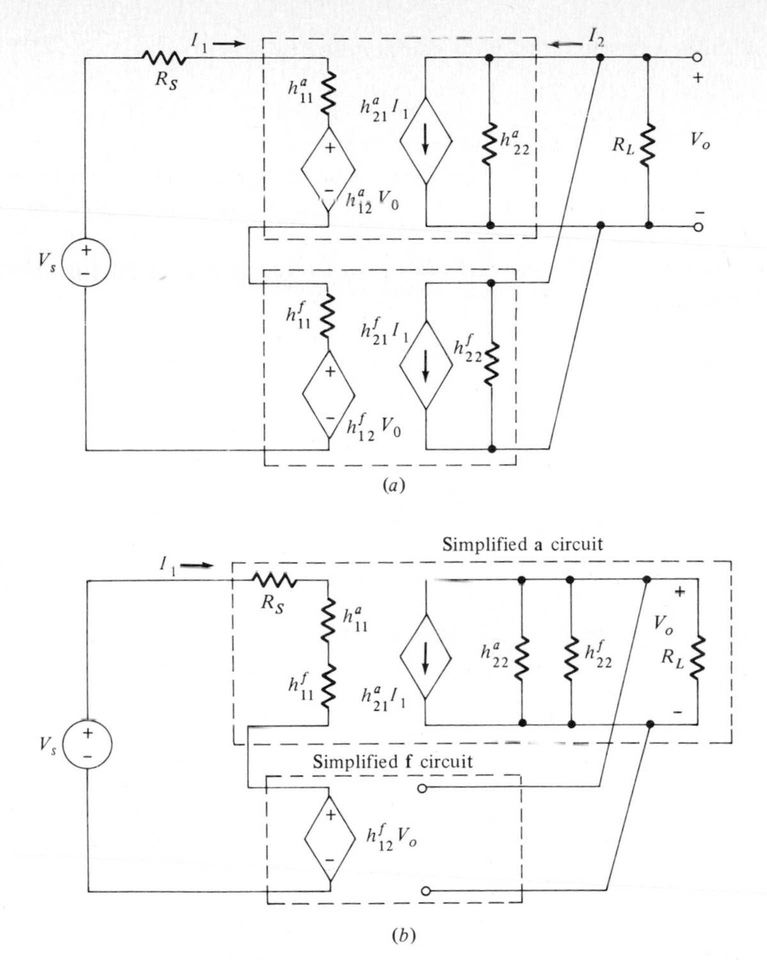

FIGURE 10.12 Series-shunt feedback configuration. (*a*) Series-shunt connected two-ports. (*b*) Redrawn and simplified using the basic approximations.

$$\mathbf{a}_v(s) = \frac{-h^a_{21}}{(h^T_{11} + R_S)(h^T_{22} + G_L)} = \left.\frac{V_o}{V_s}\right|_{f=0} \tag{10.60}$$

$$\mathbf{f}(s) = h^f_{12} \tag{10.61}$$

$$T(s) = \mathbf{a}_v\mathbf{f} = -\frac{h^a_{21}h^f_{12}}{(h^T_{11} + R_S)(h^T_{22} + G_L)} \tag{10.62}$$

Note again that in these identifications $\mathbf{a}_v(s)$ includes h^f_{11} and h^f_{22}, since $h^T_{11} = h^a_{11} + h^f_{11}$, etc., while $\mathbf{f}(s)$ depends only on the feedback network. The input and output variables are voltages; hence this configuration achieves voltage gain stabilization. The input resistance is like the series-series case, namely,

$$Z_{\text{in}} = (z^T_{11} + R_S)(1 + T) \tag{10.63}$$

The output admittance is like the shunt-shunt case, namely,

$$Y_{\text{out}} = (y_{22}^T + G_L)(1 + T) \tag{10.64}$$

Again note that $T(0) > 0$ since we have a negative-feedback amplifier.

As an application of the series-shunt feedback configuration consider the op amp circuit of Fig. 10.13. (Note that this is simply a noninverting op amp as considered in Sec. 9.2.)

The circuit is redrawn in the form of **a** and **f** circuits in Fig. 10.14*a*. From the **a** circuit we have

$$\mathbf{a}_v = \frac{V_o}{V_i} = \frac{V_o}{V_i}\frac{V_i}{V_s}$$

but

$$\frac{V_o}{V_i} = a\frac{R_{f1} + R_{f2}}{R_{f1} + R_{f2} + R_o} \simeq a \qquad \text{for } (R_{f1} + R_{f2}) \gg R_o$$

and

$$\frac{V_i}{V_s} = \frac{R_i}{R_i + R_{f1} \parallel R_{f2}} \simeq 1 \qquad \text{for } R_i \gg R_{f1} \parallel R_{f2}$$

hence

$$\mathbf{a}_v = \frac{V_o}{V_i} = a(s) = \frac{a_o}{1 + s/p_o} \tag{10.65a}$$

From the **f** circuit,

$$\mathbf{f} = h_{12}^f = \frac{R_{f1}}{R_{f1} + R_{f2}} = \mathrm{f}_o \tag{10.65b}$$

Hence from (10.65*a*) and (10.65*b*),

$$A = \frac{\mathbf{a}}{1 + \mathbf{af}} = \frac{\mathbf{a}_v}{1 + \mathbf{a}_v\mathrm{f}_o} = \frac{a_o}{(1 + s/p_o) + a_o\mathrm{f}_o} = \frac{a_o p_o}{s + p_o(1 + a_o\mathrm{f}_o)} \tag{10.66a}$$

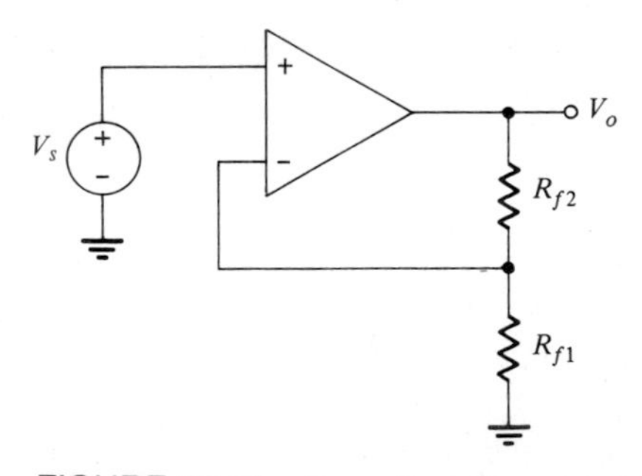

FIGURE 10.13 A noninverting op-amp circuit.

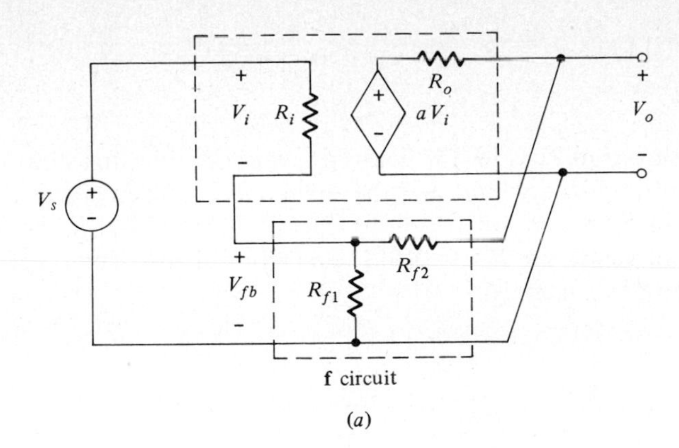

(a)

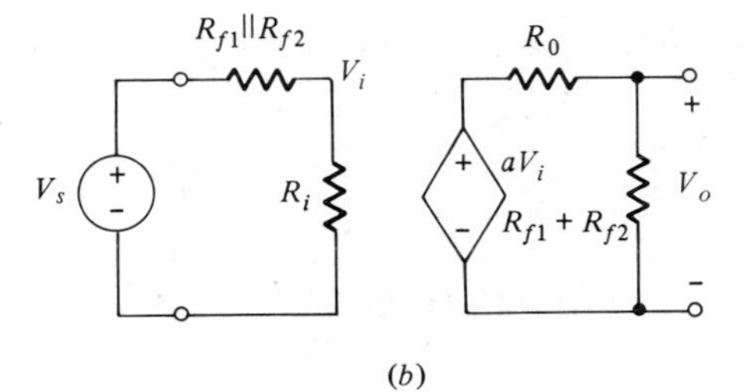

(b)

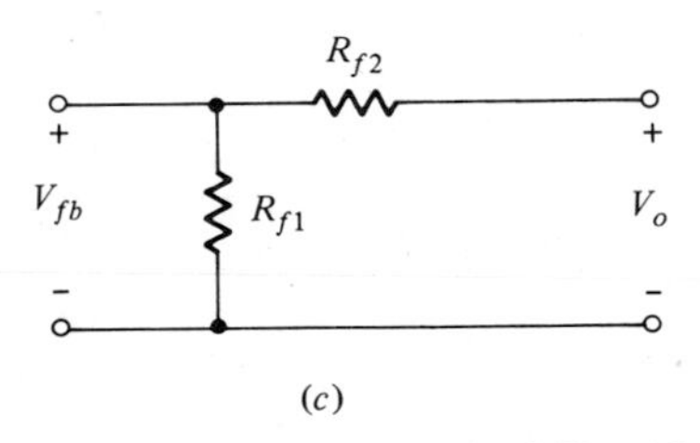

(c)

FIGURE 10.14 (*a*) Equivalent circuit of Fig. 10.13. (*b*) The **a** circuit. (*c*) The **f** circuit.

The voltage gain

$$A_v(0) \simeq \frac{1}{\mathrm{f}_o} = \frac{R_{f1} + R_{f2}}{R_{f1}} \tag{10.66b}$$

The bandwidth is determined from the pole of $A(s)$, namely,

$$\omega_{3\mathrm{dB}} = p_o(1 + a_o \mathrm{f}_o) \simeq a_o \mathrm{f}_o p_o \tag{10.67}$$

The input impedance for the circuit is

$$Z_{\mathrm{in}} = (R_i + R_{f1} \| R_{f2})(1 + T_0) \tag{10.68}$$

$$Y_{\mathrm{out}} = \left(\frac{1}{R_o} + \frac{1}{R_{f2}}\right)(1 + T_0) \tag{10.69}$$

EXAMPLE 10.4

For the op-amp circuit shown in Fig. 10.13, determine the loop transmission gain, the closed loop gain and bandwidth, and the input and output impedances. The op amp is a 741 with $a_o = 2 \times 10^5$, $p_o = 2\pi(5\text{ Hz})$, $R_i = 1\text{ M}\Omega$, $R_o = 75\ \Omega$. The circuit element values are $R_{f2} = 50\text{ k}\Omega$ and $R_{f1} = 1\text{ k}\Omega$. The open-loop gain is given by (10.65*a*), namely,

$$\mathbf{a}(s) = \frac{a_o}{1 + s/p_o}$$

From (10.65*b*)

$$\mathrm{f} = \frac{10^3}{50 \times 10^3 + 10^3} \simeq 0.02 = \mathrm{f}_o$$

$$T(0) = a(0)\mathrm{f}_o = 2 \times 10^5(0.02) = 4 \times 10^3$$

From (10.66*a*)

$$A = \frac{a_o p_o}{s + p_o(1 + a_o \mathrm{f}_o)} = \frac{2 \times 10^5[2\pi(5\text{ Hz})]}{s + 4 \times 10^3[2\pi(5\text{ Hz})]}$$

Hence $A(0) = 2 \times 10^5/4 \times 10^3 = 50$ and $\omega_{3\text{dB}} = 2\pi(20\text{ kHz})$. From (10.68) $R_{\text{in}} = 10^6(4 \times 10^3) = 4 \times 10^9\ \Omega$. From (10.69) $Y_{\text{out}} = \frac{1}{75}(4 \times 10^3)$ or $R_{\text{out}} = 75/(4 \times 10^3) = 0.02\ \Omega$.

Note that if the circuit in Fig. 10.13 is used as a unity buffer, the loop transmission will be 2×10^5, the input impedance will be of the order of $10^{11}\ \Omega$, and the output impedance will be of the order $10^{-4}\ \Omega$, which will make the circuit an *ideal buffer*.

Another configuration suitable for the series-shunt formulation is shown in Fig. 10.18 and is analyzed in Sec. 10.9.

EXERCISE 10.5

For the circuit shown in Fig. 10.13, show that

(a) $R_{\text{in}} = R_i \dfrac{a_o}{1 + R_{f2}/R_{f1}}$

(b) $R_{\text{out}} = \dfrac{R_o}{a_o}\left(1 + \dfrac{R_{f2}}{R_{f1}}\right)$

(c) $f_{3\text{dB}} = f_o \dfrac{a_o}{1 + R_{f2}/R_{f1}}$

Determine the numerical values of the above for a voltage gain of 20 and a 741 op amp.

Ans (a) $R_{in} = 10^{10}\ \Omega \rightarrow \infty$, (b) $R_o = 7.5 \times 10^{-3}\ \Omega \rightarrow 0$, (c) $f_{3dB} = 50$ kHz

10.8 SHUNT-SERIES CONFIGURATION

The shunt-series configuration is shown in Fig. 10.15. For this configuration the g-parameter description of the networks is most appropriate. The overall gain function of the circuit in Fig. 10.15a is given by

$$A_I = \frac{I_o}{I_s} = \frac{g_{21}^T}{(g_{11}^T + G_S)(g_{22}^T + R_L) - g_{12}^T g_{21}^T} \tag{10.70}$$

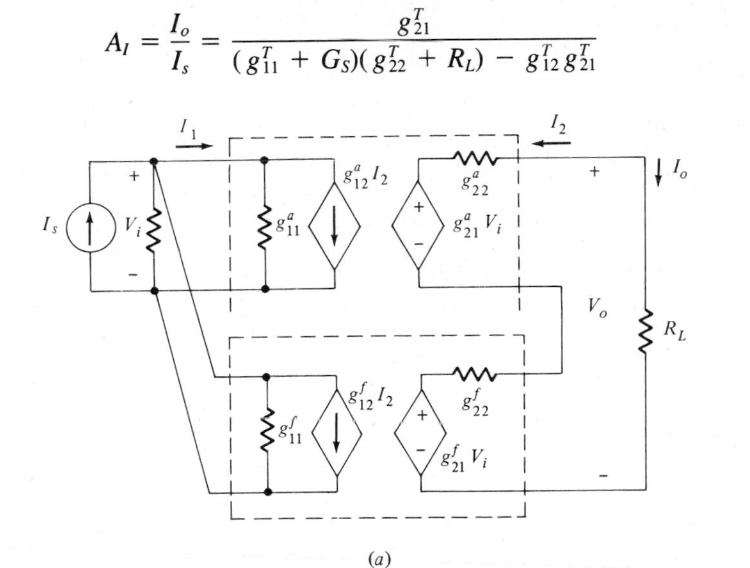

(a)

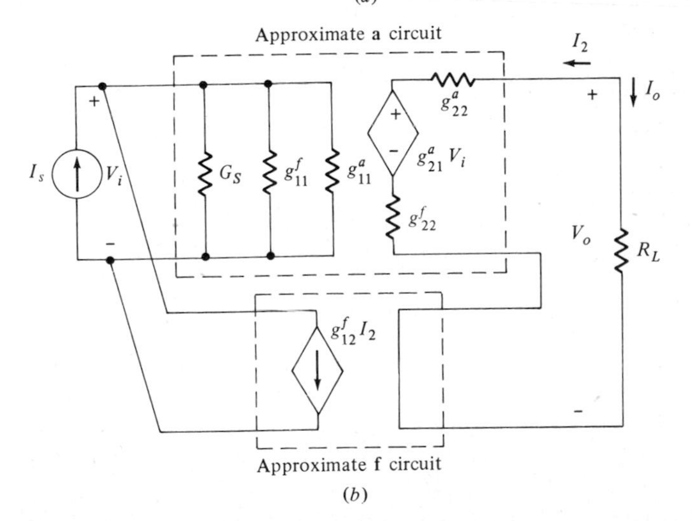

(b)

FIGURE 10.15 Shunt-series feedback configuration. (a) Shunt-series connected two ports. (b) Redrawn and simplified using the basic approximations.

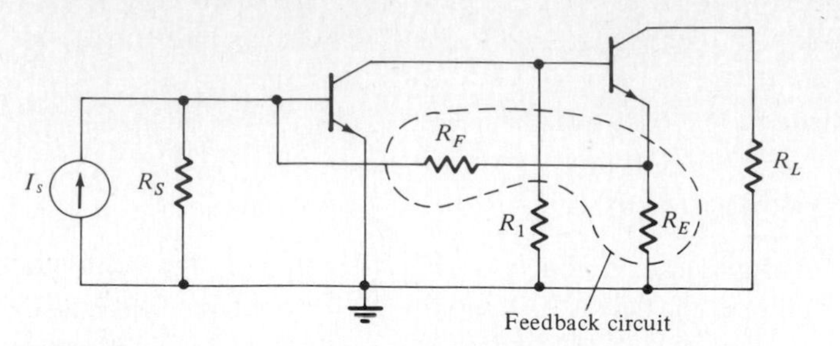

FIGURE 10.16 A shunt-series feedback pair.

In order to express (10.70) in the basic feedback equation form the two basic assumptions are $g_{21}^a \gg g_{21}^f$, and $g_{12}^a \ll g_{12}^f$. These approximations are depicted in Fig. 10.15*b*. We can rewrite (10.70) as

$$A_I \simeq \frac{g_{21}^a/(g_{11}^T + G_S)(g_{22}^T + R_L)}{1 - g_{21}^a g_{12}^f/(g_{11}^T + G_S)(g_{22}^T + R_L)} \tag{10.71}$$

Note that in this case the feedback signal is current, and we can use the subscript i for the **a** circuit. From (10.71) and (10.4) we make the following identifications:

$$\mathbf{a}_i(s) = \frac{g_{21}^a}{(g_{11}^T + G_S)(g_{22}^T + R_L)} = \left.\frac{I_o}{I_s}\right|_{\text{f}=0} \tag{10.72}$$

$$\mathbf{f}(s) = -g_{12}^f \tag{10.73}$$

$$T(s) = \mathbf{a}_i\mathbf{f} = \frac{-g_{21}^a g_{12}^f}{(g_{11}^T + G_S)(g_{22}^T + R_L)} \tag{10.74}$$

Observe that $\mathbf{a}_i(s)$ includes g_{11}^f and g_{22}^f, since $g_{11}^T = g_{11}^a + g_{11}^f$, etc., and $\mathbf{f}(s)$ depends on the feedback network alone. In this configuration the input and output variables are currents; hence it achieves current gain stabilization.

For example, consider the shunt-series feedback pair shown in Fig. 10.16. The **a** and **f** circuits are shown in Fig. 10.17*a* and *b*, respectively.

From the **f** circuit in Fig. 10.17*b*, by inspection

$$\mathbf{f} = -g_{12}^f = \left.\frac{I_{fb}}{I_o}\right|_{V_1=0} \approx \frac{R_E}{R_E + R_F} \tag{10.75a}$$

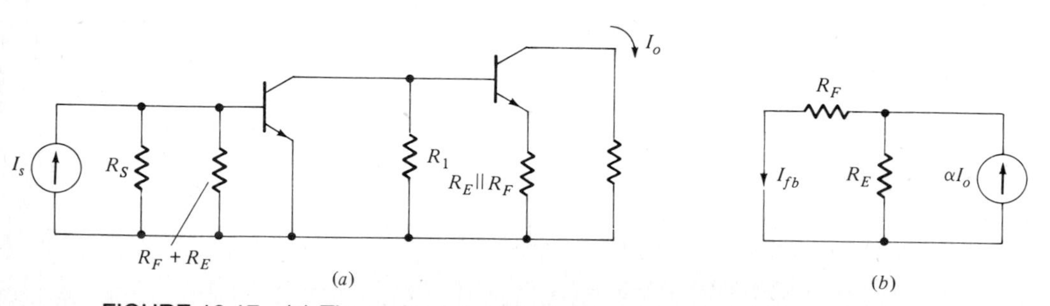

FIGURE 10.17 (*a*) The **a** circuit for the shunt-series pair. (*b*) The **f** circuit.

From the **a** circuit in Fig. 10.17a, we can readily determine the current gain I_o/I_s. Note that as mentioned earlier, we can consider the second stage simply as a modified CE stage with r_e changed to $r_e + R_E \parallel R_F$ as in Fig. 7.27b. Hence the open-loop gain of the two-stage amplifier is given by (8.5a), namely,

$$\mathbf{a}_i(0) = \beta_0^2 \frac{R_S'}{R_S' + r_b + R_i} \frac{R_1}{R_1 + r_b + R_i'} \tag{10.75b}$$

where $R_S' = R_S \parallel (R_F + R_E)$ and $R_i' = \beta_0(r_e + R_E \parallel R_F)$. Note that if $\mathbf{a}_i(0)$ is very large so that $T(0) \gg 1$, without any calculation of $\mathbf{a}_i(0)$ from (10.75a) we have

$$A_i = \frac{I_o}{I_f} \simeq \frac{1}{\mathbf{f}} = 1 + \frac{R_F}{R_E} \tag{10.75c}$$

EXAMPLE 10.5

For the circuit shown in Fig. 10.16 determine the low-frequency voltage and current gains if $R_1 = 5$ kΩ, $R_F = 5$ kΩ, $R_E = 500\ \Omega$, $R_L = R_S = 1$ kΩ. The transistor parameters are $\beta_0 = 100$, $r_e = 5\ \Omega$, $r_b = 50\ \Omega$. From (10.75a)

$$\mathbf{f} = \frac{0.5 \times 10^3}{5 \times 10^3 + 0.5 \times 10^3} = 0.091$$

To find $\mathbf{a}_i$, we first determine R_S' and R_i', namely,

$$R_S' = R_S \parallel (R_F + R_E) = (1 \times 10^3) \parallel (5.5 \times 10^3) = 0.85 \text{ k}\Omega$$

$$R_i' \simeq \beta_0(R_E \parallel R_F) = 100(0.5 \times 10^3 \parallel 5 \times 10^3) = 45.4 \text{ k}\Omega$$

From (10.75b)

$$\mathbf{a}_i = 100^2\left(\frac{0.85}{0.85 + 0.05 + 0.5}\right)\left(\frac{5}{5 + 45.4}\right) = 600$$

$$\mathbf{a}_i(0)\mathbf{f}(0) = 600(0.091) = 54.6 \gg 1$$

Hence

$$A_i = \frac{\mathbf{a}_i}{1 + \mathbf{a}_i\mathbf{f}} \simeq \frac{1}{\mathbf{f}} = 11.0$$

Since $R_S = R_L$ in this example, the voltage gain $V_o/V_S = 11.0$.

EXERCISE 10.6

For the circuit shown in Fig. 10.16, the circuit parameters are $R_1 = 2$ kΩ, $R_f = 1$ kΩ, $R_E = 100\ \Omega$, $R_S = 10$ kΩ, and $R_L = 500\ \Omega$. The transistor parameters are the same as in Example 10.5. Determine the current gain, the loop gain, and the input impedance.

Ans $A_i \simeq 11$, $T_0 = 122$, $R_{in} \simeq 4.1\ \Omega$

10.9 SUMMARY OF THE PROPERTIES OF VARIOUS CONFIGURATIONS

From the previous sections we see that feedback has an important effect on the performance of a circuit. The four basic configurations were analyzed by using the two-port theory and the basic feedback equation. Equation (10.4) is a fundamental equation for single-loop feedback systems. In the derivation of equations in the four basic configurations two simplifying basic assumptions were utilized that enabled us to identify the **f** and **a** circuits. The first assumption of $k_{12}^a \gg k_{12}^f$ (where $k_{12} = y_{12}, z_{12}, h_{12}$, or g_{12}) implies that signal transmission from the source to the load is mainly via the amplifier. The second assumption $k_{21}^f \gg k_{21}^a$ implies that reverse transmission is mainly via the feedback network. Both of these assumptions are valid, of course, and utilized to simplify analysis. The feedback function as a result became dependent only on the **f** circuit, namely $\mathbf{f} = k_{12}^f$. The **a** circuit depends on the **f** circuit and was labeled $\mathbf{a}_z$, $\mathbf{a}_i$, or $\mathbf{a}_v$ or $\mathbf{a}_y$ on the basis of the feedback **f** circuit and the signal added to the input, and these quantities were either transresistance, current gain, voltage gain, or transconductance. Again we emphasize that the product of $\mathbf{a}(s)\ \mathbf{f}(s)$ must be dimensionless, i.e., if $\mathbf{f}(s)$ is an admittance function $\mathbf{a}(s)$ *must* be an impedance function.

Table 10.1 summarizes the salient points of the previous four sections. Two of the important quantities in a single-loop feedback configuration of the types discussed so far are the loop gain function $T(= \mathbf{af})$ and **f**. If the loop gain $\mathbf{af} \gg 1$, then the closed-loop gain

$$A = \frac{\mathbf{a}}{1 + T} = \frac{\mathbf{a}}{1 + \mathbf{af}} \simeq \frac{1}{\mathbf{f}} \qquad \text{for } T \gg 1 \tag{10.76}$$

Thus by identifying the **f** circuit and the configuration type, the value of $\mathbf{f} = k_{12}^f$ (where k_{12} represents g_{12}, h_{12}, y_{12}, or z_{12}, whichever is appropriate) determines the closed-loop gain as long as $T \gg 1$, and *provided the circuit is stable*. Stability of feedback amplifiers is discussed in Sec. 10.11. We shall therefore assume for the time being that the following multistage feedback amplifiers are stable, inspite of the fact that $T_0 \gg 1$.

For example, consider the circuit shown in Fig. 10.18, which is a series-shunt feedback pair. Analysis of the circuit is similar to the one in Example 10.5. The returned voltage signal $\mathbf{f} \simeq R_{F1}/(R_{F1} + R_{F2})$, and as long as $T \gg 1$, then

$$A_v = \frac{V_o}{V_s} = \frac{1}{\mathbf{f}} = 1 + \frac{R_{F2}}{R_1} \tag{10.77}$$

Analysis of the circuit in Fig. 10.18 for the frequency response is carried out in the next section.

A series-series feedback triple is shown in Fig. 10.19. For the feedback circuit we can readily determine **f**:

$$\mathbf{f} = -z_{12}^f = \left.\frac{V_1}{I_2}\right|_{I_1=0} = \frac{V_{fb}}{I_o} = \frac{R_{E2}R_{E1}}{R_{E2} + R_{E1} + R_F} \tag{10.78}$$

TABLE 10.1 Summary of Salient Points of the Various Feedback Configurations

Feedback connection	Two-Port parameter representation	Input variable	Output variable	Transfer function stabilized	Input impedance Z_{in}	Output impedance Z_{out}	Approximations used
Shunt-shunt	y_{ij}	I_s	V_o	Transresistance $\mathbf{a}_z = \frac{V_o}{I_s}$	Low*	Low*	$y^a_{21} \gg y^f_{21}$ $y^f_{12} \gg y^a_{12}$
$T = \frac{-I_f}{I_s}$				$\mathbf{f} = y^f_{12} = \left.\frac{-I_f}{V_o}\right\vert_{V_i=0}$			
Series-series	z_{ij}	V_s	I_o	Transconductance $\mathbf{a}_y = \frac{I_o}{V_s}$	High*	High*	$z^a_{21} \gg z^f_{21}$ $z^f_{12} \gg z^a_{12}$
$T = \frac{-V_f}{V_s}$				$\mathbf{f} = -z^f_{12} = \left.\frac{-V_f}{I_o}\right\vert_{I_i=0}$			
Series-shunt	h_{ij}	V_s	V_o	Voltage gain $\mathbf{a}_v = \frac{V_o}{V_s}$	High*	Low*	$h^a_{21} \gg h^f_{21}$ $h^f_{12} \gg h^a_{12}$
$T = \frac{V_f}{V_s}$				$\mathbf{f} = h^f_{12} = \left.\frac{V_f}{V_o}\right\vert_{I_i=0}$			
Shunt-series	g_{ij}	I_s	I_o	Current gain $\mathbf{a}_i = \frac{I_o}{I_s}$	Low*	High*	$g^a_{21} \gg g^f_{21}$ $g^f_{12} \gg g^a_{12}$
$T = \frac{I_f}{I_s}$				$\mathbf{f} = -g^f_{12} = \left.\frac{I_f}{I_o}\right\vert_{V_i=0}$			

Note that in each case $T(0) > 0$ for negative-feedback amplifiers in (10.4).
*By a factor $1 + T$.

Hence

$$A_y = \frac{I_o}{V_s} = \frac{1}{\mathbf{f}} \simeq \frac{R_{E1} + R_{E2} + R_F}{R_{E2}R_{E1}} \tag{10.79}$$

The voltage gain is

$$A_v = \frac{V_o}{V_s} = R_L\left(\frac{R_{E1} + R_{E2} + R_F}{R_{E2}R_{E1}}\right) \tag{10.80a}$$

For $R_F = 0$,

$$A_v = \frac{R_L}{R_{E1} \parallel R_{E2}} \tag{10.80b}$$

Finally, the shunt-shunt feedback triple is shown in Fig. 10.20. For the feedback circuit we determine $\mathbf{f}$:

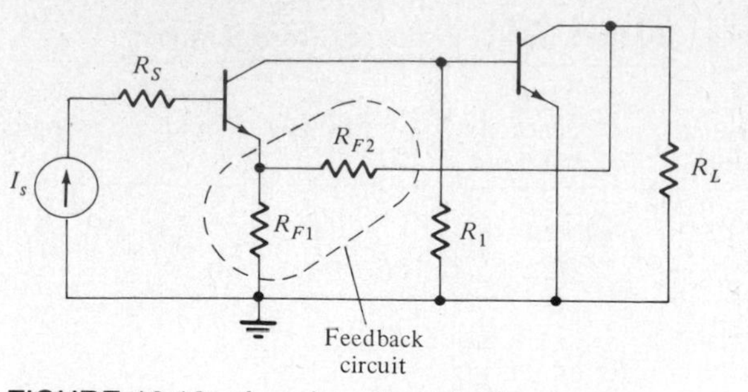

FIGURE 10.18 A series-shunt feedback pair.

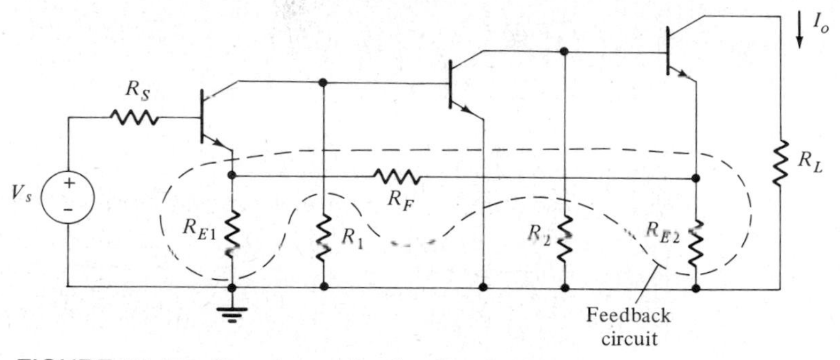

FIGURE 10.19 A series-series feedback triple circuit.

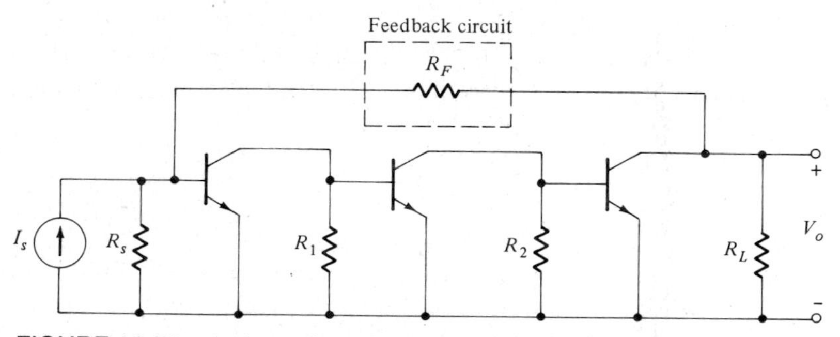

FIGURE 10.20 A shunt-shunt feedback triple circuit.

$$\mathbf{f} = y_{12}^{f} = \left.\frac{I_1}{V_2}\right|_{V_1=0} = \frac{-I_{fb}}{V_0} = -G_F = -\frac{1}{R_F} \tag{10.81}$$

Hence

$$A_z = \frac{V_o}{I_s} \simeq \frac{1}{\mathbf{f}} = -R_F \tag{10.82}$$

The current gain is

$$A_i = \frac{I_o}{I_s} = -\frac{R_F}{R_L} \tag{10.83}$$

Note the considerable simplification through the use of the feedback approach in all of the above cases.

10.10 FREQUENCY RESPONSE OF FEEDBACK AMPLIFIERS

We shall next consider the frequency response of feedback amplifiers. We have already shown in Sec. 10.4 that if the open-loop gain function has one pole, the closed-loop gain function wll also have one pole but increased by the factor $1 + a_o \mathrm{f}_o$. Specifically, if

$$\mathbf{a}(s) = \frac{a_o}{1 + s/p_o} \quad \text{and} \quad \mathbf{f} = \mathrm{f}_o \tag{10.84}$$

then

$$A(s) = \frac{\mathbf{a}}{1 + \mathbf{af}} = \frac{a_o}{s + p_o(1 + a_o \mathrm{f}_o)} = \frac{a_o}{s + p_o(1 + T_o)} \tag{10.85}$$

In other words, the bandwidth is increased by the same factor as the gain is decreased. The gain-bandwidth product remains the same. The pole locations of $\mathbf{a}(s)$ and $A(s)$ for a specific value of f_o in the s plane are shown in Fig. 10.21. Note that when $T_o = 0$, we have the open-loop pole, and as T_o is increased, the poles move away on the negative real axis. This is the so-called *root locus* plot [3].

Example 10.4 is a numerical illustration for a single-pole gain function. Suppose we have a two-pole open-loop gain function, such as would be obtained from an analysis of the series-shunt feedback pair in Fig. 10.18 or the shunt-series feedback pair of Fig. 10.16. In fact, the gain function of the circuit in Fig. 10.18 using the unilateral model for the transistor, as shown in Fig. 10.22, is given by (problem 10.13)

$$\mathbf{a}_v(s) = \frac{V_o}{V_s} = \frac{a_o p_1 p_2}{(s + p_1)(s + p_2)} \tag{10.86}$$

where $a_o = \beta_0^2 \dfrac{R_1}{R_{i1} + R_S + r_b} \dfrac{R_L}{R_{i2} + R_1 + r_{b2}}$ **(10.87a)**

$$p_1 = \frac{1}{R_{i1}C_{i1}} \frac{R_{i1} + R_S + r_b}{R_S + r_b}$$

$$p_2 = \frac{1}{R_{i2}C_{i2}} \frac{R_{i2} + R_1 + r_b}{R_1 + r_b} \tag{10.87b}$$

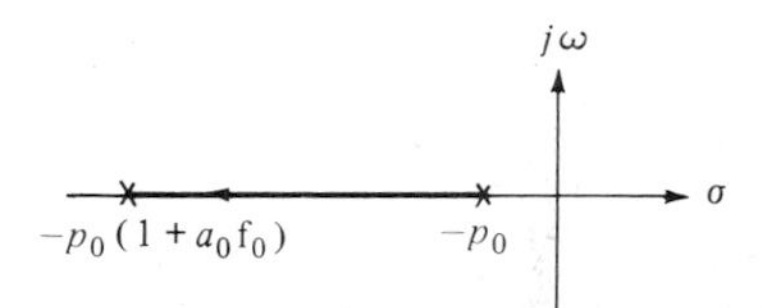

FIGURE 10.21 Pole location of **a**(s) and $A(s)$ for different values of T_o.

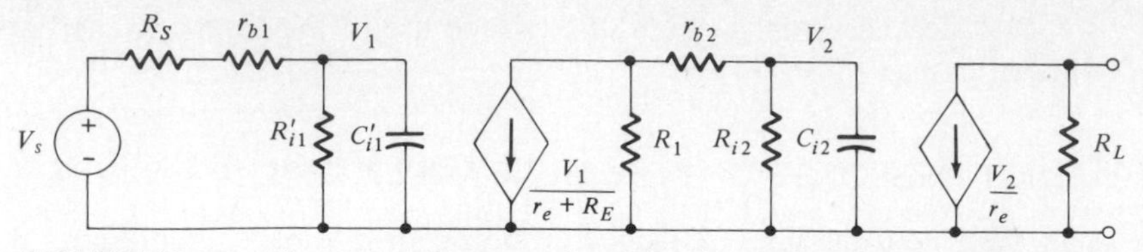

FIGURE 10.22 The approximate **a** circuit for the series-shunt feedback pair.

The capacitances, using the Miller approximation, are given by

$$C_{i1} = \frac{1 + (R_{L1} + R_E)C_c\omega_T}{(r_{e1} + R_E)\omega_T} \qquad C_{i2} = \frac{1 + R_L C_c \omega_T}{r_{e2}\omega_T}$$

where
$$R_{i1} = \beta_0(r_{e1} + R_E)$$
$$R_{i2} = \beta_0 r_{e2}$$
$$R_{L1} = R_L \,\|\, (r_b + R_{i2})$$
$$R_E = R_{F1} \,\|\, R_{F2}$$

We have already determined $\mathbf{f} = R_{F1}/(R_{F1} + R_{F2}) = \mathrm{f}_o$. Note that f_o is used to denote that it is not a function of frequency; it is a constant, but its value can be changed by selecting different values of R_{F1} and R_{F2}. The closed-loop gain, corresponding to (10.86) and dropping the subscripts, is given by

$$A = \frac{\mathbf{a}}{1 + \mathbf{af}} = \frac{a_0 p_1 p_2}{s^2 + s(p_1 + p_2) + p_1 p_2(1 + a_o \mathrm{f}_o)}$$
$$= \frac{a_o p_1 p_2}{s^2 + s(p_1 + p_2) + p_1 p_2(1 + T_0)} \qquad \textbf{(10.88)}$$

The poles of $A(s)$ can be varied by varying $T_0 = a_o \mathrm{f}_o$ as shown in Fig. 10.23. This is the root locus of the system. For a detailed discussion on root locus refer to Refs. 1 and 3. Note that by a proper choice of f_o, we can design the circuit to yield a maximally flat magnitude closed-loop response as indicated in Fig. 10.23.

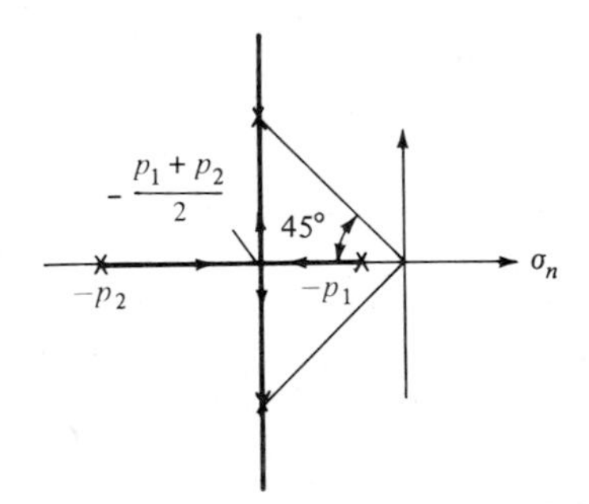

FIGURE 10.23 Pole locations (root locus) for **a**(s) with two real poles.

EXAMPLE 10.6

Consider a numerical example for the series-shunt feedback pair. Assume both transistors are biased at $V_{CE} = 5$ V and $I_C = 5$ mA. At this operating point let the transistor parameters be as follows.

$$\omega_T = 2\pi(175 \text{ MHz}) \qquad C_c = 5 \text{ pF} \qquad \beta_0 = 50 \qquad r_b = 75\ \Omega$$

$$R_S = 25\ \Omega \qquad R_L = 50\ \Omega \qquad R_{F1} = 50\ \Omega \qquad R_1 = 3.9\ \text{k}\Omega$$

Design the circuit; i.e., choose the value of R_{F2} that would yield a Butterworth response, and determine the bandwidth for this design.

Note that since the value of R_{F2} is to be determined and is unknown, we use $R_E \simeq R_{F1}$ in the preliminary design. For $A \geq 10$, $R_{F2} \geq 9R_{F1}$ and $R_E \simeq R_{F1}$.

Substitution of the numerical values in (10.86) and (10.87) yields $a_o = 40.6$, $p_1 = 21.9 \times 10^7$ rad/s, and $p_2 = 1.8 \times 10^7$ rad/s. For convenience we normalize the smaller pole to unity; i.e., a frequency normalization factor of $\Omega_0 = 1.8 \times 10^7$ is used ($s_n = s/1.8 \times 10^7$). For these normalized numerical results we have

$$\mathbf{a}(s_n) = \frac{40.6(12.2)}{(s_n + 1)(s_n + 12.2)}$$

The feedback factor $\mathrm{f} = \mathrm{f}_o$ is to be determined. The closed-loop gain function for this example is

$$A(s_n) = \frac{a}{1 + a\mathrm{f}} = \frac{40.6(12.2)}{s_n^2 + 13.2s_n + 12.2(1 + 40.6f_o)} = \frac{K}{s_n^2 + \sqrt{2}\rho s_n + \rho^2}$$

where $\rho = (\omega_{3\text{dB}})_n$. The above condition is for a Butterworth two-pole (the poles should be located at a 45° angle) (see Fig. 8.18*a*). The value of f_o that achieves this condition is $\mathrm{f}_o = 0.15$, which yields $R_{F2} = 285\ \Omega$. It should be noted that in this case the value of R_{F2} is not larger than $9R_{F1}$ and that the actual value of R_E will be $R_{F1} \parallel R_{F2} \simeq 43\ \Omega$, which is close to the value of 50 Ω; hence iteration is not necessary. In an analysis problem the values of both resistors are known, and the exact value of R_E should be used. For this example the closed-loop bandwidth is $(\omega_{3\text{dB}})_n = \sqrt{12.2(1 + 6.1)} = 9.24$ or $f_{3\text{dB}} = (9.24/2\pi)(1.8 \times 10^7) \simeq 26$ MHz.

Note that when dealing with feedback amplifiers *one cannot ignore the nondominant poles of* $\mathbf{a}(s)$ if the loop transmission gain is high. In the above example, if the second pole, which is larger by more than a factor of 10, were ignored, the open-loop gain function would have one pole and so would the closed loop, and thus no matter what the value of f_o is, the closed loop would not predict complex poles. The nondominant poles produce additional phase shift, and the contribution of the phase shift, as we shall show, cannot be ignored in a feedback system with a high amount of feedback. The models therefore should be accurate, and the *hybrid-π model should be used in an accurate analysis of feedback amplifiers, especially when the loop transmission gain is high.*

As a final example, consider the frequency response of the shunt-shunt feedback triple of Fig. 10.20. The open-loop gain function of the circuit is simply a cascade of three CE stages, which was considered in Chap. 8. Hence we can write

$$\mathbf{a}_i(s) = \frac{-a_o p_1 p_2 p_3}{(s + p_1)(s + p_2)(s + p_3)} \tag{10.89}$$

and

$$\mathbf{f}(s) = \frac{1}{R_f} = \mathrm{f}_o \tag{10.90}$$

From (10.89), (10.90), and the basic feedback equation, we have

$$A(s) = \frac{\mathbf{a}}{1 + \mathbf{af}} = \frac{-a_o p_1 p_2 p_3}{(s + p_1)(s + p_2)(s + p_3) + p_1 p_2 p_3 a_o \mathrm{f}_o} \tag{10.91}$$

In order to discuss (10.91) any further, let us consider a specific example.

EXAMPLE 10.7

Consider a numerical case for the shunt-shunt feedback triple of Fig. 10.20. Let the transistor parameters be identical, namely, $\omega_T = 10^9$ rad/s, $r_b = 50\ \Omega$, $\beta_0 = 50$, $C_c = 5$ pF, and $r_e = 5\ \Omega$. The circuit element values are $R_S = R_1 = R_2 = 100\ \Omega$, $R_L = 75\ \Omega$, and $R_F = 390$ kΩ. Determine the bandwidth of the amplifier.

We draw the **a** circuit as in Example 10.2. Calculation of $\mathbf{a}(s)$ for the circuit is very much similar to the one in Examples 8.5. We shall therefore omit the details. The result is

$$\mathbf{a}_i = \frac{-1.95 \times 10^5}{(s/p_1 + 1)^3}$$

where $p_1 = p_2 = p_3 = 3.87 \times 10^7$ rad/s.

$$\mathbf{f} = \mathrm{f}_o = -\frac{1}{390 \times 10^3} = -2.56 \times 10^{-6}$$

and

$$A(s) = \frac{1.95 \times 10^5}{s_n^3 + 3s_n^2 + 3s_n + 1.50}$$

where $s_n = s/3.87 \times 10^7$. The normalized 3-dB bandwidth as calculated from the above is $(\omega_{3\mathrm{dB}})_n = 0.88$; hence $f_{3\mathrm{dB}} = 5.4$ MHz.

For the numerical values of this example the values were chosen specifically to get identical poles. Recall that in transistors identical sections do not necessarily yield identical poles due to different D factors.

Finally, it should be noted that in determining the low-frequency response of the feedback amplifiers, we can break the complete frequency response into a separate high-frequency and low-frequency gain function, and each can be treated separately. Note that feedback also improves the low-frequency cutoff of the amplifier. For example, in an amplifier with a single pole-zero gain function as in (7.40), the low-frequency cutoff will be reduced by a factor $(1 + T_0)$, as can be readily seen from the use of the basic feedback equation.

EXERCISE 10.7

A shunt-series feedback pair has been designed to have the following open-loop gain function:

$$\mathbf{a}_i(s) = \frac{a_o}{(1 + s/p_1)^2}$$

where $a_o = 100$ and $p_1 = 10^7$ rad/s. Find the feedback factor f_o that will yield a closed-loop maximally flat magnitude function response. What is the resulting closed-loop current gain and bandwidth?

Ans $\mathrm{f}_o = 10^{-2}$, $A_i = 50$, $f_{3\mathrm{dB}} = 2.25$ MHz

10.11 STABILITY CONSIDERATIONS IN LINEAR FEEDBACK SYSTEMS

All of the advantages discussed in Sec. 10.2 can be realized only if the feedback amplifier is stable. Stability may be defined in a number of ways. Analytically, by stability we mean that the roots of the characteristic equation are in the left half-plane not including the $j\omega$ axis.

Let us consider the characteristic equation of a single-loop feedback amplifier (10.8), rewritten for convenience in the following:

$$F(s) = 1 + T(s) = 0 \quad \textbf{(10.92)}$$

To determine the stability of the system, we examine the roots of (10.92). If any of the roots of (10.92) is on the right half-plane, i.e., has a positive real part, the system is unstable. For amplifiers, we consider complex pole pairs on the $j\omega$ axis also as unstable. Analytic determination of stability may be done by the Routh-Hurwitz test. This test is discussed in detail in Ref. 3. In order to obtain a meaningful and accurate result on the stability of a feedback system, the expression used for $T(s)$ must be accurate. The unilateral model, which ignores the effect of nondominant poles, is, in general, *not* adequate for stability determination. One may use the hybrid-π model, which is unwieldy for all practical purposes. Other effects such as the excess phase shift, which is not included in the hybrid-π model, also play an important role in determining the stability of the amplifier. The inclusion of this effect further complicates the problem. Fortunately, one can ascertain the stability of a system by using the measured values of $T(j\omega)$. From (10.92) we have

$$|T(j\omega)| = 1 \qquad \arg T(j\omega) = \pm 180° \tag{10.93}$$

In other words, from a knowledge of the magnitude $|T(j\omega)|$ and the phase arg $T(j\omega) = 180°$ over a range of frequencies for which $|T(j\omega)| = 1$ and arg $T(j\omega) = 180°$, one could deduce whether the system is stable. Since the actual measured values of the gain and phase of the loop transmission is used, instead of a model, this method is accurate and preferable in feedback amplifiers.

There are basically two ways we can look at the steady-state response of $T(j\omega)$, and both yield the same results. One is the Bode plot, i.e., plots of $|T(j\omega)|$ vs frequency and arg $T(j\omega)$ vs frequency. We have already encountered Bode plots in Chap. 7. The other is the Nyquist criterion [3], which is a plot of $|T(j\omega)|$ vs. arg $T(j\omega)$ [or the imaginary part of $T(j\omega)$ vs the real part of $T(j\omega)$] with frequency as a parameter ($0 \leq \omega \leq \infty$). The Nyquist plot, in effect, combines the two Bode plots of the gain vs frequency and the phase shift vs frequency in a single plot. We shall consider the Bode plots in the following discussion.

In the Bode plot we determine the frequency ω_p at which the phase shift of $T(j\omega)$ is $-180°$. If at this frequency the gain $|T(j\omega)| \geq 1$, the amplifier is unstable.

EXAMPLE 10.8

Consider a feedback system with three identical poles, where the loop transmission function is given by

$$T(j\omega) = \mathbf{a}(s)\mathbf{f}(s)\Big|_{s=j\omega} = \frac{T_0}{(s+1)^3}\Big|_{s=j\omega} = \frac{T_0}{(1+j\omega)^3}$$

Examine the stability of the system for the three cases $T_0 = \frac{1}{2}$, $T_0 = 8$, and $T_0 = 15$. We set arg $T(j\omega) = -180°$ and determine $\omega = \omega_p = \sqrt{3}$ at $\omega_p = \sqrt{3}$; we have $|T(j\sqrt{3})| = \frac{1}{16}, 1, \frac{15}{8}$ for $T_0 = \frac{1}{2}, 8, 15$, respectively. Thus the system is unstable for $T_0 \geq 8$.

Two quantities of interest in feedback system design are the *gain margin* and the *phase margin* as shown in Fig. 10.24. The gain margin G_M is defined as the value of $[T(j\omega)]^{-1}$ in decibels at the frequency at which arg $T(j\omega) = 180°$. The radian frequency ω_p at which arg $T(j\omega) = 180°$ is called the *phase crossover* frequency. For a stable system $G_M > 0$ dB. The phase margin is defined as 180° plus arg $T(j\omega)$. The radian frequency ω_g at which $|T(j\omega)| = 1$ is called the *gain crossover* frequency. For a minimum phase network (i.e., a network with no zeros in the right half-plane),

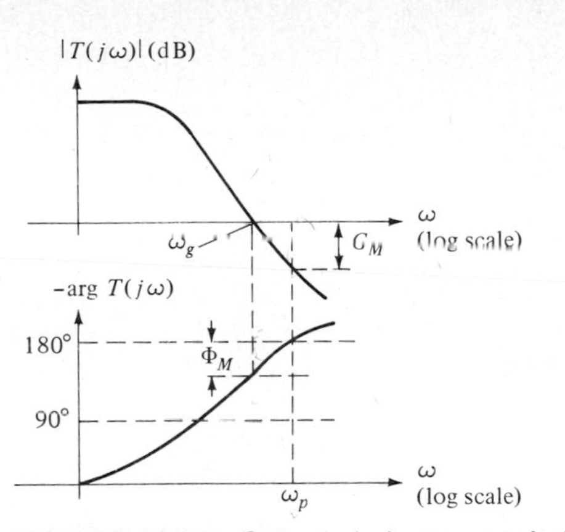

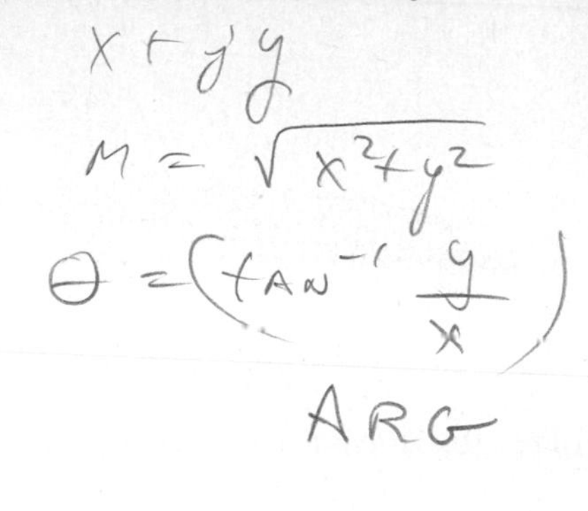

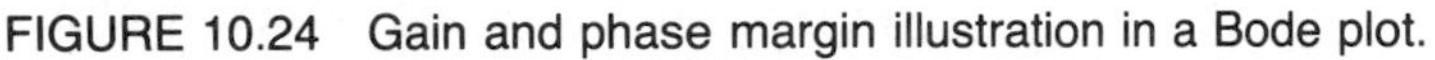

FIGURE 10.24 Gain and phase margin illustration in a Bode plot.

which is almost always used in feedback amplifiers for reasons of stability, $\Phi_M > 0$ for a stable system. As a rule of thumb, in a feedback amplifier design, gain and phase margins of at least 10 dB and 60°, respectively, are provided.

Analytically, the gain and phase margins are determined as follows: From $|T(j\omega)| = 1$ we find ω_g, and from $\arg T(j\omega) = 180°$ we find ω_p; then

$$G_M = 20 \log |T(j\omega_p)|^{-1} \qquad \text{decibels} \tag{10.94}$$

$$\Phi_M = 180° + \arg T(j\omega_g) \qquad \text{degrees} \tag{10.95}$$

EXAMPLE 10.9

If $T(s) = 4/(1 + s)^3$ determine the gain and the phase margins. From $|T(j\omega)| = 1$, we obtain $4/|1 + j\omega_g)^3 = 1$, or $\omega_g = 1.23$. From arg $T(j\omega_p) = 180°$, we obtain $3 \tan^{-1} \omega_p = 180°$ or $\omega_p = \sqrt{3}$. Hence $G_M = 20 \log T(j\sqrt{3}) = 20 \log 2 = 6$ dB and $\Phi_M = 180° + \arg T(j1.23) = 27°$. The above results, can of course, be obtained from the Bode plots of $T(j\omega)$.

The purpose of using the gain and phase margins is to make allowances for deviation in the system and to obtain a good transient response. For example, in cases where the closed-loop gain function magnitude responses include frequency peaking, the amplifier transient response will not be good (i.e., will exhibit large overshoot), and the amplifier will not be suitable for linear pulse amplification. Time-domain response is considered in Chap. 12.

EXERCISE 10.8

The characteristic equation of a certain feedback system is given by

$$1 + \frac{K_0}{(s+1)^6} = 0 \qquad K_0 \geq 0$$

For what value of K_0 will the system become unstable?

Ans 2.37

EXERCISE 10.9

If the loop transmission function of a certain feedback system is given by

$$T(s) = \frac{2}{(1+s)^4}$$

Determine the gain and phase crossover frequencies and the gain and phase margins.

Ans $\omega_g = 0.643$, $\omega_p = 1$, $G_M = 6$ dB, $\Phi_M = 50°$

10.12 OP-AMP FREQUENCY COMPENSATION

Care has to be exercised when op amps are used in feedback systems. The gain of the op amp is very high, and thus the amount of feedback applied can be *very large*. In a compensated one-pole op amp, such as the BJT 741 op amp or the BiFET LF351, where internal compensation is used ($f_{3dB} \simeq 5$ and 20 Hz, respectively), this may not pose a problem. In other IC op amps, where internal compensation is not used, one must provide for external frequency compensation.

As an example, consider the 1530-type op amp, which has a dc gain of 5000, break frequencies at $f_1 = 1$ MHz, $f_2 = 6$ MHz, and $f_3 = 20$ MHz, with f_3 being the unity gain bandwidth. The gain function can be written as

$$a(s) = \frac{5 \times 10^3}{(1 + s/p_1)(1 + s/p_2)(1 + s/p_3)} \tag{10.96}$$

where $p_i = 2\pi f_i$. The Bode plot is shown in Fig. 10.25.

This plot is typical of a number of other uncompensated IC op amps (e.g., for 702-type op amp: $a_o = 3600$, $f_1 = 1$ MHz, $f_2 = 4$ MHz, and $f_3 = 40$ MHz).

Consider the use of 1530 uncompensated as a unity gain block shown in Fig. 9.5. Because of the large amount of feedback the amplifier will be unstable. This large amount of feedback posed no problem with 741 or LF351, as they are internally compensated with break frequencies at around 5 or 20 Hz, respectively. In other words, for the compensated one-pole gain function

$$a(s) = \frac{a_o}{1 + s/p_o} \tag{10.97}$$

where $p_o = 2\pi(5\text{ Hz})$ and $a_o = 2 \times 10^5$ for 741 and $p_o = 2\pi(20\text{ Hz})$ and $a_o = 10^5$ for LF351. Recall that a one-pole gain function cannot become unstable.

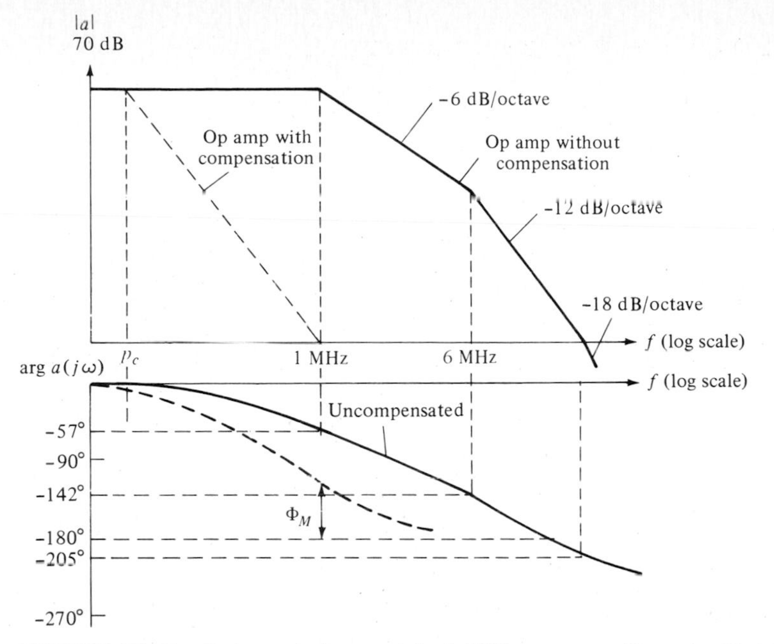

FIGURE 10.25 Gain and phase plot of 1530 op amp with and without compensation.

Thus, in order to use op amps of the type 1530 or 702, we must introduce external compensation. Op-amp manufacturers identify compensation terminals, and their data sheets give information on how to do it. The basic idea of compensation is to provide a phase margin at the unity loop transmission gain. This is shown by the heavy dashed line in Fig. 10.25. Such compensation is achieved by a narrowbanding scheme, as shown schematically in Fig. 10.26. The internal resistance R_c seen looking into this terminal and ground determines the break frequency of the narrowbanding scheme. The narrowbanding method is called a simple *lag compensation*. The simple lag-compensation method provides a pole at

$$p_c = 2\pi f_c = \frac{1}{R_c C} \tag{10.98}$$

in the loop transmission function. Either the value of R_c, which is several kiloOhms, is given by the manufacturer, or the value of C can be adjusted to get the unity-gain-

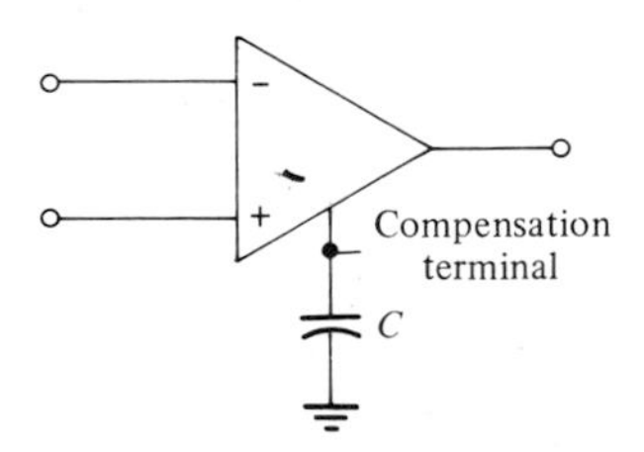

FIGURE 10.26 Simple lag compensation (narrowbanding scheme).

FIGURE 10.27 (*a*) An improved lag compensation. (*b*) Circuit representation for calculations.

bandwidth product with a phase margin of 45°. The gain function of the op amp with lag compensation is then approximately a one-pole gain function as in (10.97) with the pole value given by (10.96). Note that this method sacrifices a lot of gain-bandwidth product. In a number of applications, the designer may need considerable gain and bandwidth, and the narrowbanding scheme by the simple lag compensation may not be acceptable. For this situation one can introduce an *improved lag-compensation* scheme as shown in Fig. 10.27*a*. This compensation introduces a zero and a pole in the gain function. From Fig. 10.27*b*,

$$\frac{V_2}{V_1} = \frac{R_1}{R_1 + R_C} \frac{s + 1/R_1C}{s + 1/(R_1 + R_C)C} = K\frac{s + z}{s + p} \tag{10.99a}$$

Note that $p < z$. For maximum feedback bandwidth we set $z = p_1$, the smallest pole of the op amp (f_{3dB} of the op amp). The effect of this compensation on the frequency response (for $z = p_1$) is shown in Fig. 10.28. For $z = p_1$ we have

$$\frac{1}{R_1C} = p_1 \tag{10.99b}$$

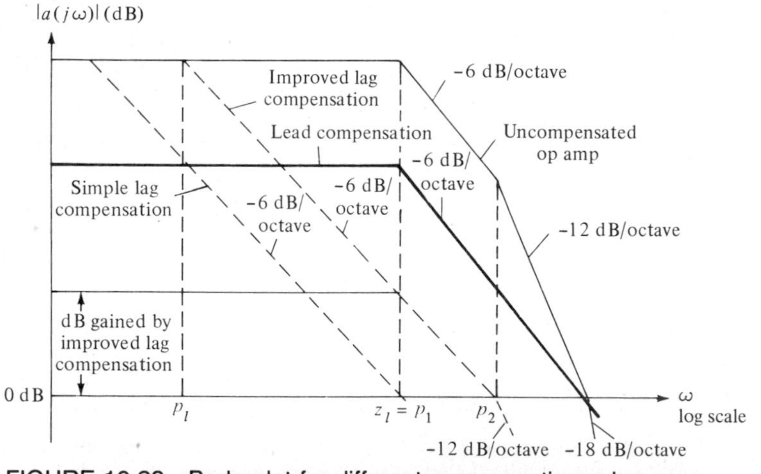

FIGURE 10.28 Bode plot for different compensation schemes.

The savings in gain bandwidth by introducing the improved lag compensation in the op amp from Fig. 10.28 is clearly in evidence. For 1530 type this amounts to 20 log $(f_2/f_1) = 15.6$ dB. The op-amp gain function for this case can still be considered as a one pole (10.97) except that the pole is at p_1 determined by the external C and R's.

Phase margin can also be provided by a *lead-compensation* network such as shown in Fig. 10.29. From Fig. 10.29 we have

$$\frac{V_2}{V_1} = \frac{s + 1/R_1C}{s + (R_2 + R_1)/R_2R_1C} = \frac{s + z}{s + p} \tag{10.100a}$$

Note that for the lead compensation $p > z$. This network provides a leading phase shift, hence the term *lead compensation*. For this case we set $z = p_2$, the second-lowest pole of the op amp, namely,

$$\frac{1}{R_1C} = p_2 \tag{10.100b}$$

the pole of the lead compensation is then designed such that it is large enough not to affect significantly the gain crossover frequency. An example below will illustrate this point.

Consider a noninverting 1530-type op amp shown in Fig. 10.30, which has a lead-compensation network. The open-loop gain function is given by (10.96). Note that the compensation is provided by the feedback circuit, namely,

$$f(s) = f_o\frac{s/z + 1}{s/p + 1} = \frac{R_{f1}}{R_{f1} + R_{f2}}\frac{s/z + 1}{s/p + 1} \tag{10.100c}$$

where $z = 1/R_{f2}C_f$ and $p = (R_{f1} + R_{f2})/R_{f1}R_{f2}C_f > z$. The zero of (10.100$c$) is set to cancel the pole $p_2 (= 2\pi \times 6 \times 10^6)$. The pole of (10.100$c$) is selected to much larger than p_3, namely, approximately $10p_3$. For 45° phase margin the Bode plot for the lead compensation is shown in Fig. 10.28. In this case the loop transmission gain is

$$T(0) \simeq \sqrt{2}(20) \simeq 29 \text{ dB}$$

Note that some low-frequency gain has been sacrificed in this compensation scheme.

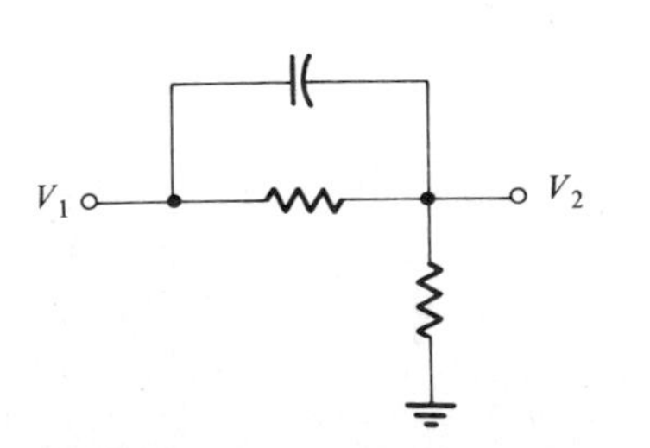

FIGURE 10.29 A lead compensation circuit.

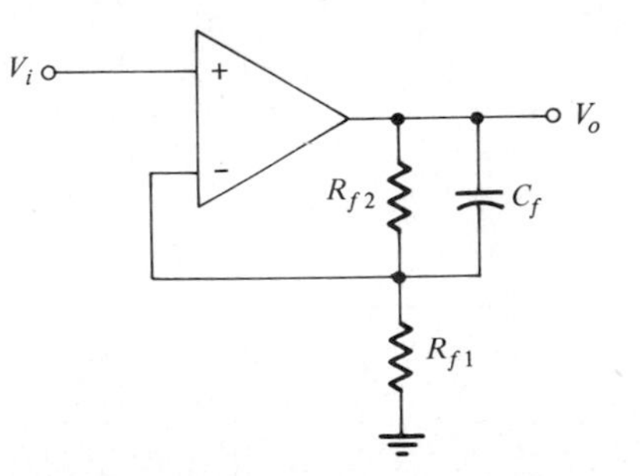

FIGURE 10.30 A noninverting op amp with lead compensation.

EXERCISE 10.10

For the circuit shown in Fig. 10.13, a 1530-type op amp is used without compensation. If $R_{f2} = (K - 1)R_{f1}$, for what value of K is the amplifier unstable?

Ans $K = 159$

EXERCISE 10.11

For the circuit shown in Fig. 10.30, a 702-type op amp is used.

(a) If $R_{f1} = 0.5\ \text{k}\Omega$, and if C_f is not present, for what value of R_{f2} will the system become unstable?

(b) What is the corresponding minimum value of the closed-loop gain?

Ans $R_{f2} \leq 32.8\ \text{k}\Omega$, $A_V = 66.7$

EXERCISE 10.12

Design the circuit of Fig. 10.30 for a voltage gain of 100, using a 702-type op amp. Determine the value of C_f, and R_{f2} if $R_{f1} = 0.5\ \text{k}\Omega$.

Ans $R_{f2} \simeq 50\ \text{k}\Omega$, $C_f = 0.8$ pF

10.13 LINEAR ANALYSIS OF SINUSOIDAL OSCILLATORS

A sinusoidal oscillator circuit is a circuit configuration that delivers a sinusoidal output waveform without an externally applied input signal. In other words, such systems supply their input from their own output without external excitation. From the feedback viewpoint, the above condition may be expressed as positive feedback with unity loop transmission, namely, $T(j\omega) = 1\ \underline{/0°}$, i.e.,

$$|T(j\omega)| = 1 \qquad \arg T(j\omega) = 0 \tag{10.101}$$

The condition in (10.101) is also referred to as the Barkhausen condition. In other words, instability is the necessary precondition for oscillation. The system is depicted in Fig. 10.31, where $\mathbf{a}(j\omega)$ provides gain and a phase reversal and $\mathbf{f}(j\omega)$ provides another phase reversal and some attenuation. If $\mathbf{a}(j\omega)$ does not have a phase reversal, then $\mathbf{f}(j\omega)$ should also have no phase reversal. If the system is stable, the steady-state output will be zero in the absence of an input. If the system is unstable, the output would theoretically indefinitely build up exponentially with time as will be predicted by a linear analysis. In an actual circuit, however, this cannot occur since due to nonlinearities of the active device, saturation occurs, which limits the amplitude of the output. Thus, a linear analysis is not strictly valid and is insufficient to characterize completely the operation of an oscillator. (Oscillation *amplitudes* are indeterminate.) Linear analysis may be used in a preliminary design, however, to establish approximately the condition of oscillation and to determine the *frequency* of oscillation.

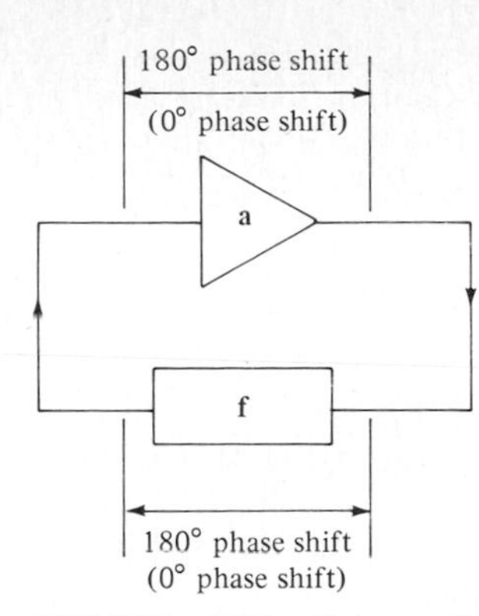

FIGURE 10.31 Schematic representation of a basic feedback oscillator.

The final design is to be refined experimentally and/or by using more accurate analytic techniques. In order to ensure oscillation, we often use $|T(j\omega)|$ slightly larger than unity so that environmental effects or replacement of the device does not cause $|T(j\omega)|$ to become less than unity. If $|T(j\omega)|$ is much larger than unity, the waveform will be very much distorted. The amplitude of oscillation in all cases will be limited by the nonlinearity of the circuit.

Note that the amplitude of oscillation cannot be predicted by a linear analysis, since the natural frequencies of a linear system do not depend on the input signal.

10.14 *RC* OSCILLATORS

RC oscillators are commonly used for audio frequencies, i.e., several hertz (Hz) to several kilohertz (kHz). The simplest practical oscillator circuits at low frequencies are the *RC* phase-shift oscillators and the Wien-bridge oscillators. These two will be considered briefly.

A. Phase-Shift Oscillator

A phase-shift oscillator is shown schematically in Fig. 10.32*a*. The amplifier gain A, which has a phase shift of 180°, may be realized by a BJT, FET, or an inverting op amp with low gain.

The operation of the circuit is simply as follows: The three *RC* sections provide a total of 180° phase shift. This phase shift with the 180° phase shift of the amplifier makes arg $T(j\omega) = 0$. For the idealized amplifier (i.e., assuming an amplifier with infinite input resistance and zero output resistance) the circuit model is shown in Fig. 10.32*b*. From the circuit the loop transmission is found by straightforward analysis. By setting the phase shift of $V_f/V_i = 180°$ and its magnitude equal to unity, we get (problem 10.27)

$$f = \frac{1}{2\pi RC\sqrt{6}} \quad \textbf{(10.102a)}$$

$$A > 29 \quad \textbf{(10.102b)}$$

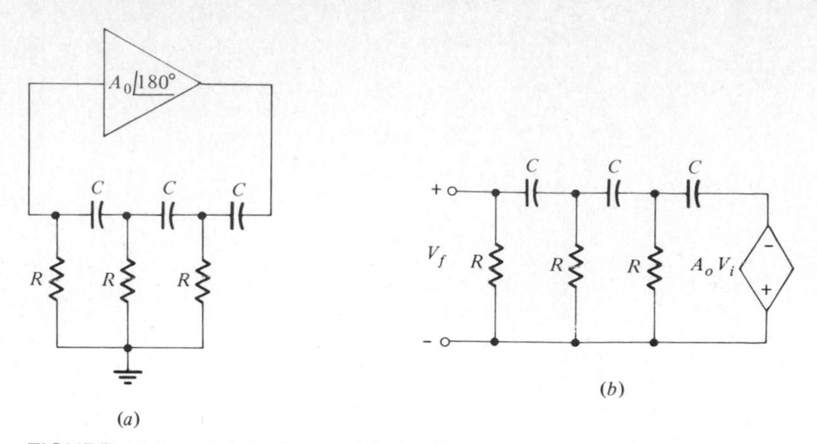

FIGURE 10.32 (*a*) A phase-shift oscillator using an ideal amplifier. (*b*) Circuit model for (*a*).

From (10.102*b*) the amplifier must have a gain of at least 29 to compensate for the loss in the feedback network.

Let us now consider other phase-shift oscillators. For example, if a bipolar transistor is utilized, the circuit is as shown in Fig. 10.33*a*. The oscillation condition may be determined by using feedback techniques and setting $|T(j\omega)| = 1$ or by loop or nodal analysis, setting the determinant equal to zero. Since *RC* oscillators are designed for low frequencies, the device model need only be the low-frequency equivalent circuit. The circuit model of Fig. 10.33*a* is shown in Fig. 10.33*b*. Note that in Fig. 10.33*b*, R_c could include the shunting effect of a large resistor value of h_{oe}^{-1}. We assume that biasing resistors R_E and $R_B = R_{B1} \| R_{B2}$ have negligible effect on the performance of the circuit so that they can be ignored in the following analysis. Further $R_c' + h_{ie} = R$, R_E is short-circuited by the bypass capacitor, and $R_B \gg h_{ie}$.

The loop transmission function is readily evaluated by opening the loop at point *b* and injecting a current I_b in the base. In order not to change the loading effect, an impedance equal to the input impedance of the amplifier is placed from *b* to ground as in Fig. 10.33*b*. The loop transmission function is then given by I_o/I_b. From Fig. 10.33*b*, either by nodal analysis or loop analysis, we obtain

$$T(s) = -\frac{I_o}{I_b} = \frac{-h_{fe}}{3 + R/R_c + (1/s)(4/RC + 6/R_cC) + (1/s^2)(1/R_c^2C^2 + 5/RR_cC^2) + (1/s^3)(1/R^2R_cC^3)} \tag{10.103}$$

The oscillation condition is $T(j\omega) = 1\underline{/0^\circ}$. Thus, from (10.103), the imaginary part of the denominator must be zero

$$-j\left[\frac{1}{\omega}\left(\frac{4}{RC} + \frac{6}{R_cC}\right) - \frac{1}{\omega^3}\frac{1}{R^2R_cC^2}\right] = 0 \tag{10.104a}$$

which simplifies to

$$\frac{1}{\omega^2 RR_cC} = 4 + 6\frac{R}{R_c} \tag{10.104b}$$

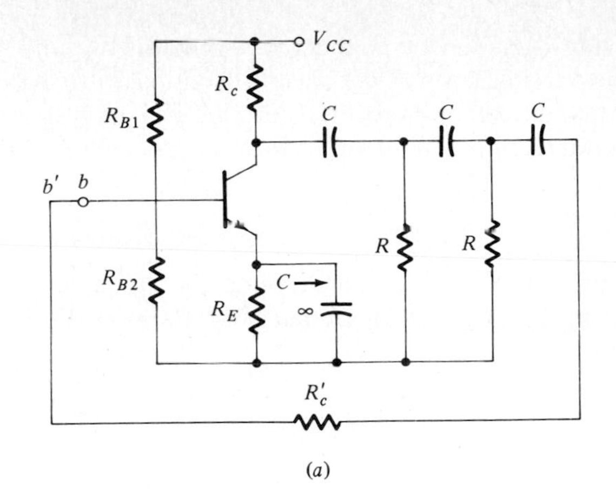

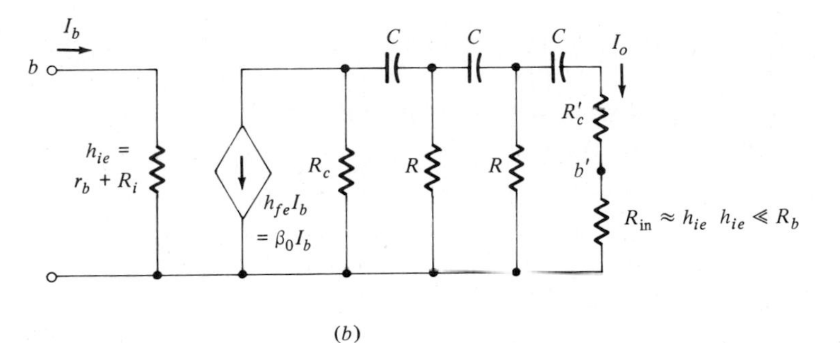

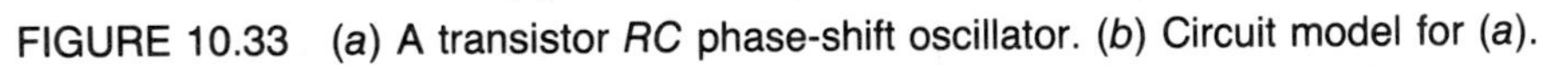

FIGURE 10.33 (*a*) A transistor *RC* phase-shift oscillator. (*b*) Circuit model for (*a*).

Similarly, setting $|T(j\omega)| = 1$ and using (10.104*b*) yields

$$\frac{-h_{fe}}{3 + R/R_c - (1/\omega^2)(1/R_c^2C^2 + 5/RR_cC^2)} = 1 \tag{10.105}$$

Substituting (10.104*b*) into (10.105) and simplifying, we get

$$h_{fe} = 23 + \frac{29}{R_c/R} + 4\frac{R_c}{R} \tag{10.106}$$

The expression on the right of (10.106) is minimum when $R_c/R = 2.7$. Since $|T(j\omega)| \geq 1$, we must have

$$h_{fe} \geq 23 + \frac{29}{2.7} + 4(2.7) = 44.5 \tag{10.107}$$

Thus, if the transistor h_{fe}, i.e., β_0, is less than 44.5, the circuit will not oscillate. If (10.107) is satisfied, the circuit will oscillate, and the frequency of oscillation will be given by (10.104*b*), namely,

$$f = \frac{1}{2\pi RC}\frac{1}{(6 + 4R_c/R)^{1/2}} \tag{10.108}$$

In practice, usually transistors have a β_0 much larger than 44.5, and a variable resistor is used for R_c. The variable resistor is adjusted such that $|T(j\omega)|$ is slightly larger than unity in order to eliminate waveform distortion. The analysis of the FET phase-shift oscillator is similar and is left as an exercise in problem 10.28.

EXERCISE 10.13

For the BJT oscillator circuit shown in Fig. 10.33, determine the frequency of oscillation and the minimum value of β_0 if $C = 0.01\ \mu\text{F}$ and $R = R_C = 1\ \text{k}\Omega$.

Ans $(\beta_0)_{\min} = 56,\ f_o = 5.04\ \text{kHz}$

B. The Wien-Bridge Oscillator

A Wien-bridge oscillator using an op amp is shown in Fig. 10.34. Note that the op amp is used in noninverting mode with a closed-loop gain equal to $1 + R_f/R_1$. The circuit can be analyzed by the loop transmission method, i.e., by finding V_f/V_o. By this method, if we let $R_f = (K - 1)R_1$, we have

$$T(s) = \left(1 + \frac{R_f}{R_1}\right)\frac{R/(sRC + 1)}{R/(sRC + 1) + 1/(R + sC)}$$

$$T(s) = \frac{KRCs}{R^2C^2s^2 + 3RCs + 1} \tag{10.109}$$

By setting $T(j\omega) = 1\underline{/0^\circ}$, we find $\omega_0 = 1/RC$ and $K = 3$; hence $R_f = 2R_1$.

Alternately, we have the method of setting the determinant equal to zero. Since there is a response for a zero input, the determinant must be equal to zero to account for it. This method is used here to familiarize the student with the alternate method. Note, however, the simplicity of the feedback approach.

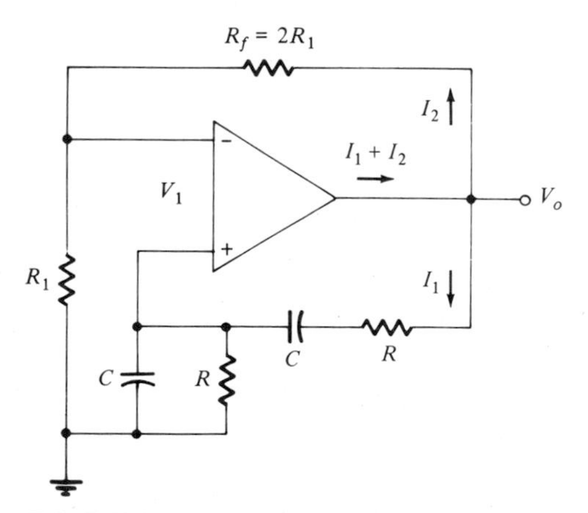

FIGURE 10.34 A Wien-bridge oscillator.

The equilibrium equations of the circuit are obtained by writing KVL around the loop as shown. The second equation is obtained by noting the $V_+ = V_-$ at the op-amp input terminals. Hence

$$\left(\frac{R}{1 + sRC} + \frac{1 + sRC}{sC}\right)I_1 + (R_1 + R_f)I_2 = 0 \tag{10.110a}$$

$$\left(\frac{R}{1 + sRC}\right)I_1 + R_1 I_2 = 0 \tag{10.110b}$$

For the circuit to oscillate the determinant of the coefficients in (10.110) are set equal to zero. Setting the determinant equal to zero, and with some manipulation, we get

$$s^2 + \frac{s}{RC}\left(2 - \frac{R_f}{R_1}\right) + \frac{1}{R^2C^2} = 0 \tag{10.111}$$

For $s = j\omega$ setting the real and imaginary parts of (10.111) equal to zero yields the conditions of oscillation, namely,

$$R_f = 2R_1 \qquad \text{and} \qquad \omega_0 = \frac{1}{RC} \tag{10.112}$$

10.15 *LC* OSCILLATORS

For radio-frequency (rf) oscillators, i.e., in the range beyond 100 kHz, up to the 100-MHz range, *LC* oscillators are usually used because of the relatively smaller sizes of the reactive elements. There are a number of such circuits; some of the classical ones are shown in Figs. 10.35 to 10.37. The circuit in Fig. 10.35 is called *Colpitts oscillator*, Fig. 10.36 is the Hartley oscillator, and Fig. 10.37 is a tuned-collector tuned-base oscillator. The inductance labeled rf choke is a large inductance having a very high impedance at the frequency of oscillation. The labeled circuit elements are

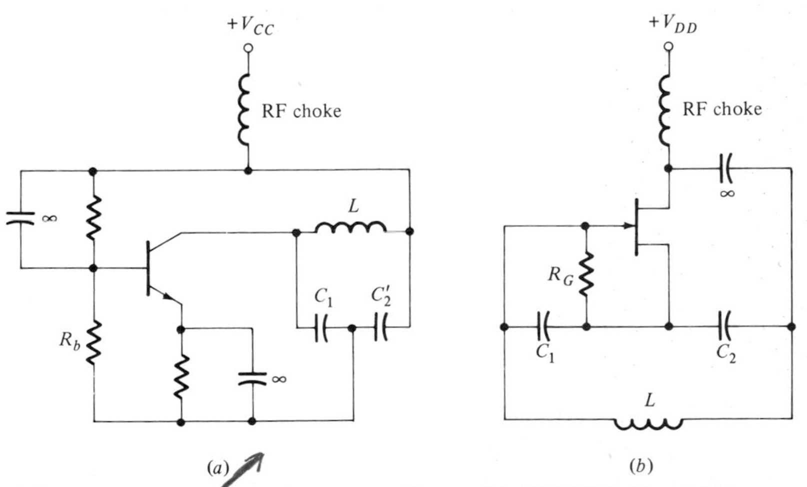

FIGURE 10.35 (*a*) BJT Colpitts oscillator. (*b*) JFET Colpitts oscillator.

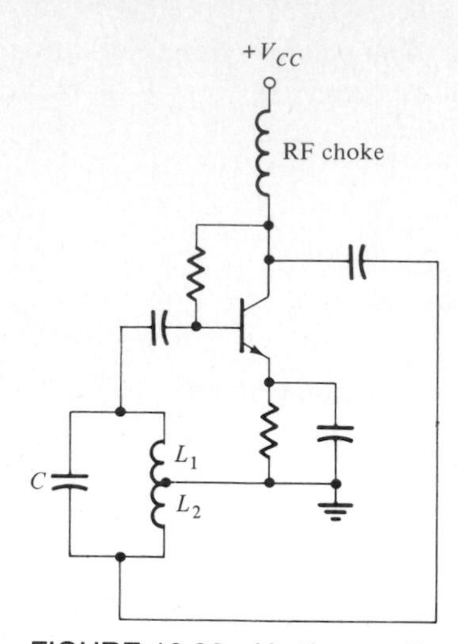

FIGURE 10.36 Hartley oscillator.

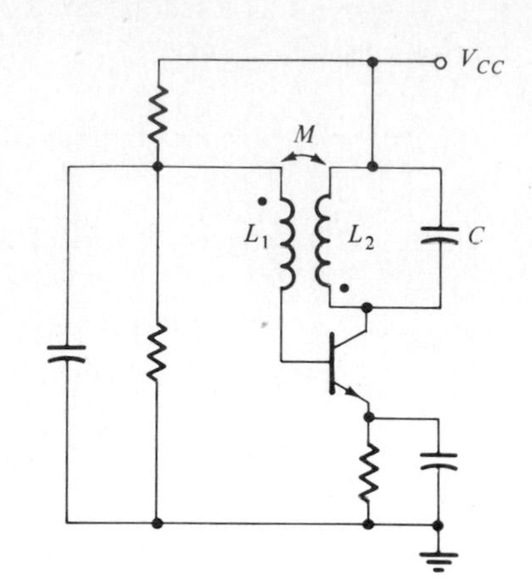

FIGURE 10.37 A tuned-circuit oscillator (tuned collector, tuned base).

the principal frequency-controlling elements. The biasing circuit-element values are such that they have a negligible effect on the operation of the circuit.

The condition of oscillation and the computation of the frequency of oscillation can be made from the feedback viewpoint by finding $T(s)$ or by setting the circuit determinant, from loop or nodal analysis, equal to zero, as was done for the Wien-bridge oscillator.

A. Colpitts Oscillator

Consider the Colpitts circuit shown in Fig. 10.35*a*. Since the operating frequency of *LC* oscillators is usually in the megahertz region, the reactive elements of the circuit models for the active device, in general, cannot be ignored. The small-signal model of the Colpitts oscillator is shown in Fig. 10.38. For simplicity, we have ignored r_b. Note that $C_i \simeq (1 + R_1 C_c \omega_T)/r_e \omega_T$.

We can now determine the return ratio function and set it equal to unity as in the previous section. The loop transmission function is found by breaking the loop open at the base and calculating the returned signal.

$$T(s) = \frac{V_f}{V_i} = \frac{-g_m R_1}{sRC_1 + (1 + sR_1C_2)(1 + s^2LC_1)} \tag{10.113a}$$

$$= \frac{-g_m R_1}{1 + s(R_1C_2 + R_1C_1) + s^2LC_1 + s^3LC_1R_1C_2} \tag{10.113b}$$

Setting (10.113) equal to $1\underline{/0^\circ}$, we obtain

$$\omega_0^2 = \frac{C_1 + C_2}{LC_1C_2} \tag{10.114a}$$

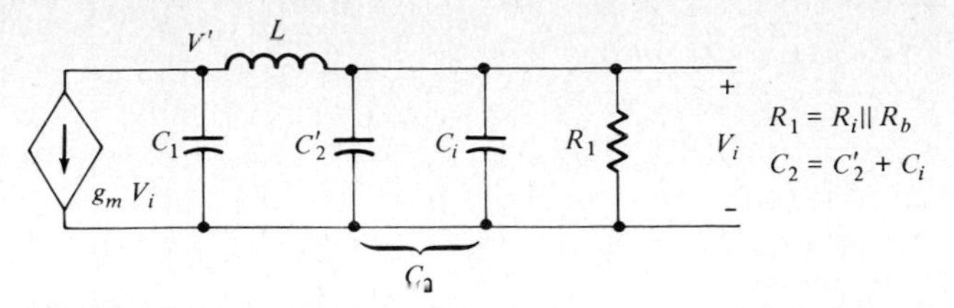

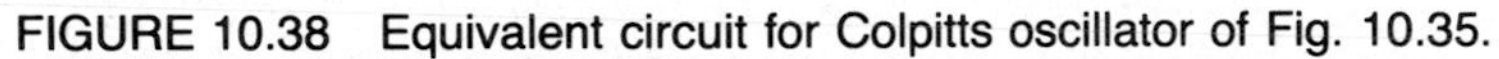

FIGURE 10.38 Equivalent circuit for Colpitts oscillator of Fig. 10.35.

and

$$\frac{-g_{m1}}{1 - \omega_0^2 LC_1} = 1 \tag{10.114b}$$

From (10.114) we obtain the condition of oscillation:

$$1 + g_m R_1 = 1 + \frac{C_1}{C_2} \tag{10.115}$$

If $R_b \gg R_i$, then $R_1 \simeq R_i = \beta_0 r_e = \beta_0/g_m$. Hence

$$\beta_0 \simeq \frac{C_1}{C_2} \tag{10.116}$$

and

$$f_o = \frac{1}{2\pi}\sqrt{\frac{C_1 + C_2}{LC_1C_2}} \tag{10.117}$$

Alternatively, we can use the nodal equation for the circuit and set the determinant equal to zero (problem 10.31) and get the same results. Analysis of the FET Colpitts oscillator is very similar and is left as an exercise in problem 10.41.

An op-amp Colpitts oscillator circuit is shown in Fig. 10.39. It is left as an exercise (problem 10.36) to show that the frequency of oscillation of this circuit is given by (10.117) with a proper choice of R_2/R_1.

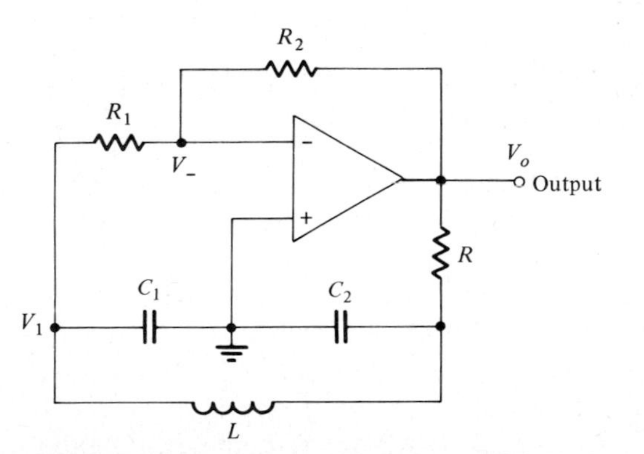

FIGURE 10.39 Op-amp Colpitts oscillator.

B. Tuned-Circuit Oscillators

There are a variety of tuned-circuit oscillators. One such example is the tuned-collector tuned-base circuit as shown in Fig. 10.37, where two separate inductively coupled coils are used. The oscillation frequency is determined mainly by the capacitor C and the inductance from the collector terminal to ground (i.e., the inductance across C).

EXERCISE 10.14

For the LC-tuned circuit shown in Fig. E10.14, determine the frequency of oscillation, and the turns ratio, n if $g_m = 4 \times 10^{-3}$ mho, $r_d = 15$ kΩ, given $L = 1$ mH, C = 1 nF and, $R = 4$ kΩ.

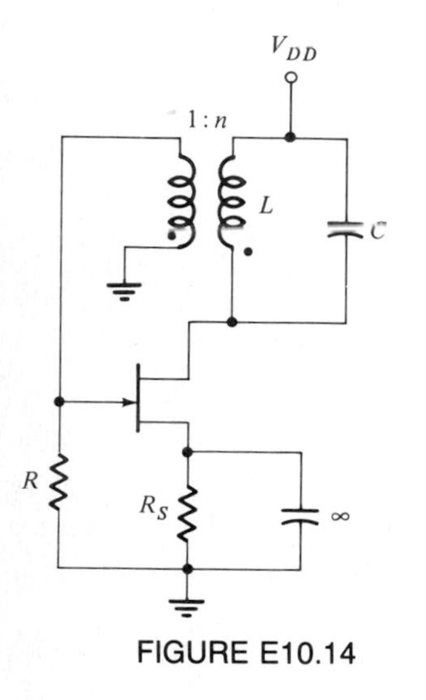

FIGURE E10.14

Ans $f_o = 0.16$ MHz and $n = 4$

10.16 CRYSTAL OSCILLATORS

For *very good frequency stability*, crystals are often used as the sole controlling element for the frequency of oscillation. The crystal oscillator is a very *important* component of the digital signal processing system. The accuracy of the total timing system is entirely dependent on the accuracy of the oscillator. The symbol and the circuit model of a vibrating piezoelectric crystal are shown in Fig. 10.40*a, b,* and *c*, respectively. The Q of a crystal can be as high as several hundred thousand. Typical values for a 10-MHz quartz crystal are $Q = 1.5 \times 10^5$, $C_0/C_1 = 300$, $L = 12$ mH, $C_1 = 10^{-11}$ F, and $R_S = 5$ Ω. Table 10.2 shows typical component values for common cuts of quartz oscillator crystals. In different cut-quartz crystals usually these parameters are specified (as in Table 10.2), or C_0, C_1, Q, and ω_0 are specified, and

TABLE 10.2 Typical Component Values for Common Cuts of Quartz Oscillator Crystals*

Frequency	32 kHz	280 kHz	525 kHz	2 MHz
Cut	*XY* bar	DT	DT	AT
R_s, Ω	40×10^3	1820	1400	82
L, H	4800	25.9	12.7	0.52
C_1, pF	0.00491	0.0126	0.00724	0.0122
C_0, pF	2.85	5.62	3.44	4.27
C_0/C_1	580	450	475	350
Q	25000	25000	30000	80000

*Courtesy of RCA.

L is determined from $L \simeq 1/C_1\omega_0^2$, where ω_0 is the resonant frequency of the crystal. R_S is determined from $R_S \simeq \omega_0 L/Q$ (Q is usually in the range of 10^4 to 10^6). A variety of crystal oscillator circuits is possible. The Colpitts-derived crystal oscillator is shown in Fig. 10.41. The analysis of the circuit is similar to that of Sec. 10.15*a* for Fig. 10.34 and will not be repeated here.

Finally, a CMOS oscillator circuit, which is often used in timing circuits, is shown in Fig. 10.42, where the basic configuration of Fig. 10.31 is utilized. The CMOS amplifier provides the gain and a phase reversal (see Sec. 5.5). The crystal provides another phase reversal at the crystal frequency.

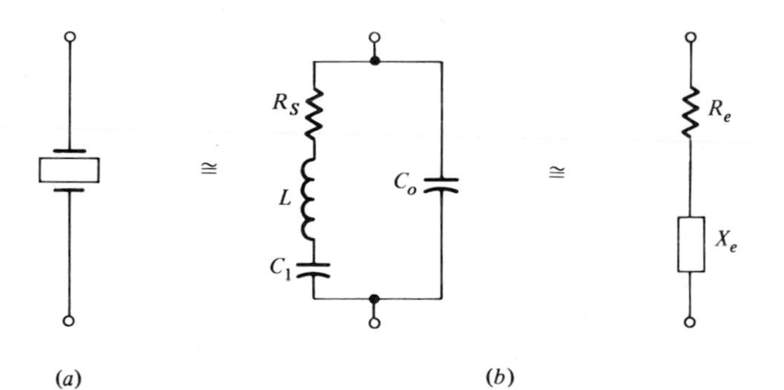

FIGURE 10.40 (*a*) Symbol for a piezoelectric crystal. (*b*) Equivalent circuit of (*a*).

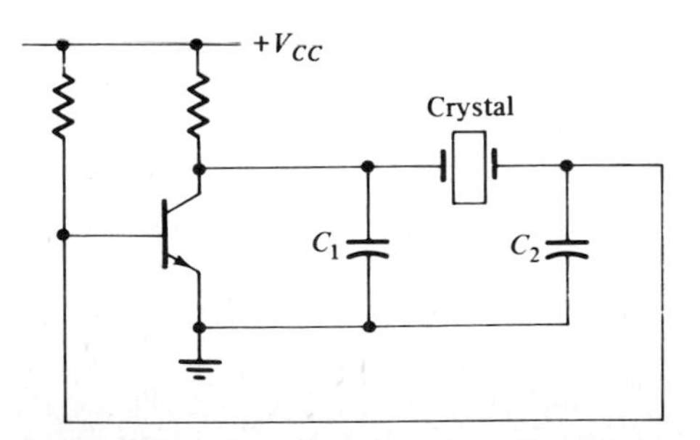

FIGURE 10.41 Colpitts crystal oscillator.

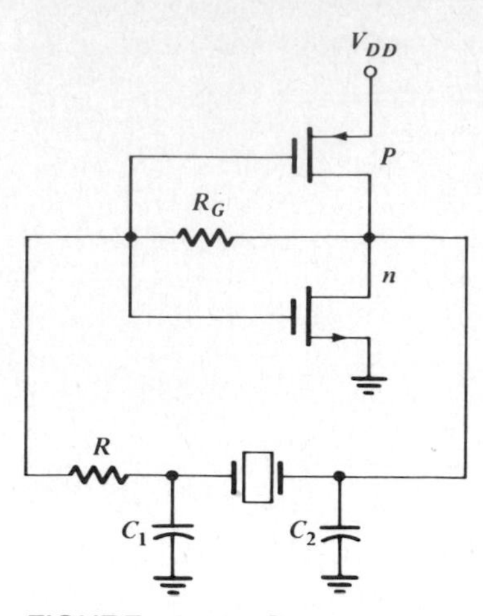

FIGURE 10.42 Colpitts crystal oscillator using a CMOS amplifier.

EXERCISE 10.15

For the crystal-controlled oscillator circuit shown in Fig. E10.15, a 32-kHz quartz crystal is used. Determine the oscillator frequency.

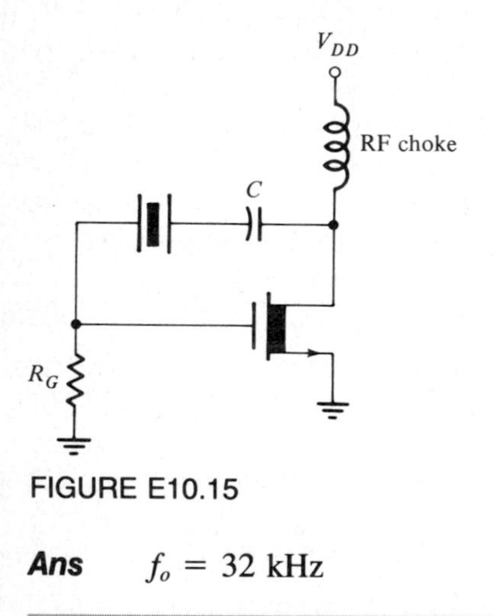

FIGURE E10.15

Ans $f_o = 32$ kHz

10.17 CONCLUDING REMARKS

In this chapter we have considered the analysis and design of feedback amplifiers by using the concept of feedback. Feedback techniques are very important in the design of electronic circuits and systems. Op amps are seldom used in open-loop configuration. Feedback is intentionally used in op amps to provide amplifying blocks of either polarity with a relatively wide bandwidth. In fact, by using the feedback concept, certain parameters can be identified by inspection. In feedback amplifiers, it is very important to ensure stability. External compensation must therefore be used if the op amp is not internally compensated. We have introduced the concepts of gain and phase margins as measures of relative stability.

We have briefly discussed sinusoidal oscillators where an active element such as an op amp, a FET, or a BJT is used in a feedback configuration with passive circuits such as *RC*, *LC*, or a crystal. For good frequency stability, crystals are often used as the sole frequency-controlling element.

REFERENCES

1. P. Gray and R. G. Meyer, *Analog Integrated Circuits,* Wiley, New York, 1977.
2. A. Sedra and K. C. Smith, *Microelectronic Circuits,* Holt, New York, 1982.
3. J. J. D'Azzo and C. H. Houpis, *Linear Control System Analysis and Design,* 2d ed., McGraw-Hill, New York, 1981.

PROBLEMS

10.1 For the single-stage, shunt-feedback amplifier shown in Fig. 10.7, the transistor parameters are as in Example 10.3. Design the circuit (i.e., find R_f) for a midband current gain of 10. What is the 3-dB bandwidth of the amplifier?

10.2 In the single-stage, shunt-feedback amplifier shown in Fig. 10.7, assume that β_0 can vary by ± 50 percent from its nominal value of 200. Design the circuit such that the effect of this variation in the midband gain is within 5 percent. What is the midband gain of the circuit? Make reasonable approximations.

10.3 For the single-stage, series-feedback amplifier shown in Fig. 10.10, assume that the transistor parameters are as in Example 10.3. If $R_S = 100\ \Omega$ and $R_L = 1\ \text{k}\Omega$, determine the value of R_f such that the midband voltage gain of the amplifier is $A_v = -10$. What is the 3-dB bandwidth of the amplifier in this case?

10.4 A single-stage FET amplifier is shown in Fig. P10.4. Determine the value of R_f for a midband voltage gain of 10. The FET parameters are $g_m = 4 \times 10^{-3}$ mho and $r_d = 50\ \text{k}\Omega$.

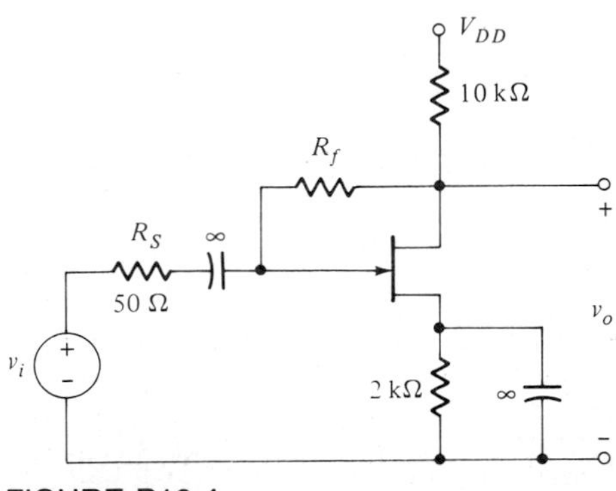

FIGURE P10.4

10.5 For the FET feedback amplifier circuit shown in Fig. P10.5, determine $\mathbf{a}(s)$, $\mathbf{f}(s)$, the voltage gain A_V, and the 3-dB bandwidth. The FET parameters are $g_m = 4 \times 10^{-3}$ mho, $r_d = 30$ kΩ, $C_{gs} = 10$ pF, and $C_{gd} = 2$ pF. The circuit parameters are $R_D = 10$ kΩ, and $R_1 = 2R_2 = 10$ kΩ.

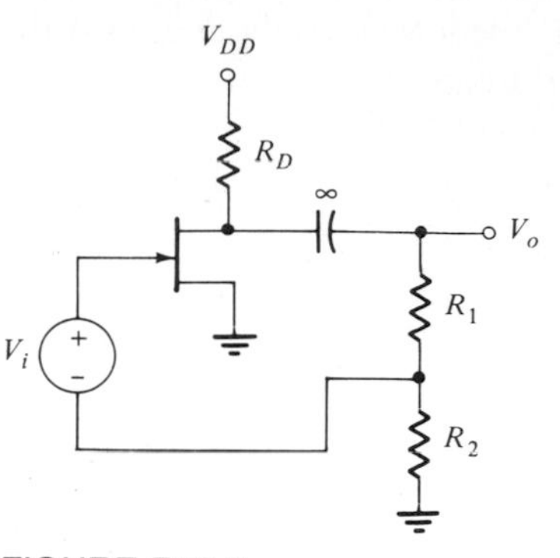

FIGURE P10.5

10.6 For the shunt-shunt, triple-feedback amplifier shown in Fig. P10.6, show the $\mathbf{a}(s)$ and $\mathbf{f}(s)$ circuits if $y_{11}^f \ll y_{11}^a$ and $y_{22}^f \ll G_L$. Write the expression for the midband $\mathbf{a}_i$, $\mathbf{f}$, and A_i.

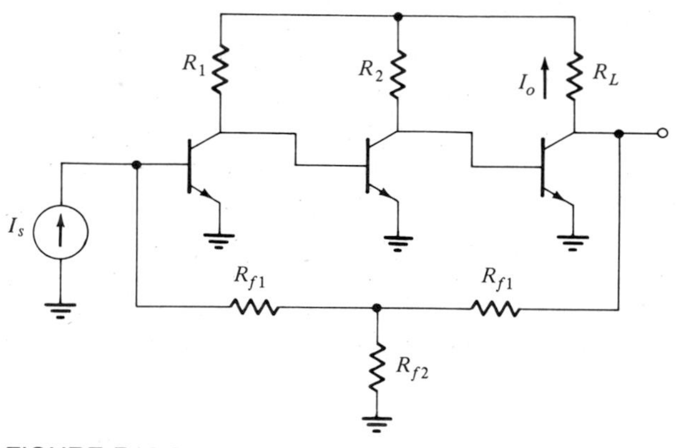

FIGURE P10.6

10.7 In Fig. P10.6 the circuit element values are as follows: $R_1 = R_2 = R_L = 50\ \Omega$, $R_{f1} = 50$ kΩ, and $R_{f2} = \infty$. The transistor parameters are as in Example 10.2. Determine the dominant pole-zero location of the closed-loop amplifier. If the circuit is stable (for simplicity, use the unilateral model), what is the gain of the amplifier?

10.8 Determine the current gain of the circuit in Fig. P10.6 if $R_1 = R_2 = 100\ \Omega$, $R_L = 500\ \Omega$, $R_{f1} = R_{f2}/2 = 50$ kΩ.

10.9 For the series-series, triple-feedback amplifier shown in Fig. P10.9, show the $\mathbf{a}_v(s)$ and $\mathbf{f}(s)$ circuits. Determine the expression for the midband $\mathbf{a}_v$, $\mathbf{f}$, and A_v.

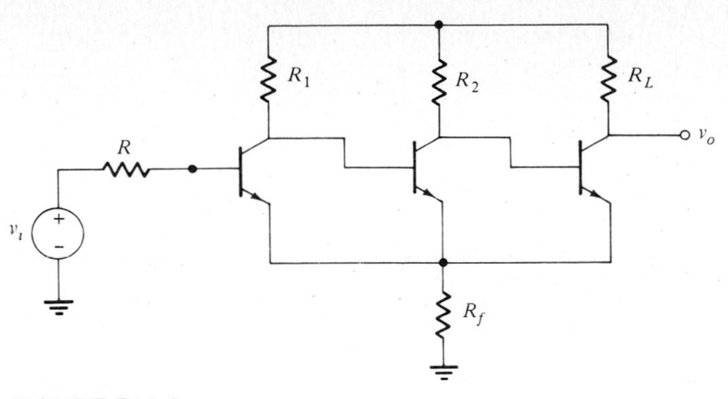

FIGURE P10.9

10.10 In Fig. P10.9 the circuit element values are as follows: $R_1 = R_2 = R_S = 50\ \Omega$, $R_f = 50\ \Omega$, and $R_L = 5\ \text{k}\Omega$. The transistor parameters are as in Example 10.2. Determine the dominant pole-zero location of the closed-loop amplifier. *Is* the circuit stable? Use the unilateral model for simplicity and make reasonable simplifications.

10.11 For the shunt-series feedback circuit shown in Fig. 10.16, $R_S = R_F = 1\ \text{k}\Omega$, $R_1 = 3\ \text{k}\Omega$, $R_L = 500\ \Omega$, and $R_E = 5\ \Omega$. The transistor parameters are $\beta_0 = 50$, $r_b = 75\ \Omega$, and $r_e = 5\ \Omega$. Determine the low-frequency values of $\mathbf{a}_i$, $\mathbf{f}$, and A_i of the circuit.

10.12 For the series-shunt circuit shown in Fig. 10.18, assume $R_{F1} = R_L = 50\ \Omega$, $R_{F2} = 300\ \Omega$, $R_S = 25\ \Omega$, and $R_1 = 3.9\ \text{k}\Omega$. The transistor parameters are as in problem 10.11. Determine the low-frequency $\mathbf{a}_v$, $\mathbf{f}$, and A_v of the circuit.

10.13 Determine the bandwidth of the series-shunt feedback pair of problem 10.12, if the transistor high-frequency parameters are $\omega_T = 10^9$ rad/s and $C_c = 5$ pF.

10.14 For the feedback circuit shown in Fig. 10.19, $R_s = R_L = 100\ \Omega$, $R_{E1} = R_{E2} = 50\ \Omega$, $R_1 = R_2 = 200\ \Omega$, and $R_F = 5\ \text{k}\Omega$. Determine the value of $\mathbf{a}_v$, $\mathbf{f}_o$, and the voltage gain of the circuit.

10.15 This problem illustrates the effectiveness of overall negative feedback as compared to the local feedback case in regards to sensitivity. Assume no interaction between the blocks in Fig. P10.15 and use differential sensitivity. Show that for an equal overall gain in Fig. P10.15

$$(S_a^A)_2 = \frac{1}{1 + af_1}(S_a^A)_1$$

Comment on the above results. Note that A_2 corresponds to Fig. P10.15*b*.

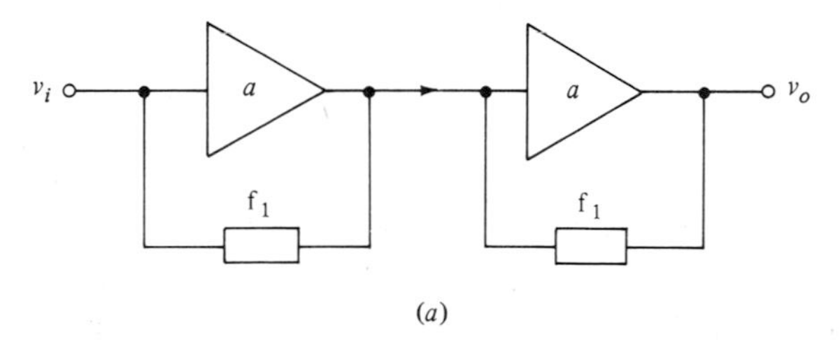

FIGURE P10.15

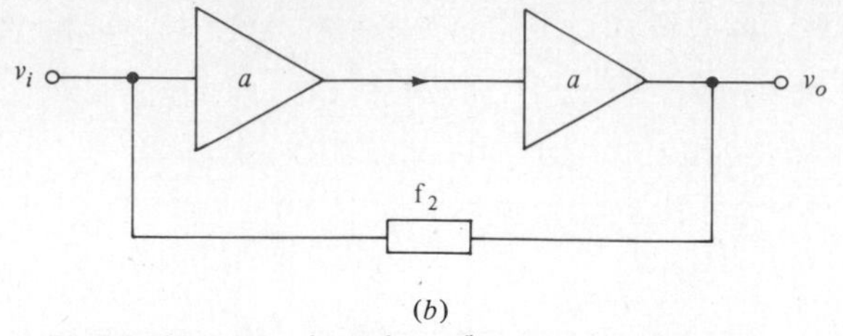

(b)

FIGURE P10.15 (*continued*)

10.16 For the FET feedback amplifier shown in Fig. P10.16, determine $\mathbf{a}(s)$, $\mathbf{f}(s)$, $A_V(s)$, and determine *roughly* whether the circuit is stable (i.e., use the unilateral model). If it is stable, what is the overall voltage gain. Use the FET parameters in problem 10.5.

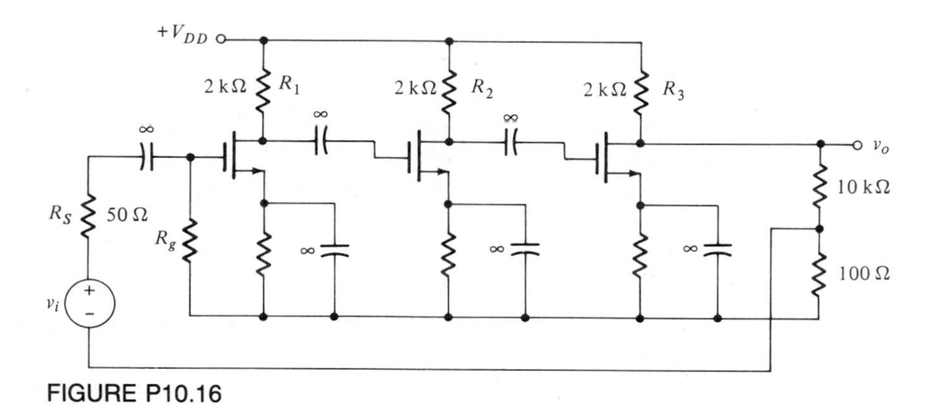

FIGURE P10.16

10.17 A feedback system is shown in Fig. P10.17, where $a(s) = 10^4/(s + 1)^2$. Determine the return ratio such that a ±20 percent variation in $\mathbf{a}(0)$ does not cause more than +1 percent variation in the closed-loop low-frequency gain of the amplifier. What is the resulting closed-loop gain? Sketch the magnitude response of the gain.

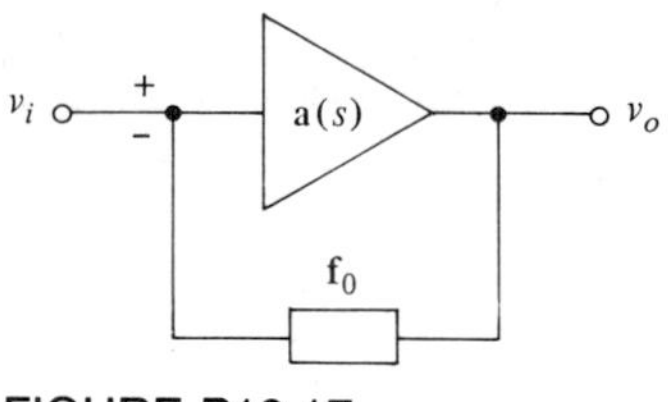

FIGURE P10.17

10.18 (*a*) For the feedback system shown in Fig. P10.15*a*, the following are given: $a_o = -200$, $f_1 = 0.1$. The closed-loop gain function $A = V_o/V_i$. Find the value of A and dA/A for $da_o/a_o = 0.05$.

(*b*) The feedback system shown in Fig. P10.15*b* is used with the following data given: $a_o = -200$, $A = 100$. Find the values of f_2 and dA/A for $da_o/a_o = 0.05$.

(*c*) If $a(s) = -200/(1 + s/10^6)$ in each case above, determine the corresponding closed-loop pole-zero locations.

10.19 The open-loop gain function of an amplifier, which includes the low- and the high-frequency responses, is given by

$$\mathbf{a}(s) = 10^8 \frac{s}{(s + 20)(s + 10^6)} \qquad \mathbf{f}(s) = f_o = 0.1$$

Determine the high-frequency and the low-frequency 3-dB bandwidths. What is the closed-loop midband gain?

10.20 For the four-stage, shunt-series feedback circuit shown in Fig. P10.20 exclusive of biasing and coupling circuitry, determine the following: $\mathbf{a}_i(0)$, $\mathbf{f}(0)$, and $A_i(0)$. Let $\beta_0 = 50$, $r_e = 10$, and $r_b = 0$.

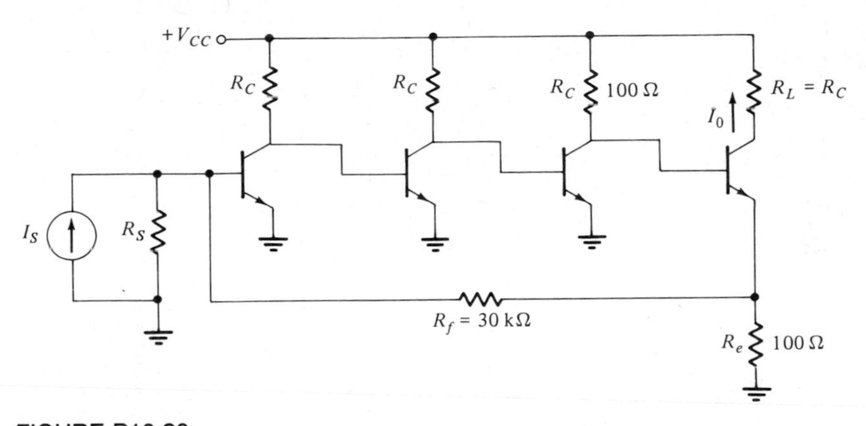

FIGURE P10.20

10.21 For the four-stage, series-shunt feedback circuit shown in Fig. P10.21 exclusive of biasing and coupling circuitry, determine the following: $\mathbf{a}_v(0)$, $\mathbf{f}(0)$, $A_v(0)$. The transistor parameters are as given in problem 10.20.

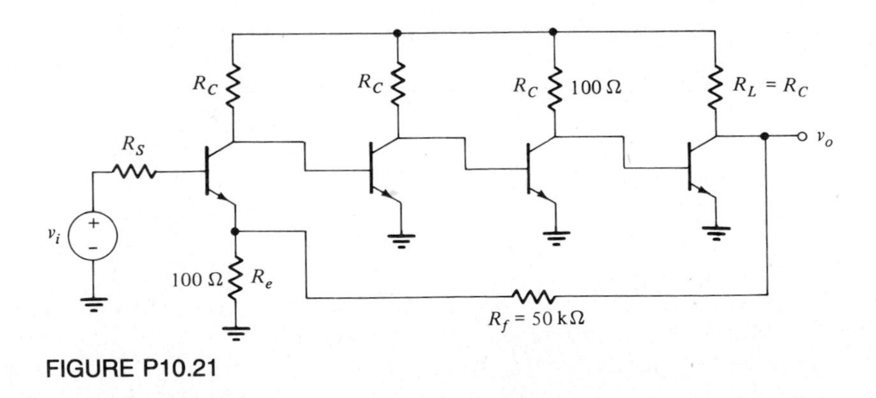

FIGURE P10.21

10.22 For the negative-feedback system shown in Fig. P10.22, $\mathbf{a}(s) = K/(s+1)^n$. Determine the maximum value of K as a function of n for which the system is stable.

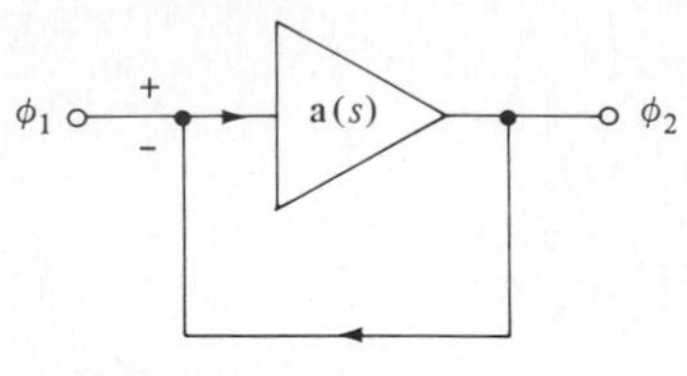

FIGURE P10.22

10.23 In a certain feedback amplifier, the following are given:

$$\mathbf{a}(s) = 5 \times 10^{14} \frac{s^2}{(s+1)(s+2)(s+10^6)(s+2\times 10^6)}$$

$$\mathbf{f}(s) = 0.002$$

Find the midband closed-loop gain and the upper and lower 3-dB cutoff frequencies of the closed-loop gain function.

10.24 Examine the stability of a feedback system if

$$\mathbf{a}(s) = \frac{100}{(1 + s/10^6)^2(1 + s/2\times 10^6)^2} \qquad \mathbf{f} = 0.1$$

Let $s_n = s/2 \times 10^6$.

10.25 The following measured data were obtained from a single-loop negative-feedback amplifier:

Frequency, MHz	$\lvert\mathbf{a}\rvert$	arg $\mathbf{a}(j\omega)$
0.01	1000	−5°
0.1	800	−20°
1	200	−80°
5	90	−120°
10	40	−200°
50	1	−260°

What is the value of the resistive feedback $\mathbf{f}_o$ that can be applied around this amplifier and provide a gain margin of 10 dB? What is the corresponding phase margin? For what values of $\mathbf{f}_o$ will the feedback amplifier become unstable?

10.26 For the FET circuit shown in Fig. P10.26, exclusive of biasing circuitry, Z_1, Z_2, and Z_3 are purely reactive elements (i.e., L or C). Set up the expression for the condition of oscillation. What type of reactive elements, i.e., L and/or C, would result in an oscillator? Use the low-frequency model for the FET.

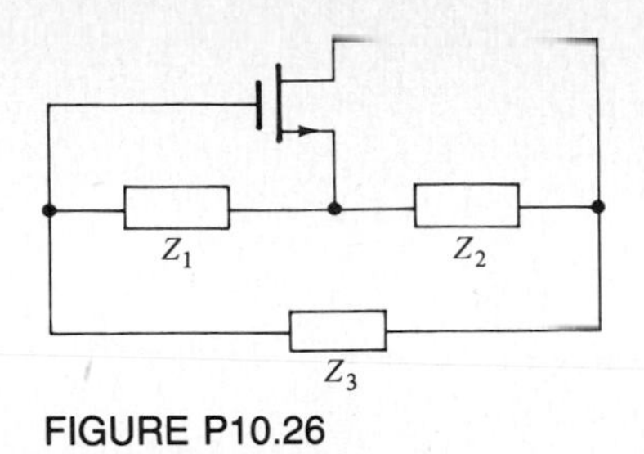

FIGURE P10.26

10.27 Derive (10.102*a*) and (10.102*b*).

10.28 For the FET phase-shift oscillator shown in Fig. P10.28, determine the condition for oscillation and the expression for the frequency of oscillation. Use the low-frequency FET parameters.

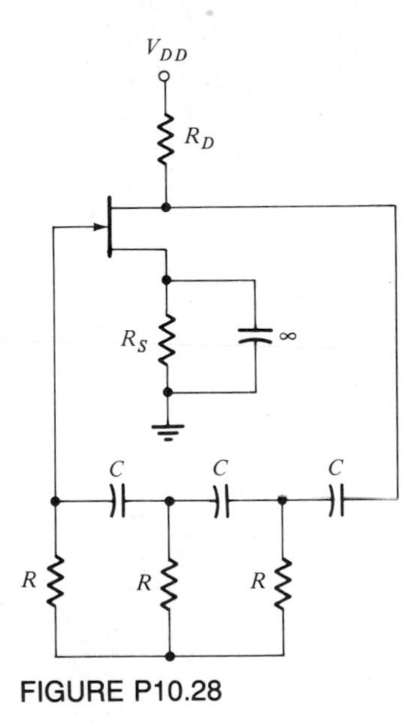

FIGURE P10.28

10.29 Design the circuit of Fig. P10.28 for $f_o = 1$ kHz if the FET parameters are $g_m = 4 \times 10^{-3}$ mho and $r_d = 40$ kΩ.

10.30 The circuit model of a tuned-collector oscillator (for $f < f_\beta$) is shown in Fig. P10.30. Derive the expression for the oscillation condition and the equation that determines the frequency of oscillation.

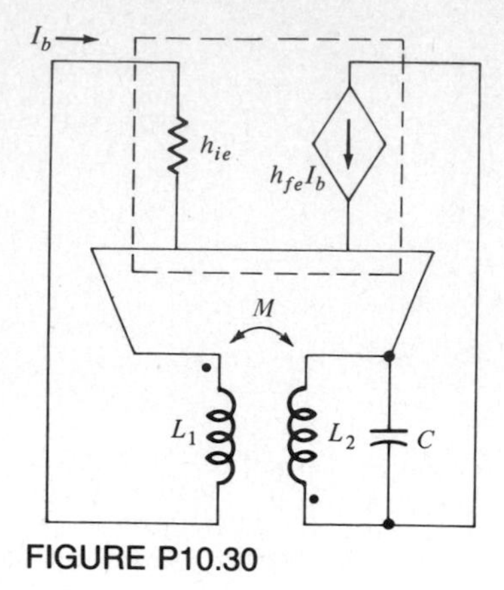

FIGURE P10.30

10.31 For the Colpitts oscillator in Fig. 10.35*a* use the nodal equations and set the determinant equal to zero, and derive (10.116) and (10.117).

10.32 For the MOSFET oscillator shown in Fig. P10.32, determine the condition for oscillation and the frequency of oscillation. Use the low-frequency model for the FET. Assume $R_{L2} \ll R$.

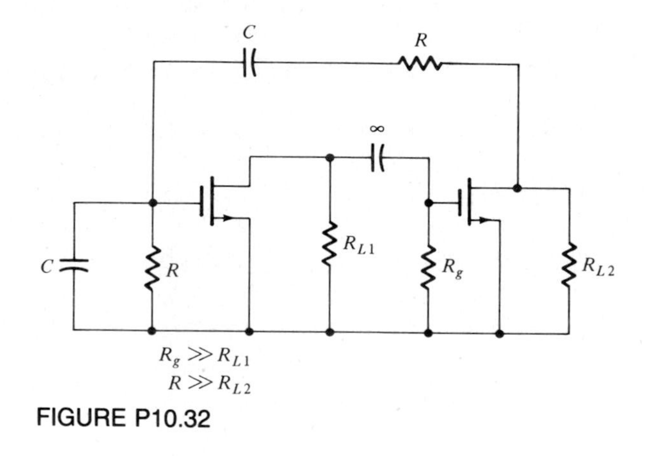

FIGURE P10.32

10.33 For the circuit shown in Fig. P10.33, R_n is a negative resistance. Determine the expression for the oscillation condition and the frequency of oscillation for the two forms of Z_1.

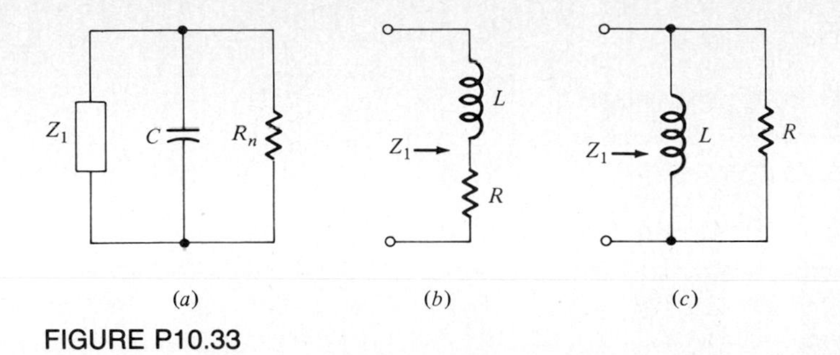

FIGURE P10.33

10.34 Determine the condition for oscillation and the frequency of oscillation for the crystal oscillator in Fig. 10.42 if a 525-kHz crystal (see Table 10.2) is used.

10.35 Determine the condition of oscillation and the frequency of oscillation for the FET Colpitts oscillator shown in Fig. 10.35*b*.

CHAPTER 11 Tuned Amplifiers and Frequency-Selective Circuits

In this chapter we consider mainly the analysis and design of tuned amplifiers. In contrast to the previous chapters where wideband amplifiers were considered, here we examine amplifiers designed to amplify a narrow band of frequencies centered around some center frequency f_o. Tuned amplifiers find many applications in telecommunications, such as in radio, radar, and television receivers, to name a few. The chapter also includes other frequency-selective circuits such as phase-locked loops. Phase-locked loops, in addition to frequency selection, are also used for frequency synchronization, frequency translation, modulation, etc.

We shall first consider general design requirements of a tuned amplifier.

11.1 GENERAL CONSIDERATIONS

In order to illustrate requirements on the tuned amplifier, consider the block diagram of a simple communication system, as shown in Fig. 11.1. The input signal is a

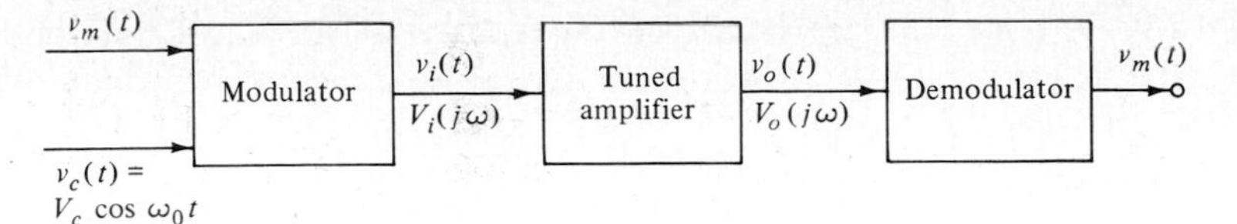

FIGURE 11.1 Simple amplitude-modulated system

message wave $v_m(t)$ that can be expressed in terms of its Fourier series and is assumed to consist of a band of low frequencies in the vicinity of dc. The modulator translates this band of frequencies to a new band of frequencies centered around the carrier frequency ω_c without disturbing the relative position of various frequency components of the signal. For example, for an amplitude-modulated (AM) signal

$$v(t) = [V_c + v_m(t)] \cos \omega_c t \tag{11.1}$$

where V_c is the amplitude of the carrier signal. The message signal and the modulated waveforms are shown in Fig. 11.2. The magnitude of $v_m(t)$ is always adjusted to be less than V_c; hence the envelope of the modulated signal never drops to zero. To illustrate, consider the special case of a single frequency. The message signal has the form $v_m(t) = V_m \cos(\omega_m t + \theta_m)$. In this case, from (11.1)

$$\begin{aligned} v(t) &= [V_c + V_m \cos(\omega_m t + \theta_m)] \cos \omega_c t \\ &= V_c[1 + m \cos(\omega_m t + \theta_m)] \cos \omega_c t \end{aligned} \tag{11.2}$$

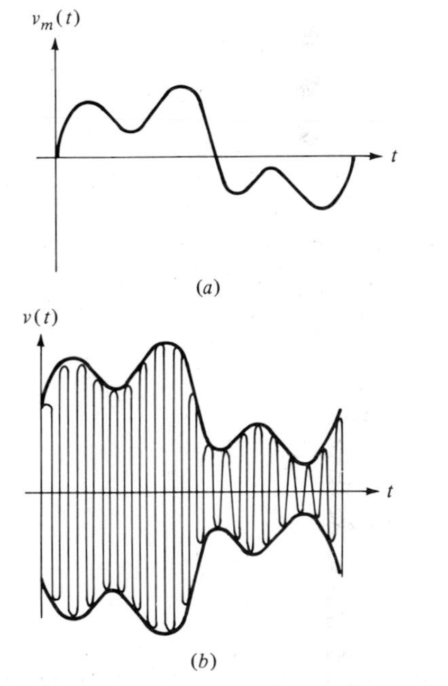

FIGURE 11.2 (*a*) Message signal. (*b*) Amplitude-modulated signal.

where $m = V_m/V_c$ is called the *modulation index* and is always less than unity and $\omega_c \gg \omega_m$. We may rewrite (11.2), using the trigonometric identity of the product of two cosines, as

$$v(t) = V_c \cos \omega_c t + \frac{m}{2} V_c \cos [(\omega_c + \omega_m)t + \theta_m] + \frac{m}{2} V_c \cos [(\omega_c - \omega_m)t - \theta_m] \quad \textbf{(11.3)}$$

Note that (11.3) consists of three frequency components, the carrier frequency and two side frequencies, located symmetrically on either side of the carrier frequency.

In the general case, there are many signal-frequency components and hence many pairs of side frequencies. These are called the *upper* and the *lower sidebands* of the AM wave. The demodulator in Fig. 11.1 converts the signal information back to its original form. The purpose of the tuned amplifiers is to perform a frequency-selective amplification. The magnitude and phase characteristics of the ideal tuned (bandpass) amplifier are shown in Fig. 11.3*a*. The corresponding characteristics for the realizable tuned amplifier are shown in Fig. 11.3*b*. The bandwidth $\omega_{3dB} = \omega_2 - \omega_1$ in a narrowband amplifier is much smaller than the center radian frequency ω_0. In other words, for a narrowband situation, $\omega_{3dB} \leq 0.1\omega_0$, and for this case considerable simplification is achieved in the design as demonstrated in later sections of this chapter.

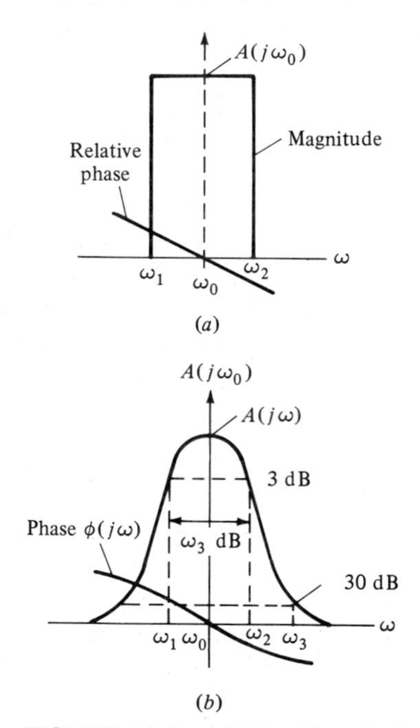

FIGURE 11.3 (*a*) Ideal bandpass characteristics. (*b*) Actual bandpass characteristics.

A measure of the effectiveness of the filtering is provided by the "skirt" of the magnitude response in Fig. 11.3*b*. Quantitatively, this measure is given by *selectivity*, which is given either by −60- to −6-dB frequency ratio or −30- to −3-dB ratio ω_3/ω_2, as shown in Fig. 11.3*b*. The design of a bandpass amplifier, as we shall soon see, has some aspects common to those of low-pass amplifier design, especially in the narrowband case. However, there are certain features of the design in a bandpass amplifier that are quite different from those of low-pass amplifiers. The chief problem in bandpass amplifier design is the possibility of oscillation, due to the reactive load and the internal feedback of the device, e.g., the C_c of a CE transistor or the C_{gd} of a common-source FET. Since C_c plays an important, undesirable role in the design and is the source of the potential instability of the device, *the unilateral model cannot, in general, be used for an inductive load and maximum power gain,* as will be shown in Sec. 11.9. For narrowband tuned amplifiers the transistor two-port parameters are measured at the center frequency and then the design is based on the approximation that these parameters are approximately the same throughout the passband of the amplifier. Since the passband extends at most to ±5 percent of the center frequency in a narrowband situation, this approximation is usually valid. For a wideband design this approximation cannot be made, and mismatching techniques are used to avoid instability and provide alignability.

11.2 SINGLE-TUNED CIRCUITS

One of the simplest bandpass amplifier interstages is the so-called single-tuned amplifier. A single-tuned circuit may utilize a FET, a BJT or an op amp. We shall consider each of these in the following sections.

A. FET Circuit

A simple single-tuned JFET stage exclusive of biasing circuitry is shown in Fig. 11.4*a*. The equivalent circuit is shown in Fig. 11.4*b*, where the device capacitance C_{ds} and resistance r_d are included in C_T and R, respectively. As discussed in Chap. 7, C_{ds} is usually very small and can be ignored. This circuit has a single resonant circuit and provides a good vehicle for introducing the analysis and design of tuned amplifiers.

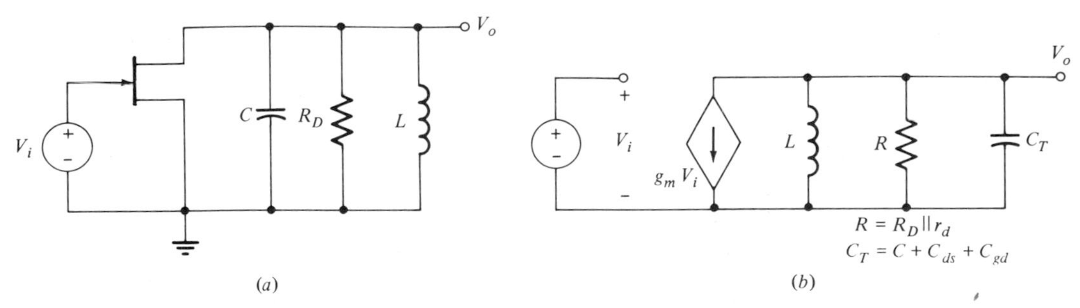

FIGURE 11.4 (*a*) FET Single-tuned circuit. (*b*) Equivalent circuit of (*a*).

The voltage gain function of the circuit is given by

$$A_v = \frac{V_o}{V_i} = -\frac{g_m}{G + sC_T + 1/sL} \tag{11.4a}$$

$$= -\frac{g_m}{C_T}\frac{s}{s^2 + (G/C_T)s + 1/LC_T} \tag{11.4b}$$

where C_T is the total interstage capacitance and $G = 1/R$ ($R = R_D \parallel r_d$). Equation (11.4) is also sometimes referred to as the resonator expression.

The magnitude of the gain function is

$$|A_v(j\omega)| = \frac{g_m}{C_T}\frac{\omega}{\sqrt{[(G/C_T)\omega]^2 + (1/LC_T - \omega^2)^2}} \tag{11.5}$$

The magnitude is maximum at the frequency where the denominator of (11.5) is minimum. This frequency is commonly called the *center* or the *resonant frequency* and is denoted by ω_0. Hence

$$\omega_0 = \frac{1}{\sqrt{LC_T}} \tag{11.6}$$

Note that at ω_0 the gain is

$$|A_v(j\omega_0)| = -\frac{g_m}{G} = -g_mR = -g_m(R_D \parallel r_d) \tag{11.7}$$

The frequencies at which the gain function is down by 3 dB from the center frequency (also called *half-power* frequencies) are determined from (11.5) and (11.7); that is,

$$\left(-\frac{g_m}{G}\right)^2\frac{1}{2} = \left(-\frac{g_m}{C_T}\right)^2\frac{\omega^2}{[(G/C_T)\omega]^2 + (\omega_0^2 - \omega^2)^2} \tag{11.8}$$

Two solutions of (11.8) are

$$\omega_2 = \frac{G}{2C_T} + \sqrt{\omega_0^2 + \frac{G^2}{4C_T^2}} \tag{11.9a}$$

$$\omega_1 = -\frac{G}{2C_T} + \sqrt{\omega_0^2 + \frac{G^2}{4C_T^2}} \tag{11.9b}$$

Note that at ω_0 the phase shift of A_v is zero, while at $\omega_{2,1}$ the relative phase shifts are $\pm 45°$, respectively. The -3-dB bandwidth (BW) of the circuit is defined as the difference between the -3-dB band-edge frequencies, namely,

$$\text{BW} = \Delta\omega_{3\text{dB}} = \omega_2 - \omega_1 = \frac{G}{C_T} \tag{11.10}$$

Also, from (11.9) and (11.10), we can solve for ω_0 in terms of the band-edge frequencies. The result is

$$\omega_0 = \sqrt{\omega_1 \omega_2} \tag{11.11}$$

i.e., ω_0 is the *geometric* mean of the upper and lower 3-dB frequencies. In the following we shall omit Δ for convenience and use ω_{3dB} only. However, it should be noted that it is the difference between two 3-dB frequencies. Most often we deal with *narrowband* tuned amplifiers, where $\omega_{3dB} \leq 0.1\omega_0$. For such cases, from (11.9) and (11.10), we obtain

$$\omega_0 \simeq \frac{\omega_1 + \omega_2}{2} \qquad \omega_{3dB} \ll \omega_0 \tag{11.12}$$

The pole-zero plot of (11.4) for a narrowband tuned amplifier is shown in Fig. 11.5.

A useful term in tuned amplifiers is the so-called *Q quality factor* of the circuit. The Q of a single-tuned circuit is defined as

$$Q = \frac{\omega_0}{\text{BW}} = \frac{\omega_0}{\omega_{3dB}} \tag{11.13}$$

From (11.6), (11.10), and (11.13) we have

$$Q = \omega_0 R C_T = \frac{R}{L\omega_0} \tag{11.14}$$

Note that a narrowband case ($\omega_{3dB} \ll \omega_0$) means a high Q (i.e., $Q \gg 1$).

Substitution of (11.6) and (11.10) into (11.4) yields a useful form for the single-tuned amplifier:

$$A(s) = -\frac{g_m}{C_T} \frac{s}{s^2 + \omega_{3dB}s + \omega_0^2} \tag{11.15a}$$

In a normalized form, using $s_n = s/\omega_0$, we have

$$A_v(s_n) = -\frac{g_m R}{Q} \frac{s_n}{s_n^2 + (1/Q)s_n + 1} \tag{11.15b}$$

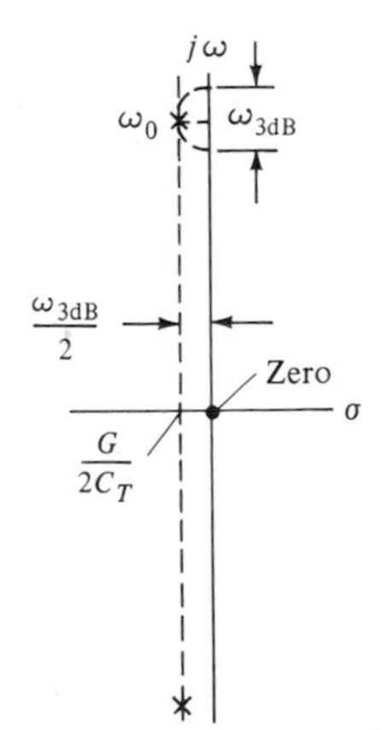

FIGURE 11.5 Pole-zero plot for a narrowband single-tuned stage ($\omega_0 \gg \omega_{3dB}$).

Note that (11.4*a*) can also be expressed as

$$A_v(j\omega) = \frac{-g_m R}{1 + j\omega RC_T - j(R/\omega L)}$$

$$= -\frac{g_m R}{1 + j\omega_0 RC_T(\omega/\omega_0 - \omega_0/\omega)} \tag{11.16a}$$

Substitution of (11.7) and (11.14) into (11.16*a*) yields

$$\frac{A_v(j\omega)}{A_v(j\omega_0)} = \frac{1}{1 + jQ(\omega/\omega_0 - \omega_0/\omega)} \tag{11.16b}$$

A plot of the magnitude and phase of (11.16*b*) for two values of Q is shown in Fig. 11.6.

We may express the response in the passband of the narrowband amplifier by letting $\omega/\omega_0 = 1 + \Delta$ with $\Delta \ll 1$. Hence (11.16*b*) reduces to

$$\frac{A_v(j\omega)}{A_v(j\omega_0)} \simeq \frac{1}{1 + j2Q\Delta} \tag{11.16c}$$

The response characteristic of (11.16*c*) is essentially the same as in Fig. 11.6. The curve of Fig. 11.6 is commonly called the universal *resonance curve*.

The gain-bandwidth product (GBP) of the single-tuned stage from (11.7) and (11.10) is given by

$$\text{GBP} = |A_v(j\omega_0)|\omega_{3\text{dB}} = \frac{g_m}{C_T} \tag{11.17}$$

We shall illustrate the above expressions by the following design example.

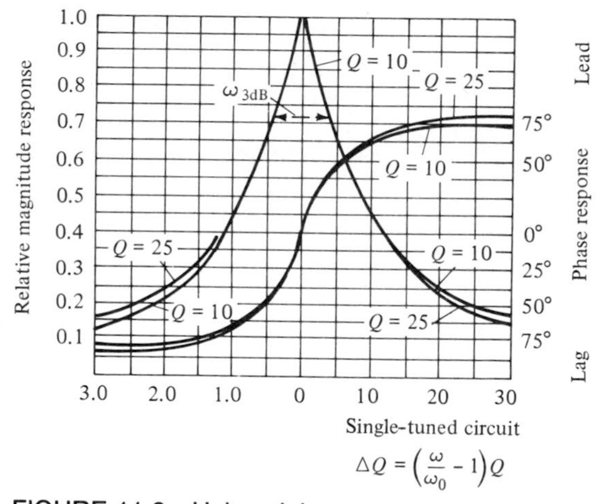

FIGURE 11.6 Universal resonance curve for a single-tuned stage.

EXAMPLE 11.1

Design the single-stage FET tuned amplifier of Fig. 11.4 for the following specifications:

$$\omega_0 = 2\pi(10^6\text{ Hz}) \qquad \omega_{3\text{dB}} = 2\pi(10^4\text{ Hz})$$

and a midband voltage gain $A_v(j\omega_0) = -10$. The FET parameters are as follows: $g_m = 4 \times 10^{-3}$ mho, $r_d = 20\text{ k}\Omega$, $C_{gs} = 40\text{ pF}$, $C_{gd} = C_{ds} = 5\text{ pF}$.

From the gain requirements, we have

$$A_v(j\omega_0) = -g_m(R_D \| r_d) = -10 \qquad R_D = 2.9\text{ k}\Omega$$

From the bandwidth specification

$$\omega_{3\text{dB}} = \frac{G}{C_T} = \frac{G_D + g_d}{C + C_{ds} + C_{gd}} = 2\pi(10^4\text{ Hz})$$

Substituting the values in the above and solving for C yields

$$C = \frac{G_D + g_d}{2\pi(10^4)} - (C_{ds} + C_{gd}) \simeq 0.0064\ \mu\text{F}$$

From the resonant frequency requirement we have $1/LC_T = \omega_0$; hence

$$L = \frac{1}{C_T\omega_0^2} = \frac{1}{6.4 \times 10^{-9}[2\pi(10^6)]^2} = 3.97\ \mu\text{H}$$

B. BJT Circuit

Consider the resistive-loaded single-tuned BJT stage shown, exclusive of biasing circuitry, in Fig. 11.7*a*. For simplicity, we assume that $r_b = 0$ and that the coil loss and the parasitic capacitance of the inductor are included in R_l and C_a, respectively (this will be discussed later).

Since the load is resistive, we can use the simplified unilateral model, which incorporates the Miller effect, as shown in Fig. 11.7*b*. Note that from (7.70) $C_\pi + C_c = 1/r_e\omega_T$; thus $C_T = C_a + C_i = C_a + (1 + R_LC_c\omega_T)/r_e\omega_T$, and $R = R_1 \| R_i$. The circuit is redrawn where all the capacitances and conductances are combined.

The current gain function of the amplifier is

$$A = \frac{I_o}{I_s} = \frac{I_o}{V'}\frac{V'}{I_s}$$

$$= \frac{-g_m}{G + sC_T + 1/sL} = \frac{-g_mR}{1 + sRC_T + R/sL} \tag{11.18a}$$

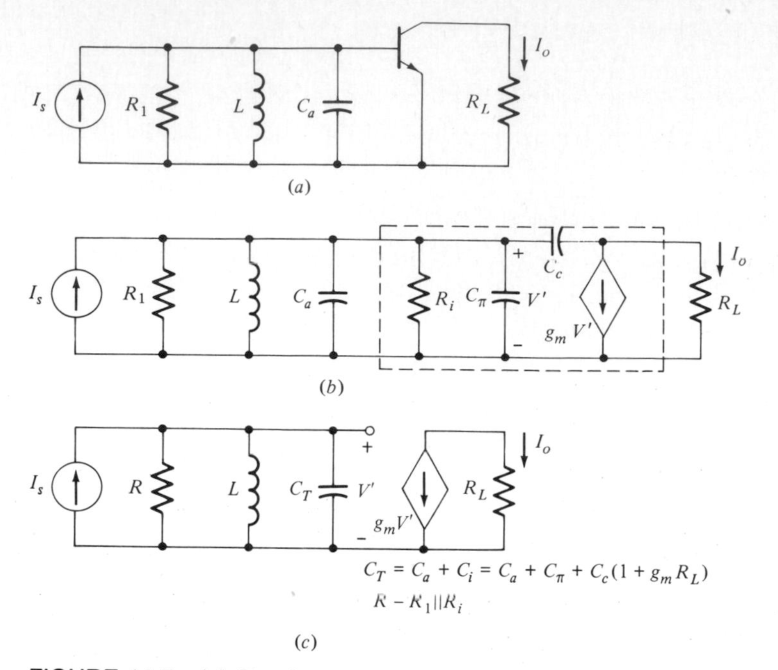

FIGURE 11.7 (*a*) Simple transistor single-tuned stage of example. (*b*) Equivalent circuit of (*a*). (*c*) Simplified equivalent circuit.

Equation (11.18*a*) is of identical form to (11.4); hence the midband current gain and the 3-dB bandwidth are given by

$$A_i(j\omega_0) = -g_m R \qquad \omega_{3\text{dB}} = \frac{1}{RC_T} \tag{11.18b}$$

To illustrate the above numerically, consider the following example.

EXAMPLE 11.2

Determine the center frequency f_0, quality factor Q, and the current gain at the center frequency $|A_i(j\omega_0)|$. The transistor parameters are given as follows:

$$f_T = 200 \text{ MHz} \qquad \beta_0 = 100 \qquad r_e = 20\ \Omega \qquad C_c = 5 \text{ pF}$$

The other circuit element values are $R_L = 200\ \Omega$, $R_1 = 2\ \text{k}\Omega$, $L = 10\ \mu\text{H}$, and $C_a = 2000$ pF. From Fig. 11.7*c*, we find

$$R = R_1 \| R_i = (2 \| 2)\ \text{k}\Omega = 1\ \text{k}\Omega$$

$$C_T = C_a + C_i = (2000 + 90)\ \text{pF} = 2.09 \times 10^{-9}\ \text{F}$$

The resonant frequency, from (11.6), is

$$f_0 = \frac{\omega_0}{2\pi} = \frac{1}{2\pi}\left[\sqrt{10^{-5}(2.09 \times 10^{-9})}\right]^{-1} = 1.1 \text{ MHz}$$

The midband gain, the 3-dB bandwidth, and the Q of the circuit are $|A_i(j\omega_0)| = -50$, $f_{3dB} = 76$ kHz, and $Q = 14.5$.

The *design* of a single-tuned BJT stage for a specified center frequency, f_0, and bandwidth, f_{3dB}, follows directly from (11.6) and (11.10), respectively, as in Example 11.1.

C. Op-Amp Circuit

The use of op amp in conjunction with *RC* elements to obtain bandpass filtering was discussed in Sec. 9.4 and will be considered again in Sec. 11.9. However, the Q of a single op-amp active *RC* filter is limited ($Q \leq 30$). For a Q larger than 30 either two or three op amps should be used for integrated circuits, or coupled coils are used as discussed later in this chapter. Tuned circuits can also be used with op amps, and the design equations are then similar to those given earlier in this section.

Many integrated-circuit tuned amplifiers are available that use the emitter-coupled circuit with *LC* resonants circuits connected externally. Such a configuration is discussed in Sec. 11.7.

D. Inductor Quality Factor (Q_L)

Often in the design of tuned amplifiers, the value of the inductor may not be practical. In other words, the inductor may not have a high Q, and the use of transformers may be necessary. For example, consider the inclusion of the inductor coil losses, as shown by a series resistance, r_s, in Fig. 11.8*a*. For convenience, the circuit may be transformed into a parallel circuit, as shown in Fig. 11.8*b*, where the equivalent parallel resistance r_p is readily determined from

$$Y(j\omega) = \frac{1}{r_s + j\omega L} = \frac{r_s - j\omega L}{r_s^2 + \omega^2 L^2} \tag{11.19a}$$

For a high-Q coil

$$Q = \frac{\omega L}{r_s} \gg 1$$

and (11.19*a*) can be written approximately as

$$Y(j\omega) \simeq \frac{r_s}{\omega^2 L^2} - j\frac{1}{\omega L} = r_p^{-1} + \frac{1}{j\omega L} \tag{11.19b}$$

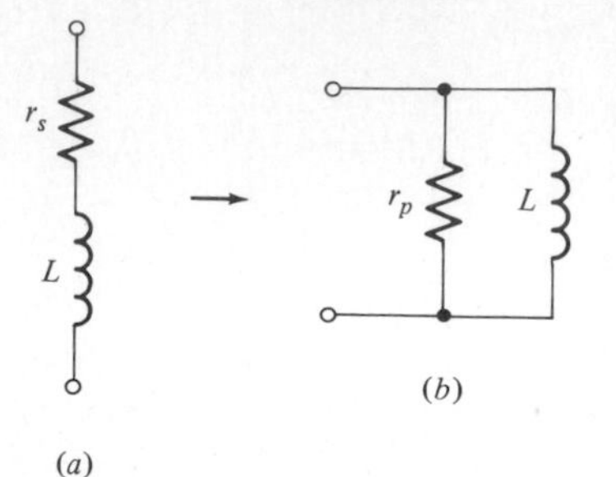

FIGURE 11.8 (*a*) Coil including the loss. (*b*) Equivalent model for a high-Q coil.

where

$$r_p = \frac{\omega^2 L^2}{r_s} = \omega L Q_L \simeq \omega_0 L Q_L \qquad \textbf{(11.19c)}$$

Thus, if the coil Q is not high enough, the value of r_p may be of the same order of magnitude as R in Fig. 11.4*b* or Fig. 11.7*c*, and the circuit Q will be lowered (i.e., ω_{3dB} will be increased). Thus, the effect of the coil loss, or coil Q_L cannot, in general, be ignored.

EXERCISE 11.1

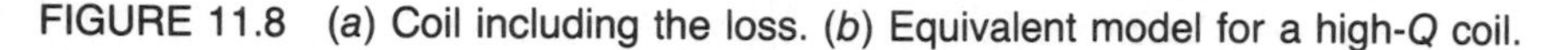

For the circuit shown in Fig. 11.4*a*, the inductor has a value of 10 μH with $Q_L = 200$ at $\omega_0 = 10^7$ rad/s. Determine the values of R_D and C for a bandwidth of $\omega_{3dB} = 2 \times 10^5$ rad/s. What is the voltage gain at the resonant frequency of $\omega_0 = 10^7$ rad/s? The FET parameters are $g_m = 5 \times 10^{-3}$ mho, $r_d = 20$ kΩ, $C_{gs} = 10$ pF, and $C_{ds} = C_{gd} = 1$ pF.

Ans $R_D = 10$ kΩ, $C = 1$ nF, $A_v = -25$.

11.3 IMPEDANCE TRANSFORMATION AND TRANSFORMER COUPLING

The circuit-element values (L and C) can be made practical by using suitable impedance-transforming networks. There are many types of impedance-transforming networks, the choice of which depends on the frequency range and the impedance levels. One such network which is particularly attractive in tuned amplifiers is the autotransformer, shown in Fig. 11.9*a*. The coils are wound on a ferrite core, and a coefficient of coupling very close to unity can be achieved. The voltage-current relationship of the ideal autotransformer is given by

$$-\frac{I_2}{I_1} = \frac{V_1}{V_2} = n > 1 \qquad \textbf{(11.20)}$$

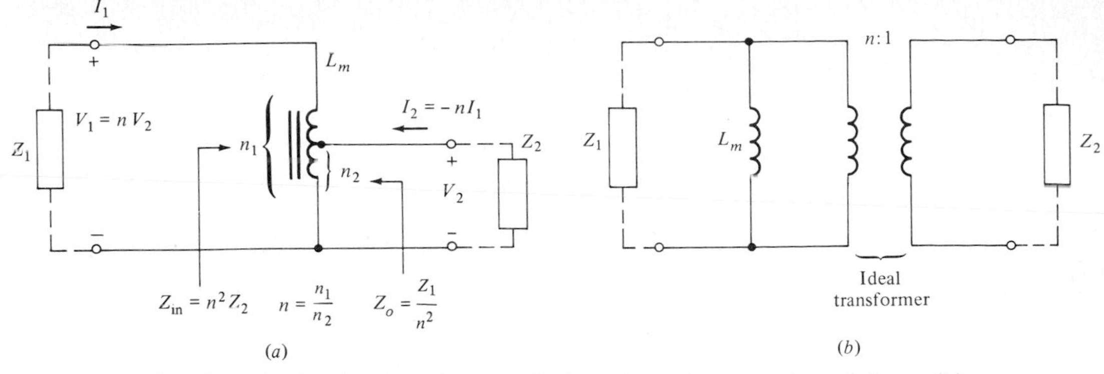

FIGURE 11.9 (*a*) Autotransformer with impedance transforming relations. (*b*) Equivalent model of (*a*) (losses are ignored).

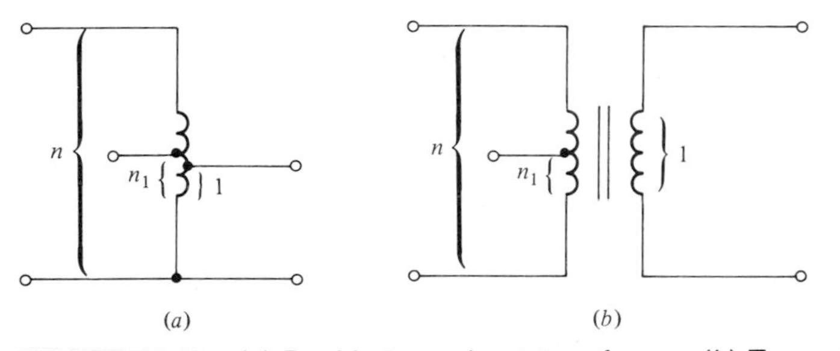

FIGURE 11.10 (*a*) Double-tapped autotransformer. (*b*) Tapped transformer.

The model for the autotransformer, neglecting the losses, is shown in Fig. 11.9*b*, where L_m is the total magnetizing inductance characterized on the primary side and n is the turns ratio. Note also that when the transformer is used, the voltage gain is reduced by a factor of $1/n$, and the current gain is increased by a factor of n in a step-down transformer. For this reason, one usually considers power gain in tuned amplifiers, which would be unaffected in a lossless transformer. Often a double-tapped inductance or a combination of tapped primary interstage transformer is used in tuned-amplifier design. These are shown in Fig. 11.10*a* and *b*, respectively. Their use is illustrated in the design examples later in this chapter.

EXERCISE 11.2

For the autotransformer circuit shown in Fig. E11.2, if $L_m = 100\ \mu\text{H}$, $n = 10$ and Z_1 is a shunt *RC* circuit (with $C = 1$ nF, $R = 100$ kΩ), determine the element values of the total impedance at the output port.

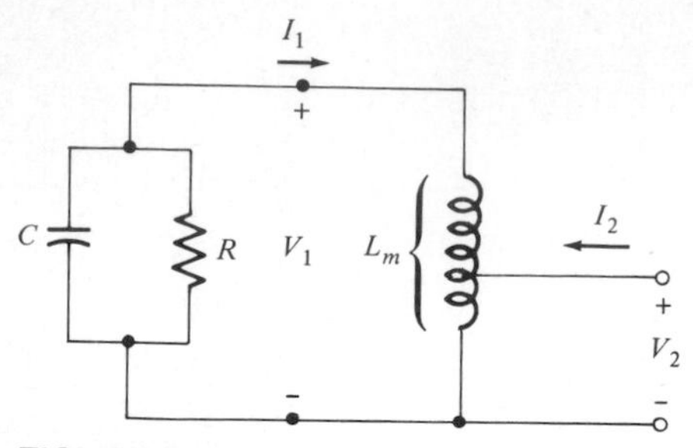

FIGURE E11.2

Ans $L_2 = 1\ \mu\text{H},\ R_2 = 1\ \text{k}\Omega,\ C_2 = 0.1\ \mu\text{F}$

11.4 TRANSISTOR SINGLE-TUNED AMPLIFIER WITH TUNED CIRCUIT AT THE OUTPUT

A single stage is seldom used in a tuned amplifier. For the sake of clarity of discussion of some basic ideas that will be encountered in multistage design, however, we consider briefly a single-stage single-tuned bipolar transistor where the tuned circuit is at the output. An interior stage of a FET with a tuned-circuit load may also be handled in a manner similar to that of the following treatment. The reason that the design of such stages can be different is because of the inductive nature of the load just under the resonant frequency. For the inductive load, the Miller approximation can lead to erroneous results. In other words, if the tuned circuit is at the output, as shown in Fig. 11.11*a*, we cannot use the unilateral model, since the output impedance of the transistor is *very high* and can be inductive or capacitive near the resonant frequency. In fact, for such a case, it is quite possible—indeed likely—for the circuit to oscillate. This fact makes the problem of a tuned-amplifier design quite different when compared to the low-pass case. The instability problem of tuned stages is discussed in Sec. 11.9.

In the majority of applications, we deal with the narrowband tuned amplifiers, i.e., $\omega_{3\text{dB}} \leq 0.1\omega_0$. This condition yields considerable simplification in the analysis and design. At the center frequency, in a high-frequency tuned amplifier, we can represent the transistor by its y parameters. This characterization at ω_0 can be very accurate and can include the parasitic and other effects. The small-signal model of the tuned amplifier is shown in Fig. 11.11*b*. In the neighborhood of the center frequency, i.e., for $\omega - \omega_0 \leq 0.05\omega_0$, the two-port parameters are approximately the same as those measured at ω_0 and can be expressed approximately as

$$\begin{aligned} y_{11} &\simeq g_{11} + j\omega b_{11} \qquad & y_{21} &\simeq g_{21} + j\omega b_{21} \\ y_{12} &\simeq g_{12} + j\omega b_{12} \qquad & y_{22} &\simeq g_{22} + j\omega b_{22} \end{aligned} \tag{11.21}$$

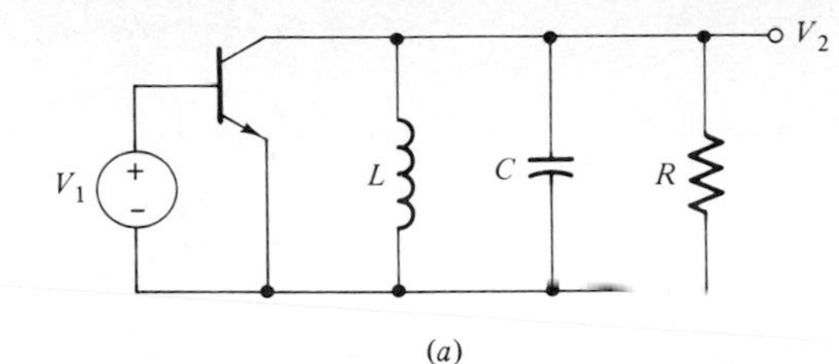

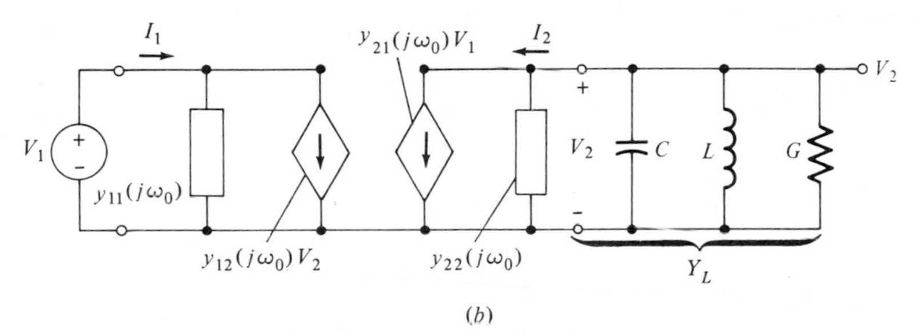

FIGURE 11.11 (a) Transistor stage with tuned circuit at the output. (b) Equivalent circuit of (a) at the center frequency.

The voltage gain of a linear two-port network with load admittance $Y_L = G + j\omega C + 1/j\omega L$, from Table C.2, is given by

$$A_v(j\omega) = \frac{V_2}{V_1} = \frac{-y_{21}}{y_{22} + Y_L} = \frac{-y_{21}}{y_{22} + G + j\omega C + 1/j\omega L}$$

$$= \frac{-y_{21}}{g_{22} + G + j(\omega C + b_{22}) - j(1/\omega L)} \tag{11.22}$$

Note that in bandwidth calculations, the effect of the real and the imaginary parts of y_{22} (especially the latter) is very important since we deal with the difference of two nearly equal quantities near resonance. However, y_{21} is not involved in the resonant part and for a narrowband, y_{21} is fairly constant over the bandwidth. Hence we can write (11.22) as

$$A_v(j\omega) = -\frac{y_{21}(j\omega_0)}{g_T + j\omega C_T + 1/j\omega L} = -\frac{g_m}{g_T + j\omega C_T + 1/j\omega L} \tag{11.23}$$

where
$$g_T = g_{22} + G$$
$$C_T = C + \frac{b_{22}}{\omega_0}$$
$$y_{21}(j\omega_0) = g_m$$

A comparison of (11.23) and (11.4*a*) shows immediately that the expressions are identical in the passband, and thus the same design equations can be used.

EXAMPLE 11.3

Consider the circuit shown in Fig. 11.11*b*. The transistor *y* parameters measured at the desired center frequency, $\omega_0 = 2\pi(10 \text{ MHz})$, are given as follows (all in mhos):

$$y_{11} = y_{ie} = (2.0 + j0.5) \times 10^{-3} \qquad y_{12} = y_{re} = -(1.0 + j5.0) \times 10^{-5}$$

$$y_{21} = y_{fe} = (20 - j5.0) \times 10^{-3} \qquad y_{22} = y_{oe} = (2.0 + j4.0) \times 10^{-5}$$

The design requires a bandwidth of $\omega_{3\text{dB}} = 2\pi(100 \text{ kHz})$ (i.e., ± 50 kHz centered at $f_0 = 10$ MHz). The stage voltage gain at ω_0 is to be approximately -50.

From the gain specifications at ω_0, we have

$$|A_v(j\omega_0)| = -\frac{|y_{21}|}{g_T} = -50$$

or

$$\frac{21 \times 10^{-3}}{2 \times 10^{-5} + G} = 50 \qquad \text{or} \qquad R = \frac{1}{G} = 2.56 \text{ k}\Omega$$

From the bandwidth specification and (11.10), we have

$$\frac{g_T}{C_T} = \frac{g_{22} + G}{b_{22}/\omega_0 + C} = 2\pi(10^5)$$

or

$$\frac{2 \times 10^{-5} + 3.9 \times 10^{-4}}{0.64 \times 10^{-12} + C} = 2\pi(10^5) \qquad \text{or} \qquad C = 652 \text{ pF}$$

The value of L is determined from the specified ω_0 and (11.6)

$$\frac{1}{LC_T} = \omega_0^2 \quad \text{namely} \quad \frac{1}{L(6.5 \times 10^{-10})} = [2\pi(10^7)]^2 \quad \text{or} \quad L = 0.385 \ \mu\text{H}$$

The element values are acceptable as an initial design. In practice, inductors with a low value of inductance have a small equivalent shunt resistor ($r_p \simeq \omega_0 L Q_L$) that will have the effect of increasing the bandwidth, thus reducing the circuit Q. A more desirable value for the inductor is obtained if an autotransformer with $n = 3$ is used. In this case

$$L_m = n^2 L = 9(0.385 \ \mu\text{H}) = 3.47 \ \mu\text{H}$$

$$C_a = \frac{C}{n^2} = \frac{652}{9} = 72.5 \text{ pF}$$

The designed circuit, exclusive of biasing circuitry, is shown in Fig. 11.12. In

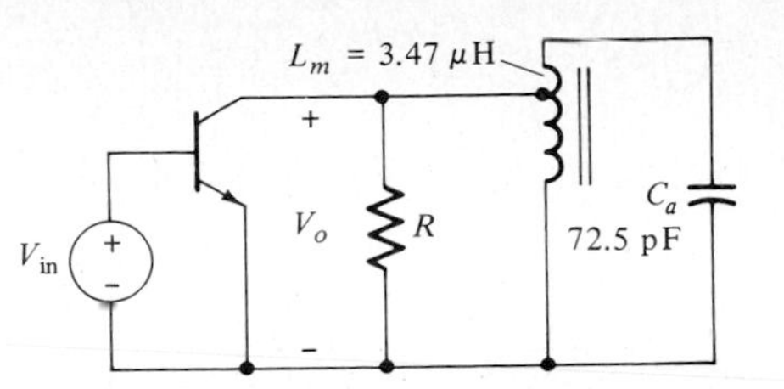

FIGURE 11.12 Tuned-amplifier circuit of the design example.

passing, we remark again that the above design method can be carried over identically to any other *nonunilateral* active device for narrowband case.

11.5 THE NARROWBAND APPROXIMATION

Narrowband tuned amplifiers play a very important role in telecommunication systems. They are used widely in the transmitter and receiver systems using modulated signals. This is fortunate, since considerable simplification can be achieved in the analysis and design of narrowband tuned amplifiers. Because of the tremendous usefulness of the narrowband approximation, we consider a brief discussion of this topic in this section.

The admittance function of a single-tuned bandpass amplifier (Fig. 11.4) is given by

$$Y(s) = G + sC + \frac{1}{sL} = G + C\left(s + \frac{\omega_0^2}{s}\right) \tag{11.24}$$

where $\omega_0^2 = 1/LC$. Now consider the admittance of the single-pole low-pass case shown in Fig. 11.13. For the low-pass case, we use the p-complex frequency plane ($p = x + jy$) so as not to confuse the low-pass and bandpass complex frequency variable planes. The admittance of Fig. 11.13 is

$$Y(p) = G + Cp \tag{11.25}$$

From (11.24) and (11.25), we introduce a *low-pass-to-bandpass transformation**:

$$p = s + \frac{\omega_0^2}{s} \tag{11.26}$$

or, solving for s in terms of p from (11.26),

$$s = \frac{p}{2} \pm \sqrt{\left(\frac{p}{2}\right)^2 - \omega_0^2} \tag{11.27}$$

---

*Other transformations that are useful in other types of filters are the following: low pass to high pass, $p = 1/s$, and low pass to band elimination, $p = s/(s^2 + \omega_0^2)$.

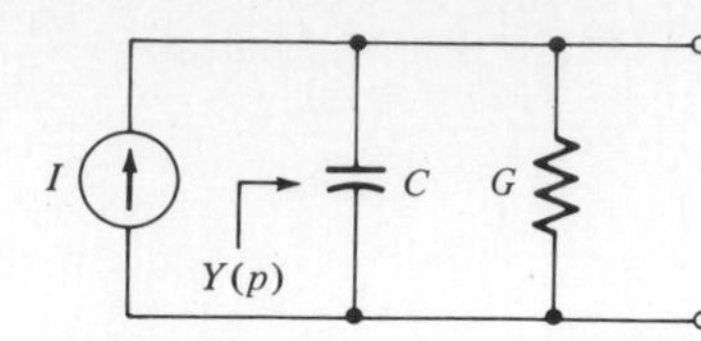

FIGURE 11.13 A low-pass single-pole admittance.

For a narrowband case, $|p/2| \ll \omega_0$, and (11.27) reduces to

$$s \simeq \frac{p}{2} \pm j\omega_0 \tag{11.28}$$

In other words, the bandpass pole-zero location is the pole-zero location of the low-pass case scaled by a factor of $\frac{1}{2}$ and translated around new origins at $\pm j\omega_0$.

For example, a single-pole low-pass pole location, under the narrowband frequency transformation, has the bandpass pole locations shown in Fig. 11.14, which are those of a single-tuned circuit. In fact, for the frequency response calculations near the center frequency, in a narrowband case, the zeros and the conjugate poles can also be ignored, as seen in the following.

Consider the pole-zero locations of a single-tuned, narrowband, bandpass amplifier shown in Fig. 11.15. The gain function corresponding to this pole-zero location is given by

$$\frac{V_o}{V_i} = -\frac{g_m}{C}\frac{s}{(s+s_1)(s+s_1^*)} \tag{11.29}$$

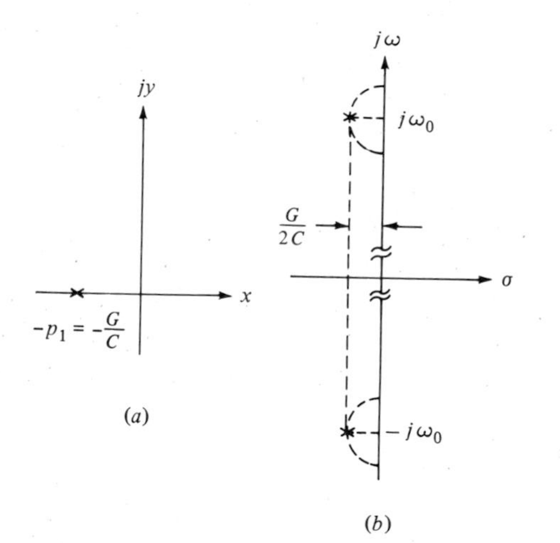

FIGURE 11.14 (*a*) Low-pass pole location in the *p* plane. (*b*) Bandpass pole location in the *s* plane.

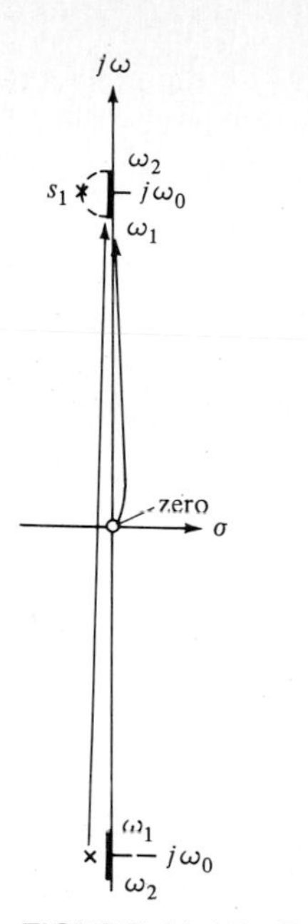

FIGURE 11.15 Pole-zero locations of a narrowband single-tuned stage.

The magnitude response is

$$\left|\frac{V_o}{V_i}(j\omega)\right| = \frac{g_m}{C}\frac{|j\omega|}{|j\omega + s_1||j\omega + s_1^*|} \tag{11.30}$$

For the frequencies near $j\omega_0$ and in the range $\omega_1 \leq \omega \leq \omega_2$, actually in the range $|\omega| > \omega_1$, the following magnitude expression (see Fig. 11.15) is approximately constant

$$\frac{|j\omega|}{|j\omega + s_1^*|} \simeq \frac{1}{2} \qquad \omega \geq \omega_1 \tag{11.31}$$

Note that for the narrowband approximation, i.e., $\omega_2 - \omega_1 \leq 0.1\omega_0$, the error in (11.31) is less than 5 percent. Therefore, from (11.31) and (11.30) we can write

$$\left|\frac{V_o}{V_i}(j\omega)\right| \simeq \frac{g_m}{2C_T}\frac{1}{|j\omega + s_1|} \qquad |\omega - \omega_0| \leq 0.05\omega_0$$

or

$$\frac{V_o}{V_i} \simeq -\frac{g_m}{2C_T}\frac{1}{s + s_1} \tag{11.32}$$

Therefore, in frequency response calculations, in the bandpass region for narrowband tuned amplifier, the zeros at the origin (or real axis) and the conjugate poles and the poles on the real axis can be ignored.

In the design of narrowband tuned amplifiers we focus on the pole-zeros in the vicinity of $j\omega_0$. The design methods of low-pass case such as maximally flat magnitude can then be directly translated into the bandpass case *around* $s = j\omega_0$, as discussed in the next section.

EXERCISE 11.3

The pole-zero plot of a tuned-amplifier design is shown in Fig. E11.3. Determine the 3-dB bandwidth and Q of the design (you may use the results in Fig. 8.10).

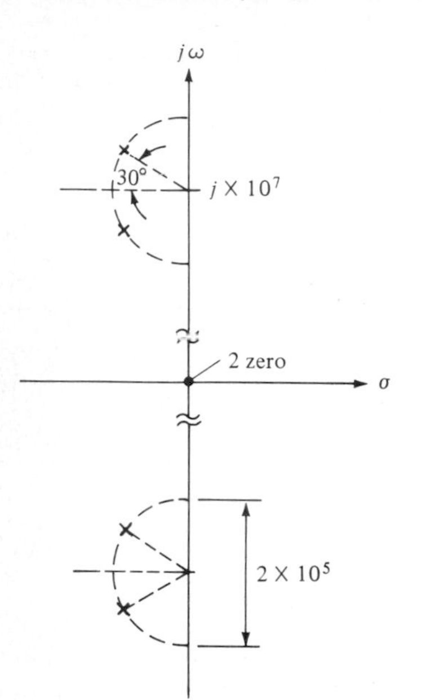

FIGURE E11.3

Ans $\omega_{3dB} = 1.56 \times 10^5$ rad/s, $Q = 64.1$

11.6 SYNCHRONOUS TUNING AND STAGGER TUNING

Synchronous tuning and stagger tuning were discussed in Sec. 8.11 in conjunction with low-pass amplifiers. In a multistage tuned amplifier, if the individual stages are unilateral or designed to have negligible interaction, and if each stage has the same center frequency and bandwidth, then the system is said to be *synchronously tuned*. For such a case, the pole-zero pattern of two synchronously tuned single-tuned stages is shown in Fig. 11.16*a*. The overall bandwidth of the two-stage design is determined (see Table 8.1) to be

$$(\omega_{3dB})_{\text{overall}} = S_2(\omega_{3dB})_i = \sqrt{2^{1/2} - 1}\,(\omega_{3dB})_i = 0.64(\omega_{3dB})_i \tag{11.33}$$

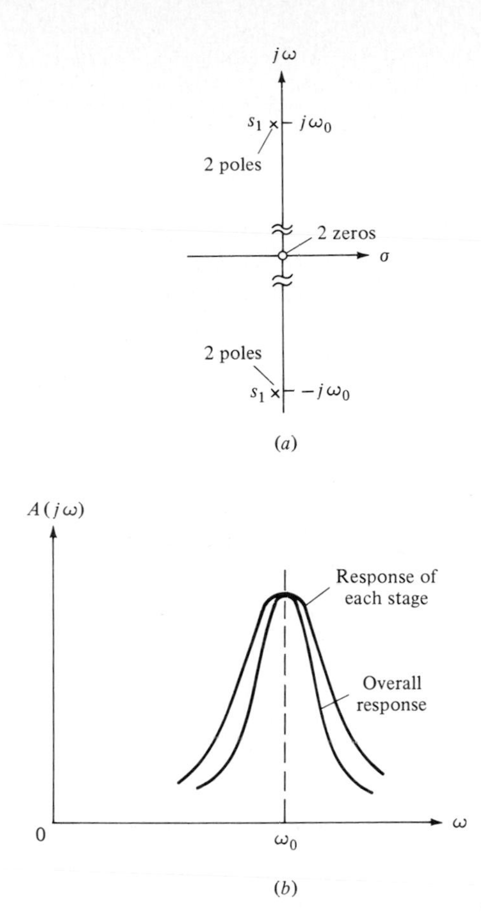

FIGURE 11.16 (*a*) Pole-zero locations of two identical single-tuned, synchronously tuned design. (*b*) Magnitude response of the individual single-tuned stage and that of a two-stage synchronously tuned design.

where S_2 is the shrinkage factor for two single-tuned stages and $(\omega_{3\text{dB}})_i$ is the bandwidth of the individual stage. The overall response and the response of individual stages for two single-tuned stages are shown in Fig. 11.16*b*. Synchronously tuned design can also be used for double-tuned stages. This will be discussed later.

The shrinkage factor can be avoided by the use of *stagger tuning*. In stagger tuning, the individual noninteracting stages do not have the same pole-zero locations, but rather they are arranged in such a manner that the overall response has the desired response shape. For example, for a narrowband case, a cascade of two noninteracting single-tuned stages can be designed to have maximally flat magnitude transmission characteristics. The pole-zero pattern for the bandpass is readily obtained from the low-pass maximally flat pole locations and the narrowband approximation. The pole-zero pattern is shown in Fig. 11.17. Thus, the individual single-tuned circuits can be designed to realize a zero and a pair of the complex poles from the pattern shown in Fig. 11.17*a*. The individual tuned-circuit response and the overall response are shown

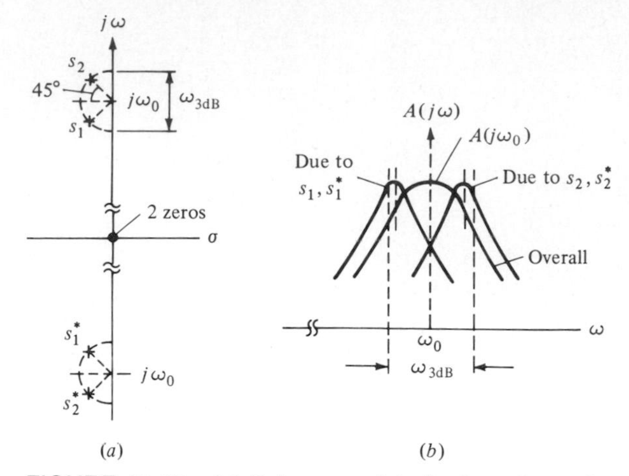

FIGURE 11.17 (*a*) Pole-zero plot of a two-stage stagger-tuned maximally flat magnitude design using two single-tuned stages. (*b*) Magnitude response of the individual single-tuned stages and of the overall stagger-tuned design.

in Fig. 11.17*b*. When many noninteracting stages are used, the poles for a stagger-tuned design may be designed to be on a circle corresponding to the Butterworth angles (see Sec. 8.10), in order to get a flat magnitude response and maximum bandwidth. Whenever a better selectivity is required, stagger tuning is often employed.

Illustrative design examples for synchronously tuned and stagger-tuned amplifiers using a cascode transistor amplifier are discussed in the next section.

11.7 THE CASCODE TUNED AMPLIFIER

The cascode circuit was discussed in Secs. 5.3 and 8.2. The cascode circuit is shown in Fig. 11.18*a*. This circuit provides excellent isolation between the input and the output circuits. In the following analysis we shall assume, for simplicity, that the transistors are identical and that $f_0 \ll f_T$, so that the internal capacitances of the CB stage can be neglected. We also assume that $r_b = 0$. The approximate equivalent circuit is shown in Fig. 11.18*b*. Note that since $h_{ib} = r_e + r_b(1 - \alpha_0) \simeq r_e$ and $h_{ib} \ll R$, the contribution of this usually very small resistance to the Miller effect is negligible. Since

$$C_i = \frac{D}{r_e \omega_T} \simeq \frac{1 + h_{ib} C_c \omega_T}{r_e \omega_T} \simeq \frac{1 + r_e C_c \omega_T}{r_e \omega_T} = \frac{1}{r_e \omega_T} + C_c \simeq \frac{1}{r_e \omega_T}$$

$$\text{for} \quad C_c \ll \frac{1}{r_e \omega_T}$$

the effect of the Miller capacitance can be ignored altogether without any significant error. The two noninteracting single-tuned circuits of Fig. 11.18*b* can be designed as

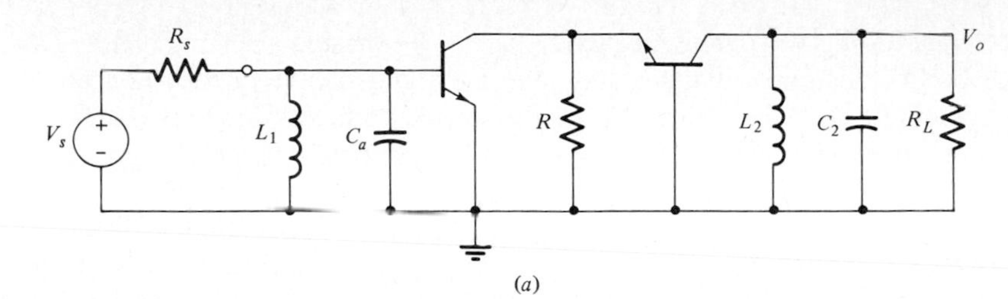

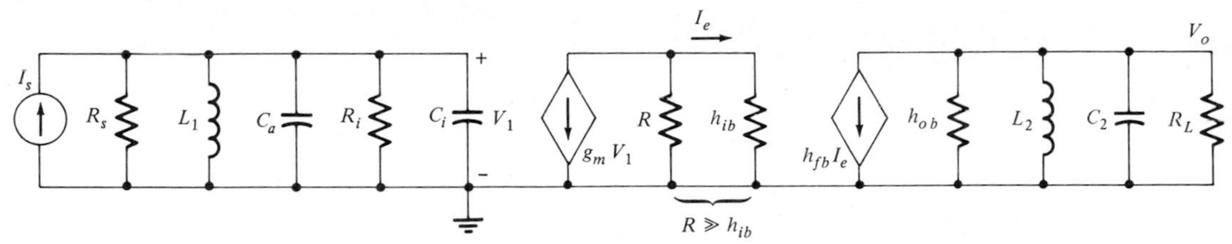

FIGURE 11.18 (*a*) Cascode amplifier with two single-tuned stages. (*b*) Equivalent circuit of (*a*).

a synchronously tuned or a stagger-tuned amplifier. We shall illustrate both of these cases in the following:

From the circuit of Fig. 11.18*b*, the gain function is given by

$$A_v = \frac{V_o}{V_s} = \frac{V_o}{I_e}\frac{I_e}{V_1}\frac{V_1}{V_s} \tag{11.34}$$

Substitution of the various expressions, as obtained from Fig. 11.18*b*, in (11.34) yields

$$\frac{V_o}{V_s} = \frac{-h_{fb}}{G_2 + sC_2 + 1/sL_2}\frac{-g_m R}{R + h_{ib}}\frac{G_s}{G_1 + sC_1 + 1/sL_1} \tag{11.35}$$

where $G_1 = G_s + G_i$

$G_2 = G_L + h_{ob} \simeq G_L$

$C_1 = C_a + C_i$

Equation (11.35) may be rewritten for convenience as

$$\frac{V_o}{V_s} = -\frac{g_m G_s}{C_1 C_2}\frac{s}{s^2 + (G_1/C_1)s + 1/L_1 C_1}\frac{s}{s^2 + (G_2/C_2)s + 1/L_2 C_2} \tag{11.36}$$

where the approximations $R \gg h_{ib}$ and $h_{fb} \simeq -\alpha_0 \simeq -1$ are used in the above.

We consider first the synchronously tuned design.

A. Synchronously Tuned Design

In a synchronously tuned design, all the single-tuned stages are identical and tuned to the same resonant frequency f_0. For this case, the pole-zero location is designed as in

Fig. 11.16. Thus, for a given center frequency, $f_0 = \omega_0/2\pi$, and a desired bandwidth, $f_{3dB} = \omega_{3dB}/2\pi$, we design *each* tuned stage to have the center frequency at f_0 and to have a bandwidth at $f_{3dB}/0.64$, where 0.64 is the shrinkage factor due to two identical poles. The design equations for (11.36) as a synchronously tuned design are as follows:

$$\frac{G_1}{C_1} = \frac{G_2}{C_2} = \frac{\omega_{3dB}}{0.64} \tag{11.37}$$

$$\frac{1}{L_1C_1} = \frac{1}{L_2C_2} = \omega_0^2 \tag{11.38}$$

The voltage gain at the center frequency from (11.36) is

$$[A_v(j\omega_0)]_{\text{sync}} = -\frac{g_m G_s}{G_1 G_2} \tag{11.39}$$

If the source and load terminations are also specified, transformers may be needed at the input and output circuits to match the source and load impedances. G_1 and G_2 are determined by the midband gain specification.

B. Stagger-Tuned Design

In a stagger-tuned design, each circuit resonates at a different frequency. In this case, the pole-zero configuration for a maximally flat magnitude corresponds to Fig. 11.17. We rewrite (11.36) in a factored form as

$$A_v(s) = -\frac{g_m G_s}{C_1 C_2} \frac{s}{(s + s_1^*)(s + s_1^*)} \frac{s}{(s + s_2^*)(s + s_2^*)} \tag{11.40}$$

For a maximally flat magnitude design with a specified center frequency f_0 and an overall bandwidth, f_{3dB}, we assign the pole location (see Sec. 8.10) as in Fig. 11.18. Thus

$$s_1, s_1^* = -\frac{\omega_{3dB}}{2\sqrt{2}} \pm j\left(\omega_0 - \frac{\omega_{3dB}}{2\sqrt{2}}\right) \tag{11.41a}$$

The resulting quadratic with roots at s_1, s_1^* for the narrowband case is given approximately by

$$s^2 + \frac{\omega_{3dB}}{\sqrt{2}}s + \omega_0^2\left(1 - \frac{\omega_{3dB}}{\sqrt{2}\,\omega_0}\right) \tag{11.41b}$$

Similarly,

$$s_2, s_2^* = -\frac{\omega_{3dB}}{2\sqrt{2}} \pm j\left(\omega_0 + \frac{\omega_{3dB}}{2\sqrt{2}}\right) \tag{11.42a}$$

The corresponding polynomial is

$$s^2 + \frac{\omega_{3dB}}{\sqrt{2}}s + \omega_0^2\left(1 + \frac{\omega_{3dB}}{\sqrt{2}\,\omega_0}\right) \tag{11.42b}$$

Hence, in a design, the corresponding coefficients of (11.41*b*) and (11.42*b*) are equated with those of (11.36), and the various quantities are thus determined.

The midband gain $|A_v(j\omega_0)|$ is readily determined by the use of the narrowband approximation. From (11.40), using the narrowband approximation, (11.29) and (11.32), we have

$$A_v(s) = -\frac{g_m G_s}{C_1 C_2}\frac{1}{2(s + s_1)}\frac{1}{2(s + s_2)} \tag{11.43}$$

Note, from Fig. 11.17 or from (11.41*a*) and (11.42*a*), that

$$|j\omega_0 + s_1| = |j\omega_0 + s_2| = \frac{\omega_{3dB}}{2} \tag{11.44}$$

Hence

$$|A_v(j\omega_0)|_{\text{stagger}} \simeq -\frac{g_m G_s}{C_1 C_2}\frac{1}{\omega_{3dB}}\frac{1}{\omega_{3dB}} \tag{11.45}$$

But

$$\omega_{3dB} = \sqrt{2}\frac{G_1}{C_1} = \sqrt{2}\frac{G_2}{C_2} \tag{11.46}$$

Therefore,

$$|A_v(j\omega_0)|_{\text{stagger}} = -\frac{g_m G_s}{2G_1 G_2} \tag{11.47}$$

It should be noted that, for the same bandwidth and capacitances, the gain of the stagger-tuned stage is larger by a factor of 2.44, since

$$\frac{|A(j\omega)|_{\text{sync}}}{|A(j\omega_0)|_{\text{stagger}}} = \frac{g_m G_s/(\omega_{3dB}/0.64)^2 C_1 C_2}{g_m G_s/2(\omega_{3dB}/\sqrt{2})^2 C_1 C_2} = \frac{0.64^2}{1} = 0.41 \tag{11.48}$$

If, in addition to the ω_0 and ω_{3dB} specifications, the source and load terminations are also specified, as in the synchronously tuned design, the use of transformers at the input and output may be required. The use of the above expressions in a design example is straightforward (problems 11.9 and 11.10) and will not be belabored any further.

Note that the cascode has negligible interaction, and thus improved *stability and alignability* result when such blocks are used in a design.

It should be pointed out that the two-transistor block using the emitter-coupled pair are also often used in tuned amplifiers in order to reduce interaction and thus

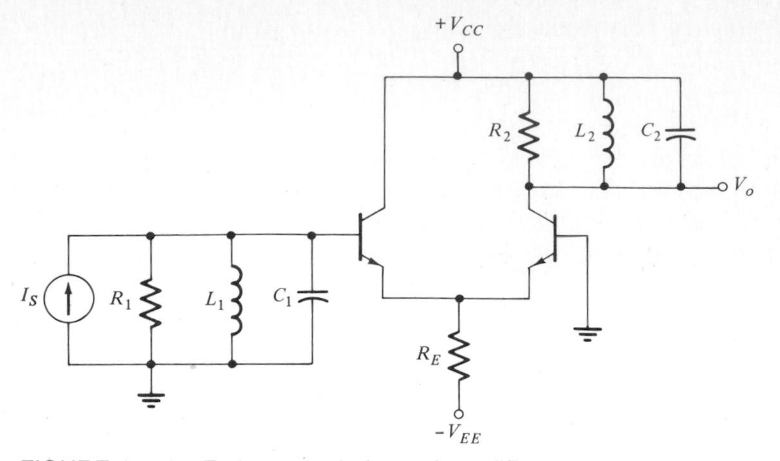

FIGURE 11.19 Emitter-coupled tuned amplifier.

achieve good stability and alignability properties. The circuit for an emitter-coupled tuned amplifier, using tuned circuits at the input and output, is shown in Fig. 11.19. The analysis of the emitter-coupled amplifier is similar to that of a cascode and is left as an exercise as follows.

EXERCISE 11.4

For the circuit shown in Fig. 11.19, the following are given: $R_1 = R_2 = R_E = 1\ \text{k}\Omega$, the transistor $\beta_0 = 100$, $r_e = 10\ \Omega$, and $r_b = 100\ \Omega$. Determine the element values of the circuit for a synchronously tuned design with the following specifications: $\omega_{3\text{dB}} = 10^4$ rad/s, $\omega_0 = 10^5$ rad/s. (Make use of the low-frequency input and output impedances of the CC and CB stages.)

Ans $L_1 = 0.81$ mH, $L_2 = 1.56$ mH, $C_1 = 0.12\ \mu$F, $C_2 = 0.064\ \mu$F

11.8 OSCILLATION POSSIBILITIES IN TUNED AMPLIFIERS

In most tuned-amplifier designs, we wish to maximize the power gain. In Appendix C, however, we have indicated that such a maximization can be achieved only if the device is inherently stable. Unfortunately, many BJTs and FETs are potentially unstable in their useful frequency range of operations. For example, consider the transistor parameters at high frequencies given in Example 11.3, repeated for convenience: at $f_0 = 10$ MHz, the parameters are

$$y_{11} = (2.0 + j0.5) \times 10^{-3} \text{ mho} \qquad y_{12} = -(1.0 + j5.0) \times 10^{-5} \text{ mho}$$

$$y_{21} = (20 - j5.0) \times 10^{-3} \text{ mho} \qquad y_{22} = (2.0 + j4.0) \times 10^{-5} \text{ mho}$$

Using the potential instability criterion (C.51), we have

$$\text{Re } y_{11} = 2.0 \times 10^{-3} > 0 \qquad \text{Re } y_{22} = 2 \times 10^{-5} > 0$$

$$\text{Re } y_{11} \text{ Re } y_{22} > \tfrac{1}{2}|y_{12}y_{21}|(1 + \cos\theta) \tag{11.49}$$

where
$$\theta = \tan^{-1}\frac{\text{Im } y_{12}y_{21}}{\text{Re } y_{12}y_{21}}$$

Re stands for the real part of y_{11}, y_{22}, and Im for the imaginary part.

$$\text{Re } y_{11} \text{ Re } y_{22} = (2 \times 10^{-3})(2 \times 10^{-5}) = 4 \times 10^{-8}$$

$$\tfrac{1}{2}|y_{12}y_{21}|(1 + \cos\theta) = \tfrac{1}{2}(106 \times 10^{-8})[1 + \cos(-114°)] = 30 \times 10^{-8}$$

Since the inequality in (11.49) is not satisfied, the device is potentially unstable. Thus, we cannot maximize the power gain, since instability arises under this condition. Note that for an ideal unilateral two-port device, $y_{12} = 0$ and (11.49) is always satisfied. A cascode configuration provides an excellent unilateral two-port circuit (see Sec. 5.2).

The possibility of oscillation can be further demonstrated from the transistor circuit shown in Fig. 11.20. Since the load of a tuned-transistor stage is inductive for frequencies below the center frequency, we consider the shunt *RL* load. For $\omega C_c \ll g_m, Y_L$ the input admittance of the circuit, using the Miller effect is

$$Y_{\text{in}} \simeq j\omega C_c(1 + g_m Z_L) = j\omega C_c\left[1 + \frac{g_m}{G - j(1/\omega L)}\right]$$

$$= -\frac{g_m C_c}{L(G^2 + 1/\omega^2 L^2)} + j\left(\omega C_c + \frac{g_m C_c G}{G^2 + 1/\omega^2 L^2}\right) \tag{11.50}$$

The appearance of the negative real part in (11.50) is the potential cause of oscillation. In fact, from Fig. 11.20 and Eq. (11.50), if

$$\text{Re } Y < \text{Re } Y_{\text{in}} = \frac{-g_m C_c}{L(G^2 + 1/\omega^2 L^2)} \tag{11.51}$$

the circuit will be unstable. In some cases circuit designers provide neutralization to cancel the effect of C_c. This fact makes the design with FETs and BJTs for wideband tuned amplifiers difficult and requires particular attention on the part of the designer.

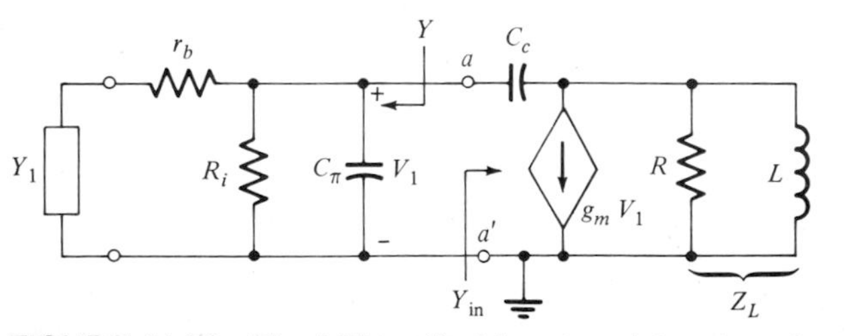

FIGURE 11.20 The Miller effect in a transistor stage for an inductive load.

The cascode (especially with a dual gate MOS, shown in Fig. 3.30) eliminates the need for neutralization which makes tunable design much easier.

Instability in tuned amplifiers can also be avoided by mismatch. By *mismatch,* we mean the condition where the source and/or load conductance is deliberately chosen to be much larger than the conductive part of the input and/or output admittance parameters of the two-port device, respectively. Mismatch achieves stability and reduces the interaction between the stages. The price paid for this is the reduction in power gain. The cascode configuration provides a mismatch, since the output impedance of a CE stage is higher by several orders of magnitude than the input impedance of a CB stage.

Commercially available IC tuned amplifiers perform a variety of functions. For example, the CA3088E can be used for amplitude modulation (AM) intermediate-frequency (IF) and radio frequency (RF) amplification up to 30 MHz. The other popular amplifier is the LM3089, which can be used for frequency modulation (FM) IF amplification and other functions.

EXERCISE 11.5

A unilateralized transistor, used for a narrowband tuned amplifier, has the following two-part parameters at the center frequency $f_0 = 59$ MHz.

$$y_{11} = (2.25 + j2.5) \times 10^{-3} \text{ mho} \qquad y_{12} = 0$$

$$y_{21} = 22.36 \times 10^{-3} \text{ mho} \qquad y_{22} = (0.09 + j0.52) \times 10^{-3} \text{ mho}$$

Is the two-port potentially unstable? If not, what is the maximum power gain at the center frequency? [*Note:* At the center frequency the circuit parameters are real and use (C.62).]

Ans No, $(G_P)_{\max} = 617$ (= 27.9 dB)

11.9 FREQUENCY SELECTION USING ACTIVE *RC* SECTIONS

We have already discussed in Sec. 9.4 the use of op amps and *RC* circuits to obtain bandpass filter sections. Note that the use of coils can altogether be avoided with active *RC* circuits. In fact, inductors can be simulated with op amp and *RC* circuits as was shown in Fig. 9.20. Active *RC* filters are practical for frequencies under 0.5 MHz. For higher frequencies the circuits discussed earlier using inductors or different schemes for integrated circuits are to be used. The circuit in Fig. 11.21*a* or *b* can be used individually to realize single-tuned circuits, and two of these sections can be used in a synchronous or stagger-tuned design. The use of these circuits are limited to $Q \leq 30$, due to sensitivity considerations as discussed in the next section. The circuit in Fig. 11.21*a* was discussed in Sec. 9.5. In Fig 11.21*b* resistors R_1 and R_2 are used to obtain a degree of freedom for the individual section gain specification. The transfer function of the circuit in Fig. 11.21*b* is given by (problem 11.15)

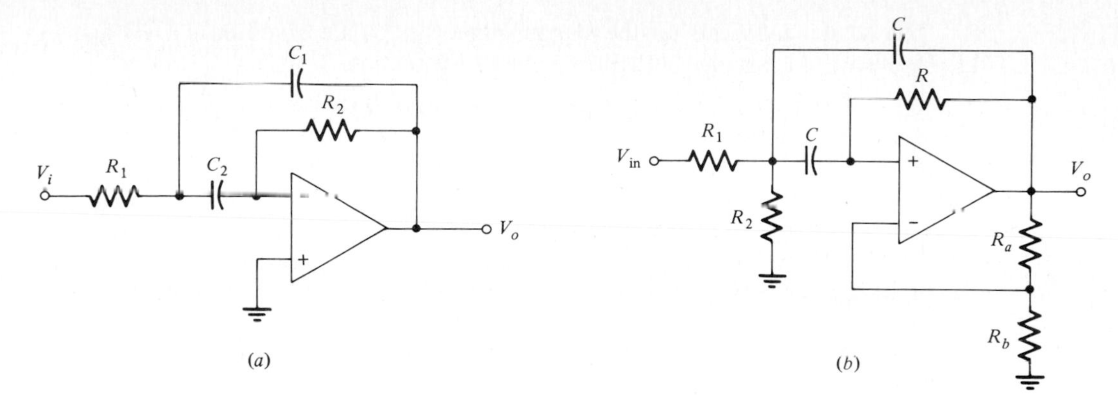

FIGURE 11.21 (*a*) A canonical active *RC* single-tuned stage. (*b*) An active *RC* single-tuned stage.

$$\frac{V_o}{V_{in}} = \frac{s[\alpha\omega_0(10.2 - 1/Q)]}{s^2 + s(\omega_0/Q) + \omega_0^2} = \frac{Ks}{s^2 + a_1 s + a_0} \tag{11.52}$$

where

$$\left.\begin{aligned} R_1 &= \frac{R}{100\alpha} \\ R_2 &= \frac{R}{100(1-\alpha)} \end{aligned}\right\} \quad \alpha = \frac{R_2}{R_1 + R_2}$$

$$R_a = \frac{50}{(1 - 5/Q)R_b} \tag{11.53}$$

$$\omega_0 = \frac{10}{RC} \qquad C_1 = C_2 = C$$

From (11.52) it is seen that the independent specifications of ω_0, Q or ω_{3dB}, and the gain at center frequency can be made.

A double-tuned frequency-selective response using two of these sections in cascade can readily be designed, as illustrated in the following example.

EXAMPLE 11.4

Consider the design of two sections of Fig. 11.21*b* cascaded to obtain a maximally flat magnitude response and the following specification:

Overall bandwidth: $\omega_{3dB} = \pi \times 10^3$ rad/s
Center frequency: $\omega_0 = 2\pi \times 10^4$ rad/s
Overall gain: $|A_v(j\omega_0)| = 100$
Each section gain $= 10$

The design equations for the stagger-tuned design are given in (11.41) and (11.42). From (11.43) and (11.69) we obtain for section 2

$$R_1 = 1.96\ \text{k}\Omega \qquad R_2 = 105\ \Omega \qquad R_a = 60.97\ \text{k}\Omega$$

$$R_b = 1\ \text{k}\Omega \qquad R = 10\ \text{k}\Omega \qquad C = 0.0162\ \mu\text{F}$$

From (11.42) and (11.52) we obtain for section 2

$$R_1 = 2.38\ \text{k}\Omega \qquad R_2 = 104\ \Omega \qquad R_a = 58.96\ \text{k}\Omega$$

$$R_b = 1\ \text{k}\Omega \qquad R = 10\ \text{k}\Omega \qquad C = 0.0137\ \mu\text{F}$$

The design is thus complete and the op amp, for example, can be a BJT(741) or a Bi-FET (LF351).

For synchronously tuned design the sections are identical, and of course, the magnitude response is not maximally flat. For a specified overall bandwidth the section center frequency is to be designed at $\omega_{01} = \omega_{02} = \omega_0/0.64$. The details of this design is left as an exercise in problem 11.17.

EXERCISE 11.6

For the circuit shown in Fig. 11.21*a* determine the transfer function for $C_1 = C_2 = C$. If $R_1 = 1\ \text{k}\Omega$, determine the values of R_2 and C for $Q = 10$ and $\omega_0 = 10^5$ rad/s.

Ans $V_o/V_i = (1/R_1C)s/[s^2 + (2/R_2C)s + 1/R_1R_2C^2]$,
$R_2 = 400\ \text{k}\Omega$, $C = 0.5$ nF

11.10 SENSITIVITY IN ACTIVE *RC* BANDPASS AMPLIFIERS

We have briefly considered the topic of sensitivity in conjunction with feedback amplifiers in Sec. 10.2. We will further explore this topic for bandpass amplifiers. In high-Q active networks, sensitivity consideration is very important [2].

The passive and active circuit elements have tolerance variation, etc. In an active network, these variations easily may be large enough to cause instability. Therefore, it is very important to have a knowledge about the degree of dependence of one quantity upon the other, hence the sensitivity of a system. We shall restrict ourselves in the sequel to variations in a single element only. However, it should be pointed out that, in any physical system, the characteristics change due to variation of many elements in the circuit, and therefore, a complete analysis must consider multiparameter sensitivity. The latter topic is beyond the scope of this book. A sensitivity measure that depends on variations of ω_0 and Q, however, is briefly discussed.

If $A(s, x)$ is the gain function and x is any parameter, *classical sensitivity* is defined as [see also (10.11)]

$$S_x^A(s, x) = \frac{dA/A}{dx/x} = \frac{d(\ln A)}{d(\ln x)} \tag{11.54}$$

In (11.54) it is assumed that the variations are incremental. If the variations are not small, a logical extension of the above definition is

$$S_x^A(s, x) = \frac{\Delta A / A}{\Delta x / x} \tag{11.55}$$

Note that S_x^A is really the linear term of the Taylor series expansion for $\Delta A/A$. Both of these definitions have the physical interpretation of percentage change in A due to percentage change in x. If the variations of x do not cause any change in A, the system has the ideal zero sensitivity. In general, $A(s)$ is the ratio of two polynomials, i.e.,

$$A(s, x) = \frac{N(s, x)}{D(s, x)} \tag{11.56}$$

From (11.55) and (11.56) we have

$$S_x^A = \frac{dA}{dx}\left(\frac{x}{A}\right) = \frac{DN' - ND'}{D^2}\left(x\frac{D}{N}\right) = x\left(\frac{N'}{N} - \frac{D'}{D}\right) \tag{11.57}$$

where $N' = \partial N/\partial x$ and $D' = \partial D/\partial x$. Consider, for example, a feedback system given by

$$A = \frac{a_1 a_2}{1 + a_1 a_2 f} \tag{11.58}$$

From (11.57) and (11.58) we have

$$S_{a_1}^A = a_1\left(\frac{a_2}{a_1 a_2} - \frac{a_2 f}{1 + a_1 a_2 f}\right) = \frac{1}{1 + a_1 a_2 f} \tag{11.59}$$

We shall now show that the real part of S_x^A corresponds to the normalized magnitude sensitivity, while the imaginary part of S_x^A corresponds to the phase sensitivity. To do this, for $s = j\omega$ we may write

$$A(j\omega) = |A(j\omega)|e^{j\phi(\omega)} \tag{11.60}$$

where $\phi(\omega)$ is arg $A(j\omega)$. Then from (11.60)

$$\ln A(j\omega) = \ln|A(j\omega)| + j\phi(\omega) \tag{11.61}$$

From (11.61) and (11.54)

$$S_x^A = \frac{d[\ln A(j\omega)]}{d(\ln x)} = \frac{d[\ln|A(j\omega)|]}{dx/x} + j\frac{d\phi(\omega)}{dx/x} \tag{11.62}$$

Since x is usually a real quantity, we have, from the above

$$\text{Magnitude sensitivity} = \text{Re } S_x^A \tag{11.63a}$$

$$\text{Phase sensitivity} = \text{Im } S_x^A \tag{11.63b}$$

where Re and Im designate the real and the imaginary parts of the term, respectively. Note that in (11.62), Re $S_{\omega_0}^A$ corresponds to the normalized change in magnitude of $A(j\omega)$ and Im $S_{\omega_0}^A$ corresponds to the change in the phase function.

EXAMPLE 11.5

Consider a bandpass transfer function of the form

$$A(s) = \frac{Ks}{s^2 + (\omega_0/Q)s + \omega_0^2}$$

Determine $S_Q^{|A|}$ and $S_{\omega_0}^{|A|}$ at $s = j\omega_0$.

We may rewrite the above in a normalized form such that $|A(j\omega_0)| = 1$, i.e.,

$$A(s) = \frac{(\omega_0/Q)s}{s^2 + (\omega_0/Q)s + \omega_0^2}$$

Then the ω_0 and Q sensitivities are determined from

$$S_Q^A = \frac{Q}{A}\frac{\partial A}{\partial Q} \qquad \text{and} \qquad S_{\omega_0}^A = \frac{\omega_0}{A}\frac{\partial A}{\partial \omega_0}$$

The results are

$$S_Q^A = \left.\frac{s^2 + \omega_0^2}{s^2 + (\omega_0/Q)s + \omega_0^2}\right|_{s=j\omega_0} = 0$$

If we let $s = j\omega_0$, then

$$S_Q^{|A(j\omega_0)|} = \text{Re } S_Q^{A(j\omega_0)} = 0$$

Similarly,

$$S_{\omega_0}^A = \left.\frac{s^2 - \omega_0^2}{s^2 + (\omega_0/Q)s + \omega_0^2}\right|_{s=j\omega_0} = -2Q$$

and

$$S_{\omega_0}^{|A(j\omega_0)|} = \text{Re } S_Q^A = -2Q$$

From the above results it is seen that for high Q the effect of variation of ω_0 on the magnitude of the gain function is much larger than that of Q. Thus in a high-Q circuit the variation of the center frequency must be kept very tight. In other words, the limiting factor on high-Q active RC networks is not the Q sensitivity but rather the center frequency sensitivity, which is always limited by RC tracking and manufacturing tolerances. In active switched-capacitor (SC) filters (Sec. 9.6) we have a different situation, and precise values of very high Q and ω_0 can be achieved without any difficulty since the value of Q is controlled by the ratio of capacitors and the value of ω_0 is controlled by a very stable crystal-controlled clock and the ratio of capacitors.

EXERCISE 11.7

For Exercise 11.6 determine the sensitivity of ω_0 and Q due to R_1, R_2, and C.

Ans $S_{R_1}^{\omega_0} = -\frac{1}{2}, S_{R_2}^{\omega_0} = -\frac{1}{2}, S_C^{\omega_0} = -1; S_{R_1}^Q = -\frac{1}{2}, S_{R_2}^Q = \frac{1}{2}, S_C^Q = 0$

EXERCISE 11.8

Repeat Exercise 11.7 for the bandwidth sensitivities.

Ans $S_{R_1}^{BW} = 0, S_{R_2}^{BW} = -1, S_C^{BW} = -1$

11.11 PHASE-LOCKED LOOPS

Phase-locked loops (PLL) are very important and are used in many applications in communication circuits that require a high degree of noise immunity and narrow bandwidth. Some of the applications are frequency synchronization, frequency multiplication and division, frequency translation, FM demodulation, and AM detection. The block diagram of a PLL is shown in Fig. 11.22. Basically the PLL consists of three parts: a phase comparator (also called *phase detector*), a low-pass filter, and a voltage-controlled oscillator (VCO).

The operation of a PLL is briefly described as follows: When no signal is applied to the system, $v_c(t)$ is zero, and VCO operates at a frequency f_0, which is called the *free-running* frequency. However, when an input signal is applied to the system, the phase detector compares the phase and the frequency of the VCO to the input and generates an error signal $v_e(t)$. The error voltage is related to the phase and the frequency difference between the input $v_i(t)$ and the output of VCO, $v_o(t)$. The error voltage is filtered, amplified, and applied to the VCO as a control voltage $v_c(t)$. The control voltage forces the VCO frequency to vary in a manner that reduces the difference between f_0 and f_i. When $f_0 = f_i$, we say that the PLL is *locked*. Under the lock condition the control voltage is a function of the net phase difference between $v_i(t)$ and $v_o(t)$. The range of frequencies over which the PLL can maintain lock with input signal is called the *lock range*. The range of frequencies over which the PLL can acquire lock with input signal is called the *capture range*. The capture range is always smaller than the lock range, as will be given shortly.

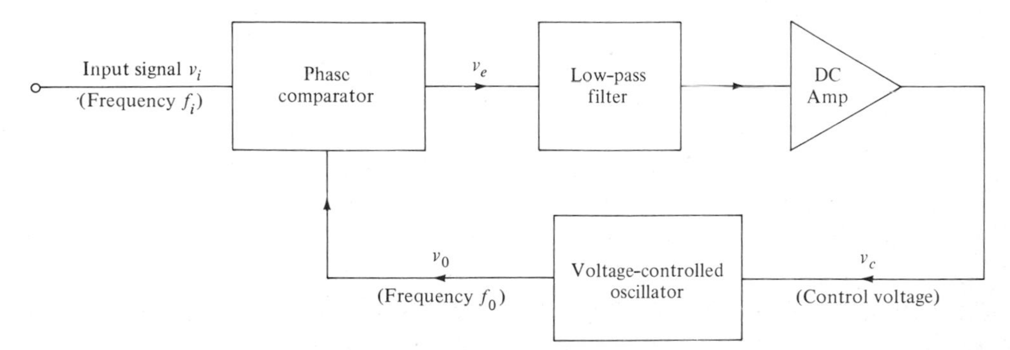

FIGURE 11.22 A phase-locked loop system.

A. Phase Detector

The phase detector is an analog multiplier as discussed in Sec. 9.4. Consider Fig. 11.22, where

$$v_i(t) = V_i \sin \omega_i t \tag{11.64}$$

$$v_o(t) = V_o \sin (\omega_0 t + \theta_e) \tag{11.65}$$

where θ_e is the phase difference or phase error. Consider the two cases now.

Unlocked State ($\omega_i \neq \omega_0$)

When the two frequencies are not the same, the loop is unlocked. In this case

$$v_e(t) = K_1 v_i v_o = \frac{K_1 V_i V_o}{2}[\cos (\omega_i - \omega_0)t - \cos (\omega_i + \omega_0)t] \tag{11.66}$$

The low-pass filter removes the high-frequency term, and thus

$$v_c(t) = AV_i V_o \cos (\omega_i - \omega_0)t \tag{11.67}$$

where $A = a_o K_1/2$ and a_o is the gain of the dc amplifier block. If ω_i and ω_0 differ substantially, there will be no effective feedback, and nothing will happen. If $\omega_i - \omega_0 = \pm\Delta\omega$ is less than some value, determined by the parameters of the loop, there will be feedback that will drive the system into lock.

Locked State ($\omega_i = \omega_0$)

When the frequencies are synchronized, that is, $\omega_i = \omega_0 = \omega$,

$$\begin{aligned} v_e &= K_1 v_i v_o = K_1 V_i V_o \sin \omega t \sin (\omega t + \theta_e) \\ &= \frac{K_1 V_i V_o}{2} [\cos \theta_e - \cos (2\omega t + \theta_e)] \end{aligned} \tag{11.68}$$

The low-pass filter removes the high-frequency ac component, and thus

$$v_c(t) = \frac{a_o K_1 V_i V_o}{2} \cos \theta_e \tag{11.69}$$

Note that $v_c(t)$ is proportional to the cosine of the phase difference θ_e. Thus, under the lock condition, v_c depends on θ_e, and for $\theta_e = \pm 90°$, $v_c(t) = 0$.

A linear model of a PLL system is shown in Fig. 11.23, where the capital letters correspond to the frequency-domain (Laplace transform) quantities. Note that the gain

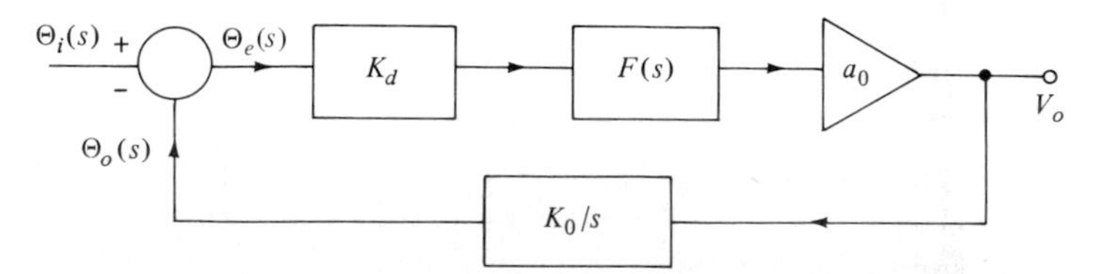

FIGURE 11.23 Block diagram of a linearized PLL system.

of the phase comparator is K_d(V/rad) of phase difference. $F(s)$ is a low-pass transfer function, and a_o is the voltage gain of the dc amplifier. The VCO gain is denoted by K_o rad/(s · V). The phase comparator is sensitive to the difference between the phase of the VCO output signal and the phase of the input signal. The phase of the VCO output is the time integral of the VCO output frequency (hence $1/s$) and K_o is in radians per second per volt. From Fig. 11.23, the closed-loop transfer function is given by

$$\frac{V_o}{\Theta_i}(s) = \frac{a_o K_d F(s)}{1 + a_o K_d K_o F(s)/s} = \frac{s a_o K_d F(s)}{s + K_d K_o a_o F(s)} \tag{11.70}$$

Recall that $F(s)$ is a low-pass transfer function. The simplest case is a single-pole RC circuit for which

$$F(s) = \frac{1}{1 + s\tau_1} \tag{11.71}$$

where $\tau_1 = RC$.

In PLL we are usually interested in the response to frequency variations, and since $\omega_i = d\theta_i/dt$ [i.e., $\Omega_i(s) = s\Theta_i(s)$], from (11.70) we have

$$\frac{V_o}{\Omega_i}(s) = \frac{a_o K_d F(s)}{s + a_o K_d K_o F(s)} = \frac{1}{K_o} \frac{K_v F(s)}{s + K_v F(s)} \tag{11.72}$$

where $K_v = a_o K_o K_d$ is the loop gain.

From (11.72) and (11.71) we obtain

$$\frac{V_o}{\Omega_i}(s) = \frac{1}{K_o} \frac{K_v}{s^2 \tau_1 + s + K_v} \tag{11.73}$$

In (11.73) it is seen that the value of K_v controls the pole locations. Usually (11.73) is designed for a maximally flat magnitude response to obtain good frequency and transient response. For the maximally flat magnitude case,

$$K_v = \frac{1}{2\tau_1} \tag{11.74}$$

and the 3-dB bandwidth is given by

$$\omega_{3\text{dB}} = \sqrt{2}\, K_v \tag{11.75}$$

The lock range $\Delta\omega_L$ is given by [3, 4]

$$\Delta\omega_L = \pm a_o K_o K_d \phi_d \tag{11.76}$$

and the capture range $\Delta\omega_C$ is given by

$$\Delta\omega_C \simeq \pm\sqrt{\frac{\Delta\omega_L}{\tau_1}} \qquad \text{for } \Delta\omega_L \tau_1 \gg 1 \tag{11.77}$$

Note that $\Delta\omega_C < \Delta\omega_L$.

Typical PLL frequency-to-voltage transfer characteristics are shown in Fig. 11.24. Note that the PLL has a frequency-selective characteristic, about the center frequency, which is set by the VCO free-running frequency ω_0. It responds to the input

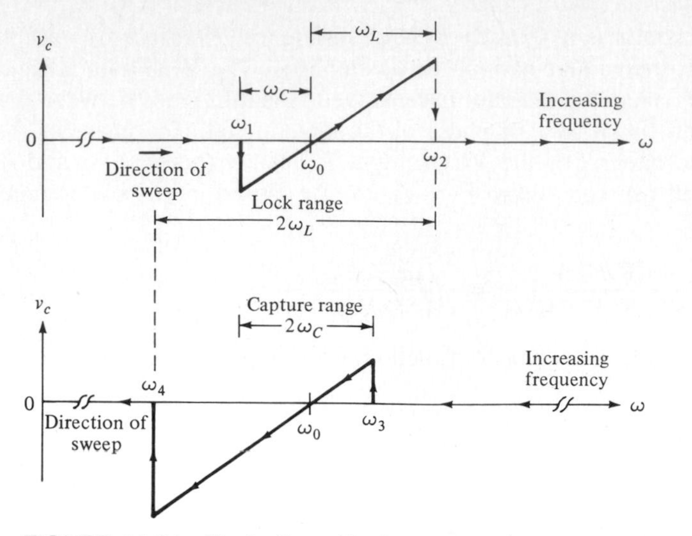

FIGURE 11.24 Illustration of lock range and capture range.

frequencies near ω_0, by less than $\Delta\omega_c$ or $\Delta\omega_L$, depending whether the loop starts with or without an initial lock condition.

B. Voltage-Controlled Oscillator

The voltage-controlled oscillator (VCO) is a very important block of the PLL. It is a circuit that provides a square wave or a triangular waveform whose frequency can be varied by a dc voltage. The VCO should have a linear voltage-to-frequency con-

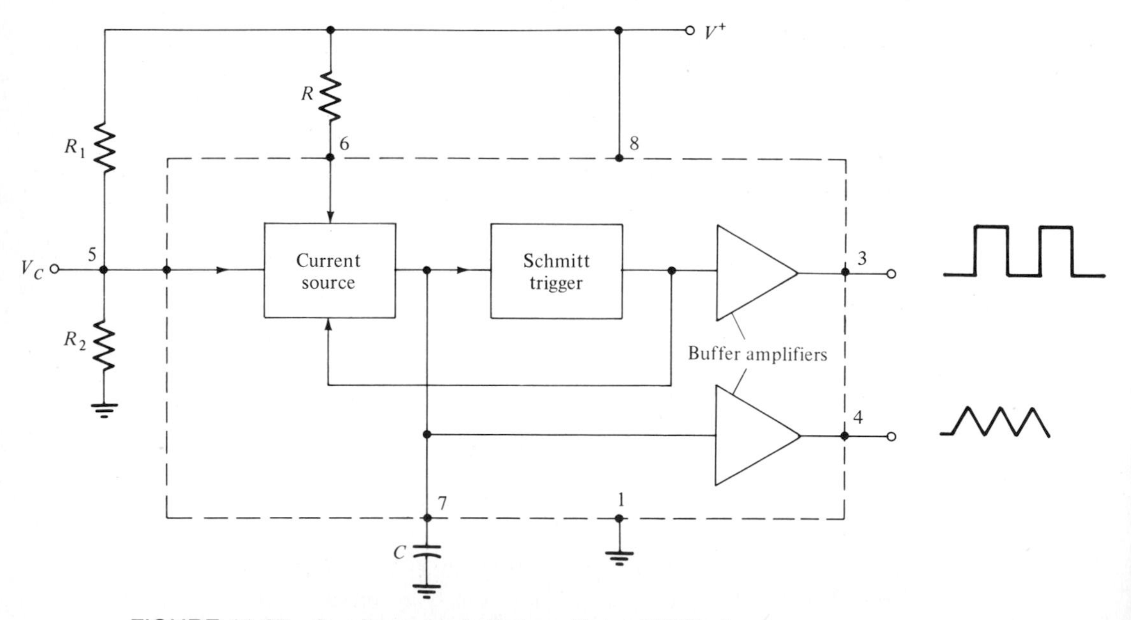

FIGURE 11.25 A voltage-controlled oscillator (VCO) circuit.

version, good frequency stability, and ease of tuning. An example of a commercial VCO is the 566 IC unit. The frequency of oscillation of the 566 unit is determined by an external resistor and capacitor and can be varied by an applied dc voltage. The 566 unit, functional block diagram with its pin connections, is shown in Fig. 11.25. Note that the output of the VCO is a square-wave periodic signal. The free-running frequency f_0 is determined by a single external capacitor C that can be continuously adjusted from a few cycles to about 1 MHz. It is determined from

$$f_0 = \frac{2}{RC}\left(1 - \frac{V_C}{V^+}\right) \tag{11.78}$$

where $\frac{3}{4} \leq V_C/V^+ \leq 1$.

EXAMPLE 11.6

Determine the free-running frequency f_0 of the 566 VCO if $V^+ = 20$ V, $R = 5$ kΩ, $C = 400$ pF, $R_2 = 4$ kΩ, and $R_1 = 1$ kΩ.

From the circuit

$$V_C = \frac{R_2}{R_1 + R_2} V^+ = \left(\frac{4}{1 + 4}\right) 20 = 16 \text{ V}$$

From (11.78)

$$f_0 = \frac{2}{5 \times 10^3 (4 \times 10^{-10})}\left(1 - \frac{16}{20}\right) = 0.2 \text{ MHz}$$

EXERCISE 11.9

The phase detector is basically an electronic switch, shown in Fig. E11.9*a*, that is in synchronism with the input. It is opened and closed by the square-wave signal of the VCO, shown in Fig. E11.9*b*. Determine the average error voltage v_e.

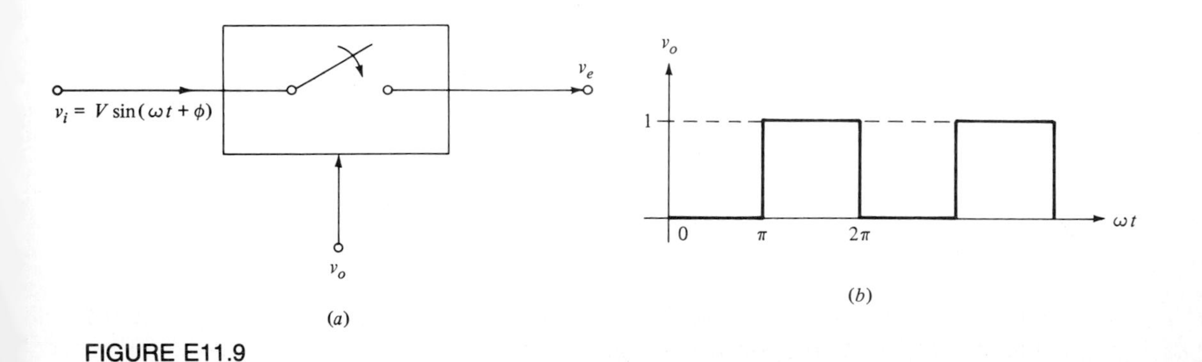

FIGURE E11.9

Ans $(V/\pi)\cos\phi$

For a more detailed study of PLL and its applications the reader is referred to Refs. 3 and 4.

11.12 CONCLUDING REMARKS

This chapter dealt with frequency-selective circuits. The design of tuned amplifiers using active elements and *LC* circuits as discussed. These circuits are used extensively in communication equipment. In integrated circuits, tuned amplifiers are available that use external *LC* tuned circuits and perform a variety of functions. Phase-locked loops, which are very important in communication circuits requiring a high degree of noise immunity and narrow bandwidth, were also briefly discussed.

REFERENCES

1. D. Schilling and C. Belove, *Electronic Circuits: Discrete and Integrated,* McGraw-Hill, New York, 1979.
2. M. S. Ghausi and K. R. Laker, *Modern Filter Design: Active RC and Switched Capacitor,* Prentice-Hall, Englewood Cliffs, N.J., 1981.
3. A. Grebene, *Analog Integrated Circuit Design,* Van Nostrand Reinhold, New York, 1972.
4. P. Gray and R. Meyer, *Analysis and Design of Analog Integrated Circuits,* Wiley, New York, 1977.

PROBLEMS

11.1 For the single-stage FET tuned amplifier shown in Fig. P11.1, determine the expression for the voltage gain function. Determine the gain, the bandwidth, and the resonant frequency if the FET parameters are $g_m = 4 \times 10^{-3}$ mho, $r_d = 20$ kΩ, $C_{iss} = 8$ pF, $C_{rss} = 2$ pF, $L = 10$ μH with a Q of 200 at ω_0, and $C_a = 500$ pF and include the stray capacitance of the inductor $R_L = 1$ kΩ.

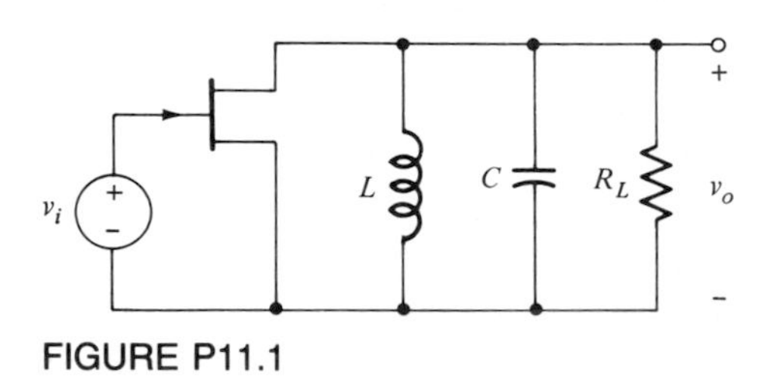

FIGURE P11.1

11.2 In the example of Fig. 11.7*a*, use the Miller effect and include r_b. Determine the expression for the gain function. Find the pole-zero location of the transfer function if $r_b = 50\ \Omega$. Determine the midband gain and the bandwidth.

11.3 Find the midband gain and the bandwidth for problem 11.2 under the following cases:
(*a*) $f_0 \gg 1/R_iC_i$, i.e., neglect R_i as compared to the shunt C_i.
(*b*) $f_0 \ll 1/R_iC_i$, i.e., neglect C_i as compared to the shunt R_i.

11.4 Design the single-stage bipolar transistor amplifier of Fig. 11.7*a* for the following specifications:

$$\omega_0 = 2\pi(500\text{ kHz}) \qquad Q = 50 \qquad A_i = -20$$

i.e., determine R_1, L, C. The load resistance $R_L = 1\text{ k}\Omega$ and the transistor parameters are the same as in Example 11.1.

11.5 A single-stage bipolar transistor has the tuned circuit at the output, as in Fig. 11.12. The transistor parameters at the resonant frequency $f_0 = 10$ MHz are given by Example 11.3. Design the circuit for a bandwidth $f_{3dB} = 500$ kHz and a midband voltage gain of -100.

11.6 A tuned amplifier has been designed such that the pole-zero pattern of the voltage gain function is as shown in Fig. P11.6. The midband voltage gain $A_v(j\omega_0) = 10^3$. Determine the 3-dB bandwidth of the amplifier using the narrowband approximation.

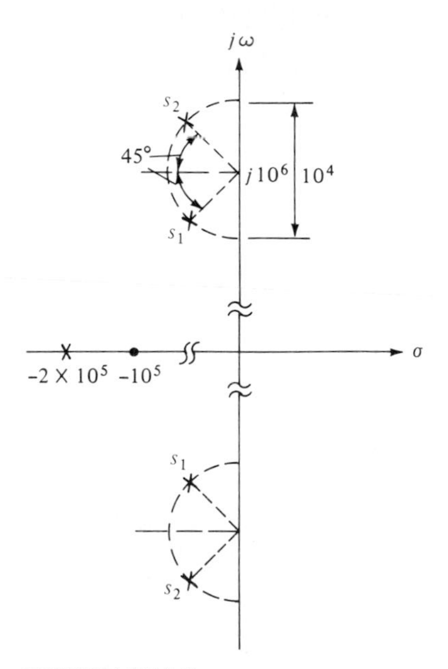

FIGURE P11.6

11.7 The circuit shown in Fig. P11.7 is to resonate at $f_0 = 20$ MHz and with $f_{3dB} = 1$ MHz. The transistor parameters are as follows:

$$\beta_0 = 50 \qquad f_T = 500\text{ MHz}$$

$$C_c = 2\text{ pF} \qquad r_e = 10\ \Omega$$

(Assume that $r_b = 0$.) Determine the values of L_m, n, n_1, and C to achieve the above specifications in addition to realizing the maximum midband current gain. What is the resulting midband current gain?

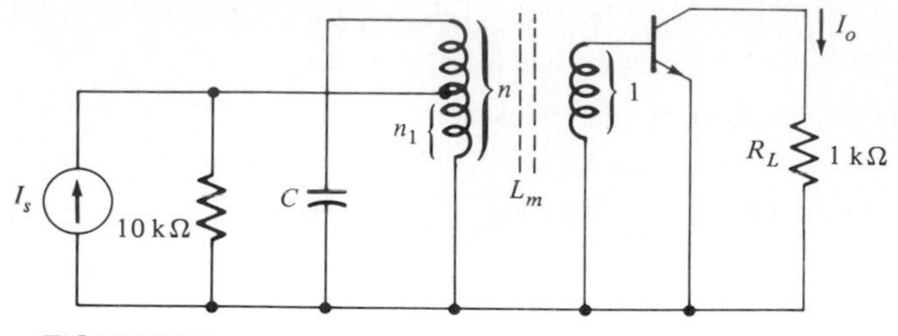

FIGURE P11.7

11.8 The y parameters of a device measured at $f = 1$ MHz are given as follows:

$$y_{11} \simeq (1 + j2) \times 10^{-3} \text{ mho} \qquad y_{12} \simeq -j(10^{-5}) \text{ mho}$$

$$y_{21} \simeq 4 \times 10^{-2} \text{ mho} \qquad y_{22} \simeq (2 + j10) \times 10^{-6} \text{ mho}$$

Is the device absolutely stable? What value of shunt conductance added at the input or the output will make the overall two-port device absolutely stable?

11.9 For the cascode amplifier shown in Fig. 11.18, the transistor parameters are $\beta_0 = 75$, $\omega_T = 8\pi \times 10^8$ rad/s, $C_c = 4$ pF, $r_e = 5\ \Omega$, $r_b = 0$, and $R_s = 100\ \Omega$. Determine the element values for a synchronously tuned design and the following specifications:

$$f_0 = 10 \text{ MHz} \qquad f_{3\text{dB}} = 1 \text{ MHz} \qquad |A_v(j\omega_0)| = -100$$

11.10 Repeat problem 11.9, but with a stagger-tuned design, corresponding to the critically coupled double-tuned case (maximally flat magnitude).

11.11 A common-base tuned circuit is shown in Fig. P11.11. The center frequency $f_0 \ll f_\alpha$, so that the internal capacitance of the CB stage may be omitted (as in Fig. 11.18b). Derive the design equations for a synchronously tuned narrowband design where ω_0 and $\omega_{3\text{dB}}$ are specified.

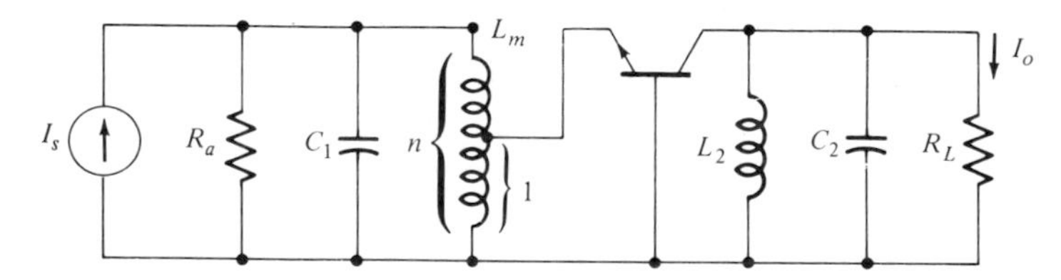

FIGURE P11.11

11.12 Repeat problem (11.11) for a stagger-tuned design with a maximally flat magnitude response characteristic.

11.13 Design the hybrid CS-CB signal circuit shown in Fig. P11.13 for the following specifications: $\omega_0 = 10^6$ rad/s, $\omega_{3\text{dB}} = 10^5$ rad/s, and a voltage gain at center frequency of 40 dB. The BJT parameters are $\beta_0 = 100$, $r_e = 10\ \Omega$, and $r_b = 0$. The FET parameters are $g_m = 4 \times 10^{-3}$ mho, $r_d = 20$ kΩ, $C_{gs} = C_{ds} = 2$ pF, and $C_{gd} = 0.5$ pF.

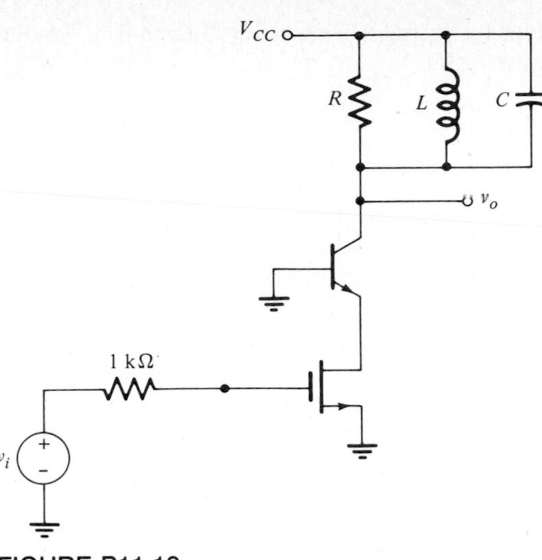

FIGURE P11.13

11.14 Design the emitter-coupled tuned amplifier shown in Fig. 11.19 for the following specifications on the transfer impedance function (V_o/I_s):

$$f_0 = 10 \text{ MHz} \qquad f_{3\text{dB}} = 500 \text{ kHz}$$

The transistor parameters are the same as in problem 11.7. Let $R_E = 100 \text{ k}\Omega$ and $R_1 \to \infty$. Use stagger-tuned design for a flat magnitude response.

11.15 Derive (11.52) and design the circuit for a voltage gain of 10, $f_{3\text{dB}} = 10^3$ Hz, and $f_0 = 10^4$ Hz. Let $C_1 = C_2 = 0.01\ \mu$F and $R_b = 1$ kΩ.

11.16 Design a frequency-selective amplifier using two circuits of Fig. 11.21*b* for the same specifications as in Example 11.4, but use a synchronously tuned design. Let $R_b = 1$ kΩ and $C = 0.001\ \mu$F.

11.17 Design the circuit of Fig. 11.21*a* for $f_{3\text{dB}} = 10^3$ Hz and $f_0 = 5 \times 10^3$ Hz. Determine the voltage gain. Assume $C_1 = C_2 = 0.001\ \mu$F.

11.18 For the single-tuned circuit shown in Fig. 11.4, determine

$$S_G^Q,\ S_L^Q, \text{ and } S_C^Q$$

11.19 If y, u, and v are single-valued differentiable functions of x and K a constant, use (11.54) to derive the following:
(*a*) $S_x^{Ky} = S_x^y$
(*b*) $S_x^{1/y} = -S_x^y$
(*c*) $S_x^{uv} = S_x^u + S_x^v$

11.20 The transfer function of a certain network is given by

$$T(s) = \frac{\omega_0^2}{s^2 + (\omega_0/Q)s + \omega_0^2}$$

Determine $S_Q^{|T|}$ and $S_{\omega_0}^{|T|}$ at $s = j\omega_0$.

11.21 For the 566 VCO unit shown in Fig. P11.21, the input ac signal v_{in} has a peak-to-peak amplitude of 2 V. If the circuit parameters are as in Example 11.6, determine the range of output frequencies of VCO.

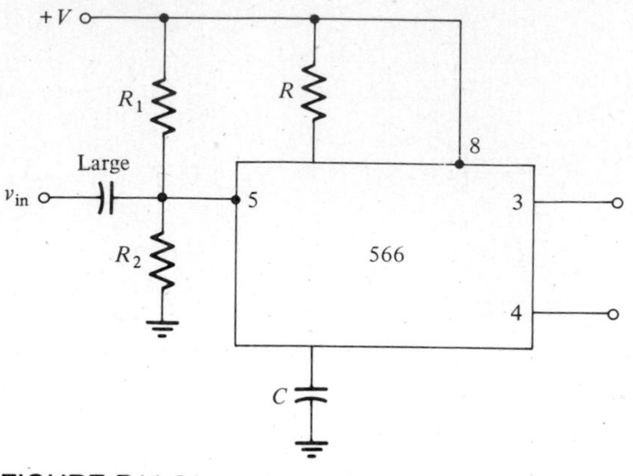

FIGURE P11.21

11.22 For the 565 IC PLL the following are given:

$$K_o = \frac{50 f_0}{V_{CC}} \text{ rad/(s} \cdot \text{V)} \qquad K_d \simeq \frac{1.4}{\pi} \text{ V/rad} \qquad a_o \simeq 1.4 \qquad \phi_d \simeq \pm\pi/2 \text{ rad}$$

Determine the lock-frequency range if the VCO frequency $f_0 = 1$ kHz for $V_{CC} = 5$ V. Determine the capture-frequency range if the time constant of the low-pass filter is $\tau_1 = 1$ ms.

CHAPTER 12

Transient Response and Switching Speed

In previous chapters we concentrated on the steady-state frequency response of linear circuits. In a linear circuit there is a relation between frequency response and time response. We shall first consider transient response of linear circuits in this chapter. Subsequently, we shall consider the switching speed of BJT and FET inverters. In linear circuits the voltage or current gain function in the frequency domain can be used to obtain the transient response. Thus, we shall use the results of the previous chapters where the small-signal models for the devices were used.

When the device is used under large-signal conditions, i.e., in the ON-OFF mode, linear analysis is not applicable. In such cases one must resort to computer-aided analysis and design. An *approximate* method of calculating the ON-OFF times of simple inverters can be made by a modified (averaging) linearized approximation, as will be done in this chapter. However, digital circuits utilize many transistors and inverters and defy simple analysis. For most calculations the propagation delay of the inverter, which is often supplied by the manufacturer, can be used, or one must resort to use a computer-aided analysis.

12.1 GENERAL TIME RESPONSE

Consider a two-port linear time-invariant circuit shown in Fig. 12.1. A lowercase letter $v_i(t)$ shall be used for time domain and an uppercase letter $V_i(s)$ for the transform frequency domain.

The input and output quantities V_i and V_o in the frequency domain are related by the transfer function $H(s)$, namely,

$$V_o(s) = H(s)V_i(s) \tag{12.1}$$

If $v_i(t)$ is a unit impulse, then $V_i(s) = 1$, and the output $v_o(t)$, or *impulse response,* is given by

$$v_o(t) = \mathcal{L}^{-1}[V_o(s)] = \mathcal{L}^{-1}[H(s)] = h(t) \tag{12.2}$$

where $\mathcal{L}^{-1}$ indicates the inverse Laplace transform operation.

In electronic circuits, such as pulse amplifiers, one is often interested in the step response of the system. Thus, if $v_i(t) = u(t)$, then $V_i(s) = 1/s$, and from (12.1)

$$V_o(s) = \frac{1}{s}H(s) \tag{12.3}$$

and

$$v_o(t) = \mathcal{L}^{-1}[V_o(s)] = \mathcal{L}^{-1}\frac{H(s)}{s} = \text{step response} \tag{12.4}$$

A typical normalized step response for network functions with complex poles is shown in Fig. 12.2. Such a response is also called the *transient response*. Key features of the transient response are *overshoot* γ, *rise time* τ_R, and *delay time* τ_D. These quantities are illustrated in Fig. 12.2.

Overshoot is the percentage of the difference between the largest peak value and the final value to the final value. *Rise time* is the difference between the time when the response first reaches 90 percent of its final value and the time when the response is 10 percent of its final value. *Delay time* is the time from $t = 0$ until the response reaches 50 percent of its final value.

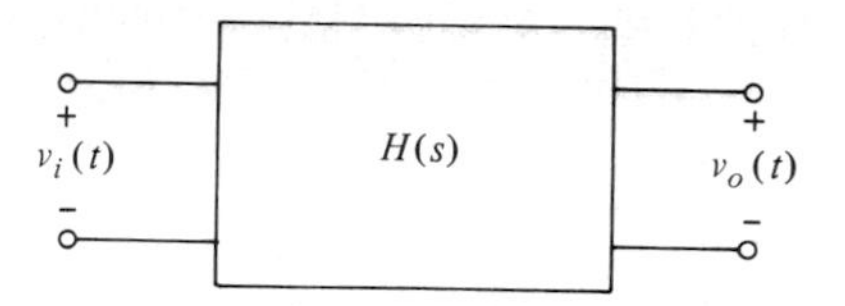

FIGURE 12.1 Schematic input-output relationship of a linear circuit.

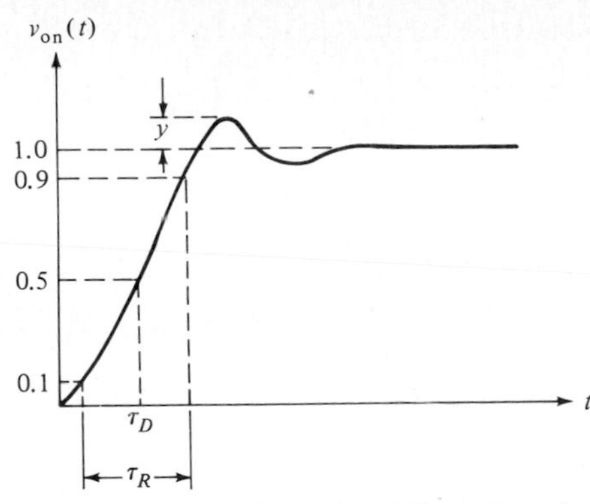

FIGURE 12.2 Normalized time-domain response for a unit-step input.

12.2 STEP RESPONSE OF AN AMPLIFIER STAGE

Consider the gain function of a simple resistively loaded amplifier stage in the midband and high-frequency regions. We have already seen this to be a low-pass function; in many cases, the approximate expression for the gain function is of the form

$$A(s) = \frac{V_o}{V_i} = \frac{A_o p_o}{s + p_o} \tag{12.5}$$

where p_o is the dominant pole of the high-frequency circuit, A_o is the midband gain, and V_o and V_i are the transforms of the output and input voltages. Note that the output and the input signals may be either currents or voltages.

If the input is a unit step voltage, $v_i(t) = u(t)$, its transform is $1/s$ and the transform of the output voltage is

$$V_o(s) = \frac{A_o p_o}{s} \frac{1}{s + p_o} \tag{12.6a}$$

The inverse Laplace transform of (12.6*a*) yields the time-domain response of the output signal $v_o(t)$. Hence

$$v_o(t) = \mathcal{L}^{-1}[V_o(s)] = A_o(1 - e^{-tp_o}) \tag{12.6b}$$

The unit step response is shown in Fig. 12.3. Note that there is no overshoot for a one-pole function. Transfer functions with all real poles also do not exhibit overshoot. However, functions with complex poles may or may not exhibit overshoot.

To find τ_D, we set $v_o(t) = 0.5A_o$ and solve for t. Thus

$$0.5A_o = A_o(1 - e^{\tau_D p_o}) \qquad \text{or} \qquad \tau_D = \frac{0.69}{p_o} \tag{12.7}$$

To determine τ_R, we set $v_o(t) = 0.1A_o$ to find t_1, and $v_o(t) = 0.9A_o$ to find t_2, and then

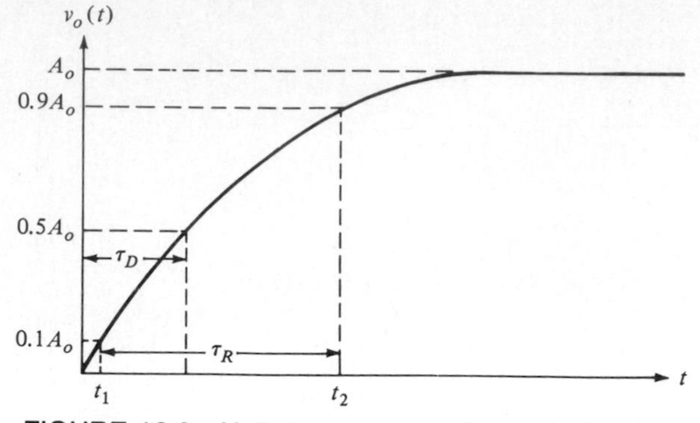

FIGURE 12.3 Unit step response for a single-pole amplifier gain function.

$\tau_R = t_2 - t_1$. Hence

$$0.1A_o = A_o(1 - e^{-t_1 p_o}) \qquad \text{or} \qquad t_1 = \frac{0.1}{p_o} \tag{12.8a}$$

$$0.9A_o = A_o(1 - e^{-t_2 p_o}) \qquad \text{or} \qquad t_2 = \frac{2.3}{p_o} \tag{12.8b}$$

The 10 to 90 percent rise time is then

$$\tau_R = t_2 - t_1 = \frac{2.2}{p_o} \tag{12.9}$$

Note that the rise time and the 3-dB bandwidth for the one-pole gain function are completely related. Since $\omega_{3dB} = p_o$, we have

$$\tau_R = \frac{2.2}{\omega_{3dB}} \qquad \text{or} \qquad \tau_R f_{3dB} = \frac{2.2}{2\pi} = 0.35 \tag{12.10}$$

Thus such a system can be completely characterized by its frequency response or its step response. The relationship in (12.10) has also been found to be empirically true for a wide class of gain functions. Specifically, if the step response is monotonically increasing vs time, the relation in (12.10) can be used. In fact, even for a non-monotonically increasing step response (12.10) can be used if the overshoot is less than approximately 5 percent. Note that the bandwidth of an oscilloscope is determined from (12.10); namely, a 10-ns scope has a bandwidth of 35 MHz.

EXAMPLE 12.1

Let us calculate the step response of an amplifier designed to have a two-pole Butterworth voltage gain function. We shall consider normalized frequency s_n and for simplicity, however, drop the subscript n in the following:

$$A_o(s) = \frac{V_o}{V_i} = \frac{A_o}{s^2 + \sqrt{2}s + 1} = \frac{A_o}{s^2 + 2\zeta s + 1}$$

For a step input, from (12.3) we have

$$V_o(s) = \frac{1}{s}\frac{A_o}{s^2 + \sqrt{2}s + 1}$$

This may be expanded into a partial fractional expansion as follows:

$$V_o(s) = \frac{1}{s}\frac{A_o}{[s + 1/\sqrt{2} + j(1/\sqrt{2})][s + 1/\sqrt{2} - j(1/\sqrt{2})]}$$

$$= \frac{A_o}{s} + \frac{A_o(\sqrt{2}/2)\angle 5\pi/4}{s + 1/\sqrt{2} + j(1/\sqrt{2})} + \frac{A_o(\sqrt{2}/2)\angle -5\pi/4}{s + 1/\sqrt{2} - j(1/\sqrt{2})}$$

where $\angle$ denotes angle in radians. From an inverse Laplace transform table we obtain

$$v_o(t) = A_o\left[1 + \sqrt{2}e^{-(1/\sqrt{2})t}\cos\left(\frac{1}{\sqrt{2}}t + \frac{5\pi}{4}\right)\right]u(t)$$

The response for $\zeta = \sqrt{2}/2 \simeq 0.71$ is plotted in Fig. 12.4 [see also (7.27)].

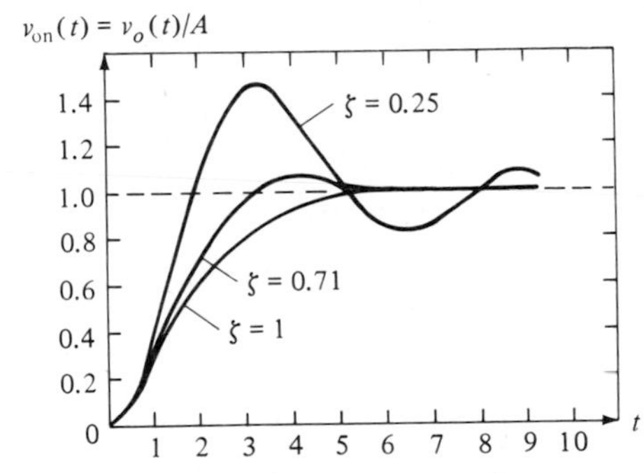

FIGURE 12.4 Step response for an amplifier with two poles ($\zeta = 0.71$ is for a Butterworth response).

The various transient response parameters for the Butterworth two-pole response are

$$\gamma \simeq 41 \text{ percent} \qquad \tau_R \simeq 2.1 \qquad \tau_D \simeq 1.4$$

Figure 12.4 also shows the step response for $\zeta = 1$ (which corresponds to two identical real poles) and $\zeta = 0.25$ (which corresponds to the complex poles having an angle of 75° with the real axis).

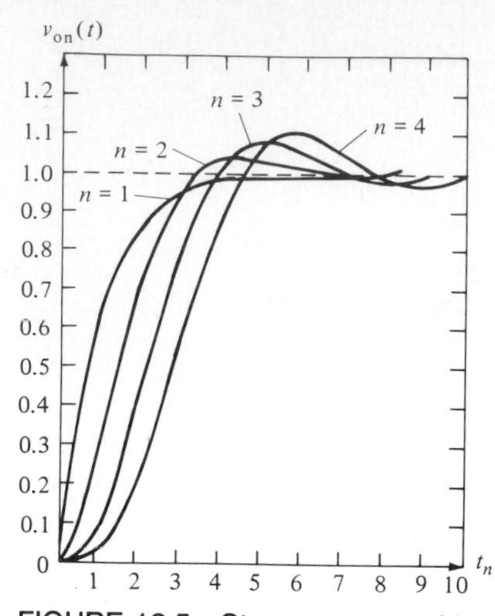

FIGURE 12.5 Step response of low-order Butterworth functions.

In pulse and video amplifiers an overshoot of higher than 5 percent is not desirable, since it causes distortion of the transient signal. Note that in the above example the normalized bandwidth $\omega_{3dB} = 1$; hence τ_R is also normalized, and thus $\tau_R\omega_{3dB} \simeq 2.1$, which is close to the relation given in (12.10). Normalized step responses for the low-order Butterworth polynomials (listed in Table 8.2) are shown in Fig. 12.5. Note that the overshoot increases for increasing order of Butterworth polynomials.

EXERCISE 12.1

The gain function of a certain amplifier is given by

$$A(s) = \frac{A_o}{(s + 10^7)(s + 2 \times 10^7)}$$

Determine the rise time of the amplifier for a step input

(a) Use the exact method.
(b) Use the approximate method.

Ans *(a)* 0.259 μs, *(b)* 0.262 μs

EXERCISE 12.2

Determine the approximate rise time, for a step input, of an amplifier that has the following gain function:

$$A(s_n) = \frac{A_o}{s_n^2 + 3s_n + 3}$$

where $s_n = s/10^6$.

Ans 1.6 μs

12.3 SIGNAL TRANSMISSION WITH MINIMUM DISTORTION

A distortionless ideal signal transmission is defined as follows: The output signal has the same waveform as the input signal (although it may be amplified and/or delayed by a constant time T_0). In other words, $v_o(t) = A_o v_i(t - T_0)$. The network function for the ideal transmission is

$$H(s) = \frac{\mathcal{L}[v_o(t)]}{\mathcal{L}[v_i(t)]} = \frac{\mathcal{L}[A_o v_i(t - T_0)}{\mathcal{L}[v_i(t)]} \tag{12.11}$$

and $\mathcal{L}[A_o v_i(t - T_0)] = A_o e^{-T_0 s} V_i(s)$. Hence

$$H(s) = \frac{A_o e^{-T_0 s} V_i(s)}{V_i(s)} = A_o e^{-T_0 s} \tag{12.12}$$

Thus, from (12.12), for ideal signal transmission

$$H(j\omega) = A_o \tag{12.13a}$$

$$\arg H(j\omega) = -T_0 \omega \tag{12.13b}$$

The ideal characteristics expressed in (12.13*a*) and (12.13*b*) are not realizable in practice, and thus we can only try to approximate them. Since lumped networks are used in the actual realizations, the approximation of $H(s)$ must be in the form of rational functions of the complex frequency. One such approximation is the so-called linear-phase response characteristic. Note that in (12.13*b*) the phase function is a linear function of frequency. The all-pole transfer function approximation of (12.12) with T_0 normalized to unity for lower-order polynomials is listed in Table 12.1. The corresponding transient responses are shown in Fig. 12.6. The transient response for

TABLE 12.1 Linear Phase Polynomials ($\tau_o = 1$)*

$s_n + 1$
$s_n^2 + 3s_n + 3$
$s_n^3 + 6s_n^2 + 15s_n + 15 = (s_n + 2.322)(s_n^2 + 3.678s_n + 6.460)$
$s_n^4 + 10s_n^3 + 45s_n^2 + 105s_n + 105 = (s_n^2 + 5.792s_n + 9.140)(s_n^2 + 4.208s_n + 11.488)$

*Note that $\tau_o = 1$ implies that the coefficient of the linear term s_n is the same as the last term constant a_o. In this case ω_{3dB} is *not* unity.

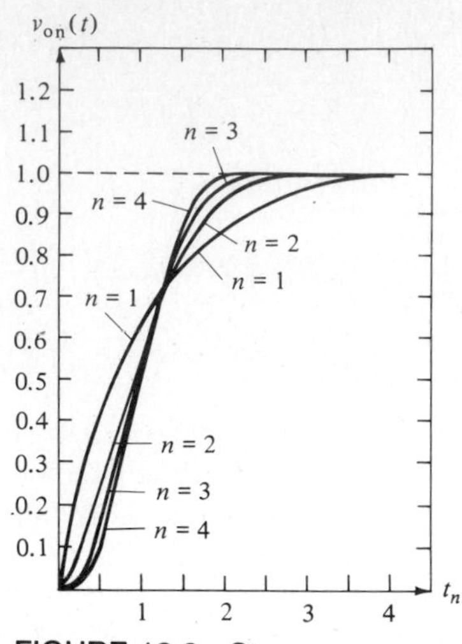

FIGURE 12.6 Step response of low-order linear phase functions.

transfer functions with linear-phase characteristics do not exhibit any overshoot. Design of a linear-phase response gain function is similar to that of Butterworth; i.e., we use the polynomials listed in Table 12.1 instead of those in Table 8.2.

12.4 TRANSIENT RESPONSE OF MULTIPLE-POLE CIRCUITS

In multistage amplifiers the gain function is comprised of many poles. We shall consider the time response of amplifier circuits, with identical poles and nonidentical real poles, respectively.

A. Identical Real Poles

At this point let us consider the time-domain response and examine the output for a unit step input for an amplifier having the voltage gain function given by

$$A(s) = \frac{V_o(s)}{V_i(s)} = \frac{A_o}{(1 + s/p_o)^n} = \frac{A_o p_o^n}{(s + p_o)^n} \tag{12.14}$$

where V_o and V_i are the transforms of the output and input voltages and n is the number of poles. If the input is a unit step voltage, i.e., $V_i(t) = u(t)$, then $V_i(s) = 1/s$ and from (12.14)

$$V_o(s) = \frac{A_o p_o^n}{s(s + p_o)^n} \tag{12.15}$$

The output voltage $v_{on}(t)$, normalized with respect to A_o [i.e., $V_{on}(s) = V_o(s)/A_o$], is given by

$$v_{on}(t) = \mathcal{L}^{-1}[V_{on}(s)] = \mathcal{L}^{-1}\frac{p_o^n}{s(s + p_o)^n} \tag{12.16}$$

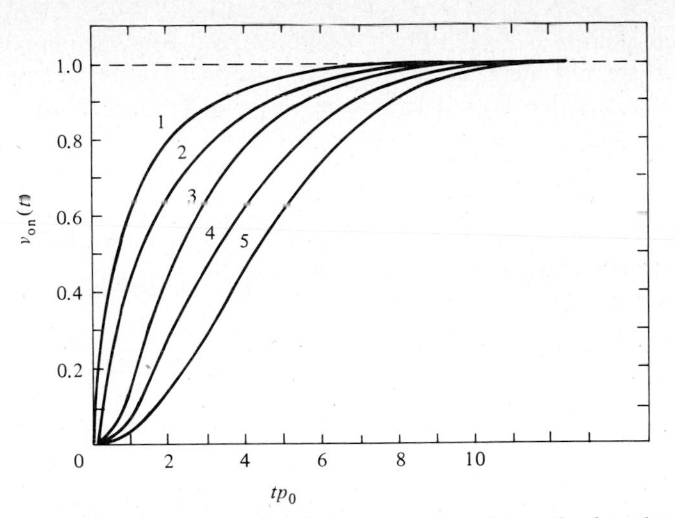

FIGURE 12.7 Normalized step response of transfer functions with n identical real poles ($n = 1$ to $n = 5$).

From a table of Laplace transforms we obtain

$$v_{on}(t) = 1 - e^{-tp_o} \sum_{k=0}^{n-1} \frac{(p_o t)^k}{k!} \tag{12.17}$$

The plot of (12.17) for n up to 5 is shown in Fig. 12.7. The 10 to 90 percent rise time τ_R is listed in Table 12.2. In Table 12.2 we have also included the product $f_{3dB}\tau_R$, where f_{3dB} is given by (12.10). Note that this product is approximately the same for each case, i.e.,

$$f_{3dB}\tau_R \simeq 0.34 \tag{12.18}$$

The result in (12.18) is also found to be true for n higher than 5. Note that this result is in agreement with the empirical relation given in (12.10).

B. Nonidentical Real Poles: Low-Pass Case

The poles of a cascaded amplifier, in general, may not be identical. Let the gain function of the amplifier be given by

$$A(s) = \frac{A_o}{(1 + s/p_1)(1 + s/p_2)(1 + s/p_3) \cdots} \tag{12.19}$$

TABLE 12.2 $p_o\tau_R$ and $f_{3dB}\tau_R$ for Identical Real Poles

n	1	2	3	4	5
$p_o\tau_R$	2.2	3.3	4.2	4.9	5.5
$f_{3dB}\tau_R$	0.35	0.34	0.34	0.34	0.34

This situation arises often in BJTs and FETs where due to the interaction of stages, the dominant poles are not identical even though identical stages are used. The frequency-domain response has been discussed in Chap. 8. From (8.23) we have

$$\frac{1}{\omega_{3\text{dB}}^2} = \left(\frac{1}{p_1}\right)^2 + \left(\frac{1}{p_2}\right)^2 + \left(\frac{1}{p_3}\right)^2 + \cdots \tag{12.20}$$

In the time domain the rise time τ_R of the output for a step input is determined by using the approximate relation from (12.10) and (12.20):

$$\tau_R^2 = \tau_{R1}^2 + \tau_{R2}^2 + \tau_{R3}^2 + \cdots = \sum_{i=1}^{N} \tau_{Ri}^2 \tag{12.21}$$

where τ_{R1}, τ_{R2}, etc., correspond to the rise time due to poles p_1, p_2, p_3, etc. In other words, $\tau_{Ri} = 2.2/p_i$.

EXAMPLE 12.2

The 3-dB bandwidth of a multistage CE transistor amplifier was determined in Example 8.6, namely, $f_{3\text{dB}} = 2.54$ MHz. Determine the approximate 10 to 90 percent rise time of the amplifier. From (12.10)

$$\tau_R = \frac{0.35}{2.54 \times 10^6} = 0.14\ \mu\text{s}$$

EXERCISE 12.3

An amplifier with identical poles has the following voltage gain function:

$$A(s) = \frac{5 \times 10^{30}}{(s + 10^7)^4}$$

Determine the exact and the approximate rise times τ_r.

Ans $\tau_r(\text{exact}) = 0.56\ \mu\text{s}$, $\tau_r(\text{approximate}) \simeq 0.50\ \mu\text{s}$

EXERCISE 12.4

Determine the approximate rise time of an amplifier, for a step input, which has the following gain function:

$$A(s_n) = \frac{A_o}{(s_n + 1)(s_n + 2)(s_n + 3)} \qquad s_n = \frac{s}{10^6}$$

Ans 2.56 μs

12.5 STEP RESPONSE OF A TYPICAL HIGH-PASS CIRCUIT AND SAG CALCULATION

The transfer function of an individual high-pass circuit, as shown in Sec. 7.4, is of the form of (7.40), rewritten as

$$A = \frac{V_o}{V_i} = A_m \frac{s + z_1}{s + p_1} \tag{12.22}$$

where $p_1 > z_1$ and both are real quantities and V_o and V_i are the transforms of the output and input signals voltages. We shall now determine the response of a typical high-pass circuit in the time domain.

The transform of the signal for a unit step input [i.e., $v_i(t) = u(t)$] is $V_i = 1/s$. Thus, the transform of the output signal is

$$V_o(s) = \frac{A_m}{s} \frac{s + z_1}{s + p_1} \tag{12.23}$$

The inverse Laplace transform of (12.23), obtained by simple partial-fraction expansion, is

$$v_o(t) = A_m \frac{z_1}{p_1}\left[1 + \left(\frac{p_1}{z_1} - 1\right)e^{-p_1 t}\right] \tag{12.24}$$

The magnitude of the output step at $t = 0$ is equal to A_m, in accordance with the initial value theorem of Laplace transforms. The output response decays with a time constant $(1/p_1)$, and the final value is $(A_m z_1/p_1)$, as shown in Fig. 12.8. For linear pulse amplification, the input pulse has a pulse duration time of, say, τ_p. We define a pulse sag δ as

$$\delta = \frac{A_m - v_o(\tau_p)}{A_m} = 1 - \frac{v_o(\tau_p)}{A_m} \tag{12.25}$$

The above quantities are illustrated in Fig. 12.9. To calculate the sag, we then use (12.25) and (12.24). Usually $p_1\tau_p \ll 1$, and we can use an approximation by retaining the linear term of the exponential expansion, i.e.,

$$e^{-p_1\tau_p} \simeq 1 - p_1\tau_p \tag{12.26}$$

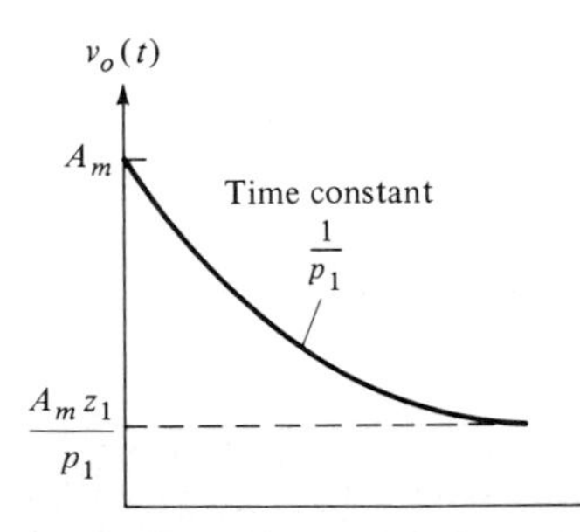

FIGURE 12.8 Unit step response of a typical high-pass function.

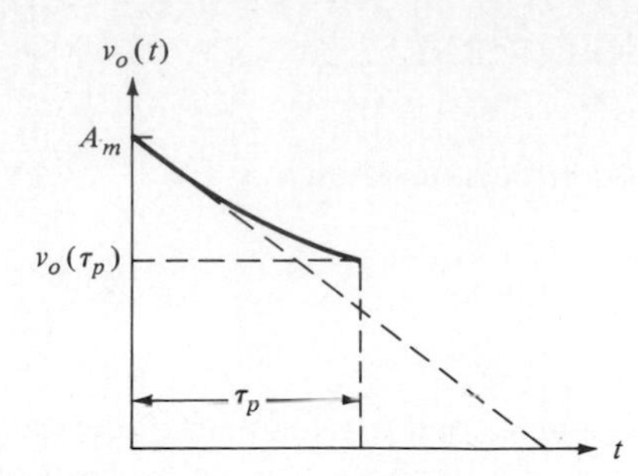

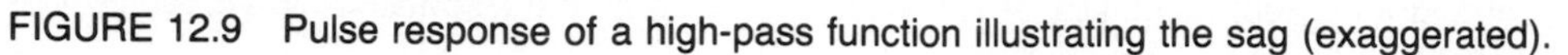
FIGURE 12.9 Pulse response of a high-pass function illustrating the sag (exaggerated).

Then from (12.26), (12.25), and (12.24) we have

$$\delta \simeq (p_1 - z_1)\,\tau_p \qquad \textbf{(12.27)}$$

Note that if we normalize $v_o(t)$ such that $v_o(0^+) = 1$ and then differentiate (12.24), we obtain

$$\frac{dv_{\text{on}}}{dt} = -(p_1 - z_1)e^{-p_1 t} \qquad \textbf{(12.28}\boldsymbol{a}\textbf{)}$$

$$\left.\frac{dv_{\text{on}}}{dt}\right|_{t=0^+} = z_1 - p_1 \qquad \textbf{(12.28}\boldsymbol{b}\textbf{)}$$

Hence the negative of the slope of the normalized output at $t = 0^+$ yields the sag per second; that is, from (12.27) and (12.28*b*)

$$\text{Sag/s} = \frac{\delta}{\tau_p} \simeq p_1 - z_1 = \left.\frac{-dv_{\text{on}}}{dt}\right|_{t=0^+} \qquad \textbf{(12.29)}$$

For many sources of sag, if the sags of the individual coupling bypass circuits are small [so that the initial slope $v_o(0^+)$ can be used to represent the sag], it can be shown that (problem 12.9)

$$\frac{d}{dt}v_o(0^+) = \sum_{i=1}^{N} (z_i - p_i) \qquad \textbf{(12.30)}$$

or alternatively stated,

$$\text{Total sag} = \tau_p \sum \text{individual sags (slopes)} \qquad \textbf{(12.31)}$$

EXAMPLE 12.3

Let a voltage pulse of amplitude A (small enough so as to operate in the linear region) and a pulse duration of 0.1 ms be applied to the FET amplifier example of Sec. 7.4. We wish to calculate the total sag.

Since $\tau_p p_1$ for each circuit (see Sec. 7.5) is much less than unity, we can use (12.27) and (12.31). The gate and drain coupling circuits each contribute a percent sag of

$$\delta_g = \delta_d = (p_1 - z_1)\tau_p = 0.1(10^{-4}) = 10^{-5}$$

The source bypass circuit contributes a sag of

$$\delta_s = (2.66 \times 10^2 - 0.66 \times 10^2)(10^{-4}) = 0.02$$

Hence the total sag is $\delta = \delta_s + \delta_d \simeq 0.02$. In other words (see Fig. 12.10), the output pulse at τ_p will be lower in amplitude than at $t = 0^+$ by 2 percent. Note that the rise time is assumed to be negligibly small and is not shown in Fig. 12.10.

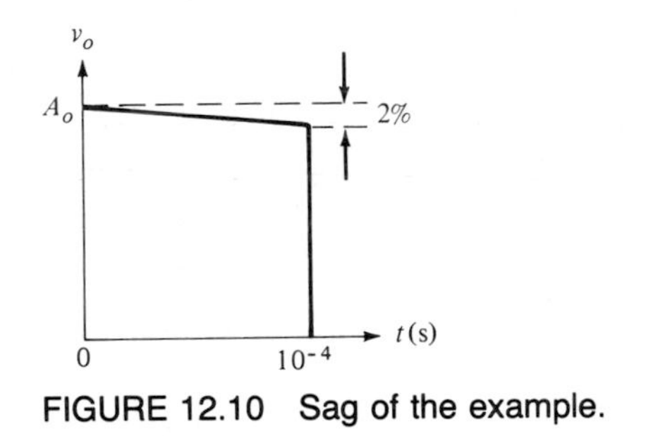

FIGURE 12.10 Sag of the example.

EXERCISE 12.5

A voltage pulse of small amplitude and 10-μs duration is applied to the BJT amplifier design example of Sec. 7.14. Determine the approximate rise time and the total sag in the output pulse.

Ans $\tau_R = 7 \times 10^{-8}$ s, $\delta = 6.3$ percent

12.6 TRANSIENT AND SLEW CONSIDERATIONS IN OP AMPS

We have already mentioned the slew rate (SR) of op amps in Sec. 9.9. In fact, the value of SR is determined by the compensating capacitor, hence the unity gain bandwidth as given in (9.74). The slew-rate limitation due to high-frequency and/or large-signal performance of an op amp will be considered in some detail. First consider a sinusoidal signal input at high frequencies. Let

$$v_i = V_m \sin \omega t \tag{12.32}$$

The maximum rate of change of the waveform is at $\omega t = 0$, namely,

$$\left.\frac{dv_i}{dt}\right|_{\max} = \omega V_m \cos \omega t \bigg|_{\omega t = 0} = \omega V_m \tag{12.33}$$

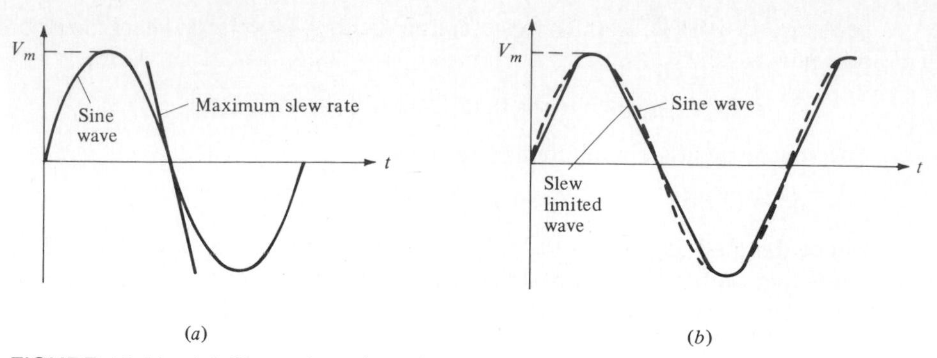

FIGURE 12.11 (*a*) Illustration of maximum slew rate. (*b*) Waveform distorted due to small value of slew rate.

The slope corresponding to (12.33) is shown in Fig. 12.11*a*. If ωV_m is larger than the slew-rate limit of the op amp, the output will not be able to follow the input as shown in Fig. 12.11*b*. This distortion is caused by the inability of capacitances in the op amp to charge or discharge fast enough to follow the signal. Special-purpose op amps with high slew rates are available (SR $\simeq$ 100 V/μs). To illustrate the SR effect, consider the following example.

EXAMPLE 12.4

The slew rate of the 741 op amp is SR $\simeq 0.5$ V/μs. The op amp is used in closed-loop inverting configuration with a gain of $-K$. For a sine-wave input $v_i = V_m \sin \omega t$ with $V_m = 0.2$ V and $\omega = 2\pi(10^5)$ rad/s, determine the maximum gain to prevent slew limiting.

The output is given by $v_o = -KV_m \sin \omega t$. From (12.33)

$$KV_m\omega \leq \text{SR} \qquad \text{or} \qquad K \leq \frac{0.5 \times 10^{-6}}{0.2(2\pi \times 10^5)} \simeq 4$$

Hence the closed-loop gain of the op amp has to be less than 4 to handle this signal without slew-limiting distortion. Note that for a given gain, K, and amplitude of the sine wave the upper bound of the frequency is readily determined from above.

As a further illustration of the slew rate limitation consider the following example.

EXAMPLE 12.5

A certain op amp is used in closed-loop inverting configuration with a gain of -4. The input signal is a square wave with a peak-to-peak voltage of 1 V and a period of 2 μs, as shown in Fig. 12.12*a*. Sketch the output waveform for the following cases:

1. An op amp with a slew rate of 20 V/μs is used.
2. A 741 op amp (slew rate = 0.5 V/μs) is used.

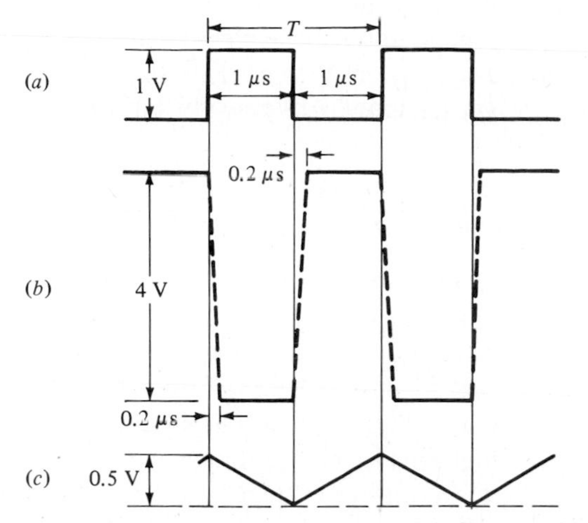

FIGURE 12.12 (*a*) Input pulse. (*b*) Output of inverting op amp with high slew rate. (*c*) Output of inverting op amp with a low value of slew rate.

For the higher-slew-rate op amp, namely, SR = 20 V/μs, we have 4 V/20 V/μs = 0.2 μs, which is much smaller than the period of the input signal. The output voltage reaches the full closed-loop gain, and the output signal has a peak-to-peak voltage of 4 V and is shown in Fig. 12.12*b*.

For the lower-SR case, namely, SR = 0.5 V/μs, we have 4 V/0.5 V/μs = 8 μs, which is larger than the period of the input signal, and the output voltage cannot reach the full peak-to-peak amplitude. In this case the amplifier can slew to $T/2$, i.e., to 1 μs, before the signal changes. Hence during this time it can reach only (0.5 V/μs)(1 μs) = 0.5 V peak to peak. The waveform is shown in Fig. 12.12*c*.

EXERCISE 12.6

The slew rate of an LF351 is 13 V/μs. Determine the maximum frequency for which the output voltage of a 5-V peak is undistorted.

Ans $f_{max} = 414$ kHz

EXERCISE 12.7

A step input of 5 V is applied to a 741 op amp in a unity gain configuration. How long will it take for the output to change by this voltage?

Ans 10 μs

We shall next consider large-signal behavior in the time domain, namely, the switching speed of the devices when it is used in the ON-OFF mode. In these cases linear-analysis methods using the frequency-domain approach are not applicable. Nonlinear analysis and computer-aided analysis must be used. In some cases modified linearized approaches can be used to get an approximate result, but caution must be exercised.

12.7 TRANSIENT TIME OF *pn*-JUNCTION DIODES

When a *pn* junction diode is driven from an ON state to an OFF state or vice versa, it takes some time for the diode to reach steady-state value. For example, consider the diode circuit shown in Fig. 12.13*a*. In the diode circuit the reverse recovery time is appreciable and cannot be ignored in fast switching circuits. When the applied input voltage is suddenly reversed, as shown in Fig. 12.13*b*, the diode current will not suddenly change to the steady-state value. This is because minority carriers are stored, which will require some time for the current to drop to zero. Hence, the diode will continue to conduct for a time period t_s (the storage time). The transition or decay time t_d depends on the reverse-biased junction capacitance discharging to 10 percent of its value from $-V_R$. Thus the total transient of a diode $t_T = t_s + t_d$. This recovery time interval can have adverse effects in BJT inverters in saturation where both diodes will be conducting simultaneously. Manufacturers usually specify this time interval as the reverse recovery time of the diode. Commercially available switching diodes have t_T

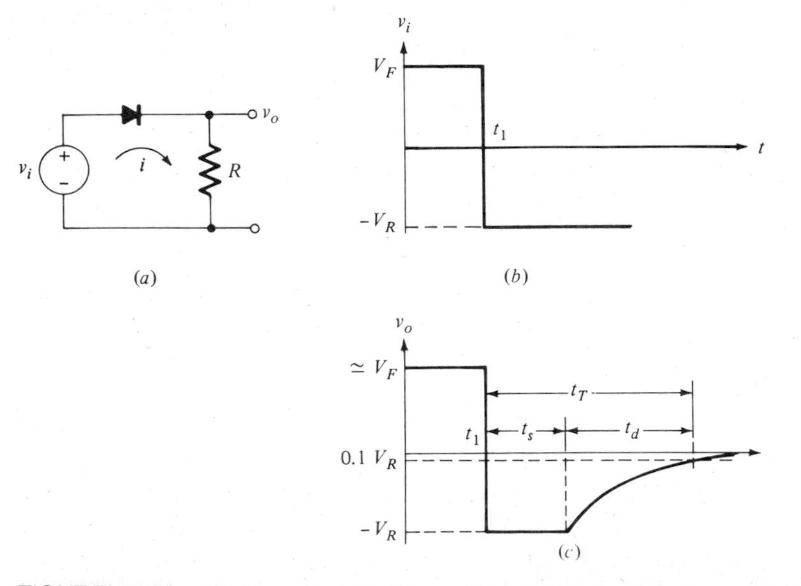

FIGURE 12.13 (*a*) A *pn* junction diode circuit. (*b*) Input waveform. (*c*) Output waveform.

from less than 1 ns up to as high as 1 μs. The high value is usually for diodes intended for switching large currents. The storage time can be almost eliminated by the use of Schottky diodes (see Appendix B), and subnanosecond recovery time can be achieved.

If we define $I_F = V_F/R$ [actually $(V_F - 0.7)/R$] and $I_R = V_R/R$, the storage time t_s, at which point all the stored charge has been removed, is given by

$$t_s = \tau_F \ln\left(1 + \frac{I_F}{I_R}\right) \tag{12.34}$$

where $\tau_F = Q_{SF}/I_F$ and Q_{SF} is the stored charge under steady-state forward-biased conditions. The decay time t_d is determined from the exponential decay equation, namely, from

$$v_o = -V_R e^{-t/RC_c} \tag{12.35a}$$

where C_c is the capacitance of the reverse model (see Fig. 1.41*b*). From (12.35*a*) setting $v_o = -0.1V_R$ determines the value of t_d, namely,

$$t_d = RC_c \ln 10 = 2.3RC_c \tag{12.35b}$$

EXAMPLE 12.5

Determine the reverse recovery time t_T of the diode in Fig. 12.13*a* if I_F = 50 mA, I_R = 10 mA, R = 500 Ω. Given τ_F = 10 ns, and C_c = 5 pF. From (12.34)

$$t_s = 10 \ln (1 + 5) = 17.9 \text{ ns}$$

From (12.35*b*)

$$t_d = 2.3(500 \times 5 \times 10^{-12}) = 5.7 \text{ ns}$$

Hence t_T = 23.6 ns.

EXERCISE 12.8

For the circuit shown in Fig. 12.13, $V_F = -V_R = 5$ V and $R = 1$ kΩ. Given $\tau_F = 20$ ns, determine the diode storage time.

Ans $t_s = 13.9$ ns

12.8 THE TRANSISTOR SWITCH: ON-OFF TIME-INTERVAL CALCULATIONS

Transistors are often used under large-signal conditions in the switching mode in digital circuits. In BJT recall that when a transistor is turned ON into the saturation region from the OFF mode, the operation transverses the active region. The ON-to-

OFF switching encounters the minority storage-time effects discussed in the previous section.

The collector current for a large input voltage pulse ($V_1/R_s > I_c/\beta_0$) is shown in Fig. 12.14. Note that the turn-on time consists of a delay time and a rise time, i.e.,

$$t_{on} = t_d + t_r \tag{12.36}$$

where t_d is defined to be the time interval between the application of base drive and the 10 percent value of collector saturation current, and t_r is defined to be the 10 to 90 percent rise time of the collector current. The turn-off time consists of a storage time and a fall time, i.e.,

$$t_{off} = t_s + t_f \tag{12.37}$$

where t_s is the storage time and t_f is defined as the 90 to 10 percent fall time of the collector current. The storage time is the time interval between the removal of the base drive ($t = T_p$) and the point where i_c just *starts* to decrease toward zero.

An accurate calculation of the time intervals can be made by a nonlinear analysis of the circuit using, say, the nonlinear model shown in Fig. 2.24. For such calculations, computer-aided analysis is a must. In fact, computer programs are available, such as SPICE,* that can handle nonlinear (as well as linear) analysis.

---

*See footnote page 43.

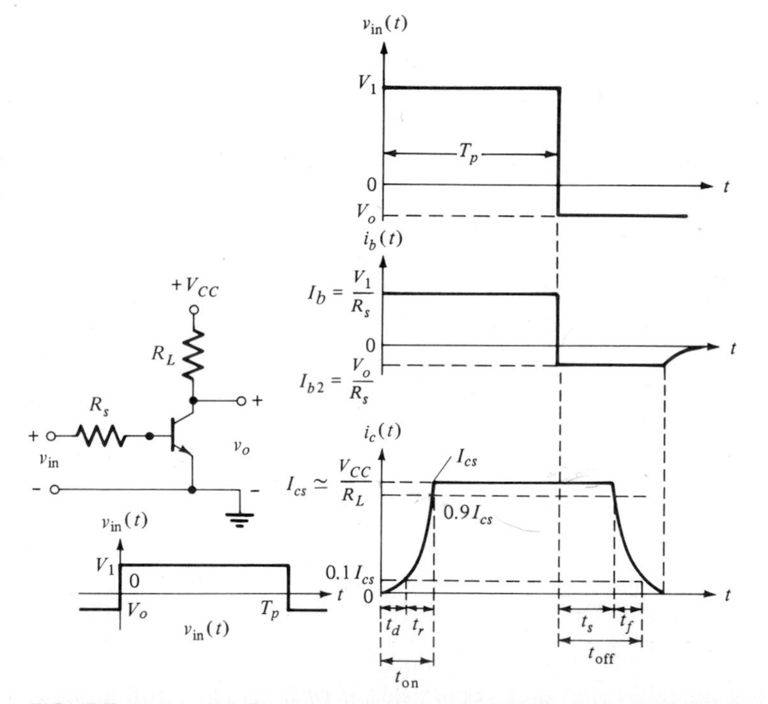

FIGURE 12.14 (*a*) Transistor invertor circuit with input voltage waveform. (*b*) Waveforms of input voltage base current and collector current.

We shall, however, use a linearized approximate method for the calculation of various time intervals. The various time intervals are calculated as follows.

A. Delay Time

To calculate t_d, we must find the time required for the emitter-base junction voltage to increase from $-V_o$ to a value that will result in a collector current equal to $0.1I_{cs}$. However, the following approximations are often made for simplicity. We find instead the time interval for v_{be} to increase from $-V_o$ to $(V_{BE})_{sat} \simeq 0.7$ V with a single time constant $\tau_d = R_s C_{ibo}$, where C_{ibo} is the value of the (reverse-biased) emitter-base junction capacitance. The value of C_{ibo} may be specified by the manufacturer. If this value is not available, a rough estimate is $C_{ibo} \simeq \overline{C}_c \simeq 2C_c$ at $-V_o$ (the bar indicates average value of C_c as the voltage changes). Usually, t_d is very small and the error in calculating t_d does not have a pronounced effect on t_{on}.

The expression for the emitter-base voltage increasing from $-V_o$ to a target voltage V_1, with a single time constant τ_d, is given by

$$v_{be} = V_1 - (V_o + V_1)e^{-t/\tau_d} \tag{12.38}$$

The delay time t_d is found by setting $v_{be} = (V_{BE})_{sat} \simeq 0.7$ V. Hence

$$0.7 = V_1 - (V_o + V_1)e^{-t_d/\tau_d} \tag{12.39a}$$

or

$$t_d = \tau_d \ln \frac{V_1 + V_o}{V_1 - 0.7} \tag{12.39b}$$

B. Rise Time

In order to calculate t_r, we can use the method of analysis discussed in Sec. 12.2, since the transistor is in the active region of operation. In other words, we use the approximate equivalent circuit of the transistor and consider a step input as shown in Fig. 12.15. An important difference between the model of Fig. 12.15 and that of Fig. 7.19*b* is that average values are used, indicated by a bar in Fig. 12.15. This must be done since the operating-point values change considerably, i.e., I_c varies from the initial off state (i.e., zero) to the saturation value $I_{cs} = V_{CC}/R_L$ and the collector junction voltage V_{CE} varies from the initial off state at V_{CC} to $(V_{CE})_{sat}$, which are different by an order of

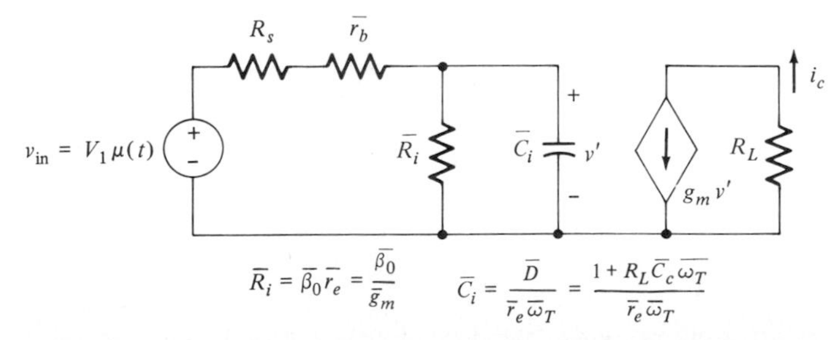

FIGURE 12.15 Transistor model in the active mode for t_r and t_f calculations.

magnitude. For these changes, I_e, β_0, C_c, r_e, and ω_T all change. As a first approximation, we may assume r_b, β_0, and ω_T to be constants. We linearize r_e and C_c. One such approximation, which has been found satisfactory, is that r_e is determined from the midpoint value of I_c, namely, the average of the initial value $I_{ci} = 0$ and the final saturation value $I_{cs} = V_{CC}/R_1$,

$$\overline{g}_m \simeq \frac{q}{kT}\frac{I_{ci} + I_{cs}}{2} \simeq \frac{q}{kT}\frac{V_{CC}}{2R_L} = \frac{1}{\overline{r}_e} \tag{12.40}$$

and the linearized collector-junction capacitance is determined from

$$\overline{C}_c \simeq 2C_c \qquad \text{evaluated (or measured) at } V_{CC} \tag{12.41}$$

For a current drive, i.e., $R_s \gg r_b + \overline{R}_i$, we have $i_b = (V_1/R_s)u(t)$.

The step response for the circuit of Fig. 12.15 is a one-pole response. For an overdriven condition ($i_c > V_{CC}/R_L$), the step response is shown in Fig. 12.16. The expression for the collector current (dashed curve in Fig. 12.16), in terms of the base drive current for the single-pole circuit of Fig. 12.15, can be written as

$$i_c(t) = I_\infty + (I_o - I_\infty)e^{-t/\tau_1} \tag{12.42}$$

where
$$\tau_1 = \frac{1}{p_1} = \frac{\overline{D}}{\omega_\beta} = \frac{\beta_0(1 + R_L\overline{C}_c\omega_T)}{\omega_T} \tag{12.43}$$

$$\begin{aligned} i_c(0) &= I_o = 0 \\ i_c(\infty) &= I_\infty = \beta_0 I_b = \beta_0 \frac{V_1}{R_s} \end{aligned} \tag{12.44}$$

The actual collector current will, of course, not reach the overdrive value but will saturate at $I_{cs} = V_{CC}/R_L$. The time required for the collector current to reach from the zero value to the saturated value T_1 is calculated from (12.42) and (12.44); that is,

$$I_{cs} = \beta_0 I_b - \beta_0 I_b e^{-T_1/\tau_1} \tag{12.45}$$

From (12.41) and (12.45) we obtain

$$T_1 = \tau_1 \ln \frac{\beta_0 I_b}{\beta_0 I_b - I_{cs}} = \frac{\overline{D}}{\omega_\beta} \ln \frac{K}{K - 1} \tag{12.46}$$

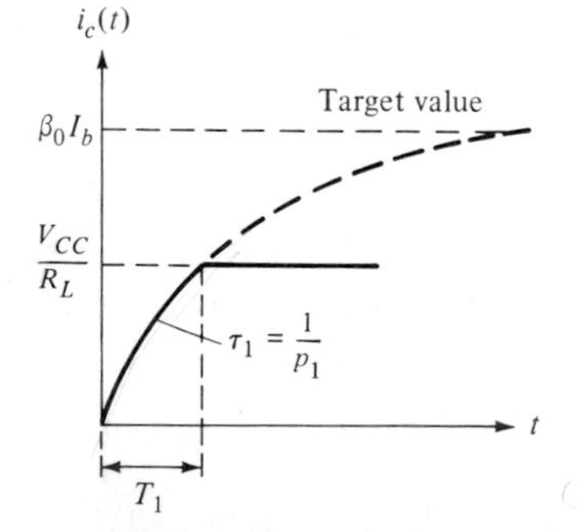

FIGURE 12.16 Collector-current waveform for rise-time calculations.

where

$$K = \frac{\beta_0 I_b}{I_{cs}} > 1 \tag{12.47}$$

K is usually called the *turn-on overdrive*. Usually (12.46) is used as the value for t_r for an overdriven stage, i.e., $t_r \simeq T_1$. However, it is a simple matter to find the 10 to 90 percent rise time t_r. From (12.45), namely,

$$0.1 I_{cs} = \beta_0 I_b - \beta_0 I_b e^{-t_1/\tau_1} \tag{12.48a}$$

$$0.91 I_{cs} = \beta_0 I_b - \beta_0 I_b e^{-t_2/\tau_1} \tag{12.48b}$$

From (12.48) and (12.47) we obtain

$$t_r = t_2 - t_1 = \tau_1 \ln \frac{1 - 0.1/K}{1 - 0.9/K} \tag{12.49}$$

For $K \gg 1$, and using the approximation

$$\ln (1 + x) = x - \frac{x^2}{2} + \cdots \qquad x \ll 1 \tag{12.50}$$

(12.49) simplifies to

$$t_r \simeq \tau_1 \frac{0.8}{K} \tag{12.51}$$

Note that for K large, t_r is much smaller than τ_1. However, a large value of K increases the storage time t_s, as we shall see below.

C. Storage Time

The transistor in saturation has a saturation charge of excess minority carriers stored in the base as we saw in the case of *pn*-junction diodes. The transistor, therefore, cannot respond to the trailing edge of the driving pulse until this excess charge has been removed. The lapse of time due to the storage delay and the time constant associated with it are discussed in detail elsewhere [1]. The storage time constant is given by

$$\tau_s \simeq \frac{\omega_\alpha + \omega_I}{\omega_\alpha \omega_I (1 - \alpha_0 \alpha_I)} \tag{12.52a}$$

$$\simeq \frac{\beta_I}{\omega_I} \qquad \text{for } \omega_\alpha \gg \omega_I,\ \beta_0 \gg 1 \tag{12.52b}$$

where ω_I and α_I are the alpha cutoff frequency and dc alpha for the inverse active region and α_0 and ω_α are for the normal active region.

Now i_c falls off from a value of I_{c1} to I_{c2}, with a single time constant τ_s. This can be written as

$$i_c(t) = I_{c2} + (I_{c1} - I_{c2}) e^{-t/\tau_s} \tag{12.53}$$

The storage time is the particular value of t, namely, t_s, when $i_c = I_{cs}$. Hence, from (12.53)

$$t_s = \tau_s \ln \frac{I_{c1} - I_{c2}}{I_{cs} - I_{c2}} \tag{12.54a}$$

$$= \tau_s \ln \frac{I_{b1} - I_{b2}}{I_{bs} - I_{b2}} \tag{12.54b}$$

where $I_{bs} = I_{cs}/\beta_0 = V_{CC}/\beta_0 R_L$. Equations (12.54) can also be written as

$$t_s = \tau_s \ln \frac{1 + K/M}{1 + 1/M} \tag{12.55}$$

where K is defined by (12.47) and $M = -I_{b2}(I_{cs}/\beta_0)$ is the turn-off overdrive factor. Note that the value of $I_b = V_1/R_s$ is used for I_{b1}. If both K, $M \gg 1$, then (12.55) simplifies to

$$t_s \simeq \tau_s \ln \left(1 + \frac{K}{M}\right) \tag{12.56}$$

Note that a large value of K increases t_s and that this is the price paid for getting a smaller t_r. Usually, either τ_s is given directly by the manufacturer, or it is found from (12.52) in terms of the parameters of the inverse and normal active region. The value of t_s can also be determined by measuring the storage time for known base-drive conditions from (12.55). Note that (12.54b) assumes positive current I_{b2}. Further, for $I_b < I_{bs}$, the transistor will come out of saturation.

D. Fall Time

The fall-time calculation is straightforward, since the transistor operates in the active region. It parallels the rise-time calculation. Hence we write the equation for fall time, for i_c to fall from $0.9I_{cs}$ to $-\beta_0 I_{b2}$ with a time constant τ_1, as shown in Fig. 12.17. Note that the negative sign of I_{b2} is included in this case. This can be written as

$$i_c(t) + \beta_0 I_{b2} = (0.9I_{cs} + \beta_0 I_{b2})e^{-t/\tau_1} \tag{12.57}$$

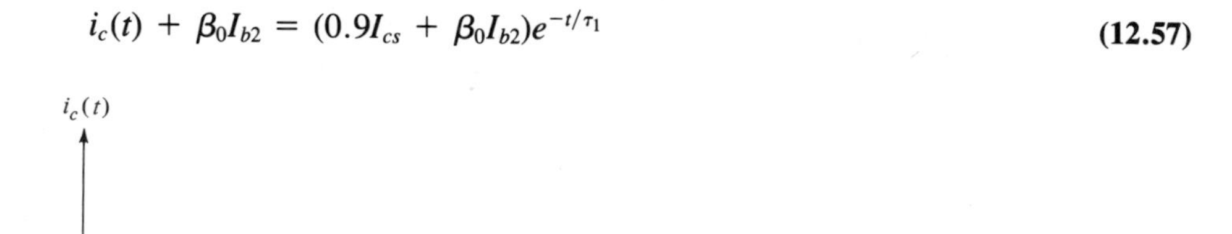

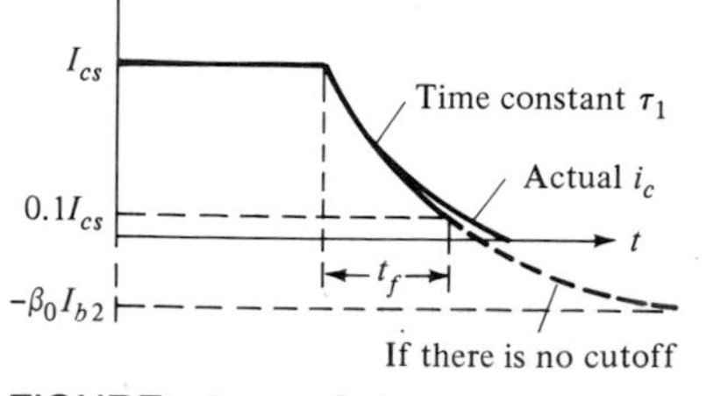

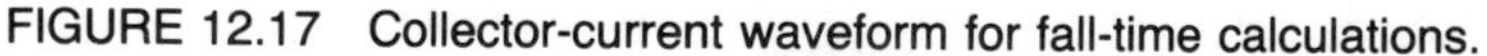

FIGURE 12.17 Collector-current waveform for fall-time calculations.

The fall time is the particular value of t, namely, t_f, when $i_c = 0.1I_{cs}$. Hence

$$t_f = \tau_1 \ln \frac{0.9I_{cs} + \beta_0 I_{b2}}{0.1I_{cs} + \beta_0 I_{b2}} \tag{12.58}$$

Equation (12.58) can also be written as

$$t_f = \tau_1 \ln \left(\frac{1 + 0.9/M}{1 + 0.1/M} \right) \tag{12.59}$$

For large turn-off overdrive, $M \gg 1$, (12.59) simplifies to

$$t_f \simeq \tau_1 \frac{0.8}{M} \tag{12.60}$$

From (12.60) it is seen that the larger the turn-off drive, the smaller the fall time for the transistor. Compare (12.64) with the rise-time result (12.54).

Finally, note that if the stage is driven by a voltage source (i.e., R_s is small) rather than a current source, the expression for I_b, namely, (12.44), will be different and indeed smaller than the current-drive stage. In order to illustrate the derived relations, we shall consider the following example.

EXAMPLE 12.7

Consider the pulse response of the circuit shown in Fig. 12.14*a*. The circuit parameters are as follows: $V_{CC} = 10$ V, $R_L = 1$ kΩ, $R_s = 10$ kΩ. The transistor parameters, which may be assumed constant in the operating range, are $\beta_0 = 100$, $f_T = 200$ MHz, and $r_b = 50$ Ω. We are also given that $C_c = 5$ pF at 10 V and that in the inverse active region $\omega_I = 10^8$ rad/s and $\beta_I = 10$. The input step is a pulse voltage of amplitude +5 V and a pulse duration of $T_p = 10$ μs. The turn-on and turn-off times of the transistor switch are to be determined.

We estimate $C_{ibo} \simeq 2C_c \simeq 10$ pF. Hence $\tau_d = R_s C_{ibo} = 10^{-7}$ s. From (12.39*b*) we calculate

$$t_d = 10^{-7} \ln \frac{5 + 0}{4.3} = 1.5 \times 10^{-8} \text{ s}$$

We next determine the average values for the active region; I_c changes from an off value of 0 mA to an on value of 10 mA. Hence

$$\bar{I}_e \simeq \frac{I_{ci} + I_{cs}}{2} = \frac{0 + 10}{2} = 5 \text{ mA} \qquad \text{and} \qquad \bar{r}_e = \frac{25}{5} = 5\ \Omega$$

Also

$$\overline{C}_c = 2C_c = 2(5 \text{ pF}) = 10 \text{ pF}$$

From (12.43)

$$\tau_1 = \frac{\overline{D}}{\omega_\beta} = \frac{\beta_0(1 + R_L\overline{C}_c\omega_T)}{\omega_T} = 1.08\ \mu\text{s}$$

From (12.49) and (12.47), with $K = 100(0.5)/10 = 5$, we have

$$t_r = 1.08 \times 10^{-6} \ln \frac{4.9}{4.1} = 0.19\ \mu\text{s}$$

Thus $t_{on} = t_d + t_r \simeq 0.21\ \mu$s.

From (12.52), since $\beta_0 \gg 1$ and $\omega_\alpha > \omega_T \gg \omega_I$, the storage time constant is

$$\tau_s \simeq \frac{\beta_I}{\omega_I} = \frac{10}{10^8} = 0.1\ \mu\text{s}$$

Hence from (12.54), with $I_{b2} = 0$ (i.e., $M = 0$), we have

$$t_s = 0.1 \times 10^{-6} \ln \frac{0.5}{0.1} = 0.16\ \mu\text{s}$$

Finally, the fall time for zero overdrive, i.e., $I_{b2} = 0$, is found from (12.58):

$$t_f = 1.08 \times 10^{-6} \ln \frac{9}{1} = 2.38\ \mu\text{s}$$

Thus $t_{off} = t_s + t_f = 2.54\ \mu$s.

[Note that if a turn-off overdrive factor of 8, i.e., $M = 8$, is used, from (12.60) $t_f \simeq 1.08 \times 10^{-6}(0.8/8) = 0.108\ \mu$s, and from (12.56) $t_s \simeq 0.1 \times 10^{-6} \ln(1 + \frac{5}{8}) = 0.05\ \mu$s.]

EXERCISE 12.9

For the circuit shown in Fig. 12.14, the following are given: $V_{CC} = 5$ V, $R_s = 10$ kΩ, $R_L = 1$ kΩ. The transistor parameters are $\beta_F = 100$, $f_T = 500$ MHz, $r_b = 0$, $C_c = 2$ pF at 5 V, $\beta_I = 10$, and $f_I = 20$ MHz. An input pulse voltage of amplitude +5 V with a pulse duration of $T_p = 10\ \mu$s is applied. Determine (*a*) the turn-on time and (*b*) the turn-off time.

Ans (*a*) $t_{on} = 0.04\ \mu$s, (*b*) $t_{off} = 1.13\ \mu$s

The turn-on and turn-off times of the transistor can be reduced by external circuitry in a number of ways. Some of these are discussed in the following. It is to be noted that fast switching transistors, with t_{off} and t_{on} under 20 ns, are commonly used in digital computer systems.

12.9 CIRCUITRY TO IMPROVE THE SWITCHING TIME OF A TRANSISTOR

Some of the simple techniques to reduce t_{on} and t_{off} of a transistor are briefly discussed in the following section.

A. Speed-up Capacitors

The turn-on time can be improved by reducing t_r. From (12.51) it is seen that t_r can also be reduced by broadbanding the stage, i.e., reducing τ_1.

From the bandwidth calculations in Chap. 7, it is obvious that when R_s is small, the bandwidth is larger, and the stage current gain lower than if R_s is large. In switching circuits it is desirable, however, to drive the stage by a current source, since it is not possible to control the base current accurately when driving it with a voltage source. We shall show in this section how to improve τ_1, hence t_r and t_f, by using a capacitor across the large value of R_s. It should be mentioned at this point that it is also possible to decrease t_r by overdriving the stage. Overdrive increases the storage time, however, as is evident in (12.56), and thus it is undesirable to increase the on overdrive in order to decrease the turn-on time. We shall show later how to eliminate the storage time by using nonsaturating circuits.

Let us consider the potential advantage of using a capacitor across R_s, as shown in Fig. 12.18*a*. The capacitor C_s is usually referred to as the *speed-up* capacitor because, as we shall show, it improves the turn-on and turn-off times. Consider the linear equivalent circuit of the base input of the transistor in the active region shown in Fig. 12.18*b*, where we have assumed $r_b \ll R_s, \overline{R}_i$. The transfer function V_{be}/V_s for this circuit is

$$\frac{V_{be}}{V_s} = \frac{Z_2}{Z_1 + Z_2} = \frac{\overline{R}_i/(1 + s\overline{R}_i\overline{C}_i)}{\overline{R}_i/(1 + s\overline{R}_i\overline{C}_i) + R_s/(1 + sR_sC_s)} \tag{12.61}$$

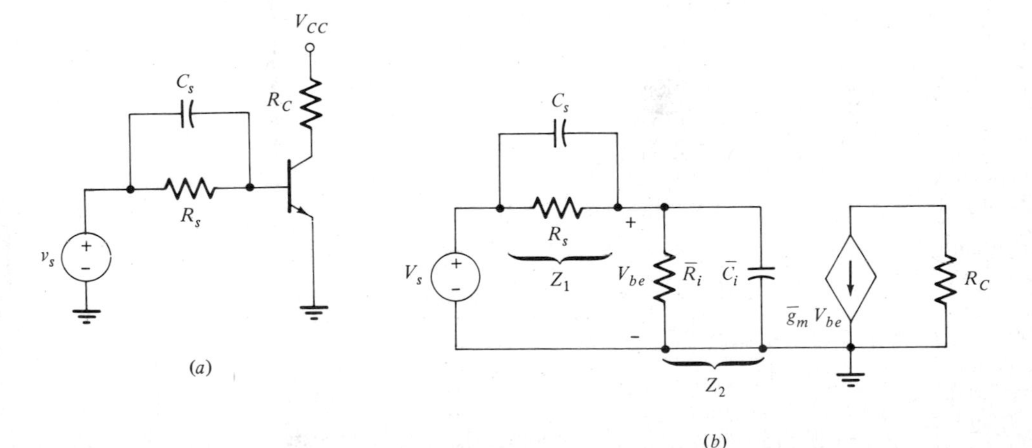

FIGURE 12.18 (*a*) Transistor invertor circuit with a speed-up capacitor. (*b*) Simplified circuit model of (*a*).

For perfect compensation, $R_sC_s = \overline{R}_i\overline{C}_i$, or in terms of the transistor parameters, the proper value of C_s is given approximately by

$$C_s \simeq \frac{\overline{R}_i\overline{C}_i}{R_s} = \frac{\beta_0\overline{r}_e(\overline{D}/\overline{r}_e\omega_T)}{R_s} = \frac{\overline{D}}{R_s\omega_\beta} \tag{12.62}$$

Note that in practice, perfect compensation is not possible. This is mainly due to the actual complicated model of the base-input circuit, also the fact that r_b is ignored in the simplified analysis in the above and that average values are used for R_i and C_i.

In practice, the value of R_s is chosen such that the base current drives the transistor at the edge of saturation, and C_s is then chosen approximately (slightly higher) by (12.62) and then adjusted experimentally. An approximate value of C_s with proper adjustments gives significant improvement in rise and fall times.

EXERCISE 12.10

For Example 12.7 determine the approximate value of C_s for compensation.

Ans 108 pF

B. Circuits That Avoid Saturation

The storage time of a transistor switch can be reduced or almost eliminated by using a Schottky transistor. The Schottky transistor is discussed in Appendix B. In the circuit shown in Fig. 12.19, a clamping Schottky diode is used to prevent saturation. The clamping Schottky diode becomes forward-biased as the transistor enters the saturation region, and thus the excess base current is shunted through the diode into the

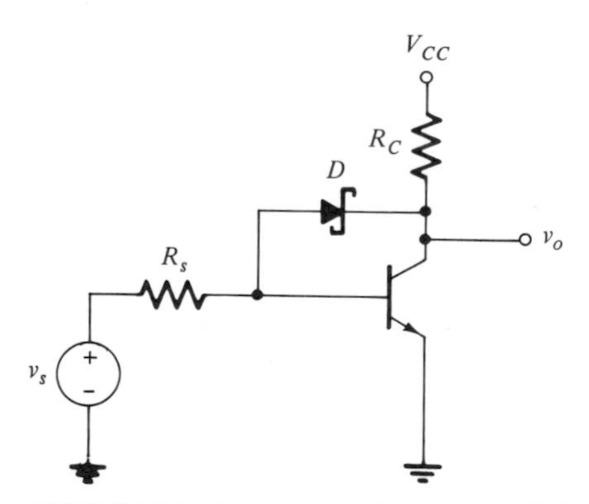

FIGURE 12.19 A Schottky-clamped diode to prevent the transistor from going into saturation.

collector circuit. An example of the Schottky transistor inverter is the advanced Schottky (TTL 74AS), discussed in Sec. 13.7.

12.10 EMITTER-COUPLED INVERTER (CURRENT-MODE CIRCUIT)

One of the most widely used circuits, in applications where fast switching time is important, is the so-called nonsaturating current-mode switch or the emitter-coupled inverter, as shown in Fig. 12.20. The transfer characteristic of the circuit was described in Fig. 2.16. The circuit is seen to be an overdriven differential amplifier as discussed in Sec. 5.11. The emitter-coupled circuit is commonly found in fast integrated-circuit logic, known as ECL (emitter-coupled logic), as discussed in Chap. 13.

The operation of the circuit may be briefly described as follows: Consider v_i to be considerably more negative than the reference voltage V_r, so that Q_1 is off and Q_2 is in the active region. As the voltage v_i is increased, at some value, Q_1 enters the active region, and both transistors are on. As v_i is further increased in the positive direction, Q_2 eventually cuts off, since the base is at a fixed reference potential. Thus, the output voltage v_o is at a value $V_{CC} - IR_c$ for large negative values of v_i, and v_o is equal to V_{CC} for large positive values of v_i. The circuit is called a *current-mode switch* because the current is switched from Q_2 to Q_1 with the total emitter current constant, namely, $I = (V_r + V_{EE} - V_{BE})/R_E$. Thus, when Q_1 is off (Q_2 on), $I_1 = 0$ and $I_2 = I$, and when Q_2 is off (Q_1 on), $I_2 = 0$ and $I_1 = I$. Note that neither of the transistors is in saturation over the entire range of operation, and thus storage time is avoided in the switch. The switching speed of ECL is the fastest and is therefore used in very-high-speed integrated circuits (VHSIC), and a switching speed of less than 1 ns is feasible.

Calculation of the switching speed of ECL requires nonlinear analysis, and even in a linearized approximate analysis all parasitic capacitances must be included. The reader is referred to Ref. 1 for such calculations.

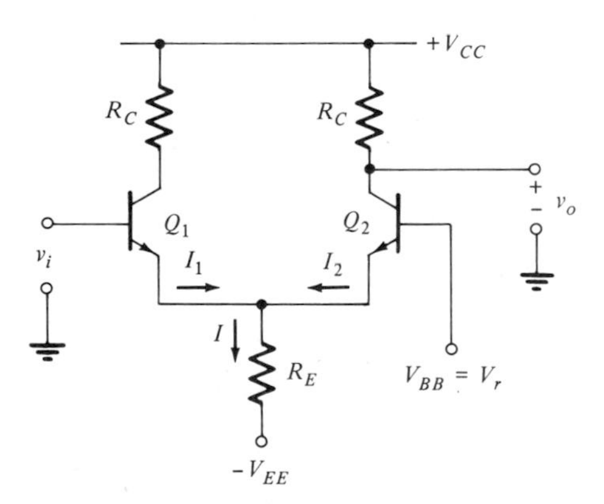

FIGURE 12.20 Emitter-coupled inverter (current-mode circuit), which prevents the transistors from going into saturation.

12.11 SWITCHING SPEED OF CMOS INVERTERS

CMOS switches drive a capacitive load in practice; hence we consider a capacitive load for the ON-OFF time calculations. The CMOS gate and its transfer characteristics are shown in Fig. 3.24. For convenience we show again the CMOS gate with a capacitive load in Fig. 12.21. We shall give the expressions for the turn-off and turn-on times of the CMOS circuit assuming PMOS and NMOS are not identical. Of course, if PMOS and NMOS are identical, i.e., the same V_T and K, then $t_{on} = t_{off}$ since one of the two identical devices is conducting current during each interval.

The turn-off time is defined as the time required for v_o to rise from zero to $0.9V_{DD}$. It is determined from [3, 4]

$$t_{off} = \frac{0.5C_L}{K_p(V_{DD} - V_{Tp})}\left(\frac{2V_{Tp}}{V_{DD} - V_{Tp}} + \ln\frac{V_{DD} - 2V_{Tp} + v_o}{V_{DD} - v_o}\right) \tag{12.63}$$

where v_o is replaced by $0.9V_{DD}$. The turn-on time is defined as the time required for v_o to drop from V_{DD} to $0.1V_{DD}$. It is determined from

$$t_{on} = \frac{0.5C_L}{K_n(V_{DD} - V_{Tn})}\left(\frac{2V_{Tn}}{V_{DD} - V_{Tn}} + \ln\frac{2V_{DD} - 2V_{Tn} - v_o}{v_o}\right) \tag{12.64}$$

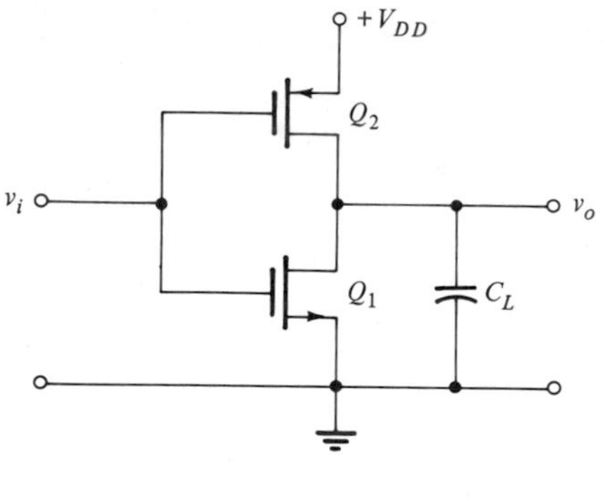

(a)

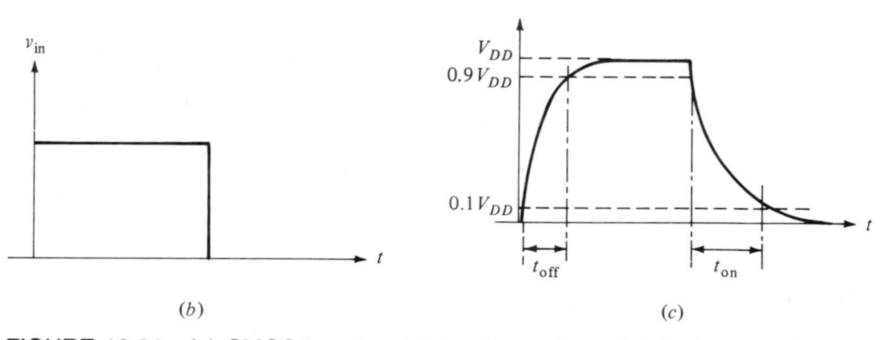

(b) (c)

FIGURE 12.21 (a) CMOS inverter. (b) Input waveform. (c) Output waveform.

where v_o is replaced by $0.1V_{DD}$. In (12.63) and (12.64) the negative sign for V_{Tp} has been incorporated; hence all quantities are positive. Note that for $K_n = K_p = K$, $V_{Tn} = |V_{Tp}| = V_T$ and $t_{on} = t_{off}$. A different approximate expression for calculating t_{on} and t_{off} for a CMOS is given in problem 12.15.

EXAMPLE 12.8

Consider a CMOS gate driving a capacitive load of $C_L = 10$ pF. For simplicity, the PMOS and CMOS are assumed identical: $K = 0.20$ mA/V^2 and $V_T = 2$ V. The supply voltage is $V_{DD} = 10$ V. Determine t_{off} of the gate.

From either (12.63) (set $v_o = 0.9V_{DD} = 9$ V) or (12.64)(set $v_o = 0.1V_{DD} = 1$ V) we get

$$t_{off} = t_{on} = \frac{0.5(10)}{0.2(8)}\left[\frac{2(2)}{8} + \ln\frac{20 - 4 - 1}{1}\right] = 10 \text{ ns}$$

The expressions for t_{on}, t_{off} for the transient calculations of NMOS with saturated and nonsaturated enhancement-type NMOS loads are given in Ref. 1 (see problems 12.16 and 12.17).

Some manufacturers supply the on-off time of their CMOS gates in a simple manner. For example, for a given supply voltage V_{DD},

$$t_{on} = \alpha_1 C_L + \tau_1 \tag{12.65a}$$

$$t_{off} = \alpha_2 C_L + \tau_2 \tag{12.65b}$$

where α_1, α_2 (ns/pF) and τ_1, τ_2 (ns) are given for the on and off times of the specific CMOS, respectively. These constants are usually different for the on and off times, since K_n of the NMOS is usually larger than K_p of the PMOS.

In digital circuits it is common to give the speed of the inverter in terms of the propagation delay. The propagation delay is discussed in Sec. 13.6.

EXERCISE 12.11

For a CMOS inverter the following are given: $V_{DD} = 5$ V, $K_n = 0.1$ mA/V^2, $V_{Tn} = 1.0$ V, $K_p = 0.07$ mA/V^2, $V_{Tp} = -1.5$ V. If the capacitive load is $C_L = 5$ pF, determine the t_{on} and t_{off} of the gate.

Ans $t_{off} = 4.18$ ns, $t_{on} = 2.0$ ns

EXERCISE 12.12

For a CMOS inverter the manufacturer gives the following data: $\alpha_1 = 0.1$ ns/pF,

$\alpha_2 = 0.2$ ns/pF, $\tau_1 = 2$ ns, and $\tau_2 = 3$ ns at $V_{DD} = 5$ V. Determine t_{on} and t_{off} of the inverter for $C_L = 5$ pF.

Ans $t_{on} = 2.5$ ns, $t_{off} = 4.0$ ns

12.13 CONCLUDING REMARKS

This chapter dealt with the time-domain response of linear amplifiers and the switching speed of BJT and FET inverters. In the linear amplifier case the transient response and the frequency response are interrelated. More specifically, if the bandwidth is specified, the rise time is also known, and vice versa. In the nonlinear case no such interrelationship exists. In the inverter case one must use the nonlinear model for the device and resort to a computer-aided analysis, such as SPICE, or use linearized approximation to estimate the switching speed of the inverter. The former approach is of course preferable. In some cases, manufacturers supply the on and off times of their inverter gates in a simple manner, for quick calculations, since a large number of inverters are usually involved.

REFERENCES

1. D. J. Hamilton and W. G. Howard, *Basic Integrated Circuit Engineering,* McGraw-Hill, New York, 1975.
2. J. Millman, *Microelectronics—Digital and Analog Circuits and Systems,* McGraw-Hill, New York, 1979.
3. C. A. Holt, *Electronic Circuits, Digital and Analog,* Wiley, New York, 1978.
4. V. Grinich and H. Jackson, *Introduction to Integrated Circuits,* McGraw-Hill, New York, 1975.

PROBLEMS

12.1 An amplifier has the following dominant-pole voltage gain function

$$\frac{V_o}{V_1}(s) = \frac{10^3}{(1 + s/10^6)[1 + s/(3 \times 10^6)]}$$

Determine the rise time and the delay time of the amplifier for a step input using the inverse Laplace transform method.

12.2 Repeat problem 12.1, but use the approximate simplified method.

12.3 The low-frequency gain function of an amplifier stage is given by

$$A_v(s) = \frac{10^3 s^2}{(s + 10)(s + 50)}$$

Sketch the output pulse response and show the sag if the input is a pulse of magnitude 1 mV and duration 1 ms.

12.4 An amplifier gain function is given as in problem 8.5, repeated in the following for convenience:

$$A_v(s) = \frac{8 \times 10^{12} s^2 (s + 3)}{(s + 5)(s + 10)(s + 200)(s + 10^5)(s + 2 \times 10^5)}$$

Sketch the output pulse waveform for an input pulse of magnitude 1 mV and duration 1 ms. Show the pertinent features such as the rise time, the sag, and the midband gain

12.5 Derive the pole locations for the second-order linear-phase transfer function, i.e., verify the result in Table 12.1. (*Hint:*

$$\tan^{-1} x = x - \frac{x^3}{3} + \frac{x^5}{5} - \cdots \qquad \text{for } |x| < 1)$$

12.6 If the transfer function of a circuit is given by

$$H(s) = \frac{s + z}{s^2 + a_1 s + a_o}$$

Show that for a linear phase $a_1^3 - 3a_1 a_o = (a_o/z)^3$, $1/z - a_1/a_o \neq 0$.

12.7 Obtain the lowest four linear-phase polynomials in Table 12.1 by using the method of

$$\frac{1}{e^s} = \frac{1}{\cosh s + \sinh s}$$

12.8 A three-stage amplifier has an identical-pole transfer function, namely,

$$A_v(s) = \frac{A(0) \times 10^{18}}{(s + 10^6)^3}$$

(*a*) Determine the exact rise time T_R of the amplifier.
(*b*) Compare your results with the one obtained by using (12.20).

12.9 Derive (12.30).

12.10 The slew rate of an op amp is given to be 1 V/μs. Sketch the output waveform if a square-wave input signal with a peak-to-peak voltage of 2 V and a period of 10 μs is applied. The op amp is used in a closed-loop inverting configuration with a gain of -20.

12.11 Consider the circuit shown in Fig. 12.14. The circuit values are $R_s = 1$ kΩ, $R_L = 1$ kΩ, and $V_{CC} = +10$ V. The input pulse is as shown in Fig. P12.14. The ON-voltage amplitude of the input pulse is 20 V, the pulse duration is 1 μs, and the OFF-voltage amplitude of the input pulse is -10 V. The transistor used is type 2N2222A (see Appendix E). Determine the approximate t_{on} and t_{off} of the transistor switch. Assume $\tau_s = 25$ ns.

12.12 Calculate the approximate switching time t_d, t_r, t_s, and t_f of the *pnp* epitaxial planar silicon transistor (type 2N3250), and compare the results with those given in the data sheets (see Appendix E). Use the parameters and circuit values of the test circuits and $\tau_S = 200$ ns.

12.13 For the circuit shown in Fig. 12.18*a*, let $C_1 = 2$ nF and let the other circuit parameters be the same as in Example 12.5. Determine t_{on} and t_{off} of the gate.

12.14 Derive (12.63), assuming that the NMOS and PMOS parameters are identical.

12.15 An approximate t_{on} (or t_{off}) for a CMOS in which both transistors have identical V_T and

K, as obtained by determining the average capacitor current I_C and $t_{on} = Q/I_C$, is given by [1]

$$t_{on} = t_{off} = \frac{3C_L V_{DD}^2}{K(V_{DD} - V_T)^2(2V_{DD} + V_T)}$$

Derive the above expression and compare the results with that of Example 12.8.

12.16 A saturated-load ($V_{GG} = V_{DD}$, i.e., V_{GG} and V_{DD} connected together) NMOS inverter is shown in Fig. P12.16. The ON-OFF times are given by [1]

$$t_{on} = \frac{3C_L(V_{DD} - V_T)}{K_1}\left[2(V_{DD} - 2V_T)^2 - (V_{DD} - V_T)^2\frac{K_2}{K_1}\right]^{-1}$$

$$t_{off} = \frac{3C_L}{K_2(V_{DD} - V_T)}$$

Determine the t_{on}, t_{off}, and the power dissipation $V_{DD}I_D$, given $K_1 = 0.5$ mA/V, $K_2 = 0.1$ mA/V, $V_T = 2$ V, $V_{DD} = 10$ V, and $C_L = 10$ pF.

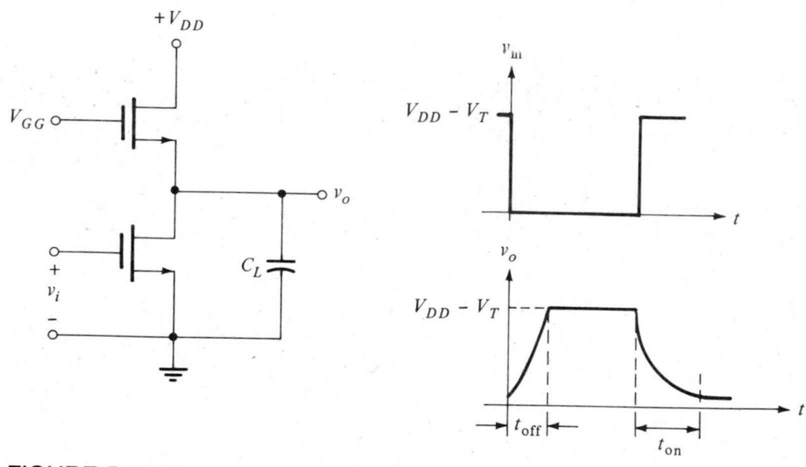

FIGURE P12.16

12.17 For the inverter with nonsaturated load ($V_{GG} - V_{DD} \geq V_T$) in Fig. P12.16, determine the power dissipation if $V_{GG} = V_{DD} + V_T$.

Introduction to Digital Circuits and Integrated-Circuit Logic Gates

This chapter will consider some of the fundamental aspects of digital circuits. Digital circuits and systems are important in electronics because of their use in computers and a wide variety of consumer products. Reliable and relatively inexpensive IC chips performing different functions are readily available as building blocks. These blocks are made from elementary circuits, called *logic gates*. In the following sections we shall consider binary numbers, logic gates, their characteristics, and interfacing different logic families.

13.1 FUNDAMENTAL CHARACTERISTICS OF DIGITAL CIRCUITS

The fundamental requirement of electronic circuits used for digital operation is that the electrical variables (current or voltages) that represent information be *discrete* as

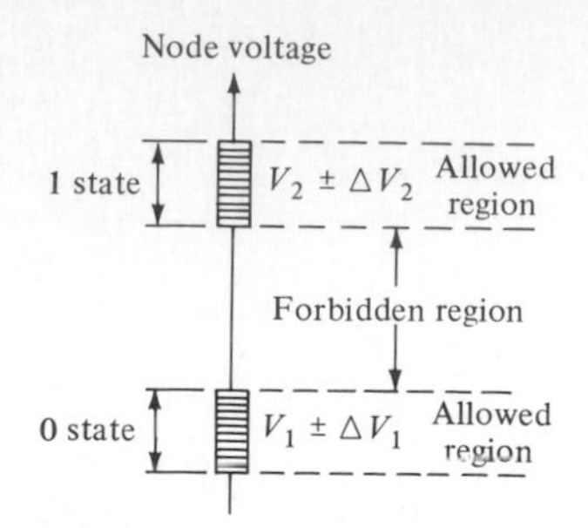

FIGURE 13.1 Node voltage levels showing allowed and forbidden ranges of values.

opposed to *continuous*. The discrete variables are most often binary, i.e., taking on only two values. Each binary digit (abbreviated bit) assumes one of two possible states, usually denoted as *zero* (**0**) and *one* (**1**). The physical states of the circuit have a permissible range, but the range for each state is nonoverlapping, so that the circuit is *unambiguously* in either one state or the other but never in both or neither of the two states.

For example, a transistor in logic circuits may operate in either the saturation or the cutoff state. Obviously, it cannot be in both saturation and cutoff at the same time, and the active mode is not allowed, so this is not much of a restriction. The transistor is not allowed to be in *neither* cutoff nor saturation, i.e., there is no linear range of operation in digital circuits. Nonsaturating circuits are also used, but again the operation is always at one or the other of two separate voltage levels. The digital-circuit designer must ensure that the signal is clearly within the range of the two allowed regions and never in the forbidden region. The two discrete states are usually designated as logic **1** and logic **0**, as shown in Fig. 13.1.

Because the two states are made widely separated, digital circuits are tolerant of much wider component variations than are linear circuits and can be made to be highly reliable.

Electronic circuits that perform complex logic functions consist of interconnections of a large number of a few basic elementary digital circuits (gates) or functional blocks. Thus, the fundamental characteristics of elementary digital circuits are

1. *Quantization:* The capability of the network to produce and preserve the two discrete states, usually represented by logic **0** and **1**
2. *Logic:* The capability of the network to implement the basic logic operations as defined in the following sections.

In addition to these two basic properties, there are other important properties that we shall consider later on.

13.2 BINARY SYSTEM

In order to understand the operation of digital circuits, we must become familiar with the binary system and its use in logic circuits. In a binary system representation of numbers only two digits, 0 and 1, are allowed. The role played by 0 and 1 in a binary system is the same as that played by 0, 1, 2, . . . , 9 in the decimal system. In a decimal system we use the base, or radix, 10. For example, the digits in the number 1984 (one thousand nine hundred eighty-four) have the following positions and meaning:

$$(1984)_{10} = 1 \times 10^3 + 9 \times 10^2 + 8 \times 10^1 + 4 \times 10^0$$

$$(64.35)_{10} = 6 \times 10^1 + 4 \times 10^0 + 3 \times 10^{-1} + 5 \times 10^{-2}$$

Note that we use numbers from 0 to 9 multiplied by powers of 10, depending on their positions. A number N to any base r can be expressed as

$$(N)_r = a_n r^n + a_{n-1} r^{n-1} + \cdots + a_1 r^1 + a_0 r^0 \tag{13.1}$$

where $n = 1, 2, 3, \ldots$; r = base (e.g., for decimal, $r = 10$; for binary, $r = 2$; for octal, $r = 8$); and a_n = numbers with values between 0 and $r - 1$.

In a binary system we use the base 2, and only the numbers 0 and 1 are used. The interpretation is, otherwise, similar. For example,

$$(1001)_2 = 1 \times 2^3 + 0 \times 2^2 + 0 \times 2^1 + 1 \times 2^0 = (9)_{10}$$

$$(101.01)_2 = 1 \times 2^2 + 0 \times 2^1 + 1 \times 2^0 + 0 \times 2^{-1} + 1 \times 2^{-2} = (5.25)_{10}$$

A decimal system of numbers can be converted to a binary system of numbers, as shown in Table 13.1. The given decimal number D to be converted to a binary system of numbers is divided by 2; the quotient C_1 and the remainder R_1 are as shown in Table 13.1. The process is repeated, i.e., divide C_1 by 2 and enter the quotient C_2 and the remainder R_2, until a quotient of 0 is obtained. The array of numbers $R_n \cdots R_1$ (which are either 0 or 1) is the binary representation. For example, consider the decimal number $(13)_{10}$. The binary representation as obtained from Table 13.1 is shown below:

0	1	3	6	← 13 (Decimal number)
1 MSB	1	0	1 LSB	Binary number

where LSB and MSB designate least significant bit and most significant bit, respectively. The result can be readily checked since $(1101)_2 = 1 \times 2^3 + 1 \times 2^2 + 0 \times 2^1 + 1 \times 2^0 = (13)_{10}$.

It may be pointed out that the same algorithm holds true for conversion to any other base, i.e., divide repeatedly by the radix r, and continue the procedure.

TABLE 13.1 Conversion of a Decimal Number to a Binary Number

Quotient of Divide by 2	$C_n = 0$	1	$\cdots$	C_2	C_1	Decimal Number D ← (D/2)
Remainder	R_n	R_{n-1}	$\cdots$	R_2	R_1	Binary Number

A decimal fraction is converted to a binary fraction by repeatedly multiplying the fraction by 2 and saving the integers. As an example, to convert $(0.79)_{10}$ to binary, we have

Decimal 0.79 fraction →	2(0.79) = 1.58	2(0.58) = 1.16	2(0.16) = 0.32	2(0.32) = 0.64	2(0.64) = 1.28	. . .
Binary fraction	1	1	0	0	1	. . .

To show how close the binary fraction 0.11001 is to the decimal 0.79,

$$(0.11001)_2 = 1 \times 2^{-1} + 1 \times 2^{-2} + 0 \times 2^{-3} + 0 \times 2^{-4} + 1 \times 2^{-5}$$

$$= 0.50 + 0.25 + 0 + 0 + 0.03125 = (0.78125)_{10}$$

Of course, carrying the process farther leads to a more accurate result. A decimal number that contains an integer and a fraction is converted to binary in two parts separately and then combined. Note that a binary point (more properly, a radix point) is used to separate a binary integer from a binary fraction in the same manner as is done for decimals.

A. The Octal and Hexadecimal Number Systems

The *octal* and *hexadecimal* number systems are related to the binary system. In the octal system the basis is *8* and the digits used are 0, 1, . . . , 7. In the hexadecimal system the base is *16* and the digits used are 0, 1, . . . , 9, A, B, C, D, E, F. Table 13.2 lists the equivalent numbers with decimal, binary, octal, and hexadecimal bases.

If we group the binary digits into sets of three, the number is converted to octal system. For example,

$$101110111100 = \underset{5}{101}\quad \underset{6}{110}\quad \underset{7}{111}\quad \underset{4}{100} = (5674)_8$$

Note that in the octal system the largest digit is 7, and we start grouping from the least significant digit.

If we group the binary digits into sets of four, the number is converted to hexadecimal system. For example,

$$\underset{\text{B}}{1011}\quad \underset{\text{E}}{1110}\quad \underset{\text{D}}{1101} = (\text{BED})_{16}$$

In computer technology 4 bits taken together is called a *nibble,* 8 bits taken together is referred to as a *byte,* while 16 or 32 bits or larger taken together is referred to as a *word.*

TABLE 13.2 Equivalent Numbers with Decimal, Binary, Octal, and Hexadecimal Bases

Decimal	Binary	Octal	Hexadecimal	Decimal	Binary	Octal	Hexadecimal
0	00000	00	00	16	10000	20	10
1	00001	01	01	17	10001	21	11
2	00010	02	02	18	10010	22	12
3	00011	03	03	19	10011	23	13
4	00100	04	04	20	10100	24	14
5	00101	05	05	21	10101	25	15
6	00110	06	06	22	10110	26	16
7	00111	07	07	23	10111	27	17
8	01000	10	08	24	11000	30	18
9	01001	11	09	25	11001	31	19
10	01010	12	0A	26	11010	32	1A
11	01011	13	0B	27	11011	33	1B
12	01100	14	0C	28	11100	34	1C
13	01101	15	0D	29	11101	35	1D
14	01110	16	0E	30	11110	36	1E
15	01111	17	0F	31	11111	37	1F

B. Binary-Coded Decimal

In the binary-coded decimal (BCD) system four binary digits A, B, C, D are used to represent a decimal digit, namely,

$$D \times 2^3 + C \times 2^2 + B \times 2^1 + A \times 2^0 \tag{13.2}$$

For example, 0110 corresponds to 6, and 1001 corresponds to 9. The use of BCD in a decoder application is given in Sec. 13.4.

EXERCISE 13.1

Convert the decimal number 3665 to a (*a*) binary, (*b*) octal, and (*c*) hexadecimal number system.

Ans (*a*) 11001010001, (*b*) $(7121)_8$, (*c*) $(E51)_{16}$

EXERCISE 13.2

Convert the binary number 110110011010 to a (*a*) decimal, (*b*) octal, and (*c*) hexadecimal number.

Ans (*a*) 3482, (*b*) $(6632)_8$, (*c*) $(D9A)_{16}$

EXERCISE 13.3

What is the decimal number corresponding to $(C2A.D8)_{16}$?

Ans 3114.656

13.3 BASIC LOGIC OPERATIONS

The three *basic* logic operations in digital circuits are designated OR, AND, and NOT. Other, more complex, logic operations are defined in terms of these three basic operations.

A. The NOT Gate

The NOT gate, shown symbolically in Fig. 13.2*a*, performs the NOT or negation operation (usually this gate is called an *inverter*). This is the simplest gate to implement and is realized by the BJT or FET inverter circuits discussed in Secs. 2.6 and 3.5.

The NOT operation is denoted by

$$B = \bar{A} \tag{13.3}$$

The table in Fig. 13.2*b*, which is called a *truth table*, completely describes the operation of the NOT gate. The truth table is a listing of the values of the dependent variable in terms of *all* the possible values of the independent variables. The truth table is used to *prove* logic equations and may also be viewed as *defining* what is meant by a logic equation such as (13.3). Note that the "bubble," in Fig. 13.2, signifies inversion, and thus if the bubble were absent we would have simply a *buffer* (e.g., a noninverting unity gain op amp).

B. The OR Gate

The OR gate, shown symbolically in Fig. 13.3*a* (for two inputs), performs the OR operation, which is given by

$$Y = A + B \tag{13.4}$$

where "+" indicates the logical OR operation. Note that the plus sign does not have the same meaning as in ordinary algebra. The algebra of binary (boolean) variables, that is, dealing with **1**s and **0**s, is called *boolean algebra*. The truth table for the OR operation is shown in Fig. 13.3*b*. Note that the output is at logical **1** if *A or B or* both are at logical **1**. A simple OR gate using diodes has already been encountered in Fig. 1.38*a*.

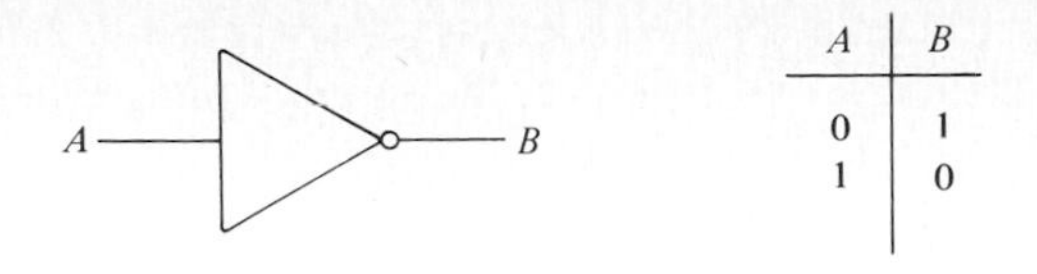

A	B
0	1
1	0

(*a*) (*b*)

FIGURE 13.2 (*a*) NOT gate. (*b*) Truth table.

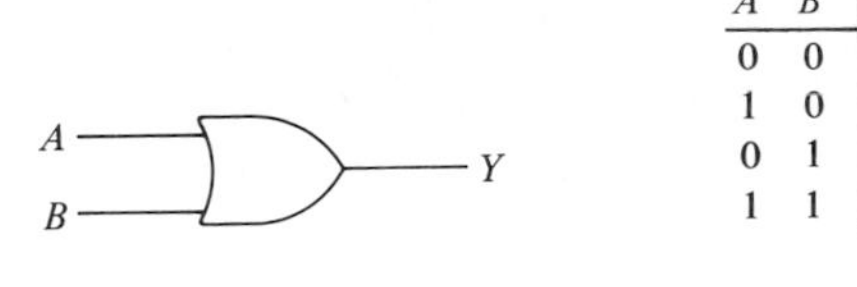

A	B	Y
0	0	0
1	0	1
0	1	1
1	1	1

(*a*) (*b*)

FIGURE 13.3 (*a*) OR gate. (*b*) Truth table.

C. The AND Gate

The AND gate, shown symbolically in Fig. 13.4*a* (for two inputs), performs the AND operation, which is given by

$$Y = A \cdot B \tag{13.5}$$

where "·" indicates the logical AND operation. The truth table for the AND operation is shown in Fig. 13.4*b*. Note that the output is at logical **1** if *A and B* are at logical **1**. A simple AND gate using diodes is shown in Fig. 1.38*b*.

Any boolean function can be implemented using NOT, AND, and OR gates. Such a set is referred to as a functionally complete set.

D. Positive and Negative Logic

The logical state of a gate is either **1** or **0**, i.e., $V(\mathbf{1})$ or $V(\mathbf{0})$. If $V(\mathbf{1}) > V(\mathbf{0})$ we have a positive logic and if $V(\mathbf{1}) < V(\mathbf{0})$ we have a negative logic. In order to avoid confusion, we shall henceforth use *positive logic,* i.e., the more positive voltage is

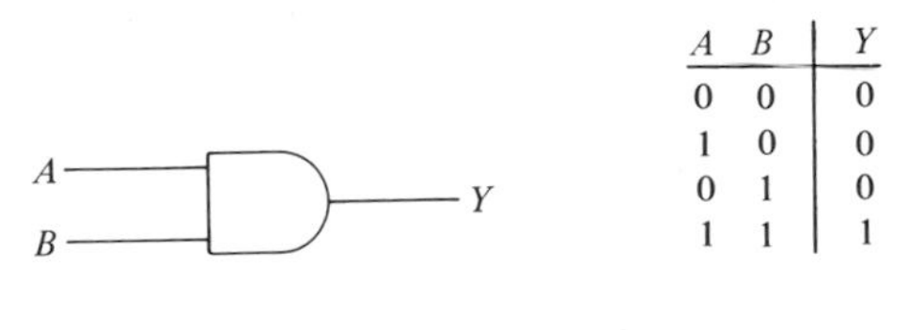

A	B	Y
0	0	0
1	0	0
0	1	0
1	1	1

(*a*) (*b*)

FIGURE 13.4 (*a*) AND gate. (*b*) Truth table.

assigned as the logical **1** state. For example for the inverters shown in Figs. 2.14 and 3.15, when the devices are OFF, the output voltage is at the supply voltage, hence at logical **1** and positive logic!

13.4 THE NOR AND NAND OPERATIONS

In the previous section we considered the basic logical operations of NOT, OR, and AND gates. Two other operations that are useful and that are obtained from the above basic operations or vice versa also exist, namely, the NOR and the NAND gates.

The NOR gate is simply an OR gate followed by an inverter as shown in Fig. 13.5*a*. The NOR operation is denoted by

$$Y = \overline{A + B} \tag{13.6}$$

The truth table for the NOR gate is given in Fig. 13.5*b*.

The NAND gate is simply an AND gate followed by an inverter, as shown in Fig. 13.6*a*. The truth table is given in Fig. 13.6*b*. The NAND operation is denoted by

$$Y = \overline{A \cdot B} \tag{13.7}$$

Many digital logic gates, as we shall see in the following sections, come basically in the NOR and NAND forms.

The following two identities, which are very useful in logic circuits, are known as *De Morgan's* theorems:

$$\overline{A + B} = \bar{A} \cdot \bar{B} \tag{13.8}$$

$$\overline{A \cdot B} = \bar{A} + \bar{B} \tag{13.9}$$

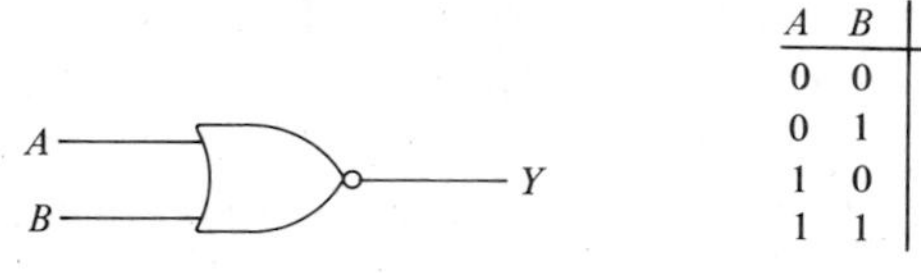

A	B	Y
0	0	1
0	1	0
1	0	0
1	1	0

(*a*) (*b*)

FIGURE 13.5 (*a*) NOR gate. (*b*) Truth table.

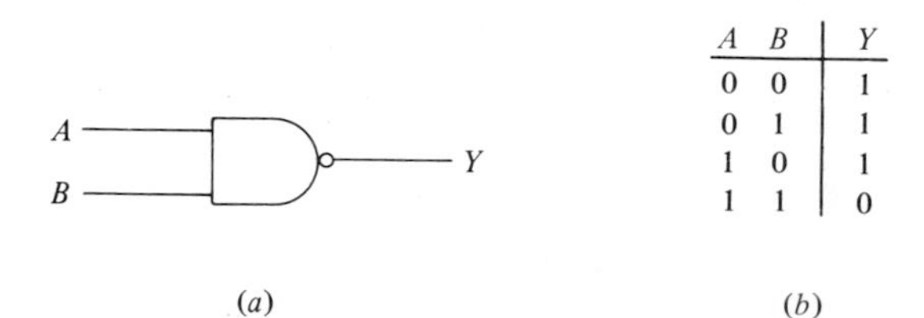

A	B	Y
0	0	1
0	1	1
1	0	1
1	1	0

(*a*) (*b*)

FIGURE 13.6 (*a*) NAND gate. (*b*) Truth table.

TABLE 13.3 Proof of De Morgan's Law: $\overline{A + B} = \bar{A} \cdot \bar{B}$

A	B	$A + B$	$\overline{A + B}$	$\bar{A}$	$\bar{B}$	$\bar{A} \cdot \bar{B}$
0	0	0	1	1	1	1
0	1	1	0	1	0	0
1	0	1	0	0	1	0
1	1	1	0	0	0	0

De Morgan's identities show that the complement of a boolean function is obtained if we change all OR symbols to ANDs and all AND symbols to ORs and complement each variable. With the help of these theorems one can also show that every logic function can be implemented with either ANDs and NOTs or ORs and NOTs.

The equivalency of (13.8) and (13.9) can be readily proved by means of a truth table. For example, Table 13.3 proves (13.8), and (13.9) can be proved similarly. De Morgan's theorems, of course, apply to more than two variables and can be proved by treating two variables at a time as in problem (13.8).

EXERCISE 13.4

Prove that the NAND and the NOR gates shown in Fig. E13.4 perform the NOT function.

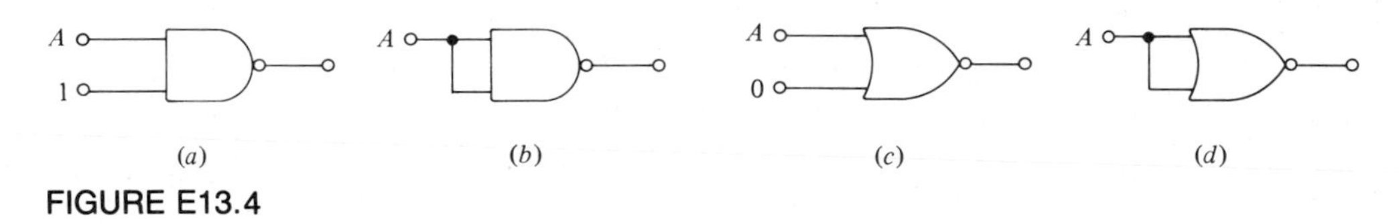

FIGURE E13.4

Ans (a) $\overline{A \cdot \mathbf{1}} = \bar{A}$, (b) $\overline{A \cdot A} = \bar{A}$, (c) $\overline{A + \mathbf{0}} = \bar{A}$, (d) $\overline{A + A} = \bar{A}$

EXERCISE 13.5

For the logic circuits shown in Fig. E13.5, determine the output function.

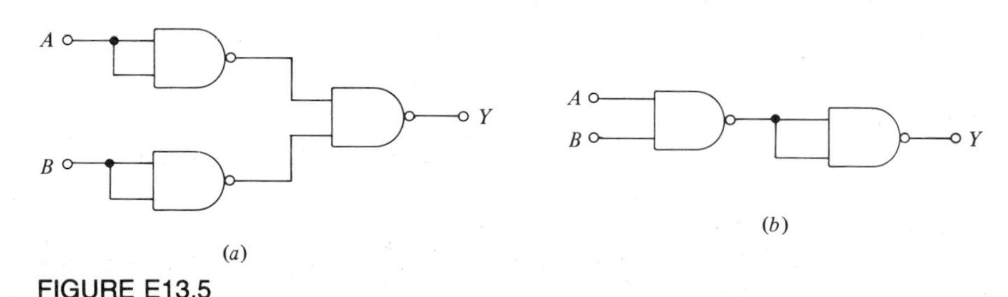

FIGURE E13.5

Ans (*a*) $Y = A + B$, (*b*) $Y = A \cdot B$

From Exercises 13.4 and 13.5 note that an arbitrary logic function can be implemented using the NAND gates alone. Similarly, it can readily be shown (problem 13.4) that arbitrary logic functions can also be realized using NOR gates alone. Finally, it should be noted that an implementation using only one or two logic functions will *not* necessarily yield the minimum number of gates.

EXAMPLE 13.1

As an application of the logic gates and boolean expressions, consider a long corridor with one light bulb and two doors with light switches at each door. Write the boolean expression for the light in the corridor and show the truth table.

We shall assume the following:

$$\text{Switch 1} = \begin{cases} \text{up} \rightarrow S_1 = \mathbf{1} \\ \text{down} \rightarrow S_1 = \mathbf{0} \end{cases}$$

$$\text{Switch 2} = \begin{cases} \text{up} \rightarrow S_2 = \mathbf{1} \\ \text{down} \rightarrow S_2 = \mathbf{0} \end{cases}$$

$$\text{Light} = \begin{cases} \text{on} \rightarrow L = \mathbf{1} \\ \text{off} \rightarrow L = \mathbf{0} \end{cases}$$

The truth table for this example is shown in Fig 13.7. Note that we have assumed the light is OFF when both switches are down. The boolean expression for the light in the corridor is

$$L = \overline{S_1} \cdot S_2 + S_1 \cdot \overline{S_2}$$

S_1	S_2	L
0	0	0
0	1	1
1	0	1
1	1	0

FIGURE 13.7 Truth table of the example.

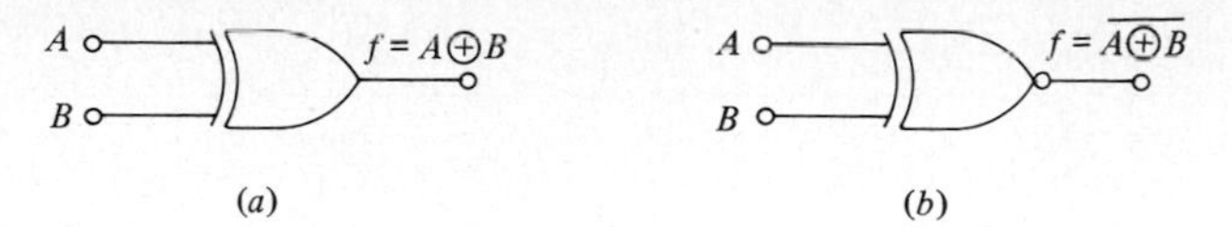

FIGURE 13.8 (*a*) Symbol for exclusive OR. (*b*) Symbol for exclusive NOR.

It is of interest to note that the boolean expression in Example 13.1 happens to be an exclusive-OR or XOR function. The exclusive-OR is an important building block in logic systems and is denoted by

$$f = A \oplus B = \bar{A} \cdot B + A \cdot \bar{B} \tag{13.10a}$$

where $\oplus$ denotes that the function is **1** if one *or* the other *but not both* of A and B equal **1**. A logic implementation for the XOR function is shown in Fig. 13.9. Figure 13.8*a* and *b* shows the logic symbols for the exclusive-OR and exclusive-NOR functions, respectively. The exclusive-NOR function is also called a *coincidence gate*. This gate can be used as an equality detector or a digital comparator. For example, for Fig. 13.8*b*,

$$f = A \odot B = \overline{A \oplus B} \tag{13.10b}$$

In other words, $f = \mathbf{1}$ only if $A = B$.

We have been using two inputs to the gates for simplicity. Of course, more than two inputs can be handled similarly. These logic functions and other more complex combinations are best described by the *boolean algebra*. For example, the truth tables for the basic functions of three boolean variables are shown in Table 13.4. Note that there are 2^n different combinations of n binary variables, e.g., for three variables we have $2^3 = 8$ different combinations.

For convenience, some of the important identities of the basic NOT, OR, and AND operations designated by ($\bar{\ }$), (+), and (·) are shown in Table 13.4. The variables are, of course, binary and the identities in the table can be proved by means of truth tables. The identities can also be used for the minimization of logic expressions.

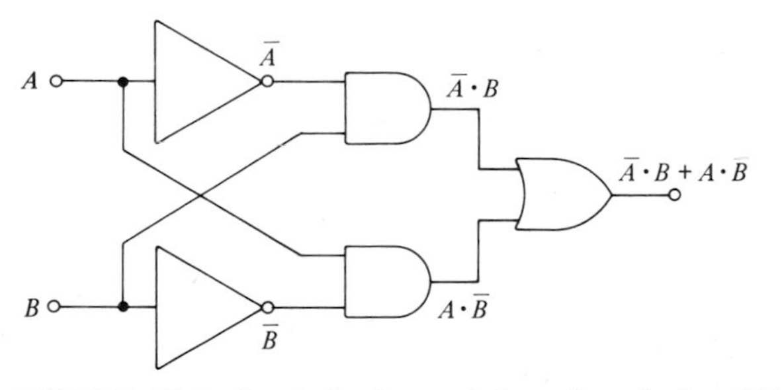

FIGURE 13.9 Logic implementation of exclusive OR.

TABLE 13.4 Truth Table for Three Boolean Variables

Variable Values ABC	OR $A + B + C$	AND $A \cdot B \cdot C$	NOR $\overline{A + B + C}$	NAND $\overline{A \cdot B \cdot C}$
000	**0**	**0**	**1**	**1**
001	**1**	**0**	**0**	**1**
010	**1**	**0**	**0**	**1**
011	**1**	**0**	**0**	**1**
100	**1**	**0**	**0**	**1**
101	**1**	**0**	**0**	**1**
110	**1**	**0**	**0**	**1**
111	**1**	**1**	**0**	**0**

EXERCISE 13.6

The exclusive-NOR gate is shown in Fig. 13.8*b*. Write the truth table for its function: $f = \overline{A \oplus B}$.

Ans

A	**0**	**0**	**1**	**1**
B	**0**	**1**	**0**	**1**
f	**1**	**0**	**0**	**1**

EXERCISE 13.7

An input data **11011** is applied to terminal A of XOR in Fig. 13.8*a*. Determine the output if terminal B is connected (*a*) to **1** and (*b*) to **0**.

Ans (*a*) **00100**, (*b*) **11011**

EXERCISE 13.8

Repeat the above if an exclusive-NOR gate of Fig. 13.8*b* is used.

Ans (*a*) **11011**, (*b*) **00100**

A. Minimization of Logic Gates

In many cases a complicated logic expression (or its gate implementation) can be simplified to obtain an implementation with minimum number of gates. The identities in Table 13.5 can be helpful for such purposes. The following example will illustrate this point.

TABLE 13.5 Binary Logic Rules

NOT operation	OR operation	AND operation
$\bar{\mathbf{1}} = \mathbf{0}$	$A + \mathbf{1} = \mathbf{1}$	$A \cdot \mathbf{0} = \mathbf{0}$
$\bar{\mathbf{0}} = \mathbf{1}$	$A + \mathbf{0} = A$	$A \cdot \mathbf{1} = A$
$\bar{\bar{A}} = A$	$A + \bar{A} = \mathbf{1}$	$A \cdot \bar{A} = \mathbf{0}$
	$A + A = A$	$A \cdot A = A$
	$A + B = B + A$	$A \cdot B = B \cdot A$
	$(A + B) + C = A + (B + C)$	$(A \cdot B) \cdot C = A \cdot (B \cdot C)$
	$A + (A \cdot B) = A$	$A \cdot (A + B) = A$
	$A + (\bar{A} \cdot B) = A + B$	$A \cdot (\bar{A} + B) = A \cdot B$
	$A + (B \cdot C) = (A + B) \cdot (A + C)$	$A \cdot (B + C) = (A \cdot B) + (A \cdot C)$
	$\overline{A + B} = \bar{A} \cdot \bar{B}$	$\overline{A \cdot B} = \bar{A} + \bar{B}$

EXAMPLE 13.2

Obtain a simplified expression for the following and show the minimum gate implementation.

$$f = A \cdot B \cdot \bar{C} \cdot \bar{D} + \bar{A} \cdot B \cdot \bar{C} \cdot \bar{D} + B \cdot \bar{C} \cdot D$$

From Table 13.3 we note that $A + \bar{A} = \mathbf{1}$ and $A \cdot \mathbf{1} = A$; hence

$$f = B \cdot \bar{C} \cdot \bar{D} \cdot (A + \bar{A}) + B \cdot \bar{C} \cdot D$$

$$= B \cdot \bar{C} \cdot \bar{D} \cdot \mathbf{1} + B \cdot \bar{C} \cdot D$$

$$= B \cdot \bar{C} \cdot \bar{D} + B \cdot \bar{C} \cdot D$$

$$= B \cdot \bar{C} \cdot (D + \bar{D}) = B \cdot \bar{C} \cdot \mathbf{1}$$

or $f = B \cdot \bar{C}$. The minimum gate implementation is shown in Fig. 13.10.

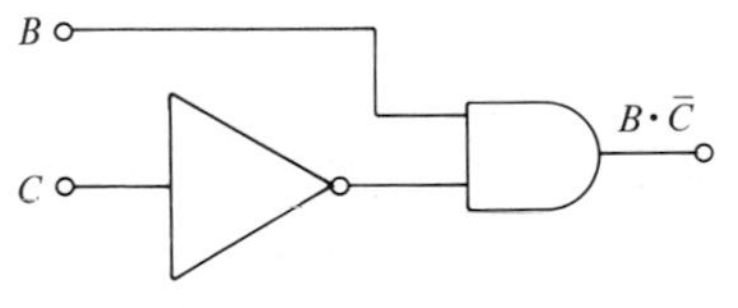

FIGURE 13.10 Logic implementation of the example.

It should be noted that there exist organized procedures for function minimization. One of the most popular and useful tools is the Karnaugh map, and the reader is referred to Ref. 1 for a detailed treatment of this topic.

EXERCISE 13.9

Simplify $f = A \cdot B \cdot \overline{C} + A \cdot \overline{B} \cdot \overline{C} + \overline{A} \cdot \overline{B} \cdot C + \overline{A} \cdot \overline{C} \cdot \overline{B} + A \cdot \overline{B}$.

Ans $A \cdot \overline{C} + \overline{B}$

EXERCISE 13.10

Simplify $f = A \cdot B \cdot \overline{C} \cdot \overline{D} + A \cdot B \cdot \overline{C} \cdot D + A \cdot B \cdot C + A \cdot B \cdot D + C \cdot D$.

Ans $A \cdot B + C \cdot D$

EXERCISE 13.11

For the logic circuit shown in Fig. E13.11, determine the simplest expression for Y as a function of A, B, and C.

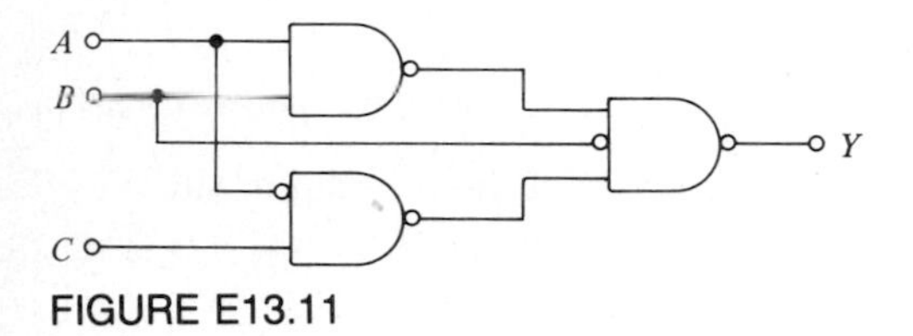

FIGURE E13.11

Ans $Y = B + C \cdot \overline{A}$

B. Decoder Application

As an application of the logic functions consider the 7447 IC decoder, which accepts a 4-bit binary-coded-decimal (BCD) as the input and uses a seven-segment light emitting diode (LED)* display for the output. Such a display is used in many applications such as meters and clocks. The basic parts, including the logic functional block diagram of a one-digit display are shown in Fig. 13.11. The decoder is driven by a decade counter, which will be described in the next chapter.

From the logic functional block and the corresponding function table note that when the input is **0110** (corresponding to decimal 6) the output of the seven segment has *a* and *b* off; hence the output is **1100000,** where **0** corresponds to the low and **1** to the high level, respectively. Note that when the gate is off, the voltage is at the

*For a discussion of LED see E. S. Yang, *Fundamentals of Semiconductor Devices,* McGraw-Hill, New York, 1978, Chap. 6.

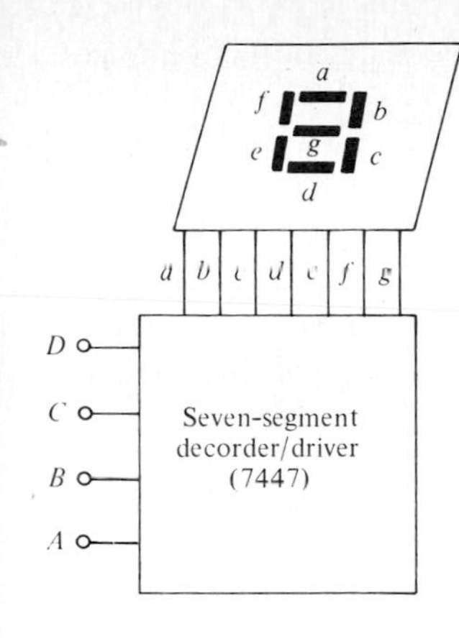

(a)

0 1 2 3 4 5 6 7 8 9

Decimal or function	Inputs				Outputs						
	D	C	B	A	a	b	c	d	e	f	g
0	L	L	L	L	ON	ON	ON	ON	ON	ON	OFF
1	L	L	L	H	OFF	ON	ON	OFF	OFF	OFF	OFF
2	L	L	H	L	ON	ON	OFF	ON	ON	OFF	ON
3	L	L	H	H	ON	ON	ON	ON	OFF	OFF	ON
4	L	H	L	L	OFF	ON	ON	OFF	OFF	ON	ON
5	L	H	L	H	ON	OFF	ON	ON	OFF	ON	ON
6	L	H	H	L	OFF	OFF	ON	ON	ON	ON	ON
7	L	H	H	H	ON	ON	ON	OFF	OFF	OFF	OFF
8	H	L	L	L	ON	ON	ON	ON	ON	ON	ON
9	H	L	L	H	ON	ON	ON	OFF	OFF	ON	ON

(b)

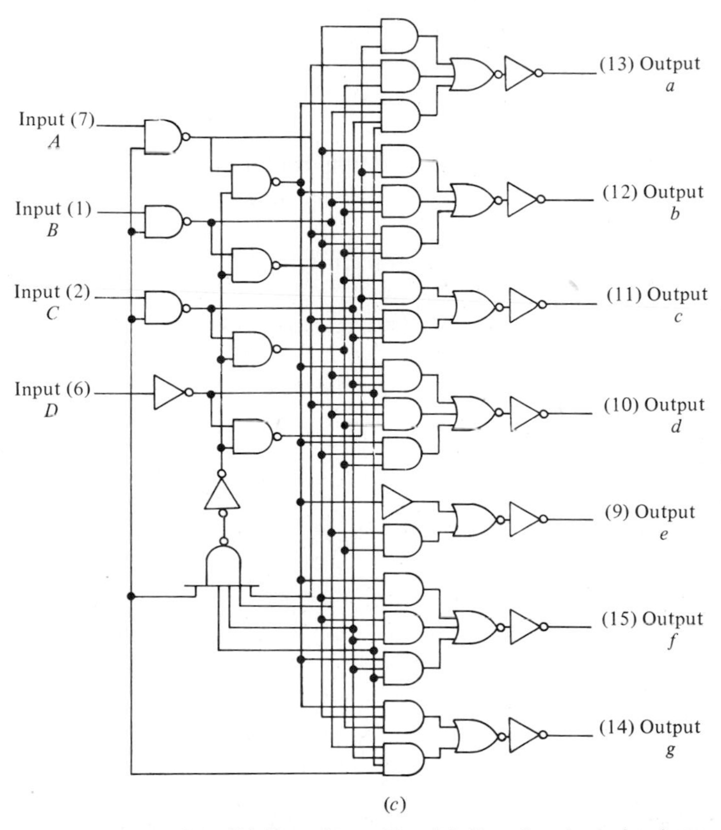

(c)

FIGURE 13.11 (a) Decoder. (b) Function table. (c) Function logic implementation. (Courtesy of Texas Instruments Inc.)

supply voltage, hence at the high level, which is logical **1**. This is a positive logic as described earlier.

13.5 INVERTER LOGIC SWING AND NOISE MARGINS

The inverter is the simplest gate and performs the NOT operation. We have already discussed the transfer function characteristics of a simple BJT inverter (Fig. 2.14), an emitter-coupled inverter (Fig. 2.16), an NMOS inverter (Fig. 3.15), and the CMOS inverter (Fig. 3.24). The transfer characteristics are all basically similar and two such characteristics are shown in Fig. 13.12*a* for a BJT inverter and in Fig. 13.12*b* for a CMOS inverter. Note that in these cases V_{OH} corresponds to logic **1** and V_{OL} corresponds to logic **0**, and since $V(\mathbf{1}) > V(\mathbf{0})$, we have a positive logic.

From Fig. 13.12 we have the following definitions:

$$\text{Logic swing} = V_{OH} - V_{OL} \tag{13.11a}$$

$$\text{Transition region} = V_{IH} - V_{IL} \tag{13.11b}$$

where O, I, H, and L designate output, input, high, and low, respectively. The high- and low-level noise margins (NM) are defined as follows:

$$\text{NM}_H = V_{OH} - V_{IH} = \mathbf{\Delta 1} \tag{13.12a}$$

$$\text{NM}_L = V_{IL} - V_{OL} = \mathbf{\Delta 0} \tag{13.12b}$$

The manufacturers usually provide the minimum values of V_{OH} and V_{IH} and the maximum values of V_{IL} and V_{OL}, which may be used in (13.12).

From (13.11) and (13.12) it is seen that the noise margin will improve, i.e., NM_H, and NM_L will be high, if either the logic swing is large or the transition region is small. Note that V_{IL} and V_{IH} are those input voltages at which the slopes of the transfer function is equal to -1. In most cases, however, this precise calculation is not necessary.

In order to illustrate the above definitions and expressions, we will consider an example in this section.

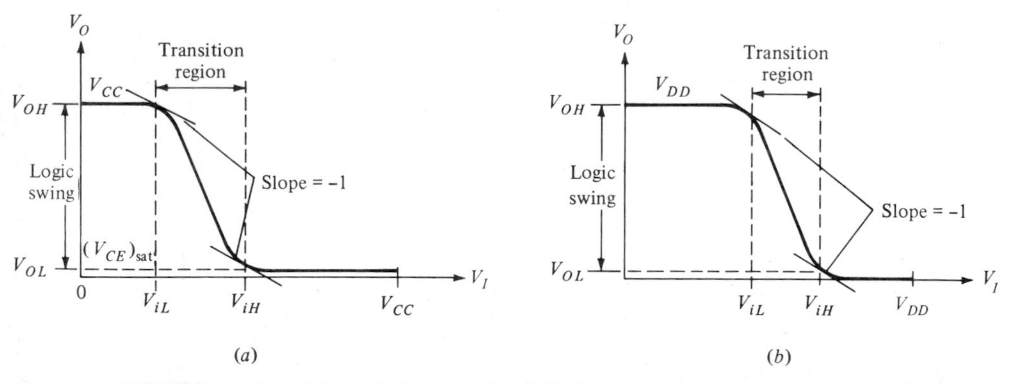

FIGURE 13.12 (*a*) Logic levels of a BJT inverter. (*b*) Logic levels of a CMOS inverter.

EXAMPLE 13.3

For the simple BJT inverter circuit shown in Fig. 13.13*a*, sketch the transfer characteristic and determine (*a*) the logic swing, (*b*) the transition region, and (*c*) the noise margins.

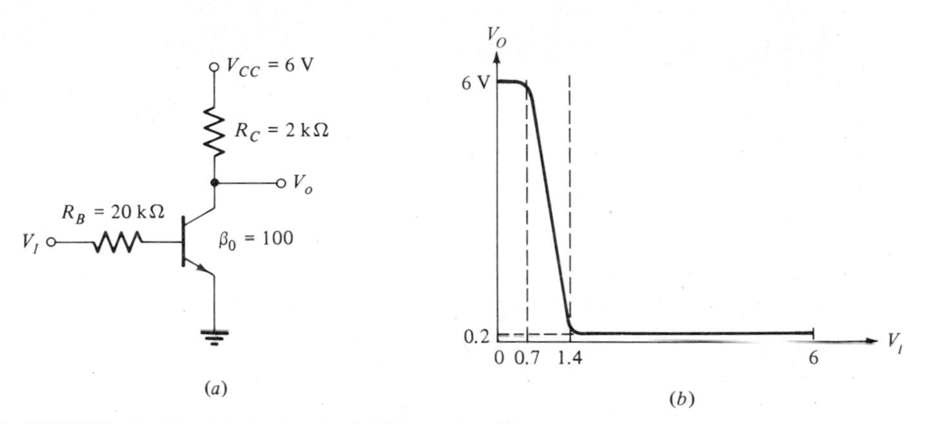

FIGURE 13.13 (*a*) A simple BJT inverter. (*b*) Its transfer characteristic.

For the circuit the transfer characteristics can readily be determined as shown in Fig. 13.13*b*. Note that $V_{OH} \simeq V_{CC}$, $V_{OL} = (V_{CE})_{sat} = 0.2$ V. If the input voltage is less than $(V_{BE})_{on} = 0.7$ V, the transistor is off; hence $V_{IL} = 0.7$ V. To find V_{IH}, we first determine the value of I_C for the transistor to enter the saturation region, namely,

$$(I_C)_{sat} = \frac{V_{CC} - (V_{CE})_{sat}}{R_C} = \frac{6.0 - 0.2}{2\ \text{k}\Omega} = 2.9\ \text{mA}$$

However, in the active region, we have $I_B = I_C/\beta_0$, and at the edge of saturation $(I_B)_{sat} = (I_C)_{sat}/\beta_0 = 2.9\ \text{mA}/100 = 29\ \mu\text{A}$. Hence $V_{IH} \simeq I_B R_B + (V_{BE})_{sat} = (29 \times 10^{-6})(2 \times 10^4) + 0.8 = 1.4$ V. From (13.11)

$$\text{Logic swing} = 6.0 - 0.2 = 5.8\ \text{V}$$

$$\text{Transition region} = 1.4 - 0.7 = 0.7\ \text{V}$$

From (13.12) the noise margins are

$$\text{NM}_H = 6.0 - 1.4 = 4.6\ \text{V}$$

$$\text{NM}_L = 0.7 - 0.2 = 0.5\ \text{V}$$

The logic gates usually drive each other, and in fact sometimes the driver gate and the load gate are of different logic families. For reliable operation, $V_{OL} < V_{IL}$ and $V_{OH} > V_{IH}$, in other words both NM_H and NM_L must be positive, as shown in Fig. 13.14. Note that if the inverter drives another similar inverter V_{OH} is reduced while V_{IH} remains the same and thus NM_H is reduced.

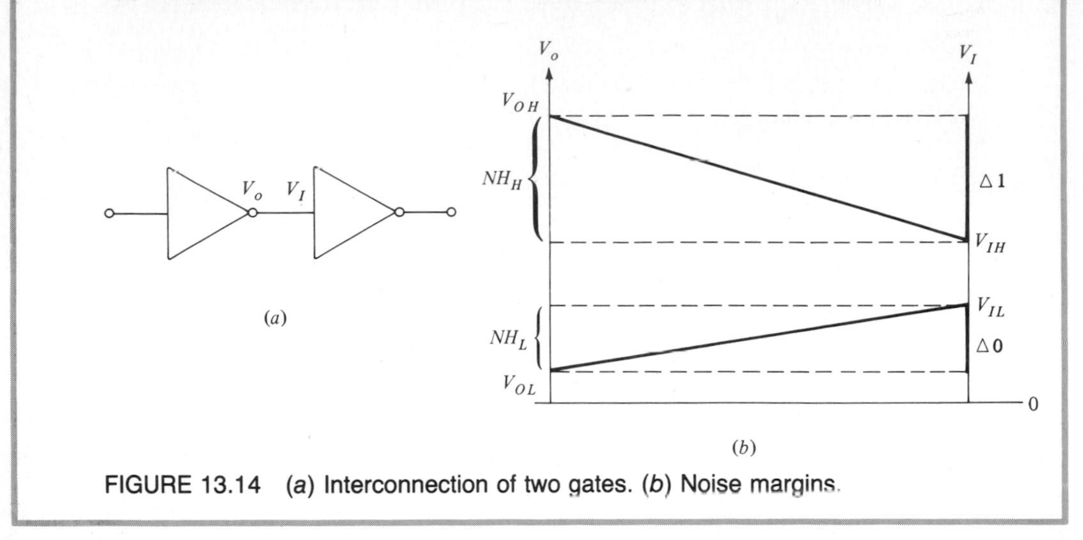

FIGURE 13.14 (*a*) Interconnection of two gates. (*b*) Noise margins.

13.6 INTEGRATED-CIRCUIT GATES

There are many different IC families that can perform the basic logic functions discussed in the previous sections. These are diode-transistor logic (DTL), transistor-transistor logic (TTL), emitter-coupled logic (ECL), *n*-channel metal-oxide semiconductor (NMOS), complementary metal-oxide semiconductor (CMOS), and integrated injection logic (IIL). Of these the most commonly used gates to date are TTL, ECL, NMOS, CMOS, and IIL (also written I^2L). We shall briefly consider each of the commonly used logic families separately in the subsequent sections.

However, before we study each logic type, there are some considerations that are important and common to all gates. We shall consider some of the key parameters since in most cases comparisons and choices of the logic families will be made based on these parameters, and, of course, cost.

A. Propagation Delay (Speed)

We have already considered the switching speed of the various inverters in the previous chapter. Integrated-circuit gates utilize many devices, and in digital circuits it is common to give the speed of the gate in terms of the *propagation delay*. The propagation delay is illustrated in Fig. 13.15. The average of the two propagation delays is defined as the propagation delay t_{PD}, namely,

$$t_{PD} = \tfrac{1}{2}(t_{PHL} + t_{PLH}) \tag{13.13}$$

where t_{PHL} and t_{PLH} are the high-low and low-high propagation delay times, defined between the 50 percent points of the input and output pulse waveforms.

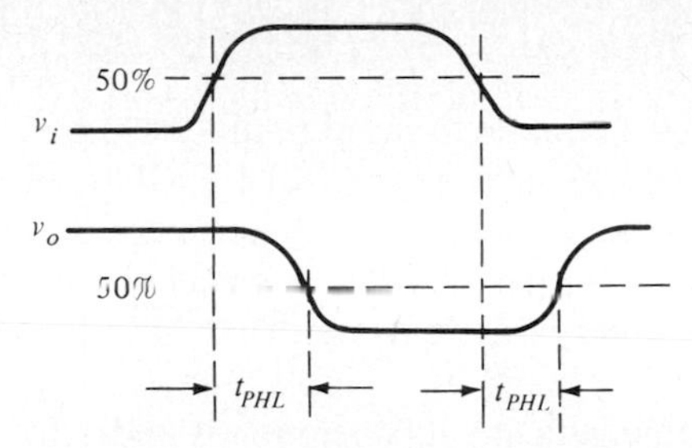

FIGURE 13.15 Illustration of propagation delay.

B. Fan-Out and Fan-In

In digital circuits the output of one gate is often connected to the input of more than one gate as shown in Fig. 13.16. *Fan-out* refers to the maximum number of gates that can be driven by a gate while remaining within the guaranteed specifications. The manufacturer usually specifies the fan-out number N. Similarly, more than one input is often connected to a gate. The number of independent nodes to a gate is called *fan-in*.

C. Power Dissipation

In large-scale integrated circuits thousands or even hundred thousands of gates are often used. Each gate dissipates a certain amount of power which we wish to minimize subject to meeting other requirements. As we shall see in the subsequent sections in general, the smaller the power dissipation, the larger the propagation delay. In other words, there is a compromise between power and delay. The product of power dissipation and the propagation delay is often used as a *figure of merit* of the gate. The power-delay product (PDP), which is on the order of picojoules (1 pJ $= 10^{-12}$ W · s), varies among the different logic families. Currently the range is from 1 to 50 pJ. We shall make a comparison of key parameters of the various logic families after covering the commonly used gates.

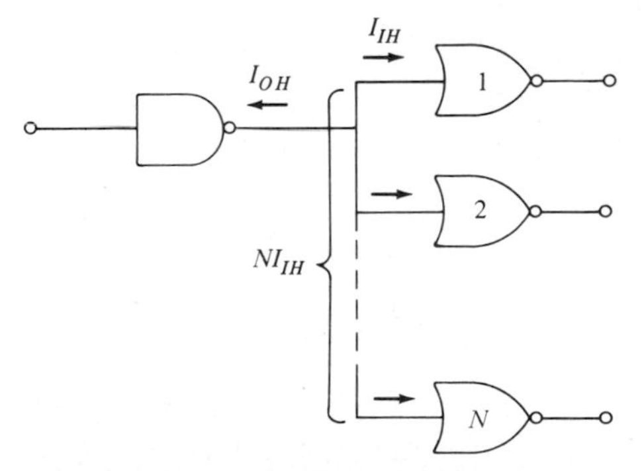

FIGURE 13.16 Illustration of fan-out.

13.7 TRANSISTOR-TRANSISTOR LOGIC (TTL)

Transistor-transistor logic (TTL) is the most commonly used family of logic gates for small-scale integration (SSI) and medium-scale integration (MSI). The basic TTL gate is shown in Fig. 13.17. The multiemitter transistor, Q_1, which performs the function of parallel connected transistors (i.e., the two base terminals and the two collector terminals connected) conserves chip area and is therefore utilized in integrated circuits.

The circuit is a two-input NAND logic gate and its operation is understood if we recall typical values for BJT (see Sec. 2.5) of $(V_{BE})_{on} = 0.7$ V, $(V_{BE})_{sat} = 0.8$ V, and $(V_{CE})_{sat} = 0.2$ V. Now for $V_a = 0.2$ V, i.e., logical **0**, the base-emitter junction of Q_1 is forward-biased ($V_{CC} \simeq 5$ V), and current will flow through R_B. The base voltage of Q_1 is then

$$V_{B1} = V_{BE} + V_a = 0.7 + 0.2 = 0.9 \text{ V}$$

The large base current drives Q_1 into saturation, and $V_{CE1} = 0.2$ V. For Q_2, since $V_{B2} = (V_{CE})_{sat} + V_a = 0.2 + 0.2 = 0.4 \text{ V} < 0.7 \text{ V}$, then Q_2 is off and $V_o \simeq V_{CC}$, i.e., logical **1**. This situation holds as long as any of the inputs is *low*. Next consider both inputs V_a *and* V_b high, i.e., $V_a = V_b = V_{CC}$ (logical **1**). For this situation the base-emitter junctions of Q_1 are reverse-biased, and no current will flow into the emitters of Q_1. On the other hand, since the base-collector junction of Q_1 is forward-biased, as will be the base-emitter junction of Q_2, large current will flow through the base of Q_1 driving Q_1 into saturation and $V_o = (V_{CE})_{sat}$, i.e., logical **0.** Hence the logic function of the gate is

$$V_o = \overline{V_a \cdot V_b} \tag{13.14}$$

A. Standard TTL

The circuit diagram of a standard two input NAND gate for the series 54/74 TTL is shown in Fig. 13.18. The corresponding voltage transfer characteristic is shown in Fig. 13.19. Series 54 is intended for operation over the temperature range 55 to 125°C, where the requirements are stringent. Series 74 is a lower-cost industrial version that

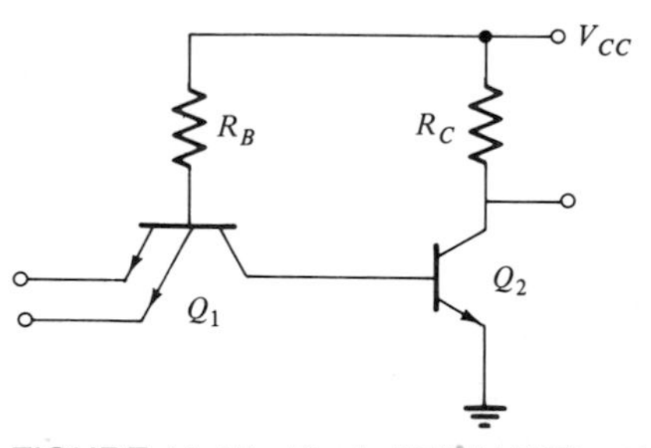

FIGURE 13.17 Basic TTL NAND gate.

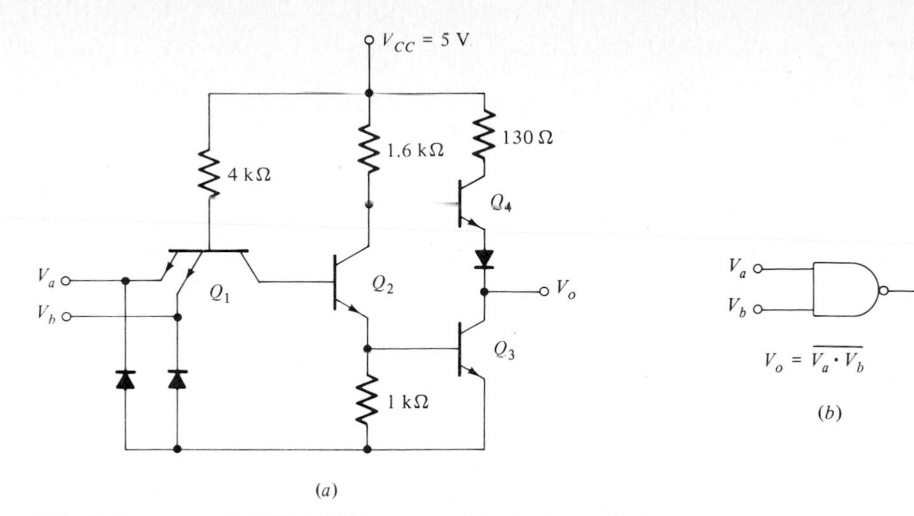

FIGURE 13.18 IC TTL NAND gate and its logic symbol.

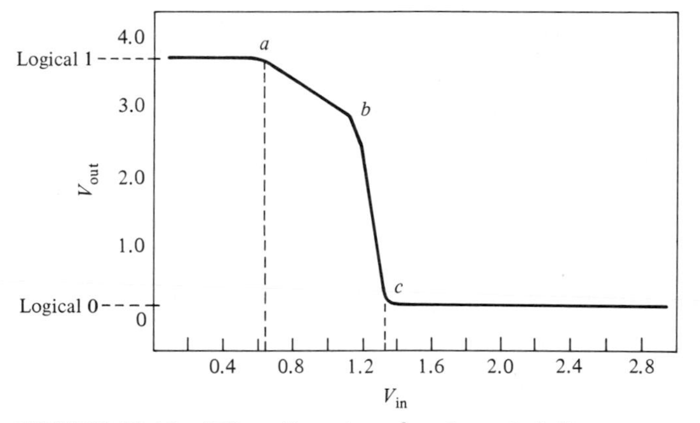

FIGURE 13.19 TTL voltage transfer characteristic.

operates over the more limited temperature range 0 to 70°C. The voltage transfer characteristics for the 54/74 series TTL NAND gate is shown in Fig. 13.20. Typical electrical characteristics of standard TTL(54/74) is given in Table 13.6.

B. Schottky-Clamped TTL

In order to improve the speed of TTL gates, Schottky-clamped diodes and Schottky transistors are used in order to prevent the diodes and the transistors from going into saturation and hence eliminating the storage time. A typical Schottky TTL NAND gate (54S/74S) circuit is shown in Fig. 13.21. The propagation delay of this circuit is 3 ns rather than 10 ns in standard TTL (54S/74S).

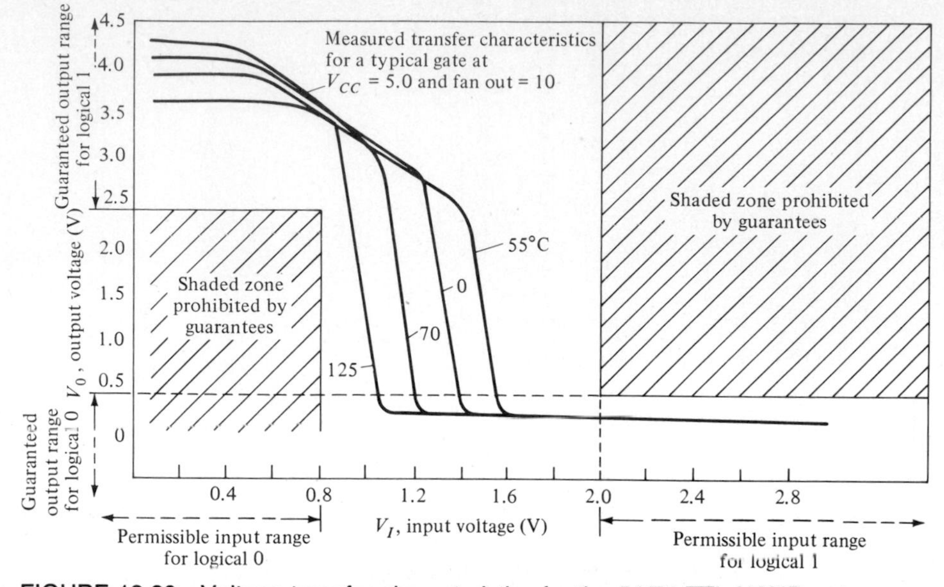

FIGURE 13.20 Voltage transfer characteristics for the 54/74 TTL NAND gate.

There are many versions of the (54/74) family commercially available. In addition to the standard-power standard 54/74 series, there are

54L/74L:	Low-power standard
54S/74S:	Standard-power Schottky
54LS/74LS:	Low-power Schottky
54AS/74AS:	Advanced Schottky
54ALS/74ALS:	Advanced low-power Schottky

The power and the delay characteristics of the various versions of characteristics of the 74 series are given in Table 13.7. For these gates the specifications are

$$\min V_{OH} = 2.4 \text{ V} \qquad \max V_{OL} = 0.5 \text{ V}$$

$$\min V_{IH} = 2.0 \text{ V} \qquad \max V_{IL} = 0.8 \text{ V}$$

TABLE 13.6 Standard Transistor-Transistor Logic (54/74 TTL) *Typical* Electrical Characteristics (T = 25°C)

V_{OH}/V_{OL}	3.4 V/0.2 V	Fan-out	10
V_{IH}/V_{IL}	1.5 V/0.5 V	Supply volts	+5.0 V
NM_H/NM_L	1.9 V/0.3 V	Power dissipation per gate	10 mW
Logic Swing	3.2 V	Propagation delay time	10 ns

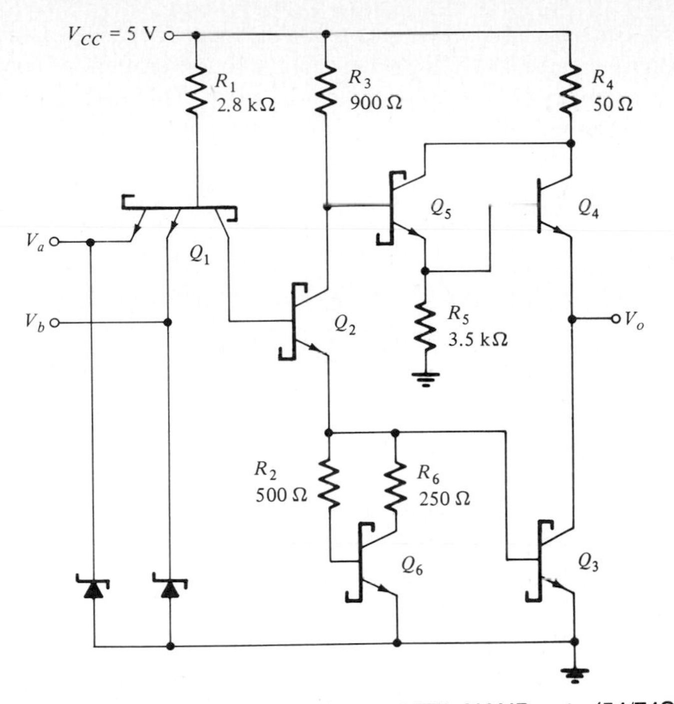

FIGURE 13.21 Schottky-clamped TTL NAND gate (54/74S).

Of the above versions the TTL low-power Schottky (LS) is the most widely used digital IC gate in SSI and MSI. The propagation delay is typically 10 ns, the same as for the standard TTL; however, the power dissipation is only 2 mW, instead of 10 mW for the standard TTL.

The propagation delay of the Schottky TTL gates increases linearly with increasing values of the load capacitance. Note that the Series 74AS has the lowest propagation delay of about 1.5 ns for a capacitive load of 15 pF in the TTL 74 family.

TABLE 13.7 Transistor-Transistor Logic Performance Characteristics (T = 25°C)

	Series 74L	Series 74S	Series 74LS	Series 74AS	Series 74ALS
Typical propagation delay time, ns	30	3	10	1.5	4
Typical power dissipation per gate, mW	1	20	2	20	1
Typical power-delay product, pJ	30	60	20	30	4

The complete data sheets for the 54LS/74LS are given in Appendix E. The data sheets are for the 74 NAND gate and the 74 NOR gate series. Fan-out information is obtained from the data approximately by $|I_{OH}/I_{IH}|$ for logical **1** and $|I_{OL}/I_{IL}|$ for logical **0**. For the 74 series the fan-out is typically 10.

EXERCISE 13.12

Determine

(a) NM_H of the inverter in Fig. 13.13 if the gate drives an identical inverter load.
(b) What is the fan-out N (the maximum number of load gates)?

Ans (*a*) 4.13 V, (*b*) $N = 76$

EXERCISE 13.13

For the 74LS logic gate in addition to the data given in this section the following are also given: min $I_{OH} = -0.4$ mA, min $I_{OL} = 8$ mA; max $I_{IH} = 20\ \mu$A, max $I_{IL} = -0.4$ mA. Determine the fan-out for logical **1** and logical **0**.

Ans 20

13.8 EMITTER-COUPLED LOGIC (ECL)

The basic ECL gate is shown in Fig. 13.22. As discussed in Chap. 12, this circuit avoids saturation of the transistor and has the highest switching speed of any gate. A propagation delay of less than 1 ns has been achieved with ECL gates. This inverter was briefly discussed for its transfer characteristics in Secs. 2.7 and 5.11. The reader is encouraged to examine Fig. 5.30.

In order to examine the logic properties *high* and *low* levels of the circuit, let us consider a numerical example of the basic gate. Let $V_{CC} = -V_{EE} = 5$ V, $R_{C1} = R_{C2} = R_C = 1$ kΩ, $I_E = 2$ mA, and $V_R = 0$. For $|V_{in}| \gg V_t$ (where $V_t = 25$ mV at room temperature) one of the transistors is on and the other is off. More specifically, if $V_{in} = +2$ V, the base voltage of transistor Q_1 is more positive than that of Q_2; hence Q_1 is on and Q_2 is off, while for $V_{in} = -2$ V the reverse is true and Q_2 is on and Q_1 is off. For $V_{in} = +2$ V,

$$V_{O1} = V_{CC} - I_{C1}R_{C1} = 5 - 2 = +3 \text{ V} \qquad V_{O2} = V_{CC} = +5 \text{ V}$$

For $V_{in} = -2$ V,

$$V_{O1} = V_{CC} = +5 \text{ V} \qquad V_{O2} = V_{CC} - I_{C2}R_{C2} = +3 \text{ V}$$

To find V_{IL} and V_{IH}, one can determine the unity slopes of the voltage transfer characteristics. However, from the diode equation or the transfer characteristic shown

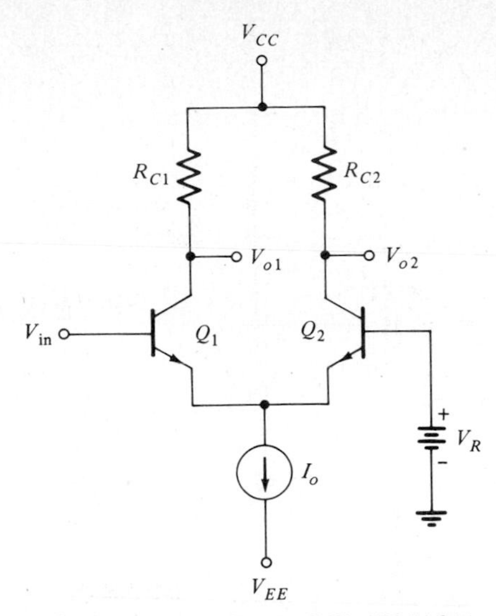

FIGURE 13.22 Basic ECL OR/NOR gate.

in Fig. 5.32, the transition width is quite small for $V_{B1} - V_{B2} \geq 5V_t \simeq 125$ mV, the current I_o will flow entirely in Q_1 or Q_2. Thus to guarantee the operation beyond the transition width, we use approximately 200 mV. Hence

$$V_{IL} \simeq V_R - 100 \text{ mV} \tag{13.15a}$$

$$V_{IH} \simeq V_R + 100 \text{ mV} \tag{13.15b}$$

In general then, we have

$$V_{OH} = V_{CC} \tag{13.16a}$$

$$V_{OL} = V_{CC} - I_C R_C = V_{CC} - \frac{R_C}{R_E}(V_R - V_{EE}) \tag{13.16b}$$

An improved IC version of the ECL gate is shown in Fig. 13.23. The OR operation is obtained as follows: When the input signal at A or B is at high level, Q_1 or Q_3 is conducting and Q_2 is off. When Q_2 is off, the emitter of Q_5 is at high level and $V_{o2} = V_A + V_B$, the OR operation. The emitter of Q_4, on the other hand, is at low level when either A or B is at high level; thus $V_{o1} = V_A + V_B$, i.e., the NOR operation.

Typical transfer characteristic of ECL(10K series) is shown in Fig. 13.24, and its electrical characteristics are listed in Table 13.8. Note that for these gates the logic swing at the output ($V_{OH} - V_{OL}$) is 0.8 V, and $\text{NM}_H = \text{NM}_L = 0.3$ V. The fan-out of these gates are typically 10. The power dissipation, which is determined from the product of V_{EE} and $(I_E)_{av}$ is typically 22 mW. The propagation delay of the circuit is proportioned to C_L and approximately equal to $0.2R_LC_L$ for Fig. 13.23. For $R_L = 1.5\ \text{k}\Omega$, $C_L = 5$ pF, $t_{pd} \simeq 1.5$ ns.

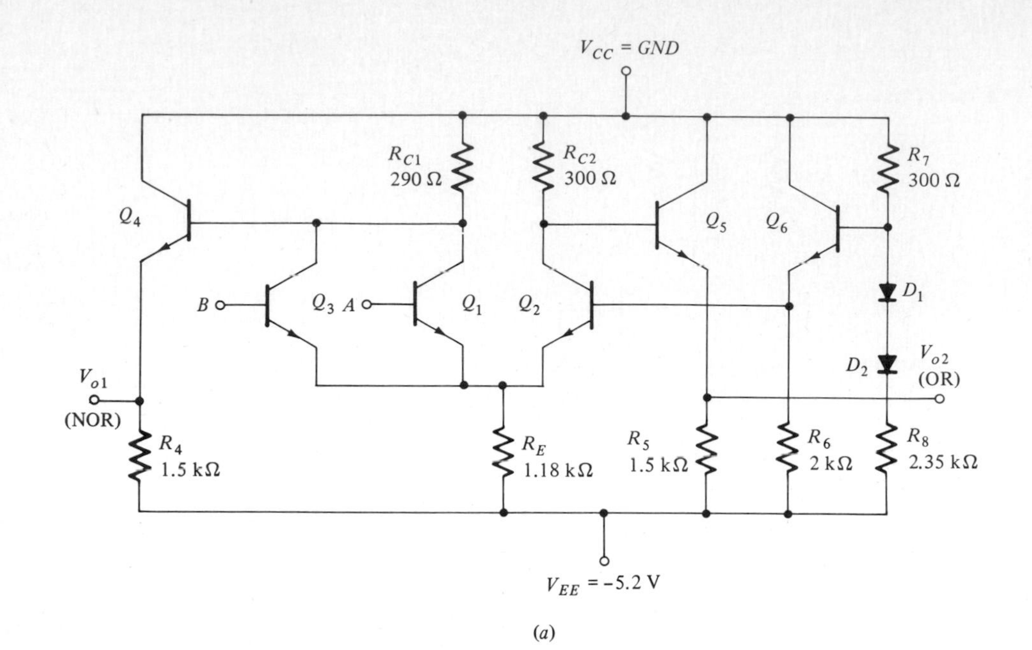

(a)

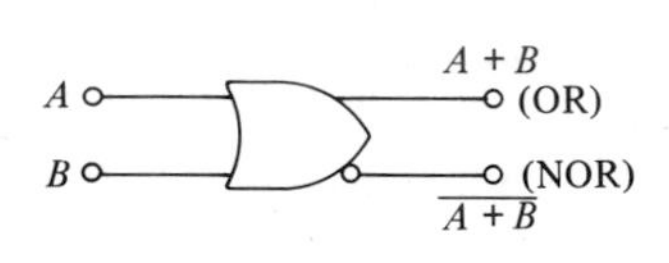

(b)

FIGURE 13.23 (a) IC ECL OR/NOR gate. (b) Logic symbol of the gate.

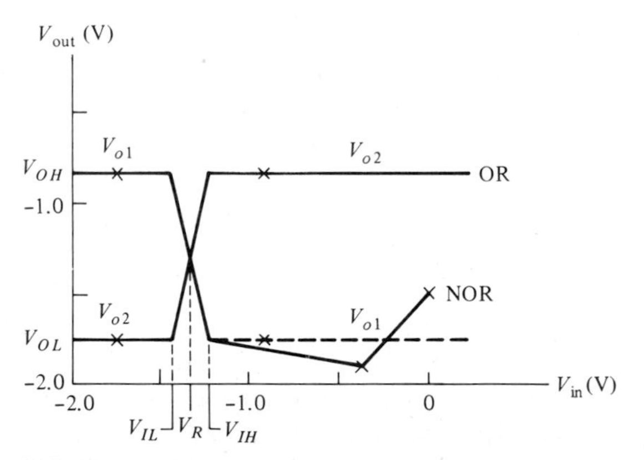

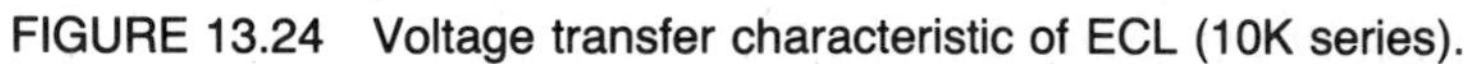

FIGURE 13.24 Voltage transfer characteristic of ECL (10K series).

TABLE 13.8 Emitter-Coupled Logic (10K) Typical Electrical Characteristics (T = 25°C)

V_{OH}/V_{OL}	−0.9 V/−1.7 V	Fan-out	10
V_{IH}/V_{IL}	−1.2 V/−1.4 V	Supply volts	−5.2 V
NM_H/NM_L	0.3 V/0.3 V	Power dissipation per gate	24 mW
Logic swing	0.8 V	Propagation delay time	2 ns

ECL and TTL gates can be interfaced with each other by using a commercially available ECL-to-TTL translator (10125 chip) and TTL-to-ECL translator (10124 chip). Finally, it should be noted that the ECL 100K series is also available. This series is an improved version of the 10K series, which is almost independent of temperature variations and of the variation in the supply voltage. The propagation delay of these gates are typically 0.75 ns with a power dissipation of about 40 mW, which results in power-delay product of 30 pJ. The data sheets for the ELC 10K series are given in Appendix E.

EXERCISE 13.14

For the basic ECL gate shown in Fig. 13.22, determine the ratio of the current I_{C1}/I_O for (*a*) $(V_{B1} - V_{B2}) = +5V_t$, and (*b*) $(V_{B1} - V_{B2}) = -5V_t$. Assume $\beta_0 \gg 1$.

Ans (*a*) 99.3 percent, (*b*) 0.67 percent

13.9 NMOS AND CMOS LOGIC GATES

The NMOS and the CMOS inverters were discussed in Secs. 3.5 and 3.6. These inverters consume much less power than the gates discussed in the previous sections occupy less real estate, and are extensively used in large-scale integration (LSI) such as in memory, microprocessors, etc.

A. NMOS Gates

The NMOS gates usually utilize inverters of the enhancement-depletion type shown in Fig. 3.19. This configuration produces an inverter with close to the ideal voltage transfer characteristic. In order to obtain a steep transfer characteristic, the W/L ratio of the enhancement-type driver is usually larger than the W/L ratio of the depletion-type load.

The basic NMOS NOR and NAND gates are shown in Fig. 13.25*a* and *b*, respectively. The NOR gate is obtained simply by connecting the driver inverters in parallel. All these driver (enhancement-type) inverters should have the same W/L

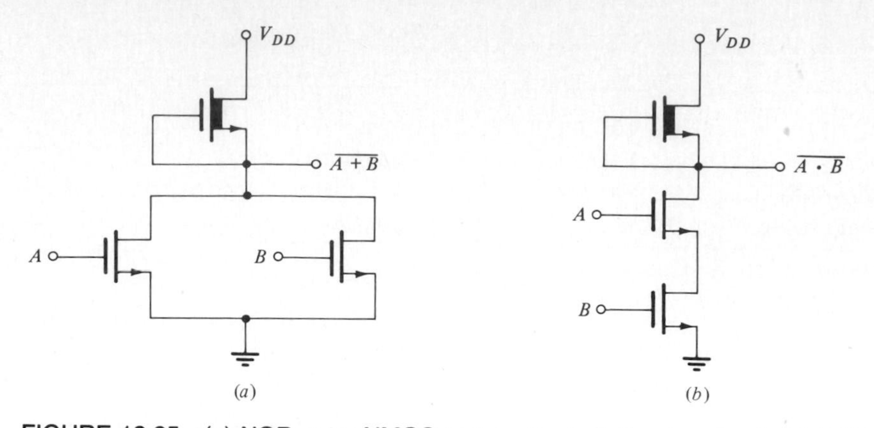

FIGURE 13.25 (*a*) NOR gate: NMOS gates with depletion-type load. (*b*) NAND gate: NMOS gates with depletion-type load.

ratio. The NAND gate is obtained by connecting all the identical driver inverters in series.

A typical transfer characteristic for an inverter using an enhancement NMOS driver and a depletion NMOS load is shown in Fig. 13.26. The key parameters of the inverter are as follows:

$V_{IL} = 1.6$ V $\qquad V_{OL} = 0.3$ V
$V_{IH} = 2.6$ V $\qquad V_{OH} = 5.0$ V
$NM_H = 2.4$ V $\qquad NM_L = 1.3$ V

The power delay product (PDP) of an NMOS gate is given by [2]

$$\text{PDP} = \frac{\gamma C_L (V_{OH} - V_{OL}) V_{DD}}{2} \tag{13.17}$$

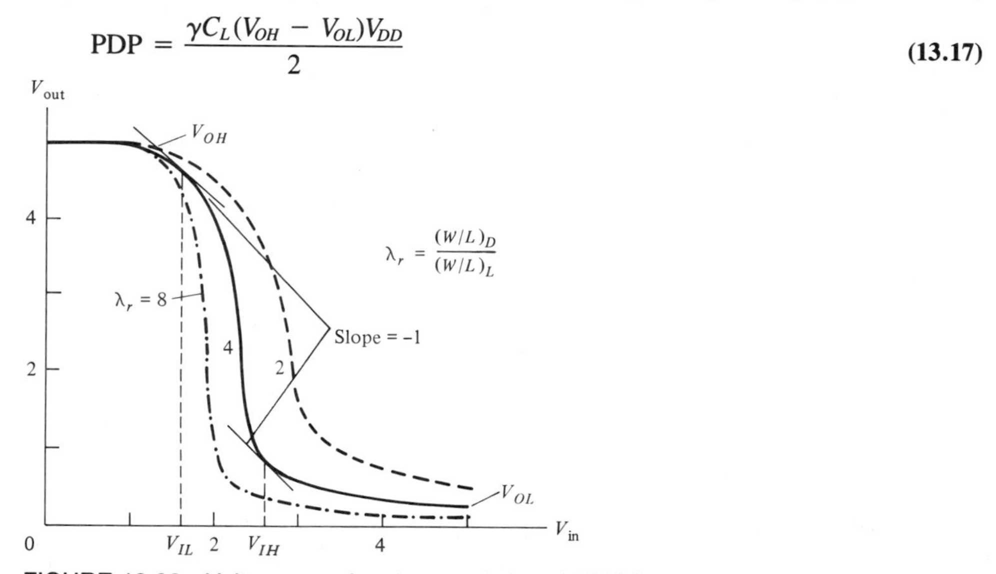

FIGURE 13.26 Voltage transfer characteristics of NMOS driver with depletion-type load.

where γ is a proportionality constant, with typical value $\gamma \simeq 0.5$. For the typical figures for $C_L = 5$ pF in the above, we have PDP $\simeq$ 25 pJ. Thus for a 0.2-mW power dissipation the propagation delay is approximately 125 ns.

B. CMOS Gates

The CMOS logic families are used in LSI applications because of their extremely low power dissipation property. In fact, the static power dissipation of a CMOS inverter is approximately 10 nW.

The basic CMOS inverter (Fig. 3.24) was discussed in Sec. 3.6. From the voltage transfer characteristic it was seen that the circuit has excellent temperature stability and nearly ideal voltage transfer characteristics. In the derivation of the results in Sec. 3.6 we assumed $K_n = K_p$. This assumption is not severe, as they can be made nearly equal by designing the W/L ratios inverse to the mobility ratios of the holes to electrons, namely,

$$\left(\frac{W}{L}\right)_{\text{PMOS}} \simeq 2.5\left(\frac{W}{L}\right)_{\text{NMOS}} \tag{13.18}$$

A NOR gate and a NAND gate in CMOS is shown in Fig. 13.27*a* and *b*, respectively. For the NOR gate note that the outputs of the PMOS devices are connected in series, whereas in the NMOS devices, they are connected in parallel. For the NAND gate the situation is reversed.

For the NOR gate note that when both the inputs V_A and V_B are at low levels the NMOS devices are off, whereas the PMOS devices are conducting and the output is at a high level, logical **1**. When V_A or V_B or both are at high level, one or both of the NMOS turn on while one or both of the PMOS turn off, resulting in a low level, logical **0** in each case. A typical voltage transfer characteristic for the basic NOR gate for three values of supply voltages V_{DD} = 5, 10, and 15 V are shown in Fig. 13.28.

For the NAND gate when both input levels are low, both NMOS devices are off,

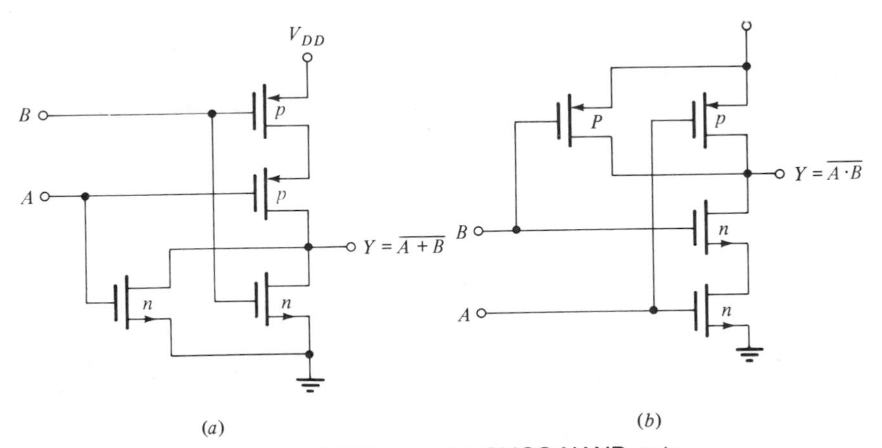

FIGURE 13.27 (*a*) CMOS NOR gate. (*b*) CMOS NAND gate.

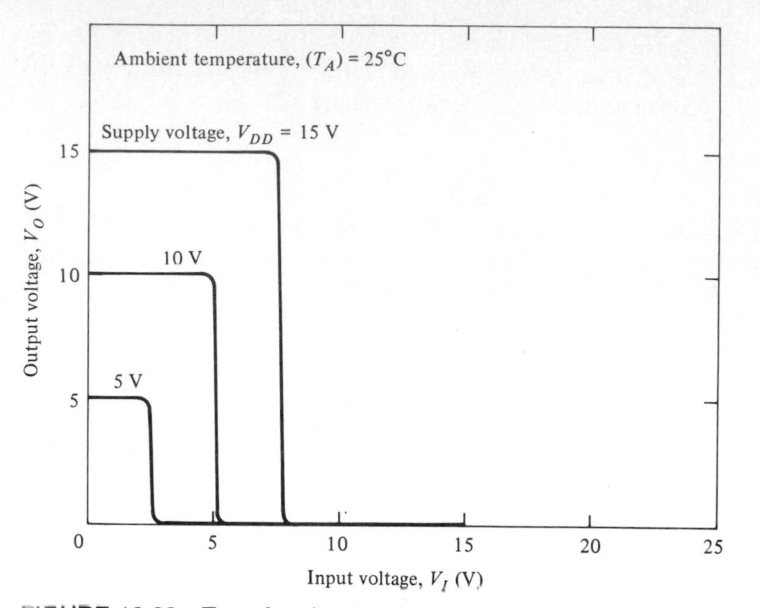

FIGURE 13.28 Transfer characteristics of 4000B CMOS gates for different supply voltages.

and PMOS devices are conducting, and thus the output level is high. When only one of the inputs is at low level, one of NMOS transistor is off, and the output remains high. When both input levels are high, the NMOS transistors are conducting, and PMOS transistors are off, and thus the output level is low.

The values of V_{IL} and V_{IH} can be determined from the transfer characteristic (3.32) by obtaining the values of the input voltage at which the slope is -1. However this is not usually done in practice. The transfer characteristic of a CMOS is nearly ideal (see Appendix E), but the mismatch between the PMOS and the NMOS does not allow the switching to occur at exactly $V_{DD}/2$. Manufacturers usually specify the two limits of the transfer characteristics as shown in Fig. 13.29. The values of V_{ILm} and V_{IHm} are taken from the minimum and maximum worst-case transfer characteristics. Normally the deviation in the output voltage is taken as $0.1V_{DD}$. For example, for a CMOS with a supply voltage $V_{DD} = 5$ V and $NM_H = NM_L = 1$ V (supplied by the manufacturer) the critical voltages are

$$V_{OL} = 0.5 \text{ V} \qquad V_{OH} = 4.5 \text{ V}$$

$$V_{ILm} = V_{OL} + NM_L = 1.5 \text{ V} \qquad V_{IHm} = V_{OH} - NM_H = 3.5 \text{ V}$$

Similarly, for $V_{DD} = 10$ V and specified $NM_H = NM_L = 2$ V, we have $V_{OL} = 1$ V, $V_{OH} = 9$ V, $V_{ILm} = 3$ V, and $V_{IHm} = 8$ V.

Power dissipation of CMOS under static condition, i.e., when not switching, is extremely small and of the order of 10 nW (dc supply current of 1 nA and $V_{DD} = 10$ V).

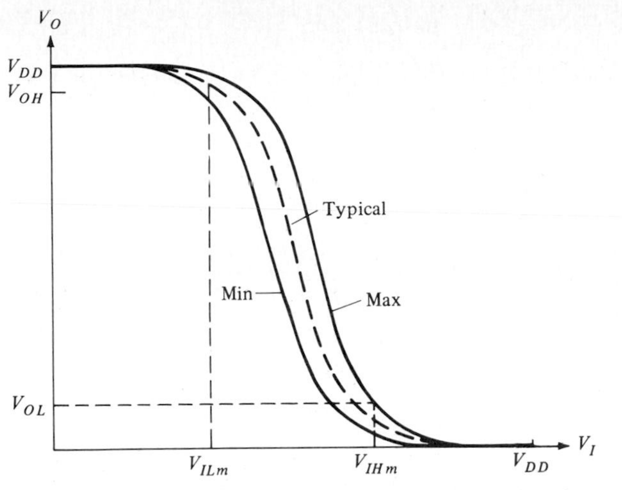

FIGURE 13.29 Maximum and minimum transfer characteristics of CMOS gates which determine V_{IL} and V_{IH}.

The dynamic power dissipation is readily determined as follows: The charge on the load capacitance rises from zero to $C_L V_{DD}$. The total energy supplied by the source is equal to $C_L V_{DD}^2$. If the switch is turned on and off periodically with a period $T = 1/f$, the average dynamic power dissipation P is then given by

$$P = C_L V_{DD}^2 f \tag{13.19}$$

For example, at a frequency of 1 MHz, $C_L = 5$ pF, and $V_{DD} = 10$ V the dynamic power dissipation is $P = 0.5$ mW.

The propagation delay of the CMOS is given approximately by

$$t_{PD} \simeq (0.7 \text{ ns/pF})C_L + 22 \text{ ns}$$

For example, for a CMOS at $V_{DD} = 10$ V and $C_L = 5$ pF the propagation delay is approximately 25 ns. Thus the power-delay product at 1 MHz is 12 pJ. The propagation delay decreases for increasing supply voltage V_{DD}, since the dynamic power dissipation increases with increasing voltage. Since the input current of the CMOS gates is extremely low, the CMOS gates have a large fan-out of typically larger than 50. Typical characteristics of the 4000B series is given in Table 13.9.

TABLE 13.9 CMOS Logic (4000 B series) Typical Electrical Characteristics ($V_{DD} = 5$ V)

V_{OH}/V_{OL}	4.5 V/0.5 V	Fan-out	50
V_{IH}/V_{IL}	3.5 V/1.5 V	Power dissipation per gate (1 MHz)	0.1 mW
NM_H/NM_L	1 V/1 V	Propagation delay	120 ns
Logic swing	4 V		

EXERCISE 13.15

For an NMOS gate the output logic swing is 4.6 V, $V_{DD} = 5$ V. Determine the PDP if $C_L = 5$ pF.

Ans 28.75 pJ

EXERCISE 13.16

For a CMOS device the following are specified for $V_{DD} = 15$ V: $V_{ILm} = 4$ V, $V_{IHm} = 11$ V. Determine the noise margins.

Ans $NM_H = NM_L = 2.5$ V

EXERCISE 13.17

For $V_{DD} = 5$ V, $C_L = 10$ pF, and $f = 100$ kHz determine the dynamic power dissipation of a CMOS gate.

Ans 25 μW

The CMOS 4000 series is the oldest CMOS logic family and is available in NAND and NOR gates. The newer version of this logic family denoted as 4000B series is now available from the manufacturers. The complete data sheet for the 4000B series is given in Appendix E. The letter B designates buffered series, which makes this family suitable for interfacing. Interfacing of the logic families is discussed in Sec. 13.11.

Another logic family of the CMOS series, known as the 74C logic families, is also available. These series are pin compatible with the TTL 74 family and are quite useful for converting TTL designs to CMOS. A high-performance newer version of 74C, designated by 74HC, is also available and can be used with TTL 74LS, with about similar overall logical performance. The low power dissipation of CMOS operated at low frequencies, however, makes it the dominant technology in LSI and VLSI among the gates considered thus far.

13.10 INTEGRATED INJECTION LOGIC (I^2L)

I^2L logic circuits are the latest development of IC gates using bipolar transistors. These are also referred to as merged transistor logic (MTL), since, as we shall see, the *npn* and *pnp* transistors are merged in one structure. These gates have certain desirable features, the foremost being the lowest power dissipation and the highest packing density that make them attractive for LSI and VLSI applications. An attractive property is the lowest delay-power product (which is constant) coupled with the ease of trading delay for power or vice versa. They are small in size, and a very large number of these devices can be put on a chip without excessive power dissipation, and therefore, they are used in LSI and VLSI circuits.

The basic I^2L digital gate is shown in Fig. 13.30*a*. The current I_o is *injected* through the *pnp* injector transistor to the common-base of the multicollector inverter transistors. The *pnp* transistor has a high current gain β_F, whereas the multicollector *npn* transistors have low current gains β_I since the roles of the emitter and collector are interchanged (compare Fig. 13.30*b* with that of Fig. 2.1). The current I_o is obtained by the external resistor and the dc voltage V_{DC}.

The power dissipation of the gate is obtained simply by the product of the current I_o and the supply voltage V_{DC}. The supply voltage is usually in the range 0.8 to 1 V. The power-delay product of the I^2L logic is essentially constant and very low in the range of 0.5 to 2 pJ. Thus power and delay can be traded by varying I_o, and power dissipation can be lowered by reducing I_o.

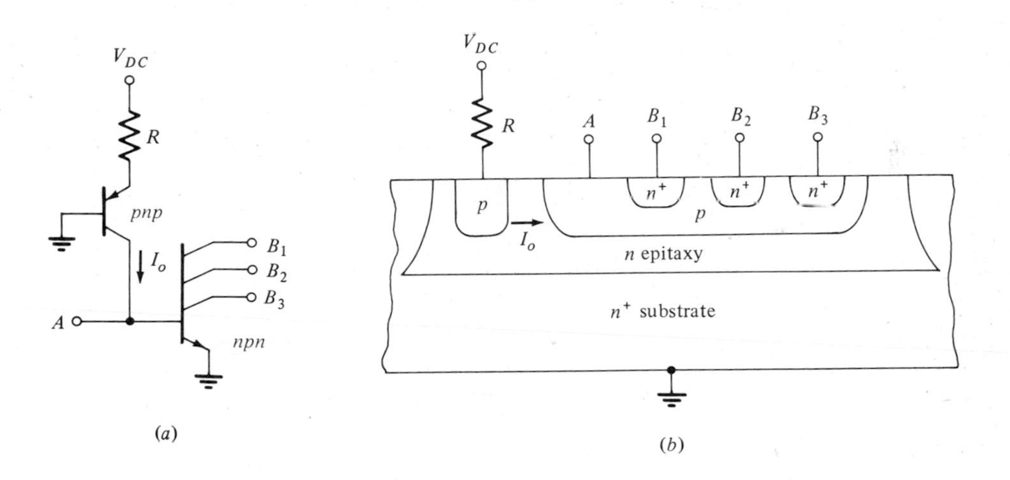

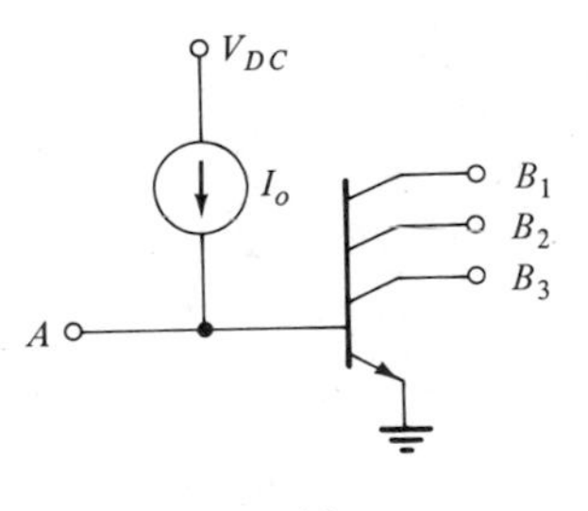

FIGURE 13.30 I^2L gate: (*a*) circuit diagram, (*b*) cross section, (*c*) equivalent circuit with a current source.

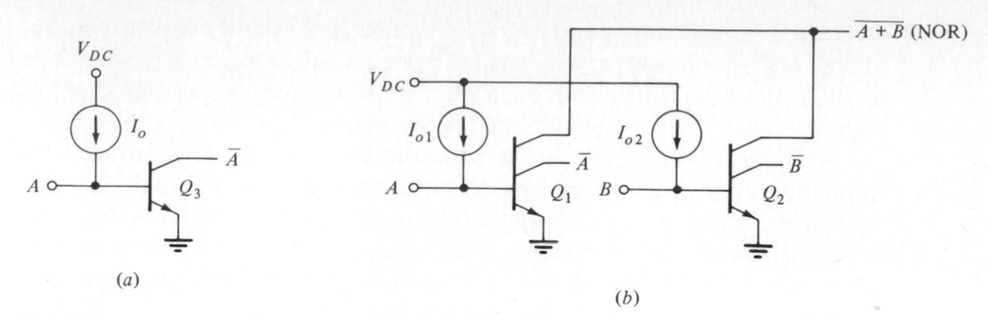

FIGURE 13.31 (*a*) I²L inverter. (*b*) I²L NOR gate.

Figure 13.31*a* and *b* shows simple I²L logic gates performing the NOT and the NOR functions, respectively. Typical values of the logic levels are

$$V_{OH} = (V_{BE})_{\text{sat}} = 0.8 \text{ V} \qquad V_{OL} = (V_{CE})_{\text{sat}} = 0.2 \text{ V}$$

$$V_{IH} = (V_{BE})_{\text{eoc}} = 0.7 \text{ V} \qquad V_{IL} = (V_{BE})_{\text{eos}} = 0.6 \text{ V}$$

where the subscripts eoc and eos denote the edge of conduction and the edge of saturation, respectively. From the above values it is seen that the logic swing and the noise margins are very small.

EXERCISE 13.18

Determine the logic swing and the noise margins of the I²L gate.

Ans LS = 0.6 V, NM_H = 0.1 V, NM_L = 0.4 V

The fan-out of the I²L gate depends on the β_I of the *npn* transistors ($\beta_I \simeq 5$). Typical values of the fan-out are therefore less than or equal to 5.

It is interesting to compare the key parameters of the I²L logic with those of TTL(LS), which are shown in Table 13.10. The basic Schottky version of I²L, shown in Fig. 13.32, is also available. The Schottky I²L achieves a lower propagation delay and a PDP of approximately 0.2 pJ, the lowest of all the gates.

TABLE 13.10 Parameters for I²L and TTL(LS)

	I²L	TTL(LS)
Power-delay product, pJ	0.5–2	10–20
Propagation delay time, ns	10–50	5–10
Supply voltage, V	1	5
Logic swing, V	0.6	2.2
Current range per gate	1 nA–1 mA	0.2–1 mA

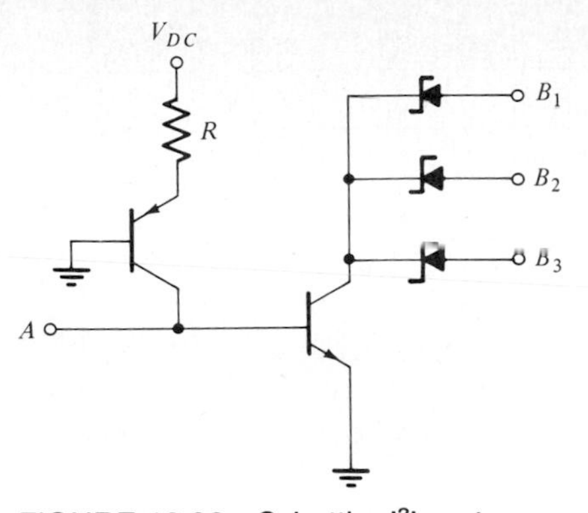

FIGURE 13.32 Schottky I²L gate.

13.11 INTERFACING

In some applications it is desirable to use a combination of different logic gates in order to take advantage of their superior properties. When this situation occurs it is necessary to interconnect or interface different logic families. When we interface one logic family to another, we must make sure that the voltage-level shifts are appropriate and also that we do exceed the fan-out capabilities of the gate.

Interfacing of ECL and TTL, i.e., interfacing ECL to TTL and TTL to ECL, is achieved by using the commercially available translators, e.g., 10125 ECL-to-TTL and 10124 TTL-to-ECL chips. The translators use nonsaturating transistors and Schottky diodes in order not to degrade the speed. Typical propagation delay of such translators are 5 ns.

In order to take advantage of the low power dissipation of CMOS and the speed of TTL, the TTL74, logic families are often interfaced with the CMOS 4000 series. Interfacing TTL and CMOS, and vice versa, is illustrated schematically in Fig. 13.33*a* and *b*, respectively. Typical values of the parameters for interfacing for TTL 74LS and

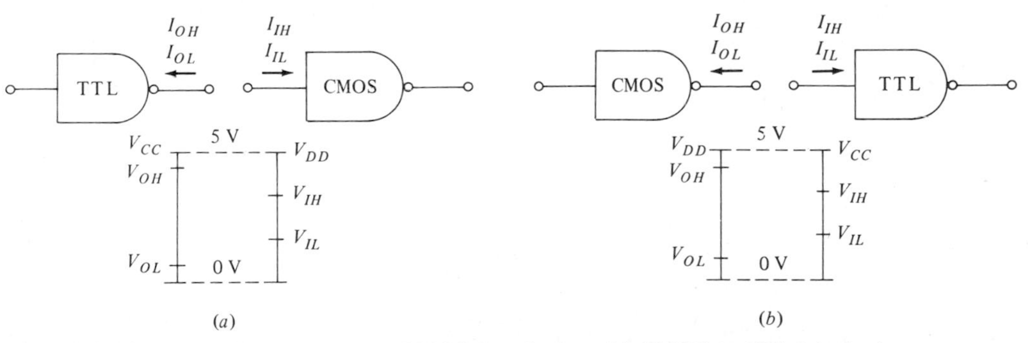

FIGURE 13.33 (*a*) TTL-to-CMOS interfacing. (*b*) CMOS-to-TTL interfacing.

TABLE 13.11 Typical Values of the Parameters for Interfacing TTL and CMOS

	V_{OH}, V	I_{OH}, mA	V_{OL}, V	I_{OL}, mA	V_{IH}, V	V_{IL}, V	I_{IH}	I_{IL}
TTL 747LS	2.7	−0.4	0.5	8	2	0.8	20 mA	−0.4 mA
CMOS 4000	4.5	−2	0.5	0.4	3.5	1.5	10 pA	−10 pA

CMOS 4000 series for a supply voltage of 5 V are given in Table 13.11. The conditions to be met for TTL to CMOS are given below (N is the fan-out):

$$V_{OH}(\text{TTL}) \geq V_{IH}(\text{CMOS}) \tag{13.20a}$$

$$V_{OL}(\text{TTL}) \leq V_{IL}(\text{CMOS}) \tag{13.20b}$$

$$-I_{OH}(\text{TTL}) \geq NI_{IH}(\text{CMOS}) \tag{13.20c}$$

$$I_{OL}(\text{TTL}) \geq -NI_{IL}(\text{CMOS}) \tag{13.20d}$$

For CMOS-to-TTL interfacing, just replace TTL by CMOS and vice versa in (13.20). In (13.20) all the inequalities are satisfied except for (13.20*a*), since $V_{OH}(\text{TTL}) \simeq 2.7$ V and $V_{IH}(\text{CMOS}) \simeq 3.5$ V. Hence the TTL circuit must be modified to make $V_{OH}(\text{TTL})$ larger than 3.5 V when TTL is the drive stage. This is done several ways and is given by the manufacturer. Similarly, when a CMOS is the driver stage, by looking at the data sheets it is found that the inequality $I_{OL}(\text{CMOS}) \geq -NI_{IL}(\text{TTL})$ is barely satisfied for $N = 1$; hence a buffer is needed to remedy this situation. It is for this reason that the buffered series is a preferred choice.

13.12 GENERAL COMPARISON OF LOGIC FAMILIES

In digital systems a large number of gates are usually available. The digital IC gates are packaged by manufacturers with several gates on a single chip. Some of the IC gates in the TTL 74 and CMOS 4000 series are listed below:

Quad two-input NAND (7400, 4011)
Quad two-input AND (7408, 4081)
Quad two-input OR (7432, 4071)
Quad two-input NOR (7402, 4001)
Quad two-input XOR (7846, 4030)
Triple three-input AND (7411, 4073)
Triple three-input NAND (7410, 4023)
Triple three-input NOR (7427, 4025)
Dual four-input NAND (7420, 4012)
Dual four-input AND (7421, 4082)
Dual four-input NOR (7425, 4002)

Quad four-input NAND (7422, 4012)
Hex inverter (7404, 4069)
Single eight-input NAND (7430, 4068)

Some of the above combinations are also available in the other logic families discussed in this section. However, the widest choice available are in the TTL and CMOS logic families. For convenience the chip numbers for the TTL 74 series with the totem-pole output and CMOS 4000B series are also listed above. The choice of a logic family depends on the requirements at hand. Some of the parameters are cost, speed of operation, (propagation delay) power dissipation, noise immunity, fan-out and fan-in.

The key performance parameters of the different gate types are for typical logic families compared in Table 13.11. Because of rapid changes in IC technology, the data in Table 13.11 are good as of the date of writing this book. From Table 13.12, it is seen that the ECL gate has the fastest speed; in fact, the series 100K has a propagation delay of 0.7 ns. However, note that the ECL gates have the highest power dissipations per gate. The CMOS and I^2L have the largest propagation delay. The constant power-delay product lines (0.1 to 100 pJ) for the logic families discussed in this chapter are shown in Fig. 13.34. It should be noted that the propagation delay of CMOS given above is based on 5 μm gate length. Scaling down the gate length to 2 μm, which is current state of the art (at the time of this writing) has reduced the propagation delay to about 2 ns. The power consumption of the 2-μm CMOS 7000 series for a capacitive load of 1 pF and a supply voltage of 5 V is 25 μW/MHz. Thus the CMOS arrays compete strongly with Schottky TTL, and the short-gate CMOS will be a strong contender in the memory area in LSI nd VLSI in the near future. The figure shows the approximate position of the various logic families with respect to PDP. Note that I^2L has the lowest PDP $\simeq$ 0.5 pJ, while TTL 54/74 has the highest PDP $\simeq$

TABLE 13.12 Basic Characteristics of Selected Typical Logic Families

	Type				
	TTL		ECL	CMOS	
Characteristic	54/74	54/74LS	(10K)	(4000)	I^2L
Positive logic function of basic gate	NAND	NAND	OR/NOR	NAND/NOR	NAND
Supply voltage, V	5	5	−5.2	5	1
Typical fan-out	10	10	20	50	5
Typical power dissipation per gate	10 mW	2 mW	25 mW	0.01 μW(static) $\simeq$ 0.5 mW(1 MHz)	10 nW–25 μW
Typical propagation delay	10 ns	10 ns	2 ns	25 ns	20 ns–50 μs
Noise performance NM_H/NM_L	2 V/0.3 V	0.7 V/0.3 V	0.3 V/0.3 V	1 V/1 V	0.1 V/0.4 V
Logic swing, V	3.3	2.2	0.8	4	0.6

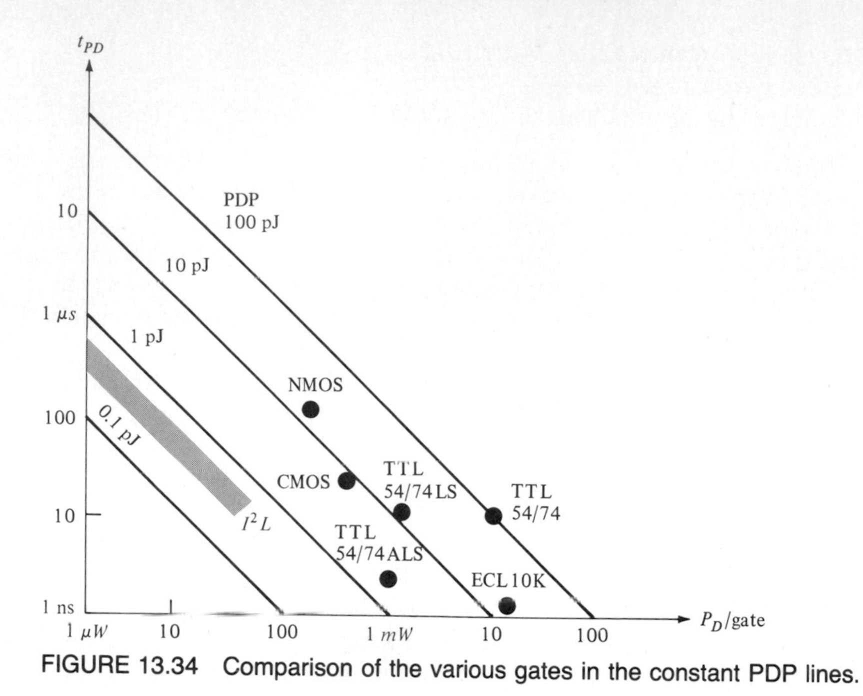

FIGURE 13.34 Comparison of the various gates in the constant PDP lines.

100 pJ. Further note that for the I²L logic speed and power can be traded, and hence it is drawn as shown. The CMOS power dissipation is considerably reduced at low-frequency switching. In other words, the power dissipation per gate increases linearly for increasing frequencies as given in (13.19).

13.13 CONCLUDING REMARKS

In this chapter we introduced the concept of binary, octal, and hexadecimal number systems, which are important in digital circuits and computers. We introduced the various logical function operations such as NOT, NAND, NOR, AND, OR, and XOR. We then showed the implementation of these gates with BJT and MOS devices. We discussed only the various logic families that are currently predominant, namely, TTL, ECL, NMOS, CMOS, and I²L logic gates. A comparison of the gates and their properties reveals the following conclusions: TTL(LS) is the most widely used logic family for SSI and MSI. For LSI applications, NMOS and CMOS gates are used as the power dissipation per gate is very low. I²L is used for VLSI applications due to their extremely low power dissipation and hence high packing density. We have also discussed interfacing, as it may be necessary or desirable in some applications to use a combination of logic families.

In Chap. 14 we shall consider higher-level building blocks using the gates discussed in this chapter.

REFERENCES

1. F. J. Hill and G. R. Peterson, *Switching Theory and Logical Design,* Wiley, New York, 1974.
2. D. A. Hodges and H. G. Jackson, *Analysis and Design of Digital Integrated Circuits,* McGraw Hill, New York, 1983.
3. R. L. Morris and J. R. Miller (eds.), *Designing with TTL Integrated Circuits,* McGraw Hill, New York, 1971.
4. W. R. Blood, Jr., *MECL System Design Handbook,* 2d ed., Motorola Semiconductor Products Inc., Phoenix, 1972.
5. W. N. Carr and J. P. Mize, *MOS/LSI Design and Applications,* McGraw-Hill, New York, 1972.
6. J. E. Smith (ed.), *Integrated Injection Electronics,* IEEE Press, New York, 1980.

PROBLEMS

13.1 Convert the decimal numbers (1) 18670.23 and (2) 2235.68 to
(*a*) Binary
(*b*) Octal
(*c*) Hexadecimal

13.2 Convert the octal numbers (1) 372.12 and (2) 420.50 to
(*a*) Decimal
(*b*) Binary
(*c*) Hexadecimal

13.3 Convert the hexadecimal numbers (1) 6ED.3C and (2) 7BA.F2 to
(*a*) Binary
(*b*) Octal
(*c*) Decimal

13.4 Show that the NOR gates alone are adequate to implement any boolean function. (Show the NOT, AND, and OR implementation corresponding to Fig. E13.5.)

13.5 Prove each of the following simplification theorems for boolean variables:
(*a*) $x \cdot y + x \cdot \bar{y} = x$
(*b*) $x + \bar{x} \cdot y = x + y$

13.6 Repeat problem 13.5 for the following:
(*a*) $x + x \cdot y = x$
(*b*) $x \cdot y + \bar{x} \cdot z + y \cdot z = x \cdot y + \bar{x} \cdot z$

13.7 Verify De Morgan's second law in (13.9) by means of a truth table.

13.8 Prove the following using De Morgan's laws. Use two variables at a time, i.e., let $Y = A + B$ and $Z = C + D$
(*a*) $\overline{A + B + C + D} = \bar{A} \cdot \bar{B} \cdot \bar{C} \cdot \bar{D}$
(*b*) $\overline{A \cdot B \cdot C \cdot D} = \bar{A} + \bar{B} + \bar{C} + \bar{D}$

13.9 Construct the truth table for a three-input NAND gate.

13.10 Repeat problem 13.9 for a three-input NOR gate.

13.11 The waveforms shown in Fig. P13.11 are the inputs to a NOR gate. Sketch the output waveform.

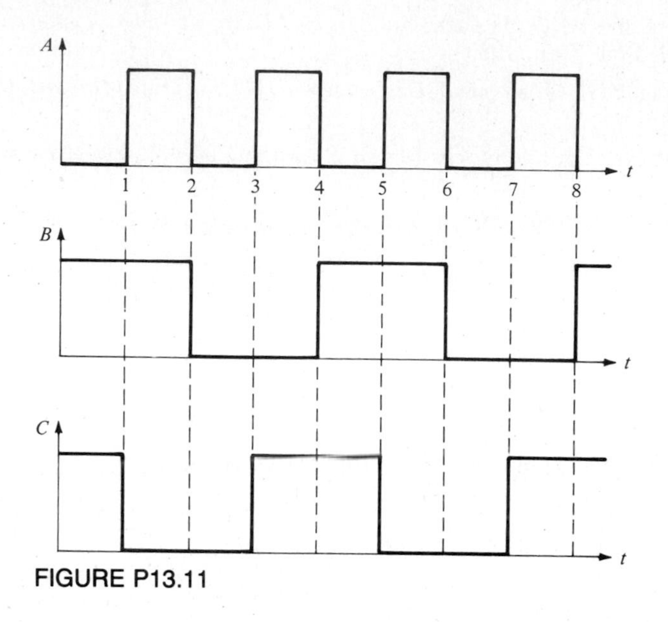

FIGURE P13.11

13.12 Repeat problem 13.11 with the condition that the waveforms are inputs to a NAND gate.

13.13 Determine the output function Y for the logic circuit shown in Fig. P13.13.

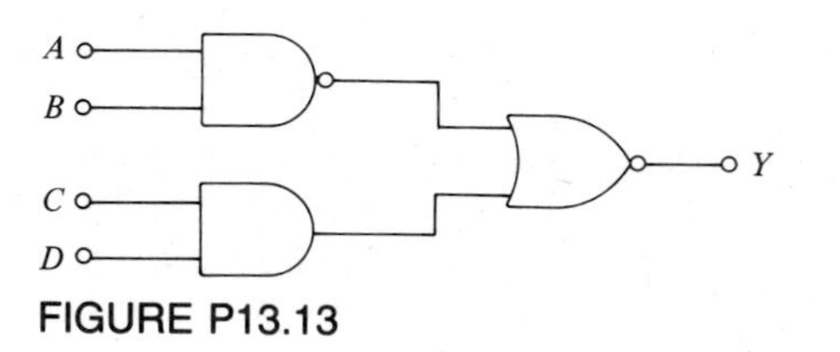

FIGURE P13.13

13.14 The exclusive-OR gate is shown in Fig. 13.8. Verify the following identities for exclusive-OR operation:

(a) $A \oplus (A + B) = \bar{A} \cdot B$

(b) $A \oplus \bar{A}B = A + B$

(c) $A \oplus (\bar{A} + B) = \overline{A \cdot B}$

13.15 Verify the following identities for the exclusive-NOR operation:

(a) $A \odot \bar{A}B = \overline{A + B}$

(b) $A \odot (\bar{A} + B) = A \cdot B$

(c) $A \odot (A + B) = \bar{A} \cdot B$

13.16 Simplify each of the following boolean expressions and draw the logic blocks for each using AND, OR, and NOT gates. All types need not be used.

(*a*) $(A + B) \cdot (\overline{A} + C) \cdot (B + C)$

(*b*) $A \cdot \overline{C} + A \cdot B \cdot C + \overline{A \cdot B \cdot (D + \overline{C})}$

(*c*) $A \cdot B + A \cdot B \cdot \overline{C} + \overline{(B + C + \overline{A})}$

13.17 Construct a truth table for each of the following boolean functions:

(*a*) $A \cdot \overline{B} + B \cdot \overline{C} + C \cdot \overline{A}$

(*b*) $\overline{A} \cdot \overline{D} + A \cdot C + \overline{B} \cdot \overline{C} \cdot \overline{D} + B \cdot C \cdot D$

13.18 Obtain a NAND-gate realization to the following function: $f = A \cdot B + B \cdot C + C \cdot A$.

13.19 Obtain a NOR-gate realization to the following function: $f = \overline{A} \cdot \overline{B} + \overline{B} \cdot \overline{C} + \overline{C} \cdot \overline{A}$.

13.20 Find the relation between the output and the input for the logic diagram shown in Fig. P13.20 and simplify.

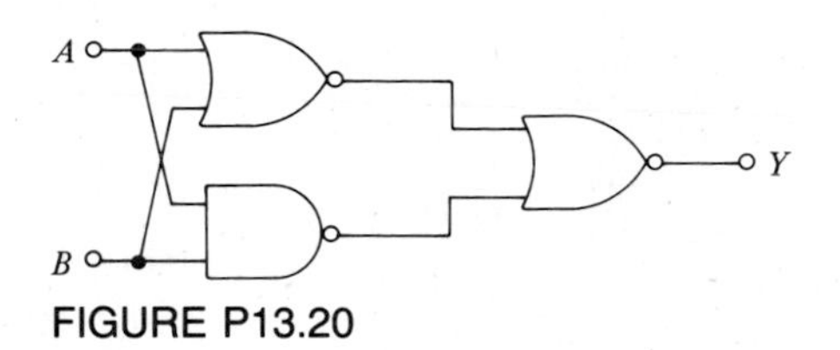

FIGURE P13.20

13.21 For the inverter shown in Fig. P13.21, determine the values of V_{OH}, V_{IH}, V_{OL}, V_{IL}, NM_H, and NM_L. Assume the following transistor parameters: $\beta_0 = 50$, $(V_{CE})_{sat} = 0.2$ V, $(V_{BE})_{sat} = 0.8$ V, $(V_{BE})_{on} = 0.7$ V.

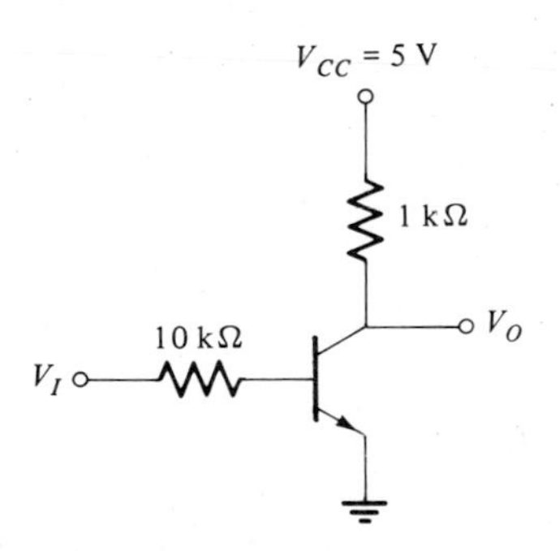

FIGURE P13.21

13.22 For the inverter circuit shown in Fig. P13.21, determine the fan-out of the gate if the inverter is loaded by N identical inverters. (*Hint:* Set $NM_H = 0$ to determine the value of N.)

13.23 A TTL inverter has the following parameters: $V_{OH} = 3.5$ V, $V_{OL} = 0.2$ V, $V_{IH} = 1.5$ V, $V_{IL} = 0.5$ V. Determine the noise margins when two TTL inverters are connected in cascade.

13.24 Repeat problem 13.23 for CMOS inverters with the following parameters: $V_{OH} = 4.5$ V, $V_{OL} = 0.5$ V, $V_{IH} = 3.5$ V, $V_{IL} = 1.5$ V.

13.25 Determine the power-delay product of an NMOS for $C_L = 10$ pF if $V_{OH} = 5$ V, $V_{OL} = 0.3$ V, and $V_{DD} = 10$ V.

13.26 Determine the dynamic power dissipation of a CMOS gate for $C_L = 10$ pF and $V_{DD} = 5$ V at $f = 1$ MHz.

13.27 For the NMOS inverter shown in Fig. P13.27, the device parameters arc $V_T = 1$ V, $K_n = 50\ \mu A/V^2$. Determine V_{OH}, V_{OL}, and NM_H and NM_L.

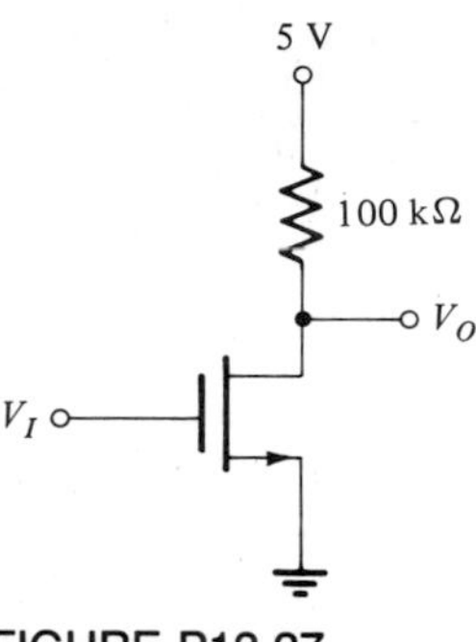

FIGURE P13.27

13.28 For the PMOS inverter shown in Fig. P13.28, determine V_{OH} and V_{OL} and the logic swing. Assume identical devices with $|V_{Tp}| = 1$ V and $K_p = 10^{-3}$ A/V^2.

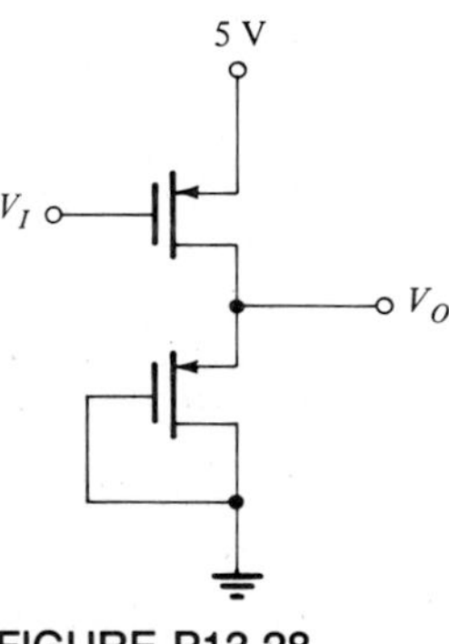

FIGURE P13.28

13.29 For the depletion load NMOS inverter shown in Fig. P13.29, determine V_{OH}, V_{OL}, NM_H, and NM_L. The following parameters are given:

For the enhancement type, $V_T = 3$ V, and $K_n = 5 \times 10^{-3}$ A/V^2.
For the depletion type, $I_{DSS} = 2$ mA, and $V_{PO} = 3$ V.

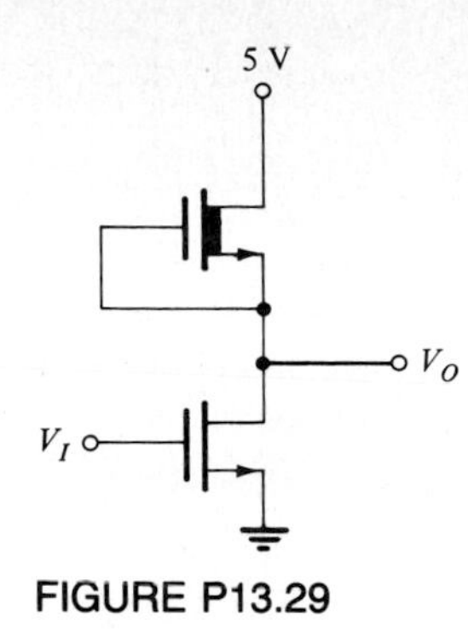

FIGURE P13.29

13.30 For the CMOS inverter shown in Fig. P13.30, determine V_{OH}, V_{OL}, NM_L, and NM_H given $K_n = K_p = 0.1$ mA/V^2 and $V_{Tn} = |V_{Tp}| = 1$ V.

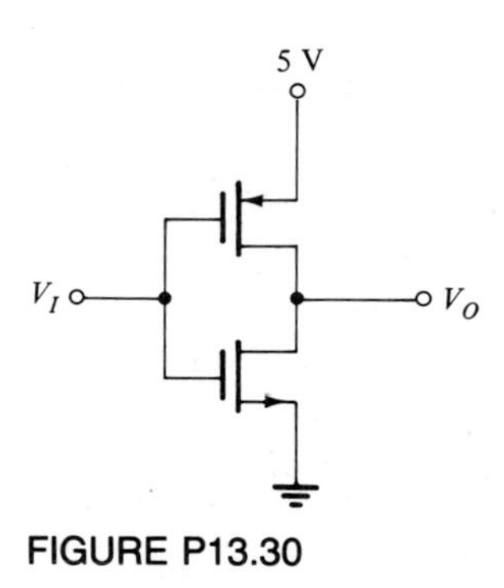

FIGURE P13.30

13.31 Determine the maximum current I_D for the CMOS circuit of problem 13.30.

13.32 The logic diagram of a half-adder circuit is shown in Fig. P13.32. Show the truth table and verify that the circuit adds 2 bits.

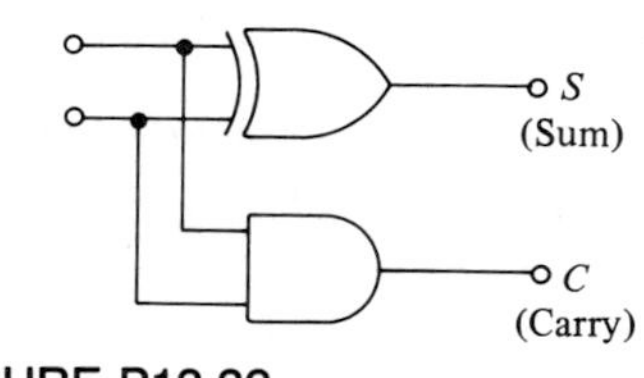

FIGURE P13.32

Flip-Flops, Multivibrators, and Digital-to-Analog and Analog-to-Digital Converters

In Chap. 13 we considered the basic logic gates and the logical relationships between the input and the output. The output signals resulted from the application of the input signals but with some propagation delay. In the digital circuits that we considered in Chap. 13 there were no feedback connections from the output to the input of any gate. These circuits are commonly called *combinational* logic circuits.

In this chapter we consider digital circuits with intentional feedback connections. The existence of this feedback can lead to a number of very important and useful functional blocks. One such element is the *memory,* which is at the heart of a digital computer. When logical elements with feedback are present in a logical circuit, the circuits are known as *sequential* logic circuits. The basic elements of a sequential circuit are bistable latches and flip-flops. We shall discuss the various types of flip-flops, multivibrators, clocks, timers, and the applications of flip-flops in shift registers and counters.

The chapter will conclude with a discussion of digital-to-analog (D/A) and analog-to-digital (A/D) converters.

14.1 THE BISTABLE LATCH

The bistable latch is the simplest form of the many types of flip-flops. It stores information temporarily or permanently. It is therefore called a *memory* element. Electronic memory capacity ranges from a few hundred bits for a pocket calculator to 10^6 bits for personal computers.

In its simplest form a memory consists of two inverters in a feedback arrangement shown in Fig. 14.1. The inverters can be simply BJT inverters or NMOS inverters such as shown in Fig. 14.2*a* and *b*, respectively, for discrete circuits. In IC circuits these are usually the NOT-connected NAND gates or the NOT-connected NOR gates, using any of the logic families discussed in Chap. 13, such as the TTL (54/74) or the CMOS (4000B) series.

Since we are considering digital circuits, the output of each inverter is either **1** or **0**, where **1** refers to high voltage levels and **0** refers to low voltage levels. In other words, we use positive logic as in Chap. 13. If Q is **1**, then $\overline{Q}$ is **0**, and vice versa. In latches both of these states are stable, and the circuit can remain indefinitely in any one of these states. Since the circuit has two stable states, it is called a *bistable circuit*. We shall, for consistency, refer to the circuit as *storing* a **1** if Q = **1** and storing a **0** if Q = **0** (where Q refers to the output of the upper inverter in Fig. 14.1). Note that when the circuit is turned ON, Q can be either **1** or **0**, but once it randomly chooses its state, it remains there indefinitely unless disturbed by a trigger or by the power supply being turned off and on. Note that in this simple latch we cannot be sure whether Q is **1** or **0**, and hence it is desirable to have a latch that can be *controlled;* i.e., if we want Q = **1**, it should be so unless we change its state purposely.

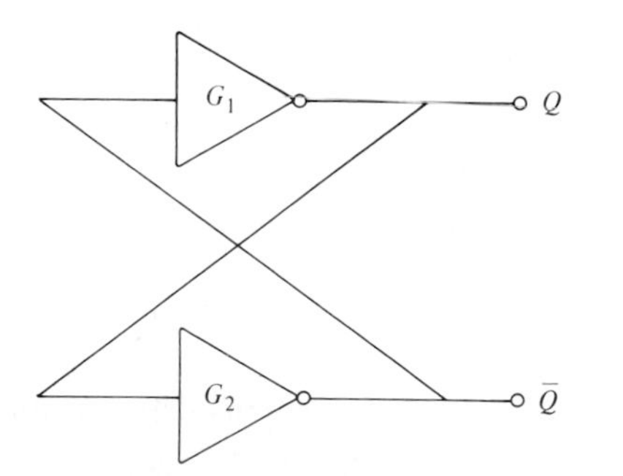

FIGURE 14.1 Basic bistable circuit.

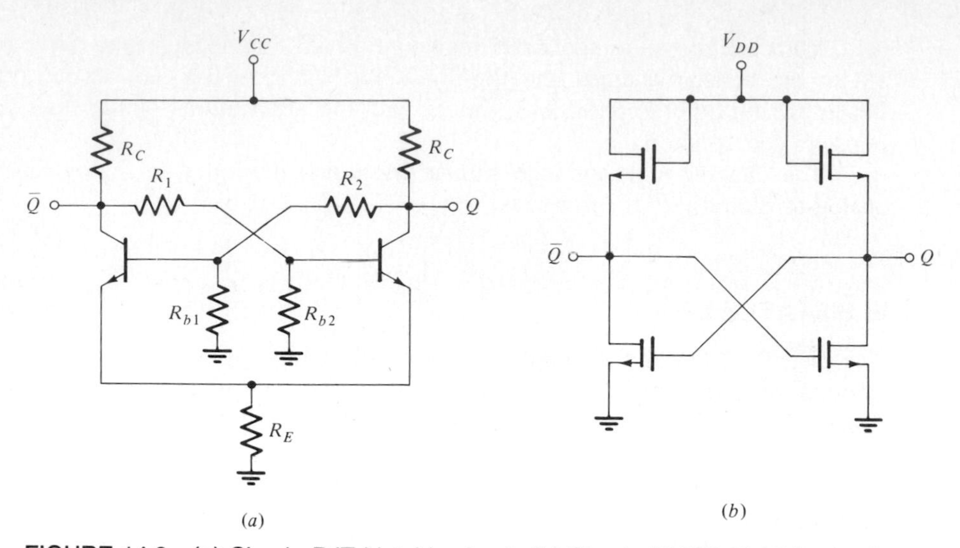

FIGURE 14.2 (*a*) Simple BJT bistable circuit. (*b*) Simple NMOS bistable circuit.

14.2 *SR* LATCH

SR latch is short for set-reset latch. This is a controllable latch as we shall see. Set means $Q = \mathbf{1}$ when $S = \mathbf{1}$; reset means $Q = \mathbf{0}$ when $R = \mathbf{1}$. *SR* latches can be obtained using either the NOR gates or the NAND gates.

Consider first the NOR-connected *SR* latch shown in Fig. 14.3*a*. The *function* or *characteristic* table, which is similar to the truth table, is shown in Fig. 14.3*b*. Note that the latch is set for $S = \mathbf{1}$, $R = \mathbf{0}$ and reset for $R = \mathbf{1}$, $S = \mathbf{0}$. The condition of $S = R = \mathbf{1}$ is not allowed since Q and $\overline{Q}$ are not consistent; i.e., if $Q = \mathbf{0}$, $\overline{Q}$ cannot be also $\mathbf{0}$. For $S = R = \mathbf{0}$ there is no change at the output. The block diagram for this latch is shown in Fig. 14.3*c*.

Next consider the NAND-connected *SR* latch shown in Fig. 14.4*a*, its function table is shown in Fig. 14.4*b*. Note again that the latch is set when $S = \mathbf{1}$ and $R = \mathbf{0}$ and reset when $R = \mathbf{1}$ and $S = \mathbf{0}$. The block diagram of the NAND-connected gates

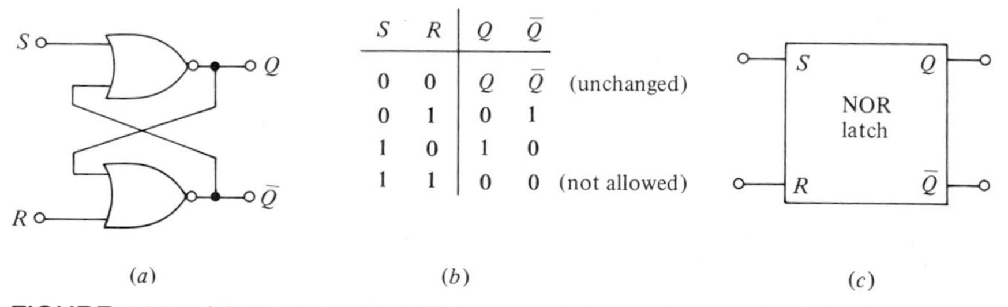

S	R	Q	$\overline{Q}$	
0	0	Q	$\overline{Q}$	(unchanged)
0	1	0	1	
1	0	1	0	
1	1	0	0	(not allowed)

FIGURE 14.3 (*a*) A latch with NOR gates. (*b*) Function table of the latch. (*c*) Block diagram of the latch.

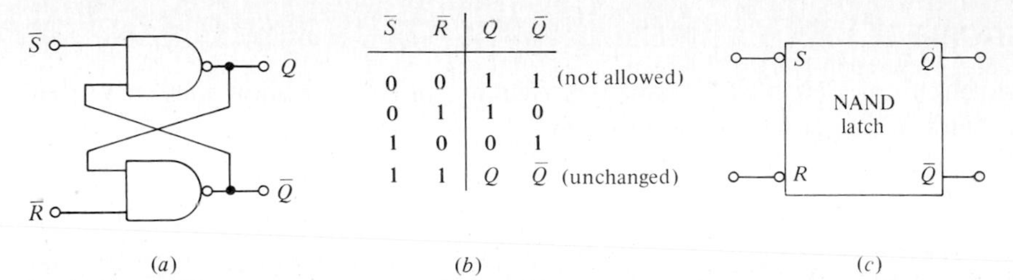

$\overline{S}$	$\overline{R}$	Q	$\overline{Q}$	
0	0	1	1	(not allowed)
0	1	1	0	
1	0	0	1	
1	1	Q	$\overline{Q}$	(unchanged)

FIGURE 14.4 (*a*) A latch with NAND gates. (*b*) Function table of the latch. (*c*) Block diagram of the latch.

is shown as in Fig. 14.4*c* and is also referred to as a *SR* latch. An example of this latch, available from the manufacturer, is the TTL 74279.

14.3 *JK* FLIP-FLOP

The ambiguity of the *SR* latches, depicted by the disallowed states in their function tables when both *S* and *R* are activated, is avoided by providing two (*J* and *K*) feedback connections. One such *JK* flip-flop using NAND gates is shown in Fig. 14.5*a*. Figure 14.5*b* and *c* is the logic symbol and the function table for the *JK* flip-flop. The clock, as discussed in Sec. 14.9, is an astable multivibrator that generates a train of square-wave pulses.

The *J* and *K* inputs are synchronized with the clock input, and thus *J* is the clocked-set, and *K* is in the clocked-reset terminals corresponding to the *SR* latch. From Fig. 14.5*c*, note that when $J = K = \mathbf{0}$, the flip-flop does not change its state after clocking. However, when $J = K = \mathbf{1}$, the flip-flop changes its state; i.e., it will *toggle* when clocked. The *JK* flip-flop that is used only for toggling is called a *T flip-flop*. Some of the important parameters in flip-flops are the power dissipation, propagation delay, and toggle frequency. These parameters are dependent on the number and types of gates used, and will be considered later in this chapter.

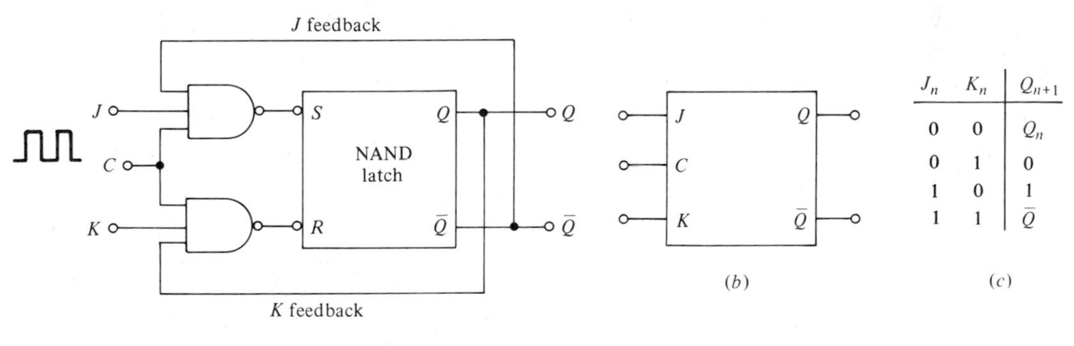

J_n	K_n	Q_{n+1}
0	0	Q_n
0	1	0
1	0	1
1	1	$\overline{Q}$

FIGURE 14.5 (*a*) *JK* flip-flop. (*b*) Block diagram. (*c*) Function table.

EXERCISE 14.1

The logic circuit of a clocked *SR* latch is shown in Fig. E14.1. Show that only when $C = \mathbf{1}$, *S* and *R* are sensed by the flip-flop.

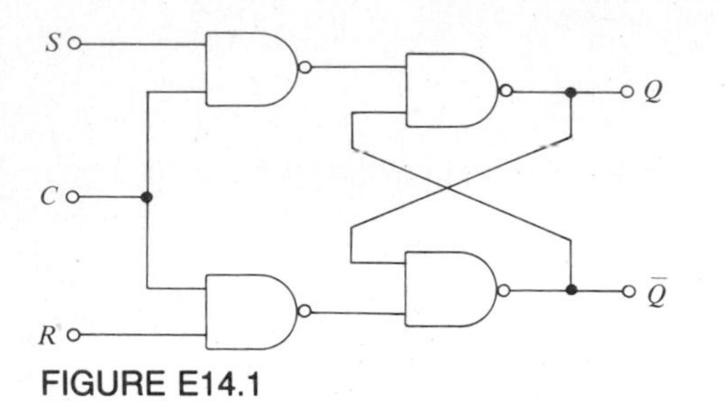

FIGURE E14.1

An *SR* latch with the same properties as of Fig. E14.1 but that uses a NOR latch and two AND gates is given in problem 14.2.

14.4 *JK* MASTER-SLAVE FLIP-FLOP

In the *JK* flip-flop there is a restriction on the pulse width of the clock, namely, that the clock pulse width must be smaller than the propagation delay of the flip-flop. This is seen as follows: Let $Q = \mathbf{0}$ and suppose *C* is high and $J = K = \mathbf{1}$; then $\overline{S} = \mathbf{0}$ and $\overline{R} = \mathbf{1}$. This will set the flip-flop. Now, $Q = \mathbf{1}$; if *C* is still high, then $\overline{S} = \mathbf{1}$ and $\overline{R} = \mathbf{0}$, which will immediately reset the flip-flop. This undesired toggeling will continue until *C* is brought low. The only way to prevent this is to make the clock pulse width smaller than the propagation delay of the flip-flop. Since the propagation delay of the IC flip-flops is very small (of the order of 10 ns), this restriction is very undesirable. To eliminate this problem, the *JK* master-slave flip-flop has been introduced. The logic-circuit diagram of a master-slave flip-flop is shown in Fig. 14.6*a*. In this case the master flip-flop is activated by the clock *C*, and the slave flip-flop is activated by $\overline{C}$.

The operation of the *JK* master-slave flip-flop is as follows: When the clock pulse rises from low to high, i.e., *C* goes from **0** to **1**, $\overline{C}$ is going from **1** to **0**. From Fig. 14.5*a* and *c* we see that when *C* goes to **1** the slave latch is disabled; i.e., its output is at Q_n when $\overline{C}$ is at **0**. However, when the clock pulse falls from high to low, i.e., *C* goes to **0**, the master latch is disabled. For $\overline{C} = \mathbf{0}$, $C = \mathbf{1}$, the slave latch is enabled and its output is at $\overline{Q}$. The output of the flip-flop is thus changed. The sequence of events in the master-slave flip-flop during a clock cycle is best illustrated in Fig. 14.6*b*. In this type of flip-flop there is no limit on the maximum width of the clock pulse, but there is a limit on the minimum pulse width, namely, it must be larger than the total propagation delay of the flip-flop. This restriction obviously is not a problem.

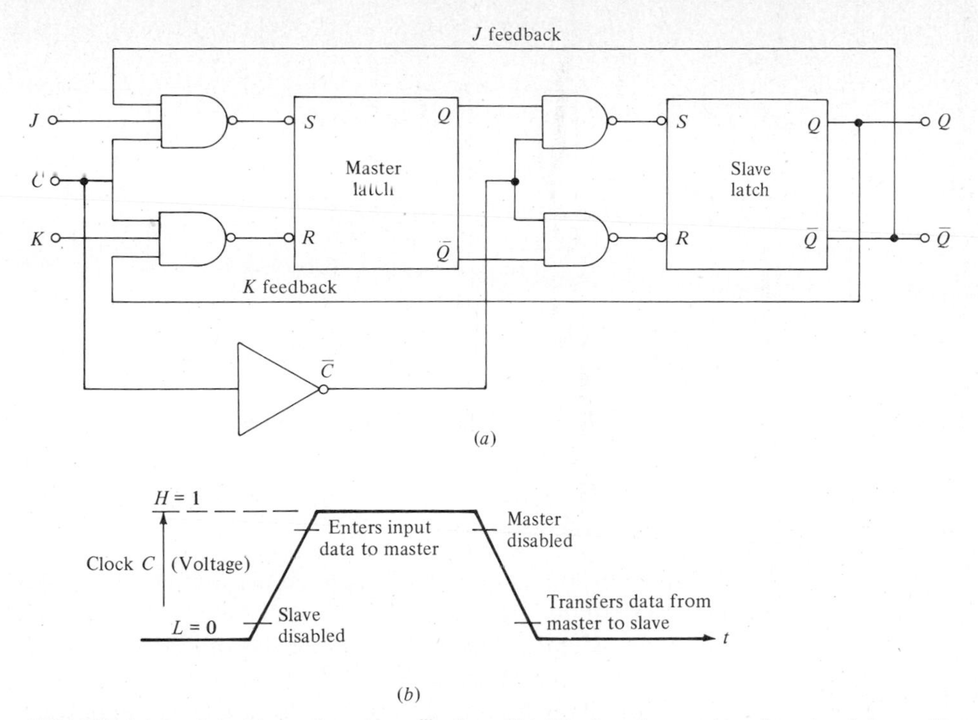

FIGURE 14.6 (*a*) *JK* master-slave flip-flop. (*b*) Clock pulse and pertinent points on the waveform.

14.5 *JK* EDGE-TRIGGERED AND *D* FLIP-FLOPS

In the *JK* master-slave flip-flop any spikes in the *J*, *K* lines can cause a problem. To see this, suppose that the slave flip-flop $Q = \mathbf{0}$; then the *K* input is locked out, while *J* is allowed to control the master. Now let *J* be low but have a small spike on the *J* line going to high just before the clock goes low. This spike will set the slave with no chance of being reset before the clock goes low since *K* is locked out. Therefore the output of the slave was determined by the *spike,* not the *level* of *J*. To avoid this problem, the *JK* edge-triggered flop-flop and the *D* flip-flop were developed.

A. Edge-Triggered *JK* Flip-Flop

In the edge-triggered flip-flops the output changes only during the transient operation by the *trailing edge* of the clock. These flip-flops are commonly referred to as *negative edge-triggered* flip-flops. The logic-circuit diagram of a *JK* edge-triggered flip-flop is shown in Fig. 14.7*a*. The logic symbol of the flip-flop is shown in Fig. 14.7*b*. The function table of the flip-flop is identical to that of Fig. 14.5*c*. For these flip-flops the propagation delays of the two input gates must be much larger than those of the other gates for the flip-flop to operate properly. The manufacturers ensure this condition in their designs.

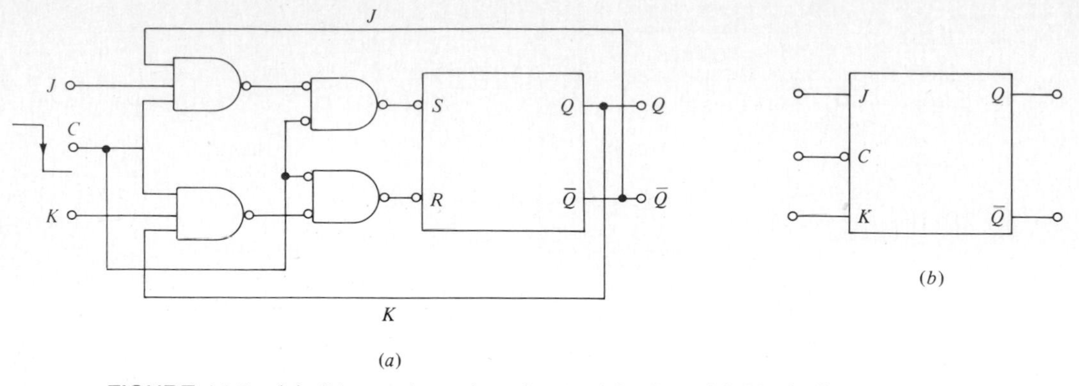

FIGURE 14.7 (*a*) *JK* negative edge-triggered flip-flop. (*b*) Block diagram.

B. The *D* Flip-Flop

The *D* flip-flop is basically a *positive edge-triggered* flip-flop. The main purpose of these flip-flops is to store and *delay* bits and is therefore often referred to as the *D* (delay) flip-flops. The *D*-type flip-flop delays the data by exactly one cycle of the clock. The logic-circuit diagram of the *D* flip-flop are shown in Fig. 14.8*a*. The logic symbol and the function table of the *D* flip-flop are shown in Fig. 14.8*b* and *c*, respectively. The *D* flip-flops are mostly used in the design of shift registers and counters, which will be discussed later in this chapter.

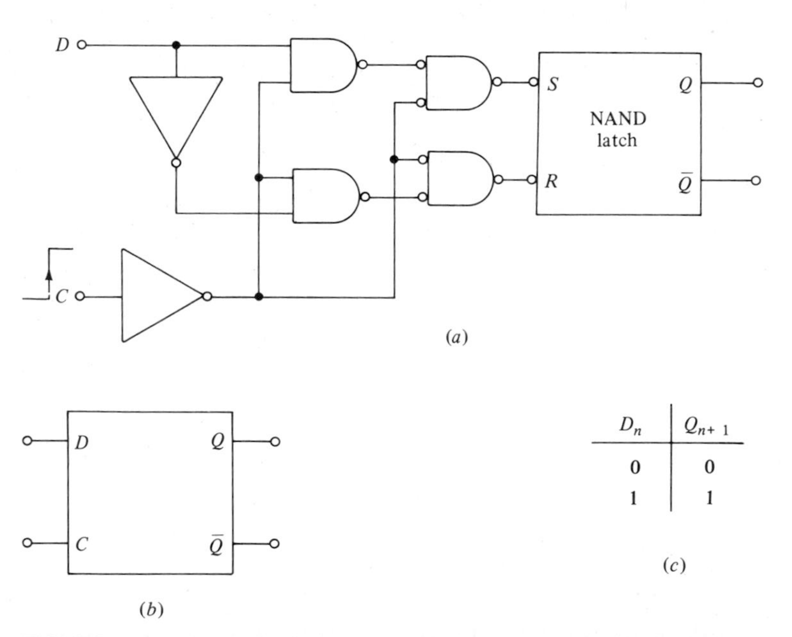

D_n	Q_{n+1}
0	0
1	1

FIGURE 14.8 (*a*) *D* flip-flop (positive edge-triggered). (*b*) Block diagram. (*c*) Function table.

EXERCISE 14.2

For the D-latch logic circuit shown in Fig. E14.2, the restriction of $R = S = \mathbf{1}$, in Fig. E14.1, is overcome by the inverter. Show that the logic behaves as a delay line, and its function table is as in Fig. 14.8c.

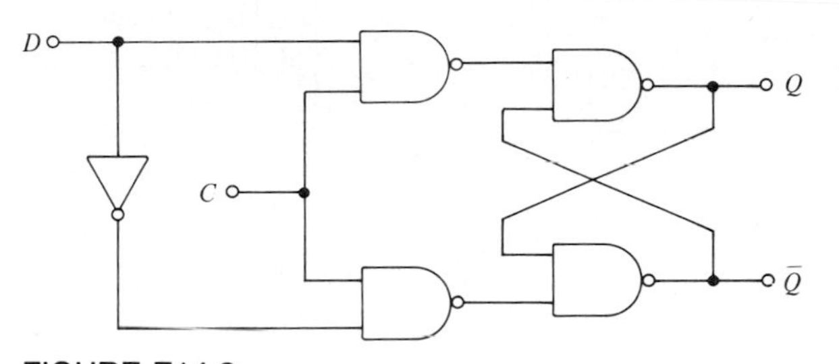

FIGURE E14.2

EXERCISE 14.3

Show that a D flip-flop can be implemented using a JK flip-flop as shown in Fig. E14.3.

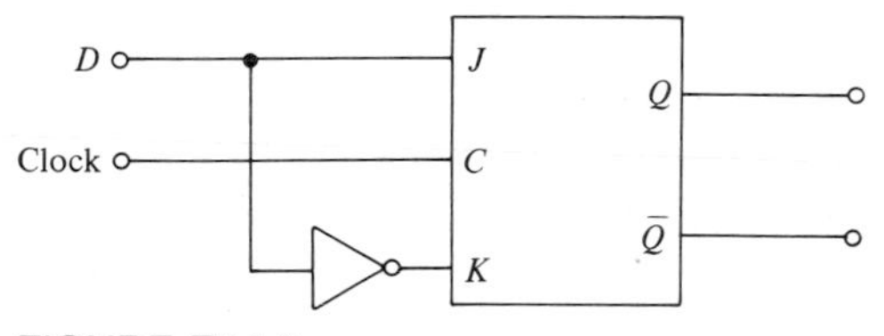

FIGURE E14.3

Ans For $D = \mathbf{1}$, $J = \mathbf{1}$ and $K = \mathbf{0}$; thus $Q = \mathbf{1}$. For $D = \mathbf{0}$, $J = \mathbf{0}$ **and** $K = \mathbf{1}$; thus $Q = \mathbf{0}$.

14.6 COMMERCIALLY AVAILABLE FLIP-FLOPS

In Chap. 13 we considered the various logic families and specifically the TTL, ECL, and CMOS gates. Flip-flops are available in all three families. For SSI these are available in dual forms; for MSI four flip-flops are in a package, and some are even available for LSI. The NMOS and I^2L families are only used for LSI design.

We will list in the following some of the currently available flip-flops from the manufacturers. Appendix E shows data sheets for the TTL (54/74) and the CMOS (4000B) series.

A. TTL

In the TTL 54/74 series, the following are available:

54/7470 *JK* positive edge-triggered
54/7472 *JK* Master-slave
54/7474 dual *D*-type positive edge-triggered

B. ECL

In the ECL 10K family, the following are available:

10135 dual *JK* master-slave
10231 dual *D* type, master-slave
10176 hex *D* type, master-slave

C. CMOS

In the CMOS 4000 series, the following are available:

4027 dual *JK* master-slave
4013 dual *D* type

In flip-flops one usually considers the toggle frequency in megahertz and the power dissipation in milliwatts. Of course, these are related to the individual logic gates from which they are made up. As we would expect, maximum toggle frequency, which is related to the propagation delay, is for the ECL flip-flops, and the lowest power dissipation is for the I^2L and CMOS flip-flops. The results are listed in Table 14.1.

TABLE 14.1 Performance Characteristics of IC Flip-Flops

	TTL		ECL		CMOS (4000)	
	74S	74LS	10K	100K	(V_{DD} = 5 V)	(V_{DD} = 10 V)
Toggle frequency, MHz						
Typical	110	45	200	550	7	16
Minimum	75	30	140	400	3.5	8
Power dissipation, mW						
Typical	75	10	230	550	10^{-4}	2×10^{-4}
Minimum	125	20	350	775	5×10^{-3}	2×10^{-2}

14.7 MULTIVIBRATOR CIRCUITS

Multivibrators are used in timing circuits. There are three classes of multivibrators, namely, *bistable, monostable,* and *astable* multivibrators. The most commonly used is the bistable multivibrator. Latches and flip-flop discussed in the previous sections are examples of these.

The bistable multivibrator has two stable states, as we have discussed already. The monostable multivibrator, also referred to as a *one-shot,* has one stable state in which it can remain indefinitely. The other state is quasi-stable and can be attained through triggering. The circuit can be held in the quasi-stable state for a predetermined duration of time, after which it returns to its stable state. These circuits are useful in generating pulses. The astable multivibrator has no stable states. Both states are quasi-stable, and the circuit changes its state without external triggering. The duration in each state can be predetermined in the design. The astable circuit is used as a clock or square-wave pulse generator.

14.8 MONOSTABLE MULTIVIBRATOR

We shall consider in the following both discrete and IC monostable multivibrators. In the discrete case we shall give the results only (and the derivations are left as exercises in the problems at the end of the chapter). For the IC case we shall derive the results and utilize the *idealized* voltage transfer characteristics of the gates.

A. Discrete Monostable Circuits

Simple discrete monostable multivibrator circuits utilizing BJT and NMOS devices are shown in Fig. 14.9*a* and *b*, respectively. In the emitter-coupled circuit of Fig.

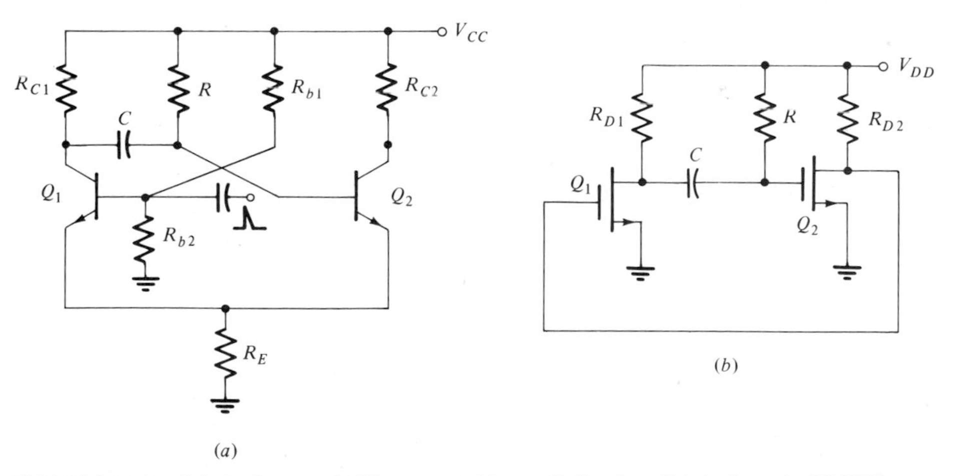

FIGURE 14.9 (*a*) A discrete BJT monostable multivibrator. (*b*) A discrete NMOS monostable multivibrator.

13.9*a* the stable state is Q_1 off and Q_2 on. When a very-short-duration negative trigger pulse is applied at the collector of Q_1, Q_1 is turned on and Q_2 is off (the quasi-stable state). However, Q_2 cannot remain off for very long. The time duration for Q_2 to change from off to on depends on the time constant RC, namely, for $R \gg R_{c1}$, it is given by

$$T \simeq RC \ln \left[\frac{1 - R_{b2}/(R_{b1} + R_{b2}) + R_{c2}/(R_E + R_{c2})}{1 - R_{b2}/(R_{b1} + R_{b2})} \right] \tag{14.1}$$

In the drain-gate coupled circuit of Fig. 14.9*b*, the stable state is Q_1 off and Q_2 on. The time duration for the quasi-stable state is given by

$$T \simeq RC \ln \left(\frac{V_{DD} - V_{on}}{V_{DD} - V_T} \right) \tag{14.2}$$

where $V_{on} = V_{DD}R_{on}/(R_{on} + R_{D2})$ and R_{on} is the drain to source on resistance of the NMOS.

For the numerical calculations consider the following exercise.

EXERCISE 14.4

For the MOS monostable multivibrator shown in Fig. 14.9*b*, the following are given: $R = 50\ \text{k}\Omega$, $C = 0.1\ \mu\text{F}$, $R_{D1} = R_{D2} = 10\ \text{k}\Omega$, $V_{DD} = 10$ V. If $V_T = 2$ V and $R_{on} = 500\ \Omega$, determine the pulse width of the circuit.

Ans 0.87 ms

B. IC Monostable Multivibrator

An IC monostable multivibrator is shown in Fig. 14.10. The circuit consists of two two-input NOR gates and an *RC* circuit. The second NOR gate with both inputs connected together is simply an inverter or a NOT gate.

For the purposes of calculations we shall assume an idealized voltage transfer characteristic for the gates as shown in Fig. 14.11. In fact, to be specific, let us consider CMOS gates whose transfer characteristics are very close to the ideal shown in Fig. 14.11 (see Fig. 13.38). Note that V_{TR} is the transition voltage of an idealized

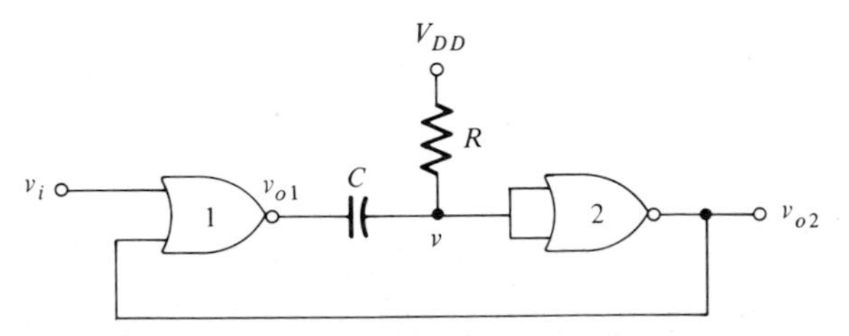

FIGURE 14.10 An IC CMOS monostable multivibrator.

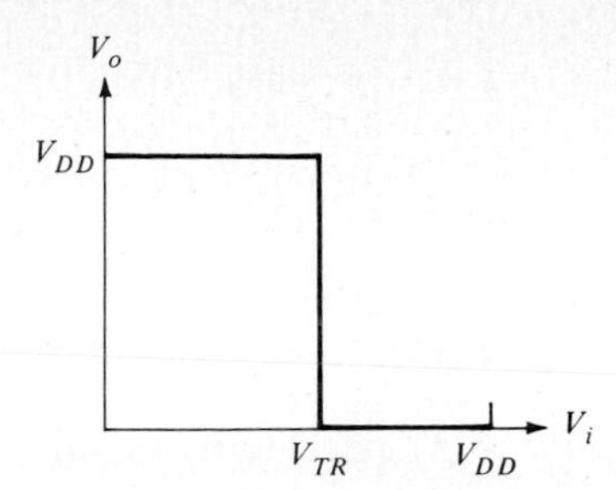

FIGURE 14.11 Idealized voltage transfer characteristic of a CMOS inverter.

CMOS inverter transfer characteristic. For a CMOS with NMOS and PMOS identical, $V_{TR} = 0.5V_{DD}$.

The expression for T is derived in the following by referring to Figs. 14.12 and 14.10. In the stable state, in Fig. 14.10, v_i in 0 V and v_{o2} is 0 V since gate 2 is inverter-connected with both inputs at high, namely, V_{DD}. After the application of a positive trigger pulse, v_{o1} goes low to 0 V, and $v(t)$ changes from zero toward V_{DD}. The charging equation is

$$v(t) = V_{DD}(1 - e^{-t/\tau}) \tag{14.3}$$

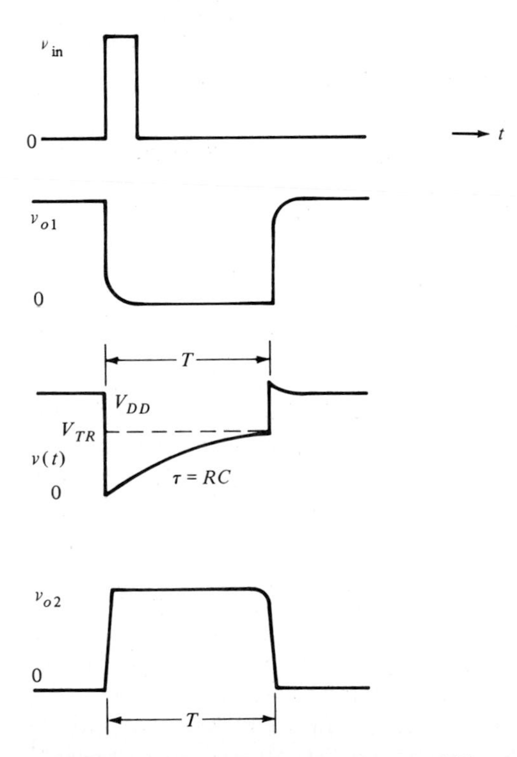

FIGURE 14.12 Voltage waveforms of Fig. 14.10.

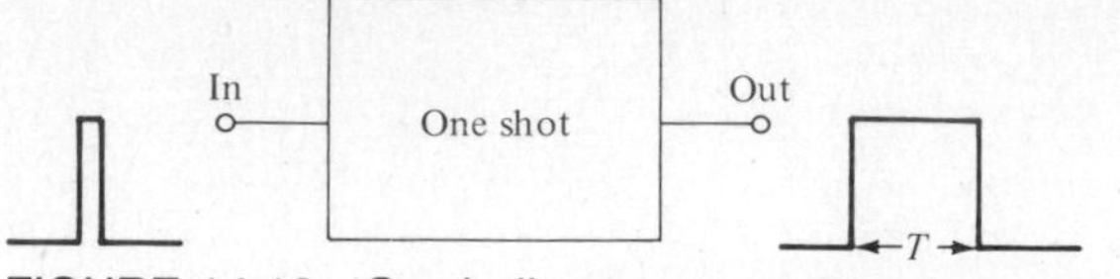

FIGURE 14.13 Symbolic representation of a monostable multivibrator.

where $\tau = RC$. However, when $v(t)$ reaches V_{TR} at $t = T$, gate 2 turns on, and v_{o2} goes to zero. Hence

$$v(T) = V_{TR} = V_{DD}(1 - e^{T/\tau}) \tag{14.4}$$

From (14.4), and with $V_{TR} = 0.5V_{DD}$, we obtain

$$T = \ln \frac{V_{DD}}{V_{DD} - V_{TR}} = \tau \ln 2 \simeq 0.7RC \tag{14.5}$$

A monostable multivibrator (one shot) is usually designated by a block diagram as shown in Fig. 14.13. The commercially available CMOS monostable multivibrator is 4047, and the TTL version of the monostable multivibrator is 74121. Monostable multivibrators are used extensively for waveshaping, timing circuits, and delay circuits.

EXERCISE 14.5

Design the IC monostable multivibrator of Fig. 14.10 for $T = 0.1$ μs, given the following CMOS parameters: $V_{TR} = 0.5V_{DD} = 2.5$ V. Use $R = 10$ kΩ.

Ans $C = 14$ pF

14.9 ASTABLE MULTIVIBRATOR (CLOCK)

An astable multivibrator, which is referred to as a clock, is commonly used for the generation of square waveforms. Clocks are commonly used in digital circuits and systems.

A. Discrete Astable Circuits

Simple discrete astable multivibrator circuits utilizing BJT and JFET devices are shown in Fig. 14.14*a* and *b*, respectively. Note that for an astable multivibrator there is no stable state, and *both* states are quasi-stable. For the circuit shown in Fig. 14.14*a*, the time duration at which each transistor is on or off is given by (problem 14.10)

$$T_1 = T_2 \simeq RC \ln 2 \simeq 0.7RC \tag{14.6}$$

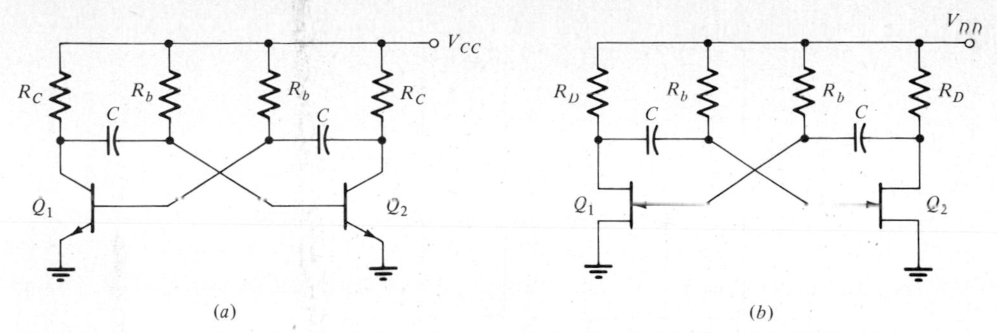

FIGURE 14.14 (*a*) A discrete BJT astable multivibrator. (*b*) A discrete JFET astable multivibrator.

The repetition or the clock frequency is

$$f = \frac{1}{T} = \frac{1}{T_1 + T_2} \tag{14.7}$$

The JFET version of the astable multivibrator in Fig. 14.14*b* is similar and performs identically to its BJT counterpart. The half-period time is given by

$$T_1 = T_2 \simeq RC \ln 2 \qquad \text{for } V_{DD} \gg V_{PO} \tag{14.8}$$

B. IC Astable Multivibrator

An IC version of the astable multivibrator type CD4047, using CMOS, is shown in Fig. 14.15. The waveforms for v_{o1}, v_{o2}, and v_{i1} are shown in Fig. 14.16. The period T of the astable multivibrator, using the idealized transfer characteristics of the CMOS, is readily derived by referring to the v_{i1} waveform in Fig. 14.16. The charging equation, with $\tau = RC$, is given by

$$v_{i1} = (V_{TR} - V_{DD}) + (2V_{DD} - V_{TR})(1 - e^{-t/\tau}) \tag{14.9}$$

At $t = T_1$, when v_{i1} reaches the value of V_{TR}, the gates change their levels, and

$$T_1 = \tau \ln \frac{2V_{DD} - V_{TR}}{V_{DD} - V_{TR}} \tag{14.10}$$

For $V_{TR} = 0.5V_{DD}$ we have

$$T_1 = \tau \ln 3 = 1.1RC \tag{14.11}$$

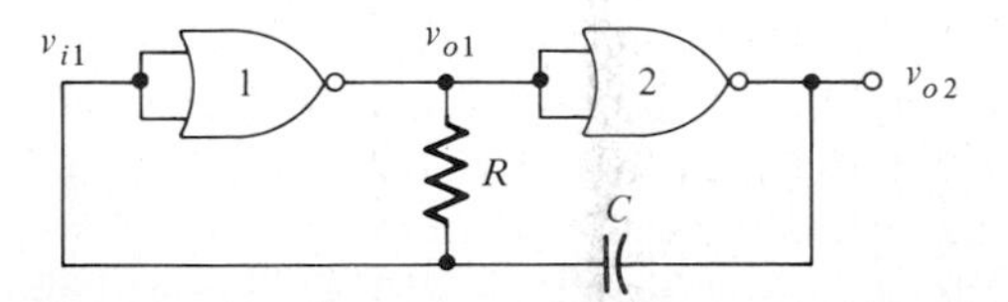

FIGURE 14.15 An IC CMOS astable multivibrator

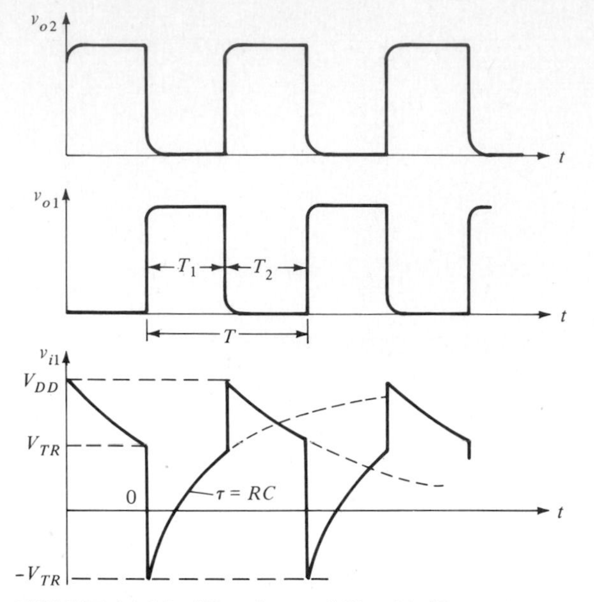

FIGURE 14.16 Waveform of Fig. 14.15.

Note that for $V_{TR} = 0.5V_{DD}$ the circuit is symmetrical; hence

$$T = T_1 + T_2 = 2T_1 \simeq 2.2RC \tag{14.12}$$

The frequency of the astable multivibrator (clock) is given by

$$f = \frac{1}{T} \simeq \frac{1}{2.2RC} \tag{14.13}$$

A commercially available astable multivibrator is the CMOS 4047.

EXERCISE 14.6

Determine the frequency of the CMOS clock if $RC = 0.1\ \mu s$.

Ans $f = 4.5$ MHz

14.10 THE 555 IC TIMER

A timing circuit that is very popular and widely used in digital circuits and systems is the so-called 555 timer. With two or three external components the circuit could be operated as either an astable multivibrator or a monostable multivibrator. The timer itself has a timing accuracy of ± 1 percent and a temperature stability of 50 ppm/°C.

The basic block diagram of the timer is shown in Fig. 14.17. The circuit consists of a latch, two comparators, an output stage, and a discharge transistor.

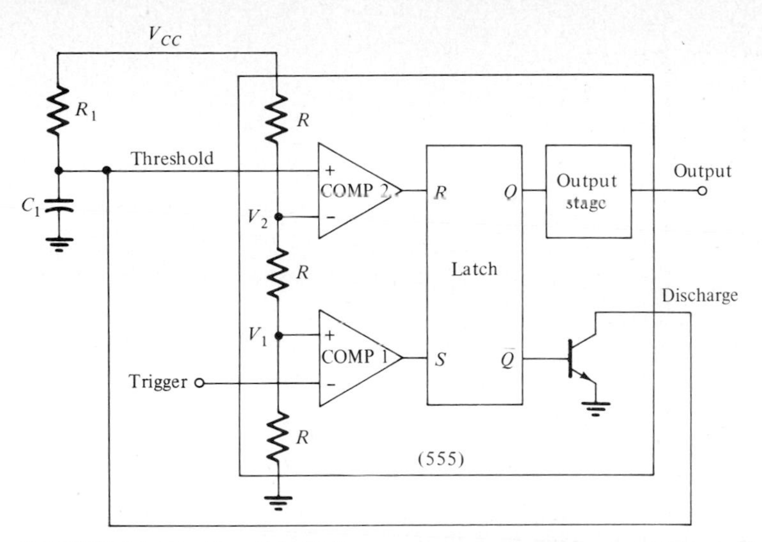

FIGURE 14.17 The 555 IC timer, connected as a monostable multivibrator.

A. Monostable Operation

Monostable operation is obtained by connecting two external elements R and C as shown by heavy lines in Fig. 14.17. First note that $V_2 = \frac{2}{3}V_{CC}$, $V_1 = \frac{1}{3}V_{CC}$. The operation of the timer is as follows: Consider the circuit is triggered at the trigger terminal with $V_t < V_1$. The first comparator sets the latch, which discharges the transistor to turn off. The timing capacitor makes the threshold voltage to charge toward V_{CC} with a time constant R_1C_1. As soon as the voltage value reaches V_2, the second comparator resets the latch, and the discharge transistor turns on, which discharges C and returns to its stable state. The pulse width is readily determined from writing the charging equation across the capacitor

$$v_{C1} = V_{CC}(1 - e^{-t/\tau}) \tag{14.14}$$

where $\tau = R_1C_1$. At $t = T$, $v_{C1} = V_2 = \frac{2}{3}V_{CC}$; hence from (14.14)

$$T = \tau \ln \frac{V_{CC}}{V_{CC} - 2V_{CC}/3} = \tau \ln 3 = 1.1R_1C_1 \tag{14.15}$$

B. Astable Operation

Astable operation is obtained by connecting the threshold and the trigger terminals and connecting external R_1, R_2, and C as shown in Fig. 14.18. In this case the capacitor charges from $V_{CC}/3$ to $2V_{CC}/3$ with a time constant $\tau_1 = (R_1 + R_2)C$. The pulse width for the charging is given by

$$T_1 \simeq 0.7(R_1 + R_2)C \tag{14.16}$$

The capacitor discharges from $2V_{CC}/3$ to $V_{CC}/3$ with a time constant $\tau_2 = R_2C$. The

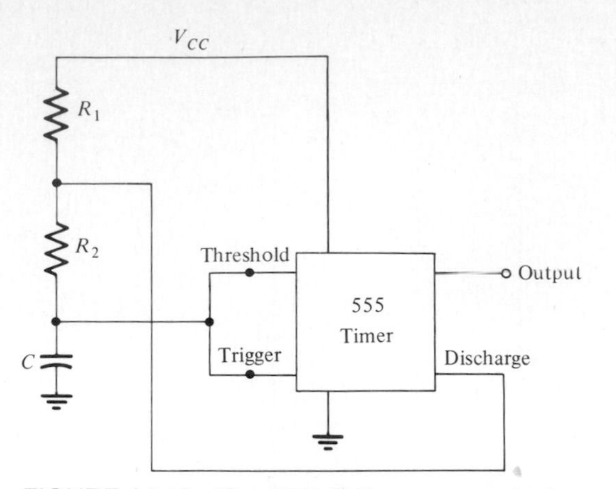

FIGURE 14.18 The 555 IC timer, connected as an astable multivibrator.

pulse width for the discharging time is given by

$$T_2 \simeq 0.7R_2C \tag{14.17}$$

Hence the frequency of the clock is

$$f = \frac{1}{T_1 + T_2} \tag{14.18}$$

The commercially available CMOS version of the 555 is 40001B.

EXERCISE 14.7

For a 555 timer used as a one shot, with $R_1 = 1\ \text{k}\Omega$, $C_1 = 0.01\ \mu\text{F}$, determine the pulse width.

Ans 0.11 μs

EXERCISE 14.8

A 555 timer is used as a clock, with $R_1 = R_2 = 10\ \text{k}\Omega$, $C = 0.01\ \mu\text{F}$, determine its frequency of oscillation.

Ans 4.76 kHz

14.11 SCHMITT TRIGGER

The Schmitt trigger, already discussed in Sec. 9.8, is used for waveshaping purposes. For example, the leading and falling edges of a waveform may be sharpened up for further data handling in digital circuits. A sinusoidal voltage may be converted into a square wave at the output.

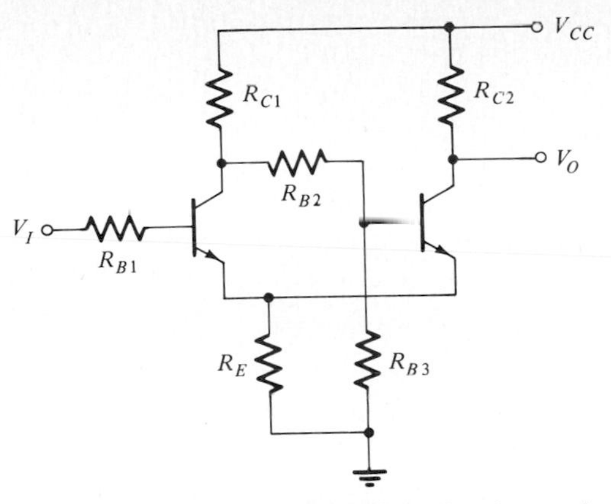

FIGURE 14.19 A discrete Schmitt trigger circuit (emitter coupled).

A simple discrete Schmitt trigger circuit is shown in Fig. 14.19. Note that the circuit is basically a bistable circuit similar to Fig. 14.2*a*. The symbolic representation for an IC Schmitt trigger and its voltage transfer characteristic is shown in Fig. 14.20*a* and *b*, respectively. The hysteresis sign in the symbol and in the transfer characteristic depicts memory of the circuit. In other words, the voltage at any moment is not determined by the value of the input voltage alone but also by the past history of the circuit. For example, an input voltage between the threshold values, i.e., $V_T^- < V_I < V_T^+$, will yield an output equal to V_o^+ if the circuit was in V_o^+ state and an output of V_o^- if the circuit was in V_o^- state. A simple example of obtaining a square wave at the output of a Schmitt trigger from a sinusoidal input is shown in Fig. 14.21.

In IC circuits several Schmitt trigger circuits are available in one package. For example, the TTL 7413 is a dual four-input, whereas the TTL 7414 is a hex Schmitt trigger. The CMOS type 4093B consists of quad two-input NAND Schmitt triggers. The logic symbol of one of the quads in 4093B is shown in Fig. 14.22*a*, with its transfer characteristic in Fig. 14.22*b*. External *RC* circuits can be added to the Schmitt trigger to obtain a monostable or an astable multivibrator as shown in Figs. 14.23 and 14.24, respectively.

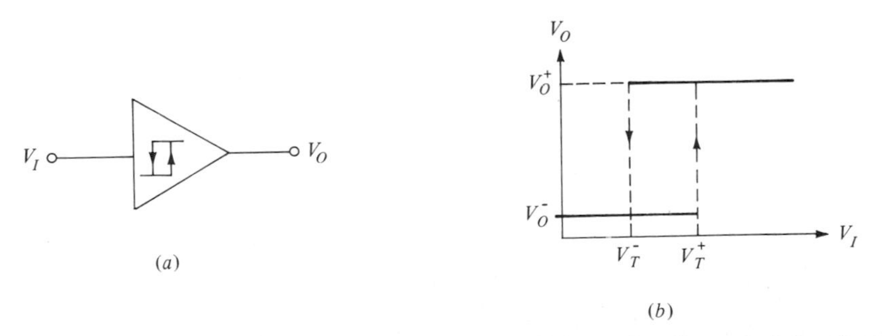

FIGURE 14.20 (*a*) Schmitt trigger symbol. (*b*) Transfer characteristic of a Schmitt trigger.

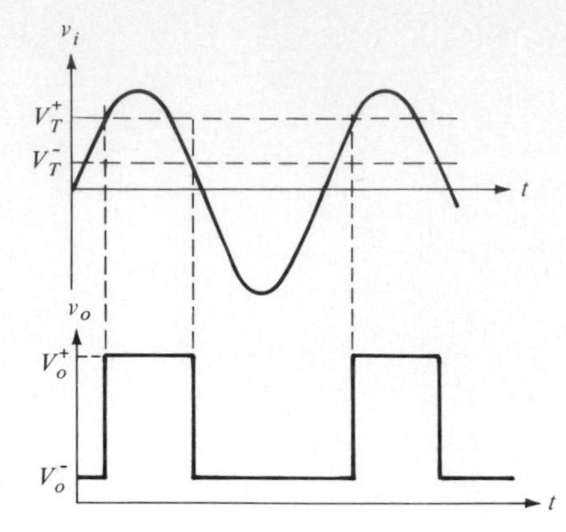

FIGURE 14.21 Schmitt trigger used to obtain a square wave.

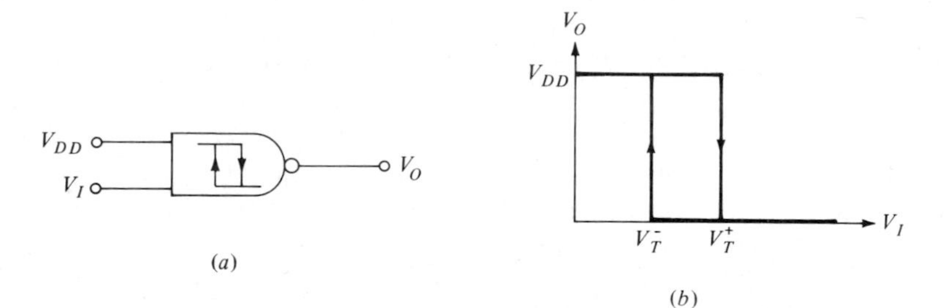

FIGURE 14.22 (*a*) IC Schmitt trigger. (*b*) Transfer characteristic.

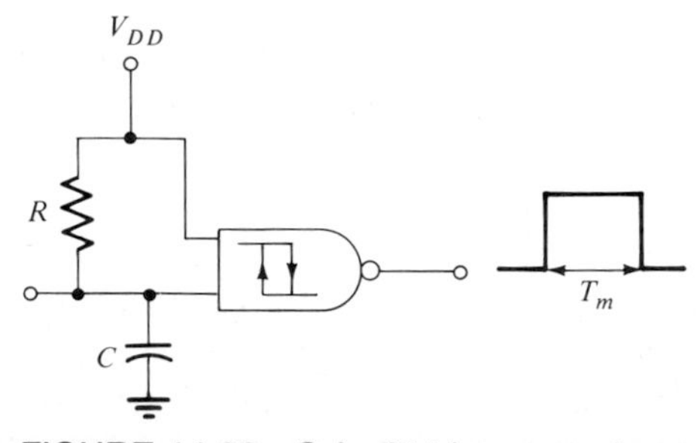

FIGURE 14.23 Schmitt trigger used as a monostable multivibrator.

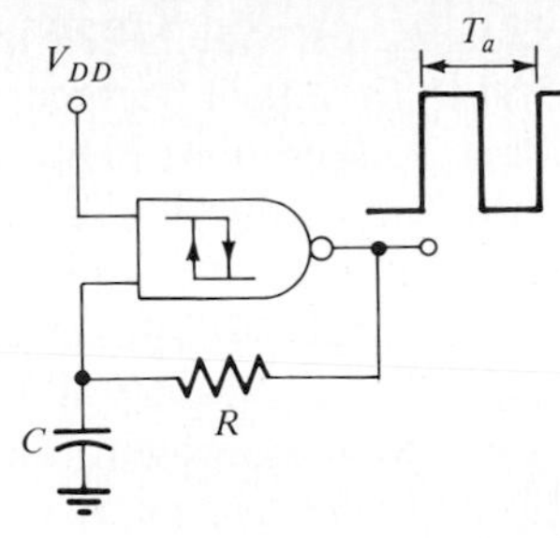

FIGURE 14.24 Schmitt trigger used as an astable multivibrator.

For the monostable multivibrator, shown in Fig. 14.23, the period T_m is given by

$$T_m = RC \ln \frac{V_{DD}}{V_{DD} - V_T^+} \tag{14.19}$$

For the astable multivibrator, shown in Fig. 14.24, the period T_a is given by

$$T_a = RC \ln \left(\frac{V_T^+}{V_T^-} \frac{V_{DD} - V_T^-}{V_{DD} - V_T^+} \right) \tag{14.20}$$

where V_T^+ and V_T^- are defined as in Fig. 14.22*b*.

EXAMPLE 14.1

A 4093B Schmitt trigger is used as an astable multivibrator as shown in Fig. 14.24. If $V_{DD} = 10$ V, $V_T^+ = 5.9$ V, $V_T^- = 3.9$ V, and $R = 100$ kΩ, determine the value of C for the astable multivibrator period to be 10 μs. From (14.20)

$$10^{-5} = 10^5 C \ln \left(\frac{5.9}{3.9} \frac{10 - 3.9}{10 - 5.9} \right) = 10^5 C(0.81) \qquad C = 123 \text{ pF}$$

14.12 SOME APPLICATIONS OF FLIP-FLOPS

There are a large variety of applications of flip-flops, and we shall consider only a few of these here. Specifically, we shall consider shift register and counters.

A. Shift Register

A register is simply an array of latches or flip-flops that are used to store a byte or word. Since each flip-flop stores one bit of data, n flip-flops are needed to register n bits. Registers are usually used for temporary storage of data for processing. One useful processing of the data is the operation of shifting bits. A shift register is capable of shifting all bits toward the right or the left. For example, if a code **11010** is shifted one place to the right, it becomes **01101**, and if shifted to the left, it becomes **10100**. An important application of shift registers is in arithmetic operation. Figure 14.25*a* shows as 4-bit shift register (shift left to right) using the *D*-type flip-flop. Other types of flip-flops such as the *JK* or the *SR* flip-flops could also be used. To illustrate the operation of the shift register, suppose we have the input datas **11010**. Assume that all the flip-flops are initially cleared; i.e., at $t = t_o$, Q_o, Q_1, Q_2, Q_3 are all **0**.

The shift register's response to these data is illustrated in the waveforms in Fig. 14.25*b*. From Fig. 14.8 recall that for the *D*-type flip-flop there is a delay of exactly one cycle of the clock and that it transfers data at the positive edge of the pulse.

There are several types of shift registers (data may be entered either in serial or in parallel): the shift-right register, the shift-left register, and the universal shift register, which can do all the above. A commercially available 4-bit register in the TTL family is the 7494, and 4-bit universal bidirectional in the CMOS family is 40104B.

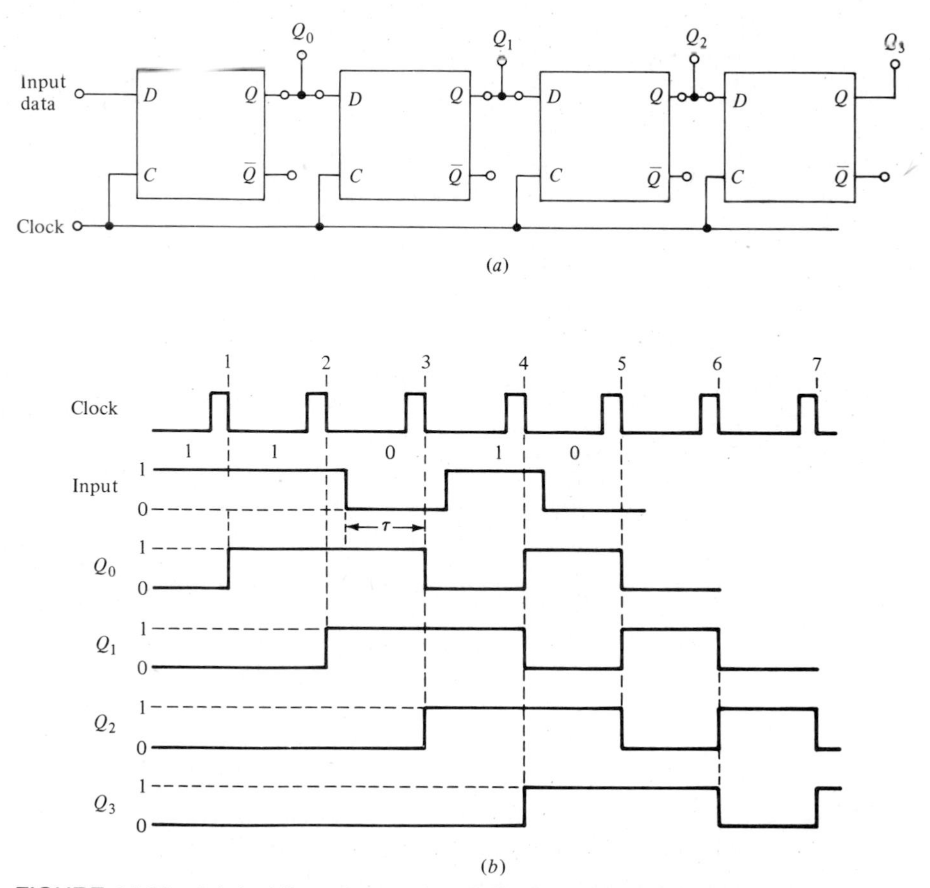

FIGURE 14.25 (*a*) A shift-register using *D* flip-flops. (*b*) Its waveforms for an input data of 11010.

EXAMPLE 14.2

Design a 2-bit shift register using *JK* flip-flops.

The register circuit is shown in Fig. 14.26. At each clock pulse, the data in the *A* flip-flop are transferred to the *B* flip-flop as follows:

If $Q_A = \mathbf{1}$, then $J_B = \mathbf{1}$ and $K_B = \mathbf{0}$; thus $Q_B = \mathbf{1}$.
If $Q_A = \mathbf{0}$, then $J_B = \mathbf{0}$ and $K_B = \mathbf{1}$; thus $Q_B = \mathbf{0}$.

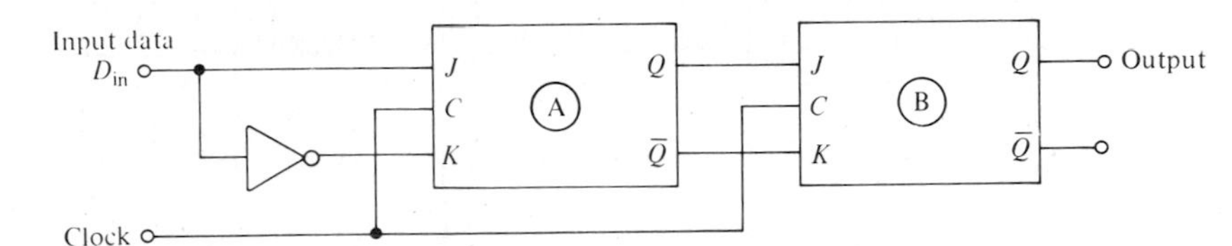

FIGURE 14.26 A 2-bit shift register example.

Also, during the same clock pulse, the input data D_{in} are transferred to the *A* flip-flop as follows:

If $D_{in} = \mathbf{1}$, then $J_A = \mathbf{1}$ and $K_A = \mathbf{0}$; thus $Q_A = \mathbf{1}$.
If $D_{in} = \mathbf{0}$, then $J_A = \mathbf{0}$ and $K_A = \mathbf{1}$; thus $Q_A = \mathbf{0}$.

B. Counters

There are basically two general types of counters. These are the synchronous counters and the asynchronous counters. In the synchronous counters the clocking signal is applied to all the flip-flops and all the flip-flops change their stages at the same time. In the asynchronous counters, the clocking signal is applied to one of the flip-flops, and the signal propagates successively. We shall now consider the ripple counter as another application of the flip-flops.

A *ripple counter* is shown in Fig. 14.27*a*, where three *JK* flip-flops are used in cascade in the toggle mode. Note that in the toggle mode *J* and *K* terminals are connected to logic **1**. One could also use the *D*-type flip-flop, where the $\overline{Q}$ output terminal is connected to the *D* input terminal. The input signal, to be counted, is connected to the clock input of the first flip-flop FF_0. The output of the first flip-flop is connected to the input of the second flip-flop and so on. The output waveform for a train of input pulses is shown in Fig. 14.27*b*. Note that the states of the flip-flops change only when the input of that flip-flop goes from high to low (**1** to **0**). The *count* *N* is determined from

$$N = Q_2 \times 2^2 + Q_1 \times 2^1 + Q_0 \times 2^0 \tag{14.21}$$

Note that $N_{min} = 0$ ($Q_i = 0$) and $N_{max} = 7$ ($Q_i = 1$), and the total count is 8. This system is referred to as scale of 8 or mod 8 since three flip-flops are used and each

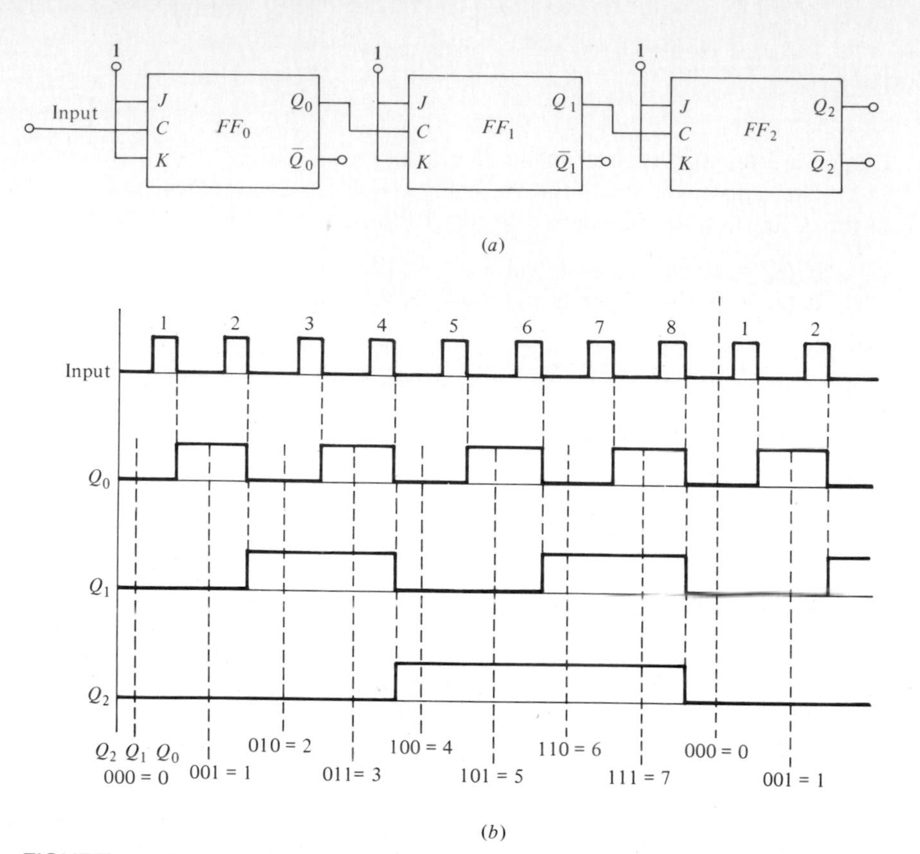

FIGURE 14.27 (*a*) A ripple counter using *JK* flip-flops. (*b*) Its waveforms.

flip-flop can be in one of two possible states ($2^3 = 8$). For four flip-flops we can obtain a count of $2^4 = 16$, i.e., from 0 to decimal 15. For a decade counter *four flip-flops* are used, and the design must eliminate the last six states.

The ripple counter shown in Fig. 14.27*a* is one of the up-count type; i.e., the sequence is $Q_2Q_1Q_0 = \mathbf{000} \rightarrow \mathbf{001} \rightarrow \mathbf{010}$, etc. One can also obtain a down-count realization by connecting each clock input to the $\bar{Q}$ of the flip-flop, except for FF_0, where the data input is applied. In this case we would achieve a down count, where

TABLE 14.2 Four-Bit Up/Down Binary Counters

	TTL		ECL	CMOS
Synchronous counter	74191	74LS191	10178	4516
Clock rate	25 MHz	25 MHz	150 MHz	8 MHz
Power dissipation	325 mW	100 mW	370 mW	200 mW

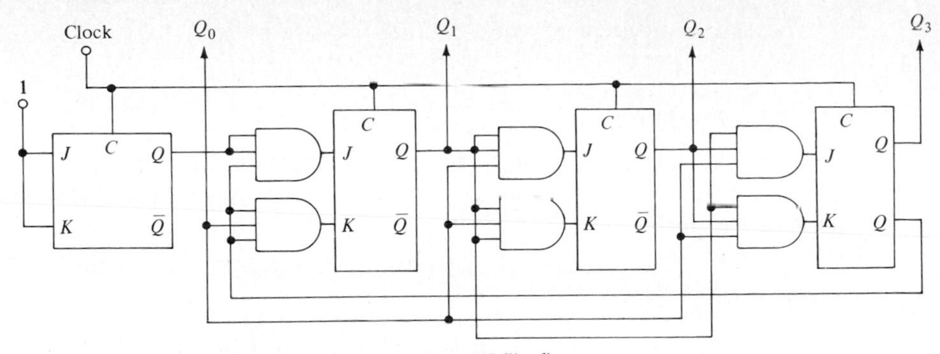

FIGURE 14.28 A decade counter using *JK* flip-flops.

the sequence would be $Q_2Q_1Q_0$ = **000** → **111** → **110**, etc. Note that negative edge-triggered flip-flops are used in Fig. 14.27*a*. For a positive edge-triggered flip-flop, such as the *D*-type, the clock input is connected to $\bar{Q}$ for up count and to Q for down count, respectively. The up/down counter is designed by incorporating two separate counters, one up counter and the other down counter.

Table 14.2 lists some of the IC 4-bit up/down counters with their maximum clock rate and their power dissipation. It should be noted that the ripple counter encounters cumulative flip-flop delays as the count progresses across the line, and thus it has a slow speed. Synchronous counters eliminate the cumulative flip-flop delays. All flip-flops in a synchronous counter are under the control of the same clock. One such counter, called a *binary divide-by-10 counter,* is shown in Fig. 14.28; the count sequence is shown in Table 14.3. Examples of commercially available decade synchronous counters are the TTL 74190 ECL 10137, and the CMOS 4017B—all of which are available as MSI products. Another popular counter is the IC 7490 unit, which can be connected as a decade counter.

TABLE 14.3 Count Sequence of Binary Divide-by-10 Counter

	Q_0	Q_1	Q_2	Q_3
0	0	0	0	0
1	1	0	0	0
2	0	1	0	0
3	1	1	0	0
4	0	0	1	0
5	1	0	1	0
6	0	1	1	0
7	1	1	1	0
8	0	0	0	1
9	1	0	0	1

The maximum frequency of a counter is given by Ref. 4:

$$f_{\max} \leq \frac{1}{Nt_{pd} + T_s} \tag{14.22}$$

where N is the number of flip-flops, t_{pd} is the propagation delay of each flip-flop, and T_s is the so-called strobe time. Strobe time is the time to gate or control a bit into a counter or register.

EXAMPLE 14.3

A four-stage binary counter using TTL flip-flops of 50-ns propagation delay time is used. If the strobe-time pulse width is $T_s = 50$ ns, determine the time required for the counter to change from **1111** to **0000**.

From (14.22)

$$T \geq 4(50) + 50 = 250 \text{ ns} \qquad f \leq 4 \text{ mHz}$$

We shall conclude the book with the topic of digital-to-analog and analog-to-digital converters.

14.13 D/A AND A/D CONVERTERS

The basic use of A/D and D/A converters in a digital signal processing system is shown schematically in Fig. 14.29. The A/D converter has its input from an analog signal and produces a digital output signal. The D/A converter does the reverse; namely, its digital input signal is converted to an analog output signal. The low-pass filter is used to remove the steps and smooth out the signal. The A/D and D/A converter thus provide an interface between analog and digital signal processing.

In converting analog signals to digital forms, one has to sample the signal. A simple circuit that performs this function is the *sample-and-hold* (S/H) circuit, as shown in Fig. 14.30*a*, where the switch S is controlled by a clock. In practice the S/H circuit shown in Fig. 9.42, which utilizes a unity gain op amp, is used in order to buffer the holding capacitor from the load or an IC version of S/H is used. The switch is closed for a duration of τ ns at which time the input is sampled. The switch is open

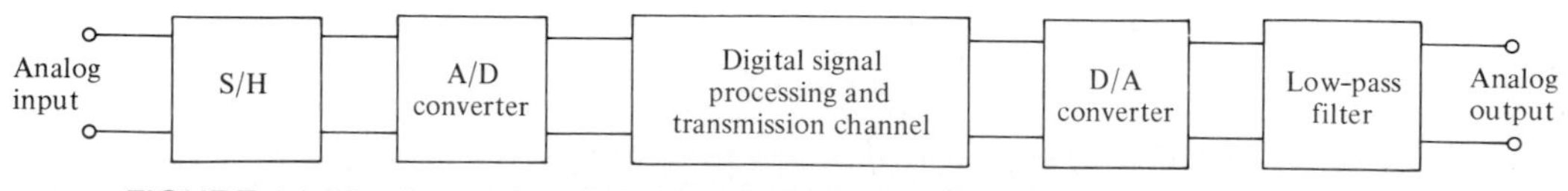

FIGURE 14.29 An analog-digital transmission system.

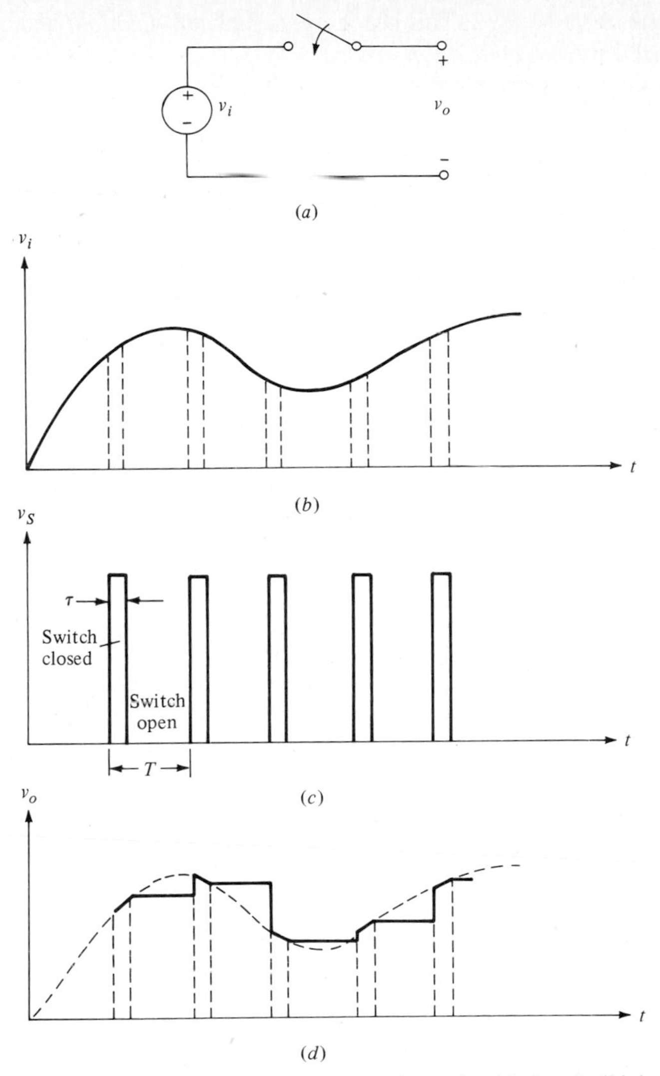

FIGURE 14.30 (*a*) A simple sample-and-hold circuit. (*b*) Input waveform. (*c*) Sampled waveform. (*d*) Output waveform of the sample and hold.

during time *T*, and the voltage is held at the same value when it was sampled. In other words, during the hold time, the capacitor *C* will have the last input value of the voltage. The process is shown in Fig. 14.30*c* and *d*.

Signal quantization is done as follows: Suppose that an analog signal has a range of values from 0 to 20 V, and we wish to convert it to a digital form with four binary digit (bit) signals. With four binary digits we have 2^4 different values (0 to 15). Hence we have an interval value or resolution of 20/15 V. Thus the analog values of 0, 4/3 V, 8/3 V, . . . , 20 V will be represented by 0000, 0001, 0010, . . . , 1111. A voltage value *within* the interval value closest to its value, e.g., 3 V, which is between

8/3 and 12/3, will be treated as 8/3. This process is called *quantization*, and the error thus caused is referred to as *quantization error*.

For a general case of n bits the resolution is given by

$$\Delta V = \frac{V_{o,\max} - V_{o,\min}}{N - 1} \quad \textbf{(14.23)}$$

where $N = 2^n$

EXERCISE 14.9

Determine the resolution of an analog signal which has a range of 0 to 10 V, using 8 bits.

Ans 39.2 mV.

A. D/A Converter Circuits

D/A converters (or DACs) are used to convert a digital signal to an analog signal. A simple circuit utilizing binary *weighing-resistors* and an op amp that can achieve this process is shown in Fig. 14.31. The switches S_i are controlled by a digital input. In practice these are digitally controlled electronic switches. In Fig. 14.31, R_o is the least significant bit (LSB), and R_3 is the most significant bit (MSB). From the circuit, which is simply a summer, we have

$$V_o = V_r R_f \left(\frac{S_3}{R_3} + \frac{S_2}{R_2} + \frac{S_1}{R_1} + \frac{S_0}{R_0} \right) \quad \textbf{(14.24)}$$

where S_i are either **1** or **0**, i.e., **1** when connected to V_r and **0** when connected to ground. From (14.24) and the values of R_i indicated in Fig. 14.31 we have

$$V_o = -\frac{V_r R_f}{R}(2^3 S_3 + 2^2 S_2 + 2^1 S_1 + 2^0 S_0) \quad \textbf{(14.25)}$$

From (14.25) it is seen that the output voltage is directly proportional to the numerical value of the input.

For example, assume $V_r = 1$ V, $R_f = R$, and a 4-bit word. If the digital input values are 0000, 0111, 1111, the corresponding analog output voltage is 0, 7, 15 V, respectively. From (14.25) it is seen that V_o is an analog voltage proportional to the digital input. From the above circuit it should be noted that for a large bit input the resistance spread will be very high. For example, for a 16-bit converter the spread in resistor values will be $2^{16} = 65536$. Hence, if $R = 1$ kΩ, $R_o = 1$ kΩ and $R_{15} = 65.536$ MΩ. An accuracy of 0.015 percent in the most significant bit corresponds to the entire value of the least significant bit. This clearly presents a problem.

A DAC circuit that avoids this problem is the so-called R-$2R$ *resistive ladder*. This circuit requires only two values of resistors R and $2R$ (hence the name R-$2R$) and

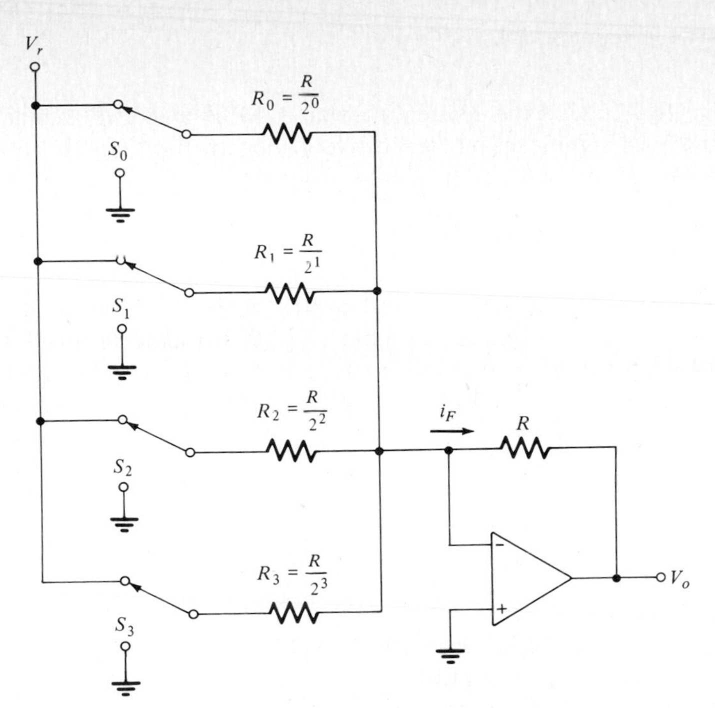

FIGURE 14.31 A digital-to-analog (D/A) converter.

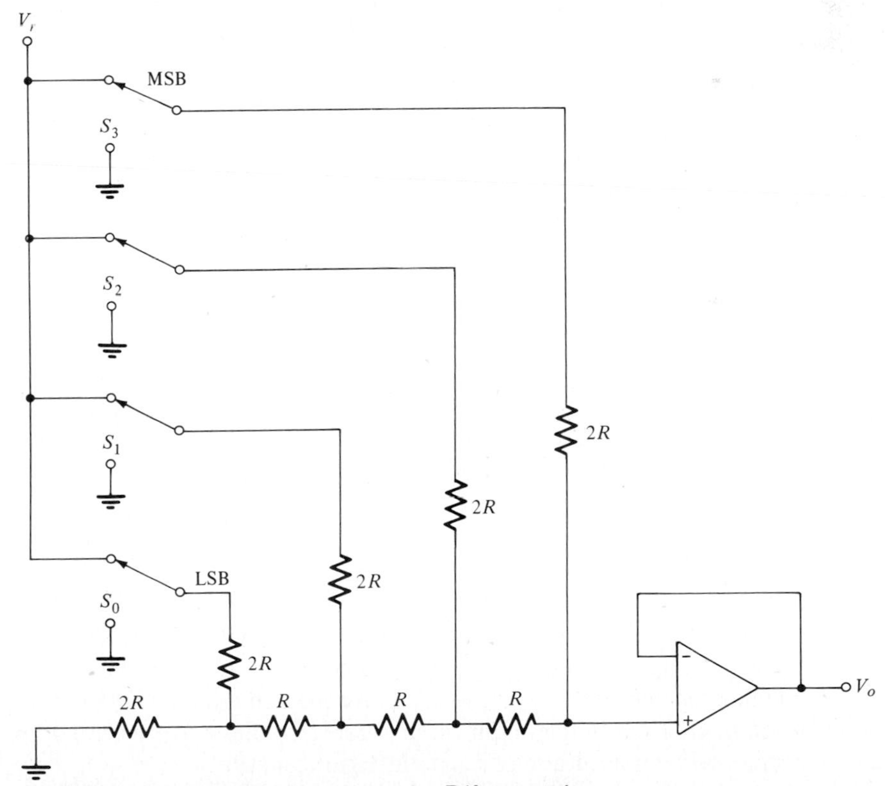

FIGURE 14.32 *R*-2*R* resistive ladder D/A converter.

is shown in Fig. 14.32. V_r is the reference voltage; MSB and LSB signify most significant bit and least significant bit, respectively. For the 4-bit circuit the output voltage is given by

$$V_o = \frac{V_r(2^3S_3 + 2^2S_2 + 2^1S_1 + 2^0S_0)}{2^4} \tag{14.26}$$

The result in (14.26) can be readily derived by the analysis of the circuit using superposition and is left as an exercise in problem 14.19. It should be noted that the circuit of Fig. 14.32 is available in a single package from the manufacturers, i.e., DAC 1200 type, which is a 12-bit D/A converter containing switches, ladder, and references.

EXAMPLE 14.4

For the circuit shown in Fig. 14.32, $R = 10\ \text{k}\Omega$ and $V_r = 10$ V. Determine the analog voltage if the digital input is 1100.

We note that $S_3 = 1$, $S_2 = 1$, $S_1 = 0$, $S_0 = 0$. Hence

$$V_o = 16\,\frac{2^3(1) + 2^2(1) + 2^1(0) + 2^0(0)}{16} = 10\ \text{V}$$

Note that greater quantization of steps can be obtained by using a larger-bit (more ladder sections) DAC. For example, for a 10-bit DAC the voltage resolution is $V_r/2^{10} = 10/1024$, i.e., approximately 10 mV vs 625 mV in this example.

B. A/D Converter

We shall assume that the signal is sampled using the sample-and-hold circuit, and thus the waveform is quantized as in Fig. 14.30*d*. The quantized signal is then an input to A/D converter. A simple 4-bit A/D converter circuit is shown in Fig. 14.33. The comparator circuit has already been discussed in Secs. 5.11 and 9.8. The comparator output is **1** when $v_a > v_b$ and **0** otherwise.

The up/down counter is a counter that can count either up or down depending on the level at the up/down control terminal. The operation of the circuit is as follows: Assume a zero count in the counter. The D/A output v_b will be zero, and since $v_a > 0$, the comparator output will be high (**1**), and the counter will be instructed to count the clock pulse in the up direction. The output v_b increases with the increase in count. The process continues until $v_b = v_a$, at which point the comparator output will be 0 and stops the counter. The counter output is, therefore, the digital equivalent of the analog input voltage with an accuracy of ± 1 least significant bit.

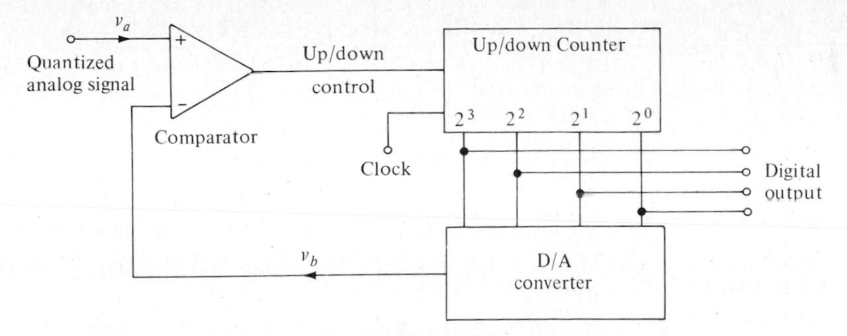

FIGURE 14.33 An analog-to-digital (A/D) converter.

A successive-approximation technique is discussed below.

EXAMPLE 14.5

Determine a 3-bit representation of an analog voltage $v_a = 5.2$ V; use $V_r = 8$ V.

We start with the first approximation, 100, which is equivalent to 4 V, i.e., $V_r/2$. Now $v_a - v_b = 1.2 \text{ V} > 0$, so we keep 1 in the MSB. We set 1 in the bit next to the MSB, and the second approximation is 110, which corresponds to 6 V. Since $v_a - v_b = -0.8 \text{ V} < 0$, we set the second MSB to zero, and the third approximation is 101, which corresponds to 5 V. The difference $v_a - v_b = 0.2 \text{ V} > 0$, and hence we keep 1; thus the digital approximation to 5.2 V is 101. The speed of the converter type shown in Fig. 14.33 is slow (10 to 100 μs) as it operates on successive approximations. Other A/D converters are also available that operate on a *parallel* conversion method with a speed of 10 to 100 ns. For more detail on this subject the reader is referred to Refs. 2 and 3.

14.14 CONCLUDING REMARKS

In this chapter we have introduced the reader to latches, flip-flops, multivibrators, and timing circuits. We then considered some applications of flip-flops such as in shift registers and counters. Digital circuits are very important and play a dominant role in electronics and are used in our everyday life. These include, of course, computers, controls, communications, consumer electronics, space, transportation, etc. The treatment of digital circuits has been purposely *very brief* in this book, and we have covered the essentials only. Many topics such as arithmetic operations, different types of memories, and microprocessors are not covered. The reader is referred to the list of books at the end of this chapter, especially Refs. 1 and 2, for further reading as well as the data books listed in Appendix E, especially their application notes.

REFERENCES

1. D. Hodges and H. G. Jackson, *Analysis and Design of Digital Integrated Circuits,* McGraw-Hill, New York, 1983.
2. H. Taub and D. L. Schilling, *Digital Integrated Electronics,* McGraw-Hill, New York, 1977.
3. A. S. Sedra and K. C. Smith, *Microelectronics Circuits,* Holt, New York, 1982.
4. R. L. Morris and John R. Miller, *Designing with TTL Integrated Circuits,* McGraw-Hill, New York, 1971.
5. *COS/MOS Integrated Circuits,* RCA Corporation, Sommerville, N.J., 1977.

PROBLEMS

14.1 For the discrete bistable circuit shown in Fig. 14.2*a*, the transistor parameters are $\beta_0 = 50$, $(V_{BE})_{\text{on}} = 0.7$ V, and $(V_{CE})_{\text{sat}} \simeq 0$. The circuit parameters are $V_{CC} = 10$ V, $R_C = 1$ kΩ, $R_1 = R_2 = 10$ kΩ, and $R_{b1} = R_{b2} = 50$ kΩ. Assume $Q = 1$. Determine all the voltages and the currents.

14.2 The circuit shown in Fig. P14.2, which is made from a NOR latch and two AND gates, is another form of clocked *SR* flip-flop. Verify that the operation is the same as in Exercise 14.1. Show that the characteristic table is given by the following:

S_n	R_n	Q_{n+1}
0	0	Q_n
0	1	0
1	0	1
1	1	Not allowed

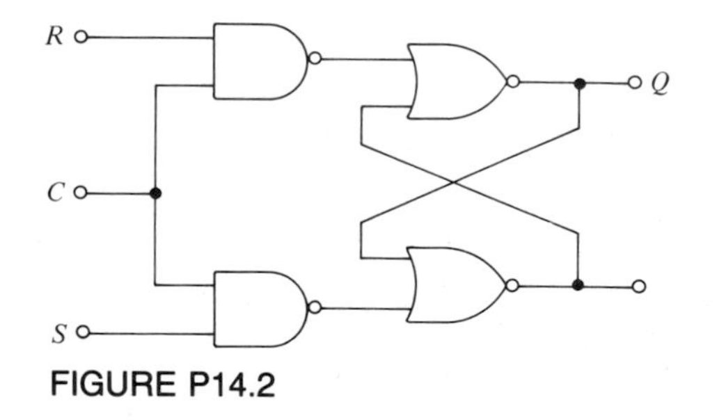

FIGURE P14.2

14.3 For the *SR* flip-flop in problem 14.2, the waveforms are given in Fig. P14.3. Sketch the waveform of Q, assume the initial state $Q = 0$.

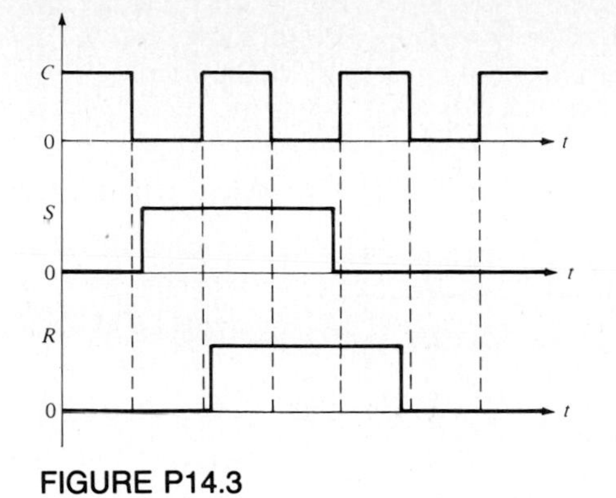

FIGURE P14.3

14.4 The timing diagram for a D flip-flop is shown in Fig. P14.4. Plot the Q timing diagram, assuming Q is 0 at $t = 0$.

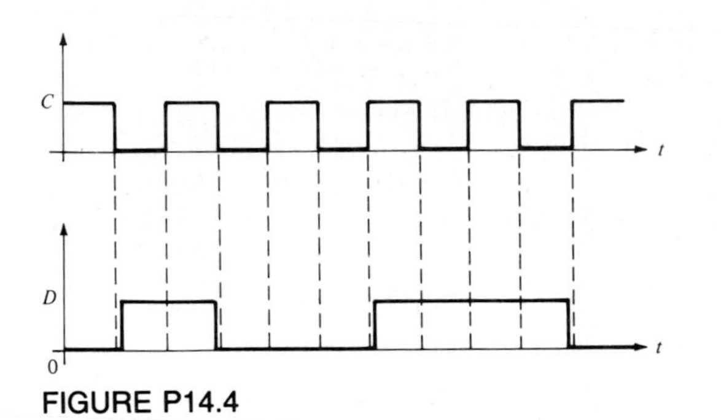

FIGURE P14.4

14.5 The timing diagram for a JK flip-flop is shown in Fig. P14.5. Complete the diagram by showing the Q output; assume the initial state of $Q = 0$.

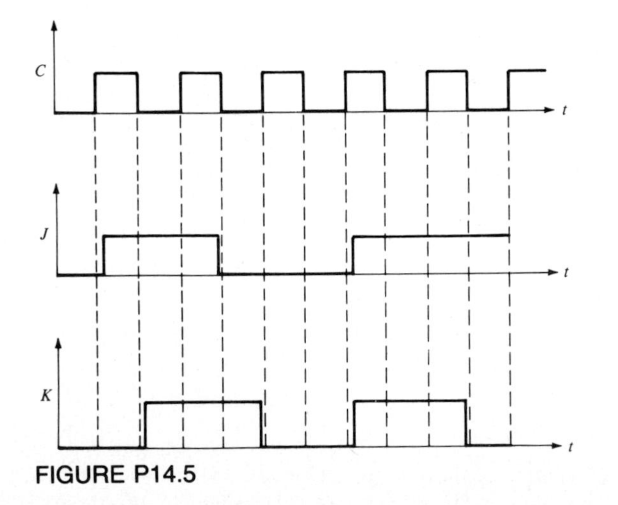

FIGURE P14.5

14.6 For the C and Q timing diagrams shown in Fig. P14.6, show a pair of J and K waveforms that will produce these waveforms in a master-slave flip-flop.

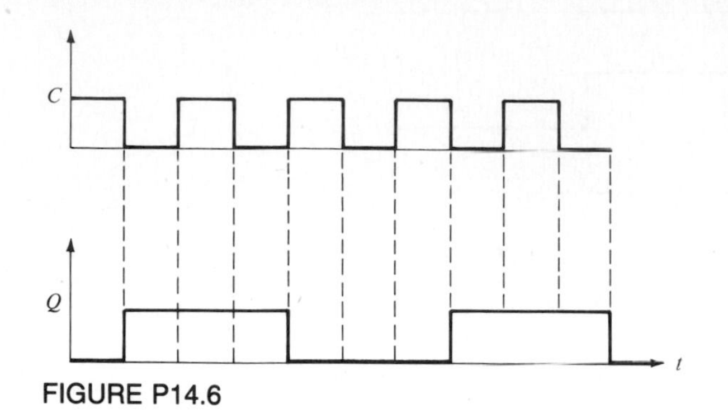

FIGURE P14.6

14.7 For the one-shot multivibrator circuit shown in Fig. 14.9*a*, let $R_E = 0$.
(*a*) Show that in the stable state Q_2 is on and Q_1 is off.
(*b*) Show that if a short trigger pulse is applied to turn Q_1 on, the pulse width (the time for Q_2 to reach turn on) is given by

$$T = -RC \ln \frac{2V_{CC}}{V_{CC}} \simeq 0.7RC$$

[*Hint:* $v_{b2}(t) \simeq V_{CC} - 2V_{CC}e^{-t/RC}$, where $(V_{BE})_{sat}$ and $(V_{CE})_{sat}$ are assumed to be zero.]

14.8 Repeat problem 14.7 for Fig. 14.9*b*.

14.9 For the astable multivibrator circuit shown in Fig. 14.14*a*, show that
(*a*) $T_{off} = T_{on} \simeq 0.7RC$
(*b*) $f = \frac{1}{T} = \frac{1}{1.4RC}$

14.10 Repeat problem 14.9 for Fig. 14.14*b*.

14.11 A CMOS monostable multivibrator circuit is shown in Fig. P14.11. Determine the pulse width if $R_x = 10$ kΩ and $C_x = 0.1$ μF. Assume $V_{TR} = 0.5\,V_{DD}$.

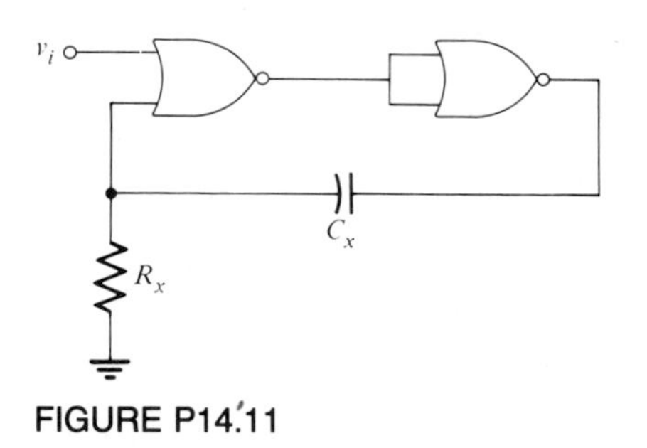

FIGURE P14.11

14.12 The duty cycle D of the 555 timer shown in Fig. 14.18 can be less than 50 percent, where $D = T_1/(T_1 + T_2)$. Show a 50 percent duty cycle astable multivibrator can be constructed by using the circuit of Fig. P14.12; more specifically, show that

$$T_1 = 0.7R_1C \qquad T_2 = \frac{R_1R_2}{R_1 + R_2} C \ln \left(\frac{R_2 - 2R_1}{2R_2 - R_1} \right)$$

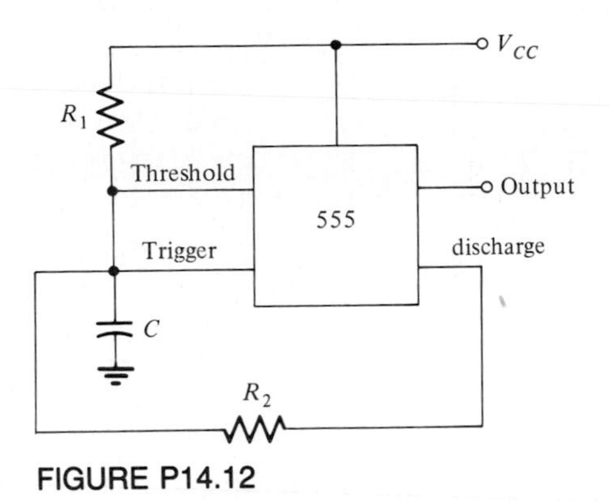

FIGURE P14.12

14.13 (*a*) Sketch the transfer characteristic of the inverting Schmitt trigger in Fig. P14.13*a*. Assume $R_1 = 50\ \text{k}\Omega$, $R_2 = 0.5\ \text{k}\Omega$, and $V_Z = 5$ V.
(*b*) Sketch the output waveform on the same scale for the input waveform given in Fig. P14.13*b*.

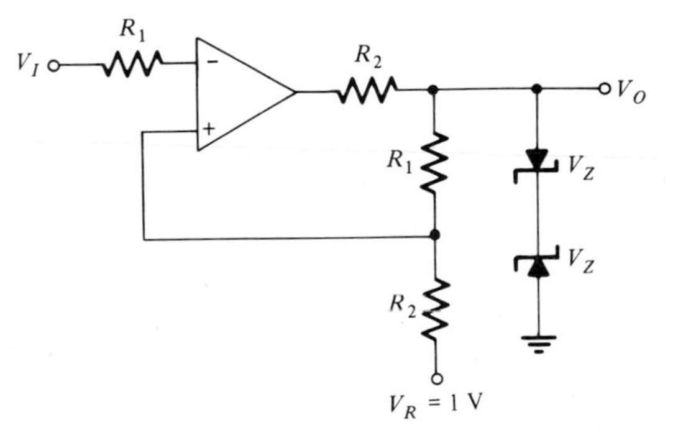

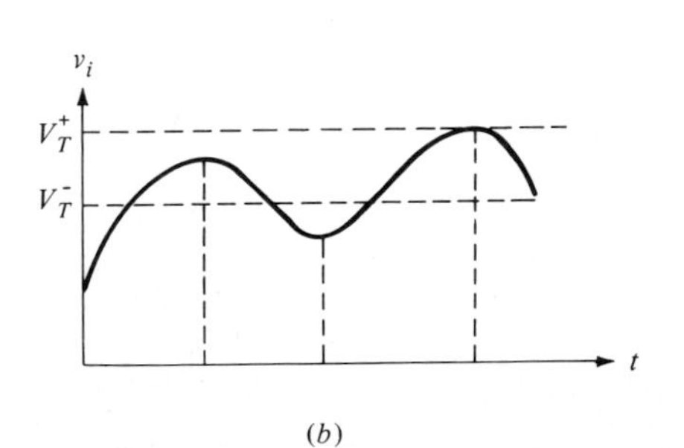

(*b*)

FIGURE P14.13

14.14 The circuit shown in Fig. P14.14 uses four *JK* FFs to perform the function of a decade counter. Show the pulse waveform or the function table for the circuit.

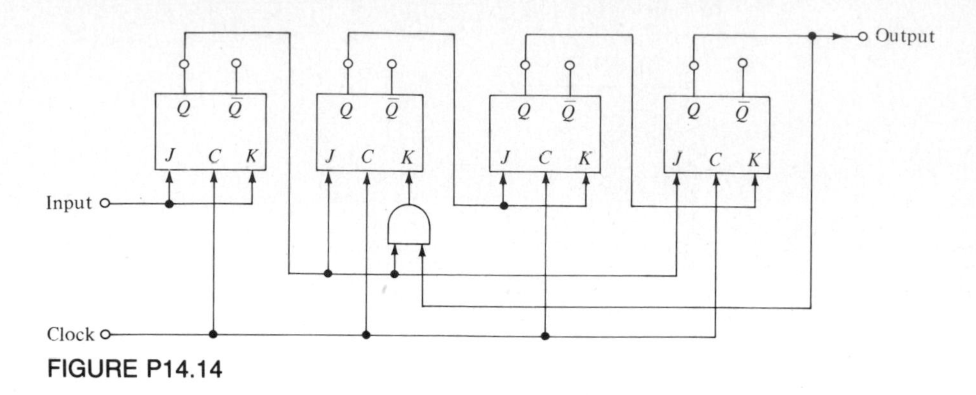

FIGURE P14.14

14.15 The circuit shown in Fig. P14.15 performs a pulse-shortening function. Describe how the circuit functions and show the output pulse. Assume a delay time $\tau = T_i/10$ for the delay block *D*.

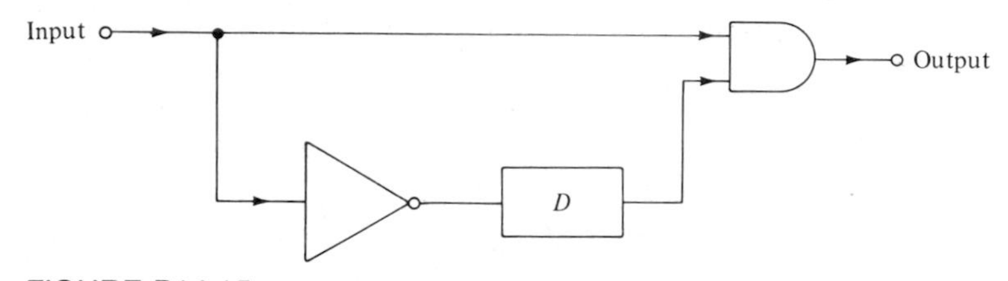

FIGURE P14.15

14.16 Show a 4-bit shift register corresponding to Fig. 14.25*a*, but use *JK* flip-flops.

14.17 The circuit shown in Fig. P14.17 utilizes three *SR* flip-flops. Show that it is a 3-bit shift register.

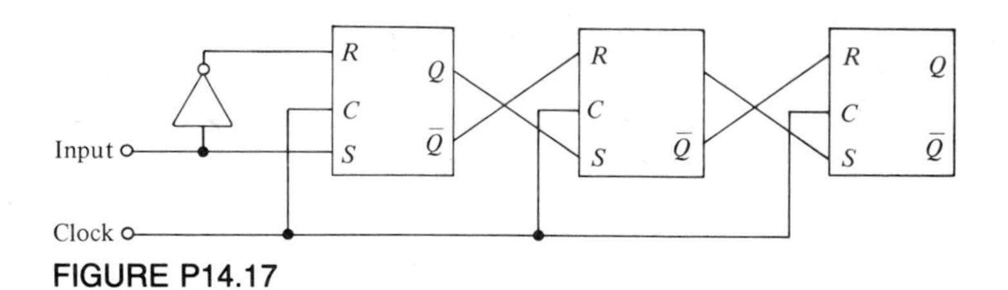

FIGURE P14.17

14.18 If three NOR gates, each having a propagation delay t_d, are cascaded in a closed loop as shown in Fig. P14.18, what would be the frequency of oscillation? If four stages were used instead of three, would you still get a square-wave oscillator? Derive an expression for the period of the oscillator when *N* gates are cascaded in a feedback loop. Is there any restriction on *N*?

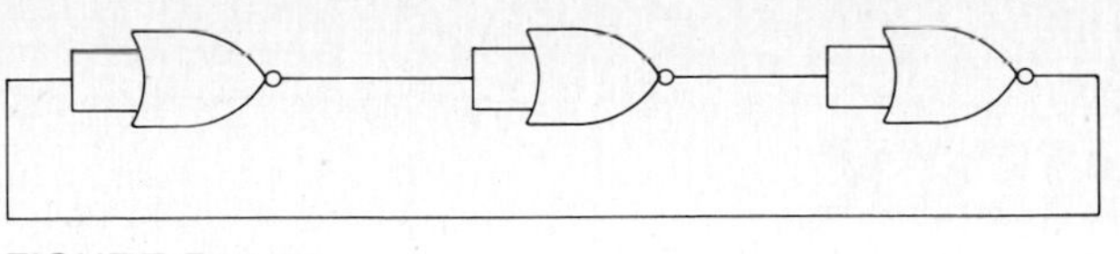

FIGURE P14.18

14.19 Derive (14.26).

14.20 A 10-bit D/A converter is used in Fig. 14.30. If the reference voltage is 10 V and the clock rate is 1 MHz, determine the conversion resolution and the maximum conversion time.

APPENDIX A

Summary of Semiconductor and *pn*-Junction Properties

The *pn* junctions are the most important and essential part of semiconductor electronic devices. They are to be found in diodes, bipolar transistors, field-effect transistors, and other integrated-circuit elements. In order to understand and appreciate the underlying principles of operation of these devices, we summarize some of the most important properties of semiconductors and *pn* junctions in this appendix. The discussion is in the form of a review, and the reader is referred to the references listed at the end of this appendix for a more complete and detailed discussion of these topics. Familiarity with this material is helpful but not necessary to understand the text material.

A.1 SEMICONDUCTORS: INTRINSIC AND EXTRINSIC

Single-crystal materials may be classified in terms of the energy levels of the outer electrons, which form energy bands. In some materials, the lower energy band (valance band) is completely filled, and the next higher energy band (conduction band) is completely empty. These bands are separated by a *forbidden energy gap* W_g of several electron volts (eV). If an electric field is applied, current cannot flow. Such materials are called *insulators*. In other materials, the valence band can partially overlap the conduction band with no forbidden energy gap in between. If an electric field is applied, current flows readily. These types of materials are called *conductors*. In still other materials, the width of the forbidden gap is less than 1.5 eV, and the material behaves as an insulator at very low temperatures. At very high temperatures, they behave as conductors. This type of material, with a conductivity level between that of insulators and that of conductors, is called a *semiconductor*. The most commonly used semiconductor in electronic devices is silicon (Si). Other semiconductor materials are also sometimes used such as germanium (Ge) and gallium arsenide (GaAs), the latter being used for high-frequency applications. The energy gap (W_g) of Si is 1.12 eV at room temperature and decreases slightly for increasing temperature.

Silicon is from group IV of the periodic table; therefore, each atom has four valence electrons with other atoms. It is a diamondlike crystal and the atoms are bound together by *covalent bonds;* i.e., each atom shares its pairs of electrons with other atoms. If the temperature of such a crystal is raised, there is an increased probability that an electron will be broken loose, or freed, from a covalent bond, and a *hole* is left behind in the bond. The free electron can participate in conduction. A hole is a missing electron in the covalent bond. Through the transfer of electrons from adjacent bonds, the hole is also free to participate in conduction and behaves very much like a positive charge $+q$ in the conduction mechanism. The process of the creation of a free electron and a hole is called *ionization*.

A pure semiconductor crystal that consists entirely of the same type of atoms is called an *intrinsic* semiconductor. A semiconductor whose conductivity is chiefly due to the presence of impurities is called an *extrinsic semiconductor*. The concentration of atoms in silicon crystal is approximately 5×10^{22} atoms/cm^3. As we shall see in Example A.1, the addition of minute impurities can cause significant effects in the conductivity of the material. In an intrinsic semiconductor, the concentrations of holes and electrons are equal, i.e.,

$$n = p = n_i \tag{A.1}$$

where the subscript i denotes intrinsic and n and p designate the concentration of electrons and holes, respectively. In the references, it is shown that the equilibrium concentration, which is strongly dependent on the temperature, is governed by the *mass-action law* and is given by

$$np = n_i^2 = AT^3 e^{-W_g/kT} = AT^3 e^{-V_g/V_t} \tag{A.2}$$

where A = proportionality constant
T = absolute temperature (K)
k = Boltzmann's constant (1.38×10^{-23} J/K = 8.62×10^{-5} eV/K
q = magnitude of electronic charge (1.6×10^{-19} C)
W_g = band-gap energy, i.e., the minimum energy required to break a covalent bond (eV), $V_g = W_g/q = 1.11$ V at 290 K.

$$V_t = \frac{kT}{q} \simeq 0.025 \text{ V at room temperature } (\simeq 290 \text{ K})$$

For Si the equilibrium intrinsic concentration is given by

$$n_i^2(T) \simeq 1.5 \times 10^{33} T^3 e^{-14{,}000/T} \text{ cm}^{-6} \qquad \textbf{(A.3a)}$$

At room temperature (300 K), the intrinsic concentration for Si is given by

$$n_i = p_i \simeq 1.5 \times 10^{10} \text{ holes or electrons/cm}^3 \qquad \textbf{(A.3b)}$$

Note that since the concentration of atoms in the Si crystal is about $5 \times 10^{22}\text{cm}^{-3}$, from (A.3*b*) one finds that roughly one atom in 10^{12} is ionized at room temperature. Note that by either thermal energy, as described earlier, or light photons of sufficient energy, we can increase the ionization in the crystal considerably.

Extrinsic, or *doped,* semiconductors are materials in which carriers of one kind (either hole or electron) predominate. The predominant carrier is called the *majority carrier*. Semiconductors in which the predominant carriers are electrons are called n-type materials. An n-type semiconductor is formed by doping (adding impurities to) the silicon crystal with elements of group V of the periodic table (phosphorus, arsenic, antimony). When a silicon atom is replaced by one of phosphorus, the impurity atom has an extra electron that is donated to the crystal. The impurity atom in this case is called a *donor* atom. The four electrons of phosphorus form covalent bonds with those of Si, and the fifth electron is left very loosely bound. The binding energy of the fifth electron in the silicon host crystal is small (about 0.04 eV) and almost independent of the type of impurities used from group V. Since the thermal energy of the crystal is of the same order of magnitude as the binding energy, all impurities are ionized at room temperature and above, and the liberated donor electrons can move freely and contribute to the conduction mechanism. The concentration of impurity atoms is usually in the range 10^{14} to 10^{18} atoms/cm^3, with 5×10^{16} a typical value. Thus, the corresponding fraction of impurities is $(5 \times 10^{16})/(5 \times 10^{22}) = 10^{-6}$, i.e., 1 ppm. From a mechanical and metallurgical point of view, the crystal is unchanged, but its electrical properties are greatly changed, as we shall presently show. A p-type semiconductor is formed by doping the silicon crystal with elements of group III of the periodic table (boron, aluminum, gallium, indium). These impurity atoms have only three valence electrons, and when complete covalent bonds are formed, the impurity atom accepts an electron. Thus, the impurity atom is called an *acceptor* atom. The electron accepted by the acceptor atom comes from a normal covalent-bonded semiconductor atom, thus producing a hole free to participate in conduction. The binding energy in this case is again small, and thus the holes can be produced readily. The conduction in the p-type crystal is then carried mainly by holes, which are the *majority carriers*. In a p-type semiconductor, the free electrons are called the *minority carriers*. In the n-type semiconductor, the electrons are the majority carriers, and the holes are the minority carriers.

The concentration of the holes and electrons are determined from $np = n_i^2$ and the charge neutrality condition $p + N_d = n + N_a$, where N_d and N_a are the donor and the acceptor concentrations, respectively. In the n-type semiconductor, the donor concentration $N_d \gg n_i$, where N_d is typically 10^{17} cm^{-3} (note that $n_i = 1.5 \times 10^{10}$ cm^{-3} in Si at room temperature). Thus, the majority and minority concentrations are given by

$$n_n \simeq N_d \qquad p_n \simeq \frac{n_i^2}{N_d} \qquad (n \text{ type}) \tag{A.4a}$$

Similarly, in a p-type semiconductor, the acceptor concentration $N_a \gg n_i$; thus the majority and minority concentrations are given by

$$p_p \simeq N_a \qquad n_p \simeq \frac{n_i^2}{N_a} \qquad (p \text{ type}) \tag{A.4b}$$

A.2 CONDUCTIVITY AND DRIFT CURRENT

There are two mechanisms by which holes and electrons move through a crystal. One of these is the *drift* current, which is caused by the application of an electric field. The other is the *diffusion* current, which is caused by the net flow of carriers from a region of high concentration to a region of lower concentration.

If an electric field $\mathscr{E}$ is applied to a semiconductor with a hole concentration p and an electron concentration n, the holes and electrons will attain an average velocity v_{dp} and v_{dn} in the direction of the field proportional to $\mathscr{E}$, namely,

$$v_{dp} = \mu_p \mathscr{E} \tag{A.5a}$$

$$v_{dn} = -\mu_n \mathscr{E} \tag{A.5b}$$

where μ_p and μ_n are the hole and electron mobilities $[cm^2/(V \cdot s)]$. The hole current density is then given by

$$J_p = qpv_d = qp\mu_p \mathscr{E} = \sigma_p \mathscr{E} \text{ A/cm}^2 \tag{A.6}$$

where

$$\sigma_p = qp\mu_p \text{ mhos/cm} \qquad \text{(hole conductivity)} \tag{A.7}$$

Similarly for the electron in the p-type material we have

$$J_n = -qnv_d = qn\mu_n \mathscr{E} = \sigma_n \mathscr{E} \tag{A.8}$$

where μ_n is the electron mobility and $\sigma_n = qn\mu_n$. Thus the total drift current density is given by

$$J = J_n + J_p = q(n\mu_n + p\mu_p)\mathscr{E} \tag{A.9}$$

Note that the electrons move in the opposite direction from the field, but the current is in the same direction as the field, so that the total currents of holes and electrons are in the same direction.

From Ohm's law, the total conductivity is

$$\sigma = q(n\mu_n + p\mu_p) \text{ (ohm} \cdot \text{cm)}^{-1} \tag{A.10}$$

For an intrinsic crystal $n = p = n_i$; thus we have

$$\sigma_i = q(\mu_n + \mu_p)n_i \qquad \text{(for intrinsic)} \tag{A.11}$$

The mobilities of Si at room temperature is given by

$$\mu_p = 500 \text{ cm}^2/\text{V}\cdot\text{s} \qquad \mu_n = 1300 \text{ cm}^2/\text{V}\cdot\text{s} \tag{A.12}$$

Thus the intrinsic conductivity of Si at room temperature is

$$\sigma_i \simeq 4.2 \times {}^{-6}(\text{ohm}\cdot\text{cm})^{-1} \tag{A.13}$$

We shall illustrate the use of the above relations by examples.

EXAMPLE A.1

Determine the conductivity of an extrinsic silicon if a donor impurity of 1 part in 10^6 silicon atoms is added.

The concentration of Si atoms is 5×10^{22} atoms/cm^3. From the impurity given, $N_D = 5 \times 10^{22}(1 \times 10^{-6}) = 5 \times 10^{16}$ atoms/cm^3. From (A.4*a*)

$$n \simeq N_D \gg p \qquad \left(p = \frac{n_i^2}{N_D} = 4.5 \times 10^3 \text{ holes/cm}^3\right)$$

Hence, from (A.10)

$$\sigma \simeq qn_n\mu = (1.6 \times 10^{-19})(5 \times 10^{16})(1.3 \times 10^3) = 0.10 \text{ (ohm}\cdot\text{cm)}^{-1}$$

Notice that the addition of 1 donor atom in 10^6 silicon atoms has increased the conductivity of silicon from 4.2×10^{-6} to 0.1, which is several orders of magnitude.

EXAMPLE A.2

A bar of Si with cross-sectional area $A = 0.001$ cm^2 and length $L = 0.2$ cm is doped with donor impurity. When a dc voltage of 10 mV is applied across the bar (at room temperature), the drift current flowing through it is found to be 1 mA. Determine the majority and the minority concentrations and the drift velocity.

The resistance of the bar, from measurement, is

$$R = \frac{V}{I} = 10\ \Omega$$

but

$$R = \frac{L}{\sigma A}$$

hence

$$\sigma = \frac{L}{AR} = \frac{0.2}{0.001(10)} = 20\ (\text{ohm} \cdot \text{cm})^{-1}$$

From the value of σ in the above and (A.10) we determine the concentrations. Note that since Si is doped with donor impurity $n \gg p$. Hence, for n type, the majority carrier concentration is

$$n \simeq \frac{\sigma}{q\mu_n} = \frac{20}{(1.6 \times 10^{-19})(1.3 \times 10^3)} = 9.61 \times 10^{16}\ \text{cm}^{-3}$$

The minority carrier concentration is found from (A.1):

$$p = \frac{n_i^2}{n} = \frac{(1.5 \times 10^{10})^2}{9.61 \times 10^{16}} = 2.34 \times 10^3\ \text{cm}^{-3}$$

Note that the approximation used above, $n\mu_n \gg p\mu_p$, is clearly justified.

The drift velocity is determined from (A.8) and (A.5):

$$v = \frac{J}{qn} = \frac{I}{Aqn} = \frac{10^{-3}}{10^{-3}(1.6 \times 10^{-19})(9.61 \times 10^{16})} = 65.0\ \text{m/s}$$

A.3 DIFFUSION CURRENT

Charge carriers are in random motion due to thermal energy whether there is any field or not. When there exists a concentration gradient, it causes carrier to flow from the high- to the low-density region. This mechanism of carrier flow in the presence of a concentration gradient is called *diffusion*. In other words, when the carrier concentration is not uniform within the crystal, there will be a net flow of carriers (even with no electric field applied), hence a diffusion current, which is proportional to the gradient of the carrier concentration. Let us assume that the hole concentration varies along the x direction and is constant in the y and z directions. Then the current density J_p in the x direction is proportional to the slope of the curve $p(x)$ and is given by

$$J_p = -qD_p\frac{dp}{dx} \qquad \text{A/cm}^2 \tag{A.14}$$

Note that the direction of hole current is from the high to the low concentration, hence the negative sign, since the slope of $p(x)$ is negative.

Similarly, for electrons, we have

$$J_n = qD_n\frac{dn}{dx} \qquad \text{A/cm}^2 \tag{A.15}$$

The proportionality constants D_p and D_n are called the *diffusion* constants and have the following values for Si:

$$D_p = 13\ \text{cm}^2/\text{s} \qquad D_n = 200\ \text{cm}^2/\text{s} \tag{A.16}$$

Note that in each case the current is given by the current density multiplied by the cross-sectional area perpendicular to the flow of current.

The mobility and diffusion constants are related, under steady-state conditions, by *Einstein's relations,* namely,

$$\frac{u_p}{D_p} = \frac{q}{kT} = V_t^{-1} \tag{A.17a}$$

$$\frac{u_n}{D_n} = \frac{q}{kT} = V_t^{-1} \tag{A.17b}$$

At room temperature (300 K), $V_t = kT/q \simeq 0.025$ V.

A.4 THE *pn* JUNCTION AND ITS PROPERTIES

A *pn* junction is formed by diffusing into an *n*-type semiconductor a *p*-type impurity, or vice versa. The transition from a *p*-type to an *n*-type doping may be abrupt, so that it may be considered a step junction, or it may be gradual, which is called the *graded junction*. Typical concentration profile of a *pn* junction is shown in Fig. A.1*a*.

The *p*-type semiconductor has many free holes contributed by the acceptor impurity atoms, while the *n*-type semiconductor has many free electrons contributed by the donor impurity atoms. Because of the nonuniform concentrations, there is a strong tendency for holes to move by diffusion from the *p*-type to the *n*-type and,

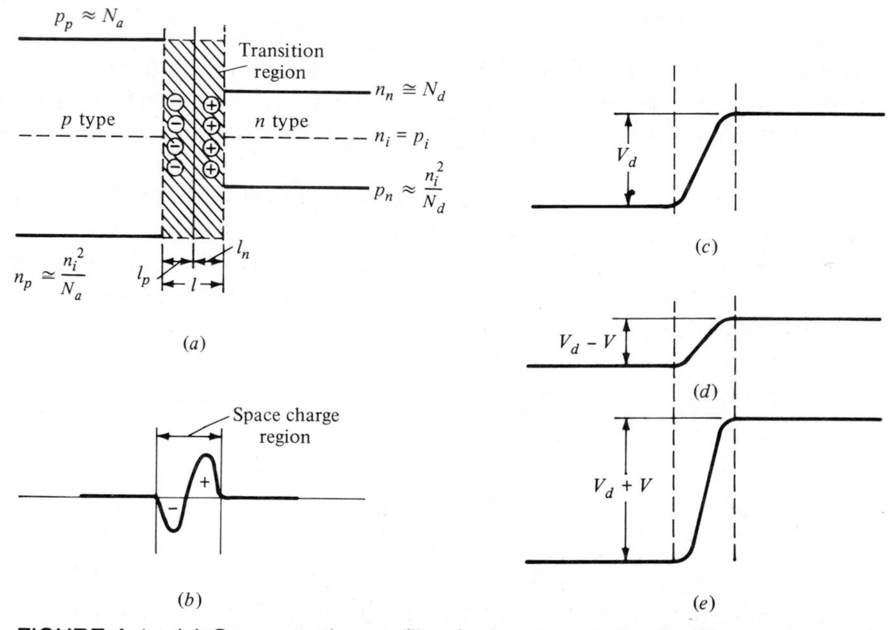

FIGURE A.1 (*a*) Concentration profile of a junction diode. (*b*) Charge density. (*c*) Contact potential with no external voltage applied. (*d*) Potential across the junction for a forward-biased applied voltage. (*e*) Potential across the junction for a reverse-biased applied voltage.

similarly, for electrons to move from the n type to the p type. However, when holes cross the junction, they find immediately abundant free electrons and a large number disappear by *recombination* very quickly. The same situation occurs for electrons passing from the n-type region to the p-type region across the junction. If such electron and hole migration took place with no restraint, then a current would flow across the junction with no external voltage applied. Since no current can flow in the pn junction without an external voltage, the current must be zero. To accomplish this a so-called *depletion region,* free from mobile charge, or equivalently, a *space-charge* region, is created. This space-charge region arises from the fact that the region near the junction plane within the n-type semiconductor is left with a net positive charge, while the region near the junction plane within the p-type semiconductor is left with a net negative charge. As a consequence of this dipole layer at the junction, an electric field is established across the junction that results in a potential difference across the junction, as shown in Fig. A.1*b*. Since, at thermal equilibrium, no net current can flow, this potential barrier V_d prevents the flow of majority carriers to the other side. The potential barrier, which is also called the *diffusion potential* or *contact potential,* is determined from the carrier concentrations by the following simple relations:

$$\frac{n_n}{n_p} = e^{(q/kT)V_d} = \frac{p_p}{p_n} \qquad \textbf{(A.18a)}$$

or

$$V_d = \frac{kT}{q}\ln\frac{n_n}{n_p} = \frac{kT}{q}\ln\frac{p_p}{p_n} \qquad \textbf{(A.18b)}$$

Equation (A.18*b*) may also be written as

$$V_d = \frac{kT}{q}\ln\frac{N_a N_d}{n_i^2} = \frac{kT}{q}\ln\frac{\sigma_n\sigma_p}{\mu_n\mu_p q^2 n_i^2} \qquad \textbf{(A.18c)}$$

For typical Si junctions the potential difference is $V_d \simeq 0.5$ to 0.8 V. (For Ge it is approximately 0.1 to 0.2 V and for GaAs it is approximately $\simeq 1.5$ V.)

Note that when an external voltage V is applied, the potential barrier will naturally be affected. In other words, the carrier concentrations near the junction in such a case is given by

$$\frac{n_n}{n_p} = e^{(q/kT)(V_d \pm V)} \qquad \textbf{(A.19a)}$$

$$\frac{p_p}{p_n} = e^{(q/kT)(V_d \pm V)} \qquad \textbf{(A.19b)}$$

where the minus sign is used for forward-biased conditions (in the forward-biased case, the plus polarity of the battery is connected to the p type), and the plus sign is used for the reverse-biased case. These are shown in Figs. A.1*d* and *e*, respectively. Observe that a forward-biased case reduces the potential hill, and thus enables many majority carriers to cross the junction by diffusion, while the reverse-biased case increases the potential hill, thus further preventing the flow of majority carriers.

A number of interesting properties of the *pn* junction such as the depletion or junction capacitance and the depletion width can be determined by solving Poisson's equation in the depletion region. *Depletion capacitance C_j is due to charge storage in the depletion region.* These are derived in the references cited. The results are

$$C_j = \frac{\epsilon A}{l} = \frac{A}{|l_p| + |l_n|} \tag{A.20}$$

where ϵ is the relative dielectric constant ($\epsilon = 12\epsilon_0$ for Si and $\epsilon_0 = 8.85 \times 10^{-12}$ F/m), A is the cross-sectional area, and l_p and l_n are the transition widths of the p and n regions. For abrupt junctions the transition widths are given by

$$l_p = \left(\frac{2\epsilon N_d}{qN_a(N_d + N_a)}\right)^{1/2} (V_d - V)^{1/2} \tag{A.21a}$$

abrupt junction

$$l_n = \left(\frac{2\epsilon N_a}{qN_d(N_d + N_a)}\right)^{1/2} (V_d - V)^{1/2} \tag{A.21b}$$

so that

$$1 = |l_p| + |l_n| = \left[\frac{2\epsilon}{q}\left(\frac{1}{N_d} + \frac{1}{N_a}\right)\right]^{1/2} (V_d - V)^{1/2} \tag{A.22}$$

For a graded junction

$$|l_p| = |l_n| = \left(\frac{3\epsilon}{2qa}\right)^{1/3} (V_d - V)^{1/3} \tag{A.23}$$

where a is the slope of the graded junction impurity profile. In general, we may express the depletion capacitance of a *pn* junction by

$$C_j = K(V_d - V)^{-m} \tag{A.24}$$

where K = constant depending on impurity concentration profile and area
m = constant depending on distribution of impurity near junction
= 0.33 or 0.5 depending on the type of junction

Note that in (A.22) to (A.24), V is a positive for a forward-biased case and negative for a reverse-biased case. The nonlinear relation $C = f(V)$ of a *pn* junction should be particularly noted as it is used in a more accurate computer-aided analysis and design. It should also be noted that in the *forward-biased case* the diode has a *diffusion capacitance C_d*, which is due to the variation of stored minority carrier charge in the neutral zone as the carriers move due to diffusion. The diffusion capacitance is proportional to the bias current and is usually larger than C_j. The diffusion capacitance is almost zero under a reverse-biased condition. The fabrication of integrated circuits based on *pn* junctions is discussed in Appendix B.

EXAMPLE A.3

Consider a *pn* junction as shown schematically in Fig. A.2. The junction is assumed to be abrupt, and the conductivities of the *p*-type and *n*-type Si crystals at room temperature are $\sigma_p = 100\ (\text{ohm}\cdot\text{cm})^{-1}$ and $\sigma_n = 1\ (\text{ohm}\cdot\text{cm})^{-1}$. We are to calculate V_d, l, and C_j at zero applied voltage with the values of l and C_j for a reverse-applied potential of -5 V and a forward-bias potential of $+0.5$ V.

The donor and acceptor concentrations are found from

$$N_d \simeq n_n \simeq \frac{\sigma_n}{q\mu_n} = 4.80 \times 10^{15}\ \text{cm}^{-3}$$

$$N_a \simeq p_p \simeq \frac{\sigma_p}{q\mu p} = 1.25 \times 10^{18}\ \text{cm}^{-3}$$

From (A.18), at room temperature we determine

$$V_d = \frac{kT}{q} \ln \frac{\sigma_n \sigma_p}{\mu_n \mu_p q^2 n_i^2} = 0.025\ (30.9) = 0.77\ \text{V}$$

at no applied voltage, $V_a = 0$, and from (A.22) we have

$$1 = \left[\frac{2\epsilon}{q}\left(\frac{1}{N_d} + \frac{1}{N_a}\right)V_d\right]^{1/2} \simeq \left(\frac{2\epsilon}{q}\frac{1}{N_d}V_d\right)^{1/2} = 4.6 \times 10^{-4}\ \text{mm}$$

Note that

$$l_n = \frac{N_a}{N_a + N_d} l \qquad \text{and} \qquad l_p \simeq \frac{N_a}{N_a + N_d} l$$

Since $N_a \gg N_d$, $l \simeq l_n \gg l_p$. From (A.20) and above we obtain

$$C_j(0) = \frac{\epsilon A}{l} = 22.7\ \text{pF}$$

At a reverse-biased applied voltage $V_a = -5$ V, the corresponding values of l and C_j from (A.21) are

$$l = 1.2 \times 10^{-3}\ \text{mm} \qquad C_j(-5\ \text{V}) = 8.3\ \text{pF}$$

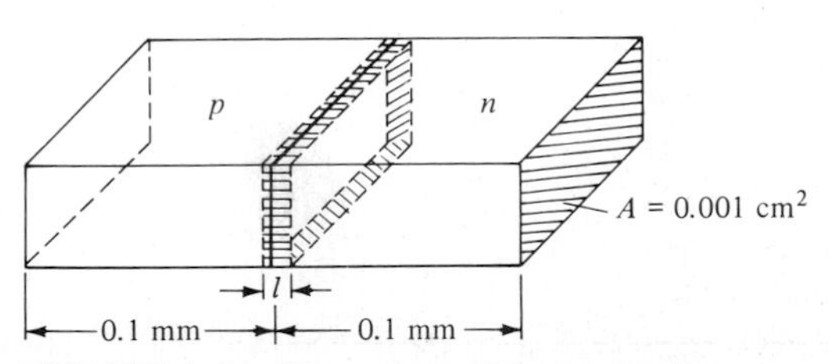

FIGURE A.2 Silicon bar of the example.

Similarly, at a forward-biased voltage $V_a = +0.5$ V, these are

$$l = 2.7 \times 10^{-4} \text{ mm} \qquad C_j(+5 \text{ V}) = 38.3 \text{ pF}$$

A.5 *V-I* CHARACTERISTIC OF *pn* JUNCTIONS

The dc voltage-current relationships of actual *pn*-junction Si and Ge diodes at room temperature is shown in Fig. A.3. The characteristics, exclusive of the breakdown region, is given by the relation

$$I = I_s(e^{\lambda qV/kT} - 1) = I_s(e^{\lambda V/V_t} - 1) \tag{A.25}$$

where $V_t = \dfrac{kT}{q} \simeq 40$ V at room temperature (20°C)

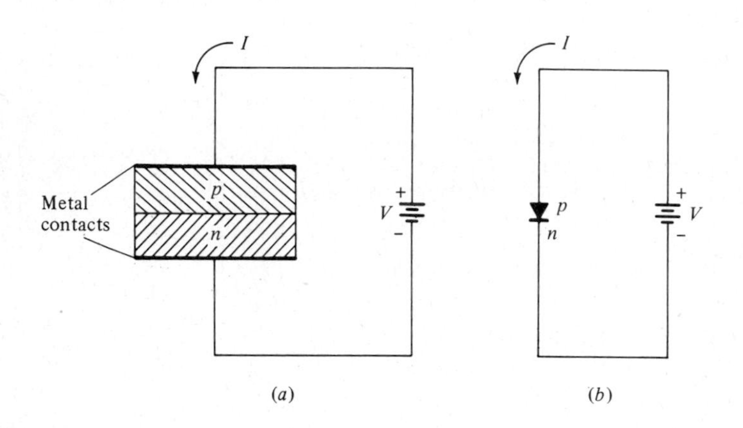

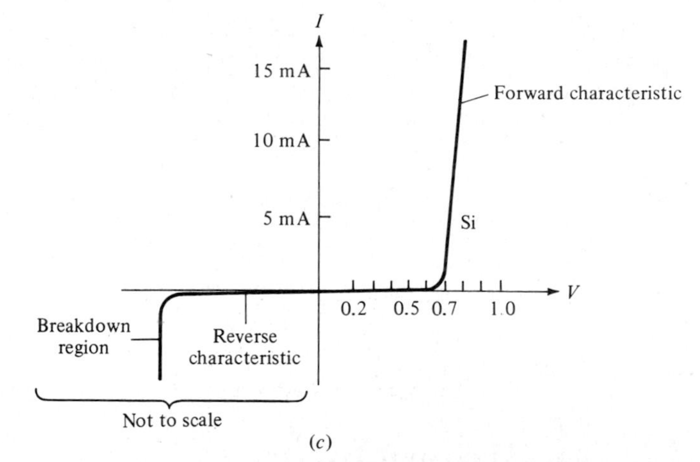

FIGURE A.3 (*a*) Schematic of a silicon diode with forward bias applied. (*b*) Graphic symbol for (*a*). (*c*) Typical *V-I* characteristics of Si diodes (reverse region not to scale).

I_s = reverse saturation current, A
λ = empirical constant, which lies between 0.5 and 1.0

We shall now show how (A.25) may be obtained for a *pn*-junction material and also show the strong dependence of I_s on temperature.

The current-voltage relationship of a *pn*-junction diode is obtained from the *continuity equation* and the appropriate boundary conditions.

The continuity equation for holes, assuming one-dimensional diffusion flow, is given by

$$\frac{\partial p(x,t)}{\partial t} = -\frac{p - p_n}{\tau_p} + D_p \frac{\partial^2 p(x,t)}{\partial x^2} \tag{A.26}$$

In other words,

Time rate of increase in volume	=	holes generated minus holes recombined per unit time per unit volume	+	difference between hole currents into and leaving volume

For dc conditions, there is no variation with time; hence $\partial p/\partial t = 0$. Equation (A.26) can be written as

$$D_p \frac{d^2 p}{dx^2} = \frac{p - p_n}{\tau_p} \tag{A.27}$$

Equation (A.27), which is referred to as the hole *diffusion* equation, may be written as

$$\frac{d^2(p - p_n)}{dx^2} = \frac{p - p_n}{L_p^2} \tag{A.28}$$

where $L_p = \sqrt{D_p \tau_p}$ is the hole diffusion length; p_n, which is constant, is the equilibrium density of holes in the *n* region far away from the junction; and τ_p is the hole lifetime in the *n* region. Solution of (A.28), subject to the following boundary conditions:

At $x = 1_n$: $\quad p = p_n e^{V/V_t}$

At $x = \infty$: $\quad p = p_n$

is given by (A.29)

$$p - p_n = p_n(e^{V/V_t} - 1)e^{(l_n - x/L_p)} \tag{A.29}$$

The hole-current density from (A.29) and (A.14) is

$$J_p(x) = q\frac{D_p p_n}{L_p}(e^{V/V_t} - 1)e^{(l_n - x)/L_p} \qquad x \geq l_n \tag{A.30}$$

Similarly, the diffusion equation for the electrons in the *p* region is

$$D_n \frac{d^2(n - n_p)}{dx^2} = \frac{n - n_p}{\tau_n} \tag{A.31}$$

The boundary conditions are

At $x = -l_p$: $\quad n = n_p e^{V/V_t}$

At $x = \infty$: $\quad p = p_n$

The electron current density, similar to the above, is

$$J_n(x) = q \frac{D_n n_p}{L_n} (e^{V/V_t} - 1) e^{(x+l_p)/L_n} \qquad x < -l_p \tag{A.32}$$

Since the transition width, l is almost always much smaller than the diffusion length, the recombination in the depletion region can be neglected, and we can write

$$J_n(-l_p) = J_n(l_n) \qquad \text{and} \qquad J_p(l_n) = J_p(-l_p) \tag{A.33}$$

The total diode current, which is constant throughout the crystal, is

$$J = J_n(l_n) + J_p(l_n) = J_n(-l_p) + J_p(l_n) \tag{A.34}$$

The total dc current density, with the cross-sectional area A, is

$$J = q\left(\frac{D_p p_n}{L_p} + \frac{D_n n_p}{L_n}\right)(e^{V/V_t} - 1) \tag{A.35}$$

The total dc current is

$$I = JA = qA\left(\frac{D_p p_n}{L_p} + \frac{D_n n_p}{L_n}\right)(e^{V/V_t} - 1) \tag{A.36}$$

We may rewrite (A.36) as

$$I = I_s(e^{V/V_t} - 1) \tag{A.37}$$

where

$$I_s = qA\left(\frac{D_p p_n}{L_p} + \frac{D_n n_p}{L_n}\right) = qA\left(\frac{D_p}{L_p N_d} + \frac{D_n}{L_n N_a}\right) n_i^2 \tag{A.38}$$

Note that n_i^2 is strongly dependent on temperature, as is evident by (A.1). Hence I_s is strongly dependent on temperature. For a real diode, an empirical factor λ, i.e., $V_t = V_t/\lambda$, where λ ranges between 0.5 and 1, is used, so as to fit the V-I characteristics with the measured data.

From (A.25) and for $\lambda V \gg V_t$, we have

$$V \simeq \frac{V_t}{\lambda} \ln \frac{I}{I_s} \tag{A.39}$$

A graph of (A.39) is shown in Fig. A.4 by the dotted straight line on the semilog plot. From such a plot the value of I_s can be determined. The actual graph of a typical *pn*-junction diode is also shown by the solid line. The deviation between the actual diode characteristic and (A.39) is due to the following: At the high voltage region

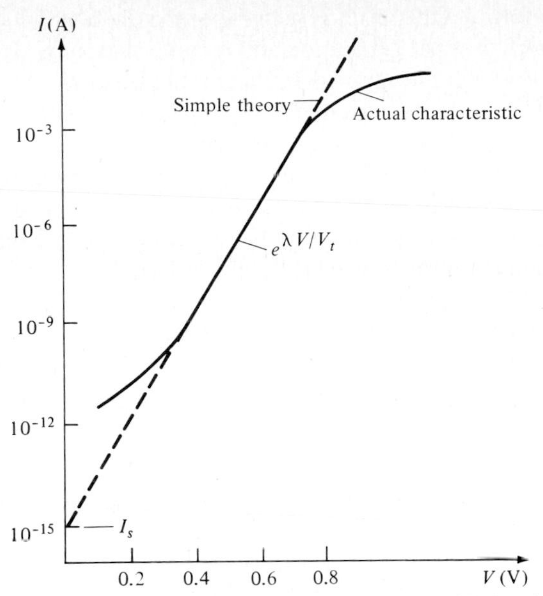

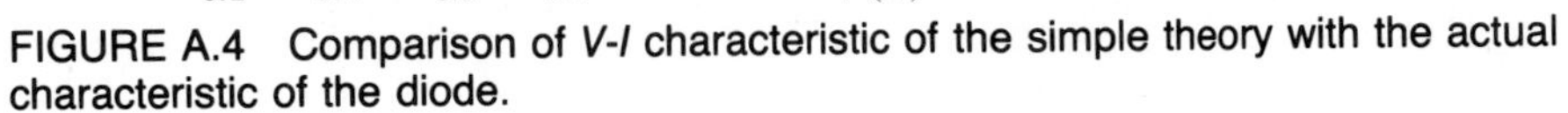

FIGURE A.4 Comparison of *V-I* characteristic of the simple theory with the actual characteristic of the diode.

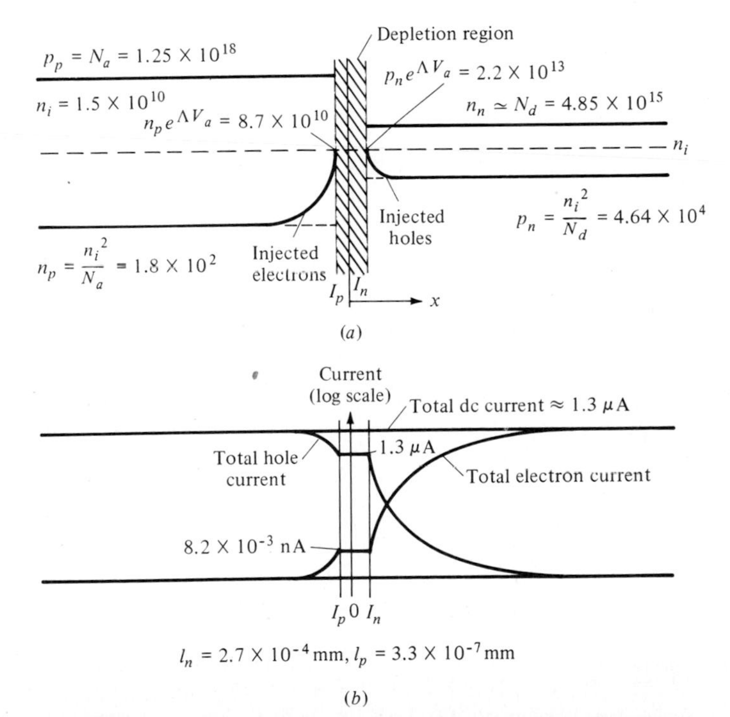

FIGURE A.5 Silicon crystal of the example with a forward bias of 0.5*V*: (*a*) concentration profile, (*b*) dc currents (not to scale).

($V \geq 0.8$), high-level injection and the series resistance of the diode are dominant. At the low current ($I \leq 0.1$ nA), thermal generation and recombination of electrons and holes are dominant; both of which are ignored in the simple theory.

The concentration profile and the current density of the silicon example of this section for a forward-bias potential of 0.5 V are shown in Fig. A.5. Note that the boundary condition concentration, the hole current, and the electron current are all shown for illustrative purposes. The saturation current of the silicon used for the example is $I_s = 2.7 \times 10^{-15}$ A. The actual value of the saturation current in a real silicon junction at room temperature, however, is much larger than this value. A typical value is 10^{-9} A. This discrepancy is due to a variety of reasons such as generation and recombination of carriers in the depletion layer, leakage current, surface effects, and high-level injection. The actual value of I_s can be obtained by measurements made on the real diode. The value of λ is also obtained by measurement to fit the experimental curve (see Fig. 1.2).

EXAMPLE A.4

For the silicon *pn*-junction diode of Example A.3, determine I_s and the applied voltage for a current of 1 mA at 300 K. Given $\tau_n = \tau_p = 1\ \mu$s,

$$L_p = \sqrt{D_p \tau_p} = 13(10^{-6}) = 3.6 \times 10^{-3}\ \text{cm}$$

$$L_n = \sqrt{D_n \tau_n} = 200(10^{-6}) = 14.1 \times 10^{-3}\ \text{cm}$$

From Example 1.3,

$$\frac{n_i^2}{N_d} = 4.7 \times 10^4\ \text{cm}^{-3} \quad \text{and} \quad \frac{n_i^2}{N_a} = 11.2\ \text{cm}^{-3}$$

Thus from (A.38)

$$I_s = (1.6 \times 10^{-19})(10^{-3})\left(\frac{13}{(3.6 \times 10^{-3})(4.8 \times 10^{15})} + \frac{200}{(14.1 \times 10^{-3})(1.25 \times 10^{18})}\right)(2.25 \times 10^{20})$$

$$= 2.73 \times 10^{-14}\ \text{A}$$

Rewriting (A.37) we have

$$V = V_t \ln\left(\frac{I}{I_s} + 1\right) \simeq 0.025 \ln \frac{10^{-3}}{2.73 \times 10^{-14}} = 0.61\ \text{V}$$

A.6 TEMPERATURE EFFECTS

The two parameters of the diode in (A.37), namely, I_s and V are both temperature dependent. The temperature dependence of the current and voltage can be found as follows: From (A.38) and (A.30) we have

$$I_s = K_1 n_i^2 = K_2 T^3 e^{-W_g/kT} \tag{A.40}$$

where K_1 and K_2 are proportionality constants.

From (A.40) we obtain

$$\frac{1}{I_s}\frac{dI_S}{dT} = \frac{1}{n_i^2}\frac{d(n_i^2)}{dT} = \frac{3}{T} + \frac{1}{T}\frac{W_g}{kT} = \frac{3}{T} + \frac{1}{T}\frac{V_g}{V_t} \tag{A.41a}$$

For Si at room temperature $V_g/V_t \simeq 44.4$; hence from (A.41*a*)

$$\frac{dI_s}{dT} = 47.4\frac{dT}{T} \tag{A.41b}$$

Thus near room temperature the percentage change in current is approximately 16 percent/K. Thus for every 5°C, I_s approximately doubles, since $(1.165)^5 = 2.1$. The actual experimental value is somewhat higher than 5, and we shall use 6°C for the doubling of I_s near room temperature, namely,

$$I_s(20°\text{C} + \Delta T) \simeq I_s(20°\text{C})\, e^{\Delta T/6°\text{C}} \tag{A.42}$$

From (A.37) for a forward-biased diode, we can neglect the unity term and obtain the following:

$$\left.\frac{dV}{dT}\right|_{I=\text{const}} = \frac{d}{dt}\left(V_t \ln\frac{I}{I_s}\right) = \frac{V}{T} - V_t\left(\frac{1}{I_s}\frac{dI_s}{d_T}\right) \tag{A.43}$$

Near the threshold voltage ($V = 0.7$ V) and at room temperature

$$\frac{dV}{dT} = \frac{700}{290} - 25(0.16) = 2.4 - 4.0 = -1.6 \text{ mV/°C} \tag{A.44}$$

The measured value is somewhat higher than the above, and we shall use −2 mV/°C; thus

$$V(20°\text{C} + \Delta T) = (0.70 - 2 \times 10^{-3}\Delta T) \tag{A.45}$$

From (A.42) and (A.45) we obtain the approximate result that the current doubles for every 6°C rise and the voltage decreases linearly with temperature increase at a rate of −2 mV/°C near room temperature.

REFERENCES

1. E. S. Yang, *Fundamentals of Semiconductor Devices*, McGraw-Hill, New York, 1978.
2. R. S. Muller and T. I. Kamins, *Device Electronics for Integrated Circuits*, Wiley, New York, 1977.
3. S. M. Sze, *Physics of Semiconductor Devices*, Wiley, New York, 1981.

A Brief Summary of Integrated-Circuit Fabrication and Some Terminology

In this appendix we will discuss briefly the technique of fabricating circuits that are orders of magnitude smaller, more reliable, and less costly than discrete circuits. *Integrated-circuit* (IC) or *microelectronics* technology has made it possible to realize large systems and complex building blocks. For example, the impact of IC technology in microprocessors, op amps, calculators, microcomputers, video games, minicomputers, and large mainframe digital computers cannot be questioned. The low cost of active elements in IC and the availability of building block modules have changed the basic ground rules of electronic-circuit design.

An IC consists of a single chip of silicon, typically 50 to 100 mil (1 mil = 0.001 in = 2.5 mm) on a side, containing both active and passive elements. There are two types of IC, namely, the monolithic and the hybrid forms. The monolithic is completely integrated, i.e., the entire circuit is fabricated on a single semiconductor substrate. The monolithic form is *small* and *inexpensive in large-quantity production*, since all components and interconnections are made by a succession of processing steps at the same time. The disadvantage of this form is the restriction imposed on the size and range of element values, and the tolerance variations of passive components in absolute values, which are typically in the range of 10 to 20 percent. Ratios of passive component values can be held as close as 3 percent. In MOS technology the ratio of capacitors can be held with a tolerance of less than 1 percent.

In the hybrid IC, various components, on separate chips, are mounted on an insulating substrate and interconnected. This form of IC is more flexible in that thin-film and diffused components can be used. Thin-film circuits consist of resistors and capacitors that are fabricated by vacuum deposition on inert substrate such as alumina or glass. These films are usually 10^{-7} to 10^{-6} m thick. Tantalum may be used to form complete thin-film circuits. A thick-film circuit consists of passive components that are formed on the substrate in the form of pastes which are fired at high temperature to form films. The manufacturing of the thick films are much simpler and economical in small production. In small-quantity production, a hybrid IC is economical than the monolithic IC. For large-quantity production, the cost of manual assembly and interconnection exceeds that of masks for monolithic IC and is, therefore, more expensive.

As an example the photomicrograph of a monolithic IC operational amplifier (741) is shown in Fig. B.1. The chip size for this circuit is about 56 × 56 mil. Capacitors and resistors usually occupy more space than active devices. For example,

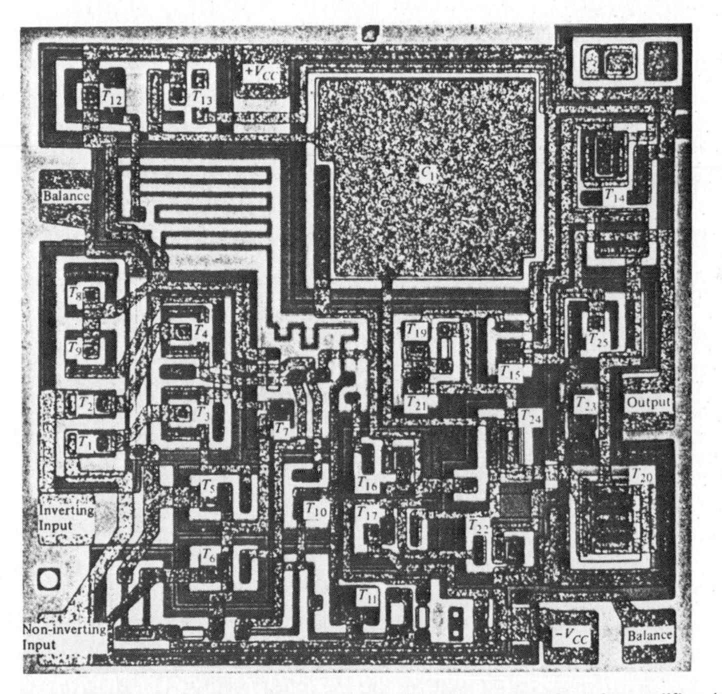

FIGURE B.1 Photomicrograph of the 741 operational amplifier (*Courtesy of Fairchild Semiconductor.*)

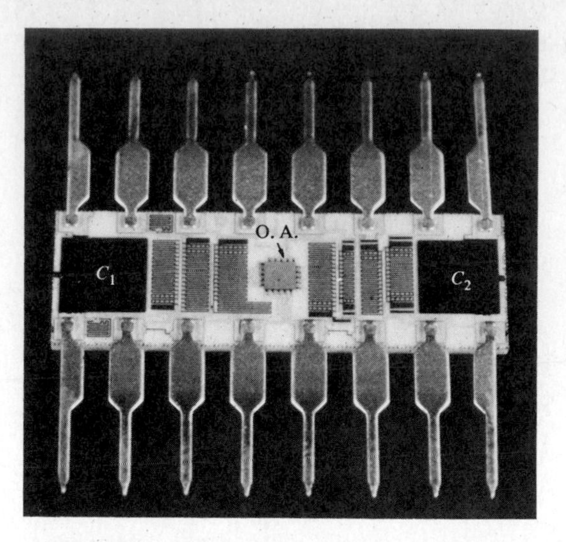

FIGURE B.2 Photomicrograph of a hybrid thin-film IC active filter module. (*Courtesy of Bell Telephone Laboratories.*)

note the large square area occupied by $C_1 = 30$ pF used for internal compensation of the op amp. The circuitry of the chip is shown in Fig. 5.25.

A hybrid thin-film general biquadratic module is shown in Fig. B.2. This module is used in active *RC* filters, and the circuit is shown in Fig. 9.31.

We shall discuss briefly some of the fabrication techniques used in an IC in order to appreciate the constraints and new problems imposed on the circuit designer.

B.1 IC TRANSISTOR FABRICATION

We shall first show the basic steps in the fabrication of an IC transistor. The main fabrication process involves epitaxial growth, oxidation, photomasking, impurity diffusion, metallization, and packaging.

A *p*-type silicon-crystal wafer with a typical resistivity of about 10 Ω/cm is first grown from a silicon bar. The Si crystal is then sawed into wafers of 0.25 to 0.4 mm in thickness. The Si wafer is then polished to a mirror finish and then used as the *substrate* material. An oxide layer is then grown, as shown in Fig. B.3*a*. The wafer surface is then coated with a photosensitive emulsion, as shown in Fig. B.3*b*. To expose the emulsion selectively, a prescribed mask for the buried-layer (n^+) diffusion is placed in contact with the SiO_2, and the masked wafer is then exposed to ultraviolet light. Note than n^+ means heavily doped n type with a doping $\geq 10^{18}$ cm^{-3}. After exposure, the emulsion is hardened, and the unexposed emulsion is removed from the buried-layer regions. The wafer is then ready for the first diffusion, as shown in Fig. B.3*c*. Arsenic or antimony is used for the buried-layer diffusion, where a heavily doped *n*-type layer is formed, as shown by the n^+ layer in Fig. B.3*d*. After the buried layer is formed, all oxide is removed by a hydrofluoric acid etchant, and an *n*-type

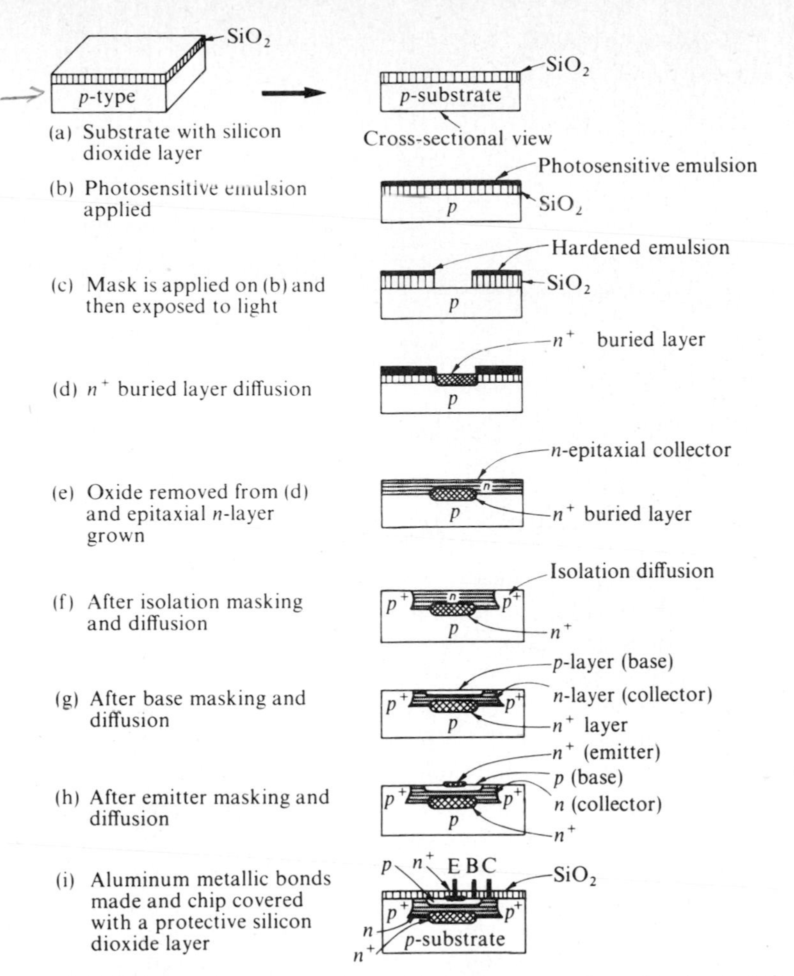

FIGURE B.3 Integrated-circuit fabrication steps for a bipolar transistor.

layer is grown epitaxially on the surface to serve as the collector region of the transistor, as shown in Fig. B.3*e*.

At this point several oxidation, photoresist, masking, cleaning, and diffusion steps take place in the following sequence: An oxide layer is grown over the surface of the *n* layer; photoresist is then applied. The isolation pattern (mask) is then applied in a manner similar to the buried-layer pattern previously described. The isolation diffusion is then completed through the *n* layer by doping the isolation region with acceptors to make it p^+, in order to provide an isolation region between this transistor and the adjacent *n* region on the chip, as shown in Fig. B.3*f*. Next we follow the same steps of oxidation, photoresist, base mask, and diffusion in forming the base and resistor regions, as shown in Fig. B.3*g*. Once more these steps are repeated for the emitter region, and a heavily doped n^+ region is formed that serves as the emitter of the transistor; a similar n^+ region is used as a topside contact to the *n*-collector region,

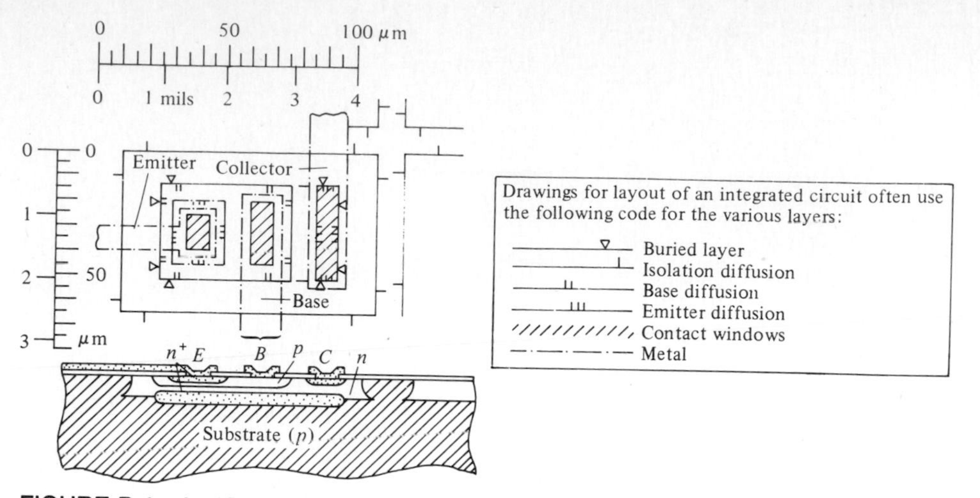

FIGURE B.4 An IC bipolar transistor.

as shown in Fig. B.3*h*. Finally, another mask is used for ohmic contacts, to which aluminum metallic bonds are made. The whole chip is then covered by a silicon dioxide for passivation purposes, as shown in Fig. B.3*i*.

Top view and cross-sectional view of a single IC transistor are shown in Fig. B.4 [3]. Notice that it is typically under 3 by 5 mil (1 mil = 25 μm). In IC because of production, yield, and economy usually only one type of transistor, namely, *npn*, is produced. One of the reasons for using *npn*, rather than *pnp*, is that the mobility of electrons is higher than that of holes, and this makes it attractive for high-frequency operation. The *npn* transistors are also compatible with NMOS devices. For special applications where *pnp* transistors are needed, two types are used; these are the *substrate* and *lateral pnp* transistors, which can be fabricated without complicating the fabrication process. The substrate and the lateral *pnp* transistors are shown in Figs. B.5 and B.6, respectively. The substrate *pnp* can be used only as an emitter follower since the collector is the substrate and must be grounded. Substrate *pnp* can be made for β_0 of about 100 and f_T to about 10 MHz. In the lateral *pnp* transistors, transistor action

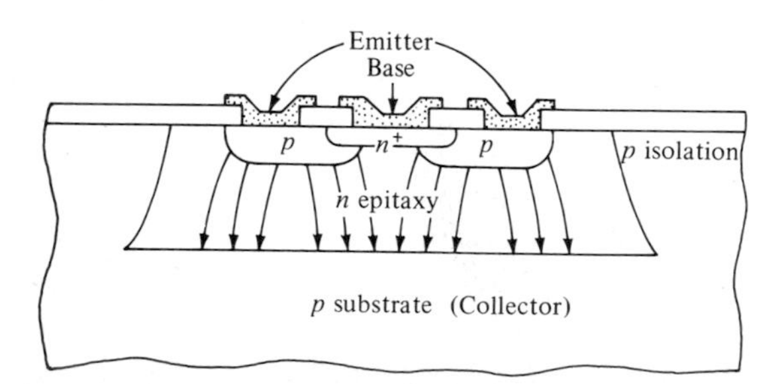

FIGURE B.5 Substrate *pnp* transistor.

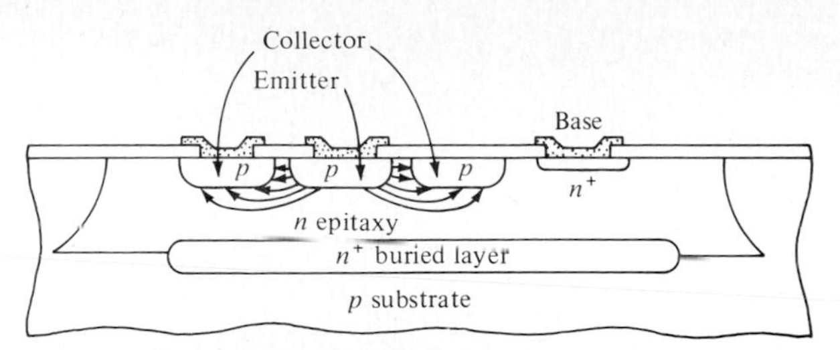

FIGURE B.6 Lateral *pnp* transistor.

takes place laterally. The gains of lateral *pnp* transistors is usually low ($\beta_0 \simeq 10$) and improved process control is needed to achieve β_0 of about 100.

In monolithic IC fabrication, the entire circuit (a large number of identical ones) is fabricated as an array of dice processed simultaneously on all chips of a wafer. A wafer is generally a circular plate about 2 to 5 in in diameter and 10 mil thick. Each chip or die (representing a circuit) is a square or rectangle of about 4000 to 40,000 mil^2. After the fabrication of the circuit is completed, the wafer is scribed and broken into individual die, each of which is a complete circuit. Figure B.7 shows various stages from grown crystal to packaged product in the processing of an integrated circuit. At the top is a silicon crystal that has been doped to provide specific electronic characteristics. The middle row shows this crystal, alongside a wafer cut from it, a wafer after polishing, and a wafer that has undergone fabrication steps similar to Fig. B.3 but for the entire circuit into hundreds of identical ICs. The bottom row shows a collection of chips separated from the wafer, followed by three common packages: a dual in-line package (DIP), a flatpack (F/P), and a TO-99 can, respectively. The number of leads in these packages vary from 3 to 22. They are available in either hermetically sealed ceramic or plastic versions. Two IC wafers of

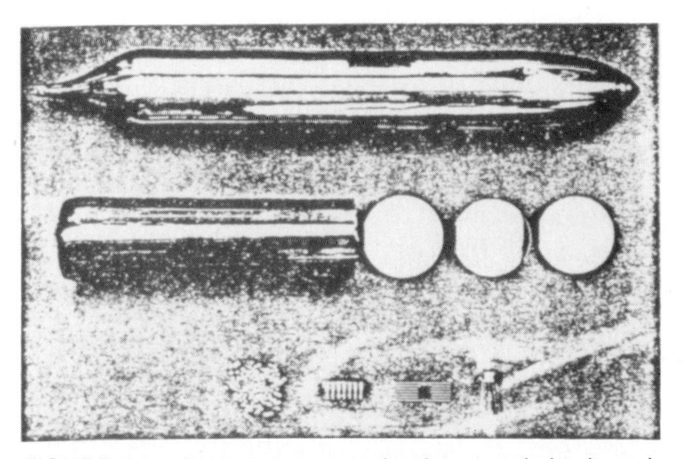

FIGURE B.7 From grown crystal to integrated circuit: various stages in processing. (*Courtesy of Fairchild Semiconductor.*)

(*a*)

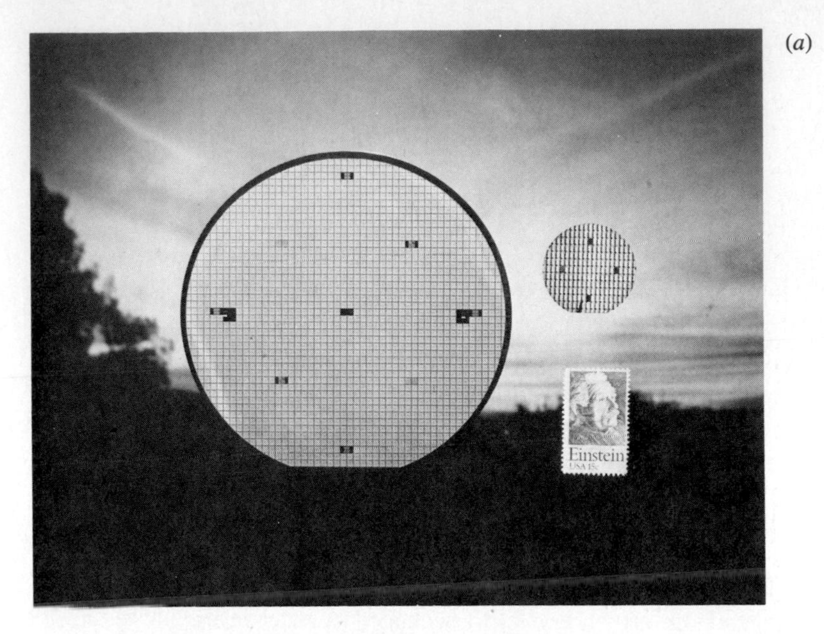

(*b*)

24-Lead Ceramic Flat Pack

TO-5 Style Package
with Straight Leads

8, 10, and 12-lead versions

Dual-In-Line Welded-Seal Ceramic Packages

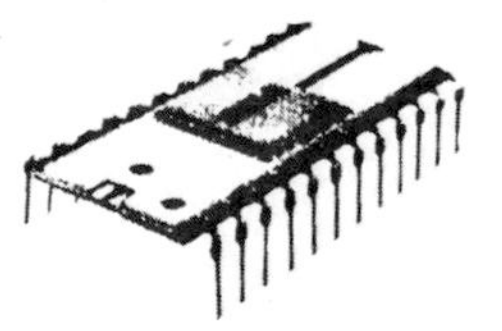

16- and 24-lead versions

Plastic Dual-In-Line Packages

16-, 18-, 22-, 24-, 28-, and 40-lead versions

FIGURE B.8 (*a*) Photograph of an IC wafer. (*Courtesy of National Semiconductor.*) (*b*) IC package configurations. (*Courtesy of RCA.*)

different size are shown in Fig. B.8*a*, note the large number of chips on each wafer. Figure B.8*b* shows the common IC package configurations just mentioned.

Note that in a 5-in wafer there can be as many as 2000 chips if the individual chips are 100 mil on a side. Assuming a yield of 75 percent there are a total of 1500 chips. The number of devices on a chip can vary, of course, and literally hundreds of thousands of MOS devices can be fabricated on a chip. Thus the cost of individual devices or chips is small, and the package cost is higher than that of the chip.

Finally, it should be noted that the base width of an IC transistor is of the order of 0.5 μm; thus, from (2.34), the cutoff frequency of an IC transistor can reach the gigahertz region and super-β transistors can be manufactured with base width approximately 0.2 μm. At this time, high-quality fast *pnp* transistors cannot be fabricated on the same chip with *npn* devices. Typical values of f_T for *pnp* devices are under 50 MHz, while the f_T of *npn* transistors is typically several hundred megahertz. With lateral and substrate *pnp*, β of 100, i.e., compatible with those of *npn*, are now achieved. Hence *pnp* transistors can be used where ultimate speed is not necessary. For very fast circuits, however, all IC transistors are invariably *npn*.

B.2 MOS DEVICES

The basic fabrication process of a MOSFET is very simple in IC technology and requires only the diffusion of two *n* regions on a *p* substrate for NMOS devices (or a *n* substrate with two *p* regions for a PMOS device), as shown in Fig. B.9. The heavily doped regions form the source and drain contacts. The insulating layer, i.e., SiO_2, Si_3N_4, or Al_2O_3, is then grown, and aluminum metallization completes the fabrication. Silicon deposited over the gate oxide in an epitaxial system is not a single crystal. It is composed of many crystals, and the layer is called *polycrystalline* silicon. Its electrical function in MOS system is similar to metal and is commonly used in MOS system. In this type of FET, no channel is actually fabricated. When the gate-source potential is positive, the so-called *enhancement mode*, an *n* channel, is induced between the source and the drain. The enhancement MOS is a normally off device.

The MOSFET, in addition to requiring a smaller number of fabrication steps, low power dissipation also requires less area on the chip than the conventional IC bipolar transistor and is, therefore, attractive in this respect as well. Typical chip area used by a MOSFET is less than 5 mil^2, which is smaller than that of a BJT by a factor of 3 to 5. In other words, MOS technology can create circuits with more than 100

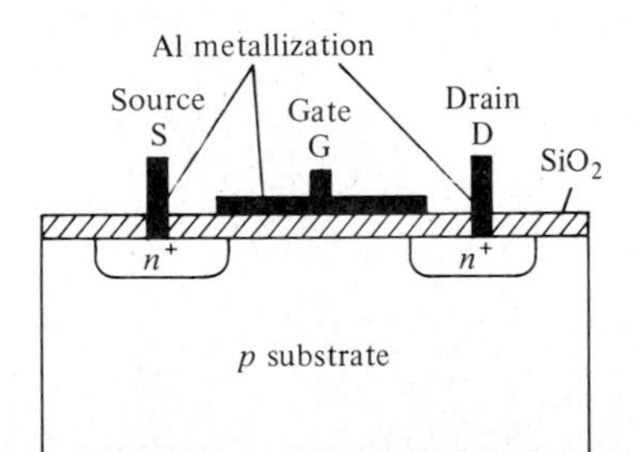

FIGURE B.9 An *n*-channel MOSFET of the enhancement type.

times as many devices per chip as the bipolar technology. Complementary MOS devices, i.e., NMOS and PMOS together, are also possible. One such arrangement, called CMOS is shown in Fig. 3.23. These devices dissipate extremely low power and have excellent temperature stability. Scaled-down versions of MOS devices (capable of high speed and low power dissipation) that operate at low supply voltages are also available under the name of high-performance MOS or HMOS.

B.3 PARASITICS IN IC

In IC analysis and design, one has to be careful with parasitic diodes and capacitances. For example, two transistors fabricated in one chip, as shown in Fig. B.10*a*, approximate the circuit model shown in Fig. B.10*b*. Note that, although the *p* substrate isolates the two transistors, the collectors of the two transistors form two back-to-back diodes with the *p* substrate. Note also that the substrate must always be connected to the most negative potential in the circuit to ensure that the isolation diodes are always back-biased, thus yielding very high impedance isolation between two transistors to maintain correct circuit operation. The presence of the low-resistivity n^+ buried layer in Fig. B.3*d* shunts the collector material and reduces the series collector resistance, r_{sc} to typically 10 Ω or less, and thus r_{sc} can, in most cases, be ignored if a buried layer is present. Because the small value of r_{sc} also reduces the transistor's apparent saturation voltage, the buried layer is almost always fabricated in an IC transistor. Finally, note the presence of the parasitic capacitances and diodes may have to be included in a complete and more accurate computer-aided analysis and design. For example, in switched capacitor filters, MOS capacitors are very small (fraction of a picofarad to several picofarads); hence parasite capacitors cannot be ignored.

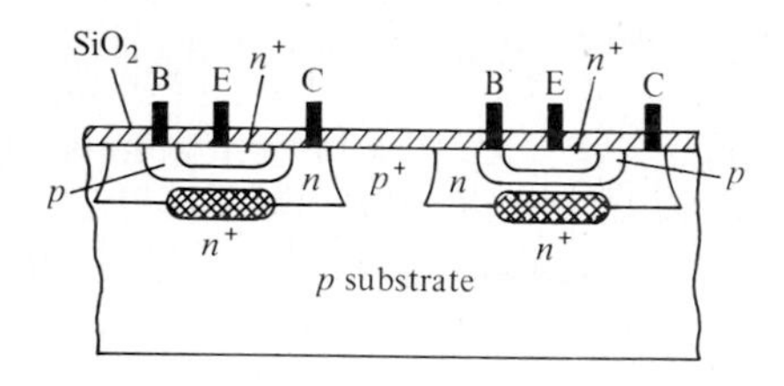

(*a*)

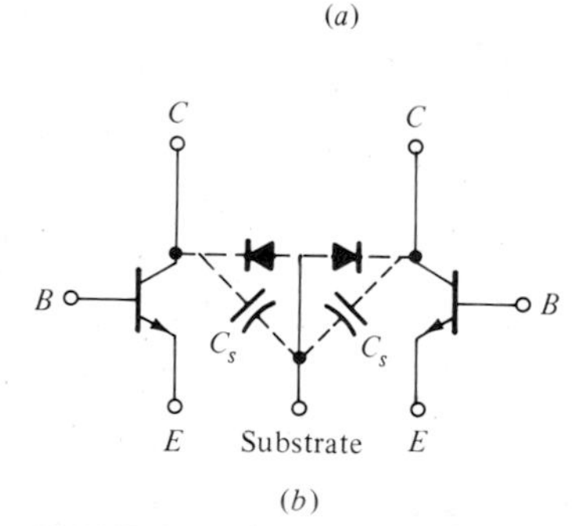

(*b*)

FIGURE B.10 (*a*) IC transistor. (*b*) Parasitics in IC transistors.

B.4 IC DIODES

IC diodes are usually constructed by proper interconnection of the IC transistor terminals. There are five combinations of diodes possible from a transistor, as shown in Fig. B.11*a–e*. Among these, the configurations in Fig. B.11*a* and *b* are commonly used. The reverse-breakdown voltage of these diodes is low (approximately 5 to 7 V). In fact, this property can be utilized to obtain a zener diode behavior in the reverse direction. A temperature-compensated zener diode is obtained as shown in Fig. B.12. The configuration in Fig. B.11*c* may be used for a higher breakdown since the breakdown voltage of the collector base is usually high ($\geq$ 15 V). It is to be noted that a collector-base diode with no emitter diffusion is also used in practice. This type of diode has a high breakdown voltage, small area, and low capacitance. It should also be noted that different diode configurations have different storage-time characteristics. The diode in Fig. B.11*a* provides the smallest storage time by an order of magnitude than the other configurations.

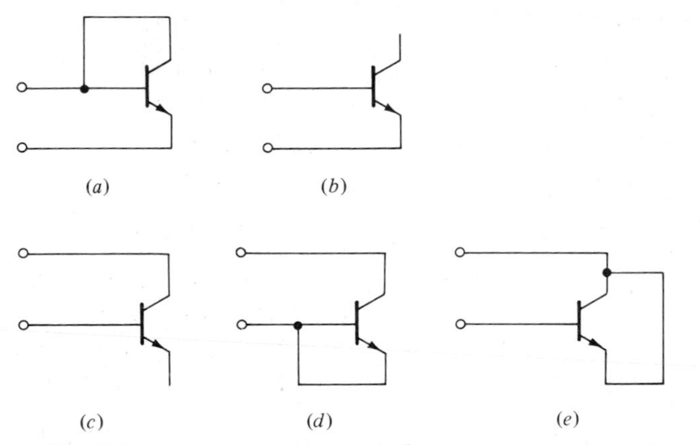

FIGURE B.11 Various forms of IC diodes as obtained from IC transistors.

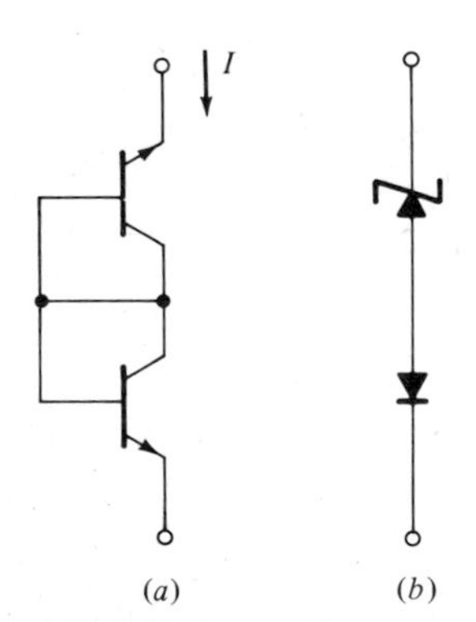

FIGURE B.12 Temperature-compensated zener diode: (*a*) IC form and (*b*) circuit symbol.

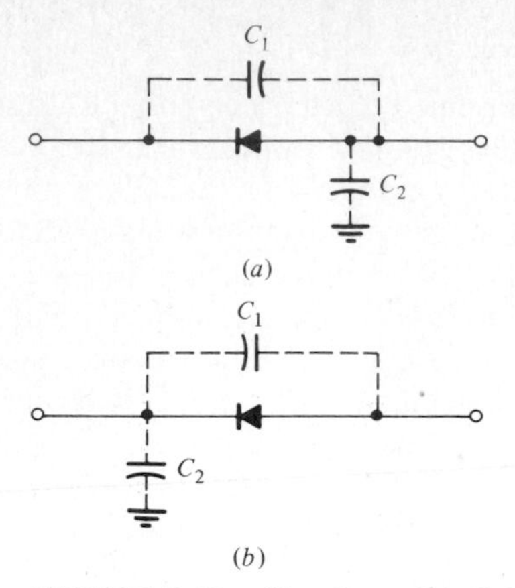

FIGURE B.13 Circuit model of an IC diode, showing parasitic capacitances (*a*) for diodes in Fig. B.11*a*, *b* and (*b*) for diodes in Fig. B.11*c–e*.

The circuit model of an IC diode, which includes the capacitance of the diode and of the reverse-biased parasitic diodes for Fig. B.11*a* and *b*, is shown in Fig. B.13*a* and in Fig. B.13*b* for the diodes in Fig. B.11*c–e*.

B.5 SCHOTTKY BARRIER DIODE, TRANSISTOR, AND FET

Schottky barrier diodes (SBD) are formed by a contact between a suitable metal and an *n*-type semiconductor. The IC structure and the symbol for SBD are shown in Fig. B.14*a* and *b*. The aluminum and the *n*-type material form a rectifying contact, and the metal and the heavily doped n^+ region form the ohmic contact. To improve yield, sometimes platinum silicide is used between the aluminum and the *n*-type semiconductor. The *V-I* characteristic of SBD is similar to the *pn*-junction diodes but with the following *two-important differences*. First, the device, under low-level injection, operates as a majority carrier, and thus the minority storage time is eliminated. Hence a faster response time can be achieved. Second, the forward voltage drop of SBD is much lower than those of *pn*-junction diodes and usually around 0.3 V instead of 0.7 V. Schottky diodes are used as clamps to speed up the switching behavior of BJT (see Sec. 12.9). The SBD clamp between the base and collector prevents the

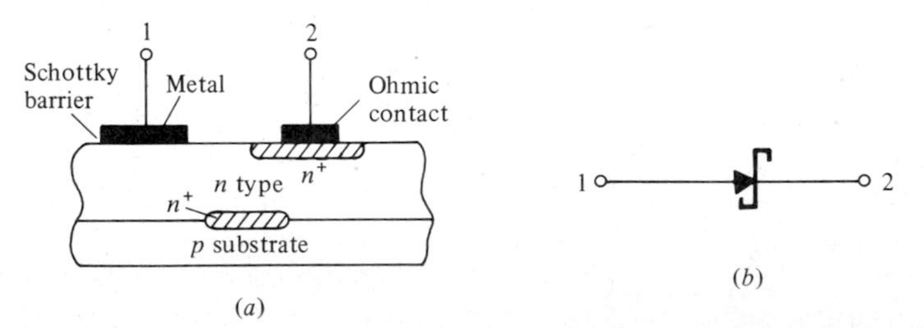

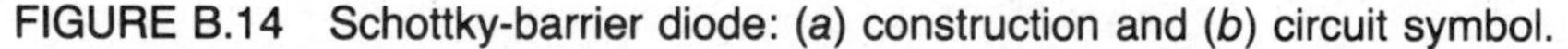
FIGURE B.14 Schottky-barrier diode: (*a*) construction and (*b*) circuit symbol.

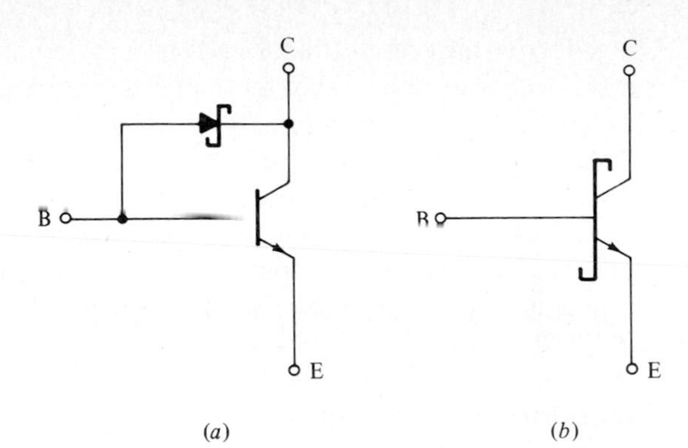

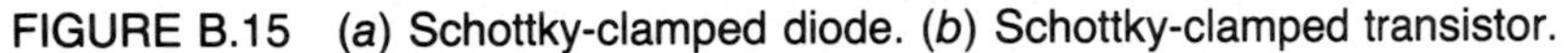
FIGURE B.15 (*a*) Schottky-clamped diode. (*b*) Schottky-clamped transistor.

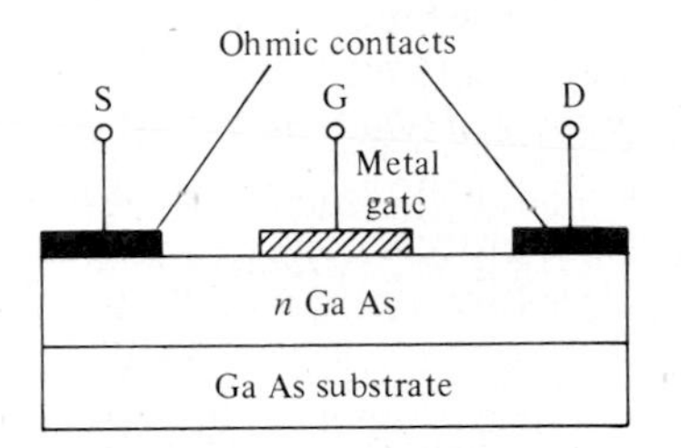

FIGURE B.16 Shottky-barrier FET (MESFET).

transistor to go into saturation. Such a circuit configuration, shown in Fig. B.15*a*, is also drawn as in Fig. B.15*b* and is referred to as the Schottky-clamped transistor.

Schottky barrier is also used to replace the *pn*-junction gate to realize a shorter channel length. From (3.44) it is seen that the frequency performance of the device is limited by the channel length and the mobility of electrons in the material. If GaAs is used as the material, which has a mobility of about 6 times that of silicon, the higher mobility and shorter channel can be used to obtain very-high-frequency performances. The Schottky barrier FET is shown in Fig. B.16, where the Schottky barrier is formed on the top surface. This structure is also called MESFET, which stands for metal semiconductor field-effect transistor. These devices can operate at frequencies up to several gigahertz.

B.6 IC CAPACITORS

There are many types of capacitors used in IC technology. Among these are the depletion capacitors of *pn* junctions and the thin-film capacitors.

The junction capacitors are the easiest to form in integrated circuits. These polarized capacitors are voltage dependent and are used primarily as decoupling and bypass capacitors. This type of capacitor has a series resistance of the bulk semiconductor resistivity associated with it that may have to be included in a careful circuit analysis. Typical values of a junction capacitor are about 300 pF/mm^2, with a tolerance of 30 percent, or higher. Thus, if an accurate capacitor of large value is required, it is made in thin-film form. Capacitors of the order of picofarads occupy larger areas in an IC than those of BJTs and FETs. For example, note the real estate occupied by the capacitor $C_1 = 30$ pF in 741 op amp as shown in Fig. B.1.

The thin-film, specifically MOS, capacitors are formed as parallel-plate capacitors by the n^+ region of the emitter and a metal film separated by the silicon dioxide dielectric, as shown in Fig. B.17. Because of the metal oxide and semiconductor layers, it is called an MOS capacitor. The value of the capacitor is proportional to its area, such as in parallel-plate capacitors. The capacitance of this device is typically between 300 to 600 pF/mm^2 (0.2 to 0.4 pF/mil^2). The MOS capacitors are *voltage-independent* and nonpolarized. The series resistance of the MOS capacitor is lower as compared to that of the junction capacitor because of the low resistivity of the n^+ emitter region forming the lower plate. MOS capacitors are also formed in metal-gate NMOS fabrication. An MOS capacitor is extremely linear and has a low temperature coefficient. Ratios of MOS capacitors can be held very accurately to within 1 percent. They are very attractive in switched-capacitor filters, among others.

Larger values of capacitors can be obtained by sandwiching the silicon dioxide with layers of aluminum to form the top and the bottom plates. These types of thin-film capacitors require additional steps in the fabrication process, but the capacitance can be as high as 2000 pF/mm^2 with the lowest power dissipation among those mentioned above.

Note that, in all cases, the maximum size of the capacitors is quite limited. The maximum value of an IC capacitor is under 200 pF, while for monolithic and MOS, IC the maximum value is under 30 pF. This is because of the area constraints in IC and the breakdown limitations. This is why direct-coupled amplifiers are used in the design of IC amplifiers for ac applications, since bypass and coupling capacitors of the order of microfarads are entirely impractical in integrated circuits.

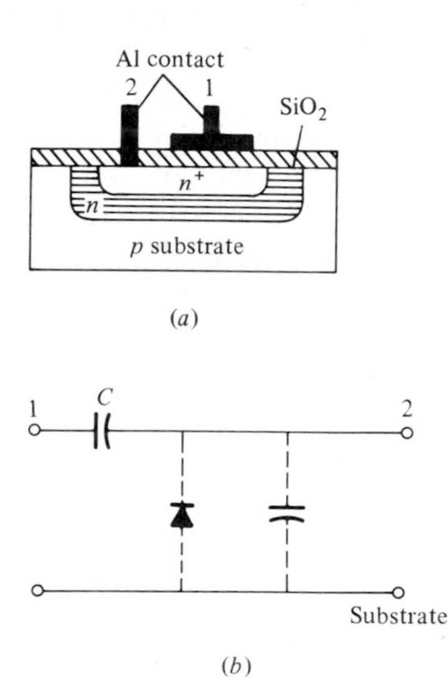

FIGURE B.17 MOS Capacitor: (*a*) construction and (*b*) circuit model.

B.7 IC RESISTORS

Similar to IC capacitors, there are many types of resistors in IC technology. Among these are MOS resistors, diffused resistors and thin-film resistors. MOS resistors shown in Fig. B.18*a* are formed in source drain diffusion.

The diffusion resistors are formed during the *p*-type base diffusion, as shown in Fig. B.18*b*. The approximate expression for the resistance, ignoring fringing effects, is

$$R = \frac{\rho l}{dw} \tag{B.1}$$

where d is the depth of the diffused p region and is usually thin. The resistance of a thin sheet of material is often specified by its sheet resistance ($R_\square$) expressed in ohms per square, namely,

$$R_\square = \frac{\rho}{d} \text{ ohms/square} \tag{B.2}$$

hence

$$R = R_\square \frac{l}{w} \text{ ohms} \tag{B.3}$$

If $l = w$, then $R = R_\square$ ohms.

Practical values of diffused resistors range from approximately 50 Ω to 20 kΩ. For the lower range, i.e., small values of resistance, the emitter diffusion ($R_\square \simeq 2$ Ω/square) instead of the base diffusion ($R_\square \simeq 200$ Ω/square) is used. The range of MOS resistors is from 100 to 200 Ω/square. Manufacturing tolerances of diffused resistors are poor (about 20 to 50 percent) in absolute value, whereas the ratios of resistors can be held to under 5 percent. This is why one should strive to

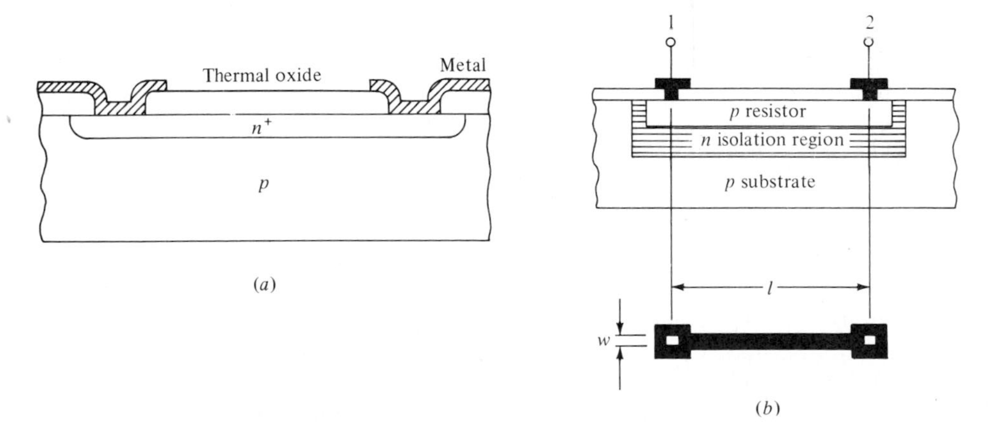

FIGURE B.18 (*a*) MOS resistor. (*b*) Diffused resistor.

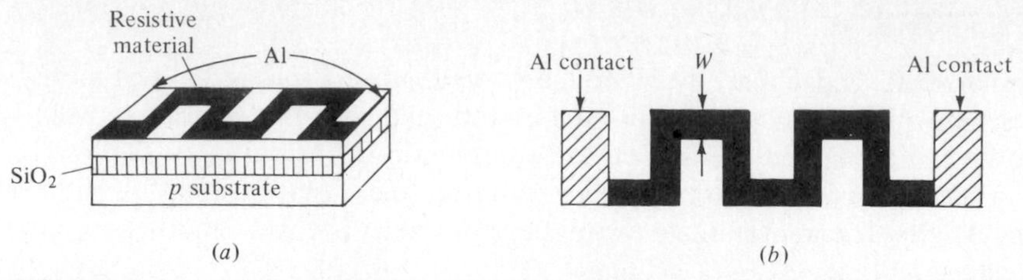

FIGURE B.19 (*a*) Construction of a large-valued resistor. (*b*) Top view of (*a*).

design the IC circuit such that the performance depends on resistor ratios rather than absolute values. Again notice the presence of the parasitic transistor formed by the *pnp* layers and the junction capacitors in Fig. B.18. Resistor ratios in diffused resistors cannot be held to the same tolerance as the ratio of MOS capacitors. Furthermore, temperature coefficients of resistors are not as good as those of MOS capacitors. Thus in MOS technology one strives to obtain a design based on ratio of capacitors.

Thin-film resistors can be fabricated in the form shown in Fig. B.19, where the resistive material may be nichrome, tin oxide, tantalum nitride, or other resistive film. Large and small values of resistance are obtained by using tin oxide and tantalum nitride, respectively. The sheet resistance of the various materials ranges from 10 to 20,000 Ω/square. Note that for a given resistive material, the large-value resistor from (B.2) requires a large number of squares, hence a large ratio of l/w. This can be achieved by reducing w and/or increasing l. The smallest value of w is limited by photographic resolution or power dissipation $w_{\min}$ is about 0.2 mil (0.0002 in); the largest value of l is limited by the area of the chip. To get the largest l, one usually designs a zigzag or serpentine pattern in order to get a large number of squares, i.e, a large l/w ratio as shown in Fig. B.19*b*.

B.8 IC SIMULATION OF INDUCTORS

No practical range of inductor values can be fabricated in IC at the present time. In integrated-circuit design, one usually tries to avoid using inductors. Apparent inductors can be obtained by active *RC* circuits such as using op amps with *RC* elements to obtain the so-called generalized impedance converter (see Sec. 9.6). If an inductor or a transformer must be used, a discrete component is connected externally to the hybrid integrated circuit.

B.9 VERY-LARGE-SCALE INTEGRATION (VLSI)

Very-large-scale integration (VLSI) represents the process of fabricating a very large number ($\geq 10^5$) of active and passive components on a silicon chip. These components usually NMOS, CMOS, or I^2L (integrated injection logic discussed in Sec. 13.11) devices are interconnected in such a way as to perform a multitude of circuit functions and thus represent a complete subsystem. The component density, i.e., the number of active and passive components per square inch on the silicon chip is very large in VLSI technology. A complete subsystem is now routinely fabricated on a chip such as memory, microcomputer, etc.

Figure B.20 shows a Texas Instrument voice-synthesis processor, a device measuring less than $\frac{1}{4}$-in. square. The chip uses digital-speech data to electronically duplicate the human vocal tract, producing a digital signal which can be converted to an analog signal and then amplified to produce synthetic speech.

Solid-state technology is improving rapidly and very complex subsystems are being made on chips with a good yield. Usually the larger the component density the lower the yield. For this reason the maximum possible component density is intentionally not used on the chip. In fact, in 1981 an NMOS computer chip with 450,000 transistors on a chip, only 0.4 cm^2 in area, was achieved commercially. Even higher density and yield in VLSI production are expected in the future. In the next couple of years 1 million device per chip is a definite possibility. The problems of analysis, design, and testing in VLSI and LSI are indeed formidable. The impact of LSI and VLSI on the development of micro-, mini-, and large computers is profound, and the use of computers in the analysis and design of VLSI and LSI is a necessity.

In integrated circuits, macromodeling and systems approach are used. In LSI, instead of focusing on the various properties of a device, one looks at the important

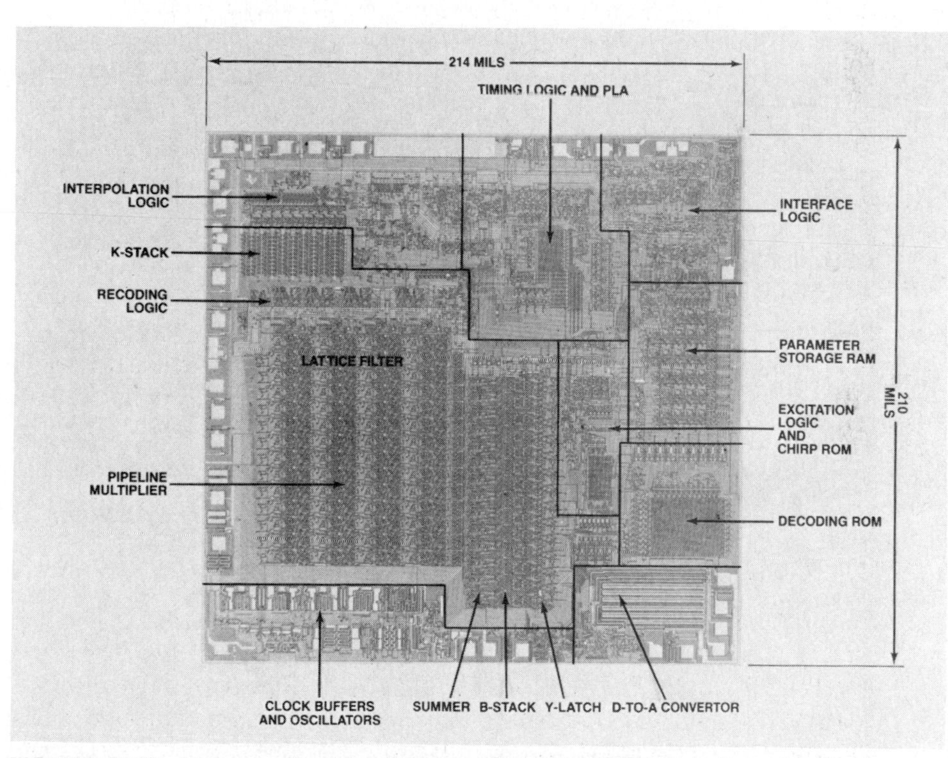

FIGURE B.20 Voice synthesis processor. (*Courtesy of Texas Instruments, Inc.*)

TABLE B.1 Important Systems Parameters in VLSI*

Parameter	1978	19xx
Minimum feature size	6 μm	0.3 μm
Transit time, τ	0.3 to 1 ns	$\simeq$0.02 ns
Switching energy per device, E_{sw}	$\simeq 10^{-12}$ J	$\simeq 2 \times 10^{-16}$ J
System clock period, T	30–50 ns	$\simeq$2–4 ns

*From Ref. 4, with permission.

system parameters. For example, Table B.1 shows the important system parameters for current technology and for future technology. The limits are, of course, imposed by physical laws. For example, the transit time τ is given by

$$\tau = \frac{(\text{distance})^2}{(\text{mobility})(\text{voltage})} \tag{B.4}$$

The parameters that govern the transit time are listed in Table B.2.

The progress of technology from microfabrication to submicrofabrication leads to smaller and smaller devices. The concept of scaling and structuring the design of VLSI to handle these was introduced by C. Mead and L. Conway [4].

Table B.3 lists the complexity and examples of small-scale integration (SSI), medium-scale integration (MSI), large-scale integration (LSI) and very-large-scale integration (VLSI) for current technology. The component counts are transistors, resistors, and capacitors. As seen in the table the higher density of components are in microcomputers, microprocessors, read-only memories (ROMs), programmable ROM (generally referred to as PROM), erasable PROM (called EPROM), random access memories (RAMs), and analog-to-digital (A/D) and digital-to-analog (D/A) converters. The density and power requirements at LSI and VLSI necessitate the use of NMOS, CMOS, and I^2L devices. For very-high-speed integrated circuits (VHSIC), bipolar transistors are used as emitter-coupled logic (ECL) gates or advanced Schottky transistor-transistor logic (TTL). For very low power dissipation, CMOS and NMOS gates are used. Integrated injection logic (I^2L) gates have the lowest power-delay

TABLE B.2 Parameters that Govern Transit Time*

Parameter	MOSFET	MESFET, JFET	BJT
Distance, μm	Channel length	Channel length	Base width
Voltage, V	$\simeq \frac{V_{DD}}{2}\left(\text{many }\frac{kT}{q}\right)$	$\simeq \frac{V_{DD}}{2}\left(\text{many }\frac{kT}{q}\right)$	$\frac{4kT}{q}$
Mobility, $cm^2/V \cdot s$	$\mu_n = 800$ (Surface mobility Si)	$\mu_n \simeq 1300$ (Bulk mobility Si)	$\mu_n \simeq 1300$ (Bulk mobility Si)

*From Ref. 4, with permission.

TABLE B.3 Complexity of Integration with Examples

Complexity	Examples
VLSI ≥ 10^5 components*	Microcomputers, ROMs, RAMs, subsystems
LSI ≥ 10^4 components	Microprocessor, A/D and D/A converters, ROMs, RAMs, subsystems
MSI ≥ 10^3 components	Adders, complex gates, ROMs, RAMs
SSI ≥ 10 components	AND, OR, NAND, NOR gates, Buffers, memory cells, op amps

*Component are transistors, resistors, and capacitors.

product (PDP) and are generally used in LSI and VLSI. The performance characteristics of the various gates are discussed in Sec. 13.12 (see Fig. 13.34).

REFERENCES

1. W. C. Till and J. T. Luxton, *Integrated Circuits: Materials, Devices and Fabrication*, Prentice-Hall, Englewood Cliffs, N.J., 1982.
2. D. J. Hamilton and W. G. Howard, *Basic Integrated Circuit Engineering*, McGraw-Hill, New York, 1975.
3. R. S. Muller and T. I. Kamins, *Device Electronics for Integrated Circuits*, Wiley, New York, 1977.
4. C. Mead and L. Conway, *Introduction to VLSI Systems*, Addison-Wesley, Reading, Mass., 1980.

APPENDIX C

Two-Port Network Properties

In this appendix we shall briefly review the basic properties of two-port networks. A two-port network is a circuit that has only *two pairs* of accessible terminals. A pair of terminals in which the current leaving one terminal is equal to the current flowing into the other terminal is called a *port*. Two-port networks constitute a very important class of networks in signal processing. We often encounter two-port systems where one port represents the input and the other the output.

The discussion and results that follow are quite general, regardless of the nature of the two-port networks, provided the networks are *linear and time invariant*.

All active electronic devices, such as bipolar transistors, field-effect transistors, semiconductor diodes, are nonlinear. Under small-signal conditions, however, nonlinear devices can be adequately approximated by linear models. In a number of applications, such as the amplification of video and audio signals, linearity of signal processing must be maintained, and in such applications, the nonlinear device is deliberately operated and analyzed under small-signal conditions.

The Laplace transform is useful and convenient for the analysis of linear time-invariant networks, and we shall, therefore, use the transform domain throughout this chapter. The transform quantities are denoted by capital letters. If the reader is not exposed to the Laplace transform by this time, it is suggested that the capacitance and inductance be replaced by an open circuit and a short circuit, respectively, for low frequencies, and thus the two-port network will be all *resistive with real parameters*.

C.1 TWO-PORT CHARACTERIZATIONS

A linear time-invariant two-port network is shown symbolically by the box marked with the conventional reference directions for the voltages and currents in Fig. C.1. There are no independent voltage or current sources within the box, and the network is in the zero state, i.e., the initial conditions are set to zero. There are four possible types of variables in such a two-port network: currents I_1 and I_2 and voltages V_1 and V_2. Among these four variables, there are only two independent variables and two linear constraints imposed by superposition. There are thus six ways of picking two quantities out of a set of four quantities; thus, six two-port descriptions are possible, depending on the choice of the dependent and independent quantities. The choice of course depends on the problem at hand. The various possibilities are as follows.

A. Short-Circuit Admittance Parameters (y_{ij})

Consider V_1 and V_2 to be the *independent* variables; i.e., the network in Fig. C.1 is excited by two independent voltage sources. We wish, therefore, to determine the currents I_1 and I_2.

For a general linear time-invariant two-port network, we can write, by superposition,

$$I_1 = y_{11}V_1 + y_{12}V_2 \tag{C.1a}$$

$$I_2 = y_{21}V_1 + y_{22}V_2 \tag{C.1b}$$

Equations (C.1*a*) and (C.1*b*) can be put in matrix form:

$$\begin{bmatrix} I_1 \\ I_2 \end{bmatrix} = \begin{bmatrix} y_{11} & y_{12} \\ y_{21} & y_{22} \end{bmatrix} \begin{bmatrix} V_1 \\ V_2 \end{bmatrix} \tag{C.1c}$$

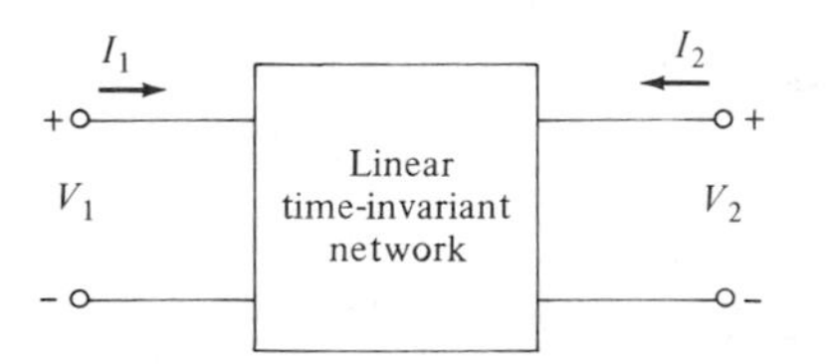

FIGURE C.1 Linear time-invariant two-port network.

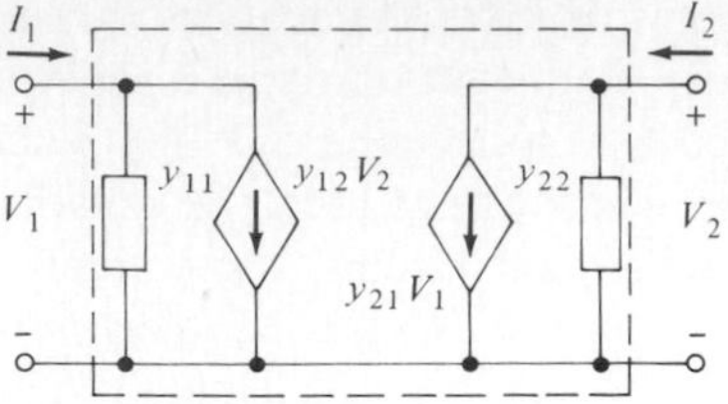

FIGURE C.2 Norton equivalent of a two-port network.

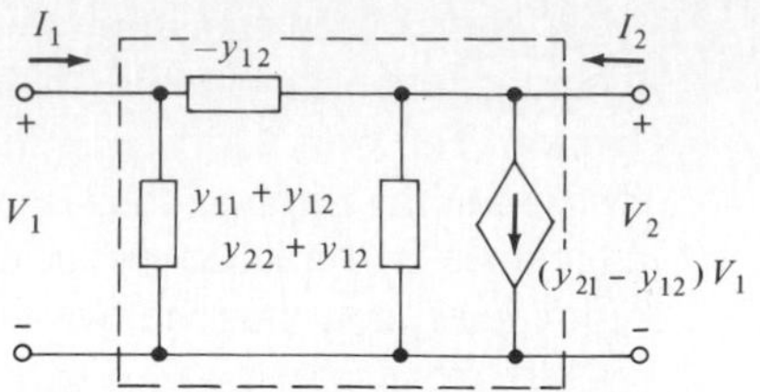

FIGURE C.3 The π equivalent of an active two-port network.

where

$$y_{11} = \left.\frac{I_1}{V_1}\right|_{V_2=0} \qquad y_{12} = \left.\frac{I_1}{V_2}\right|_{V_1=0}$$
$$y_{21} = \left.\frac{I_2}{V_1}\right|_{V_2=0} \qquad y_{22} = \left.\frac{I_2}{V_2}\right|_{V_1=0} \tag{C.2}$$

The coefficients $y_{ij}(s)$ are called the *short-circuit admittance parameters*. Notice that the short circuit is applied to the port at which the excitation is *not* being applied.

Equations (C.1) may be represented by the equivalent circuit with two dependent current sources, as shown in Fig. C.2. The circuit in Fig. C.2 is a Norton equivalent at both ports.

By adding and subtracting $y_{12}V_1$ in (C.1*b*), we obtain

$$I_2 = y_{12}V_1 + y_{22}V_2 + (y_{21} - y_{12})V_1 \tag{C.3}$$

From (C.3) and (C.1*a*), an alternate network representation with one dependent current source, usually referred to as the *π equivalent circuit,* is obtained as shown in Fig. C.3.

In circuit analysis, the y parameters of a network can sometimes be directly obtained by using Figs. C.2 or C.3.

EXAMPLE C.1

Let us calculate the short-circuit admittance parameters of the network shown in Fig. C.4. Using the definition of y_{11} from (C.2), $V_2 = 0$ means that the terminals 2-2 are short-circuited. Thus $y_{11} = G_i + s(C_\pi + C_c)$ where $G_i = 1/R_i$. Alternatively, from a comparison of Fig. C.4 and Fig. C.3, we have

$$y_{12} = -sC_c$$

$$y_{11} + y_{12} = G_i + sC_\pi$$

hence

$$y_{11} = G_i + s(C_\pi + C_c)$$

$$y_{22} + y_{12} = G_o$$

hence

$$y_{22} = G_o + sC_c$$

$$y_{21} - y_{12} = g_m$$

hence

$$y_{21} = g_m - sC_c$$

where $G_0 = R_0^{-1}$. Note that $y_{12} \neq y_{21}$ for this circuit. The network is, therefore, not reciprocal. For *passive networks* (i.e., *RLC*) we always have $y_{12} = y_{21}$.

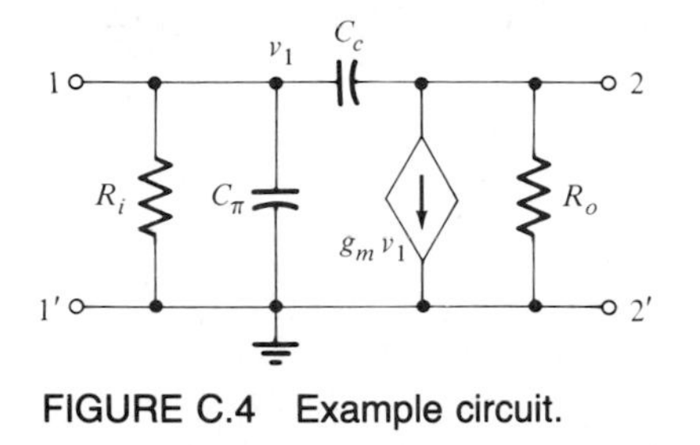

FIGURE C.4 Example circuit.

B. Open-Circuit Impedance Parameters (z_{ij})

Let us consider I_1 and I_2 as the *independent* variables (i.e., the network of Fig. C.1 excited by two independent current sources) and V_1 and V_2 as the responses to be determined. The circuit responses in terms of the excitations, again using the superposition principle, can be written in the following matrix form:

$$\begin{bmatrix} V_1 \\ V_2 \end{bmatrix} = \begin{bmatrix} z_{11} & z_{12} \\ z_{21} & z_{22} \end{bmatrix} \begin{bmatrix} I_1 \\ I_2 \end{bmatrix} \tag{C.4}$$

where z_{ij} are called the *open-circuit impedance parameters* since

$$z_{11} = \left.\frac{V_1}{I_1}\right|_{I_2=0} \qquad z_{12} = \left.\frac{V_1}{I_2}\right|_{I_1=0}$$
$$z_{21} = \left.\frac{V_2}{I_1}\right|_{I_2=0} \qquad z_{22} = \left.\frac{V_2}{I_2}\right|_{I_1=0} \tag{C.5}$$

Note that the open circuit is at the port at which excitation is *not* being applied.

Equation (C.4) may be represented by the equivalent circuit shown in Fig. C.5. The circuit has a Thevenin equivalent representation at both ports. By adding and subtracting $z_{12}I_1$ in the expression for V_2 in (C.4), we obtain

$$V_2 = z_{12}I_1 + z_{22}I_2 + (z_{21} - z_{12})I_1 \tag{C.6}$$

From (C.6) and (C.4), an alternate network representation with one dependent voltage source, referred to as the *T-equivalent* circuit of a two-port, is obtained, as shown in Fig. C.6. The use of Fig. C.6 sometimes yields the z parameters directly.

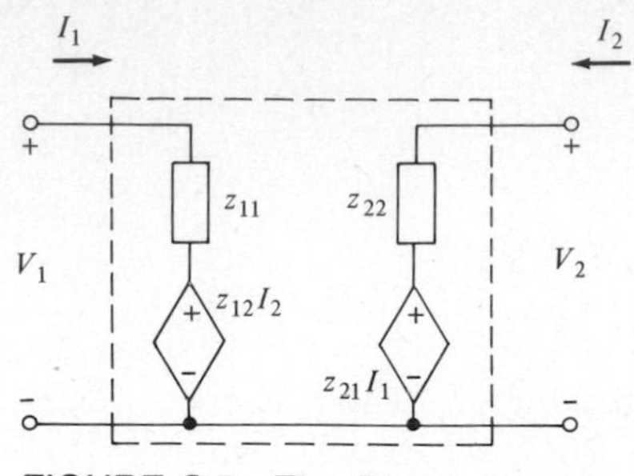

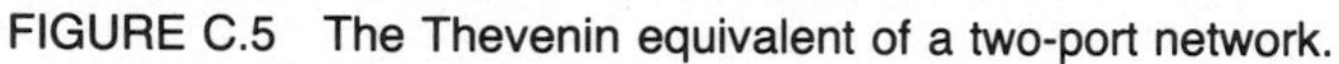
FIGURE C.5 The Thevenin equivalent of a two-port network.

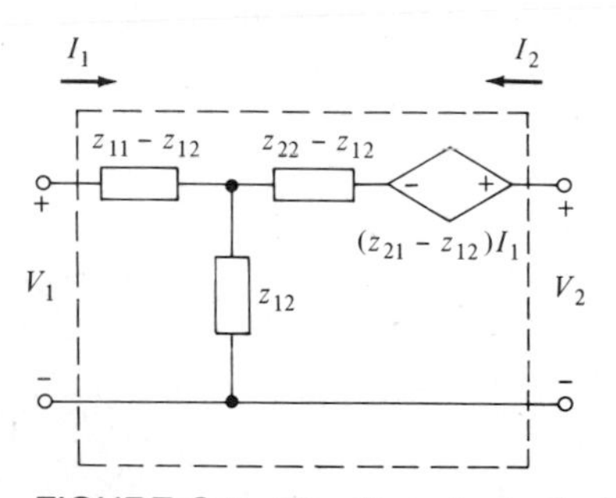

FIGURE C.6 The T equivalent of an active two-port network.

EXAMPLE C.2

Let us calculate the open-circuit impedance parameters of the circuit shown in Fig. C.7. Using the definition of z_{11} from (C.5), setting $I_2 = 0$ is equivalent to the terminals 2-2′ being open-circuited. Thus $z_{11} = r_e \parallel (1/sC_e) + r_b$, where $\parallel$ denotes "in parallel with." Alternatively, from a comparison of Fig. C.7 with Fig. C.6, we can write by inspection

$$z_{12} = r_b$$

$$z_{22} - z_{12} = Z_c$$

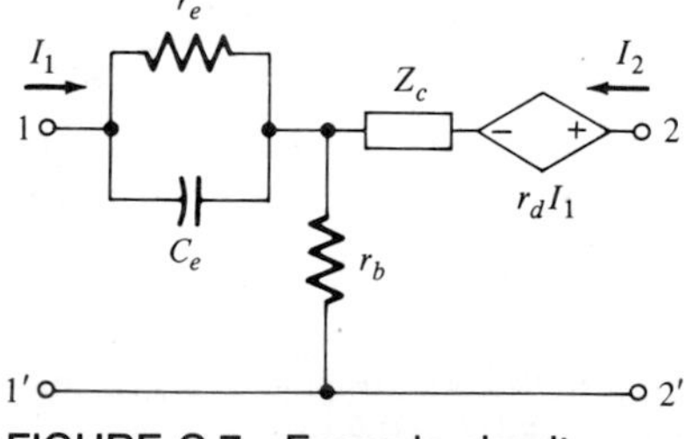

FIGURE C.7 Example circuit.

hence

$$z_{22} = r_b + Z_c$$

$$z_{21} - z_{12} = r_d$$

hence

$$z_{21} = r_d + r_b$$

$$z_{11} - z_{12} = \frac{r_e}{1 + r_e C_e s}$$

hence

$$z_{11} = \frac{r_e}{1 + r_e C_e s} + r_b$$

Note that the z-parameters can be obained directly from the y-parameters if det $[y_{ij}] \neq 0$. Since

$$\begin{bmatrix} I_1 \\ I_2 \end{bmatrix} = \begin{bmatrix} y_{11} & y_{12} \\ y_{21} & y_{22} \end{bmatrix} \begin{bmatrix} V_1 \\ V_2 \end{bmatrix} \tag{C.7}$$

we can solve for V_1 and V_2 in terms of I_1 and I_2. The solutions are, by Cramers rule [1, 2],

$$V_1 = \frac{\begin{vmatrix} I_1 & y_{12} \\ I_2 & y_{22} \end{vmatrix}}{\begin{vmatrix} y_{11} & y_{12} \\ y_{21} & y_{22} \end{vmatrix}} = \frac{y_{22}}{\Delta y} I_1 - \frac{y_{12}}{\Delta y} I_2 \tag{C.8a}$$

$$V_2 = \frac{\begin{vmatrix} y_{11} & I_1 \\ y_{21} & I_2 \end{vmatrix}}{\Delta y} = \frac{y_{11}}{\Delta y} I_2 - \frac{y_{21}}{\Delta y} I_1 \tag{C.8b}$$

where Δy is the determinant of the admittance parameters, namely,

$$\Delta y = y_{11} y_{22} - y_{12} y_{21} \neq 0 \tag{C.9}$$

Hence, from (C.4) and (C.8a)

$$\begin{aligned} z_{11} &= \frac{y_{22}}{\Delta y} \qquad & z_{12} &= -\frac{y_{12}}{\Delta y} \\ z_{22} &= \frac{y_{11}}{\Delta y} \qquad & z_{21} &= -\frac{y_{21}}{\Delta y} \end{aligned} \tag{C.10}$$

Similarly, y_{ij} may be obtained in terms of z_{ij} if det $[z_{ij}] \neq 0$. Note that $[z] = [y]^{-1}$ and the product of $\Delta y \, \Delta z = 1$.

C. The Hybrid Parameters (h_{ij} and g_{ij})

If I_1 and V_2 are chosen as the *independent* variables, then V_1 and I_2 are the dependent variables. Note that we have mixed (current and voltage) dependent and independent quantities. In other words, the excitations are an independent current source at port 1 and an independent voltage source across port 2, while the dependent responses are V_1 and I_2. The h-parameter description has been used in Sec. 2.8 for bipolar transistors. In this case, we have

$$\begin{bmatrix} V_1 \\ I_2 \end{bmatrix} = \begin{bmatrix} h_{11} & h_{12} \\ h_{21} & h_{22} \end{bmatrix} \begin{bmatrix} I_1 \\ V_2 \end{bmatrix} \tag{C.11}$$

where

$$\begin{aligned} h_{11} &= \left.\frac{V_1}{I_1}\right|_{V_2=0} \qquad & h_{21} &= \left.\frac{I_2}{I_1}\right|_{V_2=0} \\ h_{12} &= \left.\frac{V_1}{V_2}\right|_{I_1=0} \qquad & h_{22} &= \left.\frac{I_2}{V_2}\right|_{I_1=0} \end{aligned} \tag{C.12}$$

Note that h_{11} has the dimension of impedance, h_{22} has the dimension of admittance, and h_{21} and h_{12} are dimensionless since they are ratios of currents and voltages, respectively. In particular, note that h_{21} is the *short-circuit current ratio*. An equivalent circuit in terms of the hybrid h-parameters is shown in Fig. C.8. In this characterization, note that the input is a Thevenin equivalent, while the output is a Norton equivalent representation.

One could obtain the h parameters in terms of y parameters or z parameters from (C.4) or (C.7). For example, solving for V_1 and I_2 in terms of I_1 and V_2 in (C.7), we have

$$\begin{bmatrix} V_1 \\ I_2 \end{bmatrix} = \begin{bmatrix} \dfrac{1}{y_{11}} & -\dfrac{y_{12}}{y_{11}} \\ \dfrac{y_{21}}{y_{11}} & \dfrac{\Delta y}{y_{11}} \end{bmatrix} \begin{bmatrix} I_1 \\ V_2 \end{bmatrix} \tag{C.13}$$

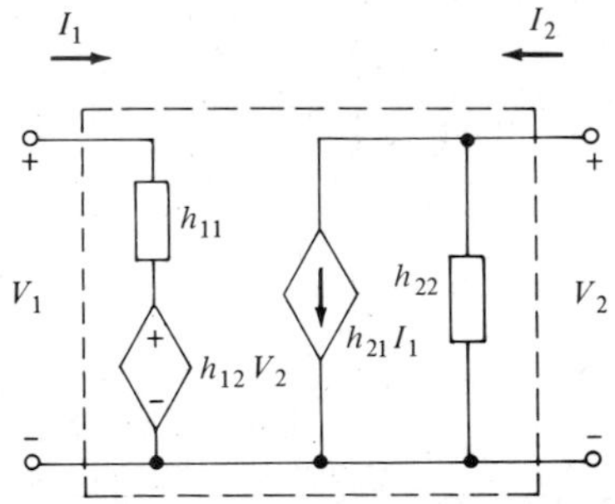

FIGURE C.8 Hybrid representation of a two-port (input Thevenin, output Norton equivalent).

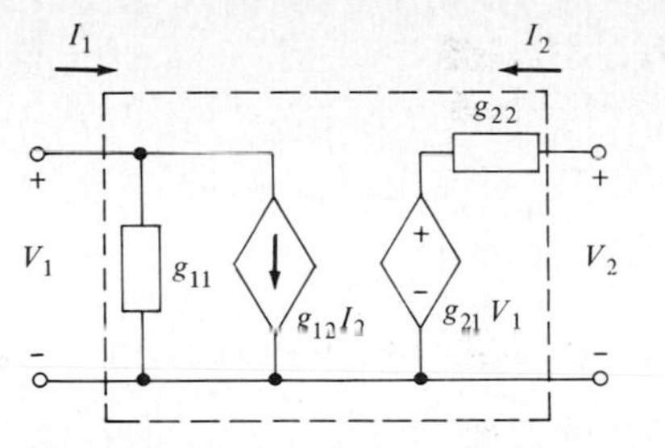

FIGURE C.9 Hybrid representation of a two-port (input Norton, output Thevenin equivalent).

hence

$$h_{11} = \frac{1}{y_{11}} \qquad h_{12} = -\frac{y_{12}}{y_{11}}$$
$$h_{21} = \frac{y_{21}}{y_{11}} \qquad h_{22} = \frac{\Delta y}{y_{11}} \tag{C.14}$$

Similarly, if V_1 and I_2 are chosen as the independent variables, then I_1 and V_2 are the dependent variables. In this case, we have the g-parameter description

$$\begin{bmatrix} I_1 \\ V_2 \end{bmatrix} = \begin{bmatrix} g_{11} & g_{12} \\ g_{21} & g_{22} \end{bmatrix} \begin{bmatrix} V_1 \\ I_2 \end{bmatrix} \tag{C.15}$$

where

$$g_{11} = \left.\frac{I_1}{V_1}\right|_{I_2=0} \qquad g_{21} = \left.\frac{V_2}{V_1}\right|_{I_2=0}$$
$$g_{12} = \left.\frac{I_1}{I_2}\right|_{V_1=0} \qquad g_{22} = \left.\frac{V_2}{I_2}\right|_{V_1=0} \tag{C.16}$$

Note that g_{11} and g_{22} have the dimensions of admittance and impedance, respectively, while g_{12} and g_{21} are dimensionless quantities. In particular, note that g_{21} is an *open-circuit voltage ratio*. An equivalent circuit, in terms of the hybrid g parameters, is shown in Fig. C.9. In this case, the input has a Norton equivalent while the output has a Thevenin equivalent circuit.

D. The Transmission (Chain) Parameters (*ABCD* and $\mathscr{ABCD}$)

The transmission parameters, also referred to as the chain parameters, relate the input and output quantities. If the independent variables are chosen as V_2 and $-I_2$, then the dependent parameters are V_1 and I_1. The minus sign preceding I_2 is used for convenience, as we shall see. The relationship between the dependent and the independent variables in this description is given by the *ABCD* parameters:

$$\begin{bmatrix} V_1 \\ I_1 \end{bmatrix} = \begin{bmatrix} A & B \\ C & D \end{bmatrix} \begin{bmatrix} V_2 \\ -I_2 \end{bmatrix} \tag{C.17}$$

where

$$A = \left.\frac{V_1}{V_2}\right|_{I_2=0} \qquad C = \left.\frac{I_1}{V_2}\right|_{I_2=0}$$
$$B = -\left.\frac{V_1}{I_2}\right|_{V_2=0} \qquad D = -\left.\frac{I_1}{I_2}\right|_{V_2=0} \tag{C. 18}$$

Note that all the preceding types of parameters (z, y, g, h) were defined as network functions, i.e., transforms of the output over the input, whereas the reciprocals of transmission parameters (e.g., A^{-1}) are network functions since in V_1/V_2, V_1 is the input and V_2 is the output.

Similarly, if the independent variables are chosen as V_1 and $-I_1$, then the dependent parameters are V_2 and I_2; this relationshp is given by $\mathscr{A}\mathscr{B}\mathscr{C}\mathscr{D}$ parameters, namely,

$$\begin{bmatrix} V_2 \\ I_2 \end{bmatrix} = \begin{bmatrix} \mathscr{A} & \mathscr{B} \\ \mathscr{C} & \mathscr{D} \end{bmatrix} \begin{bmatrix} V_1 \\ -I_1 \end{bmatrix} \tag{C.19}$$

The descriptions in (C.17) and (C.19) are *very useful in the analysis of cascaded two-ports*, and it is for this reason that the negative sign is associated with the independent current in each description, since V_2, $-I_2$ will be the input for the succeeding two-port.

EXAMPLE C.3

Consider the cascaded two-ports shown in Fig. C.10. For the two-port network N_a we have

$$\begin{bmatrix} V_1 \\ I_1 \end{bmatrix} = \begin{bmatrix} A^a & B^a \\ C^a & D^a \end{bmatrix} \begin{bmatrix} V_2 \\ -I_2 \end{bmatrix} \tag{C.20}$$

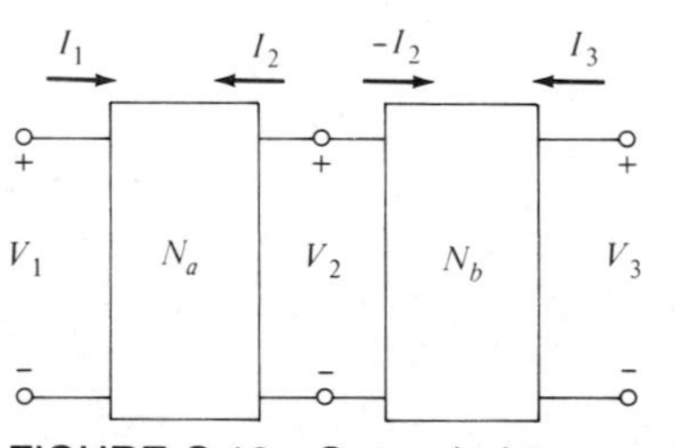

FIGURE C.10 Cascaded two-port networks.

For the two-port network N_b we have

$$\begin{bmatrix} V_2 \\ -I_2 \end{bmatrix} = \begin{bmatrix} A^b & B^b \\ C^b & D^b \end{bmatrix} \begin{bmatrix} V_3 \\ -I_3 \end{bmatrix} \tag{C.21}$$

From (C.20) and (C.21), we can immediately write

$$\begin{bmatrix} V_1 \\ I_1 \end{bmatrix} = \begin{bmatrix} A^a & B^a \\ C^a & D^a \end{bmatrix} \begin{bmatrix} A^b & B^b \\ C^b & D^b \end{bmatrix} \begin{bmatrix} V_3 \\ -I_3 \end{bmatrix} = \begin{bmatrix} A & B \\ C & D \end{bmatrix} \begin{bmatrix} V_3 \\ -I_3 \end{bmatrix} \tag{C.22}$$

$$\begin{aligned} \text{where } A &= A^aA^b + B^aC^b \\ B &= A^aB^b + B^aD^b \\ C &= C^aA^b + D^aC^b \\ D &= C^aB^b + D^aD^b \end{aligned} \tag{C.23}$$

For convenience and subsequent reference, the matrix conversions from one parameter set to another parameter set are listed in Table C.1.

The choice of parameter characterization depends on the problem at hand. In certain cases, some characterizations may not exist. For eample, the ideal transformer has no z- or y-parameter description, but the other characterizations can be used. In some other cases, a particular choice may yield a very simple representation. As an example, consider the various idealized controlled-source two-port networks.

For the voltage controlled voltage source (VCVS) in Fig. 1.8*a*, which is also referred to as the ideal voltage amplifier, the g parameters give a simple description

$$\begin{bmatrix} I_1 \\ V_2 \end{bmatrix} = \begin{bmatrix} 0 & 0 \\ \mu & 0 \end{bmatrix} \begin{bmatrix} V_1 \\ I_2 \end{bmatrix} \tag{C.24}$$

i.e., $g_{21} = \mu$, and $g_{11} = g_{22} = g_{12} = 0$. The voltage-controlled current source (VCCS) shown in Fig. 1.8*b*, which is also called the ideal transconductance amplifier, is most simply described in terms of the y parameters

$$\begin{bmatrix} I_1 \\ I_2 \end{bmatrix} = \begin{bmatrix} 0 & 0 \\ g_m & 0 \end{bmatrix} \begin{bmatrix} V_1 \\ V_2 \end{bmatrix} \tag{C.25}$$

i.e., $y_{21} = g_m$, and $y_{11} = y_{22} = y_{12} = 0$. For Fig. 1.8*c*, a current-controlled current source (CCCS), also referred to as the ideal current amplifier, the simplest description in terms of the h parameters is

$$\begin{bmatrix} V_1 \\ I_2 \end{bmatrix} = \begin{bmatrix} 0 & 0 \\ \beta & 0 \end{bmatrix} \begin{bmatrix} I_1 \\ V_2 \end{bmatrix} \tag{C.26}$$

i.e., $h_{21} = \beta$, and $h_{11} = h_{22} = h_{12} = 0$. Finally, for the current-controlled voltage source (CCVS) shown in Fig. 1.8*d*, also referred to as the ideal transresistance amplifier, the simplest description provided by the z parameters is

$$\begin{bmatrix} V_1 \\ V_2 \end{bmatrix} = \begin{bmatrix} 0 & 0 \\ r_m & 0 \end{bmatrix} \begin{bmatrix} I_1 \\ I_2 \end{bmatrix} \tag{C.27}$$

TABLE C.1 Two-Port Parameter Matrix Conversions*

	$[z_{ij}]$	$[y_{ij}]$	$[g_{ij}]$	$[h_{ij}]$	$\begin{bmatrix} A & B \\ C & D \end{bmatrix}$	$\begin{bmatrix} \mathscr{A} & \mathscr{B} \\ \mathscr{C} & \mathscr{D} \end{bmatrix}$
$[z_{ij}]$	$\begin{matrix} z_{11} & z_{12} \\ z_{21} & z_{22} \end{matrix}$	$\begin{matrix} \frac{y_{22}}{\Delta_y} & -\frac{y_{12}}{\Delta_y} \\ -\frac{y_{21}}{\Delta_y} & \frac{y_{11}}{\Delta_y} \end{matrix}$	$\begin{matrix} \frac{1}{g_{11}} & -\frac{g_{12}}{g_{11}} \\ \frac{g_{21}}{g_{11}} & \frac{\Delta_g}{g_{11}} \end{matrix}$	$\begin{matrix} \frac{\Delta_h}{h_{22}} & \frac{h_{12}}{h_{22}} \\ -\frac{h_{21}}{h_{22}} & \frac{1}{h_{22}} \end{matrix}$	$\begin{matrix} \frac{A}{C} & \frac{\Delta_A}{C} \\ \frac{1}{C} & \frac{D}{C} \end{matrix}$	$\begin{matrix} \frac{\mathscr{D}}{\mathscr{C}} & \frac{1}{\mathscr{C}} \\ \frac{\Delta_{\mathscr{A}}}{\mathscr{C}} & \frac{\mathscr{A}}{\mathscr{C}} \end{matrix}$
$[y_{ij}]$	$\begin{matrix} \frac{z_{22}}{\Delta_z} & -\frac{z_{12}}{\Delta_z} \\ -\frac{z_{21}}{\Delta_z} & \frac{z_{11}}{\Delta_z} \end{matrix}$	$\begin{matrix} y_{11} & y_{12} \\ y_{21} & y_{22} \end{matrix}$	$\begin{matrix} \frac{\Delta_g}{g_{22}} & \frac{g_{12}}{g_{22}} \\ -\frac{g_{21}}{g_{22}} & \frac{1}{g_{22}} \end{matrix}$	$\begin{matrix} \frac{1}{h_{11}} & -\frac{h_{12}}{h_{11}} \\ \frac{h_{21}}{h_{11}} & \frac{\Delta_h}{h_{11}} \end{matrix}$	$\begin{matrix} \frac{D}{B} & -\frac{\Delta_A}{B} \\ -\frac{1}{B} & \frac{A}{B} \end{matrix}$	$\begin{matrix} \frac{\mathscr{A}}{\mathscr{B}} & -\frac{1}{\mathscr{B}} \\ -\frac{\Delta_{\mathscr{A}}}{\mathscr{B}} & \frac{\mathscr{D}}{\mathscr{B}} \end{matrix}$
$[g_{ij}]$	$\begin{matrix} \frac{1}{z_{11}} & -\frac{z_{12}}{z_{11}} \\ \frac{z_{21}}{z_{11}} & \frac{\Delta_z}{z_{11}} \end{matrix}$	$\begin{matrix} \frac{\Delta_y}{y_{22}} & \frac{y_{12}}{y_{22}} \\ -\frac{y_{21}}{y_{22}} & \frac{1}{y_{22}} \end{matrix}$	$\begin{matrix} g_{11} & g_{12} \\ g_{21} & g_{22} \end{matrix}$	$\begin{matrix} \frac{h_{22}}{\Delta_h} & -\frac{h_{12}}{\Delta_h} \\ -\frac{h_{21}}{\Delta_h} & \frac{h_{11}}{\Delta_h} \end{matrix}$	$\begin{matrix} \frac{C}{A} & -\frac{\Delta_A}{A} \\ \frac{1}{A} & \frac{B}{A} \end{matrix}$	$\begin{matrix} \frac{\mathscr{C}}{\mathscr{D}} & -\frac{1}{\mathscr{D}} \\ \frac{\Delta_{\mathscr{A}}}{\mathscr{D}} & \frac{\mathscr{B}}{\mathscr{D}} \end{matrix}$
$[h_{ij}]$	$\begin{matrix} \frac{\Delta_z}{z_{22}} & \frac{z_{12}}{z_{22}} \\ -\frac{z_{21}}{z_{22}} & \frac{1}{z_{22}} \end{matrix}$	$\begin{matrix} \frac{1}{y_{11}} & -\frac{y_{12}}{y_{11}} \\ \frac{y_{21}}{y_{11}} & \frac{\Delta_y}{y_{11}} \end{matrix}$	$\begin{matrix} \frac{g_{22}}{\Delta_g} & -\frac{g_{12}}{\Delta_g} \\ -\frac{g_{21}}{\Delta_g} & \frac{g_{11}}{\Delta_g} \end{matrix}$	$\begin{matrix} h_{11} & h_{12} \\ h_{21} & h_{22} \end{matrix}$	$\begin{matrix} \frac{B}{D} & \frac{\Delta_A}{D} \\ -\frac{1}{D} & \frac{C}{D} \end{matrix}$	$\begin{matrix} \frac{\mathscr{B}}{\mathscr{A}} & \frac{1}{\mathscr{A}} \\ -\frac{\Delta_{\mathscr{A}}}{\mathscr{A}} & \frac{\mathscr{C}}{\mathscr{A}} \end{matrix}$
$\begin{bmatrix} A & B \\ C & D \end{bmatrix}$	$\begin{matrix} \frac{z_{11}}{z_{21}} & \frac{\Delta_z}{z_{21}} \\ \frac{1}{z_{21}} & \frac{z_{22}}{z_{21}} \end{matrix}$	$\begin{matrix} -\frac{y_{22}}{y_{21}} & -\frac{1}{y_{21}} \\ -\frac{\Delta_y}{y_{21}} & -\frac{y_{11}}{y_{21}} \end{matrix}$	$\begin{matrix} \frac{1}{g_{21}} & \frac{g_{22}}{g_{21}} \\ \frac{g_{12}}{g_{21}} & \frac{\Delta_y}{g_{21}} \end{matrix}$	$\begin{matrix} -\frac{\Delta_h}{h_{21}} & -\frac{h_{11}}{h_{21}} \\ -\frac{h_{22}}{h_{21}} & -\frac{1}{h_{21}} \end{matrix}$	$\begin{matrix} A & B \\ C & D \end{matrix}$	$\begin{matrix} \frac{\mathscr{D}}{\Delta_{\mathscr{A}}} & \frac{\mathscr{B}}{\Delta_{\mathscr{A}}} \\ \frac{\mathscr{C}}{\Delta_{\mathscr{A}}} & \frac{\mathscr{A}}{\Delta_{\mathscr{A}}} \end{matrix}$
$\begin{bmatrix} \mathscr{A} & \mathscr{B} \\ \mathscr{C} & \mathscr{D} \end{bmatrix}$	$\begin{matrix} \frac{z_{22}}{z_{12}} & \frac{\Delta_z}{z_{12}} \\ \frac{1}{z_{12}} & \frac{z_{11}}{z_{12}} \end{matrix}$	$\begin{matrix} -\frac{y_{11}}{y_{12}} & -\frac{1}{y_{12}} \\ -\frac{\Delta_y}{y_{12}} & -\frac{y_{22}}{y_{12}} \end{matrix}$	$\begin{matrix} -\frac{\Delta_g}{g_{12}} & -\frac{g_{22}}{g_{12}} \\ -\frac{g_{11}}{g_{12}} & -\frac{1}{g_{12}} \end{matrix}$	$\begin{matrix} \frac{1}{h_{12}} & \frac{h_{11}}{h_{12}} \\ \frac{h_{22}}{h_{12}} & \frac{\Delta_h}{h_{12}} \end{matrix}$	$\begin{matrix} \frac{D}{\Delta_A} & \frac{B}{\Delta_A} \\ \frac{C}{\Delta_A} & \frac{A}{\Delta_A} \end{matrix}$	$\begin{matrix} \mathscr{A} & \mathscr{B} \\ \mathscr{C} & \mathscr{D} \end{matrix}$

*All matrices appearing in the same row in the table are equivalent, e.g., $h_{11} = \Delta_z/z_{22} = 1/y_{11}$ and $\Delta_z = z_{11}z_{22} - z_{12}z_{21}$, etc.

i.e., $z_{21} = r_m$, and $z_{11} = z_{22} = z_{12} = 0$. In all the above ideal controlled-source cases, the only other mode of representation of the six possible uses is the *ABCD* parameters.

When two-ports are interconnected, the overall two-port parameters of the combined two-port can be obtained simply by direct addition of the respective parameters of the original two-ports (z_{ij}, y_{ij}, h_{ij}, g_{ij}) provided that the dependent variable is common to both ports and also the interconnection does not change the parameter sets. In other words, direct addition of the respective parameters is allowable if the current that enters one terminal of a port has the same value as the current that leaves the other terminal of the same port. It is usually clear in any given problem whether direct addition is permissible.

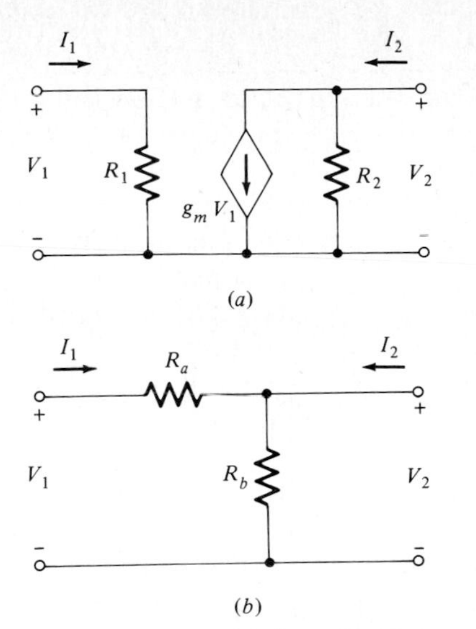

FIGURE C.11 (a) An active two port network. (b) A passive two-port network.

For example, consider the two-port networks shown in Fig. C.11. If the two circuits are connected in shunt at both ports (i.e., a shunt-shunt interconnection), the total y parameters are given by $y_{ij}^T = y_{ij}^a + y_{ij}^b$. Other permissible connections are given in Chap. 10, where the two-port theory is used extensively in the deviations. If the two circuits in Fig. C.11 are series-series interconnected $z_{ij}^T \neq z_{ij}^a + z_{ij}^b$ since the element R_a will be short-circuited in a series-series interconnection. If an ideal transformer is used to provide isolation, then in any of the series-series, shunt-shunt, series-shunt, and shunt-series interconnections, the overall tow-port parameters are always given by the respective addition of the parameters of the individual two-ports. For example, in a series-series interconnection, we may use the ideal transformer, as shown in Fig. C.12. In this case, the equalities $z_{ij}^T = z_{ij}^a + z_{ij}^b$ always hold. Hence, if N^a and N^b are the networks in Fig. C.11, we simply add each element of the [z] matrix of N^a to those of N^b to get the overall z_{ij} parameters.

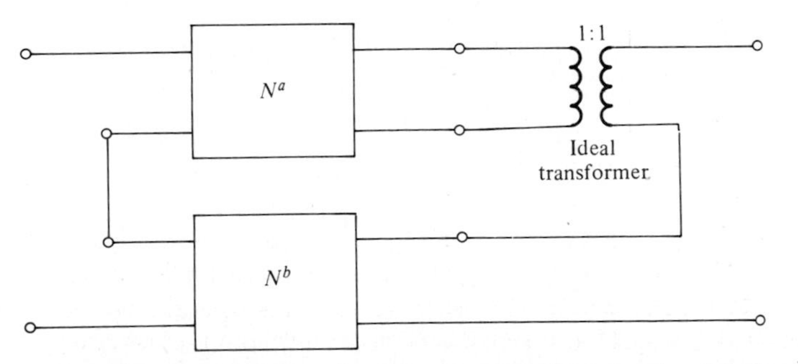

FIGURE C.12 Series-series interconnected two-port network: $z_{ij}^T = z_{ij}^a + z_{ij}^b$.

C.2 TERMINATED TWO-PORT NETWORKS

In many circuit applications, a two-port network is used to process and transmit a signal from a source to a load. The results of this section have been used extensively in Chaps. 4 and 10. Consider a linear time-invariant two-port network with initial conditions equal to zero, as represented by the box in Fig. C.13. The terminating source and load resistances are R_S and R_L, respectively. For this circuit, we are interested in determining the input impedance Z_{in}, output impedance Z_o, the voltage ratio V_2/V_1, and the current ratio I_2/I_1. The input impedance Z_{in} is the impedance seen at the input port of the two-port network when the terminating source resistance is removed. The output impedance Z_o is the impedance seen at the output port of the two-port network when the terminating load resistance is removed and the independent source input "killed."

The voltage gain V_2/V_s and the current gain $I_o/I_s = I_o/(V_s/R_S)$ can then be readily determined from the above results. The power-gain calculation is discussed in the next section. Consequently, we shall consider only the admittance parameters (y_{ij}) for the sake of clarity. The reader can carry out the steps in terms of the generalized parameters for any other parameter characterizations.*

The two-port description in terms of the y parameters is

$$I_1 = y_{11}V_1 + y_{12}V_2 \tag{C.28a}$$

$$I_2 = y_{21}V_1 + y_{22}V_2 \tag{C.28b}$$

The source and load connections yield the following additional relations:

$$-I_2 = G_L V_2 \tag{C.29}$$

$$I_1 = G_S V_S - G_S V_1 \tag{C.30}$$

Note that G_S and G_L are the reciprocals of R_S and R_L, respectively. To obtain the input admittance Y_{in}, we remove R_S and use (C.29) to eliminate I_2 in (C.28b) and solve for I_1/V_1. This yields

$$Y_{\text{in}} = \frac{I_1}{V_1} = y_{11} - \frac{y_{12}y_{21}}{y_{22} + G_L} \tag{C.31}$$

*To avoid repetition of identical steps in manipulating the equations, it is often helpful to use the generalized parameter k_{ij}, which relates the dependent and the independent variables by the following matrix equation.

$$\begin{bmatrix} \Phi_{d1} \\ \Phi_{d2} \end{bmatrix} = \begin{bmatrix} k_{11} & k_{12} \\ k_{21} & k_{22} \end{bmatrix} \begin{bmatrix} \Phi_{i1} \\ \Phi_{i2} \end{bmatrix}$$

where k_{ij} represents y_{ij}, z_{ij}, h_{ij}, or g_{ij} depending on the choice of the dependent and the independent variables. The subscripts d and i denote dependent and independent variables. For example, if Φ_{i1} and Φ_{i2} denote I_1 and V_2, and if Φ_{d1} and Φ_{d2} represent V_1 and I_2, then k_{ij} represents h_{ij}, respectively. The source will be represented by an impedance R_s and the load by an admittance G_L in this case. In order to avoid confusion, Γ_S and Γ_L should be used for the source and load terminations when the k parameters are used.

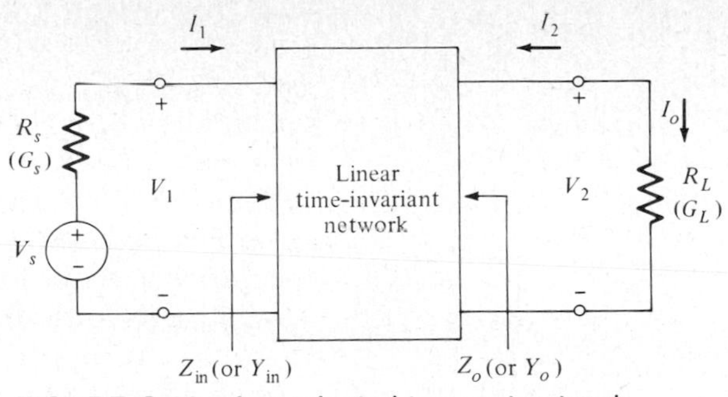

FIGURE C.13 A terminated two-port network.

Similarly, to find the output admittance we remove R_L, set $V_s = 0$, and use (C.30) to eliminate I_1 in (C.28*a*) and solve for I_2/V_2:

$$Y_0 = \frac{I_2}{V_2} = y_{22} - \frac{y_{21}y_{12}}{y_{11} + G_S} \tag{C.32}$$

The voltage ratio V_2/V_1 is found from (C.29) and (C.28*b*) by eliminating I_2. Thus

$$\frac{V_2}{V_1} = \frac{-y_{21}}{y_{22} + G_L} \tag{C.33}$$

Similarly, the current ratio is found by eliminating V_1 and V_2. The result is

$$\frac{-I_2}{I_1} = \frac{-y_{21}G_L}{\Delta y + y_{11}G_L} \tag{C.34}$$

If we are interested in finding the current gain I_o/I_s, where $I_s = V_sG_s$ and $I_o = -I_2$, we can incorporate G_S in the two-port to obtain a new two-port described by $[y]_s$ and then use (C.34). The modified two-port admittance matrix is

$$[y_{ij}]_s = \begin{bmatrix} (y_{11} + G_S) & y_{12} \\ y_{21} & y_{22} \end{bmatrix} \tag{C.35}$$

The current gain I_o/I_s can then be written directly from (C.34) and (C.37):

$$\frac{-I_2}{I_s} = \frac{-y_{21s}G_L}{\Delta y_s + y_{11s}G_L} = \frac{-y_{21}G_L}{(y_{11} + G_s)y_{22} - y_{12}y_{21} + (y_{11} + G_S)G_L}$$

hence

$$\frac{I_o}{I_s} = \frac{-y_{21}G_L}{(y_{11} + G_S)(y_{22} + G_L) - y_{12}y_{21}} \tag{C.36}$$

The voltage gain V_2/V_s is given by

$$\frac{V_2}{V_s} = \frac{V_o}{V_s} = \frac{-R_LI_2}{R_SI_s} = \frac{-y_{21}G_S}{(y_{11} + G_S)(y_{22} + G_L) - y_{12}y_{21}} \tag{C.37}$$

Similarly, if we use the z parameters, we obtain

$$Z_{\text{in}} = \frac{V_1}{I_1} = z_{11} - \frac{z_{12}z_{21}}{z_{22} + R_L} \tag{C.38}$$

$$Z_o = \frac{V_2}{I_2} = z_{22} - \frac{z_{12}z_{21}}{z_{11} + R_S} \tag{C.39}$$

$$\frac{V_2}{V_1} = \frac{z_{21}R_L}{\Delta z + z_{11}R_L} \tag{C.40}$$

$$\frac{I_2}{I_1} = -\frac{z_{21}}{z_{22} + R_L} \tag{C.41}$$

$$\frac{V_2}{V_s} = \frac{V_o}{V_s} = \frac{z_{21}R_L}{(z_{11} + R_S)(z_{22} + R_L) - z_{12}z_{21}} \tag{C.42}$$

and

$$\frac{I_o}{I_s} = \frac{-z_{21}R_S}{(z_{11} + R_S)(z_{22} + R_L) - z_{12}z_{21}} \tag{C.43}$$

Note that $I_s = V_s/R_S$. The various gain and impedance relations in terms of the other parameters are given in Table C.2. It should be noted that the voltage gain can also be obtained from

$$\frac{V_2}{V_s} = \frac{V_2}{V_1}\frac{V_1}{V_s} \tag{C.44}$$

where $V_1/V_s = Z_{\text{in}}/(Z_{\text{in}} + R_S)$. The impedance relations, the voltage ratio, and the current ratio expressions in Table C.2 are convenient for gain calculations in cascaded

TABLE C.2 Gain and Impedance Relations

	$[z_{ij}]$	$[y_{ij}]$	$[g_{ij}]$	$[h_{ij}]$	$\begin{bmatrix} A & B \\ C & D \end{bmatrix}$	$\begin{bmatrix} \mathscr{A} & \mathscr{B} \\ \mathscr{C} & \mathscr{D} \end{bmatrix}$
Input impedance, Z_i	$\frac{\Delta_z + z_{11}Z_L}{z_{22} + Z_L}$	$\frac{y_{22} + Y_L}{\Delta_y + y_{11}Y_L}$	$\frac{g_{22} + Z_L}{\Delta_g + g_{11}Z_L}$	$\frac{\Delta_h + h_{11}Y_L}{h_{22} + Y_L}$	$\frac{AZ_L + B}{CZ_L + D}$	$\frac{\mathscr{D}Z_L + \mathscr{B}}{\mathscr{C}Z_L + \mathscr{A}}$
Output impedance, Z_o	$\frac{\Delta_z + z_{22}Z_s}{z_{11} + Z_s}$	$\frac{y_{11} + Y_s}{\Delta_y + y_{22}Y_s}$	$\frac{\Delta_g + g_{22}Y_s}{g_{11} + Y_s}$	$\frac{h_{11} + Z_s}{\Delta_h + h_{22}Z_s}$	$\frac{DZ_s + B}{CZ_s + A}$	$\frac{\mathscr{A}Z_s + \mathscr{B}}{\mathscr{C}Z_s + \mathscr{D}}$
Current ratio, $-\frac{I_2}{I_1}$	$\frac{z_{21}}{z_{22} + Z_L}$	$\frac{-y_{21}Y_L}{\Delta_y + y_{11}Y_L}$	$\frac{g_{21}}{\Delta_g + g_{11}Z_L}$	$\frac{-h_{21}Y_L}{h_{22} + Y_L}$	$\frac{1}{D + CZ_L}$	$\frac{\Delta_{\mathscr{A}}}{\mathscr{A} + \mathscr{C}Z_L}$
Voltage ratio, $\frac{V_2}{V_1}$	$\frac{z_{21}Z_L}{\Delta_z + z_{11}Z_L}$	$\frac{-y_{21}}{y_{22} + Y_L}$	$\frac{g_{21}Z_L}{g_{22} + Z_L}$	$\frac{-h_{21}}{\Delta_h + h_{11}Y_L}$	$\frac{Z_L}{B + AZ_L}$	$\frac{\Delta_{\mathscr{A}}}{\mathscr{B}Y_L + \mathscr{D}}$

Voltage gain $= \frac{V_2}{V_s} = \left(\frac{Z_i}{Z_i + Z_s}\right)\frac{V_2}{V_1}$, current gain $= \frac{I_o}{I_s} = \frac{I_o}{V_s/R_s} = \left(\frac{Z_s}{Z_s + Z_i}\right)\left(-\frac{I_2}{I_1}\right)$

two-port circuits (see Chap. 4) and can simplify calculations in circuits, as demonstrated in Chap. 5.

C.3 MAXIMUM POWER GAIN IN ACTIVE TWO-PORTS

In this section, we shall consider the definition of power gain and the corresponding expression when the linear time-invariant active two-port is terminated at both ends by passive one-ports (Z_S and Z_L). It should be emphasized that the quantities defined in this section apply only to sinusoidal steady-state conditions.

Consider the network shown schematically in Fig. C.14. *Power gain G_P* is defined as the ratio of the average power delivered to the load to the average power in the active two-port network, i.e., P_o/P_i. By active two-port, we mean that the box in Fig. C.14 contains some form of controlled source such that power gain larger than unity is obtained.

The average power P across two terminals of network is given by

$$P_{av} = \tfrac{1}{2}|I|^2 \text{Re } Z(j\omega) = \tfrac{1}{2}|V|^2 \text{ Re } Y(j\omega) \tag{C.45}$$

where Re designates "real part of" and $Z(j\omega) = 1/Y(j\omega)$.

Thus by definition the power gain of a two-port network is given by

$$G_P = \frac{P_o}{P_i} = \frac{\frac{1}{2}|V_2|^2 \text{ Re } Y_L}{\frac{1}{2}|V_1|^2 \text{ Re } Y_{in}} \tag{C.46}$$

From (C.31), (C.33), and (C.46), we obtain

$$G_P = \left|\frac{y_{21}}{y_{22} + Y_L}\right|^2 \frac{\text{Re } Y_L}{\text{Re }[y_{11} - (y_{12}y_{21})/(y_{22} + Y_L)]} \tag{C.47}$$

A. Maximum Power Transfer Theorem

This theorem states that maximum power is transferred to the load if the load impedance is equal to the complex conjugate of the source impedance.

The above theorem can be readily proved by referring to the one-port network shown in Fig. C.15; let the source and load impedances be Z_S and Z_L, respectively.

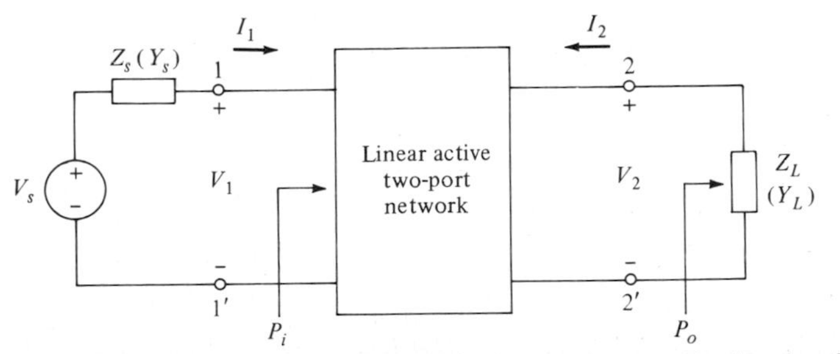

FIGURE C.14 A terminated active two-port network showing P_o and P_i.

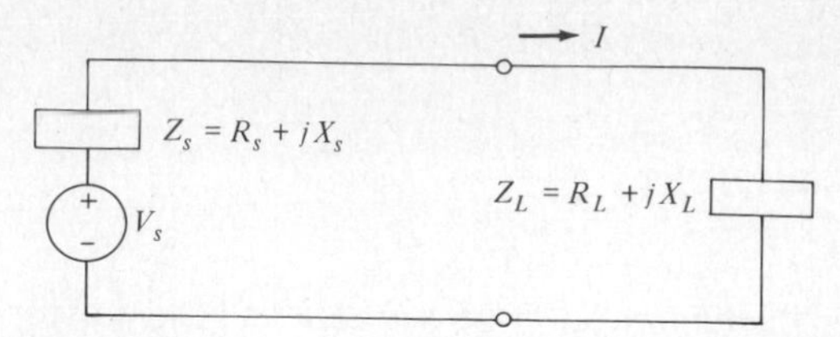

FIGURE C.15 A terminated one-port network.

At any sinusoidal source-voltage frequency the impedance can be expressed by its real and imaginary parts as shown in Fig. C.15.

The average power delivered to the load is

$$P_{av} = \tfrac{1}{2}|I|^2 \operatorname{Re} Z_L = \tfrac{1}{2}|V_s|^2 \frac{R_L}{(R_L + R_S)^2 + (X_L + X_S)^2} \tag{C.48}$$

For a given source excitation and source impedance we can find the optimum load impedance to maximize P_{av}. By inspection of (C.48), since the reactance can be negative, we choose

$$(X_L)_{opt} = -X_S \tag{C.49a}$$

From setting $\partial P_{av}/\partial R_L = 0$, we obtain

$$(R_L)_{opt} = R_S \tag{C.49b}$$

Thus

$$(Z_L)_{opt} = Z_S^* \tag{C.50a}$$

where * denotes complex conjugate. The above theories can be applied to the two-port case. If Z_i and Z_o are the input and output impedances of the two-port, maximum power gain is obtained if

$$Z_i = Z_S^* \qquad \text{and} \qquad Z_L = Z_o^* \tag{C.50b}$$

B. Potential Instability

Maximum power gain in an active two-port is meaningful only if the two-port is absolutely stable. By potential instability we mean that the two-port can be made unstable for some choice of Z_S and/or Z_L.

If the active two-port is potentially unstable, then the maximum power gain G_P is infinite and meaningless.

The conditions for potential instability in terms of generalized parameters are given below without proof.* If any of the following conditions is *not* satisfied, the two-port is potentially unstable.

---

*For proof see M. S. Ghausi, *Principles and Design of Linear Active Circuits*, McGraw-Hill, New York, 1965, Chap. 3.

$$\text{Re } k_{11} > 0 \qquad \text{Re } k_{22} > 0 \qquad \text{Re } k_{11} \text{ Re } k_{22} > \tfrac{1}{2} k_{21} k_{12} (1 + \cos \theta) \tag{C.51}$$

where

$$\theta = \tan^{-1} \frac{\text{Im } k_{12} k_{21}}{\text{Re } k_{12} k_{21}}$$

Im designates the "imaginary part" and Re the "real part."

For a potentially unstable two-port, passive terminations can be found that would make the two-port unstable, i.e., yield infinite power gain, which is meaningless.

There are several methods to find the maximum power gain for stable two-ports. Two of these methods are

1. Maximize G_P in (C.47) with respect to the real and imaginary parts of the load impedance; then select the source impedance to conjugate match the input port with Z_L optimized.
2. Simultaneous conjugate match both ports and solve for the real and imaginary parts of Z_S and Z_L.

Method (2) is quite involved if the two-port parameters are complex quantities, i.e., if there are frequency-dependent circuit elements in the box in Fig. C.14. For such cases use method 1. If, however, the parameters are real, i.e., $z_{ij} = r_{ij}$, method 2 leads to a simple solution. We shall demonstrate it in the following using r_{ij} for z_{ij}, since they are assumed real. From (C.38) and (C.39) we have

$$Z_{\text{in}} = r_{11} - \frac{r_{12} r_{21}}{r_{22} + R_L} = R_S \tag{C.52}$$

$$Z_O = r_{22} - \frac{r_{12} r_{21}}{r_{11} + R_L} = R_L \tag{C.53}$$

Solving (C.52) and (C.53) for $r_{12} r_{21}$, we have

$$(r_{11} - R_S)(r_{22} + R_L) = r_{12} r_{21} \tag{C.54a}$$

$$(r_{22} - R_L)(r_{11} + R_S) = r_{12} r_{21} \tag{C.54b}$$

Subtracting (C.54*a*) from (C.54*b*), we obtain

$$2(r_{11} R_L - r_{22} R_S) = 0 \tag{C.55}$$

From (C.55), we have

$$\frac{R_L}{r_{11}} = \frac{R_S}{r_{22}} \tag{C.56}$$

Substituting (C.56) in (C.54*a*) or (C.54*b*), we have

$$(R_S)_{\text{opt}} = \sqrt{\frac{r_{11}}{r_{22}} (r_{11} r_{22} - r_{12} r_{21})} \tag{C.57}$$

$$(R_L)_{\text{opt}} = \sqrt{\frac{r_{22}}{r_{11}} (r_{11} r_{22} - r_{12} r_{21})} \tag{C.58}$$

Since, as we note from (C.51), for an absolutely stable active two-port we must have $r_{11} > 0$, $r_{22} > 0$, and $r_{11}r_{22} - r_{12}r_{21} > 0$, the optimum terminations found in (C.57) and (C.58) are therefore positive and meaningful.

The maximum power gain obtained by using optimum terminating conditions is given by

$$(G_P)_{\max} = \frac{r_{21}^2}{(\sqrt{r_{11}r_{22} - r_{12}r_{21}} + \sqrt{r_{11}r_{22}})^2} \tag{C.59}$$

The results, *for real parameters*, are listed below in terms of the generalized parameters for convenience:

$$(\Gamma_L)_{\text{opt}} = \sqrt{\frac{k_{22}}{k_{11}}(k_{11}k_{22} - k_{12}k_{21})} \tag{C.60}$$

$$(\Gamma_S)_{\text{opt}} = \sqrt{\frac{k_{11}}{k_{22}}(k_{22}k_{11} - k_{21}k_{12})} \tag{C.61}$$

$$(G_P)_{\max} = \frac{k_{21}^2}{(\sqrt{k_{11}k_{22} - k_{12}k_{21}} + \sqrt{k_{11}k_{22}})^2} \tag{C.62}$$

where k_{ij} are the generalized parameters (i.e., any of the y_{ij}, z_{ij}, g_{ij}, or h_{ij} parameters), and all are assumed to be real quantities in this derivation such as in purely resistive networks.

$$\begin{aligned} \Gamma_L &= R_L \quad \text{for } z_{ij} \text{ and } g_{ij} \qquad \Gamma_S = R_S \quad \text{for } z_{ij} \text{ and } h_{ij} \\ \Gamma_L &= G_L \quad \text{for } y_{ij} \text{ and } h_{ij} \qquad \Gamma_S = G_S \quad \text{for } y_{ij} \text{ and } g_{ij} \end{aligned} \tag{C.63}$$

EXAMPLE C.4

Consider a transistor model whose h parameters are given by

$$h_{11} = 500\ \Omega \qquad h_{12} = 10^{-4}$$

$$h_{21} = 50 \qquad h_{22} = 10^{-4}\ \text{mho}$$

Find the optimum passive terminations (if any) and the maximum power gain. Since, from (C.51), $h_{11} > 0$, $h_{22} > 0$ and $h_{11}h_{22} - h_{12}h_{21} = 450 \times 10^{-4} > 0$, there exist optimum terminations, which are

$$(R_S)_{\text{opt}} = \sqrt{\frac{h_{11}}{h_{22}}(h_{22}h_{11} - h_{21}h_{12})} = 474\ \Omega$$

$$(G_L)_{\text{opt}} = \sqrt{\frac{h_{22}}{h_{11}}(h_{22}h_{11} - h_{21}h_{12})} = 0.95 \times 10^{-5}\ \text{mhos}$$

or

$$(R_L)_{\text{opt}} = 10.5 \text{ k}\Omega$$

$$(G_P)_{\text{max}} = \frac{50^2}{(\sqrt{450 \times 10^{-4}} + \sqrt{500 \times 10^{-4}})^2} = 1.32 \times 10^4$$

or

$$(G_P)_{\text{max}} = 10 \log (1.32 \times 10^4) = 41.2 \text{ dB}$$

REFERENCES

1. M. E. VanValkenburg, *Network Analysis,* 3d ed., Prentice-Hall, Englewood Cliffs, N.J., 1974.
2. L. S. Bobrow, *Elementary Linear Circuit Analysis,* Holt, Rinehart and Winston, New York, 1981.

APPENDIX D Noise in Transistors and Op Amps

This appendix deals with noise in diodes, bipolar transistors, field-effect transistors, and operational amplifiers.

The output signal of an amplifier, in the absence of an input signal, is called *noise*. Noise is exhibited in the amplifier output by hiss or crackle (in the case of an audio amplifier connected to a loudspeaker) when there is no input and the gain control is set at or near the maximum. The magnitude of the noise voltage generated within the circuit, in general, limits the minimum signal that can be amplified by an amplifier.

In addition to the gain and bandwidth considerations in an amplifier, as pointed out in Chap. 8, there are other factors that may have to be considered in the design. If many stages of amplification are used to obtain a very high gain, the progressive increase of signal level at the last stage may cause distortion. This is because the signal level at the input of the final stage may make large excursions and cause distortion. The distortion problem arising as such is considered in Chap. 6. The maximum signal swing is limited by the supply voltages. If, on the other hand, we use an extremely small amplitude signal at the input so that the signal level at the final stage is small and in the linear region of the characteristic curve, we have solved the distortion problem, but we may have created another problem. The signal level at the input stage may be so small (of the order of the noise voltage) that we may not be able to distinguish the signal from the noise at the output. We shall now consider how this noise voltage (or current) comes about, and what should be done to minimize this problem.

D.1 SOURCES OF NOISE

There are several sources of noise and each one is considered briefly in the following without any derivation.

A. Thermal or Johnson Noise

Noise in a resistor is due to the random motion produced by thermal agitation of the electrons. This noise is temperature dependent and is referred to as *thermal noise* or *Johnson noise*. The noise voltage in a resistor has a mean-square value given by

$$\overline{v_n^2} = 4kTR\,\Delta f \tag{D.1}$$

where T = absolute temperature, K
k = Boltzmann's constant (1.38×10^{-23} W · s)
Δf = noise bandwidth, Hz
R = resistance, Ω

The noise equivalent circuit for a resistor, as composed of a noiseless resistor, in series with a noise voltage source is shown in Fig. D.1*b*, or alternatively, the equivalent circuit for a resistor, as a noiseless resistor, in parallel with a noise current source is shown in Fig. D.1*c*. Note that polarity is not shown for the sources and that it makes no difference which way it is assigned for independent noise sources. To get an idea of the order of magnitude of the thermal-noise voltage, consider at room temperature (T = 300 K) R = 1 kΩ, and Δf = 10 kHz. Substituting these values in (D.1) and taking its square root yields an rms noise voltage v_n = 0.41 μV. Note that a wider bandwidth and/or a higher temperature will produce a larger noise voltage. Sometimes

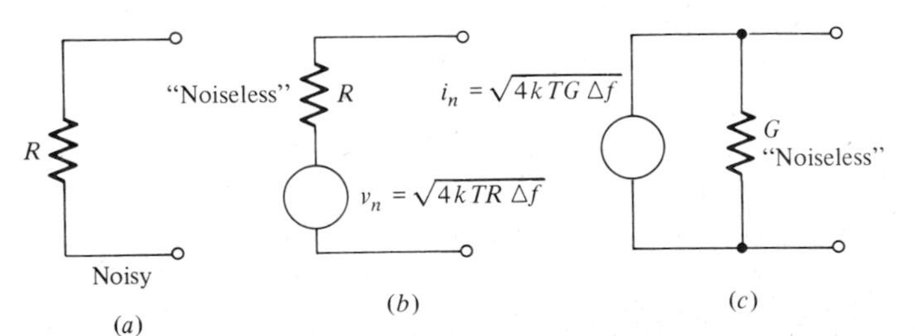

FIGURE D.1 (*a*) An actual resistor (noisy resistor). (*b*) Noiseless resistor with a noise-voltage source. (*c*) Noiseless conductance with a noise-current source.

the noise is expressed in per unit bandwidth, i.e., $\Delta f = 1$ Hz (nV/$\sqrt{\text{Hz}}$); e.g., for the above numerical case $v_n = 4$ nV/$\sqrt{\text{Hz}}$.

B. Shot Noise

The random fluctuation in the number of charge carriers when emitted from a surface or diffused across a junction is referred to as *shot noise*. For shot noise the mean-square fluctuation in the number of carriers is equal to the average number of carriers and is given by

$$\overline{i_n^2} = 2qI_o\Delta f \tag{D.2}$$

where I_o is the dc current flow in amperes, Δf is the noise bandwidth in hertz, and q is the magnitude of the electronic charge (1.6×10^{-19} C).

C. Flicker Noise

Another source of noise in a semiconductor device is the flicker noise, or $1/f$ noise, which is mainly due to surface imperfections resulting from the emission and diffusion processes. Flicker noise is always present when there is a flow of dc current. This noise is important at low frequencies and can be reduced considerably by proper fabrication processes. The flicker noise is represented by

$$\overline{i^2} = K\frac{I_o}{f}\Delta f \tag{D.3}$$

where K is a constant of the device and I_o is the dc current.

D. Generation-Recombination Noise

One source of noise in transistors is due to the random generation and recombination process in the base region of a BJT. Because of the randomness of generation and recombination, two sources of noise must be introduced to account for this phenomenon.

E. Burst Noise

Another source of low-frequency noise in discrete transistors and integrated-circuit devices is related to the presence of heavy-metal ion contamination. This noise consists typically of random pulses of variable length but about the same height. The repetition rate of the noise pulses is usually in the audio-frequency range and because of its sound this noise is also referred to as *popcorn noise*.

D.2 NOISE MODELS

Ideal capacitors and inductors have no sources of noise. Actual *LC* elements exhibit noise due to the parasitic resistors.

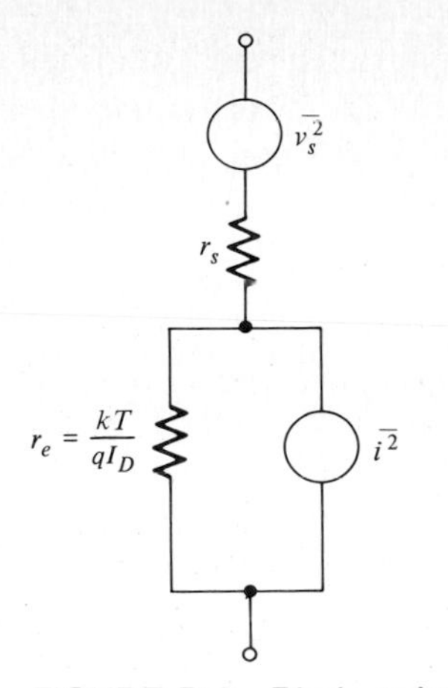

FIGURE D.2 Diode noise model.

A. Junction Diodes

The noise model for a *pn*-junction diode is shown in Fig. D.2. The noise voltage $\overline{v_s^2}$ is due to the small series resistance of the silicon material. The noise current generator is a combination of shot noise and flicker noise. It is given by

$$\overline{i^2} - 2qI_D\,\Delta f + \frac{KI_D}{f}\,\Delta f \tag{D.4}$$

where the first term is the shot noise and the second term is the flicker noise.

B. Bipolar Transistors

From a knowledge of the presence of these noise sources, a simple noise model for the bipolar junction transistor using the π model is known in Fig. D.3. The model is valid for both *npn* and *pnp* devices.

In the model of Fig. D.3, the noise generators have the following expressions:

$$\overline{v_b^2} = 4kTr_b\,\Delta f \tag{D.5a}$$

$$\overline{i_c^2} = 2qI_c\,\Delta f \tag{D.5b}$$

$$\overline{i_b^2} = 2qI_b\,\Delta f + \frac{K_1 I_b}{f}\,\Delta f \simeq 2q\frac{I_e}{\beta_0}\,\Delta f \tag{D.5c}$$

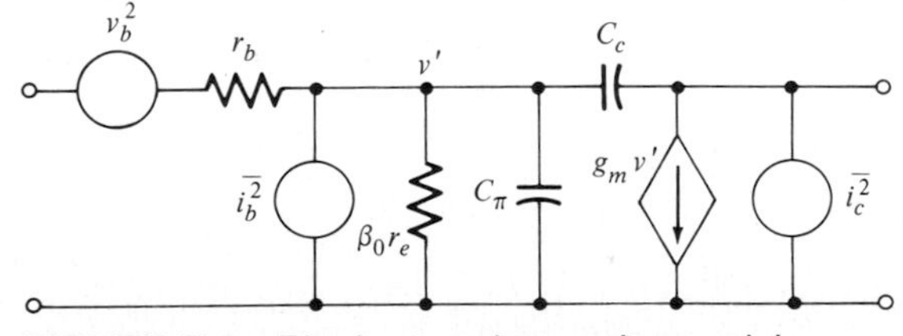

FIGURE D.3 Bipolar transistor noise model.

where at midband the $1/f$ noise can be ignored at high frequencies. We shall assume in the following that the noise sources are uncorrelated. This final assumption is found to be quite reasonable and enables one to use superposition of power in noise calculations—a factor that greatly simplifies such calculations. We shall next consider an example to illustrate noise calculation in an amplifier circuit.

EXAMPLE D.1

The amplifier circuit shown in Fig. D.4 uses a silicon transistor. We wish to calculate the midband rms noise voltage at the output of the circuit at room temperature. The bias conditions are $V_{CE} = 5$ V, $I_E = 5$ mA. For simplicity we assume that $\Delta f \simeq f_{3dB}$. In practice Δf is based on the power-spectrum considerations, or it is specified.* The transistor parameters are assumed to be $\beta_0 = 50$, $r_b = 50\ \Omega$, $C_c = 4$ pF, and $f_T = 300$ MHz. The source and load terminations are $R_S = R_L = 100\ \Omega$.

For the circuit using the unilateral model, we can readily determine the approximate 3-dB bandwidth f_{3dB}. For $R_L = 100\ \Omega$, $D = 1 + R_L C_c \omega_T = 1.7$, $C_i = D/r_e\omega_T = 1.8 \times 10^{-10}$ rad/s and the upper cutoff frequency $f_{3dB} = 1/2\pi C_i[R_i \| (r_b + R_S)] \simeq 9.3$ MHz $\simeq \Delta f$, according to our assumption. The noise model at midband frequencies is shown in Fig. D.5. Flicker noise is assumed negligible.

Since the circuit is linear and all the sources are assumed to be independent, we can separately determine the noise output due to each source and then add the results. In other words, to calculate the output noise due to the source resistance noise i_{ns}, we ignore all the other sources (short-circuit the voltage source and open-circuit the current source) and find its contribution to $\overline{v_N^2}$ alone.

*Noise bandwidth is obtained from a "brick-wall" response. Since the power gain is related to the square of its voltage transfer function, $T(j\omega)$, Δf is obtained from $T_{\max}^2\,\Delta f = \int_0^\infty |T(j\omega)|^2\,d\omega$. For a single-pole RC rolloff $\Delta f = 1.57 f_{3dB}$. For higher-order maximally flat magnitude filters $\Delta f \simeq f_{3dB}$.

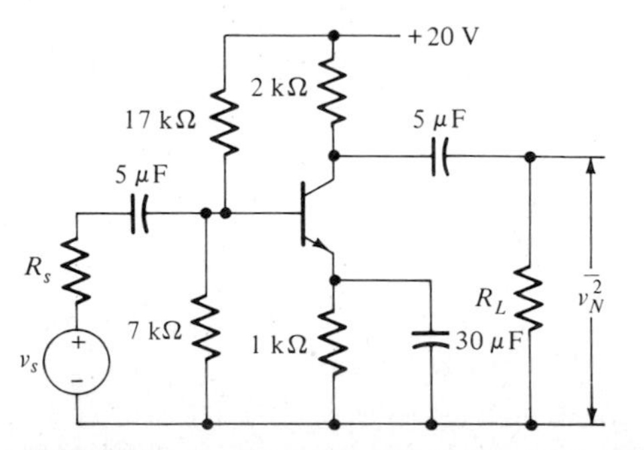

FIGURE D.4 Example circuit for noise calculations.

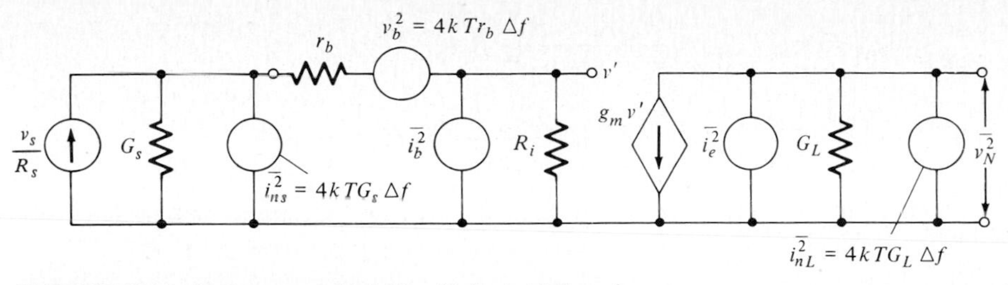

FIGURE D.5 Noise model of Fig. D.4 at midband.

We repeat the calculation for each of the other sources and then add the mean-square voltages.

Thus, the output noise contribution due to the source resistance noise $\overline{v_{ns}^2}$ is found from Fig. D.6 to be

$$v' = \frac{1}{G_S r_b G_i + G_i + G_S} i_{ns}$$

$$v_{ns} = g_m v' R_L = \frac{g_m R_L}{G_S r_b G_i + G_i + G_S} i_{ns} \tag{D.6a}$$

From (D.6*a*), and substitution of the expression for i_{ns}^2, we obtain

$$v_{ns}^2 = \left(\frac{g_m R_L}{G_S r_b G_i + G_i + G_s}\right)^2 (4kTG_S\,\Delta f) = 24.4 \times 10^{-10} \tag{D.6b}$$

Similarly, the output noise contribution due to v_b, i.e., v_{nb}^2, is found to be

$$v' = R_i \frac{v_b}{r_b + R_S + R_i}$$

$$v_{nb} = g_m v' R_L = \frac{g_m R_L R_i}{r_b + R_S + R_i} v_b \tag{D.7a}$$

$$\overline{v_{nb}^2} = \left(\frac{g_m R_L R_i}{r_b + R_S + R_i}\right)^2 4kTr_b\,\Delta f = 12.22 \times 10^{-10} \tag{D.7b}$$

The output noise contribution due to $\overline{i_b^2}$, i.e., $\overline{v_{nB}^2}$, is as follows:

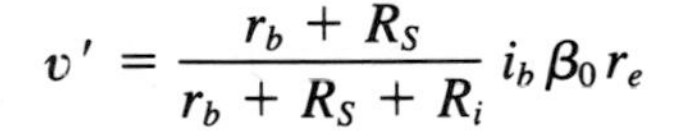

$$v' = \frac{r_b + R_S}{r_b + R_S + R_i} i_b \beta_0 r_e$$

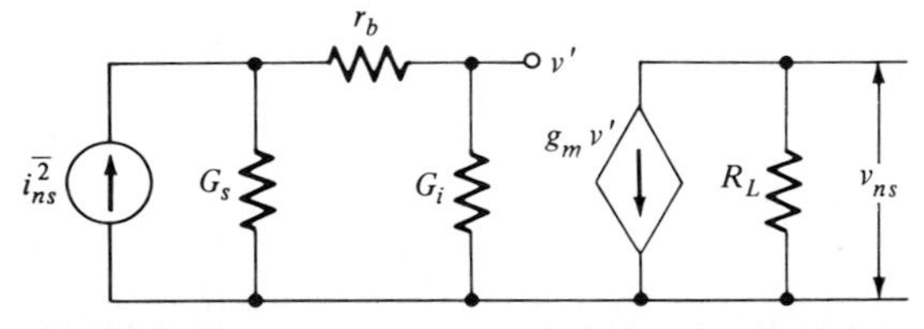

FIGURE D.6 Circuit for noise calculation due to the source resistor only.

$$v_{nB} = g_m v' R_L = \frac{R_L(r_b + R_S)}{r_b + R_S + R_i} \beta_0 i_b \tag{D.8a}$$

$$\overline{v_{nB}^2} = \left[\frac{R_L(r_b + R_S)}{r_b + R_S + R_i}\right]^2 \beta_0^2 \frac{2qI_e \, \Delta f}{\beta_0} = 10.5 \times 10^{-10} \tag{D.8b}$$

Finally, the output noise due to $\overline{i_c^2}$, i.e., $\overline{v_{nc}^2}$, and the noise due to R_L, i.e., $\overline{v_{nL}^2}$, are

$$\overline{v_{nc}^2} = R_L^2 \overline{i_c^2} = R_L^2 (2qI_c \, \Delta f) = 1.49 \times 10^{-10} \tag{D.9a}$$

and

$$\overline{v_{nL}^2} = R_L^2 \overline{i_{nL}^2} = 4kTR \, \Delta f = 0.16 \times 10^{-10} \tag{D.9b}$$

The total output noise voltage is the sum of the above quantities; that is,

$$\overline{v_N^2} = \overline{v_{ns}^2} + \overline{v_{nb}^2} + \overline{v_{nB}^2} + \overline{v_{nc}^2} + \overline{v_{nL}^2} \tag{D.10}$$

$$\overline{v_N^2} = (24.4 + 12.2 + 10.5 + 1.49 + 0.16) \times 10^{-10} = 48.8 \times 10^{-10}$$

$$v_N \simeq 7 \times 10^{-5} = 70 \ \mu\text{V}$$

In other words, in the absence of a signal, i.e., $v_s = 0$, we will have an undesired signal of 70 = μV amplitude at the output of the amplifier. In a cascaded amplifier this voltage will be further increased and amplified by the succeeding stages.

C. Field-Effect Transistors

The noise model for a field-effect transistor is shown in Fig. D.7. The independent noise generators have the values

$$\overline{i_g^2} = 2qI_G \, \Delta f \simeq 0 \tag{D.11a}$$

$$\overline{i_d^2} = 4kT\left(\frac{2}{3} g_m\right) \Delta f + \frac{K_2 I_D}{f} \Delta f \simeq 4kT\left(\frac{2}{3} g_m\right) \Delta f \tag{D.11b}$$

where at midband the second term ($1/f$ noise) can be ignored at high frequencies.

From a comparison of Fig. D.7 and Fig. D.3 it is seen that the noise model for FET devices are simpler than the BJT devices. The noise performance of FET devices

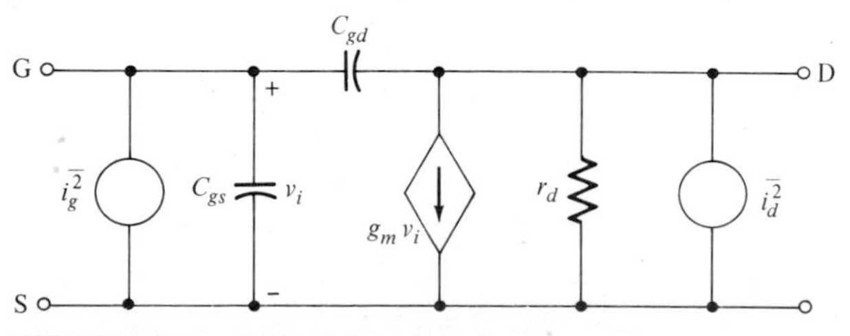

FIGURE D.7 FET noise model.

is much better at a high-input-resistance environment. Calculation of noise voltage at the output of Fig. D.7 is therefore very simple and is due to the current source $\overline{i_d^2}$ and to R_L.

D.3 EQUIVALENT INPUT NOISE GENERATORS IN OP AMPS

The noise generated by a two-port network can be represented by equivalent input noise generators and a noiseless two-port network as shown in Fig. D.8.

In Fig. D.8, $\overline{v_{ni}^2}$ and $\overline{i_{ni}^2}$ are the equivalent noise sources which are partially correlated. For the sake of simplicity we shall assume negligible correlation. Hence the values of each generator can be found individually as follows: short-circuit the input ports of both circuits and equate the open-circuit output noise voltage in each case to determine $\overline{v_{ni}^2}$. Open-circuit the input ports of both circuits and equate the short-circuit output noise currents in each case to determine $\overline{i_{ni}^2}$. The noise representation of BJT and FET are then as shown in Fig. D.9.

Noise representation of an op amp is similarly shown in Fig. D.10. The manufacturers usually provide the data for input noise voltage and input noise current as a function of frequency. A typical data for an op amp is shown in Fig. D.11.

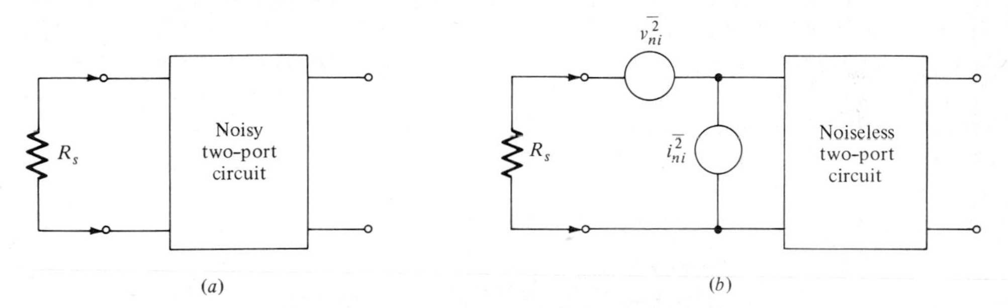

FIGURE D.8 Noise model for two-port circuits with equivalent input noise generators.

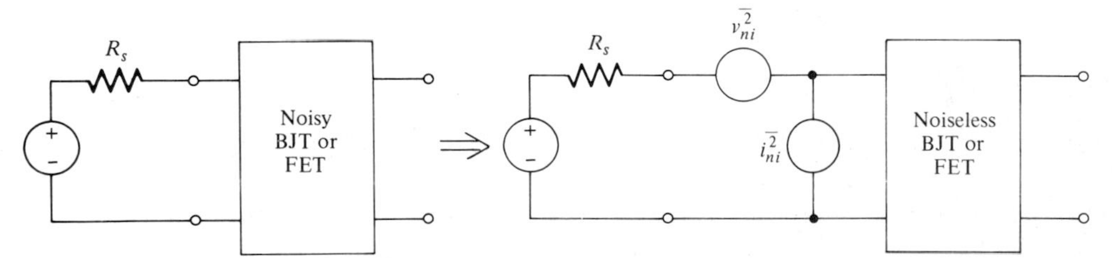

FIGURE D.9 Noise model for BJT and FET with equivalent input noise generators.

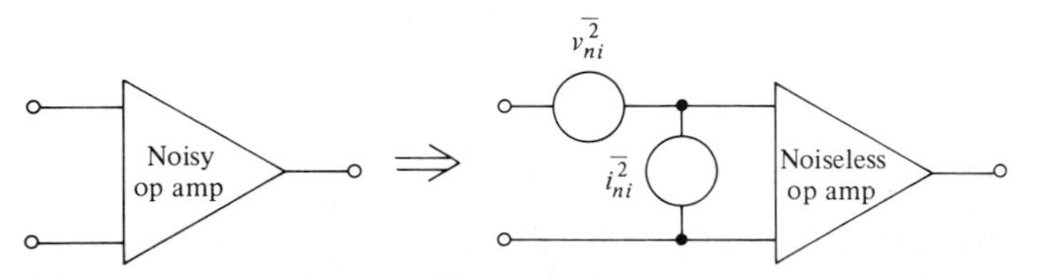

FIGURE D.10 Op-amp noise model.

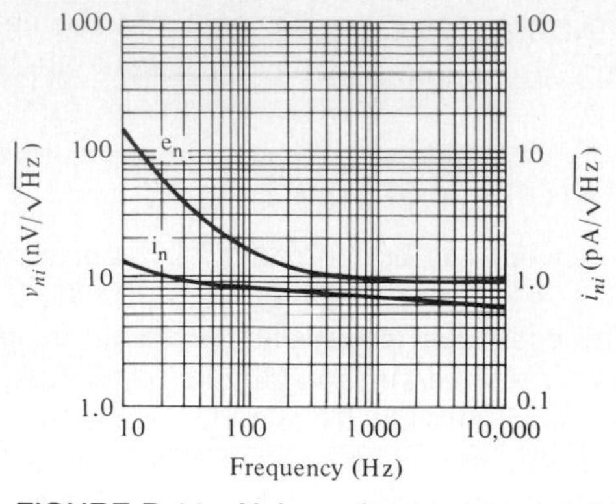

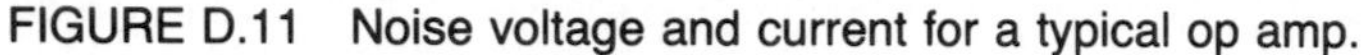

FIGURE D.11 Noise voltage and current for a typical op amp.

EXAMPLE D.2

For the noninverting op-amp circuit shown in Fig. D.12*a*, the op-amp noise voltage and current are given in Fig. D.11. If $R_S = R_2 = 1\ \text{k}\Omega$, $R_1 = 100\ \text{k}\Omega$, determine the total equivalent input noise *per unit bandwidth* at 1 kHz.

First move the current noise generator outside the feedback loop, where $\overline{i_{n2}^2} = \overline{i_{ni}^2}$ as shown in Fig. D.12*b*.

We can include the effect of the thermal noise of $R_1 \| R_2$ by considering it in series with R_S. Thus the thermal noise due the resistors $R_S + R_1 \| R_2 \simeq 2\ \text{k}\Omega$, and from (D.1) is 5.75 nV/$\sqrt{\text{Hz}}$. Reading the values of v_{ni} and i_{ni} at 1 kHz from Fig. D.11 we have $v_{ni} = 9.5$ nV/$\sqrt{\text{Hz}}$ and $i_{ni} = 0.68$ pA/$\sqrt{\text{Hz}}$. The voltage due to this current is obtained by multiplying it by 2 kΩ to obtain 1.36 nV/$\sqrt{\text{Hz}}$. Hence

$$v_N = (5.75^2 + 9.5^2 + 1.36^2)^{1/2} = 11.2\ \text{nV}/\sqrt{\text{Hz}}$$

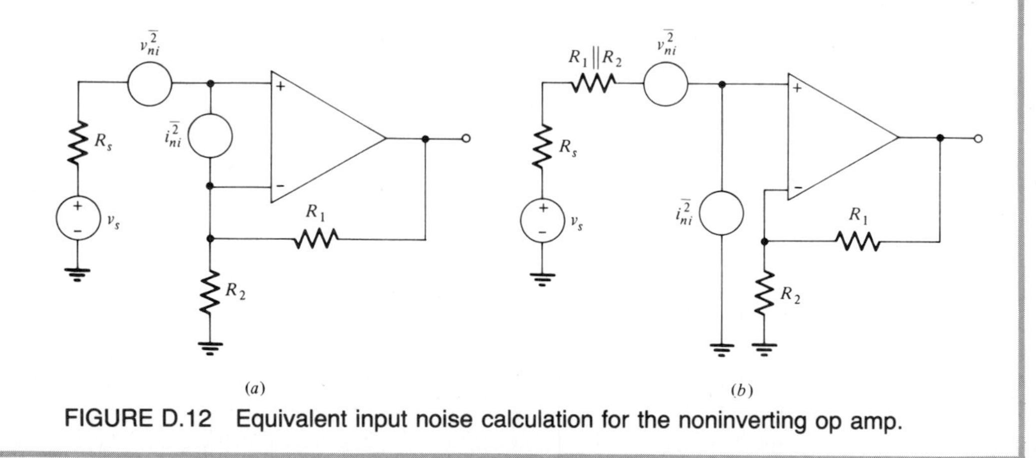

FIGURE D.12 Equivalent input noise calculation for the noninverting op amp.

EXAMPLE D.3

For the inverting op-amp circuit shown in Fig. D.13*a*, $R_1 = 10\ \mathrm{k}\Omega$, $R_2 = 100\ \mathrm{k}\Omega$. Determine the corresponding equivalent input noise per unit bandwidth at 1 kHz.

Again the noise current generator can be moved outside the feedback loop as shown in Fig. D.13*b*, where $\overline{v_{n2}^2} = \overline{v_{ni}^2} + 4kTR_1$ and $\overline{i_{n2}^2} = \overline{i_{ni}^2} + 4kT/R_2$. The thermal noise due to $R_1 = 10\ \mathrm{k}\Omega$ is 12.8 nV/$\sqrt{\mathrm{Hz}}$. The values of v_{ni} and i_{ni} from Fig. D.11 are 9.5 nV/$\sqrt{\mathrm{Hz}}$ and 0.68 pA/$\sqrt{\mathrm{Hz}}$, respectively. Hence the noise voltage due $\overline{i_{ni}^2}$ is $(10K)^2(0.68^2 + 0.40^2) = 6.2\ \mathrm{nV}/\sqrt{\mathrm{Hz}}$. Thus

$$v_N = (9.5^2 + 12.8^2 + 6.2^2)^{1/2} = 17.0\ \mathrm{nV}/\sqrt{\mathrm{Hz}}$$

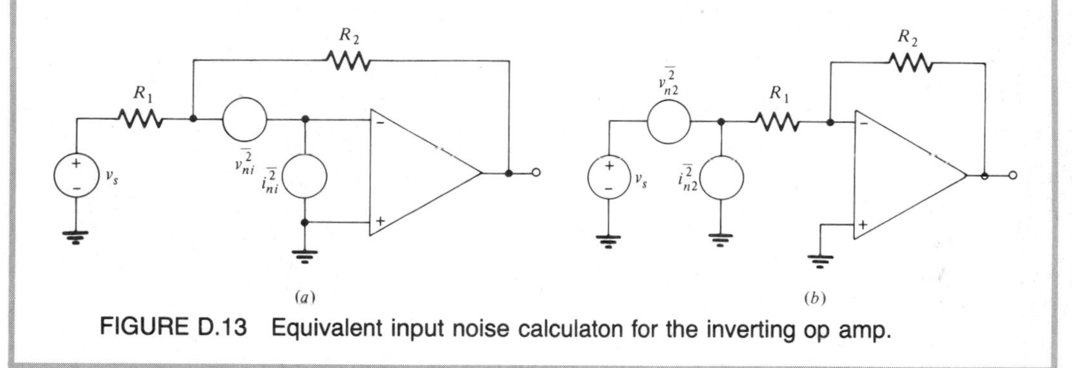

FIGURE D.13 Equivalent input noise calculaton for the inverting op amp.

D.4 NOISE FIGURE

The noise figure provides a quantitative measure of how noisy a circuit is. The noise figure F_n is defined as follows:

$$F_n = 10 \log \frac{\text{total noise power output}}{\text{noise power output due to } R_S \text{ alone}} \tag{D.12a}$$

where R_S is the source resistance. A more convenient form of (D.11*a*) is one of the following:

$$F_n = 10 \log \left. \frac{\text{total mean-square noise current at output}}{\text{mean-square noise current at output due to } R_S \text{ alone}} \right|_{R_L=0} \tag{D.12b}$$

or

$$F_n = 10 \log \left. \frac{\text{total mean-square noise voltage at output}}{\text{mean-square noise voltage at output due to } R_S \text{ alone}} \right|_{R_L=\infty} \tag{D.12c}$$

In some cases the value of R_L is chosen to match the output impedance of the amplifier in order to maximize the power gain. Furthermore, the contribution of R_L to the noise figure is usually insignificant. Note the factor of 10 for calculating the noise figure in decibels, instead of 20, because power ratios are involved, rather than voltage or current ratios.

Since the numerator is equal to the denominator plus additional terms, we may also write (D.12*a*) as

$$F_n = 1 + \frac{R_{neq}}{R_S} \tag{D.13a}$$

or

$$R_{neq} = (F_n - 1)R_S \tag{D.13b}$$

where R_{neq} is a function of kT and the circuit parameters, and its determination is explained in the following example.

EXAMPLE D.4

Determine the noise figure F_n and the equivalent noise resistor R_{neq} of the transistor amplifier shown in Fig. D.4. From (D.12), (D.10), and the results in Example (D.1) we have

$$F_n = \frac{\overline{v_N^2}}{\overline{v_{ns}^2}} = 1 + \frac{\overline{v_{nb}^2} + \overline{v_{nB}^2} + \overline{v_{nc}^2} + \overline{v_{nL}^2}}{\overline{v_{ns}^2}}$$

$$= \frac{48.8}{24.4} = 2 = 3 \text{ dB} \tag{D.14}$$

Thus the noise figure of the one-stage amplifier of the example is 3 dB. From (D.13) and (D.14) the value of $R_{neq} = R_S = 100\ \Omega$.

The expression for the noise figure in terms of the circuit parameters may easily be found by using (D.11). For the CE configuration, from (D.11*c*), (D.6*b*), (D.7*b*), (D.8*b*), and (D.9) we obtain

$$F_n = 10 \log \left[1 + \frac{r_b}{R_S} + \frac{(r_b + R_S)^2}{4kTR_S}\frac{2qI_e}{\beta_0} + \frac{2qI_e}{4kTR_S\,\beta_0}\left(\frac{r_b + R_S + \beta_0 r_e}{\beta_0}\right)^2\right] \text{ (dB)} \tag{D.15a}$$

Substituting $kT/qI_e = r_e$ in (D.15*a*), and using the approximation $\beta_0 \gg 1$, we can simplify (D.15*a*) to

$$F_n = 10 \log \left(1 + \frac{r_b}{R_S} + \frac{r_e}{2R_S} + \frac{(R_S + r_b + r_e)^2}{2\beta_0 r_e R_S}\right) \text{ decibels} \tag{D.15b}$$

From the noise expression (D.15) it is seen that F_n depends both on R_S and r_e. One can optimize F_n with respect to these two quantities, because F_n increases for high

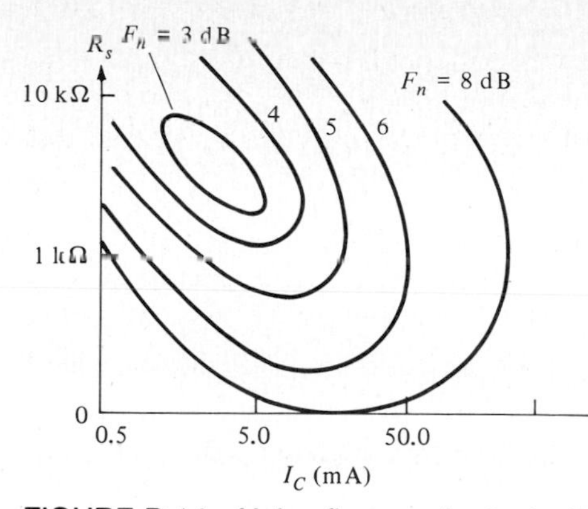

FIGURE D.14 Noise figures of a typical transistor for various source resistors and collector currents.

and very low values of r_e and R_S. If r_e is fixed, i.e., fixed I_e, then F_n can be optimized with respect to R_S. The result is

$$(R_S)_{\text{opt}} = r_b \sqrt{1 + \frac{\beta_0 r_e}{r_b}\left(2 + \frac{r_e}{r_b}\right)} \simeq r_e \sqrt{\beta_0\left(1 + \frac{2r_b}{r_e}\right)} \tag{D.16}$$

For this example $(R_S)_{\text{opt}} = 170$ and $F_{\min} = 10 \log 1.90 = 2.8$ dB. Note that the noise figure for $R_S = 100\ \Omega$ is not far off from the minimum value. Also note from (D.15) that for a low noise figure, transistors with low r_b and high β_0 are preferable. A typical plot of F_n with respect to both I_c and R_S is shown in Fig. D.14. Note that the optimum region is reasonably broad and that F_n is poor for very high and very low I_c. F_n is also high for very low or very high source resistance. Since the CE state has a moderate input impedance, it can best match the optimum source resistance for low noise performance. Thus, these considerations are to be taken into account when a minimal noise amplifier is to be designed. From the typical plot, such as in Fig. D.14, which is usually provided by the manufacturer, the selection of R_S and I_c can be made.

The noise figure of a transistor, in addition to being dependent on R_S and I_c, also depends on frequency as shown in Fig. D.15. Typical variation of F_n is fairly constant, and it increases at both the low and the high frequencies. At low frequencies the $1/f$

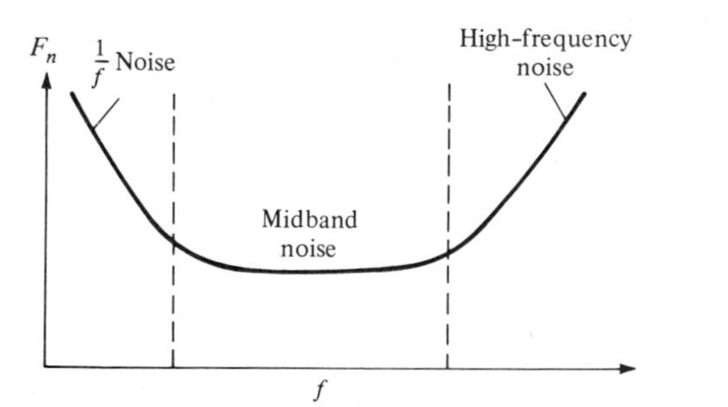

FIGURE D.15 Noise figures of a typical transistor vs frequency.

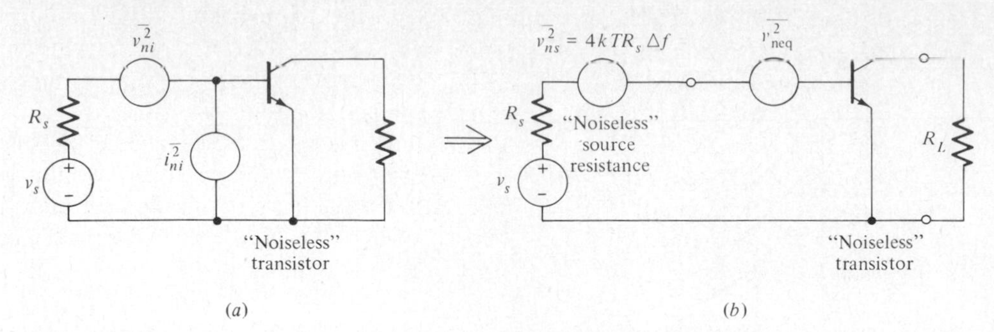

FIGURE D.16 Equivalent input noise voltage representation for BJT.

noise dominates, and it is for this reason that one must be careful in dc-coupled ac amplifiers when noise consideration is an important factor. The high-frequency increase of F_n is essentially due to the increase in $\overline{i_c^2}$ at high frequencies and also the decrease of power gain vs frequency.

The noise generated in a device can also be represented by placing an equivalent input noise voltage at the input of the device as shown for a BJT in Fig. D.16*b*. The same representation can also be made for a FET device, i.e., replace BJT by a FET. In this case, the relation between $\overline{v_{neq}^2}$ and F_n is defined as follows:

$$F_n = 1 + \frac{\overline{v_{neq}^2}}{\overline{v_{ns}^2}} \tag{D.17}$$

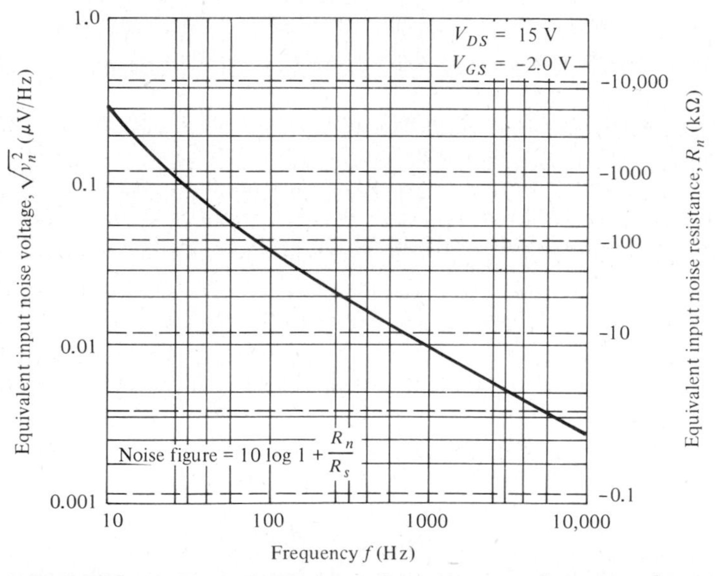

FIGURE D.17 Equivalent input noise voltage vs frequency for a typical FET. (*Courtesy of Fairchild Semiconductor.*)

From (D.17), or (D.12), the value of R_{neq}, or $\overline{v_{neq}^2}$, determines the *noise figure* F_n of the stage, and vice versa. If there are external resistors, the noise contribution due to them should be added to that due to $\overline{v_{neq}^2}$. For very large values of R_S, the noise figure of a FET is much better than that of a bipolar junction transistor, since the noise figure of a FET is small when it is connected to a very large source resistance. Furthermore, the noise figure of a FET is essentially independent of bias conditions. A typical noise figure for a FET is about 2 dB. The manufacturers of FETs usually give the values of v_{neq} or R_{neq} vs frequency. A typical plot is shown in Fig. D.17, where v_n and R_n are used for short.

EXAMPLE D.5

Determine the noise figure in Example D.2. The noise figure for the non-inverting op amp in Example D.2 is found from (D. 14):

$$F_n = \frac{\overline{v_N^2}}{\overline{v_{ns}^2}} = \frac{11.1^2}{2.88^2} = 14.86 = 11.7 \text{ dB}$$

Note that $\overline{v_{ns}^2}$ is due to R_S alone.

D.5 NOISE FIGURE OF CASCADED TWO-PORTS

For a cascade of two-stage amplifiers the overall noise figure is given by

$$F_n = F_{n1} + \frac{F_{n2} - 1}{G_A 1} \tag{D.18}$$

where F_{n1} and F_{n2} are the noise figures of the individual two-port circuits and G_{A1} is the available power gain* of circuit 1 connected to the source resistor. From (D.18) it is seen that in the overall noise figure of the cascaded amplifier, the noise contribution of the first stage is the most significant if the available power gain $G_{A1} \gg 1$. In fact, if G_A of the first stage is very high, the noise contribution due to F_{n2} will be negligible. It is for this reason, i.e., high available power, that in bipolar junction transistor circuitry, minimal noise cascaded networks use a CE stage as the input stage. The second stage, therefore, does not play an important role under the above condition and can be any configuration. However, if the input CE stage is followed by a CB stage such as in cascode circuit Fig. 11.18, the circuit will be nearly unilateral and less prone to instability, and thus the unit can be designed for an overall maximum power gain, with less instability problems than the other combinations.

**Available power gain* is defined as $G_A = P_{avo}/P_{avs}$, where P_{avs} and P_{avo} are the available power from the source and the output of the two-port, respectively. (More specifically for real parameters: $P_{avs} = |V_{in}|^2/4R_S$, i.e., the input impedance is matched to the source impedance.)

If the input source resistance is very high, a hybrid cascode circuit using a FET common source in cascade with a CB bipolar junction transistor of Fig. 5.7 may also be used to obtain a good noise performance. The best noise performance is obtained with a dual-gate MOS (see Sec. 3.8), which is a two-MOSFET cascode with a single channel.

REFERENCES

1. A. Van Der Ziel, *Noise: Sources, Characterization, Measurement,* Prentice-Hall, Englewood Cliffs, N.J., 1970.
2. P. R. Gray and R. G. Meyer, *Analysis and Design of Analog Integrated Circuits,* Wiley, New York, 1977.
3. M. Schwartz, *Information Transmission, Modulation and Noise,* 2d ed., McGraw-Hill, New York, 1980.

APPENDIX

Data Sheets

The purpose of this appendix is to present data for a few representative devices as supplied by the manufacturers. These data sheets are intended to be used as references accompanying the text in order to familiarize the reader with types of data available, and some are used with the problem assignments requiring such data. These should also provide the reader with some knowledge of the practical range of parameter values of the various devices. This appendix includes the following data sheets:

1. Zener diodes: Types 1N4728 to 1N4764
2. Bipolar transistor
   npn: types 2N2221, 2N2222
   pnp: types 2N3250, 2N3251
   General-purpose *npn* transistor array, type CA 3086
3. Field-effect transistors
   JFET: type JN51
   MOS: depletion-type MN82
   MOS: enhancement-type MN83

4. Power transistors
   npn: 2N5878
   pnp: 2N5875
5. Operational amplifiers
   BJT (741)
   BI-FET (LF351)
   Op amp guide
6. IC Gates
   TTL (54/74 series)
   ECL (10K series)
   CMOS (4000B series)
7. *JK* Flip-Flops
   TTL (54/74 series)
   CMOS (4000B series)

For complete information on a variety of discrete and IC devices and functional blocks, the reader is referred to the following representative manufacturers' data books:

Motorola Semiconductor Corporation, Data Book.
RCA Data Books: Semiconductor Devices and Integrated Circuits.
Texas Instruments Data Book: Semiconductors and Components
Fairchild Semiconductor Data Catalogs
National Semiconductor: Linear Data Book
Signetics Corporation: Digital, Linear and MOS Data Book

E.1 ZENER DIODES* (TYPES 1N4728 TO 1N4764)

MAXIMUM RATINGS

Rating	Value	Unit
DC Power Dissipation	1.0	Watt
Derating Factor	6.67	mW/°C
Junction and Storage Temperature	-65 to +200	°C

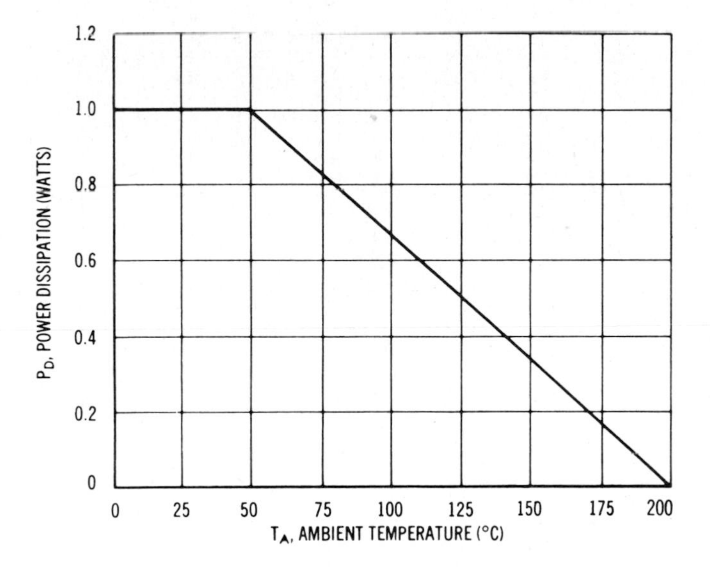

POWER RATING versus AMBIENT TEMPERATURE

*Courtesy of Motorola.

E.1 ZENER DIODES (continued) (TYPES 1N4728 THRU 1N4764)

ELECTRICAL CHARACTERISTICS (At 25°C unless otherwise specified)

$V_F = 1.5$ V max @ $I_F = 200$ mA on all types.

Type No.	Nominal Zener Voltage V_Z @ I_{ZT} Volts	Test Current I_{ZT} mA	Max Zener Impedance (Note 3)			REVERSE LEAKAGE CURRENT		Surge Current @ $T_A = 25°C$ I_R	Max DC Zener Current I_{ZM} mA
			Z_{ZT} @ I_{ZT} Ohms	Z_{ZK} @ I_{ZK} Ohms	I_{ZK} mA	I_R μA Max @	V_R Volts		
1N4728	3.3	76	10	400	1.0	100	1	1380	276
1N4729	3.6	69	10	400	1.0	100	1	1260	252
1N4730	3.9	64	9	400	1.0	50	1	1190	234
1N4731	4.3	58	9	400	1.0	10	1	1070	217
1N4732	4.7	53	8	500	1.0	10	1	970	193
1N4733	5.1	49	7	550	1.0	10	1	890	178
1N4734	5.6	45	5	600	1.0	10	2	810	162
1N4735	6.2	41	2	700	1.0	10	3	730	146
1N4736	6.8	37	3.5	700	1.0	10	4	660	133
1N4737	7.5	34	4.0	700	0.5	10	5	605	121
1N4738	8.2	31	4.5	700	0.5	10	6	550	110
1N4739	9.1	28	5.0	700	0.5	10	7	500	100
1N4740	10	25	7	700	0.25	10	7.6	454	91
1N4741	11	23	8	700	0.25	5	8.4	414	83
1N4742	12	21	9	700	0.25	5	9.1	380	76
1N4743	13	19	10	700	0.25	5	9.9	344	69
1N4744	15	17	14	700	0.25	5	11.4	304	61
1N4745	16	15.5	16	700	0.25	5	12.2	285	57
1N4746	18	14	20	750	0.25	5	13.7	250	50
1N4747	20	12.5	22	750	0.25	5	15.2	225	45
1N4748	22	11.5	23	750	0.25	5	16.7	205	41
1N4749	24	10.5	25	750	0.25	5	18.2	190	38
1N4750	27	9.5	35	750	0.25	5	20.6	170	34
1N4751	30	8.5	40	1,000	0.25	5	22.8	150	30
1N4752	33	7.5	45	1,000	0.25	5	25.1	135	27
1N4753	36	7.0	50	1,000	0.25	5	27.4	125	25
1N4754	39	6.5	60	1,000	0.25	5	29.7	115	23
1N4755	43	6.0	70	1,500	0.25	5	32.7	110	22
1N4756	47	5.5	80	1,500	0.25	5	35.8	95	19
1N4757	51	5.0	95	1,500	0.25	5	38.8	90	18
1N4758	56	4.5	110	2,000	0.25	5	42.6	80	16
1N4759	62	4.0	125	2,000	0.25	5	47.1	70	14
1N4760	68	3.7	150	2,000	0.25	5	51.7	65	13
1N4761	75	3.3	175	2,000	0.25	5	56.0	60	12
1N4762	82	3.0	200	3,000	0.25	5	62.2	55	11
1N4763	91	2.8	250	3,000	0.25	5	69.2	50	10
1N4764	100	2.5	350	3,000	0.25	5	76.0	45	9

E.2 BIPOLAR TRANSISTORS

npn: Types 2N2221, 2N2222*

***electrical characteristics at 25°C free-air temperature (unless otherwise noted)**

PARAMETER		TEST CONDITIONS		TO-5→ 2N2217		2N2218		2N2219		UNIT
				TO-18→ 2N2220		2N2221		2N2222		
				MIN	MAX	MIN	MAX	MIN	MAX	
$V_{(BR)CBO}$	Collector-Base Breakdown Voltage	$I_C = 10\ \mu A$, $I_E = 0$		60		60		60		V
$V_{(BR)CEO}$	Collector-Emitter Breakdown Voltage	$I_C = 10$ mA, $I_B = 0$,	See Note 6	30		30		30		V
$V_{(BR)EBO}$	Emitter-Base Breakdown Voltage	$I_E = 10\ \mu A$, $I_C = 0$		5		5		5		V
I_{CBO}	Collector Cutoff Current	$V_{CB} = 50$ V, $I_E = 0$			10		10		10	nA
		$V_{CB} = 50$ V, $I_E = 0$,	$T_A = 150°C$		10		10		10	μA
I_{EBO}	Emitter Cutoff Current	$V_{EB} = 3$ V, $I_C = 0$			10		10		10	nA
h_{FE}	Static Forward Current Transfer Ratio	$V_{CE} = 10$ V, $I_C = 100\ \mu A$				20		35		
		$V_{CE} = 10$ V, $I_C = 1$ mA		12		25		50		
		$V_{CE} = 10$ V, $I_C = 10$ mA	See Note 6	17		35		75		
		$V_{CE} = 10$ V, $I_C = 150$ mA		20	60	40	120	100	300	
		$V_{CE} = 10$ V, $I_C = 500$ mA				20		30		
		$V_{CE} = 1$ V, $I_C = 150$ mA		10		20		50		
V_{BE}	Base-Emitter Voltage	$I_B = 15$ mA, $I_C = 150$ mA	See Note 6		1.3		1.3		1.3	V
		$I_B = 50$ mA, $I_C = 500$ mA					2.6		2.6	
$V_{CE(sat)}$	Collector-Emitter Saturation Voltage	$I_B = 15$ mA, $I_C = 150$ mA	See Note 6		0.4		0.4		0.4	V
		$I_B = 50$ mA, $I_C = 500$ mA					1.6		1.6	
$\lvert h_{fe} \rvert$	Small-Signal Common-Emitter Forward Current Transfer Ratio	$V_{CE} = 20$ V, $I_C = 20$ mA,	f = 100 MHz	2.5		2.5		2.5		
f_T	Transition Frequency	$V_{CE} = 20$ V, $I_C = 20$ mA,	See Note 7	250		250		250		MHz
C_{obo}	Common-Base Open-Circuit Output Capacitance	$V_{CB} = 10$ V, $I_E = 0$,	f = 1 MHz		8		8		8	pF
$h_{ie(real)}$	Real Part of Small-Signal Common-Emitter Input Impedance	$V_{CE} = 20$ V, $I_C = 20$ mA,	f = 300 MHz		60		60		60	Ω

NOTES: 6. These parameters must be measured using pulse techniques. $t_w = 300\ \mu s$, duty cycle ≤ 2%.
7. To obtain f_T, the $\lvert h_{fe} \rvert$ response with frequency is extrapolated at the rate of −6 dB per octave from f = 100 MHz to the frequency at which $\lvert h_{fe} \rvert = 1$.

---

*Courtesy of Texas Instruments Inc.

E.2 BIPOLAR TRANSISTORS (continued)

*absolute maximum ratings at 25°C free-air temperature (unless otherwise noted)

	2N2217 2N2218 2N2219	2N2218A 2N2219A	2N2220 2N2221 2N2222	2N2221A 2N2222A	UNIT
Collector-Base Voltage	60	75	60	75	V
Collector-Emitter Voltage (See Note 1)	30	40	30	40	V
Emitter-Base Voltage	5	6	5	6	V
Continuous Collector Current	0.8	0.8	0.8	0.8	A
Continuous Device Dissipation at (or below) 25°C Free-Air Temperature (See Notes 2 and 3)	0.8	0.8	0.5	0.5	W
Continuous Device Dissipation at (or below) 25°C Case Temperature (See Notes 4 and 5)	3	3	1.8	1.8	W
Operating Collector Junction Temperature Range	−65 to 175				°C
Storage Temperature Range	−65 to 200				°C
Lead Temperature 1/16 Inch from Case for 10 Seconds	230				°C

NOTES: 1. These values apply between 0 and 500 mA collector current when the base-emitter diode is open-circuited.
2. Derate 2N2217, 2N2218, 2N2218A, 2N2219, and 2N2219A linearly to 175°C free-air temperature at the rate of 5.33 mW/°C.
3. Derate 2N2220, 2N2221, 2N2221A, 2N2222, and 2N2222A linearly to 175°C free-air temperature at the rate of 3.33 mW/°C.
4. Derate 2N2217, 2N2218, 2N2218A, 2N2219, and 2N2219A linearly to 175°C case temperature at the rate of 20.0 mW/°C.
5. Derate 2N2220, 2N2221, 2N2221A, 2N2222, and 2N2222A linearly to 175°C case temperature at the rate of 12.0 mW/°C.

*JEDEC registered data. This data sheet contains all applicable registered data in effect at the time of publication.

USES CHIP N24

*operating characteristics at 25°C free-air temperature

PARAMETER		TEST CONDITIONS	TO-5→ 2N2218A / TO-18→ 2N2221A MAX	TO-5→ 2N2219A / TO-18→ 2N2222A MAX	UNIT
F	Spot Noise Figure	$V_{CE} = 10$ V, $I_C = 100\ \mu A$, $R_G = 1\ k\Omega$, $f = 1$ kHz		4	dB

*switching characteristics at 25°C free-air temperature

PARAMETER		TEST CONDITIONS†	TO-5→ 2N2218A / TO-18→ 2N2221A MAX	TO-5→ 2N2219A / TO-18→ 2N2222A MAX	UNIT
t_d	Delay Time	$V_{CC} = 30$ V, $I_C = 150$ mA, $I_{B(1)} = 15$ mA, $V_{BE(off)} = -0.5$ V, See Figure 1	10	10	ns
t_r	Rise Time		25	25	ns
τ_A	Active Region Time Constant‡		2.5	2.5	ns
t_s	Storage Time	$V_{CC} = 30$ V, $I_C = 150$ mA, $I_{B(1)} = 15$ mA, $I_{B(2)} = -15$ mA, See Figure 2	225	225	ns
t_f	Fall Time		60	60	ns

†Voltage and current values shown are nominal; exact values vary slightly with transistor parameters.

‡Under the given conditions τ_A is equal to $\frac{t_r}{10}$.

*PARAMETER MEASUREMENT INFORMATION

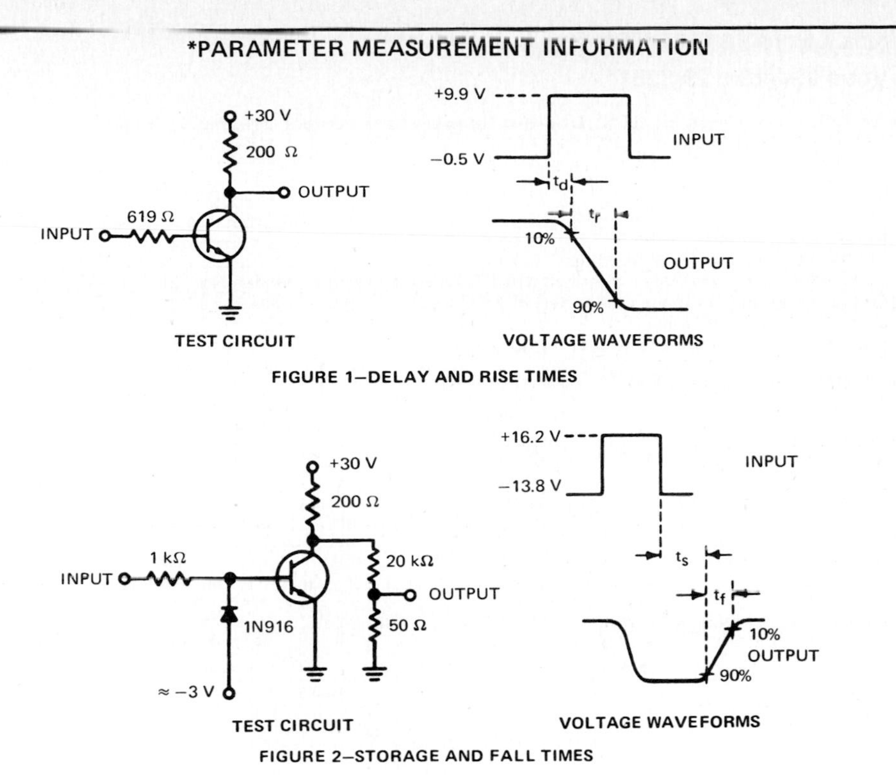

FIGURE 1—DELAY AND RISE TIMES

FIGURE 2—STORAGE AND FALL TIMES

NOTES: a. The input waveforms have the following characteristics: For Figure 1, $t_r \leqslant 2$ ns, $t_w \leqslant 200$ ns, duty cycle $\leqslant$ 2%; for Figure 2, $t_f \leqslant 5$ ns, $t_w \approx 100$ μs, duty cycle $\leqslant$ 17%.

b. All waveforms are monitored on an oscilloscope with the following characteristics: $t_r \leqslant 5$ ns, $R_{in} \geqslant 100$ kΩ, $C_{in} \leqslant 12$ pF.

*JEDEC registered data

DESIGNED FOR HIGH-SPEED, MEDIUM-POWER SWITCHING AND GENERAL PURPOSE AMPLIFIER APPLICATIONS

- h_{FE} . . . Guaranteed from 100 μA to 500 mA
- High f_T at 20 V, 20 mA . . . 300 MHz (2N2219A, 2N2222A)
  250 MHz (all others)
- 2N2218, 2N2221 for Complementary Use with 2N2904, 2N2906
- 2N2219, 2N2222 for Complementary Use with 2N2905, 2N2906

E.2 BIPOLAR TRANSISTORS (continued)
pnp: Types 2N3250, 2N3251*

***absolute maximum ratings at 25°C free-air temperature (unless otherwise noted)**

	2N3250 2N3251	2N3250A 2N3251A
Collector-Base Voltage	−50 V	−60 V
Collector-Emitter Voltage (See Note 1)	−40 V	−60 V
Emitter-Base Voltage	−5 V	−5 V
Continuous Collector Current	←−200 mA→	
Continuous Device Dissipation at (or below) 25°C Free-Air Temperature (See Note 2)	← 0.36 W →	
Continuous Device Dissipation at (or below) 25°C Case Temperature (See Note 3)	← 1.2 W →	
Storage Temperature Range	−65°C to 200°C	
Lead Temperature 1/16 Inch from Case for 60 Seconds	← 300°C →	

***electrical characteristics at 25°C free-air temperature**

PARAMETER		TEST CONDITIONS		2N3250		2N3250A		2N3251		2N3251A		UNIT
				MIN	MAX	MIN	MAX	MIN	MAX	MIN	MAX	
$V_{(BR)CBO}$	Collector-Base Breakdown Voltage	$I_C = -10\ \mu A$, $I_E = 0$		−50		−60		−50		−60		V
$V_{(BR)CEO}$	Collector-Emitter Breakdown Voltage	$I_C = -10$ mA, $I_B = 0$,	See Note 4	−40		−60		−40		−60		V
$V_{(BR)EBO}$	Emitter-Base Breakdown Voltage	$I_E = -10\ \mu A$, $I_C = 0$		−5		−5		−5		−5		V
I_{CEV}	Collector Cutoff Current	$V_{CE} = -40$ V, $V_{BE} = 3$ V			−20		−20		−20		−20	nA
I_{BEV}	Base Cutoff Current	$V_{CE} = -40$ V, $V_{BE} = 3$ V			50		50		50		50	nA
h_{FE}	Static Forward Current Transfer Ratio	$V_{CE} = -1$ V, $I_C = -0.1$ mA	See Note 4	40		40		80		80		
		$V_{CE} = -1$ V, $I_C = -1$ mA		45		45		90		90		
		$V_{CE} = -1$ V, $I_C = -10$ mA		50	150	50	150	100	300	100	300	
		$V_{CE} = -1$ V, $I_C = -50$ mA		15		15		30		30		
V_{BE}	Base-Emitter Voltage	$I_B = -1$ mA, $I_C = -10$ mA	See Note 4	−0.6	−0.9	−0.6	−0.9	−0.6	−0.9	−0.6	−0.9	V
		$I_B = -5$ mA, $I_C = -50$ mA			−1.2		−1.2		−1.2		−1.2	V
$V_{CE(sat)}$	Collector-Emitter Saturation Voltage	$I_B = -1$ mA, $I_C = -10$ mA			−0.25		−0.25		−0.25		−0.25	V
		$I_B = -5$ mA, $I_C = -50$ mA			−0.5		−0.5		−0.5		−0.5	V
h_{ie}	Small-Signal Common-Emitter Input Impedance	$V_{CE} = -10$ V, $I_C = -1$ mA, f = 1 kHz		1	6	1	6	2	12	2	12	kΩ
h_{fe}	Small-Signal Common-Emitter Forward Current Transfer Ratio			50	200	50	200	100	400	100	400	
h_{re}	Small-Signal Common-Emitter Reverse Voltage Transfer Ratio				10×10^{-4}		10×10^{-4}		20×10^{-4}		20×10^{-4}	
h_{oe}	Small-Signal Common-Emitter Output Admittance			4	40	4	40	10	60	10	60	µmho

NOTES: 1. These values apply between 0 and 200 mA collector current when the base-emitter diode is open-circuited.
2. Derate linearly to 200°C free-air temperature at the rate of 2.06 mW/deg.
3. Derate linearly to 200°C case temperature at the rate of 6.9 mW/deg.
4. These parameters must be measured using pulse techniques. $t_p = 300\ \mu s$, duty cycle $\leq 2\%$.

*Indicates JEDEC registered data

USES CHIP P23

***electrical characteristics at 25°C free-air temperature (continued)**

PARAMETER		TEST CONDITIONS	2N3250 2N3250A		2N3251 2N3251A		UNIT
			MIN	MAX	MIN	MAX	
$\|h_{fe}\|$	Small-Signal Common-Emitter Forward Current Transfer Ratio	$V_{CE} = -20$ V, $I_C = -10$ mA, f = 100 MHz	2.5		3		
f_T	Transition Frequency	$V_{CE} = -20$ V, $I_C = -10$ mA, See Note 5	250		300		MHz
C_{obo}	Common-Base Open-Circuit Output Capacitance	$V_{CB} = -10$ V, $I_E = 0$, f = 100 kHz		6		6	pF
C_{ibo}	Common-Base Open-Circuit Input Capacitance	$V_{EB} = -1$ V, $I_C = 0$, f = 100 kHz		8		8	pF
$r_b'C_c$	Collector-Base Time Constant	$V_{CE} = -20$ V, $I_C = -10$ mA, f = 31.8 MHz		250		250	ps

NOTE 5: To obtain f_T, the $|h_{fe}|$ response with frequency is extrapolated at the rate of −6 dB per octave from f = 100 MHz to the frequency at which $|h_{fe}| = 1$.

*Courtesy of Texas Instruments Inc.

*operating characteristics at 25°C free-air temperature

PARAMETER		TEST CONDITIONS	2N3250 2N3250A MAX	2N3251 2N3251A MAX	UNIT
NF	Spot Noise Figure	$V_{CE} = -5$ V, $I_C = -100$ μA, $R_G = 1$ kΩ, f = 100 Hz	6	6	dB

*switching characteristics at 25°C free-air temperature

PARAMETER		TEST CONDITIONS†	2N3250 2N3250A MAX	2N3251 2N3251A MAX	UNIT
t_d	Delay Time	$I_C = -10$ mA, $I_{B(1)} = -1$ mA, $V_{BE(off)} = 0.5$ V,	35	35	ns
t_r	Rise Time	$R_L = 275$ Ω, See Figure 1	35	35	ns
t_s	Storage Time	$I_C = -10$ mA, $I_{B(1)} = -1$ mA, $I_{B(2)} = 1$ mA,	175	200	ns
t_f	Fall Time	$R_L = 275$ Ω, See Figure 2	50	50	ns

†Voltage and current values shown are nominal; exact values vary slightly with transistor parameters. Nominal base current for delay and rise times is calculated using the minimum value of V_{BE}. Nominal base currents for storage and fall times are calculated using the maximum value of V_{BE}.

*PARAMETER MEASUREMENT INFORMATION

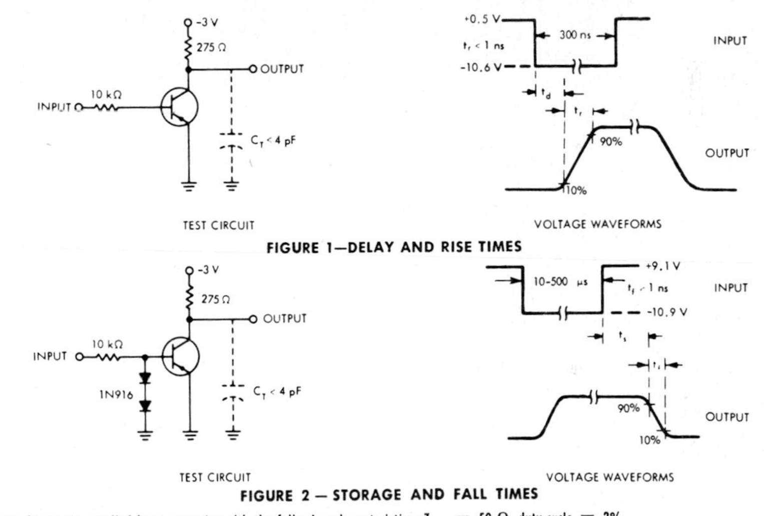

FIGURE 1—DELAY AND RISE TIMES

FIGURE 2 — STORAGE AND FALL TIMES

NOTES: a. The input waveforms are supplied by a generator with the following characteristics: $Z_{out} = 50$ Ω, duty cycle = 2%.

b. Waveforms are monitored on an oscilloscope with the following characteristics: $t_r \leq 1$ ns, $R_{in} \geq 100$ kΩ.

*Indicates JEDEC registered data

E.2 BIPOLAR TRANSISTORS (continued)
npn: CA3086* General-Purpose N-P-N Transistor Array

Three Isolated Transistors and One Differentially– Connected Transistor Pair

For Low–Power Applications from DC to 120MHz

RCA-CA3086 consists of five general-purpose silicon n-p-n transistors on a common monolithic substrate. Two of the transistors are internally connected to form a differentially-connected pair.

The transistors of the CA3086 are well suited to a wide variety of applications in low-power systems at frequencies from DC to 120 MHz. They may be used as discrete transistors in conventional circuits. However, they also provide the very significant inherent advantages unique to integrated circuits, such as compactness, ease of physical handling and thermal matching.

The CA3086 is supplied in a 14-lead dual-in line plastic package. The CA3086F is supplied in a 14-lead dual-in-line hermetic (frit-seal) ceramic package.

Applications

- **General-purpose use in signal processing systems operating in the DC to 120-MHz range**
- **Temperature compensated amplifiers**
- **See RCA Application Note, ICAN-5296 "Application of the RCA-CA3018 Integrated-Circuit Transistor Array" for suggested applications.**

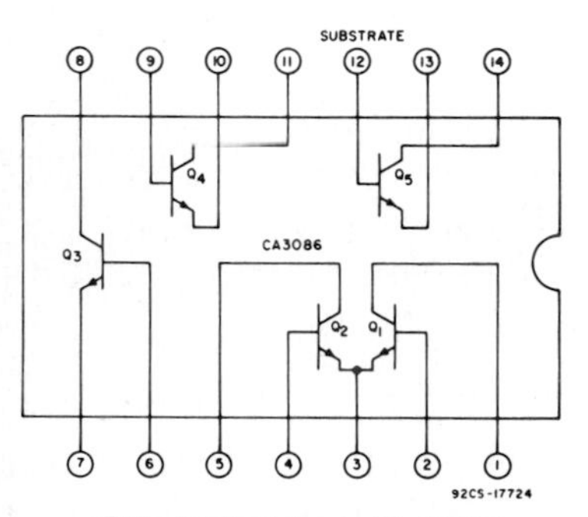

Fig.1 – Functional diagram of the CA3086.

MAXIMUM RATINGS, Absolute–Maximum Values at $T_A = 25^{\circ}C$

DISSIPATION:		
Any one transistor	300	mW
Total package up to $T_A = 55^{\circ}C$	750	mW
Above $T_A = 55^{\circ}C$	derate linearly 6.67	mW/°C
AMBIENT TEMPERATURE RANGE:		
Operating	−55 to + 125	°C
Storage	−65 to + 150	°C
LEAD TEMPERATURE (During soldering):		
At distance 1/16 ± 1/32 inch (1.59 ± 0.79mm) From case for 10 seconds max.	+ 265	°C
The following ratings apply for each transistor in the device:		
COLLECTOR-TO-EMITTER VOLTAGE, V_{CEO}	15	V
COLLECTOR-TO-BASE VOLTAGE, V_{CBO}	20	V
COLLECTOR-TO-SUBSTRATE VOLTAGE, V_{CIO}*	20	V
EMITTER-TO-BASE VOLTAGE, V_{EBO}	5	V
COLLECTOR CURRENT, I_C	50	mA

* The collector of each transistor in the CA3086 is isolated from the substrate by an integral diode. The substrate (terminal 13) must be connected to the most negative point in the external circuit to maintain isolation between transistors and to provide for normal transistor action. To avoid undesirable coupling between transistors, the substrate (terminal 13) should be maintained at either DC or signal (AC) ground. A suitable bypass capacitor can be used to establish a signal ground.

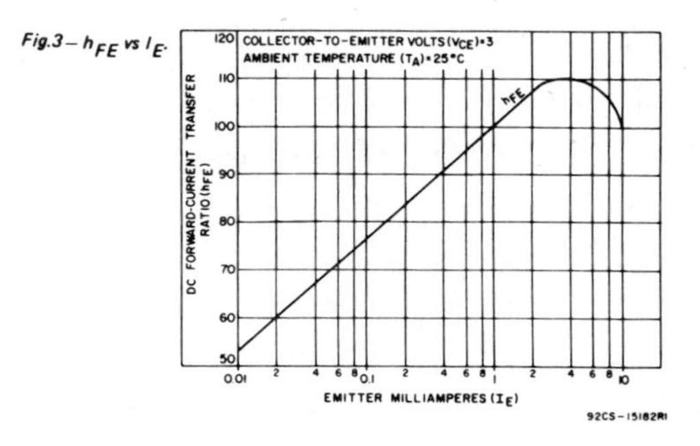

Fig.3 – h_{FE} vs I_E.

TYPICAL STATIC CHARACTERISTICS FOR EACH TRANSISTOR

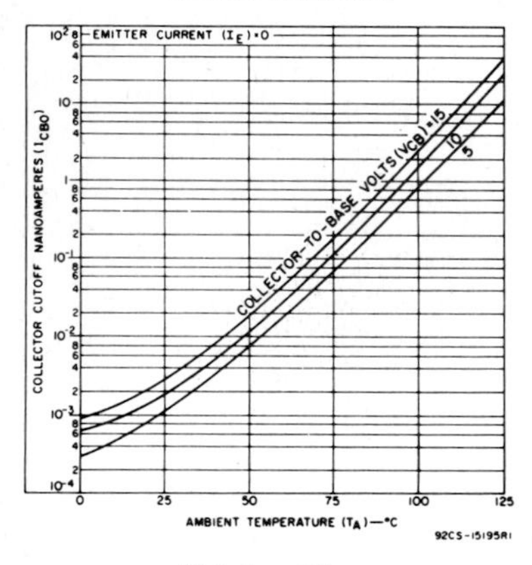

Fig.2 – I_{CBO} vs T_A.

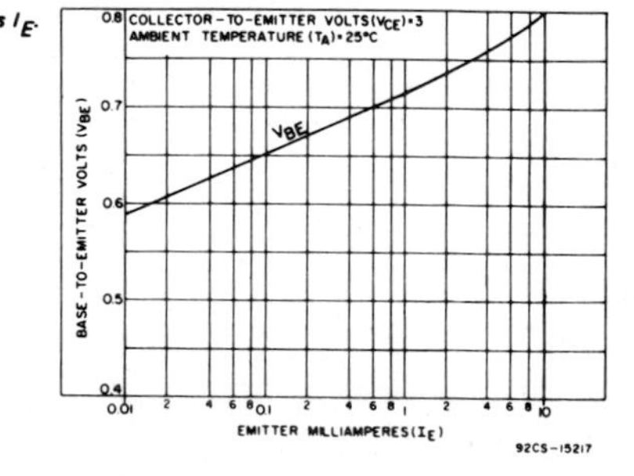

Fig.4 V_{BE} vs I_E.

*Courtesy of RCA.

ELECTRICAL CHARACTERISTICS at $T_A = 25^oC$
For Equipment Design

CHARACTERISTICS	SYMBOLS	TEST CONDITIONS	Typ. Characteristic Curves Fig. No.	LIMITS Min.	LIMITS Typ.	LIMITS Max.	UNITS
Collector-to-Base Breakdown Voltage	$V_{(BR)CBO}$	$I_C = 10\,\mu A$, $I_E = 0$	–	20	60	–	V
Collector-to-Emitter Breakdown Voltage	$V_{(BR)CEO}$	$I_C = 1mA$, $I_B = 0$	–	15	24	–	V
Collector-to-Substrate Breakdown Voltage	$V_{(BR)CIO}$	$I_C = 10\,\mu A$, $I_{CI} = 0$	–	20	60	–	V
Emitter-to-Base Breakdown Voltage	$V_{(BR)EBO}$	$I_E = 10\,\mu A$, $I_C = 0$	–	5	7	–	V
Collector-Cutoff Current	I_{CBO}	$V_{CB} = 10V$, $I_E = 0$	2	–	0.002	100	nA
Collector-Cutoff Current	I_{CEO}	$V_{CE} = 10V$, $I_B = 0$	3	—	See Curve	5	μA
DC Forward-Current Transfer Ratio	h_{FE}	$V_{CE} = 3V$, $I_C = 1mA$	4	40	100	–	

ELECTRICAL CHARACTERISTICS at $T_A = 25^oC$ Typical Values Intended Only for Design Guidance

CHARACTERISTICS	SYMBOL	TEST CONDITIONS		Typ. Characteristics Curves Fig. No.	TYPICAL VALUES	UNITS
DC Forward-Current Transfer Ratio	h_{FE}	$V_{CE} = 3V$	$I_C = 10mA$	4	100	
			$I_C = 10\,\mu A$	4	54	
Base-to-Emitter Voltage	V_{BE}	$V_{CE} = 3V$	$I_E = 1mA$	5	0.715	V
			$I_E = 10mA$	5	0.800	V
V_{BE} Temperature Coefficient	$\Delta V_{BE}/\Delta T$	$V_{CE} = 3V$, $I_C = 1mA$		6	–1.9	$mV/^oC$
Collector-to-Emitter Saturation Voltage	V_{CEsat}	$I_B = 1mA$, $I_C = 10mA$		–	0.23	V
Noise Figure (low frequency)	NF	$f = 1kHz$, $V_{CE} = 3V$, $I_C = 100\mu A$, $R_S = 1k\,\Omega$		–	3.25	dB
Low-Frequency, Small-Signal Equivalent-Circuit Characteristics:						
Forward Current-Transfer Ratio	h_{fe}	$f = 1kHz$, $V_{CE} = 3V$, $I_C = 1mA$		7	100	–
Short-Circuit Input Impedance	h_{ie}			7	3.5	$k\Omega$
Open-Circuit Output Impedance	h_{oe}			7	15.6	μmho
Open-Circuit Reverse-Voltage Transfer Ratio	h_{re}			7	1.8×10^{-4}	–
Admittance Characteristics:						
Forward Transfer Admittance	y_{fe}	$f = 1MHz$, $V_{CE} = 3V$, $I_C = 1mA$		8	31 – j1.5	mmho
Input Admittance	y_{ie}			9	0.3 + j0.04	mmho
Output Admittance	y_{oe}			10	0.001 + j0.03	mmho
Reverse Transfer Admittance	y_{re}			11	See Curve	–
Gain-Bandwidth Product	f_T	$V_{CE} = 3V$, $I_C = 3mA$		12	550	MHz
Emitter-to-Base Capacitance	C_{EBO}	$V_{EB} = 3V$, $I_E = 0$		–	0.6	pF
Collector-to-Base Capacitance	C_{CBO}	$V_{CB} = 3V$, $I_C = 0$		–	0.58	pF
Collector-to-Substrate Capacitance	C_{CIO}	$V_{CI} = 3V$, $I_C = 0$		—	2.8	pF

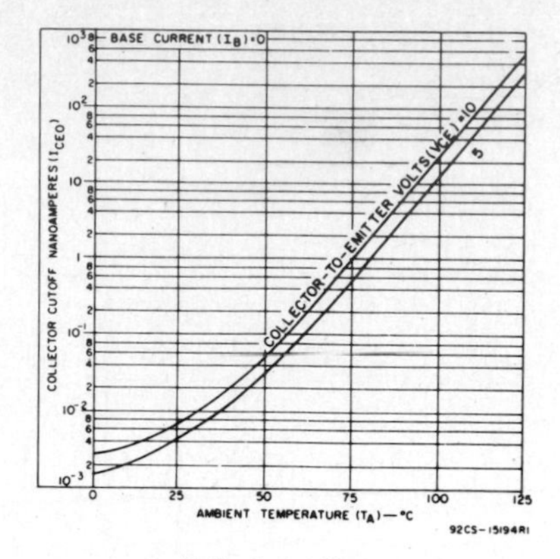

Fig.5 – I_{CEO} vs T_A.

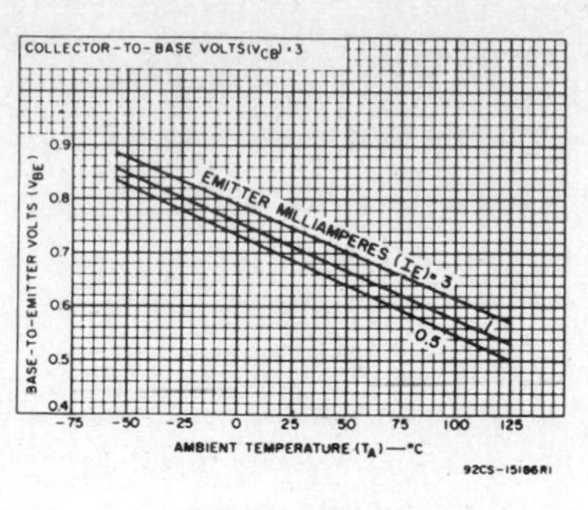

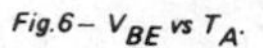

Fig.6 – V_{BE} vs T_A.

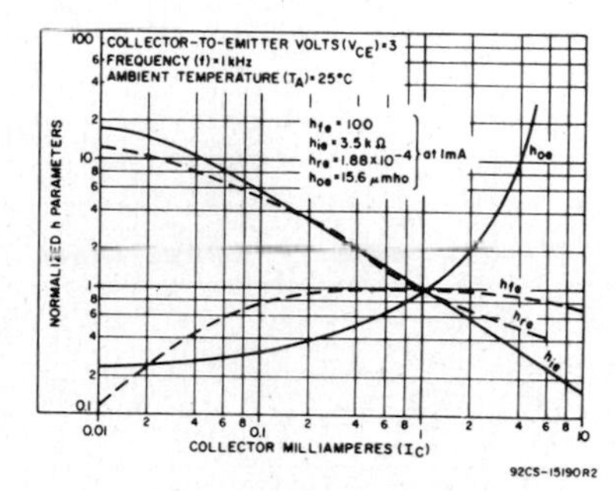

Fig.7 – Normalized h_{fe}, h_{ie}, h_{oe}, h_{re} vs I_C.

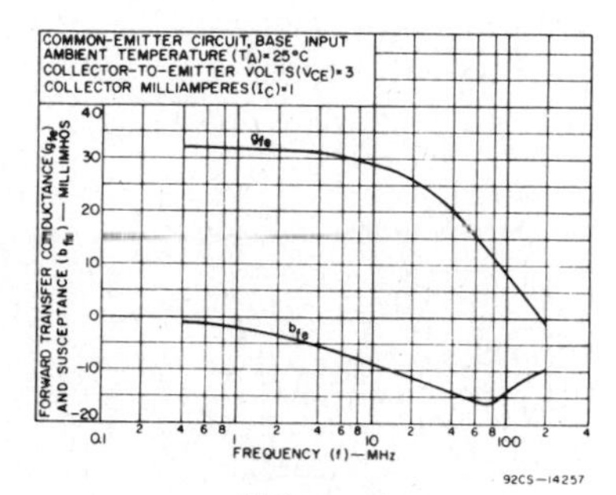

Fig.8 – y_{fe} vs f.

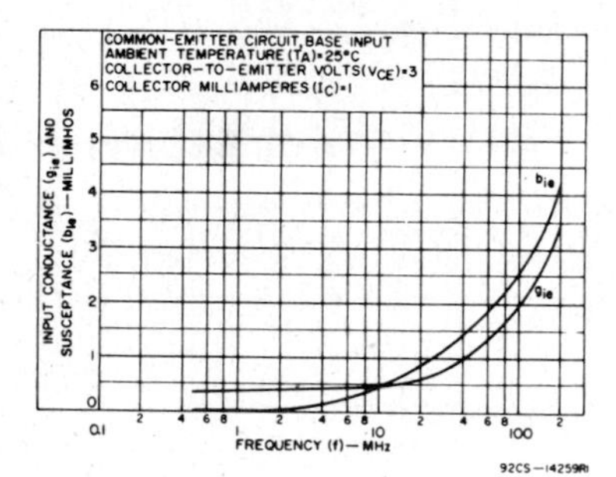

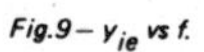

Fig.9 – y_{ie} vs f.

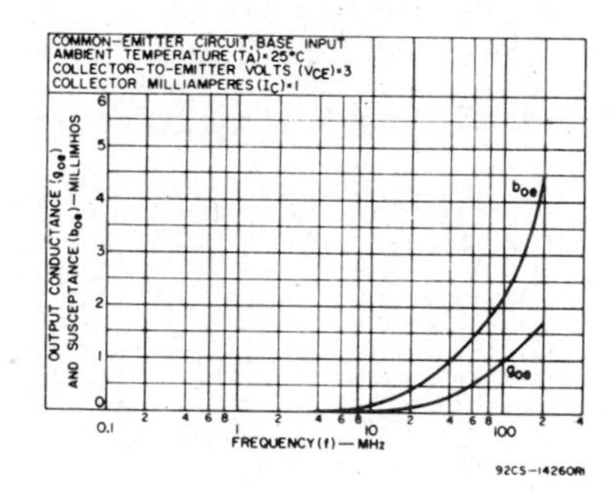

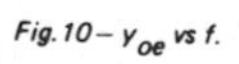

Fig.10 – y_{oe} vs f.

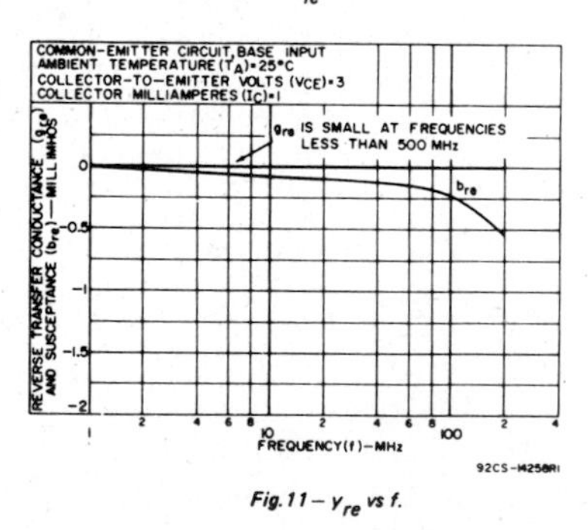

Fig.11 – y_{re} vs f.

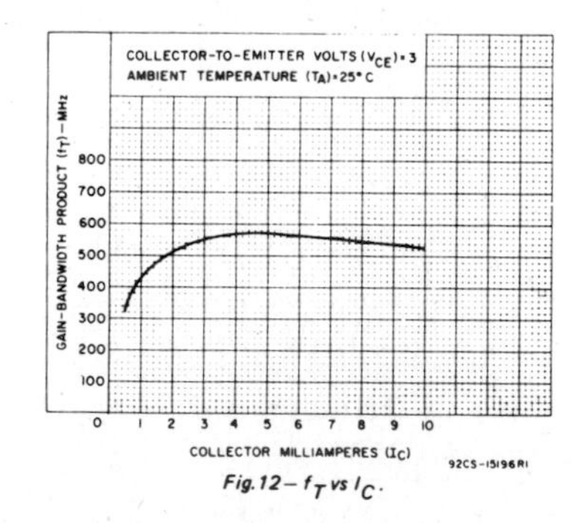

Fig.12 – f_T vs I_C.

E.3 FIELD-EFFECT TRANSISTORS

JFET: Type JN51*

electrical and operating characteristics at 25°C free-air temperature

PARAMETER		CONDITIONS			OBSERVED VALUES LOW	TYP	HIGH	UNIT
$V_{(BR)GSS}$	Gate-Source Breakdown Voltage	$I_G = -1\ \mu A$,	$V_{DS} = 0$		−30♦	−75	−100	V
I_{GSS}	Gate Reverse Current	$V_{GS} = -15$ V,	$V_{DS} = 0$			−<0.1	−2	nA
$V_{GS(off)}$	Gate-Source Cutoff Voltage	$V_{DS} = 15$ V,	$I_D = 0.5$ nA		−0.35	−3.5	−9	V
V_{GS}	Gate-Source Voltage	$V_{DS} = 15$ V,	$I_D = 100\ \mu A$		−0.25	−3	−8	V
I_{DSS}	Zero-Gate-Voltage Drain Current	$V_{DS} = 15$ V,	$V_{GS} = 0$,	See Note 1	0.5	10	24	mA
$r_{ds(on)}$	Small-Signal Drain-Source On-State Resistance	$V_{DS} = 0$,	$I_D = 0$,	f = 1 kHz	100	200	2000	Ω
$\|y_{fs}\|$	Small-Signal Common-Source Forward Transfer Admittance	$V_{DS} = 0$,	$V_{GS} = 0$,	f = 1 kHz	2	4.8	7	mmho
$\|y_{os}\|$	Small-Signal Common-Source Output Admittance					25	70	μmho
C_{iss}	Common-Source Short-Circuit Input Capacitance	$V_{DS} = 15$ V, See Note 2	$V_{GS} = 0$,	f = 1 MHz,	3.5	4.7	6	pF
C_{rss}	Common-Source Short-Circuit Reverse Transfer Capacitance				0.9	1.4	2	pF
g_{is}	Small-Signal Common-Source Input Conductance	$V_{DS} = 15$ V,	$V_{GS} = 0$,	f = 100 MHz		90	250	μmho
g_{fs}	Small-Signal Common-Source Forward Transfer Conductance				1	4	7	mmho
g_{os}	Small-Signal Common-Source Output Conductance					60	150	μmho
$\|y_{fs}\|$	Small-Signal Common-Source Forward Transfer Admittance	$V_{DS} = 15$ V,	$V_{GS} = 0$,	f = 200 MHz	2	4		mmho
g_{is}	Small-Signal Common-Source Input Conductance					0.5	1	mmho
g_{os}	Small-Signal Common-Source Output Conductance					0.15	0.3	mmho
F	Spot Noise Figure	$V_{DS} = 15$ V, $R_G = 1$ MΩ	$V_{GS} = 0$,	f = 10 Hz,		4.5	5	dB
		$V_{DS} = 15$ V, $R_G = 1$ MΩ	$V_{GS} = 0$,	f = 1 kHz,		0.2	2	
		$V_{DS} = 15$ V, $R_G = 1$ kΩ	$V_{GS} = 0$,	f = 100 MHz,		3	5	
V_n	Equivalent Input Noise Voltage	$V_{DS} = 15$ V,	$V_{GS} = 0$	f = 10 Hz		170	300	nV/$\sqrt{Hz}$
				f = 1 kHz		15	100	

†Trademark of Texas Instruments

♦This value does not modify guaranteed limits for specific devices and does not justify operation in excess of absolute maximum ratings.

NOTES: 1. This parameter was measured using pulse techniques. $t_w = 300\ \mu s$, duty cycle ⩽ 2%.

2. Capacitance measurements were made using chips mounted in *Silect* packages.

- JN51 is a 17 X 17-mil, epitaxial, planar, expanded-contact chip
- Available in TO-18, TO-71, TO-72, a short-can version of TO-78, and *Silect*† packages
- For use in low-noise amplifier, mixer, switching, and chopper circuits

*Courtesy of Texas Instruments Inc.

E.3 FIELD-EFFECT TRANSISTORS (continued)

TYPICAL CHARACTERISTICS

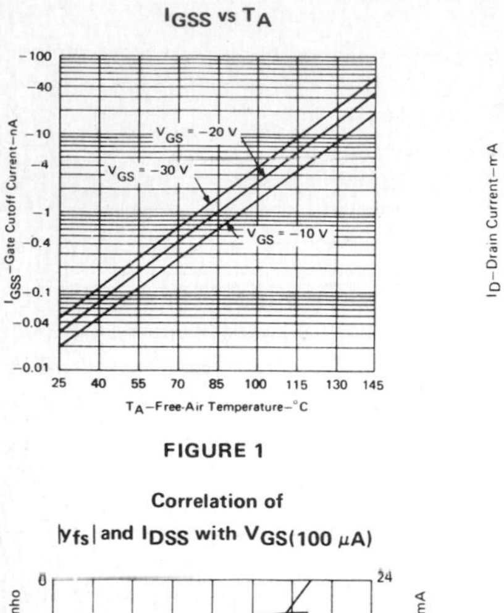

FIGURE 1

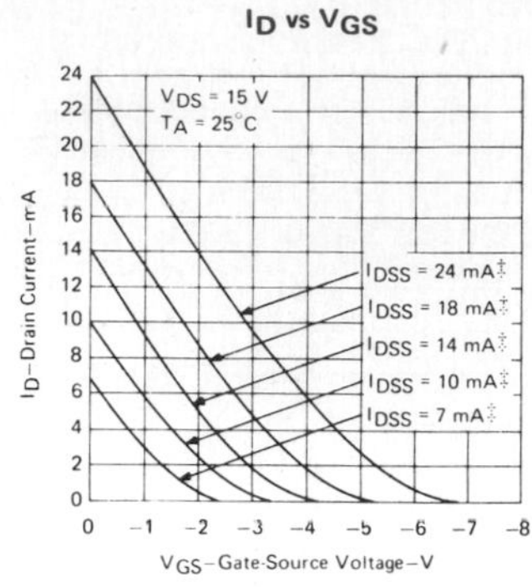

FIGURE 2

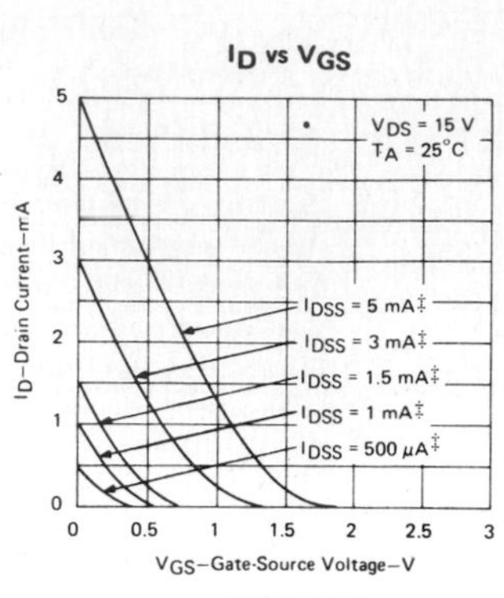

FIGURE 3

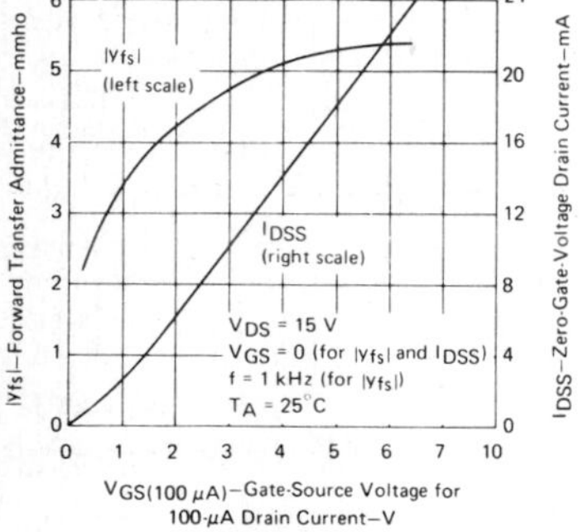

FIGURE 4

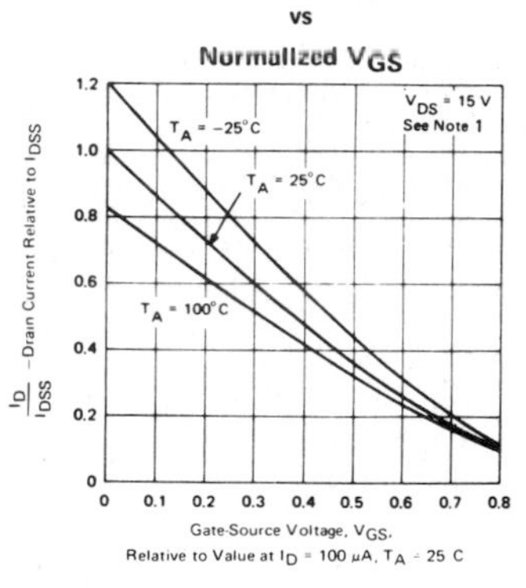

FIGURE 5

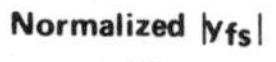

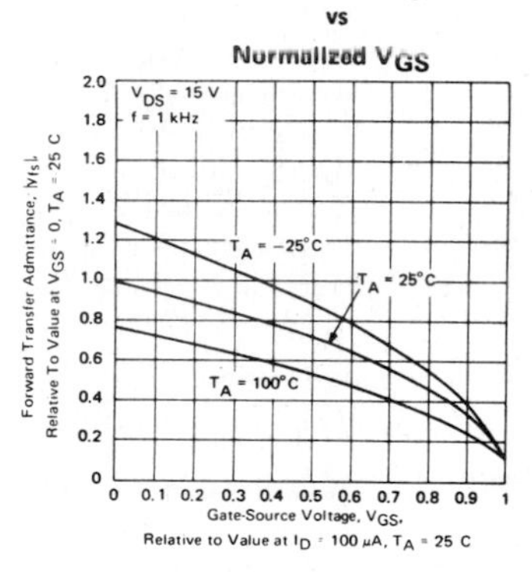

FIGURE 6

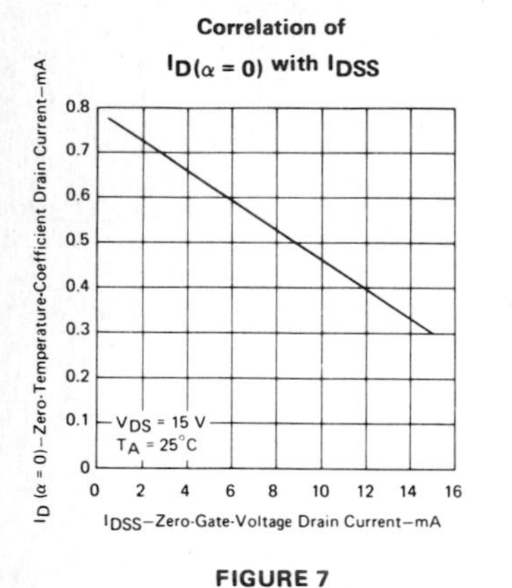

FIGURE 7

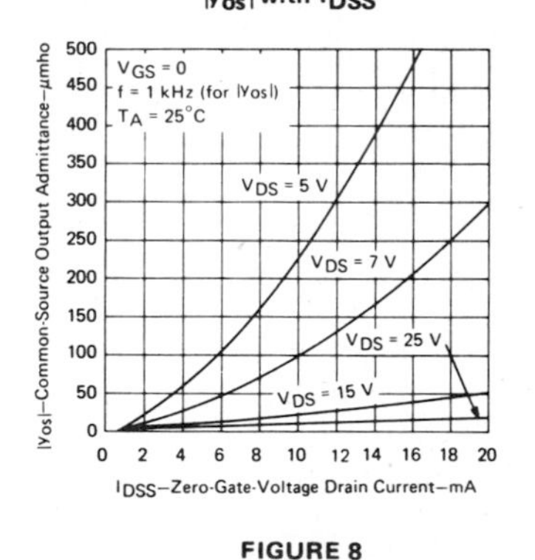

FIGURE 8

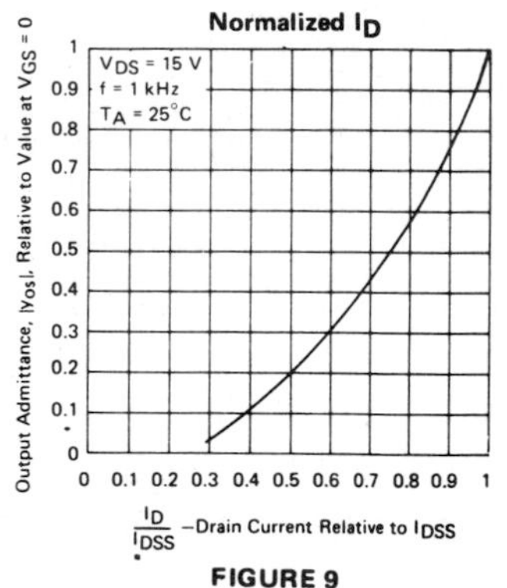

FIGURE 9

‡Data is for devices having indicated value of I_{DSS} at V_{DS} = 15 V, V_{GS} = 0, T_A = 25°C.

NOTE 1: This parameter was measured using pulse techniques. t_w = 300 μs, duty cycle ≤ 2%.

NMOS: Depletion-type MN82*

electrical and operating characteristics at 25°C free-air temperature

	PARAMETER	CONDITIONS	OBSERVED VALUES LOW	TYP	HIGH	UNIT
$V_{(BR)DSV}$	Drain-Source Breakdown Voltage	I_D = 10 μA, V_{GS} = −8 V	20♦	28		V
I_{GSSF}	Forward Gate-Terminal Current	V_{GS} = 8 V, V_{DS} = 0		<1		μA
I_{GSSR}	Reverse Gate-Terminal Current	V_{GS} = −8 V, V_{DS} = 0		−<1	−50	pA
$V_{GS(off)}$	Gate-Source Cutoff Voltage	V_{DS} = 15 V, I_D = 50 μA	−0.8	−1.5	−8	V
I_{DSS}	Zero-Gate-Voltage Drain Current	V_{DS} = 15 V, V_{GS} = 0, See Note 1	5	10	30	mA
$\lvert y_{fs} \rvert$	Small-Signal Common-Source Forward Transfer Admittance	V_{DS} = 15 V, I_D = 5 mA, f = 1 kHz	5	10	12	mmho
$\lvert y_{os} \rvert$	Small-Signal Common-Source Output Admittance			0.25		mmho
C_{iss}	Common-Source Short-Circuit Input Capacitance	V_{DS} = 15 V, I_D = 5 mA, f = 1 MHz, See Note 2		4		pF
C_{rss}	Common-Source Short-Circuit Reverse Transfer Capacitance			0.3	0.35	pF
C_{oss}	Common-Source Short-Circuit Output Capacitance			1.6		pF
g_{is}	Small-Signal Common-Source Input Conductance	V_{DS} = 15 V, I_D = 5 mA, f = 200 MHz		0.2		mmho
b_{is}	Small-Signal Common-Source Input Susceptance			4.5		
g_{fs}	Small-Signal Common-Source Forward Transfer Conductance			10		mmho
b_{fs}	Small-Signal Common-Source Forward Transfer Susceptance			−2		
g_{rs}	Small-Signal Common-Source Reverse Transfer Conductance			0.05		mmho
b_{rs}	Small-Signal Common-Source Reverse Transfer Susceptance			−0.4		
g_{os}	Small-Signal Common-Source Output Conductance			0.25		mmho
b_{os}	Small-Signal Common-Source Output Susceptance			2		
F	Spot Noise Figure	V_{DS} = 15 V, I_D = 5 mA, f = 200 MHz			5	dB

♦This value does not modify guaranteed limits for specific devices and does not justify operation in excess of absolute maximum ratings.

CAUTION: The measurement of $V_{(BR)DSV}$ may be destructive.

NOTES: 1. This parameter was measured using pulse techniques. t_w = 300 μs, duty cycle ≤ 2%.

2. Capacitance measurements were made using chips mounted in TO-72 packages.

- **MN82 is a 19 X 19-mil, epitaxial, planar, expanded-contact MOS silicon chip**
- **Available in TO-72 packages**
- **For use in VHF amplifier circuits**

---

*Courtesy of Texas Instruments Inc.

E.3 FIELD-EFFECT TRANSISTORS (continued)

TYPICAL CHARACTERISTICS

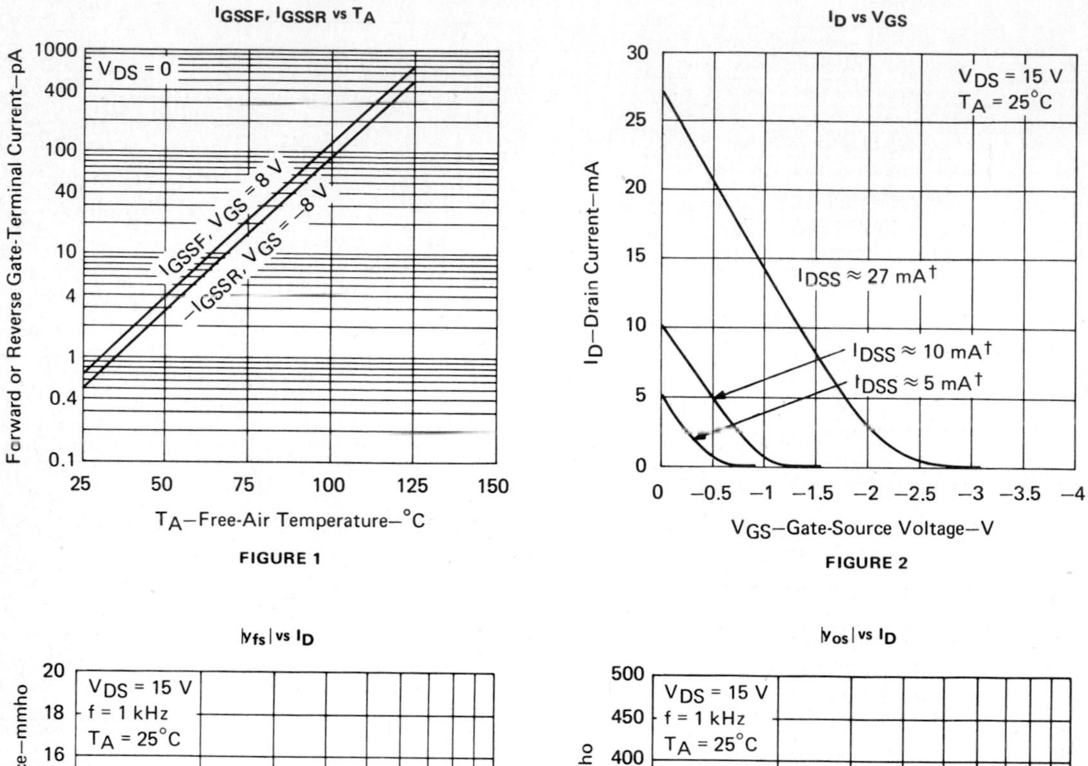

FIGURE 1

FIGURE 2

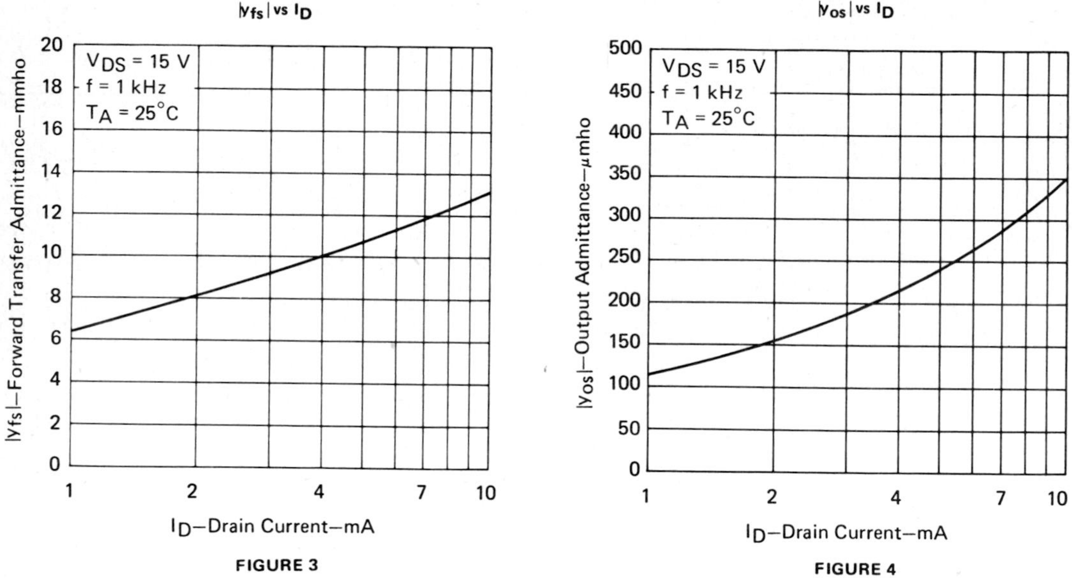

FIGURE 3

FIGURE 4

† Data is for devices having the indicated value of I_{DSS} at V_{DS} = 15 V, V_{GS} = 0, and T_A = 25°C.

NMOS: Enhancement-Type MN83*

electrical and operating characteristics at 25° C free-air temperature

PARAMETER		CONDITIONS			OBSERVED VALUES			UNIT
					LOW	TYP	HIGH	
$V_{(BR)DSS}$	Drain-Source Breakdown Voltage	$I_D = 10\ \mu A$,	$V_{GS} = 0$		25♦	40		V
I_{GSSF}	Forward Gate-Terminal Current	$V_{GS} = 35$ V,	$V_{DS} = 0$			<1	10	pA
I_{GSSR}	Reverse Gate-Terminal Current	$V_{GS} = -35$ V,	$V_{DS} = 0$			–<1	–10	pA
I_{DSS}	Zero-Gate-Voltage Drain Current	$V_{DS} = 10$ V,	$V_{GS} = 0$			<1	10	nA
$V_{GS(th)}$	Gate-Source Threshold Voltage	$V_{DS} = 10$ V,	$I_D = 10\ \mu A$		0.5	1	3	V
$I_{D(on)}$	On-State Drain Current	$V_{DS} = 10$ V,	$V_{GS} = 10$ V,	See Note 1	10	150	400	mA
$r_{ds(on)}$	Small-Signal Drain-Source On-State Resistance	$V_{GS} = 10$ V,	$I_D = 0$,	f = 1 kHz		15	200	Ω
C_{iss}	Common-Source Short-Circuit Input Capacitance	$V_{DS} = 10$ V, See Note 2	$V_{GS} = 0$,	f = 1 MHz,		4.5	6	pF
C_{rss}	Common-Source Short-Circuit Reverse Transfer Capaciatnce	$V_{DS} = 0$, See Note 2	$V_{GS} = 0$,	f = 1 MHz,		1.1	1.5	pF
$t_{d(on)}$	Turn-On Delay Time	$V_{DD} = 10$ V, $V_{GS(on)} = 10$ V,	$I_{D(on)} \approx 10$ mA, $V_{GS(off)} = 0$,	$R_L = 800\ \Omega$, Figure 1 Circuit		1		ns
t_r	Rise Time					2		ns
$t_{d(off)}$	Turn-Off Delay Time					3		ns
t_f	Fall Time					12		ns

♦This value does not modify guaranteed limits for specific devices and does not justify operation in excess of absolute maximum ratings.
CAUTION: The measurement of $V_{(BR)DSS}$ may be destructive.
NOTES: 1. This parameter was measured using pulse techniques. $t_w = 300\ \mu s$, duty cycle ⩽ 2%.
2. Capacitance measurements were made using chips mounted in TO-72 packages.

- **MN83 is a 21 X 21-mil, epitaxial, planar, expanded-contact MOS silicon chip**
- **Available in TO-72 packages**
- **For use in switching and chopper circuits**

*Courtesy of Texas Instruments Inc.

E.3 FIELD-EFFECT TRANSISTORS (continued)

PARAMETER MEASUREMENT INFORMATION

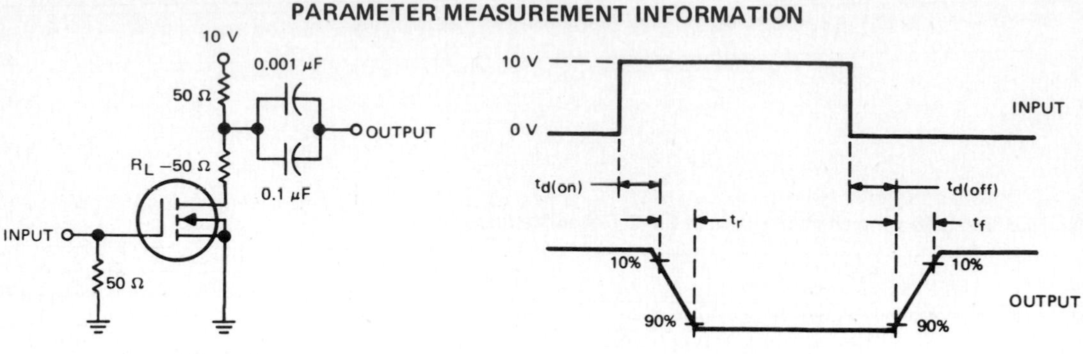

TEST CIRCUIT

VOLTAGE WAVEFORMS

NOTES: a. The input waveform is supplied by a generator with the following characteristics: Z_{out} = 50 Ω, duty cycle ⩽ 1%, t_r ⩽ 0.33 ns, t_f ⩽ 0.33 ns, t_w ≈ 100 ns.

b. Waveforms are monitored on an oscilloscope with the following characteristics: t_r ⩽ 0.4 ns, R_{in} = 50 Ω, C_{in} ⩽ 2 pF.

FIGURE 1

TYPICAL CHARACTERISTICS

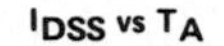

I_{DSS} vs T_A

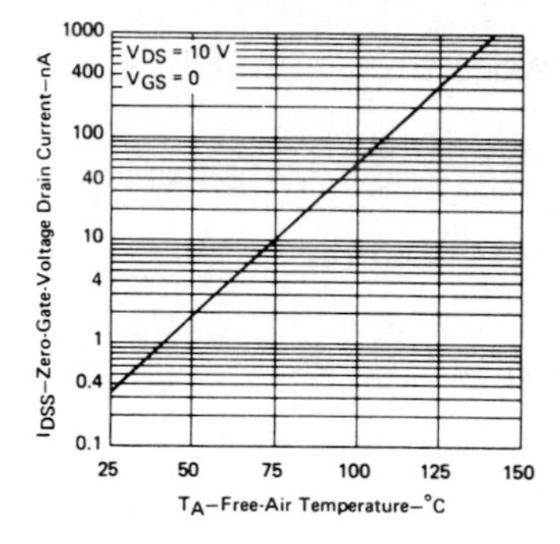

FIGURE 2

I_D vs V_{GS}

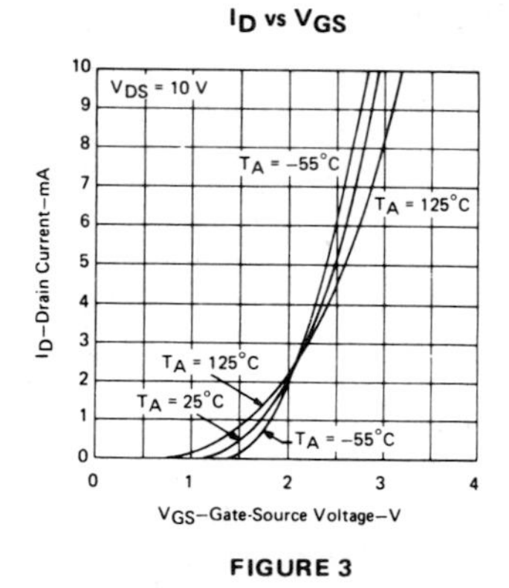

FIGURE 3

I_D vs V_{GS}

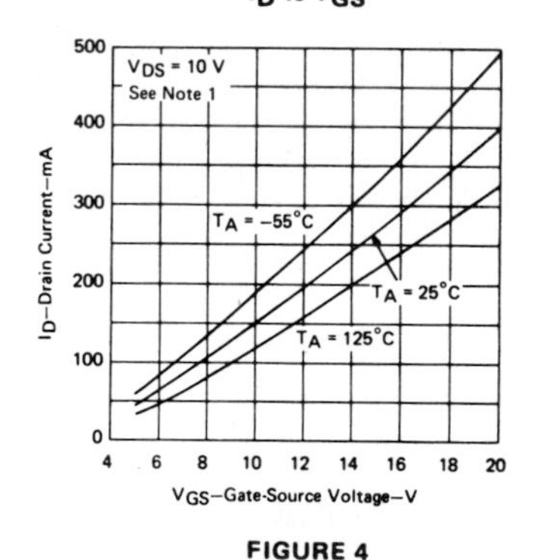

FIGURE 4

NOTE 1: This parameter was measured using pulse techniques. t_w = 300 μs, duty cycle ⩽ 2%.

E.4 POWER TRANSISTORS (*npn*: 2N5878, *pnp*: 2N5875)*

Silicon N-P-N and P-N-P Epitaxial-Base High-Power Transistors

Rugged Devices, Broadly Applicable
For Industrial and Commercial Use

The RCA-2N5875 and 2N5876 are epitaxial-base silicon p-n-p transistors featuring high gain at high current. The RCA-2N5877 and 2N5878 are epitaxial-base silicon n-p-n transistors. They may be used as complements to 2N5875 and 2N5876, respectively. These devices have a compability of 150 watts at case temperatures up to 25°C.

They differ in voltage ratings and in currents at which the parameters are controlled. All are supplied in the steel JEDEC TO-204MA hermetic package.

Features:

- **High dissipation capability**
- **Low saturation voltages**
- **Maximum safe-areas-of-operation curves**
- **Hermetically sealed JEDEC TO-3/TO-204MA package**
- **High gain at high current**
- **Thermal-cycling rating curve**

Applications:

- **Series and shunt regulators**
- **High-fidelity amplifiers**
- **Power-switching circuits**
- **Solenoid drivers**

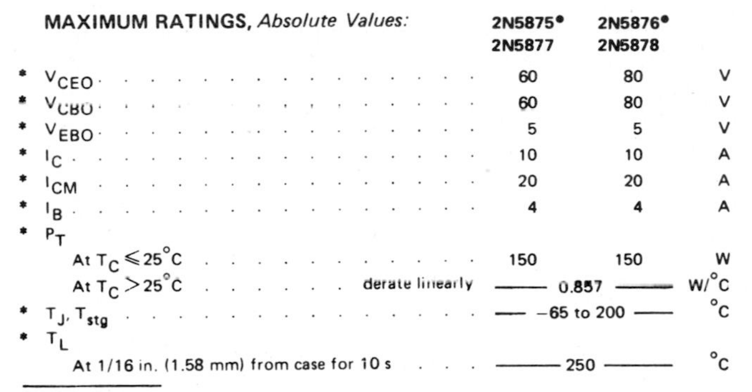

MAXIMUM RATINGS, *Absolute Values:*

		2N5875• 2N5877	2N5876• 2N5878	
*	V_{CEO}	60	80	V
*	V_{CBO}	60	80	V
*	V_{EBO}	5	5	V
*	I_C	10	10	A
*	I_{CM}	20	20	A
*	I_B	4	4	A
*	P_T			
	At $T_C \leq 25°C$	150	150	W
	At $T_C > 25°C$ derate linearly	0.857		W/°C
*	T_J, T_{stg}	−65 to 200		°C
*	T_L			
	At 1/16 in. (1.58 mm) from case for 10 s	250		°C

*In accordance with JEDEC registration data.

•For p-n-p devices, voltage & current values are negative.

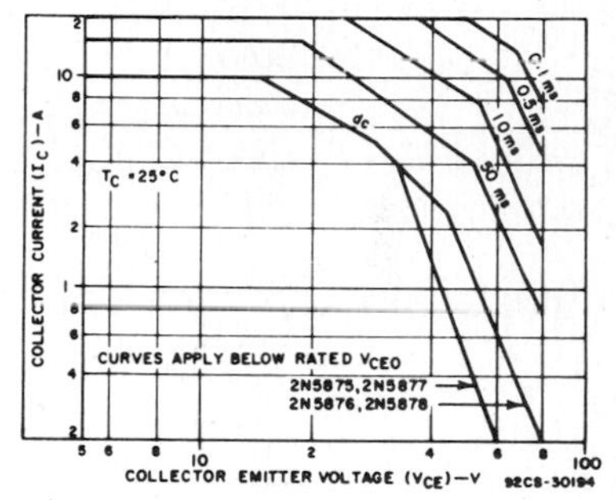

Fig. 1 – Collector-emitter voltage (V_{CE})-V

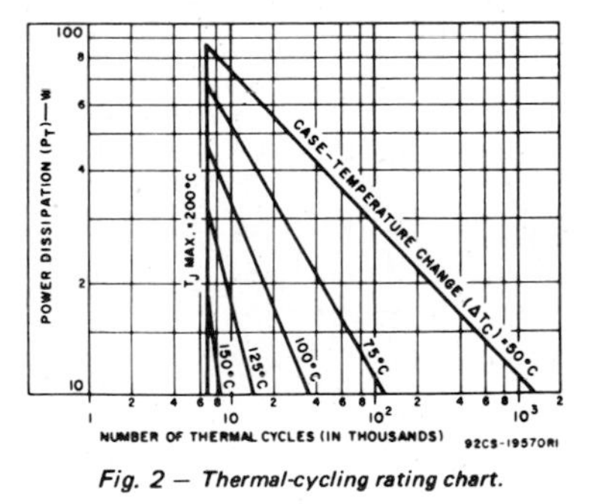

Fig. 2 – Thermal-cycling rating chart.

TERMINAL DESIGNATIONS

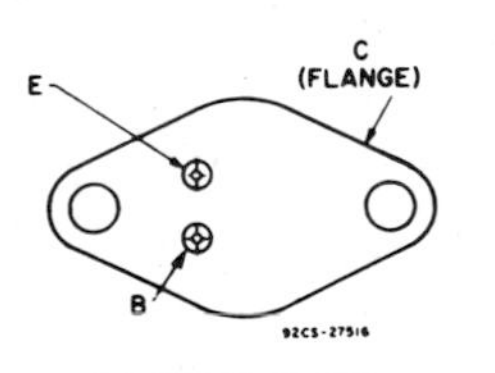

JEDEC TO-204MA
(See dimensional outline "A".)

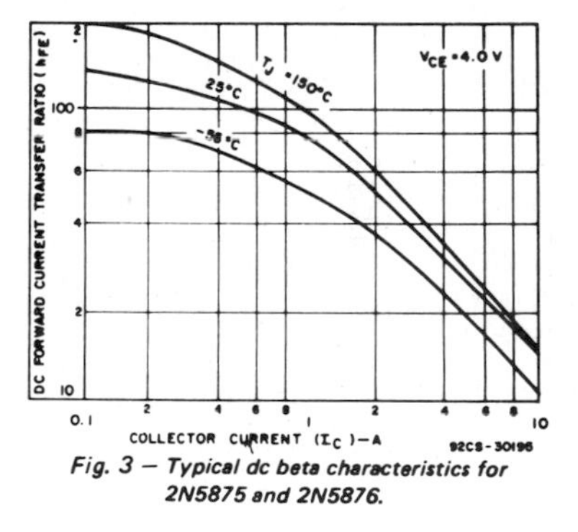

Fig. 3 – Typical dc beta characteristics for 2N5875 and 2N5876.

*Courtesy of RCA.

E.4 POWER TRANSISTORS (continued)

ELECTRICAL CHARACTERISTICS, at Case Temperature $T_C = 25°C$ Unless Otherwise Specified

CHARACTERISTIC	TEST CONDITIONS				LIMITS				UNITS
	VOLTAGE V dc		CURRENT A dc		2N5875● 2N5877		2N5876● 2N5878		
	V_{CE}	V_{BE}	I_C	I_B	MIN.	MAX.	MIN.	MAX.	
* I_{CEX}	60	1.5	–	–	–	0.5	–	–	mA
	80	1.5	–	–	–	–	–	0.5	
$T_C = 150°C$	60	1.5	–	–	–	5	–	–	
	80	1.5	–	–	–	–	–	5	
* I_{CEO}	30	–	–	0	–	1	–	–	mA
	40	–	–	0	–	–	–	1	
* I_{CBO} $I_E = 0$	60[c]	–	–	–	–	0.5	–	–	mA
	80[c]	–	–	–	–	–	–	0.5	
* I_{EBO} $I_E = 0$	–	–5	–	–	–	1	–	1	mA
* $V_{CEO(sus)}$[b]	–	–	0.2	0	60	–	80	–	V
* h_{FE}[a]	4	–	1	–	35	–	35	–	
	4	–	4	–	20	100	20	100	
	4	–	10	–	4	–	4	–	
* V_{BE}[a]	4	–	4	–	–	1.5	–	1.5	V
* $V_{BE(sat)}$[a]	–	–	10	2.5	–	2.5	–	2.5	V
* $V_{CE(sat)}$[a]	–	–	5	0.5	–	1	–	1	V
	–	–	10	2.5	–	3	–	3	
* f_T[c] f = 1 MHz	10	–	0.5	–	4	–	4	–	MHz
* h_{fe} f = 1 kHz	4	–	1	–	20	–	20	–	
* C_{ob} $V_{CB} = 10$ V 2N5875-76	–	–	–	–	–	500	–	500	pF
f = 1 MHz 2N5877-78	–	–	–	–	–	300	–	300	
* t_r $V_{CC} = 30$ V	–	–	4	0.4[d]	–	0.7	–	0.7	μs
* t_s $V_{CC} = 30$ V	–	–	4	0.4[d]	–	1.0	–	1.0	μs
* t_f $V_{CC} = 30$ V	–	–	4	0.4[d]	–	0.8	–	0.8	μs
$R_{\theta JC}$	–	–	–	–	–	1.17	–	1.17	°C/W

* In accordance with JEDEC registration data.

[a] Pulsed; pulse width ≤ 300 μs, duty cycle ≤ 2%.

[c] V_{CB}

● For p-n-p devices, voltages and current values are negative.

[b] **CAUTION:** Sustaining voltage, $V_{CEO(sus)}$, *MUST NOT* be measured on a curve tracer.

[d] $I_{B1} = -I_{B2}$

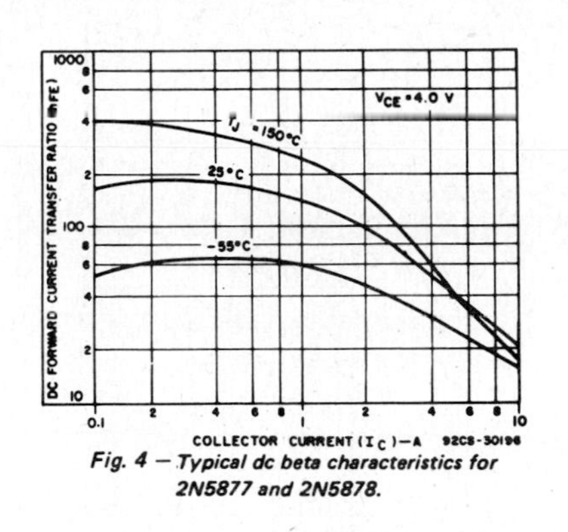

Fig. 4 — Typical dc beta characteristics for 2N5877 and 2N5878.

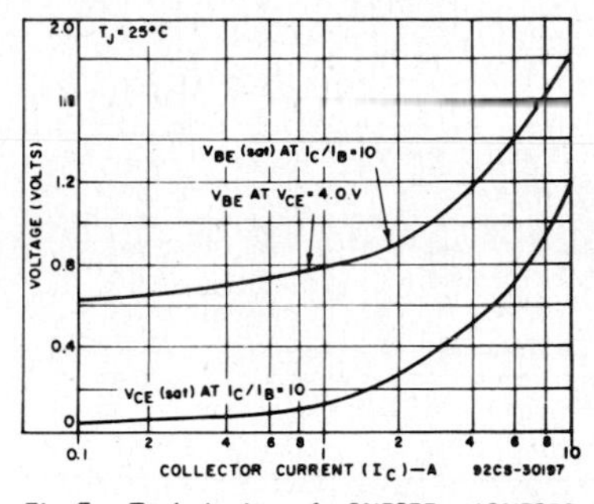

Fig. 5 — Typical voltages for 2N5875 and 2N5876.

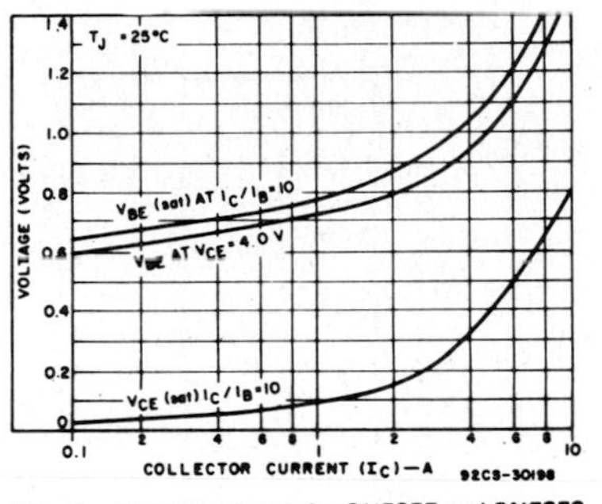

Fig. 6 — Typical voltages for 2N5877 and 2N5878.

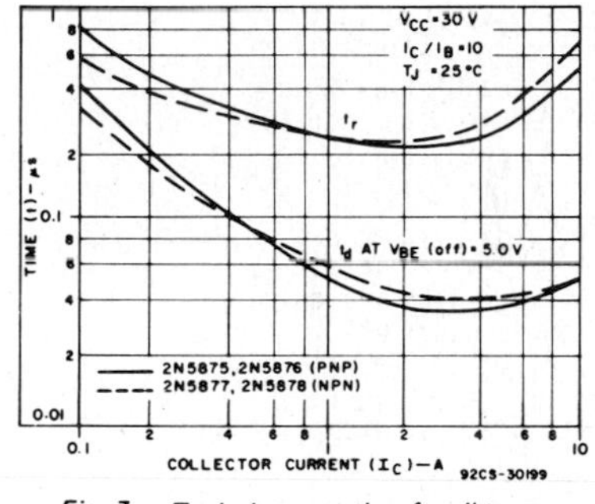
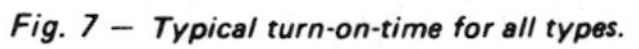

Fig. 7 — Typical turn-on-time for all types.

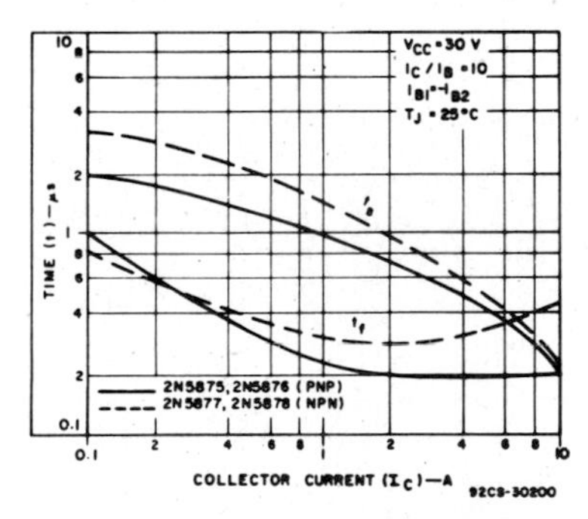

Fig. 8 — Typical turn-off-time for all types.

E.5 OPERATIONAL AMPLIFIERS*

BJT (741)*

FREQUENCY-COMPENSATED OPERATIONAL AMPLIFIER

FAIRCHILD LINEAR INTEGRATED CIRCUIT

GENERAL DESCRIPTION — The μA741 is a high performance monolithic Operational Amplifier constructed using the Fairchild Planar* epitaxial process. It is intended for a wide range of analog applications. High common mode voltage range and absence of latch-up tendencies make the μA741 ideal for use as a voltage follower. The high gain and wide range of operating voltage provides superior performance in integrator, summing amplifier, and general feedback applications. Electrical characteristics of the μA741A and E are identical to MIL-M-38510/10101.

- **NO FREQUENCY COMPENSATION REQUIRED**
- **SHORT CIRCUIT PROTECTION**
- **OFFSET VOLTAGE NULL CAPABILITY**
- **LARGE COMMON MODE AND DIFFERENTIAL VOLTAGE RANGES**
- **LOW POWER CONSUMPTION**
- **NO LATCH-UP**

ABSOLUTE MAXIMUM RATINGS

Supply Voltage	
μA741A, μA741, μA741E	±22 V
μA741C	±18 V
Internal Power Dissipation (Note 1)	
Metal Can	500 mW
Molded and Hermetic DIP	670 mW
Mini DIP	310 mW
Flatpak	570 mW
Differential Input Voltage	±30 V
Input Voltage (Note 2)	±15 V
Storage Temperature Range	
Metal Can, Hermetic DIP, and Flatpak	−65°C to +150°C
Mini DIP, Molded DIP	−55°C to +125°C
Operating Temperature Range	
Military (μA741A, μA741)	−55°C to +125°C
Commercial (μA741E, μA741C)	0°C to +70°C
Lead Temperature (Soldering)	
Metal Can, Hermetic DIPs, and Flatpak (60 s)	300°C
Molded DIPs (10 s)	260°C
Output Short Circuit Duration (Note 3)	Indefinite

CONNECTION DIAGRAMS

8-LEAD METAL CAN
(TOP VIEW)
PACKAGE OUTLINE 5B

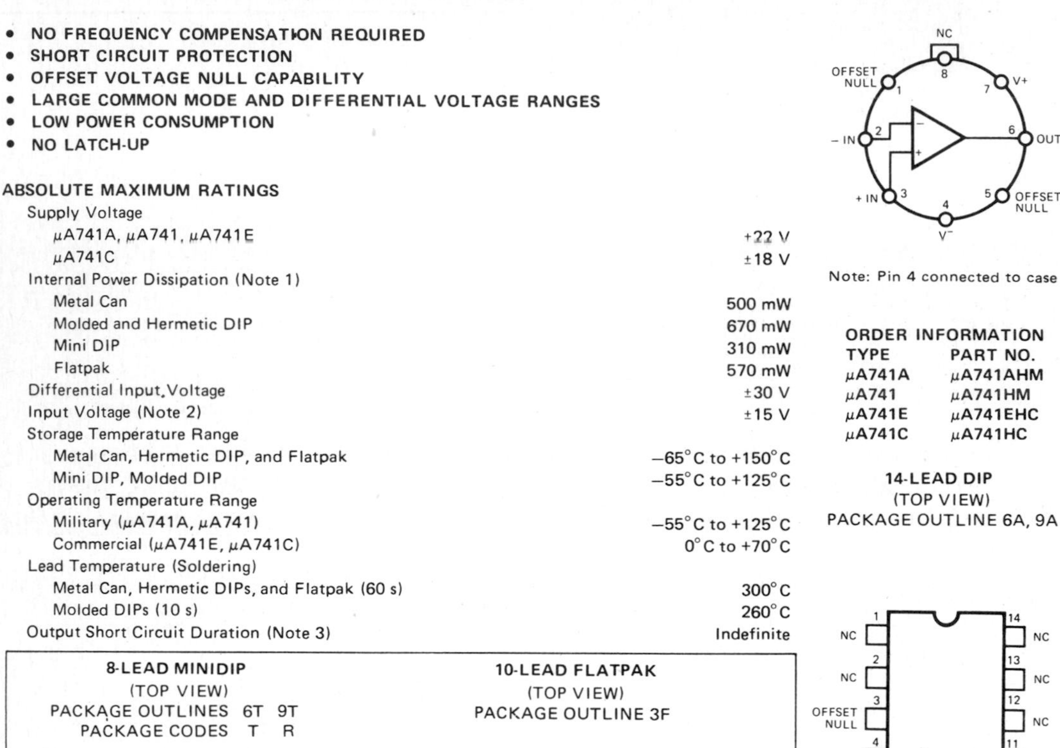

Note: Pin 4 connected to case

ORDER INFORMATION

TYPE	PART NO.
μA741A	μA741AHM
μA741	μA741HM
μA741E	μA741EHC
μA741C	μA741HC

14-LEAD DIP
(TOP VIEW)
PACKAGE OUTLINE 6A, 9A

8-LEAD MINIDIP
(TOP VIEW)
PACKAGE OUTLINES 6T 9T
PACKAGE CODES T R

10-LEAD FLATPAK
(TOP VIEW)
PACKAGE OUTLINE 3F

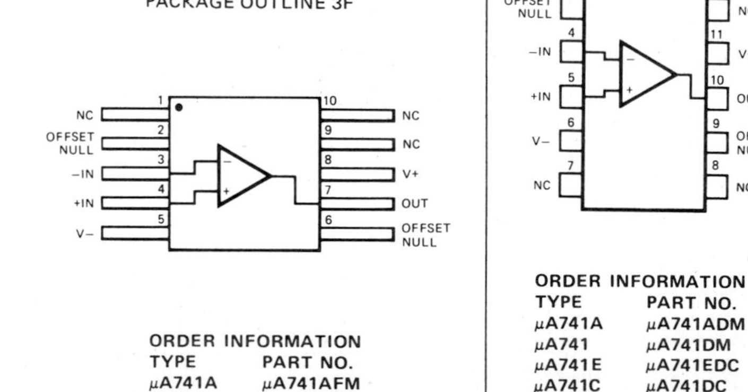

ORDER INFORMATION (8-lead minidip)

TYPE	PART NO.
μA741C	μA741TC
μA741C	μA741RC

ORDER INFORMATION (10-lead flatpak)

TYPE	PART NO.
μA741A	μA741AFM
μA741	μA741FM

ORDER INFORMATION (14-lead DIP)

TYPE	PART NO.
μA741A	μA741ADM
μA741	μA741DM
μA741E	μA741EDC
μA741C	μA741DC
μA741C	μA741PC

Notes on following pages. *Planar is a patented Fairchild process.

*Courtesy of Fairchild. For 741 circuit diagram see Fig. 5-25.

TYPICAL PERFORMANCE CURVES FOR μA741A AND μA741

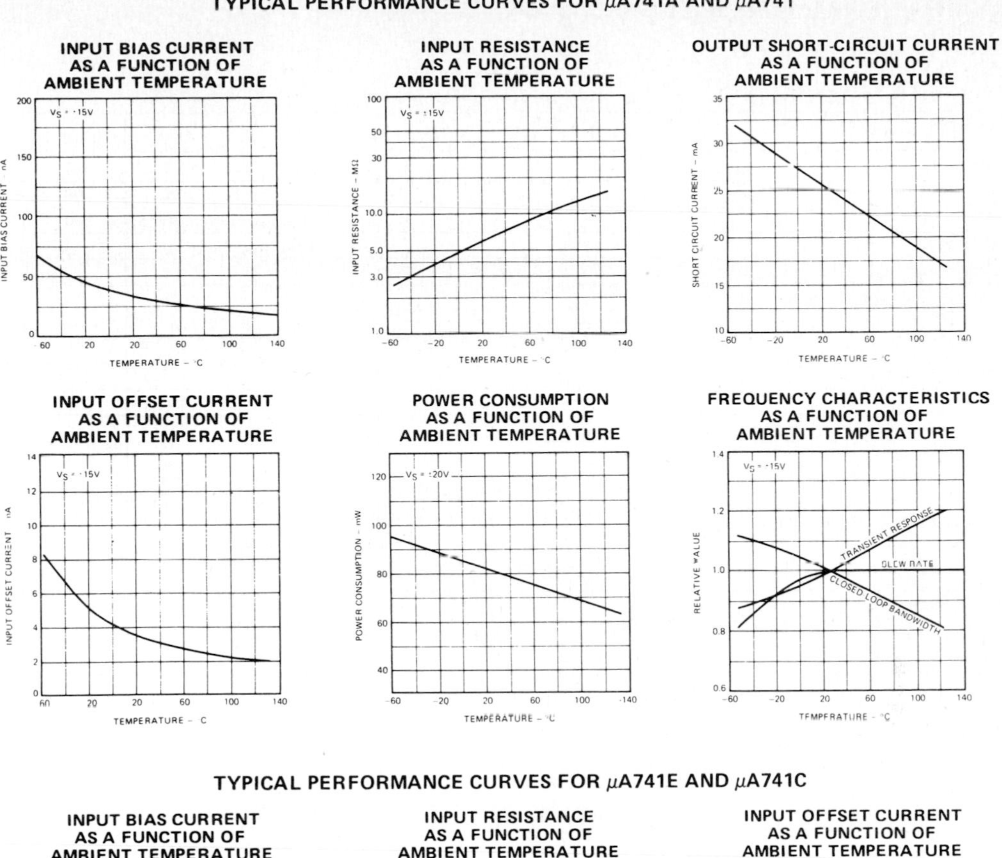

TYPICAL PERFORMANCE CURVES FOR μA741E AND μA741C

INPUT BIAS CURRENT AS A FUNCTION OF AMBIENT TEMPERATURE

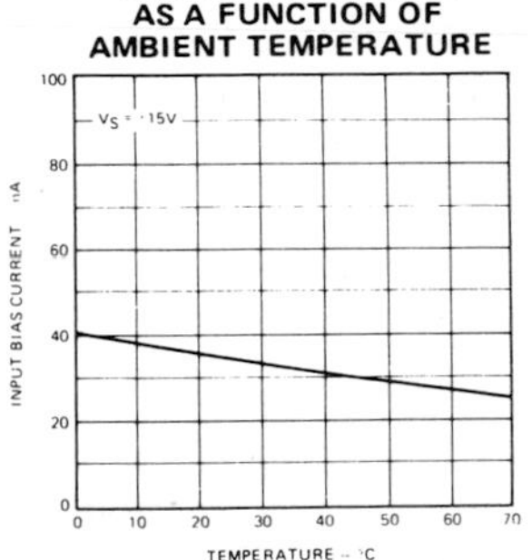

INPUT RESISTANCE AS A FUNCTION OF AMBIENT TEMPERATURE

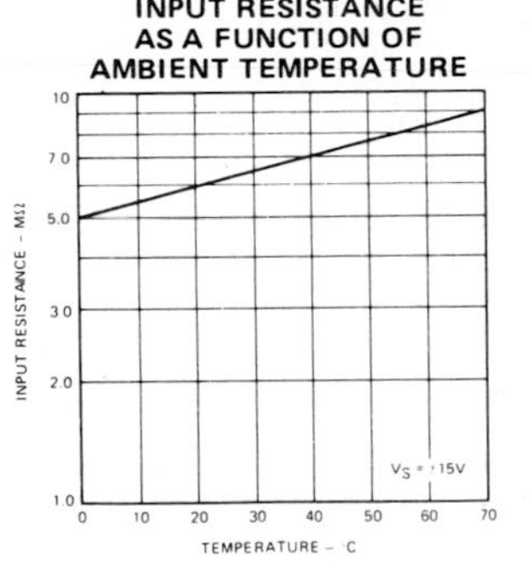

INPUT OFFSET CURRENT AS A FUNCTION OF AMBIENT TEMPERATURE

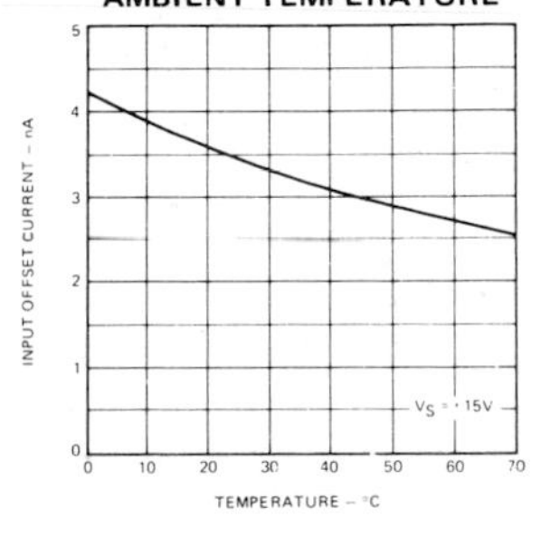

POWER CONSUMPTION AS A FUNCTION OF AMBIENT TEMPERATURE

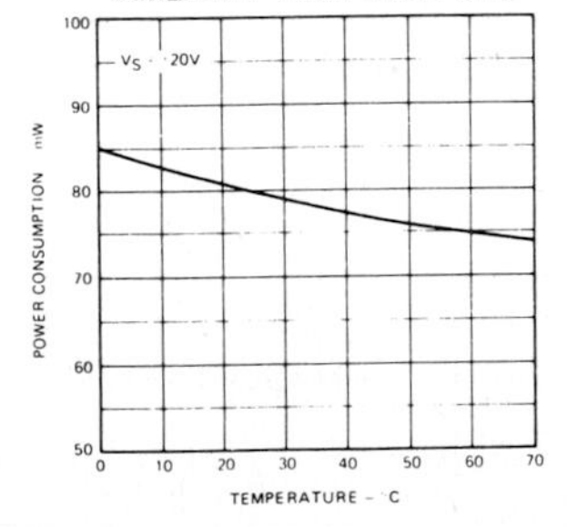

OUTPUT SHORT CIRCUIT CURRENT AS A FUNCTION OF AMBIENT TEMPERATURE

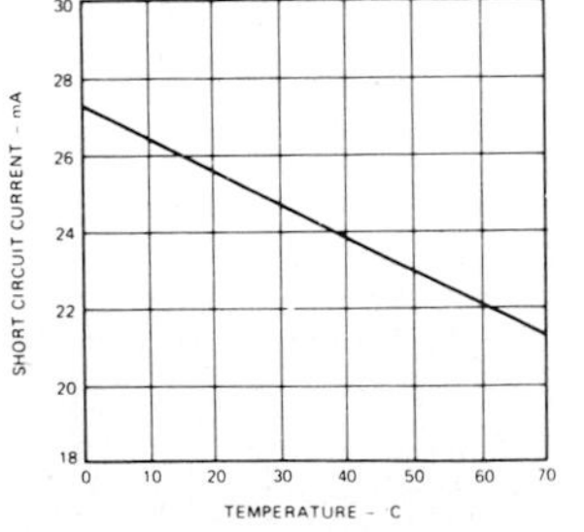

FREQUENCY CHARACTERISTICS AS A FUNCTION OF AMBIENT TEMPERATURE

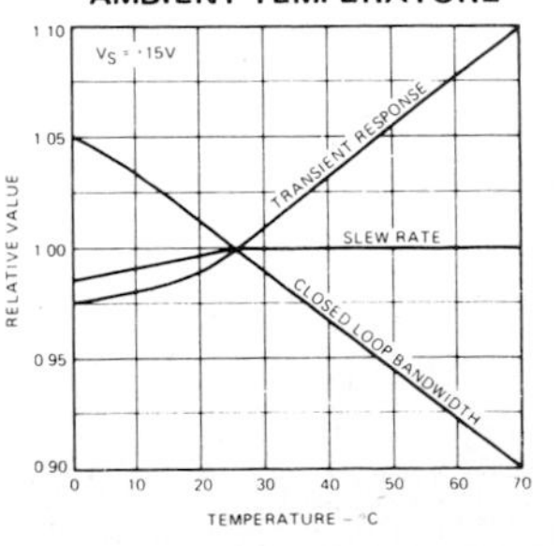

E.5 OPERATIONAL AMPLIFIERS (continued)

μA741

ELECTRICAL CHARACTERISTICS ($V_S = \pm15$ V, $T_A = 25°C$ unless otherwise specified)

PARAMETERS (see definitions)		CONDITIONS	MIN	TYP	MAX	UNITS
Input Offset Voltage		$R_S \leq 10$ kΩ		1.0	5.0	mV
Input Offset Current				20	200	nA
Input Bias Current				80	500	nA
Input Resistance			0.3	2.0		MΩ
Input Capacitance				1.4		pF
Offset Voltage Adjustment Range				±15		mV
Large Signal Voltage Gain		$R_L \geq 2$ kΩ, $V_{OUT} = \pm10$ V	50,000	200,000		
Output Resistance				75		Ω
Output Short Circuit Current				25		mA
Supply Current				1.7	2.8	mA
Power Consumption				50	85	mW
Transient Response (Unity Gain)	Rise time	$V_{IN} = 20$ mV, $R_L = 2$ kΩ, $C_L \leq 100$ pF		0.3		μs
	Overshoot			5.0		%
Slew Rate		$R_L \geq 2$ kΩ		0.5		V/μs

The following specifications apply for $-55°C \leq T_A \leq +125°C$:

PARAMETERS	CONDITIONS	MIN	TYP	MAX	UNITS
Input Offset Voltage	$R_S \leq 10$ kΩ		1.0	6.0	mV
Input Offset Current	$T_A = +125°C$		7.0	200	nA
	$T_A = -55°C$		85	500	nA
Input Bias Current	$T_A = +125°C$		0.03	0.5	μA
	$T_A = -55°C$		0.3	1.5	μA
Input Voltage Range		±12	±13		V
Common Mode Rejection Ratio	$R_S \leq 10$ kΩ	70	90		dB
Supply Voltage Rejection Ratio	$R_S \leq 10$ kΩ		30	150	μV/V
Large Signal Voltage Gain	$R_L \geq 2$ kΩ, $V_{OUT} = \pm10$ V	25,000			
Output Voltage Swing	$R_L \geq 10$ kΩ	±12	±14		V
	$R_L \geq 2$ kΩ	±10	±13		V
Supply Current	$T_A = +125°C$		1.5	2.5	mA
	$T_A = -55°C$		2.0	3.3	mA
Power Consumption	$T_A = +125°C$		45	75	mW
	$T_A = -55°C$		60	100	mW

TYPICAL PERFORMANCE CURVES FOR μA741A AND μA741

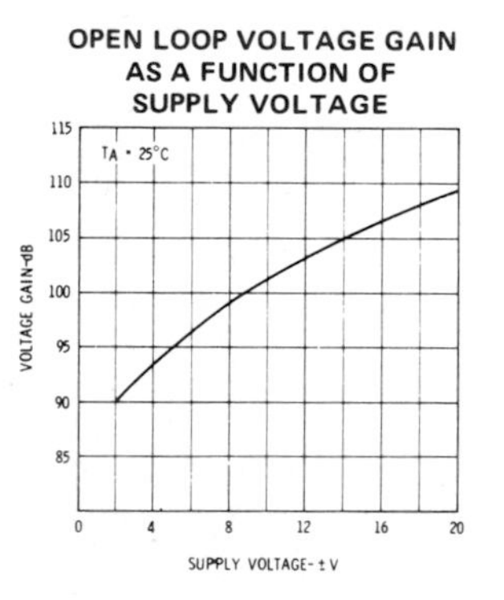

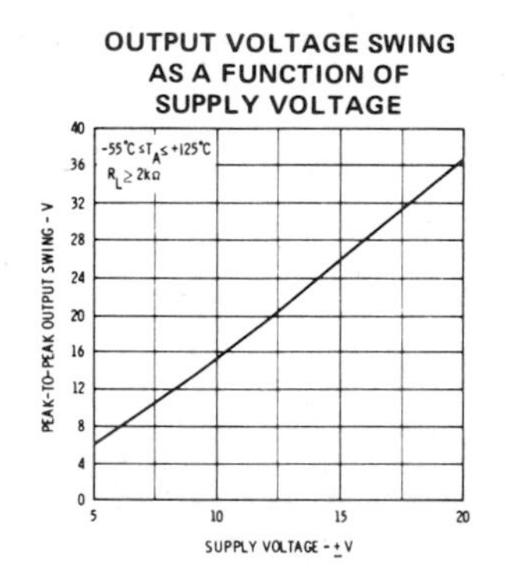

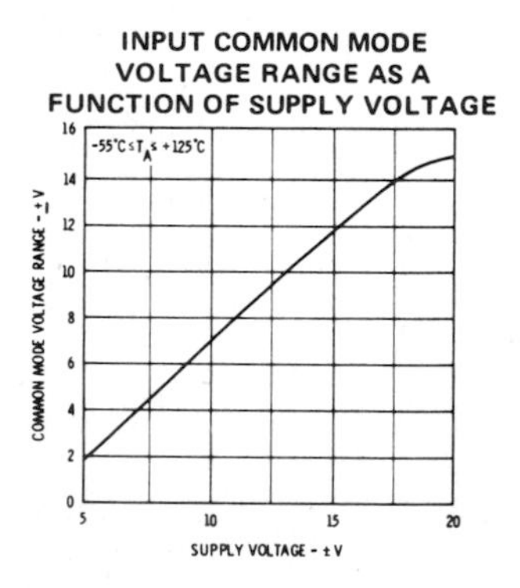

TYPICAL PERFORMANCE CURVES FOR μA741A, μA741, μA741E AND μA741C

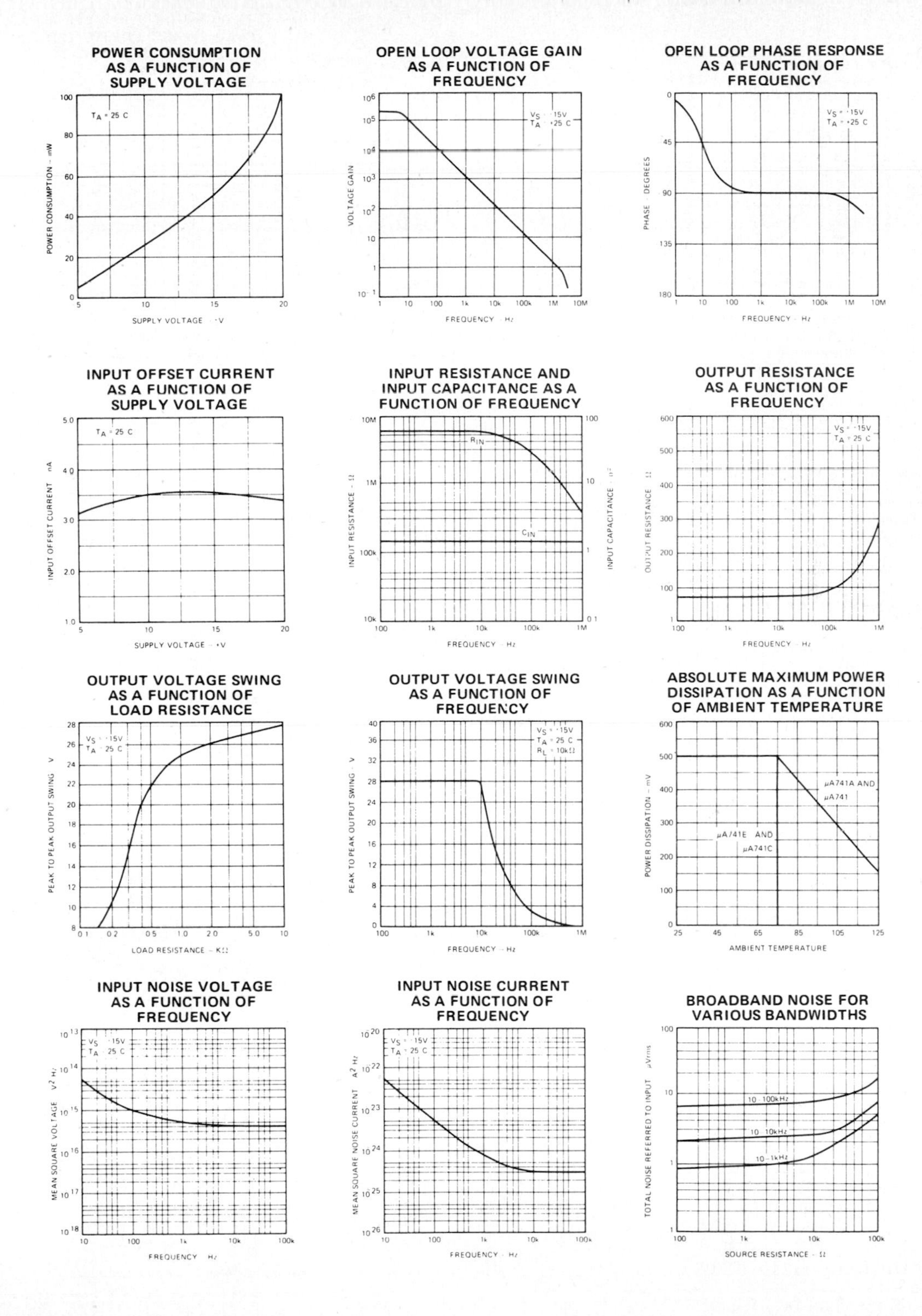

E.5 OPERATIONAL AMPLIFIERS (continued)

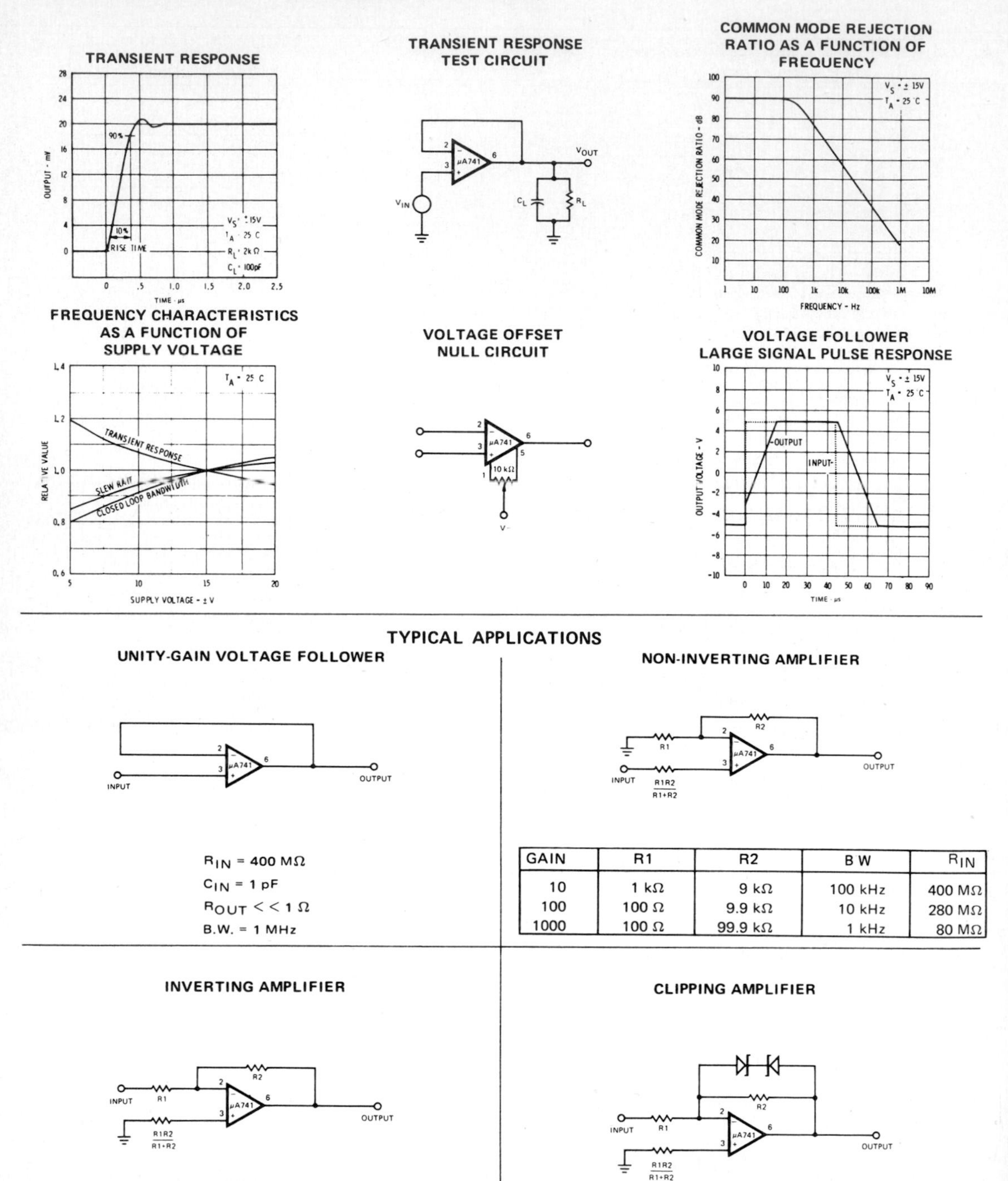

R_{IN} = 400 MΩ
C_{IN} = 1 pF
R_{OUT} << 1 Ω
B.W. = 1 MHz

GAIN	R1	R2	B W	R_{IN}
10	1 kΩ	9 kΩ	100 kHz	400 MΩ
100	100 Ω	9.9 kΩ	10 kHz	280 MΩ
1000	100 Ω	99.9 kΩ	1 kHz	80 MΩ

GAIN	R1	R2	B W	R_{IN}
1	10 kΩ	10 kΩ	1 MHz	10 kΩ
10	1 kΩ	10 kΩ	100 kHz	1 kΩ
100	1 kΩ	100 kΩ	10 kHz	1 kΩ
1000	100 Ω	100 kΩ	1 kHz	100 Ω

$$\frac{E_{OUT}}{E_{IN}} = \frac{R2}{R1} \text{ if } |E_{OUT}| \leq V_Z + 0.7\ \text{V}$$

where V_Z = Zener breakdown voltage

Bi-FET (LF351)*

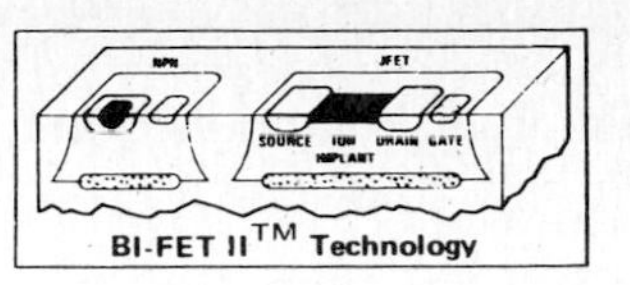

LF351 Wide Bandwidth JFET Input Operational Amplifier

General Description

The LF351 is a low cost high speed JFET input operational amplifier with an internally trimmed input offset voltage (BI-FET II™ technology). The device requires a low supply current and yet maintains a large gain bandwidth product and a fast slew rate. In addition, well matched high voltage JFET input devices provide very low input bias and offset currents. The LF351 is pin compatible with the standard LM741 and uses the same offset voltage adjustment circuitry. This feature allows designers to immediately upgrade the overall performance of existing LM741 designs.

The LF351 may be used in applications such as high speed integrators, fast D/A converters, sample-and-hold circuits and many other circuits requiring low input offset voltage, low input bias current, high input impedance, high slew rate and wide bandwidth. The device has low noise and offset voltage drift, but for applications where these requirements are critical, the LF356 is recommended. If maximum supply current is important, however, the LF351 is the better choice.

Features

- Internally trimmed offset voltage — 2 mV
- Low input bias current — 50 pA
- Low input noise voltage — 16 nV/$\sqrt{Hz}$
- Low input noise current — 0.01 pA/$\sqrt{Hz}$
- Wide gain bandwidth — 4 MHz
- High slew rate — 13 V/μs
- Low supply current — 1.8 mA
- High input impedance — $10^{12}\Omega$
- Low total harmonic distortion $A_V = 10$, $R_L = 10k$, $V_O = 20$ Vp-p, BW = 20 Hz–20 kHz — <0.02%
- Low 1/f noise corner — 50 Hz
- Fast settling time to 0.01% — 2 μs

Typical Connection

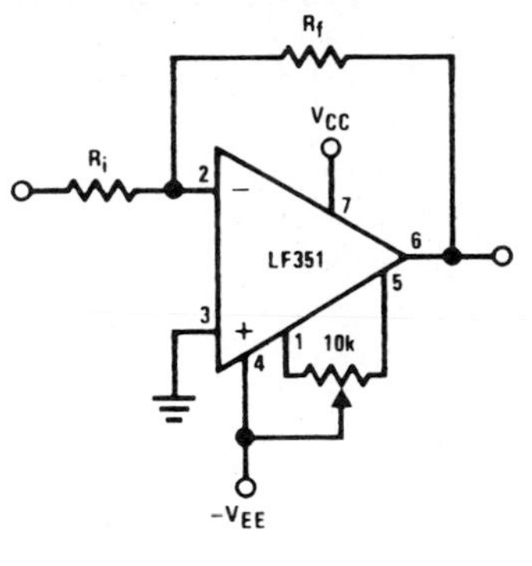

Simplified Schematic

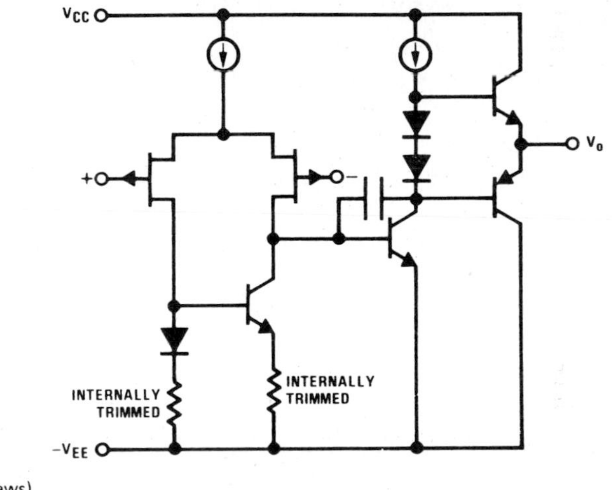

Connection Diagrams (Top Views)

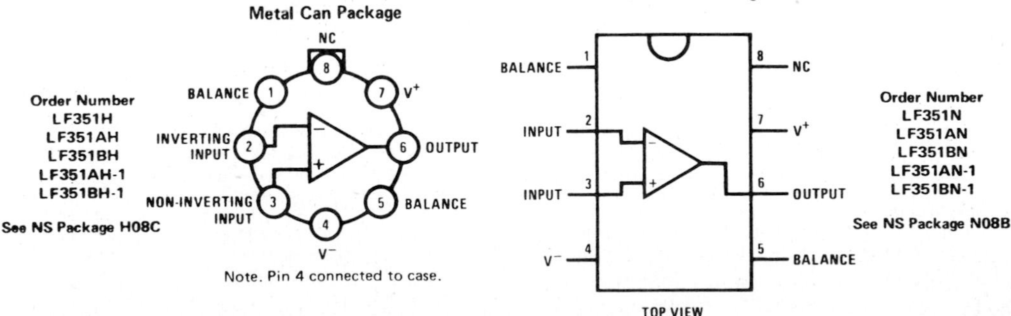

*Courtesy of National Semiconductor. For circuit diagram see Fig. 5-27.

E.5 OPERATIONAL AMPLIFIERS (continued)

Absolute Maximum Ratings

Supply Voltage	±18V
Power Dissipation (Note 1)	500 mW
Operating Temperature Range	0°C to +70°C
$T_{j(MAX)}$	115°C
Differential Input Voltage	±30V
Input Voltage Range (Note 2)	±15V
Output Short Circuit Duration	Continuous
Storage Temperature Range	−65°C to +150°C
Lead Temperature (Soldering, 10 seconds)	300°C

DC Electrical Characteristics (Note 3)

SYMBOL	PARAMETER	CONDITIONS	LF351A			LF351B			LF351			UNITS
			MIN	TYP	MAX	MIN	TYP	MAX	MIN	TYP	MAX	
V_{OS}	Input Offset Voltage	R_S = 10 kΩ, T_A = 25°C		1	2		3	5		5	10	mV
		Over Temperature			4			7			13	mV
$\Delta V_{OS}/\Delta T$	Average TC of Input Offset Voltage	R_S = 10 kΩ		10			10			10		μV/°C
		LF351A-1, LF351B-1		10	20		10	30				
I_{OS}	Input Offset Current	T_j = 25°C, (Notes 3, 4)		25	100		25	100		25	100	pA
		$T_j \leq 70°C$			2			4			4	nA
I_B	Input Bias Current	T_j = 25°C, (Notes 3, 4)		50	200		50	200		50	200	pA
		$T_j \leq 70°C$			4			8			8	nA
R_{IN}	Input Resistance	T_j = 25°C		10^{12}			10^{12}			10^{12}		Ω
A_{VOL}	Large Signal Voltage Gain	V_S = ±15V, T_A = 25°C V_O = ±10V, R_L = 2 kΩ	50	100		50	100		25	100		V/mV
		Over Temperature	25			25			15			V/mV
V_O	Output Voltage Swing	V_S = ±15V, R_L = 10 kΩ	±12	±13.5		±12	±13.5		±12	±13.5		V
V_{CM}	Input Common-Mode Voltage Range	V_S = ±15V	±11	+15 −12		±11	+15 −12		±11	+15 −12		V V
CMRR	Common-Mode Rejection Ratio	$R_S \leq 10$ kΩ	80	100		80	100		70	100		dB
PSRR	Supply Voltage Rejection Ratio	(Note 5)	80	100		80	100		70	100		dB
I_S	Supply Current			1.8	2.8		1.8	2.8		1.8	3.4	mA

AC Electrical Characteristics (Note 3)

SYMBOL	PARAMETER	CONDITIONS	LF351A			LF351B			LF351			UNITS
			MIN	TYP	MAX	MIN	TYP	MAX	MIN	TYP	MAX	
SR	Slew Rate	V_S = ±15V, T_A = 25°C	10	13			13			13		V/μs
GBW	Gain Bandwidth Product	V_S = ±15V, T_A = 25°C	3	4			4			4		MHz
e_n	Equivalent Input Noise Voltage	T_A = 25°C, R_S = 100Ω, f = 1000 Hz		16			16			16		$nV/\sqrt{Hz}$
i_n	Equivalent Input Noise Current	T_j = 25°C, f = 1000 Hz		0.01			0.01			0.01		$pA/\sqrt{Hz}$

Note 1: For operating at elevated temperature, the device must be derated based on a thermal resistance of 150°C/W junction to ambient or 45°C/W junction to case.

Note 2: Unless otherwise specified the absolute maximum negative input voltage is equal to the negative power supply voltage.

Note 3: These specifications apply for V_S = ±15V and $0°C \leq T_A \leq +70°C$. V_{OS}, I_B and I_{OS} are measured at V_{CM} = 0.

Note 4: The input bias currents are junction leakage currents which approximately double for every 10°C increase in the junction temperature, T_j. Due to limited production test time, the input bias currents measured are correlated to junction temperature. In normal operation the junction temperature rises above the ambient temperature as a result of internal power dissipation, P_D. $T_j = T_A + \Theta_{jA} P_D$ where Θ_{jA} is the thermal resistance from junction to ambient. Use of a heat sink is recommended if input bias current is to be kept to a minimum.

Note 5: Supply voltage rejection ratio is measured for both supply magnitudes increasing or decreasing simultaneously in accordance with common practice.

Typical Performance Characteristics

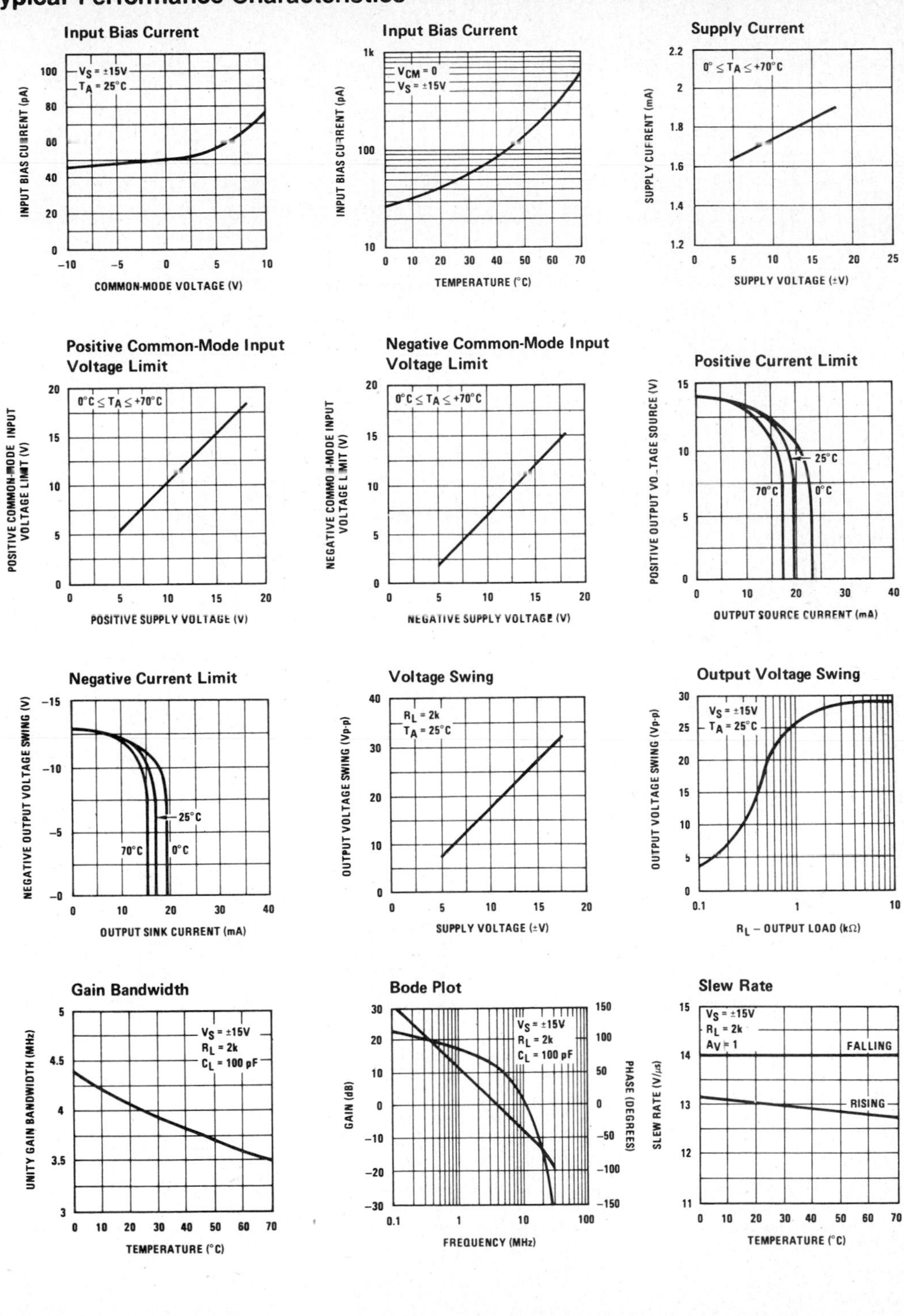

E.5 OPERATIONAL AMPLIFIERS (continued)

Typical Performance Characteristics (Continued)

Distortion vs Frequency

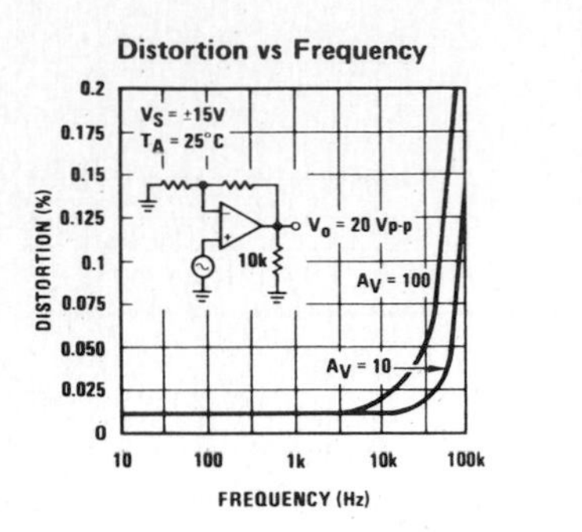

Undistorted Output Voltage Swing

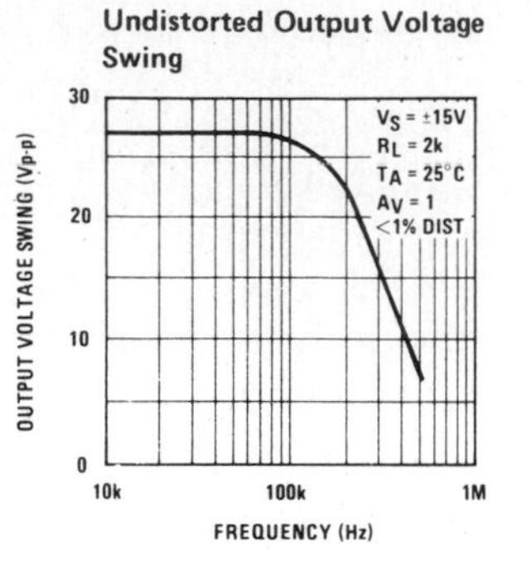

Open Loop Frequency Response

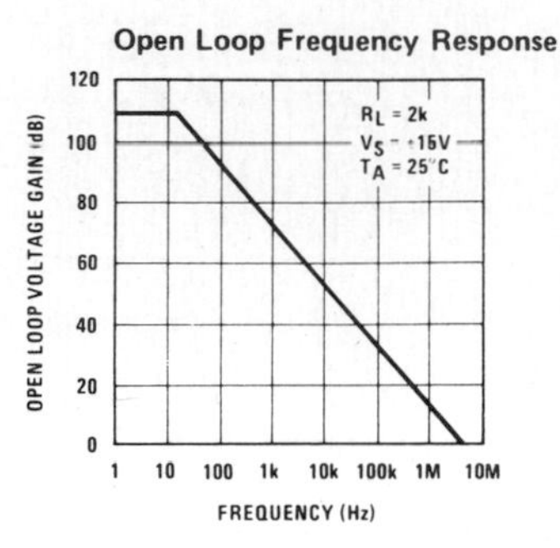

Common-Mode Rejection Ratio

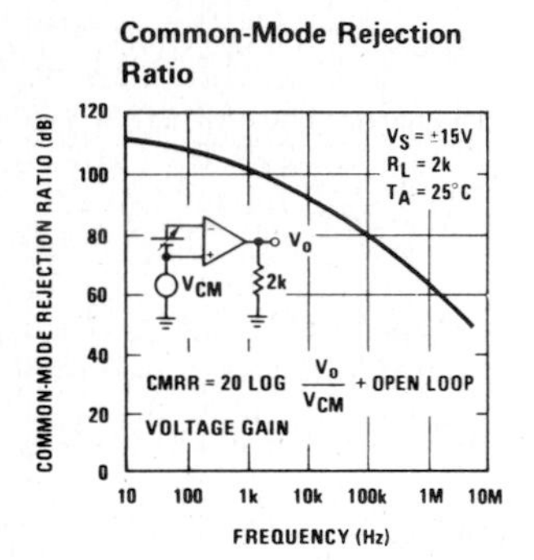

Power Supply Rejection Ratio

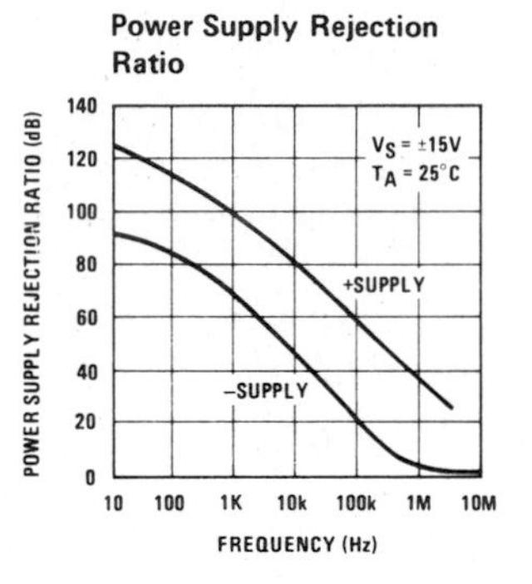

Equivalent Input Noise Voltage

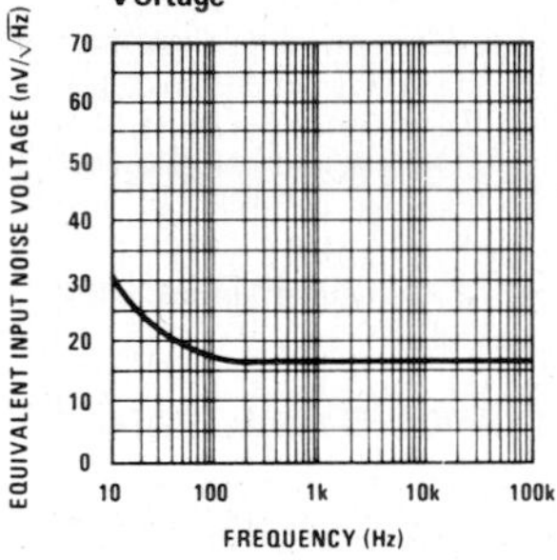

Open Loop Voltage Gain (V/V)

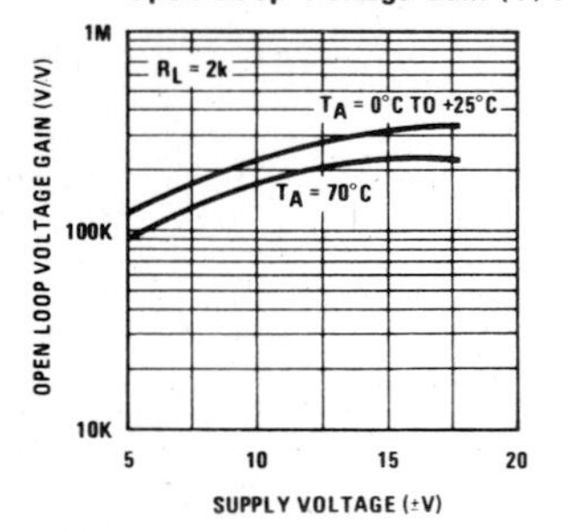

Output Impedance

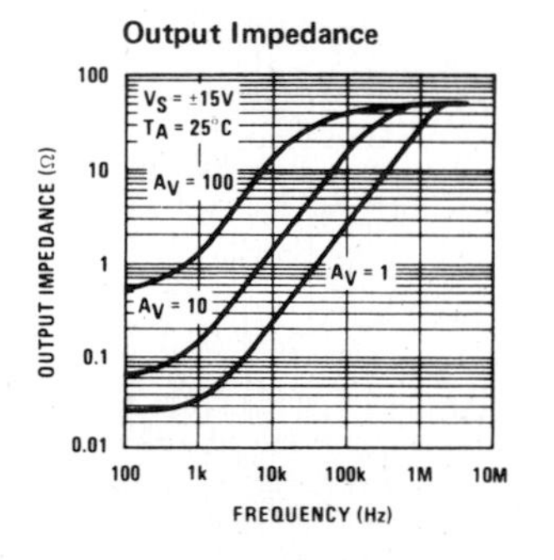

Inverter Settling Time

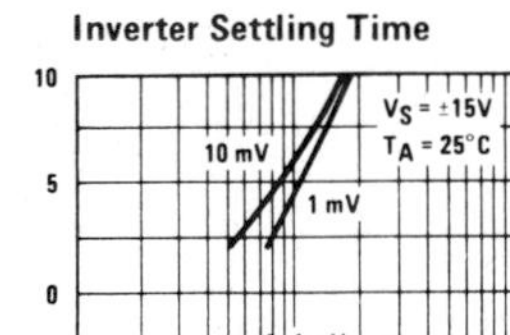

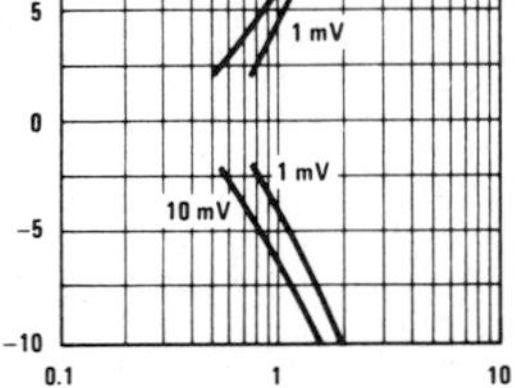

Typical Applications

Supply Current Indicator/Limiter

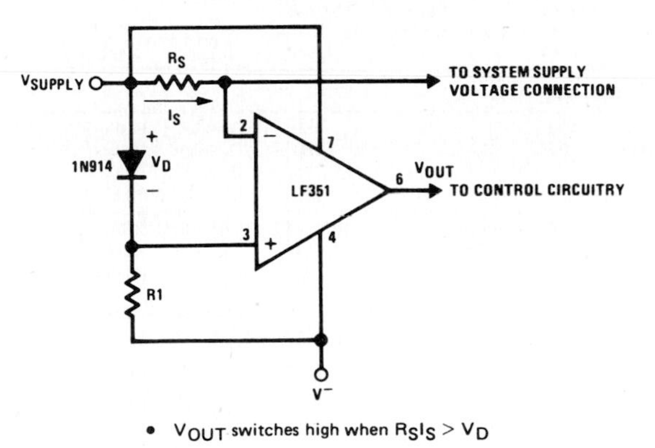

- V_{OUT} switches high when $R_S I_S > V_D$

Hi-Z_{IN} Inverting Amplifier

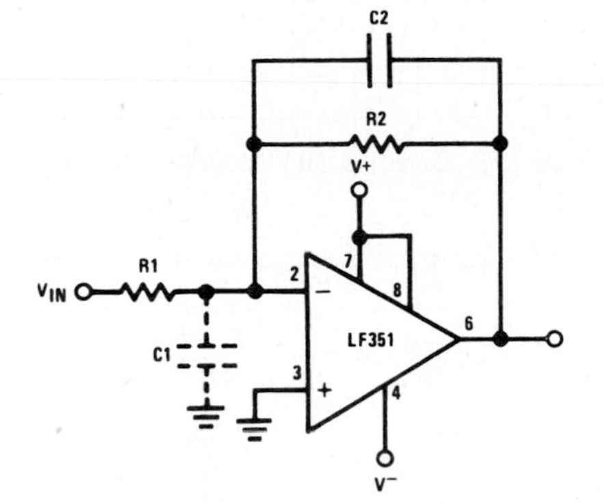

Parasitic input capacitance C1 ≅ (3 pF for LF351 plus any additional layout capacitance) interacts with feedback elements and creates undesirable high frequency pole. To compensate, add C2 such that: R2C2 ≅ R1C1.

Ultra-Low (or High) Duty Cycle Pulse Generator

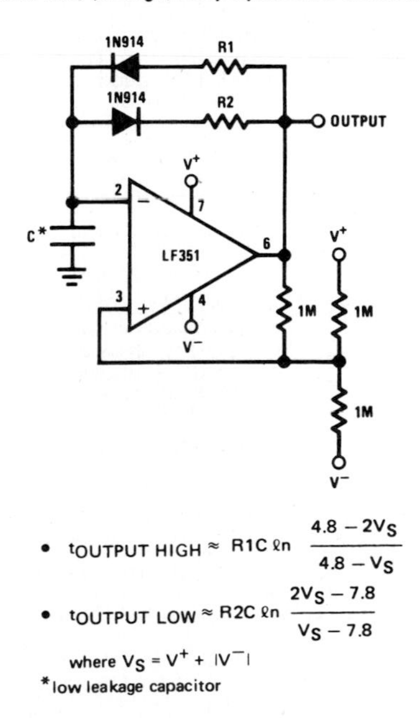

- $t_{OUTPUT\ HIGH} \approx R1C\ \ell n \dfrac{4.8 - 2V_S}{4.8 - V_S}$
- $t_{OUTPUT\ LOW} \approx R2C\ \ell n \dfrac{2V_S - 7.8}{V_S - 7.8}$

where $V_S = V^+ + |V^-|$

*low leakage capacitor

Long Time Integrator

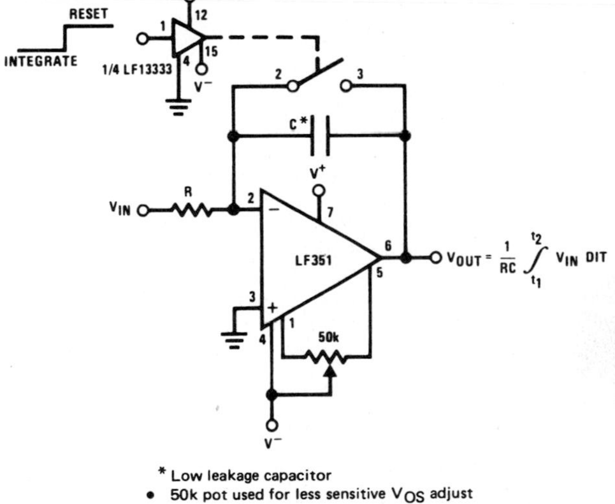

* Low leakage capacitor
- 50k pot used for less sensitive V_{OS} adjust

Op Amp Guide*

QUICK GUIDE TO BIPOLAR-FET OP AMPS

This guide does not list uncompensated versions, duals or military range versions.

Offset Voltage mV 25°C	Bias Current pA T_i = 25°C	Offset Current pA T_i = 25°C	Drift $\mu V/°C$	Unity Gain Bandwidth MHz	Slew Rate $V/\mu s$	Voltage Gain $a_o \times 10^4$	CMRR db	Supply Current per amp mA	Thermal Resistance °C/W	Model	Manufacturer
Single Units Compensated											
0.5	50	10	2T, 5M	6T, 4M	17T, 10M	100	86	4.7	150	OP-15E	PMI
0.5	50	10	2T, 5M	8T, 6M	25T, 18M	100	86	4.7	150	OP-16E	PMI
1	100	20	3T, 10M	5.7T, 3.5M	16T, 7.5M	75	86	4.7	150	OP-15F	PMI
1	100	20	3T, 10M	7.6T, 5.5M	24T, 12M	75	86	4.7	150	OP-16F	PMI
2	20	10	5T, 15M	4T	10T	100	86	15	135-145	CA3160B	RCA
2	30	10	5T	4.5T	9T	50	86	6	135-145	CR3140B	RCA
2	50	10	3T, 5M	4.5T, 4M	12T, 10M	50	85	7	150	LF356A	National*
2	50	10	3T, 5M	2.5T	5T, 3M	50	85	4	150	LF355A	National*
3	100	50	10T	5T	12T	50	80	2.8	120-210	TL081BC	Texas Instrument
3	200	50	4T, 15M	5.4T, 3M	15T, 5M	50	82	5.8	150	OP-15G	PMI
3	200	50	4T, 15M	7.2T, 5M	23T, 9M	50	82	5.8	150	OP-16G	PMI
5	30	20	6T	4T	10T	50	80	15	135-145	CA3160A	RCA
5	40	20	6T	4.5T	9T	20	70	6	135-145	CA3140A	RCA
5	100	20	8T	4T	13T	50	80	2.5	100	μAF771F	Fairchild
5	100	20	5T	5T	12T, 7.5M	50	85	7	155	LF356B	National*
6	200	100	10T	3T	12T	50	80	2.8	120-210	TL081AC	Texas Instrument
10	200	50	8T	4T	13T	25	70	2.5	100	μAF771C	Fairchild
10	200	50	10T	5T	15T	25	70	2.8	150	LF351	National
10	200	50	10T	3T	15T	25	70	2.8	120-210	TL086C	Texas Instrument
10	200	50	5T	2.5T	5T	25	80	4	150-175	LF355	National*
10	200	50	5T	5T	12T	25	80	10	150-175	LF356	National*
15	50	30	8T	4.5T	9T	20	70	6	135-145	CA3140	RCA
15	50	30	8T	4T	10T	50	70	15	135-145	CA3160	RCA
15	200	50	10T	1T	0.5T	25	70	4	150	LF13741	National
15	400	200	10T	3T	12T	25	70	2.8	120-210	TL081C	Texas Instrument
References											
0.25	2	0.3T	5M	1T	0.9T, 0.6M	100	76T	4	150	3527BM	Burr-Brown
0.5	7,000	1,000	2T, 5M	1T	0.3T	80	96	0.8	150	LM308A	National**
0.5	30,000	2,000	2T, 8M	1.3T, 0.8M	0.5T, 0.25M	100	90	2	—	OP-02E	PMI
0.5	50,000	5,000	5M	1T	0.5T	50	90	2.8	—	AD741L	Analog Devices
1	0.075	—	25M	0.35T	1T	25	66	1.5	—	AD515L	Analog Devices
6	500,000	200,000	—	1T	0.5T	20	70	2.8	—	μA741C	Fairchild**
Quadruple Units											
3	100	50	10T	3T	12T	50	80	2.8	85-97	TL084BC	Texas Instrument
5	100	20	8T	4T	13T	50	80	2.5	70	μAF774E	Fairchild
6	200	100	10T	3T	12T	50	80	2.5	85-97	TL084AC	Texas Instrument
10	200	50	8T	4T	13T	25	70	2.5	70	μAF744C	Fairchild
10	200	50	10T	5T	15T	25	70	2.8	100	LF347	National
15	200	50	—	10T, 7M	20M	25	70	2.5	100	3471	Motorola
15	400	200	10T	3T	12T	25	70	2.8	85-97	TL084C	Texas Instrument
References											
0.75	500,000	20,000	8M	2T	1T	100	90	1.5	—	OP-09/11E	PMI
3.5	300,000	100,000	2T	8T	4T	75	80	1.6	—	HA4605	Harris
6	200,000	50,000	—	1T	0.5T	25	70	1.1	—	LM348	National**
6	500,000	50,000	—	3T	1T	20	60	10	—	RC4136	Raytheon**

M—Maximum or Minimum.
T—Typical.

*Suppliers: AMD, Fairchild, Intersil, Motorola, National, PMI, Raytheon, and Texas Instruments.
**Suppliers: See IC Master.

**Electron Products* magazine, p. 53, June 1977.

E.6 IC GATES

TTL GATES (54/74 Series)*

POSITIVE-NAND GATES AND INVERTERS WITH TOTEM-POLE OUTPUTS

recommended operating conditions

	54 FAMILY / 74 FAMILY	SERIES 54 / SERIES 74 '00, '04, '10, '20, '30			SERIES 54H / SERIES 74H 'H00, 'H04, 'H10, 'H20, 'H30			SERIES 54L / SERIES 74L 'L00, 'L04, 'L10, 'L20, 'L30			SERIES 54LS / SERIES 74LS 'LS00, 'LS04, 'LS10, 'LS20, 'LS30			SERIES 54S / SERIES 74S 'S00, 'S04, 'S10, 'S20, 'S30, 'S133			UNIT
		MIN	NOM	MAX	MIN	NOM	MAX	MIN	NOM	MAX	MIN	NOM	MAX	MIN	NOM	MAX	
Supply voltage, V_{CC}	54 Family	4.5	5	5.5	4.5	5	5.5	4.5	5	5.5	4.5	5	5.5	4.5	5	5.5	V
	74 Family	4.75	5	5.25	4.75	5	5.25	4.75	5	5.25	4.75	5	5.25	4.75	5	5.25	
High-level output current, I_{OH}	54 Family			−400			−500			−100			−400			−1000	μA
	74 Family			−400			−500			−200			−400			−1000	
Low-level output current, I_{OL}	54 Family			16			20			2			4			20	mA
	74 Family			16			20			3.6			8			20	
Operating free-air temperature, T_A	54 Family	−55		125	−55		125	−55		125	−55		125	−55		125	°C
	74 Family	0		70	0		70	0		70	0		70	0		70	

electrical characteristics over recommended operating free-air temperature range (unless otherwise noted)

PARAMETER	TEST FIGURE	TEST CONDITIONS†		SERIES 54 / SERIES 74 '00, '04, '10, '20, '30			SERIES 54H / SERIES 74H 'H00, 'H04, 'H10, 'H20, 'H30			SERIES 54L / SERIES 74L 'L00, 'L04, 'L10, 'L20, 'L30			SERIES 54LS / SERIES 74LS 'LS00, 'LS04, 'LS10, 'LS20, 'LS30			SERIES 54S / SERIES 74S 'S00, 'S04, 'S10, 'S20, 'S30, 'S133			UNIT
				MIN	TYP‡	MAX	MIN	TYP‡	MAX	MIN	TYP‡	MAX	MIN	TYP‡	MAX	MIN	TYP‡	MAX	
V_{IH} High-level input voltage	1, 2			2			2			2			2			2			V
V_{IL} Low-level input voltage	1, 2		54 Family			0.8			0.8			0.7			0.7			0.8	V
			74 Family			0.8			0.8			0.7			0.8			0.8	
V_{IK} Input clamp voltage	3	V_{CC} = MIN, I_I = §				−1.5			−1.5						−1.5			−1.2	V
V_{OH} High-level output voltage	1	V_{CC} = MIN, V_{IL} = V_{IL} max, I_{OH} = MAX	54 Family	2.4	3.4		2.4	3.5		2.4	3.3		2.5	3.4		2.5	3.4		V
			74 Family	2.4	3.4		2.4	3.5		2.4	3.2		2.7	3.4		2.7	3.4		
V_{OL} Low-level output voltage	2	V_{CC} = MIN, V_{IH} = 2 V, I_{OL} = MAX	54 Family		0.2	0.4		0.2	0.4		0.15	0.3		0.25	0.4			0.5	V
			74 Family		0.2	0.4		0.2	0.4		0.2	0.4		0.25	0.5			0.5	
		V_{CC} = MIN, V_{IH} = 2 V, I_{OL} = 4 mA	Series 74LS												0.4				
I_I Input current at maximum input voltage	4	V_{CC} = MAX	V_I = 5.5 V			1			1			0.1						1	mA
			V_I = 7 V												0.1				
I_{IH} High-level input current	4	V_{CC} = MAX	V_{IH} = 2.4 V			40			50			10							μA
			V_{IH} = 2.7 V												20			50	
I_{IL} Low-level input current	5	V_{CC} = MAX	V_{IL} = 0.3 V									−0.18							mA
			V_{IL} = 0.4 V			−1.6			−2						−0.4				
			V_{IL} = 0.5 V															−2	
I_{OS} Short-circuit output current◆	6	V_{CC} = MAX	54 Family	−20		−55	−40		−100	−3		−15	−20		−100	−40		−100	mA
			74 Family	−18		−55	−40		−100	−3		−15	−20		−100	−40		−100	
I_{CC} Supply current	7	V_{CC} = MAX		See table on next page															mA

†For conditions shown as MIN or MAX, use the appropriate value specified under recommended operating conditions.
‡All typical values are at V_{CC} = 5 V, T_A = 25°C.
§I_I = −12 mA for SN54'/SN74', −8 mA for SN54H'/SN74H', and −18 mA for SN54LS'/SN74LS' and SN54S'/SN74S'.
◆Not more than one output should be shorted at a time, and for SN54H'/SN74H', SN54LS'/SN74LS', and SN54S'/SN74S', duration of short-circuit should not exceed 1 second.

*Courtesy Texas Instruments Inc.

E.6 IC GATES (continued)

POSITIVE-NAND GATES AND INVERTERS WITH TOTEM-POLE OUTPUTS

switching characteristics at $V_{CC} = 5$ V, $T_A = 25°C$

TYPE	TEST CONDITIONS#	t_{PLH} (ns) Propagation delay time, low-to-high-level output			t_{PHL} (ns) Propagation delay time, high-to-low-level output		
		MIN	TYP	MAX	MIN	TYP	MAX
'00, '10	C_L = 15 pF, R_L = 400 Ω		11	22		7	15
'04, '20			12	22		8	15
'30			13	22		8	15
'H00	C_L = 25 pF, R_L = 280 Ω		5.9	10		6.2	10
'H04			6	10		6.5	10
'H10			5.9	10		6.3	10
'H20			6	10		7	10
'H30			6.8	10		8.9	12
'L00, 'L04, 'L10, L20	C_L = 50 pF, R_L = 4 kΩ		35	60		31	60
'L30			35	60		70	100
'LS00, 'LS04 'LS10, 'LS20	C_L = 15 pF, R_L = 2 kΩ		9	15		10	15
'LS30			8	15		13	20
'S00, 'S04	C_L = 15 pF, R_L = 280 Ω		3	4.5		3	5
'S10, 'S20	C_L = 50 pF, R_L = 280 Ω		4.5			5	
'S30, 'S133	C_L = 15 pF, R_L = 280 Ω		4	6		4.5	7
	C_L = 50 pF, R_L = 280 Ω		5.5			6.5	

#Load circuits and voltage waveforms are shown on pages 3-10 and 3-11.

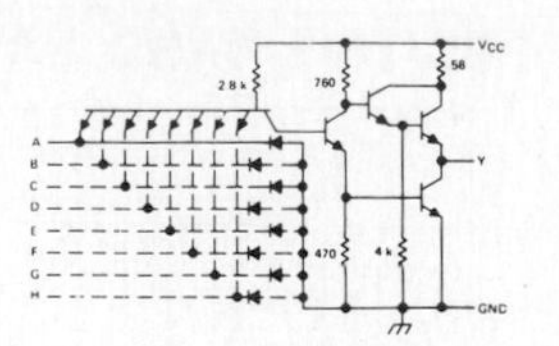

'H00, 'H04, 'H10, 'H20, 'H30 CIRCUITS

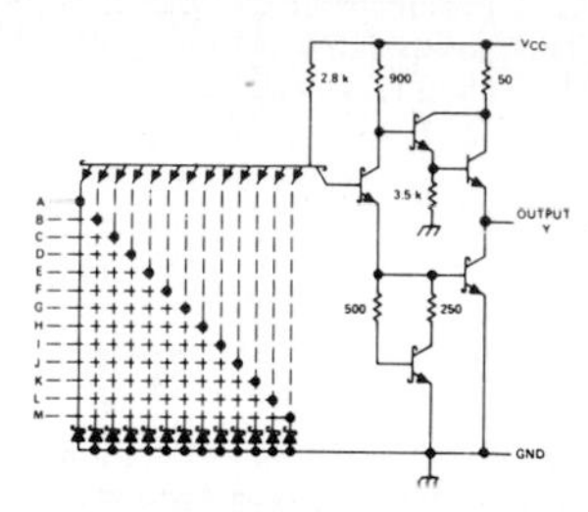

'S00, 'S04, 'S10, 'S20, 'S30, 'S133 CIRCUITS

supply current¶

TYPE	I_{CCH} (mA) Total with outputs high		I_{CCL} (mA) Total with outputs low		I_{CC} (mA) Average per gate (50% duty cycle)
	TYP	MAX	TYP	MAX	TYP
'00	4	8	12	22	2
'04	6	12	18	33	2
'10	3	6	9	16.5	2
'20	2	4	6	11	2
'30	1	2	3	6	2
'H00	10	16.8	26	40	4.5
'H04	16	26	40	58	4.5
'H10	7.5	12.6	19.5	30	4.5
'H20	5	8.4	13	20	4.5
'H30	2.5	4.2	6.5	10	4.5
'L00	0.44	0.8	1.16	2.04	0.20
'L04	0.66	1.2	1.74	3.06	0.20
'L10	0.33	0.6	0.87	1.53	0.20
'L20	0.22	0.4	0.58	1.02	0.20
SN54L30	0.11	0.33	0.29	0.51	0.20
SN74L30	0.11	0.2	0.29	0.51	0.20
'LS00	0.8	1.6	2.4	4.4	0.4
'LS04	1.2	2.4	3.6	6.6	0.4
'LS10	0.6	1.2	1.8	3.3	0.4
'LS20	0.4	0.8	1.2	2.2	0.4
'LS30	0.35	0.5	0.6	1.1	0.48
'S00	10	16	20	36	3.75
'S04	15	24	30	54	3.75
'S10	7.5	12	15	27	3.75
'S20	5	8	10	18	3.75
'S30	3	5	5.5	10	4.25
'S133	3	5	5.5	10	4.25

schematics (each gate)

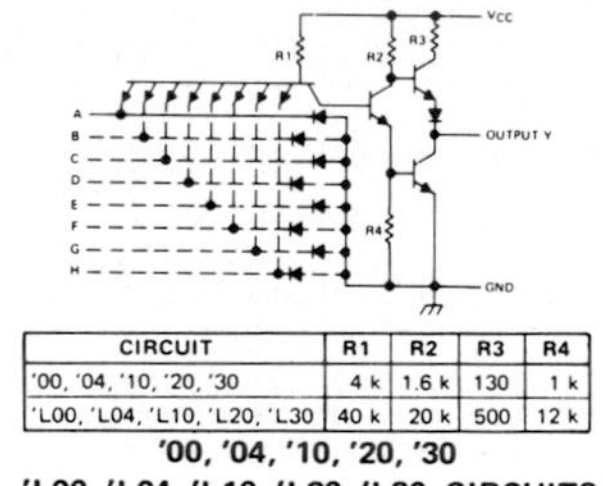

CIRCUIT	R1	R2	R3	R4
'00, '04, '10, '20, '30	4 k	1.6 k	130	1 k
'L00, 'L04, 'L10, 'L20, 'L30	40 k	20 k	500	12 k

'00, '04, '10, '20, '30
'L00, 'L04, 'L10, 'L20, 'L30, CIRCUITS
Input clamp diodes not on SN54L'/SN74L' circuits.

Resistor values shown are nominal and in ohms.

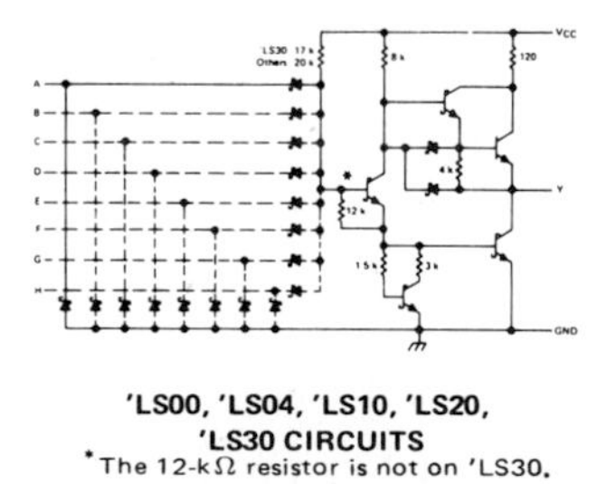

'LS00, 'LS04, 'LS10, 'LS20, 'LS30 CIRCUITS
*The 12-kΩ resistor is not on 'LS30.

¶Maximum values of I_{CC} are over the recommended operating ranges of V_{CC} and T_A; typical values are at $V_{CC} = 5$ V, $T_A = 25°C$.

POSITIVE-NOR GATES WITH TOTEM-POLE OUTPUTS

recommended operating conditions

	54 FAMILY / 74 FAMILY	SERIES 54 / SERIES 74						SERIES 54L / SERIES 74L			SERIES 54LS / SERIES 74LS			SERIES 54S / SERIES 74S			UNIT
		'02			'25, '27			'L02			'LS02, 'LS27			'S02, 'S260			
		MIN	NOM	MAX	MIN	NOM	MAX	MIN	NOM	MAX	MIN	NOM	MAX	MIN	NOM	MAX	
Supply voltage, V_{CC}	54 Family	4.5	5	5.5	4.5	5	5.5	4.5	5	5.5	4.5	5	5.5	4.5	5	5.5	V
	74 Family	4.75	5	5.25	4.75	5	5.25	4.75	5	5.25	4.75	5	5.25	4.75	5	5.25	
High-level output current, I_{OH}	54 Family			−400			−800			−100			−400			−1000	μA
	74 Family			−400			−800			−200			−400			−1000	
Low-level output current, I_{OL}	54 Family			16			16			2			4			20	mA
	74 Family			16			16			3.6			8			20	
Operating free-air temperature, T_A	54 Family	−55		125	−55		125	−55		125	−55		125	−55		125	°C
	74 Family	0		70	0		70	0		70	0		70	0		70	

electrical characteristics over recommended operating free-air temperature range (unless otherwise noted)

PARAMETER		TEST FIGURE	TEST CONDITIONS†		SERIES 54 / SERIES 74			SERIES 54L / SERIES 74L			SERIES 54LS / SERIES 74LS			SERIES 54S / SERIES 74S			UNIT
					'02, '25, '27			'L02			'LS02 'LS27			'S02, 'S260			
					MIN	TYP‡	MAX	MIN	TYP‡	MAX	MIN	TYP‡	MAX	MIN	TYP‡	MAX	
V_{IH} High-level input voltage		1, 2			2			2			2			2			V
V_{IL} Low-level input voltage		1, 2		54 Family			0.8			0.7			0.7			0.8	V
				74 Family			0.8			0.7			0.8			0.8	
V_{IK} Input clamp voltage		3	V_{CC} = MIN, I_I = §				−1.5						−1.5			−1.2	V
V_{OH} High-level output voltage		1	V_{CC} = MIN, V_{IL} = V_{IL} max, I_{OH} = MAX	54 Family	2.4	3.4		2.4	3.3		2.5	3.4		2.5	3.4		V
				74 Family	2.4	3.4		2.4	3.2		2.7	3.4		2.7	3.4		
V_{OL} Low-level output voltage		2	V_{CC} = MIN, V_{IH} = 2 V, I_{OL} = MAX	54 Family		0.2	0.4		0.15	0.3		0.25	0.4			0.5	V
				74 Family		0.2	0.4		0.2	0.4		0.35	0.5			0.5	
			I_{OL} = 4 mA	Series 74LS								0.25	0.4				
I_I Input current at maximum input voltage		4	V_{CC} = MAX	V_I = 5.5 V			1			0.1						1	mA
				V_I = 7 V									0.1				
I_{IH} High-level input current	Data inputs	4	V_{CC} = MAX	V_{IH} = 2.4 V			40			10							μA
	Strobe of '25						160										
	All inputs			V_{IH} = 2.7 V									20			50	
I_{IL} Low-level input current	All inputs	5	V_{CC} = MAX	V_{IL} = 0.3 V						−0.18							mA
	Data inputs			V_{IL} = 0.4 V			−1.6						−0.4				
	Strobe of '25						−6.4										
	All inputs			V_{IL} = 0.5 V												−2	
I_{OS} Short-circuit output current♦		6	V_{CC} = MAX	54 Family	−20		−55	−3		−15	−20		−100	−40		−100	mA
				74 Family	−18		−55	−3		−15	−20		−100	−40		−100	
I_{CC} Supply current		7	V_{CC} = MAX		See table on next page												mA

†For conditions shown as MIN or MAX, use the appropriate value specified under recommended operating conditions.

‡All typical values are at V_{CC} = 5 V, T_A = 25°C.

§I_I = −12 mA for SN54'/SN74' and −18 mA for SN54LS'/SN74LS' and SN54S/SN74S'.

♦Not more than one output should be shorted at a time, and for SN54LS'/SN74LS' and SN54S'/SN74S', duration of output short-circuit should not exceed one second.

E.6 IC GATES (continued)

POSITIVE-NOR GATES WITH TOTEM-POLE OUTPUTS

supply current¶

TYPE	I_{CCH} (mA) Total with outputs high		I_{CCL} (mA) Total with outputs low		I_{CC} (mA) Average per gate (50% duty cycle)
	TYP	MAX	TYP	MAX	TYP
'02	8	16	14	27	2.75
'25	8	16	10	19	2.25
'27	10	16	16	26	4.34
'L02	0.8	1.6	1.4	2.6	0.275
'LS02	1.6	3.2	2.8	5.4	0.55
'LS27	2.0	4	3.4	6.8	0.9
'S02	17	29	26	45	5.38
'S260	17	29	26	45	10.75

¶Maximum values of I_{CC} are over the recommended operating ranges of V_{CC} and T_A; typical values are at V_{CC} = 5 V, T_A = 25°C.

schematics (each gate)

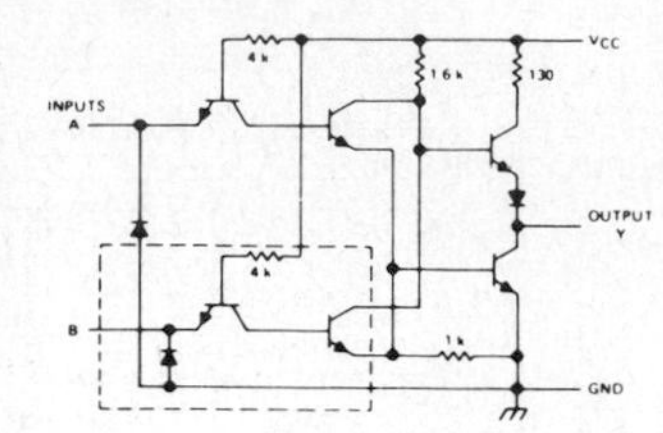

The portion of the schematic within the dashed lines is repeated for the C input of the '27.

'02, '27 CIRCUITS

switching characteristics at V_{CC} = 5 V, T_A = 25°C

TYPE	TEST CONDITIONS#	t_{PLH} (ns) Propagation delay time, low-to-high-level output			t_{PHL} (ns) Propagation delay time, high-to-low-level output		
		MIN	TYP	MAX	MIN	TYP	MAX
'02	C_L = 15 pF, R_L = 400 Ω		12	15		8	15
'25			13	22		8	15
'27			10	15		7	11
'L02	C_L = 50 pF, R_L = 4 kΩ		31	60		35	60
'LS02, 'LS27	C_L = 15 pF, R_L = 2 kΩ		10	15		10	15
'S02	C_L = 15 pF, R_L = 280 Ω		3.5	5.5		3.5	5.5
	C_L = 50 pF, R_L = 280 Ω		5			5	
'S260	C_L = 15 pF, R_L = 280 Ω		4	5.5		4	6

#Load circuit and voltage waveforms are shown on pages 3-10 and 3-11.

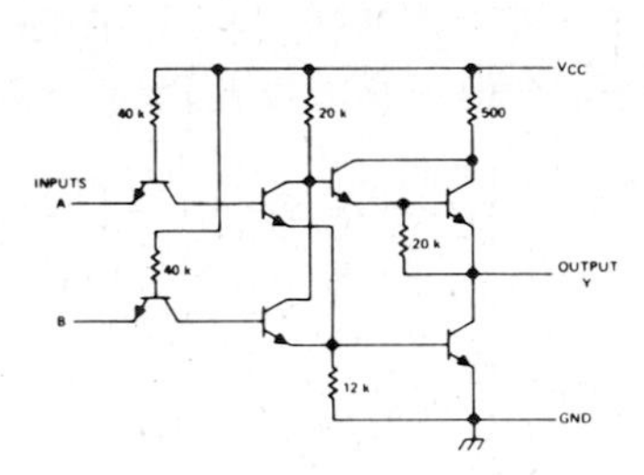

'L02 CIRCUITS

The portion of the schematic within the dashed lines applies only to the 'LS27

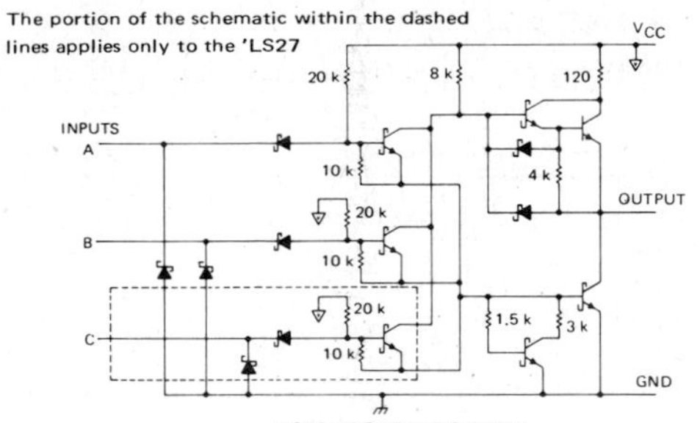

'LS02, 'LS27 CIRCUITS

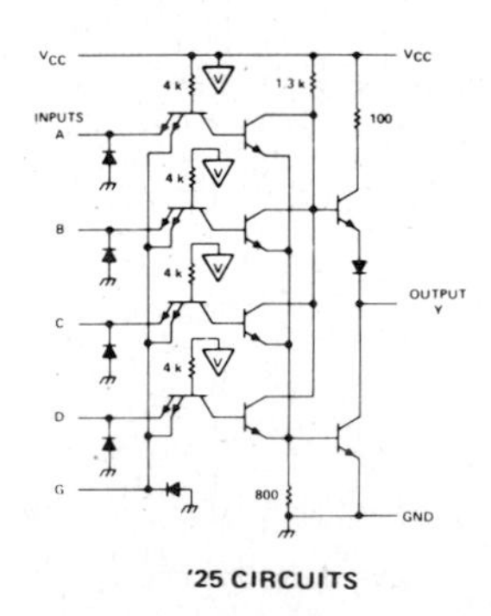

'25 CIRCUITS

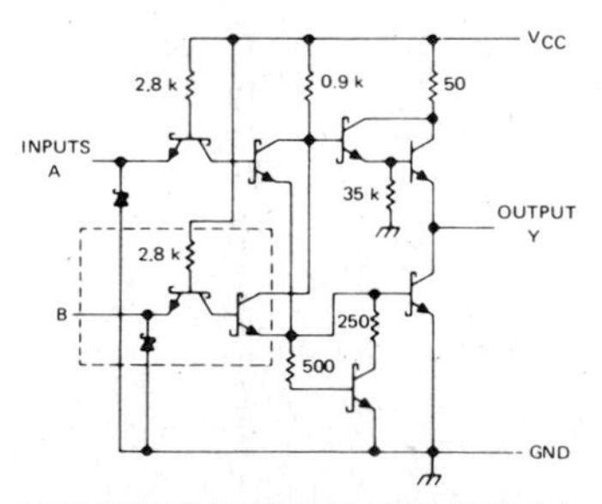

The portion of the schematic within the dashed lines is repeated for each additional input of the 'S260, and the 0.9-kΩ resistor is changed to 0.6 kΩ.

'S02, 'S260 CIRCUITS

Resistor values are nominal and in ohms.

ECL Gates (10000 Series)*

TRIPLE EXCLUSIVE OR/EXCLUSIVE NOR GATE

GX family

standard temperature range

10107

gate

The GX family of ECL silicon monolithic integrated circuits is designed for high speed central processors and digital communication systems.
With 2,0 ns typical propagation delay and only 25 mW power dissipation per gate, this family offers an excellent speed-power product and so is recommended for high speed large system design.

The GXB10107 is a three gate array designed to provide the positive EXCLUSIVE OR and NOR functions.

Input pull-down resistors (50 kΩ) allow unused inputs to be left open.

The GX family corresponds to the ECL10 000series.

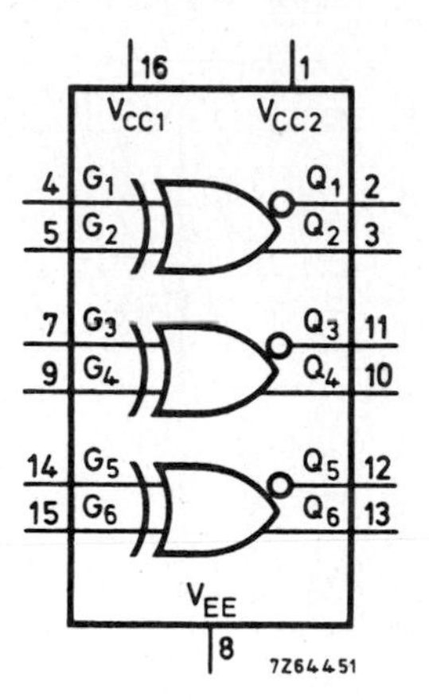

16 15 14 13 12 11 10 9
VCC1 G6 G5 Q6 Q5 Q3 Q4 G4
GXB10107
VCC2 Q1 Q2 G1 G2 n.c. G3 VEE
1 2 3 4 5 6 7 8

$V_{CC1} = V_{CC2} = 0$ V (ground)

$V_{EE} = -5,2$ V

QUICK REFERENCE DATA				
Supply voltage	V_{EE}		$-5,2 \pm 10\%$	V
Operating ambient temperature	T_{amb}		0 to +75	°C
Average propagation delay	t_{pd}	typ.	2,4	ns
Output voltage HIGH state	V_{OH}	nom.	−880	mV
LOW state	V_{OL}	nom.	−1720	mV
Power consumption per package	P_{av}	typ.	115	mW

PACKAGE OUTLINE (see General Section)
GXB10107P: plastic 16-lead dual in-line
GXB10107D: ceramic 16-lead dual in-line

*Courtesy of Signetics.

E.6 IC GATES (continued)

GX family — standard temperature range

10107 gate

CIRCUIT DIAGRAM (one gate)

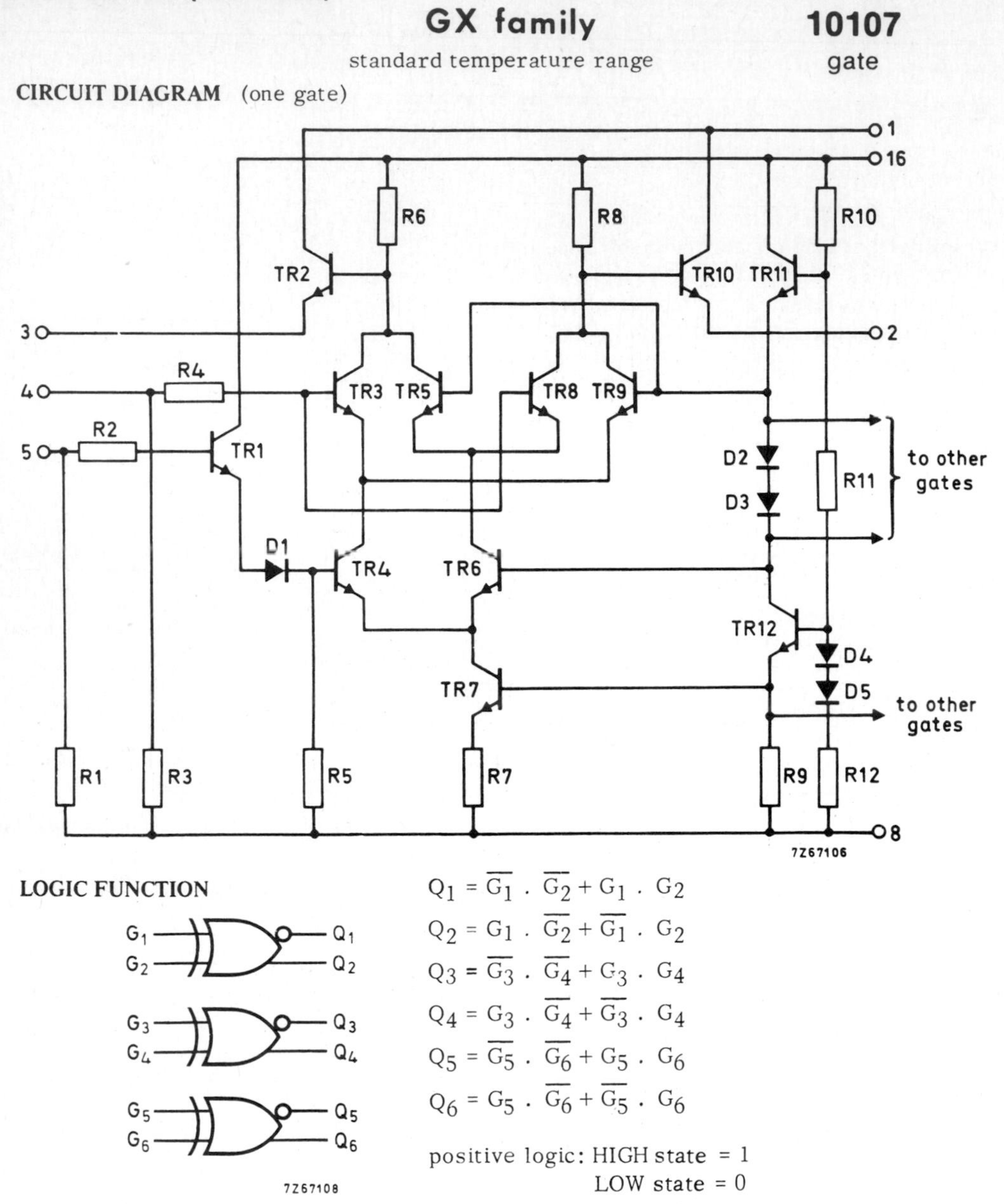

LOGIC FUNCTION

$Q_1 = \overline{G_1} \cdot \overline{G_2} + G_1 \cdot G_2$

$Q_2 = G_1 \cdot \overline{G_2} + \overline{G_1} \cdot G_2$

$Q_3 = \overline{G_3} \cdot \overline{G_4} + G_3 \cdot G_4$

$Q_4 = G_3 \cdot \overline{G_4} + \overline{G_3} \cdot G_4$

$Q_5 = \overline{G_5} \cdot \overline{G_6} + G_5 \cdot G_6$

$Q_6 = G_5 \cdot \overline{G_6} + \overline{G_5} \cdot G_6$

positive logic: HIGH state = 1
LOW state = 0

RATINGS Limiting values in accordance with the Absolute Maximum System (IEC 134)

Supply voltage (d.c.)	V_{EE}	max.	−8,0	V
Input voltage	V_I		0 to V_{EE}	
Output current	I_O	max.	50	mA
Storage temperature	T_{stg}		−55 to +125	°C
Junction temperature	T_j	max.	125	°C

GX family

standard temperature range

10107

gate

CHARACTERISTICS (d.c.) at V_{CC} = ground; V_{EE} = −5, 2 V

Each GX circuit has been designed to meet the d.c. specifications shown in the test table below, after thermal equilibrium has been established. The circuit is in a test socket or mounted on a printed circuit board and transverse air flow > 2, 5 m/s is maintained. Outputs are terminated via a 50 Ω resistor to −2, 0 V. Test values for applied conditions are given in the table and defined in the figure.

Test table

T_{amb}	0	25	75	°C
V_{IHmax}	-0, 840	-0, 810	-0, 720	V
V_{IHT}	-1, 145	-1, 105	-1, 045	V
V_{ILT}	-1, 490	-1, 475	-1, 450	V
V_{ILmin}	-1, 870	-1, 850	-1, 830	V

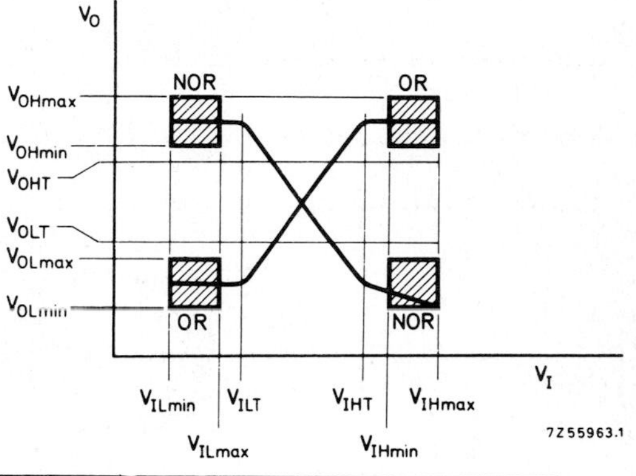

	Symbol		T_{amb} (°C) 0	25	75		Conditions
Output voltage HIGH	V_{OH}	min.	-1000	-960	-900	mV	See note 1
		typ.	–	-880	–	mV	
		max.	-840	-810	-720	mV	
Output voltage LOW	V_{OL}	min.	-1,870	-1,850	-1,830	V	See note 2
		typ.	–	-1,720	–	V	
		max.	-1,665	-1,650	-1,625	V	
Output threshold voltage HIGH	V_{OHT}	min.	-1020	-980	-920	mV	See note 3
Output threshold voltage LOW	V_{OLT}	max.	-1,645	-1,630	-1 ,605	V	See note 4

Notes

1. V_{ILmin} or V_{IHmax} on both inputs for invert outputs.
   V_{ILmin} on one input and V_{IHmax} on other input for direct outputs.
2. V_{ILmin} on one input and V_{IHmax} on other input for invert outputs.
   V_{ILmin} or V_{IHmax} on both inputs for direct outputs.
3. V_{ILT} or V_{IHT} on both inputs for invert outputs.
   V_{ILT} on one input and V_{IHT} on other input for direct outputs.
4. V_{ILT} on one input and V_{IHT} on other input for invert outputs.
   V_{ILT} or V_{IHT} on both inputs for direct outputs.

E.6 IC GATES (continued)

GX family

standard temperature range

10107
gate

CHARACTERISTICS (continued)

	Symbol		T_{amb} (^oC) 0	25	75		Conditions
Input current HIGH							
pins 4, 9, 14	I_{IH}	max.	–	350	–	μA	V_{IHmax} for input under test
pins 5, 7, 15	I_{IH}	max.	–	265	–	μA	
Input current LOW	I_{IL}	min.	–	10	–	μA	V_{ILmin} for input under test
Supply current	I_{EE}	typ.	–	23	–	mA	V_{IHmax} for all inputs
		max.	–	28	–	mA	
	$\frac{dV_{OL}}{dV_{EE}}$	typ.	–	0, 25	–		

CHARACTERISTICS (a.c.) at V_{CC} = ground; V_{EE} = –5, 2 V; T_{amb} = 25 ^oC

	Symbol	min.	typ.	max.		Conditions
Rise propagation delay times: OR output	t_{pdrOR}	1, 1	2, 0	3, 7	ns	Inputs 4, 9 or 14
		1, 1	2, 8	3, 7	ns	Inputs 5, 7 or 15
NOR output	t_{pdrNOR}	1, 1	2, 0	3, 7	ns	Inputs 4, 9 or 14
		1, 1	2, 8	3, 7	ns	Inputs 5, 7 or 15
Fall propagation delay times: OR output	t_{pdfOR}	1, 1	2, 0	3, 7	ns	Inputs 4, 9 or 14
		1, 1	2, 8	3, 7	ns	Inputs 5, 7 or 15
NOR output	t_{pdfNOR}	1, 1	2, 0	3, 7	ns	Inputs 4, 9 or 14
		1, 1	2, 8	3, 7	ns	Inputs 5, 7 or 15
Rise time	t_r	1, 1	2, 5	3, 5	ns	
Fall time	t_f	1, 1	2, 5	3, 5	ns	
Input capacitance (see note)	C_I	–	-	5	pF	reflection measurement

Note: Input resistance is positive at any frequency

CMOS Gates (4000B Series)*

CD4000B, CD4001B, CD4002B, CD4025B Types

COS/MOS NOR Gates

High-Voltage Types (20-Volt Rating)

Dual 3 Input
plus Inverter – CD4000B
Quad 2 Input – CD4001B
Dual 4 Input – CD4002B
Triple 3 Input – CD4025B

RCA-CD4000B, CD4001B, CD4002B, and CD4025B NOR gates provide the system designer with direct implementation of the NOR function and supplement the existing family of COS/MOS gates. All inputs and outputs are buffered.

The CD4000B, CD4001B, CD4002B, and CD4025B types are supplied in 14-lead hermetic dual-in-line ceramic packages (D and F suffixes), 14-lead dual-in-line plastic packages (E suffix), 14-lead ceramic flat packages (K suffix), and in chip form (H suffix).

Features:

- Propagation delay time = 60 ns (typ.) at C_L = 50 pF, V_{DD} = 10 V
- Buffered inputs and outputs
- Standardized symmetrical output characteristics
- 100% tested for maximum quiescent current at 20 V
- 5-V, 10-V, and 15-V parametric ratings
- Maximum input current of 1 μA at 18 V over full package-temperature range; 100 nA at 18 V and 25°C
- Noise margin (over full package temperature range):
  1 V at V_{DD} = 5 V
  2 V at V_{DD} = 10 V
  2.5 V at V_{DD} = 15 V
- Meets all requirements of JEDEC Tentative Standard No.13A, "Standard Specifications for Description of "B" Series CMOS Devices"

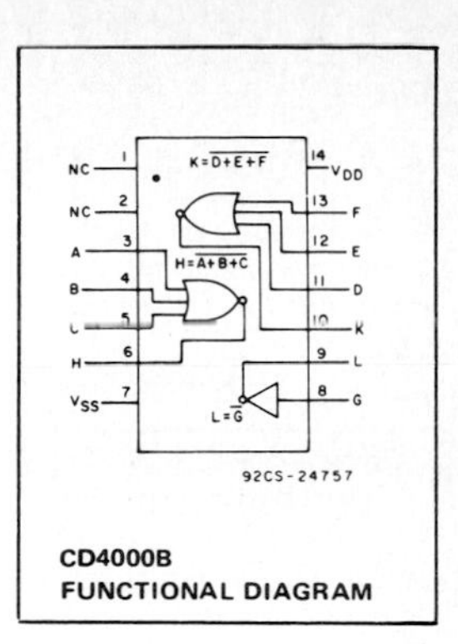

CD4000B FUNCTIONAL DIAGRAM

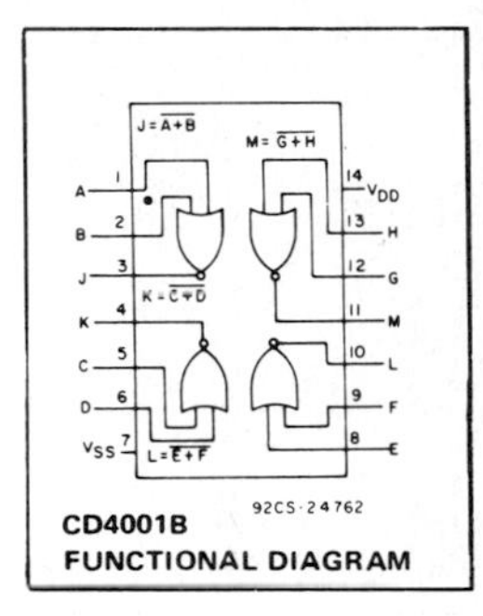

CD4001B FUNCTIONAL DIAGRAM

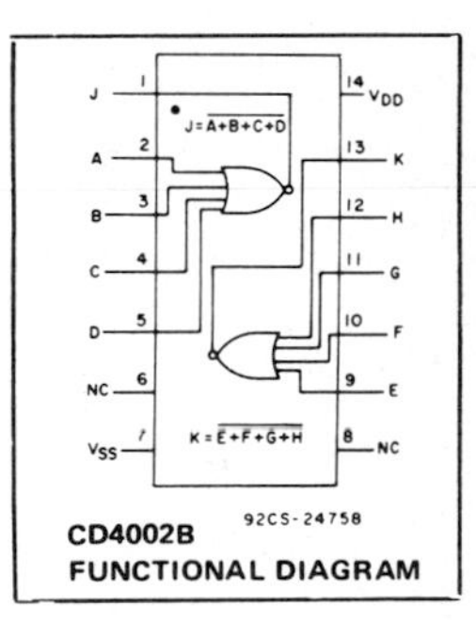

CD4002B FUNCTIONAL DIAGRAM

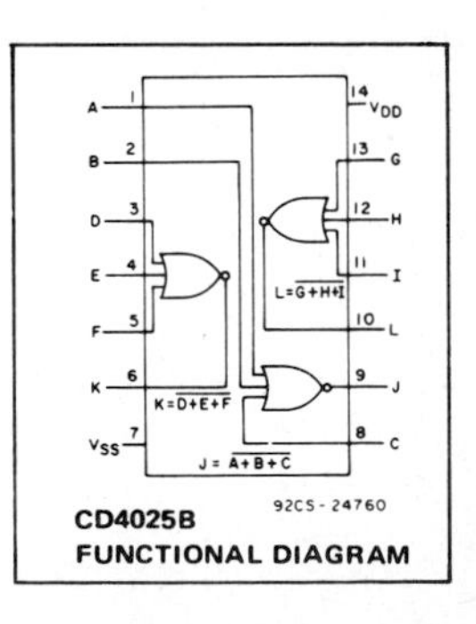

CD4025B FUNCTIONAL DIAGRAM

STATIC ELECTRICAL CHARACTERISTICS

CHARACTERISTIC	CONDITIONS			LIMITS AT INDICATED TEMPERATURES (°C) Values at –55, +25, +125 Apply to D,F,H Packages Values at –40, +25, +85 Apply to E Package							UNITS
								+25			
	V_O (V)	V_{IN} (V)	V_{DD} (V)	–55	–40	+85	+125	Min.	Typ.	Max.	
Quiescent Device Current, I_{DD} Max.	–	0,5	5	0.25	0.25	7.5	7.5	–	0.01	0.25	μA
	–	0,10	10	0.5	0.5	15	15	–	0.01	0.5	
	–	0,15	15	1	1	30	30	–	0.01	1	
	–	0,20	20	5	5	150	150	–	0.02	5	
Output Low (Sink) Current I_{OL} Min.	0.4	0,5	5	0.64	0.61	0.42	0.36	0.51	1	–	mA
	0.5	0,10	10	1.6	1.5	1.1	0.9	1.3	2.6	–	
	1.5	0,15	15	4.2	4	2.8	2.4	3.4	6.8	–	
Output High (Source) Current, I_{OH} Min.	4.6	0,5	5	–0.64	–0.61	–0.42	–0.36	–0.51	–1	–	
	2.5	0,5	5	–2	–1.8	–1.3	–1.15	–1.6	–3.2	–	
	9.5	0,10	10	–1.6	–1.5	–1.1	–0.9	–1.3	–2.6	–	
	13.5	0,15	15	–4.2	–4	–2.8	–2.4	–3.4	–6.8	–	
Output Voltage: Low-Level, V_{OL} Max.	–	0,5	5	0.05				–	0	0.05	V
	–	0,10	10	0.05				–	0	0.05	
	–	0,15	15	0.05				–	0	0.05	
Output Voltage: High-Level, V_{OH} Min.	–	0,5	5	4.95				4.95	5	–	
	–	0,10	10	9.95				9.95	10	–	
	–	0,15	15	14.95				14.95	15	–	
Input Low Voltage, V_{IL} Max.	0.5,4.5	–	5	1.5				–	–	1.5	V
	1,9	–	10	3				–	–	3	
	1.5,13.5	–	15	4				–	–	4	
Input High Voltage, V_{IH} Min.	0.5	–	5	3.5				3.5	–	–	
	1	–	10	7				7	–	–	
	1.5	–	15	11				11	–	–	
Input Current I_{IN} Max.		0,18	18	±0.1	±0.1	±1	±1	–	$\pm10^{-5}$	±0.1	μA

*Courtesy of RCA.

E.6 IC GATES (continued)

CD4000B, CD4001B, CD4002B, CD4025B Types

RECOMMENDED OPERATING CONDITIONS

For maximum reliability, nominal operating conditions should be selected so that operation is always within the following ranges:

CHARACTERISTIC	LIMITS		UNITS
	MIN.	MAX.	
Supply-Voltage Range (For T_A = Full Package Temperature Range)	3	18	V

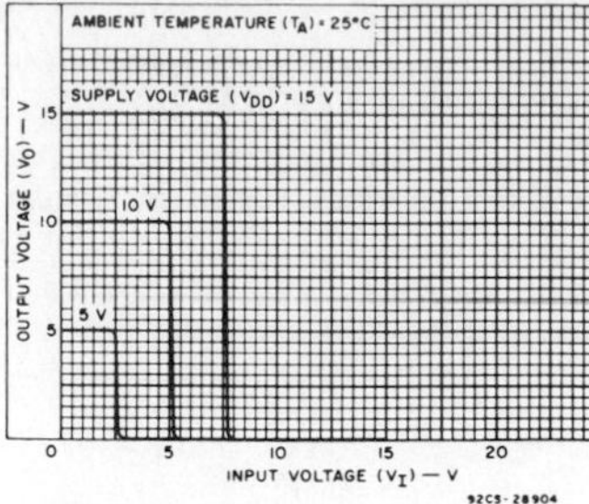

Fig. 1 — Typical voltage transfer characteristics.

MAXIMUM RATINGS, *Absolute-Maximum Values:*

DC SUPPLY-VOLTAGE RANGE, (V_{DD})
(Voltages referenced to V_{SS} Terminal) . . . . . −0.5 to +20 V
INPUT VOLTAGE RANGE, ALL INPUTS . . . . . −0.5 to V_{DD} +0.5 V
DC INPUT CURRENT, ANY ONE INPUT . . . . . ±10 mA
POWER DISSIPATION PER PACKAGE (P_D):
For T_A = −40 to +60°C (PACKAGE TYPE E) . . . . . 500 mW
For T_A = +60 to +85°C (PACKAGE TYPE E) . . . . . Derate Linearly at 12 mW/°C to 200 mW
For T_A = −55 to +100°C (PACKAGE TYPES D, F) . . . . . 500 mW
For T_A = +100 to +125°C (PACKAGE TYPES D, F) . . . . . Derate Linearly at 12 mW/°C to 200 mW
DEVICE DISSIPATION PER OUTPUT TRANSISTOR
FOR T_A = FULL PACKAGE-TEMPERATURE RANGE (All Package Types) . . . . . 100 mW
OPERATING-TEMPERATURE RANGE (T_A):
PACKAGE TYPES D, F, H . . . . . −55 to +125°C
PACKAGE TYPE E . . . . . −40 to +85°C
STORAGE TEMPERATURE RANGE (T_{stg}) . . . . . −65 to +150°C
LEAD TEMPERATURE (DURING SOLDERING):
At distance 1/16 ± 1/32 inch (1.59 ± 0.79 mm) from case for 10 s max. . . . . . +265°C

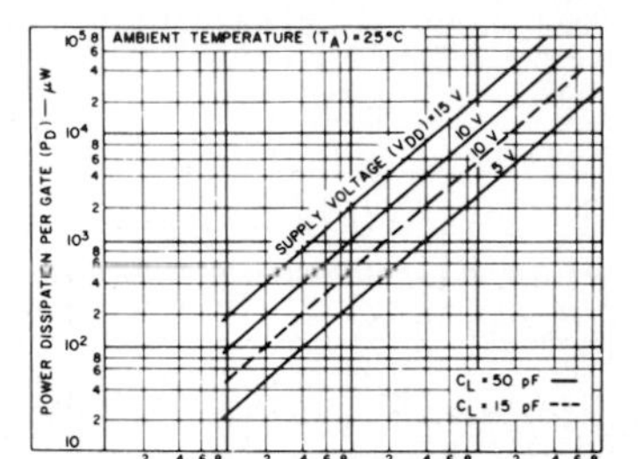

Fig.2 — Typical power dissipation vs. frequency.

DYNAMIC ELECTRICAL CHARACTERISTICS

At T_A = 25°C; Input t_r, t_f = 20 ns, C_L = 50 pF, R_L = 200kΩ

CHARACTERISTIC	TEST CONDITIONS		ALL TYPES LIMITS		UNITS
		V_{DD} VOLTS	TYP.	MAX.	
Propagation Delay Time, t_{PHL}, t_{PLH}		5	125	250	ns
		10	60	120	
		15	45	90	
Transition Time, t_{THL}, t_{TLH}		5	100	200	ns
		10	50	100	
		15	40	80	
Input Capacitance, C_{IN}	Any Input		5	7.5	pF

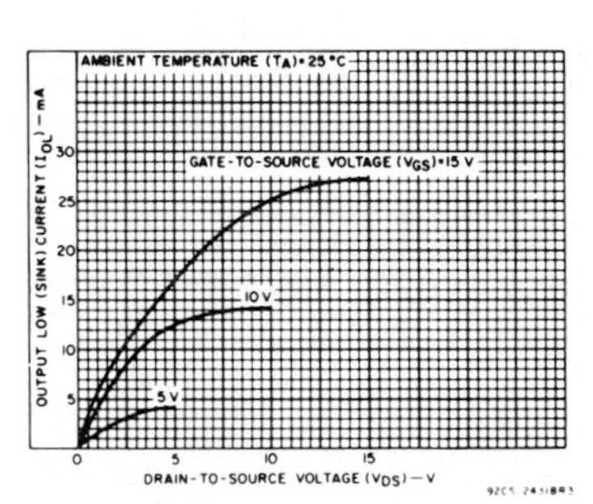

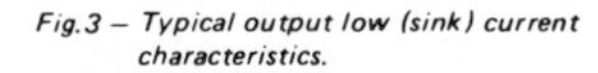

Fig.3 — Typical output low (sink) current characteristics.

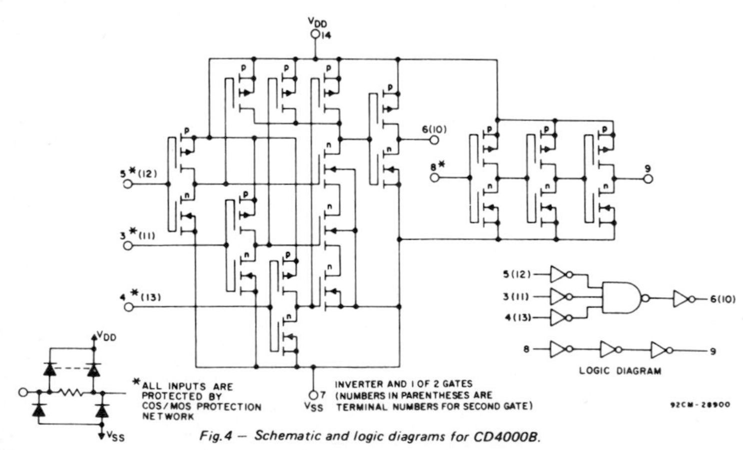

Fig.4 — Schematic and logic diagrams for CD4000B.

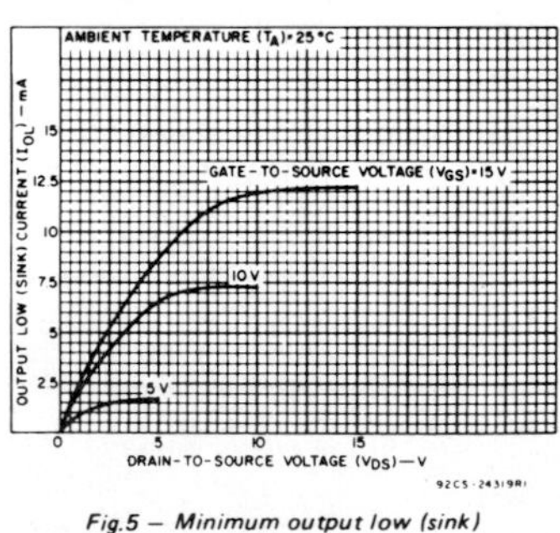

Fig.5 — Minimum output low (sink) current characteristics.

CD4000B, CD4001B, CD4002B, CD4025B Types

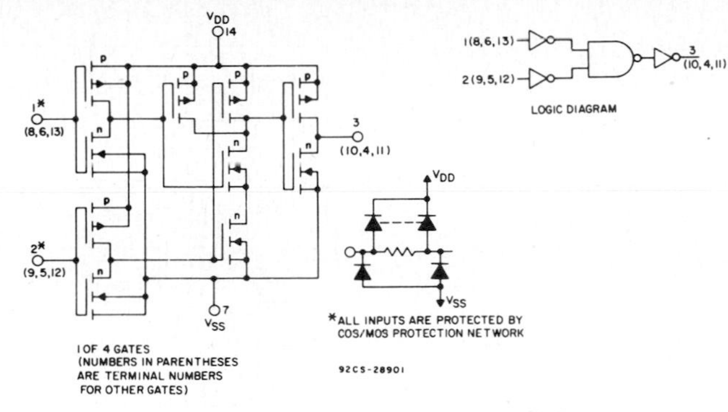

Fig.6 – Schematic and logic diagrams for CD4001B.

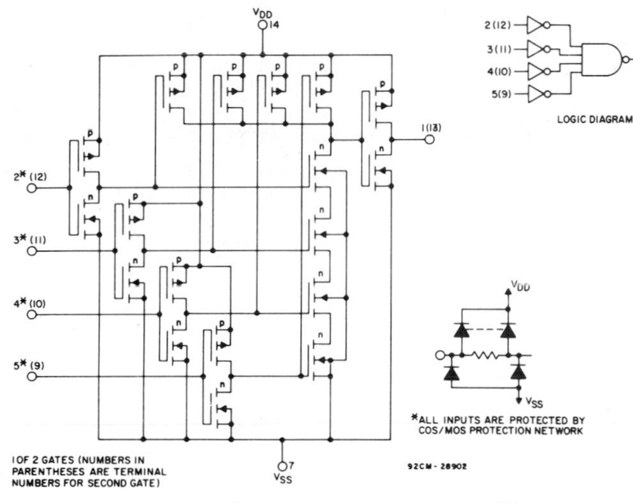

Fig.7 – Schematic and logic diagrams for CD4002B.

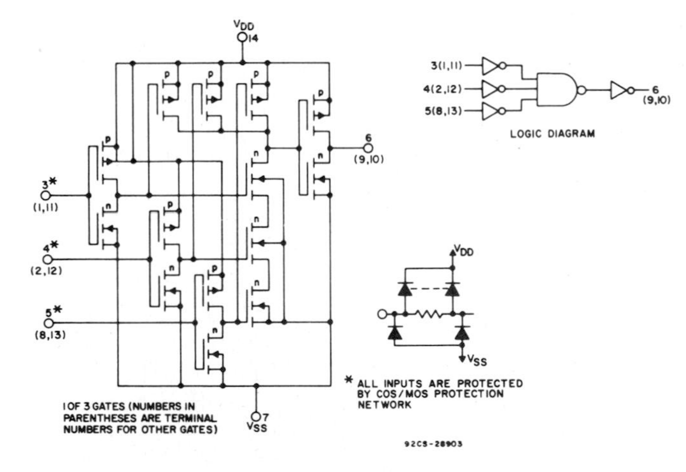

Fig.8 – Schematic and logic diagrams for CD4025B.

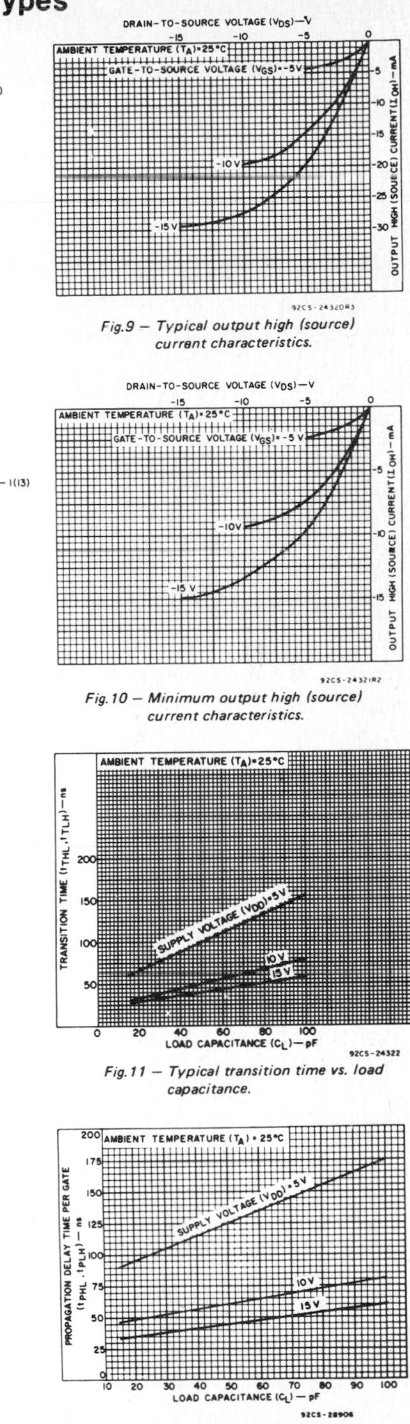

Fig.9 – Typical output high (source) current characteristics.

Fig.10 – Minimum output high (source) current characteristics.

Fig.11 – Typical transition time vs. load capacitance.

Fig.12 – Typical propagation delay time vs. load capacitance.

E.7 *JK* FLIP-FLOPS

TTL (Series 54/74)*

recommended operating conditions

	54 FAMILY / 74 FAMILY	SN54279 / SN74279 MIN	NOM	MAX	SN54LS279 / SN74LS279 MIN	NOM	MAX	UNIT
Supply voltage, V_{CC}	54 Family	4.5	5	5.5	4.5	5	5.5	V
	74 Family	4.75	5	5.25	4.5	5	5.25	
High-level output current, I_{OH}				−800			−400	μA
Low-level output current, I_{OL}	54 Family			16			4	mA
	74 Family			16			8	
Operating free-air temperature, T_A	54 Family	−55		125	−55		125	°C
	74 Family	0		70	0		70	

electrical characteristics over recommended free-air operating temperature range (unless otherwise noted)

PARAMETER		TEST CONDITIONS†			SN54279 / SN74279 MIN	TYP‡	MAX	SN54LS279 / SN74LS279 MIN	TYP‡	MAX	UNIT
V_{IH}	High-level output voltage				2			2			V
V_{IL}	Low-level output voltage			54 Family			0.8			0.7	V
				74 Family			0.8			0.8	
V_{IK}	Input clamp voltage	V_{CC} = MAX, I_I = §					−1.5			−1.5	V
V_{OH}	High-level output voltage	V_{CC} = MIN, V_{IH} = 2 V, V_{IL} = V_{IL} max, I_{OH} = MAX		54 Family	2.4	3.4		2.5	3.4		V
				74 Family	2.4	3.4		2.7	3.4		
V_{OL}	Low-level output voltage	V_{CC} = MIN, V_{IL} = V_{IL} max, V_{IH} = 2 V	I_{OL} = MAX	54 Family		0.2	0.4		0.25	0.4	V
				74 Family		0.2	0.4		0.35	0.5	
			I_{OL} = 4 mA	Series 74LS					0.35	0.4	
I_I	Input current at maximum input voltage	V_{CC} = MAX		V_I = 5.5 V			1				mA
				V_I = 7 V						0.1	
I_{IH}	High-level input current	V_{CC} = MAX		V_I = 2.4 V			40				μA
				V_I = 2.7 V						20	
I_{IL}	Low-level input current	V_{CC} = MAX, V_I = 0.4 V					−1.6			−0.4	mA
I_{OS}	Short-circuit output current♦	V_{CC} = MAX		54 Family	−18		−55	−20		−100	mA
				74 Family	−18		−57	−20		−100	
I_{CC}	Supply current	V_{CC} = MAX, See note 1				18	30		3.8	7	mA

† For conditions shown as MIN or MAX, use the appropriate value specified under recommended operating conditions.

‡ All typical values are at V_{CC} = 5 V, T_A = 25°C.

§ I_I = −12 mA for SN54'/SN74' and −18 mA for SN54LS'/SN74LS'.

♦ Not more than one output should be shorted at a time, and for SN54LS'/SN74LS', duration of the output short circuit should not exceed one second.

NOTE 1: I_{CC} is measured with all $\overline{R}$ inputs grounded, all $\overline{S}$ inputs at 4.5 V, and all outputs open.

*Courtesy Texas Instruments Inc.

switching characteristics, $V_{CC} = 5$ V, $T_A = 25°C$

PARAMETER		TEST CONDITIONS	'279 MIN	'279 TYP	'279 MAX	'LS279 MIN	'LS279 TYP	'LS279 MAX	UNIT
t_{PLH}	Propagation delay time, low-to-high-level output from $\overline{S}$ input	$C_L = 15$ pF, See Notes 2 and 3		12	22		12	22	ns
t_{PHL}	Propagation delay time, high-to-low-level output from $\overline{S}$ input			9	15		13	21	
t_{PHL}	Propagation delay time, high-to-low-level output from $\overline{R}$ input			15	27		15	27	

NOTE 2: Load circuit and voltage waveforms are shown on pages 3-10 and 3-11.

NOTE 3: $R_L = 400\ \Omega$ for '279, $R_L = 2$ kΩ for 'LS279.

schematics of inputs and outputs

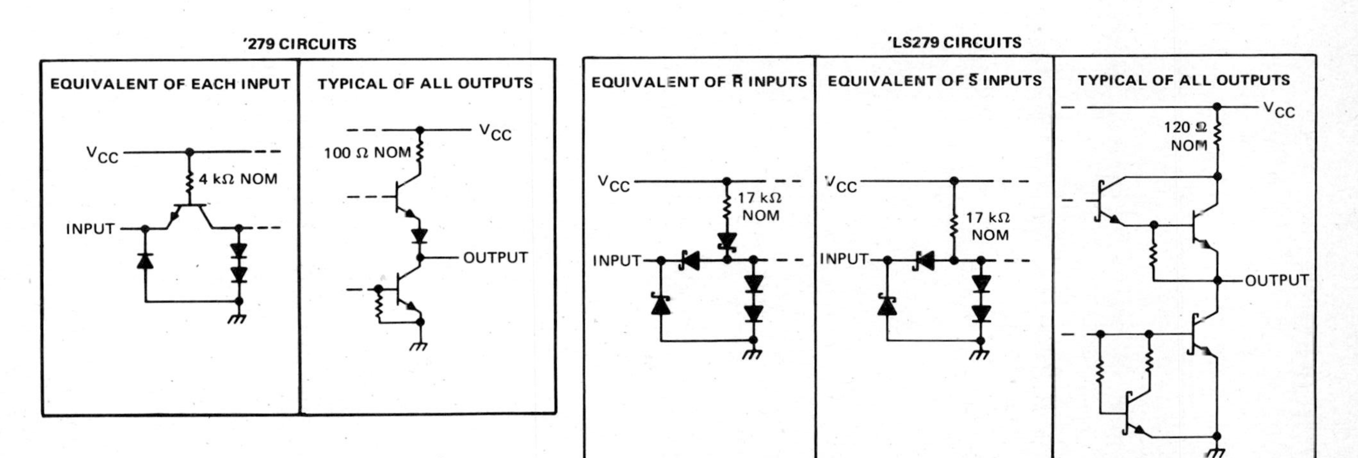

E.7 *JK* FLIP-FLOPS (continued)

SERIES 54/74 FLIP-FLOPS

functional block diagrams (continued)

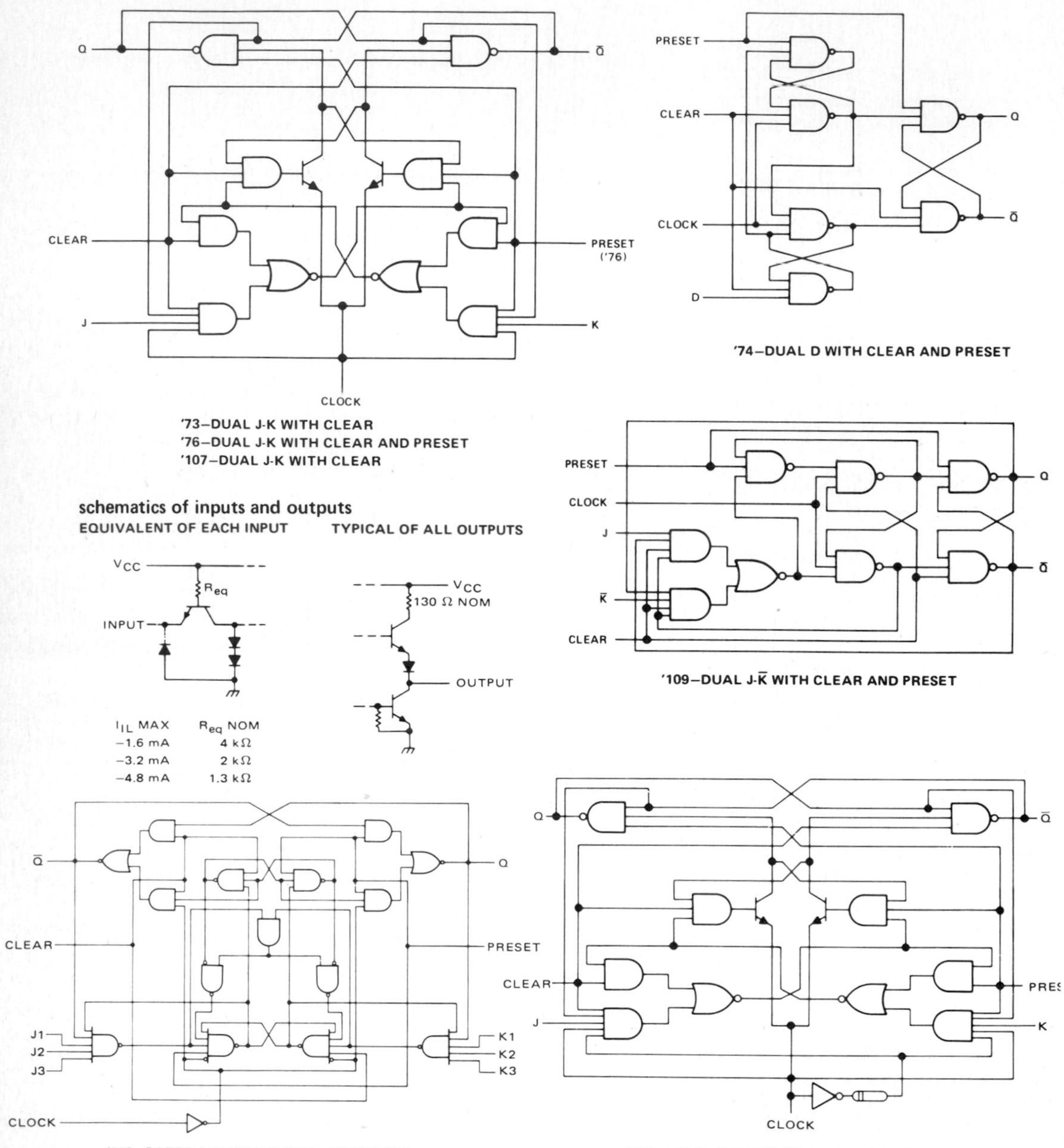

'73–DUAL J-K WITH CLEAR
'76–DUAL J-K WITH CLEAR AND PRESET
'107–DUAL J-K WITH CLEAR

'74–DUAL D WITH CLEAR AND PRESET

'109–DUAL J-K̄ WITH CLEAR AND PRESET

'110–GATED J-K WITH CLEAR AND PRESET

'111–DUAL J-K WITH CLEAR AND PRESET

SERIES 54/74 FLIP-FLOPS

recommended operating conditions

	SERIES 54/74	'70			'72, '73, '76, '107			'74			'109			'110			'111			UNIT
		MIN	NOM	MAX	MIN	NOM	MAX	MIN	NOM	MAX	MIN	NOM	MAX	MIN	NOM	MAX	MIN	NOM	MAX	
Supply voltage, V_{CC}	Series 54	4.5	5	5.5	4.5	5	5.5	4.5	5	5.5	4.5	5	5.5	4.5	5	5.5	4.5	5	5.5	V
	Series 74	4.75	5	5.25	4.75	5	5.25	4.75	5	5.25	4.75	5	5.25	4.75	5	5.25	4.75	5	5.25	
High-level output current, I_{OH}				−400			−400			−400			−800			−800			−800	μA
Low-level output current, I_{OL}				16			16			16			16			16			16	mA
Pulse width, t_w	Clock high	20			20			30			20			25			25			ns
	Clock low	30			47			37			20			25			25			
	Preset or clear low	25			25			30			20			25			25			
Input setup time, t_{su}		20$\uparrow$			0$\uparrow$			20$\uparrow$			10$\uparrow$			20$\uparrow$			0$\uparrow$			ns
Input hold time, t_h		5$\uparrow$			0$\downarrow$			5$\uparrow$			6$\uparrow$			5$\uparrow$			30$\uparrow$			ns
Operating free-air temperature, T_A	Series 54	−55		125	−55		125	−55		125	−55		125	−55		125	−55		125	°C
	Series 74	0		70	0		70	0		70	0		70	0		70	0		70	

$\uparrow\downarrow$The arrow indicates the edge of the clock pulse used for reference: $\uparrow$ for the rising edge, $\downarrow$ for the falling edge.

electrical characteristics over recommended operating free-air temperature range (unless otherwise noted)

PARAMETER		TEST CONDITIONS†	'70			'72, '73, '76, '107			'74			'109			'110			'111			UNIT
			MIN	TYP‡	MAX	MIN	TYP‡	MAX	MIN	TYP‡	MAX	MIN	TYP‡	MAX	MIN	TYP‡	MAX	MIN	TYP‡	MAX	
V_{IH} High-level input voltage			2			2			2			2			2			2			V
V_{IL} Low-level input voltage					0.8			0.8			0.8			0.8			0.8			0.8	V
V_{IK} Input clamp voltage		V_{CC} = MIN, I_I = −12 mA			−1.5			−1.5			−1.5			−1.5			−1.5			−1.5	V
V_{OH} High-level output voltage		V_{CC} = MIN, V_{IH} = 2 V, V_{IL} = 0.8 V, I_{OH} = MAX	2.4	3.4		2.4	3.4		2.4	3.4		2.4	3.4		2.4	3.4		2.4	3.4		V
V_{OL} Low-level output voltage		V_{CC} = MIN, V_{IH} = 2 V, V_{IL} = 0.8 V, I_{OL} = 16 mA		0.2	0.4		0.2	0.4		0.2	0.4		0.2	0.4		0.2	0.4		0.2	0.4	V
I_I Input current at maximum input voltage		V_{CC} = MAX, V_I = 5.5 V			1			1			1			1			1			1	mA
I_{IH} High-level input current	D, J, K, or $\overline{K}$	V_{CC} = MAX, V_I = 2.4 V			40			40			40			40			40			40	μA
	Clear				80			80			120			160			160			80	
	Preset				80			80			80			80			160			80	
	Clock				40			80			80			80			40			120	
I_{IL} Low-level input current	D, J, K, or $\overline{K}$	V_{CC} = MAX, V_I = 0.4 V			−1.6			−1.6			−1.6			−1.6			−1.6			−1.6	mA
	Clear *				−3.2			−3.2			−3.2			−4.8			−3.2			−3.2	
	Preset *				−3.2			−3.2			−1.6			−3.2			−3.2			−3.2	
	Clock				−1.6			−3.2			−3.2			−3.2			−1.6			−4.8	
I_{OS} Short-circuit output current♦	Series 54	V_{CC} = MAX	−20		−57	−20		−57	−20		−57	−30		−85	−20		−57	−20		−57	mA
	Series 74		−18		−57	−18		−57	−18		−57	−30		−85	−18		−57	−18		−57	
I_{CC} Supply current (Average per flip-flop)		V_{CC} = MAX, See Note 1		13	26		10	20		8.5	15		9	15		20	34		11	20.5	mA

†For conditions shown as MIN or MAX, use the appropriate value specified under recommended operating conditions.

‡All typical values are at V_{CC} = 5 V, T_A = 25°C.

♦Not more than one output should be shorted at a time.

*Clear is tested with preset high and preset is tested with clear high.

NOTE 1: With all outputs open, I_{CC} is measured with the Q and $\overline{Q}$ outputs high in turn. At the time of measurement, the clock input is at 4.5 V for the '70, '110, and '111; and is grounded for all the others.

E.7 *JK* FLIP-FLOPS (continued)

SERIES 54/74 FLIP-FLOPS

switching characteristics, $V_{CC} = 5$ V, $T_A = 25°C$

PARAMETER¶	FROM (INPUT)	TO (OUTPUT)	TEST CONDITIONS	'70			'72, '73 '76, '107			'74			'109			'110			'111			UNIT
				MIN	TYP	MAX	MIN	TYP	MAX	MIN	TYP	MAX	MIN	TYP	MAX	MIN	TYP	MAX	MIN	TYP	MAX	
f_{max}			$C_L = 15$ pF, $R_L = 400\ \Omega$, See Note 2	20	35		15	20		15	25		25	33		20	25		20	25		MHz
t_{PLH}	Preset (as applicable)	Q				50		16	25			25		10	15		12	20		12	18	ns
t_{PHL}		$\bar{Q}$				50		25	40			40		23	35		18	25		21	30	
t_{PLH}	Clear (as applicable)	$\bar{Q}$				50		16	25			25		10	15		12	20		12	18	ns
t_{PHL}		Q				50		25	40			40		17	25		18	25		21	30	
t_{PLH}	Clock	Q or $\bar{Q}$			27	50		16	25		14	25		10	16		20	30		12	17	ns
t_{PHL}					18	50		25	40		20	40		18	28		13	20		20	30	

¶ f_{max} ≡ maximum clock frequency; t_{PLH} ≡ propagation delay time, low-to-high-level output; t_{PHL} ≡ propagation delay time, high-to-low-level output.
NOTE 2: Load circuit and voltage waveforms are shown on page 3-10.

functional block diagrams

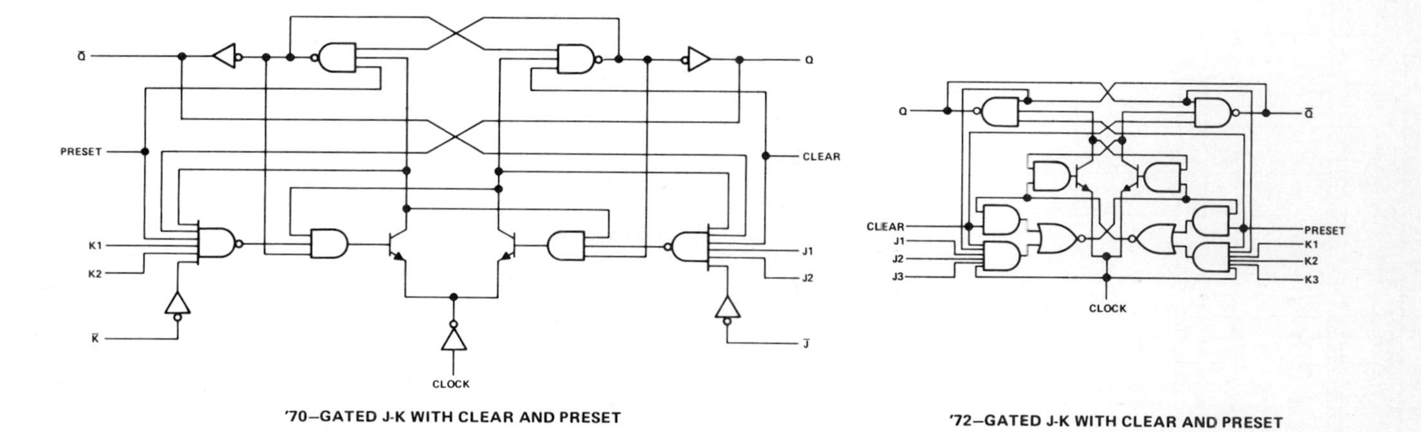

'70—GATED J-K WITH CLEAR AND PRESET

'72—GATED J-K WITH CLEAR AND PRESET

See following pages for:
'73—DUAL J-K WITH CLEAR
'74—DUAL D WITH CLEAR AND PRESET
'76—DUAL J-K WITH CLEAR AND PRESET
'107—DUAL J-K WITH CLEAR
'109—DUAL J-$\bar{K}$ WITH CLEAR AND PRESET
'110—GATED J-K WITH CLEAR AND PRESET
'111—DUAL J-K WITH CLEAR AND PRESET

CMOS (4000B Series)*

CD4095B, CD4096B Types

COS/MOS Gated J-K Master-Slave Flip-Flops

With Set-Reset Capability

High-Voltage Types (20-Volt Rating)

CD4095B Non-Inverting J and K Inputs
CD4096B Inverting and Non-Inverting J and K Inputs

The RCA-CD4095B and CD4096B are J-K Master-Slave Flip-Flops featuring separate AND gating of multiple J and K inputs. The gated J-K inputs control transfer of information into the master section during clocked operation. Information on the J-K inputs is transferred to the Q and $\overline{Q}$ outputs on the positive edge of the clock pulse. SET and RESET inputs (active high) are provided for asynchronous operation.

The CD4095B and CD4096B types are supplied in 14-lead hermetic dual-in-line ceramic packages (D and F suffixes), 14-lead dual-in-line plastic package (E suffix), 14-lead ceramic flat package (K suffix), and in chip form (H suffix).

Features:

- **16 MHz toggle rate (typ.) at $V_{DD} - V_{SS}$ = 10 V**
- **Gated inputs**
- **100% tested for quiescent current at 20 V**
- **Maximum input current of 1 μA at 18 V over full package-temperature range; 100 nA at 18 V and 25°C**
- **Noise margin over full package-temperature range: 1 V at V_{DD} = 5 V, 2 V at V_{DD} = 10 V, 2.5 V at V_{DD} = 15 V**
- **5-V, 10-V, and 15-V parametric ratings**
- **Standardized, symmetrical output characteristics**
- **Meets all requirements of JEDEC Tentative Standard No. 13A, "Standard Specifications for Description of 'B' Series CMOS Devices"**

Applications:

- **Registers** ■ **Counters** ■ **Control circuits**

MAXIMUM RATINGS, *Absolute-Maximum Values:*

DC SUPPLY-VOLTAGE RANGE, (V_{DD}) (Voltages referenced to V_{SS} Terminal)	−0.5 to +20 V
INPUT VOLTAGE RANGE, ALL INPUTS	−0.5 to V_{DD} +0.5 V
DC INPUT CURRENT, ANY ONE INPUT	±10 mA
POWER DISSIPATION PER PACKAGE (P_D):	
For T_A = −40 to +60°C (PACKAGE TYPE E)	500 mW
For T_A = +60 to +85°C (PACKAGE TYPE E)	Derate Linearly at 12 mW/°C to 200 mW
For T_A = −55 to +100°C (PACKAGE TYPES D, F)	500 mW
For T_A = +100 to +125°C (PACKAGE TYPES D, F)	Derate Linearly at 12 mW/°C to 200 mW
DEVICE DISSIPATION PER OUTPUT TRANSISTOR FOR T_A = FULL PACKAGE-TEMPERATURE RANGE (All Package Types)	100 mW
OPERATING-TEMPERATURE RANGE (T_A):	
PACKAGE TYPES D, F, H	−55 to +125°C
PACKAGE TYPE E	−40 to +85°C
STORAGE TEMPERATURE RANGE (T_{stg})	−65 to +150°C
LEAD TEMPERATURE (DURING SOLDERING): At distance 1/16 ± 1/32 inch (1.59 ± 0.79 mm) from case for 10 s max.	+265°C

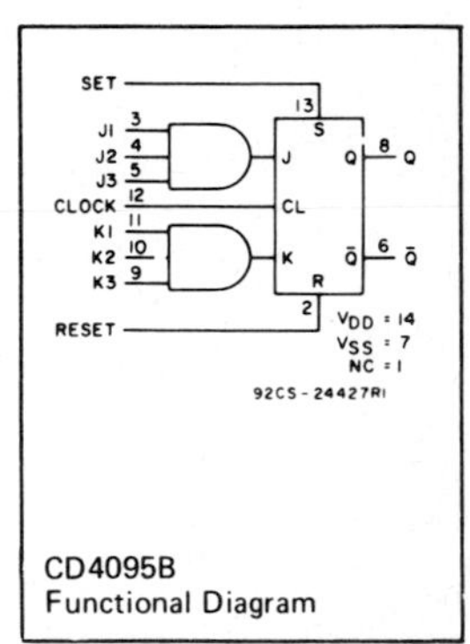

CD4095B
Functional Diagram

TRUTH TABLES

SYNCHRONOUS OPERATION (S=0 R=0)

Inputs Before Positive Clock Transition		Outputs After Positive Clock Transition	
J*	K*	Q	$\overline{Q}$
0	0	No Change	
0	1	0	1
1	0	1	0
1	1	Toggles	

* For CD4095B
J = J1 · J2 · J3
K = K1 · K2 · K3

For CD4096B
J = J1 · J2 · $\overline{J3}$
K = K1 · K2 · $\overline{K3}$

ASYNCHRONOUS OPERATION (J and K - DON'T CARE)

S	R	Q	$\overline{Q}$
0	0	No Change	
0	1	0	1
1	0	1	0
1	1	0	0

0 = V_{SS}, 1 = V_{DD}

*Courtesy of RCA.

E.7 *JK* FLIP-FLOPS (continued)

CD4095B, CD4096B Types

RECOMMENDED OPERATING CONDITIONS at $T_A = 25^\circ C$, Except as Noted. For maximum reliability, nominal operating conditions should be selected so that operation is always within the following ranges:

CHARACTERISTIC	V_{DD} (V)	LIMITS MIN.	LIMITS MAX.	UNITS
Supply-Voltage Range (For T_A = Full Package-Temperature Range)		3	18	V
Data Setup Time, t_S	5 10 15	400 160 100	– – –	ns
Clock Pulse Width, t_W	5 10 15	140 60 40	– – –	ns
Clock Input Frequency, f_{CL}	5 10 15	dc	3.5 8 12	MHz
Clock Rise and Fall Time, t_rCL, t_fCL:	5 10 15	– – –	15 5 5	μs
Set or Reset Pulse Width, t_W	5 10 15	200 100 50	– – –	ns

STATIC ELECTRICAL CHARACTERISTICS

CHARACTERISTIC	CONDITIONS V_O (V)	V_{IN} (V)	V_{DD} (V)	LIMITS AT INDICATED TEMPERATURES (°C) Values at –55, +25, +125 Apply to D,F,H Packages; Values at –40, +25, +85 Apply to E Package: –55	–40	+85	+125	+25 Min.	+25 Typ.	+25 Max.	UNITS
Quiescent Device Current, I_{DD} Max.	–	0,5	5	1	1	30	30	–	0.02	1	μA
	–	0,10	10	2	2	60	60	–	0.02	2	
	–	0,15	15	4	4	120	120	–	0.02	4	
	–	0,20	20	20	20	600	600	–	0.04	20	
Output Low (Sink) Current I_{OL} Min.	0.4	0,5	5	0.64	0.61	0.42	0.36	0.51	1	–	mA
	0.5	0,10	10	1.6	1.5	1.1	0.9	1.3	2.6	–	
	1.5	0,15	15	4.2	4	2.8	2.4	3.4	6.8	–	
Output High (Source) Current, I_{OH} Min.	4.6	0,5	5	–0.64	–0.61	–0.42	–0.36	–0.51	–1	–	
	2.5	0,5	5	–2	–1.8	–1.3	–1.15	–1.6	–3.2	–	
	9.5	0,10	10	–1.6	–1.5	–1.1	–0.9	–1.3	–2.6	–	
	13.5	0,15	15	–4.2	–4	–2.8	–2.4	–3.4	–6.8	–	
Output Voltage: Low-Level, V_{OL} Max.	–	0,5	5	0.05				–	0	0.05	V
	–	0,10	10	0.05				–	0	0.05	
	–	0,15	15	0.05				–	0	0.05	
Output Voltage: High-Level, V_{OH} Min.	–	0,5	5	4.95				4.95	5	–	
	–	0,10	10	9.95				9.95	10	–	
	–	0,15	15	14.95				14.95	15	–	
Input Low Voltage, V_{IL} Max.	0.5, 4.5	–	5	1.5				–	–	1.5	V
	1, 9	–	10	3				–	–	3	
	1.5, 13.5	–	15	4				–	–	4	
Input High Voltage, V_{IH} Min.	0.5, 4.5	–	5	3.5				3.5	–	–	
	1, 9	–	10	7				7	–	–	
	1.5, 13.5	–	15	11				11	–	–	
Input Current I_{IN} Max.		0,18	18	±0.1	±0.1	±1	±1	–	$\pm 10^{-5}$	±0.1	μA

CD4095B, CD4096B Types

DYNAMIC ELECTRICAL CHARACTERISTICS at T_A = 25 °C; Input t_r, t_f = 20 ns, C_L = 50 pF, R_L = 200 KΩ

CHARACTERISTIC	TEST CONDITIONS	V_{DD} (V)	LIMITS MIN.	TYP.	MAX.	UNITS
Propagation Delay Time: t_{PHL}, t_{PLH} Clock		5	–	250	500	ns
		10	–	100	200	
		15	–	75	150	
Set or Reset		5	–	150	300	
		10	–	75	150	
		15	–	50	100	
Transition Time, t_{THL}, t_{TLH}		5	–	100	200	ns
		10	–	50	100	
		15	–	40	80	
Maximum Clock Input Frequency, (f_{CL})*		5	3.5	7	–	MHz
		10	8	16	–	
		15	12	24	–	
Minimum Clock Pulse Width, t_W		5	–	70	140	ns
		10	–	30	60	
		15	–	20	40	
Clock Input Rise or Fall Time, t_{rcl}, t_{rcf}		5	–	–	15	μs
		10	–	–	5	
		15	–	–	5	
Minimum Set or Reset Pulse Width, t_W		5	–	100	200	ns
		10	–	50	100	
		15	–	25	50	
Minimum Data Setup Time, t_S		5	–	200	400	ns
		10	–	80	160	
		15	–	50	100	
Input Capacitance, C_{IN}	Any Input	–	–	5	7.5	pF

* t_r, t_f = 5 ns

SET
J1 3
J2 4
J3 5
CLOCK 12
K1 11
K2 10
K3 9
RESET
13 S
J
CL
K
R 2
8 Q
6 Q̄
V_{DD} = 14
V_{SS} = 7
NC = 1
92CS - 24430R1

Fig. 1 – CD4096B Functional Diagram.

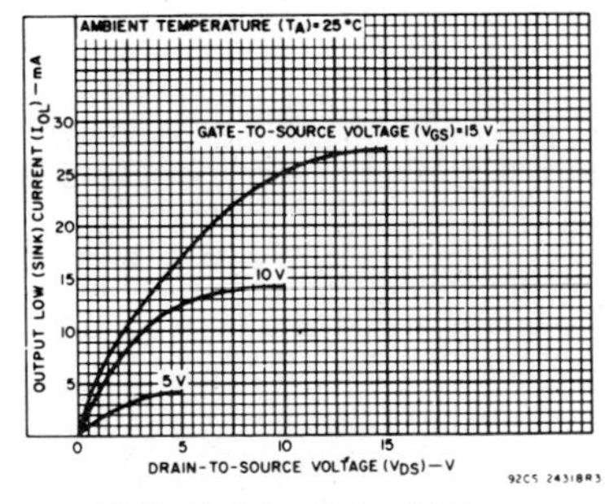

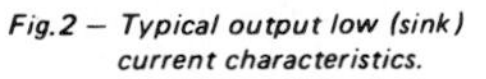

Fig.2 – Typical output low (sink) current characteristics.

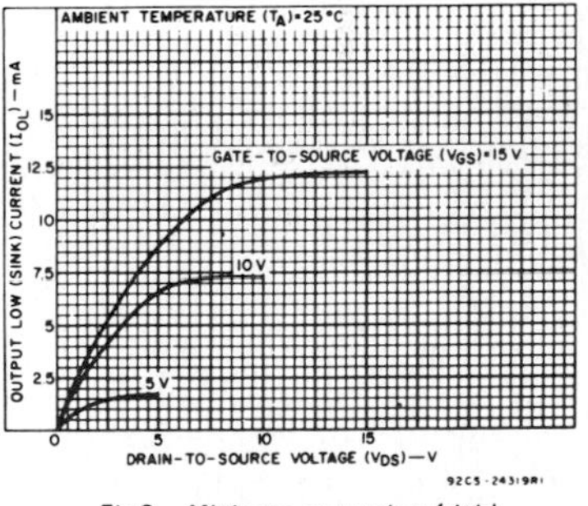

Fig.3 – Minimum output low (sink) current characteristics.

E.7 *JK* FLIP-FLOPS (continued)

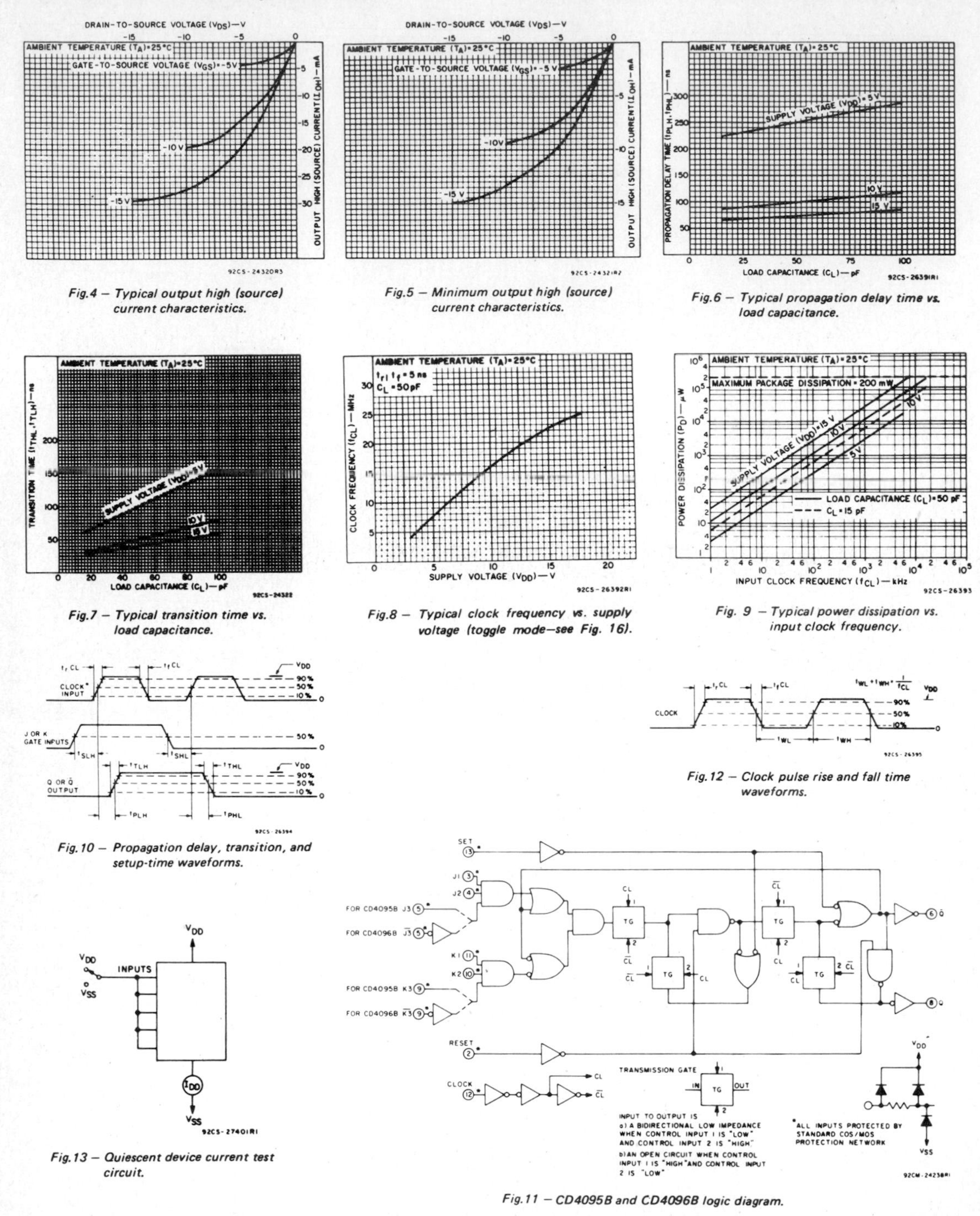

Fig.4 – Typical output high (source) current characteristics.

Fig.5 – Minimum output high (source) current characteristics.

Fig.6 – Typical propagation delay time vs. load capacitance.

Fig.7 – Typical transition time vs. load capacitance.

Fig.8 – Typical clock frequency vs. supply voltage (toggle mode–see Fig. 16).

Fig. 9 – Typical power dissipation vs. input clock frequency.

Fig.10 – Propagation delay, transition, and setup-time waveforms.

Fig.12 – Clock pulse rise and fall time waveforms.

Fig.13 – Quiescent device current test circuit.

Fig.11 – CD4095B and CD4096B logic diagram.

CD4095B, CD4096B Types

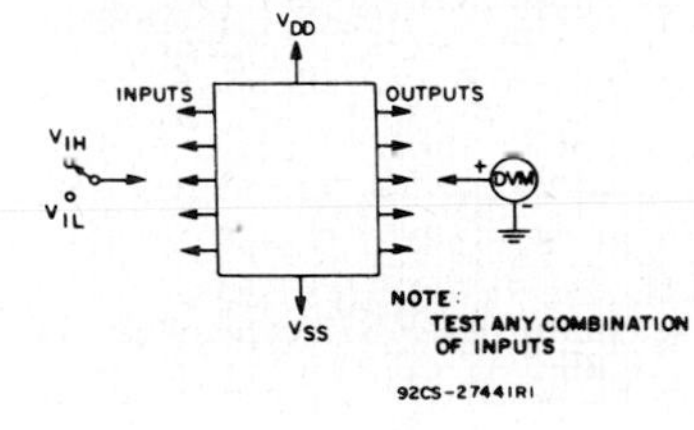

Fig.14 – Input voltage test circuit.

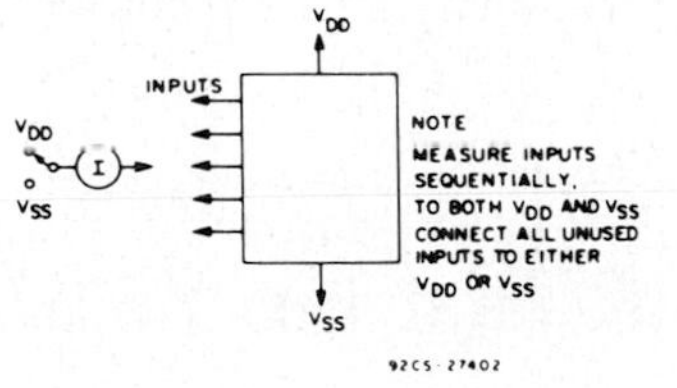

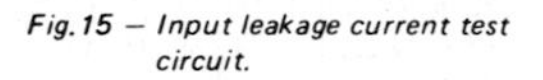

Fig.15 – Input leakage current test circuit.

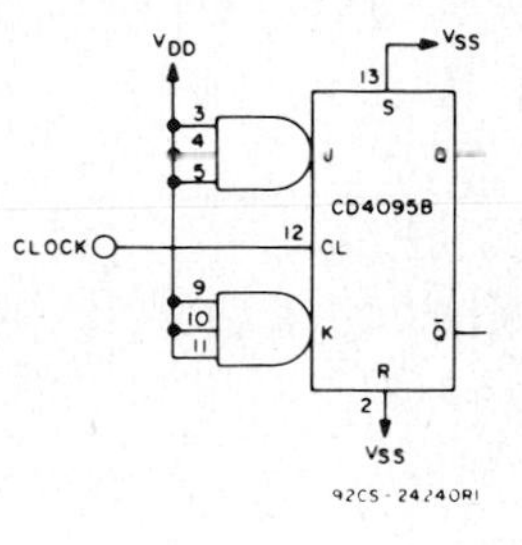

Fig.16 – CD4095B connected in toggle mode.

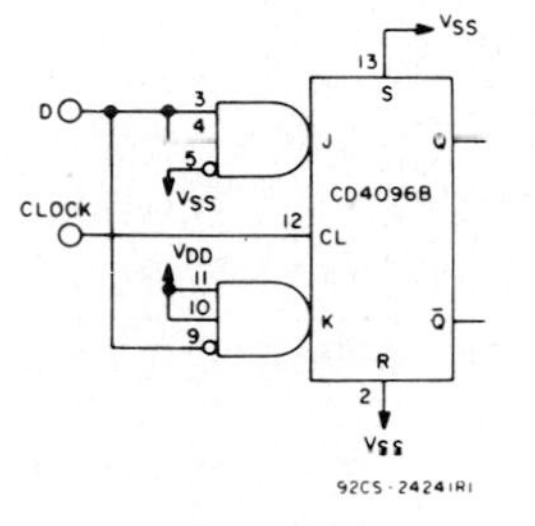

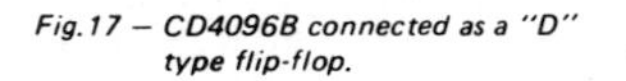

Fig.17 – CD4096B connected as a "D" type flip-flop.

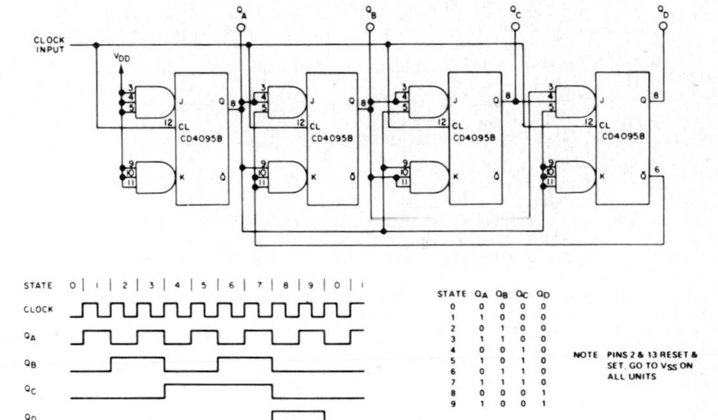

STATE	Q_A	Q_B	Q_C	Q_D
0	0	0	0	0
1	1	0	0	0
2	0	1	0	0
3	1	1	0	0
4	0	0	1	0
5	1	0	1	0
6	0	1	1	0
7	1	1	1	0
8	0	0	0	1
9	1	0	0	1

Fig. 18 – Synchronous binary divide-by-ten counter.

TERMINAL ASSIGNMENTS

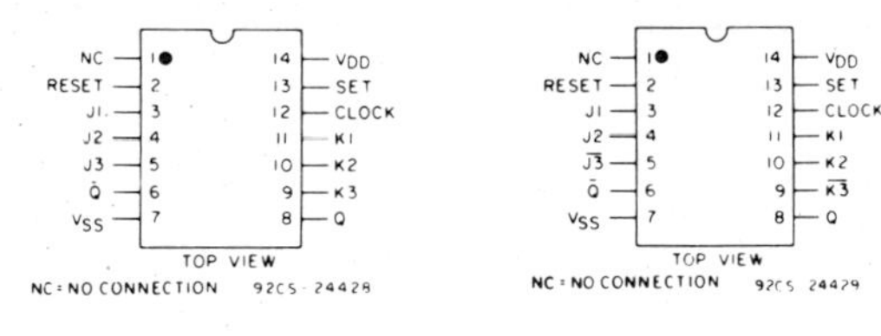

CD4095B

CD4096B

Answers to Selected Problems

CHAPTER 1

1.1 3.18 MΩ

1.4 0.56 V, 0.86 V

1.6 $V_{ab} = 0.204V_1$, $R_{eq} = 3.41\ \Omega$, $I_{eq} = 0.0588V_1$

1.9 $I_D = 50$ mA, $V_D = 0.75$ V

1.11 $I_X = 0.58$ A

1.13 $I = 7.2$ mA

1.20 $V_o = V_i$ for $V_i \leq 0.74$ V, $V_o = \frac{1}{2}(V_i + 0.74)$ for $V_i \geq 0.74$ V

1.28 $V_{DC} = 4.23$ V, $V_{rms} = 4.7$ V

1.34 $V_{DC} = V_m T/8$, $V_{rms} = V_m T/6.93$

1.42 (*a*) $\eta_r = 40.5\%$, (*b*) $\eta_r = 81\%$

1.44 $I_{Dq} = 18.3$ mA, $(V_L)_{dc} = 0$, $r_d = 2.35\ \Omega$, $i_d = 0.292$ mA, $(v_L)_{ac} = 0.07 \sin 10^4 t$

1.46 (*a*) 1.63 W, (*b*) 1.08 W

1.48 $R_{L,\min} = 111\ \Omega$, $I_{Z,\max} = 90$ mA

1.53 (*a*) $V_o = -5$ V, (*b*) $V_o = 0$, (*c*) $V_o = -5$ V

CHAPTER 2

2.1 $I_B = 2.7\ \mu A$, $I_C = 0.262$ mA, $I_E = -0.265$ mA

2.2 $I_E = 0.53$ mA, $I_C = -0.525$ mA, $I_B = 5.3\ \mu A$

2.5 (*a*) $I = 10^{-14}(e^{V_{CE}/V_t} - 1)$, (*b*) $I = 10^{-14}(e^{V_{BE}/V_t} - 1)$

2.10 $V_{BB} = 3.06$ V

2.12 $\beta_0 = 100$, $\alpha_0 = 0.99$, $(V_{CE})_{sat} = 0.2$ V, $(V_{BE})_{on} = 0.7$ V, $I_{co} \simeq 0$

2.15 $V_B = 1.7$ V, output $\simeq 10$ V, input $\simeq 1$ V

2.22 $R_{b1} = 79.4\ k\Omega$, $R_{b2} = 36.4\ k\Omega$, $\Delta I_C = 0.196$ mA, $\Delta V_{CE} = 2.92$ V

2.23 $I_C = 0.283$ mA, $V_{CE} = 19.43$ V

2.25 $z_{11} = 0.027\ \Omega$, $z_{12} = 0$, $z_{21} = 5.3\ \Omega$, $z_{22} = 1.65\ \Omega$

2.30 $I_C = 3.27$ mA, $V_{CE} = 2.97$ V, $\beta_0 \geq 16.3$

2.32 (*a*) $I_C = 3.46$ mA, $V_{CE} = 6.54$ V
(*b*) $I_C = 3.54$ mA, $V_{CE} = 6.46$ V

2.33 $R_{b1} = 15.6\ k\Omega$, $R_{b2} = 7.4\ k\Omega$, $I_{Cq} = 3$ mA, $V_{CE} = 4.5$ V

2.38 $h_{fe} = 64$, $h_{ie} \simeq 250\ \Omega$, $h_{oe} \simeq 10^{-4}$ mho; $h_{re} \simeq 0$

2.40 $g_m = 0.4$ mho, $r_b \leq 60\ \Omega$, $C_c \leq 8$ pF, $\beta_0 \simeq 60$, $R_i \simeq 150\ \Omega$, $f_T = 250$ MHz,
$C_i = 2.53$ nF, $r_c \simeq 100\ k\Omega$

2.42 $I_C = 1.07$ mA, $V_C = 4$ V

2.47 $R_{b1} = 17\ k\Omega$, $R_{b2} = 5.2\ k\Omega$

CHAPTER 3

3.1 240 Ω

3.4 7.06 kΩ

3.7 $I_D = 5$ mA, $V_{DS} = 5$ V

3.10 $I_D = 1.07$ mA, $V_{DS} = 3.58$ V

3.11 $I_D = 1.07$ mA, $V_{DS2} = 8.9$ V

3.15 (*a*) 2.5 V, (*b*) 3 V

3.18 (*a*) $R_D = 4.75\ k\Omega$, $R_S = 1.25\ k\Omega$, (*b*) $A_v = -7.9$, (*c*) $I_D = 1.77$ mA, $V_{DS} = 9.4$ V

3.20 $R_{g2} = 0.125\ M\Omega$, $R_D = 1\ k\Omega$

3.22 (*a*) $R_D = 1.25\ k\Omega$; (*b*) $R_{g1} = 1.38\ M\Omega$, $R_{g2} = 0.786\ M\Omega$;
(*c*) $I_D = 4.56$ mA, $V_{DS} = 5.05$ V

3.25 $C_{gd} = 0.3$ pF, $C_{gs} = 3.7$ pF, $g_m = 10$ mmho, $r_d = 4\ k\Omega$

3.27 (*a*) 0.4 MHz, (*b*) 62 MHz, (*c*) 2.6 MHz

CHAPTER 4

4.1 $h_{ie} \simeq 5\ \text{k}\Omega$, $h_{fe} \simeq 175$, $h_{re} \simeq 8 \times 10^{-4}$, $h_{oe} \simeq 20\ \mu\text{mho}$

4.4 $h_{ib} \simeq 25\ \Omega$, $h_{fb} \simeq -0.994$, $h_{rb} \simeq 4 \times 10^{-6}$, $h_{ob} \simeq 0.2\ \mu\text{mho}$

4.7 (a) $y_{11} = y_{12} = 0$, $y_{22} = 25\ \mu\text{mho}$, $y_{21} = 4.8$ mmho
(b) $y_{11} = y_{12} = 0$, $y_{22} = 0.25$ mmho, $y_{21} = 10$ mmho

4.10 VI $= 7 \times 10^4$

4.13 $G_p = 3.57 \times 10^4$ (45.5 dB)

4.17 $A_i = -60$, $A_v = -115$, $G_P = 38.4$ dB

4.19 $A_v = 3.3$

4.21 (a) $A_v = 6$, (b) $R_{\text{in}} \simeq 3\ \text{k}\Omega$

4.23 $R_D = 2.5\ \text{k}\Omega$, $R_{G1} = 1.3\ \text{M}\Omega$

4.25 (a) $A_v \simeq 1$, (b) $A_v = g_m(R_2 + R_3)/[1 + g_m(R_2 + R_3)]$

4.28 $-R_C/R_E$

4.30 $R_{\text{in}} \simeq \beta_0(r_e + R_E)$

CHAPTER 5

5.1 (a) $I_{C1} = 2$ mA, $V_{CE1} = 5.4$ V, $I_{C2} = 2$ mA, $V_{CE2} = 5.3$ V
(b) $A_v = 3.67 \times 10^3$

5.3 $A_v = 180$

5.5 $A_v = -87$

5.7 $A_v = -29.4$

5.9 $R_S = 500\ \text{k}\Omega$, $R_L = 5\ \text{k}\Omega$, $(G_P)_{\text{max}} = 10^4 = 40$ dB

5.11 $R_S = 250\ \text{k}\Omega$, $R_L = 10\ \text{k}\Omega$, $(G_P)_{\text{max}} = 10^6 = 60$ dB

5.13 $R_S = 2.5\ \text{k}\Omega$, $R_L = 10\ \text{k}\Omega$, $(G_P)_{\text{max}} = 80$ dB

5.15 $A_v = 45.8$

5.17 $A_v = 1.58 \times 10^4$, $A_i \simeq -2 \times 10^5$

5.19 $A_v = 2.844 \times 10^3$

5.24 (a) $R_B = 194\ \text{k}\Omega$, $I_{C2} = I_{C3} = 1.4$ mA, $I_{C1} = 4.78$ mA
(b) $A_v = -191$

5.27 CMRR = 67.5 (36.5 dB)

5.30 $A_d = 3.32$, $A_C = 0.24$, CMRR = 13.8 (22.8 dB)

5.32 $I_{C1} = I_{C2} = I_{C3} = 1.08$ mA, $V_{C2} = V_{C3} = 2.84$ V

5.34 $Z_{\text{in}} = Z_1Z_3Z_5/Z_2Z_4$

CHAPTER 6

6.1 20 W

6.3 4.8 V

6.7 $D = 2.9\%$

6.11 4.6 V

6.14 (*a*) ± 9.3 V, (*b*) same

6.16 40 W

CHAPTER 7

7.1 (*a*) 0.15 μF, (*b*) 0.12 μF

7.4 $f_l = 0.159$ Hz, $f_h = 2.72$ MHz

7.6 (*a*) $A_v = -26.6$, 44 Hz, (*b*) $R_{in} = 3.6\ \text{k}\Omega$

7.8 (*a*) $A_i = 72.7$, $\omega_{3dB} = 5 \times 10^6$ rad/s
(*b*) 45°

7.10 (*a*) exact: $\omega_{3dB} = 0.51p_o$; approximate: $\omega_{3dB} \simeq 0.58p_o$
(*b*) exact: $\omega_{3dB} = 1.96p_o$; approximate: $\omega_{3dB} \simeq 1.73p_o$

7.13 (*a*) $\omega_l = 3.16 \times 10^3$ rad/s, (*b*) $\omega_h \simeq 10^6$ rad/s, (*c*) $K = 10^{15}$

7.16 $\omega_{3dB} \simeq 1.42 \times 10^7$ rad/s

7.19 (*a*) $R_l \simeq 118\ \Omega$, $f_{3dB} = 735$ kHz
(*b*) $R_l \simeq 2.8\ \Omega$, $A_i \simeq 0.28$

7.23 (*a*) $A_v = -9.1$, $f_l = 79.6$ Hz
(*b*) $f_h = 21.7$ MHz

7.26 $A_v = 0.976$, $f_h = 322$ MHz, $f_l = 17.7$ Hz

7.27 $A_v = -13.2$, $f_h = 1$ MHz

7.30 (*a*) $A_i(0) = -11.5$, $f_{3dB} = 34.4$ MHz
(*b*) $A_i = -11.5$, $f_{3dB} = 9.55$ MHz

7.31 For $R_1 = 1\ \text{k}\Omega$, $R_2 = 50\ \text{k}\Omega$, $C = 0.002\ \mu$F

7.34 $\omega_{3dB} = 2\pi(337\ \text{MHz})$

CHAPTER 8

8.1 $A_v = 5.12 \times 10^3$, $\omega_{3dB} = 5.7 \times 10^6$ rad/s

8.4 $A_v = 0.97$, $R_{in} = 33.5\ \text{k}\Omega$, $R_{out} = 24\ \Omega$, $\omega_{3dB} = 2.1 \times 10^8$ rad/s

8.6 $A_v = 500$, $f_l = 229$ Hz, $f_h = 77$ kHz

8.7 $A_i = 500$, $\omega_l = 64$ rad/s, $\omega_h = 4.5 \times 10^5$ rad/s

8.10 $A_v = 38$, $f_{3dB} \simeq 110$ MHz

8.12 $A_v = -3.2$, $f_{3dB} = 106$ MHz

8.14 $A_v = -20$, $f_{3db} = 22.7$ MHz

8.17 (*a*) $S = 0.802$, (*b*) $f_{3dB} = 1.25$ MHz

8.19 $A_v = -40$, $\omega_l = 82$ rad/s, $\omega_h = 1.25 \times 10^7$ rad/s

8.21 $A_d = -1$, $A_c = -0.05$, CMRR = 20

CHAPTER 9

9.6 $V_o \propto \cos\phi$

9.12 $R_1 = 5\ \text{k}\Omega$, $R_2 = 15.9\ \text{k}\Omega$, $\omega_{3\text{dB}} \simeq 4 \times 10^4$ rad/s

9.14 $R_1 = 1.6\ \text{k}\Omega$, $R_2 = 3.8\ \text{k}\Omega$, $C_2 = 0.004\ \mu\text{F}$

9.16 $V_o/V_i = (K\omega_o/Q)s/[s^2 + (\omega_o/Q)s + \omega_o^2]$, where $\omega_o = \sqrt{2}/RC$,
$\omega_o/Q = (4 - K)/RC$, $K = 3.86$, $R = 2.25\ \text{k}\Omega$

9.22 $R_1 = R_2 = 15.9\ \text{k}\Omega$, $R_3 = 8\ \text{k}\Omega$, $C_3 = 0.02\ \mu\text{F}$

9.23 $f \simeq 210$ Hz

9.27 $V_o/V_i = [GB/(s + GB/K)]^2$

9.28 $R_1 = 1\ \text{M}\Omega$, $R_2 = 100\ \text{k}\Omega$, $C_1 = 10$ pF, $C_2 = 100$ pF

9.29 $C_a = 500$ pF, $C_b = 480$ pF

9.34 $V_o = -3.2$ V, 10 V if the op amp saturation voltage ≥ 10 V

CHAPTER 10

10.1 $R_f = 1\ \text{k}\Omega$, $\omega_{3\text{dB}} = 1.04 \times 10^7$ rad/s

10.3 $R_f = 100\ \Omega$, $\omega_{3\text{dB}} = 2.58 \times 10^7$ rad/s

10.7 $s_{2,3} = -(4.9 \pm j6.6) \times 10^7$ rad/s, $A_i(0) = -1319$

10.10 $s_{2,3} = (1.9 \pm j10.9) \times 10^7$ rad/s, unstable

10.12 $a_v = 38.2$, f = 0.142, $A_v = 7$

10.14 $a_v = 3300$, $f_0 = 0.5$, $A_v = 188$

10.19 $(\omega_{3\text{dB}})_h = 9.1 \times 10^6$ rad/s, $(\omega_{3\text{dB}})_l = 2.2$ rad/s, $A_o = 9.1$

10.22 $K \leq (\sec \pi/n)^n$ for $K > 0$, $K \leq 1$ for $K < 0$

10.25 $f_o = 0.0063$, $\Phi_M = 85°$, $f_o = 0.063$, $\Phi_M = 0$

10.28 $\omega_o = \dfrac{1}{\sqrt{6}RC}$, $A_V \geq 29$

10.30 $\omega_o = \dfrac{1}{\sqrt{L_2 C}}$, $h_{fe} = M/L_2$

10.32 $\omega_o = 1/RC$, $g_m^2 R_{L1} R_{L2} \simeq 3$

10.34 $\omega_o = 2\pi(525\ \text{kHz})$, $g_m \geq \frac{1}{2}(\omega_o/Q)(C_1 + C_2)$

CHAPTER 11

11.1 $A_v \simeq 4$, $\omega_{3\text{dB}} = 2.17 \times 10^6$ rad/s, $\omega_o = 1.41 \times 10^7$ rad/s

11.4 $R_1 = 500\ \Omega$, $C_a \simeq 0.04\ \mu\text{F}$, $L = 2.54\ \mu\text{H}$

11.6 $f_{3\text{dB}} = 1.59$ kHz

11.8 Potentially unstable, $R_1 < 10\ \Omega$

11.10 $C_2 = 179$ pF, $L_2 = 0.10\ \mu\text{H}$, $L_1 = 1.33\ \mu\text{H}$, $C_a = 2786$ pF, $R_L = 1.26\ \text{k}\Omega$

11.13 $R = 25\ \text{k}\Omega$, $C = 400\ \text{pF}$, $L = 2.5\ \text{mH}$

11.15 $R = 15.9\ \text{k}\Omega$, $R_1 = 1.6\ \text{k}\Omega$, $R_2 = 176\ \Omega$, $R_a = 100\ \text{k}\Omega$

11.17 $R_2 = 159\ \text{k}\Omega$, $R_1 = 6.37\ \text{k}\Omega$, $A_v = -12.5$

11.20 $S_Q^{|T|} = 1$, $S_{\omega_o}^{|T|} = 1$

11.21 $f_{oL} = 0.15\ \text{MHz}$, $f_{oH} = 0.25\ \text{MHz}$

CHAPTER 12

12.1 $T_r = 2.2\ \mu\text{s}$, $T_D = 1.1\ \mu\text{s}$

12.4 $T_r = 25\ \mu\text{s}$, $\delta = 21.5\%$, $A_{vo} = 400$

12.8 (*a*) $T_R = 4.2\ \mu\text{s}$, (*b*) $T_R = 3.8\ \mu\text{s}$

12.11 $t_{\text{on}} = 17.6\ \text{ns}$, $t_{\text{off}} = 40.4\ \text{ns}$

12.13 $t_{\text{on}} = 11.1\ \text{ns}$, $t_{\text{off}} = 27.4\ \text{ns}$

12.17 $P_D = K_L V_{DD}^3$

CHAPTER 13

13.2 (1) $(250.15625)_{10}$, 11111010.001010, $(\text{FA.28})_{\text{HEX}}$
(2) $(312.625)_{10}$, 100010000.101, $(110.\text{A})_{\text{HEX}}$

13.13 $Y = A \cdot B \cdot (\overline{C} + \overline{D})$

13.16 (*a*) $A \cdot C + \overline{A} \cdot B$, (*b*) $A \cdot (B + \overline{C})$, (*c*) $A \cdot (B + \overline{C})$

13.18 $\overline{(\overline{A \cdot B}) \cdot (\overline{B \cdot C}) \cdot (\overline{C \cdot A})}$

13.19 $\overline{A + B} + \overline{B + C} + \overline{A + C}$

13.20 $Y = A \cdot B$

13.21 $\text{NM}_H = 3.24\ \text{V}$, $\text{NM}_L = 0.5\ \text{v}$

13.22 $N = 33$

13.24 $\text{NM}_H = \text{NM}_L = 1\ \text{V}$

13.27 $V_{OH} = 5\ \text{V}$, $V_{OL} = 0.12\ \text{V}$, $\text{NM}_H = 2.9\ \text{V}$, $\text{NM}_L \simeq 1\ \text{V}$

13.28 $V_{OH} = 3.87\ \text{V}$, $V_{OL} = 1\ \text{V}$, $LS = 2.87\ \text{V}$.

13.30 $V_{OH} = 5\ \text{V}$, $V_{OL} \simeq 0\ \text{V}$, $\text{NM}_H = 4\ \text{V}$, $\text{NM}_L = 1\ \text{V}$

CHAPTER 14

14.1 $V_{C1} = V_E = 7.63\ \text{V}$, $I_{C1} = 2.37\ \text{mA}$, $V_{B2} = 6.35\ \text{V}$, $V_{BE2} = -1.28\ \text{V}$

14.11 $t_1 = 0.7\ \text{ms}$

14.18 $f = 1/T = 1/2(3t_d)$, $f = 1/2Nt_d$, $N =$ odd integer

14.20 $\Delta V = 0.0195\ \text{V}$, $8\ \mu\text{s}$

Index